Modeling in the Neurosciences

From Biological Systems to Neuromimetic Robotics

Second Edition

Modeling in the Neurosciences

From Biological Systems to Neuromimetic Robotics

Second Edition

Edited by

G. N. Reeke, R. R. Poznanski, K. A. Lindsay, J. R. Rosenberg, and O. Sporns

CRC Press is an imprint of the
Taylor & Francis Group, an **informa** business

CRC Press
Taylor & Francis Group
6000 Broken Sound Parkway NW, Suite 300
Boca Raton, FL 33487-2742

First issued in paperback 2019

ISBN-13: 978-0-415-32868-5 (hbk)
ISBN-13: 978-0-367-39317-5 (pbk)

Library of Congress Cataloging-in-Publication Data

Catalog record is available from the Library of Congress

Visit the Taylor & Francis Web site at
http://www.taylorandfrancis.com

and the CRC Press Web site at
http://www.crcpress.com

Contents

Preface to the Second Edition

The Second Edition of *Modeling in the Neurosciences* appears on the fifth anniversary of the publication of the first edition. Inspired by the wealth of new work since that date, we were determined to bring the book up to date, to discuss the most important new developments in neuronal and neuronal systems modeling, including use of robotics to test brain models.

The goal of the book is to probe beyond the realm of current research protocols toward the uncharted seas of synthetic neural modeling. The book is intended to help the reader move beyond model ingredients that have been widely accepted for their simplicity or appeal from the standpoint of mathematical tractability (e.g., bidirectional synaptic transfer, updates to synaptic weights based on nonlocal information, or mean-field theories of rate-coded point neurons), and learn how to construct more structured (integrative) models with greater biological insight than heretofore attempted. The book spans the range from gene expression, dendritic growth, and synaptic mechanics to detailed continuous-membrane modeling of single neurons and on to cell–cell interactions and signaling pathways, including nonsynaptic (ephaptic) interactions. In the final chapters, we deal with complex networks, presenting graph- and information-theoretic methods of analyzing complexity and describing in detail the use of robotic devices with synthetic model brains to test theories of brain function.

The book is neither a handbook nor an introductory volume. Some knowledge of neurobiology, including anatomy, physiology, and biochemistry is assumed, as well as familiarity with analytical methods including methods of solving differential equations. A background knowledge of elementary concepts in statistics and applied probability is also required. The book is suitable as an advanced textbook for a graduate-level course in neuronal modeling and neuroengineering. It is also suited to neurophysiologists and neuropsychologists interested in the quantitative aspects of neuroscience who want to be informed about recent developments in mathematical and computer modeling techniques. Finally, it will be valuable to researchers in neural modeling interested in learning new methods and techniques for testing their ideas by constructing rigorously realistic models based on experimental data.

The first edition of *Modeling in the Neurosciences* has contributed impressively to the analytical revolution that has so completely changed our perception of quantitative methods in neuroscience modeling. Nevertheless many different views persist regarding the most satisfactory approaches to doing theoretical neuroscience. The source of this plethora of interpretations is, perhaps, the fact that "computational" interpretations of brain function are not satisfactory, as discussed in the introductory chapter by G. Reeke. Thus, it seems appropriate in this book to attempt to obtain a more balanced picture of all the elements that must be taken into account to get a better understanding of how nervous systems in fact function in the real world where adaptive behavior is a matter of life or death. We hope the integrative viewpoint we advocate may serve as a Rosetta stone to guide the development of modern analytical foundations in the neurosciences.

To accomplish this, we have commissioned authors to provide in-depth treatments of a number of unresolved technical and conceptual issues which we consider significant in theoretical and integrative neuroscience, and its analytical foundations. We hope that this second edition will help and inspire neuroscientists to follow threads revealed herein for many years to come.

We acknowledge the support of the many colleagues who have made this book possible, in particular, the authors who have given so freely of their time to create the chapters now before you. We are also indebted to Dr. John Gillman, Publisher at Harwood Academic Publishers, Reading, UK, who authorized the commissioning of the second edition in the winter of 2001; Dr. Grant Soannes, Publisher at Taylor & Francis, London, UK, who contracted for the book in the fall of 2003; and Barbara Norwitz, Publisher at Taylor & Francis, Boca Raton, Florida, who finalized the arrangements in January of 2004. Last but not least, we thank Pat Roberson and her team for their continuing efforts in the production of a truly immaculate volume.

Sadly, one of the contributors, Aron M. Gutman, passed away soon after the launch of the first edition in the spring of 1999. Aron Gutman was born in Zhitomir, Ukraine in 1936. He received his Ph.D. (externally) from the Department of Physics of Leningrad University in 1962. From 1959 to 1999 he worked in the Neurophysiological Laboratory of Kaunas Medical Institute (now University). He was a prolific writer with more than 150 scientific publications, including two monographs (in Russian) and about 50 papers in international journals. As a distinguished biophysicist, particularly known for his work on the theory of N-dendrites, he has made a profound impact on the scientific community. In Gutman's own words: "small cells use dendritic bistability with slow rich logic."

G.N. Reeke
R.R. Poznanski
K.A. Lindsay
J.R. Rosenberg
O. Sporns

Contributors

Toshihiro Aoyama
Department of Electronic and Information Engineering
Suzuka National College of Technology
Shiroko, Suzuka-City, Mie, Japan

Armuntas Baginskas
Laboratory of Neurophysiology
Institute for Biomedical Research
Kaunas University of Medicine
Kaunas, Lithuania

Jonathan Bell
Department of Mathematics and Statistics
University of Maryland
Baltimore County
Baltimore, Maryland, USA

Helen M. Bronte-Stewart
Department of Neurology and Neurological Sciences
Stanford University Medical Center
Stanford, California, USA

Gilbert A. Chauvet
LDCI, Ecole Pratique des Hautes Etudes et CHU Angers
Paris, France

Robert Costalat
INSERM U 483
Université Pierre et Marie Curie
Paris, France

Bruno Delord
INSERM U 483
Université Pierre et Marie Curie
Paris, France

Shinji Doi
Department of Electrical Engineering
Osaka University
Suita, Osaka, Japan

Jonathan D. Evans
Department of Mathematical Sciences
University of Bath
Bath, England

Hidekazu Fukai
Department of Information Science
Faculty of Engineering
Gifu University
Yanagido, Gifu, Japan

Loyd L. Glenn
Research Informatics Unit
East Tennessee State University
Johnson City, Tennessee, USA

Mel D. Goldfinger
Department of Anatomy & Physiology
College of Science & Mathematics and School of Medicine
Wright State University
Dayton, Ohio, USA

Aron M. Gutman (deceased)
Laboratory of Neurophysiology
Institute for Biomedical Research
Kaunas University of Medicine
Kaunas, Lithuania

David Halliday
Department of Electronics
University of York
York, England

William R. Holmes
Neuroscience Program
Department of Biological Sciences
Ohio University
Athens, Ohio, USA

Jorn Hounsgaard
Laboratory of Cellular Neurophysiology
Department of Medical Physiology
Panum University of Copenhagen
Copenhagen N, Denmark

Akito Ishihara
School of Life System Science and Technology
Chukyo University
Toyota, Japan

Viktor K. Jirsa
Center for Complex Systems and Brain Sciences
Florida Atlantic University
Boca Raton, Florida, USA

Yoshimi Kamiyama
Information Science and Technology
Aichi Prefectural University
Nagakute, Aichi, Japan

Jeffrey R. Knisley
Department of Mathematics
East Tennessee State University
Johnson City, Tennessee, USA

Jeffrey L. Krichmar
The Neurosciences Institute
San Diego, California, USA

Larry S. Liebovitch
Center for Complex Systems and Brain Sciences
Florida Atlantic University
Boca Raton, Florida, USA

Kenneth A. Lindsay
Department of Mathematics
University of Glasgow
Glasgow, Scotland

George P. Moore
Department of Biomedical Engineering
University of Southern California
Los Angeles, California, USA

Taishin Nomura
Department of Mechanical Science and Bioengineering
Graduate School of Engineering Science
Osaka University
Toyonaka, Osaka, Japan

Roger Orpwood
Bath Institute of Medical Engineering
University of Bath
c/o Wolfson Centre
Royal United Hospital
Bath, England

Parag G. Patil
Neurosurgery and Neurobiology
Duke University Medical Center
Durham, North Carolina, USA

Roman R. Poznanski
Claremont Research Institute of Applied Mathematical Sciences
Claremont Graduate University
Claremont, California, USA

George N. Reeke
Laboratory of Biological Modeling
Rockefeller University
New York, USA

Jay R. Rosenberg
Division of Neuroscience and Biomedical Systems
University of Glasgow
Glasgow, Scotland

Shunsuke Sato
Department of Physical Therapy
Faculty of Nursing and Rehabilitations
Aino University
Ibaraki, Osaka, Japan

Lina A. Shehadeh
Center for Complex Systems and Brain Sciences
Florida Atlantic University
Boca Raton, Florida, USA

Gregory D. Smith
Department of Applied Science
The College of William and Mary
Williamsburg, Virginia, USA

Olaf Sporns
Department of Psychology
Indiana University
Bloomington, Indiana, USA

Natasha Svirskiene
Laboratory of Neurophysiology
Institute for Biomedical Research
Kaunas University of Medicine
Kaunas, Lithuania

Gytis Svirskis
Laboratory of Neurophysiology
Institute for Biomedical Research
Kaunas University of Medicine
Kaunas, Lithuania

Gayle Tucker
Department of Mathematics
University of Glasgow
Glasgow, Scotland

Dennis A. Turner
Neurosurgery and Neurobiology
Duke University Medical Center
Durham, North Carolina, USA

Shiro Usui
Brain Science Institute
RIKEN
Hirosawa, Wako, Japan

Harry B.M. Uylings
Netherlands Institute for Brain Research
KNAW
Amsterdam, The Netherlands

Jaap van Pelt
Netherlands Institute for Brain Research
KNAW
Amsterdam, The Netherlands

Mike West
Institute of Statistics and Decision Sciences
Duke University
Durham, North Carolina, USA

Howard V. Wheal
Southampton Neurosciences Group (SoNG)
University of Southampton
School of Biological Sciences
Southampton, England

Foreword

Numerical modeling is common, but analytical theory is rare. Fast computers and software to generate formal models have become more accessible to experimentalists at their own peril. The art of specifying and validating models has not kept pace. The complexities and nonlinearities of neuronal systems require, on the part of the modeling practitioner, an unusual degree of neurophysiological and neuroanatomical insight, as well as skill in a repertoire of numerical methods. However, a computer simulation without an underlying theory is only of heuristic value; such a model cannot be properly tested because conceptual errors cannot be distinguished from inappropriate parameter choices.

At the cellular level, numerical models that discretize the neuronal membrane suffer from an excess of degrees of freedom; it is difficult or impossible to collect enough data to constrain all the parameters of such models to unique values. In such constructs, the continuity of the neuronal membrane is sliced into pieces to form compartments (or "spiking neuron" models in the computational neuroscience literature). Much better treatments of chemical diffusion and the spatial variation of ion concentrations inside and outside the cell are badly needed. The use of continuous-membrane (partial differential equation) models to resolve some of the issues mentioned here is a strong feature of the present volume.

There are also foundational problems at the systems biology level. The next step toward the construction of a solid theoretical foundation for brain science will require a precise clarification of the subtle problems afflicting current computational neuroscience (both conceptual and epistemological), together with the development of fully integrative models across all levels of multihierarchical neural organization. It will also require a new understanding of how perceptual categorization and learning occur in the real world in the absence of preassigned category labels and task algorithms. I congratulate the editors who have produced this second edition of *Modeling in the Neurosciences*, which epitomizes a trend to move forward toward the development of more conceptual models based on an integrative approach.

Professor G.A. Chauvet, M.D., Ph.D.
Chief Editor
Journal of Integrative Neuroscience
Imperial College Press, London, U.K.

About the Editors

George Reeke was born in Green Bay, Wisconsin in 1943. He trained as a protein crystallographer with Dick Marsh at Caltech and Bill Lipscomb at Harvard, where his Ph.D. thesis contained one of the first atomic coordinate sets for any protein, that of the enzyme carboxypeptidase A. He was brought to Rockefeller University in 1970 by Gerald Edelman, with whom he collaborated in a series of studies on the three-dimensional structures of proteins of interest for understanding the function of the immune system. He developed one of the first microprocessor-controlled lab instruments, an x-ray camera, and his Fourier synthesis software was used to solve many protein structures worldwide.

In recent years, his interest has turned to problems of pattern recognition, perceptual categorization, and motor control, which he has studied primarily through computer simulations of selective neural networks and recognition automata, including brain-based robot-like devices. With Allan Coop, he has developed a computationally efficient composite method for modeling the discharge activity of single neurons, as well as a new "analytic distribution" method for assessing the interval entropy of neuronal spike trains. At present, Dr. Reeke is an Associate Professor and head of the Laboratory of Biological Modeling at The Rockefeller University and a Senior Fellow of The Neurosciences Institute. He serves as a member of the editorial board of the Journal of Integrative Neuroscience and of the Biomedical Advisory Committee of the Pittsburgh Supercomputer Center.

Roman R. Poznański was born in Warsaw, the city of Frédéric Chopin. He is an influential neuroscientist noted for his modeling work in first pinpointing the locus of retinal directional selectivity as vindicated by recent experiments.

He continues to set the stage in revolutionizing the pioneering work undertaken in the 1950s and 1960s by the NIH duo: *Richard FitzHugh* and *Wilfrid Rall* through a complete analytical treatment of the most difficult partial differential equations in classical neurophysiology, namely the Frankenhaeuser–Huxley equations. He has been instrumental in the establishment of a new generation of neural networks and their application to large-scale theories of the brain. Dr. Poznański remains a heavyweight in both neural modeling and neuropsychology with the publication of an integrative theory of cognition in the *Journal of Integrative Neuroscience* (founded with *Gilbert A. Cháuvet*). His idea is to develop integrative models to represent semantic processing based on more biologically plausible neural networks. This would naturally lead to more sophisticated artificial systems: he has coined the term "neuromimetic robotics" (with *Olaf Sporns*) to designate real-world devices beyond neuroscience-inspired technology used in today's connectionist design protocols.

Currently, Dr. Poznański holds a research faculty appointment in the Claremont Research Institute of Applied Mathematical Sciences within the School of Mathematical Sciences at Claremont Graduate University.

"*His main supervisor W R Levick, FRS was one of the most imminent neuro-psychologists of our time. And here is where Roman made his first real contribution. In fact, I was surprised but warmed to see how influential his work has been.*"

Allan W. Snyder FRS
150th Anniversary Chair, University of Sydney
Peter Karmel Chair, Australian National University

Olaf Sporns was born in Kiel, Germany, in 1963. After studying biochemistry at the University of Tübingen in Germany, he entered the Graduate Program at New York's Rockefeller University. In 1990, he received a Ph.D. in neuroscience and became a Fellow in Theoretical Neurobiology at The Neurosciences Institute in New York. Since 2000, he has held a faculty position at the Department of Psychology at Indiana University in Bloomington. Dr. Sporns' main research interest is theoretical neuroscience. A main research focus is the development of integrative and synthetic models that can be interfaced with autonomous robots and can be used to study neurobiological and cognitive functions such as perceptual categorization, sensorimotor development, and the development of neuronal receptive field properties. Another focus is the development of anatomically and physiologically detailed models of neuronal networks to investigate the large-scale dynamics of neuronal populations. This work includes the development of statistical measures for characterizing global dynamics of neuronal networks as well as methods for analyzing the structure of neuronal connectivity patterns. Dr. Sporns is a member of the AAAS, the Society for Neuroscience, the Society for Adaptive Behavior, the Cognitive Neuroscience Society and Sigma Xi. He is a member of the editorial boards of the journals *BioSystems*, *Adaptive Behavior*, the *International Journal of Humanoid Robotics*, and *Neuroinformatics*.

Kenneth Lindsay is at the Department of Mathematics, and **Jay Rosenberg** at the Division of Neuroscience and Biomedical Systems, University of Glasgow, Glasgow, Scotland.

1 Introduction to Modeling in the Neurosciences

George N. Reeke

In this book, 40 distinguished authors explore possibilities for the creation of computational models of neuronal systems that capture biologically important properties of those systems in a realistic way that increases our understanding of how such systems actually work. The authors survey the theoretical basis for believing that such studies might be worthwhile, the kinds of methods that may be employed profitably in practice, and current progress.

This field stands today at a crossroads. While computational models of cognitive function have been constructed at least since the earliest days of electronic computers, progress has been slow. This has been due partly to the inadequate performance until recently of the available computational equipment, and perhaps even more due to the overwhelming acceptance of the key postulate of what has become known as cognitive science: the idea that the brain itself is a kind of computer. Whether the field stays on this path or takes another, while continuing to draw what inspiration it can from the computer metaphor, will largely determine the kinds of results we can expect to see in the coming decades and how fast we get there.

The postulate of the computational brain has been enormously successful in taking us away from the "black box" approach of the behaviorists. It has encouraged us to look inside the brain and analyze what goes on there in terms of the logical conditions that must be satisfied in order to get from the known input to the known output. This has helped us to see, for example, that the sensory affordances spoken of by the Gibsons (1966, 1979) are the beginning, not the end, of the story. Computational science tells us that it can be productive to ask just how it might be that an affordance is transformed into an action. Thinking in computational terms has brought about a great deal of refinement in the kinds of questions that are being asked about the control of behavior and the kinds of evidence that can be brought to bear on these questions.

Nonetheless, for some time, some of us have been saying that this emphasis is holding back our understanding of the brain (Reeke and Edelman, 1988; Bickhard and Terveen, 1996; Poznanski, 2002b). Now this claim, which until recently could be argued only on rather abstract theoretical grounds, has begun to take hold as the inadequacies of the formalistic computational approach have become more evident. Here I will go briefly over this ground, to remind readers of the issues involved, then discuss briefly the kinds of facts we know about the brain that need to be taken into account to arrive at models that truly explain, rather than just emulate, cognitive processes. I will then try to show how the chapters in this book reflect work that does take these facts into account, and indicate where such work might lead to in its maturity.

Just why is it that the computational analogy is inadequate? Surely it is correct that on–off signals (neural spikes) enter the brain via multiple sensory pathways and leave via motor pathways, just as binary signals enter and leave a computer via its input/output devices. Inside, various complex transformations are applied to the input signals in order to arrive at appropriate behavioral outputs. As Marr (1982) famously proposed, these transformations (call them computations if you wish) may be analyzed at the level of their informational requirements (without certain inputs, certain outputs can never be unambiguously obtained), at the level of algorithm (by what steps the appropriate transformations can be efficiently carried out), and at the level of implementation (what kind of devices are needed to carry out the algorithms). These levels of analysis apply also to, indeed, were derived from, processes carried out in computers. Thus, what Marr has proposed is to take what has already been learned about computation, beginning with the work of Turing (1950) and von Neumann, and apply this information to understanding the brain. The problem arises when this analogy is pushed too far, to the point where one falls into the temptation of calling everything the brain does a computation. One thus arrives at the curious circularity of logic espoused by Churchland and Sejnowski (1992, p. 61), who put it this way: "Notice in particular that once we understand more about what sort of computers *nervous systems* [authors' emphasis] are, and how they do whatever it is they do, we shall have an enlarged and deeper understanding of what it is to compute and represent." In this view, the study of the brain, which is already a big enough problem, takes on the additional burden of providing a new underpinning for computer science as well. This is necessary because conventional computer science cannot adequately explain what it is that the brain does. Either computer science must be expanded, as Churchland and Sejnowski suggest, or else we must study the brain on its own terms and stop calling it a computer, as we suggest in this book.

What then, are some of the problems with the computationalist view? First and foremost, the brain is not a programmed, or even a programmable, device. There is no evidence that neuronal circuits are genetically specified and constructed during development to carry out specific algorithmic manipulations of signals, except perhaps in the simplest invertebrate brains, nor can we see how they might be, given our current understanding of the epigenetic influences that preclude perfect realization of genetically specified templates during development. (Although Chomsky [1988] and Pinker [1994] have suggested that humans have genetically specified circuits that provide innate language capabilities, they have not explained what these circuits might be or how DNA could encode the details of their construction.) Similarly, there is no mechanism in view that could explain how the brain could make use of signals derived from behavioral errors to reshape its circuitry in a way specifically directed to correct those errors. Rather, the brain is a product of evolution, constructed of neurons, glia, blood vessels, and many other components in just such a way as to have the right sort of structures to be able to generate adaptive behavior in response to environmental inputs. It has no programs written down by a programmer and entered as instructions on some sort of tape. The mystery that we must try to solve with our modeling efforts, then, is how the brain organizes itself, without the good offices of a programmer, to function in an adaptive manner. In other words, on top of the mystery of what processes the brain might be carrying out, which the computational approach investigates, there is the further mystery of how those processes construct themselves. That mystery is generally ignored by the computational approach or else solved only by biologically unrealistic mechanisms such as back-propagation of error signals (Werbos, 1974; McClelland et al., 1986; Rumelhart et al., 1986) or so-called genetic algorithms (Holland, 1975), which make a caricature of the processes of mutation and recombination that occur during evolution, applying them incorrectly to specifications of procedures rather than of structures.

A second, closely related problem, concerns the question of how neural firings in the brain come to have meaning, that is, how they come to represent objects and actions in the world, and eventually, even abstract concepts such as black holes or Gödel's proof that have no obvious referents in the sensory world around us. In ordinary Turing machine computation, of course, the signals in the machine are assigned interpretations as numbers by the machine's designer, and these numbers are in turn assigned meanings in terms of a particular problem by a programmer. A great deal of work in artificial intelligence has gone into finding ways to free machine programs from their dependence on preassigned symbol definitions and prearranged algorithms to operate upon those symbols. This work has led to systems that can prove mathematical theorems (Wos and McCune, 1991), systems that can optimize their own learning algorithms ("learn how to learn") (Laird et al., 1986), even systems that can to a limited extent answer questions posed in natural human language (Zukerman and Litman, 2001). These systems have demonstrated to everyone's satisfaction that it is possible for a formal logic system, such as a computer, to construct rich webs of symbolic representation and to reason from them to reach conclusions that were not included in their initial programming.

Nonetheless, as has been discussed perhaps most clearly by Bickhard and Terveen (1996), but also by Harnad (1990) and by Reeke and Edelman (1988), the meanings of symbols in these programs depend on derivations from one abstract symbol system to another (and ultimately on derivation from a symbol system provided by a programmer) and never on direct derivation from experience with structures in the real world. In such systems, all possible conclusions that can be deduced by logic from the propositions that are entered initially are potentially available, but new concepts cannot be derived that are directed at ("intentional" with respect to) structures in the world and their transformations. The key problem for our future brain models is how to capture within the representational systems of the brain something of the richness of the interrelationships of objects and events in the real world so that the firing of a neuronal pattern does not just signify something already known, but can invoke new associations and creative formations of concepts that are not derivable by logic within a formal system (Searle, 1984; Johnson, 1987; Lakoff, 1987).

It might be mentioned at this point that some authors have concluded that the only escape from Turing computation and Gödel incompleteness in the brain is an appeal to the uncertainties implicit in events at the submolecular quantum mechanical level (Penrose, 1989; Hameroff et al., 2002). Here, however, we take the view that the ordinary electrical and chemical fluctuations that occur in signaling at the mesoscopic scale of neuronal structures is quite enough to provide the variability necessary for unexpected behaviors to arise. It is beyond our aim here to consider the implications of these ideas for the meaning and reality of free will (Wegner, 2003). Rather, we shall attempt to show that these fluctuations, along with deviations from perfect anatomical regularity, play a fundamental role in providing the substrate for selectional processes that can remove the need for the programmer in a formal computational system.

There is one set of ideas on offer that can provide a bridge between the relational structures of the real world and those of the symbolic logic world and thereby fill in the foundational defect in the computational approach to understanding the brain. These ideas apply to the brain the same principles that Darwin so successfully applied to the origin of complex structure in biological organisms, namely, variation and selection. While proposals along these lines have been advanced by Calvin (1989), Dehaene and Changeux (2000), and others, the most complete formulation is the theory of neuronal group selection first proposed by Edelman in 1978 and spelled out in greater detail in a subsequent series of books (Edelman, 1978, 1987, 1989, 1992).

According to Edelman (my paraphrase), nervous systems are constructed during development out of components and with generic interconnection circuitry chosen by natural selection to satisfy the necessary preconditions to perform their task of generating behavior suitable to the needs of individuals of a particular species. However, these circuits do not embody in finished form all the transformations or algorithms that will be needed by that individual animal in the particular environmental niche in which it will find itself, nor could they, since the optimal configuration is a function of the unpredictable experience-history of the individual. Instead, circuits are constructed with variance around some preselected set of mean structures, thereby providing so-called primary repertoires of circuits that are suitable for the job but unfinished. Then, during life experience, that is, interaction with the environment from hatching or birth on through adulthood, the activity of these circuits generates behavioral outputs that may be more or less appropriate at each moment in time. It is this greater or lesser degree of match with the environment that provides the opportunity for selection to take place. This selection is akin to Darwinian selection during evolution, but it operates during the lifetime of an individual and does not involve the multiplication and death of cells or circuits, but rather the modification of their interconnections among themselves and with motor output pathways such that the more successful circuits in a particular situation come to dominate the behavioral output when a similar situation occurs again later. (For a discussion of whether this process can legitimately be considered a form of selection, see Reeke [2001], which is a commentary on Hull et al., [2001].) In order for this kind of selection to operate, one further postulate is necessary: the animal must have the ability to sense the consequences or adaptive value of its behavior (e.g., by sampling blood glucose levels after feeding), to generate value signals from these senses, and to apply those value signals to the control of the selective process such that outcomes that are positive for the animal lead to an increased influence of the circuits that caused that behavior, whereas negative outcomes lead to a decreased influence of those circuits. This mechanism is akin to reinforcement learning (Friston et al., 1994) and value signals are akin to what are called "error signals" in many theories of learning, but note that value signals are innate, not supplied (except in explicit training situations) by an external teacher. Value signals encode only the *ex post facto* evaluation of behavior and not any specific information relating to errors in one or another particular circuit. Because value is *ex post facto* and derived from chance interactions with the chaotic real world, learning follows a path that cannot be predicted by computation. Over time, representations in the nervous system automatically acquire a veridical relationship with the realities of the world, or else the organism would fail to survive. (In higher animals, play and parental protection must play a role in eliminating behaviors that are potentially harmful or even fatal to the individual early in the selection process.)

We note that this sketch of a theory leaves out many things that ultimately must be investigated by modeling. In particular, the actual minimal requirements for neuronal structures to be capable of undergoing selection to solve particular behavioral problems may be greater, but not less, than those derivable from algorithmic analysis. The problem of credit assignment (Minsky, 1961; Sutton and Barto, 1998) must be addressed by selectional theories of learning as much as by instructional (computational) theories. Here, however, the solution lies in the nature and timing of the value signals and just how they act to modulate selectional events at synapses, rather than in a formal computational analysis.

Supposing we accept that some form of selectional theory provides a correct picture of how the brain learns to work effectively in a particular creature in a particular environment. Nonetheless, selectionism does not specify the brain function at the algorithmic level, but rather at the level of physiological mechanisms, which may be considered analogous to the logic-gate level in computers. It is left to modeling studies, guided by psychological, psychophysical, and neurophysiological data, to sort out what processes (which may *ex post facto* be analyzed as computations) emerge from the working of these low-level mechanisms, and how they construct themselves. What do these ideas suggest to us about how to conduct our modeling studies?

The most important thing they suggest is that many of the messy biological details are likely to be important. It will no longer do to treat neurons as simply integrating on–off switches. The large variety of ion channels found in real neurons are there because they perform functions that natural selection has seen fit to retain. Most likely, these functions involve providing appropriate timings for the decays of postsynaptic and action potentials, for the interactions of dendritic inputs acting at different distances from the cell body, for the proper balance of excitatory and inhibitory inputs, for a variety of potentiation and depression phenomena, and so on. These timings obviously affect the detailed outcome of dendritic integration and of changes in synaptic efficacy and dynamics that are thought to be involved in learning and memory, as demonstrated by the recent discovery of timing-dependent synaptic plasticity (Bi and Poo, 1998).

As an example of a recently explicated system in which detailed dendritic anatomy and signal timing have been shown to play crucial roles, the directionally selective ganglion cells of the retina (as studied in the rabbit) may be cited (Barlow et al., 1964). Simplifying the story greatly, an early computational analysis (Koch et al., 1983) suggested a postsynaptic scheme whereby the ganglion cell is mainly responsible for directional selective responses, computing these responses based on inputs received from bipolar cells and amacrine cells. More recent work making use of two-photon optical recording, however, has shown that individual angular sectors of the roughly planar dendritic trees of so-called starburst amacrine cells make a significant contribution to the overall directional response (Euler et al., 2002). Dendritic calcium concentrations, but not the membrane voltages measured at the somata of these cells, are directionally selective for stimuli that move centrifugally away from the soma in the direction of the particular sector under measurement. These cells are axonless, and their processes, which are called dendrites, combine features of dendrites and axons, including the possession of synaptic output connections. Another recent study (Fried et al., 2002) shows that these dendritic outputs make presynaptic contact with synapses onto ganglion cells, where they deliver inhibitory input in advance of a stimulus progressing in the null direction. There is no asymmetry in the arrangement of these connections that would predict direction selectivity (Jeon et al., 2002). Rather, activity in individual sectors of amacrine cells separately influences ganglion cells of appropriate directional selectivity. These results were anticipated by a detailed cable model of the amacrine cell (Poznanski, 1992), which showed that dendritic responses in this cell can be confined locally, consistent with a role of presynaptic processing in the determination of the responses of the ganglion cells (see also Tukker et al. [2004]). The lesson to be drawn here is that brains are unlike computers in at least this crucial aspect: computers accomplish universal computation with the smallest possible number of different kinds of components, ideally just NAND gates alone, whereas the brain accomplishes more specific computations using whatever components evolution can come up with in any combinations that happen to work. Oversimplified ideas about computation in the brain can lead to serious oversights in understanding even something as relatively simple as the retina.

Going beyond the rather stereotyped sensory system found in the retina, the understanding of how value systems solve the credit assignment problem and how they implement theoretical notions of reinforcement learning (Sutton and Barto, 1998) and temporal difference learning (Sutton and Barto, 1990) will require modelers to distinguish the various properties of the four main modulatory systems in the brain: cholinergic, dopaminergic, serotoninergic, and noradrenergic (Doya, 2002) and eventually also the numerous poorly understood peptidergic systems, as these are the leading candidates to serve as neuronal implementations of value systems.

All of these complexities will need to be realized in large-scale network models complete enough at least to include the major sensory as well as motor pathways involved in simple behaviors. This presents an enormous challenge to the developers of modeling software, in that the basic model neuron must be complex enough to incorporate whichever of these numerous features are being investigated in a particular study, yet simple enough to yield simulation results in a reasonable length of time. Modern parallel multiprocessor computers begin to provide hope of accomplishing these goals, along with new ideas on how to implement integrate-and-fire model neurons with more realistic dynamics than earlier versions (Izhikevich, 2001) and event-driven simulation software that can provide significant speedups relative to fixed-time-step computation (Delorme and Thorpe, 2003; Makino, 2003). With these developments in place, there is no longer any excuse to include features in models that are known to be biologically incorrect, simply because they are computationally more efficient or aesthetically more attractive. Nonetheless, it will still be some time before all the known properties of real neurons can be realized in any one model. When that day comes, perhaps we will begin to see the emergence of some form of machine consciousness. In the meantime, modelers have much important science to do to work out which combinations of the known physiological properties are critical to which aspects of overall nervous system function. A bottom-up style of modeling would appear to provide the best opportunity to identify the minimum combinations of realistic features needed to realize any function of interest. For that reason, the models presented in this book are mostly of the bottom-up flavor.

Therefore, in this book, no abstract "neural network" models of brain function will be found. Instead, the editors have brought together a selection of chapters covering aspects of the problem of finding the relevant and effective mechanisms among all the biological richness of the brain, ranging from patterns of gene interactions (Chapter 2, L.S. Liebovitch, L.A. Shehadeh, and V.K. Jirsa) to the use of robotic devices to study how brain models can interact with real environments (Chapter 23, J.L. Krichmar and G.N. Reeke; Chapter 24, O. Sporns). Taken as a whole, these studies provide an early view of progress in this new style of realistic biologically based neuronal modeling and a possible road map to future developments.

The major emphasis in roughly the first half of the book is on events at the dendritic and synaptic levels. The important role of calcium in intracellular signaling is taken up in Chapter 3 by W.R. Holmes, who discusses how three generations of models have provided increasingly detailed insights into dendritic spine function, and in Chapter 13 by G.D. Smith, who describes methods for measuring and modeling intracellular changes in calcium ion concentration at a level that can encompass highly localized Ca^{2+} fluctuations. Buffered diffusion as well as sequestration and release from internal stores are major features of Ca^{2+} dynamics that are given detailed treatment in these chapters. P. Patil et al. in Chapter 4 discuss the modeling of synaptic responses beginning with the customary analysis in terms of release probability, quantal amplitude and postsynaptic site response amplitude, and their variances, and continuing with more sophisticated newer forms of analysis. Ephaptic (nonsynaptic electrostatic) interactions are covered in Chapter 14 by R. Costalat and B. Delord, who provide a full development of a partial differential equations approach to modeling these phenomena beginning with the basic physics as described by Maxwell's equations. They show how solutions to these equations may be obtained that are relevant to real neural tissue, with applications, for example, to get a better understanding of the role of ephaptic interactions in epilepsy.

Dendritic geometry and cable properties are covered in detail in several chapters. A key question in the treatment of dendritic conduction is the extent to which the level of detail incorporated in multicompartment models (Koch and Segev, 1998) will be necessary to the understanding of dendritic function in complex multicellular networks, or whether, on the other hand, simplified distillations of the results obtained from these detailed studies can be made to serve in large-scale networks. The answers to this question will clearly depend on the particular brain region and cell types being modeled; the reader may begin to discern the considerations that must be taken into account in the treatments given here. These begin with J. van Pelt and H.B.M. Uylings' presentation in Chapter 5 of a dendritic growth model that generates the observed geometry of dendritic branching patterns in various cell types through variation of a few biologically interpretable parameters. Analytical approaches to aspects of signal propagation in both passive (Chapter 6, J.D. Evans; Chapter 7, L.L. Glenn and J.R. Knisley) and active (Chapter 7; Chapter 8, R.R. Poznanski) dendritic cables are presented, with a full coverage of Green's function methods. Methods for dealing with tapered and arbitrarily branching cables are described in detail in these chapters. Evans shows how to model gap junctions as well. Many other topics are included that are not found in most treatments of the subject. In Chapter 9, J. Bell discusses questions that can be formulated as inverse problems in cable theory, such as estimating dendritic spine densities. The great simplification that is possible by constructing equivalent single cables from dendritic trees is described by K.A. Lindsay, J.R. Rosenberg, and G. Tucker in Chapter 10 — unfortunately, only passive trees can be treated in this way. These same authors go on in Chapter 11 to show how the three-dimensional structure of dendritic cables, that is, their nonzero radius, can be taken into account, thus providing a solid basis for appreciating the circumstances in which standard one-dimensional treatments are likely to fail. The curious phenomenon of dendritic bistability and its possible role in the control of neuronal responses is taken up by A. Gutman et al. (Chapter 16). Axonal dynamics, particularly the role of calcium and potassium kinetics in the propagation (or failure of propagation) of impulses across unmyelinated axonal branch points is described by M.D. Goldfinger (Chapter 18).

The treatment then moves on to the coverage of the dynamics of single cell responses. Detailed models of two specific cell types are presented: retinal cells in Chapter 12 by Y. Kamiyama et al., and neocortical pyramical cells in Chapter 15 by R.D. Orpwood. Orpwood presents both a detailed multicompartment model and a much simplified, but also much faster, model, with the aim of using the comparison to illustrate a principled approach to finding the appropriate level of simplification for a given problem. Analysis of the nonlinear dynamics of Hodgkin–Huxley spike-train generation is covered by S. Sato et al., who, in Chapter 17, apply bifurcation theory to illuminate the differing behavior and stability of neurons in different parameter regimes. As explained earlier and as is increasingly evident from the results presented in these chapters, realistic models need to be based on spiking cells rather than the rate-coded cells of an earlier generation of models. This book, and these chapters in particular, provides the necessary background for understanding and undertaking this newer style of modeling.

Techniques for the simulation of network systems without appeal to compartmental modeling are presented by Poznanski in Chapter 19. D. Halliday in Chapter 20 shows how to apply linear and nonlinear correlation analysis to find relationships between spike trains from multiple cells in both the time and frequency domains. This sets the stage for a discussion of "The Poetics of Tremor" by G.P. Moore and H. Bronte-Stewart (Chapter 21), who apply similar techniques at the behavioral level to analyze data on tremor motion from Parkinson's disease patients and others. Their work helps us understand why it is so important to consider normal and aberrant network activity beyond its purely computational function in the context of the behaving organism. O. Sporns in Chapter 22 shows how concepts derived from graph theory and information theory can be used to characterize network activity in terms of the integration of activity in segregated subsystems to generate highly complex patterns of behavior. He summarizes recent studies that he has carried out with G.M. Edelman and G. Tononi into the structural basis for informationally complex behavior in network systems, which have led to a theory of how an ever-changing "dynamic core" can provide the neuronal basis for conscious awareness in higher brains. This chapter brings the book back in full circle to the genetic mechanisms discussed in the second chapter. The results of these studies, as well as the success of the synthetic and robotic models discussed in the last two chapters by J.L. Krichmar and Reeke and by Sporns, which use realistic neuronal modeling principles to achieve learning in artificial systems without conventional programming, shed a great deal of light on the mystery of why evolution might have settled on the particular kinds of areal structures, interconnections, and dynamics found in higher brains.

2 Patterns of Genetic Interactions: Analysis of mRNA Levels from cDNA Microarrays

Larry S. Liebovitch, Lina A. Shehadeh, and Viktor K. Jirsa

CONTENTS

2.1 INTRODUCTION

2.1.1 Biological Organisms are Complex Systems

A cell in a biological organism consists of thousands of molecules that are used in various complicated biochemical processes. While a cell performs its individual functions it is also involved with other cells in the overall function of the organism. Therefore, each cell in a biological organism is a component with individual and collective properties that are essential for the survival of the system.

A biological system is a robust dynamical system that grows, evolves, and adapts to environmental changes. In order to maintain this complex system, cells in a biological organism need to interact with each other on many levels, from gene networks to brain networks (Bota et al., 2003). This chapter presents interactions at the basic genetic level.

2.1.2 Genes Interact with Each Other

Genetic information is stored as DNA (deoxyribonucleic acid). DNA is very stable as a storage molecule, a fact that is crucial because genetic information must be preserved from one generation to the next. However, DNA is also a quite passive molecule, in that it cannot easily enhance biochemical reactions between other molecules. For the information stored within the DNA to be released, the DNA must be transferred into a molecular form with a more active biochemical role. This is what is meant by "gene expression." The first step in gene expression is the production of RNA (ribonucleic acid) molecules by the process of transcription, which takes place in a cell's nucleus (in eukaryotes). RNA is chemically similar to DNA, except that it contains ribose instead of deoxyribose sugar and the base uracil (U) instead of thymine (T). RNA can perform various structural and catalytic functions within the cell, and accordingly, several types of RNA are found within a cell. The main categories are messenger RNA (mRNA), ribosomal RNA (rRNA), and transfer RNA (tRNA). Although RNA can carry out many functions, it lacks the chemical complexity that is needed for most of the jobs that keep a cell alive. Instead, mRNA often serves as a template for protein synthesis in a process called translation, which takes place in the cytoplasm. Proteins are a much more versatile kind of molecule because they can form more complex three-dimensional structures than DNA or RNA. Proteins carry out most of the catalytic and structural tasks within an organism. Thus gene expression, which ultimately controls the proteins present in a cell, is the principal determinant of an organism's nature and identity.

It was originally thought that gene expression was only a one-way street: that DNA is transcribed into RNA and RNA is translated into proteins. That paradigm was called "The Central Dogma." But we now know that living systems are much more dynamical systems where feedback plays an essential role in development and maintenance. Proteins, called transcription factors, bind to DNA and thereby regulate its expression. These proteins can modulate the transcription of their source genes into mRNA, as well as the expression of other genes, in either positive or negative feedback loops. Positive feedback provides a means for either maintaining or inducing increased gene expression. By contrast, negative feedback allows for a response that switches a gene off or reduces its expression. The switching on and off of genes, that is, inducing expression versus inhibiting expression, is a central determinant of how organisms develop. Stated simplistically, genes interact in the following way: from genes to mRNA to proteins and back to genes.

Gene expression can be measured, approximately, by the amount of mRNA transcribed. This can be analyzed using DNA microarray technology (Schena et al., 1995), which measures the mRNA and thus the expression profile of thousands of genes simultaneously. These experiments generate a flood of biological data in the form of mRNA levels. These mRNA levels are affected by systematic errors introduced by the experimental procedure (Alter et al., 2000; Miller et al., 2002), and by nonsystematic errors due to the inherent biological variability of cells (Elowitz et al., 2002; Ozbudak et al., 2002; Blake et al., 2003). Normalization methods are frequently applied to correct for systematic errors (Tseng et al., 2001; Workman et al., 2002; Yang et al., 2002, 2003a), but the cause of the corrected biases is not yet fully explored. Recent studies show that the procedure for printing the probes on the microarray can result in systematic errors in the gene-expression measurements (Balázsi et al., 2003).

There is no strict one-to-one correspondence of mRNA to protein. There is some posttranscriptional processing of mRNA, and processed mRNAs can be involved in some biochemical reactions, such as that occuring in ribosomes and in self-splicing. Further, mRNA levels can be regulated by degradation and silencing. Therefore, mRNA levels alone do not tell the full story of gene expression. Recent studies in proteomics suggest that a more complete estimate of gene expression should take into consideration the amount of proteins translated, inasmuch as mRNAs may be spliced and prevented from being translated to proteins. Therefore, the study of gene networks requires mRNA as well as protein expression information (Hatzimanikatis and Lee, 1999). Currently, proteomic analysis is developing and the technology is becoming available in the field of neuroscience (Husi and Grant, 2001). For example, recently Giot et al. (2003) have presented a protein interaction map for 4780 interactions between 4679 proteins in *Drosophila melanogaster*. They found that the *Drosophila* protein network is a small-world network that displays two levels of organization: local connectivity, representing interactions between protein complexes, and global connectivity, representing higher-order connectivity between complexes. Connectivity in this context refers to proteins that react chemically with each other. Therefore, at present, until proteomics technology is more fully developed, microarray technology using mRNA levels, despite its limitations, is still a useful way to study overall genetic interactions.

2.2 Genetic Interaction has a Special Nature

Regulated gene expression underlies cellular form and function in both development and adult life. An understanding of gene regulation is therefore fundamental to all aspects of biology. Yet, at present, our knowledge of gene regulation is still vague and the existing molecular manipulation techniques such as gene knockouts have proven that the system under study is far more complex and dynamic than had been previously assumed.

In order to understand gene regulation, we need to understand the nature of genetic interaction. From a network perspective, genes interact in a network where the nodes are genes and the links are regulatory connections between genes. A key question then arises: What is the topology of this genetic network? Is it a random network, a scale-free network, a small-world network, or a network of some other topology?

2.2.1 Network Topologies

Networks of genetic interactions, or of electric power lines, or of social interactions between people, can be pictured as nodes with connections between the nodes. For the network of genetic interactions, the nodes are genes and the connections between them are the positive and negative feedback interactions of the protein transcription factors that regulate gene expression. The structure of the network can be characterized in different ways. One important way is to determine how many nodes, $g(k)$, have connections to k other nodes. The function $g(k)$, also called the degree distribution, characterizes the connectivity pattern of a network. That is, based on the number of input or output connections of the nodes, a characteristic distribution is formed. Figure 2.1 shows the degree distribution of a random network compared with the degree distribution of a scale-free network. Both models are explained in the following sections.

A random network is a network in which each node has an equal probability of connecting to any other node. Erdos and Renyi (1959) defined a random network as a graph with N labeled nodes connected by n edges chosen randomly from the $N(N-1)/2$ possible edges. In random networks, the degree of a node follows a binomial distribution

$$P(k_i = k) = \binom{N-1}{k} p^k (1-p)^{N-1-k}, \tag{2.1}$$

where P is the probability that a given node has a connectivity k. For large N, the binomial distribution becomes a Poisson distribution,

$$P(k) \approx \mathrm{e}^{-pN} \frac{(pN)^k}{k!} = \mathrm{e}^{-\langle k \rangle} \frac{\langle k \rangle^k}{k!}. \tag{2.2}$$

A scale-free network obeys a power-law degree distribution, $g(k) = Ak^{-a}$, where $g(k)$ nodes have k links (Barabási and Albert, 1999). A scale-free network has a hierarchy of connections that is self-similar across different scales or levels of structure. In such a network, structures extend over a wide range of scales. A scale-free network is formed when new connections are added preferentially to nodes that already have many connections or when connections are broken and rejoined (Huberman and Adamic, 1999; Albert and Barabási, 2000; Bhan et al., 2002; Jeong et al., 2003). Many real-world networks are scale-free. These include the World Wide Web connected by URLs (Albert et al., 1999; Kumar et al., 1999; Broder et al., 2000), actors linked by the number of movies in which they

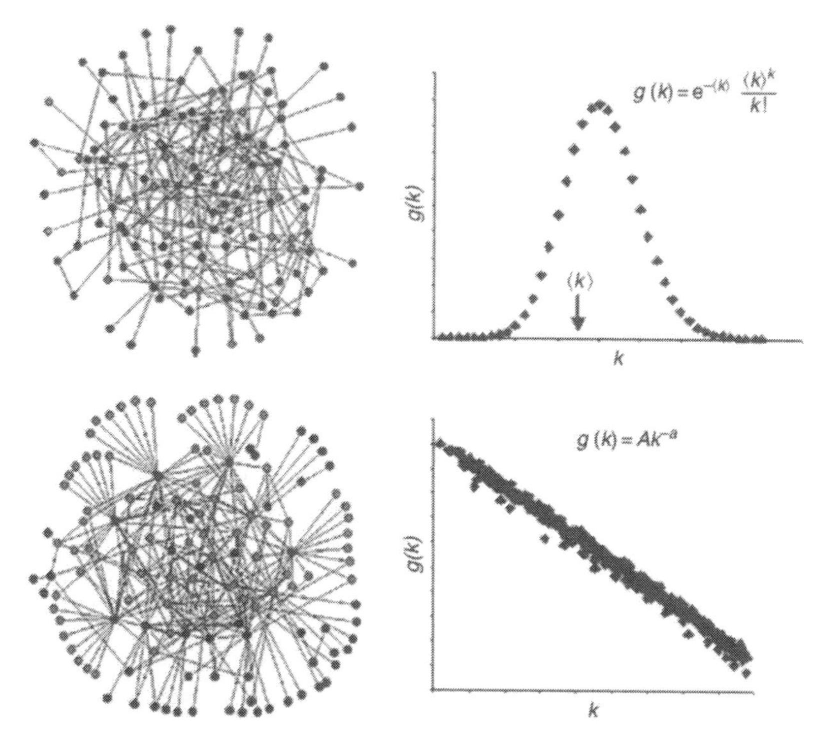

FIGURE 2.1 Degree distributions of (top) a random network and (bottom) a scale-free network. A random network generates a Poisson degree distribution. A scale-free network forms a power-law degree distribution, which appears linear on a log–log scale. (*Source:* From Barabási, A.L., *Nature* 406, 378, 2000. With permission.)

have appeared together (Barabási and Albert, 1999), substrates in metabolic pathways in 43 different organisms linked by the number of reaction pathways that separate them (Jeong et al., 2000), and proteins in *Saccharomyces cerevisiae* linked if they are found to be bound together (Jeong et al., 2001).

A small-world network is one in which few links are needed to go from one node to any other node. Two important properties of small-world networks are short characteristic path lengths and high clustering coefficients (Albert and Barabási, 2002). The average path length, l, of a network is defined as the number of edges in the shortest path between two nodes, averaged over all pairs of nodes (Watts and Strogatz, 1998). Given the same number of nodes, N, and the same average number of connections, $\langle k \rangle$, both a random network and a small-world network have a similarly small average path length, l_{path}. The clustering coefficient, C_i, of a node i that has k_i edges which connect it to k_i other nodes, is equal to the ratio between the number E_i of edges that actually exist between these k_i nodes and the total number of all possible edges $k_i(k_i - 1)/2$,

$$C_i = \frac{2E_i}{k_i(k_i - 1)}. \tag{2.3}$$

The clustering coefficient of the whole network is the average of the C_is from all the nodes (Watts and Strogatz, 1998). Given the same number of nodes, N, and average number of connections, $\langle k \rangle$, a small-world network has a higher clustering coefficient than a random network. Many real-world networks are small world. These include the neurons in the nematode worm *Caenorhabditis elegans* linked by their synapses or gap junctions, and the substations of the electrical power grid in the western United States linked by their transmission lines (Watts and Strogatz, 1998). Studies show that most scale-free networks also display a high clustering property that is independent of the number of nodes in the network (Ravasz and Barabási, 2003).

2.2.2 Genomic Networks

There is no general method for determining the topology of the network underlying the genome, that is, the network of genetic regulation between the genes. Different groups have explored different mathematical models of genetic networks and different statistical methods to analyze mRNA or protein experimental data (Smolen et al., 2000a).

Attempts to reconstruct models for gene-regulatory networks on a global, genome-wide scale have been made using ideas from genetic algorithms (Wahde and Hertz, 2000), neural networks (D'haeseleer et al., 2000), and Bayesian models (Hartemink et al., 2001). Most recently, researchers have used linear models and singular value decomposition (SVD) (D'haeseleer et al., 1999; Alter et al., 2000; Raychaudhuri et al., 2000; Dewey and Galas, 2001; Holter et al., 2001; Bhan et al., 2002; Yeung et al., 2002), to reverse-engineer the architecture of genetic networks. These models try to find the strength of the connections from one gene to another. SVD (a form of principal components analysis or Karhunen–Loeve decomposition) is a statistical method for determining the combinations of variables that contribute most to the variance of the data set (Press et al., 1992). Such an approach typically allows for an enormous dimension reduction that simplifies the analysis and visualization of the multidimensional data set. It has been frequently applied to microarray expression data. The mRNA levels of some genes go up or down together when measured under different experimental conditions. In principle, it should be possible to use these data to determine an interaction matrix that would identify which genes influence which other genes. However, the inevitable noise present in the experimental data makes this very difficult in practice. The SVD approach picks out the connections between the genes that are more significant than the noise. The advantage of this method is that it can determine the most significant connections. The disadvantage is that the number of connections found to be significant depends on how the level of significance is set. Moreover, although this method can find a few strong connections, it ignores the effects of many weak connections, whose collective effect may be important in the overall global network.

Quantitative models of genetic regulatory networks have also used Boolean networks, dynamical systems, and hybrid approaches. In Boolean models the state of each gene is either on (1) or off (0), as opposed to the models described in the previous paragraph where the strengths of the connections can take on any real values (Kauffman, 1969, 1974, 1984; Thomas and d'Ari, 1990; Boden, 1997; Liang et al., 1998; Thomas et al., 1999). In his *NK* model, Stuart Kauffman (1993) uses a random Boolean network of N nodes and K edges to model the regulatory genetic systems of cellular organisms. He shows that the system evolves to reach a state where each gene receives only a small number of inputs from other genes. At this small number of inputs, which for some models occurs at $K \sim 2$, the genetic system is at the edge of chaos (on the border between regular and random behavior), ensuring stability and the potential for progressive evolutionary improvements. Boolean models have been useful in understanding the role of feedback loops and multiple stable states of gene expression. Most of the analytic tools used in such studies use some kind of (linear or nonlinear) measure in an appropriate space to quantify the similarity between genes in terms of their expression profiles. The similarity measure compares the mRNA levels of every pair of genes in the data to quantify what is called the distance between two genes. The distance values are the entries in the distance matrix. For a given data set, each gene is described and quantified by its mRNA values (x_i) where i distinguishes either points in time or different environmental or biological conditions, such as different types or stages of cancer. All the mRNA levels of a gene are cast into a vector $X = (\ldots, x_i, \ldots)$. For any two genes, $X = (\ldots, x_i, \ldots)$ and $Y = (\ldots, y_i, \ldots)$, a commonly used similarity measure is the Euclidean distance (Michaels et al., 1998; Wen et al., 1998), which is defined as

$$d_{\mathrm{E}} = \sqrt{\sum_{i=1}^{N} (x_i - y_i)^2}. \tag{2.4}$$

Eisen et al. (1998) used another similarity measure, which captures the similarity by means of the correlation coefficient. Here the similarity score can be calculated as follows:

$$S(X,Y) = \frac{1}{N} \sum_{i=1,N} \left(\frac{X_i - X_{\text{offset}}}{\Phi_X} \right) \left(\frac{Y_i - Y_{\text{offset}}}{\Phi_Y} \right), \tag{2.5}$$

where

$$\Phi_X = \sqrt{\sum_{i=1}^{N} \frac{(x_i - x_{\text{offset}})^2}{N}} \quad \text{and} \quad \Phi_Y = \sqrt{\sum_{i=1}^{N} \frac{(y_i - y_{\text{offset}})^2}{N}}, \tag{2.6}$$

and X_{offset} and Y_{offset} are reference states of mRNA levels. When X_{offset} and Y_{offset} are set to the means of the observations, then Φ_X and Φ_Y are the standard deviations, where N is the total number of observations of a gene and $S(X, Y)$ is equal to the Pearson correlation coefficient of the observations of genes X and Y.

Dynamical systems models use real values for the strengths of the connections between the genes and are based on ordinary differential equations to describe the rates of change of the concentrations of gene products — mRNAs and proteins. These models have different mathematical properties than the Boolean models, where the strengths are only 0 and 1. The 0 and 1 thresholds introduce nonlinearities that change the system's dynamical behavior. For example, stable oscillations and steady states produced by the Boolean model may not be observed in a more realistic model that describes gene-expression rates as real variables and that can also include stochastic fluctuations in the numbers of molecules (Glass and Kauffman, 1973; Smith, 1987; Mahaffy et al., 1992; Bagley and Glass, 1996; Hlavacek and Savageau, 1996; Mestl et al., 1996; Arkin et al., 1998; Leloup and Goldberger, 1998; Gardner et al., 2000; Smolen et al., 2000b; Davidson et al., 2002). On the other hand, computer integration of the ordinary differential equation systems' models requires short time steps and therefore these simulations can require much more computer time.

Hybrid models have also been developed that combine both the Boolean and dynamical systems approaches (McAdams and Shapiro, 1995; Kerszberg and Changeux, 1998; Yuh et al., 1998). These models can help distinguish between activation or repression functions that can be modeled as logical switches versus activation or repression over a broad range of an effector concentration.

In recent studies, researchers have attempted to break down networks into their building blocks. These building blocks are called motifs, repeating significant patterns of interconnections. Networks such as transcriptional regulatory networks in the bacterium *Escherichia coli* and the yeast *S. cerevisiae* seem to have recurring subgraphs that may serve as building blocks for constructing the network and that may have specific functional roles in the network (Lee et al., 2002; Milo et al., 2002; Shen-Orr et al., 2002; Mangan and Alon, 2003; Mangan et al., 2003). This local network approach is now being used to analyze not only genetic but also protein regulatory networks, social insect networks, and communication in neural networks (Alon, 2003; Bray, 2003; Fewell, 2003; Laughlin and Sejnowski, 2003; McAdams and Shapiro, 2003).

Most of the previous studies have focused on the detailed behavior of individual genes rather than on the structure and global properties of all the genes interacting within the genome. What is the overall pattern of these genetic interactions? Perhaps genetic networks share common features with other natural networks, or, to be more specific, genetic networks may share structural features with other biological networks such as neural networks. We test the hypothesis that the global properties of genetic interactions can be characterized by the main network topologies described above. In essence, this is a top–down rather than a bottom–up approach, meaning comparison is done between behavioral phenotype and genotype in an inductive rather than deductive way using autoassociative neural networks of the Hopfield type (e.g., see Liebovitch et al., 1994; Liebovitch and Zochowski, 1997; Rolls and Stringer, 2000). In comparing gene networks with neural networks it is important to mention that behavioral phenotype is not determined by the genotype alone; interactions with nongenetic factors are often both numerous and important (Berkowitz, 1996). Furthermore, genes and behavior represent distant levels of organization and a correlation between the two may provide little understanding of the causative processes involved.

2.3 Genetic Networks in Relation to Neural Networks

In this chapter we have focused on understanding the characteristics of the pattern of gene regulation. We now describe how the methods and models used to study genetic regulatory networks are related to studying biological neural networks.

Many different systems, such as the Internet, the electrical power grid, the social connections between people, airplane route maps, genetic regulatory networks, biological neural networks, and many others, are all networks. That is, they all consist of nodes and connections between those nodes. As eloquently expressed by Barabási in his book *Linked* (Barabási, 2002), we have spent much more time in science tearing apart nature to look at the details of the pieces rather than trying to put all the pieces together to understand how they function together as a whole. Although many different physical, chemical, or biological components can form the nodes and connections of different networks, these very different networks can still share important features in their structure, dynamics, and how they were constructed. Thus, genetic regulatory networks share many common features with other networks and many of the issues we have dealt with here apply also to those other networks.

The specific issue that we have reviewed here about genetic regulatory networks is how to determine the connections between genes from the activity at each node in the network, namely, the mRNA abundance expressed by each gene. This analysis, to determine the connections in a network from the activity at the nodes, is a very general problem in the study of networks and has applicability to many different kinds of networks. For example, the Internet has grown by the independent addition of many connections and computers. Its growth is much more like that of a living organism than a centrally planned engineering project, and there is no blueprint or map of it. We must use experimental techniques to probe the Internet to determine how it is connected together. Such studies are now called Internet tomography (Coates et al., 2002).

Similarly, an essential issue in the analysis of neuronal networks is to understand how we can determine the functional connections between neurons from measurements, such as electrophysiological measurements, at each neuron. The methods that we have reviewed here, which have been used to try to determine the connectivity between different genes from the mRNA expression level at each gene, can be directly applied to try to determine the functional connections between neurons from the electrophysiological activity (e.g., spike rate) at each neuron.

Understanding genetic regulatory networks depends on understanding how the dynamics of such a network depend on the pattern of connections between the genes. Similarly, understanding biological neural networks depends on understanding how the dynamics of those networks depend on the patterns of connections between neurons. Thus, the different models presented here for genetic regulatory networks are analogous to models that have been studied for biological neural networks. We now further explore the relationships between these two types of models.

Both our approach in modeling genetic networks as well as the approach in modeling neural networks by Olaf Sporns' group (Sporns et al., 2000a; Sporns and Tononi, 2002; Sporns, 2004) use a statistically based method with an overall goal of understanding aspects of the relationship between connectivity and dynamics. Our model is structurally equivalent to Sporns' model except for the fact that Sporns introduces a transfer function that is a unimodal saturating nonlinearity such as a hyperbolic tangent. In our model, we introduce the nonlinearity by renormalizing the mRNA vector as described in the later sections to follow. Sporns includes Gaussian white noise in his model system and characterizes the dynamics of the network by means of a covariance matrix, which he interprets as a measure of functional connectivity. For linear systems, the covariance matrix COV can be computed analytically from the connection matrix $\mathbf{C}_{ij}$. Let $\mathbf{A}$ be a vector of random variables that represents the activity of the units of a given system X subject to uncorrelated noise $\mathbf{R}$ of unit magnitude. Then $\mathbf{A} = \mathbf{C}_{ij} \times \mathbf{A} + \mathbf{R}$ when the elements settle under stationary conditions. Substituting $\mathbf{Q} = [\mathbf{1} - \mathbf{C}_{ij}]^{-1}$ and averaging over the states produced by successive $\mathbf{R}$ values (denoted by $\langle\ \rangle$), the covariance matrix becomes

$$\mathrm{COV} = \langle \mathbf{A}^{\mathrm{T}} \times \mathbf{A} \rangle = \langle \mathbf{Q}^{\mathrm{T}} \times \mathbf{R}^{\mathrm{T}} \times \mathbf{R} \times \mathbf{Q} \rangle = \mathbf{Q}^{\mathrm{T}} \times \mathbf{Q}, \tag{2.7}$$

with ^{T} denoting the matrix transpose. The covariance matrix captures the second-order correlations among distinct neuronal sites and may be interpreted as *functional connectivity* (Sporns et al., 2000a; Sporns, Chapter 22, this volume). Such functional connectivity may be put in juxtaposition to the underlying anatomical connectivity of the neural networks, thereby addressing one of the oldest questions in biology, the nature of the structure–function relationship. The anatomical connection topology is referred to as *structural connectivity* and may be directly compared to the covariance matrices capturing functional connectivity. Sporns uses connection matrices with different topologies such as a random synaptic weight distribution, as well as distributions optimized for global statistical measures such as entropy or complexity. For a given system X, the entropy $H(X)$ is a measure of the system's overall degree of statistical independence. In the Gaussian case,

$$H(X) = 0.5 \cdot \ln((2\pi e)^n |\mathrm{COV}(X)|), \tag{2.8}$$

with $|\cdot|$ indicating the matrix determinant. The integration, $I(X)$, is a measure of a system's overall deviation from statistical independence, that is

$$I(X) = \sum_i H(x_i) - H(X). \tag{2.9}$$

Complexity, $C_N(X)$, captures the extent to which a system is both functionally segregated (small subsets of the system behave independently) and functionally integrated (subsets behave coherently), that is

$$C_N(X) = \sum_k \left[\frac{k}{n} I(X) - \left\langle I\left(X_j^k\right) \right\rangle \right], \tag{2.10}$$

where k stands for the subset size and n is the total number of subsets.

The corresponding network dynamics and hence functional connectivity is solved computationally and shows characteristic differences as a function of the underlying structural connectivity. In particular, connectivity structures optimized for complexity show the greatest amount of clustering in the functional connectivity. This implies that such networks have the largest capability to be, at the same time, functionally segregated and functionally integrated. Rather than using statistical covariance as a descriptor of the network dynamics, we choose to describe the statistical features of the network dynamics by means of probability density functions (PDFs). The PDFs, though non-uniquely, capture a system's connection topology and are sensitive to some connectivity parameters such as the degree of clustering of a system and the average number of connections in a system.

While Sporns' neural networks reveal "small-world" architecture (small characteristic path length and high clustering index) but no scale-free topology, our genetic networks show both small-world and scale-free topologies.

2.4 INTEGRATIVE MODELING APPROACH

Much work has been done to understand individual biochemical reactions and how these reactions form modules, that is, clusters of related chemical pathways with a well-defined physiological function. Now we want to understand the overall pattern of all these reactions and try to make sense of what the structure looks like when the modules are all connected to each other.

We use ideas from the study of other types of networks to understand the topology of genetic networks that can be deduced from experimental data. This approach may help us understand the biological structure and function of genetic networks and may have utility in clinical applications (Shehadeh et al., submitted).

First, we formulate five different types of linear models of genetic networks. These models include both random and hierarchical (small-world) models, with homogeneous, heterogeneous, and symmetric connection topologies referring to the input and output regulatory connections between genes.

Then we compute the steady-state mRNA levels from each model and evaluate the PDF, that is, the probability that the mRNA levels, x, are between x and $x + dx$, where x is the mRNA concentration and dx is the differential of mRNA concentration.

We form a dictionary whose entries are the PDFs of the mRNA expression levels of the different connectivity models. We do not claim that these PDFs are unique to the connectivity models.

Next, we determine the PDFs of many sets of mRNA expression levels from several cDNA microarray databases.

We then read our dictionary backwards, starting from the PDFs of the theoretical mRNA levels, to find the model of genetic interactions that best captures the observed statistics of the experimental data.

The strength of our approach lies in the fact that the assumptions and structures of these models can be falsified with experimental cDNA microarray data. Interestingly, our models raise specific and experimentally addressable questions that can be valuable in planning future microarray experiments. Thus, understanding a biological system becomes an integrative, iterative process, with our models feeding experimental design and the experimental data in turn feeding our models.

Our network models include global interactions as well as local repetitive interactions or motifs, connecting small parts of the network into an organized system. The integrative nature of these models come from the fact that the global properties of the network are assembled from the small-scale motifs. In the scale-free models described in Section 2.4.1, the structure of the motifs continues up to similar and larger structures at higher scales in a statistically self-similar cascade.

2.4.1 The General Model

Our genetic regulatory network model is an $N \times N$ connectivity matrix with N genes. Each element in the matrix, M_{ij}, represents the influence of the jth gene on the ith gene. $M_{ij} = 0$ if gene j has no regulatory influence on gene i and $M_{ij} \neq 0$ if gene j regulates gene i. The number of edges at a given node (gene) is referred to as the node degree. As before, the spread in the node degrees is characterized by a distribution function, $g(k)$. For example, if every gene in the network influences exactly six other genes, then $g(k) = N$ for $k = 6$, and $g(k) = 0$ for all other k. If a small number of genes regulate all the other genes, then the degree distribution of the network may be a power-law distribution where $g(k) = Ak^{-a}$, characteristic of a scale-free network. The degree distribution $g(k)$ may refer to the regulatory input from the genes, the regulatory output of the genes, or both. The degree distribution of the network, $g(k)$, characterizes the structure or topology of the network.

Having an $N \times N$ connectivity matrix M means that we limit our model to linear interactions between the genes. In a realistic interaction matrix, the nonzero elements should be either positive, representing up-regulation, or negative, representing down-regulation or inhibition. However, interaction matrices with both positive and negative values can cause a wide range of complex behaviors, including steady-state, periodic, and quasi-periodic dynamical behavior. Therefore, we are faced with a trade-off between studying less complex models that we can successfully analyze and more complex models that would be much more difficult to analyze.

In our initial studies, we therefore restrict the connectivity matrices to stimulatory or zero regulation where $M_{ij} \geq 0$ so that the interaction between the genes leads to mRNA levels that will always reach a steady state with a characteristic PDF. This function will be the goal of our analysis and can serve as a starting point for more complex models. We also consider two general kinds of matrices:

Markovian matrix. If an element M_{ij} in the matrix represents the probability of switching from the expression level of gene j to the expression level of gene i, then the matrix can be thought of as a transition matrix with Markovian probabilities (Papoulis, 1991) where $0 \leq M_{ij} \leq 1$ and the sum over i of $M_{ij} = 1$, that is, the sum of the elements, M_{ij}, in each column is equal to 1. Restricting the matrix to the well-known Markovian properties simplifies the computation of this model and helps predict the behavior of the generated mRNA levels. For example, this Markovian formulation guarantees that the transition matrix M will always have an eigenvalue of 1 and a corresponding stationary eigenvector. On the other hand, with this Markovian formulation, the model cannot include inhibition since all $M_{ij} \geq 0$. Also, with the Markovian constraints on M, we cannot independently set the weights of the elements, M_{ij}, and the degree distribution of the genetic network with $g(k)$ genes influencing k other genes. For example, since the elements in each column sum to 1, if there are m_j equal nonzero elements of M_{ij} in column j, then each nonzero element $M_{ij} = 1/m_j$. If there are different numbers of nonzero elements in different columns, then $1/m_j$, will be different in each column j and so the input from each gene cannot have the same strength.

Non-Markovian matrix. Since the Markovian formulation puts constraints on modeling arbitrary genetic interactions, we used a non-Markovian formulations for the matrices in four of the five models explained in Section 2.4.2. In these cases, we can freely choose the desired strength of the connections between the genes and represent different distributions of genetic interactions $g(k)$. The main advantage of the non-Markovian formulation is that we can model much more complex genetic networks. The disadvantage is that there is no bound on the generated mRNA expression levels; they may grow or decay exponentially as they evolve in time. However, this problem can be solved by renormalizing the mRNA levels at each time step.

2.4.2 Five Models of Genetic Networks

Considering the three main network topologies, we formulate five different models of genetic networks. In the first two models, genes interact randomly. In the last three models, genes interact in a scale-free and small-world pattern. The name of each model refers to the pattern of input regulation from other genes, output regulation of genes, or both. In the first four models, the pattern of input regulation into the genes differs from the pattern of output regulation of the genes. In the last model the patterns of input and output regulation are the same.

Model A1. Random output. The output of each gene regulates six other genes randomly chosen from the network. Each gene is therefore regulated by the input of an average of six other genes randomly placed in the network. Among the five models, this is the only matrix with Markovian properties, which means that the elements in each column sum to 1. Hence, each column has six nonzero values of $M_{ij} = \frac{1}{6}$ and all other $M_{ij} = 0$. The distribution of genetic interactions is the single-point distribution $g(k) = N$ for $k = 6$, and $g(k) = 0$ for all other k.

Model A2. Random input. Each gene is regulated by the input of six other genes randomly placed in the network. The output of each gene regulates an average of six other genes randomly chosen from the network. The elements in each row sum to 1. Hence, each row has six nonzero values of $M_{ij} = \frac{1}{6}$ and all other $M_{ij} = 0$. The distribution of genetic interactions is the single-point distribution $g(k) = N$ for $k = 6$, and $g(k) = 0$ for all other k.

Model B. Small-world homogeneous input. The rows in this matrix are similar because the pattern of input regulation into each gene is the same. Each gene i receives few inputs from the genes with low j, more inputs from the middle range of j, and many more inputs from the genes with highest j. The output pattern from each gene is a scale-free distribution, $g(k) = Ak^{-a}$, with the higher order genes regulating the entire network. The weights of the regulatory connections are the same, with $M_{ij} = 1$ for coupled ij pairs and $M_{ij} = 0$ for all other ij pairs.

Model C. Small-world heterogeneous input. The rows in this matrix are different but the columns are similar. The input pattern into each gene is a scale-free distribution, $g(k) = Ak^{-a}$, with the higher numbered genes regulated by all the other genes. The output pattern from each gene is the same. Each gene j regulates few genes with low i, more middle i genes, and many more genes with high i. The weights of the regulatory connections are the same, with $M_{ij} = 1$ for coupled ij pairs and $M_{ij} = 0$ for all other ij pairs. The matrix M of this model is the transpose of the matrix in the homogeneous input model B. The transpose, M^{T}, is formed by interchanging the rows and the columns. Therefore, both models have scale-free structure but in reversed direction.

Model D. Small-world symmetric input/output. The regulatory connections between genes are undirected. The input patterns into genes and the output patterns from genes are the same with a scale-free distribution, $g(k) = Ak^{-a}$. Here again, the weights of the regulatory connections are the same with $M_{ij} = 1$ for coupled ij pairs and $M_{ij} = 0$ for all other pairs.

In designing the power-law distributions, $g(k) = Ak^{-a}$, in models B, C, and D, we vary the slope, a, of the distribution from 1 to 6. We use a density function to choose the matrix elements, M_{ij}, to have the desired $g(k)$ (Shehadeh et al., submitted).

The connectivity matrices from these five models are presented in Figure 2.2. The nonzero elements are black dots (each of which is larger than the resolution of one row or column). The number of genes, N, in this figure is 1000, as compared, for example, to 13,600 genes in the fruitfly *D. melanogaster* and about 30,000 genes in humans. The results from the models are not sensitive to the number of genes, N, as we have seen by varying N logarithmically from 10 to 1000. In Figure 2.2, each model has exactly a total of 6000 connections. Therefore, each gene has an average of $\langle k \rangle = 6$ connections. We normalize the average connections to $\langle k \rangle = 6$ in approximate agreement with the popular notion of "6 degrees of separation" in social networks (Milgram, 1967).

In addition to being scale-free, models B, C, and D have small-world properties. These three models have high clustering coefficients, C, and short path lengths, l_{path}, when compared to random models that have the same number of nodes, N, and the same average number of connections, $\langle k \rangle$.

Genes and their corresponding proteins interact closely (in short paths) with other genes and proteins (local neighbors) to perform a local function for their subsystem. However, these subsystems may also need to interact with other less related subsystems to perform a global function for the whole biological system. In this sense, these subsystems are clusters that are connected in a small-world fashion. Also, many biological molecules perform more than one role, for example, phospholipids in the cell membrane, when hydrolyzed, can serve as "second messengers" to control other functions in the cell, such as the release of calcium. Thus, molecules that serve as structural elements also double as carriers of information to control cellular functions (Kandel et al., 2000). These separate functions are analogous to the long-range connections in the small-world models that serve to knit together seemingly different biological functions into more globally connected units.

2.4.3 Each Model Generates mRNA Levels with a Characteristic PDF

Expression levels are controlled by gene interactions in a regulatory network. The gene expression is reflected, to some degree, in the mRNA levels. We represent this system in the following model:

$$\mathbf{X}' = M \cdot \mathbf{X}, \tag{2.11}$$

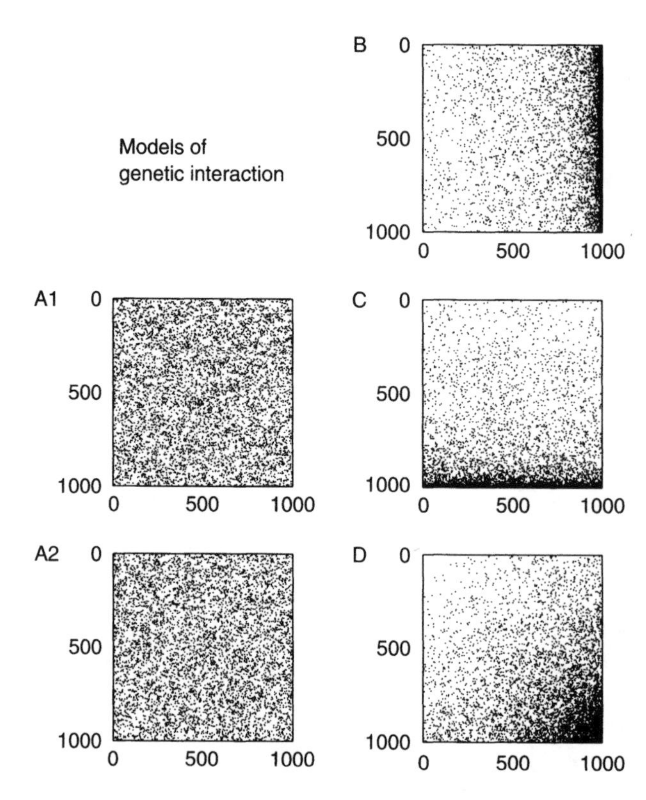

FIGURE 2.2 Models of genetic networks. Black dots represent the nonzero elements, M_{ij}, indicating regulation of the ith gene by the jth gene. (Each dot is larger than the resolution of one row or column.) The connection matrix is 1000×1000 in each model. (A1) Random output: the output of each gene regulates six other genes chosen at random. (A2) Random input: each gene is regulated by six other genes chosen at random. (B) Small-world homogeneous input: the pattern of input connections into each gene is the same. The pattern of output connections from the genes is scale-free. (C) Small-world heterogeneous input: the pattern of input connections into each gene is different and scale-free. (D) Small-world symmetric input/output: the pattern of inputs and outputs of each gene is scale-free.

where M is the connectivity matrix representing the regulatory influence of gene j on gene i, $\mathbf{X}$ is a one-dimensional column vector representing the mRNA level expressed by each gene at a given point in time, and $\mathbf{X}'$ represents the mRNA levels at the next time step. Iteration of the above equation generates the dynamics. The model starts with initial values of the elements of $\mathbf{X}$. At each subsequent time step, the connectivity matrix is multiplied by the mRNA vector to compute new elements of X_i. The equation is iterated until the mRNA levels reach a steady state. (For the random-output model [A1], the steady state $\mathbf{X}'$ can also be computed by solving an eigenvector problem. This is because the matrix M has an eigenvalue of one with the corresponding stationary eigenvector $\mathbf{X}'$.)

From the steady-state mRNA vector, we compute the statistical properties of the elements X_i by evaluating the PDF, that is, the probability that any of the mRNA levels, x, is between x and $x + \mathrm{d}x$. Since the sum of all the probabilities is equal to one, the integral of the PDF over x is equal to one. The PDF is a global property of the system and hence is the object of our analysis. We are not interested in the individual properties of the genes but rather in the global properties of all the expression levels to help understand the structure of the interacting genetic system. We compute histograms using different bin sizes for different ranges of x as described in the Appendix (Liebovitch et al., 1999).

The plots of log PDF versus log x computed from the five different models of the connectivity matrix M are shown in Figure 2.3. The PDF of the steady-state vector from the random-output model (A1) is similar to a Poisson distribution. For the random-input model (A2), the state vector converges

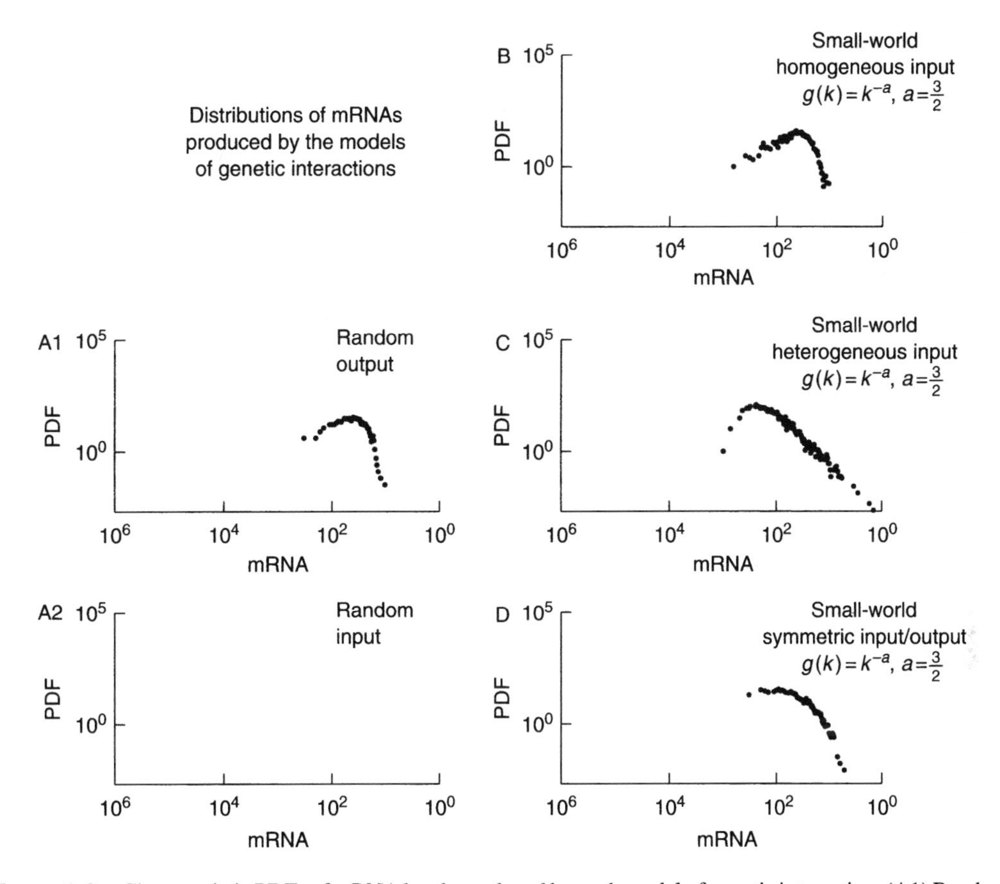

FIGURE 2.3 Characteristic PDFs of mRNA levels produced by each model of genetic interaction. (A1) Random output: the PDF resembles a Poisson distribution. (A2) Random input: the PDF is a single point distribution. (B) Small-world homogeneous input: the tail of the PDF has a power-law form. (C) Small-world heterogeneous input: the tail of the PDF has a power-law form. (D) Small-world symmetric input/output: the PDF is intermediate between models (B) and (C).

to a uniform vector, in which all the components have the same value, the average of the initial mRNA vector. This is reflected by the PDF's contracting to a delta function. For the small-world models, the PDFs are sensitive to the scaling exponent a of $g(k) = Ak^{-a}$. As a increases, the PDFs become steeper and narrower. Figure 2.3 shows the PDFs computed from the matrices with average number of connections $\langle k \rangle = 6$. We varied $\langle k \rangle$ from 1 to 3600 and observed that as $\langle k \rangle$ increases, the PDFs becomes narrower.

In summary, we found that there is a relationship between the global architecture of the genetic interactions and the statistical properties of the mRNA expression levels. Some models of the connectivity matrix M generate PDFs of the mRNA expression levels that are similar to each other while other models generate PDFs that are different from each other. Model C, Small-world heterogeneous input, is of particular interest. It generates a PDF with a power-law tail resembling the distributions of many other natural networks. Stuart et al. (2003) found that the connectivities of the networks of genes that are coexpressed across multiple organisms follows a power-law distribution with a slope of -1.5. This suggests that we may use the experimental data to differentiate between some different types of models of genetic networks. That is, this new top–down approach can be used to analyze experimental data from DNA microarrays to extract some properties of the network of genetic regulation without first needing to fully identify which specific genes regulate which other specific genes.

2.5 Biological Data

To link the models to biological reality, we evaluated the PDFs of mRNA measured by cDNA microarrays provided by Dr. Sharon Kardia and her collaborators at the University of Michigan. These researchers have searched for sets of genes (e.g., by Principle Component Analysis) that are good diagnostic markers for different tumor stages and different tumor types and their predictive value in survival probabilities for ovarian, lung, and colon cancers. An example of the data we analyzed is one where Affymetrix GeneChip cDNA microarray technology was used to develop such a molecular classification of types of lung tumor (Giordano et al., 2001; Beer et al., 2002; Schwartz et al., 2002). Figure 2.4 shows the PDFs of gene-expression profiles for primary lung adenocarcinomas from six different patients.

Thus, we determined the global statistical properties, the PDFs, of the mRNA expression levels from the theoretical models and those measured by cDNA microarray chips and then compared the results from the experiments with those from the models. The PDFs of the experimental data are reminiscent of the models in which there is an overall global pattern of interaction between the genes. The biological implication of this is that there does seem to be some pattern of overall biological structure, besides just the lower-level individual reactions and pathways from separate modules. We tested the hypothesis that different types or stages of cancers could show different PDF patterns and thus that those patterns might serve as useful clinical diagnostic tools. However, so far, the differences between different tumor types and stages seem small. We do not yet know whether the PDFs of different subsets of genes will show larger differences. Even if they do not, these results

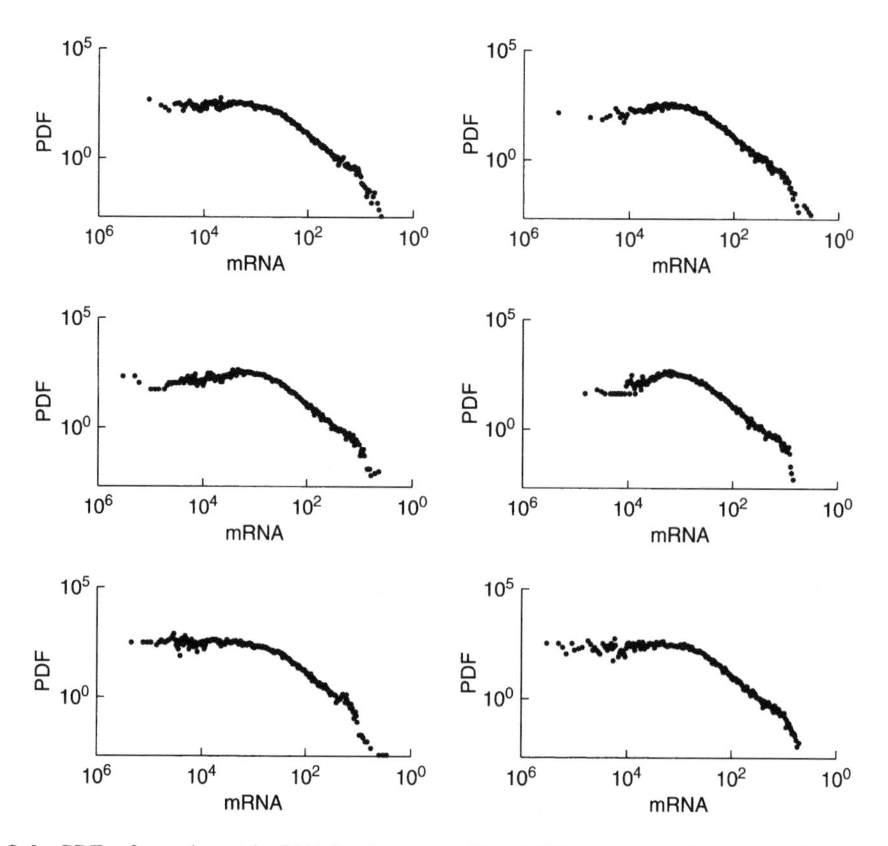

Figure 2.4 PDFs of experimental mRNA levels measured by cDNA microarrays. These data from six patients are representative of 86 primary lung adenocarcinomas (67 stage I and 19 stage III tumors), and 10 nonneoplastic lung samples.

would also be interesting from a biological standpoint, for it would mean that although different genes are active in different types and stages of cancer, still the overall pattern of interaction remains the same. For that to happen, there must be small, but important, adjustments in gene expression throughout large numbers of genes. Finding the details of these homeostatic mechanisms would give us important new insight into the global organization of organisms and may also provide new concepts for clinical intervention.

2.6 Summary

DNA is transcribed into mRNA, which is translated into proteins. These proteins can then bind back onto the DNA and increase or decrease gene expression. Nature is a circular dance of positive and negative feedback. Each gene influences the expression of other genes. New methods of analysis are now being developed to discover the nature of this network of genetic regulation. Some methods take a bottom–up approach, trying to piece together all the separate interactions between individual genes from the experimental data. Other methods take a top–down approach, trying to discover the overall organization of this network from experimental data.

In some ways, trying to understand the network of genetic regulation is analogous to trying to understand the flow of information in neural systems. The same circular dance of positive and negative feedback is also present in neural systems. Perhaps the methods described here for genetic networks will have useful applications in understanding neural networks, and vice versa.

ACKNOWLEDGMENTS

We sincerely thank Dr. Sharon Kardia and her experimental team from the Department of Epidemiology at the University of Michigan, Ann Arbor, Michigan for providing the microarray data on 86 lung adenocarcinomas and for their thorough explanation of these valuable data.

PROBLEMS

1. The global analysis of the DNA microarrays presented here depends on evaluating the PDF of a set of numbers, here the elements of $\mathbf{X}$. Determine the PDF of a set of numbers generated by a uniform random number generator.
2. In the scale-free models, $g(k)$ nodes have Ak^{-a} connections to other nodes. Generate a set of numbers whose PDF has such a power-law form.
3. Construct a connection matrix M with exactly $\langle k \rangle$ nonzero elements, chosen at random in each column. Using $\mathbf{X}' = M\mathbf{X}$, compute the PDF of the steady-state values of $\mathbf{X}$ from the matrix.
4. Vary the value of $\langle k \rangle$ in Problem 3. How does the shape of the PDF depend on $\langle k \rangle$?
5. Construct a connection matrix M in which there are $Y(R) = \{A/[(a-1)(N-R)]\}^{(1/a-1)}$ nonzero elements in each row R (or column R). Show that the number of nodes $g(k)$ that have k connections to other nodes then has the scale-free form $g(k) = Ak^{-a}$.
6. Using $\mathbf{X}' = M\mathbf{X}$, compute the PDF of the steady-state values of $\mathbf{X}$ from the matrix in Problem 5. How does the shape of the PDF depend on a?

APPENDIX: MULTI-HISTOGRAM ALGORITHM FOR DETERMINING THE PDF

To determine the PDF, we computed histograms of different bin sizes, evaluated the PDF from each histogram, and then combined those values to form the completed PDF. $N(k)$ is the number of times a given level of mRNA, $x = k\Delta x$, is produced in the range $(k-1)\Delta x < x \leq k\Delta x$. The PDF at $x = (k - \frac{1}{2})\Delta x$ is equal to $N(k)/(\Delta x N_{\mathrm{T}})$ where N_{T} is the total number of mRNA levels. From each histogram we included in the PDF the values computed from the second bin, $k = 2$, and continuing for bins $k > 2$, stopping at the first bin, $k = k_{max}$, that contained no mRNA levels, or at $k = 20$, whichever came first. We excluded from the PDF the value computed from the first bin, $k = 1$, because it includes all the levels unresolved at resolution Δx. We also excluded from the PDF the values computed from the bins $k > k_{max}$ or $k > 20$ because the mRNA levels in these bins are too sparse to give good estimates of the PDF. We used histograms of different bin size, Δx. The size of the smallest bin Δx_{min} was determined by using trial and error to find the smallest bin size for which there were mRNA levels in the first four bins. Then the procedure described above was used to compute values of the PDF from that histogram. The next histogram was formed with bin size $2\Delta x_{min}$, and the values of the PDF were computed. This procedure was iterated so that each subsequent histogram had a bin size double the size of the previous histogram. This was continued until the first bin size for which there were no mRNA levels in the second, third, or fourth bins. The complete PDF determined in this way extended over the greatest range possible because it had values computed from small bins at small mRNA levels as well as from large bins at large mRNA levels. It also included more values at mRNA levels that can be computed from overlapping bin sizes. This produces an effective weighting of the PDF, because more values are generated at mRNA levels that have more events. We found empirically that this weighting provides reliable least squares fits of PDF functions because mRNA levels with more events, where the PDF is more accurate, are weighted more in the fit. We have found that this procedure is accurate and robust in determining PDFs of different forms, such as single exponential, multiple exponential, and power laws.

3 Calcium Signaling in Dendritic Spines

William R. Holmes

CONTENTS

3.1 Introduction

The function of dendritic spines has been debated ever since they were discovered by Ramón y Cajal in 1891. Although it was widely believed that spines were important for intercellular communication, the small size of dendritic spines prevented confirmation of this function until 1959, when Gray, using the electron microscope, showed that synapses are present on spines. We now know that in many cell types the vast majority of excitatory synapses are on spines rather than on dendrites. Why should excitatory synapses be on spines rather than on dendrites? What role does spine morphology play in their function? Because spines are so small, with surface area generally less than 1 μm^2, these questions have been difficult to address experimentally. Up until about ten years ago, leading ideas about the function of spines were based on mathematical models and computer simulations. However, in recent years, the development of sophisticated imaging techniques, particularly two-photon microscopy, has allowed questions about spine function to be addressed experimentally (Denk et al., 1996).

Early theoretical studies of passive spines focused on the electrical resistance provided by the thin spine stem, R_{SS}, and suggested that changes in spine-stem (neck) diameter might be important for synaptic plasticity (Rall, 1974, 1978). Specifically, if the ratio of the spine-stem resistance to the input resistance at the dendrite at the base of the spine (R_{SS}/R_{BI}) was between 0.1 and 10 (or considering synaptic conductance, g_{syn}, between 0.1 and 10 times $[1 + 1/(g_{syn}R_{BI})]$), then a small change in spine-stem diameter could have a significant effect on the voltage in the dendrite due to input at the spine head. Later theoretical studies in the 1980s and early 1990s (reviewed by Segev and Rall, 1998) proposed that if spines possessed voltage-gated ion channels, then interactions among excitable spines could amplify synaptic input and create a number of additional interesting computational possibilities for information transfer. However, both the earlier passive and later active spine theoretical studies required R_{SS} to be 200 to 2000 MΩ for their most interesting predictions to occur. Unfortunately, when it finally became possible to estimate R_{SS} with imaging techniques in the mid-1990s, it was found that R_{SS} was 5 to 150 MΩ in many neuron types (Svoboda et al., 1996). The latest models of excitable spines reduced the required R_{SS} value to 95 MΩ by assuming kinetics for a low-threshold calcium conductance in the spine head, but low-threshold T-type calcium channels most probably do not exist on spines in the densities required by these models (Sabatini and Svoboda, 2000).

While the theoretical studies described above searched for a function for spines that depended on the *electrical* resistance of the spine neck, the spine neck might also provide a *diffusional* resistance to the flow of ions and molecules. Koch and Zador (1993) refer to the spine neck as having a small *diffusive space constant.* By restricting the flow of materials out of the spine head, the spine neck might effectively isolate the spine head and provide a localized environment where reactions specific to a particular synapse can occur. Calcium is a prime candidate for a substance that might be selectively concentrated in the spine head. Calcium is important for a large number of metabolic processes and has been shown necessary for the induction of long-term potentiation (LTP), but high concentrations of calcium can lead to cell death. Spines might provide isolated locations where high concentrations of calcium could be attained safely without disrupting other aspects of cell function (Segal, 1995). In this chapter we review theoretical models and experimental evidence suggesting that a major function of dendritic spines is to concentrate calcium. We discuss factors affecting calcium dynamics in spines and subsequent calcium-initiated reaction cascades, focusing on spines in hippocampal CA1 pyramidal cells. We do this from the point of view of developing realistic computational models to gain insights into these processes.

3.2 First-Generation Dendritic-Spine Calcium Models

First-generation dendritic-spine calcium models were developed because studies had shown that calcium was necessary for the induction of LTP (Wigstrom et al., 1979; Turner et al., 1982; Lynch et al., 1983). LTP induction requires a strong, high-frequency tetanus; a weak stimulus delivered at low to moderate frequencies can be applied as often as desired without producing potentiation. These frequency and cooperativity requirements of LTP must produce a steep nonlinearity to enable potentiation to occur. The hypothesis behind first-generation models was that there should be a steep nonlinearity in spine-head calcium concentration as a function of the number of activated synapses and the frequency of activation. Any nonlinearity in spine-head calcium concentration could be amplified further by subsequent calcium-initiated reactions.

The first-generation models, although sharing many similarities, can be divided into three groups based on the method of calcium entry into the spine head. Models by Gamble and Koch (1987) and Wickens (1988) assumed that calcium enters via voltage-dependent calcium channels. The models by Holmes and Levy (1990) and Zador et al. (1990) had calcium entering through NMDA-receptor channels, while Schiegg et al. (1995) also included calcium release from intracellular stores. Some variant of these models forms the basis of most spine models in use today. These spine models are coupled, either directly or indirectly, with a model of a cell that, for a specified input condition, calculates voltage at the spine head as a function of time. This enables calcium influx through voltage-gated calcium channels and NMDA-receptor channels to be computed appropriately.

3.2.1 Calcium Diffusion

The starting point for modeling calcium diffusion in dendritic spines is the one-dimensional diffusion equation, $\partial C/\partial t = D\partial^2 C/\partial x^2$, also called Fick's second law of diffusion. Dendritic-spine models use a discretized form of this equation in which spine geometry is represented as a series of cylindrical compartments with one or more compartments for the spine head, one or more compartments for the spine neck, and one or more compartments for the dendritic shaft. Examples are shown in Figure 3.1. It is assumed that calcium is "well-mixed" within each compartment with calcium concentration varying only between compartments. The change in calcium concentration in compartment i due to diffusion is represented mathematically as

$$\left.\frac{\mathrm{d}[\mathrm{Ca}]_i}{\mathrm{d}t}\right|_{\mathrm{diffusion}} = -\frac{D}{V_i}\left\{\left(\frac{A}{\delta}\right)_{i,i-1}([\mathrm{Ca}]_i - [\mathrm{Ca}]_{i-1}) + \left(\frac{A}{\delta}\right)_{i,i+1}([\mathrm{Ca}]_i - [\mathrm{Ca}]_{i+1})\right\}, \quad (3.1)$$

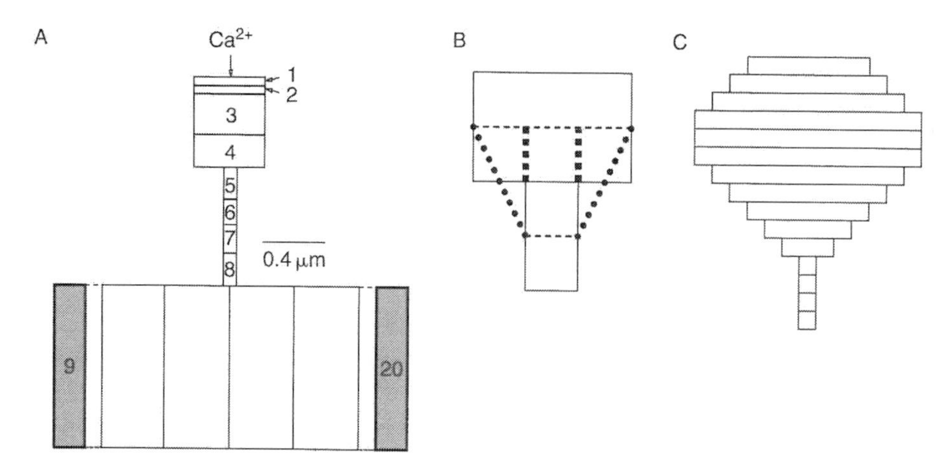

FIGURE 3.1 Compartmental representations of a spine. (A) A typical representation. (B) The boundary condition between the spine head and the spine neck can be modeled as being smooth or abrupt. Smooth coupling uses Equation (3.2) with the actual cross-sectional areas of the compartments bordering the head–neck boundary (angled heavy dotted lines). Abrupt coupling uses the neck cross-sectional area for both the neck and head compartments (vertical heavy dotted lines). (C) A representation with variable spine-head diameter.

where $[\mathrm{Ca}]_i$ is the concentration of free calcium in compartment i, D is the calcium diffusion coefficient, V_i is the volume of the ith compartment, and $(A/\delta)_{i,j}$ is the coupling coefficient between compartments i and j.

This coupling term can be defined in two different ways depending on whether the transition between compartments of unequal diameter is assumed to be smooth or abrupt. For smooth coupling, the coupling coefficient is defined as

$$\left(\frac{A}{\delta}\right)_{i,j} = 2\left(\frac{A_i\delta_i + A_j\delta_j}{(\delta_i + \delta_j)^2}\right), \tag{3.2}$$

where A_i and A_j are the cross-sectional areas and δ_i and δ_j are the thicknesses (or lengths) of compartments i and j, respectively. Abrupt coupling can be expressed with the same equation except that A_i and A_j are assumed to be equal to the cross-sectional area of the smaller compartment, typically the spine neck. Only calcium in the spine head just above the spine-neck opening is allowed to diffuse into the spine neck. The distinction is illustrated in Figure 3.1. Note, Equation (3.2) reduces to A_i/δ_i when the compartments are identical in size and to $1/\delta_i^2$ when divided by the volume, V_i.

In practice, the distinction between smooth and abrupt coupling matters little except at the transitions between spine head and neck and between spine neck and dendrite when the number of compartments is small. At these transitions one must choose compartment number and size carefully to ensure that spine shape is represented appropriately. Spine-head calcium concentration will decay significantly faster with smooth coupling and a coarse grid than with either abrupt coupling or smooth coupling with a fine grid. The use of a fine grid with abrupt coupling (as in Figure 3.1C) may be the simplest way to match specific spine morphology most accurately.

3.2.2 Calcium Buffering

Once calcium enters the cell it binds to buffers. Spines are known to possess a number of calcium-binding proteins, such as calmodulin, calcineurin, and calbindin, as well as immobile buffers and calcium stores. First-generation spine calcium models considered simple 1:1 buffering to an unspecified buffer or, in some cases, 4:1 buffering of calcium by calmodulin with buffer concentration being fixed within each compartment. Consequently, the change in calcium concentration due to buffering was modeled as

$$\left.\frac{\mathrm{d}[\mathrm{Ca}]_i}{\mathrm{d}t}\right|_{\text{buffer}} = \sum_j -k_{\mathrm{f}_j}[\mathrm{Ca}]_i[B_j]_i + k_{\mathrm{b}_j}([B_{\mathrm{t}_j}]_i - [B_j]_i), \tag{3.3}$$

where k_{f_j} and k_{b_j} are the forward and backward rate constants for binding of calcium to buffer for buffer j and $[B_j]_i$ and $[B_{\mathrm{t}_j}]_i$ are the free concentration and total concentration (bound plus unbound) of buffer j in compartment i. Equations are also required for free-buffer concentration,

$$\frac{\mathrm{d}[B_j]_i}{\mathrm{d}t} = -k_{\mathrm{f}_j}[\mathrm{Ca}]_i[B_j]_i + k_{\mathrm{b}_j}([B_{\mathrm{t}_j}]_i - [B_j]_i), \tag{3.4}$$

for each buffer j in each compartment i.

3.2.3 Calcium Pumps

There are a number of calcium pumps thought to be present in spines, including a high-affinity, low-capacity ATP-dependent pump, a low-affinity, high-capacity Na^+–Ca^{2+} exchanger, and a SERCA pump that pumps calcium into intracellular stores. These pumps are modeled either as first-order processes

$$\left.\frac{\mathrm{d}[\mathrm{Ca}]_i}{\mathrm{d}t}\right|_{\text{pump}} = \sum_j -k_{\mathrm{p}_j}\frac{A_i}{V_i}\left([\mathrm{Ca}]_i - [\mathrm{Ca}]_\mathrm{r}\right), \tag{3.5}$$

where k_{p_j}, the pump velocity for pump j, is multiplied by the surface-to-volume ratio for compartment i and $[\mathrm{Ca}]_\mathrm{r}$ is the resting calcium concentration, or as Michaelis–Menten processes

$$\left.\frac{\mathrm{d}[\mathrm{Ca}]_i}{\mathrm{d}t}\right|_{\text{pump}} = \sum_j -k_{\mathrm{m}_j}P_{\mathrm{max}_j}\frac{A_i}{V_i}\frac{[\mathrm{Ca}]}{[\mathrm{Ca}] + K_{\mathrm{d}_j}} + \frac{A_i}{V_i}J_{\text{leak}}, \tag{3.6}$$

where k_{m_j} is the maximum turnover rate for pump j, P_{max_j} is the surface density of pump sites for pump j (the product $k_\mathrm{m}P_{\mathrm{max}}$ is a measure of pump efficiency), K_{d_j} is the dissociation constant for pump j (a measure of pump affinity for calcium), and J_{leak} is the leak flux needed to maintain calcium concentration at its resting value.

3.2.4 Calcium Influx

As previously noted, first-generation spine models used different sources of calcium for calcium influx into spines. When the source of calcium is voltage-dependent calcium channels, the computation for calcium influx is straightforward. Calcium current at the spine head is computed in the neuron-level model and this current is converted into moles of calcium ions entering the volume of the most distal spine-head compartment per time interval (Gamble and Koch, 1987; Wickens, 1988). However, computation of calcium influx through NMDA-receptor channels is complicated by the fact that this current is a mixed Na/K/Ca current. Fortunately, the relative permeabilities of NMDA-receptor channels for Na, K, and Ca have been determined (Mayer and Westbrook, 1987) and computations with the constant-field equations can determine the calcium component of the NMDA-receptor channel current at a given voltage. Perhaps not surprisingly, the calcium proportion of the current through NMDA-receptor channels is relatively constant, being 8% to 12% over the physiological range of voltages. Zador et al. (1990) assumed a fixed 2% of the NMDA current was calcium, while Holmes and Levy (1990) used constant-field equation calculations to determine the calcium component.

3.2.5 Calcium from Intracellular Stores

Schiegg et al. (1995) extended the dendritic-spine model to include calcium release from internal stores. There are a number of ways in which release from stores can be modeled. Following the model of Schiegg et al. (1995), release from stores in compartment i is modeled as

$$\left.\frac{\mathrm{d}[\mathrm{Ca}]_i}{\mathrm{d}t}\right|_{\mathrm{stores}} = \rho X \left([\mathrm{Ca}_{\mathrm{store}}]_i - [\mathrm{Ca}]_i\right), \tag{3.7}$$

where ρ is the store depletion rate, X is the fraction of open channels, and $[\mathrm{Ca}_{\mathrm{store}}]_i$ is the store calcium concentration in compartment i. Store calcium concentration is modeled by

$$\frac{\mathrm{d}[\mathrm{Ca}_{\mathrm{store}}]_i}{\mathrm{d}t} = -\rho X \left([\mathrm{Ca}_{\mathrm{store}}]_i\right), \tag{3.8}$$

where store refilling is ignored for the short time course of the simulations (although refilling could be added with a term similar to the right-hand sides of Equations [3.5] or [3.6] being added to Equation [3.8]). Schiegg et al. (1995) limit the stores to one compartment within the spine head. The fraction of open channels, X, is described by

$$\frac{\mathrm{d}X}{\mathrm{d}t} = -\frac{1}{\tau_{\mathrm{store}}}(X - (RA)\mathrm{Re}(\mathrm{Ca}_i)), \tag{3.9}$$

where τ_{store} is the time constant for channel closing, RA is the probability of agonist binding to IP_3 (inositol triphosphate) or ryanodine receptors (RA is assumed to be either 0, no agonist available, or 1, saturating concentration of agonist, in the simulations), and $\mathrm{Re}(\mathrm{Ca}_i)$ expresses the calcium dependence of calcium release. $\mathrm{Re}(\mathrm{Ca}_i)$ is defined for $[\mathrm{Ca}]_i$ greater than $[\mathrm{Ca}]_\theta$ by

$$\mathrm{Re}(\mathrm{Ca}_i) = R_0 \frac{([\mathrm{Ca}]_i - [\mathrm{Ca}]_\theta)}{([\mathrm{Ca}]_{\mathrm{max}} - [\mathrm{Ca}]_\theta)} \exp\left(-\frac{([\mathrm{Ca}]_i - [\mathrm{Ca}]_\theta)}{([\mathrm{Ca}]_{\mathrm{max}} - [\mathrm{Ca}]_\theta)}\right), \tag{3.10}$$

where $[\mathrm{Ca}]_\theta$ is the threshold for calcium release and $[\mathrm{Ca}]_{\mathrm{max}}$ is the calcium concentration where the maximum release occurs. This release function is an alpha function with normalizing constant R_0. Although calcium release from stores is thought to have a bell-shaped dependence on calcium concentration, the use of an alpha function allows release via IP_3-receptor activation and ryanodine-receptor activation to be modeled with one function. This is done purely for convenience.

3.2.6 Summary

In the preceding sections, we have given various components of the differential equation for the change in calcium concentration in a spine-head compartment. All of these components must be combined into a single differential equation for the change of calcium concentration. That is,

$$\frac{\mathrm{d}[\mathrm{Ca}]_i}{\mathrm{d}t} = \left.\frac{\mathrm{d}[\mathrm{Ca}]_i}{\mathrm{d}t}\right|_{\mathrm{diffusion}} + \left.\frac{\mathrm{d}[\mathrm{Ca}]_i}{\mathrm{d}t}\right|_{\mathrm{buffer}} + \left.\frac{\mathrm{d}[\mathrm{Ca}]_i}{\mathrm{d}t}\right|_{\mathrm{pump}} + \left.\frac{\mathrm{d}[\mathrm{Ca}]_i}{\mathrm{d}t}\right|_{\mathrm{stores}} + \mathrm{influx}_i, \tag{3.11}$$

where the components due to diffusion, buffer, pumps, and stores are given by Equations (3.1), (3.3), (3.5) or (3.6), and (3.7), respectively. The influx term should be calculated in a neuron-level model, but often it is assumed to follow some simple functional form. Typically, only the outermost spine-head compartment receives calcium influx. For each compartment in the spine-head model, it will be necessary to solve Equation (3.11) for calcium concentration and Equations (3.4) and (3.8) for buffer and calcium-store concentration along with the auxiliary calcium-store equations, Equations (3.9) and (3.10). For the 20-compartment spine model shown in Figure 3.1, the system would have 100 equations, 20 each for calcium concentration, buffer concentration, and calcium-store calcium concentration, plus 40 auxiliary equations for calcium stores.

3.3 Insights from First-Generation Dendritic-Spine Calcium Models

The goal of first-generation spine-calcium models was to see whether stimulation conditions that induce LTP could create large, nonlinear increases in spine-head calcium concentration as a function of frequency and strength of the stimulus. These models found this nonlinearity and produced a number of other insights and predictions.

3.3.1 Spines Compartmentalize Calcium Concentration Changes

A brief high-frequency train of input was found to cause calcium concentration changes in the spine that were restricted to the spine head. Gamble and Koch (1987) found that spine-head calcium concentration could reach micromolar concentrations after seven spikes in 20 msec due to influx through voltage-dependent calcium channels. Holmes and Levy (1990), using eight pulses at 400 Hz, and Zador et al. (1990), using three pulses at 100 Hz paired with depolarization to −40 mV, found that calcium influx through NMDA-receptor channels could increase spine-head calcium concentration to 10 μM or more. In each of these models, the calcium transient was restricted to the spine head with very little calcium making its way through the spine neck to the dendrite. Zador et al. (1990) modeled calcium binding to calmodulin and found that compartmentalization of calmodulin fully loaded with calcium ions ($CaMCa_4$) was even more pronounced.

3.3.2 Spines Amplify Calcium Concentration Changes

When the same input stimulus that caused spine-head calcium to rise to over 10 μM was placed on the dendrite instead of the spine head, the local calcium concentration change was 1 to 2 orders of magnitude smaller. The magnitude of this difference depended on spine shape (Holmes, 1990; Zador et al., 1990). Large calcium concentrations were possible in the spine head because of its small volume and the restricted diffusion out of the spine head caused by the thin spine neck.

What was surprising was not so much that spines amplify calcium concentration changes compared to dendrites, but how few NMDA channels are needed to produce significant amplification. The NMDA conductance plots in Holmes and Levy (1990) indicate that the amplification occurred despite the average number of open NMDA-receptor channels during tetanic stimulation being much less than one (because of voltage-dependent magnesium block). This is consistent with numbers recently reported by Nimchinsky et al. (2004). Zador et al. (1990) obtained large amplification in their model assuming that the calcium component of the NMDA current was only 2% of the total, or about fivefold less than current estimates. These models found it all too easy to predict that spine-head calcium changes should be huge, even with small numbers of NMDA receptors at the synapse.

3.3.3 Spine-Head Calcium (or $CaMCa_4$) Concentration is a Good Predictor of LTP

Holmes and Levy (1990) and Zador et al. (1990) modeled spine-head calcium (or $CaMCa_4$) concentration as a function of input frequency and intensity and found that peak spine-head calcium concentration showed the same nonlinear dependence on input frequency and intensity as LTP. The results provided support for Lisman's theory (Lisman, 1989) that low to moderate levels of spine-head calcium concentration produce LTD and high levels produce LTP.

Peak spine-head calcium levels in the model were also correlated with the finding of LTP or no LTP in the relative timing of weak and strong inputs in the experiment of Levy and Steward (1983), an experimental result that foreshadowed the recent excitement about spike timing dependent plasticity (STDP). In that experiment, a weak contralateral tetanic input (eight pulse 400 Hz) was delivered just before or just after a strong ipsilateral tetanus. LTP was observed in the weak pathway if the weak input was delivered 1, 8, or 20 msec before the strong input, but LTD occurred if the weak input came after the strong input. The Holmes and Levy (1990) model found that spine-head calcium concentration was high when the weak input came before the strong input, but was low with the reverse pairing.

3.3.4 Spine Shape Plays an Important Role in the Ability of a Spine to Concentrate Calcium

While most early models assumed a spine shape resembling that of a long-thin spine, Holmes (1990) and Gold and Bear (1994) compared calcium concentration changes in long-thin, mushroom-shaped, and stubby spines. Spine dimensions for these categories of spines were taken from a study of dentate granule cells by Desmond and Levy (1985). While calcium concentration in the long-thin spine reached 28 μM in the models, levels in mushroom-shaped and stubby spines peaked at less than 2 μM. Although peak concentration was about the same in the mushroom-shaped and stubby spines, the calcium transient decayed faster in the stubby spine as calcium did not face a spine-neck diffusion barrier and rapidly diffused into the dendrite.

The quantitative numbers from these two studies can be criticized on several grounds. First, the huge change in the long-thin spine occurred when the fast buffer became saturated, while the larger spine-head volume and more buffer sites prevented buffer saturation in mushroom-shaped and stubby spines. The issue of appropriate buffer kinetics and concentrations will be discussed further in Section 3.4. Second, the input was identical for the different-shaped spines, but the larger mushroom-shaped and stubby spines are likely to have larger postsynaptic densities and hence, more NMDA receptors and more calcium influx. Nevertheless, more recent models with a slower, nonsaturating buffer and an input dependent on spine-head size also show that spine-head calcium levels are strongly dependent on spine shape. However, the differences between levels attained in long-thin and mushroom shaped spines are 2 to 5-fold rather than the 18-fold reported by the studies mentioned above.

3.4 Issues with Interpretation of First-Generation Spine-Calcium Model Results

The major problem with interpreting results from first-generation spine-calcium models is that values for many key parameters used in these models are not known in general and are not known for dendritic spines in particular. Parameter values used in the equations given above are summarized in Table 3.1.

3.4.1 Calcium Pumps

The densities, kinetics, pump velocity, and turnover rates for the various types of pumps that may exist on spines are not known. In the models, the pump parameters have a quantitative effect on results, but not a qualitative one. The pumps reduce the peak spine-head calcium concentration only a little, but do play a significant role in the final calcium decay rate. Zador et al. (1990) proposed that pumps on the spine neck might isolate the spine head from calcium concentration changes in the dendrites.

TABLE 3.1
Parameter Values for First-Generation Dendritic-Spine Models

	Description	Value	Model	Equation
D_{Ca}	Calcium diffusion coefficient	$0.6\ \mu m^2\ msec^{-1}$	Holmes, Zador, Schiegg	3.1
k_p	Pump rate constant	1.4×10^{-4} cm sec^{-1}	Holmes	3.5
k_{bf}	Forward buffer rate constant (calmodulin)	$0.5\ \mu M^{-1}\ msec^{-1}$	Holmes	3.3 and 3.4
		$0.05\ \mu M^{-1}\ msec^{-1}$	Zador	
		$0.5\ \mu M^{-1}\ msec^{-1}$	Schiegg	
k_{bb}	Backward buffer rate constant (calmodulin)	$0.5\ msec^{-1}$	Holmes	3.3 and 3.4
		$0.5\ msec^{-1}$	Zador	
		$0.5\ msec^{-1}$	Schiegg	
B_t	Total buffer concentration	$200/100\ \mu M$	Holmes	3.3 and 3.4
		$100\ \mu M$	Zador	
		$120\ \mu M$	Schiegg	
k_m	Maximum pump turnover rate	$0.2\ msec^{-1}$	Zador	3.6
$k_m P_{max}$	Pump efficiency (ATP-ase)	$1 \times 10^{-15}\ \mu mol\ msec^{-1}\ \mu m^{-2}$	Schiegg	3.6
	Pump efficiency (Na/Ca exchanger)	$5 \times 10^{-15}\ \mu mol\ msec^{-1}\mu m^{-2}$		
K_d	Pump affinity (ATP-ase)	$0.5\ \mu M$	Zador, Schiegg	3.6
	Pump affinity (Na/Ca exchanger)	$20.0\ \mu M$		
J_{leak}	Leak flux to balance pump at rest	$0.1 \times 10^{-15}\ \mu mol\ msec^{-1}\ \mu m^{-2}$ 1.25×10^{-17}	Schiegg	3.6
ρ	Store depletion rate	$(150\ msec^{-1})$	Schiegg	3.8
τ_{store}	Time constant for store channel closing	100 msec	Schiegg	3.9
RA	Probability of agonist binding to store receptor	0 or 1	Schiegg	3.9
R_0	Normalizing constant	exp(1)	Schiegg	3.10
Ca_θ	Threshold for store Ca release	150 nM	Schiegg	3.10
Ca_{max}	Concentration where maximum release occurs	250 nM	Schiegg	3.10
Ca_{store}	Store Ca concentration	25 mM	Schiegg	3.8
	Ca component of NMDA current	2%	Zador	
		10%	Schiegg	
		8% to 12% computed	Holmes	
Ca_r	Resting Ca concentration	70 nM	Holmes	3.5
		50 nM	Zador	
		50 nM	Schiegg	

Pumps on the spine neck would be in an ideal position for this role since the surface-to-volume ratio would be very high in the spine neck for a given pump density.

3.4.2 Calcium Buffers

The numbers, types, mobilities, and concentrations of calcium buffers in dendritic spines are not known. A number of different assumptions have been made about calcium buffers in spines and these can have qualitative as well as quantitative consequences for models. If the buffer has fast kinetics and is saturable, as in Holmes and Levy (1990), small changes in buffer concentration near the saturation point can have large quantitative effects on the level of spine-head calcium, as shown by Holmes (1990) and Gold and Bear (1994). Small changes in buffer concentration have less of a quantitative effect when buffer is not saturable or has low affinity or slow binding kinetics. Calmodulin, the major calcium buffer in Zador et al. (1990), is an example of a buffer with a relatively low calcium affinity, particularly for two of its four calcium-binding sites, and with calcium binding not being rapid. A difficulty with calmodulin as the primary buffer is that it is thought that only 5% of the approximately 40 μM calmodulin is free to bind calcium, at least in smooth-muscle cells (Luby-Phelps et al., 1995). A significant portion of the bound calmodulin may be released when calcium is elevated. If calmodulin is bound to neurogranin in the rest state and is released upon calcium influx (Gerendasy and Sutcliffe, 1997), then this added complication may not be a significant one.

While most model results suggest that spine-head calcium equilibrates quickly, some recent models have proposed that there is a calcium gradient within the spine head. Such a gradient cannot exist unless there is a very high concentration of fast immobile calcium buffer in the spine. Experimentally, it has been determined that equilibration of calcium in the spine head does occur and takes less than 2.5 msec (the time resolution of the optical scan) at room temperature (Majewska et al., 2000a). Because equilibration will be faster at physiological temperatures, it seems unlikely that spines have large concentrations of fast immobile buffer.

3.4.3 Calcium Source

While models consider calcium entering via voltage-gated channels or NMDA-receptor channels or from internal stores, no models to date incorporate all of these calcium sources. Holmes and Aradi (1998) modeled L, N, and T calcium channels on spine heads and found that the contribution of 5 N channels to calcium influx could be comparable to calcium influx through NMDA-receptor channels. But incorporation of all of these calcium sources poses another difficulty — calcium may rise too high in the model unless additional buffers or pumps are added.

3.5 Imaging Studies Test Model Predictions

The development and use of fluorescent calcium indicators with digital CCD camera imaging, confocal microscopy, and two-photon imaging has allowed spine calcium dynamics to be studied experimentally (Denk et al., 1996). Experimental studies have been able to test the predictions made by the models and have provided additional insights into calcium dynamics in spines and spine function.

3.5.1 Spines Compartmentalize Calcium Concentration Changes

Experiments by Muller and Connor (1991) using high-resolution fura-2 measurements suggested that spines could function as discrete compartments for calcium signaling. Svoboda et al. (1996) using two-photon imaging clearly showed that calcium can be compartmentalized in dendritic spines because of a diffusive resistance dependent on spine-neck length. What was more difficult to determine experimentally was the absolute amplitude and time course of the calcium transient in spines with different stimulation conditions. High-affinity indicators gave clear signals, but they tended to saturate at calcium concentrations above 1 μM. While low-affinity indicators could be used, the difficulty with any indicator is that the indicator itself acts as a calcium buffer, and calcium buffers, whether exogenous or endogenous, affect both the amplitude and time course of the transient (see discussion in Sabatini et al., 2001).

Despite these inherent difficulties, experimental estimates of peak spine-head calcium concentration have been made and these estimates have confirmed the predictions of the models.

Petrozzino et al. (1995) used a low-affinity indicator with CCD imaging and found that tetanic stimulation could raise spine-head calcium concentration to 20 to 40 μM. This increase in calcium was NMDA-receptor dependent, as it was blocked by APV. Yuste et al. (1999) paired five EPSPs with five action potentials and, taking into account the exogenous buffering by the indicator, estimated that calcium rose to 26 μM in the spine head (average of four spines). Sabatini et al. (2002), also taking into account buffering by the indicator, estimated that spine-head calcium reaches 12 μM for a single synaptic input when a magnesium block of NMDA receptors is relieved (by clamping to 0 mV). The extrapolation of this latter value to the physiological situation is difficult because voltage in cells is never "clamped" to 0 mV, the reversal potential for the NMDA current (although not for the calcium component of the current) is about 0 mV, and tetanic stimulation is likely to cause larger changes. Sabatini et al. (2002) also estimated the peak spine-head calcium concentration due to a single input at −70 mV to be 0.7 μM. NMDA-receptor channels are thought to be blocked at −70 mV so one would not expect any calcium influx, but it is not clear whether the voltage clamp was able to space clamp the voltage at the spine head being observed at this level.

3.5.2 Importance of Spine Geometry

Although experiments have not provided values for the peak of the spine-head calcium transient for different-shaped spines, the relative coupling between dendrite and spine has been compared for spines with different neck lengths. Volfovsky et al. (1999) used caffeine to stimulate calcium release from stores in spines with long, medium, and short neck lengths. They found that the amplitude of the spine-head calcium transient was not very different among spines with different shapes, but that the amplitude of the dendritic calcium transient was much closer in size to the spine-head transient for spines with short necks than for spines with long necks. In addition, the decay of the transient was faster in spines with short necks than in spines with long necks. The amplitudes of the spine-head transients in these experiments depended on the volume of caffeine-sensitive calcium stores in the different-shaped spines and thus depended on only one possible source of calcium influx into the spine. Majewska et al. (2000b) examined the diffusional coupling between spine and dendrite in the context of spine motility (motility is discussed further later). First they measured the calcium decay following a single action potential and found that spines with longer necks had slower decay kinetics than spines with shorter necks. Then they followed individual spines and measured changes in neck length and calcium decay at different time points over a 30-min period. They found that changes in neck length were correlated with changes in calcium decay. Holthoff et al. (2002) inferred calcium concentration changes in spines from fluorescence measurements following a backpropagating action potential. While they found a wide range of peak calcium concentration values in spines, they found no correlation between peak calcium and the diameter of the spine head suggesting that the density of calcium channels or stores compensates for changes in spine-head volume due to spine-head diameter. In agreement with the other two studies that are mentioned, they found that spines with longer necks had longer calcium decay times.

While these studies confirm the model prediction that calcium decay should be much shorter in stubby spines than in long-thin spines, the prediction about peak calcium concentration being much larger in long-thin spines than in stubby spines following synaptic or tetanic input remains to be tested.

3.6 Insights into Calcium Dynamics in Spines from Experimental Studies

Most of the problems with interpreting model results have to do with the fact that models have to make assumptions about unknown parameter values and processes. Whether the assumptions made are appropriate and realistic is often subject to debate. Recent experimental work has provided data that can be used to constrain model parameters and further refine spine calcium models. Insights from recent work have been made regarding the sources of calcium in spines, the kinetics and importance of calcium pumps in spines, and the buffer capacity of spines in different cell types.

3.6.1 Sources of Calcium in Spines

The sources for calcium in spines are well-known. Calcium enters the spine cytoplasm through NMDA-receptor channels, voltage-gated calcium channels, and from internal calcium stores. (A small component may also enter through non-NMDA glutamate channels). However, the relative importance of each calcium source is a matter of controversy — a controversy that may have arisen from the study of different subsets of highly heterogeneous spine populations.

3.6.1.1 NMDA-Receptor Channels

NMDA-receptor channels provide the major source of calcium during synaptic input in CA1 hippocampal pyramidal cells (Yuste et al., 1999; Kovalchuk et al., 2000). How much calcium enters depends on the number of NMDA receptors at a synapse and whether NMDA receptors are saturated by a vesicle of glutamate. Both models and experiments predict that the number of NMDA receptors at a synapse is small (less than 20 for long-thin spines), but whether or not a quantum of glutamate, say 2000 molecules, will saturate this small number of receptors is not known. Total calcium influx will increase with each pulse in a high-frequency tetanus if NMDA receptors are not saturated, but will not increase if they are saturated. The most convincing evidence that NMDA receptors are not saturated by a single input was provided by Mainen et al. (1999) who showed that the calcium transient for a second input, delivered 10 msec after the first, was 80% as large as the calcium transient for the first input. Given that NMDA-receptor channels have a mean open time of about 10 msec and unbind glutamate slowly, these data suggest that glutamate clearance from the synaptic cleft must be fast, or at least fast compared to glutamate binding to NMDA receptors. As for models, the data provide constraints on the size and time course of calcium influx through NMDA-receptor channels for single inputs and tetanic input. In addition, they suggest that the synaptic current through NMDA receptors should be modeled stochastically. Recent models do this (Li and Holmes, 2000; Franks et al., 2001).

3.6.1.2 Voltage-Gated Calcium Channels

Cerebellar Purkinje cells do not have NMDA receptors, so voltage-gated calcium channels are likely to be the major calcium source in these cells. In CA1 pyramidal cells, voltage-gated calcium channels provide most of the calcium influx during backpropagating action potentials (Yuste et al., 1999). Unfortunately for modelers, calcium channels come in L-, N-, T-, P/Q-, and R-types with multiple subunit combinations and possible splice variants (e.g., Tottene et al., 2000; Pietroban, 2002). So the question is: How many calcium channels and what type(s) of voltage-gated calcium channels exist on spines?

Sabatini and Svoboda (2000) report that spines on CA1 pyramidal cells contain 1 to 20 calcium channels with the larger numbers found on spines with a larger volume. They report that these channels are predominantly R-type because blockers of L-, N-, and P/Q-types of calcium channels had no effect on spine calcium transients elicited by a backpropagating action potential. The involvement of T-type calcium channels could not be ruled out, but their experiments were done in a voltage range where T-type calcium channels were thought to be inactivated. They observed calcium transients in spines evoked by backpropagating action potentials that were larger or smaller than the dendritic transients — larger because of restricted diffusion out of the spine head and smaller because of failures of spine calcium channels to open. Yuste et al. (1999) used low concentrations of Ni^{2+} to selectively block low-threshold calcium currents and found that spine calcium transients elicited by backpropagating action potentials were not affected (this concentration of Ni^{2+} would also block some of the R-type calcium current as well — the G2 or R_a and R_b subtypes [Tottene et al., 2000]).

They concluded that CA1 pyramidal cell spines have high-threshold, but not low-threshold, calcium channels. Emptage et al. (1999) also used low concentrations of Ni^{2+} to rule out low-threshold calcium channel involvement in calcium transients during single synapse activation. Schiller et al. (1998) found that calcium influx in neocortical pyramidal cells was predominantly through N-, P/Q-, and T-type calcium channels, although they did not specifically rule out R- or L-types. They report that calcium-channel subtypes and densities in spines were similar to the subtypes and densities in the neighboring dendrite but were different in apical and basilar portions of the dendritic tree. These sets of data constrain the types of channels that need to be included in models, but perhaps more importantly, they indicate that spines are not sites of calcium "hot spots" and that calcium-channel densities in spines are much lower than densities needed to produce the interesting results in excitable spine models mentioned briefly at the beginning of this chapter.

3.6.1.3 Calcium Stores

How much calcium enters CA1 pyramidal spines from calcium stores compared to other sources is an area of controversy. Emptage et al. (1999) found that single afferent stimuli produced calcium transients in the spine head that were abolished by both NMDA and calcium induced calcium release (CICR) antagonists. They hypothesized that calcium entry through NMDA-receptor channels, although not large enough by itself to be detected reliably, is necessary to trigger calcium release from stores. Release from stores then provides the bulk of the calcium in the calcium transient. In contrast Kovalchuk et al. (2000) report that calcium transients in spines following weak subthreshold stimulation are primarily due to influx through NMDA-receptor channels. CICR antagonists ryanodine or CPA reduced the amplitude of the calcium signal by about 30%, leaving a large and easily detectable signal due to influx through NMDA-receptor channels. Surprisingly, blocking AMPA receptors with CNQX had little effect on the calcium transients, suggesting that voltage-dependent magnesium block of calcium influx through NMDA channels is incomplete near the resting potential (but see Yuste et al., 1999).

This controversy could be explained if the two studies were looking at different subsets of spines. Spacek and Harris (1997) report that only 58% of immature spines and 48% of adult spines have some form of a smooth endoplasmic reticulum (SER). The presence of an SER is much more prevalent in mushroom-shaped spines than in stubby or long-thin spines. Furthermore, the spine apparatus is found in more than 80% of mushroom-shaped spines but is rare in other spine types. If the SER and spine apparatus are the calcium stores in spines, as is believed, then one would expect to see CICR only in mushroom-shaped spines and in less than half of long-thin and stubby spines. The fact that Emptage et al. (1999) consistently observed CICR suggests that the spines they studied may have been predominantly mushroom-shaped spines. Emptage et al. (1999) also found that when CICR was blocked, repetitive stimulation was needed to bring calcium influx through NMDA-receptor channels up to the levels found with CICR. This is consistent with modeling study results mentioned earlier suggesting that mushroom-shaped spines have much smaller calcium transients than long-thin spines. Conversely, Kovalchuk et al. (2000) may have studied a larger proportion of spines lacking an SER. With CICR not being possible in most spines, influx through NMDA-receptor channels dominated the calcium transient. Regardless of whether or not stores play a role in spine calcium transients, a more important role for calcium stores in CA1 pyramidal cells may be in the generation of calcium waves that propagate down the thick apical shaft toward the soma (Nakamura et al., 2002).

In contrast to CA1 pyramidal cells, almost all Purkinje cell spines have an SER. Consequently, calcium stores are a major source of calcium in these cells. Finch and Augustine (1998) report that with repetitive stimulation (15 stimuli at 60 Hz) half of the calcium transient is due to IP_3-mediated release from stores and half is due to calcium influx through voltage-gated calcium channels. The voltage-dependent influx, initiated by voltage changes produced by current through AMPA receptor channels, occurs first. IP_3 is produced following activation of metabotropic glutamate (mGlu) receptors, and induces a delayed calcium influx from stores.

3.6.1.4 Implications for Models

The experimental work elucidating the sources of calcium entry in spines has several implications for models. First, the numbers of NMDA receptors and voltage-dependent calcium channels at a spine are small and may require models to adapt a stochastic approach. Second, the number of NMDA receptors and calcium channels is correlated roughly with spine-head size. Although the type of calcium channel on spines, at least in CA1 pyramidal cells, appears to be a high-voltage-activated R-type channel, multiple R-type channels with different activation and inactivation characteristics have been identified (Tottene et al., 1996, 2000; Sochivko et al., 2002). Third, models should include calcium release from stores when spines are mushroom-shaped and probably should at least consider release from stores in non-mushroom-shaped spines.

3.6.2 Calcium Extrusion via Pumps

Work has been done to identify calcium extrusion mechanisms and their effectiveness in reducing the calcium transient. This work is difficult because fluorescent indicators buffer calcium and distort estimates of the extrusion time course. Furthermore it is difficult to separate calcium transient decay due to buffers or diffusion from decay due solely to pumps. The contribution of buffers, pumps, and diffusion to the decay of the calcium transient will be different in different shaped spines.

It is generally assumed that there are three types of pumps in spines: the SERCA pumps that pump calcium into calcium stores (SER), the plasma membrane calcium ATP pump (PMCA, Guerini and Carafoli, 1999), and the Na^{+}–Ca^{2+} exchanger (Philipson, 1999). Sabatini et al. (2002) estimated that about 30% of the calcium clearance from spines was due to SERCA pumps, although results were highly variable from spine to spine, perhaps because of the presence or absence of an SER in different spines (Spacek and Harris, 1997). They concluded that the other 70% must be due to the PMCA pump and the Na^{+}–Ca^{2+} exchanger. Blocking the SERCA pumps slowed the decay time constant of the transient by 50% (Sabatini et al., 2002) or over 100% (Majewska et al., 2000a). In addition, Majewska et al. (2000a) observed a 35% increase in the fluorescence signal when SERCA pumps were blocked, presumably because more calcium was available. Holthoff et al. (2002) found that calcium clearance scales linearly with surface area, which would be consistent with a role for plasma-membrane pumps.

When the effects of the exogenous buffer (the indicator) and temperature are taken into account in these studies of calcium clearance from spines, what is remarkable about the results is how fast extrusion mechanisms clear calcium. Majewska et al. (2000a), Sabatini et al. (2002), and Holthoff et al. (2002) all agree that without the dye, the clearance time constant is 10 to 20 msec. It should be noted that this time constant combines the effects of buffered diffusion and extrusion, which will vary from spine to spine. Given the relatively slow turnover rate of a single pump, these data imply that pump density must be extremely high or that estimates of pump turnover rate are very low (perhaps because temperature was not physiological). The three studies also agree that endogenous buffers and pumps are not saturated, since the decay time constant appears to be the same whether calcium transients are elicited once or several times at high frequency (Sabatini et al., 2002).

3.6.3 Calcium Buffers in Spines

The number, types, concentrations, and kinetics of calcium buffers in dendritic spines are still largely unknown. However, recent experimental work has revealed some characteristics of the calcium buffer in spines. Endogenous buffer capacity in spines has been estimated to be about 20 (Sabatini et al., 2002). This means that 95% of the calcium entering the spine is buffered, with 5% remaining free. This is in contrast to higher estimates of buffer capacity in CA1 pyramidal-cell dendrites which are 60 to 200 (Helmchen et al., 1996; Lee et al., 2000; Murthy et al., 2000).

Another question of interest to modelers is whether the calcium buffer is mobile or not. Cerebellar Purkinje cells have a large concentration of a mobile high-affinity buffer, thought to be calbindin (Maeda et al., 1999), along with an immobile low-affinity buffer. In contrast, the calcium buffer in CA1 pyramidal cells appears to be slowly mobile or immobile. In these cells the calcium buffer does not seem to wash out during whole cell recordings as might be expected if it were mobile, suggesting that mobile buffers do not contribute significantly to calcium buffering in spines (Sabatini et al., 2002). Murthy et al. (2000) estimate the diffusion constant for the calcium buffer to be 10 to 50 $\mu m^2 sec^{-1}$ or an order of magnitude slower than the calcium diffusion coefficient (223 $\mu m^2 sec^{-1}$, Allbritton et al., 1992). Many models use calmodulin as the major calcium buffer, but can calmodulin be considered to be an immobile buffer? Calmodulin, based strictly on size, would be expected to diffuse about an order of magnitude slower than calcium. At resting calcium levels calmodulin may be bound to neurogranin, but with calcium influx, calmodulin may be released allowing it to bind to calcium–calmodulin dependent protein kinase II (CaMKII) or calcineurin. Given that calmodulin itself may be "buffered" both at rest and during activity, calmodulin seems to have the necessary properties to be a major buffer of calcium in spines.

Finally experimental work suggests that, whatever the calcium buffer may be in spines, it is not saturated by strong input (Sabatini et al., 2002). If calmodulin were the only calcium buffer and if calmodulin concentration in spines were about 40 μM, then calmodulin itself would be capable of binding up to 160 μM of calcium. Given the size of observed calcium transients and the buffer capacity estimates for spines, it would seem that calmodulin can be a major, but not the only, calcium buffer in spines. This point is explored further in the problems presented at the end of the chapter.

3.7 Additional Insights into Spine Function from Experimental Studies

There have been a number of new insights into spine function from experimental work beyond insights into calcium dynamics. Here I shall focus on just two: spine motility and coincidence detection with backpropagating action potentials.

3.7.1 Spine Motility

Actin filaments are highly concentrated in spines and it was this high concentration of actin that led Crick to propose that calcium entry into spines might cause them to "twitch" in a way analogous to how calcium causes muscle contraction (Crick, 1982). The use of imaging techniques has allowed investigators to monitor spines in real time and it has been found that while some spines remain stable for hours or days, others are constantly changing shape over periods of seconds (Fischer et al., 1998). The highest levels of spine motility are observed in periods of development when synaptogenesis occurs (Dunaevsky et al., 1999). Spine shape tends to become more stable with age and synaptic activation, as more of the dynamic actin is replaced by stable actin. Nevertheless, spine motility has been observed at mature synapses. Morphological changes observed in development and following severe or traumatic events in the adult are dependent on activity-induced spine calcium levels. When there is a lack of activity and spine calcium levels are low, there may be a transient outgrowth of spines, but if calcium remains low, the spine may retract and be eliminated. Moderate increases of spine calcium promote spine elongation, but high levels may cause shrinkage and retraction (Segal, 2001). Such changes can occur in the normal adult, but usually they are much less pronounced.

What are the implications of spine motility for spine function? First, spine motility suggests that synaptic weights at spines are constantly changing. These changes could be subtle or significant. Shape changes can produce small electrical effects, but can also affect the amplitude and decay-time constant of the calcium transient significantly, leading to changes in calcium-initiated reaction cascades (Majewska et al., 2000a; Bonhoeffer and Yuste, 2002; Holcman et al., 2004).

Any increase or decrease in spine size may permit or be accompanied by a change in the number of postsynaptic receptors. Second, actin-based motility may position macromolecular complexes in a signal-dependent manner (Halpain, 2000). Numerous proteins make up the structure of the postsynaptic density (PSD) and how these are arranged at the synapse may depend in part on how and when they are tethered to or released from actin. Even low levels of motility could conceivably produce significant rearrangements of PSD proteins. Third, spine motility may help stabilize or destabilize the connections between the pre- and postsynaptic sides of the synapse. As the spine pushes out or retracts, the relationship between the sites of vesicle release and locations of postsynaptic receptors is tested.

3.7.2 Coincidence Detection with Backpropagating Action Potentials

Experiments in both neocortex and CA1 hippocampus have shown that pairing one or more EPSPs with a backpropagating action potential produces a supralinear summation of the calcium signals seen with either stimulus alone (Koester and Sakmann, 1998; Yuste et al., 1999). Summation was supralinear when the EPSPs preceded the action potential within a short time interval, but was sublinear with the reverse pairing. The supralinear summation is thought to occur because the backpropagating action potential relieves magnesium block of NMDA-receptor channels, allowing more calcium to enter. With the reverse pairing, the depolarization due to the action potential is largely over when the NMDA receptors are activated and no relief of magnesium block occurs. The spine "detects" the coincident pre- and postsynaptic activation by producing a larger calcium transient, but this detection only occurs when the signals are in the proper order.

Models in which spines have NMDA-receptor channels but not voltage-gated calcium channels show a notch in the spine calcium transient at the time when the action potential propagates back to the spine, and how much actual boosting of the calcium signal occurs depends on the width of the action potential (cf. figure 3A, Holmes, 2000). In most cases this boosting of the calcium signal is modest. NMDA-receptor channels are blocked very quickly by magnesium, so any relief of the magnesium block would only last as long as the action potential. To explain the supralinear summation observed experimentally, models need to incorporate appropriate types and numbers of voltage-gated calcium channels as well as NMDA-receptor channels at spines. It is possible that the relief of the magnesium block that occurs when the action potential backpropagates to the spine may be enough to cause a voltage boost sufficient to provide additional activation of voltage-gated calcium channels and, in a regenerative manner, additional NMDA-receptor channel activation, but this remains to be tested.

3.8 Second-Generation Spine Models: Reactions Leading to CaMKII Activation

While first-generation spine models searched for a nonlinearity in spine-head calcium concentration as a function of input frequency and strength, second-generation models search for this nonlinearity in CaMKII activation. Much work suggests that strong calcium signals may induce LTP by activating CaMKII (for reviews see Soderling, 1993; Lisman, 1994; Fukunaga et al., 1996; Lisman et al., 1997). When calcium enters the spine, it binds to calmodulin and the calcium–calmodulin complex ($CaMCa_4$) can then bind to individual subunits of CaMKII and activate them. Activated subunits can react with ATP to add a phosphate group at one or more positions in the subunit in a process called autophosphorylation, and this can allow the subunit to remain activated even when $CaMCa_4$ is no longer present. Second-generation models have sought to determine the stimulation conditions that lead to significant CaMKII activation, the time course of CaMKII activation, and the possible role of spine shape in CaMKII activation.

3.8.1 Modeling CaMKII Activation is Complicated

A CaMKII holoenzyme consists of two rings composed of six subunits each. Each subunit can be activated independently by $CaMCa_4$. A subunit activated by $CaMCa_4$ binding (the *bound* state) can be phosphorylated at the T^{286} position (here we consider only the α subunit type), but only if its immediate neighboring subunit in the ring is also activated. It is not clear whether this intersubunit autophosphorylation can occur only in the clockwise direction, in the counterclockwise direction, or in both directions. Once a subunit is phosphorylated at T^{286}, the rate of $CaMCa_4$ dissociation from CaMKII is extended from tens of milliseconds to tens of seconds. We say that $CaMCa_4$ is *trapped* on the subunit or that the subunit is in the trapped state, following the notation of Hanson and Schulman (1992). Eventually $CaMCa_4$ unbinds and the subunit, still phosphorylated at T^{286}, is considered *autonomous*. The subunit is still activated, but it does not require calcium to remain activated. Once $CaMCa_4$ unbinds, the calmodulin binding site in an autonomous subunit can be rapidly autophosphorylated at positions $T^{305,306}$ and the subunit is considered to be *capped*. This capping reaction may be either an intrasubunit or intersubunit autophosphorylation, although the experimental evidence is sparse for either case. The belief that the reaction is intrasubunit stems from the evidence that the slow basal autophosphorylation at T^{306} alone (to the *inhibited* state) is intrasubunit (Mukherji and Soderling, 1994); with phosphorylation at T^{286} it is thought that conformational shifts will allow both T^{305} and T^{306} to be in a better position for intrasubunit autophosphorylation. The belief that the capping reaction is intersubunit (dependent on the activation state of an immediately neighboring subunit) is based on a footnote in Mukherji and Soderling (1994) saying that they found evidence for this belief. This evidence remains unpublished.

It is thought that the bound and trapped states represent higher activity states than the autonomous and capped states. In fact the capped state is sometimes called an inhibited state because it cannot be brought back to the more highly activated bound and trapped states by calcium signals until it is dephosphorylated at $T^{305,306}$. Strictly speaking, the term "inhibited" should be reserved for the small number of subunits that spontaneously undergo basal autophosphorylation at the T^{306} site. Figure 3.2 shows kinetic reactions starting with calcium binding to calmodulin and calcium–calmodulin binding to CaMKII and calcineurin followed by transitions of activated CaMKII through the trapped, autonomous, and capped states. We have omitted the inhibited state from this figure because the reaction producing this state is slow and is likely to be reversed by phosphatase activity.

The time course of CaMKII activation will depend on the dephosphorylation action of phosphatases. Protein phosphatases 1 and 2A (PP1 and PP2A) are thought to dephosphorylate CaMKII but in different locations. PP2A dephosphorylates CaMKII primarily in the cytoplasm whereas PP1 primarily dephosphorylates CaMKII that has translocated to the postsynaptic density (Strack et al., 1997a, 1997b). PP1 activation is controlled by Inhibitor-1 which, when phosphorylated by PKA, inactivates PP1. Inhibitor-1 is dephosphorylated by calcineurin (PP2B), which is activated by $CaMCa_4$. Thus, phosphatase activity is also tightly regulated, leading many to postulate that the balance between CaMKII phosphorylation and dephosphorylation determines whether LTP is induced or not.

3.8.2 Characteristics of Second-Generation Models

Second-generation models differ in their modeling approach (deterministic or stochastic), whether or not the calcium signal is arbitrary, and whether phosphatase dynamics is included. Unfortunately, very few of these models actually consider CaMKII activation and phosphatase dynamics in the context of a dendritic spine. Those that do begin with the equations outlined earlier in Section 3.2, let calmodulin be the primary calcium buffer, and add equations for CaMKII and phosphatase dynamics as described below. If calmodulin, CaMKII, or phosphatase is allowed to diffuse, then equations analogous to Equation (3.1) are also needed for these substances.

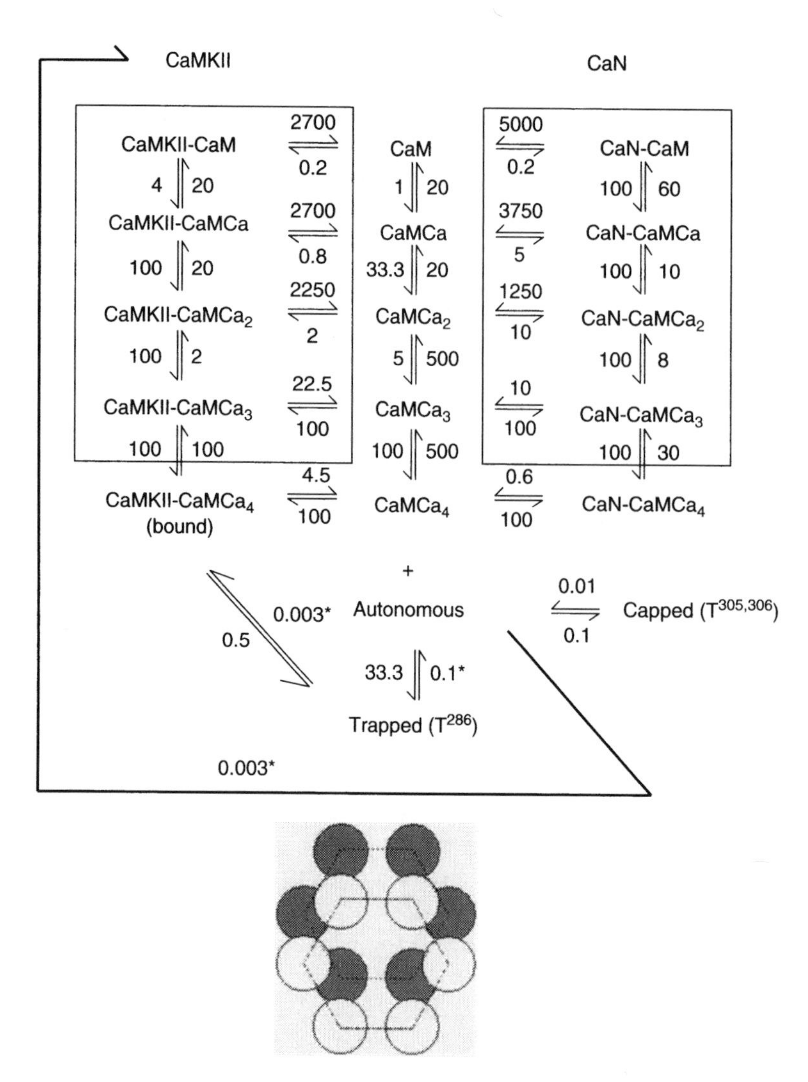

FIGURE 3.2 A summary of the reactions leading to CaMKII activation. Rate constants are those used in Holmes (2000). Units are $\mu M^{-1} sec^{-1}$ and sec^{-1}. Calcium binds to calmodulin as shown in the middle. Calmodulin with 0, 1, 2, 3, or 4 calcium ions bound binds to CaMKII or calcineurin (CaN). CaMKII with $CaMCa_4$ is considered to be in the bound state. A bound subunit may be autophosphorylated to become trapped. If $CaMCa_4$ unbinds from a trapped subunit, it becomes autonomous. An autonomous subunit can rebind $CaMCa_4$ to become trapped again. An autonomous subunit can be autophosphorylated at the calmodulin binding site to become capped. In the illustrated scheme, a capped subunit must be dephosphorylated back to the autonomous state and the autonomous state can be dephosphorylated to return the subunit to the free state. At the bottom is a cartoon of two rings of six subunits each meant to represent a CaMKII holoenzyme.

3.8.2.1 Deterministic vs. Stochastic

Models by Coomber (1998a, 1998b), Kubota and Bower (2001), Zhabotinsky (2000), and Okamoto and Ichikawa (2000a, 2000b) use differential equations to model calcium binding to calmodulin and the CaMKII activation reactions shown in Figure 3.2. Because each subunit can be in 12 different states with each holoenzyme having 8 to 10 subunits in Coomber's model, each enzyme can have 8^{12} to 10^{12} configurations. To reduce the number of equations, Coomber (1998a, 1998b) considers only four or five subunits per holoenzyme and combines states that are similar, reducing the number of equations to about 3000. Kubota and Bower (2001) also limit the number of subunits per holoenzyme to four to reduce the computational load. In contrast, Michelson and Schulman (1994) and Holmes (2000) model the CaMKII activation reactions stochastically (reactions in the lower part of Figure 3.2). This approach involves much bookkeeping. When $CaMCa_4$ binds to a CaMKII subunit, the particular holoenzyme and the particular subunit on that holoenzyme where binding occurs are assigned randomly. This is necessary because of the dependence of subsequent autophosphorylation reactions on the state of neighboring subunits. Similarly, at each time step, random numbers are chosen for each bound, trapped, autonomous, and capped subunit to determine which subunits make transitions to which neighboring states.

3.8.2.2 Calcium Signal

Second-generation models also differ with respect to the calcium signal used. Because models are interested in CaMKII dynamics and/or phosphatase dynamics, and, as we have seen above, the calcium signal in dendritic spines can be quite variable, depending on many factors, most models assume a simple form for the calcium transient. This form ranges from constant calcium concentration or regular calcium oscillations (Okamoto and Ichikawa, 2000a, 2000b), to calcium pulses (Dosemeci and Albers, 1996; Kubota and Bower, 2001; d'Alcantara et al., 2003), to step increases in calcium with exponential decay (Zhabotinsky, 2000), to exponential rise and exponential decay (Coomber, 1998a, 1998b), and to calcium influx through NMDA-receptor channels computed at a spine in a neuron-level model (Holmes, 2000) with NMDA-receptor channel openings computed either deterministically or stochastically with a synapse-level model. The use of simple calcium signals reduces computational demands of the simulations and allows insights to be obtained about what might be happening in spines, but at some point it will be necessary to model these processes within the context of actual dendritic spines to distinguish what is feasible in a model from what happens physiologically.

3.8.2.3 Calcium Binding to Calmodulin

Given the calcium signal, the next step is to model calcium binding to calmodulin. Calmodulin can bind four calcium ions, two on the N-lobe and two on the C-lobe. The binding is cooperative within a lobe. What is important for the models is that the affinity of CaMKII and calcineurin for calmodulin is very low until at least three and usually four calcium ions bind to calmodulin. Once calmodulin binds to CaMKII or calcineurin, the affinity of calmodulin for calcium increases significantly, particularly at the low-affinity binding sites. These experimental observations are built into the sample rate constants illustrated in the kinetic scheme of Figure 3.2. Consequently, the calmodulin state of primary interest for models is $CaMCa_4$.

How models compute the concentration of $CaMCa_4$ varies. Coomber (1998a), Okamoto and Ichikawa (2000a, 2000b), Zhabotinsky (2000), and d'Alcantara et al. (2003) use either a simple binding scheme where all four calcium ions bind at once or a version of the Hill equation. Kubota and Bower (2001) use an Adair–Klotz equation formulation (equation shown in Table 3.2), while Holmes (2000) uses the elaborate sequential binding scheme shown in Figure 3.2. Parameter values used in various models are given in Table 3.2. Because different investigators use different procedures to calculate $CaMCa_4$, we will not present specific differential equations here. The reader is advised to examine the individual references mentioned for specific details. However, it is straightforward to translate biochemical reactions, such as those shown in Figure 3.2, into differential equations and some examples that can be applied to calcium binding to calmodulin are given in Appendix 1.

3.8.2.4 CaMKII Autophosphorylation Reactions

Once calcium has bound to calmodulin to produce $CaMCa_4$, $CaMCa_4$ can bind to and activate CaMKII, creating a bound subunit. Although the affinity of CaMKII for $CaMCa_4$ seems quite high

TABLE 3.2
General Rate Constants Used in Second-Generation Spine Models

	Description	Value	Model
D_{Ca}	Calcium diffusion coefficient	0.6 μm^2 msec^{-1}	Holmes
		0.223 μm^2 msec^{-1}	(ref: Allbritton)
D_{CaM}	Calmodulin diffusion coefficient	0.06 μm^2 msec^{-1}	Holmes
		0.01–0.05 μm^2 msec^{-1}	(ref: Murthy)
CaM_t	Total calmodulin concentration	80 μM	Holmes
		100 μM	Coomber
		30 μM	d'Alcantara
Ca_r	Resting calcium concentration	70 nM	Holmes
k_f	Forward binding rate $CaM\text{–}Ca_x + Ca \rightarrow CaM\text{–}Ca_{x+1}$	0.5 μM^{-1} sec^{-1}	Coomber
k_b	Unbinding rate $CaM\text{–}Ca_{x+1} \rightarrow CaM\text{–}Ca_x + Ca$	0.05 sec^{-1}	Coomber
$CaMCa_4$	$= CaM_t \dfrac{0.2[Ca] + 0.32[Ca]^2 + 0.0336[Ca]^3 + 0.00196[Ca]^4}{4 + 0.8[Ca] + 0.64[Ca]^2 + 0.0448[Ca]^3 + 0.00196[Ca]^4}$		Kubota
	$= CaM/\{1 + (7/[Ca])^4\}$		d'Alcantara
	see Figure 3.2		Holmes
CaMKII	CaMKII concentration	100 holoenzymes	Holmes
		0.1 μM	Coomber (1998a)
		0.5 μM	Coomber (1998b)
		0.1–30 μM	Zhabotinsky
		0.5 μM	d'Alcantara
PP	Phosphatase concentration	0.01–0.05 μM	Coomber (1998a)
		0.1 μM	Coomber (1998b)
		0.01–1.2 μM	Zhabotinsky
		2.0 μM	d'Alcantara
CaN	Calcineurin concentration	286 molecules	Holmes
		1 μM	d'Alcantara
		1 μM	Bhalla
$CaN\text{–}CaMCa_4$	$CaMCa_4 + CaN \rightarrow CaNCaMCa_4$	1670 μM^{-1} sec^{-1}	d'Alcantara
		100 μM^{-1} sec^{-1}	Holmes
	$CaNCaMCa_4 \rightarrow CaN + CaMCa_4$	10 sec^{-1}	d'Alcantara
		0.6 sec^{-1}	Holmes
CaN activation	$= ([Ca]/K_H)^3/1 + ([Ca]/K_H)^3$	$K_H = 0.3\text{–}1.4$ μM	Zhabotinsky
PKA	Protein kinase A concentration	1 μM	
Ng	Neurogranin concentration	10–80 μM	

with a dissociation constant of about 45 nM (Meyer et al., 1992), this affinity is actually low compared to that of many calmodulin-binding proteins. For example, the dissociation constant of $CaMCa_4$ for calcineurin is about 6 nM (Meyer et al., 1992). Again, the differential equation for $CaMCa_4$ binding to CaMKII can be written directly from the biochemical equation following the procedures outlined in Appendix 1. Rate constants used in second-generation models for $CaMCa_4$ binding to CaMKII and for subsequent autophosphorylation and dephosphorylation reactions are given in Table 3.3.

The transitions from the bound state to the trapped state and subsequent reactions occur on a much slower timescale than calcium binding to calmodulin and are also handled differently by different modelers. As mentioned above, the differential equations approach for these reactions becomes quite cumbersome, whereas the stochastic approach requires bookkeeping. In the stochastic approach, when a subunit becomes bound it is assigned a random location on a random subunit from the pool of free subunits. Then in the next time interval, if the bound subunit has a bound, trapped, autonomous, or capped immediate neighbor, it becomes trapped with a certain probability. As mentioned above, it is not clear whether this intersubunit autophosphorylation reaction is unidirectional or bidirectional. Because this is an autophosphorylation and the concentration of ATP is very large compared to

TABLE 3.3
CaMKII-Related Rate Constants Used in Second-Generation Models

Reaction	Value	Model	Notes
CaMKII + $CaMCa_4$ → Bound	100 μM^{-1} sec^{-1}	Holmes	
	12 μM^{-1} sec^{-1}	Kubota	
	5 μM^{-1} sec^{-1}	Coomber	
	150 μM^{-1} sec^{-1}	d'Alcantara	
	$K_H = 4\ \mu M$	Zhabotinsky	Hill equation exp = 4
Bound → CaMKII + $CaMCa_4$	4.5 sec^{-1}	Holmes	
	$2.848([Ca] + 0.1)^{-0.1602}$ sec^{-1}	Kubota	
	100 sec^{-1}	Coomber (1998b)	
	500 sec^{-1}	Coomber (1998a)	
	2.17 sec^{-1}	d'Alcantara	
Bound → Trapped	0.5 sec^{-1}, 0.2 sec^{-1} (not B or T)	Holmes	Intersubunit
	0.5 sec^{-1}	Kubota	ATP = 1 mM
	5 sec^{-1}	Coomber (1998b)	
	20 sec^{-1}	Coomber (1998a)	
	0.5 sec^{-1}	Zhabotinsky	
	0.5 sec^{-1}	d'Alcantara	
Trapped → Bound	0.003 sec^{-1}	Holmes	Dephos
	0.05 sec^{-1}	Coomber (1998a)	For PP = 0.01 μM
	0.03 μM^{-1} sec^{-1}	d'Alcantara	Mult by active PP1
Trapped + PP ↔ Trapped-PP → Bound + PP		Kubota Coomber (1998b)	Equations also used for dephos from A and C states with same rates
Trapped + PP → Trapped-PP	0.5 μM^{-1} sec^{-1}	Kubota	
	0.5 μM^{-1} sec^{-1}	Coomber (1998b)	
Trapped-PP → Trapped + PP	0.001 sec^{-1}	Kubota	
	50 sec^{-1}	Coomber (1998b)	(?)
Trapped-PP → Bound + PP	0.5 sec^{-1}	Kubota	
	1 sec^{-1}	Coomber (1998b)	
	2 sec^{-1}	Zhabotinsky	
Trapped → Autonomous + $CaMCa_4$	$(0.00228[Ca]^{1.6919} + 9.88)^{-1}$ sec^{-1}	Holmes	
	$0.0032([Ca] + 10)^{-1.75}$ sec^{-1}	Kubota	
	0.1	Coomber (1998b)	
	0.2	Coomber (1998a)	
Autonomous + $CaMCa_4$ → Trapped	33.3 μM^{-1} sec^{-1}	Holmes	
	4 μM^{-1} sec^{-1}	Kubota	
	5 μM^{-1} sec^{-1}	Coomber	
Autonomous → Capped	0.1 sec^{-1}	Holmes	Inter
	0.1 sec^{-1}	Kubota	Intra/Inter
	20 sec^{-1}	Coomber (1998a)	ATP = 1 mM
Autonomous → Capped-1 (T^{305})	1 sec^{-1}	Coomber (1998b)	ATP = 1 mM Intra/Inter
Capped-1 (T^{305}) → Capped ($T^{305,306}$)	1 sec^{-1}	Coomber (1998b)	ATP = 1 mM Intra/Inter

TABLE 3.3
Continued

Reaction	Value	Model	Notes
Capped → Autonomous	0.01 sec^{-1}	Holmes	Dephos
Autonomous → CaMKII	0.003 sec^{-1}	Holmes	Dephos
	1 sec^{-1}	Coomber (1998a)	PP = 0.01 μM
CaMKII → Capped-2 (T^{306})	1 sec^{-1}	Coomber (1998b)	Intersubunit ATP = 1 mM
Capped-2 (T^{306}) + PP → Capped-2-PP	10 μM^{-1} sec^{-1}	Coomber (1998b)	Next rates as for trapped dephos above

CaMKII, this reaction can be considered to be a simple first-order reaction A → B. As shown in Appendix 2, the transition probability for such a reaction is easily determined. In the present case a bound subunit will remain bound in time interval Δt with probability $\exp(-k\Delta t)$ and will become trapped with probability $(1 - \exp(-k\Delta t))$. We pick a random number and if it is less than $1 - \exp(-k\Delta t)$ then the subunit becomes trapped; otherwise it remains bound.

The range of reactions considered in second-generation models differs widely. Some models consider CaMKII activation only through the trapped state (Okamoto and Ichikawa, 2000a; Zhabotinsky, 2000; d'Alcantara et al., 2003) while others include transitions to the autonomous and capped states as well (Coomber, 1998a, 1998b; Holmes, 2000; Kubota and Bower, 2001). In the stochastic approach transitions from the trapped state to the autonomous state or from the autonomous state to the capped state are handled the same way as the bound to trapped transition described above.

One additional complication that has not been considered in models to date is that calcium may unbind from calmodulin that is bound or trapped on a CaMKII subunit. Should a bound or trapped subunit that has lost a calcium ion still be regarded as a bound or trapped subunit? Models to date assume that $CaMCa_4$ must bind or unbind from the CaMKII subunit intact. This is not a large issue for binding because, as discussed above, little calmodulin will bind to CaMKII without having four calcium ions bound, but what about the loss of a calcium ion after binding? The assumption generally made is that, because calmodulin affinity for calcium is increased upon binding to CaMKII, one need only consider $CaMCa_4$ unbinding. However, it may be the case that unbinding of a calcium ion may make the unbinding of calmodulin from CaMKII more likely. Experiments have been started to measure such rate constants (Gaertner et al., 2004) and these possibilities should be included in future models.

3.8.2.5 CaMKII Dephosphorylation Reactions

Phosphatase reactions are sometimes modeled as first-order equations (Holmes, 2000) with transitions modeled similarly to the autophosphorylation reactions described above. However, if phosphatase concentration is known or modeled, representations based on enzyme kinetics equations are more appropriate. For example, d'Alcantara et al. (2003), Okamoto and Ichikawa (2000a), and Zhabotinsky (2000) use Michaelis–Menten kinetics to model dephosphorylation as described in Appendix 3. Models that include the geometry of dendritic spines can have different dephosphorylation reactions that occur in different spine compartments. CaMKII may be dephosphorylated by PP1 or PP2A depending on whether CaMKII is tethered to the surface of the synaptic membrane or is present in the cytoplasm.

Just as CaMKII is activated through a calcium-initiated cascade, the dynamics of phosphatase activation involves a number of other substances such as Inhibitor-1, calcineurin, and PKA.

Zhabotinsky (2000) and d'Alcantara et al. (2003) explicitly include equations for PP1 activation and dynamics in their models. PKA will phosphorylate Inhibitor-1 and phosphorylated Inhibitor-1 will inactivate PP1. Calcineurin (PP2B), which requires $CaMCa_4$ for activation, will dephosphorylate Inhibitor-1. The kinetic equations and rate constants for these reactions are given in Table 3.4.

3.8.2.6 Other Reactions

d'Alcantara et al. (2003) carry the analysis a step further and include equations for phosphorylation and dephosphorylation of AMPA receptors (Table 3.4). Many other substances besides CaMKII have been implicated as having a role in plasticity. Bhalla (2002) and Bhalla and Iyengar (1999)

TABLE 3.4
Other Reactions in Second-Generation Models

Reaction	Value	Model	Notes
PKC + Ca → PKC-Ca	0.6 $\mu M^{-1} sec^{-1}$	Bhalla	
PKC-Ca → PKC + Ca	0.5 sec^{-1}	Bhalla	
PKC-Ca → PKC-Ca-active	1.27 sec^{-1}	Bhalla	
PKC-Ca-active → PKC-Ca	3.5 sec^{-1}	Bhalla	
Ng + PKC-active → Ng-PKC-active			See Bhalla
Ng-PKC-active → Ng + PKC-active			
NgPKC-active → Ng-P + PKC-active			
Ng-P + $CaNCaMCa_4$ → Ng-P-$CaNCaMCa_4$			
Ng-P-$CaNCaMCa_4$ → Ng-P + $CaNCaMCa_4$			
Ng-P-$CaNCaMCa_4$ → Ng + $CaNCaMCa_4$			
I-1 + PKA → I-1-PKA	$\nu = 1\ sec^{-1}$ (applies to next 3 reactions)	Zhabotinsky	(MM like)
I-1-PKA → I-1 + PKA			= PKA activity/K_M
I-1-PKA → I-1-P + PKA			
I-1 + PKA → I-1-P	0.002 sec^{-1}	d'Alcantara	rate * [PKA]
I-1-P + $CaNCaMCa_4$ → I-1-$CaNCaMCa_4$	$\nu = 1\ sec^{-1}$ (applies to next 3 reactions)	Zhabotinsky	(MM like)
I-1-$CaNCaMCa_4$ → I-1-P + $CaNCaMCa_4$			= CaN activity/K_M
I-1-$CaNCaMCa_4$ → I-1 + $CaNCaMCa_4$			
I-1-P + $CaNCaMCa_4$ → I1	2000 $\mu M^{-1} sec^{-1}$	d'Alcantara	
CaMKII-P + PP1 → CaMKII-P-PP1			
CaMKII-P-PP1 → CaMKII-P + PP1			
CaMKII-P-PP1 → CaMKII + PP1	2.0 sec^{-1}	Zhabotinsky	K_M = 0.4–2.0 μM
PP1 + I-1-P → PP1-I-1-P (deactivated PP1)	1.0 $\mu M^{-1} sec^{-1}$	Zhabotinsky	
	30 $\mu M^{-1} sec^{-1}$	d'Alcantara	
PP1-I-1-P (deactivated PP1) → PP1 + I-1-P	0.0011 sec^{-1}	Zhabotinsky	
	0.03 sec^{-1}	d'Alcantara	
CaMKII-P + PP2A → CaMKII-P-PP2A			Similar to
CaMKII-P-PP2A → CaMKII-P + PP2A			PP1 rates ?
CaMKII-P-PP2A → CaMKII + PP2A			
GluR1 + PKA → GluR1-P-S845	0.1 sec^{-1}	d'Alcantara	rate * [PKA]
GluR1-P-S845 + PP1 → GluR1 + PP1	1.0 $\mu M^{-1} sec^{-1}$	d'Alcantara	
GluR1 + CaMKII-active → GluR1-P-S831			
GluR1-P-S845 + CaMKII-active → GluR1-P-S845-S831	10 $\mu M^{-1} sec^{-1}$	d'Alcantara	
GluR1-P-S845-S831 + PP1 → GluR1-S845	0.1 $\mu M^{-1} sec^{-1}$	d'Alcantara	
GluR1-P-S831 + PKA → GluR1-P-S831-S845			
GluR1-P-S831-S845 + PP1 → GluR1-P-S831			

include the dynamics of PKA, PKC, and MAPK activation in addition to CaMKII in their models. The rate constants for these additional reactions are often not available or else difficult to determine. Bhalla (2001) provides an excellent description of the process of parameterization one might use to integrate networks of signaling pathways into models.

3.9 Insights from Second-Generation Models

Second-generation models set out to show that stimulation conditions that lead to LTP also show high levels of CaMKII activation. The models succeeded in this regard and revealed a number of aspects of CaMKII activation that were not readily apparent.

3.9.1 Frequency Dependence of CaMKII Activation

Several models have shown that CaMKII activation is highly dependent on the frequency of the stimulation. Here we describe results from two of these models. Coomber (1998a) showed that a 5 sec 100 Hz stimulus applied to the model led to 95% of the subunits being driven to the trapped state, while a 100 sec 1 Hz stimulus led to less than 2% of the subunits being activated at any point in time. Coomber also looked at 100 pulse trains at different frequencies (Coomber, 1998b) and found that the percentage of subunits driven to the trapped state was greater than 70% for 50 to 100 Hz trains, 20–60% for 5–20 Hz trains, and less than 20% for 2 Hz trains. Holmes (2000) used eight-pulse trains of varying frequency in his model and found a steep nonlinearity in the number of trapped subunits as the frequency was raised from 20 to 100 Hz. Below 20 Hz, few subunits became trapped. Significant numbers of subunits became trapped as the frequency was raised from 20 to 100 Hz, but further increases in frequency did not cause larger numbers of trapped subunits. One difference between the two models is that while Holmes (2000) observed a high percentage of subunits in the bound state after a single high-frequency eight-pulse tetanus, the number of subunits subsequently entering the trapped state was comparatively small. This difference can be not only explained primarily by the different tetanus durations, but also by tenfold differences in the bound to trapped rate constants used in the Coomber and Holmes models (Table 3.3). In the Holmes model, large numbers of subunits in the trapped state were only obtained when the eight-pulse tetanus was repeated at 1 to 10 sec intervals, as this repetition caused the number of trapped CaMKII subunits to sum with each repetition.

In Coomber's model (Coomber, 1998b) a low-frequency stimulus caused many CaMKII subunits to enter the inhibited state (phosphorylation on T^{306} only). The low-frequency stimulus led to $CaMCa_4$ binding to only a small number of subunits — too few to permit extensive T^{286} autophosphorylation — and these activated subunits phosphorylated free CaMKII subunits at T^{306}. However, intersubunit autophosphorylation of free subunits on T^{306} is not thought to happen and is not included in other models. A basal autophosphorylation is known to occur on T^{306} but this is an intrasubunit reaction, not an intersubunit one (Mukherji and Soderling, 1994) and this transition is extremely slow — far slower than $CaMCa_4$ unbinding from CaMKII subunits (Colbran, 1993; Mukherji and Soderling, 1994). Nevertheless it would be interesting to model the effect on CaMKII activation of intrasubunit transitions to the inhibited state.

3.9.2 Different Stages of CaMKII Activation

The second-generation models showed that CaMKII activation during and after a calcium signal has different stages. Holmes (2000) found three distinct stages of activation in his model. The first was a short, highly activated stage represented by CaMKII subunits entering the bound state. This stage lasted less than 1 sec with the time course governed primarily by the dissociation rate of $CaMCa_4$ from CaMKII. The second stage was a moderately activated stage lasting up to about 40 sec. This period

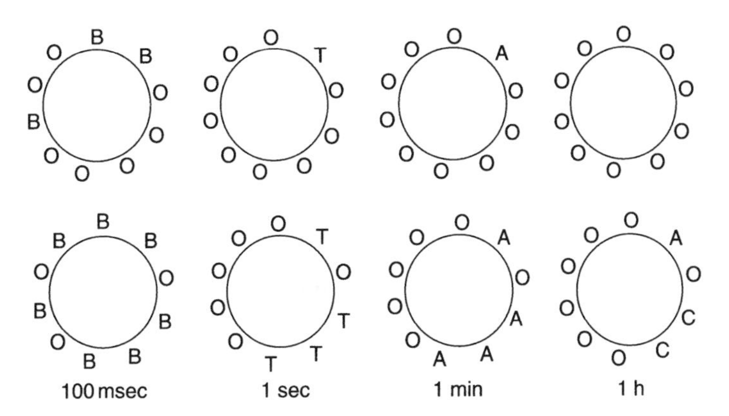

FIGURE 3.3 CaMKII subunit transitions with weak and strong calcium signals. The CaMKII holoenzyme is shown with ten subunits for illustration purposes. O = free subunit, B = bound, subunit bound with $CaMCa_4$, T = trapped, subunit autophosphorylated at T^{286} with $CaMCa_4$ trapped, A = autonomous, subunit phosphorylated at T^{286} but without $CaMCa_4$, C = capped, subunit autophosphorylated at the calmodulin binding site $T^{305,306}$ as well as at T^{286}. With a weak to moderate calcium signal, calcium binds to calmodulin and the $CaMCa_4$ complex will bind to a small number of subunits as shown on the left of the first row. Here there is only one instance where neighboring subunits are bound. This leads to one trapped subunit while $CaMCa_4$ unbinds rapidly from the other subunits. This one trapped subunit loses the $CaMCa_4$, enters the autonomous state, and later becomes dephosphorylated. The second row shows a response to a strong calcium signal where 7 of 10 subunits are bound with $CaMCa_4$. Here there are many neighboring subunits in the bound state leading to four subunits becoming autophosphorylated at the T^{286} site. The trapped $CaMCa_4$ eventually unbinds from these T subunits, but leaves these subunits still autophosphorylated at T^{286}. These autonomous subunits undergo autophosphorylation to the capped state. A rough timescale is indicated below each step. Binding of $CaMCa_4$ occurs within 100 msec, but reverses within 1 sec. Subunits that enter the trapped state hold on to $CaMCa_4$ for up to a minute. Subunits may remain in the A or C state for several minutes to several hours depending on the rate of dephosphorylation.

was governed by the length of time $CaMCa_4$ was trapped on CaMKII due to autophosphorylation at T^{286}. The final stage was a long-lasting activated stage where most subunits were in the autonomous or capped states. The duration of this final stage depended on dephosphorylation rates, but, interestingly, the decay of this stage of activation was faster from low CaMKII activation levels than from high levels. The relative timings of these stages are illustrated in Figure 3.3. Coomber's model (1998a) showed two stages of activation. As mentioned earlier, Coomber's stimulus was a long tetanus and the bound-to-trapped rate constant was tenfold faster than in the Holmes model. These factors led to a highly activated first stage composed primarily of trapped subunits that began during the 5 sec long high-frequency tetanus and lasted 5 to 10 sec after the end of the tetanus. Once the tetanus ended there was a burst of autophosphorylation at $T^{305,306}$ leading to most subunits being in the capped state. This second stage closely resembled the third stage in the Holmes model. Kubota and Bower (2001) noted transient and asymptotic stages of CaMKII activation in their model which roughly correlate with the two stages of activation found by Coomber. While it is interesting that CaMKII activation has such distinct stages, it is not clear which stage of CaMKII activation is most critical for LTP induction or whether the stages have distinct functions.

3.9.3 CaMKII Activation as a Bistable Molecular Switch

Theoretical studies have conjectured that CaMKII, because of its autophosphorylation properties, may act as a bistable molecular switch that forms the basis of learning and memory (Lisman, 1985; Lisman and Goldring,1988). There is a competition between phosphorylation and dephosphorylation of CaMKII subunits and CaMKII activation levels depend on which side wins this competition. Models have shown that CaMKII activation is a nonlinear function of tetanus frequency, as noted above, but what causes this nonlinearity?

Models have explored parameter space to determine factors that can cause switch-like behavior in CaMKII activation. Coomber (1998a) states that to get switch-like behavior in his model, the ratio of phosphatase concentration to CaMKII concentration must be small. Okamoto and Ichikawa (2000a) report that switch-like behavior occurs because phosphatase concentration is limited. Dephosphorylation is modeled as a Michaelis–Menten reaction, and low phosphatase concentration means that the maximum velocity of the dephosphorylation reaction is low. In addition, switch-like behavior depends on CaMKII concentration being large compared to the Michaelis–Menten constant of the dephosphorylation reaction. Switch-like behavior becomes more gradual as these restrictions are relaxed. In their simulations, calcium concentration was constant, but as it was stepped to higher levels, the probability that CaMKII would bind $CaMCa_4$ eventually reached a threshold beyond which almost all CaMKII entered the trapped state. Zhabotinsky (2000) used a much more realistic calcium signal and included phosphatase dynamics in his model and found bistability over a more realistic range of calcium concentrations than Okamoto and Ichikawa (2000a). Zhabotinsky (2000) also concluded that bistability was the result of the concentration of CaMKII subunits being much greater than the Michaelis–Menten constant of dephosphorylation. The region of bistability enveloped the resting calcium concentration, meaning that when calcium concentration returns to rest following a large calcium signal, CaMKII would remain fully phosphorylated. Bistability was found to be a robust phenomenon that occurred over a wide range of modeled parameter values. Interestingly, recent experiments found that CaMKII can act as a reversible switch under certain laboratory conditions (Bradshaw et al., 2003). In these experiments, CaMKII autophosphorylation alone had a switch-like dependence on calcium concentration; the presence of PP1 made this dependence steeper. It remains to be determined whether CaMKII acts as a bistable switch or as an ultrasensitive reversible switch in dendritic spines.

3.9.4 CaMKII and Bidirectional Plasticity

An interesting extension of CaMKII activation models was recently proposed by d'Alcantara et al. (2003). In this model, AMPA receptors were assumed to be in one of three states — completely dephosphorylated, phosphorylated at S^{845} (the naïve state), or phosphorylated at both S^{845} and S^{831}. In the model, $CaMCa_4$ activates both CaMKII and calcineurin, and phosphatase dynamics are included much as in the Zhabotinsky (2000) model. When calcium concentration is increased slightly from rest, phosphatase activation develops and AMPA receptors are driven to the completely dephosphorylated state, which leads to LTD. As calcium is raised above 0.5 μM, AMPA receptors are driven to the doubly phosphorylated state, which is what happens with LTP. To get this bidirectional plasticity, the phosphatase (PP1) must be activated at lower calcium concentrations than CaMKII.

3.9.5 CaMKII Activation and Spine Shape

It was discussed earlier how spine shape plays an important role in determining the amplitude and time course of the calcium transient, and so it should be no surprise that spine shape influences CaMKII activation levels. There are a number of difficulties, however, with quantifying CaMKII activation as a function of spine shape. First, different spine shapes will have different head sizes and will probably have different PSD sizes and calcium influx functions. If the PSD size and hence the number of NMDA receptors at a synapse is correlated with spine-head membrane area, as has been suggested, then models can account for this difficulty. Second, how does the number of CaMKII holoenzymes differ between spines of different shape? The simplest assumption is that CaMKII concentration is constant, meaning that the number of holoenzymes varies with head volume. How much variability exists in CaMKII concentration among spines is not known. Third, does calmodulin concentration vary with spine shape? Again the simplest assumption is that calmodulin concentration is constant, which means that the number of calmodulin molecules varies with head volume.

If the simple assumptions outlined above are made, preliminary simulations suggest that it is very difficult to get significant CaMKII activation in stubby spines (Holmes and Li, 2001). Significant CaMKII activation would require that the density of NMDA receptors at stubby spine PSDs be more than twofold higher than in long-thin spines. In addition, increasing the number of CaMKII holoenzymes to compensate for a larger spine-head volume for stubby spines has the effect of making T^{286} phosphorylation more difficult. With more subunits available, $CaMCa_4$ has more places to bind, and this reduces the probability of finding neighboring subunits in the bound state. The simulations do indicate that calcium influx and CaMKII activation are very sensitive to small changes in the number of NMDA receptors at a synapse, but only when the number of NMDA receptors exceeds a threshold number. In the model this number is less than 20 for long-thin spines, but more than 60 for stubby spines.

3.9.6 Models Predict the Need for Repetition of Short Tetanus Trains

One paradigm for inducing LTP is to apply short high-frequency trains of stimuli. Typically these short trains are repeated every 1 to 20 sec. There are two reasons why train repetition may be necessary. First, as suggested by Holmes (2000), a short train may activate 70% or more of the CaMKII subunits by taking them to the bound state, but only about 10% of these will enter the trapped state. With train repetition, the number of subunits in the trapped state will add. Subunits in the trapped state will still be in the trapped state at the time of a subsequent train and will aid in the autophosphorylation of additional subunits. After 8 to 10 repetitions a large percentage of subunits will have passed through the trapped state. Second, Li and Holmes (2000) found that if the NMDA conductance were modeled stochastically (with MCELL software, Stiles and Bartol, 2001), then the peak of the calcium transient during a short tetanus could vary threefold. The effect of repetition would be to smooth out these stochastic variations in calcium influx with individual trains and make the CaMKII activation level after 8 to 10 repetitions robust.

3.10 Future Perspectives

Obviously, second-generation spine models have not yet fully run their course, but it is time to think about what characteristics the next generation of models should include. Clearly, many of the current models do not consider the restrictions imposed by spine geometry, but it will be increasingly important to do so. Because cellular signaling mechanisms begin with calcium influx and lead to a particular synaptic change that may involve cascades of biochemical reactions that can be activated or deactivated to various degrees depending on the temporal characteristics of the calcium signal, it will be important to model the calcium signal as realistically as possible. Dozens of identified proteins play critical roles in LTP or LTD (Sanes and Lichtman, 1999) but multiprotein complexes, rather than individual molecules, may turn out to be most important for signaling (Bray, 1998; Grant and O'Dell, 2001). Kinases and phosphatases are known to bind to synaptic receptors and synaptic proteins (Gardoni et al., 1998, 1999, 2001; Leonard et al., 1999; Chan and Sucher, 2001), and their primary function may occur from this tethered state (Kennedy, 1998, 2000). At the very least, reactions that will be important to include in future models are the dynamics of phosphatase activation, CaMKII translocation, AMPA receptor phosphorylation, and AMPA receptor incorporation and removal from the PSD (a process Lisman calls "AMPAfication," Lisman, 2003).

So what will the next generation of spine models look like mathematically? First, these models will have to be completely stochastic. The deterministic approach has its obvious limitations — Coomber (1998a, 1998b) and Kubota and Bower (2001) use thousands of equations while considering only four of the subunits on a given CaMKII holoenzyme. Perhaps more importantly, the number of molecules of any given protein or ion in a dendritic spine is small (see Problem 2), and when the number of molecules is small, the deterministic approach breaks down. Second, the next generation of spine models should include realistic spine geometry with diffusional barriers as appropriate. For calcium, calmodulin, CaMKII, and other moieties to diffuse and react with each other, they may have to deal with boundaries formed by multiprotein complexes. After all, we know that the area near the PSD gets crowded (Dosemeci et al., 2000). Third, the next generation of spine models should have a stochastic calcium influx function and be able to handle many additional reactions stochastically, including those mentioned earlier.

How can diffusion and reaction in spines be modeled computationally? The major difficulty is to have a stochastic algorithm that can handle reactions between molecules that diffuse in the cytoplasm. MCELL (Stiles and Bartol, 2001) is a marvelous software program for modeling diffusion and binding to membrane-bound receptors. However, in its current incarnation MCELL cannot handle reactions among randomly diffusing molecules, but the developers are actively working on this. Another possible approach is to use a Monte Carlo lattice method as recently described by Berry (2002), where each molecule occupies a position in a three-dimensional lattice and randomly diffuses or reacts with other molecules at neighboring lattice points. Diffusion barriers are quite easily established with this approach. This approach has been used in ecological models of predator–prey interactions in heterogeneous environments. A third option that has been quite popular recently is the Gillespie algorithm (Gillespie, 1976, 1977). The major problem with the Gillespie algorithm is that it can be applied only to homogeneous volumes. As such, it can handle reactions, but not diffusion. Nevertheless, with a few modifications, the Gillespie algorithm can be applied to dendritic spines.

To see how the Gillespie algorithm might be applied to dendritic spines, it is necessary to review the Gillespie algorithm. Gillespie began with the reaction-probability function $P(\tau, \mu)$, where $P(\tau, \mu)dt$ is the probability that the next reaction in a volume will occur in the interval $(t + \tau, t + \tau + dt)$ and will be a type R_μ reaction. The function $P(\tau, \mu)$ can be decomposed into the product of two functions $P_1(\tau)$ and $P_2(\mu|\tau)$, where $P_1(\tau)d\tau$ is the probability that the next reaction will occur in the interval $(t + \tau, t + \tau + d\tau)$ and $P_2(\mu|\tau)$ is the probability that the next reaction will be of type R_μ. Gillespie showed that by drawing two random numbers r_1 and r_2 one can compute the time of the next reaction and determine which specific reaction takes place. In particular, the time of the next reaction, τ, is given by

$$\tau = \frac{1}{a} \ln\left(\frac{1}{r_1}\right), \tag{3.12}$$

where a is the sum of all of the probabilities of all possible reactions, and the specific reaction μ that occurs is given by

$$\sum_{i=1}^{\mu-1} a_i < r_2 a < \sum_{i=1}^{\mu} a_i, \tag{3.13}$$

where the a_i represents the probability of the ith reaction occurring given the number of molecules of reactants in the volume. Once the time and identity of the next reaction are known, the number of molecules of those species involved in the reaction are updated.

To extend this algorithm to spines requires two modifications. First, the dendritic spine and the associated dendrite are divided into a number of voxels that is small enough so that the homogeneity requirement of the Gillespie algorithm is satisfied within each voxel. This allows the Gillespie algorithm to be applied within each voxel. Second, diffusion between voxels is considered to be a special reaction (Stundzia and Lumsden, 1996). With these modifications, the a term in Equations (3.12) and (3.13) includes not only the probabilities of reactions, but probabilities of diffusion as well. Furthermore, the a term includes the probabilities of diffusions and reactions in all voxels. Then the first random number is used, as in Equation (3.12), to compute the time of the next reaction, but this reaction can also be a diffusion event and can occur in any voxel. The second random number is used to determine which reaction occurs in which voxel.

One obvious difficulty that the next generation of spine models will face is the large number of unknown parameter values. It will be difficult to decide which reactions are important and which can be (temporarily) ignored. As is obvious from Tables 3.1–3.4, even when a rate constant is known, it may not be known with great certainty, and concentrations and densities of buffers, pumps, and various reactants may have widespread heterogeneity. What the models can do is examine interactions among a number of different reaction pathways simultaneously, an advantage that experimentalists with the ability to manipulate one variable at a time do not share. But to gain insight into the mechanisms being studied, computational experiments will have to be designed with considerable care.

3.11 Summary

Numerous theories have been proposed to explain the function of dendritic spines. Beginning in the 1970s, theories focused on the electrical resistance of the spine neck and showed how changes in neck diameter might play a role in plasticity. More recently, spines have been viewed as isolated biochemical compartments where calcium can be safely concentrated. The amplitude and temporal characteristics of spine-head calcium concentration provide the signal to initiate specific calcium-dependent reaction cascades that lead to changes in synaptic strength. In this chapter we discussed three generations of spine models and related experimental results that provide insight into spine function. First-generation spine models demonstrated that: (a) spines can concentrate and amplify calcium concentration, (b) spine-head calcium concentration generated by various stimulation conditions is a good predictor of LTP, and (c) spine shape plays an important role in the amplitude and time course of the calcium signal. The development and use of fluorescent calcium indicators with digital CCD camera imaging, confocal microscopy, and two-photon imaging has allowed many of the predictions of first-generation spine models to be tested and verified. These new experimental techniques have provided additional insights into spine calcium dynamics, particularly with regard to the roles of calcium pumps, buffers, and stores, as well as new insights into spine function, including the role of spine motility and backpropagating action potentials for coincidence detection. Second-generation spine models study reaction cascades following calcium entry. These models have shown that: (a) activation of CaMKII generated by various stimulation conditions is a good predictor of LTP, (b) CaMKII activation has specific stages with unique levels and time courses, and (c) CaMKII may function as a bistable molecular switch. To gain further insights into spine function, third-generation spine models will have to model diffusion and reaction in spines stochastically while taking into account appropriate spine geometry and barriers and realistic calcium influx.

PROBLEMS

1. Older theoretical models proposed a function for spines that depended on the electrical properties of the spine, in particular on the spine-stem resistance, R_{SS}. Peters and Kaiserman-Abramoff (1970) measured spine dimensions in cortical neurons and categorized spines as long-thin, mushroom-shaped, or stubby. They found average spine-neck dimensions to be approximately 0.1×1.1 μm (long-thin), 0.2×0.8 μm (mushroom-shaped), and 0.6×1.0 μm (stubby).
   (a) Compute R_{SS} for these spine necks assuming an axial resistivity, R_a, of 100 Ω cm.
   (b) Harris and Stevens (1989) report that in hippocampal area CA1, spine-neck diameters range from 0.038 to 0.46 μm and spine length ranges from 0.16 to 2.13 μm. Given these data, compute upper and lower bounds for R_{SS} values in CA1 hippocampal neurons.

2. Resting calcium concentration is thought to be about 70 nM inside the cell. Let a spine head be represented as a sphere with diameter of 0.6 μm.
   (a) Compute the number of free calcium ions in the spine head at rest.
   (b) If calcium rises to 1 μM, how many free calcium ions are present in the spine head?
   (c) If 95% of the calcium that enters is immediately buffered, how many calcium ions have to enter the spine head to raise calcium concentration to 10 μM?
   (d) Repeat the calculations for a spherical spine head with a 1.0 μm diameter.
   (e) Suppose an NMDA-receptor channel with a single-channel conductance of 50 pS and reversal potential of 0 mV were open for 10 msec. Assume resting potential is −65 mV and that 10% of the NMDA current is calcium. How many calcium ions enter the spine through this one channel? Suppose magnesium blocked this NMDA channel 60% or 80% of the time it was open. How many calcium ions would enter now? Discuss with regard to the answers to earlier parts of this question.
3. Use Equations (3.1) to (3.6) and values in Table 3.1 to create a model of a spine similar to that pictured in Figure 3.1A. Assume that calcium current into the outermost spine-head compartment is 0.6 pA for 100 msec (you may want to adjust these values depending on the shape of your modeled spine). Compare the calcium transients in the spine head, spine neck, and dendrite for the cases of smooth and abrupt boundary conditions at the spine-head–spine-neck junction and at the spine-neck–dendrite junction (see Figure 3.1B and the discussion in the text). How different are the decay transients with the different boundary conditions?
4. Using the model created in Problem 3, compare calcium transients for long-thin spines, mushroom-shaped spines, and stubby spines.
   (a) How much larger must you make the stubby spine calcium current to enable the stubby spine to have a peak calcium concentration of the same amplitude as that found with the long-thin spine?
   (b) Modify the spine model to include release of calcium from stores (Equations [3.7] to [3.10]). How does this affect the amplitude and time course of the calcium transient at the stubby spine?
5. Several second-generation models find that CaMKII can act as a bistable molecular switch. The major requirement for this to occur is that CaMKII concentration must be much larger than the Michaelis–Menten constant of dephosphorylation. Develop a second-generation model along the lines of Zhabotinsky (2000) or Okamoto and Ichikawa (2000a) and explore the dependence of switching characteristics on this condition, paying particular attention to the slope of the transition between low CaMKII activation and high CaMKII activation.
   (a) For fixed K_m of dephosphorylation (0.1–15 μm), at what CaMKII concentration does the slope of the "switch" change from being less than 1 to being greater than 1?
   (b) For fixed CaMKII concentration (1–40 μm), at what K_m of dephosphorylation does the slope of the "switch" change from being less than 1 to being greater than 1?
   (c) Repeat (a) and (b) to find values that give a slope greater than 5.
   (d) Okamoto and Ichikawa (2000a) find that phosphatase concentration must also be low. Vary phosphatase concentration and determine the effect this has on switching characteristics.
6. Third-generation models will be completely stochastic. Develop a third-generation model following one of the approaches outlined in the text.
   (a) As with models of presynaptic calcium influx, one would expect there to be calcium microdomains near NMDA-receptor channels that would dissipate very quickly. Experimental work shows that calcium within the spine head is uniform within the time resolution of the imaging study (2.5 msec). Can gradients of calcium concentration be established within the spine head? If so, how fast do they dissipate? If not, how can model parameters be changed to create a gradient?
   (b) What effect does CaMKII localization within the spine head have on whether or not it gets activated? Compare CaMKII activation levels in simulations where CaMKII is distributed

randomly in the spine with simulations where CaMKII is localized in or adjacent to the postsynaptic density.

(c) It is not known whether calmodulin concentration is a limiting factor in CaMKII activation. The free concentration of calmodulin in spines is thought to be small, because calmodulin may be bound to various calmodulin-binding proteins. Okamoto and Ichikawa (2000b) suggest that free calmodulin may diffuse from neighboring spines. Compare simulation results in which (a) calmodulin concentration is low, (b) calmodulin is readily diffusible, (c) calmodulin is immobile, and (d) calmodulin is released from calmodulin-binding proteins upon calcium influx, to determine conditions where calmodulin may and may not be limiting.

APPENDIX 1. TRANSLATING BIOCHEMICAL REACTION EQUATIONS TO DIFFERENTIAL EQUATIONS

Translating word problems into differential equations is something many people find difficult. Fortunately there are many excellent sources that describe how to do this on a very basic level (e.g., Braun et al., 1983; Keen and Spain, 1992). Translating biochemical reaction equations into differential equations is particularly simple. We illustrate this with a simple example.

Consider the reaction of reactant A combining with reactant B to form a product C:

$$A + B \underset{k_{-1}}{\overset{k_1}{\rightleftarrows}} C. \tag{3.A1}$$

To form a differential equation for this reaction, we need to think about how the concentrations of A, B, and C change in time. We do this first for reactant A.

The concentration of A will decrease as A and B combine to form C. The rate of this decrease will be proportional to the current concentrations of both A and B. The proportionality constant is characteristic for a particular reaction and here we call it k_1. The constant k_1 is usually called the forward rate constant and since there are two reactants (second-order reaction), it has units of concentration^{-1} time^{-1} or, as typically given in the tables in this chapter, μM^{-1} sec^{-1}. Therefore the change in the concentration of A over time due to its reaction with B will be $-k_1AB$, where the minus sign indicates that the change is negative, and the k_1AB indicates that the change is proportional to the concentrations of both A and B.

In addition, the concentration of A will increase as C breaks down to form A and B. The rate of this increase is proportional to the concentration of C with proportionality constant k_{-1}. This constant is called the backward rate constant, and since C does not combine with anything else for this reaction, k_{-1} has units of time^{-1} or, as typically given in the tables above, sec^{-1}. Therefore the change in the concentration of A over time due to the breakdown of C will be $+k_{-1}C$, where the plus sign indicates that the change is positive, and the $k_{-1}C$ indicates that the change is proportional to the concentration of C.

We indicate the change in the concentration of A over time with the differential dA/dt or the change in A with respect to a change in time. Putting all of this together we have:

$$\frac{dA}{dt} = -k_1AB + k_{-1}C. \tag{3.A2}$$

If we analyze the changes in B and C in the same manner, we find that the differential equation for B is the same as that for A, and the differential equation for C is the negative of that for A or B. This makes sense because A and B must decrease at the same rate as they combine to form C and they must increase at the same rate as C breaks down. Similarly the change in C must be exactly opposite that of A and B. Consequently the differential equations for B and C are:

$$\frac{\mathrm{dB}}{\mathrm{d}t} = -k_1\mathrm{AB} + k_{-1}\mathrm{C}, \qquad \frac{\mathrm{dC}}{\mathrm{d}t} = +k_1\mathrm{AB} - k_{-1}\mathrm{C}. \tag{3.A3}$$

Often the individual forward and backward rate constants are unavailable. What may be available is the ratio of these rate constants, given, for example, as a dissociation constant K_d. The dissociation constant comes from a steady-state measurement. To go from the differential equations to the steady state is quite simple. In the steady state, all derivatives with respect to time are zero. Consequently, from Equation (3.A2) we have

$$0 = -k_1\mathrm{AB} + k_{-1}\mathrm{C}, \tag{3.A4}$$

which implies

$$\frac{k_{-1}}{k_1} = \frac{[\mathrm{A}][\mathrm{B}]}{[\mathrm{C}]} = K_\mathrm{d}, \tag{3.A5}$$

where the units of K_d are time^{-1}/(concentration^{-1} time^{-1}) or concentration. When only the K_d value is provided in an experiment, the modeler must choose what seem to be reasonable k_1 and k_{-1} values that are consistent with the K_d. This is what was done for many of the forward and backward rate constants shown in Figure 3.2.

As a second example, consider

$$4\mathrm{A} + \mathrm{B} \underset{k_{-1}}{\overset{k_1}{\rightleftarrows}} \mathrm{C}. \tag{3.A6}$$

We might have this situation if we modeled the formation of $\mathrm{CaMCa_4}$ as the binding of four calcium ions to calmodulin all at once. The analysis of this case is similar to that in the previous example except that now four molecules of A are consumed for every one of B and the rate of the forward reaction is proportional to $\mathrm{A}^4\mathrm{B}$ instead of A. Thus,

$$\frac{\mathrm{dA}}{\mathrm{d}t} = 4(-k_1\mathrm{A}^4\mathrm{B} + k_{-1}\mathrm{C}), \tag{3.A7}$$

$\mathrm{d}B/\mathrm{d}t = \frac{1}{4}\,\mathrm{dA}/\mathrm{d}t$, and $\mathrm{dC}/\mathrm{d}t = -\mathrm{dB}/\mathrm{d}t$.

APPENDIX 2. STOCHASTIC RATE TRANSITIONS

To model transitions between two states stochastically, we need to have an expression for the transition probability. In most articles, this transition probability is just given without explanation inasmuch as the probabilistic derivation is well-known. Rather than repeat that derivation here, we provide an intuitive understanding of the transition probability by considering analytic solutions to the differential equations for a simple first-order reaction.

Consider the simple case of one reactant being converted irreversibly into one product or

$$\mathrm{A} \xrightarrow{k} \mathrm{B}. \tag{3.A8}$$

This reaction could represent the autophosphorylation reaction for the transition of a bound subunit to a trapped subunit or an autonomous subunit becoming capped. For this reaction the differential equations for A and B are:

$$\frac{\mathrm{dA}}{\mathrm{d}t} = -k\mathrm{A}, \qquad \frac{\mathrm{dB}}{\mathrm{d}t} = +k\mathrm{A}. \tag{3.A9}$$

If A_0 and B_0 are the initial concentrations of A and B, we can solve these equations analytically to obtain:

$$\mathrm{A}(t) = \mathrm{A}_0 \exp(-kt), \qquad \mathrm{B}(t) = \mathrm{A}_0(1 - \exp(-kt)) + \mathrm{B}_0. \tag{3.A10}$$

The first term on the right-hand side of the equation for B represents the amount of B that has come from conversion of A by time t and equals the initial concentration of A minus the current concentration of A.

While Equation (3.A10) gives expressions for the whole population of A and B molecules, what happens to an individual A molecule? Equation (3.A10) tells us that at time t the proportion of the original number of A molecules that are still A molecules is $\exp(-kt)$ and the proportion that have become B molecules is $1-\exp(-kt)$. This will be true regardless of the initial number of A molecules. We can infer from this that in time interval Δt, an individual A molecule will remain an A molecule with probability $\exp(-k\Delta t)$ and will make the transition to B with probability $1 - \exp(-k\Delta t)$.

Stochastic models choose a random number between 0 and 1 for each A molecule at each Δt. If the random number is less than $1 - \exp(-k\Delta t)$ then a transition to B has been made; otherwise there is no transition. Because k and Δt are usually small and exp is a computationally expensive function, many models approximate $1 - \exp(-k\Delta t)$ with $k\Delta t$.

One additional note should be made regarding random number generators. It is important to test a random number generator before using it in a simulation. In particular, if the rate constant k and the time interval Δt are small, then the transition probability, $k\Delta t$, will be extremely small. In this case, the random number generator must be able to return extremely small numbers. If the transition probability is smaller than the smallest random number that can be returned by the random number generator, then the transition will never occur. One might have to get a random number generator that can produce very small numbers or else rescale the units of the rate constants and time so that the transition probabilities are not extremely small. Alternatively, one might separate extremely fast and extremely slow reactions.

APPENDIX 3. USE OF MICHAELIS–MENTEN KINETICS IN DEPHOSPHORYLATION REACTIONS

While converting regular biochemical reaction equations to differential equations, as shown in Appendix 1, is relatively simple, the use of the Michaelis–Menten equation to model dephosphorylation reactions is a bit more complicated. Consider the Michaelis–Menten type reaction often used to model the dephosphorylation reactions mentioned in the text:

$$\mathrm{E} + \mathrm{S} \underset{k_{-1}}{\overset{k_1}{\rightleftarrows}} \mathrm{ES} \underset{k_{-2}}{\overset{k_2}{\rightleftarrows}} \mathrm{E} + \mathrm{P}. \tag{3.A11}$$

Usually we ignore the second backward reaction (let $k_{-2} = 0$), which is reasonable for most enzymes. Here E is the phosphatase concentration, S could be phosphorylated CaMKII, and P could be dephosphorylated CaMKII.

Following the procedure outlined in Appendix 1, we observe that E will decrease as E and S combine to form ES with proportionality constant k_1, and E will increase as ES breaks down either to form E and P with rate constant k_2 or E and S with rate constant k_{-1}. S will decrease as E and S combine to form ES and will increase as ES breaks down to form E and S. P can only increase and does so as ES breaks down to form E and P with rate constant k_2. Finally ES increases as E and S combine and decreases as it breaks down to either E and S or E and P. This knowledge allows us to write the following differential equations directly:

$$\begin{aligned}
\frac{\mathrm{d[E]}}{\mathrm{d}t} &= -k_1[\mathrm{E}][\mathrm{S}] + (k_{-1} + k_2)[\mathrm{ES}], \\
\frac{\mathrm{d[S]}}{\mathrm{d}t} &= -k_1[\mathrm{E}][\mathrm{S}] + k_{-1}[\mathrm{ES}], \\
\frac{\mathrm{d[ES]}}{\mathrm{d}t} &= k_1[\mathrm{E}][\mathrm{S}] - (k_{-1} + k_2)[\mathrm{ES}], \\
\frac{\mathrm{d[P]}}{\mathrm{d}t} &= k_2[\mathrm{ES}].
\end{aligned} \tag{3.A12}$$

(We have chosen the subscript notation for the rate constants to allow the reader to interpret the rate constants used in Zhabotinsky [2000] correctly, and we have used brackets to indicate concentrations to distinguish between [E][S] and [ES] in the equations.)

We can solve the equations as given in Equation (3.A12) directly or we can make some simplifications. In the model we may make an assumption about the total concentration of a particular phosphatase. If we know the total concentration $[\mathrm{E_t}]$, then we can eliminate the need to solve $\mathrm{d[E]}/\mathrm{d}t$ by substituting $[\mathrm{E_t}] - [\mathrm{ES}]$ wherever [E] appears in the other equations. This gives us:

$$\begin{aligned}
\frac{\mathrm{d[S]}}{\mathrm{d}t} &= -k_1([\mathrm{E_t}] - [\mathrm{ES}])[\mathrm{S}] + k_{-1}[\mathrm{ES}], \\
\frac{\mathrm{d[ES]}}{\mathrm{d}t} &= k_1([\mathrm{E_t}] - [\mathrm{ES}])[\mathrm{S}] - (k_{-1} + k_2)[\mathrm{ES}], \\
\frac{\mathrm{d[P]}}{\mathrm{d}t} &= k_2[\mathrm{ES}].
\end{aligned} \tag{3.A13}$$

Then if we want to know [E], we merely compute $[\mathrm{E_t}] - [\mathrm{ES}]$ where $[\mathrm{E_t}]$ is known and [ES] is computed from the solution of Equation (3.A13).

While we can solve Equations (3.A13) directly, an alternative is to make the Michaelis–Menten assumption that the formation and breakdown of ES is in a quasi-steady state. This implies that [ES] is not changing or equivalently, $\mathrm{d[ES]}/\mathrm{d}t = 0$. Then solving for [ES] gives

$$[\mathrm{ES}] = \frac{[\mathrm{E_t}][\mathrm{S}]}{[\mathrm{S}] + (k_1 + k_2)/k_{-1}} = \frac{[\mathrm{E_t}][\mathrm{S}]}{[\mathrm{S}] + K_\mathrm{m}}, \tag{3.A14}$$

where K_m is the Michaelis–Menten constant for the reaction. Given this value for [ES] one can simplify the differential equations for [S] and [P] (algebra left to the reader as an exercise) to get:

$$\frac{d[S]}{dt} = -\frac{k_2[E_t][S]}{[S]+K_m},$$
$$\frac{d[P]}{dt} = +\frac{k_2[E_t][S]}{[S]+K_m}. \tag{3.A15}$$

The product $k_2[E_t]$ is usually called V_{max} (the maximum velocity of the breakdown of ES, which would occur if all of the enzyme were in the form ES, that is, $[ES] = [E_t]$). Both Zhabotinsky (2000) and d'Alcantara et al. (2003) use a version of Equation (3.A15) to model dephosphorylation in their calculations.

4 Physiological and Statistical Approaches to Modeling of Synaptic Responses

Parag G. Patil, Mike West, Howard V. Wheal, and Dennis A. Turner

CONTENTS

4.1 Introduction

4.1.1 Modeling Synaptic Function in the CNS

Transmission of signals across synapses is the central component of nervous system function (Bennett and Kearns, 2000; Manwani and Koch, 2000). Primarily, such synaptic transmission is mediated by chemical neurotransmitter substances that are released into the synapse by the presynaptic neuron and detected by receptors located upon the postsynaptic neuron. In many areas of the nervous system

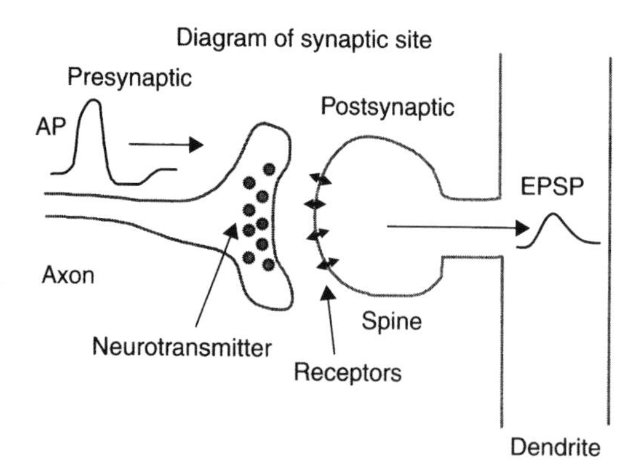

FIGURE 4.1 Diagram of synaptic site. This diagram shows the presynaptic axon (to the left) and the presynaptic terminal containing vesicles and neurotransmitter. The presynaptic action potential (AP) conducted along the axon leads to calcium release and vesicle fusion and release occurs, though as a stochastic function. This uncertainty leads to a release probability (π), which is simply the number of releases occurring in comparison to the number of presynaptic action potentials. The neurotransmitter then binds to receptors on the postsynaptic side of the synaptic cleft (in the hippocampus usually a dendritic spine that is attached to a dendrite), resulting in ionic channel opening and a postsynaptic current or voltage (EPSP) being generated in the dendrite, with peak amplitude μ. On any given neuron there may be multiple such release sites (k), acting in concert, but likely with different parameters. The EPSP signals are summated in time and space from these multiple synapses to determine the postsynaptic activity of the neuron.

(outside of the hippocampus and neocortex) there may also be dendritic neurotransmitter release, which appears to behave in a very different fashion from the more traditional axonal neurotransmitter release (Ludwig and Pittman, 2003). However, because synaptic structures are microscopic and inaccessible, direct measurement of synaptic properties is not often feasible experimentally. As a result, our understanding of the mechanisms underlying synaptic transmission and modulation of function derives from statistical inferences, which are made from observed effects on populations of synapses. Synapses in the central nervous system (CNS) have the distinct property of being unreliable and probabilistic, hence a statistical framework is critical to define function. A schematic of a typical synapse is shown in Figure 4.1, indicating that the information transfer is between a relatively secure presynaptic action potential (on the left) and a highly insecure neurotransmitter release and subthreshold postsynaptic response. For most CNS neurons, spatial and temporal integration within a neuron is an additional complication, often requiring hundreds of synaptic responses summing together to reach the action potential threshold.

Typically, studies of synaptic function have made productive use of statistical methodologies and conceptual frameworks developed for a prototypical (but very unusual) model synapse, the neuromuscular junction (NMJ), which is described in a classic work by Del Castillo and Katz (1954). At the NMJ, an action potential in the presynaptic axon arrives at the synaptic terminal and, with a finite probability of occurrence π, triggers the release of neurotransmitter. Following this probabilistic event, the neurotransmitter diffuses across the synaptic cleft to bind postsynaptic neurotransmitter receptors. The binding of postsynaptic neurotransmitter receptors produces electrophysiological responses in the postsynaptic neuron, namely, a change of magnitude, μ, in the transmembrane current or potential. Because synapses at the NMJ contain multiple sites, which release neurotransmitter in concert following a single stimulus to produce a single end-plate potential, analysis of the activity and parameters of individual release sites requires statistical inference. The NMJ is unusual in comparison with most CNS synapses in that the overall response (a postsynaptic muscle twitch) is highly reliable, so the inherent insecurity of synaptic release is overcome by multiple synaptic sites located in very close proximity. In contrast, synapses are separated on the surface of CNS neurons and are much less potent, creating a large capability for synaptic interactions and summation (Craig and Boudin, 2001).

At synapses such as the NMJ that are designed for security rather than capacity for computational functions, the amount of released neurotransmitter may saturate postsynaptic receptors. Such saturation ensures that presynaptic action potentials are reliably transduced into postsynaptic depolarizations. As the necessary consequence of such reliability, presynaptic neurotransmitter release at such synapses is relatively information poor, since variation in the magnitude of release only minimally affects the magnitude of postsynaptic depolarization. By contrast, at central synapses (Figure 4.1), the concurrent activation of hundreds of synapses may be required to produce a postsynaptic response. The requirement for postsynaptic signal integration enables substantial information transfer. Together with variation in the numbers of simultaneously activated presynaptic terminals, the modulation of presynaptic release at individual terminals and the modulation of postsynaptic responses convey information between CNS neurons. Synaptic modulation may thereby encode highly complex cognitive processes, including sensation, cognition, learning, and memory (Manwani and Koch, 2000). Since multiple CNS synapses may be concurrently activated between individual neurons, the examination of synaptic release and neuronal response characteristics often requires a complex statistical model.

The goal of this chapter is to review available statistical models, their applicability to CNS synapses typically studied, as well as the criteria under which the models may be evaluated by physiological assessment. Complexities associated with synaptic heterogeneity and synaptic plasticity, which statistical models must address, are discussed first. We then explore the qualities of electrophysiological measurements that enable meaningful statistical inferences to be drawn from experimental data. After a discussion of classical statistical models of central synapses, we present and evaluate the Bayesian approach used in our laboratory to study mechanisms of synaptic plasticity in CNS neurons, in comparison with other types of models available (Bennett and Kearns, 2000).

4.1.2 Complexity Introduced by Synaptic Heterogeneity and Plasticity

If central synapses were static and homogeneous, simple mathematical models would adequately describe synaptic transmission. For example, end-plate potentials at the NMJ might be assumed to be produced by an arbitrary but fixed number of independent release sites, n, which release their quanta with average probability, p, in response to presynaptic stimulation. Such release characteristics would produce a binomial distribution of end-plate potential magnitudes. Given such release characteristics, the parameters, n and p, could be estimated from a histogram summarizing the magnitude of responses to presynaptic stimulation (Katz, 1969). The computational requirements of CNS neurons necessitate much greater complexity, and an important objective of the statistical modeling of CNS synapses is to reveal the physiological manifestations of this complexity through its effects on synaptic release parameters.

Central synapses exhibit both heterogeneity and plasticity — their properties differ both between synaptic sites (intersite variability) and over time (intrasite variability). Such complexity arises from multiple factors including variable packaging of neurotransmitter, neurotransmitter dynamics in the synaptic cleft, agonist–receptor interactions at the synapse, channel opening properties, and postsynaptic modification of receptor proteins (Redman, 1990; Korn and Faber, 1991; Faber et al., 1992; Buhl et al., 1994; Bennett and Kearns, 2000; Harris and Sultan, 1995; Craig and Boudin, 2001). In addition, the dendritic location of postsynaptic synapses, voltage-dependent dendritic properties, and inhibition of dendritic signaling may modulate the effects of synaptic events upon the summation sites of postsynaptic neurons (Turner, 1984a, 1984b; Spruston et al., 1993; Magee and Johnston, 1997). The existence of such location-dependent heterogeneity is supported by experimental data (Buhl et al., 1994; Bolshakov and Siegelbaum, 1995; Harris and Sultan, 1995; Turner et al., 1997).

There are many sources of variability inherent to synaptic activation. Release is dependent on calcium entry into the presynaptic terminal, often resulting in failures of release (Glavinovic and Rabie, 2001). Most neurotransmitters require some uptake mechanism for recycling, as well as a synthetic cycle to redistribute them, resulting in variable neurotransmitter packaging in synaptic vesicles that changes with development (Mozhayeva et al., 2002). Once a vesicle is primed and docked for potential release, the duration of docking and the amount of neurotransmitter released per evoked action potential are unknown, and particularly whether multiple cycles of release can occur from a single vesicle (Bennett and Kearns, 2000). Once neurotransmitter is released into the synaptic cleft, a receptor–ligand mismatch may result in a variable postsynaptic response from trial to trial by stochastic diffusion and binding (Faber et al., 1992). Depending on the postsynaptic density of ligand-gated receptors and the type of receptor, a variable postsynaptic response at the synaptic site may be elicited. The typical postsynaptic response may be further altered by variable dendritic conduction pathways to the soma, and by inhibition or enhancing mechanisms on the conduction path (Turner, 1984b; Magee and Johnston, 1997). The large chain of events from presynaptic action potential to postsynaptic response at a soma or summating output site includes many sources of variability. Further, all finite samples of data are subject to sampling variability as well as measurement error, which complicate interpretations.

Synapses are not static; they evolve over time. For example, systematic changes in the glutamate receptor phenotype (from NMDA to non-NMDA), probability of release, and degree of potentiation have been suggested to occur in the hippocampus during development (Bolshakov and Siegelbaum, 1995; Durand et al., 1996; Craig and Boudin, 2001; Hanse and Gustafsson, 2001). This change in receptor phenotype has also been observed at "new" synapses induced with long-term potentiation (LTP), by observing normally "silent" NMDA synapses (at resting potential). These silent synapses can be transformed into active non-NMDA synapses by a combination of postsynaptic depolarization and synaptic activation, presumably with receptor modification induced by the calcium influx associated with the NMDA receptor activation (Isaac et al., 1995). The overall implication of this induction of "new" synapses is an increased efficacy of synaptic function, but the specific parameters underlying this enhancement remain unresolved. Thus, synaptic release parameters at individual sites vary over time and as a function of development, as well as between individual sites; this variability results in a very complex set of responses, even when reduced to a small number of connections, as noted with unitary (single presynaptic to single postsynaptic) responses (Debanne et al., 1999; Pavlidis and Madison, 1999; Kraushaar and Jonas, 2000).

Other critical instances in which statistical models may help to clarify the physiology underlying synaptic function include the phenomena of short- and long-term plasticity (Isaac et al., 1995; Stricker et al., 1996b; Debanne et al., 1999). Short-term facilitation is considered to primarily involve changes in presynaptic factors, particularly probability of release, since these changes occur over a few milliseconds (Redman, 1990; Turner et al., 1997). However, LTP may exhibit many different mechanisms of plasticity, particularly structural changes if the duration is longer than 10 to 30 min. Due to the popularity of the study of LTP and the wide variety of animal ages and preparations in which this phenomenon is studied, multiple and controversial mechanisms have been suggested (Bolshakov and Siegelbaum, 1995; Durand et al., 1996; Stricker et al., 1996b; Debanne et al., 1999; Sokolov et al., 2002). Since the evoked response histograms change radically after the induction of LTP, various interpretations have included both enhanced probability of release and also increased amplitude of synaptic site responses. Few of these physiological studies have shown detailed statistical evaluation of the physiological signals that would allow for extensive variation between synaptic release sites, other than a compound binomial approach (Stricker et al., 1996b). Thus, current statistical approaches in the evaluation of LTP may be premature in that variation between synaptic release sites (and the possible addition of release sites with potentiation) has been insufficiently studied.

There are many physiological situations where inferring the underlying statistical nature of synaptic release may lead to understanding the mechanism of critical processes, for example, memory formation and retention. Thus, statistical interpretation of synaptic responses has a clear rationale in physiological interpretation. However, the ambiguity of statistical interpretation, at least to a physiologist's perceived need for an unambiguous result, has lent an esoteric slant to the field of statistical analysis of synaptic signals and modeling.

4.1.3 Complexity Associated with Physiological Recordings

Statistical interpretation of data arising from studies of synaptic transmission may be obscured by ambient noise associated with either the neuron or the recording conditions. Korn and Faber (1991) and Redman (1990) have previously defined the appropriate physiological circumstances for a proper (and reliable) analysis of underlying synaptic response parameters. These requirements focus on two-cell recordings (either using sharp or patch electrodes) where clear, unitary (but usually multisite) responses can be identified and include:

1. The stimulation of a single presynaptic spike and control of the presynaptic element.
2. Optimization of the noise level to isolate miniature events to compare with evoked events, with miniature events occurring from the same synaptic sites as those from which evoked events emanate.
3. Resolution of single site amplitude, with minimum cable or electrotonic filtering (usually requiring a perisomatic or proximal dendritic synaptic location).
4. Direct data on variations in quantal amplitude at a single site.
5. Morphological identification of the number of active sites involved and their dendritic locations.
6. Independent evidence on uniformity of amplitude and probability between sites.

These fairly rigid criteria are primarily designed for two-cell recording and in circumstances where both elements can be stained and then anatomically analyzed. Korn and Faber have extensively performed this type of analysis on goldfish Mauthner cells, which have a glycinergic inhibitory synapse which fulfills these criteria (Korn and Faber, 1991). However, in most other CNS cells, particularly in mammalian systems, unitary recordings are much more difficult to obtain, particularly with evaluation of both physiological responses and the anatomical details of the number and location of synaptic sites between two neurons. Recently, however, the hippocampal slice culture has shown considerable promise, with Debanne et al. showing dual recordings between CA3 and CA1 neurons with a small number of connections and interesting potentiation properties (Debanne et al., 1999; Pavlidis and Madison, 1999). This same connection in acute *in vitro* slices has shown much less promise due to the small percentage of CA1 and CA3 cells showing excitatory synaptic connections (Redman, 1990; Sayer et al., 1990; Stricker et al., 1996a). Other situations which have been explored include Ia afferent synapses onto motoneurons in the spinal cord (Redman, 1990) and isolation of axonal afferents onto CA1 pyramidal cells using microstimulation techniques (Bolshakov and Siegelbaum, 1995; Isaac et al., 1995; Turner et al., 1997). All of these excitatory synaptic types, however, show primarily dendritic synapses, which are at some electrotonic distance from the soma and the recording electrode and generally higher noise levels are prevalent, obscuring individual responses. Since there are thousands (25,000) of synapses onto these cells there usually cannot be a rigorous comparison of evoked and spontaneous synaptic currents since the relationship between the origin of the two separate events is unknown.

Another approach is to identify inhibitory connections, which are usually perisomatic, have a much larger signal-to-noise ratio, but which may also have multiple synaptic release sites per individual synapse (Staley, 1999; Kraushaar and Jonas, 2000). However, these connections may be identified using two-cell recordings, since cells are close to each other and have a high frequency of synaptic interconnection.

Yet another critical characteristic of electrophysiological data is a low level of signal noise. This is achievable using whole-cell patch electrode techniques, where the baseline noise standard deviation may be less than 1 pA, producing a signal-to-noise ratio of 4:1 or better for many excitatory synaptic ensembles. Having full physiological control of a presynaptic cell or axon is critical, so responses in the postsynaptic cell may be clearly linked to the activity of the presynaptic response, which occurs either spontaneously or may be triggered (by intracellular current or axonal stimulation). Voltage-clamp simulations demonstrate that most dendritic synaptic responses are characterized by poor voltage clamp, and that there may be decreased measurement error by using current-clamp modalities with whole-cell patch electrodes (Spruston et al., 1993).

To minimize analytic complexity, experimental preparations should be developed to limit stimulation to as few synapses as possible (generally 8 to 10). Such stimulation control allows for better differentiation of individual synapses within the patch-clamp recordings. We have detailed the conditions important for capture and analysis of excitatory postsynaptic current (EPSC) data in Turner et al. (1997). Briefly, these conditions include limitation of the synaptic signals to a single glutamate receptor type (non-NMDA), careful activation of as few synapses as possible through microstimulation, and restriction of the analysis to carefully defined data sets in terms of noise and stationarity. In hippocampal slice cultures the number of synaptic connections between CA3 and CA1 pyramidal neurons is enhanced by reinnervation, so that a few synapses may be obtained, even in cells at a significant distance from each other (Debanne et al., 1999).

Statistical analysis of synaptic transmission requires that the experimental preparation be stable over time. All of the statistical techniques require a certain number of evoked response samples to be obtained for one to be able to perform the analysis satisfactorily. For example, in the initial least mean squares optimization technique employed by Redman (1990) to analyze synaptic plasticity, the minimal critical sample size was approximately 750 responses, assuming a signal-to-noise ratio of 3 to 1 or better. These samples need to be obtained at a rate that is stable for the preparation and that does not in itself lead to plasticity (either depression or facilitation). This rate varies from 2 Hz for motoneurons to 0.5 to 0.1 Hz for many hippocampal preparations. The duration of acquiring such a sample, which is presumably not changing during the acquisition and fulfills the requirements for stationarity, is extensive (up to 40 min at a rate of 0.2 Hz). This property reflecting the stability of the fluctuations during the period recorded is termed stationarity, and it implies both the absence of a trend and that the fluctuations during the period are not changing in any essential manner. Stationarity can be partly assessed by a trend plot and also partly by analyzing sequential subgroups of approximately 50 responses for net significant changes (Turner et al., 1997). A Bayesian time-series analysis and formal trend analysis has also been developed to statistically assess stationarity (West and Harrison, 1997).

Finally, skepticism about the value of statistical inference from synaptic responses is widespread among physiologists because of the difficulty in assessing whether a solution is either unique or good in an absolute sense and the indirect nature of the analysis, often requiring numerous assumptions and the lack of validation of the statistical analysis schemes.

4.1.4 Classical Statistical Models

Chemical synapses involve common release mechanisms, though the specific parameters of release, type of neurotransmitter elaborated, and pre- and postsynaptic receptors involved differ considerably. Such synapses are highly probabilistic with responses from a few synapses showing considerable fluctuation from trial to trial (Del Castillo and Katz, 1954; Redman, 1990; Korn and Faber, 1991; Faber et al., 1992; Harris and Sultan, 1995; Bennett and Kearns, 2000). Statistical models of synaptic release mechanisms have been developed to explain the probabilistic release of neurotransmitter and to infer critical changes at individual synaptic release sites when recordings are usually performed from multiple sites at once (Del Castillo and Katz, 1954). These models have been adapted to synapses in the hippocampus for understanding the development, neural plasticity, and potentiation.

The appropriateness of a statistical model used to describe what is occurring at a single synapse depends very much upon the validity of the simplifying assumptions of the model and whether there are independent physiological methods to confirm these assumptions.

Historically, statistical models at the NMJ centered on binomial models, since it was assumed that vesicles emptied near-identical quantities of neurotransmitter, multiple vesicles were released with each impulse, and each vesicle possessed the same likelihood of release (Del Castillo and Katz, 1954; Katz, 1969; Bennett and Kearns, 2000). With pharmacological manipulation the number of vesicles released could be significantly reduced and the limit of the binomial distribution, a Poisson distribution, could be reached. With the Poisson distribution the binomial parameters n (number) and p (probability) are not separately defined but are merged into a single value, m, applicable only when p is very small compared to n. The adaptation of this framework to central neurons proved to be difficult because of the much higher noise level associated with recording using sharp electrodes, requiring techniques to remove the effects of the noise (Redman, 1990).

Maximum likelihood estimation (MLE) approaches were developed to optimize searches for multiple parameters in light of this contaminating noise, using a binomial statistical model extended to a compound binomial allowing differences in probability of release (Bennett and Kearns, 2000). In this model it is assumed that a certain number of synaptic sites mix with ambient noise, summate with each other according to the respective probabilities of the sites, and form a set of responses that represent the sum of these different levels over time. The statistical inference problem is thereby to analyze the contributions of individual sites in this mixture, particularly to determine the number of sites and the individual parameters of each site. The MLE approach optimizes a likelihood function over a set of possible synaptic site parameters. The optimization is performed by fixing one group of parameters and allowing only one or a few critical parameters to vary, such as the number of sites, n. The distribution calculated from the optimized number of sites is then convolved with the recorded noise and the result is compared to the original histogram. If the result is not optimal, according to a chosen measure, further iterations are performed. MLE analysis therefore does not necessarily provide a unique solution, as many "optimal" solutions may exist. Recent MLE approaches have also added estimation of quantal variance and comparison to nonbinomial models, though these are based on a smoothed function approximation to the original data histogram (Stricker et al., 1996a).

Other approaches have included histogram-based smoothing and "peakiness" functions, using the histogram to identify the summation of site responses (termed components) rather than analyzing the primary data. This approach is severely limited by how the histogram is constructed (bin width, etc.) and very similar data may show highly different histogram appearance randomly (Walmsley, 1995). Thus, approaches based on analyzing the exact, original data are preferred over superficial histogram appearance.

Guidelines for the suitability of physiological data sets have also been established, based on simulation approaches (Redman, 1990). Minimal sample sizes are in the range of 500 to 750 data points, each from an evoked synaptic response, and a signal-to-noise ratio of at least 3 to 1 is required for optimal extraction of component locations. Stationarity of the degree of fluctuation of the signal is also required since if the basic synaptic parameters vary a reliable analysis cannot be performed (Redman, 1990). Stationarity can usually be assessed by trends of the entire data set or analysis of successive subgroups (Turner et al., 1997). The noise itself can be approximated by either a single or dual Gaussian normal function (σ_{noise}), rather than by use of the raw data histogram. The results of the MLE analysis are limited in that particular algorithms often cannot identify a local minimum for parameters and may reach different conclusions based on the starting point used for the initial parameters. Thus, multiple initial parameters must be tried and whether the final result is "optimal" or "good" in some absolute sense cannot usually be ascertained. Identifying the underlying distribution is a much more difficult task, particularly with a single finite sample, since random samples from any distribution of a finite size may be very unrepresentative in terms of their histogram appearance (Walmsley, 1995). An additional limitation of the current MLE methods is that variability between site amplitudes cannot be detected, even though variability in site probability is allowed with a compound binomial distribution. Together, these limitations severely curtail the number of data sets that can be analyzed to a small fraction of those that are derived from experiments.

However, many of the basic assumptions of such quantal models have not been tested in CNS neurons. These assumptions include near uniformity of synaptic release amplitudes between sites (low intersite variability) and limited variation in release amplitudes over time (low intrasite variability). Together, such variability in synaptic strength is termed "quantal variance." Recently however, considerable interest has arisen regarding both forms of quantal variance (Bolshakov and Siegelbaum, 1995; Turner et al., 1997). Such heterogeneity requires more complex statistical models than a simple or compound binomial model to interpret parameter changes at individual synapses.

One approach has been to identify or presume a single or unitary connection between cells whenever possible, and, using either the rate of failures or the average success (potency), to identify critical changes in release parameter values (Bolshakov and Siegelbaum, 1995; Isaac et al., 1995; Liu and Tsien, 1995; Stricker et al., 1996a, 1996b; Debanne et al., 1999). This approach simplifies analysis, at the risk of misidentification of the physiological processes underlying the change. A more detailed alternative approach would be to define an open statistical framework without such assumptions, and then to allow the experimental data to define the parameter changes underlying synaptic plasticity on long- and short-term timescales.

As a detailed example, quantitative parameters describing synaptic release sites include the number of activated sites on any one trial (k), probability of release (π), and amplitude (μ) and electrotonic conduction from the site of origin to the summation site. The goal of quantal analysis is to infer these parameters at single synapses but usually in the physiological context in which a number of release sites may be activated concurrently (Redman, 1990; Korn and Faber, 1991). However, a high level of background synaptic and recording noise often occurs during recordings, obscuring much of the detail and so requiring statistical inference to deduce the content of the original signal. Though a simple quantal–binomial model may hold for inhibitory synapses onto the Mauthner cell (Korn and Faber, 1991) a compound (or complex) binomial distribution has been suggested to be more applicable in the mammalian hippocampus, with varying probability of release (and likely varying amplitude) between required synaptic sites (Sayer et al., 1990; Turner et al., 1997). Recent studies of small EPSCs (Bolshakov and Siegelbaum, 1995; Turner et al., 1997) have suggested considerable heterogeneity over time at synaptic sites and between synapses. Thus, techniques to analyze parameters at individual synapses have been limited by adapted statistical frameworks and newer models, including MLE models (Korn and Faber, 1991; Stricker et al., 1996a, 1996b), Bayesian component models (Turner and West, 1993; West and Turner, 1994; Escobar and West, 1995; Cao and West, 1997), and site-directed models (Turner et al., 1997; West, 1997) may provide improved analytical capabilities. Newer approaches have also indicated that within-site and between-site variance may be systematically underestimated, requiring different approaches for estimation (Frerking and Wilson, 1999; Uteshev et al., 2000).

4.2 Nontraditional Models of Synaptic Transmission

4.2.1 Introduction to the Bayesian Model and Comparison to Classical Models

Understanding the fundamental processes underlying synaptic function rests upon an accurate description of critical parameters including probability of release, the strength of presynaptic neurotransmitter release, the postsynaptic response, and the relationship between the postsynaptic response and experimental measurements. Classical statistical models of synaptic release generally focus upon changes observed in ensemble histograms of synaptic responses. In such models, the statistical distribution underlying the synaptic responses is assumed from the model. The uncertainty surrounding this assumption is often difficult to assess. In addition, the numbers and nature of the statistical parameters are often incorporated into the model rather than being determined from the data.

We developed Bayesian site analysis in order to directly represent synaptic structure in terms of measurable parameters and unobserved variables that describe individual synaptic sites and experimental characteristics (Turner et al., 1997; West, 1997). Bayesian analysis is based upon entire data sets rather than binned histograms or smoothed data set distributions. Our approach performs a global optimization procedure to estimate (and provide bounds for uncertainty) both for the number of synaptic sites and for the parameters of each site (Turner and West, 1993; West and Turner, 1994; Escobar and West, 1995; Cao and West, 1997; West, 1997). Critical assumptions in our approach are that individual synaptic sites retain characteristic amplitudes, since sites are ordered during the analysis according to magnitude, and that the underlying ambient noise can be characterized by a smooth Gaussian function that is independent of synaptic signals. A similar type of unrestrained analysis, with a different approach and optimization technique, has also been suggested by Uteshev et al. (2000).

By comparison with MLE and other pure likelihood-based approaches, Bayesian inference aims to explore completely the regions of parameter space supported and indicated by the data, and formally summarize relevant parameter ranges and sets using summary probability distributions rather than just point estimates and error bars. Current standard techniques involve repeated simulation of parameter values jointly from such posterior distributions. Notice that this subsumes and formally manages problems of local maxima in likelihood surfaces, and other such technical problems that often invalidate MLE approaches; the Bayesian strategy is to map out more thoroughly the parameter surfaces via informed navigation of the surface in regions identified as relevant probabilistically. An initial statistical outline appears in West (1997), though our work has since developed to address estimation of intrinsic variation over time more fully (Turner et al., 1997).

Bayesian site analysis begins with initial assumptions about the data, and then performs a multidimensional optimization using the likelihood function. Because the resulting analytic function is complex and is without analytic solution, Monte Carlo simulation methods must be used to estimate the solution, termed the predictive probability distribution. The strengths of the Bayesian approach are that it has much stronger theoretical justification that a global solution can be obtained than with the MLE approach. In addition, error bounds can be provided on all of the estimated parameters. These error bounds reflect the uncertainty surrounding both the parameters and the overall solution.

4.2.2 Bayesian Site Analysis

Bayesian site analysis directly represents individual synaptic site characteristics in terms of amplitude, μ, and release probability, π, parameters, together with an unknown degree of intrinsic, intrasite "quantal," variation over time, τ. Such quantal variation can be divided into two components: that due to intersite heterogeneity, which is directly analyzed, and that due to intrasite variation over time, which is estimated from the analysis. Analysis is via Bayesian inference, providing point and interval estimates of all unknown parameters, together with full probabilistic descriptions of uncertainty (Turner and West, 1993; West and Turner, 1994). The model represents the interaction of a number of synaptic sites, with each site possessing its representative set of parameters. Importantly, the model does not assume equality of either probability or amplitude between sites, and includes offset of the failures component, an estimation of intrinsic standard deviation, and a variable degree of adherence to the initial data-set noise standard deviation. Hence the Bayesian format uses a highly unconstrained model.

The statistical model requires initial specification to define the unknown parameters. First, an arbitrary value of k is chosen as an upper bound on the number of sites, depending on the raw data histogram. If there are fewer than the specified k sites, then each of the extra sites may simply assume a zero value for μ (see Appendix). A marginal probability is generated that indicates how likely k sites will fully explain the data set, termed $P(k|D)$, or the probability of k sites underlying the data set, D.

This marginal probability allows inference on the true number of active sites, since both too few and too many sites are clearly identified by low values for $P(k|D)$. Second, the amplitudes of individual active sites, μ_j, are allowed to take either equal or different values, allowing for tests of equality between subsets of the synaptic sites that take on common values. Third, the release probabilities for the active sites, π_j, are allowed to take on any values between zero and unity. As with amplitudes, the site analysis allows for exactly equal values among the π_j, to assess for equivalence of this parameter within the subsets of sites. Further, prior specification is required to describe the initial uncertainty about the noise variance; the measured noise variance of the recorded data may be used to provide initial estimates and ranges.

Statistical site analysis uses methods of stochastic simulation, namely Monte Carlo iterations, to successively generate sampled values of all model parameters, including values for each of the μ_j and π_j pairs, together with estimates of quantal variation and signal noise. These sampled parameter vectors from the data-based Bayesian posterior probability distribution collectively summarize the information in the data and the corresponding uncertainties, generating the predictive probability density function (pdf) and confidence ranges for all parameters.

We define this approach further below with specific details of both simulations and experimental data sets. Statistical inference is inherent to the Bayesian approach and all parameters are given with appropriate error ranges, including the number of synaptic release sites. Detailed discussion of the model and its theoretical foundations may be found in the references (Cao and West, 1997; Turner et al., 1997; West, 1997).

4.2.3 Application of the Bayesian Site Model to Simulated Data

Simulations were performed to accurately guide the definition of the prior parameters under particular conditions. The signal-to-noise ratio, and location and number of sites were variously selected but compared to the actual data after the analyses. The simulated data sets were in general similar to those expected from companion physiological studies, except that the underlying distribution from which the random sample was drawn was well defined for the simulation ensembles. Once these data sets were generated the Bayesian analysis was performed in a manner identical to that used for the physiological data sets. Once optimized, the parameter values (and error ranges) were compared between the results of the analysis and those of the original distribution. If the original distribution parameter values were bracketed by the 95% error ranges of the simulation results, they were counted as correct parameter values, and as errors if the analysis values were outside the predicted ranges. The errors in numbers of sites and parameters were accumulated and calculated as a percentage of the total number of ensembles. This definition of error used the full error estimates available from the Bayesian estimation.

Random ensembles from known distributions, composed of 1 to 5 sites, were analyzed using varying signal-to-noise and ensemble size values. The raw histograms varied extensively depending on the random sample obtained from the underlying distribution, confirming that histogram appearance may be very misleading (Walmsley, 1995), so the underlying data were used for the analysis rather than the histograms. Numerous simulations were performed using random ensembles drawn from distributions with a variety of site numbers, relative amplitudes (both equal and unequal), signal-to-noise values, quantal standard deviations, and sample sizes to estimate the errors associated with the Bayesian analysis. Each set of simulations included 100 to 200 separate simulated ensembles, each with $n = 200$ to $n = 750$ data points, which underwent analysis and comparison to the underlying distributions in the same way as for the experimental data sets. Errors in the analysis were defined in two separate ways:

1. The relative error of the standard deviation normalized to the value of the parameter (analyzed as SD/parameter value), similar to that suggested by Stricker et al. (1996a).

2. The percentage of errors per site in which the number of sites, underlying site amplitude, or probability did not fall within the 95% confidence interval established for that data set.

This latter definition shows a more usable definition of error, inasmuch as one can estimate how often the analysis will be incorrect taking into account both the limited sample size of the experiment (including sampling errors) and the errors of the analysis. The 95% confidence interval around the median was chosen because the distributions were often asymmetric about the median.

The number of sites (k) derived from the Bayesian analysis partly reflects the prior values, particularly the adherence of the noise in the analysis to the estimated noise standard deviation (σ_{noise}) value. For weak adherence with only a few degrees of freedom (d.f. $= 10$) the noise value itself can fluctuate away from the provisional value, leading to fewer sites than in the distribution. Therefore the adherence of the Bayesian analysis to the noise was usually set to a tighter value (d.f. $= n - 1$), reflecting a strong faith in the preliminary noise estimate from the experiment. The number of sites with the noise d.f. $= n - 1$ evaluated to be either the correct number or larger, in effect partially overestimating the number of sites to ensure that the site parameters showed the least possible error. Thus, with the noise set at this level in the analysis, the number of sites was highly likely to be at least as large as the number in the simulations. The correct number of sites from the underlying (known) distributions was compared to the predicted number with a cumulative probability of at least 0.5 ($P[k|D] > 0.5$). The summary results of some of the simulations are further defined in Turner et al. (1997). Results show that with irregular sites a much larger sample size is needed to resolve the differences between sites. The overlap between sites also increases the difficulty in correctly resolving the actual number of sites.

4.2.4 Application of the Bayesian Model to Recorded Data

Physiological data sets were acquired as described in detail in Turner et al. (1997), using minimal stimulation to assure that only a small, stable number of synaptic sites was activated. The postsynaptic responses were recorded as EPSCs using low-noise whole-cell patch electrodes, and the ensembles were carefully assessed for stationarity. Figure 4.2 shows a physiological average for a minimal stimulation EPSC ensemble (U2N1), recorded from a CA1 pyramidal neuron under highly controlled conditions, using minimal stimulation just above threshold for the presence of a response with averaging. The peak average was 8.87 ± 7.91 pA ($n = 244$) and $\sigma_{noise} = 1.22$ pA. Figure 4.2B shows that the noise (closed circles) and signal (open circles) peak data showed neither a trend nor an irregular pattern over the course of the ensemble. The data set was then analyzed using noise d.f. $= 243$, with a resulting probability favoring a single site [$P(1|D) = 0.93$], and less than 0.07 likelihood showing that an additional site was needed. The parameters for the single site (Figure 4.2D) showed $\mu = 10.2$ pA, $\pi = 0.87$, and a median intrinsic variance of 40%, with the variation around these parameters illustrated as error bars in Figure 4.2D. The raw histogram and calculated Bayesian predictive pdf are shown in Figure 4.2C along with error limits on the pdf as dashed lines. Note the excellent fit of the raw histogram and the pdf. However, the raw histogram was not in any way used in the analysis and is shown only for comparison to the predictive pdf, which was calculated directly from the data. For both of these physiological examples 20,000 Monte Carlo draws were used in the estimation of the parameters.

Figure 4.3 shows another EPSC ensemble (U4N1), again obtained from the CA1 region during a whole-cell patch-clamp recording from a CA1 pyramidal neuron (Turner et al., 1997). Figure 4.3A illustrates an averaged EPSC trace ($n = 386$) and Figure 4.3B shows a trend analysis, with no net changes over the ensemble period. Figure 4.3C shows the comparison of the raw data histogram and the predictive pdf (smooth line). Figure 4.3D shows the two sites which were extracted using the Bayesian analysis: $P(1|D) = 0.01$, $P(2|D) = 0.99$, and $P(3|D) = 0$, with parameters $\mu_1 = 7.63$ pA ($\pi_1 = 0.11$), $\mu_2 = 19.0$ pA ($\pi_2 = 0.32$), and an intrinsic variance of 22%. Note that the two sites do

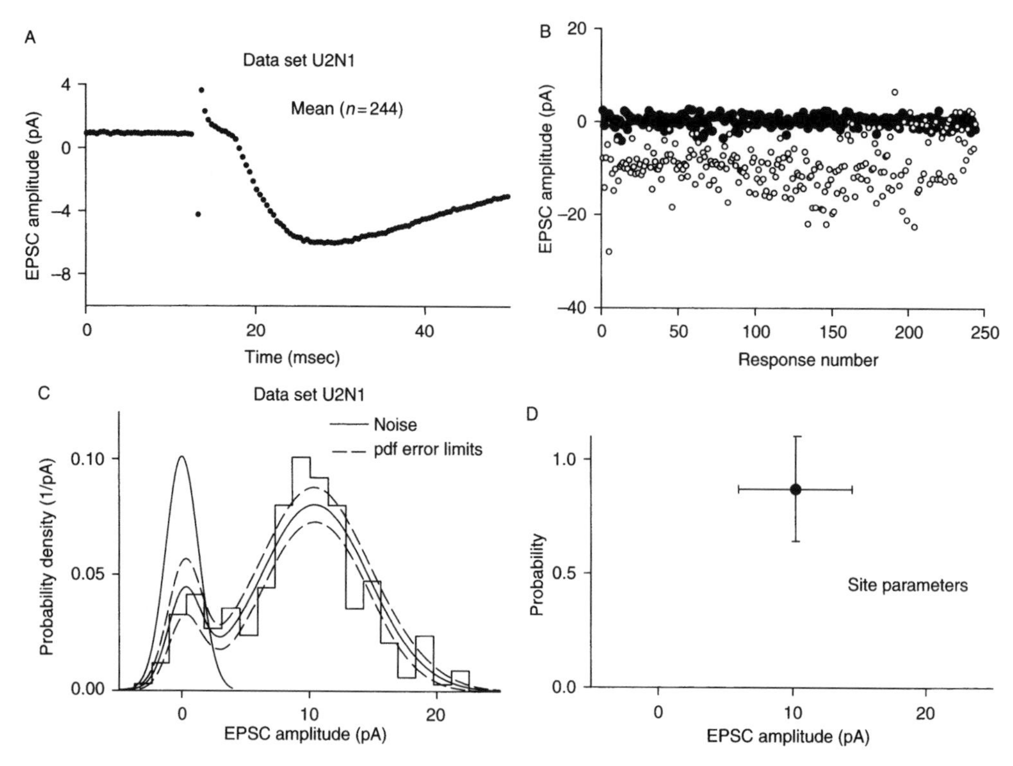

FIGURE 4.2 Data set U2N1 with 1 site. (A) The figure shows the mean trace from the data set U2N1 ($n = 244$ responses). (B) The scatter plot for stationarity is shown, indicating no change in the degree of fluctuation over the course of the experiment. The open circles show the responses and the closed circles the noise samples over the course of the ensemble. (C) The response histograms and Bayesian analysis results are illustrated here. The Gaussian peak centered at 0 shows the noise function used in the analysis, overlapping the failure peak. The smooth line with two peaks indicates the predictive Bayesian pdf and the dashed lines the error ranges around the pdf. (D) The individual synaptic site parameters, with a dual plot comparing site amplitude on the abscissa versus site probability on the ordinate, are shown. Note the single site at approximately 10 pA, corresponding well with the histogram in (C). This ensemble shows an example of a simple response with either failures or occurrences.

not overlap and both the amplitudes and probabilities are different, suggesting that this ensemble is a complex response, inasmuch as neither the simple binomial nor the compound binomial would be a suitable fit. Many further simple and more complex responses are shown in Turner et al. (1997), along with the effects of paired-pulse plasticity on site parameters.

The conclusions from the analysis of these data focus on the considerable between-site heterogeneity as well as the moderate level of within-site (intrinsic or quantal) variance observed. Less than half of the previously identified and analyzed data sets could be fit by a traditional statistical model, indicating that most responses recorded in the CA1 pyramidal neuron showed unequal amplitudes and probabilities, requiring the more complex model for an adequate fit. Informal assessment of convergence was performed as discussed in Section 4.2. These summaries reflect the inherent uncertainties in estimating the number of active sites. Additionally, it was not possible in the analysis to indicate the degree of ambiguity in the physiological "solutions," since there were always alternative explanations that could also explain the data. This form of data analysis will clearly require an alternative approach for unambiguous interpretation. The optimal correlative approach would be to identify each postsynaptic site unambiguously (such as with a marker visible with a confocal microscope system) at the time of activation. This method of optical identification could be used to verify which sites were active, and on which trial, to clarify the physiologically recorded responses.

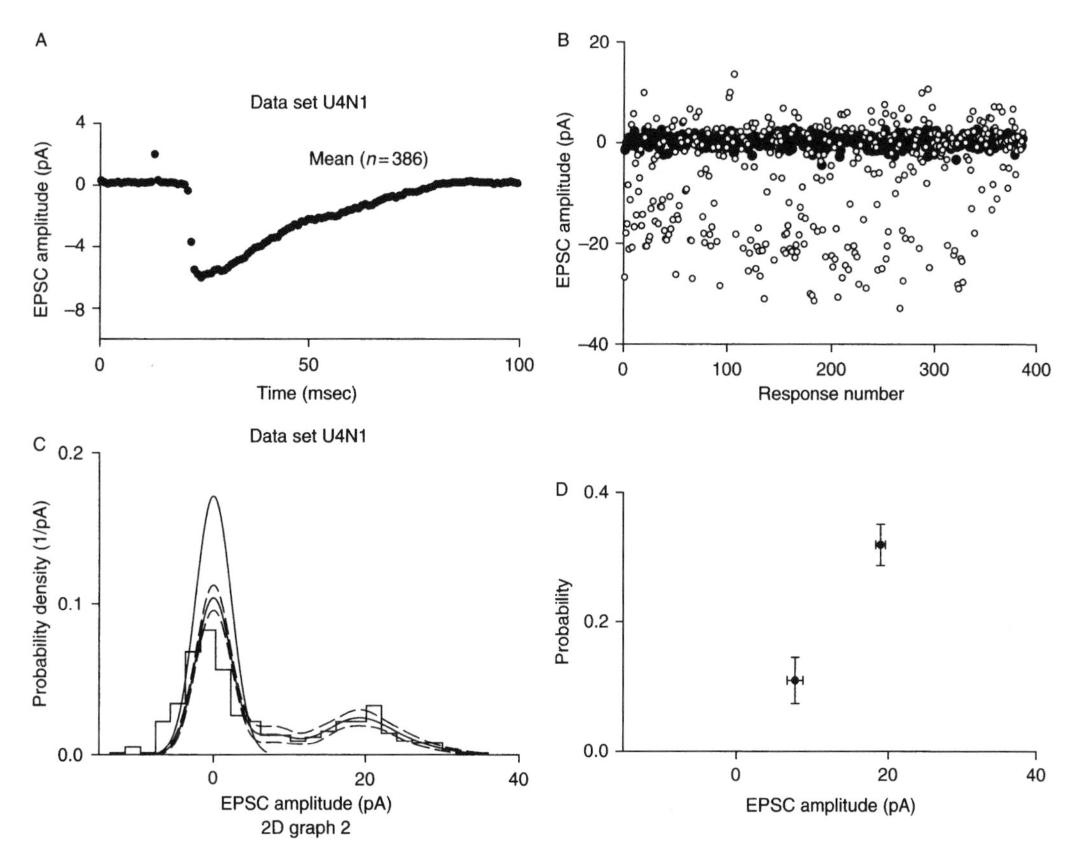

FIGURE 4.3 Data set U4N1 with 2 sites. (A) The figure shows the mean trace from the data set U4N1 ($n = 386$ responses). (B) The scatter plot for stationarity is shown, indicating no change in the degree of fluctuation over the course of the experiment. The open circles show the responses and the closed circles the noise samples over the course of the ensemble. (C) The response histograms and Bayesian analysis results are illustrated here. The Gaussian peak centered at 0 shows the noise function used in the analysis, overlapping the failure peak. The smooth line with two peaks indicates the predictive Bayesian pdf and the dashed lines the error ranges around the pdf. (D) The individual synaptic site parameters, with a dual plot comparing site amplitude on the abscissa versus site probability on the ordinate, are shown. Note the dual sites, with the first at approximately 4 pA and the second at approximately 18 pA, corresponding well with the histogram in (C).

4.3 Discussion

4.3.1 Comparison of Simulations and Physiological Data Sets

The simulation data sets were purposefully designed to emulate many of the physiological conditions observed, particularly the level of noise, the signal-to-noise ratio, the level of probability of release, and the amplitudes. The variations on signal-to-noise and sample size were also solidly established to be close to the physiological values. For most of these realistic parameter values the Bayesian site analysis performed well, though with low signal to noise (less than 2:1), small sample sizes (particularly less than 300), and irregular sites, the extraction of parameters was more difficult. The number of parameter errors allows a quantitative estimate of the likelihood of a single random sample leading to the definition of the underlying distribution from which the sample was drawn. This likelihood includes both the effects of a small sample size (particularly when less than 500 to 750) and the randomness of the subsequent draw and the ability of the analysis technique to resolve the underlying distribution. Since parameter errors have not been estimated in this practical context previously for neurophysiological applications, it is difficult to compare this technique to other currently available statistical analysis formats, particularly the bootstrap technique for the MLE algorithm and the compound binomial approach of Stricker et al. (1996a). Further simulations, particularly with more sites, may help define the limits of the Bayesian site analysis. Currently the analysis will work with up to 8–10 synaptic release sites, but beyond 4–5 sites practical problems arise with identifying or clustering the individual sites and more site overlap prevents such effective clustering. Thus, analysis of data sets with 1–3 sites may be more robust than those with more than 4–5 sites, and we are particularly cautious regarding a still larger number of sites, as yet.

Virtually all of the analysis techniques discussed here have limited availability in terms of distribution, and are highly specialized in their application to particular data sets. Korn and Faber's binomial analysis program is only applicable under strict conditions and has not become widely available (Korn and Faber, 1991). The Bayesian components (Turner and West, 1993; West and Turner, 1994) and site-directed programs (Turner et al., 1997; West, 1997) are likewise not yet sufficiently robust to become generally available, and require the use of several ranges of parameters to estimate the best solution. In particular, the Bayesian site program does not work well when there are likely multiple sites (greater than 5 to 6) and there are very few stable data sets currently available that fit the range of limited applicability of this analysis program. Limited usefulness of each type of program has made it difficult to compare the efficacy of parameter and distribution detection; it may be better to compare a standard suite of simulation data between laboratories rather than share such highly specialized programs. However, this comparison is critical and should be performed. This lack of interoperability between various approaches and difficulty in comparison hampers any validation of any particular approach. Likewise, independent methods (such as optical methods for direct quantal analysis) to evaluate validity will be critical to underscore the reliability of any particular method (Oertner, 2002; Oertner et al., 2002; Emptage et al., 2003).

4.3.2 Analysis of Components in Contrast to Sites

The analysis of sites is a more robust approach than using a histogram approximation procedure, regardless of the particular model and optimization approach utilized. For example, Korn and Faber (1991) have used binomial models, which are site directed, thus ensuring that a low probability tail is adequately represented in the analysis. However, Redman and his group have used a component-based approach, beginning with a smoothed histogram but fitting peaks in this smoothed representation of the data (Redman, 1990; Stricker et al., 1996a). The component-based approach fits histogram peaks rather than the natural sums and products of underlying sites, and thus the peaks may or may not represent sums of site contributions appropriately. The differences between the two approaches represent an ongoing debate as to their relative usefulness, but clearly the histograms are sums of individual synaptic site contributions with their own respective amplitudes and probabilities. Whenever there is more than one site, the statistical prediction is that the appropriate sums of the sites (with the probability being a product of the individual values) should be adequately represented, even though on an individual draw the histogram may not be accurate or representative for the distribution (Walmsley, 1995). To study these different approaches we have worked through both types of analysis.

Bayesian data analysis has been previously used for analysis of components (peaks in histograms representing sums of synaptic site contributions) (Turner and West, 1993; West and Turner, 1994) and data are presented here showing the enhanced usefulness of a site-directed approach. If all synaptic sites are assumed equal then a binomial (or in an extreme limiting case, a Poisson) distribution may be appropriate, and the components may be decomposed into their underlying sites. For a binomial distribution, this requires estimation of several variables: n, the number of sites, p, the probability of release at each site, and q, the amplitude at each site. The histogram is normalized by what is felt to represent the unit amplitude (q) and then integers representing the sums of the individual site contributions are used to assess the probability and number of sites. As shown previously, there is often no unique solution, particularly with the most likely situation that there is contaminating noise and where the first peak may be offset by an arbitrary value (Stricker et al., 1996a). Estimation of n may be particularly difficult. The situation is highly complicated by the additional likelihood that the probability is different between sites, leading to a more complex compound binomial distribution (but which still requires equal site amplitudes). When the ambient noise becomes significant, removal of the noise effect also requires an optimization routine, usually one based on deconvolution (Redman, 1990). Because of the multiple inferences possible with this form of deconvolution procedure, the results are difficult to confirm, particularly the likelihood that the solution is "good" or "optimal" in any way.

We have developed the Bayesian site method of analysis to overcome many of these problems. By starting with the possibility that sites may not be similar in terms of either amplitude or probability and by not assuming a fixed number of sites, a more flexible and global analysis format is possible. Additionally, as shown in Figure 4.2 and Figure 4.3, the analysis gives error ranges for the predictive pdf and in particular the site parameters, as well as the likelihood for how many sites are required. Though the analysis uses fitted Gaussians for the noise, this is only a starting point, and depending on the degrees of freedom accompanying the noise value, this starting point is more or less strictly used in the analysis. The Bayesian analysis also gives a posterior distribution for the noise variance, indicating how strict this adherence was for a particular data set. Likewise, the Bayesian format is general, and plans for development include making the prior distribution for the noise much more flexible, particularly by using a sampled noise distribution rather than any form of fitted smooth distribution. However, this format of analysis begins to give some idea as to the "goodness" of the analyzed fit to the raw data, though of course it does not in any way guarantee that the fit is unique or optimal. However, the Bayesian approach uses all of the data for the analysis, as opposed to the more conventional MLE and compound binomial (Stricker et al., 1996a) procedures, which utilize optimized histogram data. Additionally, the Bayesian approach provides a more rigorous and extensive investigation of the data-supported regions of the parameter space than the more restricted MLE solution algorithm (West and Turner, 1994). The direct Bayesian site format may eventually prove to be very helpful to identify changes directly with plasticity and experience rather than the more "blind" approaches which have conventionally been used.

4.3.3 Analysis of Physiological Data Sets

The results of the two data sets shown here and the larger group of data sets from Turner et al. (1997) explicitly highlight the considerable heterogeneity of synaptic site parameters in hippocampal neurons during paired-pulse plasticity (Bolshakov and Siegelbaum, 1995; Hanse and Gustafsson, 2001). The heterogeneity noted between sites (62% of the mean site amplitude) of non-NMDA responses may be attributable to differences in presynaptic neurotransmitter release, postsynaptic receptor density, and electrotonic filtering, but is clearly larger than the identified range of intra-site or intrinsic variability (36% of the mean site amplitude after correction), defining the relative importance of significant alterations in presynaptic release over time or ligand–receptor mismatch (Faber et al., 1992). One strength of the Bayesian site analysis is to resolve these two separate aspects of quantal variance, that due to either differences at a single site over time (here termed intrinsic variance) or that due to differences between site parameters at a single occurrence. The resolution achieved by this analysis suggests that intersite heterogeneity is clearly the rule in CA1 pyramidal neurons, for a variety of expected and important reasons, whereas intrinsic variability at a single site over time is also important and should be explicitly considered. Together these two sources of variability will lead to difficulty with histogram-based approaches, since both tend to merge "peaky" histograms into much smoother and more poorly defined histograms (Turner et al., 1997). This effect of smoothing can easily lead to the two-component histograms defined by earlier studies, in which only a "failure" and a "success" component can be identified. This result confirms that histogram-based approaches may also mislead interpretations regarding underlying sites due to this smoothing effect and also random histogram variations arising from sampling effects (Walmsley, 1995).

A small subset of histograms showed inflections suggesting approximately equal spacing of synaptic site contributions, but the rule (as observed previously in most reports) is that the histogram appearance was not usually "peaky" and histograms may be unreliable in pointing to the underlying statistical distribution (Walmsley, 1995). Using a paired-pulse stimulation protocol the histogram distributions clearly changed in many ensembles, but there were also data sets that did not exhibit a significant paired-pulse potentiation but rather showed depression (Turner et al., 1997). Paired-pulse plasticity appeared to arise primarily from changes in probability of release, as described by many other authors (Redman, 1990). However, the degree of changes in probability appeared to vary between sites, suggesting that past history of activation and/or other factors may be important in the underlying process leading to the short-term changes. The degree and direction of the paired-pulse plasticity response did correlate with the initial probability of release (for the critical synaptic site underlying most of the change), suggesting that enhanced plasticity may be present if the response starts from a lower initial probability. However, other factors may also be important in the determination of the direction and degree of short-term plasticity, since different sites often appear to respond in varying ways to the same conditioning impulse. Thus, the presence of a response on the conditioning pulse may lead to enhanced plasticity compared to responses which show a failure on the first response. In contrast, Debanne et al. suggest that an initial response may lead to depression on the test pulse rather than potentiation (Debanne et al., 1999), possibly due to neurotransmitter depletion.

We have evaluated the hypothesis that different release sites over the dendritic surface lead to amplitude and waveform shape differences due to electrotonic filtering of events (Turner, 1984a, 1984b; Williams and Stuart, 2003). EPSCs from various data sets were found to contain events with different rise times, which clearly reflected differences in electrotonic filtering. This electrotonic filtering likely is another source of postsynaptic heterogeneity. These data also lead us to suggest that minimal stimulation may lead to activation of only a few release sites onto individual CA1 cells, often located at different electrotonic locations, and that each site may possess different release probabilities and synaptic site amplitudes. Such variation has also been suggested by anatomical studies showing large variation in synaptic shape and presynaptic densities (Harris and Sultan, 1995). Thus, there is evidence for variable amplitudes of synaptic contributions (as recorded at the soma) due to dendritic spatial filtering, and possibly also due to intrinsic differences between synaptic sites located at similar spatial locations. Additionally, the data presented above indicate that the probability of release may vary extensively between synapses, and there does not appear to be a strong rationale for assuming that probability is uniform among any mammalian CNS synapses (Redman, 1990). These findings argue for significant intersite heterogeneity due to spatial and perhaps inherent site differences as well as a certain contribution of intrinsic quantal variability due to changes in release over time. The intrinsic variability can arise from such sources as variable neurotransmitter packaging in vesicles and variable ligand–receptor binding in the synaptic cleft (Faber et al., 1992). The heterogeneity noted at both intrasite and intersite levels of analysis suggests strongly that changes with plasticity may also vary considerably between synapses, possibly as a function of the history of use of each synapse. The important role of intrinsic variance is similar to that described by Bolshakov and Siegelbaum (1995; approximately 20% to 35% in their analyzed data sets) but more than that suggested by Stricker et al. (1996a, 1996b; less than 10% in the analyzable data sets).

4.3.4 Conclusions and Future Perspectives

Depending on the suspected number of synaptic release sites there are several applicable statistical models. Because of suspected significant heterogeneity in the hippocampal CA1 neurons, as discussed before, we have applied Bayesian analysis because it is the most general and unconstrained in terms of site parameter freedom. As described in Turner et al. (1997) the Bayesian complex framework was required for interpretation of the ensemble in a little over half of the data sets, and so clearly allows extension of the number of data sets that can be analyzed. However, limitations include the potential loss of identifiability of data sets with larger numbers of synaptic sites (possibly greater than 5 to 6 sites), the current lack of realistic noise modeling for complex noise, and the lack of any online feedback during recordings to show the number of sites activated on a trial-by-trial basis. Thus, further anticipated developments include simulations to model additional numbers of sites, improved noise characteristics, better characterization of data set validity and stationarity, addition of a dynamic updating to allow some on-line analysis during an experiment, and placing the data stream from individual synaptic sites into a time-series format of analysis. The latter will allow historical effects to be included, such as potentiation with time, in a dynamic format, and may be more predictive of future trends. Addition of dendritic models to include conduction from dendritic sites to the soma may also enhance the link between synaptic inputs and cell-firing capabilities and cell behavior in general (Williams and Stuart, 2003).

In addition to the scientific and statistical issues arising in connection with investigations of intrinsic variability, several other areas are of current interest and are under initial investigation. As presented, our models make no explicit allowance for outliers in signal recordings, nor for systematic features that arise when synaptic tissue exhibits inhibitory as well as excitatory responses to stimuli. An example of the latter is discussed in West and Turner (1994). Perhaps the most needed generalization, currently, is to relax the assumption of normally distributed experimental noise. This is currently in development, using noise models based on work in Bayesian density estimation following Escobar and West (1995).

One budding area of research involves issues of dependence in release characteristics over time and across sites (addressed in a different manner by Uteshev et al., 2000). On the first, synaptic responses in general may be viewed as streams where, at each site, the response may depend on previous occurrences. Current analyses assume independence. The very nature of the paired-pulse potentiation experiments, however, explicitly recognizes dependence effects, both in terms of response/no response and in terms of response levels, and of the importance of the time elapsed between successive stimuli generating the responses. Statistical developments of time-series issues are of interest here to explore and, ultimately, to model time variation in patterns of synaptic transmission. Interest in transmission dependencies also extends across sites, that is, to investigations of "connectivities" between sites that may have physiological interpretation and importance, though serious inroads into study of this issue are some way in the future.

It will be apparent that the technical aspects of simulation-based analysis in the novel mixture models here bear further study from a practical perspective. It is certainly desirable to investigate variations on the basic Gibbs sampling schemes to enhance and improve convergence characteristics. It is possible that hybrid schemes based, in part, on the use of auxiliary variables (e.g., Besag and Green, 1993) will prove useful here; the need for improved computational tools will be more strongly felt in more complex models incorporating nonnormal noise components. That said, the current approach described and illustrated here provides a useful basis for exploring and assessing synaptic data in attempts to more directly evaluate the structure and stochastic characteristics of synaptic transmission as it is currently viewed in the neurophysiological community. Our current work on the applications side is focused on refining this basic framework and in exploring ranges of synaptic data sets to isolate and summarize the diversity of observed response mechanisms in mammalian CNSs.

Clearly, physiological methods are needed to clarify the ambiguity associated with the various parameters at even a single synapse. Optical quantal analysis allows the investigation of single synaptic sites and correlation with physiology and statistical measures (Oertner, 2002). As these newer physiological approaches narrow the degree of possible solutions, the statistical methods can be more sharply aligned and more realistic and specific.

ACKNOWLEDGMENTS

Supported by the National Science Foundation (DAT, MW), and Department of Veterans Affairs Merit Review (DAT).

APPENDIX: MATHEMATICAL DERIVATION OF THE MODEL

Ensembles of maximum synaptic signal levels (specifically excitatory postsynaptic potentials [EPSPs], excitatory postsynaptic currents [EPSCs], inhibitory postsynaptic currents [IPSCs], etc.) obtained in response to stimulation are assumed to be independently generated according to an underlying stochastic, synaptic site-specific mechanism described as follows. At the identified synaptic connection, the experiment either isolates both the pre- and postsynaptic neurons for complete control of all interactions between the two cells (as shown diagrammatically in Figure 4.1), or a region on the axon branch of a presynaptic neuron leading to activation of only a small number of postsynaptic sites. Stimulation applied to the isolated presynaptic element leads to chemical neurotransmitter release at a random number of available synaptic transmission sites, similar to those shown in Figure 4.1. If release occurs, then each individual site contributes a small electrical signal, which then sums across sites to induce the overall resulting electrical signal in the postsynaptic neuron, which is measured physiologically, though usually in concert with interfering noise. Each of the $s > 0$ release sites transmits to a site-specific maximum level independently of all other sites. There is no identification of individual sites (e.g., such as might arise were we able to physically mark sites in the tissue and identify the location of individual transmissions) so we arbitrarily label the sites $1, \ldots, s$. Site by site we assume that, on any trial:

- Sites release independently, with individual, site-specific chances of release occurring on any occasion.
- A transmitting site produces a site-specific, fixed packet, or quantum of neurotransmitter leading to a characteristic amplitude response at the synaptic site.
- Recorded maximum synaptic signal levels represent the sums of signals induced by the various sites randomly transmitting, with additive synaptic and experimental noise.

In any experiment we assume individual transmitter release probabilities and transmission levels to be fixed and with only slight perturbations over time, but no net trend. Changes in these characteristics over time due to forced experimental and environmental changes is of later interest, and evaluating such changes is one of the guiding motivations in developing these models. Symbolically, we represent recorded signal levels as $y = \{y_1, \ldots, y_n\}$ where y_i is the level on trial i; the sample size n is typically a few hundreds. Given s sites, site j transmits on any trial with probability π_j, independently of other sites and independently across trials. The level of transmission on occurrence is the site-specific quantity μ_j. Physically, signal levels are measured via either sharp intracellular probes or low-noise patch electrodes as electrical signals induced on cell bodies, and responses are in units of either current (pA) or potential (mV). By convention, we change the readings to predominantly positive values so that the μ_j must be nonnegative values. The recordings are subject to background synaptic and experimental noise, including both electronic distortions in recording the cellular potential and errors induced in computing the estimated maximum potential levels. These noise sources combine to produce additive errors, assumed to be approximately normally distributed independently across trials, as discussed in Section 4.1; write v for the variance of these normal errors. Concomitant noise measurements are available to assess these experimental errors and to provide prior information relevant to v; blank signals measured without experimental stimulus provide these measurements. It is stressed that the noise affecting signal recordings has the same distribution, of variance v, as the raw noise recordings obtained without a signal, there being no scientific or experimental reason to assume otherwise. We also model a systematic bias in the signal recordings;

although the associated noise measures are typically close to zero mean, the signal recordings are often subject to a small but nonnegligible baseline offset induced by the experimental procedures; call this shift m. We then have the following basic signal model.

For $i = 1, \ldots, n$, the signal observations are conditionally independent,

$$y_i \sim N(y_i|\theta_i, v) \quad \text{with } \theta_i = m + \sum_{j=1}^{s} z_{ij}\mu_j$$

and where the z_{ij} are latent, unobserved, site transmission indicators,

$$z_{ij} = \begin{cases} 1, & \text{if site } j \text{ transmits on trial } i, \\ 0, & \text{otherwise.} \end{cases}$$

Under the model assumptions, these indicators are conditionally independent Bernoulli quantities with $P(z_{ij} = 1) = \pi_j$ for each site j and all trials i.

Thus signal data are drawn from a discrete mixture of (at most) $k = 2^s$ normal components of common variance, induced by averaging normals determined by all possible combinations of the z_{ij} and weighted by the corresponding chances of those combinations. On any trial, the normal component selected is determined by the column $s-$vector $z_i = (z_{i1}, \ldots, z_{is})'$ realized on that trial by individual sites releasing or not; the selected component has mean $\theta_i = m + z_i'\mu$ where $\mu' = (\mu_1, \ldots, \mu_s)$, and is selected with chance $\prod_{j=1}^{s} \pi_j^{z_{ij}}(1-\pi_j)^{1-z_{ij}}$; specifically, and conditional on all model assumptions and parameters, the y_i are conditionally independently drawn from the distribution with density

$$p(y_i|\mu, \pi, m, v) = \sum_{z_i} p(z_i|\pi)p(y_i|\mu, z_i, m, v) = \sum_{z_i} \left\{ \prod_{j=1}^{s} \pi_j^{z_{ij}}(1-\pi_j)^{1-z_{ij}} \right\} N(y_i|m + z_i'\mu, v), \tag{4.A1}$$

where the $s-$vector z_i ranges over all 2^s possible values. If there are common values among the site levels μ_j, then the mixture distribution function will be equivalent to one with fewer than 2^s components. Also, if either $\mu_j = 0$ or $\pi_j = 0$ then site j disappears from the model and the distribution reduces to a mixture of at most 2^{s-1} components. Additional cases of special interest include:

1. So-called compound binomial models, in which the site response levels are equal, all at a basic "quantum" level μ_0, but release probabilities differ. The aggregate response levels then run between m and $m + s\mu_0$ and the mixture of 2^s normal components effectively reduces to a mixture of just $s + 1$, with associated weights determined by the discrete compound binomial resulting from the distinct chances $\pi_1, \ldots, \pi_s$, across sites. These kinds of special cases have received considerable attention in the literature (Stricker et al., 1996a).
2. Precise quantal–binomial models, in which the site levels are equal and the release probabilities are constant too, $\pi_j = \pi_0$. Now the mixture reduces to one with distinct normal means $m + j\mu_0$ ($j = 0, \ldots, s$), and corresponding binomial weights $\binom{s}{j}\pi_0^j(1-\pi_0)^{s-j}$ (Redman, 1990).

Our previous work, following others, modeled synaptic data in the framework of "standard" mixture models for density estimation, from a particular Bayesian viewpoint; see West and Turner (1994), for example, and Escobar and West (1995) for statistical background. In these models, the component means and weights are essentially unrestricted, rather than being modeled directly as functions of the underlying synaptic parameters μ and π. A significant drawback is that it is then, in general, very difficult to translate posterior inferences about unrestricted normal mixture model parameters to the underlying μ and π parameters of scientific interest, especially in the context of uncertainty about s. One simple example in West and Turner (1994) shows how this can be done; that is a rare example in which the data appear to be consistent with the full quantal hypothesis, so that the convoluted process of backtracking from the mixture model to the underlying site-specific parameters is accessible. However, we have encountered very few data sets in which this is the case, and this inversion process is generally difficult. Hence the direct approach to inferring the neural parameters has been developed.

4.A1 Prior Distributions for Synaptic Parameters

Model completion requires specification of classes of prior distributions for determining the parameters μ, π, m, and v for any given s. Assessment of reasonable values of s, as well as values of the μ and π quantities for any given s, is part of the statistical inference problem. From a technical viewpoint, uncertainty about s can be formally included in the prior distribution, so as to provide posterior assessment of plausible values. The models below allow this. From the viewpoint of scientific interpretation, however, inferences about the site parameters are best made conditional on posited values of s, and then issues of sensitivity to the number of sites explored arise.

Begin by conditioning on a supposed number of sites s. Current implementations assume priors for all model parameters that, though rather intricately structured in certain dimensions to represent key qualitative features of the scientific context, are nevertheless inherently uniform in appropriate ways, providing reference initial distributions. Note that other classes of priors may be used, and some obvious variations are mentioned below; however, those used here are believed to be appropriately vague or uninformative in order that resulting posterior distributions provide benchmark or reference inferences that may be directly compared with supposedly objective non-Bayesian approaches (Redman, 1990; comparative analyses will be reported elsewhere in a further article).

In addressing uncertainty about s alone, we make the observation that models with fewer than s sites are implicitly nested within a model having s sites. To see this, constrain some $r > 0$ (arbitrarily labeled) site levels μ_j to be zero; then Equation (4.A1) reduces to precisely the same form based on $s - r$ sites, whatever the value of the π_j corresponding to the zeroed μ_j. Zeroing some μ_j simply confounds those sites with the noise component in the mixture — a site transmitting a zero level, with any probability, simply cannot be identified. Hence assessing whether or not one or more of the site levels are zero provides assessment of whether or not the data support fewer sites than assumed. This provides a natural approach to inference of the number of active sites, assuming that the model value s is chosen as an upper bound.

Note that a similar conclusion arises by considering $\pi_j = 0$ for some indices j; that is, an inactive site may have a zero release probability rather than (or as well as) a zero release level. However, the zeroing of a site probability induces a degeneracy in structure of the model and obviates its use as a technical device for inducing a nesting of models with fewer than s sites in the overall model. Hence our models define inactive sites through zeros among the μ_j, restricting the π_j to nonzero values, however small. Thus $\mu_j = 0$ for one or more sites j is the one and only way that the number of active sites may be smaller than the specific s.

4.A1.1 General Structure

For a specified s, analyses reported below are based on priors with the following structure. Quantities μ, π, (m, v) are mutually independent. We will describe classes of marginal priors for μ and π separately; each will involve certain hyperparameters that will themselves be subject to uncertainty described through hyperpriors, in a typical hierarchical modeling framework, and these hyperparameters will also be assumed to be mutually independent. To anticipate development below, hyperparameters denoted q and a are associated with the prior for μ, a single quantity b determines the prior for π, and the joint prior is then of the form

$$p(\mu, \pi, m, v, q, a, b) = p(\mu, \pi, m, v|q, a, b)p(q, a, b) = p(\mu|q, a)p(\pi|b)p(m, v)p(q)p(a)p(b). \tag{4.A2}$$

The component densities here are now described in detail. Two comments on notation: first, conditioning statements in density functions include only those quantities that are required to determine the density, implicitly indicating conditional independence of omitted quantities; second, for any vector of h quantities $x = (x_1, \dots, x_h)$, for any $j \leq h$ the notation x_{-j} represents the vector x with x_j removed, that is, $x_{-j} = x - \{x_j\} = (x_1, \dots, x_{j-1}, x_{j+1}, \dots, x_h)$.

4.A1.2 Priors for μ and Associated Hyperparameters q, a

We develop a general classes of priors for μ, and comment on various special cases. The class has the following features:

1. A component baseline uniform distribution for each μ_j over a prespecified range $(0, u)$, with a specified upper bound u.
2. Components introducing positive prior probabilities at zero for each of the μ_j in order to permit assessment of hypotheses that fewer than the chosen (upper bound) s are actually nonzero, and hence to infer values of the number of active sites.
3. Components permitting exact common values among the elements of μ to allow for the various special cases of quantal transmission, and specifically the questions of whether or not pairs or subsets of sites share essentially the same quantal transmission level.

These are developed as follows. Let $F(\cdot)$ be a distribution on $(0, u)$, having density $f(\cdot)$ for some specified upper bound u. Write $\delta_0(x)$ for the Dirac delta function at $x = 0$, and $U(\cdot|a, b)$ for the continuous uniform density over (a, b). Then suppose the μ_j are *conditionally independently* drawn from the model

$$(\mu_j|F, q) \sim q\delta_0(\mu_j) + (1 - q)f(\mu_j)$$

for some probability q. Under this prior, the number h of nonzero values among the μ_j — the number of active sites — is binomial (Bn), $(h|s, q) \sim \text{Bn}(s, 1 - q)$, with mean $s(1 - q)$, independently of F. If F were uniform, say $U(\cdot|0, u)$, this prior neatly embodies the first two desirable features 1. and 2. described above. Note the two distinct cases:

- Setting $q = 0$ implies that $h = s$ is the assumed number of active sites, so then inference proceeds conditional on $\mu_j > 0$ for each j.
- Otherwise, restricting to $q > 0$ allows for assessment of the number of active sites, subject to the specified upper bound s.

In practice we will assign a hyperprior to q in the case $q > 0$. The class of beta (Be) distributions is conditionally conjugate, and the uniform prior suggests itself as a reference, $q \sim U(q|0, 1)$. One immediate, and nice, consequence of a uniform prior is that the resulting prior for the number of nonzero values among the μ_j has (averaging the binomial with respect to q) a discrete uniform prior over $0, 1, \dots, s$. This is a suitably vague and unbiased initial viewpoint with respect to the number of active sites. Other beta priors may be explored, of course. One specific choice we use in current work is $(q|s) \sim \text{Be}(s - 1, 1)$; note the explicit recognition of dependence on the specified values of s. The reasoning behind this choice is as follows. First, we are currently focused on experiments designed to isolate rather small numbers of sites, down to just a few, say 1 to 4 from the viewpoint of scientific intent and expectation. So s values up to 7 or 8 may be explored, but lower values are typical. Whatever value of s is chosen, h is expected to be in the low integers, so the guiding choice of the prior for q is to induce a prior for h favoring smaller values. Consider values of s in the relevant range $3 \leq s \leq 8$, or so. Then, integrating $p(h|s, q)$ with respect to the specific prior $(q|s) \sim \text{Be}(s-1, 1)$ we obtain a distribution $p(h|s)$ that is almost completely insensitive to s, having a diffuse and decreasing form as h increases, and with $E(h|s) = 1$ for any such s. Thus, this specific beta prior for q has the attractive feature of consistency with a scientifically plausible prior $p(h|s) \approx p(h)$, incorporating the scientific view of a likely small numbers of active sites, almost independently of the specified upper bound s.

This structure provides a baseline uniform prior for release levels, together with the option for allowing a smaller number of sites than the s specified. So far, however, there is no explicit recognition of the special status, in the scientific area, of quantal hypotheses as represented through common values among the μ_j. If $F(\cdot)$ is a continuous distribution, then the prior implies that the nonzero μ_j are distinct. Though this might allow arbitrarily close μ_j values, it is desirable to have the opportunity to directly assess questions about common values, and perhaps subgroups of common values, in terms of posterior probabilities. There is also a significant technical reason, discussed below in connection with issues of parameter identification that calls for a prior that gives positive probability to exact equality of collections of the nonzero μ_j. We therefore extend the prior structure so far discussed to provide this. We do this using a standard Dirichlet model. Specifically, a Dirichlet process prior for F induces a discrete structure that gives positive prior probability to essentially arbitrary groupings of the set of nonzero μ_j into subsets of common values; this is a general framework that permits varying degrees of partial quantal structure, from the one extreme of completely distinct values to the other of one common value. This structure is most easily appreciated through the resulting set of complete conditional posterior distributions for each of the μ_j given μ_{-j}. These are defined by

$$p(\mu_j|\mu_{-j}, q, a) = q\delta_0(\mu_j) + (1-q)\left\{r_j U(\mu_j|0, u) + (1-r_j)h_j^{-1}\sum_{i \in N_j} \delta_{\mu_i}(\mu_j)\right\}, \tag{4.A3}$$

where h_j is the number of nonzero elements of μ_{-j} and N_j is the corresponding set of indices, $N_j = \{i|\mu_i > 0,\ i = 1, \dots, s;\ i \neq j\}$, and $r_j = a/(a + h_j)$. The hyperparameter a is subject to uncertainty and is included in the analysis using existing approaches for inference on precision parameters in Dirichlet models, developed in West (1997) and illustrated in Escobar and West (1995). As shown there, gamma priors for a, or mixtures of gamma priors, are natural choices, and our application currently uses diffuse gamma models.

In summary, Equation (4.A3) shows explicitly how site level μ_j may be zero, implying an inactive site, or take a new nonzero value, or be equal to one of the nonzero values of other sites. The roles of hyperparameters q and a are evident in this equation. With this structure we have defined prior components $p(\mu|q, a)p(q)p(a)$ of the full joint prior in Equation (4.A2).

4.A1.3 Priors for π and Associated Hyperparameter b

The structure of the prior for π parallels, in part, is that of μ in allowing common values. The detailed development for μ can be followed through with the same reasoning about a baseline prior and a structure to induce positive probabilities over subsets of common values. We restrict this analysis to positive release probabilities, as discussed earlier, so that the prior structure for the π_j is simpler in this respect. Specifically, we assume that π_j is independently drawn from a distribution on $(0, 1)$ that is assigned a Dirichlet process prior. We take Dirichlet precision $b > 0$ and base measure to be $bU(\cdot|0, 1)$. Then, as in the development for μ, the full joint prior $p(\pi|b)$ is defined by its conditionals

$$(\pi_j|\pi_{-j}, b) \sim wU(\pi_j|0, 1) + (1 - w)(s - 1)^{-1} \sum_{i=1, i\neq j}^{s} \delta_{\pi_i}(\pi_j), \qquad (4.A4)$$

where $w = b/(b + s - 1)$ for each $j = 1, \ldots, s$.

As for the site levels, the induced posteriors will now allow inference on which sites may have common release probabilities, as well as on the precise values of such probabilities.

Nonuniform beta distributions might replace the uniform baseline in application, particularly with expected low levels of release probabilities. Our analyses reported below retain the uniform prior for illustration and so avoid questions of overtly biasing toward lower or higher values. Note also that we explicitly exclude the possibility of nonzero release probabilities, though some or all may be very small, and so rule out the use of $\pi_j = 0$ as a device for reducing the number of active sites; that is completely determined by zero values among the μ_j, as discussed earlier. As with a in the model for the site levels, the hyperparameter b will be assigned a prior, and again a diffuse gamma prior is natural. With this structure, we have defined the additional prior components $p(\pi|b)p(b)$ of the full joint prior in Equation (4.A2).

4.A1.4 Priors for Noise Moments *m* and *v*

In most experiments we anticipate a possible systematic bias m in both noise and signal recordings, induced by the direct measurement of cell membrane potential. This is typically very small relative to induced signal levels, and the raw noise recordings provide data to assess this. In addition, the noise measurements inform on background variability v, that is, they are drawn from the $N(\cdot|m, v)$ distribution. In our analyses, our prior for m and v as input to the signal analysis is simply the posterior from a reference analysis of the noise data alone, that is, a standard conjugate normal inverse gamma distribution based on the noise sample mean, variance, and sample size. With this structure, we have defined the final component $p(v, m)$ of the full joint prior in Equation (4.A2).

4.A2 Posterior Distributions and Computation

Calculation of posterior distributions is feasible, as might be expected, via variants of Gibbs sampling (e.g., Gelfand and Smith, 1990; Smith and Roberts, 1993). The specific collection and sequence of conditional posterior distributions used to develop simulation algorithms are briefly summarized here. A key step is to augment the data with the latent site transmission indicators z; thus the sampling model is expanded as

$$\begin{aligned} p(y, z|\mu, \pi, m, v, q, a, b) &= p(y|\mu, z, m, v)p(z|\pi) \\ &= \left\{\prod_{i=1}^{n} N(y_i|m + z_i'\mu, v)\right\} \left\{\prod_{j=1}^{s}\prod_{i=1}^{n} \pi_j^{z_{ij}}(1 - \pi_j)^{1-z_{ij}}\right\}, \end{aligned}$$

providing various conditional likelihood functions that are neatly factorized into simple components. The full joint posterior density, for all model parameters together with the uncertain indicators z, has the product form

$$p(\mu, \pi, z, m, v, q, a, b|y) \propto p(\mu|q, a)p(q)p(a)p(z|\pi)p(\pi|b)p(b)p(m, v)p(y|\mu, z, m, v), \qquad (4.A5)$$

where the component conditional prior terms are as described in the previous section.

4.A3 Conditional Posteriors and Markov Chain Monte Carlo Model

The posterior Equation (4.A5) yields a tractable set of complete conditional distributions characterizing the joint posterior, and hence leads to implementation of Gibbs sampling based on this structure. Each iteration of this Markov Chain Monte Carlo Model (MCMC) sampling scheme draws a new set of all parameters and latent variables by sequencing through the conditionals now noted:

1. Sampling site levels μ proceeds by sequencing through $j = 1, \ldots, s$, at each step generating a new value of μ_j given the latest sampled values of μ_{-j} and all other conditioning quantities. For each j, this involves sampling a posterior that is a simple mixture of several point masses with a truncated normal distribution. Sampling efficiency is improved using variations of so-called configuration sampling for the discrete components.
2. Sampling site release probabilities π similarly proceeds by sequencing through $j = 1, \ldots, s$, at each step generating a new value of π_j given the latest sampled values of π_{-j} and all other conditioning quantities. For each j, this involves sampling a mixture of discrete components with a beta component. Again, configuration sampling improves simulation efficiency.
3. Sampling transmission indicator z involves a set of n independent draws from conditional multinomial posteriors for the n individual binary 2^s − vectors z_i, $i = 1, \ldots, n$. These simulations are easily performed.
4. Sampling the systematic bias quantity m involves a simple normal draw.
5. Sampling the noise variance v involves sampling an inverse gamma posterior.
6. Sampling the hyperparameter q is also trivial, simply involving a draw from an appropriate beta distribution.
7. Sampling the hyperparameters a and b follows West (1997), and Escobar and West (1995), and involves a minor augmentation of the parameter space, with simple beta and gamma variate generations.

Iterating this process produces sequences of simulated values of the full set of quantities $\phi \stackrel{\text{def}}{=} \{\mu, \pi, z, m, v, q, a, b\}$ that represent realizations of a Markov chain in ϕ space whose stationary distribution is the joint posterior Equation (4.A5) characterized by the conditionals. The precise sequence Equations (4.A1) to (4.A7) of conditional posteriors detailed above is the sequence currently used in simulation; at each step, new values of the quantities in question are sampled based on current values of all quantities required to determine the conditional in question. Convergence is assured by appeal to rather general results, such as Tierney (1994, especially theorem 1 and corollary 2) relevant to the models here; this ensures that successive realizations of $\phi = \{\mu, \pi, z, v, q, a, b\}$ generated by this Gibbs sampling setup eventually resemble samples from the exact posterior in Equation (4.A5). Irreducibility of the resulting chain is a consequence of the fact that all full conditionals defining the chain are positive everywhere. Practical issues of convergence are addressed by experiments with several short runs from various starting values. Other important issues of convergence are discussed in the next section.

4.A4 Parameter Identification and Relabeling Transmission Sites

The model structure as described is not identifiable in the traditional sense of parameter identification especially associated with mixture models. This is obvious in view of the lack of any physical identification of the neural transmission sites we have arbitrarily labeled $j = 1, \ldots, s$; for example,

π_1 is the release probability for the first site in our labeling order, but this could be any of the actual release sites. As a result, the model is invariant under arbitrary permutations of the labels on sites, and likelihood functions are similarly invariant under permutations of the indices of the sets of parameters (μ_j, π_j). Also, as the priors for the μ_j and π_j separately are exchangeable, corresponding posteriors are also invariant under labeling permutations.

To impose identification we need a physically meaningful restriction on the parameters. Perhaps the simplest, and most obvious, constraint is to order either site levels or site release chances. We can do this with the site release levels. In posterior inferences, relabel so that site j corresponds to release level μ_j, the jth largest value in μ. As the priors, hence posteriors, include positive probabilities on common values, we break ties by imposing a subsidiary ordering on the release probabilities for (ordered) sites with common release levels. Thus, if μ_3, μ_4, and μ_5 happened to be equal, the sites relabeled 3 to 5 would be chosen in order of increasing values of their chances π_j; were these equal too, as permitted under the priors, and hence posteriors here, then the labeling is arbitrary and we have (in this example) three indistinguishable sites with common release levels and common release probabilities. The posterior analysis required to compute corresponding posterior inferences is extremely simple in the context of posterior simulations; each iteration of the Gibbs sampling analysis produces a draw of (μ, π) (among other things). Given a draw, we first create a separate vector μ^* containing the *ordered* values of μ, and a second vector π^* containing the elements of π rearranged to correspond to the ordering of μ to μ^*. If subsets of contiguous values in μ^* are common, the corresponding elements in π^* are then rearranged in increasing order themselves. Through iterations, repeat draws of (μ^*, π^*) are saved, thereby building up samples from the required posterior distribution that incorporates the identification of sites.

Note that we could alternatively identify the model by a primary ordering on the π_j, rather than the μ_j. In data analyses we routinely explore posterior summaries identified each way — that is, ordered by μ_j and ordered by π_j — as each is only a partial summary of the full posterior. This is relevant particularly in cases when it appears that the data are in conformity, at least partially, with quantal structure.

In connection with this ordering issue, the quantal structure of the priors for μ and π is very relevant, as we alluded to earlier. Consider an application in which the data support common, or very close, values among the μ_j. Suppose that the prior is simply uniform, with no chance that consecutive values of the ordered μ_j are equal. Then imposing the ordering for identification results in a distortion of inferences — values we should judge to be close, or equal, are "pushed apart" by the imposed ordering. This undesirable effect is ameliorated by the priors we use, allowing exact equality of neighboring levels. The same applies to the π_j.

Note that we retain the original, unidentified parametrization for model specification, and the simulation analysis operates on the unconstrained posteriors. This has a theoretical benefit in allowing easy and direct application of standard convergence results (Tierney, 1994). More practically, it has been our experience that imposing an identifying ordering on parameters directly through the prior distribution hinders convergence of the resulting MCMC on the constrained parameter space. This arises naturally as the ordering induces a highly structured dependence between parameters, relative to the unconstrained (albeit unidentified) parameter space. In addition, one rather intriguing practical payoff from operating the MCMC on the unrestricted space is an aid in assessing convergence of the simulation. This is due to the exact symmetries in posterior distributions resulting from permutation invariance, partly evidenced through the fact that marginal posteriors $p(\mu_j, \pi_j|y)$ are the same for all j. Hence summaries of marginal posteriors for, say, the μ_j should be indicative of the same margins for all j; the same is true of the π_j.

However, in many applications convergence will be slow. One reason for this is that, in some models and with some observed signal data sets, the posterior will heavily concentrate in widely separate regions of the parameter space, possibly being multimodal with well-separated modes. Such posteriors are notorious in the MCMC literature, and the Gibbs sampling routine may get "stuck" in iterations in just one (or typically several) region of high posterior probability, or around a subset of modes. There is certainly need for further algorithmic research to provide faster alternatives to the raw Gibbs sampling developed here (possibly based on some of the methods mentioned in Besag and Green [1993], or Smith and Roberts [1993], for example). However, this is not a problem as far as the ordered site levels, and the correspondingly reordered release probabilities, are concerned, due to the symmetries in the unconstrained posterior. This is because posterior samples from one region of the (μ, π) space may be reflected to other regions by permuting site indices, while the resulting ordered values remain unchanged. Hence the unconstrained sampling algorithm may converge slowly as far as sampling the posterior for (μ, π), but the derived sequence for (μ^*, π^*) converges much faster.

Due to the lack of identification, actual output streams from MCMC simulations exhibit switching effects, as parameter draws jump between the modes representing the identification issue. For example, in a study supporting two distinct site release levels with values near 1 and 2 mV, respectively, the simulated series of the unidentified μ_1 (μ_2) will tend to vary around 1 mV (2 mV) for a while, then randomly switch to near 2 mV (1 mV) for a while, and so on. Our experience is that, across several analyses of different data sets, convergence is generally "clean" in the sense that, following what is usually a rapid initial burn-in period, the output streams remain stable and apparently stationary between points of switching between these posterior modes. Informal diagnostics based on a small number of short repeat runs from different starting values verify summarized analyses from one final, longer run. In addition, the theoretical convergence diagnostic based on the symmetry of the unidentified posterior is used in the final analysis.

4.A5 Incorporation of Intrinsic Variability into the Model

Recent attention has focused on questions of additional variability in synaptic signal outcomes due to so-called *intrinsic variability* in release levels of individual synapses (Turner et al., 1997); we describe this concept here, and define an elaborated class of models incorporating it. This concept gives rise to model modifications in which components of the normal mixture representation have variances that increase with level, and this leads to considerable complications, both substantive and technical. The technical complications have to do with developing appropriate model extensions, and associated MCMC techniques to analyze the resulting models; a brief account of the development process is mentioned here with examples. The substantive complication is essentially that competing models with and without intrinsic variance components cannot be readily distinguished on the basis of the observed data alone; an observed data configuration might arise from just one or two sites with significant intrinsic variance, or it might arise from a greater number of sites with low or zero intrinsic variance. In such cases, and especially when inferences about site characteristics are heavily dependent on the number of sites and levels of intrinsic variance, we are left reliant on the opinions of expert neurophysiologists to judge between the models. Unfortunately, in its current state, the field is represented by widely varying expert opinions, from the one extreme of complete disregard of the notion of intrinsic variability, to the other of belief in high levels of intrinsic variability as the norm. In some of the examples that we have studied, including the data sets shown here, this issue is relatively benign, as inferences about release levels and probabilities are relatively insensitive to intrinsic variances. In other cases, it is evidently highly relevant. Further collaborative research to refine knowledge of intrinsic variability effects is part of the current frontiers of the field. Here we give a short discussion of the notion and our current approach to modeling intrinsic variability.

Intrinsic variability refers to variation in levels of neurotransmitter release at specific sites. As developed above, site j has a fixed release level μ_j, and, while these levels may differ across sites, they are assumed fixed for the duration of the experiment under the controlled conditions. However, the mechanism of electrochemical transmission suggests that this may be an oversimplification. A site transmits by releasing a packet of (many) molecules of a chemical transmitter, and these molecules move across the synaptic cleft to eventually bind to the postsynaptic receptors. The induced signal response is proportional to the number of binding molecules. So, the assumption of fixed μ_j implies that (a) the number of molecules transmitted is constant across occasions, and (b) all ejected molecules are bound to receptors on the postsynaptic cell. Each of these is questionable, and the issue has given rise to the notion of intrinsic variance, that is, variability in the site-specific level of release across occasions (Faber et al., 1992). Intrinsic variance may also arise when two sites are close to each other and cannot be separated in any statistical sense. Then, it will appear that one site may be fluctuating. Whatever the actual structure of variability, it is apparent in data sets through mixture components having higher variance at higher release levels.

The basic extension of our models to admit the possibility of intrinsic variability (I) involves remodeling the data y_i as coming from the conditional normal model

$$y_i \sim N\left(y_i|m + \sum_{j=1}^{s} z_{ij}\gamma_{ij}, v\right), \tag{4.A6}$$

where the γ_{ij} are new site- and case-specific release levels and v is the noise variance. Retaining independence across sites and trials, we assume that the γ_{ij} are distributed about an underlying expected release level μ_j — the same site-specific level as before, but now representing underlying average levels about which releases vary. Then the form of the distribution of the γ_{ij} about μ_j represents the intrinsic variability in induced responses due to variation in amounts of neurotransmitter released by site j and also to variation in the success rate in moving the transmitter across the synaptic cleft. Various parametric forms might be considered; our preliminary work, to date, is based on exploration of models in which

$$p(\gamma_{ij}|\mu_j, \tau_j) \propto N(\gamma_{ij}|\mu_j, \tau_j^2\mu_j^2)I, \quad 0 < \gamma_{ij} < u \tag{4.A7}$$

for all trials i and each site j; here u is the earlier specific upper bound on release levels. The new parameters $\tau_j > 0$ measure intrinsic variability of sites $j = 1, \ldots, s$; ignoring the truncation in Equation (4.A7), τ_j is effectively a constant (though site-specific) coefficient of variation in release levels about the underlying μ_j. This model has been implemented, extending the prior structure detailed in Section 4.2.2, to incorporate the full set of site- and trial-specific release levels $\{\gamma_{ij}\}$ together with the new parameters $\tau_1, \ldots, \tau_s$. Note that, as our model allows $\mu_j = 0$ for inactive sites, we use Equation (4.A7) only for $\mu_j > 0$; otherwise, $\mu_j = 0$ implies $\gamma_{ij} = 0$ for each $i = 1, \ldots, n$.

We can see the effects of quantal variability by integrating the data density in Equation (4.A6) with respect to Equation (4.A7). This is complicated due to the truncation to positive values; as an approximation, for the purposes of illustrating structure here, assume this is not binding, that is, that μ_j and τ_j are such that the mass of the basic normal distribution in Equation (4.A7) lies well within the interval $(0, u)$. Then Equations (4.A6) and (4.A7) combine and marginalize over γ_{ij} to give

$$y_i \sim N\left(y_i|m + \sum_{j=1}^{s} z_{ij}\mu_j, v + \sum_{j=1}^{s} z_{ij}\tau_j^2\mu_j^2\right). \tag{4.A8}$$

The mean here is as in the original formulation, Equation (4.A1), with active sites contributing the expected release levels μ_j. But now the variance of y_i is not simply the noise variance v; it is inflated by adding in factors $\tau_j^2\mu_j^2$ for each of the active sites. Hence the feature relevant to the modeling shows the expected increased spread of the data configuration at higher response levels.

It should be clear that this extended model can be managed computationally with direct extensions of the MCMC algorithms discussed so far. We are now interested in the full posterior $p(\mu, \pi, z, \gamma, \tau, m, v, q, a, b|y)$, extending the original posterior in Equation (4.A5) to include the new quantities $\gamma = \{\gamma_{ij}, i = 1, \ldots, n; j = 1, \ldots, s\}$ and $\tau = \{\tau_1, \ldots, \tau_s\}$. There are difficulties in the posterior MCMC analysis due to the complicated form of Equation (4.A7) as a function of μ_j, and also due to the truncation of the basic normal model. These issues destroy part of the nice, conditionally conjugate sampling structure, and are currently handled using some direct analytic approximations and Metropolis–Hasting accept/reject steps in the extended simulation analysis. It is beyond the scope of the current chapter to develop the technical aspects of this fully here, but our discussion would be incomplete without raising the issues of intrinsic variability, currently becoming vogue in the field, and without providing some exploratory data analysis. After further technical refinements and practical experience, full modeling and technical details will be reported elsewhere (West, 1997).

5 Natural Variability in the Geometry of Dendritic Branching Patterns

Jaap van Pelt and Harry B.M. Uylings

CONTENTS

5.1 INTRODUCTION

Dendritic branching patterns are complex and show a large degree of variation in their shapes, within as well as between different cell types and species. This variation is found in typical shape parameters, such as the number, length, and connectivity pattern (topological structure) of the segments in the dendritic tree, the curved nature of these segments, and the embedding of the dendrite in three-dimensional (3D) space (Uylings and Van Pelt, 2002). Dendritic structure plays an important role in the spatial and temporal integration of postsynaptic potentials, both metrically (e.g., Rall et al., 1992; Rapp et al., 1994) and topologically (Van Pelt and Schierwagen, 1994, 1995; Van Ooyen et al., 2002), and is consequently an important determinant of the characteristics of neuronal firing patterns (e.g., Mason and Larkman, 1990; Mainen and Sejnowski, 1996; Sheasby and Fohlmeister, 1999; Washington et al., 2000; Bastian and Nguyenkim, 2001; Schaefer et al., 2003). A good account of the extent of dendritic morphological variation between individual dendritic trees is needed in order to explore its implications for neuronal signal integration. Dendritic morphological variation can be expressed by means of distribution functions for specific shape parameters, but functional properties need to be determined for complete dendritic trees. When original reconstructions are not available in sufficient number, one may use representative dendrites synthesized using models of dendritic structures.

Among the algorithmic approaches to synthesize dendritic trees with geometrical characteristics and variation similar to those of observed dendrites, two different classes can roughly be distinguished. In the *reconstruction model approach*, dendritic shapes are reconstructed using empirical distribution functions for the shape parameters. Representative dendrites are obtained by random sampling of these distributions. For instance, Burke et al. (1992) used empirically obtained distribution functions for the lengths and diameters of dendritic branches and for the diameter relations at bifurcation points. Random dendrites were generated by a repeated process of random sampling of these distributions for deciding whether an elongating neurite should branch and for obtaining the diameters of the daughter branches. The modeled dendrites obtained in this way conform to the original distribution functions for shape characteristics. These algorithms were further elaborated and extended to include orientation in 3D space and environmental influences (Ascoli et al., 2001; Ascoli, 2002a, 2002b). An important assumption in this approach is that the shape parameters are independent of each other. Kliemann (1987) considered the segments at a given centrifugal order as individuals of a generation that may give rise to individuals in a next generation by producing a bifurcation point with two daughter segments. Mathematically, such a dendritic reconstruction could be described by a Galton–Watson process, based on empirically obtained splitting probabilities for defining *per generation* whether or not a segment will be a terminal or an intermediate one. Also in this example it is assumed that the successive choices in the reconstruction of a dendrite are independent. The model's ability to predict branching patterns was greatly enhanced by including a systematic component additional to the random one (Carriquiry et al., 1991). Applications of this method to dendritic growth *in vitro* can be found in Uemura et al. (1995). Tamori (1993) used a set of six "fundamental parameters" to describe dendritic morphology and introduced a principle of least effective volume to derive dendritic branch angles.

In contrast to the reconstruction model approach, the *growth model approach* is based on modeling dendritic structure from principles of dendritic development. Neurons grow out by a process of elongation and branching of their neuritic extensions. These processes are governed by the dynamic behavior of growth cones, highly motile structures at the tips of outgrowing neurites. Neurite elongation requires the polymerization of tubulin into cytoskeletal microtubules. Branching occurs when a growth cone, including its actin cytoskeletal meshwork, splits into two parts while partitioning the number of microtubules, and each daughter growth cone continues to grow on its own (e.g., Black, 1994; Kater et al., 1994; Letourneau et al., 1994; Kater and Rehder, 1995). Growth cones may also retract by microtubule depolymerization, or may redirect their orientations. The dynamic behavior of a growth cone results from the integrated outcome of many intracellular mechanisms and interactions with its local environment. These include, for instance, the exploratory interactions of filopodia with a variety of extracellular (e.g., guidance, chemorepellant, chemoattractant) molecules in the local environment (e.g., Kater et al., 1994; McAllister, 2002; Whitford et al., 2002), receptor-mediated transmembrane signaling to cytoplasmic regulatory systems (e.g., Letourneau et al., 1994), intracellular regulatory genetic and molecular signaling pathways (e.g., Song and Poo, 2001), and electrical activity (e.g., Cline, 1999; Ramakers et al., 1998, 2001; Zhang and Poo, 2001). Electrical activity and neurotransmitters have an especially strong modulatory influence on neurite outgrowth via their effects on intracellular calcium levels and on gene expression, by which many different proteins become involved, such as neurotrophins, GAP43, CaMKII, and CPG15 (reviewed in Cline, 1999). Further regulatory influences are exerted by the phosphorylation state of microtubule associated proteins (MAPs) on the stabilization of the microtubule cytoskeleton (modeled in Van Ooyen and Van Pelt, 2002). There is increasing evidence that the Rho family of small GTPases, including Rho, Rac, and Cdc42 play an important role in neuronal development. These molecules act as molecular switches in signaling pathways down to the actin cytoskeleton within the growth cone. Their role in neuronal development may be further stressed by noting that mutations in their genes are especially involved in the origin of mental retardation, suggesting the development of abnormal neuronal network structures (Ramakers, 2002).

In addition to these regulatory mechanisms, the dendritic outgrowth process may also be subjected to basic constraints which limit its responsiveness to modulation. For instance, neurite elongation proceeds by the growth of microtubules, which requires the production, transport, and polymerization of tubulin cytoskeletal elements. The production rate of cytoskeletal elements sets an upper limit to the average total rate of increase of the length of the dendrite. In addition, the division of flow of cytoskeletal elements at a bifurcation modulates the elongation rate of the daughter branches. Earlier model studies of microtubule polymerization (neurite elongation) in relation to tubulin production and transport demonstrated how limited supply conditions may lead to competition between growth cones for tubulin, resulting in alternating advance and immobilization (Van Ooyen et al., 2001). Empirical evidence for such competitive behavior has subsequently been obtained from time-lapse studies of outgrowing neurons in tissue culture by Ramakers (see Costa et al., 2002). Limiting resources may also lead to competitive phenomena between axons, where target-derived neurotrophins are required for the maintenance and stabilization of synaptic axonal endings on target dendrites. Modeling studies have shown that the precise manner in which neurotrophins regulate the growth of axons determines what patterns of target innervation can develop (Van Ooyen and Willshaw, 1999). Taking a spatial dimension into account, these authors show that distance between axonal endings on the target dendritic tree mitigates competition and permits the coexistence of axons (Van Ooyen and Willshaw, 2000). These examples illustrate the complexity of the outgrowth process in terms of involved molecules, interactions, signaling pathways, and constraints. The integrated action and the details of all these intracellular and extracellular mechanisms finally form the basis for dendritic morphological characteristics, for morphological differentiation between cell types, and for the diversity between neurons of the same type (e.g., Acebes and Ferrus, 2000).

Modeling neurite outgrowth at the level of the details mentioned is an immense task, being only in its early phase and proceeding in a step by step fashion by focusing on well manageable subsystems, as shown in the examples. An overview of modeling studies of neural development can be found in Van Ooyen (2003). A phenomenological approach in modeling dendritic outgrowth, on the contrary, aims at finding an algorithmic framework to describe the behavior of growth cones directly in terms of elongation and branching rates. This approach is very powerful for obtaining an understanding of how dendritic shape characteristics and variations arise from the details in the elongation and branching rates. Because of the large number of processes involved in the actual behavior of growth cones, it is a reasonable assumption to model elongation and branching as outcomes of stochastic processes. Although the probability functions underlying these processes may be very complex, it is an interesting question for which minimal assumptions are required to obtain model dendrites that compare closely in their morphological properties and variations to the observed ones. Longitudinal studies on dendritic development *in vivo* are difficult to perform, because the tissue needs to be processed in order to enable morphological quantification. Tissue-culture studies might form an alternative, although, especially in dissociated cultures, neurons develop in an environment different from their natural one.

The *growth model approach* has been applied in the past decades by several investigators (e.g., Smit et al., 1972; Berry and Bradley, 1976; Van Pelt and Verwer, 1983; Ireland et al., 1985; Horsfield et al., 1987; Nowakowski et al., 1992). One of the growth model approaches will be described in more detail in this chapter. It is based on the assumption that growth actions are outcomes of a stochastic process. The branching and elongation probabilities allow a direct interpretation in terms of the underlying developmental biological process. Additionally, the dependence of these probabilities on developmental time and on the growing structure itself are implicitly accounted for. It will be shown that the assumptions of random branching and random elongation of terminal segments (or "growth cones") are sufficient for generating dendrites with realistic variations in their topological and metrical shape parameters. Because of the structure of the model, the variables are directly related to the dynamic behavior of growth cones, which allows for empirical verification. This also makes the model important for developmental studies, because it provides a tool for analyzing reconstructed dendrites in terms of hypothetical developmental histories, which themselves are often not accessible by empirical techniques.

In the following sections, dendritic shape parameters are introduced and examples are given for the extent of their variations as found in observed dendritic trees. The dendritic growth model is reviewed and discussed, following the modular extensions that have been incorporated in the course of time. The functional implications of dendritic morphological variation are briefly discussed.

5.2 Dendritic Shape Parameters

For the characterization of their shapes, dendrites are generally simplified to regular geometric structures, which allow easy quantification. An example of such a simplification is given in Figure 5.1 for a cat superior colliculus neuron, in which the segments in the dendrites are stretched and represented by cylinders with a given length and diameter (Schierwagen, 1986).

The embedding in 3D space and the irregularity and curvature of the branches is thereby lost. The simplified dendrite is subsequently characterized by the number, length, diameter, and connectivity pattern (*topological structure* or *tree type*) of its segments, that is, the branches between successive bifurcation points and/or terminal tips. *Intermediate segments* (ending in a bifurcation point) and *terminal segments* are distinguished (see Figure 5.2). A given segment can be labeled by its *degree* (denoting the number of terminal segments in its subtree) and/or by its *centrifugal order* (denoting the number of bifurcation points on the path from the root up to that segment). Bifurcation points

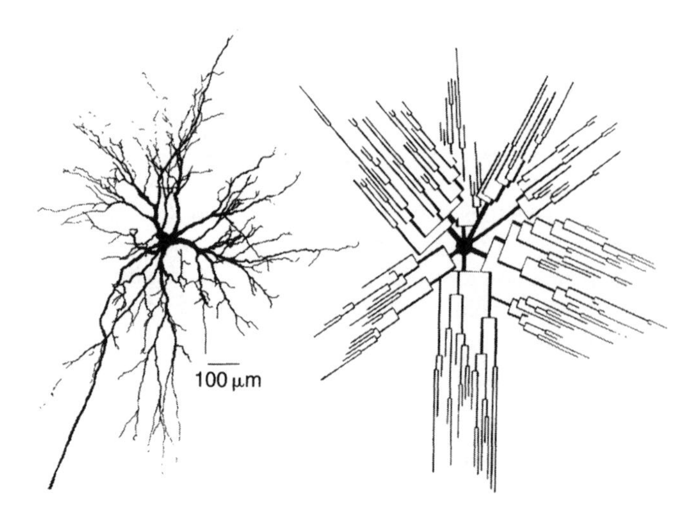

FIGURE 5.1 (Left) Plane reconstruction of the soma-dendritic profile of a tecto-reticulo-spinal neuron of cat superior colliculus. (Right) Approximation of neuronal arborizations by regular geometric bodies. (*Source:* From Schierwagen, A.K., *J. Hirnforsch.* 27, 679–690, 1986. With permission.)

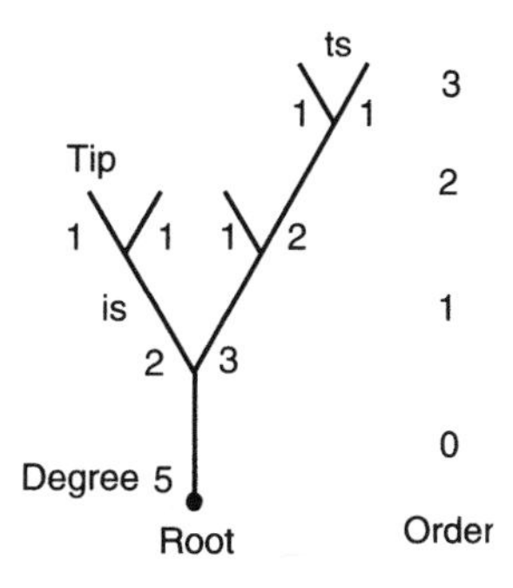

FIGURE 5.2 Elements of a topological tree, with a distinction of intermediate (is) and terminal (ts) segments. Segments are labeled according to the number of tips in their subtrees (*degree*) or distance from the root (*centrifugal order*).

are labeled by their *partitions*, that is, the *degrees* of the two subtrees arising from that bifurcation. Dendritic morphological variation is expressed in terms of variations in these shape parameters.

Dendritic shape parameters have been reviewed earlier by Uylings et al. (1989), Verwer et al. (1992), and Uylings and Van Pelt (2002).

5.2.1 Dendritic Topology

5.2.1.1 Connectivity Pattern

Topological variation arises because of the finite number of *tree types* that are possible for a given number of segments. A summary of all different tree types with up to and including eight terminal segments is given in Figure 5.3 (taken from Van Pelt and Verwer, 1983). The trees are labeled by a rank number according to the ranking scheme described in Van Pelt and Verwer (1983).

For 3D trees there is no distinction between left and right side at bifurcation points. Therefore, the trees can be displayed in a standardized way, such that at each bifurcation point the subtree with the highest rank number is drawn at the right side. The number of 3D-tree types with n terminal segments, N_α^n (α denoting the set of 3D-tree types), is given by the iterative equation

$$N_\alpha^n = \frac{1}{2}\left(\sum_{r=1}^{n-1} N_\alpha^r N_\alpha^{n-r} + (1-\varepsilon(n))N_\alpha^{n/2}\right) \quad \text{and} \quad N_\alpha^1 = 1 \tag{5.1}$$

(Harding, 1971), with

$$\varepsilon(n) = 0 \text{ for even } n \quad \text{and} \quad \varepsilon(n) = 1 \text{ for odd } n. \tag{5.2}$$

The number of tree types increases rapidly with n and a reasonable approximation for the order of magnitude is given by $N_\alpha^n \approx 2.4^n$ (Van Pelt et al., 1992). Note that, for 2D branching patterns, as for rivers, the left–right distinction is meaningful and results in a larger number of 2D-tree types N_τ^n (with τ denoting the set of 2D-tree types) that relates to the number of terminal segments n by

$$N_\tau^n = \frac{1}{2n-1} \cdot \binom{2n-1}{n} \tag{5.3}$$

(Caley, 1859; cf. Shreve, 1966).

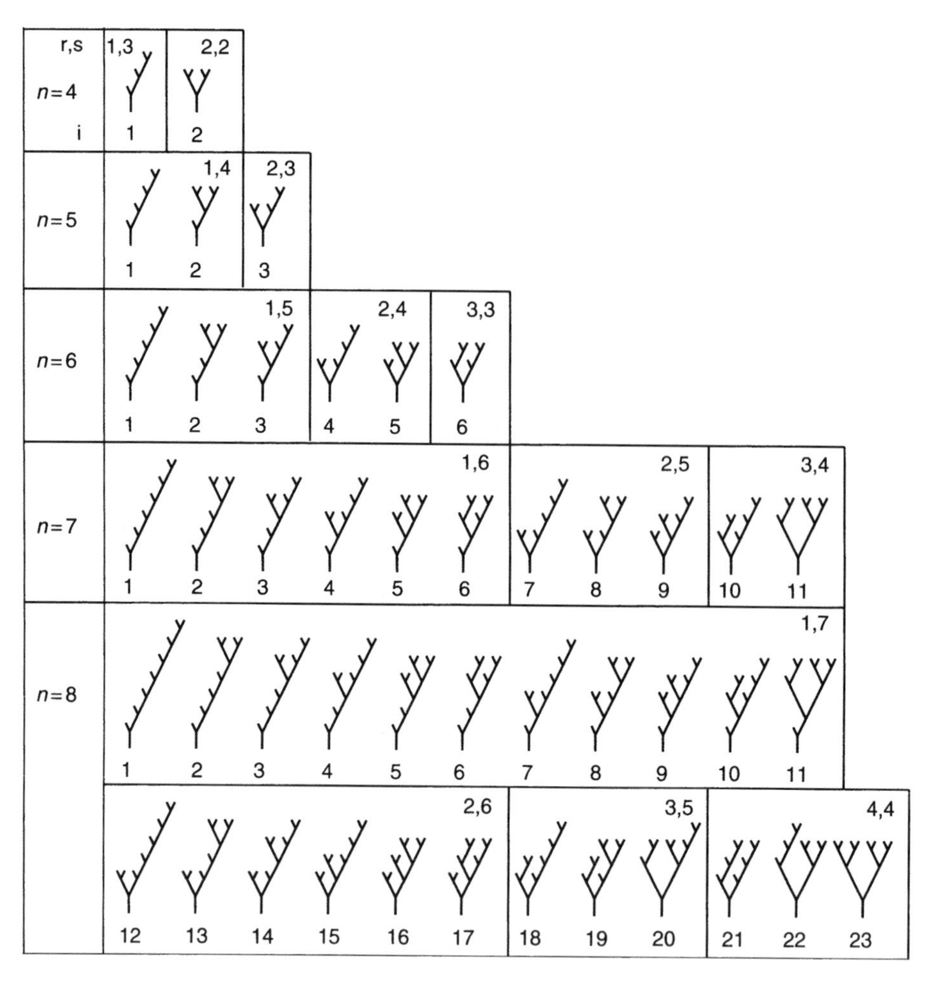

FIGURE 5.3 Table of tree types of degree $n = 4$ to 8. The trees are numbered according to a ranking scheme described in Van Pelt and Verwer (1983). The number pairs (r, s) above groups of tree types indicate the degrees of the first-order subtrees. The tree types are plotted in a standardized way with the subtree with the highest rank number plotted to the right side at each bifurcation.

5.2.1.2 Tree Asymmetry

Tree-type frequency distributions become unmanageably large with larger n. A numerical index for the topological structure of a tree would therefore greatly facilitate the quantitative analysis of topological variation within a set of dendritic trees. Among the topological indices used, the *tree asymmetry index*, A_t, has proven to be the most discriminative one, that is, the one which is best able to distinguish most (but not all) of the possible tree types (Van Pelt et al., 1989). The tree asymmetry index A_t of a given tree α^n with n terminal segments (and thus with $n - 1$ bifurcation points) is defined by

$$A_t(\alpha^n) = \frac{1}{n-1} \sum_{j=1}^{n-1} A_p(r_j, s_j), \tag{5.4}$$

which is the mean value of all the $n - 1$ partition asymmetries, $A_p(r_j, s_j)$, in the tree, which, at each of the $n - 1$ bifurcation points, indicate the relative difference in the number of bifurcation points $r_j - 1$ and $s_j - 1$ in the two subtrees emerging from the jth bifurcation point. The partition asymmetry

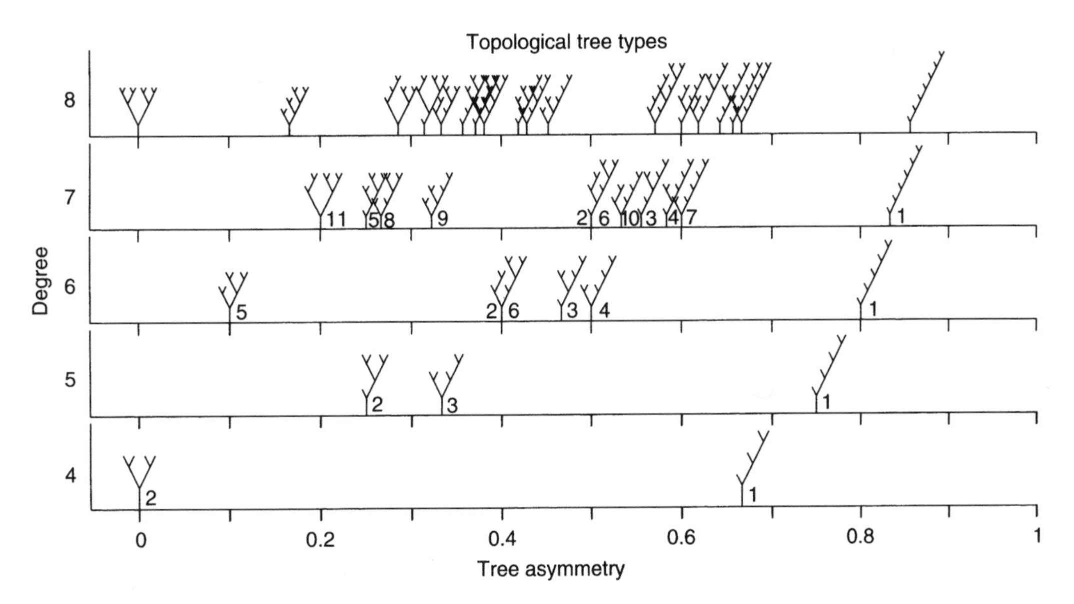

FIGURE 5.4 Tree types of degree 4 to 8, plotted versus their tree asymmetry value. The asymmetry values range from zero for fully symmetric trees (only possible when the degree is a power of two) to a maximum value for fully asymmetric trees, which value approaches one when the asymmetric tree becomes larger. The tree numbers correspond to those in Figure 5.3. Some trees have equal values for the asymmetry index and are plotted at the same position, for instance tree numbers 2 and 6 of degree 6.

A_p at a bifurcation is defined as

$$A_p(r, s) = \frac{|r - s|}{r + s - 2} \tag{5.5}$$

for $r+s > 2$, with r and s denoting the number of terminal segments in the two subtrees. $A_p(1, 1) = 0$ by definition. The discrete values of the tree asymmetry index for trees up to and including degree 8 are shown in Figure 5.4.

Two different coding schemes can be used to represent a topological tree by a linear string, viz. the *label array* and the *branching code*. The *label array* representation is based on a standardized procedure to trace a tree along its bifurcation points and terminal tips, labeled by 1 and 0, respectively. The tree is uniquely represented by the sequence of labels (see also Overdijk et al., 1978). The sequence starts at the root segment and, at bifurcations, first follows the segment at the right side and then that at the left. A label "1" is used for an intermediate segment and a label "0" for a terminal segment. For instance, tree number 11 of degree 7 in Figure 5.3 is represented by the array 1110010011000. For a tree of degree n the number of labels equals $2n - 1$ with $n - 1$ "1"s and n "0"s. Many of the $\binom{2n-1}{n}$ different permutations of n labels do not represent trees. Actually, from each subset of permutations that transform into each other by rotation, thus containing $2n - 1$ permutations, only one represents a tree. This makes the total number of different 2D-tree types of degree n equal to $\binom{2n-1}{n}/(2n - 1)$, which explains Equation (5.3). The *branching code* representation of a tree indicates recursively at each bifurcation the number of terminal segments in both subtrees within brackets. The code starts with the number of terminal segments in the complete tree. For instance, the above mentioned tree is represented by the branching code 7(3(1 2(1 1)) 4(2(1 1) 2(1 1))). The *label arrays* and *branching codes* of the tree types with up to and including eight terminal segments are given in Table 5.1. Note that the ranking scheme used (see also Van Pelt and Verwer, 1983) is based on recurrency in the succession of trees. For instance, the sequence of trees of degree 6 is repeated as first-order subtrees in trees of degree 7.

TABLE 5.1
List of *Label Arrays, Branching Codes,* and *Tree Asymmetry* Values of Tree Types with up to and Including Eight Terminal Segments

Degree	No.	Label array	Branching code	Asymmetry index
1	1	0	1	0
2	1	100	2(1 1)	0
3	1	11000	3(1 2(1 1))	0.500
4	1	1110000	4(1 3)	0.667
	2	1100100	4(2 2)	0
5	1	111100000	5(1 4(1 3))	0.750
	2	111001000	5(1 4(2 2))	0.250
	3	111000100	5(2 3)	0.333
6	1	11111000000	6(1 5(1 4(1 3)))	0.800
	2	11110010000	6(1 5(1 4(2 2)))	0.400
	3	11110001000	6(1 5(2 3))	0.467
	4	11110000100	6(2 4(1 3))	0.500
	5	11100100100	6(2 4(2 2))	0.100
	6	11100011000	6(3 3)	0.400
7	1	1111110000000	7(1 6(1 5(1 4(1 3))))	0.833
	2	1111100100000	7(1 6(1 5(1 4(2 2))))	0.500
	3	1111100010000	7(1 6(1 5(2 3)))	0.556
	4	1111100001000	7(1 6(2 4(1 3)))	0.583
	5	1111001001000	7(1 6(2 4(2 2)))	0.250
	6	1111000110000	7(1 6(3 3))	0.500
	7	1111100000100	7(2 5(1 4(1 3)))	0.600
	8	1111001000100	7(2 5(1 4(2 2)))	0.267
	9	1111000100100	7(2 5(2 3))	0.322
	10	1111000011000	7(3 4(1 3))	0.533
	11	1110010011000	7(3 4(2 2))	0.200
8	1	111111100000000	8(1 7(1 6(1 5(1 4(1 3)))))	0.857
	2	111111001000000	8(1 7(1 6(1 5(1 4(2 2)))))	0.571
	3	111111000100000	8(1 7(1 6(1 5(2 3))))	0.619
	4	111111000010000	8(1 7(1 6(2 4(1 3))))	0.643
	5	111110010010000	8(1 7(1 6(2 4(2 2))))	0.357
	6	111110001100000	8(1 7(1 6(3 3)))	0.571
	7	111111000001000	8(1 7(2 5(1 4(1 3))))	0.657
	8	111110010001000	8(1 7(2 5(1 4(2 2))))	0.371
	9	111110001001000	8(1 7(2 5(2 3)))	0.419
	10	111110000110000	8(1 7(3 4(1 3)))	0.600
	11	111100100110000	8(1 7(3 4(2 2)))	0.314
	12	111111000000100	8(2 6(1 5(1 4(1 3))))	0.667
	13	111110010000100	8(2 6(1 5(1 4(2 2))))	0.381
	14	111110001000100	8(2 6(1 5(2 3)))	0.429
	15	111110000100100	8(2 6(2 4(1 3)))	0.452
	16	111100100100100	8(2 6(2 4(2 2)))	0.167
	17	111100011000100	8(2 6(3 3))	0.381
	18	111110000011000	8(3 5(1 4(1 3)))	0.619
	19	111100100011000	8(3 5(1 4(2 2)))	0.333
	20	111100010011000	8(3 5(2 3))	0.381
	21	111100001110000	8(4(1 3) 4(1 3))	0.571
	22	111001001110000	8(4(1 3) 4(2 2))	0.286
	23	111001001100100	8(4(2 2) 4(2 2))	0

Note: Trees of equal degree are given a rank number, as displayed in the second column (see also Figure 5.3). The asymmetry index was introduced by Van Pelt et al. (1992) and the values given are calculated according to option 1 for the asymmetry index definition in that paper (the values given by Uylings et al. [1989] have been calculated using option 2 and thus differ from the values listed below). Note that the branching codes of (sub)trees of degrees 2 and 3 are expanded in the first two trees only.

5.2.2 Dendritic Metrics

The quantification of the metrical properties of dendritic trees (e.g., Uylings and Van Pelt, 2002) depends on the method of reconstruction. With present manual reconstruction techniques, the branches in the dendrite are approximated by one or more straight lines or cylinders. The metrical properties of such simplified dendritic trees are determined by the lengths and diameters of their segments. Path lengths are defined by the summed lengths of the segments between two characteristic points, for example, those between the root point and a terminal tip. The radial distance between two points in the original structure is determined by the length of the straight connecting line. The difference between path length and radial distance is an indication of the irregularity and curvature of the branches in the original dendritic structure. New reconstruction techniques, which are now under development, are based on digitized images of dendrites, for instance, obtained via confocal microscopy. The digitized representations allow powerful image analysis and differential geometry techniques to be used to quantify (automatically) the dendritic shape with a much more powerful repertoire of 3D shape descriptors, including multiscale fractal approaches (e.g., Costa et al., 2002; Streekstra and Van Pelt, 2002). Recent reviews of reconstruction approaches can be found in Jaeger (2000), Van Pelt et al. (2001c), and Ascoli (2002a).

5.2.2.1 Segment Length and Diameter

In the conventional manual reconstruction approach, curved dendritic branches are approximated by a series of straight segments with given lengths and diameters. Dendritic segments, however, may be covered with spines, making the diameter an ill-defined quantity. Even in the case of smooth dendrites, the diameter may not be constant along the segment. One speaks of tapering if the diameter gradually decreases. Practically, in the case of irregularities the diameter is taken as an average value over several measuring points.

5.3 Observed Variability in Dendritic Shape Parameters

5.3.1 Variation in Topological Structure

The topological variation in a set of dendrites is expressed by the frequencies of the different tree types. Such an analysis is given in Figure 5.5 for basal dendrites of rat cortical pyramidal cells (Uylings et al., 1990). Each panel displays the tree-type frequencies for a subgroup with a given number of terminal segments. The frequency distributions appear to be highly nonuniform, with some tree types abundantly present while others hardly occur. Especially for larger trees, many possible tree types will not occur in a data set resulting in tree-type frequency distributions with many empty classes.

Rather than dealing with distributions of tree-type numbers, topological variation can also be expressed by the mean and standard deviation (SD) of the tree asymmetry index. Examples of observed values for dendrites of different cell types are given in Table 5.2. These data show that both rat and guinea pig Purkinje cells have highest asymmetry values of about 0.5, with a small standard deviation because of the large number of bifurcations (partitions) in the trees.

5.3.2 Variation in the Number of Dendritic Segments

The numbers of segments in dendritic trees show large variations. In a study on terminal-segment number distributions (Van Pelt et al., 1997) it was shown that the number of terminal segments per dendrite ranges from 1 to about 50 in motoneurons in cat, rat, or frog and from 1 to about 15 in both rat pyramidal, multipolar nonpyramidal, and in human dentate granule cells. The shape of the terminal-segment number distributions is highly characteristic for the different cell types

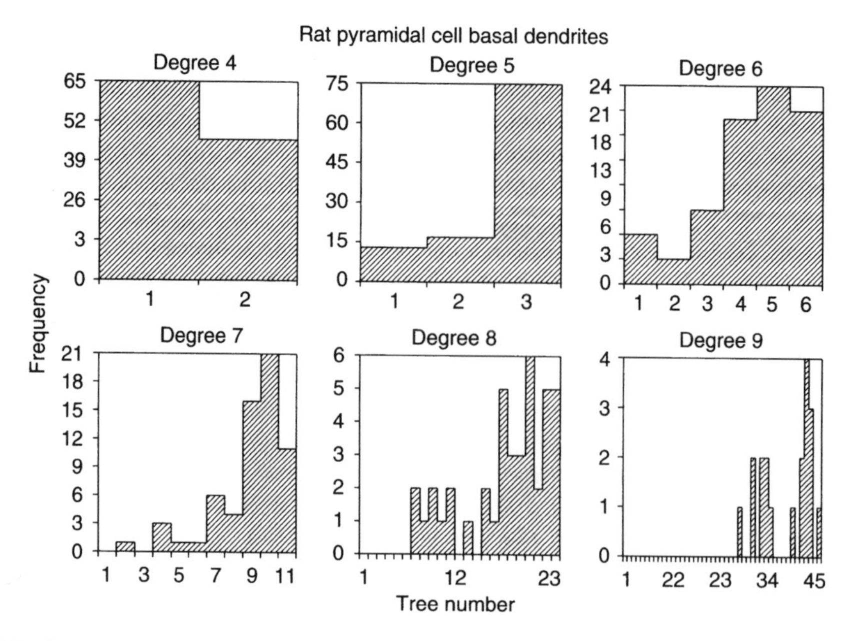

FIGURE 5.5 Frequency distributions of tree types observed in a set of basal dendrites of rat cortical pyramidal neurons. The tree numbers correspond to those in Figure 5.3. Note that the distributions are highly nonuniform.

TABLE 5.2
Mean and Standard Deviation of the Tree Asymmetry Index of Dendritic Trees from Different Cell Types

Cell type	Tree asymmetry index mean (SD)	Reference
Rat cortical pyramidal basal	0.38 (0.22)	Van Pelt et al., 1992
Rat multipolar nonpyramidal	0.43 (0.26)	Van Pelt et al., 1992
Rat cerebellar Purkinje	0.492 (0.020)	Van Pelt et al., 1992
Spinal motoneurons (rat, cat, frog)	0.42–0.47	Dityatev et al., 1995
Cat superior colliculus (deep layer)	0.41 (0.15)	Van Pelt et al., 2001a
S1-rat cortical layer 2,3 pyramidal	0.41 (0.24)	Van Pelt et al., 2001b
Guinea pig cerebellar Purkinje	0.50 (0.01)	Van Pelt et al., 2001b

(see the figures in Van Pelt et al., 1997). Examples of such distributions for dendrites of 150-day old rat pyramidal and multipolar nonpyramidal cells, and of aged human fascia dentata granule cells, are given in Figure 5.6.

5.3.3 Variation in Segment Length

A large variation has been observed in the lengths of dendritic segments. A general observation is that terminal segments can be substantially longer than intermediate ones, as is shown in pyramidal cell basal dendrites of layer V with $\bar{l}_t = 117$ (SD $= 33$) μm for terminal segments and $\bar{l}_i(median) = 11$ μm for intermediate segments (Larkman, 1991a), in S1-rat cortical layer 2,3 pyramidal cell basal dendrites with $\bar{l}_t = 110.7$ (SD $= 45.2$) μm for terminal segments and $\bar{l}_i = 22.0$ (SD $= 17.9$) μm for intermediate segments (Van Pelt et al., 2001b) (see also Figure 5.7), in superior colliculus neurons with $\bar{l}_t = 115$ (SD $= 83.3$) μm and $\bar{l}_i = 80.3$ (SD $= 75.5$) μm (Schierwagen and Grantyn, 1986),

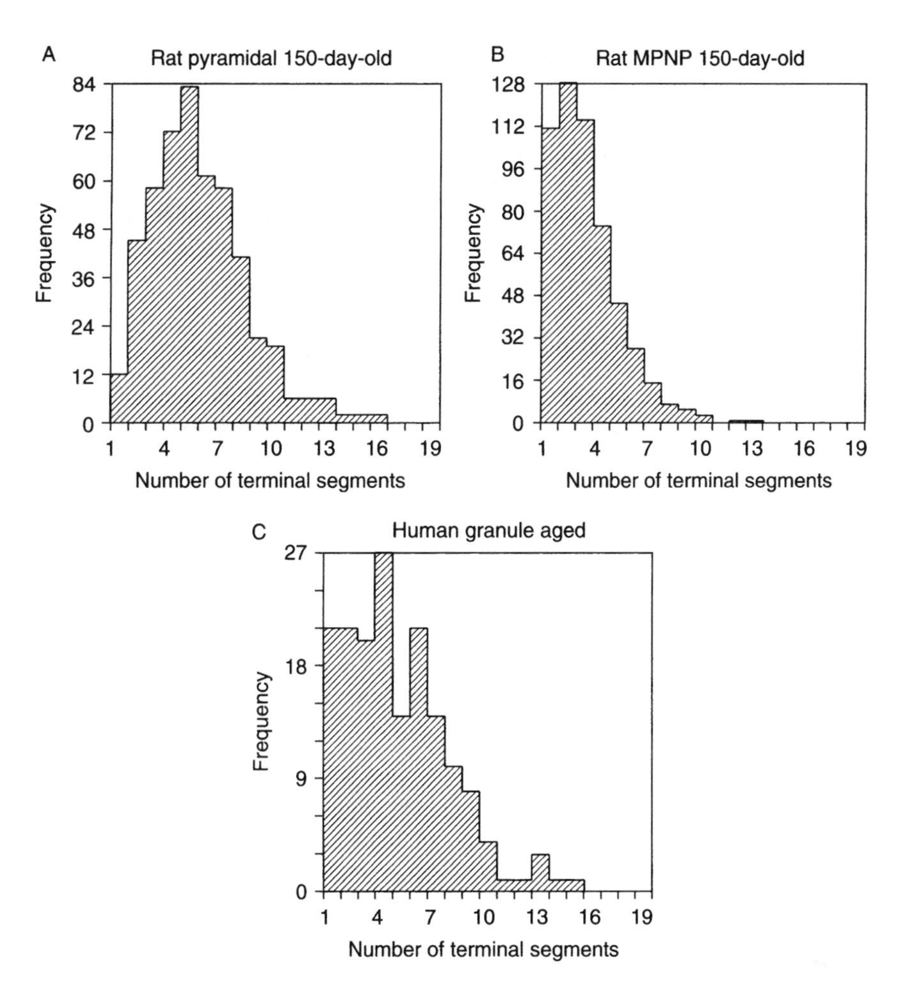

FIGURE 5.6 Frequency distributions of the numbers of terminal segments per dendritic tree for (A) 150-day old rat cortical pyramidal and (B) multipolar nonpyramidal (MPNP) cells (see Uylings et al., 1990) and (C) human fascia dentata granule cells (De Ruiter and Uylings, 1987).

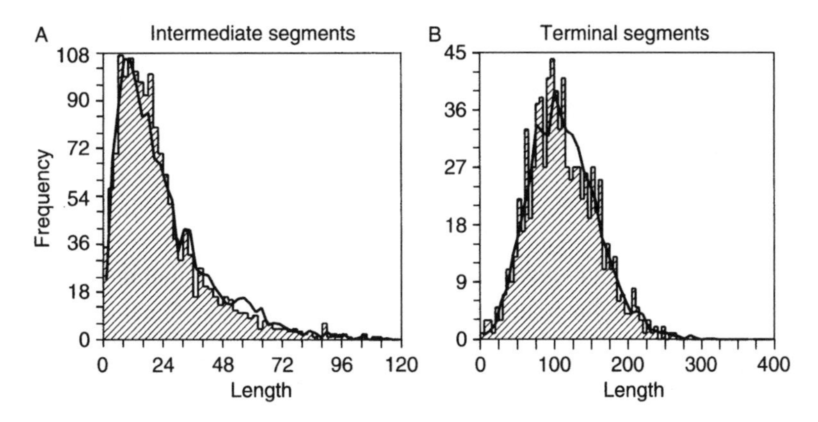

FIGURE 5.7 Frequency distributions of the length of (A) intermediate and (B) terminal segments of S1-rat cortical layer 2,3 pyramidal cell basal dendrites (Van Pelt et al., 2001b).

and in rat Purkinje neurons with terminal segments having a length of about 13 μm or 8 μm and intermediate segments having a length of about 5 μm (Woldenberg et al., 1993). An extensive review of intermediate- and terminal-segment lengths in dendrites of different cell types and species is given by Uylings et al. (1986, table II). Additionally, terminal-segment length may decrease strongly with increasing centrifugal order as is shown, for instance, by Uylings et al. (1978b). Van Veen and Van Pelt (1993) showed that such decreases in length as well as differences between intermediate- and terminal-segment lengths arise when the number and the positions of the branch points along the path from the soma to a terminal tip are determined by a Poisson process.

5.3.4 Variation in Dendritic Diameter

Dendritic segments have small diameters, down to the resolution of the optical microscope. The observed variation in segment diameters may therefore be partly attributed to measurement uncertainties. Nevertheless, a substantial part of the variation is induced by the different positions of the segments in the tree, which are caused by the correlation that is found between the diameter of a segment and its *degree* (Hillman, 1988). The neuritic cytoskeleton is believed to be a major factor determining the diameter of dendritic segments. Hillman (1979, 1988) assumed a positive correlation between segment diameter and the number of microtubules. This correlation also results in a relation found between the diameters of a parent (d_{p}) and its daughter segments (d_1 and d_2) at a bifurcation, formulated by Rall (1959) as a power-law relation

$$d_{\mathrm{p}}^{e} = d_1^{e} + d_2^{e} \tag{5.6}$$

with e being the branch power parameter. Segment diameters thus decrease across bifurcation points with terminal segments having the smallest diameters. Terminal segments were found to have mean diameters of 1.1 μm for Purkinje cells (Hillman, 1979), 0.7 μm for pyramidal neurons (Hillman, 1979; Larkman, 1991a), 0.7 (SD = 0.3) μm in cat phrenic motoneurons (Cameron et al., 1985), and about 1 μm for cat superior colliculus neurons (Schierwagen and Grantyn, 1986). Values for the branch power were reported of $e = 2$ for rat Purkinje and neocortical pyramidal neurons (Hillman, 1979), of $e = 1.47$ (SD = 0.3) (Schierwagen and Grantyn, 1986) for dendrites of cat superior colliculus neurons, of $1.5 < e < 2$ for rat visual cortex pyramidal neurons (Larkman et al., 1992), and of $\frac{3}{2}$ for the distal parts of rat dentate granule cells (Desmond and Levy, 1984). Cullheim et al. (1987a) found a value of 1.13 (SD = 0.34) for the mean ratio of $d_{\mathrm{p}}^{1.5}/(d_1^{1.5} + d_2^{1.5})$ in cat α motoneurons.

5.4 Modeling Dendritic Branching Patterns

In the *growth model approach* it is assumed that dendritic growth proceeds by branching and elongation of terminal segments. At a branching event a terminal segment is replaced by an intermediate one ending in a bifurcation point from which two new daughter segments emerge. Randomness is implemented by assigning branching probabilities and elongation rates to each terminal segment. The major challenge in this approach is to find (minimal) schemes for branching and elongation that result in model trees with geometrical properties similar to their observed counterparts.

The model was constructed step by step in the course of time, first, concentrating on the topological variation between dendritic trees with equal number of terminal segments, second, on the variation in the number of terminal segments per dendritic tree, and finally on the variation in segment length. Each step has been accompanied by a thorough validation of the model outcomes with experimental data. The final model thus has a modular structure in which the different parameters can be optimized for their corresponding shape properties in dendritic trees.

The first step concerned the branching process only, aiming at describing the observed frequency distribution of tree types with a given number of terminal segments (topological variation), as shown in Figure 5.5. The second step aimed at describing the observed distribution of the number of terminal segments, as shown in Figure 5.6. The third step concerned the joint branching and elongation process, aiming at describing segment length distributions, as shown in Figure 5.7. These successive steps will be described in detail in the subsequent sections.

5.4.1 Modeling Topological Variation (QS Model)

Topological variation arises when branching events variably occur at different segments in the growing tree, such that each growth sequence, after a given number of branching events, may end in a different tree type. In the so-called QS model, branching events were initially assumed to occur at both intermediate and terminal segments, with the selection probability of a segment for branching taken to depend on the type of the segment and on its centrifugal order. The selection probability p_{term} for branching of a terminal segment at centrifugal order γ is defined by $p_{\text{term}} = C_1 2^{-S\gamma}$, with parameter S modulating the dependence on centrifugal order and C_1 being a normalization constant to make the sum of the branching probabilities of all the segments in the tree equal to one. For $S = 0$, all terminal segments have the same probability of being selected for branching. For $S = 1$, the branching probability of a terminal segment decreases by a factor of two for each increment in centrifugal order. The selection probability of an intermediate segment, p_{int}, for branching relates to that of a terminal segment of the same order via $p_{\text{int}} = (Q/(1-Q))p_{\text{term}}$, with Q a parameter having values between 0 and 1. The parameter Q roughly indicates the total branching probability for all intermediate segments in a tree, with branching of terminal segments only for $Q = 0$, and branching of intermediate segments only for $Q = 1$. With the two parameters Q and S, a range of growth modes can be described, including the well-known random-terminal mode of growth (*rtg*) with $(Q, S) = (0, 0)$ and the random-segmental mode of growth (*rsg*) with $(Q, S) = (0.5, 0)$. An accurate description of the topological variability in dendrites from several neuron types and species could be obtained by assuming branching to occur at terminal segments only (i.e., $Q = 0$), with possibly a slight dependence of the selection probability on the centrifugal order (small or zero value for S) (Van Pelt and Verwer, 1986; Van Pelt et al., 1992; Dityatev et al., 1995). The QS model reduces to the S model for $Q = 0$ and to the Q model for $S = 0$. The probabilities of occurrence of tree types appear to depend strongly on the mode of growth. This is clearly shown in Figure 5.8, which displays the probabilities of trees of degree 8 for four different modes of growth, including *rtg* and *rsg*.

Analytical expressions for the probabilities and expected values of some shape parameters could be obtained for the Q model only. Dendritic growth according to the Q model results in partition probabilities $p(r, n-r|Q)$ (i.e., probabilities for a tree of degree n to have first-order subtrees of degrees r and $n-r$) given by

$$p(r, n-r|Q) = 2^{1-\delta(r,n-r)} \left\{1 + Q\left(\frac{n(n-1)}{2r(n-r)} - 2\right)\right\} \frac{1}{n-1-Q} \prod_{i=1}^{r-1} \frac{1-Q/i}{1-Q/(i+n-r-1)}, \tag{5.7}$$

$$p(1, n-1|Q) = \frac{2+Q(n-4)}{n-1-Q}, \tag{5.8}$$

$$\lim_{n\to\infty} p(1, n-1|Q) = Q, \tag{5.9}$$

$$p(r, n-r|\text{rtg}) = \frac{2^{1-\delta(r,n-r)}}{n-1}, \tag{5.10}$$

$$p(r, n-r|\text{rsg}) = \frac{N_\tau^r N_\tau^{n-r}}{N_\tau^n} 2^{1-\delta(r,n-r)} \tag{5.11}$$

(Van Pelt and Verwer, 1985), with δ denoting the Kronecker δ.

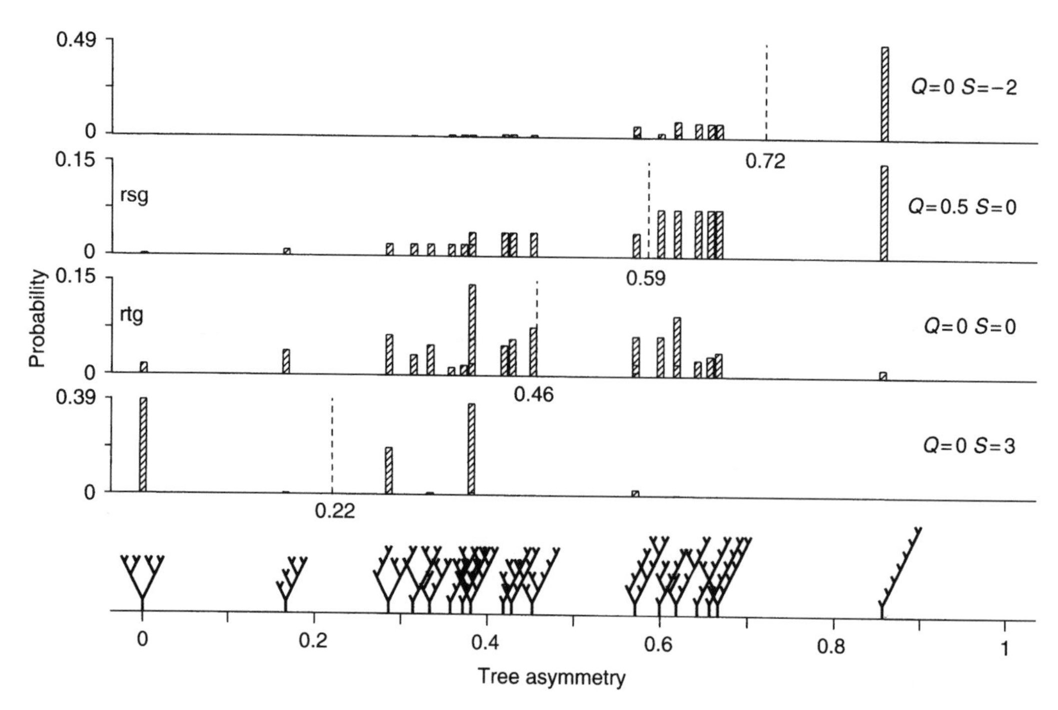

FIGURE 5.8 Plot of probabilities of occurrence of tree types of degree 8. The trees are plotted versus the tree asymmetry values in the bottom row. The probabilities have been calculated for four different modes of dendritic growth, namely, for $(Q, S) = (0, -2)$ in the top row, the *rsg* $(Q, S) = (0.5, 0)$ in the second row, the *rtg* $(Q, S) = (0, 0)$ in the third row, and for $(Q, S) = (0, 3)$ in the fourth row. The mean value of the tree asymmetry index for a given mode of growth is indicated by a dotted line. The figures demonstrate that the tree-type probabilities and the tree asymmetry expectations depend strongly on the mode of growth. Note that probability bars of trees with equal asymmetry index do overlap.

The tree asymmetry expectation for the *rtg* mode of growth is given by

$$E\{A_t^n|\text{rtg}\} = \frac{2n}{3(n-1)}\left\{\frac{2-3n_e/n}{4(n_e-1)} - \frac{2}{n_e} + \sum_{k=n_e/2}^{n_e}\frac{1}{k}\right\}, \tag{5.12}$$

with $n_e = n - \varepsilon(n)$ and $\varepsilon(n)$ defined in Equation (5.2). In the limit of large n

$$\lim_{n\to\infty} E\{A_t^n|\text{rtg}\} = -\frac{1}{3} + \sum_{k=1}^{\infty}\frac{1}{(k+1)(2k-1)} = \frac{2}{3}\ln 2 = 0.4621 \tag{5.13}$$

(Van Pelt et al., 1992).

For the *rsg* mode of growth the tree asymmetry expectation is given by

$$E\{A_t^n|\text{rsg}\} = \frac{2}{(n-1)N_\tau^n}\sum_{m=2}^{n}\frac{2(n-m)-1}{m-2}N_\tau^{n-m}\sum_{r=1}^{m/2}\{2-\delta(r,m-r)\}(m-2r)N_\tau^r N_\tau^{m-r} \tag{5.14}$$

(Van Pelt et al., 1992). These expectations (mean values) are indicated in Figure 5.8 by the dotted lines for the displayed modes of growth.

The QS model has been used to analyze dendrites from a variety of neuronal cell types and species (Van Pelt et al., 1992; Dityatev et al., 1995). In all cases, optimized parameter values could be found that accurately reproduced the observed topological variation. A given value for the mean asymmetry index, however, is not uniquely associated with a particular combination of QS values, but with a contour in QS space. For values of the asymmetry index smaller than about 0.5, the contours remain close to the S axis. All contours start at the S axis, implying that the S axis ($Q = 0$) is able to generate a full range of asymmetry expectations (Van Pelt et al., 1992). At present, all reported asymmetry values for neuronal dendrites are smaller than or equal to 0.5 (see also Table 5.2). Therefore, we assume in the following that $Q = 0$, that is, that branching occurs at terminal segments only, while the probability for a terminal segment to be selected may depend slightly on its position in the tree (i.e., its centrifugal order).

5.4.2 Modeling the Variation in the Number of Terminal Segments per Dendrite (BE, BES, and BEST Models)

Variation in the number of segments per dendritic tree emerges when trees experience a variable number of branching events during a particular period of outgrowth. In the so-called BE growth model (Van Pelt et al., 1997), it is assumed that branching events occur at random points in time and at terminal segments only. To this end, the developmental time period is divided into a number, N, of time bins with not necessarily equal durations. In each time bin, i, a terminal segment in the tree may branch with a probability given by

$$p_i = B/Nn_i^E, \tag{5.15}$$

with the parameter B denoting the expected number of branching events at an isolated segment in the full period, and the parameter E denoting the dependence of the branching probability of a terminal segment on the total number of terminal segments, n_i, in the growing tree. Equation (5.15) can also be written as $p_i = Dn_i^{-E}$ with $D = B/N$ denoting the branching probability per time bin of an isolated terminal segment. The duration of the time bins is taken to be sufficiently small (i.e., the number of time bins sufficiently large) to make the branching probabilities p_i much smaller than one, thus making the probability for more than one branching event per time bin in the tree negligibly small. For $E = 0$, the branching probability is constant, independent of the number of terminal segments in the tree.

Examples of growth sequences are given in Figure 5.9 for parameter values $B = 3$ and $E = 0$. The two sequences show the random occurrences of branching events in time as well as the difference in the number of terminal segments in the final trees. Additionally, they show that the number of branching events in the full period can be (much) larger than the value of $B = 3$ because of the proliferation in the number of terminal segments during growth. The variation in terminal-segment number is clearly seen in Figure 5.10, in which the distribution functions are displayed of the number of terminal segments per dendritic tree in sets of trees obtained for $B = 4$ and for different values of the growth parameter E. For $E = 0$, the distribution is monotonously decreasing with increasing number of terminal segments and has a long tail. For $E > 0$, the distributions become unimodal and increasingly narrower while long tails disappear for increasing values of E.

The distribution of the number of terminal segments in dendritic trees after a period of growth can be calculated by means of the recursive expression

$$P(n,i) = \sum_{j=0}^{n/2} P(n-j, i-1) \binom{n-j}{j} [p(n-j)]^j [1 - p(n-j)]^{n-2j} \tag{5.16}$$

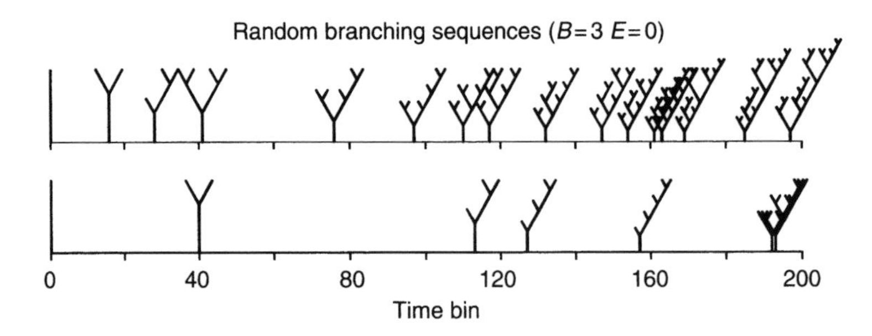

FIGURE 5.9 Two examples illustrating the growth of a branching pattern for the parameter values $B = 3$ and $E = 0$. The full time period is divided into $N = 200$ time bins. At each time bin, any of the terminal segments in the growing tree may branch with a constant probability. The trees are plotted at the time bin where a branching event has occurred.

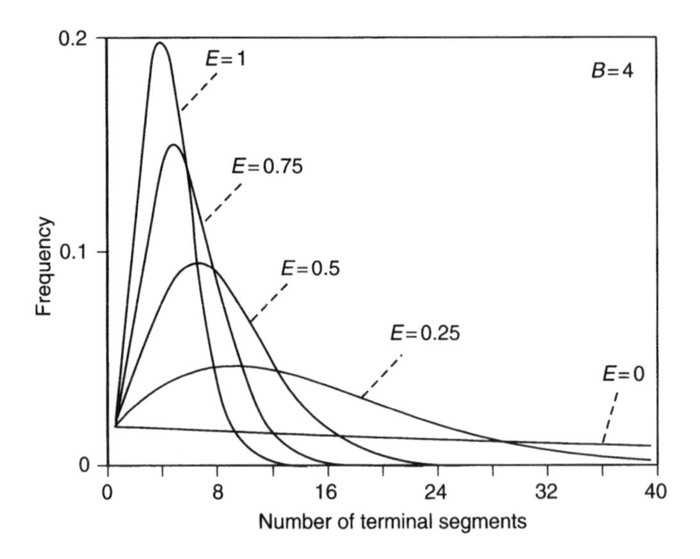

FIGURE 5.10 Degree distributions of trees obtained by growth modes with $B = 4$ and different values for the parameter E, that is, with different dependencies of the branching probability of a terminal segment on the total number of terminal segments in the tree. For $E = 0$, the distribution is monotonically decreasing and has a long tail. For increasing values of E, the distributions get a mode and become increasingly narrower.

(Van Pelt et al., 1997), with $P(n, i)$ denoting the probability of a tree of degree n at time bin i with $P(1, 1) = 1$, and $p(n)$ denoting the branching probability per time bin of a terminal segment in a tree of degree n, with $p(n) = Bn^{-E}/N$. A tree of degree n at time bin i emerges when j branching events occur at time bin $i - 1$ in a tree of degree $n - j$. The recursive equation expresses the probabilities of all these possible contributions from $j = 0, \ldots, n/2$. The last two terms express the probability that, in a tree of degree $n-j$, j terminal segments will branch while the remaining $n-2j$ terminal segments will not do so. The combinatorial coefficient $\binom{n-j}{j}$ expresses the number of possible ways of selecting j terminal segments from the existing $n - j$ ones. A continuous time implementation of this recurrent equation has recently been described by Van Pelt and Uylings (2002). The BE model has been shown to be able to reproduce accurately the shape of terminal-segment number distributions from dendrites of different cell types and species (Van Pelt et al., 1997). Optimal parameter values were found of $(\bar{B}(\text{sem}), \bar{E}(\text{sem})) = (3.55(0.17), 0.68(0.02))$ for rat pyramidal cell basal dendrites, of (3.92 (0.4), 0.29 (0.04)) for the combined group of motoneurons from cat, rat, and frog, of $(\bar{B}, \bar{E}) = (2.36, 0.38)$

for human dentate granule cells, of $(\bar{B}, \bar{E}) = (21.49, 0.42)$ for rat multipolar nonpyramidal cells, and of (50, 0.54) for one-month-old rat cerebellar Purkinje cells.

5.4.2.1 Including Topological Variation in the BE Model (BES Model)

In the BE model, all terminal segments have equal probability for branching and the topological variation produced by the BE model is similar to that produced by the *rtg* mode of growth ($Q = 0$, $S = 0$). An account of the topological variability can now be given in a combined BES model by taking the branching probability of a terminal segment per time bin, p_i, to be also dependent on the centrifugal order of the segment, as in the S model, such that

$$p_i = C2^{-S\gamma}B/Nn_i^E, \tag{5.17}$$

with γ denoting the centrifugal order of the terminal segment and $C = n/\sum_{j=1}^{n} 2^{-S\gamma_j}$ being a normalization constant, with the summation over all n terminal segments. The normalization ensures that the summed branching probability per time bin of all the terminal segments in the tree is independent of the value of S.

5.4.2.2 Modeling Branching Processes in Real Time (BEST Model)

The time bins have been introduced without specifying their durations. To calculate the branching process in real time, the time-bin scale has to be mapped onto the real timescale. The branching probability per time bin, $p_i = Dn_i^{-E}$, then transforms into a branching probability per unit of time $p(t) = Dn_i^{-E}/\Delta T_i$, with ΔT_i being the duration of time bin i. A linear mapping involves time bins of equal duration. A typical example of a growth curve of the terminal-segment number per dendrite is given in Figure 5.11. It shows how the mean and standard deviation increase with time. Also, the growth rate increases with the highest value at the end of the period. The branching probability per unit of time, however, declines with time because of the increasing number of terminal segments in the growing tree.

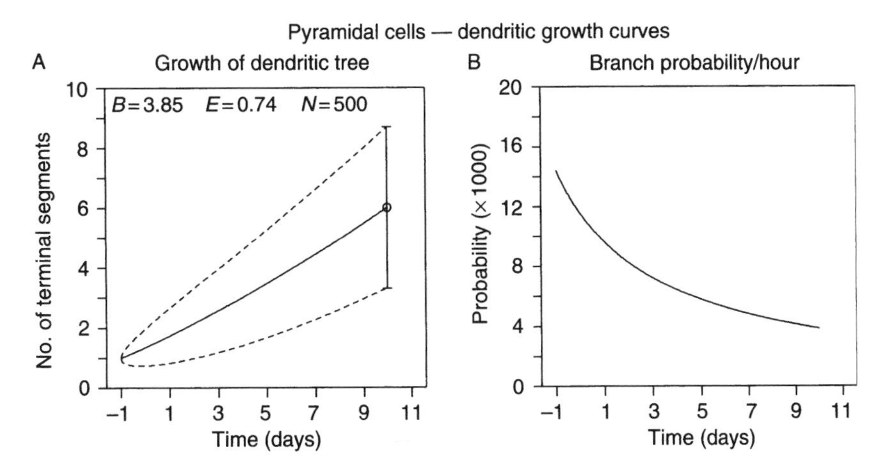

FIGURE 5.11 Growth curve of the mean terminal-segment number in dendritic trees for the growth mode $B = 3.85$, $E = 0.74$. The dashed lines indicate the standard deviation intervals around the mean. The right panel indicates how the branching probability per hour declines with age because of the increasing number of terminal segments and the nonzero value of E.

A nonlinear mapping of the time bins to real time modifies the shape of the growth curve, but, importantly, does not change the relation between mean and standard deviation. Therefore, it offers the possibility of adapting the growth curve to a set of observed values at different points in time for the mean and standard deviation (see Van Pelt et al., 1997 for an example of Purkinje cell development). A nonlinear mapping of the time-bin scale has as a consequence that the constant D transforms into a nonlinear function of time, $D(t)$, also called the baseline branching-rate function. Recently, the dependence of the dendritic-growth function on the baseline branching-rate function, $D(t)$, has been described in detail by Van Pelt and Uylings (2002). Using a developmental data set of Wistar rat layer IV multipolar nonpyramidal neurons, it was shown that an exponentially decreasing baseline branching-rate function with a decay time constant of $\tau = 3.7$ days resulted in a terminal-segment-number growth function that matched the observed data very well.

5.4.3 Modeling the Variation in the Length of Dendritic Segments (BESTL Model and Simulation Procedure)

5.4.3.1 BESTL Model

Segment lengths are determined by the rates of both elongation and branching of terminal segments. In the previous section it was shown that the parameters of the branching process can be estimated from the dendritic-segment-number distribution. Segment elongation thus needs to be included in the branching process for studying segment lengths, as illustrated for the growth of dendrites of large layer V rat cortical pyramidal neurons. The relevant empirical findings for these neurons are summarized in Table 5.3. Note that growth starts one day before birth and continues with branching up to day 10, while the elongation of segments continues up to day 18 (Uylings et al., 1994).

The topological variation was well described by the model with parameters $Q = 0$, $S = 0.87$ (Van Pelt et al., 1992). The terminal-segment-number distribution was well reproduced by the branching-process parameters $B = 3.85$ and $E = 0.74$ (Van Pelt et al., 1997) (see also Table 5.4).

TABLE 5.3
Developmental and Shape Characteristics of Basal Dendrites of Large Layer V Rat Cortical Pyramidal Neurons

Onset of branching/elongation	≈ Day 1 (24 h) *an*	Uylings (unpublished observations)
Stop of branching	Day 10 (240 h) *pn*	Uylings et al., 1994
Stop of elongation	Day 18 (432 h) *pn*	Uylings et al., 1994
Number of terminal segments per dendrite	6.0 (SD = 2.7)	Larkman, 1991a
Intermediate-segment length	11 (median) μm	Larkman, 1991a
Terminal-segment length	117 (SD = 33) μm	Larkman, 1991a
Path length to tips	156 (SD = 29) μm	Larkman, 1991a
Total dendritic length	777.6 μm	[a]
Terminal-segment diameter	0.8 (SD = 0.2) μm	Larkman, 1991a
Branch power	$1.5 < e < 2$	Larkman et al., 1992
Tree asymmetry	0.38 (SD = 0.22)	Van Pelt et al., 1992

Note: Abbreviations: *an* before birth, *pn* after birth.

[a] The mean total length per dendritic tree of 777.6 μm is estimated from the total basal dendritic length per pyramidal neuron of 4.510 mm (Larkman and Mason, 1990) and the mean number of basal dendrites per pyramidal neuron of 5.8 (SD = 1.8) (Larkman, 1991a).

TABLE 5.4
Parameter Values Used for Modeling the Growth of Basal Dendrites of Large Layer V Rat Cortical Pyramidal Neurons

Parameter	Use	Optimization on
$B = 3.85$	Free	Degree mean, SD
$E = 0.74$	Free	Degree mean, SD
$V_{el} = 0.51$ μm/h	Free	Mean terminal-segment length
CV in prop. rate = 0.28	Free	SD in path length
$V_{br} = 0.22$ μm/h	Calculated	a
$T_{onset} = -24$ h	Observed	
$T_{bstop} = 240$ h	Observed	
$T_{lstop} = 432$ h	Observed	

Note: The second column indicates which parameters were used for optimizing the shape parameters in the third column to the observed values (indicated as free parameters), and which ones were directly derived from observed data. T_{onset}, T_{bstop}, and T_{lstop} denote the time of onset of branching and elongation, the stop time of branching, and the stop time of elongation, respectively.

[a] The mean elongation rate during the branching phase, V_{br}, is calculated from the optimized value for the elongation rate during the elongation phase, V_{el}, the durations of the branching phase, $T_{br} = T_{bstop} - T_{onset} = 264$ h, and elongation phase, $T_{el} = T_{bstop} - T_{bstop} = 192$ h, and the mean path length $L_p = 156$ μm, by means of $V_{br} = (L_p - V_{el} \times T_{el})/T_{br}$.

Equal time bins, or, similarly, a constant function D, will be assumed. For the elongation rate of terminal segments we initially assume a fixed value of 0.34 μm h^{-1}, estimated from the mean path length of 156 μm traversed in the total period of growth of 456 h. With these model parameters, and including different stop times for the branching and elongation processes, dendrites were produced with mean values for their intermediate- and terminal-segment lengths of 23.6 (SD = 18.9; median = 18.8) and 96.0 (SD = 22.1) μm, respectively. These values are longer, respectively, shorter than the observed values of 11 (median) and 117 (SD = 33) μm, indicating that the assumption of constant segment elongation gives incorrect results. Branching terminates at 240 h postnatal. Terminal segments become longer when the growth cones propagate faster during the elongation phase. To maintain the same mean elongation rate during the total developmental period this implies a slower rate during the branching phase. Next, elongation rates of 0.22 μm h^{-1} during the branching phase and of 0.51 μm h^{-1} during the elongation phase were taken, still without variation in these rates. These runs resulted in dendrites with mean (±SD) values for the intermediate- and terminal-segment lengths of 15.2 (±12.2) μm (median = 12.1) and 117.4 (±14.2) μm, respectively, and path lengths of 156 μm without variation. The mean values are now in good agreement with the observed values. Note that the standard deviations in the segment lengths are solely due to randomness in the branching process and are smaller than the observed values. In the final simulation run, randomness was incorporated in the elongation rate by assigning, at the time of the birth of a growth cone after a branching event, a random value from a normal distribution with mean 0.22 μm h^{-1} during the branching phase and 0.51 μm h^{-1} during the elongation phase, and with a coefficient of variation (CV) of 0.28. With this CV value it was found that the model-predicted standard deviation in the path-length distribution optimally matches the observed value of 29 μm. A summary of the parameter values is given in Table 5.4.

An example of a growing dendrite is given in Figure 5.12. The dendrite is displayed at successive days of its development. Note that branching terminates 11 days after onset and is followed by a period of elongation only. Typical examples of random full-grown dendritic trees, at the end of their developmental period, are shown in Figure 5.13.

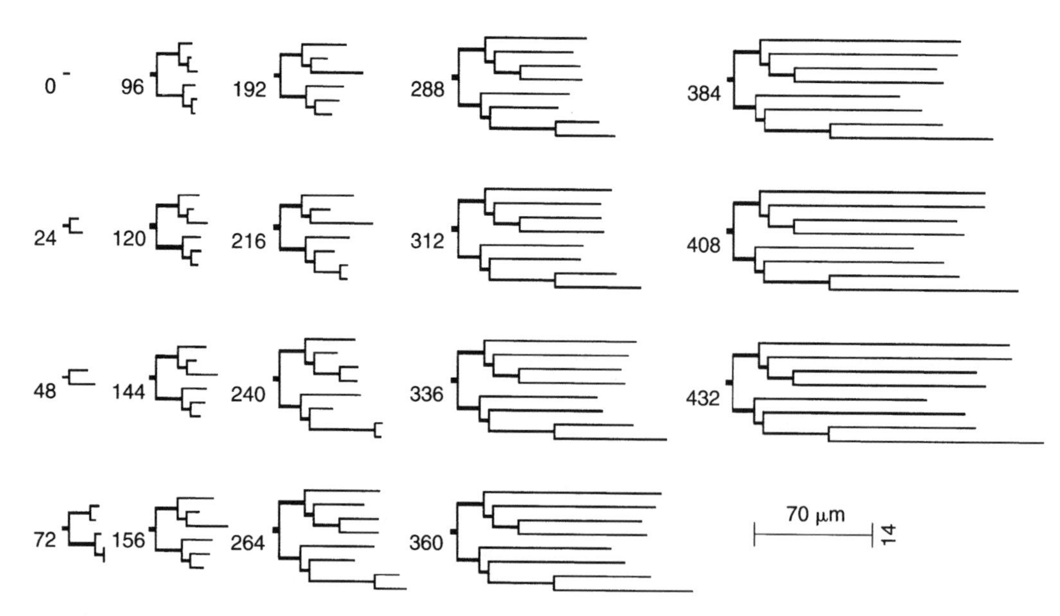

FIGURE 5.12 Plot of a model dendritic tree at successive days of its development (the numbers at the root of each plot denote the postnatal time in hours). The growth parameters used were optimized for basal dendrites of large layer V rat cortical pyramidal neurons and are given in Table 5.4. A period of branching and elongation from 1 day before birth up to 10 days after birth (denoted as branching phase) is followed by a period of elongation only up to day 18 (denoted as elongation phase). Different elongation rates are assumed of 0.22 $\mu m\,h^{-1}$ in the branching phase and 0.51 $\mu m\,h^{-1}$ in the elongation phase.

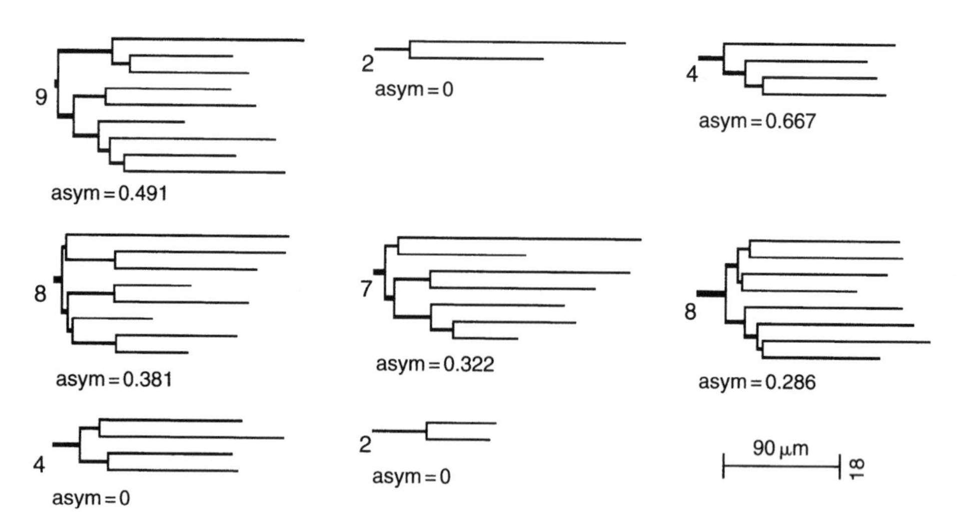

FIGURE 5.13 Examples of full-grown model dendrites obtained for model parameters given in Table 5.4, which were optimized for basal dendrites of large layer V rat cortical pyramidal neurons. For each tree, the number of terminal segments is given as well as the tree asymmetry index (asym).

The statistical shape properties of the model dendrites are summarized in Table 5.5 and are in good agreement with the observed data both in the mean and in the standard deviation. The last column indicates whether the agreement is the result of the optimization of the model parameters or an emergent outcome of the model ("prediction"). The agreement between the predicted and the observed values underscores the validity of the model and its underlying assumptions.

TABLE 5.5
Shape Properties of Model Dendrites Grown by Means of the Growth Model for the Parameter Values Given in Table 5.4

Shape variables		Obs.	Model	Outcomes
Degree	Mean	6.0	6.0	Optimized
	SD	2.7	2.7	Optimized
Asymmetry	Mean	0.38	0.36	Optimized
	SD	0.22	0.20	Prediction
Centrifugal order	Mean		2.26	Prediction
	SD		1.24	Prediction
Total length	Mean	777.6	774.6	Prediction
	SD		342.9	Prediction
Terminal length	Mean	117	117.1	Optimized
	SD	33	31.4	Prediction
Intermediate length	Mean		15.4	Prediction
	SD		13.4	Prediction
	Median	11	11.6	Prediction
Path length	Mean	156	156.2	Estimated
	SD	29	29.2	Optimized

Note: The last column indicates which shape variables were matched to the observed values by optimizing the model parameters (see Table 5.4) and which shape variables are just outcomes of the growth model and can be considered as predictions for further empirical verification.

SD = Standard deviation.

The distributions obtained for several shape parameters are shown in Figure 5.14. The terminal-segment-number distribution closely corresponds to the observed one, as shown in Van Pelt et al. (1997). Unfortunately, experimental distributions for the other parameters were not available. The distribution of the tree asymmetry values shows the irregular pattern characteristics for the discrete tree types and a mean value of 0.36 that is typical for the $S = 0.87$ mode of growth (i.e., random branching of terminal segments with a slight preference for the proximal ones). The simulations without and with variation in elongation rates make clear which part of the variation in segment lengths and total dendritic lengths is due to randomness in branching and which part is due to randomness in elongation. Note that the optimized value of 0.28 for the CV in the elongation rate is larger than the value of 0.186 ($=\frac{29}{156}$) for the observed path-length distribution. The reason is that path lengths in a dendritic tree are not independent from each other because paths to terminal tips share common intermediate segments (for instance, all paths have the root segment in common).

The intermediate-segment-length distribution of the model trees has an exponential shape (Figure 5.14). Experimental distributions of other cell types with higher resolution, however, show clear-cut model shapes (e.g., Uylings et al., 1978a, 1994; Nowakowski et al., 1992; also shown in Figure 5.7). Apparently, short intermediate segments in dendritic reconstructions occur less frequently than expected. Nowakowski et al. (1992), who first noticed this finding, assumed a transient suppression of branching immediately after a branching event. To account for this observation without interfering with the branching scheme, we have alternatively assumed that newly formed segments after a branching event have an initial length when stabilized and are able to form subsequent branches. This additional assumption resulted in accurate reproductions of the observed intermediate-segment-length distributions (Van Pelt et al., 2001a, 2001b, 2003). Additionally, the baseline branching-rate function, $D(t)$, may not be a constant but rather a decreasing function of time, as was recently shown for rat cortical multipolar nonpyramidal dendrites (Van Pelt and Uylings, 2002). Such a function will

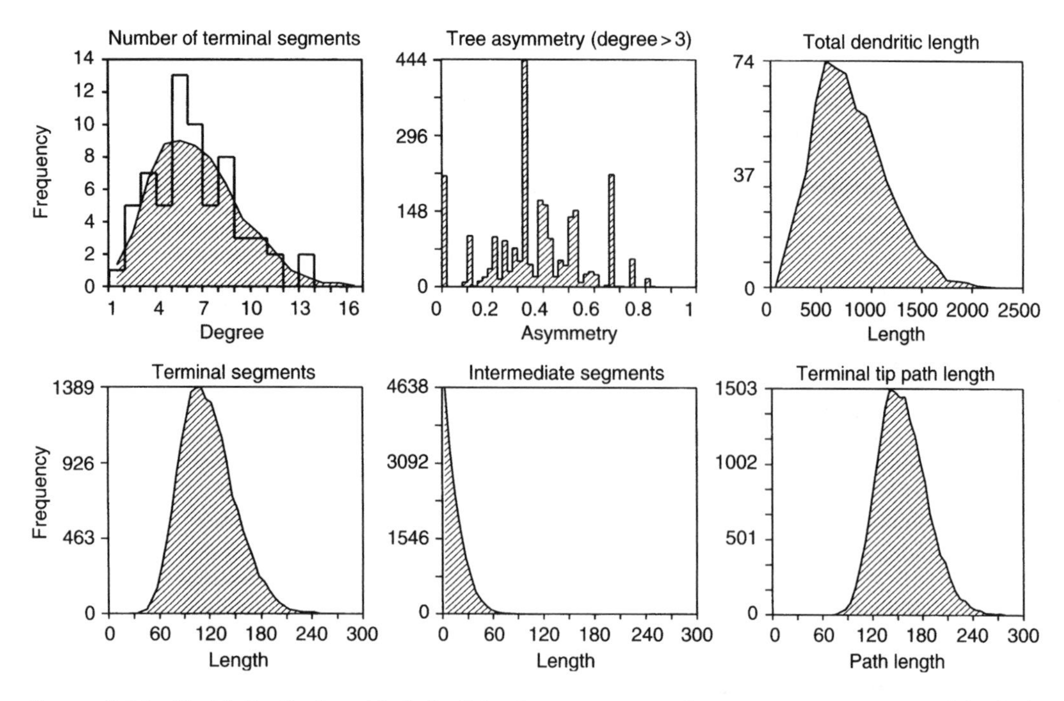

FIGURE 5.14 Model distributions (shaded) of the shape parameters degree, tree asymmetry, total dendritic length, terminal-segment length, intermediate-segment length, and terminal tip path length of model dendrites generated with parameter values given in Table 5.4. The values of the mean and standard deviation are given in Table 5.5. The first panel also includes the observed distribution of the number of terminal segments.

also have implications for the segment-length distributions. Unfortunately, neither the initial length nor the nonlinear function, $D(t)$, could be incorporated in this analysis of the large layer V cortical pyramidal neurons because the experimental data set for this cell group (Larkman, 1991a) did not include experimental segment-length distributions.

The implication of a nonlinear baseline branching-rate function for the metrical development of dendritic trees has recently been studied by Van Pelt and Uylings (2003). Using a developmental data set of Wistar rat layer IV multipolar nonpyramidal neurons, it was shown that an exponentially decreasing baseline branching-rate function with a decay time constant of $\tau = 3.7$ days resulted in a total tree-length growth function that matched the observed data very well.

Agreement between experimental and modeled distributions of dendritic shape parameters has now been obtained for a variety of cells types and species, viz., rat cortical layer 5 pyramidal neuron basal dendrites (Van Pelt and Uylings, 1999), S1 rat cortical layer 2,3 pyramidal neurons (Van Pelt et al., 2001b), guinea pig cerebellar Purkinje cell dendritic trees (Van Pelt et al., 2001b), cat superior colliculus neurons (Van Pelt et al., 2001a; Van Pelt and Schierwagen, 2004), Wistar rat visual cortex layer IV multipolar nonpyramidal neurons (Van Pelt et al., 2003), Wistar rat layer 2,3 associative pyramidal neurons (Granato and Van Pelt, 2003), and for terminal-segment-number distributions only (Van Pelt et al., 1997). Optimized model parameter values are summarized in Van Pelt et al. (2001c).

5.4.3.2 Simulation Procedure

The simulation of the growth process for the BESTL growth model includes the following steps:

1. Parameter S is estimated from the topological variation in the observed trees (see Van Pelt et al., 1992).

2. Parameters B and E are estimated from the observed degree of distribution (see Van Pelt et al., 1997).
3. The baseline branching-rate function, $D(t)$, has to be estimated. When experimental data are available for the number of terminal segments at several developmental time points, one may follow two approaches. The first one is to find a (non)linear mapping of the time bins onto real time (see Van Pelt et al., 1997; Van Pelt and Uylings, 2002), such that the model growth curve predicted for constant D transforms into a curve that matches the observed data. Then, using the bin-to-time mapping, each time bin corresponds to a time period for which the function $D(t) = B/\Delta T(i)N$ can be calculated. In the second approach, an analytical baseline branching-rate function is assumed (e.g., an exponential one) and its parameters are subsequently optimized by fitting the predicted terminal-segment-number growth function to the observed one (detailed examples of these different approaches are given in Van Pelt and Uylings, 2002, 2003). When the observed dendrites originate from only one point in time, the shape of the baseline branching-rate function has to be assumed (e.g., a linear or exponential one).
4. Trees are now generated according to the following iterative algorithm: for a given tree at a given point in time t, the branching probabilities are calculated for all of the $n(t)$ terminal segments with, for $S \neq 0$, the centrifugal order γ being taken into account for each of them by use of $p(t) = C(t)2^{-S\gamma}D(t)n(t)^{-E}$. The normalization constant $C(t)$ is obtained via $C(t) = n(t)/\sum_{j=1}^{n(t)} 2^{-S\gamma_j}$. Then, using a random number between 0 and 1 from a uniform distribution, it is decided for each terminal segment whether a branching event actually occurs in the present time step (i.e., a branching event occurs when the random number is smaller than or equal to the branching probability for that segment). A branching event implies that the terminal segment becomes an intermediate segment ending in a bifurcation with two daughter terminal segments, each of which is given an initial length and elongation rate. These values are obtained by random sampling from gamma distributions with means and standard deviations as given for the initial length and for the elongation rate for that point in time, respectively. All terminal segments (including the new ones) are subsequently elongated according to their individual rates. The process starts at the time of onset with a single (root) segment and stops at the last point in the specified time (see also Van Pelt et al., 2001b).

5.4.4 Modeling the Variation in Segment Diameter

No developmental rules have so far been used to generate the diameters of segments. The segment diameters in the displayed dendrograms have therefore been estimated by assuming a power-law relation. By the power-law relation, the diameter of an intermediate segment (d_s) is determined by the number of terminal segments, n_s, in its subtree according to $d_s^e = n_s d_t^e$ or $d_s = d_t n_s^{1/e}$. Accordingly, the diameter of a segment in a growing tree is expected to increase with the increasing number of terminal segments. The segment diameters in a tree have been estimated by means of the following procedure. First, terminal-segment diameters are estimated by random sampling from the observed terminal-segment diameter distribution (or a normal distribution based on the observed mean and SD values). At each bifurcation, the diameter of the parent segment is estimated by using a branch power value obtained by random sampling from the observed branch power distribution. For the dendrograms in Figure 5.12 and Figure 5.13, the terminal-segment diameters were randomly sampled from a normal distribution with mean 0.8 μm and standard deviation 0.2 μm (Larkman, 1991a). The branch power distribution was assumed to be a normal one with mean 1.6 μm and standard deviation 0.2 μm (values estimated from Larkman et al., 1992).

5.5 Discussion

The growth model for dendritic trees discussed here is based on a stochastic description of branching and elongation of terminal segments. The model is able to reproduce dendritic complexity with respect to number, connectivity, and length of dendritic segments. It should be noted that both branching and elongation determine the lengths of segments, such that their distributions can only be studied by taking both processes into account. Agreement between model outcomes and empirical data was obtained for a larger number of shape parameters than the number of optimized model parameters, giving further support to the assumptions made in the model. Five aspects have been distinguished in dendritic growth, viz. (a) the baseline branching-rate function $D(t)$ of individual growth cones; (b) the proliferation of the number of growth cones in the growing tree $n(t)$; (c) the modulation of the branching probability by the increasing number of growth cones (determined by the parameter E) and the position of the segment in the tree (via parameter S); (d) the elongation of segments by the migration of growth cones; and (e) the assignment of initial lengths to newly formed segments after branching.

Because of the structure of the model, both its parameters (i.e., the rates of branching and migration of growth cones) and its outcomes (i.e., the generated dendrites and the distributions of their shape parameters) can directly be associated with the biological processes and structures. This enables the empirical validation of its assumptions and outcomes. The model has been shown to be able to accurately reproduce dendritic geometrical properties of a variety of neuronal cell types. Dendritic growth is a complex process, mediated by the dynamic behavior of growth cones showing periods of forward migration but also of retraction. On a short timescale, these actions are implicitly accounted for by the stochasticity of the model. On a longer timescale, changes in growth-cone behavior can be described by specifying the time profiles of the baseline branching-rate function and of the migration rates. Such specification, however, requires the availability of experimental data on dendritic shapes at more than one point in time. For instance, in the present example, different segment-elongation rates were assumed during the two phases of development. Dendrites may also experience phases of remodeling, when the number of segments substantially declines. Dendritic regression is not included in the present model, but may be incorporated with an unavoidable increase in complexity of the model. In a study on the impact of pruning on the topological variation of dendritic trees, Van Pelt (1997) showed that under uniform random pruning schemes the topological variance of the pruned trees remained consistent with the mode of initial outgrowth.

The progress in modeling dendritic structure makes it possible to study systematically the functional implications of realistic morphological variations. Since the pioneering work on cable theory of Rall (e.g., see Rall et al., 1992; Rall, 1995b), many methods have been developed for studying the processing of electrical signals in passive and active dendritic trees. Both the branching-cable and the compartmental models (e.g., Koch and Segev, 1989; see also Figure 5.15) have made it clear how signal processing depends on dendritic structure and membrane properties. The impact of topology on input conductance and signal attenuations in passive dendritic trees was first studied systematically by Van Pelt and Schierwagen (1994, 1995). The use of one single vector equation for the whole dendritic tree, including the Laplace-transformed cable equations of all the individual segments, greatly simplified and accelerated the electrotonic calculations (Van Pelt, 1992). These studies demonstrated how dendrites with realistic topological structures differ in their passive electrotonic properties from symmetrical trees, and how the natural topological variation in dendritic morphologies contributes to the variation in functional properties. Regarding the influence of morphology on dendrites with active membrane, it was shown by Duijnhouwer et al. (2001) that dendritic shape and size differentiate the firing patterns from regular firing to bursting when the dendrite is synaptically stimulated randomly in space and time. Similar effects of dendritic shape under conditions of somatic stimulation with fixed current injections were reported earlier by Mainen and Sejnowski (1996). Systematic studies of the effect of dendritic topology on neuronal firing have shown that firing frequency correlates with the mean dendritic path length when the neuron is stimulated at

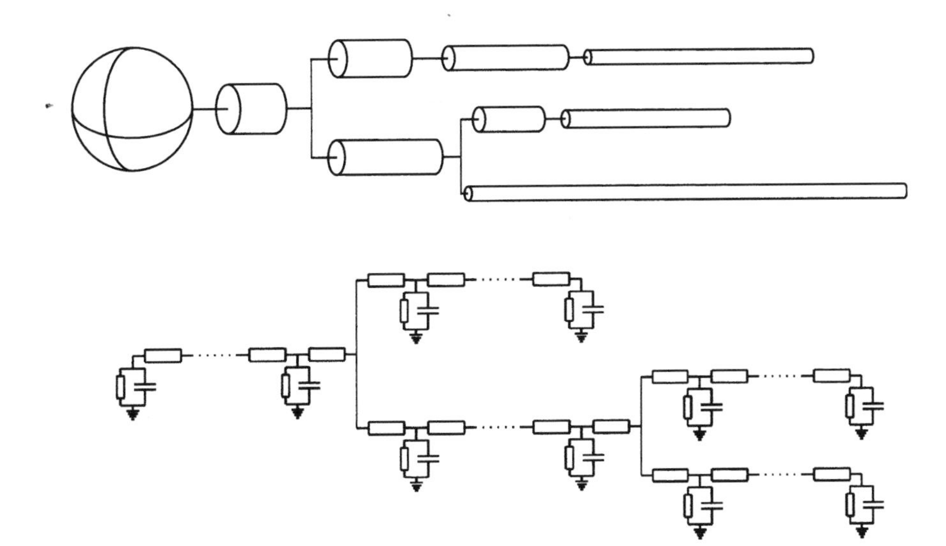

FIGURE 5.15 (Top) Example of a branching cable model in which each segment is modeled as one or more cylindrical membranes. (Bottom) A compartmental model in which each segment is modeled as a series of isopotential R–C compartments. (*Source:* Adapted from Koch and Segev, *Methods in Neuronal Modeling*, MIT Press, figure 3.3, 93–136, 1989.)

the soma (Van Ooyen et al., 2002). In these topological studies all the dendrites were given the same metrical and membrane properties. Schaefer et al. (2003) found that the detailed geometry of proximal and distal oblique dendrites originating from the main apical dendrite in thick tufted neocortical layer five pyramidal neurons determines the degree of coupling between proximal and distal action-potential initiation zones. Coupling is important for coincidence detection insomuch as it regulates the integration of synaptic inputs from different cortical layers when coincident within a certain time window. They conclude that dendritic variation may even have a stronger effect on coupling than variations in active membrane properties.

All these studies indicate that morphological variability of neurons may be an important contributor to their functional differentiation, keeping in mind that such approaches reflect a caricature of the natural branching structure (see Figure 5.15). Further insight into the subtleties of active neuronal processing of electrical signals and information is needed to fully clarify the functional implications of morphological details and variations. However, it is clear that integration in the brain is one such subtlety or "bug" that arises from intricate biological variability, irrespective of the computational capabilities a neuron is thought to possess (cf. Vetter et al., 2001; Schaefer et al., 2003).

Dendritic shapes have been studied in this chapter by focusing on the metrics and connectivity of the individual segments (see Section 5.2). Clearly, dendrites are 3D structures with a particular embedding in 3D space, and further shape descriptors are needed for their full 3D characterization, such as the angles between segments at bifurcation points, the curvature of the segments, and multiscale fractal measures (e.g., Costa et al., 2002). Recent developments in modeling 3D dendritic geometry now also concern dendritic orientation in 3D and environmental influences (Ascoli et al., 2001; Ascoli, 2002a). The 3D dendritic geometry acquires functional implications when the extracellular space participates in the electrical circuitry in an inhomogeneous manner and/or when dendrites are connected to each other via electrical contacts (for instance via gap junctions) (e.g., Poznanski et al., 1995). It is expected that the proximity of dendritic branches becomes an important aspect in dendritic 3D geometry, for which the topological structure and the angles between branches at bifurcation points are crucial parameters. The present equivalent cylinder, branching-cable, and compartmental models, however, do not incorporate these 3D aspects.

5.6 CONCLUSIONS AND FUTURE PERSPECTIVES

The mathematical modeling of dendritic growth using a stochastic description of growth-cone behavior has brought many dendritic shape parameters and their variations into a coherent framework. It allows for analysis of dendritic shapes in terms of their developmental histories and for predictions of growth-cone branching and elongation. Importantly, the model provides a useful tool for generating sets of randomly varying dendritic trees for use in studies of the functional implications of morphological variations. Both metrical and topological aspects of dendritic shapes have to be incorporated in assessing how the dendrite integrates in space and time the trains of incoming postsynaptic potentials resulting in particular series of generated action potentials. Clearly, membrane properties are also subjected to variations as for instance caused by the types and the spatial distributions of the membrane channels, which contribute to the individual functional properties of the particular dendrite. Additionally, a neuron is a highly complex dynamic system and many structural aspects of the system have time-dependent properties with large ranges of time constants, such as, for example, activity-dependent influences on membrane conductances (e.g., Abbott and LeMasson, 1993) and neurite outgrowth (e.g., Van Ooyen, 1994). The integrative properties and firing characteristics are therefore expected to depend on the dynamic history of the neuron in a very complex manner.

It is to be expected that structural cable models will become increasingly important in future studies of dendritic structure–function relationships. Structural cable models will also need to describe dendritic morphologies with increasing natural detail in order to get a better understanding of the rich dynamic repertoire of dendritic operation. A typical example of greater realism is the 3D dendritic geometry and its embedding in the neuropil. For the modeling of 3D geometry, the angles between the segments at bifurcation points and the curvature of the segments should therefore also be taken into account. For the growth model it would require the migration of growth cones to proceed in 3D as well as a qualification of the 3D environment of the outgrowing neurites. The growth model presented in this chapter for generating branching patterns with realistic topological and metrical variations contributes to this development and allows such future extensions quite naturally.

5.7 SUMMARY

Dendritic branching patterns are complex and show a large degree of variation in their shapes, within, as well as between different neuronal cell types and species. Dendritic structure plays an important role in the spatial and temporal integration of postsynaptic potentials and is consequently an important determinant for the characteristics of neuronal firing patterns. How dendritic shape influences neuronal spiking is indispensable for systematic studies on neuronal structure–function relationships. In this chapter, a quantitative description of the geometry of dendritic branching patterns is presented for the generation of model dendrites with realistic shape characteristics and variability. Several measures are introduced for the quantification of both metrical and topological properties of dendritic shape. The dendritic growth model is based on principles of neuronal development and its parameters are biologically interpretable directly. The outcomes of the dendritic growth model are validated by comparison with large sets of experimentally reconstructed 3D neurons through optimization of the model parameters. These model parameters appear to be powerful discriminators between different cell types, and thus concise descriptors of the neuronal developmental process.

ACKNOWLEDGMENTS

We are very grateful to Dr. M.A. Corner for his critical comments and improvements of the manuscript. Part of this work was supported by NATO grant CRG 930426 (JVP).

PROBLEMS

1. Real dendrites are structures embedded in 3D space having irregular branches and orientations. How can these morphological aspects be best built into the growth model? The growth process needs to be formulated in 3D space, with specific assumptions for the direction of outgrowth of growth cones. New possibilities are then offered for validating these assumptions using observed orientation and irregular properties of dendritic branches.
2. The growth-cone branching probability has been formulated as a function of the total number of growth cones, a position-dependent term, and a baseline branching rate representing all other factors determining the branching process. The baseline branching rate is a function of time that can be estimated when dendritic data at several developmental time points are available. When data at only one time point are available, what is the best choice for the baseline branching-rate function?
3. A migration rate is assigned to a growth cone at the time of its birth (i.e., after a branching event) by random sampling from a given distribution, and remains constant for its life span (i.e., up to its next branching event). Other ways of implementing stochasticity in growth-cone migration, however, are also possible (such as stochasticity in the actual migration at each time step). Are there other, or even better ways to capture the stochasticity of the underlying biological process, and how would you test them?
4. A basic assumption in the model is independence in the branching and migration of growth cones. To what extent do possible correlations in the occurrences of branching and elongation events result in altered variations in dendritic shape parameters?
5. A thorough experimental test of the model is possible when growth cone behavior is quantified in a longitudinal *in vivo* study of identified neurons in development, and the neurons are morphologically reconstructed afterwards. What might be the best experimental approach to this end?
6. The power of the growth model approach is that dendritic geometrical properties and their variations can possibly be captured by a limited set of parameters describing the underlying dendritic growth process. This set of parameters has to be determined for many different neuron types. Such an inventory will give insight in the similarity and dissimilarity of growth actions in the development of different neuron types.

6 Multicylinder Models for Synaptic and Gap-Junctional Integration

Jonathan D. Evans

CONTENTS

6.1 INTRODUCTION

One objective of neuronal modeling is to construct a mathematical model of synaptic spread and summation which agrees with existing measurements to within a specified accuracy and can be used with confidence to predict future observations and behaviour. Thus, to pursue neuronal modeling successfully (as in other areas of mathematical modeling) two contrasting but complimentary skills are needed, namely, the ability to formulate the problem in appropriate mathematical terms, and the possession of sufficient knowledge (or techniques) to obtain useful information from that mathematical model. The skill in formulation lies in finding a model that is simple enough to give useful information easily, but that is complex enough to give all the information with sufficient accuracy. In attempting to construct a model, judgment has to be made about which features to include and which to neglect. If an important feature is omitted, then the model will not describe the observed phenomenon accurately enough or the model may not be self-consistent. If unnecessary features are included then the model will be more difficult to solve because of its increased complexity. Thus it is advisable to adopt a simple approach with a minimum of features included at the first attempt, and additional features added, if necessary, one by one. It should then be possible to estimate the effect of each additional feature on the results of the original model. It is this bottom-up (rather than top-down) approach that has been followed in the development of neuron models, where the passive properties of neurons need to be investigated first before proceeding to models that include active properties.

In this chapter, we discuss both passive and active neuron models and concentrate on the recent development of analytical solutions for these models rather than their numerical solutions by, for example, segmental cable analysis or compartmental analysis. Analytical results are of value not only in their own right to reveal trends and provide simple rules of thumb in certain situations, but also as benchmarks for confirming the results from numerical approaches, which are often of unknown accuracy.

As remarked by Rall (1990), there is no one model for dendritic neurons, but rather a variety that have been used to address different problems. Indeed, a model may be appropriate in one set of circumstances but of no value in others because it is either over-complicated or too simple. This is certainly the case with the range of passive cable models that have been developed to investigate, for example, the roles of dendritic branching, tapering, and spines on the passive voltage spread within the cell. These models are reviewed mostly in the discussion section while the main text is concerned with the analysis of the multicylinder model (Evans et al., 1992) together with the effects of dendritic ion channels, tapering, and coupling. This model is the many equivalent cylinder version of Rall's single equivalent cylinder model (Rall, 1962a, 1962b), reviews for which are given in Jack et al. (1983), Tuckwell (1988a), and Rall (1989). Figure 6.1 summarizes the main transformations in constructing the Rall model.

6.2 The Multicylinder Model

6.2.1 The Mathematical Problem

The multicylinder model with N equivalent cylinders emanating from the soma with a shunt resistance is illustrated in Figure 6.2, the dimensional mathematical statement for which is the following,

$$\text{in } 0 < x_j < \ell_{\text{cyl}j}, \quad t > 0 \qquad v_j + \tau_{\text{m}}\frac{\partial v_j}{\partial t} - \lambda_j^2\frac{\partial^2 v_j}{\partial x_j^2} = r_{\text{m}j}i_j(x_j, t; v_j); \tag{6.1}$$

$$\text{at } x_j = \ell_{\text{cyl}j} \qquad \frac{\partial v_j}{\partial x_j} = 0; \tag{6.2}$$

$$\text{at } x_j = 0 \qquad v_j = v_{\text{s}}; \tag{6.3}$$

$$\text{and} \qquad v_{\text{s}} + \tau_{\text{s}}\frac{\text{d}v_{\text{s}}}{\text{d}t} - \sum_{j=1}^{N}\frac{R_{\text{s}}}{r_{\text{a}j}}\frac{\partial v_j}{\partial x_j} = R_{\text{s}}i_{\text{s}}(t); \tag{6.4}$$

$$\text{at } t = 0, \quad 0 \le x_j \le \ell_{\text{cyl}j} \qquad v_j = h_j(x_j); \tag{6.5}$$

for $j = 1, \ldots, N$, where $v_j(x_j, t)$ is the transmembrane potential in volts at a distance x_j (cm) along the jth equivalent cylinder from the soma ($x_j = 0$) and at time t (sec); v_{s} is the soma potential in volts; $i_j(x_j, t; v_j)$ is the current density term in amperes per centimeter for the jth equivalent cylinder; $i_{\text{s}}(t)$ is an applied current in amperes at the soma; and $h_j(x_j)$ is the initial voltage distribution in volts in the jth equivalent cylinder. The current density term $i_j(x_j, t; v_j)$ as expressed in Equation (6.1) mimics current applied through a pipette (additional contributions are added in Equation [6.28]), generating

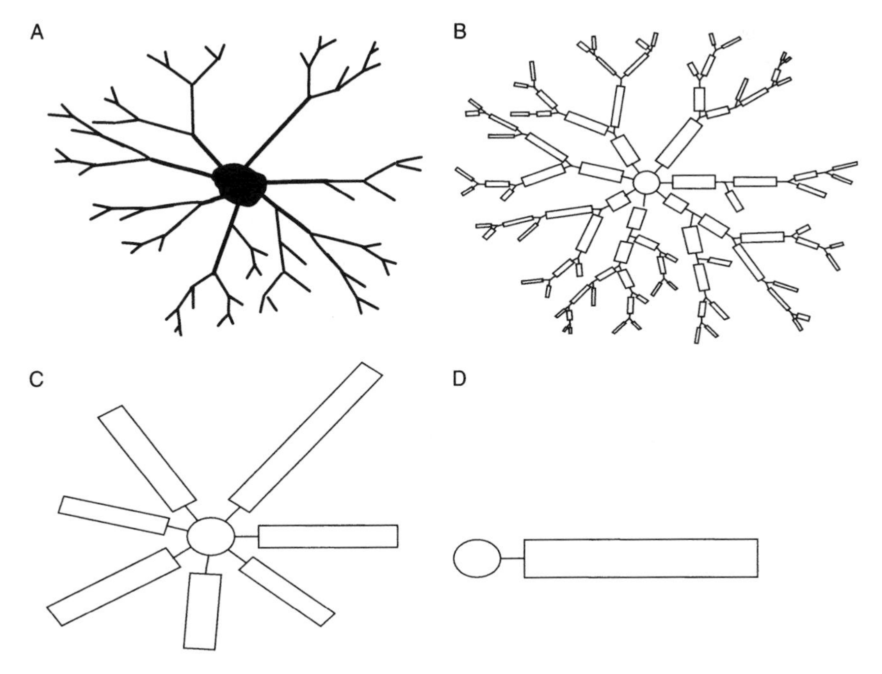

FIGURE 6.1 Transformations made in the development of models of differing branching complexity. The nerve cell (A) is represented by a model (B) with isopotential soma and each dendritic segment as a cylinder of uniform passive membrane. Under certain symmetry assumptions (e.g., see Tuckwell, 1988a), each dendritic tree can be represented by an equivalent cylinder giving the equivalent multicylinder model (C). To obtain the Rall model, it is assumed that the individual equivalent cylinders can be combined into a single equivalent cylinder (D).

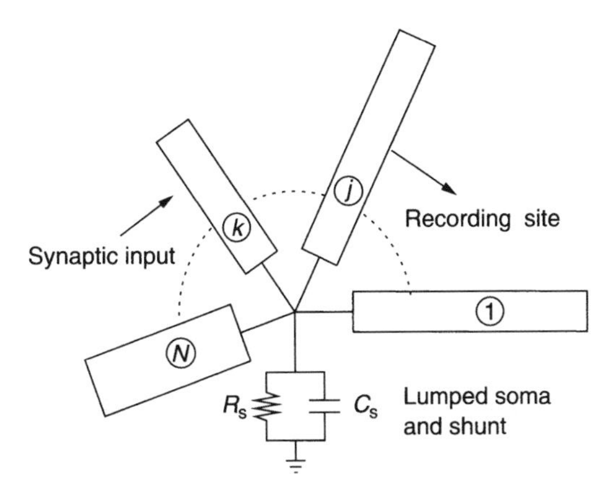

FIGURE 6.2 Illustration of the multicylinder model with N equivalent cylinders emanating from a common soma. (Adapted from Evans, J.D. et al., *Biophys. J.* 63, 350–365, 1992. With permission.)

outward current (since by convention inward current is negative and outward current is positive). The index j has been introduced as a suffix to the variables and parameters for the jth equivalent cylinder where j can range between 1 and N, inclusive. It is now worth emphasizing the *core* dimensional parameters,

R_m membrane specific resistance ($\Omega\,\text{cm}^2$),

C_m membrane specific capacitance ($\text{F}\,\text{cm}^2$),

R_i axial resistivity ($\Omega\,\text{cm}$),

d_{cylj} physical diameter of the jth equivalent cylinder (cm),
ℓ_{cylj} physical length of the jth equivalent cylinder (cm),
A_s area of soma (cm^2),
R_{sh} somatic shunt resistance (Ω),

in terms of which the other more commonly used dimensional parameters can be defined such as,

R_s resistance of soma (Ω), ($1/R_s = A_s/R_m + 1/R_{sh}$)
$C_s = A_s C_m$ capacitance of soma (F),
$\tau_m = R_m C_m$ membrane time constant of the equivalent cylinder (sec),
$\tau_s = R_s C_s$ membrane time constant of the soma (sec),
$r_{aj} = 4R_i/\pi {d_{cylj}}^2$ axial resistance per unit length of the jth equivalent cylinder ($\Omega\, cm^{-1}$),
$r_{mj} = R_m/\pi d_{cylj}$ membrane resistance of a unit length of the jth equivalent cylinder (Ω cm),
$\lambda_j = (r_{mj}/r_{aj})^{1/2}$ space constant of the equivalent cylinder (cm),
$c_{mj} = \pi d_{cylj} C_m$ membrane capacity per unit length of the jth equivalent cylinder ($F\, cm^{-1}$).

A feature worth emphasizing for this model is that each equivalent cylinder can represent one or more dendritic trees (provided that the dendritic trees fulfill the necessary conditions for reduction and in particular have the same electrotonic length). The physical diameter d_{cylj} of the jth equivalent cylinder is defined in terms of the diameters of the primary dendrites of the dendritic trees that it represents as

$$d_{cylj} = \left(\sum_{st \in stemsj} d_{st}^{3/2} \right)^{2/3}, \tag{6.6}$$

where d_{st} is the diameter of a stem segment and the summation is taken over the set stems$j$ of all such stem segments emanating from the soma to be represented by the jth equivalent cylinder. Similarly, the physical length ℓ_{cylj} of the jth equivalent cylinder is defined as

$$\ell_{cylj} = d_{cylj}^{1/2} \sum_{i \in pathj} \frac{\ell_i}{d_i^{1/2}}, \tag{6.7}$$

where ℓ_i and d_i are respectively the length and diameter of a dendritic segment indexed by i, and the summation is over the set path$j$ of all dendritic segments that, for any chosen dendritic process being represented by the jth equivalent cylinder, compose the path from the soma to its termination (by assumption all such paths in the dendritic trees being represented by the jth equivalent cylinder have the same electrotonic length). The result (Equation [6.7]) is derived from the electrotonic length of the jth equivalent cylinder being the same as that of any one of the dendritic trees or processes that it represents, with the electrical parameters R_i and R_m having been removed, and leaving the result in terms of the effective length, which for each segment is $\ell_i/d_i^{1/2}$ and for the jth equivalent cylinder is $\ell_{cylj}/d_{cylj}^{1/2}$. The conditions for the reduction of a dendritic tree to an equivalent cylinder can be analogously stated using effective length in place of electrotonic length and is sometimes used (e.g., Clements, 1984; Stratford et al., 1989; Moore and Appenteng, 1991; Rall et al., 1992), the former quantity only depending on morphological parameters.

This model was considered in Evans et al. (1992) with further analysis of the solution behavior being given in Evans et al. (1995). It involves solving the standard cable Equation (6.1) in each equivalent cylinder for appropriate boundary conditions at the ends of the cylinders (Equation [6.2]) together with conditions of voltage continuity (Equation [6.3]) and current conservation at the soma (Equation [6.4]). Indeed, it is the current conservation condition at the soma (Equation [6.4]) that enforces discontinuity in the derivative of the voltage at the focal point of the equivalent cylinders which makes the problem interesting mathematically. It should be mentioned that Redman (1973), Rinzel and Rall (1974), and Holmes et al. (1992) have obtained analytic or semianalytic solutions for a multicylinder model with a point soma, which is a particular case of the above-mentioned model.

6.2.2 Problem Normalization and General Solution

The dimensional mathematical equations for the model have been presented in Section 6.2.1, where the nerve cell is represented by N equivalent cylinders emanating from a uniformly polarized soma.

We nondimensionalize Equations (6.1) to (6.5) with

$$\begin{aligned} x_j &= \lambda_j X_j, \qquad t = \tau_m T, \qquad v_j = \bar{V} V_j, \qquad v_s = \bar{V} V_s, \\ r_{mj} i_j &= \bar{V} I_j, \qquad R_s i_s = \bar{V} I_s, \qquad h_j = \bar{V} H_j, \end{aligned} \tag{6.8}$$

where $\bar{V}$ is an appropriate voltage scaling determined by the type of input. For example, for a point charge input of Q_0 coulombs, we may take $\bar{V} = Q_0 R_s / \tau_m$, while Example 1 illustrates the choice when synaptic reversal potentials are present. It should be emphasized that the role of the scaling $\bar{V}$ is to normalize the voltage scale in the case of single inputs and to introduce a meaningful relative voltage comparison when more than one input type is present. We introduce the nondimensional parameters,

$$\epsilon = \frac{\tau_s}{\tau_m}, \qquad \gamma_j = \frac{R_s}{\lambda_j r_{aj}}, \qquad L_j = \frac{\ell_{cylj}}{\lambda_j}, \tag{6.9}$$

for $j = 1, \ldots, N$ which gives the nondimensional mathematical model,

$$\text{in } 0 < X_j < L_j, \quad T > 0 \qquad V_j + \frac{\partial V_j}{\partial T} - \frac{\partial^2 V_j}{\partial X_j^2} = I_j(X_j, T; V_j); \tag{6.10}$$

$$\text{at } X_j = L_j \qquad \frac{\partial V_j}{\partial X_j} = 0; \tag{6.11}$$

$$\text{at } X_j = 0 \qquad V_j = V_s; \tag{6.12}$$

$$\text{and} \qquad V_s + \epsilon \frac{dV_s}{dT} - \sum_{j=1}^{N} \gamma_j \frac{\partial V_j}{\partial X_j} = I_s(T); \tag{6.13}$$

$$\text{at } T = 0, \quad 0 \le X_j \le L_j \qquad V_j = H_j(X_j); \tag{6.14}$$

for $j = 1, \ldots, N$, illustrated in Figure 6.3.

The solution to the general problem, Equations (6.10) to (6.14), may be expressed as a convolution integral of the unit impulse solution $G_j^k(X_j, Y_k, T)$ with the appropriate input currents and initial

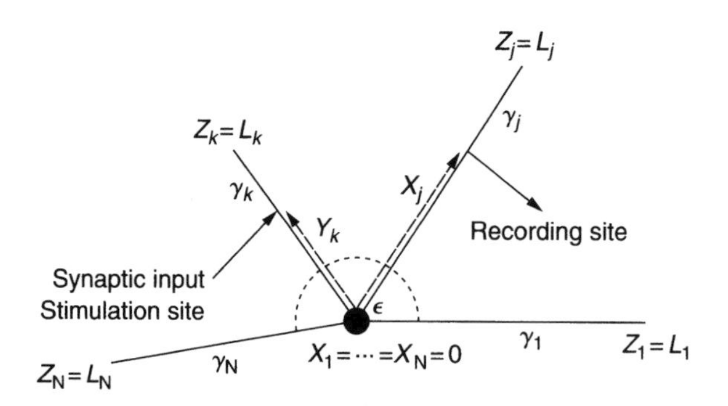

FIGURE 6.3 Illustration of the domain and dimensionless parameters for the nondimensional mathematical model of the multicylinder model. (Adapted from Evans, J.D. et al., *Biophys. J.* 63, 350–365, 1992. With permission.)

voltage distribution in the form

$$
\begin{aligned}
V_j(X_j, T) = &\sum_{k=1}^{N} \int_0^{L_k} \int_0^{T} G_j^k(X_j, \xi_k, T-u) I_k(\xi_k, u; V_k(\xi_k, u))\, \mathrm{d}u\, \mathrm{d}\xi_k \\
&+ \sum_{k=1}^{N} \int_0^{L_k} G_j^k(X_j, \xi_k, T) H_k(\xi_k)\, \mathrm{d}\xi_k + \frac{1}{\gamma_j} G_j^j(0, X_j, T) H_j(0) \\
&+ \frac{1}{\gamma_j} \int_0^{T} G_j^j(0, X_j, T-u) I_s(u)\, \mathrm{d}u,
\end{aligned}
\tag{6.15}
$$

for $j = 1, \ldots, N$. When I_k is voltage dependent, Equation (6.15) represents a system of integral equations for the voltage that has to be solved simultaneously for each cylinder. The system of integral equations is nonlinear if I_i is a nonlinear function of the voltage. It is worth remarking that as an approximation to the full Hodgkin–Huxley equations for modeling active ion channels, of particular interest are cubic voltage dependencies for I_i as described in Jack et al. (1983). These may be related to the reduced FitzHugh–Nagumo equations (e.g., see Tuckwell, 1988b) in which there is no recovery variable. When I_k is voltage independent, the system reduces with Equation (6.15) then giving the explicit solution. In all cases, the unit impulse solution (or Green's function) $G_j^k(X_j, Y_k, T)$ is important in the solution construction. It is defined as the solution of Equations (6.10) to (6.14) with

$$
\begin{aligned}
&I_k = \delta(T)\delta(X_k - Y_k), \quad I_j = 0, \qquad j = 1, \ldots, N, \quad j \neq k, \\
&I_s = 0, \qquad H_j = 0, \qquad j = 1, \ldots, N;
\end{aligned}
$$

which represents a unit impulse input at a (nondimensional) distance Y_k from the soma in the kth equivalent cylinder. The unit impulse solution $G_j^k(X_j, Y_k, T)$ so defined represents the response at a point X_j in the jth equivalent cylinder to a unit impulse input at a point Y_k along the kth cylinder. The index k emphasizes that the input or stimulating site Y_k is in the kth cylinder while the index j emphasizes that the recording site X_j is in the jth cylinder. Both indexes j and k lie in the range $1, \ldots, N$ and may or may not be equal. For example, $G_2^1(0.5, 0.75, T)$ represents the unit impulse solution recorded at the point $X_2 = 0.5$ in equivalent cylinder 2 for the unit impulse input at $Y_1 = 0.75$ in equivalent cylinder 1. Necessarily, $X_j \in [0, L_j]$ and $Y_k \in [0, L_k]$ and our notation extends to the case where $X_j = 0$ and $Y_k = 0$, that is, recording and stimulation is at the soma.

Two useful expressions for the solution G_j^k can be derived, one in the form of an eigenfunction expansion particularly appropriate for $T = O(1)$ while another is useful for $T = o(1)$.

The unit impulse solution G_j^k may be expressed as an eigenfunction expansion in the form

$$G_j^k(X_j, Y_k, T) = \sum_{n=0}^{\infty} E_{kn}\psi_{jn}(X_j)\psi_{kn}(Y_k)\,\mathrm{e}^{-(1+\alpha_n^2)T}, \tag{6.16}$$

where the spatial eigenfunctions are given by

$$\psi_{jn}(X_j) = \frac{\cos[\alpha_n(L_j - X_j)]}{\cos[\alpha_n L_j]}, \qquad \psi_{kn}(Y_k) = \frac{\cos[\alpha_n(L_k - Y_k)]}{\cos[\alpha_n L_k]}, \tag{6.17}$$

the eigenvalues $\alpha_n, n = 0, 1, 2, \ldots,$ are the (nonnegative) roots of the transcendental equation

$$1 - \epsilon(1 + \alpha^2) = \alpha \sum_{j=1}^{N} \gamma_j \tan[\alpha L_j], \tag{6.18}$$

and the coefficients E_{kn} are defined as

$$E_{kn} = \frac{2\gamma_k}{2\epsilon + \sum_{j=1}^{N} \gamma_j L_j((\tan[\alpha_n L_j]/\alpha_n L_j) + \sec^2[\alpha_n L_j])}. \tag{6.19}$$

The coefficient E_{k0} needs particular care in the case $\epsilon = 1$ since the first root of the transcendental equation $\alpha_0 = 0$, and now Equation (6.19) is replaced with the correct limit $E_{k0} = 2\gamma_k/[1 + \sum_{j=1}^{N} \gamma_j L_j]$.

The derivation of this result is discussed in detail in Evans et al. (1992) by two methods; one by the application of the Laplace transform method with complex residues following Bluman and Tuckwell (1987) for the single cylinder case and another by the method of separation of variables with a modified orthogonality condition for this more general equivalent multicylinder problem. The former is the preferred method, mainly because of its ease of generalization but also because the Laplace transform solution gives explicit information (in the form of its poles) regarding the eigenvalues of the problem. One of the limitations of the solution expressed in the form of an eigenfunction expansion as in Equation (6.16) is that the eigenvalues α_n for $n = 0, 1, \ldots,$ must form a complete set, which is more difficult to prove for the N-cylinder problem than for the single cylinder case. Generally this is the case, however, in certain circumstances they do not, and a second set of eigenvalues must also be considered. The special circumstances are when the electrotonic lengths of the input and recording cylinders are precisely odd integer multiples of one another, for example, $L_k = L_j, L_k = 3L_j, 5L_k = L_j$, etc. The particular case in which the recording cylinder is the input cylinder is also included, $L_k = L_j$ since $j = k$. Indeed, if we denote this second set of eigenvalues by $\{\beta_n\}_{n=0}^{\infty}$ then they arise only when

$$\cos(\beta_n L_k) = 0 \quad \text{and} \quad \cos(\beta_n L_j) = 0,$$

that is, L_k and L_j are in odd integer ratios. In this case the solution (Equation [6.16]) is augmented to,

$$G_j^k(X_j, Y_k, T) = \sum_{n=0}^{\infty} E_{kn}\psi_{jn}(X_j)\psi_{kn}(Y_k)\,\mathrm{e}^{-(1+\alpha_n^2)T} + \sum_{n=0}^{\infty} D_n\phi_{jn}(X_j)\phi_{kn}(Y_k)\,\mathrm{e}^{-(1+\beta_n^2)T},$$

with

$$\phi_{jn}(X_j) = \cos[\beta_n(L_j - X_j)], \qquad \phi_{kn}(Y_k) = \cos[\beta_n(L_k - Y_k)],$$

and the coefficients D_n are defined as

$$D_n = \begin{cases} -\dfrac{2\gamma_k \operatorname{cosec}[\beta_n L_k] \operatorname{cosec}[\beta_n L_j]}{L_k L_j(\gamma_k/L_k + \sum_{i\in S_n}(\gamma_i/L_i))} & j \neq k \\ -\dfrac{2\gamma_k}{L_k^2(\gamma_k/L_k + \sum_{i\in S_n}(\gamma_i/L_i))} + \dfrac{2\sin(\beta_n Y_k)\operatorname{cosec}[\beta_n L_k]}{L_k \cos[\beta_n(L_k - Y_k)]}, & j = k, Y_k < X_k \end{cases}$$

with the expression for $X_k < Y_k$ in the $j = k$ case being obtained by simply interchanging X_k and Y_k in this last expression. The index set S_n is defined as

$$S_n = \{i \in (1, \ldots, N), i \neq k : \cos(\beta_n L_i) = 0\},$$

and thus includes any cylinder, other than the input cylinder, whose electrotonic length satisfies $\cos(\beta_n L_i) = 0$ and thus must be an odd integer multiple of the electrotonic length of the input cylinder. Note that the members of this set may change with n, the eigenvalue index. The case in which the recording cylinder is not the input cylinder (i.e., $j \neq k$) is straightforward, with the set S_n necessarily containing the index j for each admissible n and also any other cylinder satisfying the given criterion $\cos(\beta_n L_i) = 0$. The case in which the recording and input cylinder are the same (i.e., $j = k$) needs discussion. In this case, the eigenvalues $\{\beta_n\}_{n=0}^{\infty}$ are given by $\beta_n = (2n+1)\pi/2L_k$, with the set S_n containing any cylinder, other than the input cylinder, whose electrotonic length satisfies $\cos(\beta_n L_i) = 0$. Note that if this set is empty for any particular n then $D_n = 0$ (which can be obtained from the above expressions for D_n), and importantly this means that when the electrotonic length of the input cylinder is not an odd integer multiple of the electrotonic lengths of any of the other cylinders then $D_n = 0$ for each n.

The full solution has been stated for completeness and is easily derived from the Laplace transform solution (Evans et al., 1992). The two sets of eigenvalues $\{\alpha_n\}_{n=0}^{\infty}$ and $\{\beta_n\}_{n=0}^{\infty}$ arise naturally as the roots ($x \geq 0$) of the transcendental equation,

$$\cos[xL_j]\cos[xL_k]\left(1 - \epsilon(1 + x^2) - x\sum_{i=1}^{N}\gamma_i \tan[xL_i]\right) = 0.$$

The second set of eigenvalues $\{\beta_n\}_{n=0}^{\infty}$ only occur in the certain cases already mentioned and are removed by small adjustments, for example, $O(10^{-5})$ to the electrotonic lengths of the cylinders involved, with no discernible effects on the overall resulting waveforms. As such they are less important and the solution (Equation [6.16]) is adequate, provided these adjustments are made when the electrotonic lengths of the input cylinder and any other cylinder are in odd integer ratios. This point is discussed more fully in Evans et al. (1992) and is a feature that does not arise in the single cylinder case. For the particular case of $\epsilon = 1$, Rall (1969a) obtained an expression similar to Equation (6.18) for the eigenvalues in this N-cylinder case, while Holmes et al. (1992) obtained expressions similar to Equation (6.16) for the particular case of a point soma at which a point charge input occurs.

The case in which electrotonic lengths are equal also needs special mention, since it can be dealt with as just discussed (by small adjustments to the electrotonic lengths) or by combining the cylinders into one. If we assume that the electrotonic lengths of all the cylinders are equal, so that $L_1 = L_2 = \cdots = L_N = L$ then the Green's function $G(X, Y, T)$ for the single-cylinder problem satisfies

$$\text{in } 0 < X < L, \quad T > 0 \qquad G + \frac{\partial G}{\partial T} - \frac{\partial^2 G}{\partial X^2} = \delta(X - Y); \tag{6.20}$$

$$\text{at } X = L \qquad \frac{\partial G}{\partial X} = 0; \tag{6.21}$$

$$\text{at } X = 0 \qquad G = G_s; \tag{6.22}$$

$$\text{and} \qquad G_s + \epsilon \frac{dG_s}{dT} - \gamma \frac{\partial G}{\partial X} = 0; \tag{6.23}$$

$$\text{at } T = 0, \quad 0 \le X \le L \qquad G = 0; \tag{6.24}$$

where $\gamma = \sum_{j=1}^{N} \gamma_j$, $Y = Y_k$, $X = X_j$, and G is related to the Green's function of the N-cylinder problem by

$$G(X, Y, T) = \frac{\sum_{j=1}^{N} G_j^k(X_j, Y_k, T)}{\sum_{j=1}^{N} \gamma_j}. \tag{6.25}$$

An analogous expression to Equation (6.25) holds when fewer than N cylinders, but all of equal electrotonic length, are combined; the summations now being taken only over the cylinders involved.

While the expression Equation (6.16) is valid for $T \ge 0$, its convergence is slow for $T \ll 1$ and the following asymptotic expansion may be obtained (Evans et al., 1992),

$$G_j^k \sim \frac{\gamma_k}{\epsilon} \sum_{i=1}^{4} e^{(h\chi_i + h^2 T)} \operatorname{erfc}\left(\frac{\chi_i}{2\sqrt{T}} + h\sqrt{T}\right) + \delta_{jk} \frac{1}{2\sqrt{\pi T}} \sum_{i=1}^{4} (-1)^{i+1} e^{-(\Omega_i)^2/4T} \quad \text{as } T \to 0, \tag{6.26}$$

where

$$\begin{aligned}
\chi_1 &= 2L_k - Y_k + X_j, & \Omega_1 &= X_k - Y_k, \\
\chi_2 &= Y_k + X_j, & \Omega_2 &= 2L_k - X_k - Y_k, \\
\chi_3 &= 2L_k - Y_k + 2L_j - X_j, & \Omega_3 &= L_k + Y_k + X_k, \\
\chi_4 &= Y_k + 2L_j - X_j, & \Omega_4 &= 2L_k + Y_k - X_k,
\end{aligned} \tag{6.27}$$

with

$$h = \frac{\gamma}{\epsilon}, \qquad \gamma = \sum_{j=1}^{N} \gamma_j.$$

The Kronecker delta δ_{jk} has been introduced for notational conciseness and is defined as,

$$\delta_{jk} = \begin{cases} 1 & \text{if } j = k, \\ 0 & \text{if } j \ne k. \end{cases}$$

The expression (6.26) holds when $j = k$ for $Y_k < X_k$ with the corresponding expression for $X_k < Y_k$ being obtained by interchanging X_k and Y_k in Equation (6.26). The expansion (Equation [6.26]) takes on simplifying forms for the two cases $h = O(1)$ and $h \gg 1$, which are discussed in Evans et al. (1995). The limit $h \to \infty$ (which arises if $\epsilon \to 0$ or $\gamma \to \infty$) can be obtained from Equation (6.26) but needs care. It is worth mentioning that Abbott et al. (1991), Abbott (1992), and Cao and Abbott (1993) present a trip-based method for determining the Green's function for no soma or shunt in the equivalent multicylinder case. This can alternatively be derived as is shown in Evans et al. (1995) by retaining higher order terms in the expansion (Equation [6.26]), with the interpretation of the terms in Equation (6.27) and others of higher order being image or reflection terms. It is noted that in the case of no soma and shunt, more terms in the expansions can be obtained, allowing the solution to hold for arbitrarily long times. However, this is not the case once the soma and shunt are introduced, suggesting that in this case the trip-based method would not produce a solution valid for arbitrarily large times.

The unit impulse solution may be evaluated in the following manner, as discussed in Evans et al. (1995). The leading order terms are kept for the small time solution (Equation [6.26]) and the number of terms in the approximation (Equation [6.16]) is increased until reasonable agreement is obtained between the two results in the form of an overlap region for a range of small T values. A bound on the error involved in truncating the series (Equation [6.16]) for its numerical calculation can be obtained and is derived in Evans et al. (1995). The error bound explicitly illustrates the fact that as $T \to 0$ more terms of the series need to be retained to achieve a specified accuracy.

Determination of the Green's function allows the voltage response to more general current inputs through use of the convolution integrals in Equation (6.15). These have been explicitly evaluated in Evans et al. (1995) to obtain expressions for the voltage response to short and long time step-current input and for smooth input current time courses in the form of an alpha function and a multiexponential function. Other types of current inputs may be considered as well as multiple input sites.

It should be noted that a numerical solution of the system of Equations (6.1) to (6.5) may be obtained by using, for example, finite difference methods. Suitable schemes appropriate to parabolic governing equations are discussed in Tuckwell (1988b). While these provide numerical output for the system of equations and are often relatively fast to solve, the analytical solution still provides a benchmark against which to compare such numerical solutions as well as other approximations such as suitable asymptotic solutions. Numerical solution methods are often more general and provide solutions in circumstances where an analytical solution cannot be obtained. However, a better understanding of a problem is often achieved when these are used in conjunction with analytical solutions, even if the latter are only available for certain reductions of the full problem and particularly if these provide asymptotic approximations in regimes in which numerical methods encounter difficulties (e.g., many singular perturbation problems).

6.2.3 Synaptic Reversal Potentials and Quasi-Active Ionic Currents

The representation of synaptic input as a conductance change, rather than current injection, is generally more realistic and allows the effects of synaptic excitation and inhibition on the soma–dendritic surface to be included in the model. The derivation of the cable theory with synaptic reversal potentials can be found in Rall (1977) and leads to voltage-dependent current input terms in Equation (6.1). In addition, we allow quasi-active dendritic properties to be included for subthreshold behavior through ionic currents as described in Poznanski and Bell (2000a) and Poznanski (2001a). Inclusion of such terms then gives dimensional current input terms of the general functional form

$$i_j(x_j, t; v_j) = -i_{\text{ion}j}(x_j, t; v_j) - i_{\text{syn}j}(x_j, t; v_j) + i_{\text{app}j}(x_j, t), \tag{6.28}$$

for the jth equivalent cylinder. Here, we only consider a single ionic species (although the extension to include more is straightforward). We consider M_j synaptic input terms of the form

$$i_{\text{syn}j}(x_j, t; v_j) = -\sum_{m=1}^{M_j} g^m_{\text{syn}j}(x_j, t)(v^m_{\text{rev}j} - v_j),$$

with constant reversal potentials $v^m_{\text{rev}j}$ and conductance changes per unit length $g^m_{\text{syn}j}(x_j, t)$ ($\Omega^{-1}\text{cm}^{-1}$) (the time variations of which are usually taken in the form of alpha functions). For synaptic reversal potentials only, the single equivalent cylinder problem ($N = 1$) has been considered by Poznanski (1987) for a single synaptic input ($M_1 = 1$) and in Poznanski (1990) for the tapering case with two synaptic inputs ($M_1 = 2$). $i_{\text{app}j}$ (A cm^{-1}) represents an applied current density. Following Poznanski and Bell (2000a), we consider linearized ionic input terms of the form

$$i_{\text{ion}j}(x_j, t; v_j) = g_{\text{ion}j}(x_j, t)v_j,$$

where the conductance term $g_{\text{ion}j}(x_j, t)$ usually incorporates a quasi-active voltage dependence through activation and inactivation variables. Such linearizations can provide reasonable approximations for voltage differences in the subthreshold range (see Poznanski and Bell, 2000a). Implicit here is the assumption that an ion pump counters the resting ionic fluxes to maintain the membrane potential at equilibrium.

The expression (6.28) may be alternatively written as

$$i_j(x_j, t; v_j) = \hat{g}_j(x_j, t)v_j(x_j, t) + i_{Aj}(x_j, t), \tag{6.29}$$

where

$$\hat{g}_j(x_j, t) = -g_{\text{ion}j}(x_j, t) - \sum_{m=1}^{M_j} g^m_{\text{syn}j}(x_j, t), \qquad i_{Aj}(x_j, t) = i_{\text{app}j}(x_j, t) + \sum_{m=1}^{M_j} g^m_{\text{syn}j}(x_j, t)v^m_{\text{rev}j}.$$

The terms $g^m_{\text{syn}j}(x_j, t)v^m_{\text{rev}j}$ have been effectively absorbed into a general applied current density term i_{Aj}.

Nondimensionalization follows Equation (6.8) and introducing

$$r_{\text{m}j}\hat{g}_j = g_j, \qquad r_{\text{m}j}i_{Aj} = \bar{V}I_{Aj} \tag{6.30}$$

gives the dimensionless current terms of the form

$$I_j(X_j, T; V_j) = g_j(X_j, T)V_j(X_j, T) + I_{Aj}(X_j, T). \tag{6.31}$$

The scaling $\bar{V}$ is based on a characteristic value for the general applied current density term i_{Aj} (Example 1 below explains this further). Different resistance scales may be appropriate for different terms in i_{Aj}. It is worth remarking that if $i_{\text{ion}j} = i_{\text{app}j} = 0$ but $i_{\text{syn}j} \neq 0$, then the characteristic value for $\bar{V}$ may be based on reversal potentials. Assuming a zero initial voltage distribution but retaining the somatic current input, Equation (6.15) becomes

$$\begin{aligned} V_j(X_j, T) = &\sum_{k=1}^{N} \int_0^{L_k} \int_0^T G^k_j(X_j, \xi_k, T-u)g_k(\xi_k, u)V_k(\xi_k, u)\,\text{d}u\,\text{d}\xi_k \\ &+ \sum_{k=1}^{N} \int_0^{L_k} \int_0^T G^k_j(X_j, \xi_k, T-u)I_{Ak}(\xi_k, u)\,\text{d}u\,\text{d}\xi_k \\ &+ \frac{1}{\gamma_j}\int_0^T G^j_j(0, X_j, T-u)I_{\text{s}}(u)\,\text{d}u, \end{aligned} \tag{6.32}$$

for $j = 1, \ldots, N$. This system may be conveniently expressed using matrix and integral operator notation as

$$\text{V} = \text{f} + \mathcal{A}\text{V}, \tag{6.33}$$

where Equation (6.32) represents the jth entry of Equation (6.33), explicitly

$$
\begin{aligned}
(\mathrm{V})_j &= V_j(X_j, T), \\
(\mathrm{f})_j &= f_j(X_j, T) = \sum_{k=1}^{N} \int_0^{L_k} \int_0^T G_j^k(X_j, \xi_k, T-u) I_{Ak}(\xi_k, u)\, \mathrm{d}u\, \mathrm{d}\xi_k \\
&\quad + \frac{1}{\gamma_j} \int_0^T G_j^j(0, X_j, T-u) I_{\mathrm{s}}(u)\, \mathrm{d}u, \\
(\mathcal{A}\mathrm{V})_j &= \sum_{k=1}^{N} (\mathcal{A}_{jk} V_k)(X_j, T).
\end{aligned}
\tag{6.34}
$$

Here, $\mathcal{A}_{jk}$ denotes the linear integral operator

$$
\mathcal{A}_{jk} = \int_0^{L_k} \int_0^T G_j^k(X_j, \xi_k, T-u) g_k(\xi_k, u)\, \mathrm{d}u\, \mathrm{d}\xi_k, \tag{6.35}
$$

which represents the jkth entry in matrix integral operator $\mathcal{A}$ such that

$$
(\mathcal{A}_{jk} V_k)(X_j, T) = \int_0^{L_k} \int_0^T G_j^k(X_j, \xi_k, T-u) g_k(\xi_k, u) V_k(\xi_k, u)\, \mathrm{d}u\, \mathrm{d}\xi_k.
$$

The summation convention has not been used in Equation (6.34) in order to aid clarity. Proceeding formally, the Neumann series solution to Equation (6.33) can be expressed as

$$
\mathrm{V} = \mathrm{f} + \mathcal{R}\mathrm{f} \tag{6.36}
$$

where $\mathcal{R}$ is the resolvent matrix integral operator

$$
\mathcal{R} = \sum_{n=1}^{\infty} \mathcal{A}^n \tag{6.37}
$$

with associated kernel termed the resolvent kernel. Thus, explicitly,

$$
V_j(X_j, T) = f_j(X_j, T) + \sum_{k=1}^{N} (\mathcal{R}_{jk} f_k)(X_j, T), \tag{6.38}
$$

where

$$
(\mathcal{R}_{jk} f_k)(X_j, T) = \sum_{n=1}^{\infty} \int_0^{L_k} \int_0^T \mathcal{K}^n(X_j, \xi_k; T, u) f_k(\xi_k, u)\, \mathrm{d}u\, \mathrm{d}\xi_k, \tag{6.39}
$$

and the iterated kernels are defined recursively by

$$
\mathcal{K}^n(X_j, \xi_k; T, u) = \sum_{i=1}^{N} \int_0^{L_i} \int_u^T G_j^i(X_j, Y_i, T-s) g_i(Y_i, s) \mathcal{K}^{n-1}(Y_i, \xi_k; s, u)\, \mathrm{d}s\, \mathrm{d}Y_i, \tag{6.40}
$$

with

$$
\mathcal{K}^1(X_j, \xi_k; T, u) = G_j^k(X_j, \xi_k, T-u) g_k(\xi_k, u), \tag{6.41}
$$

that is, $\mathcal{K}^0(Y_i, \xi_k; s, u) = \delta(Y_i - \xi_k)\delta(s-u)$ (the identity). $\mathcal{K}^n$ is the kernel associated with the operator $\mathcal{A}^n$. In unfolding the iterates (Equation [6.40]) it should be remembered that $G_j^k(X_j, \xi_k, \tau) = 0$

for $\tau < 0$, which places suitable limits on the time ranges of integration. Explicitly, Equation (6.40) may be written for $n \geq 2$ as

$$\mathcal{K}^n(X_j, \xi_k; T, u) = \sum_{i_1=1}^{N} \int_{Y_{i_1}=0}^{L_{i_1}} \int_{s_{i_1}=u}^{T} \sum_{i_2=1}^{N} \int_{Y_{i_2}=0}^{L_{i_2}} \int_{s_{i_2}=u}^{s_{i_1}} \cdots \sum_{i_{n-1}=1}^{N} \int_{Y_{i_{n-1}}=0}^{L_{i_{n-1}}} \int_{s_{i_{n-1}}=u}^{s_{i_{n-2}}} \mathcal{K}^1(X_j, Y_{i_1}; T, s_{i_1}) \mathcal{K}^1(Y_{i_1}, Y_{i_2}; s_{i_1}, s_{i_2}) \mathcal{K}^1(Y_{i_2}, Y_{i_3}; s_{i_2}, s_{i_3}) \cdots \times \mathcal{K}^1(Y_{i_{n-1}}, \xi_k; s_{i_{n-1}}, u) \mathrm{d}Y_{i_1}\, \mathrm{d}Y_{i_2} \cdots \mathrm{d}Y_{i_{n-1}}\, \mathrm{d}s_{i_1}\, \mathrm{d}s_{i_2} \cdots \mathrm{d}s_{i_{n-1}}. \tag{6.42}$$

Under suitable restrictions on the synaptic conductances (i.e., boundedness conditions), it is straightforward to show that the Neumann series converges absolutely and uniformly. Assuming for each j the uniform bounds

$$|g_j(X_j, T)| < M \qquad \text{and} \qquad |f_j(X_j, T)| < m, \tag{6.43}$$

where m and M are both constants, then Equation (6.41) gives

$$\left| \int_0^{L_k} \int_0^T \mathcal{K}^1(X_j, \xi_k; T, u) f_k(\xi_k, u) \mathrm{d}u \mathrm{d}\xi_k \right| \leq mM \int_0^{L_k} \int_0^T G_j^k(X_j, \xi_k, T-u) \mathrm{d}\xi_k \mathrm{d}_u \leq mMT,$$

on using $G_j^k(X_j, \xi_k, T) \geq 0$ and $\int_0^{L_j} G_j^k(X_j, \xi_k, u)\mathrm{d}X_j \leq 1$ as well $\int_0^{L_k} \int_0^T G_j^k(X_j, \xi_k, T-u)\mathrm{d}\xi_k \leq 1$ (due to the definition of G_j^k as the response to a unit-impulse input and the symmetry of the Green's function; in fact, it may be shown directly from the boundary-value problem satisfied by G_j^k that $Q = \sum_{j=1}^{N} \int_0^{L_j} \gamma_j G_j^k(X_j, Y_k, T)\mathrm{d}X_j + G_j^k(0, Y_k, T) = \gamma_k e^{-T}$). Similarly Equation (6.42) gives

$$\left| \int_0^{L_k} \int_0^T \mathcal{K}^n(X_j, \xi_k; T, u) f_k(\xi_k, u) \mathrm{d}u \mathrm{d}\xi_k \right| \leq mM^n N^{n-1} \frac{T^n}{n!}, \tag{6.44}$$

after using the integral mean value theorem, bounding the spatial integrals by unity and then evaluating and bounding the time integrals. It follows that

$$\left| \sum_{n=1}^{\infty} \int_0^{L_k} \int_0^T \mathcal{K}^n(X_j, \xi_k; T, u) f_k(\xi_k, u) \mathrm{d}\xi_k \mathrm{d}u \right| \leq \sum_{n=1}^{\infty} \int_0^{L_k} \int_0^T |\mathcal{K}^n(X_j, \xi_k; T, u)| |f_k(\xi_k, u)| \mathrm{d}\xi_k \mathrm{d}u < mMTe^{MNT} \tag{6.45}$$

as required.

From a computational perspective, it is useful to obtain bounds on the error made in approximating Equation (6.36) by truncating the resolvent kernel after a certain number of terms. Denoting

$$V_j^{n_{\max}}(X_j, T) = f_j(X_j, T) + \sum_{k=1}^{N} \int_0^{L_k} \int_0^T \sum_{n=1}^{n_{\max}} \mathcal{K}^n(X_j, \xi_k; T, u) f_k(\xi_k, u)\, \mathrm{d}u\, \mathrm{d}\xi_k, \tag{6.46}$$

then

$$|V_j(X_j, T) - V_j^{n_{\max}}(X_j, T)| = \left| \sum_{k=1}^{N} \int_0^{L_k} \int_0^T \sum_{n_{\max}+1}^{\infty} \mathcal{K}^n(X_j, \xi_k; T, u) f_k(\xi_k, u)\, \mathrm{d}u\, \mathrm{d}\xi_k \right| < m \frac{(MNT)^{n_{\max}+1}}{(n_{\max}+1)!} \mathrm{e}^{MNT}. \tag{6.47}$$

Thus $|V_j(X_j,T) - V_j^{n_{\max}}(X_j,T)| \to 0$ as $n_{\max} \to \infty$ and hence Equation (6.47) gives an indication of the number of terms $n_{\max}$ required in order to achieve a specified accuracy.

It is worth remarking on the structure of the solution (6.36). The solution is explicit with the first term on the right-hand side representing the solution to the problem with voltage-independent forcing terms. The effect of the voltage-dependent forcing terms is represented by the resolvent kernel in the second term. This latter term determines whether the resulting waveform is accentuated or reduced relative to the voltage-independent forcing case.

We now describe two relevant situations in which the above expressions simplify.

1. If the synaptic reversal potential or ionic input is limited to a single equivalent cylinder then the system (Equation [6.32]) decouples. For example, suppose that the synaptic reversal potential or ionic input is confined to cylinder 1 only (i.e., $g_j(X_j,T) = 0$ unless $j = 1$); then Equation (6.32) gives the following integral equation for V_1,

$$V_1(X_1,T) = f_1(X_1,T) + \int_0^{L_1}\int_0^T G_1^1(X_1,\xi_1,T-u)g_1(\xi_1,u)V_1(\xi_1,u)\,\mathrm{d}u\,\mathrm{d}\xi_1. \tag{6.48}$$

The potential in the remaining cylinders $j \neq 1$ is then given by

$$V_j(X_j,T) = f_j(X_j,T) + \int_0^{L_1}\int_0^T G_j^1(X_j,\xi_1,T-u)g_1(\xi_1,u)V_1(\xi_1,u)\,\mathrm{d}u\,\mathrm{d}\xi_1, \tag{6.49}$$

using the solution from Equation (6.48), that is, the potential in cylinder 1 now enters as an additional forcing term in the solution expression for the other cylinders. Since $g_j(X_j,T) = 0$ for $j \neq 1$ then Equations (6.40) and (6.41) reduce to

$$\mathcal{K}^1(X_j,\xi_k;T,u) = G_j^1(X_j,\xi_1,T-u)g_1(\xi_1,u)\delta_{k1}, \tag{6.50}$$

$$\mathcal{K}^n(X_j,\xi_k;T,u) = \int_0^{L_1}\int_u^T G_j^1(X_j,Y_1,T-s)g_1(Y_1,s)\mathcal{K}^{n-1}(Y_1,\xi_1;s,u)\,\mathrm{d}s\,\mathrm{d}Y_1\delta_{k1}, \tag{6.51}$$

the Kronecker delta emphasizing that these expressions are zero when $k \neq 1$. The explicit expression (6.42) reduces to

$$\begin{aligned}\mathcal{K}^n(X_j,\xi_1;T,u) = {} & \int_{Y_1=0}^{L_1}\int_{s_1=u}^{T}\int_{Y_2=0}^{L_1}\int_{s_2=u}^{s_1}\cdots\int_{Y_{n-1}=0}^{L_1}\int_{s_{n-1}=u}^{s_{n-2}} \\ & \mathcal{K}^1(X_j,Y_1;T,s_1)\mathcal{K}^1(Y_1,Y_2;s_1,s_2)\mathcal{K}^1(Y_2,Y_3;s_2,s_3)\cdots\mathcal{K}^1(Y_{n-1},\xi_1;s_{n-1},u) \\ & \times \mathrm{d}Y_1\,\mathrm{d}Y_2\cdots\mathrm{d}Y_{n-1}\,\mathrm{d}s_1\,\mathrm{d}s_2\cdots\mathrm{d}s_{n-1}\end{aligned} \tag{6.52}$$

and is zero for $k \neq 1$. The solution (6.36) becomes

$$V_j(X_j,T) = f_j(X_j,T) + \sum_{n=1}^{\infty}\int_0^{L_1}\int_0^T \mathcal{K}^n(X_j,\xi_1;T,u)f_1(\xi_1,u)\,\mathrm{d}u\,\mathrm{d}\xi_1. \tag{6.53}$$

It is worth remarking that for $j \neq 1$, Equation (6.53) is consistent with Equation (6.49) since Equation (6.51) gives

$$\mathcal{K}^n(X_j,\xi_1;T,u) = \int_{Y_1=0}^{L_1}\int_{s=u}^{T}\mathcal{K}^1(X_j,Y_1;T,s)\mathcal{K}^{n-1}(Y_1,\xi_1;s,u)\,\mathrm{d}s\,\mathrm{d}Y_1, \tag{6.54}$$

and thus

$$
\begin{aligned}
V_j(X_j,T) =& f_j(X_j,T) + \int_{\xi_1=0}^{L_1}\int_{u=0}^{T}\sum_{n=1}^{\infty}\mathcal{K}^n(X_j,\xi_1;T,u)f_1(\xi_1,u)\,\mathrm{d}u\,\mathrm{d}\xi_1 \\
=& f_j(X_j,T) + \int_{Y_1=0}^{L_1}\int_{s=0}^{T}\mathcal{K}^1(X_j,Y_1;T,s)\int_{\xi_1=0}^{L_1}\int_{u=0}^{s}\sum_{n=1}^{\infty}\mathcal{K}^{n-1}(Y_1,\xi_1;s,u) \\
&\times f_1(\xi_1,u)\,\mathrm{d}u\,\mathrm{d}\xi_1\,\mathrm{d}s\,\mathrm{d}Y_1 \\
=& f_j(X_j,T) + \int_{Y_1=0}^{L_1}\int_{s=u}^{T}\mathcal{K}^1(X_j,Y_1;T,s)V_1(Y_1,s)\,\mathrm{d}s\,\mathrm{d}Y_1,
\end{aligned}
\tag{6.55}
$$

remembering that $\mathcal{K}^0$ is a product of delta functions and interchange of summation and integration is permissible since the series for the resolvent kernel is absolutely (and uniformly) convergent.

2. Simplification also occurs for synaptic inputs that are discretely distributed, where the integral equations reduce to those of Volterra type. For example, following Poznanski and Bell (2000a), we may assume for the jth equivalent cylinder synaptic conductances of the functional form

$$
g_j(X_j,T) = \sum_{i=1}^{N_j} g_j^i(T)\delta(X_j - X_{0j}^i), \tag{6.56}
$$

where N_j denotes the number of synaptic sites located at $\{X_{0j}^i\}_{i=1}^{N_j}$ along the jth cylinder. For generality, the time course $g_j^i(T)$ has been allowed to differ at each location X_{0j}^i. The same authors suggest taking a time course at each synaptic location of the form

$$
g_j^i(T) = c_j^i + b_j^i\,\mathrm{e}^{-a_j^i T} \tag{6.57}
$$

with a_j^i, b_j^i, c_j^i as suitable constants. Such a functional dependence may be derived from a linearization of the ion activation or inactivation variables.

The system (6.32) then reduces to

$$
V_j(X_j,T) = f_j(X_j,T) + \sum_{k=1}^{N}\sum_{i_k=1}^{N_k}\int_0^T G_j^k(X_j,X_{0k}^{i_k},T-u)g_k^{i_k}(u)V_k(X_{0k}^{i_k},u)\,\mathrm{d}u, \tag{6.58}
$$

where (6.35) becomes the Volterra-type operator

$$
\mathcal{A}_{jk} = \sum_{i_k=1}^{N_k}\int_0^T G_j^k(X_j,X_{0k}^{i_k},T-u)g_k^{i_k}(u)\,\mathrm{d}u. \tag{6.59}
$$

The solution (6.36) becomes

$$
V_j(X_j,T) = f_j(X_j,T) + \sum_{k=1}^{N}\sum_{p_k=1}^{N_k}\int_0^T\sum_{n=1}^{\infty}\mathcal{K}^n(X_j,X_{0k}^{p_k};T,u)f_k(X_{0k}^{p_k},u)\,\mathrm{d}u, \tag{6.60}
$$

and the iterated kernels are defined recursively by

$$\mathcal{K}^n(X_j, X_{0k}^{p_k}; T, u) = \sum_{i=1}^{N} \sum_{p_i=1}^{N_i} \int_u^T G_j^i(X_j, X_{0i}^{p_i}, T - s) g_i^{p_i}(s) \mathcal{K}^{n-1}(X_{0i}^{p_i}, X_{0k}^{p_k}; s, u)\, \mathrm{d}s \tag{6.61}$$

with

$$\mathcal{K}^1(X_j, X_{0k}^{p_k}; T, u) = G_j^k(X_j, X_{0k}^{p_k}, T - u) g_k^{p_k}(u), \tag{6.62}$$

that is, $\mathcal{K}^0(X_j, X_{0k}^{p_k}; s, u) = \delta(X_j - X_{0k}^{p_k})\delta(s - u)$. For $n \geq 2$, Equation (6.42) becomes

$$\begin{aligned}
\mathcal{K}^n(X_j, X_{0k}^{p_k}; T, u) = {} & \sum_{i_1=1}^{N} \sum_{p_{i_1}=1}^{N_{i_1}} \int_{s_{i_1}=u}^{T} \sum_{i_2=1}^{N} \sum_{p_{i_2}=1}^{N_{i_2}} \int_{s_{i_2}=u}^{s_{i_1}} \cdots \sum_{i_{n-1}=1}^{N} \sum_{p_{i_{n-1}}=1}^{N_{i_{n-1}}} \int_{s_{i_{n-1}}=u}^{s_{i_{n-2}}} \mathcal{K}^1(X_j, X_{0i_1}^{p_{i_1}}; T, s_{i_1}) \\
& \times \mathcal{K}^1(X_{0i_1}^{p_{i_1}}, X_{0i_2}^{p_{i_2}}; s_{i_1}, s_{i_2}) \mathcal{K}^1(X_{0i_2}^{p_{i_2}}, X_{0i_3}^{p_{i_3}}; s_{i_2}, s_{i_3}) \cdots \\
& \times \mathcal{K}^1(X_{0i_{n-1}}^{p_{i_{n-1}}}, X_{0k}^{p_k}; s_{i_{n-1}}, u)\, \mathrm{d}s_{i_1}\, \mathrm{d}s_{i_2} \cdots \mathrm{d}s_{i_{n-1}}.
\end{aligned} \tag{6.63}$$

In the single-cylinder case, Equation (6.60) reduces to

$$V_1(X_1, T) = f_1(X_1, T) + \sum_{p_1=1}^{N_1} \int_0^T \sum_{n=1}^{\infty} \mathcal{K}^n(X_1, X_{01}^{p_1}; T, u) f_1(X_{01}^{p_1}, u)\, \mathrm{d}u \tag{6.64}$$

with

$$\mathcal{K}^1(X_1, X_{01}^{p_1}; T, u) = G_1^1(X_1, X_{01}^{p_1}, T - u) g_1^{p_1}(u) \tag{6.65}$$

and for $n \geq 2$,

$$\begin{aligned}
\mathcal{K}^n(X_1, X_{01}^{p_k}; T, u) = {} & \sum_{p_1=1}^{N_1} \int_{s_1=u}^{T} \sum_{p_2=1}^{N_1} \int_{s_2=u}^{s_1} \cdots \sum_{p_{n-1}=1}^{N_1} \int_{s_{n-1}=u}^{s_{n-2}} \mathcal{K}^1(X_1, X_{01}^{p_1}; T, s_1) \\
& \times \mathcal{K}^1(X_{01}^{p_1}, X_{01}^{p_2}; s_1, s_2) \mathcal{K}^1(X_{01}^{p_2}, X_{01}^{p_3}; s_2, s_3) \cdots \mathcal{K}^1(X_{01}^{p_{n-1}}, X_{01}^{p_k}; s_{n-1}, u) \\
& \times \mathrm{d}s_1\, \mathrm{d}s_2 \cdots \mathrm{d}s_{n-1}.
\end{aligned} \tag{6.66}$$

This expression for the iterated kernels differs from that given in Poznanski and Bell (2000a), by the inclusion here of repeated summation over the discrete locations of the synaptic input sites. However, it should be noted that this does not change the subsequent computational results in their paper since they only employed a single kernel approximation.

6.2.3.1 Example 1

As an example of the above theory and to illustrate nondimensionalization, we consider a nerve cell represented by a two-cylinder model with dimensional parameters

$$\begin{aligned}
& R_{\mathrm{m}} = 50{,}000\,\Omega\,\mathrm{cm}^2, \quad C_{\mathrm{m}} = 1\,\mu\mathrm{F\,cm}^{-2}, \quad R_i = 200\,\Omega\,\mathrm{cm}, \quad 1/R_{\mathrm{sh}} = 0, \\
& A_{\mathrm{s}} = 2 \times 10^{-4}\,\mathrm{cm}^2, \quad d_{\mathrm{cyl1}} = 4\,\mu\mathrm{m}, \quad \ell_{\mathrm{cyl1}} = 0.237\,\mathrm{cm}, \quad d_{\mathrm{cyl2}} = 4\,\mu\mathrm{m}, \quad \ell_{\mathrm{cyl2}} = 0.158\,\mathrm{cm}.
\end{aligned}$$

Thus $\lambda_1 = \lambda_2 = 0.158$ cm, $r_{a1} = r_{a2} = 1.59155 \times 10^9\ \Omega\,\text{cm}^{-1}$, $r_{m1} = r_{m2} = 3.97887 \times 10^7\ \Omega\,\text{cm}$, $\tau_m = \tau_s = 50$ msec.

Reverting to the representation of synaptic stimuli as currents, synaptic stimulus is applied to cylinder 1 at the location $x_1 = y_1$ cm in the form of the following alpha function:

$$i_{\text{app1}}(x_1, t) = \hat{b}t\,\mathrm{e}^{-\hat{a}t}\delta(x_1 - y_1), \qquad i_{\text{app2}}(x_2, t) = 0,$$

where $\hat{a}$ (sec^{-1}) is a parameter that controls the breadth of the time course and $\hat{b}/\hat{a}^2 = \int_0^\infty \hat{b}t\,\mathrm{e}^{-\hat{a}t}\,\mathrm{d}t$ represents the total charge delivered. Consequently, we take $i_{\text{syn}j} = 0$. It should be noted that the dimensions of $\delta(x_1 - y_1)$ are cm^{-1} (since $\delta(\lambda X) = \delta(X)/\lambda$ is one of the delta function's properties for scalar λ). We also consider the case of a discrete distribution of persistent sodium (Na^+P) channels located throughout both cylinders so that

$$\hat{g}_j(x_j, t) = -g_{\text{ion}j}(x_j, t) = -\sum_{i=1}^{N_j} \hat{g}_j^i(t)\delta(x_j - x_{0j}^i),$$

for $j = 1, 2$. The persistent-conductance time courses are taken in the form

$$\hat{g}_j^i(t) = \hat{c}_j - \hat{b}_j\,\mathrm{e}^{-\hat{a}_j t},$$

where $\hat{b}_j$ [S], $\hat{c}_j$ [S], and $\hat{a}_j$ (msec^{-1}) are constants associated with linearizing the first-order reaction equation satisfied by the ion activation variable at each location x_{0j}^i (termed a "hot spot"). Here, we have assumed that these constants may take different values in the two cylinders, although within each cylinder they are assumed the same. Based on similar values used by Poznanski and Bell (2000a), we take

$$\hat{b} = 1.24 \times 10^{-2}\ \mu\text{A/msec}, \quad \hat{a} = 0.25/\text{msec}, \quad \hat{a}_j = 1/\text{msec},$$

$$\hat{b}_j = 2.7\hat{g}_s^* \frac{N_j^*}{\pi d_{\text{cyl}j}}, \quad \hat{c}_j = \frac{0.056}{2.7}\hat{b}_j,$$

where $\hat{g}_s^* = 18\,pS$ is the maximum conductance of a single ion channel and N_j^* is the number of ion channels for each hot spot in cylinder j, that is, $N_j^*/\pi d_{\text{cyl}j}$ gives the number of persistent ion channels per unit circumference length on the membrane surface. Typical ranges for N_j^* are of the order 5 to 20. First, nondimensionalization follows Equation (6.8). Using the above values, Equation (6.9) gives the dimensionless parameters as:

cylinder 1	$\gamma_1 = 1,$	$L_1 = 1.5,$
cylinder 2	$\gamma_2 = 1,$	$L_1 = 1,$
soma	$\epsilon = 1.$	

The alpha-function input is used to provide the characteristic voltage scaling $\bar{V} = r_{m1}\hat{b}\tau_m/(\lambda_1\alpha^2)$, the dimensionless applied current being

$$I_{A1}(X_1, T) = \alpha^2 T\,\mathrm{e}^{-\alpha T}\delta(X_1 - Y_1), \qquad I_{A2}(X_2, T) = 0,$$

where

$$y_1 = \lambda_1 Y_1, \qquad \alpha = \hat{a}\tau_m.$$

If the above typical values are adopted then $\alpha = 12.5$ and $\bar{V} = 10^3$ mV. (The dimensionless factor α^2 has been introduced into the dimensionless applied current for convenience and implies that the total dimensionless current delivered is unity. The size of $\hat{b}$ was conveniently chosen so that $\bar{V} = O(10^3 \text{ mV})$.) Using Equation (6.30), the dimensionless synaptic conductances are of the form (6.56), where

$$g_j^i(T) = N_j^*(c - b\,\mathrm{e}^{-\beta_j T}), \tag{6.67}$$

(the same for each i), introducing the additional dimensionless parameters

$$\beta_j = \hat{a}_j \tau_{\mathrm{m}}, \qquad N_j^* b = \frac{r_{\mathrm{m}j}\hat{b}_j}{\lambda_j}, \qquad N_j^* c = \frac{r_{\mathrm{m}j}\hat{c}_j}{\lambda_j}.$$

On using the above typical values we obtain $\beta_j = 50,\ b = 10,\ c = 0.2$. It is worth remarking that $\beta_j/\alpha = 4$. We are interested in quantifying the effect of the persistent sodium channels (their number N_j, distribution X_{0j}^i, and strength N_j^*) on the resulting EPSPs recorded at the soma.

The solution (6.60) gives

$$V_1(X_1, T) = f_1(X_1, T) + \sum_{k=1}^{2}\sum_{p_k=1}^{N_k}\int_0^T \sum_{n=1}^{\infty} \mathcal{K}^n(X_1, X_{0k}^{p_k}; T, u) f_k(X_{0k}^{p_k}, u)\,\mathrm{d}u, \tag{6.68}$$

$$V_2(X_2, T) = f_2(X_2, T) + \sum_{k=1}^{2}\sum_{p_k=1}^{N_k}\int_0^T \sum_{n=1}^{\infty} \mathcal{K}^n(X_2, X_{0k}^{p_k}; T, u) f_k(X_{0k}^{p_k}, u)\,\mathrm{d}u, \tag{6.69}$$

where

$$f_1(X_1, T) = \alpha^2 \int_0^T G_1^1(X_1, Y_1, T-u) u\,\mathrm{e}^{-\alpha u}\,\mathrm{d}u, \qquad f_2(X_2, T) = \alpha^2 \int_0^T G_2^1(X_2, Y_1, T-u) u\,\mathrm{e}^{-\alpha u}\,\mathrm{d}u. \tag{6.70}$$

The kernels are given by

$$\mathcal{K}^1(X_j, X_{0k}^{p_k}; T, u) = G_j^k(X_j, X_{0k}^{p_k}, T-u) g_k^{p_k}(u) \tag{6.71}$$

with the $g_k^{p_k}$ given by Equation (6.67) and for $n \geq 2$,

$$\begin{aligned}\mathcal{K}^n(X_j, X_{0k}^{p_k}; T, u) = {} & \sum_{i_1=1}^{2}\sum_{p_{i_1}=1}^{N_{i_1}}\int_{s_{i_1}=u}^{T} \sum_{i_2=1}^{2}\sum_{p_{i_2}=1}^{N_{i_2}}\int_{s_{i_2}=u}^{s_{i_1}} \cdots \sum_{i_{n-1}=1}^{2}\sum_{p_{i_{n-1}}=1}^{N_{i_{n-1}}}\int_{s_{i_{n-1}}=u}^{s_{i_{n-2}}} \\ & \times \mathcal{K}^1(X_j, X_{0i_1}^{p_{i_1}}; T, s_{i_1}) \mathcal{K}^1(X_{0i_1}^{p_{i_1}}, X_{0i_2}^{p_{i_2}}; s_{i_1}, s_{i_2}) \mathcal{K}^1(X_{0i_2}^{p_{i_2}}, X_{0i_3}^{p_{i_3}}; s_{i_2}, s_{i_3}) \cdots \\ & \times \mathcal{K}^1(X_{0i_{n-1}}^{p_{i_{n-1}}}, X_{0k}^{p_k}; s_{i_{n-1}}, u)\,\mathrm{d}s_{i_1}\,\mathrm{d}s_{i_2} \cdots \mathrm{d}s_{i_{n-1}}.\end{aligned} \tag{6.72}$$

The terms in Equation (6.70) may be explicitly evaluated using the Green's function to give

$$f_j(X_j, T) = \sum_{n=0}^{\infty} E_{1n}\psi_{jn}(X_j)\psi_{1n}(Y_1)\left(\frac{\alpha}{1+\alpha_n^2-\alpha}\right)^2 (\mathrm{e}^{-(1+\alpha_n^2)T} + ((1+\alpha_n^2-\alpha)T - 1)\,\mathrm{e}^{-\alpha T})$$

a necessary restriction being $\alpha \neq 1 + \alpha_n^2$.

For illustrative purposes, we evaluate the solution at soma $X_1 = X_2 = 0$ and consider the hot spots either uniformly distributed throughout the cylinder $\{X^i_{0j} = iL_j/N_j\}_{i=1}^{N_j}$ or located proximally $\{X^i_{0j} = 0.2iL_j/N_j\}_{i=1}^{N_j}$ near to the soma. The number of persistent sodium channels per hot spot N^*_j is kept constant at 20. Figure 6.4 shows the results for a proximally placed input $Y_1 = 0.1L_1$ in Figure 6.4A and C and distally placed input at $Y_1 = 0.9L_1$ in Figure 6.4B and D. Figure 6.4A and B have the number of hot spots N_j fixed at 5 while Figure 6.4C and D consider double the number of hot spots $N_j = 10$. In each of the figures the hot-spot locations were taken in only one of the cylinders to clarify their effect. The total number of persistent sodium channels $\sum_{j=1}^{N} N_j N^*_j$ is fixed at 100 in Figure 6.4A and B and 200 in C and D. Figure 6.5 shows the effect of hot spots located in both cylinders.

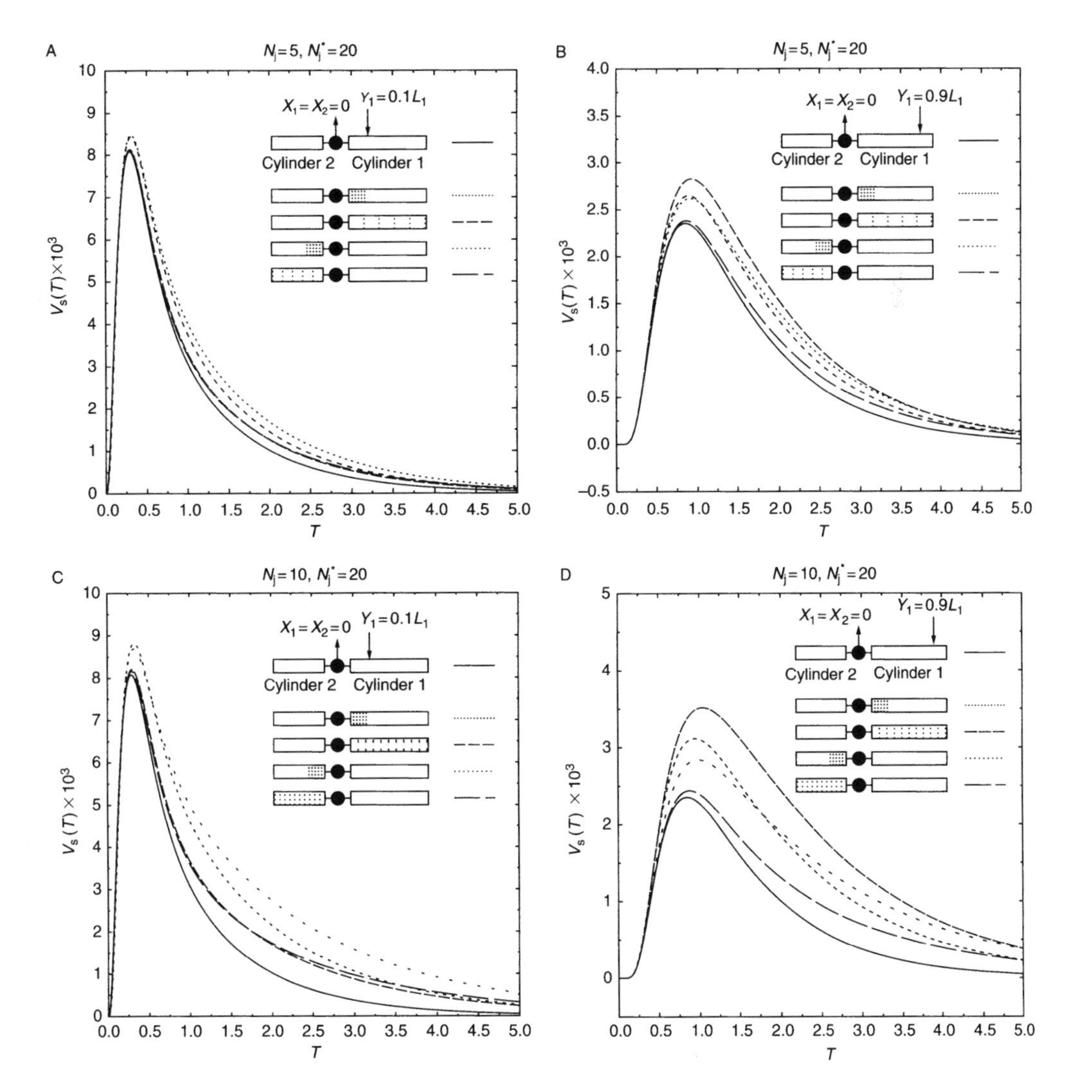

FIGURE 6.4 The effect of persistent sodium hot spots on the amplification of EPSPs for the two-cylinder model of Example 1. The voltage at the soma $V_S(T) = V_1(0,T) = V_2(0,T)$ is shown for an alpha-function input located in cylinder 1 at $Y_1 = 0.1L_1$ in (A) and (C) and at $Y_1 = 0.9L_1$ in (B) and (D). Each figure shows the solution with and without persistent sodium ion channels present, the location and distribution being indicated in each by the inset cylinder model schematics. (A) and (B) have five hot-spot locations $N_j = 5$, with $N^*_j = 20$ the number of persistent sodium channels per hot spot. (C) and (D) give the corresponding results for ten hot-spot locations $N_j = 10$ with the same number, $N^*_j = 20$, of persistent sodium channels per hot spot.

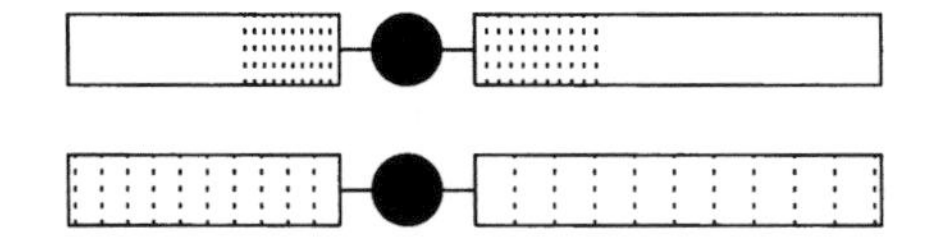

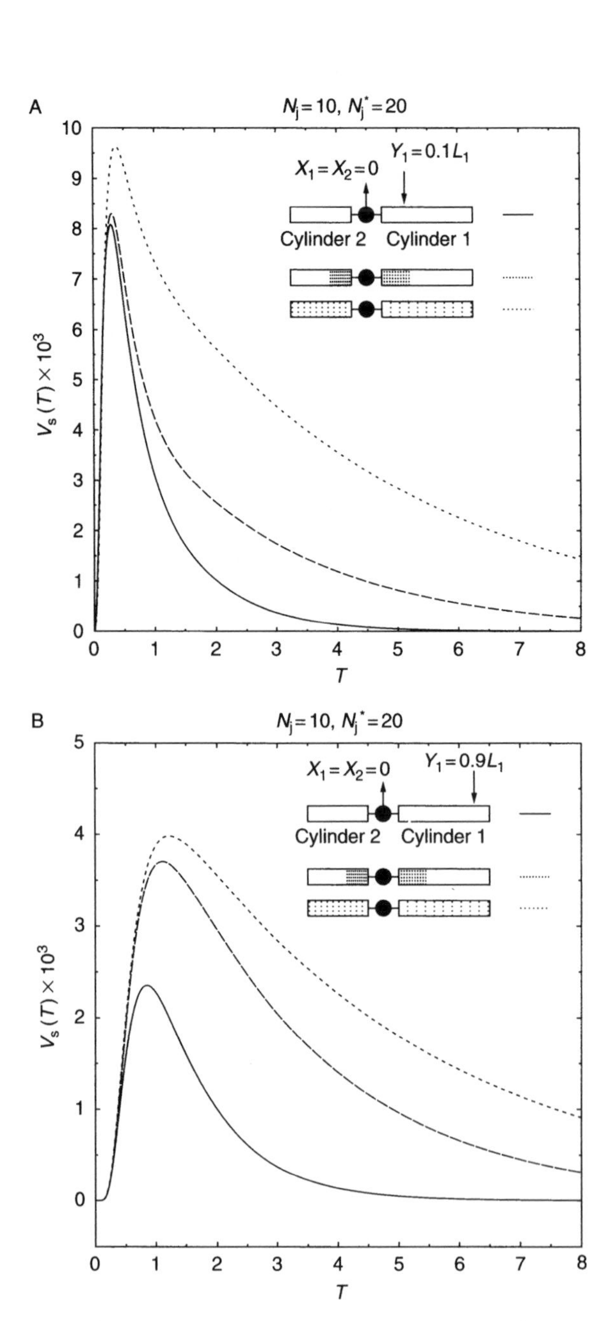

FIGURE 6.5 The effect of persistent sodium hot spots on the amplification of EPSPs for the two-cylinder model of Example 1. The voltage at the soma $V_S(T) = V_1(0, T) = V_2(0, T)$ is shown for an alpha-function input located in cylinder 1 at $Y_1 = 0.1L_1$ in (A) and at $Y_1 = 0.9L_1$ in (B). Each figure shows the solution with and without persistent sodium ion channels present, the location and distribution being indicated in each by the inset cylinder model schematics. The total number of hot-spot locations $N_j = 10$ is fixed for each cylinder with $N_j^* = 20$ the number of persistent sodium channels per hot spot (the total number of persistent sodium channels over the two cylinders is fixed at 400).

6.3 The Multicylinder Model with Taper

6.3.1 The Mathematical Problem

We consider a neuron composed of a soma and N' dendritic trees and assume that each dendritic tree may either be reduced to an equivalent cylinder with distinct taper or combined with other dendritic trees and then reduced to an equivalent cylinder with distinct taper (see Rall, 1962b; Jack et al., 1983). As in Section 6.2, the nerve cell is represented by N (where $1 \leq N \leq N'$) equivalent tapering cylinders not necessarily of the same physical or electrotonic length emanating from a uniformly polarized soma.

The jth equivalent tapering cylinder, now of physical length ℓ_j rather than $l_{\text{cyl}j}$ as used earlier, represents one or more dendritic structures in which $r_j = r_j(x_j)$ is the common branch radius at distance x_j and $n_j = n_j(x_j)$ denotes the number of branches at distance x_j. Accordingly, the transmembrane potential $v_j(x_j, t)$ satisfies the following dimensional governing equation (cf. Jack et al., 1983; Poznanski, 1988),

$$v_j + \tau_{\text{m}} \frac{\partial v_j}{\partial t} = \lambda_{\text{taper}j}^2 \left[\frac{\partial^2 v_j}{\partial x_j^2} + \frac{\partial v_j}{\partial x_j} \frac{\text{d}}{\text{d}x_j} [\ell n(r_j^2 n_j)] \right], \tag{6.73}$$

where $\lambda_{\text{taper}j}$ is a continuously changing characteristic length function defined as

$$\lambda_{\text{taper}j} = \lambda_{0j} \left(\frac{r_j}{r_{0j}} \right)^{1/2} \left(\frac{\text{d}s_j}{\text{d}x_j} \right)^{-1/2}, \tag{6.74}$$

$$\lambda_{0j} = \left(\frac{R_{\text{m}} r_{0j}}{2R_i} \right)^{1/2}, \tag{6.75}$$

is the characteristic length of a cylinder with radius r_{0j} $(= r_j(0)$, the initial radius at $x_j = 0)$ and

$$\frac{\text{d}s_j}{\text{d}x_j} = \left[1 + \left(\frac{\text{d}r_j}{\text{d}x_j} \right)^2 \right]^{1/2}. \tag{6.76}$$

The membrane time constant τ_{m}, axial resistivity R_i, and membrane specific resistance R_{m} are as defined in Section 6.2.

A convenient change of variable is given by

$$\frac{\text{d}Z_j}{\text{d}x_j} = \frac{1}{\lambda_{\text{taper}j}}, \tag{6.77}$$

with $Z_j = 0$ at $x_j = 0$, for which the governing Equation (6.73) becomes

$$v_j + \tau_{\text{m}} \frac{\partial v_j}{\partial t} = \left[\frac{\partial^2 v_j}{\partial Z_j^2} + \frac{\partial v_j}{\partial Z_j} \frac{\text{d}}{\text{d}Z_j} \ln \left(r_j^{3/2} n_j \left(\frac{\text{d}s_j}{\text{d}x_j} \right)^{1/2} \right) \right], \tag{6.78}$$

and the electrotonic length of the jth equivalent cylinder is given by

$$L_j = \int_0^{\ell j} \frac{1}{\lambda_{\text{taper}j}(x_j)} \text{d}x_j. \tag{6.79}$$

We must prescribe r_j and n_j as suitable functions of x_j in order to specify the type of tapering and branching to be considered. For the jth equivalent tapering cylinder we write,

$$r_j^{3/2} n_j \left(\frac{\mathrm{d}s_j}{\mathrm{d}x_j}\right)^{1/2} = r_{0j}^{3/2} n_{0j}\, F_j(Z_j), \tag{6.80}$$

where $F_j(Z_j)$ is termed the geometric ratio function for the jth equivalent tapering cylinder; it imposes a taper on the equivalent cylinder. We will discuss later the forms of $F_j(Z_j)$ that will be considered; here it suffices to say that strictly tapering, strictly flaring, or a combination of both tapering and flaring can be accommodated. Now consider the case in which each dendritic branch is represented by a cylinder of uniform cross-section ($\mathrm{d}r_j/\mathrm{d}x_j = 0$) or taper at a small rate $(\mathrm{d}r_j/\mathrm{d}x_j)^2 \ll 1$), then using diameter instead of radius ($d_{0j} = 2r_{0j}$ and $d_j = 2r_j$), Equation (6.80) becomes

$$n_j d_j^{3/2} / n_{0j} d_{0j}^{3/2} = F_j(Z_j),$$

where $n_j d_j^{3/2}$ is termed the combined dendritic trunk parameter and $n_j d_j^{3/2}/n_{0j} d_{0j}^{3/2}$ the geometric ratio factor (GRF_j) for the jth dendritic structure.

As a note of practical application, we remark that for the jth dendritic structure,

$$n_j(x_j) d_j(x_j)^{3/2} = \sum_{k=1}^{n_j(x_j)} d_{j,k}(x_j)^{3/2},$$

and thus Equation (6.80) becomes

$$\sum_{k=1}^{n_j(x_j)} d_{j,k}(x_j)^{3/2} = \sum_{k=1}^{n_j(0)} d_{j,k}(0)^{3/2} F_j(Z_j),$$

where $d_{j,k}(x_j)$ is the diameter of the kth branch at a distance x_j from the soma (which is measured along branches between successive branch points). In our assumptions, we have that $d_{j,k}(x_j) = d_j(x_j)$ for each branch k at the distance x_j, that is, the branches have a common diameter.

When the $\mathrm{GRF}_j > 1$, the branching pattern for the dendritic tree or trees represented by the jth equivalent cylinder exhibits a wide range of profuseness, while if $\mathrm{GRF}_j < 1$, there is relative paucity of branching as discussed by Jack et al. (1983). If $\mathrm{GRF}_j = 1$ then the combined dendritic trunk parameter is constant, permitting the dendritic structure to be transformed into an equivalent cylinder that has diameter D_{0j}^* given by

$$D_{0j}^* = \left(\sum_{i=1}^{n_{0j}} d_{j,i}(0)^{3/2}\right)^{2/3},$$

and characteristic length parameter λ_{0j}^* given by

$$\lambda_{0j}^* = \left(\frac{R_\mathrm{m}}{4R_i} D_{0j}^*\right)^{1/2}.$$

When $\mathrm{GRF}_j \neq 1$, the diameter of the tapering equivalent cylinder, which changes continuously with electrotonic distance from the soma, is

$$D_j^* = \left(\sum_{i=1}^{n_j} d_{j,i}(x_j)^{3/2} \right)^{2/3} = D_{0j}^* F_j(Z_j)^{2/3},$$

and the characteristic length parameter is

$$\lambda_j^* = \left(\frac{R_\mathrm{m}}{4R_i} D_j^* \right)^{1/2} = \lambda_{0j}^* F_j(Z_j)^{1/3}.$$

Under the preceding assumptions, the mathematical statement for N general tapering equivalent cylinders subject to current input, sealed end conditions, and a uniformly polarized soma may be presented as follows:

$$\text{in } 0 < Z_j < L_j, \quad t > 0 \qquad v_j + \tau_\mathrm{m} \frac{\partial v_j}{\partial t} - \left(\frac{\partial^2 v_j}{\partial Z_j^2} + \frac{1}{F_j} \frac{\mathrm{d}F_j}{\mathrm{d}Z_j} \frac{\partial v_j}{\partial Z_j} \right) = \frac{R_\mathrm{m}}{\pi D_j^* \lambda_j^*} i_j(Z_j, t; v_j); \tag{6.81}$$

$$\text{at } Z_j = L_j \qquad \frac{\partial v_j}{\partial Z_j} = 0; \tag{6.82}$$

$$\text{at } Z_j = 0 \qquad v_j = v_\mathrm{s}; \tag{6.83}$$

$$\text{and} \qquad v_\mathrm{s} + \tau_\mathrm{s} \frac{\mathrm{d}v_\mathrm{s}}{\mathrm{d}t} - \sum_{j=1}^{N} \frac{R_\mathrm{s}}{\lambda_{0j}^* r_{a0j}^*} \frac{\partial v_j}{\partial Z_j} = R_\mathrm{s} i_\mathrm{s}(t); \tag{6.84}$$

$$\text{at } t = 0, \quad 0 \le Z_j \le L_j \qquad v_j = h_j(Z_j); \tag{6.85}$$

for $j = 1, \ldots, N$ with $r_{\mathrm{a}j}^* = 4R_i/\pi D_j^{*2} = r_{\mathrm{a}0j}^* F_j^{-4/3}$ the axial resistance per unit length of the jth equivalent tapering cylinder ($\Omega\,\mathrm{cm}^{-1}$), $r_{\mathrm{a}0j}^* = r_{\mathrm{a}j}^*(0) = 4R_i/\pi D_{0j}^{*2}$ the axial resistance per unit length of a uniform equivalent cylinder of diameter D_{0j}^*. The somatic time constant τ_s and soma resistance R_s are as defined in Section 6.2 and again i_s is an applied current to the soma, $h_j(Z_j)$ is the initial voltage in the jth cylinder, while $i_j(Z_j, t; v_j)$ denotes a current term for the jth cylinder, which allows for applied currents and synaptic reversal potentials. The model is illustrated in Figure 6.6.

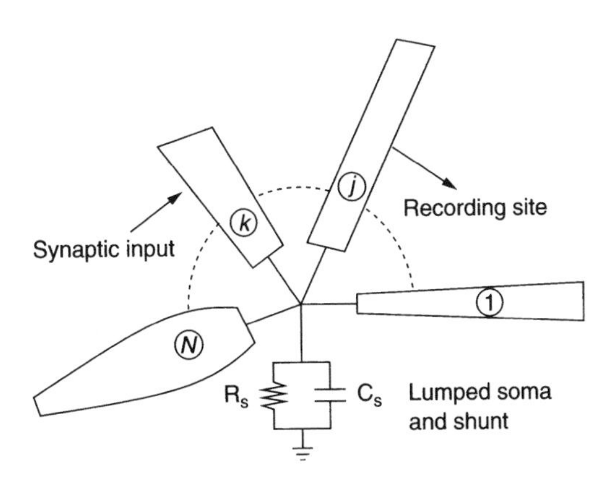

FIGURE 6.6 Illustration of the multicylinder model with N tapering equivalent cylinders emanating from a common soma. (Adapted from Evans, J.D. *IMA J. Math. Appl. Med. Biol.* 17, 347–377, 2000. With permission.)

6.3.2 Problem Normalization and General Solution

In addition to the dimensionless variables Z_j, the dimensionless parameters L_j, and the dimensionless functions $F_j(Z_j)$, we complete the nondimensionalization of Equations (6.81) to (6.85) with the scalings

$$t = \tau_m T, \quad v_j = \bar{I}\bar{R}V_j, \quad v_s = \bar{I}\bar{R}V_s, \quad h_j = \bar{I}\bar{R}H_j, \quad i_j = \bar{I}I_j, \quad i_s = \bar{I}I_s, \tag{6.86}$$

where $\bar{I}$ is an appropriate current scaling and $\bar{R}$ an appropriate resistance scaling. The choice for the current scaling $\bar{I}$ is determined by the type of input. For example, for a point charge input of Q_0 coulombs, we could take $\bar{I} = Q_0/\tau_m$, while for a constant current input of magnitude I, $\bar{I} = I$ is a possibility. There are several choices for $\bar{R}$, namely R_s or one of the $\{\lambda_{0j}^* r_{a0j}^*\}_{j=1}^N$. Here, we will take $\bar{R}$ to be $\lambda_{01}^* r_{a01}^*$, for definiteness. This then introduces the $(N+1)$ dimensionless parameters

$$\epsilon = \frac{\tau_s}{\tau_m}, \qquad \sigma_s = \frac{\lambda_{01}^* r_{a01}^*}{R_s}, \qquad \sigma_j = \frac{\lambda_{01}^* r_{a01}^*}{\lambda_{0j}^* r_{a0j}^*} = \left(\frac{D_{0j}^*}{D_{01}^*}\right)^{3/2}, \tag{6.87}$$

for $j = 1, \ldots, N$, where it should be noted that $\sigma_1 = 1$. The advantage of taking these scalings is that the case of a point soma may now be obtained by the continuous limit $\sigma_s \to 0$. In the previous section, $\bar{R} = R_s$ was taken as the appropriate resistance scaling, which gave rise to the N dimensionless parameters $\gamma_j = R_s/\lambda_{0j}^* r_{a0j}^* = \sigma_j/\sigma_s$. However, in this case the limit of a point soma requires care since the input currents now need appropriate scaling.

The system (6.81) to (6.85) becomes

$$\text{in } 0 < Z_j < L_j, \quad T > 0 \qquad V_j + \frac{\partial V_j}{\partial T} - \frac{\partial^2 V_j}{\partial Z_j^2} - \frac{1}{F_j}\frac{dF_j}{dZ_j}\frac{\partial V_j}{\partial Z_j} = \frac{1}{\sigma_j F_j(Z_j)} I_j(Z_j, T; V_j); \tag{6.88}$$

$$\text{at } Z_j = L_j \qquad \frac{\partial V_j}{\partial Z_j} = 0; \tag{6.89}$$

$$\text{at } Z_j = 0 \qquad V_j = V_s, \tag{6.90}$$

$$\text{and} \qquad \sigma_s\left(V_s + \epsilon\frac{dV_s}{dT}\right) - \sum_{j=1}^{N} \sigma_j \frac{\partial V_j}{\partial Z_j} = I_s(T); \tag{6.91}$$

$$\text{at } T = 0, \quad 0 \le Z_j \le L_j \qquad V_j = H_j(Z_j); \tag{6.92}$$

for $j = 1, \ldots, N$ and is illustrated in Figure 6.7. We now introduce the transformation

$$\hat{V}_j(Z_j, T) = F_j(Z_j)^{1/2} V_j(Z_j, T), \tag{6.93}$$

for the jth cylinder, noting that $F_j(0) = 1$ and $F_j(Z_j) \neq 0$, under which the system (6.88) to (6.92) then becomes

$$\text{in } 0 < Z_j < L_j, \quad T > 0 \qquad a_j\hat{V}_j + \frac{\partial \hat{V}_j}{\partial T} - \frac{\partial^2 \hat{V}_j}{\partial Z_j^2} = \frac{1}{\sigma_j F_j(Z_j)^{1/2}} I_j(Z_j, T; \hat{V}_j); \tag{6.94}$$

$$\text{at } Z_j = L_j \qquad \frac{\partial \hat{V}_j}{\partial Z_j} = b_j\hat{V}_j; \tag{6.95}$$

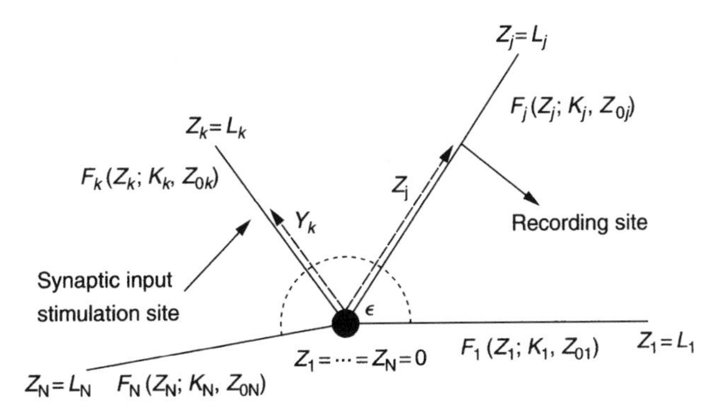

FIGURE 6.7 Illustration of the domain and dimensionless parameters for the nondimensional mathematical model of the multicylinder model with taper. (Adapted from Evans, J.D. *IMA J. Math. Appl. Med. Biol.* 17, 347–377, 2000. With permission.)

$$\text{at } Z_j = 0 \qquad \hat{V}_j = \hat{V}_s, \tag{6.96}$$

$$\text{and} \qquad \sigma_s\left((1+c)\hat{V}_s + \epsilon\frac{d\hat{V}_s}{dT}\right) - \sum_{j=1}^{N}\sigma_j\frac{\partial \hat{V}_j}{\partial Z_j} = I_s(T); \tag{6.97}$$

$$\text{at } T = 0, \quad 0 \leq Z_j \leq L_j \qquad \hat{V}_j = F_j(Z_j)^{1/2}H_j(Z_j); \tag{6.98}$$

for $j = 1, \ldots, N$, where

$$b_j = \frac{1}{2F_j(L_j)}\frac{dF_j(L_j)}{dZ_j}, \quad c_j = \frac{1}{2}\frac{dF_j(0)}{dZ_j}, \quad c = \sum_{j=1}^{N}\frac{\sigma_j}{\sigma_s}c_j, \quad A_j(Z_j) = F_j(Z_j)^{-1/2}, \tag{6.99}$$

and

$$a_j = 1 + \frac{1}{F_j^{1/2}}\frac{d^2F_j^{1/2}}{dZ_j^2}. \tag{6.100}$$

The analysis so far holds for general geometric ratio functions F_j, the only restriction being $F_j \neq 0$. Now, we consider only those F_j for which a_j defined in Equation (6.100) is constant, in which case the functions $e^{(a_j-1)T}\hat{V}_j$ satisfy the standard cable equation. In this case, Equation (6.100) is easily integrated to yield six distinct solution types that are listed in Table 6.1 together with the corresponding values of the constants a_j, b_j, and c_j. This class of functions was initially derived by Kelly and Ghausi (1965). It is emphasized that the geometric ratio function $F_j(Z_j; K_j, Z_{0j})$ is normally prescribed with the two parameters K_j and Z_{0j} (arising as constants of integration from Equation [6.100]) and usually determined by "fitting" to anatomical data (e.g., see Poznanski, 1988, 1990, 1994a, 1994b, 1996; Poznanski and Glenn, 1994). The dimensionless parameters K_j and Z_{0j} in some taper types have restrictions, which are also given in Table 6.1. It should be noted that the exponential (EXP) geometric ratio function can be obtained from the hyperbolic cosine-squared (HCS) taper function by letting $Z_{0j} \to \pm\infty$ to obtain the $\pm K_j$ cases, respectively. The square (SQU) geometric ratio function can be obtained from the hyperbolic sine-squared (HSS) by letting $K_j \to 0$. Figure 6.8 shows the profiles that these taper types generate.

Thus, with the use of the appropriate values for the constants a_j, b_j, c_j, and the function $A_j(Z_j)$ corresponding to the type of geometric ratio function chosen for the jth branch, the solution to Equations (6.88) to (6.92) represents a multicylinder model in which any of the six distinct types of

TABLE 6.1
A List of the Class of Six Taper Types with the Expressions for their Geometric Ratio Functions F_j

Type of taper	Parameter range	Geometric ratio factor, $F_j(Z_j)$	a_j	b_j	c_j
Hyperbolic cosine-squared (HCS)	$0 \leq K_j < +\infty$ $-\infty < Z_{0j} < +\infty$	$\dfrac{\cosh^2[K_j(Z_j + Z_{0j})]}{\cosh^2[K_j Z_{0j}]}$	$1 + K_j^2$	$K_j \tanh[K_j(Z_{0j} + L_j)]$	$K_j \tanh[K_j Z_{0j}]$
Exponential (EXP)	$-\infty < K_j < +\infty$	$e^{2K_j Z_j}$	$1 + K_j^2$	K_j	K_j
Uniform (UNI)	—	1	1	0	0
Hyperbolic sine-squared (HSS)	$0 \leq K_j < +\infty$, $-\infty < Z_{0j} < -L_j$ or $Z_{0j} > 0$	$\dfrac{\sinh^2[K_j(Z_j + Z_{0j})]}{\sinh^2[K_j Z_{0j}]}$	$1 + K_j^2$	$K_j \coth[K_j(L_j + Z_{0j})]$	$K_j \coth[K_j Z_{0j}]$
Square (SQU)	$-\infty < Z_{0j} < -L_j$ or $Z_{0j} > 0$	$\left(1 + \dfrac{Z_j}{Z_{0j}}\right)^2$	1	$\dfrac{1}{L_j + Z_{0j}}$	$\dfrac{1}{Z_{0j}}$
Trigonometric cosine-squared (TCS)	$0 \leq K_j < +\infty$, $0 < L_j < \dfrac{\pi}{K_j}$ $-\dfrac{\pi}{2K_j} < Z_{0j} < \dfrac{\pi}{2K_j} - L_j$	$\dfrac{\cos^2[K_j(Z_j + Z_{0j})]}{\cos^2[K_j Z_{0j}]}$	$1 - K_j^2$	$-K_j \tan[K_j(L_j + Z_{0j})]$	$-K_j \tan[K_j Z_{0j}]$

Notes: The parameter ranges for each function are also stated together with evaluation of the constants a_j, b_j, and c_j defined in Equations (6.99) and (6.100). The tapering profiles that these functions represent are illustrated in Figure 6.8.

Source: From Evans, J.D. *IMA J. Math. Appl. Med. Biol.* 17, 347–377, 2000. With permission.

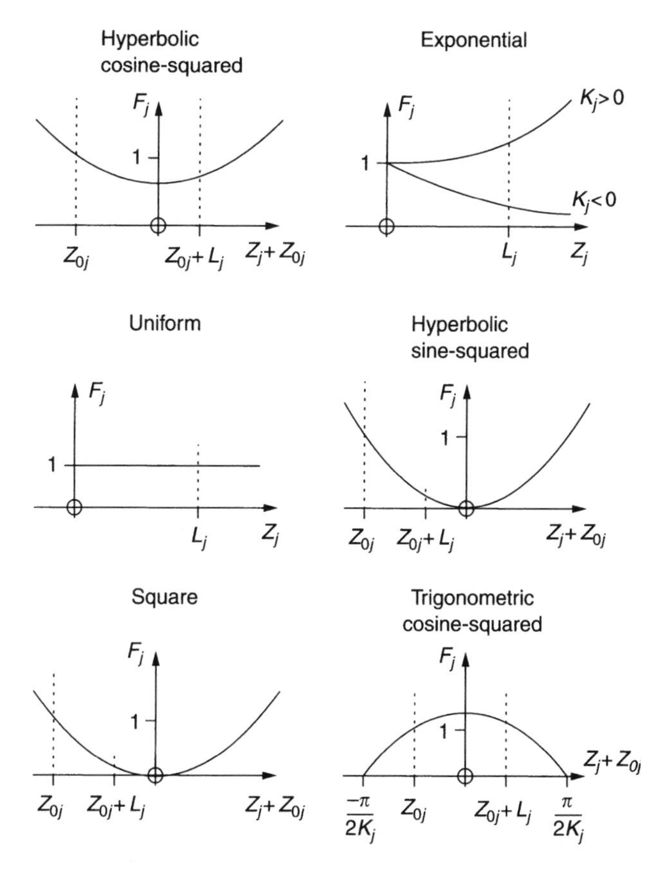

FIGURE 6.8 Schematic illustration of the six geometric ratio functions F_j listed in Table 6.1. Each taper function is shown as a function of the variable $Z_j + Z_{0j}$ for convenience and hence $Z_j + Z_{0j} = Z_{0j}$ denotes the position of the start of the equivalent cylinder and $Z_j + Z_{0j} = Z_{0j} + L_j$ the endpoint. The functions F_j show the type of tapering profiles the equivalent cylinder can attain. Note that all the functions F_j are normalized at $Z_j = 0$, that is, $F_j = 1$. Depending on the taper function, the choice of the parameters Z_{0j}, K_j determines whether the function F_j represents taper, flare, or a combination of both and the degree to which these occur. The functions F_j represent the possible tapering profiles for the jth equivalent cylinder. For any of these specific taper types, the two parameters Z_{0j}, K_j are chosen so that the loss in dendritic trunk parameter for the dendritic tree or trees represented by the jth equivalent cylinder is suitably approximated by the GRF F_j. (*Source:* From Evans, J.D., *IMA J. Math. Appl. Med. Biol.* 17, 347–377, 2000. With permission.)

tapers (hyperbolic cosine-squared, exponential, uniform, hyperbolic sine-squared, square, and trigonometric cosine-squared) may be chosen for the tapering equivalent cylinders. The dimensionless model now has up to $(4N + 1)$ nondimensional parameters

$$\epsilon, \quad \sigma_s, \quad \{\sigma_j\}_{j=1}^N, \quad \{L_j\}_{j=1}^N, \quad \{K_j\}_{j=1}^N, \quad \{Z_{0j}\}_{j=1}^N,$$

depending upon the type of geometric ratio functions F_j chosen and where we note that $\sigma_1 = 1$.

The parameter ranges are as follows: $\{\sigma_j, L_j\} > 0$, $0 \le \epsilon \le 1$, with the ranges for the parameters Z_{0j} and K_j arising in the geometric ratio functions being given in Table 6.1. Typical values are $\gamma_j = \sigma_j/\sigma_s \in (1, 20)$, $L_j \in (0.5, 2)$, $\epsilon \in (0.1, 1)$. Poznanski (1988) reports $K_j \in (-0.5, -2)$ for an exponential (EXP) taper geometric ratio function fitting the data of Turner and Schwartzkroin (1980, 1983). Poznanski (1990) suggests $Z_{0j} = -0.5$ and $K_j = 2$ for a trigonometric cosine-squared (TCS) geometric ratio function that approximates data for cat retinal delta-ganglion cells. The values of these parameters may vary considerably beyond those just stated depending on the type of neuron being considered (e.g., see Strain and Brockman, 1975; Jack et al., 1983; Poznanski, 1988; Rose and Dagum, 1988; Tuckwell, 1988a; Holmes and Rall, 1992a, 1992b; Holmes et al., 1992; Rall et al., 1992, among many others).

The system (6.88) to (6.92) may be reduced to the system of integral equations given by

$$V_j(Z_j,T) = \sum_{k=1}^{N} \int_0^{L_k} \int_0^{T} G_j^k(Z_j,\xi_k,T-u) I_k(\xi_k,u;V_k(\xi_k,u))\,du\,d\xi_k$$
$$+ \sum_{k=1}^{N} \int_0^{L_k} G_j^k(Z_j,\xi_k,T)H_k(\xi_k)\,d\xi_k + G_j^k(0,0,T)H_s + \int_0^{T} G_j^k(Z_j,0,T-u)I_s(u)\,du, \tag{6.101}$$

for $j = 1, \ldots, N$, where $H_s = H_j(0)$ is the initial somatic polarization. When $I_j(Z_j,T;V_j) = I_j(Z_j,T)$ is independent of V_j, then Equation (6.101) gives the general solution. If $I_j(Z_j,T;V_j)$ depends nonlinearly on V_j, then Equation (6.101) is a system of nonlinear integral equations. Explicit analytical progress is possible when $I_j(Z_j,T;V_j)$ depends linearly on V_j with the analysis of Section 6.2.3 carrying over, allowing for notational differences such as X_j being replaced with Z_j. Specifically, for

$$I_j(Z_j,T;V_j) = g_j(Z_j,T)V_j(Z_j,T) + I_{Aj}(Z_j,T), \tag{6.102}$$

(which allows for synaptic reversal potentials and quasi-active ionic currents), Equation (6.101) can be written in the form of Equation (6.33) with the general solution given by Equation (6.36) or explicitly in component form,

$$V_j(Z_j,T) = f_j(Z_j,T) + \sum_{k=1}^{N} \int_0^{L_k} \int_0^{T} \sum_{n=1}^{\infty} \mathcal{K}^n(Z_j,\xi_k;T,u) f_k(\xi_k,u)\,du\,d\xi_k, \tag{6.103}$$

where

$$f_j(Z_j,T) = \sum_{k=1}^{N} \int_0^{L_k} \int_0^{T} G_j^k(Z_j,\xi_k,T-u) I_{Ak}(\xi_k,u)\,du\,d\xi_k$$
$$+ \sum_{k=1}^{N} \int_0^{L_k} G_j^k(Z_j,\xi_k,T)H_k(\xi_k)\,d\xi_k + G_j^k(0,0,T)H_s + \int_0^{T} G_j^k(Z_j,0,T-u)I_s(u)\,du.$$

The iterated kernels are

$$\mathcal{K}^1(Z_j,\xi_k;T,u) = G_j^k(Z_j,\xi_k,T-u)g_k(\xi_k,u), \tag{6.104}$$

and for $n \geq 2$

$$\mathcal{K}^n(Z_j,\xi_k;T,u) = \sum_{i_1=1}^{N} \int_{Y_{i_1}=0}^{L_{i_1}} \int_{s_{i_1}=u}^{T} \sum_{i_2=1}^{N} \int_{Y_{i_2}=0}^{L_{i_2}} \int_{s_{i_2}=u}^{s_{i_1}} \cdots \sum_{i_{n-1}=1}^{N} \int_{Y_{i_{n-1}}=0}^{L_{i_{n-1}}} \int_{s_{i_{n-1}}=u}^{s_{i_{n-2}}}$$
$$\mathcal{K}^1(Z_j,Y_{i_1};T,s_{i_1})\mathcal{K}^1(Y_{i_1},Y_{i_2};s_{i_1},s_{i_2})\mathcal{K}^1(Y_{i_2},Y_{i_3};s_{i_2},s_{i_3})\cdots$$
$$\times \mathcal{K}^1(Y_{i_{n-1}},\xi_k;s_{i_{n-1}},u)\,dY_{i_1}\,dY_{i_2}\cdots dY_{i_{n-1}}\,ds_{i_1}\,ds_{i_2}\cdots ds_{i_{n-1}}. \tag{6.105}$$

However, the Green's function G_j^k is now the tapering impulse solution to Equations (6.88) to (6.92), with

$$\begin{aligned} &I_k = \delta(T)\delta(Z_k - Y_k), \quad I_j = 0, \qquad j = 1, \ldots, N, \quad j \neq k, \\ &I_s = 0, \quad H_j = 0, \qquad j = 1, \ldots, N, \end{aligned} \tag{6.106}$$

which represents a unit impulse to the kth branch at a nondimensional distance Y_k from the soma $Z_j = 0$ ($j = 1, \ldots, N$).

$G_j^k(Z_j, 0, T)$ is the solution to a unit impulse at the soma (i.e., the solution of Equation [6.88] to Equation [6.92] when $I_j \equiv 0, H_j \equiv 0$ for $j = 1, \ldots, N$ and $I_s = \delta(T)$). Explicitly,

$$G_j^k(Z_j, 0, T) = \lim_{Y_k \to 0} G_j^k(Z_j, Y_k, T) = A_j(Z_j) \sum_{n=0}^{\infty} E_n \psi_{jn}(Z_j)\, \mathrm{e}^{-(1+\alpha_n^2)T},$$

since $A_k(0) = 1$ and $\psi_{kn}(0) = 1$. Similarly, $G_j^k(0, 0, T) = \lim_{Z_j \to 0} G_j^k(Z_j, 0, T)$.

An explicit expression for the Green's function may be derived by Laplace transforms, which are detailed in Evans (2000). As an eigenfunction expansion, the expression takes the form

$$G_j^k(Z_j, Y_k, T) = A_j(Z_j)A_k(Y_k) \sum_{n=0}^{\infty} E_n \psi_{kn}(Y_k)\psi_{jn}(Z_j)\, \mathrm{e}^{-(1+\alpha_n^2)T}, \tag{6.107}$$

where

$$\psi_{jn}(Z_j) = \frac{\cos s_{jn}(L_j - Z_j) - (b_j/s_{jn}) \sin s_{jn}(L_j - Z_j)}{\cos s_{jn}L_j - (b_j/s_{jn}) \sin s_{jn}L_j}, \tag{6.108}$$

$$s_{jn} = \sqrt{1 - a_j + \alpha_n^2}, \tag{6.109}$$

$$\begin{aligned} E_n = 2 \Bigg/ \Bigg[& 2\sigma_s\epsilon + \sum_{j=1}^{N} \frac{\sigma_j}{s_{jn}} \left\{ \frac{\tan s_{jn}L_j + b_j/s_{jn}}{1 - (b_j/s_{jn}) \tan s_{jn}L_j} \right\} \\ & + \sum_{j=1}^{N} \sigma_j \left(L_j \left(1 + \frac{b_j^2}{s_{jn}^2} \right) - \frac{b_j}{s_{jn}^2} \right) \frac{\sec^2 s_{jn}L_j}{(1 - (b_j/s_{jn}) \tan s_{jn}L_j)^2} \Bigg], \end{aligned} \tag{6.110}$$

for $j = 1, \ldots, N$ and including $j = k$. The $\alpha_n, n = 0, 1, 2, \ldots,$ are defined as the roots ($\alpha = \alpha_n$) of the transcendental equation

$$1 + c - \epsilon(1 + \alpha^2) = \sum_{j=1}^{N} \frac{\sigma_j}{\sigma_s} s_j \left\{ \frac{\tan s_jL_j + b_j/s_j}{1 - (b_j/s_j) \tan s_jL_j} \right\}, \tag{6.111}$$

where

$$s_j = \sqrt{1 - a_j + \alpha^2}, \tag{6.112}$$

and the coefficients a_j, b_j, c, and A_k are defined in Equation (6.99). An additional set of eigenvalues exist if, for any two distinct branches, say the m_1th and the m_2th, the solutions of

$$\cosh q_{m_1} L_{m_1} = \frac{b_{m_1}}{q_{m_1}} \sinh q_{m_1} L_{m_1},$$

and

$$\cosh q_{m_2} L_{m_2} = \frac{b_{m_2}}{q_{m_2}} \sinh q_{m_2} L_{m_2},$$

satisfy

$$q_{m_1}^2 - a_{m_1} = q_{m_2}^2 - a_{m_2}.$$

The contribution to the solution of these eigenvalues has been omitted since as in the nontapering case, small adjustments to the electrotonic lengths, for example, L_{m_1} or L_{m_2} on the order of $\pm 10^{-6}$, will remove the corresponding eigenfunctions with no discernible effect on the solution.

Again, we note the symmetry inherent in the solution when input and recording sites are interchanged, that is, $G_j^k = G_k^j$.

The analytical solution (6.107) is dependent upon the determination of the (nonnegative) roots of the transcendental Equation (6.111), a detailed analysis of which is given in Evans (2000). Although this solution is valid for all time $T \geq 0$, again it converges increasingly slowly for small time, for which an alternative expression (analogous to Equation [6.26]) can be derived using the Laplace transform solution.

6.3.3 Synaptic Reversal Potentials

Here, we consider the effect of taper on the transient voltage response to excitatory and inhibitory synaptic inputs. Proceeding with the general case first, we consider M_j synaptic reversal potentials in cylinder j of the form

$$i_j(Z_j, t; v_j) = \sum_{n=1}^{M_j} g_{\mathrm{syn}j}^n (Z_j, t)(v_{\mathrm{rev}j}^n - v_j) \tag{6.113}$$

with constant reversal potentials $v_{\mathrm{rev}j}^n$ and conductance changes $g_{\mathrm{syn}j}^n(Z_j, t)$ $[\Omega^{-1}]$. Of particular interest are functions of the form

$$g_{\mathrm{syn}j}^n (Z_j, t) = g_{\mathrm{syn}j,\max}^n \alpha_j^n (t/\tau_{\mathrm{m}})\, \mathrm{e}^{1-\alpha_j^n (t/\tau_{\mathrm{m}})} \delta(Z_j - Z_{0j}^n), \tag{6.114}$$

representing a synaptic conductance change at the location Z_{0j}^n with maximum conductance $g_{\mathrm{syn}j,\max}^n$ and an alpha-function time course with $\alpha_j^n/\tau_{\mathrm{m}}$ being the dimensional time to peak. In the expression (6.114), it is assumed that the synaptic inputs are synchronously activated, although it is straightforward to incorporate delays by changing the time origin. If nondimensionalization follows Equation (6.86), then

$$I_j(Z_j, T; V_j) = \sum_{n=1}^{M_j} \Gamma_{\mathrm{syn}j}^n (Z_j, T)(V_{\mathrm{rev}j}^n - V_j), \tag{6.115}$$

where

$$\Gamma^n_{\text{syn}j} = \bar{R} g^n_{\text{syn}j}, \qquad V^n_{\text{rev}j} = \frac{v^n_{\text{rev}j}}{\bar{I}\bar{R}}.$$

For $g^n_{\text{syn}j}$ given by Equation (6.114) then

$$\Gamma^n_{\text{syn}j}(Z_j, T) = \Gamma^n_{\text{syn}j,\text{max}} \alpha^n_j T \, \mathrm{e}^{1-\alpha^n_j T} \delta(Z_j - Z^n_{0j}), \tag{6.116}$$

where

$$\Gamma^n_{\text{syn}j,\text{max}} = \bar{R} g^n_{\text{syn}j,\text{max}}, \tag{6.117}$$

are the dimensionless maximum conductance changes. The characteristic resistance scaling was specified after Equation (6.86) as $\bar{R} = \lambda^*_{01} r^*_{\text{a}01}$. Here, in the absence of voltage-independent applied currents, a suitable current scaling $\bar{I}$ may be chosen based on the reversal potentials. For definiteness, we take the term with the largest constant reversal potential so that $\bar{I} = \max(|v^n_{\text{rev}j}|)/\bar{R}$ (where the maximum is taken over all j and n).

The dimensionless current terms of Equation (6.115) are of the form Equation (6.102), where

$$g_j(Z_j, T) = -\sum_{n=1}^{M_j} \Gamma^n_{\text{syn}j}(Z_j, T), \qquad I_{Aj}(Z_j, T) = \sum_{n=1}^{M_j} \Gamma^n_{\text{syn}j}(Z_j, T) V^n_{\text{rev}j}.$$

The general solution is then given by Equation (6.103).

6.3.3.1 Example 2

As a specific example, we consider a two-cylinder model with an excitatory synaptic input in cylinder 1 and an inhibitory input that may be in either cylinder. The current input terms may be conveniently stated as follows

$$i_j(Z_j, T; v_j) = \delta_{j1} \, g^{\text{ex}}_{\text{syn}1}(Z_1, t)(v^{\text{ex}}_{\text{rev}1} - v_1) + \delta_{jq} \, g^{\text{in}}_{\text{syn}q}(Z_q, t)(v^{\text{in}}_{\text{rev}q} - v_q),$$

for $j = 1$ and 2, where the superscripts "in" and "ex" label the terms as inhibitory and excitatory, respectively. The index q may be 1 or 2 depending on the location of the inhibitory input; the Kronecker delta terms summarize the cylinder location of the inputs. The synaptic conductance changes are taken in the form of Equation (6.114), where the same parameter values are used for inhibitory input of Equation (6.114) independent of its location, that is, $v^{\text{in}}_{\text{rev}1} = v^{\text{in}}_{\text{rev}2}$, $g^{\text{in}}_{\text{syn}1,\text{max}} = g^{\text{in}}_{\text{syn}2,\text{max}}$, $\alpha^{\text{in}}_1 = \alpha^{\text{in}}_2$. Thus

$$\begin{aligned} I_j(Z_j, T; V_j) =& \delta_{j1} \Gamma^{\text{ex}}_{\text{syn}1,\text{max}} \alpha^{\text{ex}}_1 T \, \mathrm{e}^{1-\alpha^{\text{ex}}_1 T} \delta(Z_1 - Z^{\text{ex}}_{01})(V^{\text{ex}}_{\text{rev}1} - V_1) \\ &+ \delta_{jq} \Gamma^{\text{in}}_{\text{syn}q,\text{max}} \alpha^{\text{in}}_q T \, \mathrm{e}^{1-\alpha^{\text{in}}_q T} \delta(Z_q - Z^{\text{in}}_{0q})(V^{\text{in}}_{\text{rev}q} - V_q). \end{aligned} \tag{6.118}$$

Considering the dimensional parameter values

$$\begin{aligned} v^{\text{ex}}_{\text{rev}1} &= 90\,\text{mV}, \quad g^{\text{ex}}_{\text{syn}1,\text{max}} = 2\,\text{nS}, \quad \alpha^{\text{ex}}_1 = 1.5, \\ v^{\text{in}}_{\text{rev}q} &= -9\,\text{mV}, \quad g^{\text{in}}_{\text{syn}q,\text{max}} = 20\,\text{nS}, \quad \alpha^{\text{in}}_q = 1.25, \end{aligned} \tag{6.119}$$

(the inhibitory parameters being the same for $q = 1$ or 2), then $\bar{I} = v^{\text{ex}}_{\text{rev1}}/\bar{R}$, and if we assume for simplicity that $\bar{R} = \lambda^*_{01} r^*_{\text{a01}} = 500\,\text{M}\Omega$ then $\bar{I} = 0.18\,\text{nA}$ and

$$\Gamma^{\text{ex}}_{\text{syn1,max}} = 1, \qquad \Gamma^{\text{in}}_{\text{syn}q\text{,max}} = 10, \qquad V^{\text{ex}}_{\text{rev1}} = 1, \qquad V^{\text{in}}_{\text{rev}q} = -0.1.$$

(In the investigation that follows we will consider the solution sensitivity to $\Gamma^{\text{in}}_{\text{synq,max}}$ and allow it to take different values.) For calculation of the Green's function, we adopt the following dimensionless parameters

cylinder 1	$\sigma_1 = 1,$	$L_1 = 1.5,$	GRF = EXP,	$K_1,$
cylinder 2	$\sigma_2 = 1,$	$L_2 = 1,$	GRF = UNI,	
soma	$\sigma_s = 1,$	$\epsilon = 1.0.$		

Cylinder 1 will be taken with exponential GRF to illustrate the effects of tapering, while cylinder 2 will be uniform.

Figure 6.9A shows the soma potential ($V_s(T) = V_1(0, T) = V_2(0, T)$) for excitatory and inhibitory inputs located at the same site at the midpoint ($Z^{\text{ex}}_{01} = Z^{\text{in}}_{01} = 0.5L_1$) of cylinder 1 which is taken as uniform $K_1 = 0$. The strength of the inhibitory conductance change is varied, illustrating its effect and sensitivity on the potential at the soma. Surprisingly large values of $\Gamma^{\text{in}}_{\text{syn1,max}} = O(10)$ are required before the potential sign reversal is obtained.

Figure 6.9B to D illustrates the effect of taper. The excitatory input is kept fixed in each figure at the given locations, midpoint in B, proximal in C, and distal in D. In each of these figures, the location of the inhibitory input is varied throughout the two cylinders. The maximum strength of the inhibitory inputs are fixed with $\Gamma^{\text{in}}_{\text{syn1,max}} = \Gamma^{\text{in}}_{\text{syn2,max}} = 2$ being twice as strong as the excitatory input. Plotted in each figure is the percentage reduction in the peak soma potential defined as

$$\frac{\max V^{\text{ex}}_{\text{s}}(T) - \max V^{\text{ex,in}}_{\text{s}}(T)}{\max V^{\text{ex}}_{\text{s}}(T)} \times 100\%, \tag{6.120}$$

where $V^{\text{ex}}_{\text{s}}(T)$ denotes the soma potential with only the excitatory input and $V^{\text{ex,in}}_{\text{s}}(T)$ denotes the soma potential with both excitatory and inhibitory inputs. In each figure, three values of the tapering parameter, $K_1 = 0, -0.5, -1$ are taken.

As expected, Figure 6.9B to D illustrates that the inhibitory input is most effective when located at the recording site (the soma) with greater effects on distal excitatory inputs than proximally located ones. However, when comparing equal electrotonic distances from the soma, the figures also suggest that the inhibitory input is more effective in cylinder 2, that is, the cylinder with the shorter electrotonic length. As the taper of the cylinder containing the excitatory input increases, an inhibitory input in cylinder 2 (i.e., remote from the excitatory input and tapered cylinder) has an increased effect irrespective of the location of the excitatory input in the tapered cylinder. However, when located in the tapered cylinder containing the excitatory input, increased taper has less effect when located between the excitatory input and the cylinder terminal endpoint, while increased taper has more effect when located between the excitatory input and the soma. These transitions are most clearly seen in Figure 6.9B and C. This may not be unexpected since increased taper reduces the strength of the synaptic inputs. In Figure 6.9D, the attainment of 100% reduction indicates that the peak potential has reversed sign, which appears possible for distally located excitatory inputs in a cylinder with sufficient taper.

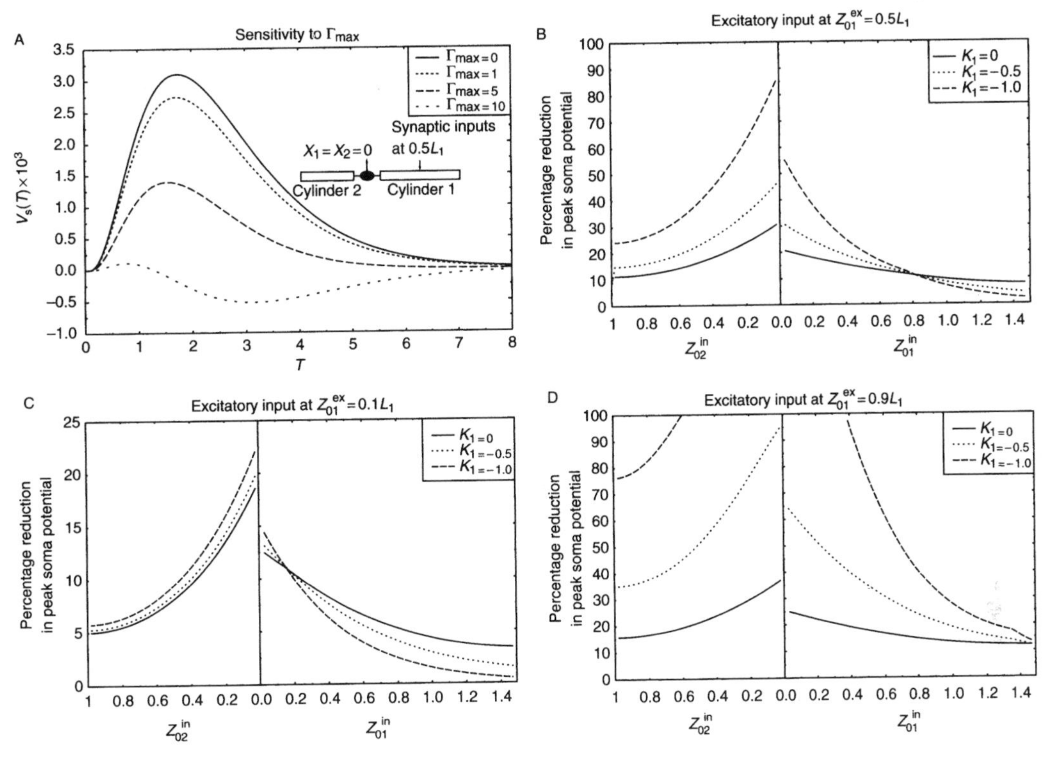

Figure 6.9 The effects of taper on synaptic inputs for Example 2. (A) shows first the effect of the strength of the inhibitory conductance change $\Gamma^{\text{in}}_{\text{syn1,max}}$ on the soma potential ($V_S(T) = V_1(0,T) = V_2(0,T)$) for excitatory and inhibitory inputs located at the same site at the midpoint of cylinder 1, which is taken as uniform $K_1 = 0$. (B) to (D) illustrate the effect of taper. The excitatory input is kept fixed in each figure, midpoint in (B), proximal in (C), and distal in (D). In each of (B) to (D), the location Z^{in}_{0i} of the inhibitory input is varied throughout the two cylinders. The maximum strength of the inhibitory inputs is fixed with $\Gamma^{\text{in}}_{\text{syn}q,\text{max}} = 2$ for $(q = 1, 2)$. Plotted in each figure is the percentage reduction in the peak soma potential defined as Equation (6.120) for the three selected tapering parameters $K_1 = 0, -0.5, -1$.

6.4 Two Gap-Junctionally Coupled Multicylinder Models with Taper

Evidence for neuronal coupling within the hippocampal subfields CA3, CA1, and dentate gyrus has been given by MacVicar and Dudek (1980a, 1980b; 1981), MacVicar et al. (1982), and Schuster (1992). Schmalbruch and Jahnsen (1981) have observed gap junctions between CA3 pyramidal neurons, and it is these that are thought to mediate the electrical coupling. However, such coupling has not been taken into account when attempts have been made to estimate the passive electrical parameters of such neurons and may cause significant distortions in the results (Getting, 1974; Rall, 1981; Skrzypek, 1984; Publicover, 1989). Recently, Amitai et al. (2002) have determined that gap junctions contribute approximately one-half of the input conductance in GABAergic neurons of the neocortex. The dynamics of neurons interacting via chemical synapses have been studied extensively (e.g., see Golomb et al., 2001). In comparison, only a few theoretical studies have addressed the dynamics of neurons coupled by electrical synapses. Pfeuty et al. (2003) have investigated a network of active-conductance-based neurons, in particular illustrating that synchrony between neurons is promoted by potassium currents and impeded by persistent sodium currents. The neurons are represented as isopotential units without spatial structure. Poznanski and Umino (1997) incorporate spatial properties of each cell in a network in the form of a leaky cable with passive membrane properties.

As a first step in the cable analysis of the effects of coupling on synaptic potentials and the passive spread of electrical signals, Poznanski et al. (1995) considered two exponentially tapered equivalent cylinders that were coupled through their isopotential somas. The analysis of the model investigated the effect of the coupling resistance and somatic shunt parameter on the potential response at the two somas for synaptic input at selected points along one of the coupled neurons. The mathematical analysis was restricted to determining the potential at the two somas with partial analytic expressions being obtained and completion of the solution being given by numerical techniques. Here, we provide a complete analytical solution for the model and extend the model to include synaptic reversal potentials as well as allowing for more general taper types (as described in Section 6.3), of which the exponential is one particular type. Again, the motivation for obtaining an analytical solution for the passive case is twofold. First, when the solution is expressed as a weighted series of time-dependent exponentials, it gives the explicit dependence of the voltage amplitude terms and time constants on the model parameters and hence information on their relative effect and importance as well as aiding the practical determination and extraction of such parameters. (Poznanski [1998] investigates the effect of electrotonic coupling on passive electrical constants, approximating soma and dendritic coupling by a leak resistance at the soma and terminal end of an equivalent cylinder, respectively.) Second, it provides a reference solution against which numerical and analytical solutions of extended models can be compared.

6.4.1 Soma–Somatic Coupling

We consider two neurons, each represented by a single tapering equivalent cylinder and coupled at their isopotential somas through a gap junction. The model for the tapering equivalent cylinder is described in Section 6.3, where for the case of interest here, the number of equivalent cylinders N representing each neuron is taken to be one ($N = 1$) and the taper types are restricted to the same class of six taper functions, namely the hyperbolic cosine-squared (HCS), hyperbolic sine-squared, trigonometric cosine-squared (TCS), exponential (EXP), square (SQU), and uniform (UNI). A schematic illustration of the model is shown in Figure 6.10A.

Denoting by $v_i(Z_i, t)$ the potential at tapering electrotonic distance Z_i and time t in cylinder $i = 1, 2$, the mathematical statement of the model is as follows:

$$\text{in } 0 < Z_1 < L_1, \quad t > 0 \qquad v_1 + \tau_m \frac{\partial v_1}{\partial t} - \left(\frac{\partial^2 v_1}{\partial Z_1^2} + \frac{1}{F_1} \frac{dF_1}{dZ_1} \frac{\partial v_1}{\partial Z_1} \right) = \lambda_1^* r_{a1}^* \hat{I}_1(Z_1, t; v_1); \tag{6.121}$$

$$\text{in } 0 < Z_2 < L_2, \quad t > 0 \qquad v_2 + \tau_m \frac{\partial v_2}{\partial t} - \left(\frac{\partial^2 v_2}{\partial Z_2^2} + \frac{1}{F_2} \frac{dF_2}{dZ_2} \frac{\partial v_2}{\partial Z_2} \right) = \lambda_2^* r_{a2}^* \hat{I}_2(Z_2, t; v_2); \tag{6.122}$$

$$\text{at } Z_1 = L_1 \qquad \frac{\partial v_1}{\partial Z_1} = 0; \tag{6.123}$$

$$\text{at } Z_2 = L_2 \qquad \frac{\partial v_2}{\partial Z_2} = 0; \tag{6.124}$$

$$\text{at } Z_1 = Z_2 = 0 \qquad \frac{1}{R_{s1}} \left(v_1 + \tau_{s1} \frac{\partial v_1}{\partial t} \right) - \frac{1}{\lambda_{01}^* r_{a01}^*} \frac{\partial v_1}{\partial Z_1} = \hat{I}_{s1}(t) + \frac{(v_2 - v_1)}{R_{junc}}; \tag{6.125}$$

$$\text{and} \qquad \frac{1}{R_{s2}} \left(v_2 + \tau_{s2} \frac{\partial v_2}{\partial t} \right) - \frac{1}{\lambda_{02}^* r_{a02}^*} \frac{\partial v_2}{\partial Z_2} = \hat{I}_{s2}(t) - \frac{(v_2 - v_1)}{R_{junc}}; \tag{6.126}$$

$$\text{at } t = 0, \qquad v_1 = h_1(Z_1) \quad v_2 = h_2(Z_2). \tag{6.127}$$

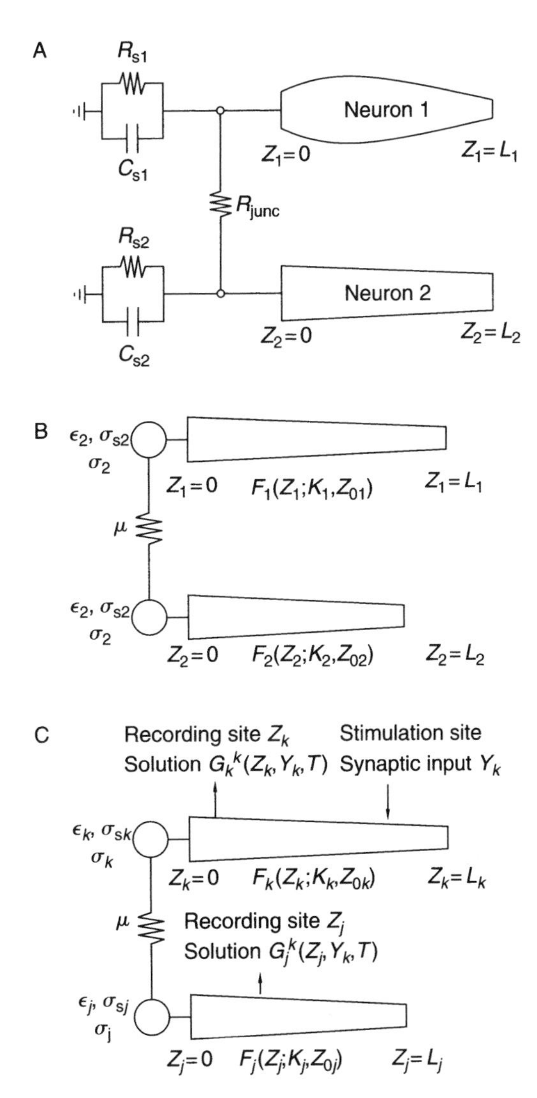

FIGURE 6.10 (A) Gives a schematic illustration for a coupled model of two tapered equivalent cylinders (representing the branched structures of two neurones). The neurones are coupled at their somas by a junctional resistance R_{junc} representing electrical coupling. (B) Illustrates the nondimensional coupled model with the dimensionless model parameters. (C) Demonstrates the notation used for the Green's function introduced by Equation (6.146), distinguishing between the cases when the recording site is in the same cylinder or in a different cylinder from that of the synaptic/stimulating site.

In addition to the previously defined parameters, given in Sections 6.2 and 6.3, we have for $i = 1$ or 2 that

$\tau_{si} = R_{si}C_{si}$	is the somatic time constant (sec) of soma i,
R_{si}	is the lumped resistance (Ω) for soma i, which can include a shunt resistance,
C_{si}	is the lumped capacitance (F) for soma i,
$\hat{I}_{si}$	is an applied current to the soma of neuron i,
$\hat{I}_i(Z_i, t)$	is an applied current to equivalent cylinder i,
$h_i(Z_i)$	is the initial voltage in cylinder i,
R_{junc}	denotes the junctional resistance (Ω) between the somas of the two neurons.

The axial current through the gap-junction resistance is $(v_2(0,t)-v_1(0,t))/R_{\text{junc}}$ and differing somatic time constants and soma resistances have been assumed for both neurons. Sealed-end termination boundary conditions are imposed for both neurons and the geometric ratio functions $F_i(Z_i)$, $i = 1, 2$, are assumed to belong to the six types listed in Table 6.1 in Section 6.3. The boundary conditions (6.125) and (6.126) follow from the conservation-of-current conditions at the somas.

It is noted that for $i = 1, 2$, the variables Z_i, parameters L_i, and geometric functions $F_i(Z_i)$ are dimensionless. Nondimensionalization of the equations is completed with the scalings

$$t = \tau_m T, \qquad v_i = \bar{I}\bar{R}V_i, \qquad h_i = \bar{I}\bar{R}H_i, \qquad \hat{I}_i = \bar{I}I_i, \qquad \hat{I}_{si} = \bar{I}I_{si},$$

for $i = 1, 2$, where $\bar{I}$ is an appropriate current scaling and $\bar{R}$ an appropriate resistance scaling. The choice for $\bar{I}$ would be guided by the type of current input, for example, $\bar{I} = Q_0/\tau_m$ for a point charge input of Q_0 coulombs, or $\bar{I} = I$ for a constant current input of magnitude I(A). There are several choices for $\bar{R}$, namely R_{s1}, R_{s2}, or one of the $\lambda^*_{0i}r^*_{a0i}$ for $i = 1, 2$. Here, we will take $\bar{R} = \lambda^*_{01}r^*_{a01}$. This then introduces the dimensionless parameters

$$\begin{aligned} &\epsilon_1 = \frac{\tau_{s1}}{\tau_m}, \qquad \epsilon_2 = \frac{\tau_{s2}}{\tau_m}, \qquad \sigma_{s1} = \frac{\lambda^*_{01}r^*_{a01}}{R_{s1}}, \qquad \sigma_{s2} = \frac{\lambda^*_{01}r^*_{a01}}{R_{s2}}, \\ &\sigma_1 = 1, \qquad \sigma_2 = \frac{\lambda^*_{01}r^*_{a01}}{\lambda^*_{02}r^*_{a02}} = \left(\frac{D^*_{02}}{D^*_{01}}\right)^{3/2}, \qquad \mu = \frac{\lambda^*_{01}r^*_{a01}}{R_{\text{junc}}}. \end{aligned} \tag{6.128}$$

The system (6.121) to (6.127) then becomes

$$\text{in } 0 < Z_1 < L_1, \quad T > 0 \qquad V_1 + \frac{\partial V_1}{\partial T} - \frac{\partial^2 V_1}{\partial Z_1^2} - \frac{1}{F_1}\frac{dF_1}{dZ_1}\frac{\partial V_1}{\partial Z_1} = \frac{1}{\sigma_1 F_1(Z_1)} I_1(Z_1, T; V_1); \tag{6.129}$$

$$\text{in } 0 < Z_2 < L_2, \quad T > 0 \qquad V_2 + \frac{\partial V_2}{\partial T} - \frac{\partial^2 V_2}{\partial Z_2^2} - \frac{1}{F_2}\frac{dF_2}{dZ_2}\frac{\partial V_2}{\partial Z_2} = \frac{1}{\sigma_2 F_2(Z_2)} I_2(Z_2, T; V_2); \tag{6.130}$$

$$\text{at } Z_1 = L_1 \qquad \frac{\partial V_1}{\partial Z_1} = 0; \tag{6.131}$$

$$\text{at } Z_2 = L_2 \qquad \frac{\partial V_2}{\partial Z_2} = 0; \tag{6.132}$$

$$\text{at } Z_1 = Z_2 = 0 \qquad \sigma_{s1}\left(V_1 + \epsilon_1\frac{\partial V_1}{\partial T}\right) - \sigma_1\frac{\partial V_1}{\partial Z_1} = I_{s1}(T) + \mu(V_2 - V_1); \tag{6.133}$$

$$\text{and} \qquad \sigma_{s2}\left(V_2 + \epsilon_2\frac{\partial V_2}{\partial T}\right) - \sigma_2\frac{\partial V_2}{\partial Z_2} = I_{s2}(T) - \mu(V_2 - V_1); \tag{6.134}$$

$$\text{at } T = 0, \quad 0 \le Z_j \le L_j \qquad V_1 = H_1(Z_1), \quad V_2 = H_2(Z_2). \tag{6.135}$$

Although $\sigma_1 = 1$, we retain this parameter in order to maintain the symmetry in the equations. We now introduce the transformation

$$\hat{V}_i(Z_i, T) = F_i(Z_i)^{1/2} V_i(Z_i, T), \tag{6.136}$$

for $i = 1, 2$, noting that $F_i(0) = 1$ and $F_i(Z_i) \neq 0$, under which the system (6.129) to (6.135) becomes

$$\text{in } 0 < Z_1 < L_1, \quad T > 0 \qquad a_1\hat{V}_1 + \frac{\partial \hat{V}_1}{\partial T} - \frac{\partial^2 \hat{V}_1}{\partial Z_1^2} = \frac{1}{\sigma_1 F_1(Z_1)^{1/2}} I_1(Z_1, T; \hat{V}_1); \tag{6.137}$$

$$\text{in } 0 < Z_2 < L_2, \quad T > 0 \qquad a_2\hat{V}_2 + \frac{\partial \hat{V}_2}{\partial T} - \frac{\partial^2 \hat{V}_2}{\partial Z_2^2} = \frac{1}{\sigma_2 F_2(Z_2)^{1/2}} I_2(Z_2, T; \hat{V}_2); \tag{6.138}$$

$$\text{at } Z_1 = L_1 \qquad \frac{\partial \hat{V}_1}{\partial Z_1} = b_1\hat{V}_1; \tag{6.139}$$

$$\text{at } Z_2 = L_2 \qquad \frac{\partial \hat{V}_2}{\partial Z_2} = b_2\hat{V}_2; \tag{6.140}$$

$$\text{at } Z_1 = Z_2 = 0 \qquad \sigma_{s1}\left(\hat{V}_1 + \epsilon_1 \frac{\partial \hat{V}_1}{\partial T}\right) + c_1\sigma_1\hat{V}_1 - \sigma_1 \frac{\partial \hat{V}_1}{\partial Z_1} = I_{s1} + \mu(\hat{V}_2 - \hat{V}_1), \tag{6.141}$$

and

$$\sigma_{s2}\left(\hat{V}_2 + \epsilon_2 \frac{\partial \hat{V}_2}{\partial T}\right) + c_2\sigma_2\hat{V}_2 - \sigma_2 \frac{\partial \hat{V}_2}{\partial Z_2} = I_{s2} - \mu(\hat{V}_2 - \hat{V}_1); \tag{6.142}$$

$$\text{at } T = 0, \quad 0 \leq Z_i \leq L_i \qquad \hat{V}_1 = F_1(Z_1)^{1/2} H_1(Z_1), \quad \hat{V}_2 = F_2(Z_2)^{1/2} H_2(Z_2); \tag{6.143}$$

where for $i = 1, 2$, a_i, b_i, c_i, and A_i are defined as

$$b_i = \frac{F_i'(L_i)}{2F_i(L_i)}, \qquad c_i = \frac{F_i'(0)}{2}, \qquad A_i(Z_i) = F_i(Z_i)^{-1/2}, \tag{6.144}$$

with $'$ denoting $\mathrm{d}/\mathrm{d}Z_i$ and

$$a_i = 1 + \frac{1}{F_i^{1/2}} \frac{\mathrm{d}^2 F_i^{1/2}}{\mathrm{d}Z_i^2}. \tag{6.145}$$

Figure 6.10B illustrates the nondimensional coupled model. The dimensionless model now has the eight nondimensional parameters,

$$\epsilon_1, \quad \epsilon_2, \quad \sigma_{s1}, \quad \sigma_{s2}, \quad \sigma_2, \quad \mu, \quad L_1, \quad L_2,$$

since $\sigma_1 = 1$ and up to four dimensionless tapering parameters,

$$K_1, \quad K_2, \quad Z_{01}, \quad Z_{02},$$

depending upon the type of geometric ratio functions, F_1 and F_2, chosen. Suitable parameter ranges have been discussed at the end of Section 6.3.1 (allowing for notational differences).

6.4.1.1 The Green's Function

The Green's function is defined to be the solution of Equations (6.129) to (6.135) with a unit impulse input into one of the cylinders, denoted as cylinder k (where $k = 1$ or 2), while evaluated in either the same cylinder k or the other cylinder, denoted by j, where $j = 1$ or 2, but $j \neq k$. For example, input could be into cylinder 1, in which case $k = 1$ and hence $j = 2$, but note that recording could take place in cylinder 1 or cylinder 2. Thus we consider,

$$\begin{aligned} I_k &= \delta(T)\delta(Z_k - Y_k), \qquad I_j = 0 \quad \text{for } j \neq k, \\ I_{s1} &= I_{s2} = 0, \qquad H_1 = H_2 = 0, \end{aligned} \tag{6.146}$$

which represents a unit impulse to cylinder k (where $k = 1$ or 2) at a nondimensional distance Y_k from the soma of cylinder k located at $Z_k = 0$. We denote the solution by $G_k^k(Z_k, Y_k, T)$ when recording in cylinder k at a nondimensional distance Z_k from the soma (located at $Z_k = 0$), and by $G_j^k(Z_j, Y_k, T)$ the solution when recording in cylinder $j \neq k$ at a distance Z_j from the soma in cylinder j (which is located at $Z_j = 0$). This notation is illustrated in Figure 6.10C. Using the transformation of Equation (6.136), we obtain

$$G_i^k(Z_i, Y_k, T) = A_i(Z_i)U_i^k(Z_i, Y_k, T), \qquad \text{for } i = j \text{ or } k, \tag{6.147}$$

where $U_i^k(Z_i, Y_k, T)$ with $i = j$ or k is the corresponding Green's function for Equations (6.137) to (6.143) and satisfies,

$$\text{in } 0 < Z_k < L_k, \quad T > 0 \qquad a_k U_k^k + \frac{\partial U_k^k}{\partial T} - \frac{\partial^2 U_k^k}{\partial Z_k^2} = \frac{A_k(Z_k)}{\sigma_k}\delta(Z_k - Y_k)\delta(T); \tag{6.148}$$

$$\text{in } 0 < Z_j < L_j, \quad T > 0 \qquad a_j U_j^k + \frac{\partial U_j^k}{\partial T} - \frac{\partial^2 U_j^k}{\partial Z_j^2} = 0; \tag{6.149}$$

$$\text{at } Z_k = L_k \qquad \frac{\partial U_k^k}{\partial Z_k} = b_k U_k^k; \tag{6.150}$$

$$\text{at } Z_j = L_j \qquad \frac{\partial U_j^k}{\partial Z_j} = b_j U_j^k; \tag{6.151}$$

$$\text{at } Z_k = Z_j = 0 \qquad \sigma_{sk}\left(U_k^k + \epsilon_k \frac{\partial U_k^k}{\partial T}\right) + c_k \sigma_k U_k^k - \sigma_k \frac{\partial U_k^k}{\partial Z_k} = \mu(U_j^k - U_k^k); \tag{6.152}$$

$$\text{and} \qquad \sigma_{sj}\left(U_j^k + \epsilon_j \frac{\partial U_j^k}{\partial T}\right) + c_j \sigma_j U_j^k - \sigma_j \frac{\partial U_j^k}{\partial Z_j} = \mu(U_k^k - U_j^k); \tag{6.153}$$

$$\text{at } T = 0, \qquad U_k^k = U_j^k = 0; \tag{6.154}$$

for $j = 1$ or 2, $k = 1$ or 2 but $j \neq k$.

This problem for U_k^k and U_j^k now includes mixed boundary conditions at the ends of the tapering cylinders as given by Equations (6.150) and (6.151).

The Laplace transform of U_i^k,

$$\bar{U}_i^k(Z_i, Y_k, p) = \int_0^\infty \mathrm{e}^{-pT} U_i^k(Z_i, Y_k, T)\,\mathrm{d}T,$$

where $i = k$ or j satisfies

$$\text{in } 0 < Z_k < L_k \qquad (a_k + p)\bar{U}_k^k - \frac{\partial^2 \bar{U}_k^k}{\partial Z_k^2} = \frac{A_k}{\sigma_k}\delta(Z_k - Y_k); \tag{6.155}$$

$$\text{in } 0 < Z_j < L_j \qquad (a_j + p)\bar{U}_j^k - \frac{\partial^2 \bar{U}_j^k}{\partial Z_j^2} = 0; \tag{6.156}$$

$$\text{at } Z_k = L_k \qquad \frac{\partial \bar{U}_k^k}{\partial Z_k} = b_k \bar{U}_k^k; \tag{6.157}$$

$$\text{at } Z_j = L_j \qquad \frac{\partial \bar{U}_j^k}{\partial Z_j} = b_j \bar{U}_j^k; \tag{6.158}$$

$$\text{at } Z_k = Z_j = 0 \qquad (\sigma_{sk}(1 + \epsilon_k p) + c_k\sigma_k + \mu)\bar{U}_k^k - \sigma_k \frac{\partial \bar{U}_k^k}{\partial Z_k} = \mu \bar{U}_j^k; \tag{6.159}$$

$$\text{and} \qquad (\sigma_{sj}(1 + \epsilon_j p) + c_j\sigma_j + \mu)\bar{U}_j^k - \sigma_j \frac{\partial \bar{U}_j^k}{\partial Z_j} = \mu \bar{U}_k^k. \tag{6.160}$$

The solution to Equations (6.155) to (6.160) may be conveniently stated as follows,

$$\bar{U}_j^k = \frac{\mu}{\Delta} A_k(Y_k)\bar{\psi}_k(Y_k)\bar{\psi}_j(Z_j), \tag{6.161}$$

and

$$\bar{U}_k^k = \begin{cases} A_k(Y_k)\bar{\psi}_k(Z_k)\left[\dfrac{(\bar{w}_j + \mu)}{\Delta}\bar{\psi}_k(Y_k) + \dfrac{\sinh q_k Y_k}{\sigma_k q_k}\right] & \text{if } Z_k > Y_k, \\ A_k(Y_k)\bar{\psi}_k(Y_k)\left[\dfrac{(\bar{w}_j + \mu)}{\Delta}\bar{\psi}_k(Z_k) + \dfrac{\sinh q_k Z_k}{\sigma_k q_k}\right] & \text{if } Z_k < Y_k, \end{cases} \tag{6.162}$$

where we define

$$q_j = \sqrt{a_j + p}, \qquad q_k = \sqrt{a_k + p}, \qquad \Delta = \bar{w}_j\bar{w}_k + \mu(\bar{w}_j + \bar{w}_k),$$

with

$$\bar{w}_k = \sigma_{sk}(1 + \epsilon_k p) + c_k\sigma_k + \sigma_k q_k \tanh(q_k L_k - \hat{\beta}_k), \qquad \tanh \hat{\beta}_k = \frac{b_k}{q_k},$$

$$\bar{w}_j = \sigma_{sj}(1 + \epsilon_j p) + c_j\sigma_j + \sigma_j q_j \tanh(q_j L_j - \hat{\beta}_j), \qquad \tanh \hat{\beta}_j = \frac{b_j}{q_j},$$

and for $i = k$ or j,

$$\bar{\psi}_i(Z_i) = \frac{(\cosh q_i(L_i - Z_i) - (b_i/q_i)\sinh q_i(L_i - Z_i))}{(\cosh q_i L_i - (b_i/q_i)\sinh q_i L_i)},$$

where $\bar{\psi}_i(Y_i)$ can be obtained from this last expression by replacing Z_i with Y_i.

In the limit $\mu \to 0$ the equations decouple and we recover the result for the single cylinder case, namely $\bar{U}_j^k = 0$ and

$$\bar{U}_k^k = \begin{cases} A_k(Y_k)\bar{\psi}_k(Z_k)\left[\dfrac{1}{\bar{w}_k}\bar{\psi}_k(Y_k) + \dfrac{\sinh q_k Y_k}{\sigma_k q_k}\right] & \text{if } Z_k > Y_k, \\ A_k(Y_k)\bar{\psi}_k(Y_k)\left[\dfrac{1}{\bar{w}_k}\bar{\psi}_k(Z_k) + \dfrac{\sinh q_k Z_k}{\sigma_k q_k}\right] & \text{if } Z_k < Y_k, \end{cases}$$

which agrees with the solution stated in Evans (2000) for the particular case of a single cylinder ($N = 1$).

The other limit of interest is $\mu \to \infty$, where the somas effectively fuse and in which case we obtain

$$\bar{U}_j^k = \frac{A_k(Y_k)}{\bar{w}_j + \bar{w}_k}\bar{\psi}_k(Y_k)\bar{\psi}_j(Z_j),$$

and

$$\bar{U}_k^k = \begin{cases} A_k(Y_k)\bar{\psi}_k(Z_k)\left[\dfrac{1}{(\bar{w}_j + \bar{w}_k)}\bar{\psi}_k(Y_k) + \dfrac{\sinh q_k Y_k}{\sigma_k q_k}\right] & \text{if } Z_k > Y_k, \\ A_k(Y_k)\bar{\psi}_k(Y_k)\left[\dfrac{1}{(\bar{w}_j + \bar{w}_k)}\bar{\psi}_k(Z_k) + \dfrac{\sinh q_k Z_k}{\sigma_k q_k}\right] & \text{if } Z_k < Y_k, \end{cases}$$

which is the two-cylinder solution described in Evans (2000) upon making the identification $\sigma_s = \sigma_{sk} + \sigma_{sj} = \sigma_{s1} + \sigma_{s2}$ and $\epsilon_1 = \epsilon_2 = \epsilon$.

The solution for U_j^k and U_k^k may be obtained by inverting Equations (6.161) and (6.162), which gives

$$U_j^k(Z_j, Y_k, T) = A_k(Y_k)\sum_{n=0}^{\infty} E_{jn}\psi_{kn}(Y_k)\psi_{jn}(Z_j)\,\mathrm{e}^{-(1+\alpha_n^2)T}, \tag{6.163}$$

and

$$U_k^k(Z_k, Y_k, T) = A_k(Y_k)\sum_{n=0}^{\infty} E_{kn}\psi_{kn}(Y_k)\psi_{kn}(Z_k)\,\mathrm{e}^{-(1+\alpha_n^2)T}, \tag{6.164}$$

where

$$\psi_{kn}(Y_k) = \frac{\cos s_{kn}(L_k - Y_k) - (b_k/s_{kn})\sin s_{kn}(L_k - Y_k)}{\cos s_{in}L_k - (b_k/s_{kn})\sin s_{kn}L_k},$$

$$\psi_{jn}(Z_j) = \frac{\cos s_{jn}(L_j - Z_j) - (b_j/s_{jn})\sin s_{jn}(L_j - Z_j)}{\cos s_{jn}L_j - (b_j/s_{jn})\sin s_{jn}L_j},$$

and $\psi_{kn}(Z_k)$ can be obtained from $\psi_{kn}(Y_k)$ by replacing Y_k with Z_k. Also, for $i = j$ or k,

$$s_{in} = \sqrt{1 - a_i + \alpha_n^2},$$

and

$$E_{jn} = \frac{\mu}{w'_{jn}(w_{kn} + \mu) + w'_{kn}(w_{jn} + \mu)},$$

$$E_{kn} = \frac{w_{jn} + \mu}{w'_{jn}(w_{kn} + \mu) + w'_{kn}(w_{jn} + \mu)},$$

with again for $i = j$ or k,

$$w_{in} = \sigma_{si}(1 - \epsilon_i(1 + \alpha_n^2)) + c_i\sigma_i - \sigma_i s_{in} \tan(s_{in}L_i + \beta_{in}),$$

$$w'_{in} = \sigma_{si}\epsilon_i + \frac{\sigma_i}{2}\left(\frac{\tan(s_{in}L_i + \beta_{in})}{s_{in}} + \left(L_i - \frac{b_i}{b_i^2 + s_{in}^2}\right)\sec^2(s_{in}L_i + \beta_{in})\right),$$

and $\tan\beta_{in} = b_i/s_{in}$.

The $\alpha_n, n = 0, 1, 2, \ldots$, are defined as the roots ($\alpha = \alpha_n$) of the transcendental equation

$$(w_k + \mu)(w_j + \mu) = \mu^2, \tag{6.165}$$

where for $i = j$ or k,

$$w_i = \sigma_{si}(1 - \epsilon_i(1 + \alpha^2)) + c_i\sigma_i - \sigma_i s_i \tan(s_iL_i + \beta_i), \tag{6.166}$$

with

$$s_i = \sqrt{1 - a_i + \alpha^2}, \qquad \tan\beta_i = \frac{b_i}{s_i}.$$

The solution in (6.163) and (6.164) represents the contribution from the simple poles $p = -(1 + \alpha_n^2)$, $n = 0, 1, 2, \ldots$.

Finally, the Green's function G_i^k ($i = j$ or k), corresponding to Equations (6.129) to (6.135), is obtained from U_i^k ($i = j$ or k) by the transformation (6.136) and is given explicitly as follows,

$$G_j^k(Z_j, Y_k, T) = A_j(Z_j)A_k(Y_k)\sum_{n=0}^{\infty} E_{jn}\psi_{kn}(Y_k)\psi_{jn}(Z_j)\,\mathrm{e}^{-(1+\alpha_n^2)T}, \tag{6.167}$$

and

$$G_k^k(Z_k, Y_k, T) = A_k(Z_k)A_k(Y_k)\sum_{n=0}^{\infty} E_{kn}\psi_{kn}(Y_k)\psi_{kn}(Z_k)\,\mathrm{e}^{-(1+\alpha_n^2)T}. \tag{6.168}$$

It is of interest to note that this solution is not necessarily symmetric when the input and recording sites are interchanged.

The series in Equations (6.167) and (6.168) converge rapidly for time $T = O(1)$ and may be used to numerically calculate G_j^k and G_k^k by truncating the series after a suitable number of terms. This solution is dependent upon the determination of the (nonnegative) roots of the transcendental Equation (6.165); an important observation for which is that the unbounded points of the functions $y_i = y_i(\alpha) = w_i + \mu$ for $i = j$ or k, with w_i defined in Equation (6.166), provide consecutive nonoverlapping intervals in which to search for the roots (where none or more than one may occur). The slow convergence of the series for small time (as $T \to 0$) requires successively more terms to be retained in order to obtain suitable accuracy. While this places little restriction on the practical evaluation of the solution, we should note that more appropriate expressions valid for small times exist. These may be found by expanding the Laplace transform solutions (6.161) and (6.162) for $(p, q_j, q_k) \gg 1$ and inverting. The leading order behavior is recorded below for completeness assuming $\epsilon_j\epsilon_k\sigma_{sj}\sigma_{sk} \neq 0$.

$$G_j^k \sim \frac{\mu}{\epsilon_j\epsilon_k\sigma_{sj}\sigma_{sk}} A_k(Y_k)A_j(Z_j) \times \sum_{i=1}^{4} r_i \left[\left(T + \frac{\chi_i^2}{2}\right)\operatorname{erfc}\left(\frac{\chi_i}{2\sqrt{T}}\right) - \chi_i\left(\frac{T}{\pi}\right)^{1/2} e^{(\chi_i^2/4T)}\right] \quad \text{as } T \to 0, \tag{6.169}$$

where

$$r_1 = 1, \quad r_2 = \kappa_j, \quad r_3 = \kappa_k, \quad r_4 = \kappa_j\kappa_k,$$

$$\kappa_j = \begin{cases} 1 & \text{if } b_j = O(1), \\ -1 & \text{if } b_j = \infty, \end{cases} \qquad \kappa_k = \begin{cases} 1 & \text{if } b_k = O(1), \\ -1 & \text{if } b_k = \infty, \end{cases}$$

and

$$\begin{aligned} \chi_1 &= Y_k + Z_j, & \chi_2 &= 2L_j - Z_j + Y_k, \\ \chi_3 &= 2L_k - Y_k + Z_j, & \chi_4 &= 2L_k - Y_k + 2L_j - Z_j. \end{aligned} \tag{6.170}$$

Similarly, for $Z_k > Y_k$ we have

$$G_k^k \sim A_k(Y_k)A_k(Z_k)\frac{1}{2\sigma_k\sqrt{\pi T}} \sum_{i=1}^{4}(-1)^{i-1} e^{-(\Omega_i)^2/4T} + A_k(Y_k)A_k(Z_k)\frac{1}{\epsilon_k\sigma_{sk}} \sum_{i=1}^{4} r_i \left[\operatorname{erfc}\left(\frac{\chi_i}{2\sqrt{T}}\right)\right] \quad \text{as } T \to 0, \tag{6.171}$$

where

$$\begin{aligned} \Omega_1 &= Z_k - Y_k, & \Omega_2 &= Z_k + Y_k, \\ \Omega_3 &= 2L_k - Z_k - Y_k, & \Omega_4 &= 2L_k + Y_k - Z_k, \end{aligned} \tag{6.172}$$

and the χ_i are defined in Equation (6.170) with j set to k.

The required expansions in the case $Z_k < Y_k$ may be obtained from Equation (6.171) by simply interchanging Z_k and Y_k in the expressions (6.172) for the Ω_i and (6.170) for the χ_i.

6.4.1.2 The General Solution

The solution to the general problem, Equations (6.129) to (6.135), is given by convolution of the unit impulse solution G_k^k and G_j^k with appropriate input currents and an initial voltage distribution.

The general solution, in this case, being

$$V_i(Z_i,T) = \sum_{k=1}^{2}\int_0^{L_k}\int_0^T G_i^k(Z_i,Y_k,T-u)I_k(Y_k,u;V_k)\,du\,dY_k + \sum_{k=1}^{2}\int_0^{L_k} G_i^k(Z_i,Y_k,T)H_k(Y_k)\,dY_k + \sum_{k=1}^{2}\int_0^T G_i^k(Z_i,0,T-u)I_{sk}(u)\,du, \quad (6.173)$$

where $i = 1$ or 2. $G_i^k(Z_i,0,T)$ is the solution to a unit impulse at the soma of neuron k ($k = 1$ or 2), that is, the solution of Equations (6.129) to (6.135) with $I_1 = I_2 = 0$, $H_1 = H_2 = 0$, and $I_{sk} = \delta(T)$, $I_{sj} = 0$ ($j \neq k$). Explicitly,

$$G_i^k(Z_i,0,T) = \lim_{Y_k\to 0^+} G_i^k(Z_i,Y_k,T) = A_i(Z_i)\sum_{n=0}^{\infty} E_{in}\psi_{in}(Z_i)\,e^{-(1+\alpha_n^2)T},$$

since $A_k(0) = 1$ and $\psi_{kn}(0) = 1$.

If $I_i(Z_i,T;V_i) = I_i(Z_i,T)$ is independent of V_i then Equation (6.173) gives the explicit general solution. When $I_i(Z_i,T;V_i)$ depends linearly on V_i (for synaptic reversal potentials and quasi-active ionic currents) in the form

$$I_i(Z_i,T;V_i) = g_i(Z_i,T)V_i(Z_i,T) + I_{Ai}(Z_i,T), \quad (6.174)$$

then the analysis of Section 6.2.3 carries straight over. The general solution is

$$V_i(Z_i,T) = f_i(Z_i,T) + \sum_{k=1}^{2}\int_0^{L_k}\int_0^T \sum_{n=1}^{\infty} \mathcal{K}^n(Z_i,\xi_k;T,u)f_k(\xi_k,u)\,du\,d\xi_k, \quad (6.175)$$

where

$$f_i(Z_i,T) = \sum_{k=1}^{2}\int_0^{L_k}\int_0^T G_i^k(Z_j,\xi_k,T-u)I_{Ak}(\xi_k,u)\,du\,d\xi_k + \sum_{k=1}^{2}\int_0^{L_k} G_i^k(Z_i,Y_k,T)H_k(Y_k)\,dY_k + \sum_{k=1}^{2}\int_0^T G_i^k(Z_i,0,T-u)I_{sk}(u)\,du,$$

and the iterated kernels are as defined in Equations (6.104) and (6.105) with $N = 2$ and Z_j replaced with Z_i.

6.4.1.3 Example 3

The dependence of the solution upon the coupling parameter μ is shown here through the response to excitatory and inhibitory synaptic inputs at discrete locations with alpha-function conductance changes. We consider the following model parameters,

cylinder 1	$\sigma_1 = 1$,	$L_1 = 1.5$,	GRF = UNI,
soma 1	$\epsilon_1 = 1.0$,	$\sigma_{s1} = 1$,	
cylinder 2	$\sigma_2 = 1$,	$L_2 = 1$,	GRF = UNI,
soma 2	$\epsilon_2 = 1.0$,	$\sigma_{s2} = 1$,	

which are fixed in the following simulations. The coupling parameter μ will be taken as 0, 1, 5, 10, 50 to illustrate its effects. Cylinders 1 and 2 have uniform taper type (UNI), the effect of tapering not being explicitly considered here.

In Figure 6.11 and Figure 6.12, we consider excitatory and inhibitory synaptic reversal potential inputs in the form

$$I_i(Z_i, T_i, V_i) = \delta_{i1}\,\Gamma^{\text{ex}}_{\text{syn1,max}}\,\alpha_1^{\text{ex}} T\, \mathrm{e}^{1-\alpha_1^{\text{ex}}T}\,\delta(Z_1 - Z_{01}^{\text{ex}})(V_{\text{rev1}}^{\text{ex}} - V_1)$$
$$+ \delta_{iq}\Gamma^{\text{in}}_{\text{syn}q,\text{max}}\alpha_q^{\text{in}} T\, \mathrm{e}^{1-\alpha_q^{\text{in}}T}\,\delta(Z_q - Z_{0q}^{\text{in}})(V_{\text{rev}q}^{\text{in}} - V_q),$$

where the excitatory input is kept fixed and located at $Z_{01}^{\text{ex}} = 0.5L_1$, the midpoint of cylinder 1, while the inhibitory input is located at Z_{0q}^{in} where q is taken to be 1 or 2 depending on its location in

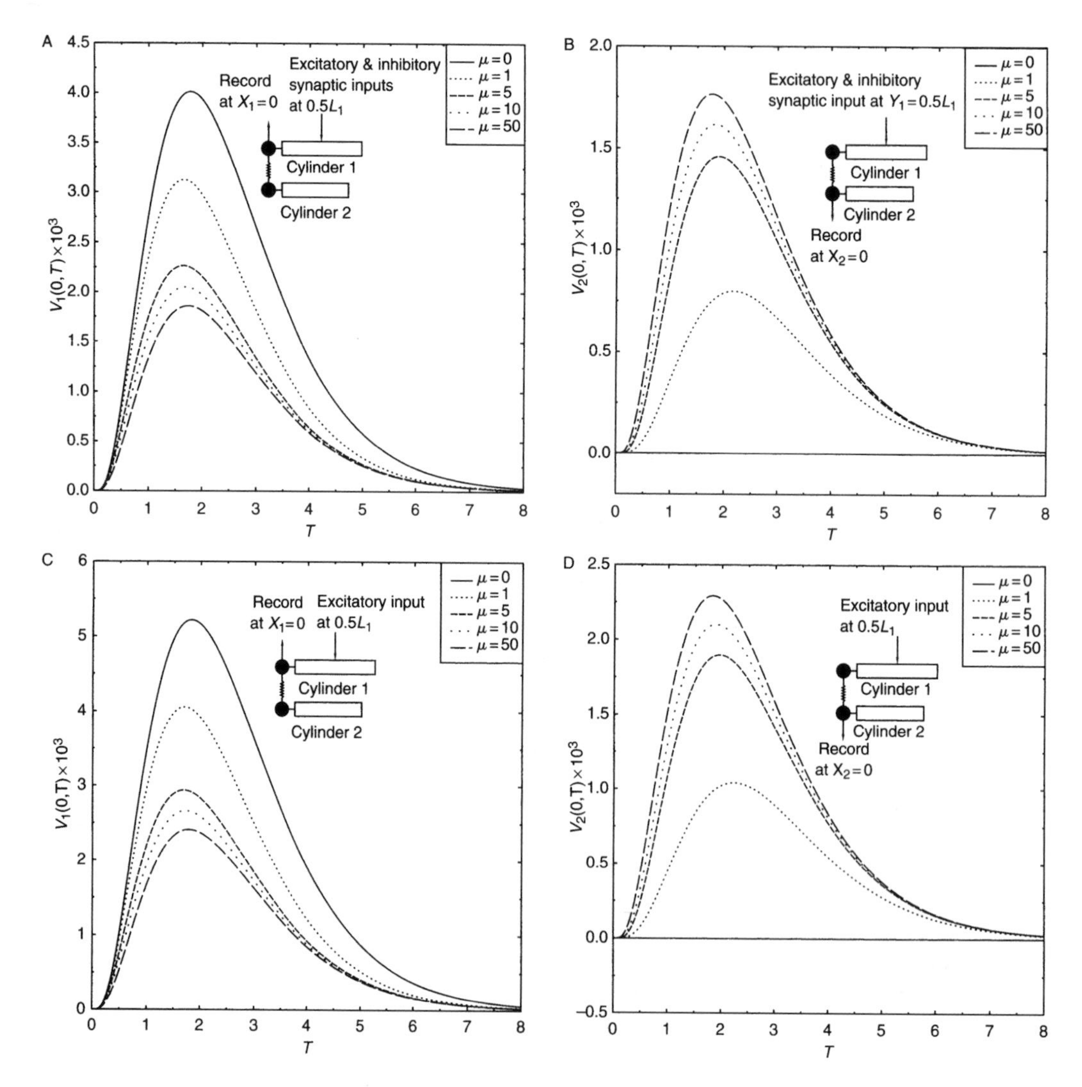

FIGURE 6.11 Illustration of the potential at the two somas for excitatory and inhibitory synaptic inputs of Example 3. (A) and (B) show the results for an excitatory and inhibitory inputs at the same location (midpoint of cylinder 1), while (C) and (D) are for an excitatory input only (also at the midpoint of cylinder). Parameter values are stated in the text.

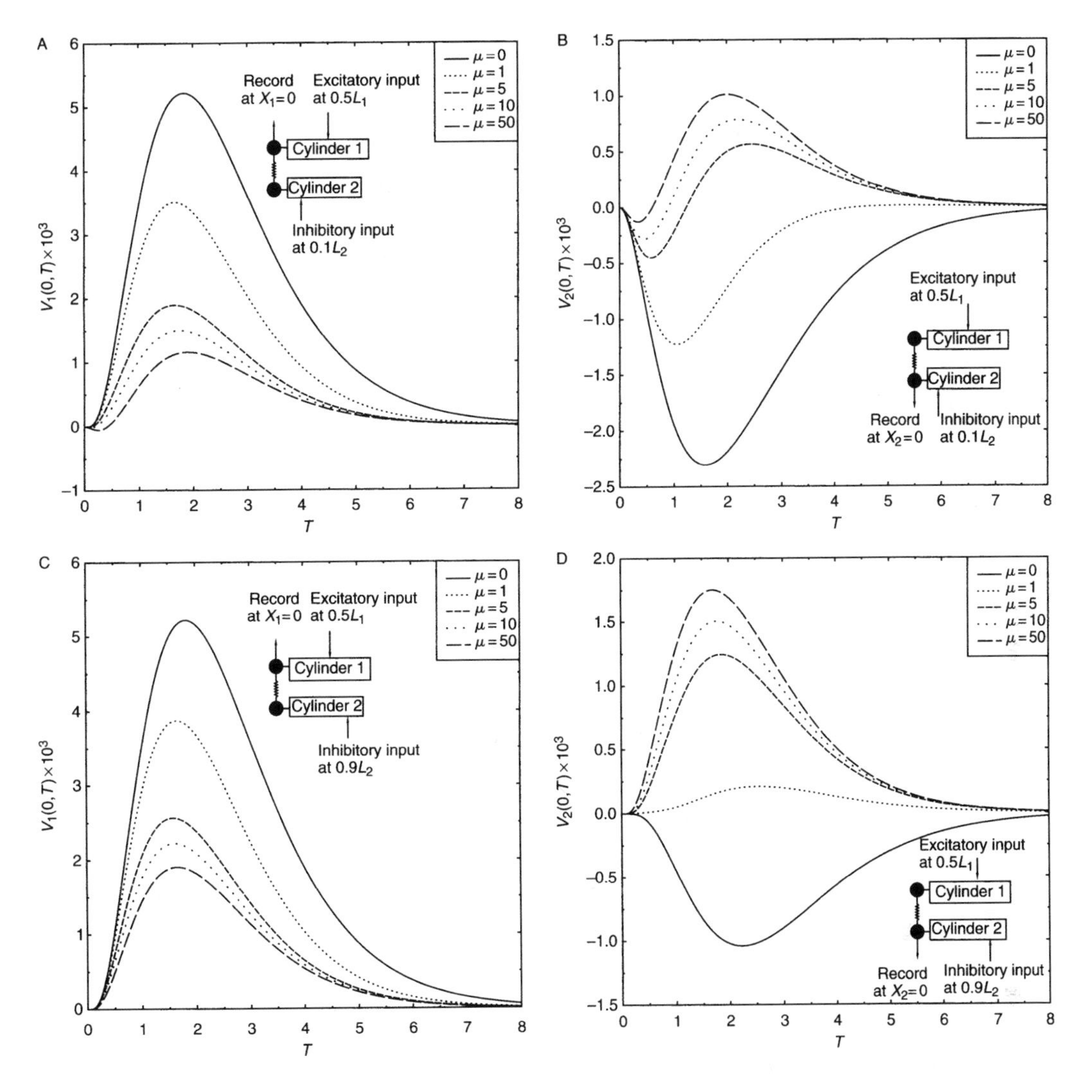

FIGURE 6.12 Illustration of the potential at the two somas for excitatory and inhibitory synaptic inputs in different cylinders. Parameter values are as given in Example 3. The excitatory input is kept fixed at the midpoint of cylinder 1, while the inhibitory input is located proximally in cylinder 2 for (A) and (B) and distally for (C) and (D).

cylinder 1 or 2, respectively. The parameter values adopted are,

$$\Gamma^{\text{ex}}_{\text{syn1,max}} = 1, \quad \Gamma^{\text{in}}_{\text{syn}q,\text{max}} = 2, \quad \alpha^{\text{ex}}_1 = 1.5, \quad \alpha^{\text{in}}_q = 1.25, \quad V^{\text{ex}}_{\text{rev1}} = 1, \quad V^{\text{in}}_{\text{rev}q} = -0.1,$$

the inhibitory maximum conductance change being twice as strong as that for the excitatory input and the inhibitory time course being slightly slower, $\alpha^{\text{in}}_q < \alpha^{\text{ex}}_1$. Figure 6.11A and B illustrates the effect of μ on the potentials at the somas when the excitatory and inhibitory inputs are at the same location ($Z^{\text{in}}_{01} = 0.5L_1$). Figure 6.11C and D shows the equivalent responses for just the excitatory input (the inhibitory input removed). The Green's function solution at the soma is expected to show similar sensitivity to changes in μ. Figure 6.11C and D illustrates how relatively large potentials can form in the soma of the remote cylinder, even for small values of the coupling parameter.

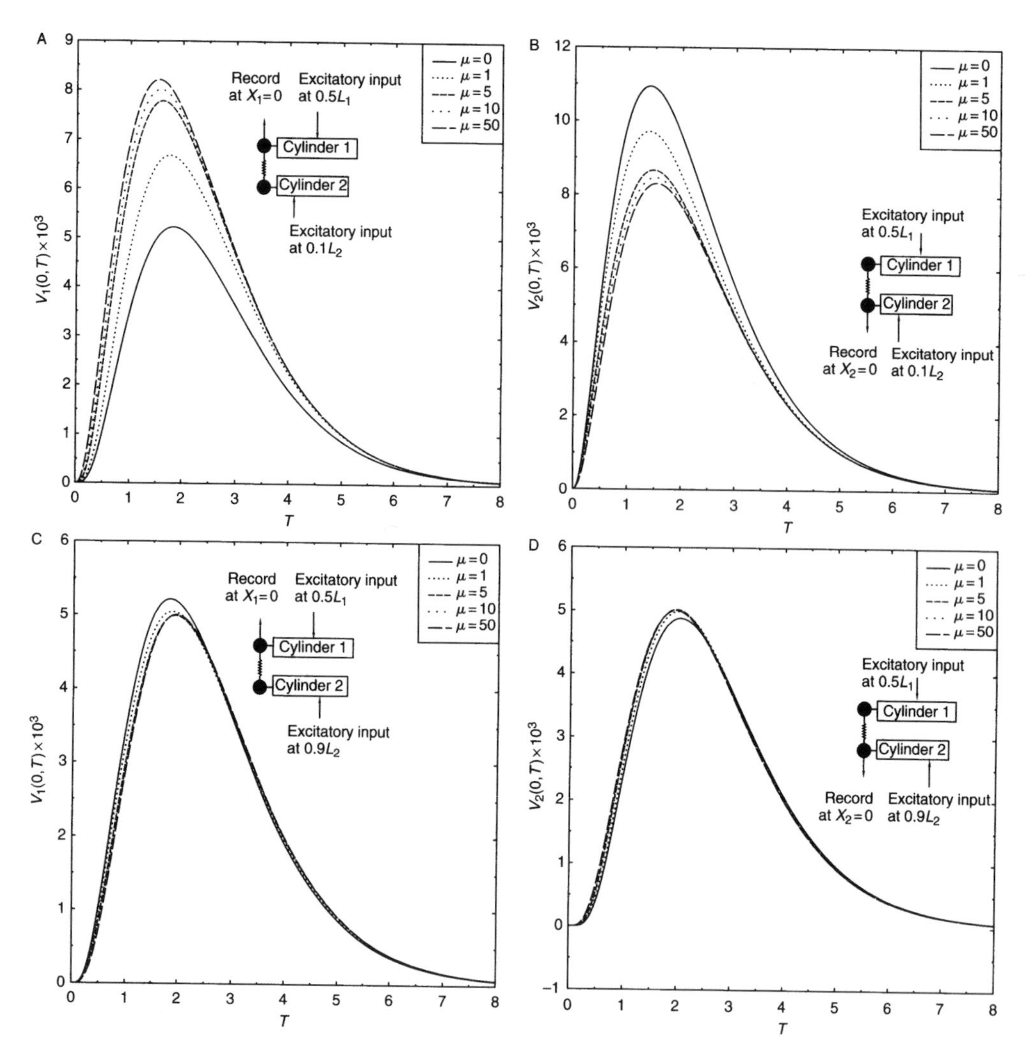

FIGURE 6.13 Illustration of the potential at the two somas for excitatory synaptic inputs in different cylinders. Parameter values are as given in Example 3. The excitatory input in cylinder 1 is kept fixed at its midpoint, while the excitatory input in cylinder 2 is located proximally for (A) and (B) and distally for (C) and (D).

Figure 6.12 shows the potentials of the somas when the inhibitory input is located in cylinder 2, remote from the excitatory input in cylinder 1. The inhibitory input is taken at a proximal location $Z_{02}^{\text{in}} = 0.1L_2$ in Figure 6.12A and B and a distal location $Z_{02}^{\text{in}} = 0.9L_2$ in Figure 6.12C and D. Figure 6.12A and C illustrate that for small μ the location of the remote inhibitory input is less significant than for larger values of μ. Figure 6.12B and D illustrates that the remotely placed excitatory input has far greater impact on the soma waveform not only as μ increases but as the location of the inhibitory input from the soma increases. Again, the impact even for small μ is significant.

Figure 6.13 shows the effect of two excitatory inputs with the same strength and time course but located in separate cylinders. The input is taken in the form,

$$I_i(Z_i, T; V_i) = \delta_{i1}\Gamma^1_{\text{syn1,max}}\,\alpha_1^1 T\,\mathrm{e}^{1-\alpha_1^1 T}\delta(Z_1 - Z_{01}^{\text{ex}})(V^1_{\text{rev1}} - V_1)$$
$$+\,\delta_{i2}\Gamma^1_{\text{syn2,max}}\alpha_2^1 T\,\mathrm{e}^{1-\alpha_2^1 T}\delta(Z_2 - Z_{02}^{\text{ex}})(V^1_{\text{rev2}} - V_2),$$

where

$$\Gamma^1_{\text{syn1,max}} = \Gamma^1_{\text{syn2,max}} = 1, \qquad \alpha^1_1 = \alpha^1_2 = 1.5, \qquad V^1_{\text{rev1}} = V^1_{\text{rev2}} = 1.$$

The excitatory input in cylinder 1 is fixed with $Z^{\text{ex}}_{01} = 0.5L_1$ as before, and for the input in cylinder 2, a proximal and a distal location, $Z^{\text{ex}}_{02} = 0.1L_2$ in Figure 6.13A and B and $Z^{\text{ex}}_{02} = 0.9L_2$ in Figure 6.13C and D is taken, respectively. Figure 6.13A and B illustrates the relative large sensitivity of the soma potentials to proximally placed inputs, while Figure 6.13C and D show the relative insensitivity to proximally placed excitatory inputs. The relative lack of sensitivity in Figure 6.13C and D is unexpected.

Importantly, the model allows quantification of the effect of the coupling parameter on voltage time course with regard to both time to peak and decay. Single tapering cylinders were chosen here to represent each neuron, but in principle, tapering multicylinders could be used. Further extensions include coupling between several neurons, all of which admit closed form analytical solutions in the passive or quasi-active cases.

6.4.2 Dendro–Dendritic Coupling

Here we consider two neurons, each represented by a single tapering equivalent cylinder and isopotential soma, coupled at their terminations through a gap junction. Following the notation of Section 6.4.1, the mathematical statement of the dimensional equations for this model is as follows,

$$\text{in } 0 < Z_1 < L_1, \quad t > 0 \qquad v_1 + \tau_m \frac{\partial v_1}{\partial t} - \left(\frac{\partial^2 v_1}{\partial Z_1^2} + \frac{1}{F_1}\frac{dF_1}{dZ_1}\frac{\partial v_1}{\partial Z_1} \right) = \lambda_1^* r_{a1}^* \hat{I}_1(Z_1, t; v_1); \tag{6.176}$$

$$\text{in } 0 < Z_2 < L_2, \quad t > 0 \qquad v_2 + \tau_m \frac{\partial v_2}{\partial t} - \left(\frac{\partial^2 v_2}{\partial Z_2^2} + \frac{1}{F_2}\frac{dF_2}{dZ_2}\frac{\partial v_2}{\partial Z_2} \right) = \lambda_2^* r_{a2}^* \hat{I}_2(Z_2, t; v_2); \tag{6.177}$$

$$\text{at } Z_1 = L_1 \text{ and } Z_2 = L_2 \qquad -\frac{1}{\lambda_1^* r_{a1}^*}\frac{\partial v_1}{\partial Z_1} = \frac{v_1 - v_2}{R_c} = \frac{1}{\lambda_2^* r_{a2}^*}\frac{\partial v_2}{\partial Z_2}; \tag{6.178}$$

$$\text{at } Z_1 = 0 \qquad \frac{1}{R_{s1}}\left(v_1 + \tau_{s1}\frac{\partial v_1}{\partial t} \right) - \frac{1}{\lambda_{01}^* r_{a01}^*}\frac{\partial v_1}{\partial Z_1} = \hat{I}_{s1}(t); \tag{6.179}$$

$$\text{at } Z_2 = 0 \qquad \frac{1}{R_{s2}}\left(v_2 + \tau_{s2}\frac{\partial v_2}{\partial t} \right) - \frac{1}{\lambda_{02}^* r_{a02}^*}\frac{\partial v_2}{\partial Z_2} = \hat{I}_{s2}(t); \tag{6.180}$$

$$\text{at } t = 0, \qquad v_1 = h_1(Z_1), \qquad v_2 = h_2(Z_2). \tag{6.181}$$

In this model, the axial current through the gap-junction resistance is given by $(v_2(L_2, t) - v_1(L_1, t))/R_c$, with the boundary conditions (6.178) following from the conservation of current at the cylinder terminations. We have also allowed for differing somatic time constants and resistances for both neurons. The variables Z_i, parameters L_i, and geometric functions $F_i(Z_i)$ are dimensionless and nondimensionalization of the equations is completed with the scalings

$$t = \tau_m T, \qquad v_i = \bar{I}\bar{R}V_i, \qquad h_i = \bar{I}\bar{R}H_i, \qquad \hat{I}_i = \bar{I}I_i, \qquad \hat{I}_{si} = \bar{I}I_{si},$$

where we take $\bar{R} = \lambda_{01}^* r_{a01}^*$, and $\bar{I}$ is an appropriate current scaling (the choice being determined by the type of input). This introduces the dimensionless parameters,

$$\epsilon_1 = \frac{\tau_{s1}}{\tau_m}, \quad \epsilon_2 = \frac{\tau_{s2}}{\tau_m}, \quad \sigma_{s1} = \frac{\lambda_{01}^* r_{a01}^*}{R_{s1}}, \quad \sigma_{s2} = \frac{\lambda_{01}^* r_{a01}^*}{R_{s2}},$$
$$\sigma_1 = 1, \quad \sigma_2 = \frac{\lambda_{01}^* r_{a01}^*}{\lambda_{02}^* r_{a02}^*} = \left(\frac{D_{02}^*}{D_{01}^*}\right)^{3/2}, \quad \mu = \frac{\lambda_{01}^* r_{a01}^*}{R_c}. \tag{6.182}$$

The system (6.176) to (6.181) then becomes

$$\text{in } 0 < Z_1 < L_1, \quad T > 0 \qquad V_1 + \frac{\partial V_1}{\partial T} - \frac{\partial^2 V_1}{\partial Z_1^2} - \frac{1}{F_1}\frac{dF_1}{dZ_1}\frac{\partial V_1}{\partial Z_1} = \frac{1}{\sigma_1 F_1(Z_1)} I_1(Z_1, T; V_1); \tag{6.183}$$

$$\text{in } 0 < Z_2 < L_2, \quad T > 0 \qquad V_2 + \frac{\partial V_2}{\partial T} - \frac{\partial^2 V_2}{\partial Z_2^2} - \frac{1}{F_2}\frac{dF_2}{dZ_2}\frac{\partial V_2}{\partial Z_2} = \frac{1}{\sigma_2 F_2(Z_2)} I_2(Z_2, T; V_2); \tag{6.184}$$

$$\text{at } Z_1 = L_1 \text{ and } Z_2 = L_2 \qquad \sigma_1 F_1(L_1)\frac{\partial V_1}{\partial Z_1} = \mu(V_2 - V_1) = -\sigma_2 F_2(L_2)\frac{\partial V_2}{\partial Z_2}; \tag{6.185}$$

$$\text{at } Z_1 = 0 \qquad \sigma_{s1}\left(V_1 + \epsilon_1 \frac{\partial V_1}{\partial T}\right) - \sigma_1 \frac{\partial V_1}{\partial Z_1} = I_{s1}(T); \tag{6.186}$$

$$\text{at } Z_2 = 0 \qquad \sigma_{s2}\left(V_2 + \epsilon_2 \frac{\partial V_2}{\partial T}\right) - \sigma_2 \frac{\partial V_2}{\partial Z_2} = I_{s2}(T); \tag{6.187}$$

$$\text{at } T = 0 \qquad V_1 = H_1(Z_1), \quad V_2 = H_2(Z_2). \tag{6.188}$$

Although $\sigma_1 = 1$, this parameter is retained to maintain the symmetry in the equations. The transformation Equation (6.136) gives

$$\text{in } 0 < Z_1 < L_1, \quad T > 0 \qquad a_1\hat{V}_1 + \frac{\partial \hat{V}_1}{\partial T} - \frac{\partial^2 \hat{V}_1}{\partial Z_1^2} = \frac{1}{\sigma_1 F_1(Z_1)^{1/2}} I_1(Z_1, T; \hat{V}_1); \tag{6.189}$$

$$\text{in } 0 < Z_2 < L_2, \quad T > 0 \qquad a_2\hat{V}_2 + \frac{\partial \hat{V}_2}{\partial T} - \frac{\partial^2 \hat{V}_2}{\partial Z_2^2} = \frac{1}{\sigma_2 F_2(Z_2)^{1/2}} I_2(Z_2, T; \hat{V}_2); \tag{6.190}$$

$$\text{at } Z_1 = L_1 \text{ and } Z_2 = L_2 \qquad \frac{\partial \hat{V}_1}{\partial Z_1} + b_1\hat{V}_1 = d_1\hat{V}_2, \tag{6.191}$$

$$\frac{\partial \hat{V}_2}{\partial Z_2} + b_2\hat{V}_2 = d_2\hat{V}_1; \tag{6.192}$$

$$\text{at } Z_1 = 0 \qquad \sigma_{s1}\left(\hat{V}_1 + \epsilon_1 \frac{\partial \hat{V}_1}{\partial T}\right) + c_1\sigma_1\hat{V}_1 - \sigma_1 \frac{\partial \hat{V}_1}{\partial Z_1} = I_{s1}; \tag{6.193}$$

$$\text{at } Z_1 = 0 \qquad \sigma_{s2}\left(\hat{V}_2 + \epsilon_2 \frac{\partial \hat{V}_2}{\partial T}\right) + c_2\sigma_2\hat{V}_2 - \sigma_2 \frac{\partial \hat{V}_2}{\partial Z_2} = I_{s2}; \tag{6.194}$$

$$\text{at } T = 0, \qquad \hat{V}_1 = F_1(Z_1)^{1/2} H_1(Z_1), \quad \hat{V}_2 = F_2(Z_2)^{1/2} H_2(Z_2), \tag{6.195}$$

where for $i = 1, 2$,

$$b_i = \frac{2\mu - \sigma_i F_i'(L_i)}{2\sigma_i F_i(L_i)}, \qquad c_i = \frac{1}{2}F_i'(0), \qquad d_i = \frac{\mu}{\sigma_i (F_1(L_1)F_2(L_2))^{1/2}}, \qquad A_i(Z_i) = F_i(Z_i)^{-1/2}, \tag{6.196}$$

with $'$ denoting $\mathrm{d}/\mathrm{d}Z_i$ and a_i given by Equation (6.145).

The dimensionless model has the eight nondimensional parameters, $\epsilon_i, \sigma_{si}, \sigma_i, L_i, \mu$ (since $\sigma_1 = 1$), and up to four dimensionless tapering parameters, K_i, Z_{0i} for $i = 1, 2$.

6.4.2.1 The Green's Function

The Green's function is defined to be the solution of Equations (6.183) to (6.188) with a unit impulse input into one of the cylinders, denoted as cylinder k (where $k = 1$ or 2). The solution can be evaluated in either the same cylinder k or the other cylinder, denoted by j where $j = 1$ or 2 but $j \neq k$. Thus we consider,

$$\begin{aligned} I_k &= \delta(T)\delta(Z_k - Y_k), \qquad I_j = 0, \quad j \neq k, \\ I_{s1} &= I_{s2} = 0, \qquad H_1 = H_2 = 0, \end{aligned} \tag{6.197}$$

which represents a unit impulse to cylinder k (where $k = 1$ or 2) at a nondimensional distance Y_k from the soma of cylinder k located at $Z_k = 0$. We denote the solution by $G_k^k(Z_k, Y_k, T)$ when recording in cylinder k at a nondimensional distance Z_k from the soma at $Z_k = 0$ and by $G_j^k(Z_j, Y_k, T)$ the solution when recording in cylinder $j \neq k$ at a distance Z_j from the soma in cylinder j located at $Z_j = 0$. Using the transformation Equation (6.136), we obtain

$$G_i^k(Z_i, Y_k, T) = A_i(Z_i)U_i^k(Z_i, Y_k, T), \quad \text{for } i = j \text{ or } k, \tag{6.198}$$

where $U_i^k(Z_i, Y_k, T)$ with $i = j$ or k is the corresponding Green's function for Equations (6.189) to (6.195) and satisfies

$$\text{in } 0 < Z_k < L_k, \quad T > 0 \qquad a_k U_k^k + \frac{\partial U_k^k}{\partial T} - \frac{\partial^2 U_k^k}{\partial Z_k^2} = \frac{A_k(Z_k)}{\sigma_k}\delta(Z_k - Y_k)\delta(T); \tag{6.199}$$

$$\text{in } 0 < Z_j < L_j, \quad T > 0 \qquad a_j U_j^k + \frac{\partial U_j^k}{\partial T} - \frac{\partial^2 U_j^k}{\partial Z_j^2} = 0; \tag{6.200}$$

$$\text{at } Z_k = L_k \text{ and } Z_j = L_j \qquad \frac{\partial U_k^k}{\partial Z_k} + b_k U_k^k = d_k U_j^k, \quad \frac{\partial U_j^k}{\partial Z_j} + b_j U_j^k = d_j U_k^k; \tag{6.201}$$

$$\text{at } Z_k = 0 \qquad \sigma_{sk}\left(U_k^k + \epsilon_k \frac{\partial U_k^k}{\partial T}\right) + c_k \sigma_k U_k^k - \sigma_k \frac{\partial U_k^k}{\partial Z_k} = 0; \tag{6.202}$$

$$\text{at } Z_j = 0 \qquad \sigma_{sj}\left(U_j^k + \epsilon_j \frac{\partial U_j^k}{\partial T}\right) + c_j \sigma_j U_j^k - \sigma_j \frac{\partial U_j^k}{\partial Z_j} = 0; \tag{6.203}$$

$$\text{at } T = 0, \qquad U_k^k = U_j^k = 0, \tag{6.204}$$

for j and k taking the values 1 or 2, but $j \neq k$.

The Laplace transform of U_i^k is

$$\bar{U}_i^k(Z_i, Y_k, p) = \int_0^\infty e^{-pT} U_i^k(Z_i, Y_k, T)\, dT,$$

where $i = k$ or j satisfies

$$\text{in } 0 < Z_k < L_k \qquad (a_k + p)\bar{U}_k^k - \frac{\partial^2 \bar{U}_k^k}{\partial Z_k^2} = \frac{A_k}{\sigma_k}\delta(Z_k - Y_k); \tag{6.205}$$

$$\text{in } 0 < Z_j < L_j \qquad (a_j + p)\bar{U}_j^k - \frac{\partial^2 \bar{U}_j^k}{\partial Z_j^2} = 0; \tag{6.206}$$

$$\text{at } Z_k = L_k \quad \text{and} \quad Z_j = L_j \qquad \frac{\partial \bar{U}_k^k}{\partial Z_k} + b_k \bar{U}_k^k = d_k \bar{U}_j^k, \quad \frac{\partial \bar{U}_j^k}{\partial Z_j} + b_j \bar{U}_j^k = d_j \bar{U}_k^k; \tag{6.207}$$

$$\text{at } Z_k = 0 \qquad (\sigma_{sk}(1 + \epsilon_k p) + c_k\sigma_k)\, \bar{U}_k^k - \sigma_k \frac{\partial \bar{U}_k^k}{\partial Z_k} = 0; \tag{6.208}$$

$$\text{at } Z_j = 0 \qquad \left(\sigma_{sj}(1 + \epsilon_j p) + c_j\sigma_j\right) \bar{U}_j^k - \sigma_j \frac{\partial \bar{U}_j^k}{\partial Z_j} = 0. \tag{6.209}$$

The solution to Equations (6.205) to (6.209) is,

$$\bar{U}_j^k = \frac{d_j}{\Delta} \frac{A_k(Y_k)}{\sigma_k} \bar{\psi}_k(Y_k)\bar{\psi}_j(Z_j), \tag{6.210}$$

and

$$\bar{U}_k^k = \begin{cases} \dfrac{A_k(Y_k)}{\sigma_k} \bar{\psi}_k(Y_k) \left[\dfrac{(b_j + \bar{w}_j q_j)}{\Delta} \bar{\psi}_k(Z_k) + \dfrac{\sinh q_k(L_k - Z_k)}{q_k} \right] & \text{if } Z_k > Y_k, \\ \dfrac{A_k(Y_k)}{\sigma_k} \bar{\psi}_k(Z_k) \left[\dfrac{(b_j + \bar{w}_j q_j)}{\Delta} \bar{\psi}_k(Y_k) + \dfrac{\sinh q_k(L_k - Y_k)}{q_k} \right] & \text{if } Z_k < Y_k, \end{cases} \tag{6.211}$$

where

$$q_j = \sqrt{a_j + p}, \qquad q_k = \sqrt{a_k + p}, \qquad \Delta = (b_k + \bar{w}_k q_k)(b_j + \bar{w}_j q_j) - d_j d_k, \tag{6.212}$$

and for $i = k$ or j,

$$\bar{w}_i = \frac{(\sigma_{si}(1 + \epsilon_i p) + c_i\sigma_i)\cosh(q_i L_i) + \sigma_i q_i \sinh(q_i L_i)}{(\sigma_{si}(1 + \epsilon_i p) + c_i\sigma_i)\sinh(q_i L_i) + \sigma_i q_i \cosh(q_i L_i)},$$

$$\bar{\psi}_i(Z_i) = \frac{(\sigma_{si}(1 + \epsilon_i p) + c_i\sigma_i)\sinh(q_i Z_i) + \sigma_i q_i \cosh(q_i Z_i)}{(\sigma_{si}(1 + \epsilon_i p) + c_i\sigma_i)\sinh(q_i L_i) + \sigma_i q_i \cosh(q_i L_i)},$$

with $\bar{\psi}_i(Y_i)$ being obtained from this last expression by replacing Z_i with Y_i.

The limit $R_c \to \infty$ corresponds to $\mu \to 0$ and hence $d_k \to 0$ and $d_j \to 0$. In this case, the equations decouple and we recover the single cylinder result, namely that $\bar{U}_j^k = 0$ and

$$\bar{U}_k^k = \begin{cases} A_k(Y_k)\bar{\psi}_k^*(Z_k)\left[\dfrac{\bar{\psi}_k^*(Y_k)}{\sigma} + \dfrac{\sinh q_k Y_k}{\sigma_k q_k}\right] & \text{if } Z_k > Y_k, \\ A_k(Y_k)\bar{\psi}_k^*(Y_k)\left[\dfrac{\bar{\psi}_k^*(Z_k)}{\sigma} + \dfrac{\sinh q_k Z_k}{\sigma_k q_k}\right] & \text{if } Z_k < Y_k, \end{cases} \tag{6.213}$$

where

$$\sigma = \sigma_{sk}(1 + \epsilon_k p) + c_k\sigma_k + \sigma_k q_k \left\{\frac{\tanh q_k L_k - (b_k^*/q_k)}{1 - (b_k^*/q_k)\tanh q_k L_k}\right\},$$

and

$$\bar{\psi}_k^*(Z_k) = \frac{\cosh q_k(L_k - Z_k) - (b_k^*/q_k)\sinh q_k(L_k - Z_k)}{\cosh q_k L_k - (b_k^*/q_k)\sinh q_k L_k}, \quad \text{with } b_k^* = \frac{F_k'(L_k)}{2F_k(L_k)},$$

which agrees with Equation (6.40) in Evans (2000) (when interpreted for the particular case of a single cylinder [$N = 1$]).

The other limit of interest is $R_c \to 0$, which corresponds to $\mu_j \to \infty$ and $\mu_k \to \infty$. In this case the two neurons effectively fuse, and if we require our tapering profiles to be smooth, so that $F_k(L_k) = F_j(L_j)$ and $F_k'(L_k) = F_j'(L_j)$, then we obtain

$$\bar{U}_j^k = \frac{A_k(Y_k)}{\sigma_k(\bar{w}_j q_j + \bar{w}_k q_k)}\bar{\psi}_k(Y_k)\bar{\psi}_j(Z_j), \tag{6.214}$$

and

$$\bar{U}_k^k = \begin{cases} \dfrac{A_k(Y_k)}{\sigma_k}\bar{\psi}_k(Y_k)\left[\dfrac{1}{(\bar{w}_j q_j + \bar{w}_k q_k)}\bar{\psi}_k(Z_k) + \dfrac{\sinh q_k(L_k - Z_k)}{q_k}\right] & \text{if } Z_k > Y_k, \\ \dfrac{A_k(Y_k)}{\sigma_k}\bar{\psi}_k(Z_k)\left[\dfrac{1}{(\bar{w}_j q_j + \bar{w}_k q_k)}\bar{\psi}_k(Y_k) + \dfrac{\sinh q_k(L_k - Y_k)}{q_k}\right] & \text{if } Z_k < Y_k. \end{cases} \tag{6.215}$$

This solution corresponds to that of the single tapered cylinder case with a soma located at either end.

The solutions for U_j^k and U_k^k may be obtained by inverting Equations (6.210) and (6.211) (through evaluating the residues at the poles of the functions), which gives

$$U_j^k(Z_j, Y_k, T) = A_k(Y_k)\sum_{n=0}^{\infty} E_{jn}\psi_{kn}(Y_k)\psi_{jn}(Z_j)\,\mathrm{e}^{-(1+\alpha_n^2)T}, \tag{6.216}$$

and

$$U_k^k(Z_k, Y_k, T) = A_k(Y_k)\sum_{n=0}^{\infty} E_{kn}\psi_{kn}(Y_k)\psi_{kn}(Z_k)\,\mathrm{e}^{-(1+\alpha_n^2)T}, \tag{6.217}$$

where

$$\psi_{jn}(Z_j) = \frac{(\sigma_{sj}(1-\epsilon_j(1+\alpha_n^2)) + c_j\sigma_j)\sin(s_{jn}Z_j) + \sigma_j s_{jn}\cos(s_{jn}Z_j)}{(\sigma_{sj}(1-\epsilon_j(1+\alpha_n^2)) + c_j\sigma_j)\sin(s_{jn}L_j) + \sigma_j s_{jn}\cos(s_{jn}L_j)},$$

$$\psi_{kn}(Y_k) = \frac{(\sigma_{sk}(1-\epsilon_k(1+\alpha_n^2)) + c_k\sigma_k)\sin(s_{kn}Y_k) + \sigma_k s_{kn}\cos(s_{kn}Y_k)}{(\sigma_{sk}(1-\epsilon_k(1+\alpha_n^2)) + c_k\sigma_k)\sin(s_{kn}L_k) + \sigma_k s_{kn}\cos(s_{kn}L_k)},$$

and $\psi_{kn}(Z_k)$ can be obtained from $\psi_{kn}(Y_k)$ by replacing Y_k with Z_k, for $i = j$ or k,

$$s_{in} = \sqrt{1 - a_i + \alpha_n^2}, \tag{6.218}$$

$$E_{jn} = \frac{2d_j}{\sigma_k[(b_k - w_{kn}s_{kn})(w'_{jn} + (w_{jn}/s_{jn})) + (b_j - w_{jn}s_{jn})(w'_{kn} + (w_{kn}/s_{kn}))]},$$

and

$$E_{kn} = \frac{2(b_j - w_{jn}s_{jn})}{\sigma_k[(b_k - w_{kn}s_{kn})(w'_{jn} + (w_{jn}/s_{jn})) + (b_j - w_{jn}s_{jn})(w'_{kn} + (w_{kn}/s_{kn}))]},$$

with again for $i = j$ or k,

$$w_{in} = \tan(s_{in}L_i - \eta_{in}), \tag{6.219}$$

$$w'_{in} = \sec^2(s_{in}L_i - \eta_{in})\left[L_i + \left(\frac{2\sigma_{si}\epsilon_i}{\sigma_i} + \frac{\tan\eta_{in}}{s_{in}}\right)\cos^2\eta_{in}\right], \tag{6.220}$$

where

$$\tan\eta_{in} = \frac{\sigma_{si}(1-\epsilon_i(1+\alpha_n^2)) + c_i\sigma_i}{\sigma_i s_{in}}. \tag{6.221}$$

The $\alpha_n, n = 0, 1, 2, \ldots,$ are defined as the roots ($\alpha = \alpha_n$) of the transcendental equation

$$d_j d_k = (b_k - w_k s_k)(b_j - w_j s_j), \tag{6.222}$$

where for $i = j$ or k,

$$w_i = \tan(s_i L_i - \eta_i), \tag{6.223}$$

with

$$s_i = \sqrt{1 - a_i + \alpha^2}, \qquad \tan\eta_i = \frac{\sigma_{si}(1-\epsilon_i(1+\alpha^2)) + c_i\sigma_i}{\sigma_i s_i}. \tag{6.224}$$

The solution (6.163) represents the contribution from the simple poles $p = -(1+\alpha_n^2), n = 0, 1, 2, \ldots$.

Finally, the Green's function G_i^k ($i = j$ or k) corresponding to the Equations (6.183) to (6.188) is obtained from U_i^k ($i = j$ or k) by transformation (Equation [6.93]) and is given explicitly as follows,

$$G_j^k(Z_j, Y_k, T) = A_j(Z_j)A_k(Y_k)\sum_{n=0}^{\infty} E_{jn}\psi_{kn}(Y_k)\psi_{jn}(Z_j)\,\mathrm{e}^{-(1+\alpha_n^2)T}, \tag{6.225}$$

and

$$G_k^k(Z_k, Y_k, T) = A_k(Z_k)A_k(Y_k)\sum_{n=0}^{\infty} E_{kn}\psi_{kn}(Y_k)\psi_{kn}(Z_k)\,\mathrm{e}^{-(1+\alpha_n^2)T}. \tag{6.226}$$

6.4.2.2 The General Solution

The Green's functions just determined may be used to express Equations (6.183) to (6.188) in the form of (6.173) with the results of Section 6.4.1.2 carrying straight over. For inputs $I_i(Z_i, T)$ that are voltage independent, Equation (6.173) gives the explicit general solution, while when $I_i(Z_i, T; V_i)$ depends linearly on V_i (for synaptic reversal potentials and quasi-active ionic currents) then Equation (6.175) gives the general solution. Otherwise, Equation (6.173) gives a nonlinear system of integral equations.

6.4.2.3 Example 4

The dependence of the solution upon the coupling parameter μ is shown here again through the response to excitatory and inhibitory synaptic inputs at discrete locations with alpha-function conductance changes. We consider the following model parameters:

cylinder 1	$\sigma_1 = 1$,	$L_1 = 1.5$,	GRF = UNI,
soma 1	$\epsilon_1 = 1.0$,	$\sigma_{s1} = 1$,	
cylinder 2	$\sigma_2 = 1$,	$L_2 = 1$,	GRF = UNI,
soma 2	$\epsilon_2 = 1.0$,	$\sigma_{s2} = 1$,	

which are fixed in the following simulations. The coupling parameter μ will be taken as 0, 1, 5, 10 to illustrate its effects. Cylinders 1 and 2 have uniform taper type (UNI), the effect of tapering not being considered.

Figure 6.14 considers a single excitatory input of the form

$$I_i(Z_i, T; V_i) = \delta_{i1}\,\Gamma^1_{\mathrm{syn1,max}}\,\alpha_1^1 T\,\mathrm{e}^{1-\alpha_1^1 T}\delta(Z_1 - Z_{01}^1)(V_{\mathrm{rev1}}^1 - V_1) \tag{6.227}$$

with

$$\Gamma^1_{\mathrm{syn1,max}} = 1, \qquad \alpha_1^1 = 1.5, \qquad V_{\mathrm{rev1}}^1 = 1. \tag{6.228}$$

The times-to-peak of the potentials do not appear to be as significantly changed by the coupling as the peak amplitude and decay times.

Figure 6.15 considers a synchronously activated excitatory input in cylinder 1 and inhibitory input in cylinder 2 of the form

$$\begin{aligned} I_i(Z_i, T; V_i) = {} & \delta_{i1}\,\Gamma^1_{\mathrm{syn1,max}}\,\alpha_1^1 T\,\mathrm{e}^{1-\alpha_1^1 T}\,\delta(Z_1 - Z_{01}^1)(V_{\mathrm{rev1}}^1 - V_1) \\ & + \delta_{i2}\,\Gamma^1_{\mathrm{syn2,max}}\,\alpha_2^1 T\,\mathrm{e}^{1-\alpha_2^1 T}\delta(Z_2 - Z_{02}^1)(V_{\mathrm{rev2}}^1 - V_2), \end{aligned} \tag{6.229}$$

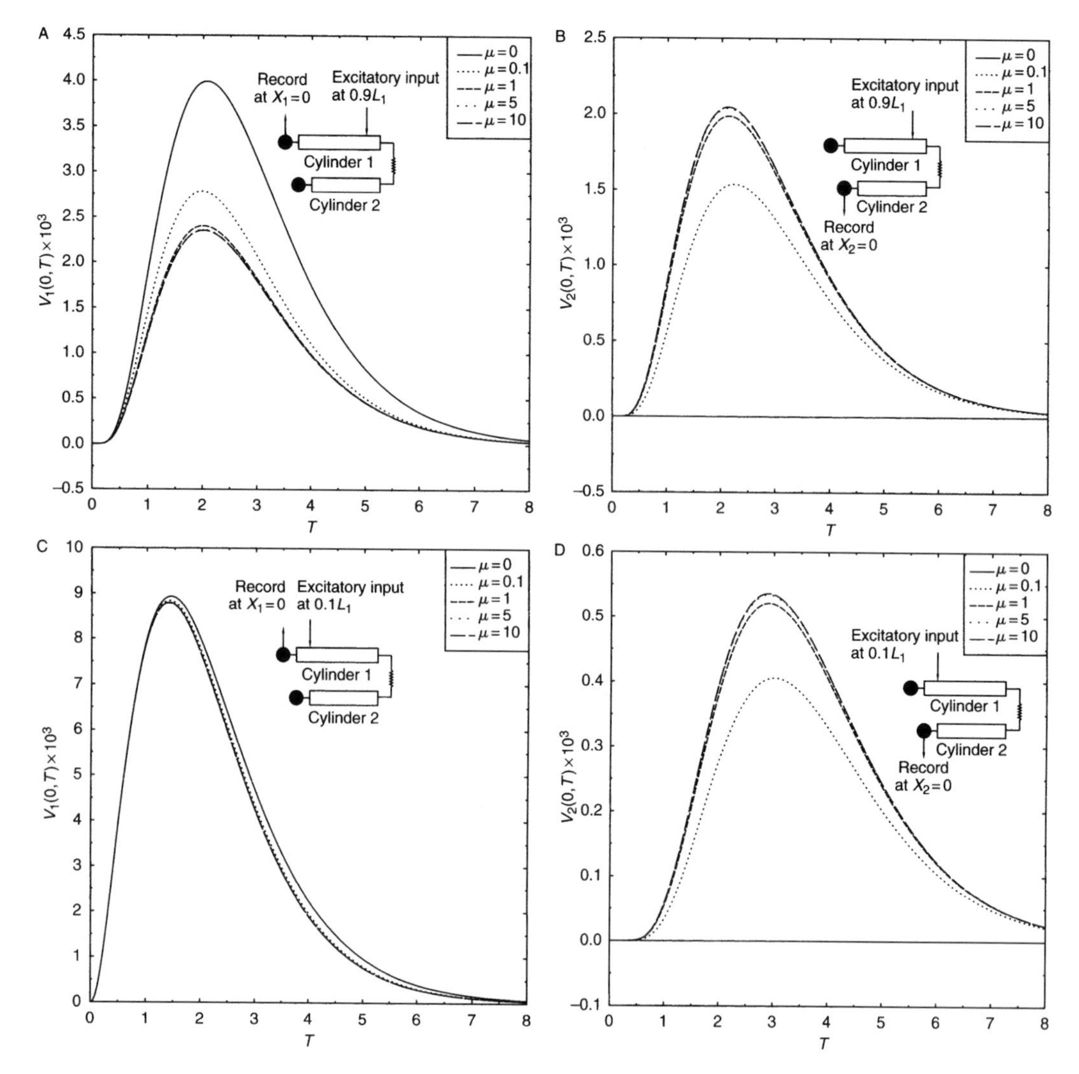

FIGURE 6.14 Illustration of the effect of the coupling parameter on the potential at the two somas for the excitatory synaptic input in cylinder 1 of Example 4. The input is located distally in (A) and (B) and proximally in (C) and (D).

where the parameters for the excitatory input are as given in Equation (6.228) and for the inhibitory input are

$$\Gamma^1_{\text{syn2,max}} = 2, \qquad \alpha^1_2 = 1.25, \qquad V^1_{\text{rev2}} = -0.1. \tag{6.230}$$

The inhibitory reversal potential is taken as small relative to that for excitatory input. The excitatory input is fixed at $Z^1_{01} = 0.5L_1$ while proximal at $Z^1_{02} = 0.1L_2$ and distal at $Z^1_{02} = 0.9L_2$ locations are taken for the inhibitory input. Figure 6.15A and C indicates that the coupling strength has a more significant impact than the location of the inhibitory input in the remote cylinder. However, this is not the case regarding inputs in the cylinder with recording at its soma. Figure 6.15B and D illustrates that the soma potential for distally placed inhibitory inputs is significantly affected by the remote excitatory input even for weak coupling. The effect of the remote excitatory input declines

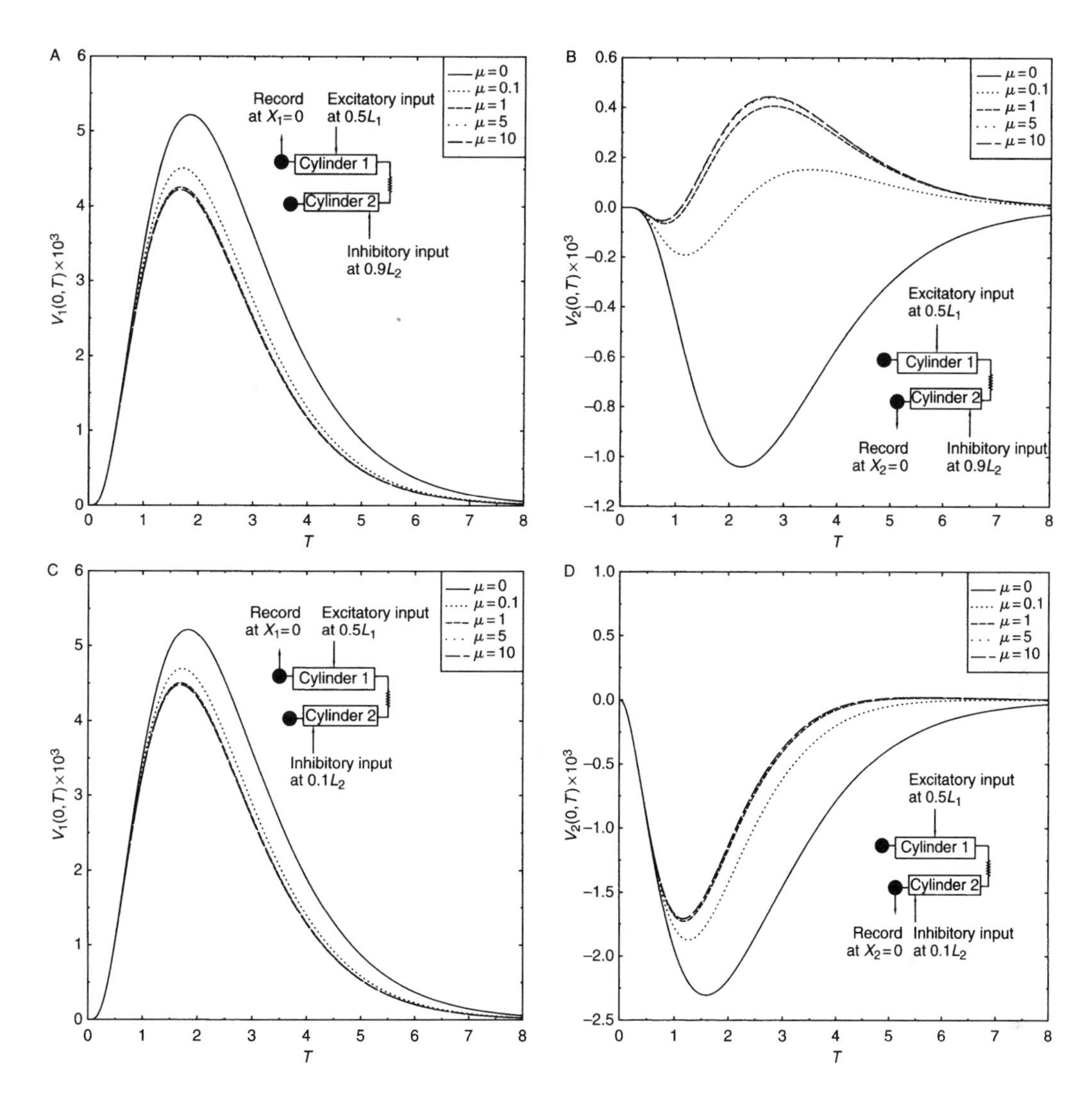

FIGURE 6.15 Illustration of the potential at the two somas for excitatory and inhibitory synaptic inputs in different cylinders. Parameter values are as given in Example 4. The excitatory input is kept fixed at the midpoint of cylinder 1, while the inhibitory input is located distally in cylinder 2 for (A) and (B) and proximally for (C) and (D).

for inhibitory inputs closer to the recording site at the soma. The sensitivity of the potential at the soma to the coupling parameter declines as the inhibitory input is located closer to the soma.

Figure 6.16 shows the effect when the inhibitory input for Figure 6.15 is replaced with an excitatory input (of the same strength and time course as the excitatory input in cylinder 1). The input is taken in the form of Equation (6.229) with the parameters for the input in cylinder 1 as given by Equation (6.228), while Equation (6.230) is replaced with

$$\Gamma^1_{\text{syn2,max}} = 1, \qquad \alpha^1_2 = 1.5, \qquad V^1_{\text{rev2}} = 1.0. \tag{6.231}$$

Figure 6.16A and C reinforces the earlier observation of the soma potential sensitivity to the coupling parameter being more important than the location of the remote excitatory input. Figure 6.16B and D

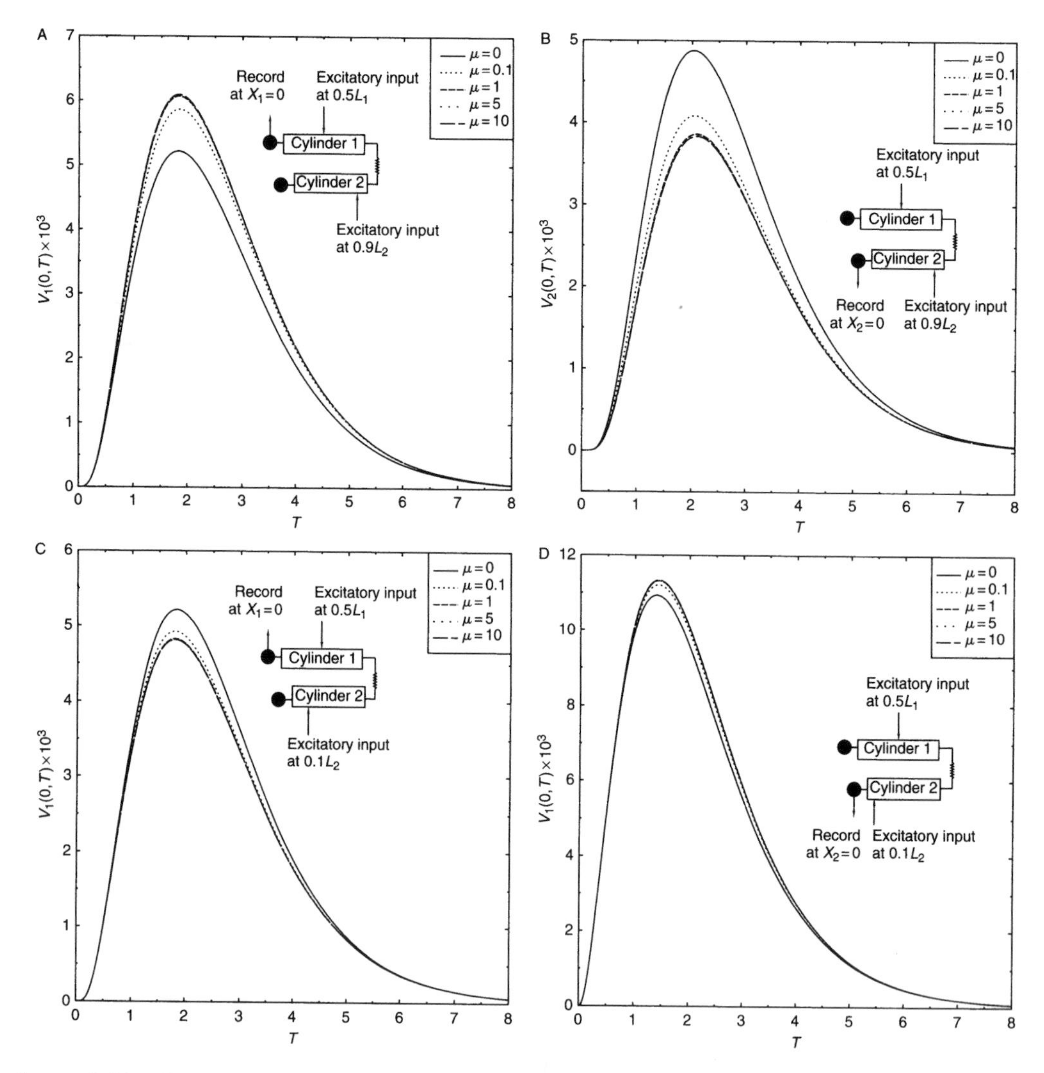

FIGURE 6.16 Illustration of the potential at the two somas for excitatory synaptic inputs in different cylinders. Parameter values are as given in Example 4. The excitatory input in cylinder 1 is kept fixed at its midpoint, while the excitatory input in cylinder 2 is located distally for (A) and (B) and proximally for (C) and (D).

suggests that remote excitatory inputs are relatively less significant on the soma potential for proximal than for distal excitatory inputs.

6.5 DISCUSSION

The equivalent multicylinder model involves one or more equivalent cylinders emanating from a common soma, retaining the powerful representation of a dendritic tree by an equivalent cylinder, as first introduced in the Rall model. Some of the benefits of considering and understanding the solution to this model in detail are (a) that it provides a reference case for other more involved multicylinder models that attempt to accommodate the additional effects of tapering of dendritic segments, nonuniform electrical parameter values, and dendritic spines; (b) that the solution in this basic geometry can be used to obtain the solution to arbitrary branching cylinder models by observing that the entire dendritic structure is composed of a multicylinder problem at the soma and at each branch point. Importantly, the solution of the multicylinder model is characterized by the $(2N + 1)$ dimensionless parameters $(\epsilon, \gamma_j, L_j)$ for $j = 1, \ldots, N$, where N is the number of equivalent cylinders emanating from the common soma. This set of dimensionless parameters quantifies two important properties in the multicylinder model: (a) the (N) electrotonic lengths L_j, $j = 1, \ldots, N$, which are the ratio of the actual length to the space constants for each equivalent cylinder; (b) the passage of current between cylinders is moderated by the soma where conservation of current introduces the dimensionless parameter ϵ (the ratio of somatic to dendritic membrane time constants) and the N dimensionless parameters γ_j, $j = 1, \ldots, N$ (the ratios of dendritic and somatic conductances for each equivalent cylinder). In the single cylinder case $(N = 1)$, we obtain $(\epsilon, \gamma_1, L_1)$ as the three controlling dimensionless parameters of the single cylinder somatic shunt model and (γ_1, L_1) as the two dimensionless parameters of the Rall equivalent cylinder model. These dimensionless parameters are typically $O(1)$, although the effect on the solution of certain physically based limits of them can be considered, which allows an understanding of their effect in the model. Simplification of the model by reduction in the number of cylinders by neglect and collapse can also be investigated and criteria developed for systematic application of each method that also allows an estimate of the error involved in the procedure.

Here, we have addressed the analysis of the forward problem which is well posed and has a well defined solution available in closed analytical form. However, practitioners are generally interested in the solution of an inverse problem; the search for model parameters that best describe the physical system being studied. The methodology presented here is in part preparatory to the study of the inverse problem, since a primary requirement before solving an inverse problem is ensuring that it has a unique solution (among other properties), so that it is mathematically well posed. The search for the solution of the inverse problem may be carried out using dimensional parameters (the dimensional inverse problem) or dimensionless parameters (the dimensionless inverse problem). The process of problem normalization allows the identification of the minimum number of fundamental dimensionless parameters for the model and in terms of which the solution is unique. This number together with the scaling parameters, which are needed to transform the data set to which the model is to be fitted into the required form for working with the dimensionless inverse problem, then sets exactly the number of parameters (dimensional or dimensionless) that can be determined by fitting of the model to the data set. If more than this number of parameters are to be fitted, then the inverse problem is ill-posed but can be made well posed if the values of the additional parameters can be prescribed (usually by independent means, for example, morphological measurements).

It is straightforward to extend the model to consider inhomogeneous mixed boundary conditions of the form,

$$\text{at } x_j = \ell_{\text{cyl}j} \qquad a_j v_j + b_j \frac{\partial v_j}{\partial x_j} = c_j\,, \tag{6.232}$$

for $j = 1, \ldots, N$ with a_j, b_j, c_j constants, in place of the sealed-end conditions (6.2). These may arise when representing a leakage current between the interior and exterior media (Rall, 1969a) or during the analysis for synaptic input located at the terminal end with the synaptic conductance represented by a rectangular pulse (Poznanski, 1987). This system of equations can be considered with little trouble using the same techniques, the solution for mixed homogeneous boundary conditions being a particular case arising in the transformation of the independent variable for the solution of the modified cable equation representing a tapering core as discussed in Evans (2000).

Other terminal boundary conditions include the so called "natural" boundary conditions (Tuckwell, 1988a) representing the sealing of the end of a nerve cylinder with the nerve membrane itself. These takes the form,

$$\text{at } x_j = \ell_{\text{cyl}j} \qquad v_j + \tau_{ej} \frac{\partial v_j}{\partial t} + \frac{R_{ej}}{r_{aj}} \frac{\partial v_j}{\partial x_j} = 0\,, \tag{6.233}$$

for the jth equivalent cylinder, where R_{ej} is the terminal swelling resistance (Ω), C_{ej} is the terminal swelling capacitance (F), and $\tau_{ej} = R_{ej}C_{ej}$. These may be considered appropriate for representing dendritic swellings, as discussed in the "dumbell" model of Jackson (1993) and more recently by Krzyzanski et al. (2002). Again, these boundary conditions pose little difficulty within our framework.

The multicylinder model has been extended in several ways to incorporate and investigate more realistic variations in geometrical and electrical nonuniformities. These extensions are now briefly reviewed.

1. *Multicylinder models with nonuniform electrical parameters*: Variations in the specific membrane resistivity (R_m), the cytoplasmic resistivity (R_i), and the specific membrane capacitance (C_m) in the cell can occur naturally, from microelectrodes and micropipettes behaving as additional processes when inserted into a cell, or from simplified representations that introduce nonuniformly adjusted values, all of which are discussed in detail in Major and Evans (1994). Consequently, variations in any one of these parameters leads to differing membrane time constants within the dendritic structure. The multicylinder model may be extended to allow differing membrane time constants for each equivalent cylinder, enabling the investigation of the effect of a variation of the dimensional parameters R_m, R_i, or C_m between (but not within) dendritic trees of the neuron. For this case, the cable Equation (6.1) becomes,

$$v_j + \tau_{mj}\frac{\partial v_j}{\partial t} - \lambda_j^2\frac{\partial^2 v_j}{\partial x_j^2} = \lambda_j r_{aj} i_j(x_j, t), \tag{6.234}$$

where τ_{mj} emphasizes that the membrane time constants can take different (constant) values in each equivalent cylinder. The model is presented and solved in detail in Evans and Kember (1994), using analogous techniques introduced for the solution of the multicylinder model.

2. *Multicylinder models with dendritic spines*: It is usual to incorporate dendritic spines into the dendritic membrane by a suitable rescaling of either the physical length and diameter or the electrical parameters of the membrane cylinder representing the dendritic segment with its spines (Shelton, 1985; Stratford et al., 1989; Larkman, 1991b; Douglas and Martin, 1992; Rapp et al., 1992). However, in general, a more careful treatment of dendritic spines is required particularly for those which receive synaptic input, so that the effects of spine neck resistances, synaptic conductance changes (Koch et al., 1982; Jack et al., 1983; Koch and Poggio, 1983; Brown et al., 1988) and possible active conductances in spine heads (Miller et al., 1985; Perkel and Perkel, 1985) can be investigated.

The first mathematical models of spines (Rall and Rinzel, 1971a, 1971b) and later Wilson (1984) were limited to discrete, passive spine(s) and generalized the original Rall model. Recently, a continuum theory for active spines (i.e., a density of spines with voltage-dependent channels), was advanced by Baer and Rinzel (1991). Their continuum model for spines was studied mostly numerically with emphasis on exploring the behavior of active spines with some analysis of the steady-state, single branch passive neuron model with passive spines. Following Baer and Rinzel (1991), a model for passive spines as a continuum in the multicylinder case was presented and analyzed in Kember and Evans (1995). The essential features of the model are that it couples the cable equation for voltage spread in the dendritic shaft with another representing isopotentiality of the spine head. The standard cable Equation (6.1) is replaced with,

$$\begin{aligned} &\text{(dendrite)} \quad v_j + \tau_m\frac{\partial v_j}{\partial t} - \lambda_j^2\frac{\partial^2 v_j}{\partial x_j^2} + \frac{\bar{N}_j r_{mj}}{R_{ss,j}}(w_j - v_j) = \lambda_j r_{aj} i_{vj}, \\ &\text{(spinehead)} \quad \tau_{sp}\frac{\partial w_j}{\partial t} + w_j + \frac{R_{sp}}{R_{ss,j}}(v_j - w_j) = R_{sp} i_{wj}, \end{aligned} \tag{6.235}$$

where for the jth equivalent cylinder, $v_j(x_j, t)$ is the dendritic membrane potential, $w_j(x_j, t)$ the spine head membrane potential, and i_{vj}, i_{wj} are applied currents to the jth cylinder and to the spine head membrane in the jth cylinder, respectively. The dimensional parameters introduced for the spines are the spine density $\bar{N}_j$ spines/unit-length (assumed constant), a spine-stem resistance $R_{ss,j}$ (Ω), the spine head membrane time constant $\tau_{sp} = R_{sp}C_{sp}$ (secs), where $R_{sp}(\Omega)$ and $C_{sp}(\mu F)$ are the spine head resistance and capacitance, respectively. The spine-stem resistance can be expressed as $R_{ss,j} = 4R_i\ell_{ss,j}/\pi d_{ss,j}^2$, where $\ell_{ss,j}$ and $d_{ss,j}$ are the length and diameter of the spine stems, respectively. The other dimensional parameters remain unchanged from their definition in the multicylinder model, Equations (6.1) to (6.5). The current-conservation condition at the soma can also be modified to include spines.

The Equation (6.235) models the dendritic spine system as a continuum, but the spines are electrically independent from each other. The equation for the spinehead in Equation (6.235) contains no term for direct coupling between neighboring spines; voltage spread along the passive dendrite to which they are attached is the only mechanism for spine interaction. For the multicylinder model, the spines on each dendritic tree are assumed to have uniform properties, while the spine density, spine length, and spine stem diameter can all vary between dendritic trees. The general solution and its behavior for this system is presented in Kember and Evans (1995).

3. *Arbitrarily branching cables*: Dendritic trees in general do not satisfy all of the constraints needed for the formation of the single equivalent cylinder model of Rall nor those for the multicylinder models. As remarked by Rall (1989), these constraints on dendritic branching are not presented as laws of nature but are used to define mathematical idealizations that provide useful and valuable reference cases. The tapering multicylinder models were developed to accommodate the fact that dendritic trees do not (or only partly, up to a certain electrotonic distance) satisfy the 3/2 power law for their branch diameters mainly due to branches terminating at different electrotonic lengths. The formulation of the tapering multicylinder models in Evans (2000) allows for the mathematical description of a much wider class of actual dendritic trees than those given by the nontapering models, and to which real neurones may be fitted with increased accuracy through suitable choices of the geometric ratio functions. However, for a more accurate description of the branching pattern that allows the consideration of branches and segments within dendritic trees and not just between them, branching cylinder models need to be considered.

The system of equations for an arbitrary branching geometry have been considered and solved analytically in Major et al. (1993a, 1993b), the solutions being extensions of those obtained in the multicylinder case. There are simplified representations of the branching structure that are intermediary between the full branching structure and the multicylinder models, where dendritic trees not satisfying the equivalent cylinder criteria are simplified using equivalent dendrite collapses (Stratford et al., 1989). Branching cylinder models allow segments and branches within dendritic trees to be considered and are not restricted to comparisons between dendritic trees as in the multicylinder models. The effects on the voltage response of nonuniform values of the dimensional parameters R_m, C_m, R_i that differ within dendritic trees as well as between them, was considered in Major and Evans (1994). The flexibility of the branching models can be seen by their ability to treat tapering branch segments as a series of uniform but different-diameter cylinders placed end to end.

However, multicylinder models may be perfectly adequate (particularly those with tapering equivalent cylinders) in providing a powerful representation of the passive properties of a nerve cell without the need to complicate the model with complex branching patterns. Indeed, it is these types of intermediary models, which are simplified enough to allow understanding of their behavior and yet may be complex enough to adequately describe the system they are simulating, which should be developed and investigated fully.

6.6 Conclusions and Future Perspectives

The methodology presented here for analysis of the multicylinder model can be applied to linear cable models with arbitrary branching structures and thus can be studied in the same framework.

In particular, perturbation techniques may be used to investigate cylinder collapse (e.g., see, Evans, 2000), which can be used to establish the error involved for collapsing dendritic segments in fully branched structures that do not exactly satisfy the criteria for collapse to an equivalent cylinder. A qualitative description may then be obtained of branching structures that do not satisfy the reduction to equivalent cylinders but nevertheless may be represented by them to within an acceptable error margin (which can be estimated).

The fitting of these models to the experimental data necessarily involves parameter identification and thus consideration of the inverse problem. This problem has received a lot more attention in recent years, as difficulties such as ill-posedness and ill-conditionedness have been encountered. Stratford et al. (1989), Holmes et al. (1992), and Major et al. (1994) have advocated examining problems associated with parameter estimation. In fact, Rall (1990) stated that modeling the complexity of single neurons with multicompartmental models results in too many degrees of freedom and one way to reduce the degree of freedom is to construct simpler models. The inclusion of tapering may allow for construction of more realistic models within the equivalent cylinder framework.

An important step in the evolution of these models is the incorporation of nonuniform membrane properties, by introducing time- and voltage-dependent nonlinearities as originally described by Hodgkin and Huxley (1952b). Accurate modeling requires a description of densities, spatial distributions, and kinetics of time- and voltage-dependent conductances. Several recent theoretical studies extending the passive cable work presented here are worth mentioning. Ohme and Schierwagen (1998) have numerically solved a nonuniform equivalent cable model for dendritic trees with active membrane properties, while Coombes and Bressloff (2000) have analyzed the continuum active spine model of Baer and Rinzel (1991). Poznanski and Bell (2000a, 2000b) formulate a model in which the densities of voltage-dependent ionic channels are discretely imposed at specific locations (hot spots) along an equivalent cylinder. We have discussed this model within our framework in Section 6.2.3. Although the voltage-dependent current is linearized in this work, thus allowing analytical progress, the model does seem to give a suitable alternative to the continuously distributed case and is worth pursuing in a more general context.

ACKNOWLEDGMENT

I am especially grateful to Dr. R.R. Poznanski for providing very helpful suggestions and information on the ionic cable equation, synaptic reversal potentials, and gap junctions.

PROBLEMS

1. Extend the multicylinder model, Equations (6.1) to (6.5), to consider inhomogeneous mixed boundary conditions of the form given by Equation (6.232). What are the physical applications of boundary conditions of this form? Nondimensionalize the model and show that the solution $V_j(X_j, T)$ satisfies Equations (6.10) to (6.14) with the boundary condition of Equation (6.11) replaced by

$$\text{at } X_j = L_j \qquad A_j V_j + \frac{\partial V_j}{\partial X_j} = B_j;$$

for suitable constants A_j and B_j. Show that the solution can be written in the form

$$V_j(X_j, T) = U_j(X_j, T) + W_j(X_j),$$

where $W_j(X_j)$ may be interpreted as the steady-state solution satisfying the inhomogeneous boundary conditions while $U_j(X_j, T)$ is the transient solution satisfying a suitable subproblem of Equations (6.10) to (6.14) with homogeneous mixed boundary conditions at $X_j = L_j$ and an appropriate initial condition. Obtain expressions for the functions W_j and U_j.

2. Consider synaptic inputs represented as conductance changes in the form suggested by Equation (6.31) with the time courses given by standard alpha functions. Derive a set of integral equations for the voltage response and investigate their properties.
3. Discuss the inverse and parameter identification problems for the multicylinder problem mentioned briefly in the Section 6.5. The following references may be of interest: D'Aguanno et al. (1986), Fu et al. (1989), and White et al. (1992).
4. Formulate and explore suitable nonlinear extensions to the linear cable models discussed in this chapter suitable for active membranes. In particular, consider polynomial membrane current–voltage relations as suggested in Jack et al. (1983). A suitable cubic function may be related to the reduced FitzHugh–Nagumo model with no recovery variable.
5. Experimental suggestion: As discussed in Section 6.3 and Evans (2000), multicylinder models can be formulated with geometric ratio functions F_j that belong to a class of six specific tapering forms. Determine which functional forms and their parameters from this class best describe the dendritic trunk parameter for different neuron types (that should be obtained by measurements). The number of equivalent cylinders to be used in the models should also be investigated.
6. Extend the multicylinder model to include dendrodendritic coupling on several cylinders to a second multicylinder model. Formulate the model equations and investigate their mathematical properties.

7 Voltage Transients in Branching Multipolar Neurons with Tapering Dendrites and Sodium Channels

Loyd L. Glenn and Jeffrey R. Knisley

CONTENTS

7.1 Introduction

Over a decade ago Major and colleagues obtained analytical solutions for arbitrarily branching cables (Major, 1993; Major et al., 1993a, 1993b; Major and Evans, 1994). The importance of arbitrarily branching cable models is that the full dendritic arbor can be modeled rather than a collapsed version of the tree, the latter of which is known as an *equivalent* model. These solutions for transients in arbitrarily branching passive dendrites represented the most advanced breakthrough obtained in the history of neuron modeling, but they did not incorporate two important physical characteristics.

The first characteristic that was not incorporated was smooth tapering of the branches. In fact, taper was only considered through a progressively smaller diameter chain of short cables. Such a method approximates taper with stepwise changes to successive cable segments and is often referred to as the segmental cable approach (e.g., see Butz and Cowan, 1974; Horwitz, 1983; Turner, 1984a; Holmes, 1986; Abbott et al., 1991; Abbott, 1992; Cao and Abbott, 1993; Steinberg, 1996).

Unfortunately the stepwise reduction at branch points in a model with untapered branches further discretizes each of the cable segments to numerical approximations involving compartments, each having a unique diameter (e.g., see Pongracz et al., 1993; Surkis et al., 1996; De Schutter and Steuber, 2000a). The biggest disadvantage of both the segmental cable model and the compartmental method is that they do not provide a continuous description of tapering dendrites, but rather a discrete approximation. Thus, the model described in the present chapter includes both branching and tapering dendrites.

The second characteristic that has not yet been incorporated into full arbor models is active channels. The seminal studies describing analytical solutions for transients in arbitrarily branching dendrites as in the series of papers by Major and colleagues, or the series of papers by Abbott and colleagues, never considered the treatment of voltage-dependent conductances. This is a serious detriment of their practical usefulness as it is now known from experimental studies that dendrites are not passive, but contain a variety of different types of ionic channels embedded in the membrane at spatially localized positions (for reviews see Reyes, 2001; Migliore and Shepherd, 2002). In this chapter, we treat linearized sodium conductances in the full arbor model with smoothly tapering branches by using an alternative method to that proposed by Evans (see Chapter 6). Moveover, two types of active conductances are considered: persistent sodium channels with only an activation parameter and Hodgkin–Huxley sodium channels with both an activation and an inactivation parameter. It is envisaged that such explorations will eventually bring us closer to the ultimate analytical solutions for transients in arbitrary branching active dendrites with active conductances possessing arbitrary kinetic properties.

The problem of impedance mismatches between the large membrane surface of a single cable and the small membrane surface of fine distal dendrites necessitates choosing lower densities of ionic channels in such reduced models in order to match realistic voltage transients. Taper introduced into the reduced model is expected to remove this problem, circumventing the above criticism made by De Schutter and Steuber (2000a). Unfortunately, no analytical theory exists to predict whether propagation will be successful given a dendritic morphology and a set of realistic channel densities and kinetics, so instead, Vetter et al. (2001) have shown with a single tapering compartmental model that action potential propagation in active dendrites via compartmental models depends purely on branching regardless of whether taper is present in the reduced compartmental model.

In addition to providing derivations and integral solutions for the full arbor model with active conductances, we also consider four issues that stem from this analytic model: (a) the relationship between tapers in dimensional and dimensionless units, (b) the relationship between a tapering, unbranched, multipolar model and a tapering equivalent cylinder model, (c) physical insights from an analysis of the intermediary equations, and (d) theoretical points that ultimately lead to the inadequacy of the compartmental model.

7.2 Solutions for Transients in a Full Arbor Model with Tapered Branches and Branch-Specific Membrane Resistivity

7.2.1 Definition of the System

The multipolar models of Glenn (1988), Holmes et al. (1992), and Major et al. (1993a, 1993b) have the simplifying assumption of constant diameter branches. This is a potentially important limitation because neither dendrites nor equivalent cylinder representations of complete dendritic trees have constant diameters. Consequently, a model that combines tapering along with the other three basic types of nonuniformity or nonlinearity (diameter, length, time constant) would appear to be at least useful, and may be argued to be the minimal model that should be used to analyze experimental charging curves from neurons. Evans and Kember (1998) and Evans (2000) developed models of this type. Extensions of these models are described in this chapter. In particular, a fully arborized model incorporating active properties in the form of a persistent inward current is introduced. Although a one-dimensional cable is assumed in this approach, it should be noted that Lindsay, Rosenberg, and Tucker (2004) derived solutions for a three-dimensional cable, albeit in a single unbranched cable.

The term tapering has been used to describe at least two different morphological traits — the tapering of individual dendritic branches, hereafter called branch tapering, and the occurrence of electrotonic mismatches at branch points, hereafter called step changes in the equivalent diameter. These step changes are caused by decreases in summed diameter to the three-halves power at branch points (Rall, 1962a).

Tapering may refer either to variations in a single dendritic branch or to variations in the summed diameters for whole neurons after the dendrites have been collapsed into a single cone-like structure called an equivalent cylinder. Dendritic trees are collapsed by summing the diameters (to the three-halves power) at increasing distances from the soma along the dendritic paths. The value of the sum at any particular path distance from the soma is called the dendritic trunk parameter. If this sum is taken to the $\frac{2}{3}$ power, it is called the equivalent diameter. In most neurons studied to date, branch tapering, step changes in net diameter, and termination of dendrites all contribute to the (whole neuron) tapering of the equivalent diameter (Ulfhake and Kellerth, 1984; Cullheim et al., 1987a; Clements and Redman, 1989), with dendritic termination predominating.

The equivalent cylinders of Clements and Redman (1989) or Bush and Sejnowski (1993) have nonuniform, discontinuous dendritic profiles because they are implemented in computational simulations via the compartmental modeling approach, and they constitute an even greater simplification than the continuous equivalent profile because the compartmental modeling approach is a less subtle approach than the continuous arbitrarily branched cable (Major et al., 1993a), or multiple equivalent cylinder approaches (Evans and Kember, 1998; Evans, 2000). This is because it replaces the continuous membrane of the cell by a set of interconnected compartments and therefore abandons the intuitive relationship between distance and signal attenuation that is of paramount importance in modeling realistic neurons with continuous membrane.

Finally, and most importantly, all the above equivalent cylinders are ill-suited for dendritic stimulation, since they were formulated to represent an approximate electrical load of the dendrites when viewed from a micropipette inserted at the soma.

In contrast to the limited number of studies on the tapering properties of individual branches, there are a large number of studies on whole neuron tapering — that is, tapering of the dendritic trunk parameter. Such studies were motivated by a desire to approximate the whole neuron by a single cable model, or to rule out the possibility that a single constant-diameter cable can be used to model the passive voltage–current responses of the neuron. Two of the earlier studies were by Lux et al. (1970), who used radioactive amino acids as a label, and by Barrett and Crill (1974), who used fluorescent compounds to label motoneurons and plot the whole neuron electrotonic profile. Neither of these studies allows full visualization of the entire dendritic tree, but with the advent of biocytin and horseradish peroxidase as intracellular labels, complete staining of dendritic trees became possible in many types of neurons. Using these techniques, the profile of the equivalent cylinder was determined in spinal motoneurons (Cameron et al., 1981; Ulfhake and Kellerth, 1984; Cullheim et al., 1987b; Fleshman et al., 1988; Clements and Redman, 1989), brainstem motoneurons (Bras et al., 1987; Nitzan et al., 1990; Moore and Appenteng, 1991), and hippocampal pyramidal cells (Turner and Schwartzkroin, 1980). Limitations to these techniques have been quantified by Szilagi and De Schutter (2004).

A picture of the relationship between the diameter, $D_{\text{taper}} = (\sum d^{3/2})^{2/3}$, and the electrotonic distance, Z, from the soma has emerged from these studies. For hippocampal pyramidal cells, the D_{taper} as a function of Z is similar in appearance to an exponential decay. Figure 7.1 illustrates an example of the dendritic profile for a chosen CA3 hippocampal cell with a decline that is essentially exponential. The major physical limitation in single tapering cable models is the reality that the

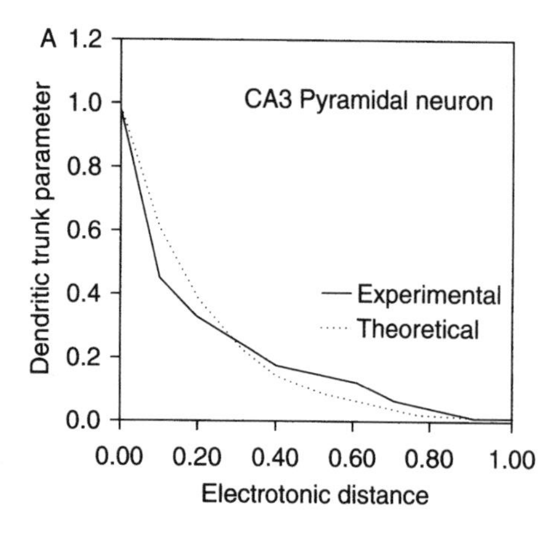

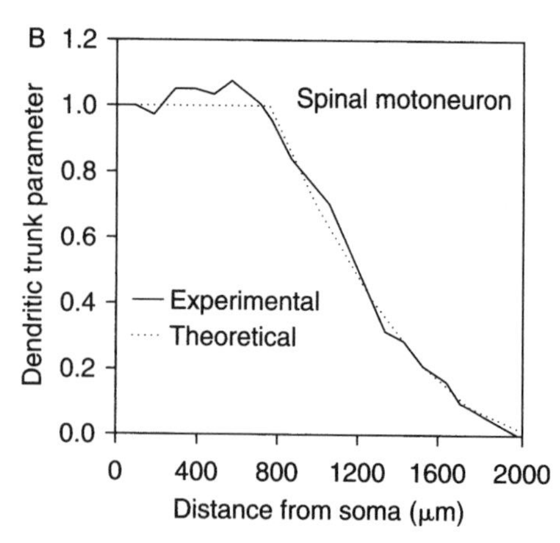

FIGURE 7.1 Taper of equivalent cylinders. (A) The equivalent taper for the CA3 hippocampal pyramidal neuron. Normalized $\sum d^{3/2}$ (dendritic trunk parameter) plotted against Z (electrotonic distance). Note that the taper is well approximated by an exponential taper in electrotonic coordinates (dashed curve). The taper rate for the dashed line is $K = -4.75$. Replotted from Poznanski (1994a). (B) Curve fit of a model taper with the morphometric taper of a spinal motoneuron whose dendrites are collapsed into a single cable by the $d^{3/2}$ rule. The taper is quadratic in conventional dimensioned units (x), which corresponds to an exponential taper in electrotonic units (Z).

dendritic profile decreases in discrete steps at each branch point as governed by the number of branches in the ith segment, n_i, rather than continuously with electrotonic distance (X or Z) as illustrated in Figure 7.1 (see Jack et al., 1975, for a discussion on this point). Yet, a clear distinction needs to be made between the various types of discrete equivalent profiles, as they do not render to the same physical principles.

Figure 7.1 also shows a plot of the dendritic trunk parameter of spinal motoneurons at different distances from the soma. The diameters of a small proportion of spinal motoneurons can be approximated by a linear taper (Clements and Redman, 1989), but curvilinear tapers are necessary for most (Ulfhake and Kellerth, 1984; Cullheim et al., 1987b). The taper begins several hundred micrometers from the soma rather than at the soma itself as would be required for a linear or exponential taper. Indeed, for the first few hundred micrometers, there is a slight flaring of the profile for spinal motoneurons (Figure 7.1). The figure shows that spinal motoneurons collapsed by the $d^{3/2}$ rule should be approximated by at least two cables: (a) a short cable emanating from the soma with a slight flare continued by (b) a longer cable with a curvilinear taper of diameter with distance. Brainstem motoneurons have shown both of the above types of taper, with jaw-closer motoneurons showing the spinal motoneuron tapering profile (Moore and Appenteng, 1991) and vagal motoneurons showing a profile similar to hippocampal pyramidal neurons (Nitzan et al., 1990).

The curve used for the spinal motoneuron in Figure 7.1 to model the taper is a quadratic function of distance. The quadratic curve approximates both the proximal flare and the distal taper of spinal motoneurons. If the curve is converted to dimensionless units, specifically to $d^{3/2}$ by electrotonic distance, the curve corresponds exactly to that of Poznanski (1991) and Goldstein and Rall (1974) — a fact that motivates and justifies the study of the models described in the remainder of this chapter. The relation between the dimensional taper k and the dimensionless taper rate K is simply $K = 3\lambda k$. Problem 4 at the end of this chapter concerns the identity of exponential tapers in K and Z to quadratic tapers in k and x.

The difference in the relationship between the form of a taper in electrotonic or dimensionless space (usually denoted by the variable Z) and the form of the taper in metric or dimensional space (usually indicated by the variable x) has not been widely appreciated. Some have mistakenly supposed that the form of the taper in dimensionless space is the same as the form of the taper in dimensional space. Moreover, the problem of determining the form of the taper in one space given the taper in the other space is not trivial. However, if morphological data are to be applied to solutions that are expressed only in dimensionless variables, it is essential that more of these types of results should be discovered.

Now, if we write the equation for the rule of a taper as

$$D_{\text{taper}} = D_0(1 + kx)^2,$$

where D_0 is the initial diameter and x is the distance from the point at which the initial diameter is measured, then the taper rate k can be calculated from morphometric measurements as shown for the motoneuron in Figure 7.1. The procedure has three major steps. First, the diameter next to the origin of a dendritic branch is measured. Next, the diameter is measured at a given distance (x) from the origin. All measurements are converted to centimeters using 1 μm = 0.0001 cm and then the three values are substituted into the equation for D_{taper}, after which we solve for k. If the dendritic branch gets narrower with longitudinal distance, k will be negative (i.e., a taper).

For theoretical purposes, it is convenient to convert physical distance from the central side of a single branch (x) into a generalized, dimensionless, electrotonic variable, Z, following the notation and approach of others (such as Rall, 1962; Jack et al., 1975; Poznanski, 1988; Major et al., 1993). The relation between Z and x is

$$Z = \int \frac{dx}{\lambda(x)},$$

$$\lambda(x) = \lambda_0 \left(\frac{D(x)}{D_0}\right)^{1/2},$$

where λ_0 is the electrotonic length characteristic at the central end of the cable ($Z = x = 0$). The electrical potential along each branch satisfies the standard one-dimensional tapering cable equation (Rall, 1962; Jack et al., 1975; Poznanski, 1988). Although either the dimensional k or the dimensionless K could be used in the derivations in this chapter, K is used to simplify the solutions.

7.2.2 Relationship between Tapering Multicylinder Model and Tapering Single Cylinder

Before we consider the fully arborized model, one of the questions that arises is the degree to which individual dendrites are well approximated by a single tapered equivalent cylinder. Although a lot of work has been published on collapsing the *entire* dendritic tree of an entire neuron into a single equivalent cylinder, there is much less on collapsing *single* dendrites, defined as all branches stemming from a *single* dendritic trunk that is attached to a soma. We tested this by comparing the responses of branching trees with a single lumped soma. Ten trees were created by the full stochastic growth algorithm (Model II with grandparent branch correction) of Burke et al. (1992). The average tree had 11.0 ± 7.8 SD terminal branches with a range of 2 to 26. The number of parent branches was 10.0 ± 7.9 SD parent branches with a range of 1 to 25, for a total of 21 branches per dendritic tree. The average diameter of the trunk was 7.0 ± 2.4 SD μm with a range of 3.0 to 11.1 μm, and the total cumulative area was 48,503 ± 26,110 SD μm^2 with a range of 12,730 to 90,320 μm^2. The average path length from soma to terminal was 236 ± 170 μm, and the average maximum path length for each tree was 1,891 ± 332 μm. The correlation between the trunk diameter and total membrane area was 0.82. The tree structure compares closely with that of spinal motoneurons (Cullheim et al., 1987a) with some relatively minor discrepancies (see Burke et al., 1992).

The responses of these nonequivalent trees (considered in isolation from the soma using a sealed-end boundary condition) were compared to that of a single tapered (equivalent) cable using a program we developed for modeling branched neurons (ArborVT). The results are shown in Table 7.1. The key parameters that required matching were the k and l, which were randomly varied until a best

TABLE 7.1
Best Fit Relation Between Tapered Equivalent Cylinder and Nonequivalent Branching Dendritic Tree Model

	Minimum	Maximum	Mean	SD
Length (μm)[a]	2,200	6,800	3,490	1,546
Taper (k)[a]	−5.67	2.62	−2.40	2.38
Membrane area (μm$^{\theta}$)[b]	4,041	23,771	13,276	6,787
Max path distance (μm)[b]	1.386	2,574	1,892	333
Number of branches[b]	3	51	21	16
Mean length (μm)[b]	236	727	454	170
Mean length/max length[b]	0.14	0.55	0.25	0.13

[a] From single tapering equivalent cylinder model.
[b] From nonequivalent branching tree model.

fit chi-squared was obtained between the waveforms. The best fit was obtained on average when $k = -2.4$. The fact that the k is meaningful is indicated by the fact that it correlated with both the mean path distance ($r = 0.76$, $p < 0.02$) and with the ratio of the mean path distance to the maximum path distance of the branched dendritic trees ($r = 0.88$, $p < 0.002$). The best fit for the maximum dendritic path-to-terminal distance was about $l = 3490$ μm, corresponding to the actual distance of 1891 μm for the fully branched tree model. The poor correspondence between the maximum path length of the tree and the l of the single tapered cable reflects the relatively rapid effective taper of the tree. As l approaches $-k^{-1}$, which is the case in motoneurons, taper can vary over a wide range without producing any change in the voltage response to a current pulse because the summed $d^{3/2}$ tapers to a negligible magnitude.

We conclude by addressing the estimated error caused by the assumption that dendrites or equivalent cylinder representations of whole neurons have a uniform diameter. A 12-cable model that fits the morphometric data of Ulfhake and Kellerth (1983) and the electrotonic data of Fleshman et al. (1988) was tested. Each cable had a unique length and diameter and represented a single collapsed dendritic tree. When the cables were assumed to be untapered, the time constant for the first (slowest) exponential component (τ_0) was 10.0 msec. In comparison, τ_0 was 7.43 msec in the physiological case. The time required for the signal to decay to 90% of its initial value correlates with the amplitude of the fastest exponential components in a signal. It was 0.30 msec in the untapered case, but 0.41 msec in the tapered case. The time required for a 50% decay was 5.6 msec in the untapered case and 3.5 msec in the tapered case. The reduction in both cases is attributed to the lower average L when tapering is incorporated into the model. This elementary change was also noted by Poznanski (1988) and Holmes and Rall (1992a) and it was observed in the present model with multiple dendrites and nonuniform tapering.

The consequences of these changes for the estimation of the electrotonic parameters of experimentally derived transients from neurons is beyond the scope of this chapter. However, our results in quadriceps (spinal) motoneurons indicate that the average L is between 0.4 and 0.8 for all of the alternative models tested (L.L. Glenn and J.R. Knisley, unpublished findings). The incorporation of some form of tapering, whether by tapering a single cable or varying the lengths of a nontapering multipolar model (Poznanski and Glenn, 1994), would appear to be essential to obtain accurate estimates of electrotonic parameters.

Now that the importance of tapering — the relation between metric (dimensional) and electrotonic (dimensionless) spaces, the $d^{3/2}$ rule — and the limitations of collapsing into equivalent representations have been reviewed, we will turn our attention to the generation of the full arbor model with active sodium conductances.

7.2.3 Separation of Variables Solution

The derivation is based on a mixture of established mathematical methods following that of Bluman and Tuckwell (1987), Kawato (1984), and Major et al. (1993). The voltage response to a current stimulus in a membrane cable is formed from the sum of an infinite series of exponential waveforms. The configuration of the model determines the initial value and boundary conditions, which in turn determine the amplitude, C, and time constants, β, for each of the exponential waveforms. A particular solution is obtained when all of the amplitudes and rates for the exponential series have been specified. To obtain the solution for the model in Figure 7.2, the separation of variables method, Parseval's theorem, and the modified orthogonality relation (Churchill, 1942) are applied to the cable equation (Rall, 1977). Now, from Poznanski (1988), for any branch in a dendritic arbor,

$$\frac{\partial^2 V_{j_p}}{\partial Z_{j_p}^2} + K_{j_p}\frac{\partial V_{j_p}}{\partial Z_{j_p}} - \tau_{j_p}\frac{\partial V_{j_p}}{\partial t} - V_{j_p} = 0,$$

where the set notation of Figure 7.2 is used, and j_p refers to a particular dendrite of the pth order.

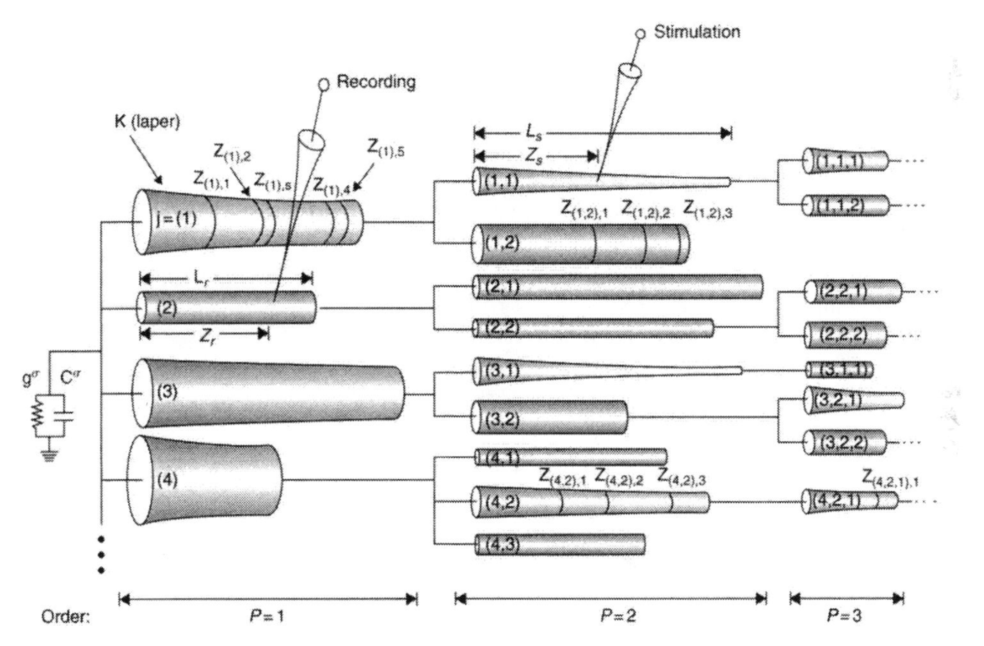

FIGURE 7.2 Schematic diagram of the full arbor model for which the analytic solution is obtained in the present chapter. Note the branching, tapering multipolar structure. Any given cable can give rise to any number of descendants, including zero. The number of branch points per dendrite is arbitrary. Dendritic branches up to the third order are depicted, but the model allows branches of any order ($p = 1, 2, 3, \ldots, N$), so that some or all of the dendrites can continue indefinitely arborizing to the right of the diagram. Voltage-dependent channels (hot spots) are shown as dark rings along the cylinder, labeled with a subscripted Z. Any number of channels (total number is denoted as M_j) are allowed on any branch at electrotonic distances of $Z_{j,1}, Z_{j,2}, Z_{j,3}, \ldots, Z_{j,M_j}$. The number of stem ($p = 1$) cables connected to the soma (σ) is arbitrary (indicated by ellipsis points). The ordered set notation used to identify cables is indicated on the cables in parentheses. Each cable has an arbitrary L_j, Z_j, τ_j, K_j, and g_j^{∞} ($\approx$ diameter) where j indicates a particular cable such as $j = (1, 1, 2)$. If j is subscripted (j_p), the subscript indicates the order of the cable, or all cables of a particular order. Stimulation and recording sites can be arbitrarily selected, but branches (1, 1) and (2) are shown in this example, respectively. Recording or stimulation on the soma is obtained by setting $Z_j = 0$, where $j = j_1$. A very short cylinder (j_0) can be substituted for the lumped RC soma. Note that each cylinder segment has a quadratic taper as a function of x, which is exponential in electrotonic units (Z).

A different approach was used for the separation of variables than was used in previous approaches in the literature (Rall, 1959; Jack and Redman, 1971; Durand, 1984; Kawato, 1984; Ashida, 1985). The primary difference is that τ is on the spatial side of the partial differential equation rather than the temporal side, as is customary. In our hands, this method was required to obtain the solution to the two models (Section 7.2 and Section 7.3) in this chapter. After separation (where α is the separation constant) and application of the completeness relation to the eigenfunctions, by the methods of Tuckwell (1988a) and Major et al. (1993), the solution was found to have the following form:

$$V_r = \sum_{i=-\infty}^{\infty} C_i \, \mathrm{e}^{-t/\beta_i}, \qquad \beta = \sqrt{\alpha^2 \tau_m - 1}, \tag{7.1}$$

$$C_i = \beta^{\delta} I \frac{\phi_{ir}(Z_r)\phi_{is}(Z_s)}{D_i}, \tag{7.2}$$

$$\phi_{ij_p}(Z_{j_p}) = \frac{1}{A_{ij_p}} \left[B_{ij_p} \sin \sqrt{\frac{\tau_{j_p}}{\beta_i} - \frac{K_{j_p}^2}{4} - 1}(L_{j_p} - Z_{j_p}) + \cos \sqrt{\frac{\tau_{j_p}}{\beta_i} - \frac{K_{j_p}^2}{4} - 1}(L_{j_p} - Z_{j_p}) \right]. \tag{7.3}$$

Each cable in the model is described by one ϕ_{ij} where $j = 1, 2, 3, \ldots, n$ and n is the total number of membrane cables comprising a given model. The n depends on the number of stem dendrites, branch points, and dendrites per branch point. The subscripts r and s each refer to any particular cable from $j = 1$ to n, where $j = r$ is the dendrite from which a recording is made and $j = s$ is the dendrite stimulated (by an impulse function). The I refers to the amplitude of a current step or instantaneous pulse, and $\beta^{\delta} = 1$ for an instantaneous pulse (Green's function) and $\beta^{\delta} = \beta_i$ for a current step. The β^{δ} for response to the current was obtained by integrating the response to the impulse function (7.1) over time, and this is the response to a current step (Jack et al., 1975; Rall, 1977).

The summation range on the separation constant from $-\infty$ to $+\infty$, rather than from 0 to $+\infty$, is convenient. The reason is that summation over $-\infty$ to $+\infty$ simplifies the derivation of, and expression for, the C_i. Specifically, if the eigenfunctions are orthogonal (i.e., no lumped soma), and summation is over 0 to $+\infty$, then the C_i for $i > 0$ is double that for $i = 0$ (i.e., C_0). When summing over $-\infty$ to $+\infty$, note that there is only one C_0, but two C_i for every $i \neq 0$. The two C_i, one for i negative and one for i positive, sum together to produce the voltage response, effectively doubling it for all $C_{i \neq 0}$. This prevents not only the need for a separate expression for C_0 but also the need for a separate derivation of C_0 (as in Jack et al., 1975; Rall, 1977).

The A_{ij}, B_{ij}, and D_i are coefficients determined by the boundary conditions and are independent of Z_i. The A_{ij} and B_{ij} arise from the space-dependent term of the separated cable equation whereas D_i arises from the time-dependent term. Consequently, the A_{ij} and B_{ij} vary from cable to cable (reflected by subscript j), whereas D_i does not vary across cables. The quantity $\phi_{is}D_i$ is simply the equivalent cell capacitance at the soma for the ith exponential (V^{σ}). The A_{ij} physically represent the steady-state ratio voltage at the end of the jth cable for the ith exponential waveform. The ratio B_{ij} is a dimensionless ratio of the conductance of the distal cable branches to that of a semiinfinite extension of cable j, corresponding to the B values of the steady-state case in Rall (1959, 1977). The separation constant is α_i. The equation structure above was used to gain analytic insight into the models. The β_i are the roots of the solution. In the form of the equation used above, the β_i are time constants for the exponential series of Equation (7.1), which simplifies the usual practice of determining time constants by a separate expression, using $\tau_i = \tau_m/(1 + \alpha_i^2)$ (Rall, 1977; Durand et al., 1983; Glenn, 1988; Major et al., 1993). The variable n^{σ} is the number of first-order dendrites, and n_j is the number of second-order dendrites connected to the peripheral end of any given dendrite j.

The next step is to derive Green's function, $G(Z_r, Z_s, t)$, for the system, which is the solution to the homogenous equation with the initial condition

$$G(Z_r, Z_s, 0) = \delta(Zr - Zs),$$

which is the voltage response at a particular distance on a particular dendrite where the membrane potential is initially zero. From Equation (7.1), the form of the solution is

$$G(Z_r, Z_s, t) = \sum_{i=-\infty}^{\infty} \beta^{\delta} I \frac{\phi_{ir}(Z_r)\phi_{is}(Z_s)}{D_i} e^{-t/\beta_i}. \tag{7.4}$$

The formula for axial current at Z_r is

$$I_{x_r}^{\text{axial}}(Z_r, Z_s, t) = \frac{-1}{r_r^{\text{axial}}} \frac{\partial V(Z_r, Z_s, t)}{\partial Z_r} = \sum_{i=-\infty}^{\infty} \beta^{\delta} I g_r^{\infty} b_r \frac{\phi'_{ir}(Z_r)\phi_{is}(Z_s)}{D_i} e^{-t/\beta_i}, \tag{7.5}$$

$$\phi'_{ir}(Z_r) = \frac{g_r^{\infty}}{A_{ir}} b_r [B_{ir} \cos b_r(L_r - Z_r) - \sin b_r(L_r - Z_r)], \tag{7.6}$$

where the D_i and β_i are later defined by Equations (7.12) and (7.21).

When all $K = 0$, and all τ are equal, Equations (7.7) and (7.8) reduce to Equations (78) and (84) in the model of Major et al. (1993). When $K = 0$, but $\tau^s \neq \tau_1 \neq \tau_2 \neq \cdots \neq \tau_m$, the model reduces to a nontapering, branching model with nonuniform membrane resistivity, derived in Appendix I of Major et al. (1994).

For the special case of multiple unbranched cables, $Y_{ij_2} = 0$ and $j = j_1$, resulting in the equation

$$D_i = g^{\sigma}\tau^{\sigma} + \frac{1}{2}\sum_{j=1}^{n} g_j^{\infty}{}_j\tau_j \left[\left(1 - \frac{\tau_j}{\beta_i}\right) \frac{L_j \sec^2 b_j L_j + (\cot b_j L_j / b_j)}{(-(K_j/2) + b_j \cot b_j L_j)^2} + \frac{2}{-(K_j/2) + b_j \cot b_j L_j} \right]. \tag{7.7}$$

The eigenvalue function in this special case, Equation (7.7), reduces to

$$0 = g^{\sigma}\left(\frac{\tau^{\sigma}}{\beta^{\sigma}} + 1\right) + \sum_{j=-\infty}^{\infty} \frac{g_j^{\infty}(1 - (\tau_j/\beta_i))}{-(K_j/2) + b_j \cot b_j L_j}, \tag{7.8}$$

where g^{σ} is the lumped soma conductance and g_j^{∞} is the input conductance for the cylinder of semiinfinite length. The branched model of Equations (7.7) and (7.8) reduces to that of the unbranched multipolar model of Evans and Kember (1998) and Evans (2000) when $\tau = \tau_e = \tau_e = \cdots = \tau_m$. When the number of dendrites is one ($j = 1$), and the final value theorem is applied to the Laplace transform of Equation (7.7), the steady-state equation for a single tapered cable (see Equation [6.6] of Poznanski, 1988) is obtained.

The ϕ functions are the only terms that relate to the site of stimulation and recording; all other terms would be identical regardless of the stimulation and recording sites chosen on a particular model. Time (t) appears only in the exponential eigenfunction term, whereas τ appears in all terms of the eigenfunction. The τ are time constants, and so are parameters of time. The inclusion of τ in the spatial part of the equation does not change the idea that time (t) and space (Z or x) variables are unseparated; it does, however, change the idea that C, often called a *spatial eigenfunction*, is independent of τ. The inclusion of τ in the spatial part of the equation shows that there is an interdependency between τ and l in determining the voltage response (see below)

The form of the general solution used in most other studies (Rall, 1977; Glenn, 1988), based on expansions in $e^{-[1+\alpha^2]/\tau}$, could not be successfully used to solve the current problem, because previous models contained the dendritic τ, which is implicitly uniform. An alternative form of the solutions in the present report can be obtained by substituting $\beta'\tau_{ref}$ for β in the equations, where the β' are a related set of eigenvalues and τ_{ref} is any time constant (dendritic or somatic) that will serve as the reference time constant. If the soma time constant is chosen as the reference ($\tau^\sigma = \tau_{ref}$) and the dendritic τ are uniform, the eigenvalues β' provide the time constants for the exponentials (Equation [7.1]) in terms of their relation to the soma time constant. Under these circumstances, the formulas could also be simplified by introducing a parameter such as $\theta = \tau_j/\tau^\sigma$ of Kawato (1984) or $\varepsilon = \tau^\sigma/\tau_j$ of Durand (1984).

When $\beta_i = \tau_m/(1+\alpha_i^2)$ and all the τ_j are constant ($\tau_j = \tau_m$), the equation reduces to that of Equation (37) of Major et al. (1993), Equation (A11) of Holmes et al. (1992), and to the solutions in Glenn et al. (1988) and Glenn (1988).

7.2.3.1 The Lumped Soma and the Somatic Shunt Solution

In most neuron models, the soma is assumed to be isopotential and is modeled by an RC circuit. Often the time constant τ_0 for the soma is assumed smaller than the other time constants, in which case the model is said to have a shunt soma and is called a somatic shunt model (Durand, 1984). Such models have had some use in explaining experimental results, but the instability of the inverse problem has limited their usefulness (White et al., 1992).

The somatic shunt model is a limiting case of the multipolar models, as we now illustrate. Let us assume that the somatic cable has a taper of 0. Then we can write

$$\hat{H}(s) = \left(-g_0^\infty\sqrt{s\tau_0+1}\tanh(L_0\sqrt{s\tau_0+1}) + \sum_{j=1}^{n}\frac{g_j^\infty(1+s\tau_j)}{K_j/2-\hat{b}_j\coth(\hat{b}_jL_j)}\right)^{-1}.$$

We define the constant $\rho = G_0^\infty L_0$ so that

$$\hat{H}(s) = \left(-\rho\sqrt{s\tau_0+1}\frac{\tanh(L_0\sqrt{s\tau_0+1})}{L_0} + \sum_{j=1}^{n}\frac{g_j^\infty(1+s\tau_j)}{K_j/2-\hat{b}_j\coth(\hat{b}_jL_j)}\right)^{-1}.$$

If L_0 is arbitrarily small (i.e., if we take the limit as $L_0 \to 0$) then

$$\hat{H}(s) = \left(-\rho(s\tau_0+1) + \sum_{j=1}^{n}\frac{g_j^\infty(1+s\tau_j)}{K_j/2-\hat{b}_j\coth(\hat{b}_jL_j)}\right)^{-1}.$$

Correspondingly, the eigenvalue equation reduces to

$$\sum_{j=1}^{n}\frac{g_j^\infty(\alpha^2\tau_j-1)}{K_j/2-\sqrt{\alpha^2\tau_j-K_j^2/4-1}\cot(L_j\sqrt{\alpha^2\tau_j-K_j^2/4-1})} = \rho(1-\tau_0\alpha^2). \tag{7.9}$$

If each of the tapers is identically zero and if $\tau_j = \tau_m$ for all $j = 1, \ldots, N$, then Equation (7.9) reduces to

$$\rho \frac{1 - \varepsilon(1 + \beta^2)}{\beta} = \sum_{j=1}^{n} g_j^{\infty} \tan(\beta L_j), \tag{7.10}$$

where $\beta = \sqrt{\alpha^2 \tau_m - 1}$ and $\varepsilon = \tau_0/\tau_m$. This is the same as Equation (6.23) in Major et al. (1993a). This shows that a very short cable approaches the formula for shunts. Very short cables have advantageous analytic properties because the eigenfunctions are *orthogonal*. If an RC shunt is used in a model, the eigenfunctions will not be orthogonal, even if the model is as simple as a single, nontapering, equivalent cylinder (Durand, 1984). The lack of orthogonality has been shown to cause instability in the important inverse problem of parameter identification (White et al., 1992). Consequently, given that the alternative of modeling the soma as a short dendrite is asymptotically identical, we would recommend against the incorporation of shunt somas into models.

7.2.3.2 Eigenvalues and Decay Rates

The solution for the β_i requires finding roots (eigenvalues) for an equation in the form of a sum of transcendental functions. For the multipolar branching model of Figure 7.2, nodes can be divided into two types: the central node at the soma and peripheral nodes at the branch points. The nodes (the wires between cables in Figure 7.2) directly connect one cable to one or more other cables by a zero-resistance bridge. Boundary conditions for both voltage continuity and current continuity apply at each node. The boundary condition for the dendritic terminals is zero axial current at the distal terminal, corresponding to an insulated end of infinite electrical resistance (sealed-end condition), that is, $A_{ij} = 0$. At the central node, the current continuity condition is combined with voltage continuity condition to obtain

$$\left. \frac{\sum_{j_1=(1)}^{(n^\sigma)} -g_{j_1}^{\text{axial}} (\partial V_{j_1}/\partial Z_{j_1})}{V_{j_1}} \right|_{Z_{j_1}=0} + \frac{g^\sigma V^\sigma + C^\sigma (\partial V^\sigma/\partial t)}{V^\sigma} = 0, \tag{7.11}$$

$$f(\beta^{-1}) = g^\sigma \left(1 - \frac{\tau^\sigma}{\beta_i}\right) + \sum_{j_1=(1)}^{(n^\sigma)} Y_{j_1} = 0, \tag{7.12}$$

$$Y_j = g_j^{\infty} b_j \frac{(B_{ij}(K_j/2b_j) - 1) \tan b_j L_j + ((K_j/2b_j) + B_{ij})}{1 + B_{ij} \tan b_j L_j}, \tag{7.13}$$

$$b = \sqrt{\frac{\tau_j}{\beta_i} - \frac{K_j^2}{4} - 1}, \tag{7.14}$$

where Y is the rate-dependent (i.e., b-dependent) input conductance at the soma, or concisely the input admittance. When the dendritic time constants are all equal, and $\beta = \tau/(1 + \alpha^2)$, then this solution reduces to Equation (9) of Glenn (1988) and Equation (22) of Major et al. (1993).

7.2.3.3 Eigenfunctions and Amplitude Coefficients

At the peripheral nodes, the boundary conditions give

$$B_{ij} = (b_j)^{-1} \left(\frac{-K_j^2}{4} - \frac{1}{g_j^{\infty}} \sum_{j_{p+1}=(j_1,\ldots,j_p,1)}^{(j_1,\ldots,j_p,n_j)} Y_{j_{p+1}} \right),$$

where the summation is over all dendrites attached to the peripheral end of dendrite j. The B and Y are recursively related in that $Y_1 = f(B_1), B_1 = f(Y_2), B_2 = f(Y_3)$, and so on where the subscript refers to the order of branching in a line of descendants. The B_{ij} for the first-order dendrites are determined by the properties of those second-order dendrites that are their descendants (Equation [7.12]), the B_{ij} of which are in turn determined by the properties of the third-order descendants (Equation [7.13]), and so forth. For a branch with a sealed end, such as a terminal branch, the input admittance reduces to $Y_{ij} = 0$, and

$$B_{ij} = \frac{-K_j}{2b_{ij}},$$

which is a nonzero quantity that represents an adjustment for the taper between two ends of a given cable. This contrasts with untapered and unbranched models, where the sealed-end condition is simply $B_{ij} = 0$ (Rall, 1959). The continuity-of-voltage condition for dendrites attached in series requires the A_{ij} (ratio of voltage at distal cable end to that of the soma) to be related by

$$\begin{aligned} A_{ij_p} &= e^{-K_{jp}L_{jp}/2}(\cos L_{j_p} + B_{j_p} \sin b_{j_p} L_{j_p}) A_{ij_{p-1}} \quad \text{for } p > 0, \\ A_{ij_0} &= 1. \end{aligned} \tag{7.15}$$

Importantly, if a given b is imaginary, then both B and $\sin bL$ are imaginary. Consequently, A is always a real number in this model. The summation in the nonrecursive form is, simply stated, over all dendrites attached to the peripheral of the pth-order dendrite $j = (j_1, j_2, \ldots, j_p)$.

The next step is to obtain the amplitude coefficients (Equation [7.2]) for an instantaneous pulse. These will be obtained by the application of the same boundary conditions to the steady-state equation in Laplace space (Jack and Redman, 1971; Butz and Cowan, 1974; Holmes, 1986; Tuckwell, 1988a) and by obtaining the inverse transform for the series of coefficients by sequential evaluation of the complex residues (Kawato, 1984; Bluman and Tuckwell, 1987; Major et al., 1993). The three terms comprising C can be represented in both the transformed and inverse transformed equations, providing for analytic insights.

Rearranging (Equation [7.1]) and letting $\beta^\delta = 1$,

$$G(Z_r, Z_s, t) - \sum_{i=-\infty}^{\infty} \frac{\phi_{ir}(Z_r)\phi_{is}(Z_s)}{D_i} e^{-t/\beta_i} = 0, \tag{7.16}$$

and separating any single amplitude coefficient $i = k$,

$$\frac{\phi_{hr}(Z_r)\phi_{hs}(Z_s)}{D_h} = \lim_{t\to\infty} e^{-t/\beta_h} \left[G(Z_r, Z_s, t) - \sum_{i=-(h-1)}^{h-1} \frac{\phi_{ir}(Z_r)\phi_{is}(Z_s)}{D_i} e^{-t/\beta_i} \right], \tag{7.17}$$

a formula is found for the mth amplitude coefficient. The Laplace transform is

$$\begin{aligned} \frac{\phi_{hr}(Z_r)\phi_{hs}(Z_s)}{D_h} &= \lim_{p\to 0} \left[p\bar{V}_r^\delta \left(Z_r, Z_s, p - \frac{1}{\beta_h} \right) - p \frac{\sum_{i=-(h-1)}^{h-1} \phi_{ir}(Z_r)\phi_{is}(Z_s)}{((1/\beta_i) - (1/\beta_h))D_i} \right] \\ &= \lim_{p\to 0} p\bar{V}_r^\delta \left(Z_r, Z_s, p - \frac{1}{\beta_h} \right). \end{aligned} \tag{7.18}$$

The values of the amplitude coefficients can be obtained by the above procedure, which is a variant of the final value method of Kawato (1984) and the residue method of Bluman and Tuckwell (1987). The formula is applied to the equation for $V_j(Z_r, Z_s, p)$ in Laplace space using the same boundary conditions above. The equations in Laplace space are found by taking the Laplace transform of p in Equations (7.2) and (7.14), which are

$$\bar{V}_j^{\delta}(Z_r, Z_s, p) = \frac{\bar{\phi}_r(Z_r)\bar{\phi}_s(Z_s)}{\bar{D}_i}\bar{I}(Z_s, p), \tag{7.19}$$

$$\phi_j(Z_j) = \frac{1}{\bar{A}_j}[\bar{B}_j \sinh\sqrt{\tau_j p + 1}(L_j - Z_j) + \cosh\sqrt{\tau_j p + 1}(L_j - Z_j)], \tag{7.20}$$

where the boundary conditions are applied, except now as $f(p)$ rather than $f(t)$. Procedures for the formulation of Equation (7.4) for particular membrane properties and branching patterns can be found in Tuckwell (1988), Butz and Cowan (1974), Jack et al. (1975), and Holmes (1986).

Substitution of the boundary conditions for Equations (7.11) through (7.14) into (7.15), followed by application of the inverse transform formula in Equation (7.20), we find the expression required for a full solution to Green's function.

$$D_i = \frac{df(\beta^{-1})}{d(-\beta^{-1})} = g^{\sigma}\tau^{\sigma} + \sum_{j_1=(1)}^{(n^{\sigma})} \frac{dY_{j_1}}{d(-(1/\beta_i))}, \tag{7.21}$$

$$\frac{dY_{j_p}}{d(-(1/\beta_i))} = \sum_{j_p=(j_1,j_2,\ldots,1)}^{(j_1,j_2,\ldots,n_j)} \left(\frac{\tau_{j_p} Y_{j_p}}{2b_{j_p}^2} + \frac{g_{j_p}^{\infty}\tau_{j_p} K_{j_p}}{4b_{j_p}^3} + \frac{g_{j_p}^{\infty}\tau_{j_p} L_{j_p}(1 - B_{ij_p}^2)}{2A_{ij_p}^2} \right.$$
$$\left. + A_{j_p}^{-1}\left[-\frac{g_{j_p}^{\infty}\tau_{j_p} K_{j_p}}{2b_{j_p}^2} - \frac{\tau_{j_p}}{b_{j_p}^2} \sum_{j_{p+1}=(j_1,j_2,\ldots,j_p,1)}^{(j_1,j_2,\ldots,j_p,n_{j(p+1)})} Y_{j_{p+1}} + 2 \sum_{j_{p+1}=(j_1,j_2,\ldots,j_p,1)}^{(j_1,j_2,\ldots,j_p,n_{j(p+1)})} \frac{dY_{j_{p+1}}}{d(-(1/\beta_i))} \right] \right). \tag{7.22}$$

The eigenfunctions represented by the D_i can be shown to be orthogonal. To reiterate, the terminal branch has an admittance of $dY/d\beta^{-1} = Y = 0$, thus $B = -K/2b$, which is homologous to the condition of $B = 0$ in untapered, single cable models (Rall, 1962, 1977).

From Equations (7.21) and (7.22), the change in admittance with decay rate for terminal branches is

$$\frac{dY_{j_p}}{d(\beta_i^{-1})} = \sum_{j_p=(j_1,j_2,\ldots,1)}^{(j_1,j_2,\ldots,n_j)} \frac{g_{j_p}^{\infty}\tau_{j_p}}{4b_{ij_p}^2} \left(\frac{2Y_{j_p}}{G_{j_p}^{\infty}} - K_{j_p} + \frac{-2b_{j_p}(b_{ij_p}L_{j_p}(1 + (K_{j_p}^2/4b_{ij_p}^2)) - (K_{j_p}/2b_{ij_p}))}{(\cos b_{ij_p}L_{j_p} - (K_{j_p}/2b_{ij_p})\sin b_{ij_p}L_{i_p})^2} \right).$$

And the quantity dY is real whether b is imaginary or real.

7.3 Solutions for Transients in a Full Arbor Model with Linear Embedding of Sodium Channels

7.3.1 Definition of the System

In Section 7.1, it was pointed out that the experimental evidence has led to the conclusion that models without active properties are unrealistic. In this section, we show how to incorporate any number of sodium channels at any set of loci on the soma or dendritic arbor.

For the model developed in Section 7.2, let the j_pth branch have M_{j_p} ion channels (also known as *hot spots*) at electrotonic distances of $Z_{j_pi}, i = 1, 2, \ldots, M_{j_p}$. Let the soma be modeled by an arbitrarily short cable corresponding to $j = 0$ with no hot spots, but with an external current stimulus $I^{\text{in}}(t)$ at point Z_{01}. We pointed out the advantages of using a short cylinder over an RC shunt in Section 7.2.3.

The potential $V_{j_p}(Z_{j_p}, t)$ along the j_pth branch segment satisfies

$$\frac{\partial^2 V_{j_p}}{\partial Z_{j_p}^2} + K_{j_p}\frac{\partial V_{j_p}}{\partial Z_{j_p}} - \tau_{j_p}\frac{\partial V_{j_p}}{\partial t} - V_{j_p} = \sum_{i=1}^{M_{j_p}} I_{\text{ion}}(Z_{j_pi}, t; V_{j_p})\,\delta(Z_{j_p} - Z_{j_pi}), \tag{7.23}$$

where $I_{\text{ion}}(Z_{j_pi}, t; V_{j_p})$ is the ionic current density at time t at the hot spot located at $Z_{j_pi}, i = 1, \ldots, M_{j_p}$ (Poznanski and Bell, 2000a). Accordingly, the short cable modeling the soma satisfies the soma criterion (see Section 7.2.3)

$$\frac{\partial^2 V_0}{\partial Z_0^2} + K_0\frac{\partial V_0}{\partial Z_0} - \tau_0\frac{\partial V_0}{\partial t} - V_0 = I^{\text{in}}(t)\delta(Z_0 - Z_{01}), \tag{7.24}$$

which is itself of the form of Equation (7.23) if we identify $I_{\text{ion}}(Z_{01}, t; V_{01}) = I^{\text{in}}(t)$ and $M_0 = 1$. Given Green's function and a linear system, the solution voltage response to any input current can be obtained by a convolution integral. Specifically, the solution for I_{ion} and for the Green's function in Equations (7.3), (7.4), (7.21), and (7.22) is

$$V_r(Z_r, t) = \sum_{p=1}^{N}\sum_{j_p=1}^{n_p}\sum_{i_{j_p}=1}^{M_{j_p}} \int_0^t G(Z_r, Z_{j_p}; t - s) I_{\text{ion}}(Z_{j_pi_l}, s; V_{j_pi_{j_p}})\,\mathrm{d}s, \tag{7.25}$$

where $V_{j_pi_{j_p}} = V_{j_p}(Z_{j_pi_{j_p}}, s)$. Here, p takes the values of $1, 2, \ldots, N$ representing N arborizations. For each p there are n_p branches and each branch (labeled as j_p) has M_{j_p} hot spots.

7.3.2 Approximate Solutions for Persistent (Na^+P) Sodium Channels

Like axonal persistent sodium channels, dendrosomatic persistent sodium channels are opened by depolarization of the membrane at 10 to 15 mV below threshold of the classical Na^+ current (Crill, 1996; Clay, 2003). Nonlinear behavior, such as this, poses challenges to obtaining an analytic solution. The convolution Equation (7.25) can be used to produce a solution to Equations (7.3), (7.4), (7.21), and (7.22) when $I_{\text{ion}}(Z_{j_pi}, t; V_{j_pi})$ is not a linear function of $V_{j_pi}(t) = V_{j_p}(Z_{j_pi}, t)$. Specifically, we assume that at Z_{j_pi}, $I_{\text{ion}}(Z_{j_pi}, t; V_{j_pi})$ induces an equal-magnitude membrane capacitative current

$$C_m\frac{\mathrm{d}V_{j_pi}}{\mathrm{d}t} = I_{\text{ion}}(Z_{j_pi}, t; V_{j_pi}).$$

Using the notation for the uncollapsed, full arbor model in Section 7.2, the persistent sodium channels can be modeled by

$$C_m \frac{\mathrm{d}V_{j_p i}}{\mathrm{d}t} = \gamma_{j_p}[m_{j_p}(V_{j_p i})(V_{j_p i} - V_{\mathrm{NaP}})], \tag{7.26}$$

$$\frac{\mathrm{d}m_{j_p}}{\mathrm{d}t} = \alpha_{j_p}(V_{j_p i})(1 - m_{j_p}) - \beta_{j_p}(V_{j_p i})m_{j_p}, \tag{7.27}$$

where the rate coefficients are

$$\alpha_{j_p} = \frac{-1.74(V_{j_p} - 81)}{\mathrm{e}^{(V_{jp}-81)/12.94} - 1} \quad \text{and} \quad \beta_{j_p} = \frac{0.06(V_{j_p} - 75.9)}{\mathrm{e}^{(V_{jp}-75.9)/4.47} - 1}. \tag{7.28}$$

The coefficients incorporate the resting membrane voltage of -70 mV. Moreover, let us further assume that for positive integers R and S

$$\alpha_{j_p} \cong A_0 + A_1 V_{j_p i} + A_2 V_{j_p i}^2 + \cdots + A_R V_{j_p i}^R,$$

$$\beta_{j_p} \cong B_0 + B_1 V_{j_p i} + B_2 V_{j_p i}^2 + \cdots + B_R V_{j_p i}^R.$$

In other words, the rate coefficients are well approximated by polynomials in V_{jl}.

To simplify, let

$$j = j_p,$$

where the subscript j implies recursive branch-by-branch summation on j_p exactly as in Equation (7.22). Now apply a linear embedding scheme to Equations (7.26) and (7.27) by the methods of Kowalski and Steeb (1991) and Montroll and Helleman (1976). For k and n nonnegative integers, let

$$U_{k,n} = V_{ji}^k m_j^n.$$

Then the product rule implies that

$$\frac{\mathrm{d}U_{k,n}}{\mathrm{d}t} = kV_{ji}^{k-1} m_j^n \frac{\mathrm{d}V_{ji}}{\mathrm{d}t} + nV_{ji}^k m_j^{n-1} \frac{\mathrm{d}m_j}{\mathrm{d}t}.$$

Substitution using Equation (7.26) then leads to:

$$\frac{\mathrm{d}U_{k,n}}{\mathrm{d}t} = kV_{ji}^{k-1} m_j^n [\gamma_j m_j (V_{ji} - V_{\mathrm{NaP}})] + nV_{ji}^k m_j^{n-1} [\alpha_j(V_{ji})(1 - m_j) - \beta_j(V_{ji})m_j],$$

and expansion then leads to

$$\begin{aligned} \frac{\mathrm{d}U_{k,n}}{\mathrm{d}t} &= k\gamma_j V_{ji}^k m_j^{n+1} - k\gamma_j V_{\mathrm{NaP}} V_{ji}^{k-1} m_j^{n+1} \\ &\quad + nV_{ji}^k m_j^{n-1} \alpha_j(V_{ji}) - nV_{ji}^k m_j^n \alpha_j(V_{ji}) - nV_{ji}^k m_j^n \beta_j(V_{ji}). \end{aligned}$$

If we now use polynomial approximations of the rate coefficients, we obtain

$$\frac{\mathrm{d}U_{k,n}}{\mathrm{d}t} = k\gamma_j V_{ji}^k m_j^{n+1} - k\gamma_j V_{\mathrm{NaP}} V_{ji}^{k-1} m_j^{n+1} + n\sum_{\mu=0}^{R} A_\mu V_{ji}^{k+\mu} m_j^{n-1} - n\sum_{\mu=0}^{R} A_\mu V_{ji}^{k+\mu} m_j^n - n\sum_{\mu=0}^{R} B_\nu V_{ji}^{k+\nu} m_j^n.$$

Finally, the definition of $U_{k,n}$ implies that

$$\frac{\mathrm{d}U_{k,n}}{\mathrm{d}t} = k\gamma_j U_{k,n+1} - k\gamma_j V_{\mathrm{NaP}} U_{k-1,n+1} + n\sum_{\mu=0}^{R} A_\mu U_{k+\mu,n-1} - n\sum_{\mu=0}^{R} A_\mu U_{k+\mu,n} - n\sum_{\nu=0}^{R} B_\nu U_{k+\nu,n}. \tag{7.29}$$

The system of equations (Equation [7.29]) is an infinite-dimensional *linear* system of first-order differential equations. It has been shown (Gaude, 2001) that if the infinite-dimensional linear system is truncated to $k < P$ and $n < Q$ for positive integers P and Q, then the solution of the resulting truncated system is a good approximation of the actual system as in Equation (7.29) and thus provides an accurate approximation of $I_{\mathrm{ion}}(Z_{ji}, t; V_{ji})$. This approximation of $I_{\mathrm{ion}}(Z_{ji}, t; V_{ji})$ can then be used in Equation (7.25) to produce a final solution to the tapered, active, full arbor model of Equations (7.3), (7.4), (7.21), and (7.22).

However, the values of P and Q often must be quite large (a range of 500 to 1500 is typical). Thus, implementation of the truncation of Equation (7.29) requires the use of a computer algebra system or other mathematical software. When the number of dendrites is one ($j = 1$) and when that dendrite is untapered ($K = 0$), then the above solution reduces to that of Equation (12) through Equation (14) of Poznanski and Bell (2000a). In conclusion, Carleman embedding is an excellent alternative to other linearization methods for obtaining analytic solutions to multipolar tapered models.

7.3.3 Approximate Solutions for Transient (Na^+) Sodium Channels

Let the sodium ion channels now satisfy Hodgkin and Huxley (1952a) kinetics, where m represents the opening or activation variable and h is the closing or inactivation variable.

$$C_m \frac{\mathrm{d}V_{ji}}{\mathrm{d}t} = \gamma_j [m_j^3 h_j (V_{ji} - V_{\mathrm{Na}})], \tag{7.30}$$

$$\frac{\mathrm{d}m_j}{\mathrm{d}t} = \alpha_{m_j}(1 - m_j) - \beta_{m_j} m_j, \qquad \frac{\mathrm{d}h_j}{\mathrm{d}t} = \alpha_{h_j}(1 - h_j) - \beta_{h_j} h_j. \tag{7.31}$$

Now, following the previous section, the polynomials must approximate both channel activation and inactivation.

$$\begin{aligned}
\alpha_{m_j} &\cong A_{m0} + A_{m_1} V_{ji} + A_{m_2} V_{ji}^2 + \cdots + A_{m_R} V_{ji}^R, \\
\beta_{m_j} &\cong B_{m0} + B_{m_1} V_{ji} + B_{m_2} V_{ji}^2 + \cdots + B_{m_R} V_{ji}^R, \\
\alpha_{h_j} &\cong A_{h0} + A_{h_1} V_{ji} + A_{h_2} V_{ji}^2 + \cdots + A_{h_R} V_{ji}^R, \\
\beta_{h_j} &\cong B_{h0} + B_{h_1} V_{ji} + B_{h_2} V_{ji}^2 + \cdots + B_{h_R} V_{ji}^R.
\end{aligned} \tag{7.32}$$

For β_m, the coefficients B_R satisfy

$$B_R = \frac{4}{(18)^R R!}.$$

For α_h, the coefficients A_R satisfy

$$A_R = \frac{0.07}{(20)^R R!}.$$

At first glance, it may seem that the use of polynomials instead of kinetic equations defeats the mechanistic value of the kinetic equations (i.e., the rate of opening and closing of gates). The fact is Hodgkin and Huxley (1952a) fit the sodium I–V curve to equations of the form of Equation (7.31) for reasons that were somewhat arbitrary. After describing the two possibilities that they favored from among many for modeling the I–V curves, they selected Equation (7.31) on this basis: "The second alternative was chosen since it was simpler to apply to the experimental results" (p. 512 of Hodgkin and Huxley, 1952a). Accordingly, we chose Equation (7.32) because it simplified an analytic solution for the full arbor model.

For k, n, and o nonnegative integers, let

$$U_{k,n,o} = V_{ji}^k m_j^n h_j^o. \tag{7.33}$$

Then by the product rule

$$\frac{\mathrm{d}U_{k,n,o}}{\mathrm{d}t} = V_{ji}^k m_j^n h_j^{o-1} \frac{\mathrm{d}h_j}{\mathrm{d}t} + m_j^n h_j^o k V_{ji}^{k-1} \frac{\mathrm{d}V_j}{\mathrm{d}t} + V_{ji}^k h_j^o n m_j^{n-1} \frac{\mathrm{d}m_j}{\mathrm{d}t}. \tag{7.34}$$

Expanding after substitution using Equations (7.30) and (7.31),

$$\begin{aligned}
\frac{\mathrm{d}U_{k,n,o}}{\mathrm{d}t} = {} & V_{ji}^k m_j^n h_j^{o-1} \alpha_{h_j} - V_{ji}^k m_j^n h_j^o \alpha_{h_j j} - V_{ji}^k m_j^n h_j^o \beta_{h_j} + \gamma_j m_j^{n+3} h_j^{o+1} k V_{ji}^k \\
& - \gamma_j m_j^{n+3} h_j^{o+1} k V_{ji}^{k-1} V_{\mathrm{NaP}} + \gamma_j m_j^n h_j^o k V_{ji}^k \eta_j - \gamma_j m_j^n h_j^o k V_{ji}^{k-1} \eta_j V_{\mathrm{KP}} \\
& + V_{ji}^k h_j^o n m_j^{n-1} \alpha_{m_j} - V_{ji}^k h_j^o n m_j^n \alpha_{m_j} - V_{ji}^k h_j^o n m_j^n \beta_{m_j}.
\end{aligned}$$

Applying the polynomial approximations, we get

$$\frac{dU_{k,n,o}}{dt} = \gamma_j m_j^{n+3} h_j^{o+1} k V_{ji}^k - \gamma_j m_j^{n+3} h_j^{o+1} k V_{ji}^{k-1} V_{\text{NaP}} + \gamma_j m_j^n h_j^o k V_{ji}^k \eta_j - \gamma_j m_j^n h_j^o k V_{ji}^{k-1} \eta_j V_{\text{KP}}$$
$$+ n \sum_{\mu=0}^{R} A_{m_\mu} V_{ji}^k m_j^{n-1} h_j^o - n \sum_{\mu=0}^{R} A_{m_\mu} V_{ji}^k m_j^n h_j^o + \sum_{\mu=0}^{R} A_{h_\mu} V_{ji}^k m_j^n h_j^{o-1} - \sum_{\mu=0}^{R} A_{h_\mu} V_{ji}^k m_j^n h_j^o$$
$$- \sum_{\nu=0}^{R} B_{h_\nu} V_{ji}^k m_j^n h_j^o - n \sum_{\nu=0}^{R} B_{h_\nu} V_{ji}^k h_j^o m_j^n.$$

Finally, from the definition of $U_{k,n,o}$,

$$\frac{dU_{k,n,o}}{dt} = \gamma_j k U_{k,n+3,o+1} - \gamma_j k V_{\text{NaP}} U_{k-1,n+3,o+1} + \gamma_j k \eta_j U_{k,n,o} - \gamma_j k \eta_j V_{\text{KP}} U_{k-1,n,o}$$
$$+ n \sum_{\mu=0}^{R} A_{m_\mu} U_{k,n-1,o} - n \sum_{\mu=0}^{R} A_{m_\mu} U_{k,n,o} + \sum_{\mu=0}^{R} A_{h_\mu} U_{k,n,o-1} - \sum_{\mu=0}^{R} A_{h_\mu} U_{k,n,o}$$
$$- \sum_{\nu=0}^{R} B_{h_\nu} U_{k,n,o} - n \sum_{\nu=0}^{R} B_{h_\nu} U_{k,n,o}. \tag{7.35}$$

As in the previous section with persistent sodium channels, Equation (7.35) is used to obtain a solution for any number of Hodgkin and Huxley (m^3h) sodium channels at any location on the dendritic arbor or soma of a fully arborized tree with arbitrary branching, branch tapering rates, and branch membrane resistivity.

7.4 Discussion

In the present chapter, we formulate analytic models that range from the simple to high complexity and realism. The simpler models are single (equivalent) models in which an entire neuron is collapsed by the $d^{3/2}$ rule. The complex models are fully arborized models with branch tapering and active conductances, in which the $d^{3/2}$ assumption is not used, and each of the thousands of cables represent all dendritic branches (i.e., the segments between adjacent branch points). The question addressed now is how to decide among these different models for any given experimental approach.

Uncollapsed Models. The fully arborized models above are best suited to neurons for which the complete dendritic architecture (length, diameter, and connectivity of branches) is known. Generally, this knowledge comes from filling neurons with a dye or chromagen-producing substance such as horseradish peroxidase or biotin, and meticulously reconstructing and measuring the branches in a series of sections. These models are the most architecturally detailed to date, but also require the most time and work. The models with active conductances also require either assumptions about kinetics or voltage-clamp data on channel kinetics.

Partially Collapsed Models. These models were derived and described by Evans and Kember (1998) and Evans (2000), and are most suited for neurons in which the soma diameter and stem diameters of all dendritic trees arc known, but little is known about the branching and tapering architecture of individual trees. In this case, each dendritic tree can be assigned a diameter based on direct morphometric measurement, with the length and taper estimated from the soma diameter. Such estimates require that the relation between dendritic diameter, length, and taper for the neuron under study is represented by neurons in which such relations have been determined (Ulfhake and

Kellerth, 1983, 1984; Cullheim et al., 1987a, 1987b). Limited architectural data is common in morphological studies and arises from the difficulty of obtaining high-contrast, complete fills of arbors with chromagen-producing agents.

Fully Collapsed Models. Models of the single equivalent cylinder type, particularly the tapering type developed by Poznanski, are optimal when little or nothing is known about the architecture of the experimentally recorded neurons and when all stimulation and recording is in the soma (i.e., antidromic). It is far more difficult, expensive, and time-consuming to obtain reliable passive response recordings from neurons with a dye-filled microelectrode than from plain microelectrodes. The difficulty level is compounded in smaller neurons. Consequently, there will always be experimental problems for which such fully collapsed models are the best choice.

The primary goal of analytic models of neural activity, as opposed to computational approaches, is an understanding of the physical mechanisms implied by the equations and their solutions. Computational methods, such as finite element modeling, treat the neuron as a black box where the black box is adjusted until the inputs and outputs match the neuron. Analytic models are like glass cages, where the mechanistic relations between input and output are visible and inherent aspects of the models. The analytic multipolar-branching, tapered model with hot spots in the present chapter is a new model that has not been solved before, despite its obvious importance to understanding the theory of transient conduction in neurons and in interpreting experimental data. The model can be applied to interpret the voltage responses of neurons to current steps or synaptic currents in almost any type of multipolar neuron, including those with sodium channels of nearly arbitrary kinetics.

Nearly six decades have passed since Hodgkin and Huxley (1952a) developed their unparalleled mathematical model of the axonal action potential. Despite mounting evidence that dendrites have active properties, and with the realization that these properties are vital to brain function, very few inroads have been made in these six decades in the incorporation of Hodgkin–Huxley-style kinetics into passive models, particularly the more realistic passive models that include tapering, nonuniform resistivity or branching. The reason is that the equations were intractable, leading many theoreticians to abandon efforts in frustration in this important area. The recent work of Poznanski and Bell (2000) along with the proliferation of linear embedding methods through mathematical communities, has led to solutions in series form, and with that, renewed hopes for anatomically realistic active models of dendritic propagation that use analytic solutions. The present study is clearly only the beginning. In the opinion of the authors, we are over the hump. Now may be time to expand research on active fully arborized models and make headway. Indeed, the linear embedding method has revolutionized many areas in the physical and natural sciences. It can do the same for the neurosciences by enabling solutions to important and sophisticated models, which when combined with experimental data, can elucidate the mechanisms of active dendritic integration in the central nervous system.

7.5 Summary

A new analytic model is considered with the following characteristics: voltage-dependent sodium conductances, full dendritic arbors of arbitrary size, arbitrary branching structure, and unique branch properties. The unique branch properties include diameter, taper rate, and membrane resistivity. The taper form is quadratic, which is an exponential taper in dimensionless electrotonic units. The voltage-dependent conductances are either persistent sodium channels with activation terms only (m) or transient sodium channels with both activation and inactivation terms (m^3h). Any number of channels can be incorporated into the model, at any point on any branch or soma in the model. The analytic solutions for the nonlinearities in this model are obtained with the use of Carleman linear embedding. Carleman embedding can be used to provide linearized solutions in open form for any type of channel that has nonlinearities which follow first-order kinetics. The structure of the solution and the intermediary equations provide new physical insights on the relation between dendritic structure and the propagation of transients. A decision tree can be used to assist experimentalists in the selection of an analytic model according to the depth and breadth of experimental data available. The analytic model considered in this chapter is in the class that requires the greatest anatomical and physiological detail.

ACKNOWLEDGMENT

Thanks to Scott La Voie for preparing the Mathematica Notebook in Problem 6.6.

PROBLEMS

1. Find the complete solution in closed form of Equations (7.1) through (7.3) when $I_{\text{ion}}(Z_{ji}, t) \equiv 0$ and $I^{\text{in}}(t) = I^{\text{stim}} t\, \mathrm{e}^{-\gamma t}$, where I^{stim} and γ are positive constants.
2. In a morphological study, principle neurons were filled with stain to the dendritic tips. Large intrinsic neurons were only partially filled, but the soma and all first-order dendrites were elucidated. Small intrinsic neurons were unfillable, however, at least the soma diameters could be determined from silver stains. Decide which category of model in Section 7.4 should be selected for each of the three types of neurons, given the morphological data available.
3. Using Equations (7.25) to (7.28) and the linearization method of Poznanski and Bell (2000a), instead of Carleman linear embedding, show that the following is the formal solution for linearized I_{NaP} hot spots in a collapsed multipolar model:

$$V_j(Z_j, t) \approx \int_0^t G_j^{01}(Z_j, t-s) I^{\text{in}}(s)\, \mathrm{d}s + \sum_{l=1}^{N} \sum_{i=1}^{M_j} \gamma_j \int_0^t G_j^{li}(Z_j, t-s)\rho(s) \times \int_0^s G_j^{01}(Z_{ij}, s-u) I^{\text{in}}(u)\, \mathrm{d}u.$$

4. Goldstein and Rall (1974) and Poznanski (1991) showed that when the radius of the branch is small relative to unity, then

$$D_{\text{taper}} = D_0\, \mathrm{e}^{2KZ/3}$$

and that

$$\frac{\mathrm{d}Z}{\mathrm{d}x} = \frac{1}{\lambda}\sqrt{\frac{D_0}{D_{\text{taper}}}}.$$

Use this to derive

$$D_{\text{taper}} = D_0(1 + kx)^2.$$

(Hint: $k = K/3\lambda$).

5. Let us suppose that we define a *state* of the multipolar model to be, $\mathbf{F} = \langle F_0(Z_0), F_1(Z_1), \ldots, F_N(Z_N)\rangle$ where $F_j(Z_j)$ is the voltage distribution in the jth cable, and let us suppose that we define the inner product of two states $\mathbf{E}$ and $\mathbf{F}$ to be

$$\langle \mathbf{E}, \mathbf{F}\rangle = \sum_{j=0}^{n} G_j^{\infty} \tau_j \int_0^{L_j} E_j(Z_j) F_j(Z_j)\, \mathrm{d}Z_j.$$

Under what conditions on the parameters of the model are two different solutions of Equations (7.14) through (7.18) orthogonal?

6. Use *Maple*, *Mathematica*, *Matlab*, or some other mathematical programming environment to implement Equation (7.24) for some large positive integers P and Q. A starting example in *Mathematica* for a single hot spot on a single cable can be found at http://faculty.etsu.edu/knisleyj/carleman.htm.
7. The eigenvalue equation for the unbranched multipolar model is

$$0 = G^{\sigma}\left(\frac{\tau^{\sigma}}{\beta^{\sigma}} + 1\right) + \sum_{j=-\infty}^{\infty} \frac{G_j(1 - (\tau_j/\beta_i))}{-(K_j/2) + b_j \cot b_j L_j},$$

where

$$b = \sqrt{\frac{\tau_j}{\beta_i} - \frac{K_j^2}{4} - 1}.$$

Use *Maple*, *Mathematica*, *Matlab*, or some other mathematical programming environment to develop a method of estimating the first five eigenvalues.

APPENDIX: CARLEMAN LINEARIZATION

When a differential equation has a polynomial nonlinearity, the technique of *linear embedding* is often an attractive alternative to linearization and perturbation techniques. In particular, the embedding method of Carleman has been used successfully to study a wide range of nonlinear equations (Kowalski and Steeb, 1991). Pilot studies in equivalent cylinder models indicated that the method of linear embedding can lead to solutions for dendritic models with hot spots in open form (Knisley, 2003). Carleman embedding has four basic steps, which are summarized in this Appendix.

Let $V(Z,t)$ be the solution to the tapered equivalent cylinder model

$$\frac{\partial^2 V}{\partial Z^2} + K\frac{\partial V}{\partial Z} - \tau\frac{\partial V}{\partial t} - V = \gamma\delta(Z-p)I(Z,t;V), \tag{7.36}$$

where γ is constant and where $I(Z,t;V)$ is the ionic current at the "hot spot" located at $Z = p$. Given suitable boundary conditions, the Green's function $G(Z,p;t)$ is unique and satisfies

$$\frac{\partial^2 G}{\partial Z^2} + K\frac{\partial G}{\partial Z} - \tau\frac{\partial G}{\partial t} - G = 0,$$

$$G(Z,p;0) = \gamma\delta(Z-p).$$

It follows that the solution to Equation (7.36) is a convolution of the form

$$V(Z,t) = \int_0^t G(Z,p;t-s)I(Z,s;V)\,\mathrm{d}s, \tag{7.37}$$

which implies that we need only determine $I(Z,t;V)$ in order to complete the solution to Equation (7.36).

In their seminal work, Poznanski and Bell (2000a) modeled persistent ionic currents using a linear approximation to the I–V relationship. Carleman linear embedding allows the use of a polynomial approximation, which is more physiologically realistic than linear. In fact, it allows optimal curvilinear fits, in the sense that the least-squares sums of the deviations are minimized.

Suppose now that the ionic current satisfies a polynomial nonlinear differential equation of the form

$$\frac{dI}{dt} = a_0(t) + a_1(t)I + a_2(t)I^2 + \cdots + a_n(t)I^n, \tag{7.38}$$

where $a_0(t), \ldots, a_n(t)$ are sufficiently smooth. Such polynomial nonlinear models follow from considering the hot spot to be an "arbitrarily short voltage clamp" (see Equation [8.38] of Tuckwell, 1988b). Equation (7.38) cannot in general be solved in closed form. However, we can estimate the solution to Equation (7.38) by using what is known as embedding coefficients (U_m) by letting

$$U_m(t) = [I(Z,t;V)]^m.$$

Equation (7.38) then implies that

$$\begin{aligned}
\frac{dU_m}{dt} &= mI^{m-1}\frac{dI}{dt} \\
&= mI^{m-1}(a_0 + a_1 I + a_2 I^2 + \cdots + a_n I^n) \\
&= ma_0 I^{m-1} + ma_1 I^m + ma_2 I^{m+1} + \cdots + ma_n I^{m+n-1} \\
&= ma_0 U_{m-1} + ma_1 U_m + ma_2 U_{m+1} + \cdots + ma_n U_{m+n-1}.
\end{aligned}$$

The above system of equations can be solved either in closed form, or numerically as an ordinary differential equation. The result is an infinite-dimensional *linear* system of differential equations.

$$\frac{d}{dt}\begin{bmatrix} 1 \\ U_1 \\ U_2 \\ U_3 \\ U_4 \\ \vdots \end{bmatrix} = \begin{bmatrix} 0 & 0 & \ldots & 0 & 0 & \ldots & & & & & \\ 0 & a_0 & a_1 & \ldots & a_n & 0 & \ldots & & & & \\ 0 & 0 & 2a_0 & 2a_1 & \ldots & 2a_n & 0 & \ldots & & & \\ 0 & 0 & 0 & 3a_0 & 3a_1 & \ldots & 3a_n & 0 & \ldots & & \\ 0 & 0 & 0 & 0 & 4a_0 & 4a_1 & \ldots & 4a_n & 0 & \ldots \\ \vdots & \vdots & \vdots & \vdots & & \ddots & & \ddots & & \ddots \end{bmatrix} \begin{bmatrix} 1 \\ U_1 \\ U_2 \\ U_3 \\ U_4 \\ \vdots \end{bmatrix}.$$

The ionic current $I(Z,t;V)$ can then be estimated by truncating the system after a (large) number of terms, N, and then solving the resulting (large) $N \times N$ system of linear equations. The first term in the solution is

$$U_1(t) = I(V,t),$$

which thus serves as a good approximation of the actual ionic current (Gaude, 2001). If Equation (7.38) is nonlinear, then it may be necessary to compute a large number of terms before they become negligible. However, if Equation (7.38) is linear, then Carleman linearization converges very quickly. In any case, substituting the estimate of $I(Z,t;V)$ into the convolution Equation (7.37) subsequently produces a good approximation to the solution to Equation (7.36). Other aspects of linear embedding are treated in the works of Kowalski and Steeb (1991), Gaude (2001), and Montroll and Helleman (1976).

8 Analytical Solutions of the Frankenhaeuser–Huxley Equations Modified for Dendritic Backpropagation of a Single Sodium Spike

Roman R. Poznanski

CONTENTS

8.1 Introduction

The Hodgkin and Huxley (1952b) equations are a system of nonlinear partial differential equations describing nerve impulse (action potential or spike) conduction along the unmyelinated giant axon of the squid *Loligo*. The original Hodgkin–Huxley (H–H) equations are a fourth-order system comprising variables for membrane potential (or voltage), sodium (Na^+) activation and inactivation, and potassium (K^+) activation. The gating variables satisfy first-order kinetic equations with voltage-dependent rate coefficients. The Na^+ activation and inactivation have been experimentally shown to be coupled (Goldman and Schauf, 1972, 1973) in *Myxicola* giant axons, and therefore are described by two coupled first-order equations that are equivalent to a second-order equation (Goldman, 1975).

Subsequent experimental results suggested modification of the H–H equations when applied to saltatory transmission (i.e., action-potential propagation) between nodes of Ranvier along the myelinated axon from *Xenopus* (Frankenhaeuser and Huxley, 1964) and from rabbit sciatic nerve (Chui et al., 1979). A variation on saltatory conduction is present in the dendrites of some neurons as was first advocated by Rodolfo Llinás (see Llinás et al., 1968, 1969; Llinás and Nicholson, 1971; Traub and Llinás, 1979). In particular, Llinás and Nicholson (1971) were the first to show by computer simulation how active propagation on dendrites might occur by saltatory conduction between patches of excitable membrane or "hot spots" containing unusual clusterings of ion channels (Shepherd, 1994). These patches are separated from the site of impulse initiation in the cell body and axon by a leaky dendritic membrane. Saltatory transmission of action potentials is governed by the Frankenhaeuser–Huxley (F–H) equations, with the exception that postsynaptic somatic "dendritic spikes" may be transmitted decrementally in the somatopetal direction.

Sodium-based action potentials have been shown to initiate at the axon hillock and backpropagate antidromically into the dendritic trees of several different varieties of mammalian neurons with variable amounts of attenuation (e.g., Stuart and Häusser, 1994; Stuart and Sakmann, 1994; Buzsaki et al., 1996; Stuart et al., 1997b; Williams and Stuart, 2000; Golding et al., 2001). Furthermore, backpropagating action potentials (bAPs) are inevitably regulated by an abundant number of synaptic and extrasynaptic receptors, juxtaposed with voltage-dependent ion channels (Yamada et al., 2002; Antic, 2003; Goldberg et al., 2003). Indeed, dendrites show a high density of ligand-gated channels (especially those gated by extracellular ligands at both synaptic and extrasynaptic sites, and intracellular ligands such as calcium and second messengers) that may invoke large-conductance leaks in the dendrites from neuromodulators, especially if the dendrites are subjected to an intense synaptic bombardment in intact networks and are in a "high-conductance" state (Destexhe et al., 2003).

The *ad hoc* application of computational modeling to excitable dendrites (exhibiting a variety of exotic membrane currents) has underestimated the epistemological limitations of the H–H equations (Strassberg and DeFelice, 1993; Meunier and Segev, 2002). In order to develop realistic input/output functions of dendritic integration in a biophysical model that includes all the complexities introduced by active dendrites (for reviews see Johnston et al., 1996; Reyes, 2001), it is vital to introduce a heterogeneous membrane with voltage-dependent ion channels, and in particular, Na^+ channels, as discrete macromolecules, sparsely (in comparison with densities found along the axonal membrane) distributed along dendrites (Hille, 2001). Evidence for a relatively sparse density of transient Na^+ channels comes indirectly from the observed decrement in the amplitude of bAPs, which is a consequence of cable properties, including the distribution and locus of Na^+ channels (Caldwell et al., 2000) (which on average are approximately six times more frequent at synapses than at other sites along the plasma membrane as found using immunocytochemical localization [Hanson et al., 2004]), their prolonged inactivation (Jung et al., 1997), slow inactivation (Callaway and Ross, 1995; Colbert et al., 1997; Jung et al., 1997; Mickus et al., 1999), frequency-dependent dendritic Na^+ spike broadening (Lemon and Turner, 2000), and a distance-dependent modifiable threshold (Bernard and Johnston, 2003). However, clustering of Na^+ channels in extrasynaptic membrane remains undetermined.

The modeling of such dendrites requires the enforcement of a "sparsely" excitable membrane rather than a "weakly" excitable membrane; the former can be ascribed to a discrete spatial distribution of channels along the dendrites, while the latter can be ascribed to a low density yet continuous distribution of ionic channels. Indeed, in earlier modeling of active nerve fibers, a low or diminished excitability (i.e., a low density of channels) was achieved through scaling of conductances without invoking a discrete spatial distribution of channels (e.g., see Cooley and Dodge, 1966; Leibovic and

Sabah, 1969; Sabah and Leibovic, 1972; Dodge, 1979; Rapp et al., 1996; Horikawa, 1998). This has led to a modification of the H–H equations so as to give decremental (graded) pulses or *g-pulses*, for short (Leibovic, 1972). In an attempt to model such sparsely excitable dendrites under current-clamp conditions for small voltage perturbations from rest, Poznanski and Bell (2000a, 2000b) have provided a fruitful way to solve nonlinear cable problems analytically through so-called "ionic cable theory." The gist of this theory is to include density distributions of specific channel proteins into passive cable models of neural processes. These channels are divided into two classes, carriers and pores (ion channels), by distributing their specific voltage-dependent currents discretely.

Earlier analytical work dealing with (V, m, h) *reduced systems* with only the K^+ activation held constant at the resting value produced an action potential having an infinite plateau potential (FitzHugh, 1960). In other words, this model failed to reproduce the pulse recovery necessary for modeling the attenuation of bAPs (e.g., Migliore, 1996). Given that excitable membranes without voltage-gated K^+ channels show, and are able to maintain, recovery in the presence of an electrogenic sodium pump simply through possessing a large leak conductance like that found in mammalian myelinated axons (Chiu et al., 1979; Brismar, 1980), it may be conceivable that the sparse distribution of transient Na^+ channels, rather than K^+ channels, is responsible for the attenuation of bAPs in some, but not all, neurons. For example, inactivation of dendritic A-type K^+ channels was shown not to be responsible for the amplification in peak amplitude of bAPs in neocortical layer five pyramidal cells (Stuart and Häusser, 2001), suggesting that bAPs may possibly decay by leakage through the membrane.

The analysis presented here entails a modification to the F–H equations, applicable for dendrites in an all Na^+ system. However, unlike the Poznanski–Bell model, where hot spots act as point sources of *outward* current as a consequence of current density mimicking current applied through a pipette, in the present model *discrete* loci of Na^+ channels or hot spots act as point sources of *inward* current. This current represents the macroscopic behavior of voltage-dependent ionic channels to the extent of the validity of the constant-field assumption inherent in the Nernst–Planck electrodiffusion theory (cf. Attwell and Jack, 1978; Cooper et al., 1985). The clusters of Na^+ channels or hot spots are imposed on a homogeneous (nonsegmented) leaky cable structure with a high density of large-conductance leak channels (these may fall into various categories, including K^+, Cl^-, and various neuromodulators, including Na^+P channels) in dendrites, dampening the amplitude of bAPs needed for action-potential repolarization. Each hot spot is assumed to occupy an infinitesimal region containing a cluster of fast pores (i.e., Na^+ channels) in the presence of ion fluxes due to active transport processes, such as a Na^+ pump (i.e., proteins specialized as slow carriers). The discrete positioning of Na^+ channels along the dendrite allows for a complete analytical solution of the problem without requiring the simplified assumptions inherent in the kinetics of gating currents used in reduced models that assume Na^+ activation to be sufficiently fast to be adiabatically eliminated (e.g., Krinskii and Kokoz, 1973; Casten et al., 1975; Rinzel, 1985; Kepler et al., 1992; Meunier, 1992; Wilson, 1999a; Penney and Britton, 2002).

The H–H equations and their precursors form the most widely accepted model of nerve impulse propagation, but because of analytic complexity in finding closed-form solutions, most work on the models has been numerical (e.g., Cole et al., 1955; FitzHugh, 1962; Cooley et al., 1965; Cooley and Dodge, 1966; Goldman and Albus, 1968; Kashef and Bellman, 1974; Moore and Ramon, 1974; Adelman and FitzHugh, 1975; Moore et al., 1978; Hines, 1984, 1989, 1993; Halter and Clark, 1991; Hines and Carnevale, 1997; Bower and Beeman, 1998; Zhou and Chiu, 2001). Furthermore, the nonlinearity inherent in the H–H equations gives unstable solutions for certain changes to the original parameters. For example, changes to the K^+channel density in the H–H membrane can often lead to small-amplitude oscillations as investigated by Taylor et al. (1995), as well as to large-amplitude periodic solutions that reflect a repetitive discharge of action potentials as investigated by Troy (1976). Rinzel and Miller (1980) have introduced numerical methods that could be applied to evaluate the unstable periodic solutions that join the stable large- and small-amplitude solutions. Furthermore, Holden (1980) has shown that specific changes in K^+ channel density can lead to autorhythmicity or oscillatory discharges as found in pyramidal dendrites in the electrosensory lateral-line lobe of weakly electric fish (Turner et al., 1994). However, a discrete distribution of Na^+ with or without K^+ channels can alleviate such problems because of a quasilinear rather than nonlinear nature of the membrane.

A large number of papers have been produced (mostly by mathematicians) concerning the existence and uniqueness of solutions (e.g., Evans and Shenk, 1970; Evans, 1972; Troy, 1976; Mascagni, 1989b) and in particular, the existence of traveling wave solutions (Hastings, 1976; Bell and Cook, 1979; Engelbrecht, 1981; Muratov, 2000). Such approaches rely on a "similarity transformation" of the independent variables (Hodgkin and Huxley, 1952b; Leibovic, 1972):

$$x \to vt,$$

where v is the constant propagation velocity of the action potential. This reduces the cable equation (a partial differential equation) to a discretized compartmental equation (an ordinary differential equation). Unfortunately, the transformation commonly used to solve such nonlinear problems (see Wilson, 1999b for a review) is invalid for dendrites because of their nonuniform conduction velocities (Poznanski, 2001d). In other words, the resulting system of equations is not translationally invariant (cf. Keener, 2000). Therefore, leading-edge approximations of the H–H (or the F–H) equations will not yield analytical solutions as had been previously explored (e.g., Rissman, 1977; Cronin, 1987; Scott, 2002). Indeed, the H–H equations and their modified variants were once believed to be analytically intractable, with only a few studies obtaining approximate analytical solutions using perturbation methods based on simplified channel dynamics (e.g., Noldus, 1973; Casten et al., 1975). Apart from a few simplified models that are analytically tractable (e.g., Bell, 1984, 1986; Bell and Cosner, 1986), there are no formal analytical solutions of the F–H equations.

In the last decade we have seen a rapid rise in computational models without recourse to further development of nonlinear cable theory. Nonlinear dendrites are modeled through the development and extension of ionic cable theory for active propagation in dendrites, integrating the macroscopic current formalism of the gating variables in H–H equations with microscopic models of single channel activity.

8.2 The Cable Equation for Discretely Imposed Na^+ Channels

Let V be the depolarization (i.e., membrane potential less than the resting potential) in mV, and $I_{\text{Na}}\delta(x - x_i)$ be the inward Na^+ current density in the ith hot spot per membrane surface of dendritic cable in A/cm^2. We consider the membrane conductivity to be continuous, reflecting a high density of large-conductance leak channels resulting from various unknown neuromodulators impinging on the surface of the dendrites. We neglect the more realistic discrete nature of leakage channels in biological membranes because the error in doing so has been shown to be marginal (see Mozrzymas and Bartoszkiewicz, 1993). The voltage response (or depolarization) along a leaky cable representation of a cylindrical leaky dendritic segment with Na^+ channels occurring at discrete points along the cable (see Figure 8.1) satisfies the following cable equation:

$$C_{\text{m}}V_t = (d/4R_i)V_{xx} - g_{\text{L}}(V - V_{\text{L}}) - \sum_{i=1}^{N} I_{\text{Na}}(x, t; V)\delta(x - x_i) - \sum_{z=1}^{N} I_{\text{Na(pump)}}\delta(x - x_z), \quad t > 0, \tag{8.1}$$

where x is the distance in cm, t is the time in sec, d is the diameter of the cable in cm, C_{m} $(= c_{\text{m}}/\pi d)$ is the membrane capacitance (F/cm^2), g_{L} is the leak conductance (S/cm^2),

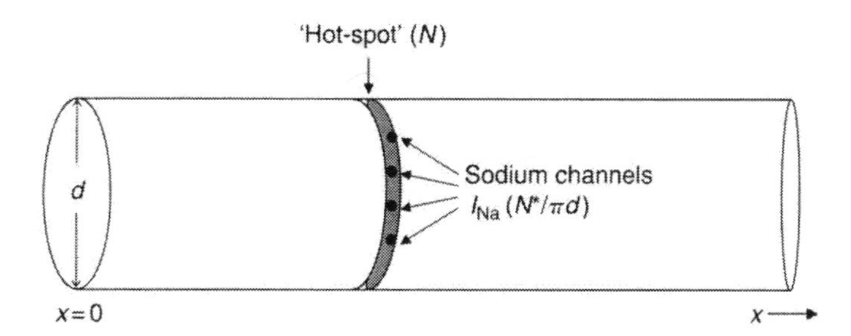

FIGURE 8.1 A schematic illustration of a dendrite of diameter d (μm) studded with clusters or hot spots of Na^+ ionic channels. Each hot spot reflects the notion of I_{Na} representing a point source of inward current imposed on an infinitesimal area on the cable. N denotes the number of hot spots and N^* denotes the number of Na^+ channels in each hot spot per membrane surface of cable, represented schematically (not to scale) as dark spots. The total number of Na^+ channels is given as NN^*. (*Source:* From Poznanski, R.R., *J. Integr. Neurosci.*, 3, 267–299, 2004. With permission.)

$V_L = E_L - E_R$ is the reversal potential (mV) where $E_L \approx -70$ mV is the leak equilibrium potential and $E_R = -70$ mV is the resting membrane potential (Frankenhaeuser and Huxley, 1964), $R_i (= r_i \pi d^2/4)$ is the cytoplasmic resistivity (Ω cm), N is the number of clusters of Na^+ channels or hot spots (dimensionless), $I_{Na(pump)}\delta(x - x_z)$ is the sodium pump current density associated with the zth hot spot per membrane surface of dendritic cable in A/cm^2. Note that subscripts x and t indicate partial derivatives with respect to these variables and δ is the Dirac delta function reflecting the axial positions along the cable where the ionic currents are positioned (see FitzHugh, 1973) with the suprathreshold (current) injection location at $x = 0$, the pump location at $x = x_z$, and ionic channel locations at $x = x_i$.

Equation (8.1) can be cast in terms of nondimensional space and time variables, $X = x/\lambda$ and $T = t/t_m$, where $R_m (= r_m \pi d) = 1/g_L$ is the membrane resistivity (kΩ cm^2), and $\lambda = (R_m d/4R_i)^{1/2}$ and $\tau_m = R_m C_m$ are, respectively, the space and time constants in centimeters and seconds. λ and τ_m are cable parameters defined in terms of passive cable properties; by definition, they remain unaffected by the permeability changes associated with active propagation (Goldstein and Rall, 1974). Thus Equation (8.1) becomes

$$V_T = V_{XX} - V - (R_m/\lambda)\sum_{i=1}^{N} I_{Na}(X,T;V)\delta(X - X_i) - (R_m/\lambda)\sum_{z=1}^{N} I_{Na(pump)}\delta(X - X_z), \quad T > 0, \tag{8.2}$$

where the $X_i = x_i/\lambda$ represent loci along the cable of ionic current, expressed in terms of the dendritic space constant, and I_{Na} is the nonlinear transient Na^+ inward current density in each hot spot per membrane surface of cable (μA/cm) (expressed as a sink of current, since by convention inward current is negative and outward is positive) based upon the constant-field equation (Dodge and Frankenhaeuser, 1959; Frankenhaeuser and Huxley, 1964):

$$I_{Na}(X,T;V) = P_{Na}(V)\lambda[Na^+]_e(VF^2/RT)\{(\exp[(V - V_{Na})F/RT] - 1)/(\exp[VF/RT] - 1)\}, \tag{8.3}$$

where F, R, and T are the Faraday constant ($F = 96{,}480$ C/mol), the gas constant ($R = 8.315$ J/K/mol), and the absolute temperature ($T = 273.15$ K), P_{Na} is the permeability of Na^+ ions (cm/sec), $[Na^+]_e$ is the external Na^+ concentration in mM (μ moles/cm^3), $V_{Na} = E_{Na} - E_R$, $E_R = -70$ mV is the resting membrane potential, and $E_{Na} = 65$ mV is the Na^+ equilibrium potential (mV). The constant-field model of ion permeation assumes a steady state, so the concentration is a function of distance only and in Equation (8.3) the presence of the space constant (λ) is the result of conversion from a dimensional to a nondimensional space variable of ionic flux due to diffusion in accordance with Fick's principle. In addition to a barrier, there are "membrane gates" controlling the flow of Na^+ionic current, with the time-dependent gating requiring a lower power for activation than in the standard H–H gate formalism (Dodge and Frankenhaeuser, 1959; Frankenhaeuser and Huxley, 1964; Chiu et al., 1979):

$$P_{Na}(V) = m^2(V)h(V)P_{Na}(0).$$

If conversion of molar flux to charge flux is adopted, that is, flux is expressed in terms of electrical current, then permeability is usually represented by a conductance. This conductance is similar to but not equivalent to the permeability, that is,

$$FP_{Na}(0)\lambda[Na^+]_e(F/RT) \rightarrow g_{Na}, \tag{8.4}$$

where $P_{Na}(0) = 8 \times 10^{-3}$ cm/sec is the maximum permeability of Na^+ ions (Frankenhaeuser and Huxley, 1964), $[Na^+]_e = 114.5$ mM (Frankenhaeuser and Huxley, 1964), and g_{Na} denotes the maximum Na^+ conductance in each hot spot per membrane surface of cable (μS/cm) given by (Hodgkin, 1975):

$$g_{Na} = g^*_{Na}N^*. \tag{8.5}$$

$N^* = \theta/\pi d$ is the number (θ) of transient Na^+ channels in each hot spot per membrane surface of cable per cm, and $g^*_{Na} = 18$ pS is the maximum attainable conductance of a single Na^+ channel (Sigworth and Neher, 1980; Stuhmer et al., 1987). Thus, $g_{Na} = 0.01430\theta$ μS/cm. A possible nonuniform distribution of channels can be determined from Equation (8.5) by redefining the number of channels as a function of space, that is, $g_{Na}(X) = g^*_{Na}N^*(X)$ so that at location $X = X_i$ there will be $N^*(X_i) = \theta(X_i)/\pi d$ Na^+ channels, where $\theta(X_i) \equiv \theta_i$ is the number of Na^+ channels at the ith hot spot, assumed to be a nonuniform function of distance along the cable.

8.3 Scaling the Macroscopic Na^+ Current Density

A space clamp of the membrane allowed Hodgkin and Huxley to set and hold the membrane potential at particular values in order to determine the underlying macroscopic ionic currents. Such currents are macroscopic because they reflect an ensemble average of many localized Na^+ and K^+ channels. They resemble the results seen in intracellular and whole-cell patch recordings.

The macroscopic Na^+ current across the axon membrane of the giant squid was shown by Hodgkin and Huxley (1952b) to follow a simple linear Ohmic relationship expressed through a driving term for the ionic current ($V - V_{Na}$) under voltage-clamp conditions:

$$I_{Na} = g_{Na}m^3(V)h(V)[V - V_{Na}], \tag{8.6}$$

where $m(V, t)$ refers to the onset of Na^+ channel openings, referred to as *activation* of the Na^+ channels, and $h(V, t)$ to the decay from the peak or *inactivation* of the Na^+ channels. If the membrane potential is fixed and the instantaneous transmembrane current density per unit membrane surface of cable is represented through a linear relationship (i.e., Equation [8.6]); then one can readily solve for the state variables (Hodgkin and Huxley, 1952b):

$$m(V,T) = m_\infty(V) + \{m_0 - m_\infty(V)\}\exp[-T\tau_\mathrm{m}/\tau_{\mu(V)}],$$

$$h(V,T) = h_\infty(V) + \{h_0 - h_\infty(V)\}\exp[-T\tau_\mathrm{m}/\tau_{h(V)}],$$

where $m_0 = \alpha_\mathrm{m}(0)/[\alpha_\mathrm{m}(0) + \beta_\mathrm{m}(0)]$ and $h_0 = \alpha_h(0)/[\alpha_h(0) + \beta_h(0)]$ are the initial values of $m(V,T)$ and $h(V,T)$ at $T = 0$, respectively, $m_\infty = \alpha_\mathrm{m}(V)/[\alpha_\mathrm{m}(V) + \beta_\mathrm{m}(V)]$ and $h_\infty = \alpha_h(V)/[\alpha_h(V) + \beta_h(V)]$ are the steady-state values of the activation $m(V,T)$ and inactivation $h(V,T)$, respectively, and $\tau_\mu = 1/(\alpha_\mathrm{m} + \beta_\mathrm{m})$ and $\tau_h = 1/(\alpha_h + \beta_h)$ are time constants (sec).

The empirically derived (at 6.3°C) rate constants (sec^{-1}) are dependent on the instantaneous value of membrane potential at some specified time and so are voltage-dependent but not time-dependent: $\alpha_\mathrm{m} = 0.1(25 - V)/\{\exp[(25 - V)/10] - 1\}$, $\beta_\mathrm{m} = 4\exp(-V/18)$, $\alpha_h = 0.07\exp(-V/20)$, and $\beta_h = 1/\{\exp[(30 - V)/10] + 1\}$ where the αs determine the rate of transfer from outside to inside of the membrane and should increase when the membrane is depolarized, while the βs determine the rate of transfer from inside to outside of the membrane and should decrease when the membrane is depolarized.

Unfortunately, Equation (8.6) does not necessarily apply to all preparations. If the electrodiffusion of Na^+ current across the dendritic membrane barrier is assumed to be nonlinear (cf. Pickard, 1974), then the driving forces due to concentration gradients needs to be replaced with the constant-field equations, which take into account the effect of the ion concentrations in the membrane (Vandenberg and Bezanilla, 1991a). This is especially prevalent at hot spots in preparations of small diameter where spatial ionic concentration changes are expected to be large (Qian and Sejnowski, 1989). Thus the Na^+ transmembrane current density in each hot spot per membrane surface of cable (μA/cm) represented by Equation (8.3) and expressed in terms of a conductance (Equation [8.4]) can be approximated as follows:

$$I_\mathrm{Na}(V) \approx \varepsilon g_\mathrm{Na}\mu^2(V)\eta(V)V\{(\exp[(V - V_\mathrm{Na})F/RT] - 1)/(\exp[VF/RT] - 1)\} < 0, \tag{8.7}$$

where $\varepsilon \ll 1$ is a parameter scaling the "whole-cell" macroscopic transmembrane current density into spatially discrete clusters of Na^+ channels, and $\mu(V)$, $\eta(V)$ represent ensemble averages of the activation and inactivation gating variables, respectively, of the mesoscopic current density in each hot spot per length of cable (dimensionless).

It should be noted that the macroscopic current density is a good representation of reality for large θ_i ($\gg$100), in comparison with the mesoscopic current density which is a good representation for moderate values of θ_i ($\gg$1). For example, the density of the Na^+ channels at the mammalian node of Ranvier is about $\theta = 10{,}000/\mu\mathrm{m}^2$ (McGeer et al., 1987) and Hodgkin (1975) estimated an optimum density of Na^+ channels to be about $\theta = 500/\mu\mathrm{m}^2$ in a 0.1 μm diameter squid axon, while Magee and Johnston (1995) estimated the number of channels per patch to average $\theta = 10/\mu\mathrm{m}^2$ in the dendrites of rat hippocampal CA1 pyramidal neurons. It is expected that spatial ionic concentration changes in these clusters ($\theta \gg 100/\mu\mathrm{m}^2$) would be sufficiently large to take into account the effect of ion concentration in the membrane, and $\epsilon \approx 0.001$ in order to reduce the Na^+ value in myelinated axons to those found in the dendrites.

8.4 Reformulation of the Cable Equation as a Nonlinear Integral Equation

This model does not display threshold behavior, but instead we consider $X = 0$ to be a point which is current clamped to an action-potential response triggered as a result of a high-density aggregation of transient sodium channels having a functional representation of a single nerve impulse (cf. Chiang, 1978):

$$V(0,T) \equiv U(T) = U_0\{150\exp(-2T) - 100\exp(-4T) - 50\exp(-1.2T)\}, \tag{8.8}$$

where U_0 is a predetermined fixed value of the membrane potential (mV). It is important to remember that the point $X = 0$ is part of the dendritic cable and not the soma, and so its properties are those of the dendritic membrane. This empirical approach of including a functional representation of an action potential based on published data from recent studies of neocortical pyramidal neurons to yield a typical action-potential response profile (e.g., see Orpwood, Chapter15, this volume) of course excludes such subtleties as shifts in rate constants caused by temperature (FitzHugh, 1966; Cole et al., 1970).

Equation (8.8) yields the time course of the somatically generated action potential at a single point, namely $X = 0$. Our focus is to determine the time course of active bAP in nonspace-clamped cables. This is accomplished by solving the following equation found by inserting Equation (8.7) into Equation (8.2) and adding Equation (8.8):

$$\begin{aligned} V_T &= V_{XX} - V - \varepsilon(g_{\mathrm{Na}}R_{\mathrm{m}}/\lambda)\sum_{i=1}^{N}\mu^2(V)\eta(V) \\ &\quad \times V\{(\exp[(V - V_{\mathrm{Na}})\Lambda] - 1)/(\exp[V\Lambda] - 1)\}\delta(X - X_i) \\ &\quad + U(T)H(T)\delta(X) - (R_{\mathrm{m}}/\lambda)\sum_{z=1}^{N} I_{\mathrm{Na(pump)}}\delta(X - X_z), \end{aligned} \tag{8.9}$$

where $\Lambda = F/RT$ $(\mathrm{mV})^{-1}$. Although Equation (8.9) is expressed in terms of nondimensional space and time variables, we have still to remove dimensions from the membrane potential by use of some characteristic value of the membrane potential. A reasonable value would be the peak membrane potential of the initial spike U_{peak}. Hence, the membrane potential expressed in dimensional terms is normalized via

$$\Phi \to V/U_{\mathrm{peak}}, \tag{8.10}$$

where U_{peak} is the peak value of the membrane potential in mV. Since $\mu(\cdot)$ and $\eta(\cdot)$ are already dimensionless, they can be equated to the corresponding dimensionless potential Φ, and therefore Equation (8.9) becomes

$$\begin{aligned} \Phi_T &= \Phi_{XX} - \Phi - \varepsilon(g_{\mathrm{Na}}R_{\mathrm{m}}/\lambda)\sum_{i=1}^{N}\mu^2(\Phi)\eta(\Phi) \\ &\quad \times \Phi\{(\exp[(\Phi - \Phi_{\mathrm{Na}})U_{\mathrm{peak}}\Lambda] - 1)/(\exp[\Phi U_{\mathrm{peak}}\Lambda] - 1)\}\delta(X - X_i) \\ &\quad + [U(T)/U_{\mathrm{peak}}]H(T)\delta(X) - [R_{\mathrm{m}}/(\lambda U_{\mathrm{peak}})]\sum_{z=1}^{N} I_{\mathrm{Na(pump)}}\delta(X - X_z). \end{aligned} \tag{8.11}$$

The current balance equation at the site of the hot spot reads at rest (0 mV): $I_{\mathrm{Na}} + I_{\mathrm{Na(pump)}} = 0$ where $I_{\mathrm{Na(pump)}}$ is the outward sodium pump current density. If the outward sodium pump current density is positive (by convention) and the resting potential $\Phi_{\mathrm{R}} = \Phi(0,0) = 0$ then, upon using

L'Hospital's rule:

$$I_{\mathrm{Na(pump)}} = \varepsilon g_{\mathrm{Na}} \mu_0^2 \eta_0 (1 - \exp[-\Phi_{\mathrm{Na}} U_{\mathrm{peak}} \Lambda])/\Lambda > 0, \tag{8.12}$$

where $\mu_0 \equiv \mu_0(0)$ and $\eta_0 \equiv \eta_0(0)$.

Assuming the hot spot location $X = X_i$ and the sodium pump location $X = X_z$ are juxtaposed (McGeer et al., 1987), inserting Equation (8.12) into Equation (8.11) yields

$$\begin{aligned}\Phi_T =& \Phi_{XX} - \Phi - \varepsilon(g_{\mathrm{Na}} R_{\mathrm{m}}/\lambda) \sum_{i=1}^{N} \{\mu^2(\Phi)\eta(\Phi) \\ &\times \Phi(\exp[(\Phi - \Phi_{\mathrm{Na}})U_{\mathrm{peak}}\Lambda] - 1)/(\exp[\Phi U_{\mathrm{peak}}\Lambda] - 1) + [\mu_0^2 \eta_0/(U_{\mathrm{peak}}\Lambda)] \\ &\times (1 - \exp[-\Phi_{\mathrm{Na}} U_{\mathrm{peak}}\Lambda])\}\delta(X - X_i) + [U(T)/U_{\mathrm{peak}}]\mathrm{H}(T)\delta(X).\end{aligned} \tag{8.13}$$

Analytical solutions of such nonlinear reaction–diffusion systems appear not to have been given in the mathematical literature (cf. Britton, 1986). By applying the Green's function method of solution (see Appendix), Equation (8.13) can be expressed in terms of a Green's function G, viz.

$$\begin{aligned}\Phi_{\mathrm{p}}(T) =& \int_0^T \Bigg[(U(s)/U_{\mathrm{peak}})H(s)G_{\mathrm{p}0}(T-s) - \varepsilon(g_{\mathrm{Na}} R_{\mathrm{m}}/\lambda) \sum_{i=1}^{N} \{\mu^2[\Phi_i(s)]\eta[\Phi_i(s)] \\ &\times \Phi_i(s)[\exp[(\Phi - \Phi_{\mathrm{Na}})U_{\mathrm{peak}}\Lambda] - 1]/[\exp(\Phi U_{\mathrm{peak}}\Lambda) - 1]G_{\mathrm{p}i}(T-s)\}\Bigg]\,\mathrm{d}s \\ &- \varepsilon R_{\mathrm{m}} g_{\mathrm{Na}} \mu_0^2 \eta_0 (1 - \exp[-\Phi_{\mathrm{Na}} U_{\mathrm{peak}}\Lambda])/(\lambda U_{\mathrm{peak}}\Lambda) \sum_{i=1}^{N} G_{\mathrm{p}i}(T),\end{aligned} \tag{8.14}$$

where the subscripts correspond to the voltage response at location $X = X_{\mathrm{p}}$ in the presence of a bAP at $X = X_0$ and hot spot locations at $X = X_i$.

Let

$$f(\Phi) = -\mu^2[\Phi]\eta[\Phi]\Phi(\exp[(\Phi - \Phi_{\mathrm{Na}})U_{\mathrm{peak}}\Lambda] - 1)/(\exp[\Phi U_{\mathrm{peak}}\Lambda] - 1) \tag{8.15}$$

and

$$\begin{aligned}\Psi_{\mathrm{p}}(t) =& \int_0^T (U(s)/U_{\mathrm{peak}})H(s)G_{\mathrm{p}0}(T-s)\,\mathrm{d}s \\ &- \varepsilon R_{\mathrm{m}} g_{\mathrm{Na}} \mu_0^2 \eta_0 (1 - \exp[-\Phi_{\mathrm{Na}} U_{\mathrm{peak}}\Lambda])/(\lambda U_{\mathrm{peak}}\Lambda) \sum_{i=1}^{N} G_{\mathrm{p}i}(T),\end{aligned} \tag{8.16}$$

then Equation (8.14) becomes

$$\Phi_{\mathrm{p}}(T) = \Psi_{\mathrm{p}}(T) + \varepsilon(g_{\mathrm{Na}} R_{\mathrm{m}}/\lambda) \int_0^T \sum_{i=1}^{N} f[\Phi_i(s)]G_{\mathrm{p}i}(T-s)\,\mathrm{d}s. \tag{8.17}$$

Equation (8.17) is a nonlinear Volterra integral equation in the membrane potential (Φ), which will be solved using a perturbation expansion.

8.5 A Perturbative Expansion of the Membrane Potential

Let the depolarization be represented as a membrane potential perturbation from the passive bAP in the form of a regular expansion:

$$\Phi_{\mathrm{p}}(T) = \Phi_{0\mathrm{p}}(T) + \sum_{\upsilon=1}^{\infty} \varepsilon^{\upsilon} \Phi_{\upsilon\mathrm{p}}(T), \tag{8.18}$$

where $\Phi_{\upsilon\mathrm{p}}(T)$ is the perturbed voltage from the leading (zeroth-order in ε) term $\Phi_{0\mathrm{p}}(T)$ given in the Appendix. The convergence of such expansions has been investigated by Marcus (1974) for random reaction–diffusion equations (see also Tuckwell and Feng, 2004).

On substituting Equation (8.18) into Equation (8.17), and equating coefficients of powers of ε, a sequence of equations governing the nonlinear perturbations of the voltage from the passive cable voltage response (Φ_0) are obtained via a Taylor expansion of $f(\cdot)$ to $O(\varepsilon^4)$. Hence the analytical solutions are found from Equation (8.18) to yield

$$\Phi_{\mathrm{p}}(T) = \Phi_{0\mathrm{p}}(T) + \varepsilon\Phi_{1\mathrm{p}}(T) + \varepsilon^2\Phi_{2\mathrm{p}}(T) + \varepsilon^3\Phi_{3\mathrm{p}}(T) + O(\varepsilon^4), \tag{8.19a}$$

where

$$\Phi_{1\mathrm{p}}(T) = g_{\mathrm{Na}}(R_{\mathrm{m}}/\lambda) \sum_{i=1}^{N} \left\{ \int_0^T G_{\mathrm{p}i}(T-s) f[\Phi_{0i}(s)]\,\mathrm{d}s \right.$$
$$\left. + \mu_0^2\eta_0(1 - \exp[-\Phi_{\mathrm{Na}} U_{\mathrm{peak}} \Lambda])/(\Lambda U_{\mathrm{peak}}) G_{\mathrm{p}i}(T) \right\}, \tag{8.19b}$$

$$\Phi_{2\mathrm{p}}(T) = g_{\mathrm{Na}}(R_{\mathrm{m}}/\lambda) \sum_{i=1}^{N} \int_0^T G_{\mathrm{p}i}(T-s) f'[\Phi_{0i}(s)]\Phi_{1i}(s)\,\mathrm{d}s, \tag{8.19c}$$

$$\Phi_{3\mathrm{p}}(T) = g_{\mathrm{Na}}(R_{\mathrm{m}}/\lambda) \sum_{i=1}^{N} \int_0^T G_{\mathrm{p}i}(T-s)\{f'[\Phi_{0i}(s)]\Phi_{2i}(s) + f''[\Phi_{0i}(s)]\Phi_{1i}^2(s)/2!\}\,\mathrm{d}s, \tag{8.19d}$$

where the Green's function (G) is derived in the Appendix, and the first three derivatives with respect to Φ_0 of Equation (8.15) can be easily evaluated symbolically with *Mathematica*®.

8.6 Voltage-Dependent Activation of the Sodium Channels

The dimensionless activation $\mu(\Phi)$ reflects the openings of a single cluster of Na^+ channels and is governed by a first-order reaction equation (Hodgkin and Huxley, 1952b):

$$\sigma(1/\tau_{\mathrm{m}})\partial\mu/\partial T = \alpha_{\mathrm{m}}(1-\mu) - \beta_{\mathrm{m}}\mu, \tag{8.20}$$

where μ represents the proportion of activating Na^+ ions on the inside of the membrane, $(1-\mu)$ represents the proportion that are outside of the membrane, and σ is a scalar (dimensionless) factor.

The rate constants α_m and β_m (m/sec) are Boltzmann-type functions of the form:

$$\alpha_m = a_\alpha(\Phi U_{peak} + E_R + b_\alpha)/c_\alpha/\{1 - \exp[-(b_\alpha + E_R + \Phi U_{peak})/c_\alpha]\}, \quad (8.21a)$$

$$\beta_m = -a_\beta(b_\beta + E_R + \Phi U_{peak})/c_\beta/\{1 - \exp[(b_\beta + E_R + \Phi U_{peak})/c_\beta]\}, \quad (8.21b)$$

where a_α, b_α, c_α, a_β, b_β, and c_β are all positive constants of the system. In particular, a_α and a_β are the maximum rates (sec^{-1}), b_α and b_β are the voltages at which the function equals half the maximum rate (mV), and c_α and c_β are inversely proportional to the steepness of the voltage dependence (mV).

We obtain the following result upon integrating Equation (8.20) (Evans and Shenk, 1970; Mascagni, 1989b):

$$\mu(X,T) = (\tau_m/\sigma)\int_0^T \{\alpha_m[\Phi(X,s)] - (\alpha_m[\Phi(X,s)] + \beta_m[\Phi(X,s)])\}\mu(X,s)\,ds, \quad T > 0. \quad (8.22)$$

The Volterra integral equation of the second type expressed by Equation (8.22) is linear and can be solved analytically by rewriting it in the form corresponding to the voltage response at location $X = X_p$:

$$\mu_p(T) = \zeta_p(T) + (-\tau_m/\sigma)\int_0^T [M_p(T,s)]\mu_p(s)\,ds, \quad T > 0, \quad (8.23)$$

where $\zeta_p(T) = (\tau_m/\sigma)\int_0^T \alpha_m(\Phi_p)\,ds$, and $M_p(T,s) = \alpha_m[\Phi_p(s)] + \beta_m[\Phi_p(s)]$.

The Liouville–Neumann series or Neumann series solution of the Volterra integral Equation (8.23) is given as a series of integrals:

$$\mu_p(T) = \zeta_p(T) + (-\tau_m/\sigma)\int_0^T \mathfrak{R}_p(T,s)\zeta_p(s)\,ds, \quad T > 0, \quad (8.24)$$

where the resolvent kernels $\mathfrak{R}_p$ are the sums of the uniformly convergent series

$$\mathfrak{R}_p(T,s) = M_p(T,s) + \sum_{j=1}^{\infty}(-\tau_m/\sigma)^j M_p^{(j)}(T,s)$$

and the j-fold iterated kernels $M_p^{(j)}$ are defined inductively by the relations

$$M_p^{(j)}(T,s) = \int_s^T M_p(T,\xi)M_p^{(j-1)}(\xi,s)\,d\xi, \qquad j = 1,2,\ldots \quad \text{and} \quad M_p^0 \equiv M_p.$$

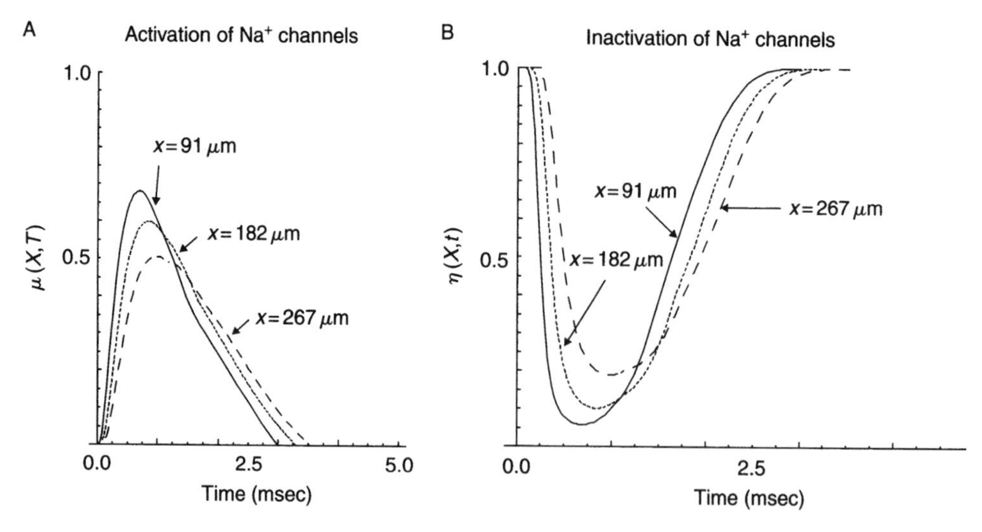

FIGURE 8.2 (A) The activation and (B) the inactivation of the Na^+ current. The abscissa is time in msec and the ordinate is in dimensionless units. Numerical quadrature was employed to evaluate the convolution integrals with *Mathematica*®. (*Source:* From Poznanski, R.R., *J. Integr. Neurosci.*, 3, 267–299, 2004. With permission.)

By writing Equation (8.24) in terms of the rate functions explicitly, we can show with the help of the Dirichlet transformation that to a second-order approximation

$$\mu_p(T) \approx (\tau_m/\sigma) \int_0^T \alpha_m[\Phi_p(s)]\,ds - (\tau_m/\sigma)^2 \int_0^T \left\{ [\alpha_m[\Phi_p(s)] + \beta_m[\Phi_p(s)]] \times \int_0^s \alpha_m[\Phi_p(\xi)]\,d\xi \right\} ds + (\tau_m/\sigma)^3 \int_0^T \{\alpha_m[\Phi_p(\xi)] + \beta_m[\Phi_p(\xi)]\} \times \left(\int_0^\xi \{\alpha_m[\Phi_p(s)] + \beta_m[\Phi_p(s)]\} \left\{ \int_0^s \alpha_m[\Phi_p(\zeta)]\,d\zeta \right\} ds \right) d\xi, \quad T > 0, \tag{8.25}$$

with $\mu_p(0) = 0.0005$ (Frankenhaeuser and Huxley, 1964). Equation (8.25) is valid if the leakage conductance is large so that terms $o(\tau_m^4)$ can be ignored. We chose $\sigma = 10$ in order to take into consideration the fast activation time of Na^+ channels.

A large leak conductance will affect the activation gating variable (m) in such a way that it will not reach the expected value of unity (Hodgkin and Huxley, 1952b; Frankenhaeuser and Huxley, 1964; Chiu et al., 1979). However, a clear broadening in the time course as observed by Chiu et al. (1979) is clearly evident in Figure 8.2. Note that $d\mu_p/d\Phi_{0p} = \tau_m[\alpha_m - (\alpha_m + \beta_m)\mu_p]\,d\Phi_{0p}/dt$ and $d^2\mu_p/d\Phi_{0p}^2 = d/dt[d\mu_p/d\Phi_{0p}]/d\Phi_{0p}/dt$ when substituted in Equation (8.19).

8.7 STEADY-STATE INACTIVATION OF THE SODIUM CHANNEL

The dimensionless inactivation $\eta(\Phi)$ reflects the closings of a single cluster of Na^+ channels and therefore the repolarization (i.e., return to rest) of the membrane potential. It is governed by the first-order reaction equation (Hodgkin and Huxley, 1952b):

$$(1/\tau_m)\partial\eta/\partial T = \alpha_h(1-\eta) - \beta_h\eta, \tag{8.26}$$

where η represents the proportion of inactivating Na^+ ions on the outside of the membrane and $(1-\eta)$ represents the proportion on the inside of the membrane. The rate constants α_h and β_h (sec^{-1}) are modified from the H–H formalism to take into account single Na^+ channel description (cf. Patlak, 1991):

$$\alpha_h = \gamma_0(1 - \mu) \tag{8.27a}$$

$$\beta_h = \gamma_1\mu^3 + 3\gamma_2\mu^2(1 - \mu) + 3\gamma_3\mu(1 - \mu)^2, \tag{8.27b}$$

with $\gamma_0 = 0.8$, $\gamma_1 = 1.0$, $\gamma_2 = 0.43$, and $\gamma_3 = 0.25\,\text{sec}^{-1}$ (Marom and Abbott, 1994). Note that the inactivation of single Na^+ channel is voltage independent and dependent on the activation (Armstrong and Bezanilla, 1977; Bezanilla and Armstrong, 1977; Aldrich et al., 1983). Goldman (1975) combined the activation and inactivation into two coupled first-order equations equivalent to a second-order equation, while Bell and Cook (1978, 1979) further developed a third-order equation by coupling the Na^+ activation and inactivation variables into a single variable. Hoyt (1984) has also proposed that the Na^+ conductance can be simulated by two related and simultaneous first-order equations. However, the decoupling of the Na^+activation from the Na^+ inactivation is chosen here because it has been possible to block inactivation by the application of the enzyme pronase without altering Na^+ activation.

We obtain the following result upon letting $\partial\eta/\partial T = 0$ in Equation (8.26):

$$\eta_p(T) = \alpha_h[\Phi_p(T)]/(\alpha_h[\Phi_p(T)] + \beta_h[\Phi_p(T)]), \quad T > 0, \tag{8.28}$$

with $\eta_p(0) = 0.8249$ (Frankenhaeuser and Huxley, 1964). The steady-state inactivation is shown in Figure 8.2. Note that $d\eta_{0p}/d\Phi_{0p} = d^2\eta_{0p}/d\Phi_{0p}^2 = 0$ when substituted in Equation (8.19).

8.8 Results

8.8.1 Electrotonic Spread of bAP without Na^+ Ion Channels

The first question of our investigation was to see how the leakage conductance affects the passive bAP. In Figure 8.3 we illustrate the passive bAP as a result of a spike input at the point $X = 0$. It is clear that the bAP peak amplitude attenuates with distance and tends to zero. During its passive backpropagation, the amplitude of the action potential decreases to half-amplitude in $\approx 0.34\lambda$ and its shape becomes rounded as was shown numerically in earlier work by Dodge (1979) (see also Vermeulen et al., 1996).

The results also show that in a passive cable, a decreased conductance decreases the electrotonic spread of bAP evoked by a voltage spike input at the point $X = 0$. The reduced electrotonic spread of bAP may therefore suggest a decreased leak conductance as the mechanism of experimentally observed bAPs. However, notwithstanding the experimental demonstration of bAP being reduced in amplitude by the blockade of Na^+ channels with bath application of tetrodotoxin (TTX) (Stuart and Sakmann, 1994), a clear indication that leakage conductance decrease does not govern bAPs is the observed latency (measured as the time taken for the voltage to rise 10% of the peak value). It is observed that passive bAPs have a greater latency than experimentally observed bAPs. Thus, reduced electrotonic spread of bAPs, although caused by a decrease in leakage conductance, is not the mechanism involved in active bAPs. Furthermore, passive bAPs have a clear "undershoot" in the afterpotential (also referred to as after-hyperpolarization), whereas almost all experimentally observed bAPs show either negligible or no undershoot, but a slowly decaying afterpotential that lasts for >50 msec (not shown in Figure 8.4 to Figure 8.6, but see Bernard and Johnston, 2003). This prolonged afterpotential is the result of a very slow decline of the K^+ channel conductance mimicked here by a large leakage conductance in the absence of K^+ channels. It is also interesting

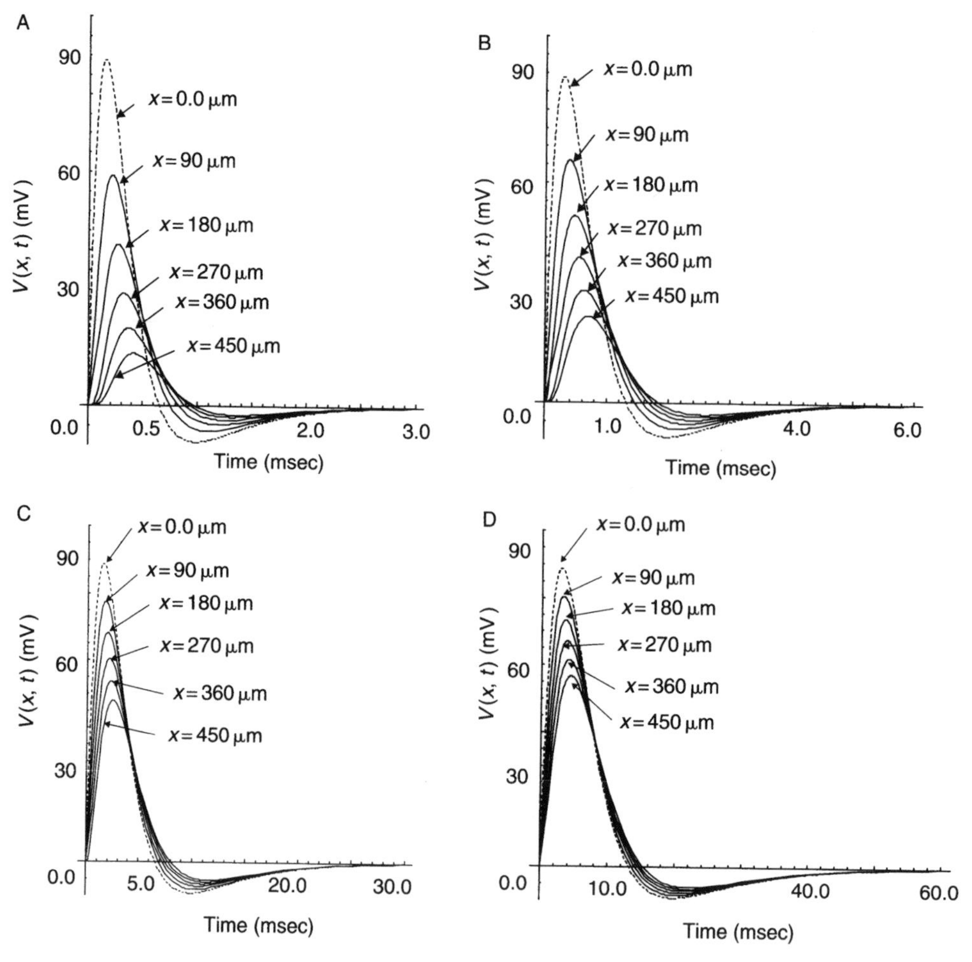

FIGURE 8.3 The attenuation of passive bAPs measured at three distinct locations mediated by an input functional representation of an action potential at the point $X = 0$ (denoting a point close to the soma) with varying leakage conductance values: (A) $R_m = 500\,\Omega\,\text{cm}^2$, (B) $R_m = 1,000\,\Omega\,\text{cm}^2$, (C) $R_m = 5,000\,\Omega\,\text{cm}^2$, and (D) $R_m = 10,000\,\Omega\,\text{cm}^2$. The following other parameters were used: $R_i = 70\,\Omega\,\text{cm}$, $C_m = 1\,\mu\text{F/cm}^2$, and $d = 4\,\mu\text{m}$. Note the change in time scales as a result of the nonconstant R_m. The abscissa is time in msec and the ordinate is voltage in mV. (*Source:* From Poznanski, R.R., *J. Integr. Neurosci.*, 3, 267–299, 2004. With permission.)

to observe that the larger the membrane time constant, the longer it takes for the depolarization at any point to rise. As well, the larger the space constant, the longer will be the depolarization at any time at a given point along the cable.

8.8.2 How the Location of Hot Spots with Identical Strengths of Na^+ Ion Channels Affects the bAP

We investigated the question of how the local density distribution of I_{Na} affects the bAP by considering two variations in the spatial distribution of hot spots along a semi-infinite cable of diameter $d = 4$ mm. The first variation (Figure 8.4A) is a nonuniform distribution of hot spots located distal to the point $X = 0$ at length intervals of $0.1\,j/N$, where $j = 1, 2, \ldots, N$. The second variation (Figure 8.4B) is a uniform distribution of hot spots along the entire cable at length intervals of j/N, where $j = 1, 2, \ldots, N$. In each case, an identical number of sodium channels per hot spot ($N^* = 10/\pi d$) is assumed.

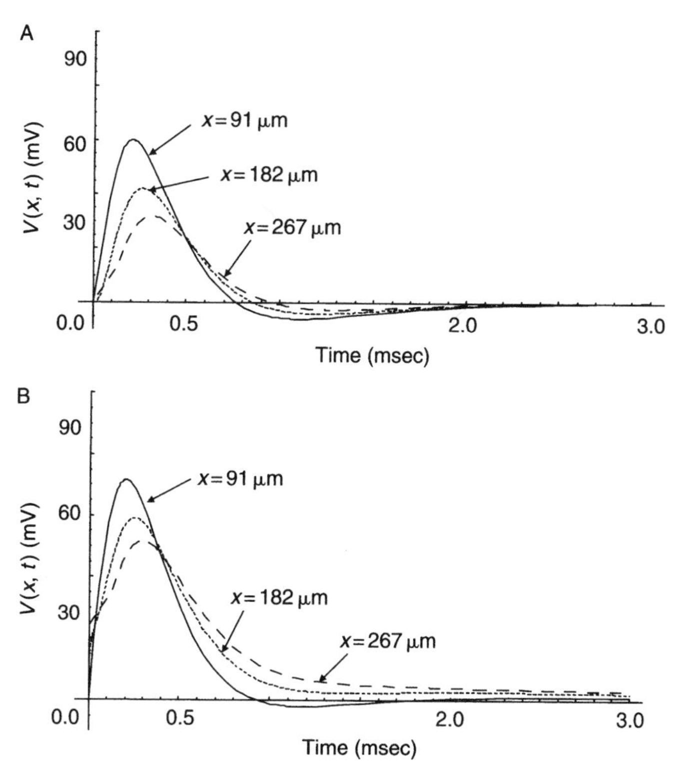

FIGURE 8.4 The attenuation of active bAPs measured at 91, 182, and 267 μm from the location of the activated action potential mediated by Na^+ hot spots distributed: (A) near the $x = 0$ region (up to 27 μm from the location of the input site), and (B) uniformly for 267 μm from the location of the input site. The number of hot spots is assumed to be $N = 10$ in each case. The following other parameters were used: $R_i = 70\,\Omega\,\text{cm}$, $C_m = 1\,\mu\text{F/cm}^2$, $R_m = 500\,\Omega\,\text{cm}^2$, $d = 4\,\mu\text{m}$, $\lambda = 0.0267\,\text{cm}$, $g_{Na} = 0.143\,\mu\text{S/cm}$. The value of g_{Na} was based on the number of sodium channels per hot spot of $N^* = 10/\pi d$. The abscissa is time in msec and the ordinate is voltage in mV. (*Source:* From Poznanski, R.R., *J. Integr. Neurosci.*, 3, 267–299, 2004. With permission.)

It is evident from Figure 8.4 that a local distribution of Na^+ channels appears to produce a faster time to decay of the amplification for more distally recorded bAPs in comparison with a uniform distribution. Indeed, this difference can be especially apparent when the recording point is further away from the location of the local density of Na^+ channels. Furthermore, for a proximal density distribution (i.e., at 0.34λ) it can be seen from Figure 8.4 that the local distribution yields a significantly smaller amplification of bAP in comparison with a uniform distribution of hot spots along the cable. These results are compatible with the concept that a more distributed density of Na^+ channels should produce a greater amplification in the voltage response in comparison to a local density distribution of Na^+ channels at non-local positions along the cable.

In the case of proximally placed Na^+ channels as shown in Figure 8.4A, the amplification of bAP at the point $x = 91\,\mu\text{m}$ is relatively close to the enhancement observed for regional distributions of dendritic hot spots as shown in Figure 8.4B. Therefore, it seems that the "local density," defined as *the density of* Na^+ *channels positioned in close proximity to the recording site*, does not have a significantly greater amplifying effect in comparison to nonlocal densities. However, locally placed Na^+ channels (although having a similar amplifying effect when recorded locally) generate a significantly smaller amplification of the bAP for distally recorded locations along the dendritic cable compared to regionally distributed Na^+ channels. This confirms the simulation results of Jung et al. (1997) that locally distributed I_{Na} alone has a small effect on propagation of bAP. Hence the results show that the local density of Na^+ channels is not the main factor for the enhancement of bAP, but it is the regional distribution of dendritic hot spots that matters.

8.8.3 How the Number (N) of Uniformly Distributed Hot Spots with Identical Strengths of Na^+ Ion Channel Densities Affects the bAP

The next problem was to investigate how the hot spot number (N) affects the bAP, assuming uniform distribution and identical strength of Na^+ ion channel densities along the dendritic cable. This was done by considering a cable that has a diameter of 4 μm and a regional (uniform) distribution of hot spots located at length intervals of j/N, where $j = 1, 2, \ldots, N$. All hot spots were assumed to have identical numbers of Na^+ channels (i.e., $N^* = 10/\pi d$) based on an average density of Na^+ channels per patch (Magee and Johnston, 1995), assuming that each hot spot has surface area similar to a patch-clamp pipette. We selected a wide range of numbers of hot spots, $N = 5$, 20, and 40, and investigated the response by computing Equation (8.18) for a cable of diameter $d = 4\,\mu\text{m}$, again assuming a uniform distribution of hot spots located at length intervals of j/N. It is important to stress that the evaluation using *Mathematica*® (version 3) software requires several days of computing time for large values of N on a standard PC. As a significantly greater amount of computing time is required if more than $N = 5$ hot spots are assumed and/or greater numbers of Na^+ channels per hot spot are assumed, only first-order approximations were used in Equation (8.18).

In Figure 8.5, all bAPs are measured at the electrotonic distances of $X = 0.34$ ($x \approx 91\,\mu\text{m}$), 0.68 ($x \approx 182\,\mu\text{m}$), and 1.0 ($x \approx 267\,\mu\text{m}$) in the presence of an outward current or sodium pump. The results in Figure 8.5 clearly show that a greater number of hot spots results in greater amplification of the bAP. What is particularly interesting is that for a greater number of Na^+ channels regionally distributed along the cable, the peak amplitude of the bAP is not significantly increased, but rather the time course of the bAP is broadened with a clear absence of an undershoot or hyperpolarizing potential range (cf. Figure 8.5A and Figure 8.5C). Furthermore, I_{Na} need not be distributed along the entire length of the cable to produce significant broadening of the response. As well, prolonged broadening in the bAP time course is observed in some neurons to be frequency dependent (Lemon and Turner, 2000), which may indicate the presence of other rapidly gating channels in the nerve cell membrane (e.g., see Ma and Koester, 1996; Shao et al., 1999).

A question of concern is whether there is a saturation point where further increases in the number of hot spots result in no further amplification of the bAP, that is, is there an optimal number of hot spots that produces the greatest amplification of bAP? It was found that increasing the number of hot spots to $N = 40$ resulted in the amplification not saturating, but further increasing the bAP to a degree that it quenched the more proximal responses (cf. Figure 8.5D). The F–H dynamics associated with the I_{Na}–V curve expressed through Equation (8.7) is nonlinear, and so saturation in the amplification is to be expected. The amplitude of the functional representation of the integrand $f[V_0(\lambda, t)]G(x, \lambda; t)$ for the first-order perturbation term in Equation (8.19) is small (a few microvolts) [see Figure 8.5B] and varies from 0.75 and 2.0 msec, and so the reason for not observing a plateau could be that a large number of hot spots is required to activate the succession in the amplification. The utilization of a greater number of hot spots and higher than first-order approximation is ongoing and will be presented elsewhere.

8.8.4 How the Conductance Strength of Na^+ ion Channel Densities with Identical Regional Distribution of Hot Spots Affects the bAP

The next question of concern was to know how the number of Na^+ channels per hot spot affects the amplification of the bAP assuming a regional distribution of a fixed number of hot spots. We assumed

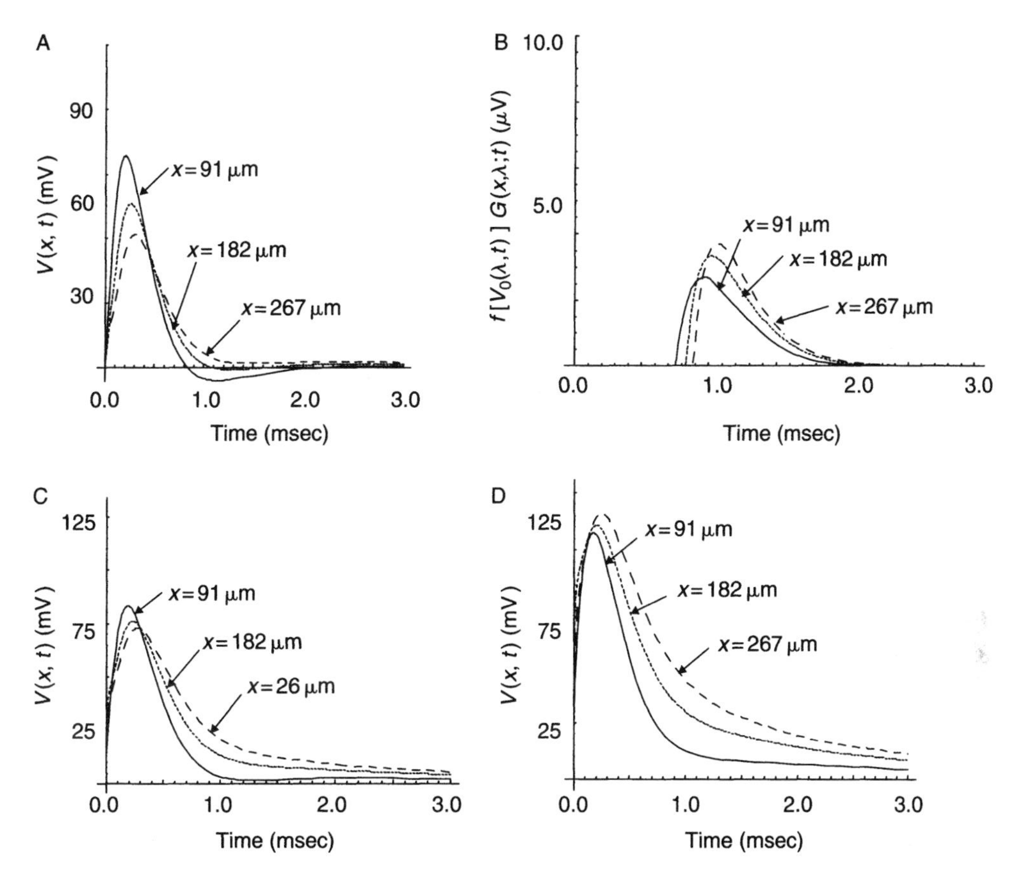

FIGURE 8.5 The attenuation of active bAPs measured at 91, 182, and 267 μm from the location of the input site mediated by Na^+ hot spots distributed uniformly for three different numbers of hot spots: (A) $N = 5$, (C) $N = 20$, and (D) $N = 40$. The abscissa is time in msec and the ordinate is voltage in mV. The same parameters as those in Figure 8.4 were used. In (B) the nonlinear functional $f[V_0(\lambda, t)]G(x, \lambda; t)$ is shown representing the integrand in the convolution integral of the first-order perturbation. Note that in (B) the ordinate has dimension of μV. (*Source:* From Poznanski, R.R., *J. Integr. Neurosci.*, 3, 267–299, 2004. With permission.)

the number of Na^+ channels per hot spot to take on values of $\theta = 6$, 12, 24, and 36, based on an approximate density range of Na^+ channels per patch (Magee and Johnston, 1995) assuming that each hot spot has a similar surface area to a patch-clamp pipette. We investigated the response by computing Equation (8.18) for a cable of diameter $d = 4\,\mu$m, again assuming a uniform regional distribution of hot spots located at length intervals of j/N. The results are presented in Figure 8.6 for $N = 5$ hot spots. It is interesting to observe that greater numbers of Na^+ channels per hot spot produce a greater amplification of the bAP, in agreement with experimental results (Jung et al., 1997).

It is interesting to see whether an optimum number of Na^+ channels can be found for maximum amplification of the bAP. When the number of Na^+ channels per hot spot is increased via the conductance g_{Na}, a relatively small number of hot spots suffice to determine whether the response saturates and there is an optimal number of Na^+ channels. A saturation period of the response is defined for values of θ where no significant change occurs in the amplification of the bAP. The results for $N = 5$ and a selection of θ values are presented in Figure 8.6. Here there is clearly no optimum number of Na^+ channels per hot spot because the Na^+ conductance is not represented in the functional as shown in Figure 8.5B, but is only a parameter multiplying the nonfunctional in Equation (8.17). Although the basis of a standard H–H model (with m^2h kinetics) and a constant-field

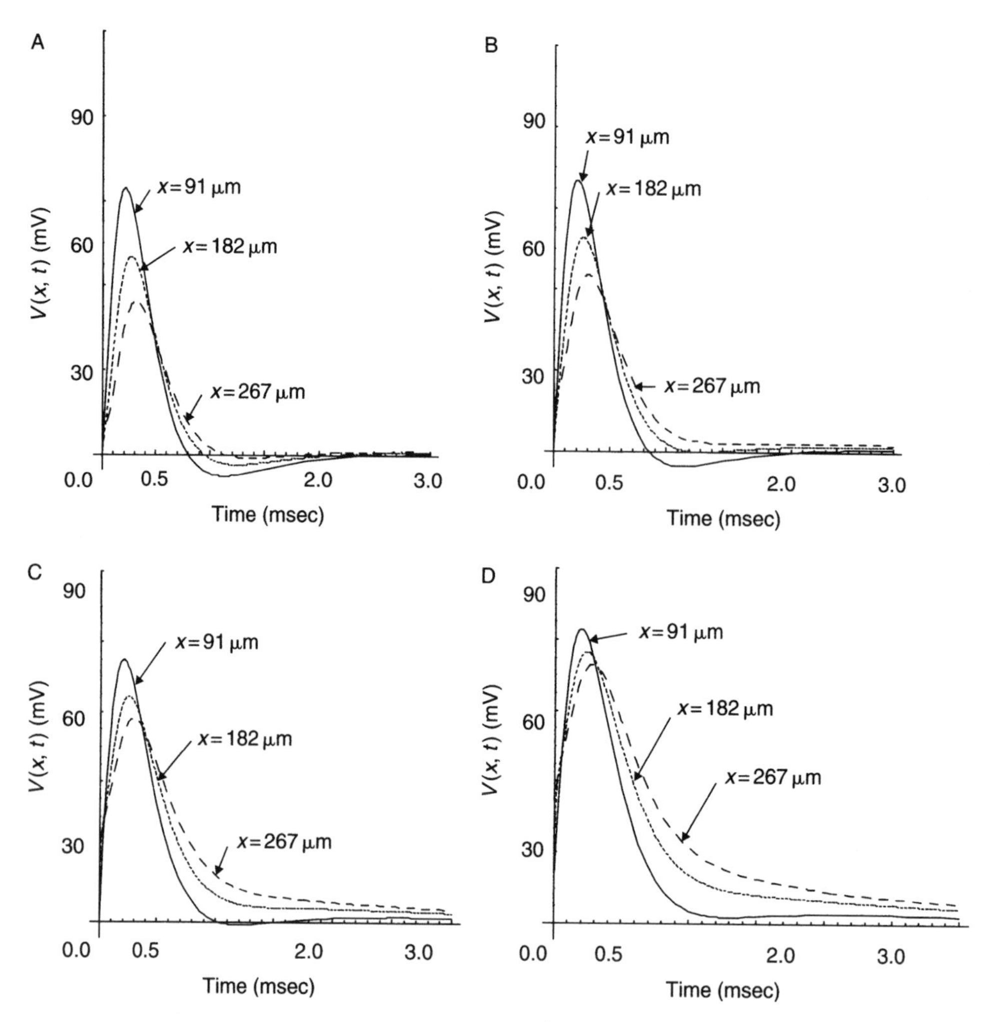

FIGURE 8.6 The attenuation of active bAPs measured at 91, 182, and 267 μm from the location of the input site mediated by Na^+ hot spots distributed uniformly with varying densities. The number of Na^+ channels per hot spot is assumed to be equal with (A) $N^* = 6/\pi d$ (i.e., $g_{Na} = 0.0858\,\mu S/cm$), (B) $N^* = 12/\pi d$ (i.e., $g_{Na} = 0.1716\,\mu S/cm$), (C) $N^* = 24/\pi d$ (i.e., $g_{Na} = 0.2288\,\mu S/cm$), and (D) $N^* = 36/\pi d$ (i.e., $g_{Na} = 0.2860\,\mu S/cm$). The abscissa is time in msec and the ordinate is voltage in mV. The same parameters are used as in Figure 8.4 but the number of hot spots is $N = 5$. (*Source:* From Poznanski, R.R., *J. Integr. Neurosci.*, 3, 267–299, 2004. With permission.)

approximation of the current density is a valid one, the scaling of the current density corresponding to a mesoscopic-level description of Na^+ current density needs to be looked into with more detail.

8.9 DISCUSSION

An *a priori* assumption of the theory is that Na^+ channels along the dendritic axis occur in far less abundance than those found along the somatoaxonal axis and moreover in discrete patches resembling hot spots (but this assumption can be made as general as we like by reducing the distance between hot spots in order to approximate a continuous layer of ionic channels). As yet there is no experimental verification of the validity of this assumption, because determination of the spatial distribution of channels in dendrites is a difficult if not impossible task, and in most instances an average estimate is found based on recording from only a few dendritic patches (e.g., Magee and Johnston, 1995). Physiologically, dendrites serve a different function from axons. The former are covered with synaptic receptors (absent along the axon membrane), and this makes it highly improbable that Na^+ channels can occupy continuous positions along the dendritic membrane as is inherent in the H–H model of the unmyelinated axon. Therefore, the assumption that Na^+ channels and other voltage-dependent ionic channels are found distributed in discrete patches or hot spots at extrasynaptic sites along the plasma membrane is a viable assumption which, however, needs further verification using immunohistological and freeze-fracture techniques.

The leaky-membrane time constant $\tau_m = R_m C_m$ is often calculated from estimates of the membrane resistivity (R_m), given that the membrane capacitance (C_m) is nearly a fixed constant of $1\ \mu F/cm^2$. However, it is incorrect not to increase the membrane capacitance when expressing τ_m in terms of the Na^+ conductance rather than R_m. This can lead to erroneous conclusions regarding the speed of processes involving the membrane potential, with values estimated as low as $\tau_m \approx 0.01$ msec (Muratov, 2000), faster than the closing of Na^+ channels ($\sim$0.2 msec). Furthermore, given that synaptic receptors are located peripherally in the dendrites of some neurons, it is probable that R_m is low in distal dendrites of these neurons because of the conductance of the ligand-gated synaptic and extrasynaptic receptor channels. Therefore, the concept of a large-conductance leak, as described in the present model, as the sole mechanism responsible for repolarization also needs further experimental verification.

A favorable fit of R_m to data for neocortical pyramidal dendrites has been achieved in a compartmental model with a sigmoidally decreasing function to a final value of $R_m = 5357\ \Omega\, cm^2$ (Stuart and Spruston, 1998). However, notwithstanding the problems inherent in such neural modeling protocols, when active properties of the dendritic membrane are included it is unlikely that values from *in vitro* preparations will be appropriate models for values encountered *in vivo*. The utilization of even a lower $R_m (\approx 500\ \Omega\, cm^2)$ used by Chiu et al. (1979) for myelinated axons in comparison to $R_m \approx 3300\ \Omega\, cm^2$ (or a leak conductance of $0.3\ mS/cm^2$) used by Hodgkin and Huxley (1952b) for an unmyelinated axon membrane is in line with the value used by earlier investigators interested in the passive dendritic properties of CNS neurons.

A metabolically operated Na^+ pump was included in the model to counter the resting ionic influx of Na^+ and maintain equilibrium, so that Na^+ conductances at each hot spot are at their resting values and the membrane potential returns to the resting state after bAP. There is experimental evidence of the independence of the Na^+ pump from the Na^+ channels (McGeer et al., 1987). The rate of pumping appears to depend on the internal concentration of Na^+ and in the giant squid axon the currents are always negligible (Moore and Adelman, 1961). In small dendrites, however, with much smaller volume-to-surface ratios than for squid axon (at least 5000 times smaller for an $0.1\ \mu m$ diameter dendrite [Hodgkin, 1975]), the metabolic pump activity is significantly greater because only a few bAPs are enough to affect the internal ionic composition. In this case the Na^+ pump plays a significant role in maintaining the ionic fluxes of the action potentials.

The ionic cable equation is quasilinear (i.e., weakly nonlinear), and it does not exhibit the same unstable behavior as the original fully nonlinear H–H equations: from complex chaotic attractors and resonances, to simpler stationary states, periodic orbits, solitary waves, and limit cycles; complex behavior, in particular, can be expected when periodic forcing terms are present. The nonlinearity is probably too small to produce the unstable behavior exhibited by the H–H equations, nevertheless Poznanski and Bell (2000b) found previously that saturation in the amount of amplification is possible with a weakly nonlinear cable equation. The contradistinction between earlier results and our present findings rests on the fact that the amount of nonlinearity per hot spot used previously was far in excess of the estimates used here.

When K^+ is reduced or blocked, autorhythmicity occurs, however, a sparse density distribution of channels does not itself lead to autorhythmicity, and suprathreshold inputs lead to decremental propagation as observed experimentally, that is, the amplitude and width of the pulse can also depend on the magnitude of the leakage conductance. For example, an increase in the leakage conductance could allow an increased flow of negatively charged leakage ions to accompany the influx of Na^+ ions on the upstroke of the action potentials, thus reducing the depolarizing effect of the Na^+ ions. This gives a strong impetus for the application and extension of the present model to dendrites with sparse distributions of fast TTX-sensitive Na^+ channels. Of particular interest are the directionally selective starburst amacrine cells in the rabbit retina (Fried et al., 2002). These cells are known to produce fast Na^+-based spikes (Cohen, 2001), yet it is still unclear how these Na^+-based spikes invoke the directionally selective Ca^{2+} influxes observed in experiments (Euler et al., 2002). A recent experimental study showing local Ca^{2+} mobilization by a spatially asymmetric distribution of chloride (Cl^-) cotransporters (Gavrikov et al., 2003) may provide the missing ingredient.

The K^+ current plays an important role in repolarization, as revealed by the low g_L/g_K ratio of about 0.008 (Hodgkin and Huxley, 1952b), and by the experimental observation that bAPs are attenuated by voltage-gated K^+ channels in hippocampal dendrites (Hoffman et al., 1997). However, we assumed here $g_L/g_K \approx 1$, implying that repolarization is accomplished by a high density of large-conductance leak channels rather than voltage-gated K^+ channels (Bekkers, 2000). Dendrites show a high density of ligand-gated channels that can produce a large-conductance "composite" leak throughout the entire cell membrane, so our assumption is plausible for dendritic backpropagation as opposed to orthodromic propagation. In future work, it would be interesting to consider synaptic input as a conductance change expressed in terms of an alpha function and explicitly include voltage-gated K^+ currents (e.g., I_K, I_A, I_h) at their proper densities to elucidate their relevance in spike trains and repolarization during bAP activation in cortical and hippocampal pyramidal neurons (Migliore et al., 1999; Bekkers, 2000).

8.10 SUMMARY

Hodgkin and Huxley's ionic theory of the nerve impulse embodies principles that are applicable also to the impulses in vertebrate nerve fibers, as demonstrated by Bernhard Frankenhaeuser and Andrew Huxley 40 years ago. Frankenhaeuser and Huxley reformulated the classical H–H equations in terms of electrodiffusion theory and computed action potentials specifically for saltatory conduction in myelinated axons. In this chapter, we found approximate analytical solutions of the Frankenhaeuser–Huxley equations modified for a model of *sparsely* excitable, nonlinear dendrites with clusters of transiently activating TTX-sensitive Na^+ channels, discretely distributed as point sources of inward current along a continuous (nonsegmented) leaky cable structure. Each cluster or hot spot, corresponding to a mesoscopic-level description of Na^+ ion channels, included known cumulative inactivation kinetics observed at the microscopic level. In such a third-order system, the "recovery" variable is an electrogenic sodium pump imbedded in the passive membrane, and the system is stabilized by the presence of a large leak conductance mediated by a composite number of ligand-gated channels permeable to monovalent cations Na^+ and K^+. In order to reproduce antidromic propagation and the attenuation of action potentials in the presence of suprathreshold input, a nonlinear integral equation was solved along with a constant-field electrodiffusion equation at each hot spot with membrane gates controlling the flow of current. A perturbative expansion of the nondimensional membrane potential (Φ) was utilized to obtain time-dependent analytical solutions involving voltage-dependent Na^+ activation (μ) and state-dependent inactivation (η) gating variables. It was shown through the model that action potentials attenuate in amplitude in accordance with experimental findings, and that the spatial density distribution of transient Na^+ channels along a long dendrite contributes significantly to their discharge patterns.

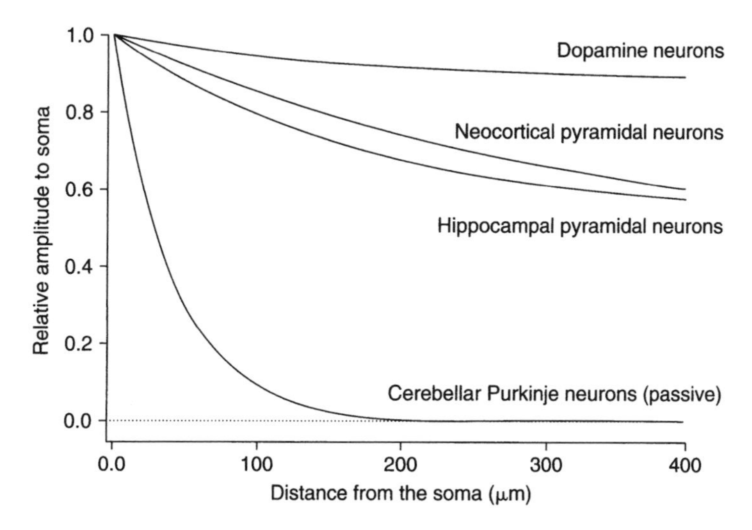

FIGURE 8.7 A summary of experimental results showing relative peak amplitude of bAPs with distance from the soma in several mammalian neurons. (*Source:* From Stuart, G.J., Spruston, N., Sakmann, B., and Hausser, M., *Trends Neurosci.* 20, 125–131, 1997. With permission.)

8.11 CONCLUSIONS

Analytical solutions of the modified F–H equations for suprathreshold input representation of a bAP at the origin were found for a model of leaky dendrite containing clusters of transient Na^+ channels spatially distributed at discrete locations along a segment of the dendritic tree. The nonlinear phenomena characterizing signal backpropagation were investigated to predict qualitatively the experimentally proven attenuation of Na^+-based bAPs found in the majority of principal neurons by Stuart et al. (1997b), with results summarized and presented in Figure 8.7.

In conclusion, it is clear from this chapter that bAPs in comparison to standard H–H action potentials in unmyelinated axons have a significantly faster upstroke and the absence of an undershoot or after-hyperpolarization, replaced by a more slowly decaying after-depolarization attributable to a high level of leak current. The depolarization of the bAP occupies only about 0.5 msec (submillisecond), which may be of importance for signal integration in dendrites, where they could be propagated over long distances without complete attenuation and summated more flexibly as was first envisaged by Leibovic (1972).

ACKNOWLEDGMENTS

I thank S. Hidaka, G.N. Reeke, J. Rubinstein, and D.A. Turner for discussions and for providing helpful suggestions for improvement. I am indebted to Imperial College Press for permission to reproduce earlier work published in the Journal of Integrative Neuroscience.

APPENDIX

By rewriting Equation (8.16):

$$\Psi(X,T) = \int_0^T [U(T-s)/U_{\text{peak}}]H(T-s)G(X,0;s)\,\mathrm{d}s$$
$$+ \{\varepsilon R_{\text{m}} g_{\text{Na}} \mu_0^2 \eta_0 (1 - \exp[-\Phi_{\text{Na}} U_{\text{peak}} \Lambda])/(\Lambda\lambda U_{\text{peak}})\} \sum_{i=1}^{N} G(X, X_i; T), \quad T > 0, \tag{8.A1}$$

where $U(T)$ is the time course of the bAPs expressed through Equation (8.8) and $G(X, 0; T)$ corresponds to the Green's function (dimensionless), that is, response at time T at location X to a unit impulse at location $X = 0$ at time $T = 0$. G is given by the solution of the following initial-value problem:

$$G_T(X, 0; T) = G_{XX}(X, 0; T) - G(X, 0; T) + \delta(X)\delta(T), \quad T > 0,$$

$$G(X, 0; 0) = 0.$$

In the case of a single semi-infinite cable with boundary condition $G(0, 0; T) = 0$ and $G(\infty, 0; T) = 0$, $G(X, 0; T)$ can be shown to be given by:

$$G(X, 0; T) = \exp(-T)/(2\sqrt{\pi T^3})X \exp(-X^2/4T).$$

The Green's function $G(X, X_i; T)$ corresponds to the response at time T and location X to a unit impulse at location $X_i \neq 0$ and time $T = 0$, and is given by the solution of the initial-value problem:

$$G_T(X, X_i; T) = G_{XX}(X, X_i; T) - G(X, X_i; T) + \delta(X - X_i)\delta(T),$$

$$G(X, X_i; 0) = 0.$$

In the case of a single semi-infinite cable with boundary conditions $G(0, X_i; T) = 0$ and $G(\infty, X_i; T) = 0$, G can be shown to be given by (Tuckwell, 1988a):

$$G(X, X_i; T) = \exp(-T)/(2\sqrt{\pi T})\{\exp[-(X - X_i)^2/4T] - \exp[-(X + X_i)^2/4T]\}.$$

Note: The expression outside of the brackets can be approximated, viz. $\exp(-T)/(2\sqrt{\pi T}) \approx K_0(T)/\sqrt{2}$, where K_0 is the modified Bessel function of 2nd kind of order zero.

The evaluation of Equation (8.A1) gives the following time course for the first component of the nonlinear Volterra integral equation solution governed by Equation (8.17):

$$\Psi(X, T) = \Phi_0(X, T) + \{\varepsilon R_m g_{Na} \mu_0^2 \eta_0 (1 - \exp[-\Phi_{Na} U_{peak} \Lambda])/(U_{peak} \lambda \Lambda)\} \sum_{i=1}^{N} G(X, X_i; T),$$

where Φ_0 is the solution of the linear diffusion equation corresponding to the passive bAP (zero-order perturbation in Equation [8.A1]) and is found easily by quadrature:

$$\begin{aligned}
\Phi_0(X, T) &= \int_0^T [U(T - s)/U_{peak}] H(T - s) G(X, 0; s)\, ds \\
&= (U_0/2\sqrt{\pi}\, U_{peak}) \int_0^T (150 \exp(-2T) X s^{3/2} \exp[s - X^2/(4s)] \\
&\quad - 100 \exp(-4T) X s^{3/2} \exp[3s - X^2/(4s)] \\
&\quad - 50 \exp(-1.2T) X s^{3/2} \exp[0.2s - X^2/(4s)])\, ds. \qquad (8.A2)
\end{aligned}$$

We need to solve the integral of the form:

$$I = \int_0^T (\zeta^{\upsilon} \exp[\zeta(\alpha - 1) - X^2/(4\zeta)] \, d\zeta.$$

Put $\zeta = X^2/4z$, $d\zeta = -X^2/4z^2 \, dz$

$$\begin{aligned} I &= (X^2/4)^{\upsilon+1} \int_{\infty}^{X^2/4T} (z^{-\upsilon-2} \exp[X^2/4z(\alpha - 1) - z])(-dz) \\ &= (X^2/4)^{\upsilon+1} \int_{X^2/4T}^{\infty} (z^{-\upsilon-2} \exp[X^2/4z(\alpha - 1) - z]) \, dz. \end{aligned}$$

But $\exp[X^2/4z(\alpha - 1)] = \sum_{n=0}^{\infty} (\alpha - 1)^n/n!(X^2/4z)^n$ and therefore

$$I = (X^2/4)^{\upsilon+1} \sum_{n=0}^{\infty} (\alpha - 1)^n/n!(X^2/4)^n \int_{X^2/4T}^{\infty} (z^{-\upsilon-2-n} \exp(-z)) \, dz.$$

(Remark: The order of summation and integration has been interchanged, as the series is uniformly convergent.) But

$$\int_{X^2/4T}^{\infty} z^{-\upsilon-2-n} \exp(-z) \, dz = \Gamma(-n - \upsilon - 1; X^2/4T),$$

where Γ is the incomplete gamma function. Hence

$$I = (X^2/4T)^{\upsilon+1} T^{\upsilon+1} \sum_{n=0}^{\infty} [(\alpha - 1)T]^n/n!(X^2/4T)^n \Gamma(-n - \upsilon - 1; X^2/4T). \quad (8.A3)$$

Substituting (8.A3) into (8.A2) yields

$$\begin{aligned} \Phi_0(X, T) = &(50U_0/\sqrt{\pi}\, U_{\text{peak}}) \sum_{n=0}^{\infty} X^{2n}/n! \Gamma(-n + 1/2; X^2/4T) \\ &\times \{3(0.25)^n \exp(-2T) - 2(0.75)^n \exp(-4T) - (0.05)^n \exp(-1.2T)\}. \end{aligned} \quad (8.A4)$$

The evaluation of Equation (8.A4) for $n = 0, 1$ yields the time course of the passively propagating action potential. It is useful because numerical evaluation of Equation (8.A1) in the perturbation expansion of the active propagating action potential using *Mathematica*® becomes too demanding on the memory capacity of standard computer workstations. A further simplification can be made upon substituting into Equation (8.A4) the following identities (Gradshteyn and Ryzhik, 1980):

$$\Gamma(1/2, X^2/4T) = \sqrt{\pi} \operatorname{erfc}[X/(2\sqrt{T})],$$

and

$$\Gamma(-1/2, X^2/4T) = 4\sqrt{T}/X \exp(-X^2/4T) 2\sqrt{\pi} \operatorname{erfc}[X/(2\sqrt{T})].$$

Hence,

$$\begin{aligned}\Phi_0(X,T) =&(50U_0/\sqrt{\pi}\,U_{\text{peak}})\{4X\sqrt{T}\exp(-X^2/4T)[0.75\exp(-2T) - 1.5\exp(-4T)\\ &- (0.05)\exp(1.2T)] + \sqrt{\pi}\,\text{erfc}[X/2\sqrt{T})][(3-1.5X^2)\exp(-2T)\\ &- (2-3X^2)\exp(-4T) - (1-0.1X^2)\exp(-1.2T)]\}, \end{aligned}\quad (8.\text{A5})$$

where erfc is the complementary error function.

PROBLEMS

1. To analyze repetitive activity in response to suprathreshold currents consider a spike train triggered at the point $X = 0$ corresponding to a lumped soma with a functional representation:

$$\begin{aligned}U(T) = U_0\sum_{k=0}^{\infty}&\{\exp(-0.2T)\sin[(2\pi/5)T] + 150\exp(-2T) - 100\exp(-4T)\\ &- 50\exp(-1.2T)\}H(T-k\tau_{\text{m}}/\omega),\end{aligned}$$

   where U_0 is a fixed value of the membrane potential (mV), ω is the interspike interval of the spike train (sec), and $H(\cdot)$ is the Heaviside step function.
   Show that the solution Φ_0 of the linear cable equation corresponding to the passive bAP in a semi-infinite cable satisfies the following convolution integral:

$$\Phi_0(X,T) = \int_0^T\sum_{k=0}^{\infty}[U(s)/U_0]H(s-k\tau_{\text{m}}/\omega)G(X,0;T-s)\,\text{d}s,$$

   where $G(X,0;T) = \rho_\infty\exp[\rho_\infty X - (1-\rho_\infty^2)T]\,\text{erfc}\,(\rho_\infty\sqrt{T} + X/2\sqrt{T})$ is the Green's function, and $\rho_\infty = (\pi/2)\{d^{3/2}/S\}(\sqrt{R_{\text{m}}}/\sqrt{R_i})$ is the dendritic-somatic conductance ratio (dimensionless) with S being the surface area of the soma (cm^2).
   (a) Evaluate the convolution integral analytically to obtain the time course of passively bAPs. Compare your results with those of Dodge (1979) and others. (b) Repeat the exercise now with F–H dynamics to investigate bAP trains. Compare your results with those of Leibovic (1972) to investigate the susceptibility of bAPs to modification by a second stimulus applied after the first one.
2. Clustering may be a way of distributing Na^+ channels to optimize bAP propagation because sparsely excitable dendrites show extremely low and nonconstant conduction velocities (Poznanski, 2001). At present there is no theory of saltatory nerve conduction that successfully treats the dependence of conduction velocity on cable and membrane current properties. Show that the ionic cable equation has a solution that resembles a traveling wave by introducing the traveling coordinates $\xi = x - \phi(t)$, and $\tau = t$. Use the method of averaging (Keener, 2000) to calculate the average velocity. Compare the result with the average conduction velocities obtained by Poznanski (2001).
3. Include synaptic input current along a finite cable of electrotonic length L. Investigate the interaction between a bAP and the synaptic potentials evoked by an excitatory input at a single point along the cable satisfying F–H dynamics. Compare your results with those found by Stuart and Häusser (2001) for neocortical pyramidal cells.

{Hint: The Green's function takes on the following form:

$$
\begin{aligned}
G(X,X_i;T) = {} & \exp(-T)/(2\sqrt{\pi T})[H(X_i - X)\{\exp[-(-X - X_i + 2L)^2/4T] \\
& + \exp[-(-X + X_i)^2/4T] - \exp[-(X - X_i + 2L)^2/4T] \\
& - \exp[-(X + X_i)^2/4T] - \exp[-(-X - X_i + 4L)^2/4T] \\
& - \exp[-(-X + X_i + 2L)^2/4T] + \exp[-(X - X_i + 4L)^2/4T] \\
& + \exp[-(X + X_i + 2L)^2/4T] - \exp[-(-X + X_i + 2L)^2/4T] \\
& - \exp[-(-X + 3X_i)^2/4T] + \exp[-(X + X_i + 2L)^2/4T] \\
& + \exp[-(X + 3X_i)^2/4T] + \exp[-(-X - X_i + 6L)^2/4T] \\
& + \exp[-(-X + X_i + 4L)^2/4T] - \exp[-(X - X_i + 6L)^2/4T] \\
& - \exp[-(X + X_i + 4L)^2/4T] + 2\exp[-(-X + X_i + 4L)^2/4T] \\
& + 2\exp[-(-X + 3X_i + 2L)^2/4T] - 2\exp[-(X + X_i + 4L)^2/4T] \\
& - 2\exp[-(X + 3X_i + 2L)^2/4T] + \exp[-(-X + 3X_i + 2L)^2/4T] \\
& + \exp[-(-X + 5X_i)^2/4T] - \exp[-(X + 7X_i + 2L)^2/4T] \\
& - \exp[-(X - 7X_i)^2/4T]\} + H(X - X_i)\{\exp[-(-X - X_i + 2L)^2/4T] \\
& + \exp[-(X - X_i)^2/4T] - \exp[-(-X + X_i + 2L)^2/4T] \\
& - \exp[-(X + X_i)^2/4T] - \exp[-(-X - X_i + 4L)^2/4T] \\
& - \exp[-(X - X_i + 2L)^2/4T] + \exp[-(-X + X_i + 4L)^2/4T] \\
& + \exp[-(X + X_i + 2L)^2/4T] - \exp[-(X - X_i + 2L)^2/4T] \\
& - \exp[-(3X - X_i)^2/4T] + \exp[-(X + X_i + 2L)^2/4T] \\
& + \exp[-(3X + X_i)^2/4T] + \exp[-(-X - X_i + 6L)^2/4T] \\
& + \exp[-(X - X_i + 4L)^2/4T] - \exp[-(-X + X_i + 6L)^2/4T] \\
& - \exp[-(X + X_i + 4L)^2/4T] + 2\exp[-(X - X_i + 4L)^2/4T] \\
& + 2\exp[-(3X - X_i + 2L)^2/4T] - 2\exp[-(X + X_i + 4L)^2/4T] \\
& - 2\exp[-(3X + X_i + 2L)^2/4T] + \exp[-(3X - X_i + 2L)^2/4T] \\
& + \exp[-(5X - X_i)^2/4T] - \exp[-(7X + X_i + 2L)^2/4T] \\
& - \exp[-(-7X + X_i)^2/4T]\}],
\end{aligned}
$$

where X_i is the location of the ith hot spot.}

4. De Schutter and Steuber (2000a) criticize reduced morphologies for active models, in particular the so-called "ball and stick" model, because the densities on the "equivalent" dendrite may not reflect the true values found in real dendrites. Extend the solutions of the F–H equations presented herein for a dendritic tree with a single branch point as shown schematically below (Figure 8.8). Show to what extent dendritic morphology influences the control of bAPs by comparing the result to a single cable representation. What densities of Na^+ channels yield similar bAPs? Hint: Apply Laplace transforms to the following infinite-dimensional nonlinear

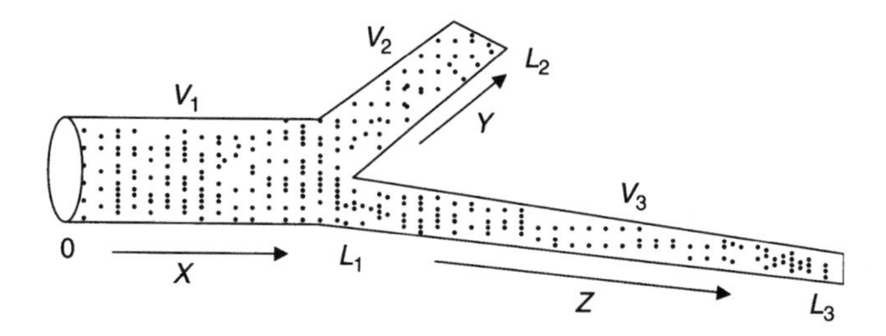

FIGURE 8.8 A schematic illustration of a dendritic tree with a single branch point.

dynamical system:

$$C_{1m}V_{1t} = (d_1/4R_{1i})V_{1xx} - g_{2L}V_1 - \sum_{i=1}^{N} f(x,t;V_1)\delta(x-x_i) - \sum_{n=0}^{\infty} U_n(t)H(t-nt)\delta(x), \quad t>0,$$

$$C_{2m}V_{2t} = (d_2/4R_{2i})V_{2yy} - g_{2L}V_2 - \sum_{i=1}^{N} f(y,t;V_2)\delta(y-y_i), \quad t>0,$$

$$C_{3m}V_{3t} = (d_3/4R_{3i})V_{3zz} - g_{3L}V_3 - \sum_{i=1}^{N} f(z,t;V_3)\delta(z-z_i), \quad t>0.$$

Subject to initial conditions:

$$V_1(x,0) = V_2(y,0) = V_3(z,0) = 0,$$

subject to boundary conditions:

$$V_{1x}(0,t) = V_{2y}(L_2,t) = V_{3z}(L_3,t) = 0,$$

subject to conservation of charge:

$$(1/R_{1i})V_{1x}(L_1,t) = (1/R_{2i})V_{2y}(0,t) + (1/R_{3i})V_{3z}(0,t),$$

and subject to continuity of potential:

$$V_1(L_1,t) = V_2(0,t) = V_3(0,t).$$

9 Inverse Problems for Some Cable Models of Dendrites

Jonathan Bell

CONTENTS

9.1 Introduction

It has been more than a century since it was established that the neuron is an independent functioning unit and that neurons, in most cases, have extensive dendritic processes. As staining techniques and other methodologies became more precise, more complex morphology was identified with these branching structures, along with substructures like dendritic spines. But dendrites are small and delicate objects to study. They are generally less than 3 μm in diameter and rapidly decrease in size the more distal they are from the cell body (soma). Until recently they were not accessible to direct measurement and electrical recordings had to be done at the soma, concealing many properties the dendrites could possess. But with carefully designed experiments and some good modeling, more information was gleaned about dendrites without direct measurements (e.g., see Hausser et al., 2000). Direct measurements are now possible because of such tools as infrared differential interference contrast video microscopy, patch-pipette recordings, two-photon microscopy, etc. (Johnston et al., 1996; Reyes, 2001). The development of calcium-dependent dyes, confocal microscopy, and more recently two-photon microscopy has enabled optical imaging of calcium dynamics in individual dendritic spines in response to synaptic inputs (Denk et al., 1996). This has allowed estimation of electrical properties of dendritic spines whose length is on the order of 0.1 to 1 μm. Direct observations have confirmed earlier results of Rodolfo Llinás (1968), who suggested that action potentials might actively propagate in the dendrites. Indeed, Segev and London (2003) write of being in the midst of the "dendritic revolution," where systematic and intimate electrical and optical experiments become possible on dendrites. The old perception of dendrites as electrical cables that carry information unidirectionally from the numerous dendritic synapses to the soma and on to the output axon has undergone significant revision. We now know that, for example, dendrites in many central nervous system neurons have active propagation of action potentials "backward" from axon to the dendrites. These backpropagating action potentials trigger plastic changes in the dendritic synapses and these changes are believed to underlie learning and memory in the brain. It appears there is nothing static about dendrites, that is, they dynamically change in an activity-dependent way. New experiments highlight the nonlinear nature of signal processing in dendrites and confirm earlier theoretical exploration of the biophysical consequences of such nonlinearities. Coupling molecular methods (Crick, 1999) with direct measurements of electrical properties, one can mark and manipulate specific membrane proteins in dendrites to identify the type and distribution of various receptors and ion channels in the dendritic membrane. When combined with sophisticated analytic and numerical models, these data are beginning to give us a functional picture of dendrites. For more detailed information, consult Stuart et al. (1999a), Hausser et al. (2000), Segev and London (2000), Euler and Denk (2001), and other chapters in this volume.

To move this program forward requires incorporating more physiological features in mathematical models, including spatially nonuniform morphological and electrical properties. This requires obtaining more microscale details about distributions of such properties from experiments, and this means considering distributed parameter problems at the cellular level. Our interest in this chapter is to discuss the recovery of certain spatially distributed parameters and to point out a number of other inverse problems along the way, all in the context of cable theory. In the next section we review the model basis for work discussed in succeeding sections. In Section 9.3 we discuss ion-channel distributions occurring in central nervous system neurons. Although there is much experimental and theoretical work to be done in this area, we narrow our focus to an algorithmic approach to recover a nonuniform distribution of a single channel density from microelectrode measurements for a passive cable. Then we extend this approach to estimate the potassium channel density in a segment of active dendrite. Here we utilize the Morris–Lecar model for convenience (Morris and Lecar, 1981), but the Hodgkin–Huxley model, or any other conductance-based model with nonlinear conductances can be substituted. In Section 9.4 we discuss a specific theory of a dendrite with spines, and again narrow our focus to recover the spatially distributed density function, after highlighting some interesting direct problems to consider further. We then complete the section with a discussion of a theory for recovering a density of cytoplasmic extensions, called filopodia, in a model for the primary sense-of-touch receptor, the Pacinian corpuscle. Finally, we end the chapter with a Saint Venant-type result motivated by behavior of the numerical method, and give some perspectives about other problems.

9.2 Conductance-Based Modeling

The cable theory model for the transmembrane potential of a cell, $v(x, t)$, is generally derived from an electrical circuit model of the cell's excitable membrane (e.g., see Jack et al., 1975; Tuckwell, 1988a). Thus, it represents a current balance expression given by

$$C_{\mathrm{m}}\frac{\partial v}{\partial t} + I_{\mathrm{ion}}(v) = \frac{a}{2R_{\mathrm{i}}}\frac{\partial^2 v}{\partial x^2} + I_{\mathrm{ext/syn}}. \tag{9.1}$$

That is, capacitive current density (with C_{m} being membrane capacitance per unit area) plus resistive (i.e. ionic) current density makes up membrane current (with R_{i} being axial resistivity and a being the constant cable radius) up to any synaptic or external current sources given by $I_{\mathrm{ext/syn}}$. The current–voltage relation I_{ion} is the sum of ion current densities, where we follow the classical Hodgkin–Huxley formulation in terms of conductances; thus,

$$I_{\mathrm{ion}}(v, \ldots) = \sum_j g_j(v - v_j) = \left(\sum_j g_j\right)v - \left(\sum_j v_j g_j\right). \tag{9.2}$$

If there are no active conductances associated with I_{ion}, we have a passive current–voltage relation, namely $I_{ion}(v) = v/R_m$, where R_m is the membrane resistivity. For each ion, v_j is the reversal potential, and from the Nernst equation (Fall et al., 2002), it is proportional to the log of the ratio of intracellular concentration to extracellular concentration of the specific ion. We consider these values fixed and known, but interesting, sometimes pathological, behavior happens when this is not the case. The most studied case involves major changes in the concentration of calcium ions in cells (e.g., see Keener and Sneyd, 1998).

In expression (9.2) for I_{ion}, we indicated that there could be other dependencies. For active conductances, and specifically ones gated by voltage, the individual conductances, g_j, can be written in terms of gating variables, that is, $g_j = \bar{g}_j r^\alpha s^\beta$, where α, β are nonnegative integers, $\bar{g}_j$ is the maximal conductance parameter for the specific jth conductance, r is an activating variable, and s is an inactivating variable associated with that conductance. Unless there is evidence to the contrary, these processes are associated with two-state gates (one open state, one closed state), which means that the dynamics for r and s is given by

$$\frac{\partial y}{\partial t} = \{y_\infty(v) - y\}/\tau_y(v), \quad y = r, s. \tag{9.3}$$

Thus, the extra dependencies in I_{ion} are with regard to all the gating variables. Indicated by this representation is y_∞ (the quasi-steady state for y) and τ_y (the timescale), both of which can be functions of potential, and in more complicated cases, such as ligand-gated channels (Fall et al., 2002), can be functions of ion concentrations, temperature, or membrane stress. In the simplest case where the fiber has constant radius and there are no active conductances, that is, the cable is passive, and without external current sources, the model given in Equation (9.1) reduces to

$$C_m \frac{\partial v}{\partial t} + g(v - v_l) = \frac{a}{2R_i} \frac{\partial^2 v}{\partial x^2}. \tag{9.4}$$

Considering the domain to be $0 < x < L, 0 < t < T$, the appropriate boundary conditions of interest to us in the rest of the chapter are

$$(\partial v/\partial x)(0,t) = -(R_i/\pi a^2) i_0(t), \qquad (\partial v/\partial x)(L,t) = 0. \tag{9.5}$$

That is, we want to inject a controlling current into the cell at $x = 0$, while the second condition (a "sealed-end" condition) expresses the situation that no longitudinal current escapes at $x = L$. If the soma is considered the terminal end of the fiber, then the boundary condition is sometimes represented by a pair of extra parameters, C_s representing the somatic membrane capacitance, and g_s representing the somatic conductance. With the soma having surface area A_s, to balance currents at $x = 0$ the left boundary condition is replaced by condition

$$(\pi a^2)/(R_i)(\partial v/\partial x)(0,t) - A_s[g_s v(0,t) + C_s(\partial v/\partial t)(0,t)] = -i_0(t).$$

The only practical complication this "oblique" boundary condition adds to the analysis is that in the associated eigenvalue problem the eigenfunctions are not orthogonal. In Krzyzanski and Bell (1999), for modeling reasons, we had both boundaries having similar oblique boundary conditions, but numerically it made very little difference from using Neumann conditions. We can assume from

Equation (9.5) that the fiber terminates at $x = L$, or experimentally we can set up *in vitro* to block current flow at that end. With an initial condition specified, and certain compatibility conditions given, the problem can be solved by elementary means. This assumes that all parameters are known and specified. But if the passive electrical parameters are spatially distributed, and particularly when they depend on an ionic-channel density that is *a priori* unknown, even the passive cable models pose significant analytic challenges. This will be discussed in Section 9.3.

9.3 Nonuniform Ion-Channel Densities

Ionic channels play a major role in characterizing the types of excitable responses expected of the cell type. Dendritic and axonal processes, along with the soma itself, have a variety of spatially distributed densities of ionic channels. These densities are usually represented as constant parameters in neural models because of the difficulty in estimating them experimentally. However, through microelectrode measurements and selective ion staining techniques, it is known that ion channels are spatially distributed nonuniformly. Further motivation for considering such a problem in this context comes, for example, from Johnston et al. (1996), Magee and Johnston (1995), Magee (1998), Safronov (1999), and Traub et al. (1992). Johnston et al. (1996) discussed studies of low-voltage-activated, and moderate high-voltage-activated conductance calcium channel types along the pyramidal dendrites in the mammalian hippocampus. Magee (1998) investigated hyperpolarization-activated (I_h) channels in CA1 hippocampal dendrites. I_h plays a variety of important roles in neuronal cell types (see Pape, 1996 for a review). Magee found that the density of these I_h channels increased sixfold from soma to distal dendrites. Traub et al. (1991), based on experimental work from several labs, modeled the CA3 pyramidal cell dendrites by 19 compartments with 6 different active ionic conductances that have spatially varying distributions. Poolos and Johnston (1999) examined the Ca^{2+}-activated K^+ current in hippocampal CA1 pyramidal neurons in relation to regional differences in neuronal excitability (see Johnston et al. [1996] for a review of the nonuniform distribution of Na^+ and various Ca^{2+} channel subtypes in the same pyramidal cells, and Safronov [1999] for a review of Na^+ and K^+ channel distributions in spinal dorsal horn neurons). The aim in this section is to outline an approach to estimate a nonuniformly distributed channel density parameter from electrophysiological measurements.

9.3.1 Recovering a Density in Passive Cable

There has been a fair amount of work in this setting in estimating model parameters, starting with the work of Rall in the 1960s (Rall, 1960, 1962b). Rall and colleagues developed "peeling" methods for single fibers and "constrained optimization" methods for coupled compartmental models for multiple neuron cases (Rall, 1977; Holmes and Rall, 1992b; Rall et al., 1992). Other methods have subsequently been described; among them are those by Jack and Redman (1971), Brown et al. (1981), Durand et al. (1983), Kawato (1984), D'Aguanno et al. (1986), Schierwagen (1990), and Cox (1998). White et al. (1992) discuss the effect of noisy data on the estimates and propose a technique that improves the robustness of the inverse mapping problem. Cox (Cox, 1998; Cox and Ji, 2000, 2001; Cox and Griffith, 2001; Cox and Raol, 2004) developed a moment method that he calls an "input impedance" method to recover (uniquely) constant parameters R_i, C_m, g, L, and thereby the cell's membrane time constant and electrotonic length constant; see also D'Aguanno et al. (1986), Tawfik and Durand (1994), Vanier and Bower (1999). That is, along with the boundary conditions stated above and the initial condition, it is sufficient to have $v(0, t)$ as the overspecified data for the inverse problem (constant-parameter case), and to compute a certain number of integrals of $v(0, t)$ and $i_0(t)$. Through a recording electrode one can inject current and record voltage at the same location, for example, at the soma. The same boundary data play a crucial role for our methodology given here. (Note that all these approaches only deal with estimating constant parameters.)

First consider the linear model of Equations (9.4) and (9.5) with known parameters except for the single distributed channel density parameter $N = N(x)$. Consider also the boundary conditions of Equation (9.5). In the notation of Hodgkin (1975), g and $C_{\rm m}$ can be represented as $g = g_0 N$, $C_{\rm m} = C_0 N + C_1$, where g_0, C_0, C_1 are, respectively, the appropriate conductance per channel, the capacitance per channel, and the capacitance per unit area of membrane in the absence of channels. We are not concerned with the estimation of these parameters here, but rather focus on the methodology for recovering $N(x)$. However, see Hille (2001) for some values of these parameters that have been found from experiments. With a recording electrode we can measure the voltage at $x = 0$, which we take as the location of the soma, while injecting current there. We substitute the above expressions for g and $C_{\rm m}$ into the equation, then transform to dimensionless variables via

$$\tilde{v}(\tilde{x},\tilde{t}) = [v(x,t)/v_1] - 1, \qquad \tilde{x} = x\lambda, \quad \tilde{t} = tg_0/C_0,$$
$$q(\tilde{x}) = C_0 N(x/\lambda)/C_1, \quad \text{where } \lambda = \sqrt{2R_{\rm i}C_1 g_0/aC_0}.$$

After dropping the tilde notation, we have the problem

$$(1+q(x))\frac{\partial v}{\partial t} + q(x)v = \frac{\partial^2 v}{\partial x^2}, \quad 0 < x < L_0, \ 0 < t < T \tag{9.6}$$
$$v(x,0) = 0, \quad 0 < x < L_0 \tag{9.7}$$
$$(\partial v/\partial x)(0,t) = -i(t), \qquad (\partial v/\partial x)(L_0,t) = 0, \tag{9.8}$$

along with the extra measurements

$$v(0,t) = f(t). \tag{9.9}$$

Now the mathematical problem is to recover $q(x)$, which is proportional to $N(x)$, on $[0, L_0]$, given $i(t)$ and $f(t)$ on $(0, T)$.

While there is considerable literature now on inverse problems for parabolic equations (e.g., see Anger [1990], Engl and Rundell [1995], Isakov [1998]), much of the work deals with recovery of source terms, or diffusion coefficients, or time-dependent only coefficients (and, in some cases, initial conditions). The challenge of the above problem is to recover a spatially distributed coefficient from overspecified temporal data. To elaborate further, previous work did not have neurobiological applications in mind, and though inverse problems are pervasive in neuroscience, analytical work on these problems has been limited.

From the theory of parabolic initial boundary value problems, it can be shown that there exists a unique solution to the above scaled problem for a given $q \in L^2[0, L_0]$. What we want for our problem is that, for a given $i(t)$, we have uniqueness of q for given $v(0,t)$. (For a uniqueness result, see Problem 6.) In Bell and Cracium (2003), we developed a numerical procedure to solve the type of inverse problems we described above in Equations (9.6) to (9.9). For a basic explanation of the procedure, discretize Equation (9.6) using an explicit method. Suppose we know the initial condition $v(x,0)$ for $0 < x < L_0$, and boundary conditions $v(0,t)$, $(\partial v/\partial x)(0,t)$ for $0 < t < t_{\max}$. For a fixed integer n, consider an $n \times n$ uniform grid for the domain $[0, L_0] \times [0, t_{\max}]$. Denote by v_i^m and q_i the values of $v(x,t)$ and $q(x)$ on the discrete grid, where $1 \le m, i \le n$, $\Delta x = L_0/n$, $\Delta t = t_{\max}/n$. The simplest explicit discretization of the equation is

$$(1+q_i)(v_i^{m+1} - v_i^m)/\Delta t + q_i v_i^m = (v_{i+1}^m - 2v_i^m + v_{i-1}^m)/\Delta x^2.$$

This can be solved for q_i; write this symbolically as

$$q_i = F(v_{i-1}^m, v_i^m, v_{i+1}^m, v_i^{m+1}) \tag{9.10}$$

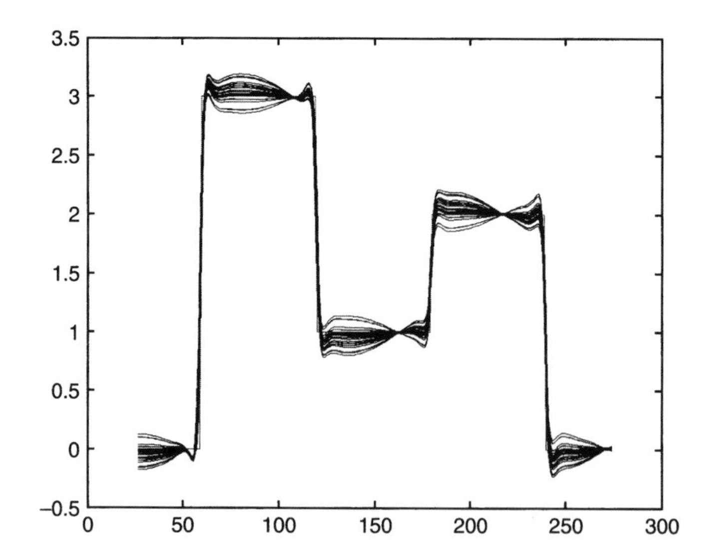

FIGURE 9.1 Recovery of $q(x)$ using the Crank–Nicholson numerical method for our distributed parameter identification procedure on the problem in Equations (9.6) to (9.9). Shown here are 20 computed versions of q, each with 0.01% random noise in left-hand boundary conditions. Vertical axis in all figures is $q(x)$. Horizontal axis in all figures is the space coordinate in nondimensional units.

(assuming $v_i^m \neq 0$). Alternatively, the difference equation can also be solved for

$$v_{i+1}^m = G(v_{i-1}^m, v_i^m, v_i^{m+1}, q_i). \tag{9.11}$$

Because of our knowledge of the boundary data, we have the values of $\{v_i^1, v_1^m, v_2^m\}$ for all $1 \leq m$, $i \leq n$. The algorithm to compute $q_2, q_3, \ldots, q_{n-1}$ is then: compute q_2 from Equation (9.10), since we know v_1^1, v_2^1, v_3^1, and v_2^2; then use Equation (9.11) and the value of q_2 just computed to compute v_3^m for $0 \leq m \leq n-1$. Next compute q_3 and v_4^m for $2 \leq m \leq n-2$, again by employing Equations (9.10) and (9.11); continue to iterate on i. In Bell and Cracium (2003) we used implicit as well as explicit methods and tested the approach on a number of problems with various (x, t) values. We added 0.1 to 1.0% noise to the boundary data for purposes of representing errors in the measurement of $v(0, t)$ while considering the injection of stimulus current $\partial v/\partial x$ at $x = 0$ as being known exactly. Boundary function $i(t)$ was treated as being "small" and monotone increasing. In Figure 9.1 we show the effectiveness of our method.

9.3.2 Recovering a Density in Active Cable

We wanted a method, however, that was applicable to cable models with active (nonlinear) conductances. Morris and Lecar (1981) developed a model for the barnacle giant muscle that became a popular two-variable nonlinear model to compute certain prototypical behavior, such as trains of propagating action potentials. While their model is not directly applicable to investigations of dendritic behavior, it provides a good nonlinear test model for our procedure. Their model of the membrane potential incorporates a noninactivating fast calcium current, a noninactivating slower potassium current, and a small (ohmic) leak current. The simulations of their model provided reasonable agreement with their experimental measurements. Because the calcium current is activated on a fast timescale, it has become standard to replace the calcium activation variable with its quasi-steady-state parameter. In Bell and Cracium (2003) we considered, rather arbitrarily, only the potassium conductance to be nonuniformly distributed spatially.

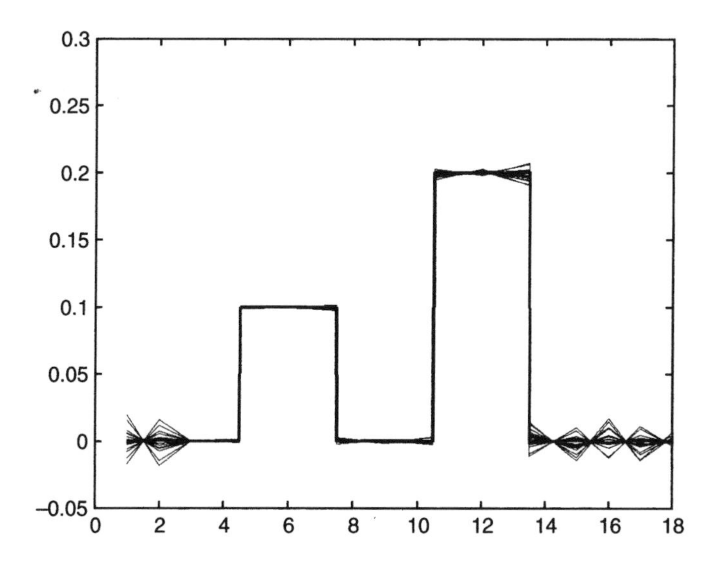

FIGURE 9.2 Results of using our scheme on the Morris–Lecar model to recover the potassium ion density, with 1% noise introduced on the left-hand boundary conditions. Twenty computed versions of q are shown by thin lines, while the thick line is the exact $q(x)$.

After nondimensionalizing the problem, using the parameter values given in Fall et al. (2002), the model becomes

$$(1+q(x))\frac{\partial v}{\partial t}+g_{\mathrm{Ca}}m_{\infty}(v)(v-v_{\mathrm{Ca}})+q(x)w(v-v_{\mathrm{K}})+g_{\mathrm{leak}}(v-v_{\mathrm{leak}})=\frac{\partial^2 v}{\partial x^2} \quad (9.12)$$

$$\frac{\partial w}{\partial t}=\{w_{\infty}(v)-w\}/\tau_w(v), \quad (9.13)$$

where w is the gating variable for the potassium conductance. As in the passive-cable case we impose initial and boundary data, considering input current on the left end of the segment, and a sealed-end boundary condition on the right end of the fiber. We next choose a potassium channel density $q(x)$ and run a forward problem to obtain the extra measurements, then apply our inverse problem procedure to see if we can recover the main features of q. One such case is given in Figure 9.2.

9.3.3 Numerical Results

One sample of the numerical results for the passive cable is given in Figure 9.1. Given the sparse conductance measurements of nerve fiber studies, which lead to histograms of the channel density distributions, considering a piecewise constant q is appropriate. But we do not have an exact solution of the problem in Equations (9.6) to (9.9) to compare with, so we have to generate the extra boundary data through solving a forward problem. The Crank–Nicholson scheme is used on the diffusion operator, with 0.1% boundary noise. The exact $q(x)$ is used (the fine line barely visible in Figure 9.1). The particular $q(x)$ chosen is rather arbitrary. With the same level of boundary noise added, errors are comparable whether using a smooth $q(x)$ or the piecewise constant $q(x)$. We found every run to represent the distribution closely, but we could do better by running the simulation many times (in Figure 9.1 there were 20 computed versions of $q(x)$), and then averaging the results. The scheme is computationally very fast.

For the Morris–Lecar model, Equations (9.12) and (9.13), with initial data, Equation (9.8), measured data, Equation (9.9), and adjusting the algorithm in the obvious way, Figure 9.2 shows a simulation where again a number of runs were taken and averaged, here with 1% noise added to the left-hand boundary condition. Again the agreement is excellent, but deteriorates quickly with increasing amounts of noise.

9.4 Dendrites with Excitable Appendages

Taking the approach of Baer and Rinzel (1991), the assumptions behind their continuum theory of dendrites with spines are: (a) the spine heads are very small so that the membrane potential for the spine head located at x, that is, $v_{\text{sh}}(x, t)$, may be considered a point source of potential for the nerve; (b) any spine head is just resistively coupled to the dendritic cylinder with no other electrical connections to neighboring spines, so that the spine-stem current is given by $I_{\text{ss}} = (v_{\text{sh}} - v_{\text{d}})/R_{\text{ss}}$, where v_{d} is the dendritic potential; (c) by current conservation principles, the excitable membrane may be modeled as a capacitive current in parallel with a resistive (ionic) current, so that

$$C_{\text{sh}}\frac{\partial v_{\text{sh}}}{\partial t} + I_{\text{ion}}(v_{\text{sh}}, \ldots) + I_{\text{ss}} + I_{\text{syn}} = 0.$$

Here I_{syn} is the synaptic current source, which is often represented by an alpha-function form of the conductance, namely $I_{\text{syn}} = g_{\text{syn}}(x, t)(v - v_{\text{syn}})$, where $g_{\text{syn}} = \bar{g}_{\text{syn}}(x)(t/\tau_{\text{syn}})\exp[1 - t/\tau_{\text{syn}}]$. We also have the passive-cable assumption for the dendrite, so that

$$C_{\text{m}}\frac{\partial v_{\text{d}}}{\partial t} + G_{\text{m}}v_{\text{d}} = \frac{a}{2R_{\text{i}}}\frac{\partial^2 v_{\text{d}}}{\partial x^2} + n(x)I_{\text{ss}}.$$

The parameter $n(x)$ is associated with the density of spine heads per unit length of membrane centered at x.

Baer and Rinzel (1991) employed Hodgkin–Huxley dynamics for the active spine-head membrane (and uniform radius). In their numerical investigations, and in our studies (Zhou and Bell, 1994), the principle parameters of concern were the spine density $n(x)$ and the spine-stem resistance R_{ss}. Baer and Rinzel obtained a number of interesting results, including showing that for R_{ss} sufficiently large (but still within the physiologically acceptable range), signal propagation was prevented. Also, they showed that successful transmission depends on the distribution of spines. In particular, they simulated the model dendrite near $x = 0$ for a fiber of length 2 mm with a fixed number (24) of spines. A threshold stimulus was given at $x = 0$, and transmission was considered successful if the spine-head potential was superthreshold at $x = 2$. Three cases were given. In the first case, the spines are distributed uniformly (so $n = 12$); in this case transmission fails. However, if the spines are clustered into eight groups of three spines each (cluster width being 0.04 and distance between clusters 0.24, so that $n = 75$), then transmission is successful. Hence, the clusters are close enough, and each cluster of spines has enough regenerative capacity to sustain propagation of a signal in this case. If, however, the spines are clustered into four groups of six spines each (cluster width being 0.08 and distance between clusters 0.42, then again $n = 75$), transmission in this case also fails. It would be interesting to formulate an analytically tractable problem and investigate this scenario further.

Zhou and Bell (1994) investigated conduction properties of the Baer–Rinzel model, with Morris–Lecar dynamics instead of Hodgkin–Huxley dynamics, in the presence of jumps and taper in the passive dendrite. They showed that for uniform spine density there is a bounded set in the n, R_{ss} parameter space where the model supports propagating action potentials. With a finite range of $\rho = 1/R_{\text{ss}}$ bounded away from 0, this is consistent with observations made by Baer and Rinzel for Hodgkin–Huxley dynamics.

9.4.1 Recovering a Spatially Distributed Spine Density

Consider the dendrite problem with a nonuniform distribution of spines along a dendritic segment that is proximal to the cell body. Using a recording microelectrode at the soma, where we can control the current injected into the cell and record potential with the same electrode, can we recover the spine density? The real question is whether the algorithmic approach discussed in Section 9.3 is applicable in this case. In Bell and Cracium (2003) we employed the Baer–Rinzel model with Morris–Lecar dynamics at the spine heads and considered only the spine density as being nonuniformly distributed. After nondimensionalizing the model, the problem takes the form

$$\frac{\partial v_{\mathrm{d}}}{\partial t} + v_{\mathrm{d}} = \frac{\partial^2 v_{\mathrm{d}}}{\partial x^2} + \rho n(x)(v_{\mathrm{sh}} - v_{\mathrm{d}}) \tag{9.14}$$

$$\delta \frac{\partial v_{\mathrm{sh}}}{\partial t} + i_{\mathrm{ion}}(v_{\mathrm{sh}}, w) = \rho(v_{\mathrm{d}} - v_{\mathrm{sh}}) \tag{9.15}$$

$$\frac{\partial w}{\partial t} = \varphi\{w_{\infty}(v_{\mathrm{sh}}) - w\}/\tau_w(v_{\mathrm{sh}}), \tag{9.16}$$

where δ is the ratio of time constants, φ is a temperature scale factor, ρ is the reciprocal of stem resistance, and i_{ion} is the scaled Morris–Lecar ionic current density. Again we use the sealed-end condition on the right ($\partial v_{\mathrm{d}}/\partial x = 0$ at $x = L$), known current injection at the soma ($\partial v_{\mathrm{d}}/\partial x = -i(t)$ at $x = 0$), and measured values of voltage ($v_{\mathrm{d}}(0, t) = v_{\mathrm{soma}}(t)$). The problem here, given the analogous boundary conditions to those in the Morris–Lecar example in Section 9.3, is to recover the (scaled) spine density function $n = n(x)$.

We can take what we highlighted in Section 9.3 and apply the same algorithm to solve for $n_2, n_3, \ldots, n_{k-1}$, given initial conditions $v_{\mathrm{d}}(x, 0), v_{\mathrm{sh}}(x, 0), w(x, 0)$, and boundary conditions $v_{\mathrm{d}}(0, t), \partial v_{\mathrm{d}}/\partial x(0, t)$. Then we input the initial data $\{v^1_{\mathrm{d},i}, v^1_{\mathrm{sh},i}, w^1_i\}$ and boundary data $v^j_{\mathrm{d},1}, v^j_{\mathrm{d},2}$ into our algorithm to recover $n_2, n_3, \ldots, n_{k-1}$. One case of this approach is given in Figure 9.3. Note that we add independent noise terms to all the input data, with the only restriction being that we keep the difference $v^j_{\mathrm{d},2} - v^j_{\mathrm{d},1}$ noise free.

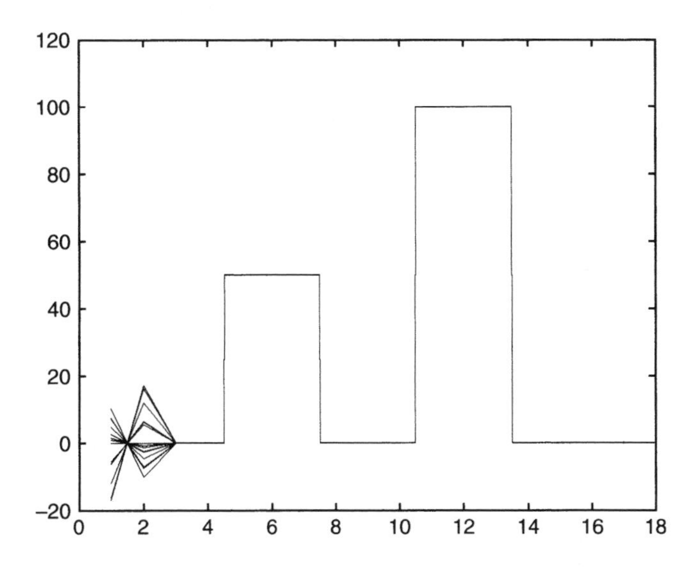

FIGURE 9.3 One result of adapting our procedure to find the spine density $n(x)$ for the Baer–Rinzel model.

along with measuring

$$u(L,t) = g(t) \tag{9.21}$$

(on $0 < t < T$). Since recovering $c(x)$ is tantamount to recovering the strain, the question is, in this simpler framework, whether this can be done knowing just h and g. A uniqueness result can be developed along the lines of the dendritic problem mentioned in Problem 6. The issue we wish to mention here is existence, and in particular whether an equation for the coefficient $c(x)$ can be found. The transformation

$$u(\cdot,t) = \frac{2}{\sqrt{2\pi t}} \int_0^\infty e^{-\tau^2/4t} u^*(\cdot,\tau)d\tau \tag{9.22}$$

is a useful tool in this case. This allows us to consider the wave equation

$$\frac{\partial^2 u^*}{\partial \tau^2} - \frac{\partial^2 u^*}{\partial x^2} + c(x)u^* = 0, \quad 0 < x < L, \ 0 < \tau < T^* \tag{9.23}$$

$$u^*(x,0) = 0 = \frac{\partial u^*}{\partial x}(x,0), \quad 0 < x < L \tag{9.24}$$

$$\frac{\partial u^*}{\partial x}(0,\tau) = 0, \quad \frac{\partial u^*}{\partial x}(L,\tau) = h^*(\tau), \quad 0 < \tau < T^* \tag{9.25}$$

with

$$u^*(L,\tau) = g^*(\tau). \tag{9.26}$$

The strategy here is: if $\{u, h, g\}$ are defined through the transformation Equation (9.22) and u^* solves the problem in Equations (9.23) through (9.26), then one can show that u satisfies the problem in Equations (9.18) through (9.21). The approach comes from Isakov (1998), who employed the technique in another parabolic problem, and who points out that the sensitive part of the process is in transforming h to h^*. But we have the power of hyperbolic theory to work with in developing an equation for $c(x)$. The idea is to look at Equation (9.23) in the strip $\Omega = \{x < x_0, 0 < t < T^*\}$ and take advantage of the finite speed of propagation of information. By a change of variables an integral equation for the solution can be given using d'Alembert's formula. Call the solution U. A solution can be established by a contraction mapping argument. Using this solution, and various growth estimates, we can obtain the following result:

Theorem. *Let $g^* \in C^2[0,T^*], h^* \in C^1[0,T^*]$ with $g^*(0) = 0, g^{*\prime}(0) = h^*(0)$:*

1. *The (hyperbolic) inverse problem is equivalent to solving the following Volterra-type integral equation for c on $0 < \tau < T^*/2$:*

$$\int_0^\tau c(s + x_0 - \tau)\{h^*(s) + U_t(s + x_0 - \tau, s + \tau)\}ds = -g^{*\prime\prime}(2\tau) + h^{*\prime}(2\tau). \tag{9.27}$$

2. *A solution $c(x)$ to Equation (9.27) is unique on $(0, T^*/2)$. For $T^* > 0$ small, there exists a solution to the inverse problem.*

When the semi-infinite strip problem is translated over to the problem in Equations (9.23) through (9.26), we have the determination of $c(x)$ on the interval $(L/2, L)$. To obtain $c(x)$ on the whole interval we return to the u Equations (9.18) through (9.20). By reflection we can extend the determination to the whole interval. This approach avoids determination of the full spectrum for the original problem, but forces us to deal with a coupled system of nonlinear Volterra integral equations. So, though it provides an argument for existence (e.g., see Pogorzelski [1966] for solution methodology), it is not a practical representation to estimate $c(x)$. Also μ, hence also c, in actuality depend on both x and t. This makes for a much harder inverse problem to solve.

9.5 Discussion

The method described here leaves many issues unresolved. We are dealing with an ill-posed problem and we did not incorporate any regularization mechanism, so the sensitivity of the method needs further consideration. We also do not have any convergence results. We make little use of the right-hand boundary data since we march across the domain from left to right computing successive estimates of $q(x)$ at nodal points. This needs to be studied further, but perhaps there is at work a diffusive Saint-Venant type decay of the influence of the right-hand boundary conditions (e.g., see Ignaczak, 1998). For one such example consider our passive cable Equation (9.6) on $\{-\infty < x < L_0, 0 < t\}$, that is, consider the local behavior near the right boundary $x = L_0$, and consider the left boundary "very far" away. It is a bit easier to analyze to make a change of variables by letting $y = L_0 - x, r(y) = 1 + q(L_0 - y)$, and $s(y) = r(y) - 1$, so that Equation (9.6) is transformed to

$$r(y)\frac{\partial u}{\partial t} + s(y)u = \frac{\partial^2 u}{\partial y^2}, \qquad y > 0, \quad t > 0.$$

Then one can show the following for a given nonnegative q.

Lemma. *Let $u(y,t)$ be any classical solution to the above equation on domain $\{y > 0, t > 0\}$, with $u(y,0) = 0$ for $y > 0$, arbitrary boundary data at $y = 0$ ($x = L_0$), and $u\,\partial u/\partial x = o(1)$ as $y \to \infty$. Let*

$$E(y,t) = \int_0^t \int_y^\infty \left\{ \frac{1}{2} r(w)[u(w,\tau)]^2 + G(w,u(w,\tau)) + \int_0^\tau \left(\frac{\partial u}{\partial w}(w,s)\right)^2 \mathrm{d}s \right\} \mathrm{d}w\,\mathrm{d}\tau$$

be the total energy associated with $u(y,t)$ stored in $\{(z,t)|t > 0, z > y\}$ from "heating" by the boundary condition at $y = 0$, where $G(w,u)$ satisfies $\partial G/\partial\tau = s(w)[u(w,\tau)]^2$. Then, for a ρ satisfying $0 < \rho < r(y)$ for $y > 0$,

$$E(y,t) \le E(0,t)\exp[-y\sqrt{2\rho/t}] \qquad \text{for } y > 0, t > 0.$$

Without going into details here, one can show that E satisfies $E(y,0) = 0$, $E(y,t) > 0$ for $t > 0$, and, through an application of Schwartz inequality, $\partial E/\partial y + \sqrt{2\rho/t}E \le 0$. From Gronwall's inequality, the result follows.

Use of the two boundary conditions on the left endpoint is reminiscent of the Cauchy problem for the heat equation (Canon, 1984), a very ill-posed problem. Canon considers the classical one-dimensional heat equation, but it is a straightforward generalization, though computationally messy, to consider

$$(1+q)\frac{\partial u}{\partial t} + qu = \frac{\partial^2 u}{\partial x^2} \quad \text{on } \{0 < x < L, 0 < t < T\}, \qquad u(0,t) = f(t), \quad \frac{\partial u}{\partial x}(0,t) = g(t). \tag{9.28}$$

Here q is a fixed, positive constant. This problem does not necessarily have a solution, but when it does, the solution is unique, and the pair $\{f, g\}$ is defined as a *compatible pair*. Canon gives an instance where this is the case. He lets $u(x, t)$ be a power series $\sum_j a_j(t)x^j$ and defines a Holmgren class of functions $H = H(L, T, C)$ to be the set of infinitely differentiable functions h on $0 < t < T$ that satisfy derivative conditions $|h^{(i)}(t)| \leq C(2i)!/L^{2i}$, $i > 0$. Then it turns out that if $f, g \in H$, the power series for u converges uniformly and absolutely for $|x| < r < L$, and u is a solution to the Cauchy problem, Equation (9.28). If q is a polynomial, Canon's result should generalize, and with our use of rather simple boundary data, we may have inadvertently used a form of compatible data corresponding to the piecewise constant q that is used.

While the algorithm outlined above is a nonoptimization approach to recovering a channel density distributed parameter, there are powerful PDE-constrained optimization approaches, such as versions of the augmented Lagrangian or the reduced sequential quadratic programming methods (e.g., see Kupfer and Sachs, 1992) that should be able to handle the problem in a stable manner, and these are being considered as an alternative approach.

Very early in the chapter it was pointed out that the kind and distribution of ion channels have a major role of determining the electrical behavior of nerve cells, and that ion channels are not uniformly distributed over the dendritic or somatic membrane. An important element in the "dendritic revolution" is about refining our understanding of how distributions of channel types and synaptic inputs influence signal processing in various cell types. From a theoretical standpoint we will have to deal more with the multi-timescale aspects of these neural processes. The local dynamics of ion concentrations like calcium and chloride play a crucial role in long-term plasticity and stability considerations of the neural network. From a dynamicist's standpoint, we have layers of fast–slow systems to consider. But ion channels and receptors are constantly experiencing turnover and synthesis via endocytotic mechanisms. The metabolic turnover rate varies depending on location and innervation (Hille, 2001). For example, in embryonic muscle cells the nACH receptors have a mean lifetime of 17 to 23 h. So while we discuss recovering channel densities, they may be (slowly) changing. Just as in the case of the membrane-strain function above, considering $q(x)$ also as a function of time significantly increases the difficulty of getting reliable data along with developing an effective algorithmic approach to solving an inverse problem for $q(x, t)$.

In considering q, hence N, time independent in this chapter, we assumed that ion-channel densities are completely controlled by transcriptional and translational processes. If one takes the viewpoint that during neuronal development the individual cell's identity, and therefore its activity levels, is established, then it is possible that the cell regulates the channel distributions to control its activity levels. The idea of activity-dependent regulation of neuronal conductances was investigated in LeMasson et al. (1993), Liu et al. (1998), Golowasch et al. (1999b), and Soto-Trevino et al. (2001). For a good review of possible mechanisms, see Marder and Prinz (2002). The general scheme introduced in these papers, within the framework of single compartment neuronal models, is that the maximal conductance parameters in Section 9.2, $\bar{g}_j$, have a slow timescale dependence on intracellular calcium concentration in such a way that there is a stabilizing negative feedback loop that controls the neuronal input/output relationship for the cell (and for local networks). The possibility of these biochemical pathways and this form of plasticity is an intriguing concept both from an experimental and from a theoretical modeling standpoint, and it is worth further exploration in the future. Tied into this dynamic ion conductance topic is the fact that dendritic spines are also dynamic, both in terms of their distribution and the fact that individual spines seem to move dynamically during electrical activity (Matus, 2000).

9.6 Conclusions and Future Perspectives

A major question in neurobiology is how much information processing is done at the individual cellular level. To understand this better, one needs to be involved in modeling and simulation along with experimental studies. Because nonuniform properties are the norm in cellular neurobiology, it is desirable to incorporate heterogeneities into mathematical models and study emergent properties that arise that can help explain the functionality of the cell. To incorporate such properties as the spatial distribution of channels, or equivalently conductances, we need finer resolution of such properties from experimental investigations. Effective methods to estimate spatially distributed physical parameters associated with delicate dendritic processes will go hand in hand with those investigations. In this chapter we have outlined one such approach that directly uses current and voltage measurements and is a fast estimator.

Our method is designed to recover the piecewise-continuous distribution of an ion channel on a continuous cylindrical cable of finite length. To our knowledge it is the first study within neuronal cable theory to consider recovering a spatially nonuniform channel density. One can also ask about recovering an approximation to the density $q(x)$ when it is known that the specific ion channels are sparse enough to be considered discrete and are spatially distributed nonuniformly. Of course, ion channels are always discrete objects and there is now the capability to make single channel recordings. From the standpoint of estimating channel densities for use in cable models, the length scale of single channels is much smaller than the length-scale phenomena usually of interest when using cable theory to model specific neural behavior. Nevertheless, one can imagine the case where $q(x)$, up to scaling, is a sum of Dirac delta functions representing the sparse channel locations. A useful device has been to replace the delta functions with a small step function so that $q(x)/\bar{q} = \sum_{j=1}^{m} \delta(x - x_j) \approx \sum_{j=1}^{m} g_\varepsilon(x - x_j)$, or more generally $q(x) \approx \sum_1^m \bar{q}_j g_\varepsilon(x - x_j)$. In this case $q(x)$ is approximated by a piecewise constant function that vanishes on some subintervals of our spatial domain and is positive and constant on other subintervals. The main issue here with the method given in this chapter is that very small step sizes must be used, and since our algorithm deals with a square matrix, that implies a need to use very small time steps as well. This amount of computation may cause a problem with error accumulation. We leave it for a future study to investigate this problem numerically and analytically.

It is a wonder that any part of the Hodgkin–Huxley theory remains in the theoretical study of dendritic processes in the presence of branching, local differences in ion channels and their densities, nonuniformities in electrical and geometric properties, dendritic spines of different classes and densities, and ion channels gated by many kinds of stimuli (ligand binding, voltage, temperature, and mechanical stress). Yet various aspects of nonlinear cable theory continue to give us the basis for modeling and analyzing numerous neural behaviors, and ultimately keep us inspired by the diversity that we observe.

9.7 Summary

By coupling molecular, imaging, and microscopic experimental methodologies, neuroscientists are beginning to obtain a functional picture of dendritic activities. Further understanding must also include sophisticated neural modeling and simulation efforts, including incorporation of more spatially nonuniform morphological and electrical properties. This requires obtaining more microscale details about spatially distributed parameters from experiments at the cellular level. To understand how some details can be recovered from experiment, we consider a few inverse problems associated with cable theory models of dendrites. In particular, we focus on estimating dendritic spine densities from "spiny" dendrites, and single ion-channel density functions from microelectrode experiments.

ACKNOWLEDGMENTS

Conversations with Steve Cox motivated me to look more seriously at approaches to estimating ion-channel densities in neuronal cable models, and Gheorghe Craciun was a collaborator in much of the work of this chapter. I also thank the Mathematical Biosciences Institute, Ohio State University, for support during my visit and chance to meet Gheorghe.

PROBLEMS

1. Consider a passive cable model for a dendritic segment with sealed-end boundary conditions. Suppose there is a single location along the segment where a known current source, $i_0(t)\delta(x-x_0)$, is given. Assume all model parameters are known, but the location, x_0, of the current source is unknown. Derive a way to find x_0. One way is to use the method of Cox (1998). A more interesting case is to find the current source location, x_0, when the input represents a synaptic source $i_{syn}(x,t) = g_{syn}(t)(v(x,t) - v_{syn})\delta(x - x_0)$. For recent work on this, see Cox (2004).
2. In Section 9.4 we discussed a result in a paper of Baer and Rinzel (1991) about the distribution of dendritic spines on a cable segment allowing or preventing propagation of spikes. For some choice of voltage dynamics, and for a given threshold stimulus on the left boundary, establish conditions of the spine density function, $n(x)$, for attaining threshold in finite time on the right side of the cable segment. For one way of reducing the complexity of the model, see Coombes and Bressloff (2000).
3. We applied a certain direct method in Sections 9.3 and 9.4 for the recovery of a single ion-channel density (or, equivalently, a single spatially distributed conductance). More work needs to be done on robust methods to handle such problems. Develop another effective and efficient method to handle the ill-posedness of the problem. Can the method determine more than one spatially distributed conductance simultaneously? What constraints are needed? Can a convergence result be proved?
4. At the beginning of Section 9.5 we discussed one (rather restrictive) set of conditions for the boundary data to be a compatible pair for the linear cable Cauchy problem. Develop another approach to the problem that arrives at alternative conditions for the set of boundary data to be compatible.
5. After consulting the papers on activity-dependent regulation discussed in Section 9.5, extend those authors' single compartment approach to modulating $N(x,t)$ to the case of a cylindrical cable model, and investigate the dynamics of this intracellular calcium sensor mechanism.
6. Consider the following result:

   Lemma. *Let v_1, v_2 be solutions to Equations (9.6) through (9.8) corresponding to q_1, $q_2 \in L^2[0, L_0]$, for the same stimulus $i(t)$, where $i \in C^2(0,T)$ with $i(0) = 0, di(0)/dt \neq 0$. If $v_1(0,t) = v_2(0,t)$ on $(0,T)$, then $q_1(x) = q_2(x)$ on $[0, L_0]$.*

   This lemma is an application of inverse Sturm–Liouville eigenvalue problem theory. The proof can be adapted from results in Kirsch (1995). What is really needed is a result that states that if $|v_1(0,t) - v_2(0,t)|$ (or $|v_1(0,t) - v_2(0,t)| + |v_1(L,t) - v_2(L,t)|$) remains small on $[0, T]$, then $|q_1(x) - q_2(x)|$ remains small on $[0, L]$. This may be a difficult result to prove if even true, but it would be the type of result that would put a foundation to the problem area. If not true, then give a counterexample.

10 Equivalent Cables — Analysis and Construction

Kenneth A. Lindsay, Jay R. Rosenberg, and Gayle Tucker

CONTENTS

10.1 INTRODUCTION

The observed complexity and variety of dendritic geometries raise important questions concerning the role played by morphology in neuronal signal processing. Cajal (1952) recognized this diversity of shape over 100 years ago, assumed that it had purpose, and lamented that he was unable to discern this purpose. Much of our present insight into the properties of dendritic trees has been obtained through the development and application of cable theory (e.g., Tuckwell, 1988a, 1988b; Segev et al., 1995). However, the mathematical complexity introduced when cable theory is extended to branched dendrites is such that the role of specific neuronal morphology in conditioning the behavior of the dendrite still remains obscure. The complexity of the biophysical dendrite resides largely in its intricate branched structure; the input structure is simple. To appreciate the role played by morphology in shaping the electrical properties of a dendrite, a simpler representation of dendritic morphology is required, presumably at the cost of a more intricate input structure. The *equivalent cable* provides one possible framework for understanding how dendritic geometry shapes neuronal function. It enables the electrical properties of a branched dendrite connected to a parent structure at $\mathcal{P}$ to be replaced by those of an unbranched dendrite (cable) with electrical properties that are indistinguishable from those of the branched dendrite it replaces. Equivalence requires that every configuration of electrical activity on the branched dendrite corresponds to a unique configuration of electrical activity on the cable and vice versa. This requirement fixes the length of the equivalent cable and allows the effect of interactions among inputs, conditioned by the branch structure, to be studied as interactions among inputs on a simpler, but equivalent, unbranched structure.

The equivalent cable approach to the understanding of neuronal morphology is inspired by the success of Rall's equivalent cylinder (Rall, 1962b) in quantifying the role played by passive dendrites in neuronal function. Rall showed that the effect on the soma of a passive dendritic tree can be replaced by that of a single uniform *equivalent cylinder* attached to the soma whenever the tree satisfies certain very restrictive geometrical conditions. However the Rall equivalent cylinder is incomplete (Lindsay et al., 2001a). Although the input structure on the original branched dendrite defines that on the equivalent cylinder, the converse, as recognized by Rall, is false. This is, in part, a consequence of the inadequate length of the equivalent cylinder. Rall's work raises two important questions: first, can Rall's equivalent cylinder be extended to an equivalent cable, and assuming that this is possible, second, to what extent can Rall's restrictive conditions be relaxed but still guarantee the existence of the equivalent cable for a branched dendrite?

Whitehead and Rosenberg (1993) were the first to provide a partial answer to this question by demonstrating that a dendritic Y-junction with uniform limbs of different electrotonic lengths also has an equivalent cable, which in this instance is not uniform. They used the Lanczos tridiagonalization algorithm, applied numerically or analytically, to show that this cable has two independent components, only one of which is connected to the parent structure. The procedure simultaneously generates both components of the cable and the unique association between configurations of inputs on the branched dendrite and these components. More recent work has demonstrated that branched dendrites with passive membrane properties and nonuniform limbs also have equivalent cables (Lindsay et al., 1999, 2001b). The extent to which equivalent cables can be constructed for branched dendrites with active membrane properties remains an open question.

The purpose of this article is, first, to demonstrate why arbitrarily branched dendrites with passive membranes have equivalent cables and, second, to illustrate how these cables can be constructed both analytically and numerically.

10.2 CONSTRUCTION OF THE CONTINUOUS MODEL DENDRITE

A branched dendrite may be regarded as a collection of dendritic segments connected together so as to preserve continuity of transmembrane potential and conservation of core current at internal branch points. The basic building block of the mathematical model of a branched dendrite is therefore a mathematical model of a dendritic segment together with sufficient connectivity and boundary conditions to connect these components into a model dendrite.

The mathematical model of a dendritic segment is expressed in terms of the departure of the transmembrane potential V from its resting value (assumed to be $V = 0$). If x measures distance along the segment in centimetres, t measures time intervals in milliseconds, and $A(x)$, $P(x)$ are, respectively, the segment cross-sectional area and perimeter at x, then the transmembrane potential $V(x,t)$ on a segment with a passive membrane satisfies the *cable equation*

$$P(x)\left(c_{\mathrm{m}}\frac{\partial V}{\partial t}+g_{\mathrm{m}}V\right)+\mathcal{I}(x,t)=\frac{\partial}{\partial x}\left(g_{\mathrm{a}}A(x)\frac{\partial V}{\partial x}\right). \tag{10.1}$$

The biophysical constants c_{m} (μF/cm^2), g_{m} (mS/cm^2), and g_{a} (mS/cm) are, respectively, the specific capacitance and specific conductance of the segment membrane and the conductance of the segment core. Finally, $\mathcal{I}(x,t)$ (μa/cm) is the exogenous linear density of transmembrane current. The core current $I(x,t)$ along the segment is calculated from the definition

$$I(x,t)=-g_{\mathrm{a}}A(x)\frac{\partial V(x,t)}{\partial x}. \tag{10.2}$$

The solution of Equation (10.1) at the endpoints of each dendritic segment must either satisfy a boundary condition if the segment is terminal or maintain continuity of membrane potential and conservation of axial current if the endpoint is a branch point or point of connection to a parent structure. Equation (10.1) can be reduced to canonical form using the nondimensional time s and electrotonic length z defined by

$$s=t\frac{g_{\mathrm{m}}}{c_{\mathrm{m}}},\qquad z=\int\sqrt{\frac{P(x)\,g_{\mathrm{m}}}{g_{\mathrm{a}}A(x)}}\,\mathrm{d}x=\int\frac{\mathrm{d}x}{\lambda(x)}, \tag{10.3}$$

where s represents nondimensional time and z measures distance in electrotonic units. This change of variables reduces Equation (10.1) to the nondimensional form

$$c(z)\left(\frac{\partial V(z,s)}{\partial s}+V(z,s)\right)+J(z,s)=\frac{\partial}{\partial z}\left(c(z)\frac{\partial V(z,s)}{\partial z}\right), \tag{10.4}$$

where the characteristic segment conductance $c(z)$ and the nondimensional exogenous linear density $J(z,s)$ of transmembrane current are defined by

$$c(z)=\sqrt{Pg_{\mathrm{m}}\,g_{\mathrm{a}}A},\qquad J(z,s)=\mathcal{I}(x,t)\,\frac{c_{\mathrm{m}}}{g_{\mathrm{m}}}\sqrt{\frac{g_{\mathrm{a}}A}{Pg_{\mathrm{m}}}}. \tag{10.5}$$

The derivation of $J(z,s)$ from $\mathcal{I}(x,t)$ is based on the observation that the charge injected during the time interval $(t,t+\mathrm{d}t)$ across the segment membrane in $(x,x+\mathrm{d}x)$, namely $\mathcal{I}(x,t)\,\mathrm{d}x\,\mathrm{d}t$, must equal the charge injected during the nondimensional time interval $(s,s+\mathrm{d}s)$ across the equivalent segment membrane in $(z,z+\mathrm{d}z)$, namely $J(z,s)\,\mathrm{d}z\,\mathrm{d}s$. Similarly, the nondimensional expression for the core current (10.2) is

$$I(z,s)=-c(z)\,\frac{\partial V(z,s)}{\partial z}. \tag{10.6}$$

Equation (10.4) and definition (10.6) define, respectively, the nondimensional canonical representations of the cable equation and the core current for a nonuniform dendritic segment.

To summarize, a model of a branched dendrite consists of a family of cable Equations (10.4), one for each segment. These equations are connected together by requiring continuity of membrane potential and conservation of core current at branch points. At dendritic tips, either the membrane potential or the core current is prescribed while continuity of membrane potential and core current is enforced at the connection between the dendrite and its parent structure.

10.2.1 Construction of the Discrete Model Dendrite

Suppose that a branched dendrite with n dendritic segments of lengths $L_1, \ldots, L_n$ is transformed by the nondimensionalization (10.3) into a branched dendrite with segments of electrotonic length $l_1, \ldots, l_n$, respectively. As a result of this transformation, the equations for each segment of the transformed dendrite take the form (10.4) in which individual segments are distinguished by their characteristic conductance $c(z)$ and their electrotonic length.

The total electrotonic length of the branched dendrite over which exogenous current is active is $l_1 + l_2 + \cdots + l_n$. The construction of the equivalent cable requires that every configuration of electrical activity on the branched dendrite correspond to a unique configuration of electrical activity on the equivalent cable, and vice versa. Therefore, whenever the equivalent cable exists, it will necessarily have electrotonic length $l_1 + l_2 + \cdots + l_n$ and an expression for $c(z)$ in Equation (10.4) to be determined. Incidentally, this argument explains why the Rall equivalent cylinder is an incomplete description of a branched dendrite as are all empirical cables inspired by the success of the Rall equivalent cylinder (Fleshman et al., 1988; Clements and Redman, 1989).

The practical construction of the equivalent cable begins by subdividing the continuous dendrite into subunits of *fixed electrotonic length, h*. In the discretization procedure, each segment is assigned an electrotonic length that is the integer multiple of h closest to the segment's exact electrotonic length. The mis-specification in the length of each dendritic segment behaves like a uniformly distributed random variable in the interval $[-h/2, h/2)$ — it has an expected value zero and variance $h^2/12$. In practice, the total electrotonic length of the discretized dendrite therefore behaves as a normal deviate with an expected value which is the length of the original dendrite and a standard deviation $h\sqrt{n/12}$. In particular, the limit of the electrotonic length of the discretized dendrite is that of the continuous dendrite as $h \to 0^{(+)}$.

Since the discretization procedure alters the electrotonic length of a segment, $c(z)$ must be modified to preserve the total membrane conductance of that segment, defined by

$$\int_0^L P(x) g_{\mathrm{m}} \, \mathrm{d}x = \int_0^l \sqrt{P(x)A(x)g_{\mathrm{m}}g_{\mathrm{a}}} \, \mathrm{d}z = \int_0^l c(z) \, \mathrm{d}z.$$

Suppose that the discretized segment has total electrotonic length, mh, then in order to ensure that the discretized piecewise uniform segment and the continuous segment have the same total membrane conductance, it is enough to impose the constraint

$$d_k = \frac{1}{h} \int_{(k-1)l/m}^{kl/m} c(z) \, \mathrm{d}z = \frac{l}{mh} \, c\left(\frac{(2k-1)l}{2m}\right) + O(h^2), \quad k = 1, \ldots, m, \tag{10.7}$$

where d_k represents the corrected characteristic conductance of the discretized cable. The conservation of total membrane conductance for a segment guarantees that the charge crossing the membrane of that segment in a fixed interval of time for a given constant potential difference is the same for the original and electrotonic representations of the segment. In practice, the characteristic conductance of each section of a discretized segment is obtained from Equation (10.7) by ignoring the $O(h^2)$ remainder.

10.3 Mathematical Model of a Uniform Cable

The construction of the discretized dendrite from the continuous dendrite is the only stage in the development of the equivalent cable at which an approximation is made; otherwise the analysis of the discretized tree is exact. From a biophysical perspective, current input to a dendrite is usually associated with synaptic points of contact on its membrane. Consequently, one development of the equivalent cable assumes point input of synaptic current as opposed to continuous distributions. With this description of the membrane current, the mathematical equation satisfied by the membrane potential in each discretized section of length h is obtained from the general Equation (10.4) by setting $J(z, s) = 0$ and treating $c(z)$ as a constant on that section. The simplified form for Equation (10.4) is therefore

$$\frac{\partial V(z,s)}{\partial s} + V(z,s) = \frac{\partial^2 V(z,s)}{\partial z^2}. \tag{10.8}$$

In this description, exogenous membrane current enters the mathematical formulation of the problem through the conservation of core current at section boundaries.

The treatment of Equation (10.8) proceeds through the development of an identity connecting the Laplace transforms of the membrane potential V and core current I at each end of a uniform section. This identity is constructed without any loss of generality on the basis of zero initial values for the membrane potential. The Laplace transform of Equation (10.8) gives

$$\frac{\mathrm{d}^2\widetilde{V}(z,p)}{\mathrm{d}z^2} - \omega^2\widetilde{V}(z,p) = 0, \tag{10.9}$$

where $\omega^2 = p + 1$ and $\widetilde{V}(z,p)$ is the Laplace transform of $V(z,s)$, defined by

$$\widetilde{V}(z,p) = \int_0^\infty V(z,s)\,\mathrm{e}^{-ps}\,\mathrm{d}s. \tag{10.10}$$

It is easy to verify that the solution of Equation (10.9) is

$$\widetilde{V}(z,p) = \widetilde{V}_{\mathrm{L}}(p)\cosh\omega z - \widetilde{I}_{\mathrm{L}}(p)\,\frac{\sinh\omega z}{c\,\omega}, \tag{10.11}$$

in which $\widetilde{V}_{\mathrm{L}}(p)$ and $\widetilde{I}_{\mathrm{L}}(p)$ are the Laplace transforms of the potential and core current respectively at $z = 0$, the left-hand endpoint of the section, and c is the characteristic conductance (constant) of the section. Furthermore, the potential and core current at $z = h$, the right-hand endpoint of this section, are respectively

$$\begin{aligned}
\widetilde{V}_{\mathrm{R}}(p) &= \widetilde{V}(h,p) = \widetilde{V}_{\mathrm{L}}(p)\cosh\omega h - \widetilde{I}_{\mathrm{L}}(p)\,\frac{\sinh\omega h}{c\,\omega}, \\
\widetilde{I}_{\mathrm{R}}(p) &= -c\,\frac{\mathrm{d}\widetilde{V}(h,p)}{\mathrm{d}z} = -c\,\omega\widetilde{V}_{\mathrm{L}}(p)\sinh\omega h + \widetilde{I}_{\mathrm{L}}(p)\cosh\omega h.
\end{aligned} \tag{10.12}$$

Equations (10.12) are now solved simultaneously for $\widetilde{I}_L(p)$ and $\widetilde{I}_R(p)$ in terms of $\widetilde{V}_L(p)$ and $\widetilde{V}_R(p)$ to obtain

$$\frac{\sinh \omega h}{\omega}\widetilde{I}_L = c\,\widetilde{V}_L \cosh \omega h - c\,\widetilde{V}_R, \qquad \frac{\sinh \omega h}{\omega}\widetilde{I}_R = c\,\widetilde{V}_L - c\,\widetilde{V}_R \cosh \omega h. \tag{10.13}$$

10.3.1 Notation

Equations (10.13) will be applied in what follows to tree-like and cable-like structures. Although the former consists of a collection of cables, points on tree-like structures cannot be enumerated sequentially, unlike those on a cable. This feature makes the mathematical description of a tree-like structure quite different from that of a cable. To avoid ambiguities relating to objects that are defined simultaneously for a tree and a cable, calligraphic symbols are used to refer to the cable and roman symbols to the tree. Where no ambiguity exists, a roman symbol will be used. This notation is defined in Table 10.1.

10.3.2 Application to a Cable

Figure 10.1 illustrates a cable of electrotonic length nh subdivided into n uniform sections each of length h and characteristic conductance d_k by the points $P_0, P_1, \ldots, P_n$. Suppose further that current $\mathcal{I}_k(s)$ is injected at point P_k at potential $\mathcal{V}_k(s)$.

TABLE 10.1
Mathematical Notation for the Matrices and Vectors Appearing in the Description of Branched and Unbranched Structures

Description	Cable	Tree	No distinction
Cable matrix	$\mathcal{A}$	A	
Symmetrizing matrix	$\mathcal{S}$	S	
Diagonal matrix	$\mathcal{D}$	D	
Tridiagonal matrix			T
Householder matrix			H
Injected current	$\mathcal{I}$	I	
Membrane voltage	$\mathcal{V}$	V	

Note: Roman characters are used when no distinction exists between tree and cable properties.

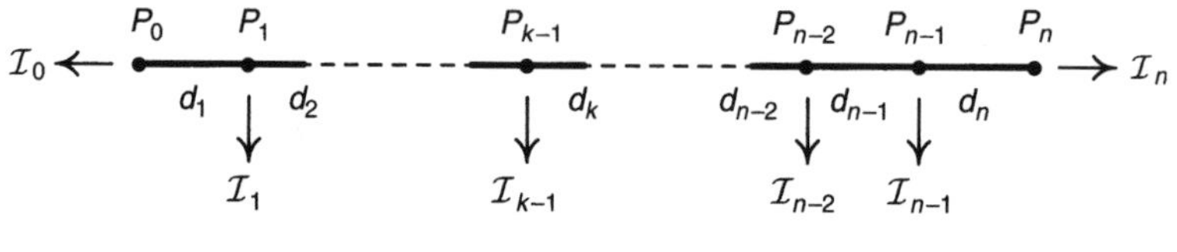

FIGURE 10.1 A piecewise uniform cable of electrotonic length nh is subdivided into n regions of length h such that the kth region has characteristic conductances d_k.

Formulae (10.13), particularized to the kth section of the piecewise uniform cable, give

$$\frac{\sinh \omega h}{\omega}\,\widetilde{I}_{\mathrm{L}}^{(k)} = d_k\,\widetilde{\mathcal{V}}_{k-1}\cosh \omega h - d_k\,\widetilde{\mathcal{V}}_k, \tag{10.14}$$

$$\frac{\sinh \omega h}{\omega}\,\widetilde{I}_{\mathrm{R}}^{(k)} = d_k\,\widetilde{\mathcal{V}}_{k-1} - d_k\,\widetilde{\mathcal{V}}_k\cosh \omega h, \tag{10.15}$$

where $\widetilde{\mathcal{V}}_k$ is the Laplace transform of the membrane potential $\mathcal{V}_k(s)$, and $\widetilde{I}_{\mathrm{L}}^{(k)}$ and $\widetilde{I}_{\mathrm{R}}^{(k)}$ are respectively the Laplace transforms of the core currents at the left- and right-hand endpoints of the kth section of the cable. Continuity of membrane potential is guaranteed by construction. Consequently, the mathematical description of the cable is based on conservation of core current at points $P_0, P_1, \dots, P_n$. This requires that

$$\begin{aligned} &I_{\mathrm{L}}^{(1)}(s) = -\mathcal{I}_0(s), \qquad I_{\mathrm{R}}^{(k)}(s) = \mathcal{I}_k(s) + I_{\mathrm{L}}^{(k+1)}(s), \quad k = 1,\dots,n-1, \\ &I_{\mathrm{R}}^{(n)}(s) = \mathcal{I}_n(s). \end{aligned} \tag{10.16}$$

In Laplace transform space, Equations (10.16) take the form

$$\widetilde{I}_{\mathrm{L}}^{(1)} = -\widetilde{\mathcal{I}}_0, \qquad \widetilde{I}_{\mathrm{R}}^{(k)} = \widetilde{\mathcal{I}}_k + \widetilde{I}_{\mathrm{L}}^{(k+1)}, \quad k = 1,\dots,n-1, \qquad \widetilde{I}_{\mathrm{R}}^{(n)} = \widetilde{\mathcal{I}}_n, \tag{10.17}$$

where it should be borne in mind that the uniqueness of the Laplace transform representation guarantees that Equations (10.16) and (10.17) are equivalent. The equations to be satisfied by $\widetilde{\mathcal{V}}_0, \dots, \widetilde{\mathcal{V}}_n$ are constructed from Equations (10.17) by replacing $\widetilde{I}_{\mathrm{L}}^{(k)}$ with expression (10.14) and $\widetilde{I}_{\mathrm{R}}^{(k)}$ with expression (10.15). The result of these substitutions for the cable in Figure 10.1 is

$$\begin{aligned} -\widetilde{\mathcal{V}}_0\cosh \omega h + \widetilde{\mathcal{V}}_1 &= \frac{\sinh \omega h}{d_1\,\omega}\,\widetilde{\mathcal{I}}_0, \\ \frac{d_k}{d_k + d_{k+1}}\,\widetilde{\mathcal{V}}_{k-1} - \widetilde{\mathcal{V}}_k\cosh \omega h + \frac{d_{k+1}}{d_k + d_{k+1}}\,\widetilde{\mathcal{V}}_{k+1} &= \frac{\sinh \omega h}{(d_k + d_{k+1})\,\omega}\,\widetilde{\mathcal{I}}_k, \\ \widetilde{\mathcal{V}}_{n-1} - \widetilde{\mathcal{V}}_n\cosh \omega h &= \frac{\sinh \omega h}{d_n\,\omega}\,\widetilde{\mathcal{I}}_n. \end{aligned} \tag{10.18}$$

Let $\mathcal{D}$ be the diagonal $(n+1)\times(n+1)$ matrix

$$\mathcal{D} = \mathrm{diag}[d_1, (d_1 + d_2), \dots, (d_k + d_{k+1}), \dots, (d_{n-1} + d_n), d_n], \tag{10.19}$$

and let

$$\mathcal{I}(s) = [\mathcal{I}_0(s), \mathcal{I}_1(s), \dots, \mathcal{I}_k(s), \dots, \mathcal{I}_n(s)]^{\mathrm{T}} \tag{10.20}$$

denote the vector of injected currents; then the system of Equations (10.18) describing the piecewise uniform cable has the matrix representation of

$$\mathcal{A}\widetilde{\mathcal{V}} = \frac{\sinh \omega h}{\omega}\,\mathcal{D}^{-1}\,\widetilde{\mathcal{I}} \tag{10.21}$$

in which $\mathcal{A}$, called the *cable matrix*, is the $(n+1)\times(n+1)$ tridiagonal matrix with entries

$$\mathcal{A}_{k,k-1} = \frac{d_k}{d_k + d_{k+1}}, \qquad \mathcal{A}_{k,k} = -\cosh \omega h, \qquad \mathcal{A}_{k,k+1} = \frac{d_{k+1}}{d_k + d_{k+1}}, \tag{10.22}$$

where $d_0 = d_{n+1} = 0$ for the purpose of this definition. To solve the system of Equations (10.21), it is necessary to specify either injected current or membrane potential at each node $P_0, \ldots, P_n$. Once this is done, Equations (10.21) divide into two sets. The first is written and solved for the unknown membrane potentials in terms of known injected currents, and the second determines the unknown injected currents from the now completely determined membrane potentials. Conversely, when a mathematical representation of a cable is given, it may be incomplete in the sense that equations contributed by nodes at which the potential is known do not appear in the description. These equations must be added in a way that is consistent with the representation (10.21) for a cable matrix. They serve to determine the injected current required to sustain the known potentials. The examples in Subsection 10.3.4 and the discussion in Subsection 10.5.3 illustrate how this is achieved.

10.3.3 Symmetrizing a Cable Matrix

Given a cable matrix $\mathcal{A}$ and a nonsingular diagonal matrix $\mathcal{S} = \text{diag}(1, s_1, \ldots, s_n)$, then matrix $T = \mathcal{S}^{-1}\mathcal{A}\mathcal{S}$ is tridiagonal with nonzero entries

$$T_{k,k} = \mathcal{A}_{k,k}, \qquad T_{k,k+1} = \frac{s_k}{s_{k+1}} \mathcal{A}_{k+1,k}, \qquad T_{k+1,k} = \frac{s_{k+1}}{s_k} \mathcal{A}_{k,k+1}, \tag{10.23}$$

whenever the indices conform to the dimensions of T, $\mathcal{A}$, and $\mathcal{S}$. Furthermore, T will be a symmetric matrix provided

$$s_{k+1} = \pm s_k \sqrt{\frac{\mathcal{A}_{k+1,k}}{\mathcal{A}_{k,k+1}}}, \qquad s_0 = 1, \tag{10.24}$$

and in this instance,

$$\mathcal{A}_{k+1,k} = \frac{T_{k,k+1}^2}{\mathcal{A}_{k,k+1}} \tag{10.25}$$

for suitable values of k. The choice of algebraic sign in Equation (10.24) is made to ensure that $\mathcal{A}_{k,k+1}$ is positive whatever the value assigned to $T_{k,k+1}$. Note that the existence of $\mathcal{S}$ in Equation (10.24) relies on the fact that all off-diagonal entries of a cable matrix have the same algebraic sign (positive). By premultiplying Equation (10.21) by $\mathcal{S}^{-1}$, the equation for a cable can be reexpressed in the symmetrized form

$$T(\mathcal{S}^{-1}\widetilde{\mathcal{V}}) = \frac{\sinh \omega h}{\omega} (\mathcal{D}\mathcal{S})^{-1} \widetilde{\mathcal{I}}. \tag{10.26}$$

Suppose on the other hand that a symmetric matrix T is given and it is required to find $\mathcal{S}$ and $\mathcal{A}$. Since the sub- and super-diagonals of T are identical while those of $\mathcal{A}$ are not, it is clear that Equations (10.24) and (10.25) are inadequate to determine $\mathcal{A}$ and $\mathcal{S}$ without further information. This shortfall is met by the requirement that the sum of the off-diagonal entries in each row of a cable matrix is unity, that is,

$$\mathcal{A}_{k,k-1} + \mathcal{A}_{k,k+1} = 1 \tag{10.27}$$

for suitable values of k. Furthermore, a cable matrix $\mathcal{A}$ corresponding to a piecewise uniform cable with n sections has dimension $(n+1) \times (n+1)$ and satisfies $\mathcal{A}_{0,1} = \mathcal{A}_{n,n-1} = 1$.

10.3.3.1 Practical Considerations

In subsequent numerical work, the equivalent cable will be extracted from the symmetric tridiagonal matrix T taking account of the restrictions imposed by the structure of the cable matrix. In order to avoid rounding error in the repeated calculation of $\mathcal{A}_{k,k+1}$ from $\mathcal{A}_{k,k-1}$ via the formula $\mathcal{A}_{k,k+1} = 1 - \mathcal{A}_{k,k+1}$, it is beneficial to satisfy this condition identically by the representation

$$\mathcal{A}_{k,k-1} = \cos^2\theta_k, \qquad \mathcal{A}_{k,k+1} = \sin^2\theta_k. \tag{10.28}$$

With this representation of the entries of the cable matrix, iteration begins with $\theta_0 = \pi/2$ ($\mathcal{A}_{0,1} = 1$) and ends when $\theta_n = 0$ ($\mathcal{A}_{n,n-1} = 1$). Of course, in a numerical calculation the cable section will be deemed to be complete whenever $\theta_n < \epsilon$ where ϵ is a user-supplied small number. The cable itself is constructed from the condition (10.25), expressed in the iterative form

$$\theta_{k+1} = \cos^{-1}\left(\frac{|T_{k,k+1}|}{\sin\theta_k}\right), \qquad \theta_0 = \frac{\pi}{2}. \tag{10.29}$$

The characteristic conductances of the individual cable sections are determined from the definition of $\mathcal{A}_{k,k-1}$ by the iterative formula

$$d_{k+1} = d_k \tan^2\theta_k, \quad d_1 \text{ given.} \tag{10.30}$$

10.3.4 The Simple Y-Junction

Figure 10.2 illustrates a simple Y-junction in which two limbs of characteristic conductances c_1 and c_2 and electrotonic lengths h terminate at P_1 and P_2. The Y-junction is connected to its parent structure at P_0 with current I_0 flowing from the Y-junction to the parent structure.

Let $V_0(s)$, $V_1(s)$, and $V_2(s)$ be the membrane potentials at points P_0, P_1, and P_2, respectively, on the Y-junction of Figure 10.2, then the Laplace transforms of V_0, V_1, and V_2 satisfy the equations

$$\begin{aligned}
-\widetilde{V}_0\cosh\omega h + \frac{c_1}{c_1+c_2}\widetilde{V}_1 + \frac{c_2}{c_1+c_2}\widetilde{V}_2 &= \frac{\sinh\omega h}{(c_1+c_2)\omega}\widetilde{I}_0,\\
\widetilde{V}_0 - \widetilde{V}_1\cosh\omega h &= \frac{\sinh\omega h}{c_1\omega}\widetilde{I}_1,\\
\widetilde{V}_0 - \widetilde{V}_2\cosh\omega h &= \frac{\sinh\omega h}{c_2\omega}\widetilde{I}_2.
\end{aligned} \tag{10.31}$$

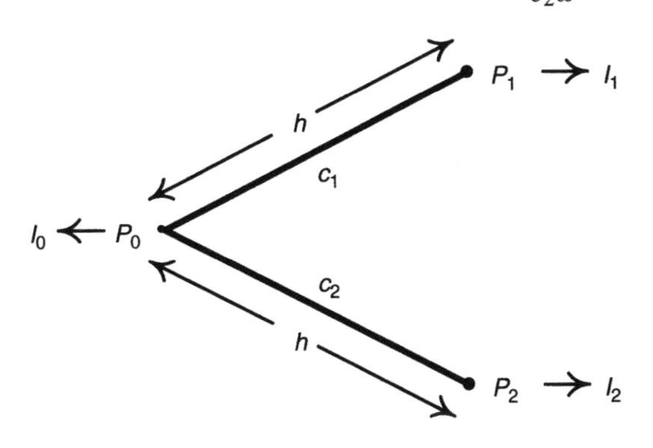

FIGURE 10.2 A simple Y-junction with limbs of equal electrotonic length h but different characteristic conductances c_1 and c_2.

The first of Equations (10.31) is based on conservation of current at the branch point P_0, that is, $I_0(s) + I_L^{(1)}(s) + I_L^{(2)}(s) = 0$ where $I_L^{(1)}$ and $I_L^{(2)}$ are the core currents at the proximal ends of the two limbs. Each is described by a particularized form of Equation (10.14). Similarly, the second and third Equations in (10.31) are particularized forms of Equation (10.15). Equations (10.31) have a matrix form

$$\begin{bmatrix} -\cosh \omega h & \dfrac{c_1}{c_1+c_2} & \dfrac{c_2}{c_1+c_2} \\ 1 & -\cosh \omega h & 0 \\ 1 & 0 & -\cosh \omega h \end{bmatrix} \begin{bmatrix} \widetilde{V}_0 \\ \widetilde{V}_1 \\ \widetilde{V}_2 \end{bmatrix} = \frac{\sinh \omega h}{\omega} \begin{bmatrix} \dfrac{\widetilde{I}_0}{c_1+c_2} \\ \dfrac{\widetilde{I}_1}{c_1} \\ \dfrac{\widetilde{I}_2}{c_2} \end{bmatrix}. \tag{10.32}$$

The purpose of this example is to provide a simple instance of a branched structure with an equivalent cable representation. This example also illustrates how the procedure for constructing equivalent cables may generate cables that are mathematically legitimate but not realized in practice. When Equation (10.32) is premultiplied by the 3×3 matrix

$$\begin{bmatrix} 1 & 0 & 0 \\ 0 & \dfrac{c_1}{c_1+c_2} & \dfrac{c_2}{c_1+c_2} \\ 0 & \dfrac{\sqrt{c_1c_2}}{c_1+c_2} & -\dfrac{\sqrt{c_1c_2}}{c_1+c_2} \end{bmatrix}, \tag{10.33}$$

the resulting system of equation can be expressed in the form

$$\begin{bmatrix} -\cosh \omega h & 1 & 0 \\ 1 & -\cosh \omega h & 0 \\ 0 & 0 & -\cosh \omega h \end{bmatrix} \begin{bmatrix} \widetilde{V}_0 \\ \dfrac{c_1\widetilde{V}_1 + c_2\widetilde{V}_2}{c_1+c_2} \\ \dfrac{\sqrt{c_1c_2}(\widetilde{V}_1 - \widetilde{V}_2)}{c_1+c_2} \end{bmatrix} = \frac{\sinh \omega h}{\omega} \begin{bmatrix} \dfrac{\widetilde{I}_0}{c_1+c_2} \\ \dfrac{\widetilde{I}_1 + \widetilde{I}_2}{c_1+c_2} \\ \dfrac{c_2\widetilde{I}_1 - c_1\widetilde{I}_2}{\sqrt{c_1c_2}\,(c_1+c_2)} \end{bmatrix}. \tag{10.34}$$

Equation (10.34) is the symmetrized form of the original Y-junction. To justify the claim that the Y-junction can be represented by a cable, it is necessary to show that Equation (10.34) can be associated with a cable matrix, and then extract its characteristic conductances and the mapping between injected currents on the original Y-junction and those on the cable.

Note that Equations (10.34) divide naturally into the 2×2 system

$$\begin{bmatrix} -\cosh \omega h & 1 \\ 1 & -\cosh \omega h \end{bmatrix} \begin{bmatrix} \widetilde{V}_0 \\ \dfrac{c_1\widetilde{V}_1 + c_2\widetilde{V}_2}{c_1+c_2} \end{bmatrix} = \frac{\sinh \omega h}{\omega} \begin{bmatrix} \dfrac{\widetilde{I}_0}{c_1+c_2} \\ \dfrac{\widetilde{I}_1 + \widetilde{I}_2}{c_1+c_2} \end{bmatrix} \tag{10.35}$$

and the single equation

$$-\frac{\sqrt{c_1 c_2}(\widetilde{V}_1 - \widetilde{V}_2)}{c_1 + c_2} \cosh \omega h = \frac{\sinh \omega h}{\omega}\left[\frac{c_2 \widetilde{I}_1 - c_1 \widetilde{I}_2}{\sqrt{c_1 c_2}\,(c_1 + c_2)}\right]. \tag{10.36}$$

The process of extracting the equivalent cable can be achieved in this instance by comparing Equations (10.35) and (10.36) in turn with the known form for a cable of length h, namely,

$$\begin{bmatrix} -\cosh \omega h & 1 \\ 1 & -\cosh \omega h \end{bmatrix} \begin{bmatrix} \widetilde{\mathcal{V}}_0 \\ \widetilde{\mathcal{V}}_1 \end{bmatrix} = \frac{\sinh \omega h}{\omega} \begin{bmatrix} \dfrac{\widetilde{\mathcal{I}}_0}{d_1} \\[2mm] \dfrac{\widetilde{\mathcal{I}}_1}{d_1} \end{bmatrix}. \tag{10.37}$$

The identifications

$$\begin{aligned} \mathcal{V}_0(s) &= V_0(s), \qquad & d_1 &= c_1 + c_2, \\ \mathcal{I}_0(s) &= I_0(s), \qquad & \mathcal{I}_1(s) &= I_1(s) + I_2(s) \end{aligned} \tag{10.38}$$

render Equation (10.35) structurally identical to (10.37). The first pair of conditions in (10.38) ensures continuity of membrane potential and conservation of core current between the first section of the equivalent cable and its parent structure. The second pair in (10.38) determines the characteristic conductance of the first section of the equivalent cable and the injected current at its distal end.

When the single Equation (10.36) is compared with the first equation in the generic representation (10.37) of a cable of length h, they are identical provided

$$\begin{aligned} \mathcal{V}_0(s) &= \frac{\sqrt{c_1 c_2}\,(V_1(s) - V_2(s))}{c_1 + c_2}, \qquad & d_1 &= c_1 + c_2, \\ \mathcal{I}_0(s) &= \sqrt{\frac{c_2}{c_1}}(I_1(s) - I_2(s)), \qquad & \mathcal{V}_1(s) &= 0. \end{aligned} \tag{10.39}$$

With this identification, the second equation of (10.37) now specifies the current to be injected at P_1 to maintain zero potential at P_1, the distal end of the cable. Unlike the first cable, however, this cable is not unique. To recognize this nonuniqueness, it is enough to express Equation (10.36) in the equivalent mathematical form

$$-(\widetilde{V}_1 - \widetilde{V}_2) \cosh \omega h = \frac{\sinh \omega h}{\omega\, c_2}\left[\frac{c_2}{c_1}\widetilde{I}_1 - \widetilde{I}_2\right]. \tag{10.40}$$

Direct comparison of this equation with the first equation in the representation (10.37) leads, in this instance, to the identification

$$\begin{aligned} \mathcal{V}_0(s) &= V_1(s) - V_2(s), \qquad & d_1 &= c_2, \\ \mathcal{I}_0(s) &= \frac{c_2}{c_1} I_1(s) - I_2(s), \qquad & \mathcal{V}_1(s) &= 0. \end{aligned} \tag{10.41}$$

Evidently Equation (10.41) is obtained by rescaling Equation (10.39). This means that the characteristic conductance of the second cable is arbitrary, but once it is given a value, the membrane potentials and injected currents on the second section are determined uniquely. This initial non-uniqueness, however, presents no dilemma since the first cable is complete and the second cable is detached from it, and is therefore disconnected from the parent structure.

In conclusion, it has been demonstrated that a Y-junction with limbs of electrotonic length h and current-injected dendritic tips has an equivalent cable of electrotonic length $2h$ consisting of two independent cables, only one of which is connected to the parent structure. If c_1 and c_2 are respectively the characteristic conductances of the limbs of the Y-junction and $I_1(s)$ and $I_2(s)$ are the currents injected at its terminal ends, then the Y-junction has the equivalent cable

$$\begin{aligned} &\text{Section 1} \quad \begin{bmatrix} d_1 = c_1 + c_2 \\ \mathcal{I}_1(s) = I_1(s) + I_2(s), \end{bmatrix} \\ &\text{Section 2} \quad \begin{bmatrix} d_2 = c_1 + c_2 \\ \mathcal{I}_2(s) = (c_1 + c_2)\left[\dfrac{I_1(s)}{c_1} - \dfrac{I_2(s)}{c_2}\right]. \end{bmatrix} \end{aligned} \tag{10.42}$$

This cable is equivalent to the original Y-junction, first, because it preserves continuity of membrane potential and conservation of core current at its point of connection to the parent structure, and second, any configuration of injected currents on the Y-junction defines a unique configuration of injected currents on the cable, and vice versa. While this example identifies all the ingredients of an equivalent cable, its successful construction depended on knowing the matrix (10.33).

10.3.5 Application to a Branched Dendrite

The construction of the model equations for a complex branched dendrite is facilitated by the notion of "parent" and "child" segments. A segment is a parent of all the segments to which it is connected at its distal end, whereas the same segment is a child of the parent segment to which it is connected at its proximal end. The model equations for a branched dendrite are constructed from those of a segment by requiring continuity of potential and conservation of core current at branch points and the point of connection to the parent structure (which may be the soma or a branched dendrite). Continuity of membrane potential between parent and child segments is enforced by ensuring that both have the same potential at their point of contact. The model equation associated with any point is formed by equating the contributions to core current from all the segments meeting at the point to the exogenous current injected at that point. The equation for conservation of core current at any point in a dendrite is a special case of the equation for a branch point.

10.3.5.1 Branch Point

If current $I_B(s)$ is injected into a branch point, then conservation of core current requires that

$$I_B(s) = I_P(s) - \sum I_C(s), \tag{10.43}$$

where $I_P(s)$ and $I_C(s)$ denote respectively the core current in the parent and child segments that meet at the branch point, and summation is taken over all child segments. The branch point condition is constructed from the Laplace transform of Equation (10.43), namely

$$\widetilde{I}_P - \sum \widetilde{I}_C = \widetilde{I}_B = \int_0^\infty I_B(s)\, e^{-sp}\, ds. \tag{10.44}$$

The currents $\widetilde{I}_P$ and $\widetilde{I}_C$ in Equation (10.44) are replaced by

$$\begin{aligned} \frac{\sinh \omega h}{\omega} \widetilde{I}_C &= c^C \widetilde{V}_B \cosh \omega h - c^C \widetilde{V}_C, \\ \frac{\sinh \omega h}{\omega} \widetilde{I}_P &= c^P \widetilde{V}_P - c^P \widetilde{V}_B \cosh \omega h, \end{aligned} \tag{10.45}$$

where V_B is the membrane potential at the branch point, V_C is the membrane potential at the distal end of the first section of a child segment, and V_P is the membrane potential at the proximal end of the last section of the parent segment. The formulae for $\widetilde{I}_C$ and $\widetilde{I}_P$ are particularized forms of Equation (10.14) for $\widetilde{I}_L$ and Equation (10.15) for $\widetilde{I}_R$, respectively. In expressions (10.45), c^C is the characteristic conductance of the first section of the child segment, while c^P is the characteristic conductance of the last section of the parent segment. Substitution of Equations (10.45) into Equation (10.44) gives

$$c^P \widetilde{V}_P - \left(c^P + \sum c^C\right) \widetilde{V}_B \cosh \omega h + \sum c^C \widetilde{V}_C = \frac{\sinh \omega h}{\omega} \widetilde{I}_B, \tag{10.46}$$

where all summations in (10.46) are again taken over the child segments. Equation (10.46) is divided by the sum of the characteristic conductances of all segments meeting at the branch point to give the standardized branch point equation

$$\frac{c^P \widetilde{V}_P}{(c^P + \sum c^C)} - \widetilde{V}_B \cosh \omega h + \frac{\sum c^C \widetilde{V}_C}{(c^P + \sum c^C)} = \frac{\sinh \omega h}{\omega (c^P + \sum c^C)} \widetilde{I}_B. \tag{10.47}$$

10.3.5.2 Connection to Parent Structure

The equation describing the connection of the dendritic tree to its parent structure may be determined directly from the branch-point condition by ignoring all contributions from the parent segment and replacing the injected current I_B by the current I_0 flowing from the dendritic tree into the parent structure. The result of this operation is the equation

$$-\widetilde{V}_0 \cosh \omega h + \frac{\sum c^C \widetilde{V}_C}{\sum c^C} = \frac{\sinh \omega h}{\omega \sum c^C} \widetilde{I}_0, \tag{10.48}$$

where in this instance summation is taken over all segments meeting at P_0, the point of connection of the dendrite to its parent structure.

10.4 Structure of Tree Matrices

The successful construction of equivalent cables for dendritic structures depends critically on the fact that any dendritic structure described by $(n+1)$ nodes has a tree matrix A comprising $(n+1)$ nonzero diagonal entries, one for each node, and $2n$ positive off-diagonal entries distributed symmetrically about the main diagonal of A, giving a total of $3n+1$ nonzero entries. The structural symmetry of A arises from the observation that if node k is connected to node j, then node j is connected to node k. The number of nonzero off-diagonal entries of A is now established by a recursive counting argument that takes advantage of the self-similarity inherent in a branched structure.

10.4.1 Self-Similarity Argument

Any point on a dendritic tree is either an internal point of a segment, a branch point, a dendritic terminal, or the point at which the tree connects to its parent structure. The self-similarity argument starts at the tip of a dendrite and moves towards the point of connection to the parent structure, counting in the process the deficit in off-diagonal entries in A with respect to two entries per point. Internal points of segments generate no deficit whereas each dendritic terminal generates a deficit of one. If N terminal segments and a parent segment meet at a branch point, then the row of A corresponding to that branch point contains $(N + 1)$ off-diagonal entries giving a surplus of $(N - 1)$. Consequently the branch point reduces the total deficit to one and behaves like a dendritic terminal with respect to further counting. The deficit of one is conserved until the point of connection to the parent structure is reached. The parent structure, however, unlike a parent segment, contributes only one point thereby increasing the deficit to two for the entire tree. Thus A contains exactly n pairs of nonzero off-diagonal entries.

10.4.2 Node Numbering

Prior to node numbering, each node is designated as a node at which membrane potential or injected current is to be specified. Without loss of generality, nodes at which injected current is known and the potential is to be found are enumerated first. Nodes at which the potential is known and the injected current is to be found are enumerated subsequently. The point of connection to the parent structure is taken to be P_0. The injected current at P_0 and the core current supplied by the parent structure are identical by conservation of current. Thereafter nodes on any path from P_0 to a dendritic terminal are numbered sequentially, omitting any nodes at which the potential is known. The enumeration scheme jumps to a second path which starts with a node that has a connection to a node on the first path (generally to an existing path) and continues this sequential numbering, omitting nodes at which the potential is known, until a dendritic tip is reached. This procedure is repeated until all paths are exhausted. Only nodes at which the membrane potential is unknown have been enumerated so far. The enumeration scheme is now repeated, this time numbering the nodes at which the potential is known and the injected current is to be determined.

Figure 10.3 illustrates the node-numbering procedure for a simple dendrite when the injected current is known at all nodes and the membrane potential is to be determined. The system of equations

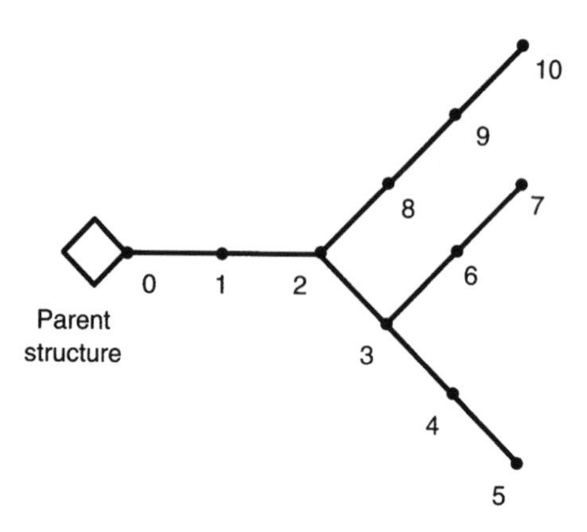

FIGURE 10.3 A dendritic tree with known injected current at all nodes and the membrane potential to be determined.

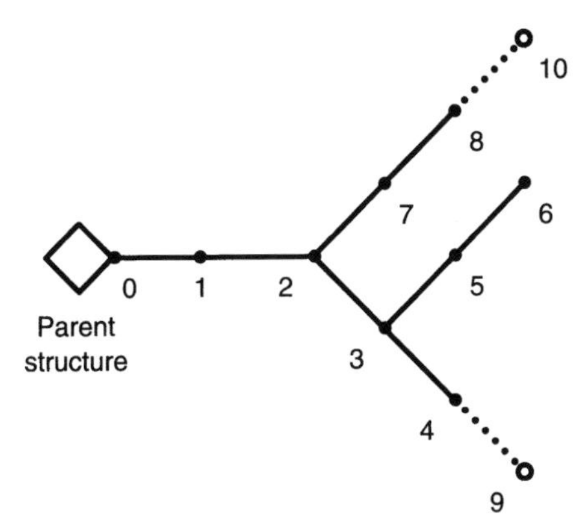

FIGURE 10.4 A dendritic tree with known injected current at all nodes except the two dendritic terminals at which the membrane potential is specified.

associated with this node-numbering scheme is

$$\begin{bmatrix} \blacksquare & \blacksquare & 0 & 0 & 0 & 0 & 0 & 0 & 0 & 0 & 0 \\ \blacksquare & \blacksquare & \blacksquare & 0 & 0 & 0 & 0 & 0 & 0 & 0 & 0 \\ 0 & \blacksquare & \blacksquare & \blacksquare & 0 & 0 & 0 & 0 & \blacksquare & 0 & 0 \\ 0 & 0 & \blacksquare & \blacksquare & \blacksquare & 0 & \blacksquare & 0 & 0 & 0 & 0 \\ 0 & 0 & 0 & \blacksquare & \blacksquare & \blacksquare & 0 & 0 & 0 & 0 & 0 \\ 0 & 0 & 0 & 0 & \blacksquare & \blacksquare & 0 & 0 & 0 & 0 & 0 \\ 0 & 0 & 0 & \blacksquare & 0 & 0 & \blacksquare & \blacksquare & 0 & 0 & 0 \\ 0 & 0 & 0 & 0 & 0 & 0 & \blacksquare & \blacksquare & 0 & 0 & 0 \\ 0 & 0 & \blacksquare & 0 & 0 & 0 & 0 & 0 & \blacksquare & \blacksquare & 0 \\ 0 & 0 & 0 & 0 & 0 & 0 & 0 & 0 & \blacksquare & \blacksquare & \blacksquare \\ 0 & 0 & 0 & 0 & 0 & 0 & 0 & 0 & 0 & \blacksquare & \blacksquare \end{bmatrix} \begin{bmatrix} \mathcal{V}_0 \\ \mathcal{V}_1 \\ \mathcal{V}_2 \\ \mathcal{V}_3 \\ \mathcal{V}_4 \\ \mathcal{V}_5 \\ \mathcal{V}_6 \\ \mathcal{V}_7 \\ \mathcal{V}_8 \\ \mathcal{V}_9 \\ \mathcal{V}_{10} \end{bmatrix} = \frac{\sinh \omega h}{\omega} \begin{bmatrix} \mathcal{I}_0/\mathcal{D}_{0,0} \\ \mathcal{I}_1/\mathcal{D}_{1,1} \\ \mathcal{I}_2/\mathcal{D}_{2,2} \\ \mathcal{I}_3/\mathcal{D}_{3,3} \\ \mathcal{I}_4/\mathcal{D}_{4,4} \\ \mathcal{I}_5/\mathcal{D}_{5,5} \\ \mathcal{I}_6/\mathcal{D}_{6,6} \\ \mathcal{I}_7/\mathcal{D}_{7,7} \\ \mathcal{I}_8/\mathcal{D}_{8,8} \\ \mathcal{I}_9/\mathcal{D}_{9,9} \\ \mathcal{I}_{10}/\mathcal{D}_{10,10} \end{bmatrix}.$$

Figure 10.4 illustrates a node-numbering scheme for the dendrite in Figure 10.3 when terminals 9 and 10 are held at a known potential.

The system of equations associated with this node-numbering scheme is

$$\left[\begin{array}{ccccccccc|cc} \blacksquare & \blacksquare & 0 & 0 & 0 & 0 & 0 & 0 & 0 & 0 & 0 \\ \blacksquare & \blacksquare & \blacksquare & 0 & 0 & 0 & 0 & 0 & 0 & 0 & 0 \\ 0 & \blacksquare & \blacksquare & \blacksquare & 0 & 0 & 0 & \blacksquare & 0 & 0 & 0 \\ 0 & 0 & \blacksquare & \blacksquare & \blacksquare & \blacksquare & 0 & 0 & 0 & 0 & 0 \\ 0 & 0 & 0 & \blacksquare & \blacksquare & 0 & 0 & 0 & 0 & \blacksquare & 0 \\ 0 & 0 & 0 & \blacksquare & 0 & \blacksquare & \blacksquare & 0 & 0 & 0 & 0 \\ 0 & 0 & 0 & 0 & 0 & \blacksquare & \blacksquare & 0 & 0 & 0 & 0 \\ 0 & 0 & \blacksquare & 0 & 0 & 0 & 0 & \blacksquare & \blacksquare & 0 & 0 \\ 0 & 0 & 0 & 0 & 0 & 0 & 0 & \blacksquare & \blacksquare & 0 & \blacksquare \\ \hline 0 & 0 & 0 & 0 & \blacksquare & 0 & 0 & 0 & 0 & \blacksquare & 0 \\ 0 & 0 & 0 & 0 & 0 & 0 & 0 & 0 & \blacksquare & 0 & \blacksquare \end{array}\right] \left[\begin{array}{c} \mathcal{V}_0 \\ \mathcal{V}_1 \\ \mathcal{V}_2 \\ \mathcal{V}_3 \\ \mathcal{V}_4 \\ \mathcal{V}_5 \\ \mathcal{V}_6 \\ \mathcal{V}_7 \\ \mathcal{V}_8 \\ \hline \mathcal{V}_9 \\ \mathcal{V}_{10} \end{array}\right] = \frac{\sinh \omega h}{\omega} \left[\begin{array}{c} \mathcal{I}_0/\mathcal{D}_{0,0} \\ \mathcal{I}_1/\mathcal{D}_{1,1} \\ \mathcal{I}_2/\mathcal{D}_{2,2} \\ \mathcal{I}_3/\mathcal{D}_{3,3} \\ \mathcal{I}_4/\mathcal{D}_{4,4} \\ \mathcal{I}_5/\mathcal{D}_{5,5} \\ \mathcal{I}_6/\mathcal{D}_{6,6} \\ \mathcal{I}_7/\mathcal{D}_{7,7} \\ \mathcal{I}_8/\mathcal{D}_{8,8} \\ \hline \mathcal{I}_9/\mathcal{D}_{9,9} \\ \mathcal{I}_{10}/\mathcal{D}_{10,10} \end{array}\right].$$

In this case, the system of equations described by the upper 9×9 matrix determines the potentials at the nine nodes at which the injected current is given and the potential is unknown, while the last two equations determine the injected currents necessary to sustain the prescribed potentials at nodes 9 and 10.

10.4.3 Symmetrizing the Tree Matrix

It has been shown that a general tree matrix A of dimension $(n+1) \times (n+1)$ has $(n+1)$ diagonal entries and $2n$ positive off-diagonal entries such that $A_{kj} \neq 0$ if and only if $A_{jk} \neq 0$ where $j \neq k$. Given any cable matrix $\mathcal{A}$, recall that it is possible to find a nonsingular diagonal matrix $\mathcal{S}$ such that $\mathcal{S}^{-1}\mathcal{A}\mathcal{S}$ is a symmetric matrix. It is now demonstrated that tree matrices can be symmetrized in the same way. Let $S = \text{diag}(1, s_1, \ldots, s_n)$ be a nonsingular $(n+1) \times (n+1)$ diagonal matrix, then $S^{-1}AS$ is the $(n+1) \times (n+1)$ matrix with entries

$$[S^{-1}AS]_{j,k} = \frac{A_{jk}s_k}{s_j}, \qquad [S^{-1}AS]_{k,j} = \frac{A_{kj}s_j}{s_k}. \tag{10.49}$$

The matrix $S^{-1}AS$ will be symmetric provided there is a matrix S such that the entries of $S^{-1}AS$ satisfy $[S^{-1}AS]_{j,k} = [S^{-1}AS]_{k,j}$ for all $j \neq k$ and $A_{jk} \neq 0$, $A_{kj} \neq 0$. This set of n equations in n unknowns has solution

$$s_k = s_j \sqrt{\frac{A_{kj}}{A_{jk}}}, \tag{10.50}$$

and with this choice of S,

$$[S^{-1}AS]_{j,k} = [S^{-1}AS]_{k,j} = \sqrt{A_{j,k}A_{k,j}}. \tag{10.51}$$

To appreciate that Equation (10.50) determines all the entries of S, it is enough to observe that each point of the tree is connected to at least one other point and so there is no entry of S which does not appear at least in one of Equation (10.50). Since $s_0 = 1$, Equation (10.50) determines uniquely $s_1, \ldots, s_n$ and so every tree matrix can be symmetrized by an appropriate choice of nonsingular diagonal matrix $S = \text{diag}(1, s_1, \ldots, s_n)$.

10.4.4 Concept of the Equivalent Cable

The concept of an equivalent cable comes from the observation that under certain conditions, a symmetric tridiagonal matrix may be interpreted as a cable matrix. The process of constructing an equivalent cable begins by recognizing that for a given tree matrix A, there is a diagonal matrix S such that $S^{-1}AS$ is symmetric, but of course, not tridiagonal. However, it is possible to reduce every symmetric matrix to a symmetric tridiagonal matrix T using the Householder or Lanczos procedures, leading to the possibility that A may be associated with a cable through the interpretation of T as a symmetrized cable matrix. This cable will be equivalent to the original dendritic tree, and for this reason is called the *equivalent cable*. Moreover, the operations which transform the original tree matrix to a cable matrix also indicate how injected current on the tree and cable are related. This construction of equivalent cables is now demonstrated using several examples of trees with simple geometry for which the algebraic manipulations are feasible. For larger trees, however, analytical work is not feasible and so the construction procedure must be implemented numerically.

10.5 Illustrative Examples

For any branched dendrite, the construction of the equivalent cable consists of a sequence of three operations. The tree matrix A is first symmetrized by a suitable choice of S, the symmetrized tree matrix $S^{-1}AS$ is then transformed to a symmetric tridiagonal matrix T, and finally, T is associated with a piecewise uniform cable matrix $\mathcal{A}$ and symmetrizing mapping $\mathcal{S}$. Irrespective of the complexity of the dendritic geometry, the process by which a branched dendrite is transformed to an equivalent cable is exact.

10.5.1 A Simple Asymmetric Y-Junction

Figure 10.5 illustrates an asymmetric Y-junction in which two limbs of electrotonic lengths h and $2h$ meet at P_0. Exogenous currents I_1, I_2, and I_3 are injected at points P_1, P_2, and P_3, while core current I_0 flows from the Y-junction to its parent structure at P_0.

The application of Equation (10.47) at P_0 and Equation (10.17) with particularized forms of Equations (10.14) and (10.15) at P_1, P_2, and P_3 lead to the algebraic equations

$$\begin{aligned}
-\widetilde{V}_0 \cosh \omega h + \frac{c_1}{c_1+c_3}\widetilde{V}_1 + \frac{c_2}{c_1+c_3}\widetilde{V}_3 &= \frac{\sinh \omega h}{\omega\,(c_1+c_3)}\widetilde{I}_0,\\
\frac{c_1}{c_1+c_2}\widetilde{V}_0 - \widetilde{V}_1 \cosh \omega h + \frac{c_2}{c_1+c_2}\widetilde{V}_2 &= \frac{\sinh \omega h}{\omega\,(c_1+c_2)}\widetilde{I}_1,\\
\widetilde{V}_1 - \widetilde{V}_2 \cosh \omega h &= \frac{\sinh \omega h}{\omega\, c_2}\widetilde{I}_2,\\
\widetilde{V}_0 - \widetilde{V}_3 \cosh \omega h &= \frac{\sinh \omega h}{\omega\, c_3}\widetilde{I}_3,
\end{aligned} \tag{10.52}$$

connecting the Laplace transforms of V_0, V_1, V_2, and V_3, the respective membrane potentials at P_0, P_1, P_2, and P_3. These equations have matrix representation

$$A\widetilde{V} = \frac{\sinh \omega h}{\omega} D^{-1}\widetilde{I}, \tag{10.53}$$

where $\widetilde{V}$ is the column vector whose components are the Laplace transforms of the membrane potentials at P_0, P_1, P_2, and P_3, respectively, and $\widetilde{I}$ is the corresponding column vector of Laplace

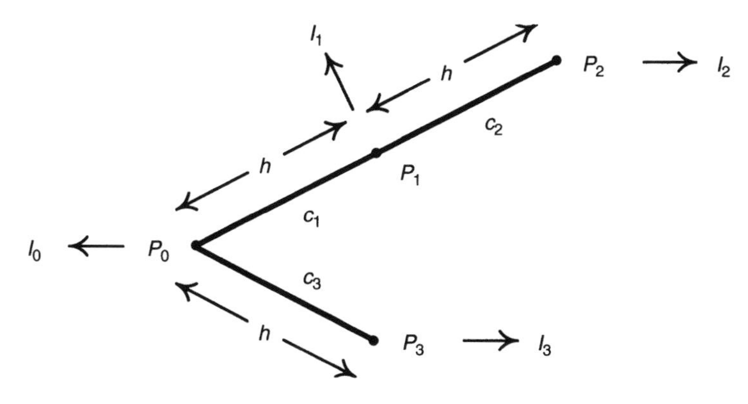

FIGURE 10.5 A Y-junction with limbs of unequal electrotonic length. The sections joining P_0 to P_1, P_1 to P_2, and P_0 to P_3 each have length h and have characteristic conductances c_1, c_2, and c_3, respectively.

transforms of the injected currents. The tree matrix is

$$A = \begin{bmatrix} -\cosh \omega h & \dfrac{c_1}{c_1+c_3} & 0 & \dfrac{c_3}{c_1+c_3} \\ \dfrac{c_1}{c_1+c_2} & -\cosh \omega h & \dfrac{c_2}{c_1+c_2} & 0 \\ 0 & 1 & -\cosh \omega h & 0 \\ 1 & 0 & 0 & -\cosh \omega h \end{bmatrix} \quad (10.54)$$

and D is the diagonal matrix whose (k, k) entry is the sum of the characteristic conductances of the sections which meet at the kth node. In this instance,

$$D = \text{diag}[c_1 + c_3,\ c_1 + c_2,\ c_2,\ c_3]. \quad (10.55)$$

When the procedure described in Section 10.4.3 is applied to the tree matrix A, it can be demonstrated that the diagonal matrix

$$S = \text{diag}\left[1, \sqrt{\frac{c_1+c_3}{c_1+c_2}}, \sqrt{\frac{c_1+c_3}{c_2}}, \sqrt{\frac{c_1+c_3}{c_3}}\right]$$

reduces the tree matrix A to the symmetric matrix

$$S^{-1}AS = \begin{bmatrix} -\cosh \omega h & p & 0 & q \\ p & -\cosh \omega h & r & 0 \\ 0 & r & -\cosh \omega h & 0 \\ q & 0 & 0 & -\cosh \omega h \end{bmatrix}, \quad (10.56)$$

where p, q, and r represent the expressions

$$p = \frac{c_1}{\sqrt{(c_1+c_2)(c_1+c_3)}}, \qquad q = \sqrt{\frac{c_3}{c_1+c_3}}, \qquad r = \sqrt{\frac{c_2}{c_1+c_2}}. \quad (10.57)$$

The first operation of the construction procedure is completed by premultiplying Equation (10.53) with S^{-1} to obtain

$$(S^{-1}AS)S^{-1}\widetilde{V} = \frac{\sinh \omega h}{\omega}(DS)^{-1}\widetilde{I}. \quad (10.58)$$

The second operation requires $S^{-1}AS$ to be transformed into a symmetric tridiagonal matrix. This task is achieved in this example by observing that

$$H = \begin{bmatrix} 1 & 0 & 0 & 0 \\ 0 & \dfrac{p}{\sqrt{p^2+q^2}} & 0 & \dfrac{q}{\sqrt{p^2+q^2}} \\ 0 & 0 & 1 & 0 \\ 0 & \dfrac{q}{\sqrt{p^2+q^2}} & 0 & -\dfrac{p}{\sqrt{p^2+q^2}} \end{bmatrix} \quad (10.59)$$

is an orthogonal symmetric matrix satisfying the property

$$H^{-1}(S^{-1}AS)H = T = \begin{bmatrix} -\cosh\omega h & \sqrt{p^2+q^2} & 0 & 0 \\ \sqrt{p^2+q^2} & -\cosh\omega h & \dfrac{pr}{\sqrt{p^2+q^2}} & 0 \\ 0 & \dfrac{pr}{\sqrt{p^2+q^2}} & -\cosh\omega h & \dfrac{qr}{\sqrt{p^2+q^2}} \\ 0 & 0 & \dfrac{qr}{\sqrt{p^2+q^2}} & -\cosh\omega h \end{bmatrix}. \tag{10.60}$$

For complex branched dendrites, H can be derived in a systematic way as a product of a finite sequence of Householder transformations (see, Golub and Van Loan, 1989; Lindsay et al., 1999), and will be described in a subsequent section. Since T is a symmetric tridiagonal matrix, it may be interpreted as the symmetric form of a cable matrix. The second operation is completed by premultiplying Equation (10.58) by H^{-1} to obtain

$$T(SH)^{-1}\widetilde{V} = \frac{\sinh\omega h}{\omega}(DSH)^{-1}\widetilde{I}. \tag{10.61}$$

Recall from Equation (10.26) that the canonical form for a cable in its symmetric form is

$$T(\mathcal{S}^{-1}\widetilde{\mathcal{V}}) = \frac{\sinh\omega h}{\omega}(\mathcal{DS})^{-1}\widetilde{\mathcal{I}}. \tag{10.62}$$

Equations (10.61) and (10.62) are identical provided membrane potentials and injected currents on the unbranched and branched dendrites are connected by the formulae

$$\mathcal{S}^{-1}\widetilde{\mathcal{V}} = (SH)^{-1}\widetilde{V}, \qquad (\mathcal{DS})^{-1}\widetilde{\mathcal{I}} = (DSH)^{-1}\widetilde{I}. \tag{10.63}$$

Formulae (10.63) relate potentials and injected current on the equivalent cable to those on the branched dendrite. By inversion of the corresponding Laplace transforms, the potentials and injected currents on the cable and tree satisfy the general relations

$$\begin{aligned} \mathcal{V}(s) &= \Psi_V\, V(s), & \Psi_V &= \mathcal{S}HS^{-1}, \\ \mathcal{I}(s) &= \Psi_C\, I(s), & \Psi_C &= \mathcal{D}(\mathcal{S}HS^{-1})D^{-1} = \mathcal{D}\,\Psi_V\, D^{-1}. \end{aligned} \tag{10.64}$$

The matrices Ψ_V and Ψ_C are respectively the voltage and current electrogeometric projection (EGP) matrices. To finalize the construction of the equivalent cable, it remains to calculate $\mathcal{S}$, the equivalent cable matrix $\mathcal{A}$, and the characteristic conductance of each cable section.

10.5.1.1 Extracting the Equivalent Cable

The off-diagonal entries of the equivalent cable matrix $\mathcal{A}$ and the symmetrizing matrix $\mathcal{S}$ which transforms $\mathcal{A}$ into $T = \mathcal{S}^{-1}\mathcal{AS}$ are extracted from T using the algorithm

$$\mathcal{A}_{0,1} = 1, \qquad \mathcal{A}_{k+1,k} = \frac{T^2_{k,k+1}}{\mathcal{A}_{k,k+1}}, \qquad \mathcal{A}_{k,k-1} + \mathcal{A}_{k,k+1} = 1, \tag{10.65}$$

$$s_0 = 1, \qquad s_{k+1} = s_k\sqrt{\frac{\mathcal{A}_{k+1,k}}{\mathcal{A}_{k,k+1}}}, \tag{10.66}$$

whenever the indices conform to the dimensions of T, $\mathcal{A}$, and $\mathcal{S}$. The calculation advances through each row of T as follows:

$$
\begin{aligned}
&\text{Row 1} \quad && \mathcal{A}_{1,0} = \frac{T_{0,1}^2}{\mathcal{A}_{0,1}} = \frac{c_1^2 + c_1c_3 + c_2c_3}{(c_1+c_2)(c_1+c_3)}, \\
& && \mathcal{A}_{1,2} = 1 - \mathcal{A}_{1,0} = \frac{c_1c_2}{(c_1+c_2)(c_1+c_3)}, \\
&\text{Row 2} \quad && \mathcal{A}_{2,1} = \frac{T_{1,2}^2}{\mathcal{A}_{1,2}} = \frac{c_1(c_1+c_3)}{c_1^2 + c_1c_3 + c_2c_3}, \\
& && \mathcal{A}_{2,3} = 1 - \mathcal{A}_{2,1} = \frac{c_2c_3}{c_1^2 + c_1c_3 + c_2c_3}, \\
&\text{Row 3} \quad && \mathcal{A}_{3,2} = \frac{T_{2,3}^2}{\mathcal{A}_{2,3}} = 1.
\end{aligned}
\tag{10.67}
$$

Since $\mathcal{A}_{3,2} = 1$, the last row of $\mathcal{A}$ confirms that the equivalent cable ends on a current-injected terminal. Furthermore, the matrix $\mathcal{S}$ transforming the equivalent cable into its symmetrized representation is

$$
\mathcal{S} = \operatorname{diag}\left[1, \sqrt{\frac{c_1^2 + c_1c_3 + c_2c_3}{(c_1+c_2)(c_1+c_3)}}, \sqrt{\frac{c_1+c_3}{c_2}}, \frac{(c_1+c_3)}{c_2}\sqrt{\frac{c_1^2 + c_1c_3 + c_2c_3}{c_3(c_1+c_3)}}\right]. \tag{10.68}
$$

Now that $\mathcal{A}$ is determined, it is straightforward matrix algebra to demonstrate that the voltage EGP matrix is

$$
\Psi_V = \begin{bmatrix} 1 & 0 & 0 & 0 \\ 0 & \dfrac{c_1}{c_1+c_3} & 0 & \dfrac{c_3}{c_1+c_3} \\ 0 & 0 & 1 & 0 \\ 0 & \dfrac{c_1+c_2}{c_2} & 0 & -\dfrac{c_1}{c_2} \end{bmatrix} \tag{10.69}
$$

and the current EGP matrix is

$$
\Psi_C = \begin{bmatrix} \dfrac{d_1}{c_1+c_3} & 0 & 0 & 0 \\ 0 & \dfrac{c_1(d_1+d_2)}{(c_1+c_3)(c_1+c_2)} & 0 & \dfrac{d_1+d_2}{c_1+c_3} \\ 0 & 0 & \dfrac{d_2+d_3}{c_2} & 0 \\ 0 & \dfrac{d_3}{c_2} & 0 & -\dfrac{c_1d_3}{c_2c_3} \end{bmatrix}. \tag{10.70}
$$

A necessary condition for the equivalence of the branched and unbranched dendrites is that $\mathcal{I}_0(s) = I_0(s)$, that is, both dendrites have the same current flowing into the parent structure. It therefore follows immediately from the first row of equation $\mathcal{I}(s) = \Psi_C I(s)$ that $d_1 = c_1 + c_3$. The definition (10.22) of the cable matrix $\mathcal{A}$ in terms of the characteristic conductances of the cable sections gives

$$\frac{d_1}{d_1+d_2} = \frac{c_1^2+c_1c_3+c_2c_3}{(c_1+c_2)(c_1+c_3)}, \qquad \frac{d_2}{d_2+d_3} = \frac{c_1(c_1+c_3)}{c_1^2+c_1c_3+c_2c_3}, \tag{10.71}$$

from which it follows by straightforward algebra that

$$d_1 = c_1+c_3, \qquad d_2 = \frac{c_1c_2(c_1+c_3)}{c_1^2+c_1c_3+c_2c_3}, \qquad d_3 = \frac{c_2^2c_3}{c_1^2+c_1c_3+c_2c_3}. \tag{10.72}$$

Now that d_1, d_2, and d_3 are known, the matrix Ψ_C can be expressed in terms of the characteristic conductances of the branched dendrite to give

$$\Psi_C = \begin{bmatrix} 1 & 0 & 0 & 0 \\ 0 & \dfrac{c_1(c_1+c_3)}{c_1^2+c_1c_3+c_2c_3} & 0 & \dfrac{(c_1+c_2)(c_1+c_3)}{c_1^2+c_1c_3+c_2c_3} \\ 0 & 0 & 1 & 0 \\ 0 & \dfrac{c_2c_3}{c_1^2+c_1c_3+c_2c_3} & 0 & -\dfrac{c_1c_2}{c_1^2+c_1c_3+c_2c_3} \end{bmatrix}. \tag{10.73}$$

Thus the Y-junction in Figure 10.5 with limbs of electrotonic length $2h$ and h has an equivalent cable of electrotonic length $3h$ consisting of three piecewise uniform sections with characteristic conductances and current distributions

$$\text{Section 1} \quad \left[\begin{array}{l} d_1 = c_1+c_3 \\ \mathcal{I}_1(s) = \dfrac{c_1+c_3}{c_1^2+c_1c_3+c_2c_3}[c_1I_1(s)+(c_1+c_2)I_3(s)], \end{array}\right.$$

$$\text{Section 2} \quad \left[\begin{array}{l} d_2 = \dfrac{c_1c_2(c_1+c_3)}{c_1^2+c_1c_3+c_2c_3} \\ \mathcal{I}_2(s) = I_2(s), \end{array}\right.$$

$$\text{Section 3} \quad \left[\begin{array}{l} d_3 = \dfrac{c_2^2c_3}{c_1^2+c_1c_3+c_2c_3} \\ \mathcal{I}_3(s) = \dfrac{c_2}{c_1^2+c_1c_3+c_2c_3}[c_3I_1(s)-c_1I_3(s)]. \end{array}\right.$$

10.5.2 A Symmetric Y-Junction

In this example, the treatment of the simple symmetric Y-junction considered in Section 10.3.4 is extended to the Y-junction illustrated in Figure 10.6. The two limbs of this dendrite each have electrotonic length $2h$ and meet at point P_0. Currents $I_0(s), \ldots, I_4(s)$ are injected into the branched dendrite at the points $P_0, \ldots, P_4$ respectively.

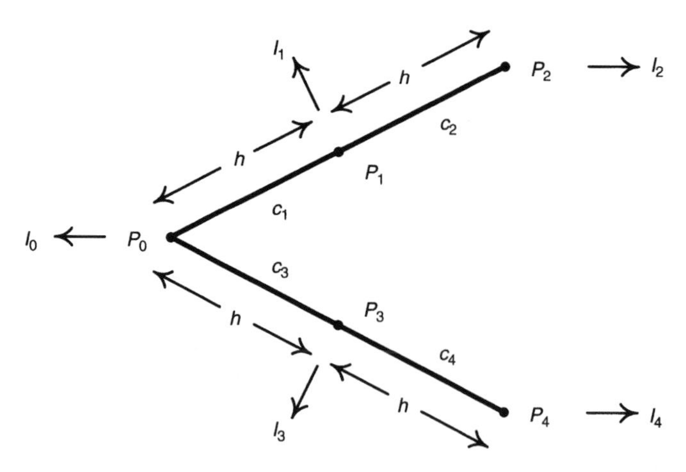

FIGURE 10.6 A Y-junction with limbs of electrotonic length $2h$. The sections joining P_0 to P_1, P_1 to P_2, P_0 to P_3, and P_3 to P_4 each have length h and characteristic conductances c_1, c_2, c_3, and c_4 respectively.

The Laplace transforms of the membrane potentials $V_0(s), \ldots, V_4(s)$ at the points $P_0, \ldots, P_4$ respectively in Figure 10.6 satisfy the algebraic equations

$$
\begin{aligned}
-\widetilde{V}_0 \cosh \omega h + \frac{c_1}{c_1+c_3}\widetilde{V}_1 + \frac{c_2}{c_1+c_3}\widetilde{V}_3 &= \frac{\sinh \omega h}{\omega(c_1+c_3)}\widetilde{I}_0, \\
\frac{c_1}{c_1+c_2}\widetilde{V}_0 - \widetilde{V}_1 \cosh \omega h + \frac{c_2}{c_1+c_2}\widetilde{V}_2 &= \frac{\sinh \omega h}{\omega(c_1+c_2)}\widetilde{I}_1, \\
\widetilde{V}_1 - \widetilde{V}_2 \cosh \omega h &= \frac{\sinh \omega h}{\omega c_2}\widetilde{I}_2, \\
\frac{c_3}{c_3+c_4}\widetilde{V}_0 - \widetilde{V}_3 \cosh \omega h + \frac{c_4}{c_3+c_4}\widetilde{V}_4 &= \frac{\sinh \omega h}{\omega(c_3+c_4)}\widetilde{I}_3, \\
\widetilde{V}_3 - \widetilde{V}_4 \cosh \omega h &= \frac{\sinh \omega h}{\omega c_4}\widetilde{I}_4
\end{aligned}
\tag{10.74}
$$

with matrix representation

$$
A\widetilde{V} = \frac{\sinh \omega h}{\omega} D^{-1}\widetilde{I}, \tag{10.75}
$$

in which $\widetilde{V}$ is the vector of Laplace transforms of the membrane potentials, $\widetilde{I}$ is the vector of Laplace transforms of the injected currents, and the tree matrix A is

$$
A = \begin{bmatrix}
-\cosh \omega h & \dfrac{c_1}{c_1+c_3} & 0 & \dfrac{c_3}{c_1+c_3} & 0 \\
\dfrac{c_1}{c_1+c_2} & -\cosh \omega h & \dfrac{c_2}{c_1+c_2} & 0 & 0 \\
0 & 1 & -\cosh \omega h & 0 & 0 \\
\dfrac{c_3}{c_3+c_4} & 0 & 0 & -\cosh \omega h & \dfrac{c_4}{c_3+c_4} \\
0 & 0 & 0 & 1 & -\cosh \omega h
\end{bmatrix}. \tag{10.76}
$$

The entries of the diagonal matrix D are again the sums of characteristic conductances of sections meeting at each point of the piecewise uniform dendrite. As in the previous example, it can be shown that the diagonal matrix

$$S = \text{diag}\left[1, \sqrt{\frac{c_1+c_3}{c_1+c_2}}, \sqrt{\frac{c_1+c_3}{c_2}}, \sqrt{\frac{c_1+c_3}{c_3+c_4}}, \sqrt{\frac{c_1+c_3}{c_4}}\right] \tag{10.77}$$

has the property

$$S^{-1}AS = \begin{bmatrix} -\cosh\omega h & p & 0 & q & 0 \\ p & -\cosh\omega h & r & 0 & 0 \\ 0 & r & -\cosh\omega h & 0 & 0 \\ q & 0 & 0 & -\cosh\omega h & w \\ 0 & 0 & 0 & w & -\cosh\omega h \end{bmatrix} \tag{10.78}$$

where p, q, r, and w represent the expressions

$$p = \frac{c_1}{\sqrt{(c_1+c_2)(c_1+c_3)}}, \quad q = \frac{c_3}{\sqrt{(c_1+c_3)(c_3+c_4)}}, \quad r = \sqrt{\frac{c_2}{c_1+c_2}}, \quad w = \sqrt{\frac{c_4}{c_3+c_4}}. \tag{10.79}$$

Equation (10.75) is now premultiplied by S^{-1} to reduce it to the symmetric form

$$(S^{-1}AS)S^{-1}\widetilde{V} = \frac{\sinh\omega h}{\omega}(DS)^{-1}\widetilde{I}. \tag{10.80}$$

Since $S^{-1}AS$ is symmetric but not tridiagonal, the construction of the equivalent cable proceeds by tridiagonalizing $S^{-1}AS$. This is achieved by using the symmetric orthogonal matrix

$$H = \begin{bmatrix} 1 & 0 & 0 & 0 & 0 \\ 0 & \frac{p}{\sqrt{p^2+q^2}} & 0 & \frac{q}{\sqrt{p^2+q^2}} & 0 \\ 0 & 0 & \frac{pr}{\sqrt{p^2r^2+q^2w^2}} & 0 & \frac{qw}{\sqrt{p^2r^2+q^2w^2}} \\ 0 & \frac{q}{\sqrt{p^2+q^2}} & 0 & -\frac{p}{\sqrt{p^2+q^2}} & 0 \\ 0 & 0 & \frac{qw}{\sqrt{p^2r^2+q^2w^2}} & 0 & -\frac{pr}{\sqrt{p^2r^2+q^2w^2}} \end{bmatrix}. \tag{10.81}$$

When Equation (10.80) is premultiplied by H^{-1}, the result is

$$T(SH)^{-1}\widetilde{V} = \frac{\sinh\omega h}{\omega}(DSH)^{-1}\widetilde{I}. \tag{10.82}$$

where $T = (SH)^{-1}A(SH)$ is a symmetric tridiagonal matrix

$$\begin{bmatrix} -\cosh \omega h & \sqrt{p^2+q^2} & 0 & 0 & 0 \\ \sqrt{p^2+q^2} & -\cosh \omega h & \sqrt{\dfrac{p^2r^2+q^2w^2}{p^2+q^2}} & 0 & 0 \\ 0 & \sqrt{\dfrac{p^2r^2+q^2w^2}{p^2+q^2}} & -\cosh \omega h & \dfrac{pq(r^2-w^2)}{\sqrt{p^2+q^2}\sqrt{p^2r^2+q^2w^2}} & 0 \\ 0 & 0 & \dfrac{pq(r^2-w^2)}{\sqrt{p^2+q^2}\sqrt{p^2r^2+q^2w^2}} & -\cosh \omega h & \dfrac{rw\sqrt{p^2+q^2}}{\sqrt{p^2r^2+q^2w^2}} \\ 0 & 0 & 0 & \dfrac{rw\sqrt{p^2+q^2}}{\sqrt{p^2r^2+q^2w^2}} & -\cosh \omega h \end{bmatrix}. \tag{10.83}$$

In general, the ratio of successive elements of $\mathcal{S}$, say s_i and s_{i+1}, takes the algebraic sign of $T_{i,i+1}$. In this example, it will be assumed that $r^2 > w^2$ (or equivalently $c_2c_3 - c_1c_4 > 0$) so that the off-diagonal entries of T are all positive and consequently each element of $\mathcal{S}$ is positive. It is clear from expression (10.83) that the structure of T when $r = w$ is significantly different from that when $r \neq w$. In the former, T decomposes into a 3×3 block matrix and a 2×2 block matrix, thereby giving an equivalent cable consisting of two separate cables, only one of which is connected to the parent structure. When $r \neq w$, the equivalent cable consists of a single connected cable.

10.5.2.1 Extracting the Equivalent Cable

The off-diagonal entries of the equivalent cable matrix $\mathcal{A}$ and the symmetrizing matrix $\mathcal{S}$ for which $T = \mathcal{S}^{-1}\mathcal{A}\mathcal{S}$ are extracted from T using the algorithm described in Equations (10.65). The calculation now advances through each row of T on the assumption that $c_1c_4 - c_2c_3 > 0$. The entries of the cable matrix $\mathcal{A}$ are calculated algebraically to obtain

$$\begin{aligned}
&\text{Row 1} \quad \mathcal{A}_{1,0} = \frac{T_{0,1}^2}{\mathcal{A}_{0,1}} = \frac{c_1^2(c_3+c_4)+c_3^2(c_1+c_2)}{(c_1+c_2)(c_1+c_3)(c_3+c_4)}, \\
&\qquad\quad \mathcal{A}_{1,2} = 1 - \mathcal{A}_{1,0} = \frac{c_1c_2(c_3+c_4)+c_3c_4(c_1+c_2)}{(c_1+c_2)(c_1+c_3)(c_3+c_4)}, \\
&\text{Row 2} \quad \mathcal{A}_{2,1} = \frac{T_{1,2}^2}{\mathcal{A}_{1,2}} = \frac{(c_1+c_3)[c_1^2c_2(c_3+c_4)^2+c_3^2c_4(c_1+c_2)^2]}{[c_1^2(c_3+c_4)+c_3^2(c_1+c_2)][c_1c_2(c_3+c_4)+c_3c_4(c_1+c_2)]}, \\
&\qquad\quad \mathcal{A}_{2,3} = 1 - \mathcal{A}_{2,1} = \frac{c_1c_3(c_2c_3-c_1c_4)^2}{[c_1^2(c_3+c_4)+c_3^2(c_1+c_2)][c_1c_2(c_3+c_4)+c_3c_4(c_1+c_2)]}, \\
&\text{Row 3} \quad \mathcal{A}_{3,2} = \frac{T_{2,3}^2}{\mathcal{A}_{2,3}} = \frac{c_1c_3[c_1c_2(c_3+c_4)+c_3c_4(c_1+c_2)]}{c_1^2c_2(c_3+c_4)^2+c_3^2c_4(c_1+c_2)^2}, \\
&\qquad\quad \mathcal{A}_{3,4} = 1 - \mathcal{A}_{3,2} = \frac{c_2c_4[c_1^2(c_3+c_4)+c_3^2(c_1+c_2)]}{c_1^2c_2(c_3+c_4)^2+c_3^2c_4(c_1+c_2)^2}, \\
&\text{Row 4} \quad \mathcal{A}_{4,3} = \frac{T_{3,4}^2}{\mathcal{A}_{3,4}} = 1.
\end{aligned} \tag{10.84}$$

In this case the equivalent cable consists of four sections and terminates in a current injected terminal. The matrix $\mathcal{S}$ mapping the equivalent cable into the symmetrized Y-junction is determined from Equation (10.66) and takes the value

$$\mathcal{S} = \text{diag}\left[1, \sqrt{\frac{c_1^2(c_3+c_4)+c_3^2(c_1+c_2)}{(c_1+c_2)(c_1+c_3)(c_3+c_4)}}, \sqrt{\frac{[c_1+c_3][c_1^2c_2(c_3+c_4)^2+c_3^2c_4(c_1+c_2)^2]}{[c_1c_2(c_3+c_4)+c_3c_4(c_1+c_2)]^2}},\right.$$
$$\left.\sqrt{\frac{[c_1+c_3][c_1^2(c_3+c_4)+c_3^2(c_1+c_2)]}{[c_2c_3-c_1c_4]^2}}, \sqrt{\frac{[c_1+c_3][c_1^2c_2(c_3+c_4)^2+c_3^2c_4(c_1+c_2)^2]}{c_2c_4[c_2c_3-c_1c_4]^2}}\right]. \tag{10.85}$$

Given $\mathcal{S}$, it is straightforward matrix algebra to show that the voltage EGP matrix is

$$\Psi_V = \begin{bmatrix} 1 & 0 & 0 & 0 & 0 \\ 0 & \dfrac{c_1}{c_1+c_3} & 0 & \dfrac{c_3}{c_1+c_3} & 0 \\ 0 & 0 & \dfrac{c_1c_2(c_3+c_4)}{\eta} & 0 & \dfrac{c_3c_4(c_1+c_2)}{\eta} \\ 0 & \dfrac{c_3(c_1+c_2)}{c_2c_3-c_1c_4} & 0 & -\dfrac{c_1(c_3+c_4)}{c_2c_3-c_1c_4} & 0 \\ 0 & 0 & \dfrac{c_3(c_1+c_2)}{c_2c_3-c_1c_4} & 0 & -\dfrac{c_1(c_3+c_4)}{c_2c_3-c_1c_4} \end{bmatrix}, \tag{10.86}$$

where $\eta = c_1c_2(c_3+c_4)+c_3c_4(c_1+c_2)$. The characteristic conductances of the sections of the equivalent cable are determined directly from $\mathcal{A}$ and take the values

$$\begin{aligned} d_1 &= c_1 + c_3, \\ d_2 &= (c_1+c_3)\left[\frac{c_1c_2(c_3+c_4)+c_3c_4(c_1+c_2)}{c_1^2(c_3+c_4)+c_3^2(c_1+c_2)}\right], \\ d_3 &= \frac{c_1c_3\,[c_2c_3-c_1c_4]^2\,[c_1c_2(c_3+c_4)+c_3c_4(c_1+c_2)]}{[c_1^2(c_3+c_4)+c_3^2(c_1+c_2)]\,[c_1^2c_2(c_3+c_4)^2+c_3^2c_4(c_1+c_2)^2]}, \\ d_4 &= \frac{c_2c_4[c_2c_3-c_1c_4]^2}{c_1^2c_2(c_3+c_4)^2+c_3^2c_4(c_1+c_2)^2}. \end{aligned} \tag{10.87}$$

The current EGP matrix $\Psi_C = \mathcal{D}\,\Psi_V\,\mathcal{D}^{-1}$ may be computed from Ψ_V and expressions (10.87).

10.5.3 Special Case: $c_1c_4 = c_2c_3$

In the special case in which $r = w$, or equivalently $c_1c_4 = c_2c_3$, the tridiagonal matrix T in equations

$$T(SH)^{-1}\widetilde{V} = \frac{\sinh \omega h}{\omega}\,(DSH)^{-1}\widetilde{I} \tag{10.88}$$

takes the particularly simple form

$$\left[\begin{array}{ccc|cc} -\cosh\omega h & \sqrt{p^2+q^2} & 0 & 0 & 0 \\ \sqrt{p^2+q^2} & -\cosh\omega h & r & 0 & 0 \\ 0 & r & -\cosh\omega h & 0 & 0 \\ \hline 0 & 0 & 0 & -\cosh\omega h & r \\ 0 & 0 & 0 & r & -\cosh\omega h \end{array}\right]. \tag{10.89}$$

The block diagonal form of T forces the construction of the equivalent cable to proceed in two stages, the first dealing with the tridiagonal matrix T_1 defined by the upper 3×3 block matrix in T, and the second dealing with the tridiagonal matrix T_2 defined by the lower 2×2 matrix in T. Thus Equation (10.88) decomposes into the independent sets of equations

$$T_1 M_1 \widetilde{V} = \frac{\sinh\omega h}{\omega} R_1 \widetilde{I}, \qquad T_2 M_2 \widetilde{V} = \frac{\sinh\omega h}{\omega} R_2 \widetilde{I}, \tag{10.90}$$

in which M_1 and M_2 are respectively the 3×5 and 2×5 matrices

$$\begin{gathered} M_1 = \begin{bmatrix} 1 & 0 & 0 & 0 & 0 \\ 0 & \dfrac{\sqrt{c_1(c_1+c_2)}}{c_1+c_3} & 0 & \dfrac{\sqrt{c_3(c_3+c_4)}}{c_1+c_3} & 0 \\ 0 & 0 & \dfrac{\sqrt{c_1c_2}}{c_1+c_3} & 0 & \dfrac{\sqrt{c_3c_4}}{c_1+c_3} \end{bmatrix}, \\ M_2 = \begin{bmatrix} 0 & \dfrac{\sqrt{c_3(c_1+c_2)}}{c_1+c_3} & 0 & -\dfrac{\sqrt{c_1(c_3+c_4)}}{c_1+c_3} & 0 \\ 0 & 0 & \dfrac{\sqrt{c_2c_3}}{c_1+c_3} & 0 & -\dfrac{\sqrt{c_1c_4}}{c_1+c_3} \end{bmatrix} \end{gathered} \tag{10.91}$$

and R_1 and R_2 are respectively the 3×5 and 2×5 matrices

$$\begin{gathered} R_1 = \frac{1}{c_1+c_3}\begin{bmatrix} 1 & 0 & 0 & 0 & 0 \\ 0 & \sqrt{\dfrac{c_1}{c_1+c_2}} & 0 & \sqrt{\dfrac{c_3}{c_3+c_4}} & 0 \\ 0 & 0 & \sqrt{\dfrac{c_1}{c_2}} & 0 & \sqrt{\dfrac{c_3}{c_4}} \end{bmatrix}, \\ R_2 = \frac{1}{c_1+c_3}\begin{bmatrix} 0 & \sqrt{\dfrac{c_3}{c_1+c_2}} & 0 & -\sqrt{\dfrac{c_1}{c_3+c_4}} & 0 \\ 0 & 0 & \sqrt{\dfrac{c_3}{c_2}} & 0 & -\sqrt{\dfrac{c_1}{c_4}} \end{bmatrix}. \end{gathered} \tag{10.92}$$

The matrices M_1 and M_2 are formed from the first three rows and last two rows respectively of $(SH)^{-1}$, while R_1 and R_2 are likewise formed from the first three rows and last two rows respectively of $(DSH)^{-1}$. Each set of equations in (10.90) represents a different component of the equivalent cable, only one of which is connected to the parent dendrite.

10.5.3.1 Connected Cable

The first component of Equation (10.90) is now examined. The entries of the corresponding cable matrix $\mathcal{A}$ are calculated sequentially to obtain

$$\begin{aligned}
\text{Row 1} \qquad & \mathcal{A}_{1,0} = \frac{T_{0,1}^2}{\mathcal{A}_{0,1}} = \frac{c_1}{c_1+c_2} = \frac{c_3}{c_3+c_4}, \\
& \mathcal{A}_{1,2} = 1 - \mathcal{A}_{1,0} = \frac{c_2}{c_1+c_2} = \frac{c_4}{c_3+c_4}, \\
\text{Row 2} \qquad & \mathcal{A}_{2,1} = \frac{T_{1,2}^2}{\mathcal{A}_{1,2}} = 1,
\end{aligned} \tag{10.93}$$

where $T = T_1$ and the transformation from $\mathcal{A}$ to T is effected with the diagonal matrix

$$\mathcal{S}_1 = \left[1, \sqrt{\frac{c_3}{c_3+c_4}}, \sqrt{\frac{c_3}{c_4}}\right]. \tag{10.94}$$

In this instance the expressions for the voltage and current EGP matrices corresponding to the formulae (10.64) are respectively

$$\begin{aligned}
\Psi_V = \mathcal{S}_1 M_1 &= \begin{bmatrix} 1 & 0 & 0 & 0 & 0 \\ 0 & \dfrac{c_1}{c_1+c_3} & 0 & \dfrac{c_3}{c_1+c_3} & 0 \\ 0 & 0 & \dfrac{c_1}{c_1+c_3} & 0 & \dfrac{c_3}{c_1+c_3} \end{bmatrix}, \\
\Psi_C = \mathcal{D}\mathcal{S}_1 R_1 &= \frac{d_1}{c_1+c_3} \begin{bmatrix} 1 & 0 & 0 & 0 & 0 \\ 0 & 1 & 0 & 1 & 0 \\ 0 & 0 & 1 & 0 & 1 \end{bmatrix}.
\end{aligned} \tag{10.95}$$

Since $\mathcal{I}_0 = I_0$ then $d_1 = c_1 + c_3$ and the expression for $\mathcal{A}_{0,1}$ leads to $d_2 = c_2(c_1+c_3)/c_1$. Cable potentials and injected currents $\mathcal{V}$ and $\mathcal{I}$ are related to tree potentials and currents V and I by the respective formulae $\mathcal{V} = \Psi_V V$ and $\mathcal{I} = \Psi_C I$. These relationships have component form

$$\begin{aligned}
& \mathcal{V}_0 = V_0, && \mathcal{I}_0 = I_0, \\
& \mathcal{V}_1 = \frac{c_1 V_1 + c_3 V_3}{c_1+c_3}, && \mathcal{I}_1 = I_1 + I_3, \\
& \mathcal{V}_2 = \frac{c_1 V_2 + c_3 V_4}{c_1+c_3}, && \mathcal{I}_2 = I_2 + I_4.
\end{aligned} \tag{10.96}$$

10.5.3.2 Detached Cable

The second component of Equation (10.90) is now examined. The entries of the associated cable matrix $\mathcal{A}$ are calculated sequentially to obtain

$$\begin{aligned}
\text{Row 1} \qquad & \mathcal{A}_{1,0} = \frac{T_{0,1}^2}{\mathcal{A}_{0,1}} = \frac{c_2}{c_1+c_2} = \frac{c_4}{c_3+c_4}, \\
& \mathcal{A}_{1,2} = 1 - \mathcal{A}_{1,0} = \frac{c_1}{c_1+c_2} = \frac{c_3}{c_3+c_4},
\end{aligned} \tag{10.97}$$

where $T = T_2$ and the transformation from $\mathcal{A}$ to T is effected with the diagonal matrix

$$\mathcal{S}_2 = \left[1, \sqrt{\frac{c_2}{c_1 + c_2}}\right]. \tag{10.98}$$

In this instance the expressions for the voltage and current EGP matrices corresponding to the formulae (10.64) are respectively

$$\begin{aligned}
\Psi_V = \mathcal{S}_2 M_2 &= \frac{\sqrt{c_3(c_1+c_2)}}{c_1+c_3}\begin{bmatrix} 0 & 1 & 0 & -1 & 0 \\ 0 & 0 & \dfrac{c_2}{c_1+c_2} & 0 & -\dfrac{c_2}{c_1+c_2}\end{bmatrix}, \\
\Psi_C = \mathcal{D}\mathcal{S}_2 R_2 &= \frac{d_1}{c_1+c_3}\frac{\sqrt{c_3(c_1+c_2)}}{c_2}\begin{bmatrix} 0 & \dfrac{c_2}{c_1+c_2} & 0 & -\dfrac{c_2}{c_3+c_4} & 0 \\ 0 & 0 & 1 & 0 & -\dfrac{c_1}{c_3}\end{bmatrix}.
\end{aligned} \tag{10.99}$$

Note that Ψ_V and Ψ_C only give the potential and injected currents at the first two nodes of the detached cable. Specifically,

$$\begin{aligned}
\mathcal{V}_0 &= \frac{\sqrt{c_3(c_1+c_2)}}{c_1+c_3}[V_1 - V_3], \\
\mathcal{V}_1 &= \frac{c_2}{c_1+c_3}\sqrt{\frac{c_3}{c_1+c_2}}[V_2 - V_4], \\
\mathcal{I}_0 &= \frac{d_1}{c_1+c_3}\frac{\sqrt{c_3(c_1+c_2)}}{c_2}\left[\frac{c_2 I_1}{c_1+c_2} - \frac{c_2 I_3}{c_3+c_4}\right], \\
\mathcal{I}_1 &= \frac{d_1}{c_1+c_3}\frac{\sqrt{c_3(c_1+c_2)}}{c_2}\left[I_2 - \frac{c_1}{c_3}I_4\right].
\end{aligned} \tag{10.100}$$

In the expressions for $\mathcal{I}_0$ and $\mathcal{I}_1$, the characteristic conductance of the first section of the detached cable is indeterminate by contrast with the first section of the connected cable for which there is an explicit expression for d_1 determined by the conservation of core current between the parent dendrite and the first section of the equivalent cable.

It is clear from formulae (10.100) that the current and potential at the third node of the detached cable are not given by the EGP matrices Ψ_V and Ψ_C. It can be demonstrated that $\mathcal{V}_2 = 0$ and that the equation contributed by the third node on the detached cable determines the injected current required to maintain $\mathcal{V}_2 = 0$. As previously discussed, this third equation does not appear in the mathematical formulation of the cable based on the determination of unknown potentials. To summarize, although the mathematical description of the detached cable contains three nodes and three unknown functions, only two of these are unknown potentials; the third unknown is an injected current which, of course, does not feature in a matrix representation of the detached cable based on unknown potentials.

The detached cable, in combination with the connected cable constructed in the previous subsection, completes the mapping of injected currents from the branched structure to injected currents on the equivalent cable. In particular, injected current input to the branched structure falling on the detached section of the equivalent cable does not influence the behavior of the parent structure.

10.6 Householder Transformations

Householder's procedure is used in the construction of the equivalent cable to convert the symmetric tree matrix into a symmetric tridiagonal matrix (which is subsequently interpreted as a symmetrized cable matrix). Previous developments of the equivalent cable have been based on the Lanczos procedure. The Householder procedure, on the other hand, enjoys two important benefits over the Lanczos procedure. First, it is numerically stable by contrast with the latter, which is well known to suffer from rounding errors (Golub and Van Loan, 1989). Second, the Lanczos procedure often fails to develop all the components of the symmetric tridiagonal matrix in a single operation by contrast with the Householder procedure, which always develops the complete symmetric tridiagonal matrix.

10.6.1 Householder Matrices

Given any unit column vector U of dimension n, the Householder matrix H (see Golub and Van Loan, 1989) is defined by

$$H = I - 2UU^{\mathrm{T}}, \qquad (10.101)$$

where I is the $n \times n$ identity matrix. By construction, the matrix H is symmetric and orthogonal. While the symmetry of H is obvious, the orthogonality property follows from the calculation

$$\begin{aligned} H^2 &= (I - 2UU^{\mathrm{T}})(I - 2UU^{\mathrm{T}}) \\ &= I - 4UU^{\mathrm{T}} + 4U(U^{\mathrm{T}}U)U^{\mathrm{T}} \\ &= I - 4UU^{\mathrm{T}} + 4UU^{\mathrm{T}} = I. \end{aligned}$$

Thus $H = H^{\mathrm{T}} = H^{-1}$. Given any symmetric $(n+1) \times (n+1)$ tree matrix $S^{-1}AS$, there is a sequence of $(n-1)$ orthogonal matrices $Q_1, Q_2, \ldots, Q_{n-1}$ such that

$$Q_{n-1}^{-1} \cdots Q_1^{-1}(S^{-1}AS)Q_1 \cdots Q_{n-1} = T, \qquad (10.102)$$

where T is a symmetric tridiagonal matrix (see Golub and Van Loan, 1989). To interpret the final tridiagonal form of the tree matrix as a cable attached to the parent structure, it is essential for the Householder procedure to start with the row of the symmetrized tree matrix corresponding to the point of connection (node zero in this analysis) to the parent structure.

Let the orthogonal matrix Q_1 and the symmetric tree matrix $W = S^{-1}AS$ have respective block matrix forms

$$W = \begin{bmatrix} w_{00} & Y^{\mathrm{T}} \\ Y & Z \end{bmatrix}, \qquad Q_1 = \begin{bmatrix} I_1 & 0 \\ 0 & H_1 \end{bmatrix}, \qquad (10.103)$$

where Y is a column vector of dimension n, Z is a symmetric $n \times n$ matrix and H_1 is an $n \times n$ Householder matrix constructed from a unit vector U. Assuming that the first row and column of W are not already in tridiagonal form, the specification of U in the construction of H_1 is motivated by the result

$$Q_1^{-1}WQ_1 = \begin{bmatrix} w_{00} & (H_1Y)^{\mathrm{T}} \\ H_1Y & H_1^{\mathrm{T}}ZH_1 \end{bmatrix}.$$

The vector U is chosen to ensure that all the elements of the column vector H_1Y are zero except the first element. If this is possible, the first row and column of $Q_1^{-1}WQ_1$ will form the first row and column of a tridiagonal matrix. Furthermore, $H_1^{\mathrm{T}} Z H_1$ is itself an $n \times n$ symmetric matrix which assumes the role of W in the next step of the Householder procedure. This algorithm proceeds iteratively for $(n-1)$ steps, finally generating a 2×2 matrix $H_{n-1}^{\mathrm{T}} Z H_{n-1}$ on the last iteration. It can be shown that the choice

$$U = \frac{Y + \alpha\,|Y| E_1}{\sqrt{2\,|Y|\,(|Y| + \alpha\,Y_1)}}, \qquad \begin{aligned} E_1 &= [1, 0, \ldots\ (n-1) \text{ times } \cdots\ ]^{\mathrm{T}}, \\ Y_1 &= Y^{\mathrm{T}} E_1 \end{aligned} \tag{10.104}$$

with $\alpha^2 = 1$ defines an H_1 with the property that $H_1 Y = -\alpha |Y| E_1$, that is, the entries of $H_1 Y$ are all zero except the first entry. This property of H_1 can be established by elementary matrix algebra. The stability of the Householder procedure is guaranteed by setting $\alpha = 1$ if $Y_1 \geq 0$ and $\alpha = -1$ if $Y_1 < 0$, that is, α is conventionally chosen to make αY_1 nonnegative.

Once H_1 is known, the symmetric $n \times n$ matrix $H_1^{\mathrm{T}} Z H_1$ is computed and the entire procedure repeated using the $(n+1) \times (n+1)$ orthogonal matrix Q_2 with block form

$$Q_2 = \begin{bmatrix} I_2 & 0 \\ 0 & H_2 \end{bmatrix},$$

in which H_2 is an $(n-1) \times (n-1)$ Householder matrix. Continued repetition of this procedure generates a sequence of orthogonal matrices $Q_1, Q_2, \ldots, Q_{n-1}$ such that

$$(Q_{n-1} \cdots Q_k \cdots Q_1)(S^{-1} A S)(Q_1 \cdots Q_k \cdots Q_{n-1}) = T, \tag{10.105}$$

where T is a symmetric tridiagonal matrix to be interpreted as the symmetrized matrix of an equivalent cable. In order to construct the mapping of injected current on the branched dendrite to its equivalent cable and vice versa, it is necessary to know the orthogonal matrix $Q = Q_1, Q_2, \ldots, Q_{n-1}$. In practice, this matrix can be computed efficiently by recognizing that the original symmetrized tree matrix can be systematically overwritten as Q is constructed provided the calculation is performed backwards, that is, the calculation begins with Q_{n-1} and ends with Q_1. Using this strategy, it is never necessary to store the Householder matrices $H_1, \ldots, H_{n-1}$.

10.7 Examples of Equivalent Cables

The equivalent cables of four cholinergic interneurons located in laminae III/IV of the dorsal horn of the spinal cord are generated numerically using the Householder procedure. These interneurons are thought to be last-order interneurons involved in presynaptic inhibition (Jankowska, 1992). Whether or not they receive myelinated and/or unmyelinated primary afferent axons is part of a larger study (Olave et al., 2002). The Neurolucida files for our sample were kindly provided by David Maxwell. A full description of these neurons, including the distribution of input, can be found in Olave et al. (2002). The following are the first examples of equivalent cables constructed from real neurons by contrast with those constructed by piecewise empirical methods (e.g., Segev and Burke, 1989).

Figure 10.7 and Figure 10.8 show a spatial representation of the neuron (right) and its electrotonic dendrogram (left) and equivalent cable (below). The equivalent cable of each cell is terminated whenever its diameter first falls below 0.05 μm.

10.7.1 Interneurons Receiving Unmyelinated Afferent Input

Figure 10.7 illustrates two cholinergic interneurons that receive unmyelinated primary afferent input. Cell (A) has total electrotonic length 3.92 eu given by the sum of the electrotonic lengths of the

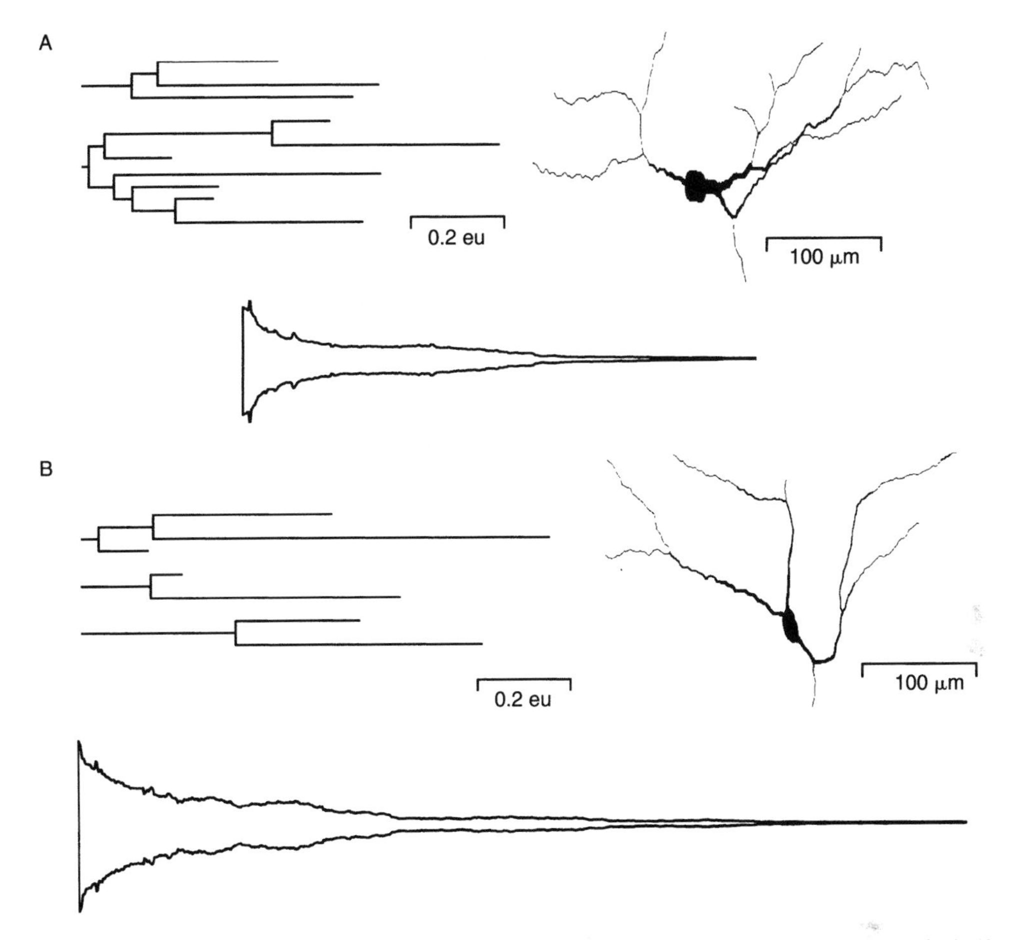

FIGURE 10.7 Two examples of cholinergic interneurons (A) and (B) with their associated electrotonic dendrograms (left) and equivalent cables (below). The dendrograms and equivalent cables are drawn to the same electrotonic scale.

segments in its dendrogram. Its equivalent cable has initial diameter 7.47 μm and electrotonic length 3.92 eu of which the first 1.40 eu is shown in Figure 10.7A. Cell (B) has total electrotonic length 3.36 eu. Its equivalent cable has initial diameter 12.20 μm and electrotonic length 3.36 eu of which the first 1.96 eu is shown in Figure 10.7B.

10.7.2 Neurons Receiving Myelinated Afferent Input

Figure 10.8 illustrates two cholinergic interneurons which receive myelinated primary afferent input. Cell (A) has total electrotonic length 2.08 eu. Its equivalent cable has initial diameter 8.12 μm and electrotonic length 2.08 eu of which the first 1.10 eu is shown in Figure 10.8A. Cell (B) has total electrotonic length 2.95 eu. Its equivalent cable has initial diameter 5.92 μm and electrotonic length 2.95 eu of which the first 2.01 eu is shown in Figure 10.8B. Of the two classes of cells examined, those receiving myelinated afferent input appear to have equivalent cables that are more complex.

These examples illustrate how the presence of structural symmetry in a branched dendrite can have a significant impact on the form of its equivalent cable. The equivalent cables illustrated in Figure 10.7 and Figure 10.8 are structurally different. Vetter et al. (2001) use such differences to predict which cells are most likely to support backpropagated action potentials, and conclude that the equivalent cable gives the best predictor of whether or not a cell supports backpropagation.

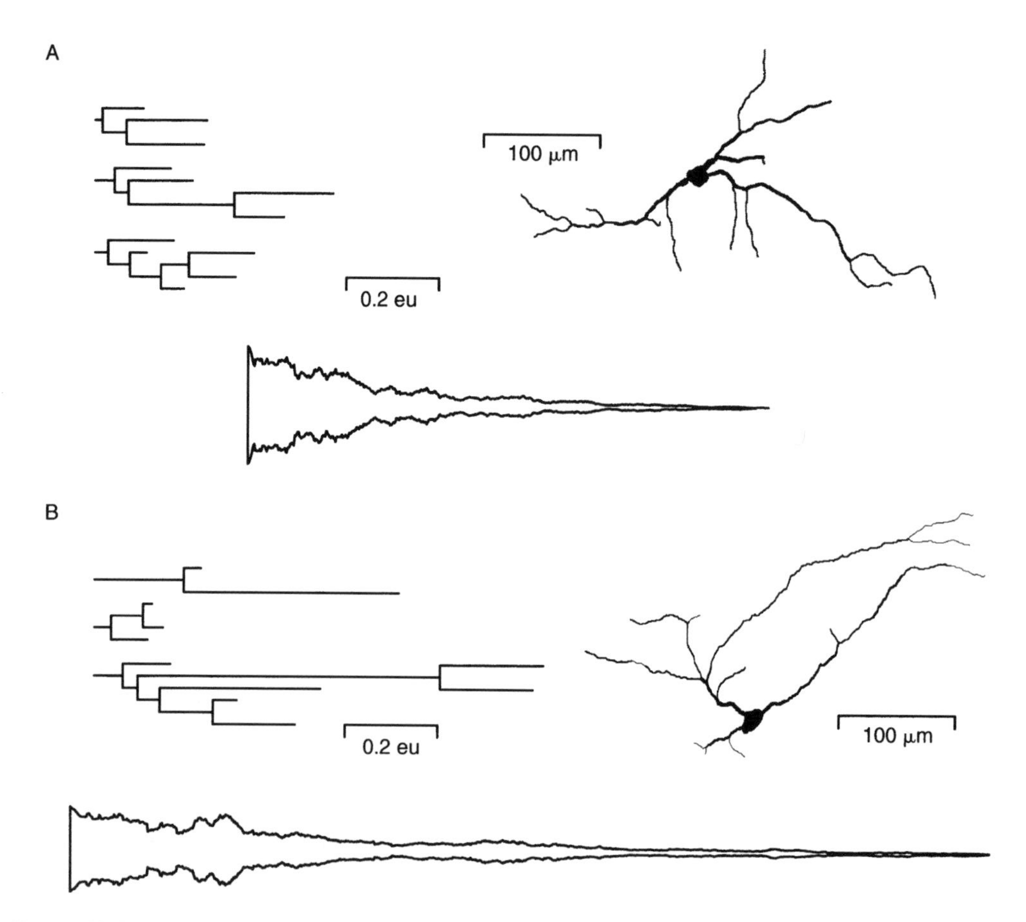

Figure 10.8 Two examples of cholinergic interneurons (A) and (B) with their associated electrotonic dendrograms (left) and equivalent cables (below). The dendrograms and equivalent cables are drawn to the same electrotonic scale.

10.8 Discussion

We have developed a novel analytical procedure that transforms a piecewise-uniform representation of an arbitrarily branched passive dendrite into an unbranched electrically equivalent cable. This equivalent representation of the dendrite is generated about its point of connection to a parent structure, usually the soma of the cell. The equivalent cable consists of a cable connected to the parent structure and often additional electrically isolated cables. Configurations of current on the branched dendrite that map solely onto an isolated cable have no influence on the potential of the soma. Conversely, the electrically isolated sections of the equivalent cable can be used to identify configurations of input on the branched dendrite that exert no influence at the soma. The electrotonic length of the equivalent cable equals the total electrotonic length of the branched dendrite. Isolated cables, when they exist, necessarily have at least one end with known voltage. The shape of the connected component of the equivalent cable is described by its characteristic conductance which, in turn, is determined uniquely by the way in which the morphology and biophysical properties of the original dendrite are combined in the characteristic conductances of its segments.

The construction of equivalent cables for branched dendrites with active membranes require special treatment. For example, an approximate equivalent cable may be constructed for branched dendrites under large-scale synaptic activity (also see Poznanski [2001a]) for conditions under which active dendrites might be incorporated into the equivalent cable framework). To appreciate how this occurs, it is important to recognize that the effects of synaptic activity can be partitioned into several conceptually different components, only one of which is nonlinear and makes a small contribution to the effect of synaptic input relative to the contribution of the other components. Presynaptic spike trains are generally treated as stationary stochastic processes. This implies that the ensuing synaptic conductance changes can be treated as random variables, say $g(t)$, whose means $\bar{g}$ are constant. If V_S is the equilibrium potential associated with a particular synapse and V_M is the resting membrane potential, then the synaptic current at membrane potential V is modeled by $g(t)(V - V_S)$. The partitioning of the synaptic current into linear and nonlinear components can be achieved using the decomposition

$$g(t)(V - V_S) = \underbrace{(g(t) - \bar{g})(V - V_M)}_{\text{(a)}} + \underbrace{\bar{g}(V - V_M)}_{\text{(b)}} + \underbrace{g(t)(V_M - V_S)}_{\text{(c)}}. \qquad (10.106)$$

Component (a) in this partition represents the nonlinear contribution to synaptic current, whereas components (b) and (c) are both linear and can be incorporated exactly into the equivalent cable framework. Each of these contributions to the large-scale synaptic current is now discussed in detail.

The mean value of $(g(t) - \bar{g})$ is zero by definition. Experimental evidence suggests that the mean value of $(V - V_M)$, sampled as a random variable in time, is small since the membrane potential largely fluctuates about V_M, the equilibrium potential (Bernander et al., 1991; Raastad et al., 1998). For example, the data of Bernander et al. (1991, figure 1c) give approximately one millivolt for the expected value of $(V - V_M)$ in the presence of large-scale synaptic input. Both Bernander et al. (1991) and Raastad et al. (1998) indicate that extreme deviations of V from V_M, corresponding to the full nonlinear contribution of the membrane currents, are short lived. For the cell types studied by Bernander et al. (1991, figure 4b) and Raastad et al. (1998, figure 1a), the membrane potential was well away from its average value at most 5% of the time in the presence of large-scale synaptic activity. In addition, the random variables $(g(t) - \bar{g})$ and $(V - V_M)$ are likely to be almost independent, since the latter depends on the global activity in the dendrite while the former is local. Consequently, the mean value of component (a) is expected to be small.

The effect of synaptic input is therefore largely expressed through components (b) and (c). Component (b) describes how the presence of synaptic activity increases membrane conductance. On the basis of this partitioning, the passive model can be expected to give a valid interpretation of dendritic behavior in the presence of large-scale synaptic activity provided membrane conductances are readjusted to take account of the synaptically induced conductance changes. Figure 10.9 illustrates how increasing levels of uniform synaptic activity leave the shape of the equivalent cable unchanged while at the same time increasing its total electrotonic length and characteristic conductance (and consequently its time constant).

These model predictions concerning the effect of large-scale synaptic activity are consistent with the experimental observations of Bernander et al. (1991) and Raastad et al. (1998).

Finally, component (c) behaves as an externally supplied stochastic current and will emphasize the timing of the synaptic inputs. For particular ionic channels (e.g., sodium), the factor $(V_M - V_S)$ may be very large and consequently the timing of inputs may be very important as has been suggested by recent work (Softky, 1994).

The contention of this argument is that a major component of the effect of synaptic input is well described by linear processes which in turn can be incorporated exactly into the framework of the equivalent cable. Furthermore, the procedure for constructing the equivalent cable incorporates the effect of synaptically induced conductance changes in a nonlinear way into both the structure of the cable and the mapping from it to the dendritic tree. Lindsay et al. (2003) describe how the properties of the mapping can be used to quantify the distribution of synaptic contacts on branched dendrites. The equivalent cable can therefore be used to investigate synaptic integration and how

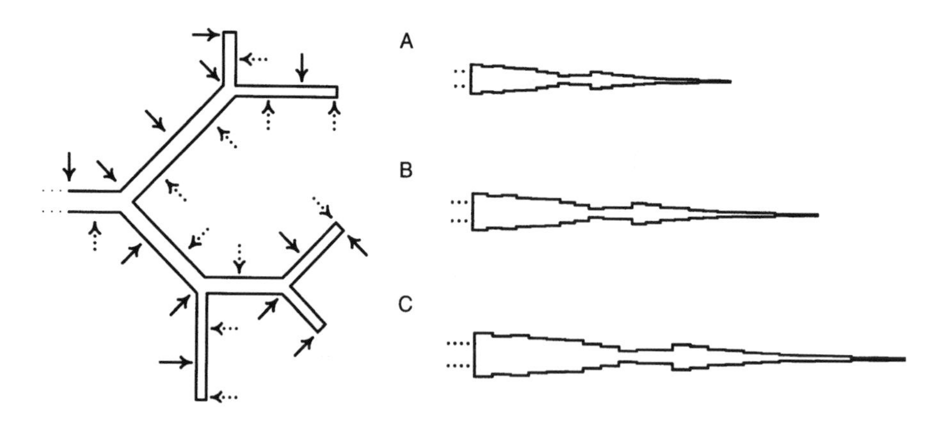

FIGURE 10.9 The effect on the equivalent cable of variable levels of uniform synaptic input on a branched dendrite. The equivalent cable of this dendrite is illustrated (A) in the absence of synaptic input, (B) in response to uniform synaptic input (solid arrows), and (C) in response to a further increase in synaptic activity (dotted arrows). The shape of the equivalent cable is independent of the level of uniform synaptic activity, but its conductance and electrotonic length increase as the level of this activity increases.

dendritic geometry interacts with large-scale synaptic activity to shape the effect of synaptic input at the soma.

ACKNOWLEDGMENTS

David Maxwell's work is supported by a Wellcome Trust project grant (No. 05462). Gayle Tucker is funded by the Future and Emerging Technologies programme (IST-FET) of the European Community, under grant IST-2000-28027 (POETIC). The information provided is the sole responsibility of the authors and does not reflect the Community's opinion. The Community is not responsible for any use that might be made of data appearing in this publication.

PROBLEMS

1. Construct two different dendritic trees which, when reduced to their equivalent cables, have identical connected sections. (Hint: Start with one tree and reduce some of its distal branches to a cable which can then be replaced by an equivalent Rall tree.)
2. Given that it is possible to have two trees with identical connected sections, does this mean that the influence of both trees on the soma is likewise identical? Justify your answer.

   If now you are told that both trees also have the same electrotonic length with isolated sections of the same length, would you wish to change your answer?
3. Start with a Y-junction with uniform limbs of equal electrotonic length, say nh where n is an integer. Now increase the electrotonic length of one limb of the Y-junction by h and decrease the other limb by h (thereby conserving the electrotonic length of the equivalent cable). Describe the changes which take place in the shape of the equivalent cable.
4. Repeat the process of the previous question with mh, where m takes integer values between $m = 2$ and $m = n - 1$, and observe the form of the equivalent cable. How does the presence of isolated sections depend on the relative values of m and n?
5. Show that when the nodes of a branched dendrite are numbered such that the branch point node is the last to be numbered in any section of the tree (Cuthill-McKee enumeration), the structure matrix of that dendrite has an LU factorization in which the nonzero elements of the lower

triangular matrix L fall in the positions of the nonzero elements of the connectivity matrix below the main diagonal, with a like property for the nonzero elements of the upper triangular matrix U.

6. Show how cut terminals in the original dendrite may be incorporated into the construction of the equivalent cable. Repeat Problems 3 and 4 when one or both terminals of the Y-junction are cut.

APPENDIX

The C function `house` performs Householder transformations on a symmetric $n \times n$ matrix `a[ ][ ]` reducing it to a symmetric tridiagonal matrix T whose main diagonal is stored in `d[ ]` and whose super-diagonal (identical to the subdiagonal) is stored in the last $(n - 1)$ entries of `e[ ]`. After tridiagonalization of `a[ ][ ]` is complete, the orthogonal matrix mapping from `a[ ][ ]` into its tridiagonal form T is stored in `a[ ][ ]`. The program further requires three vectors each of length n in addition to the memory storage required for `a[ ][ ]`, `d[ ]`, and `e[ ]`. In particular, the underlying Householder matrices are never stored explicitly. The final mapping from tree matrix to tridiagonal form is constructed by backward multiplication of reduced matrices to minimize memory requirements.

```
void house( int n, long double **a, long double *d, long double *e)
{
    int i, j, k;
    long double beta, g, s, sum, *q, *u, *w;

/*  Allocate two working vectors each of length n */
    q = (long double *) malloc( n*sizeof(long double) );
    u = (long double *) malloc( n*sizeof(long double) );
    w = (long double *) malloc( n*sizeof(long double) );

/*  A total of (n-2) householder steps are required - start on the
    first row of a[ ][ ] and progress to the third last row - the
    last 2 rows already conform to the tridiagonal structure */
    e[0] = 0.0;
    for ( i=0 ; i<n-2 ; i++ ) {
        d[i] = a[i][i];

/*  Determine the magnitude of the working row */
        for ( s=0.0,j=i+1 ; j<n ; j++ ) s += a[i][j]*a[i][j];
        s = sqrt(s);
        if ( a[i][i+1] < 0.0 ) s = -s;
        e[i+1] = -s;
        g = s+a[i][i+1];
        if ( s == 0.0 ) {
            a[i][i] = 1.0;
        } else {
            beta = 1.0/(s*g);
            u[i+1] = g;
            for ( j=i+2 ; j<n ; j++ ) u[j] = a[i][j];
            for ( j=i+1 ; j<n ; j++ ) {
                for ( sum=0.0,k=i+1 ; k<n ; k++ ) sum += a[j][k]*u[k];
                w[j] = sum*beta;
            }
            for ( sum=0.0,j=i+1 ; j<n ; j++ ) sum += u[j]*w[j];
            sum *= 0.5*beta;
            for ( j=i+1 ; j<n ; j++ ) q[j] = w[j]-sum*u[j];
```

```
            for ( j=i+1 ; j<n ; j++ ) {
                for ( k=i+1 ; k<n ; k++ ) a[j][k] -= (q[j]*u[k]+u[j]*q[k]);
            }

/*  Store vector to generate orthogonal matrix */
            a[i][i] = beta;
            for ( j=i+1 ; j<n ; j++ ) a[i][j] = u[j];
        }
    }
    d[n-2] = a[n-2][n-2];
    d[n-1] = a[n-1][n-1];
    e[n-1] = a[n-2][n-1];

/*  Restructure a[ ][ ] to hold product of Householder matrices */
    a[n-2][n-1] = a[n-1][n-2] = 0.0;
    a[n-2][n-2] = a[n-1][n-1] = 1.0;
    for ( i=n-3 ; i>=0 ; i-- ) {
        beta = a[i][i];
        for ( j=i+1 ; j<n ; j++ ) u[j] = a[i][j];
        a[i][i] = 1.0;
        for ( j=i+1 ; j<n ; j++ ) a[i][j] = a[j][i] = 0.0;
        for ( j=i+1 ; j<n ; j++ ) {
            for ( sum=0.0,k=i+1 ; k<n ; k++ ) sum += a[j][k]*u[k];
            w[j] = sum*beta;
        }
        for ( j=i+1 ; j<n ; j++ ) {
            for ( k=i+1 ; k<n ; k++ ) a[j][k] -= u[j]*w[k];
        }
    }
    free(q);
    free(u);
    free(w);
    return;
}
```

11 The Representation of Three-Dimensional Dendritic Structure by a One-Dimensional Model — The Conventional Cable Equation as the First Member of a Hierarchy of Equations

Kenneth A. Lindsay, Jay R. Rosenberg, and Gayle Tucker

CONTENTS

11.1 INTRODUCTION

Dendrites are recognized to be complex three-dimensional structures that convert input spike trains arriving at different locations on the membrane of the dendrite into an output spike train conditioned by the geometry and biophysical properties of the dendrite. Neuroanatomical data of dendritic structure provide the position in space and diameter at all the sampled points of a dendrite. Although these data are three-dimensional, the traditional mathematical model characterizing the spatial and temporal evolution of the transmembrane potential is one-dimensional; that is, the membrane potential is assumed to vary only along the length of the dendrite and not radially. Nevertheless, dendritic current flow must be both axial and radial, the latter being largely associated with transmembrane current.

The basis for the one-dimensional view of a dendrite is motivated by the fact that the length of a typical dendritic section far exceeds its diameter (e.g., see Rall, 1969a), and therefore changes in the transmembrane potential largely take place on the axial length scale and not on the radial length scale. While this is undoubtedly true, currents are proportional to gradients of potential, and small changes of intracellular potential in the radial direction may contribute appreciable currents inasmuch as the radial dimension is small. The primary aim of this chapter is to show how radial variation in intracellular potential can be accommodated within a cable-like mathematical model of a dendritic section. To achieve this it is first necessary to formulate a full three-dimensional model of the evolution of the transmembrane potential, and then to derive from this model an appropriate one-dimensional model for the behavior of the transmembrane potential. Our approach contrasts with the conventional methodology in which the starting point is the one-dimensional model of the dendrite derived by analogy with a transmission line composed of elemental circuits (e.g., see Jack et al., 1988). In this context dendritic taper, for example, is incorporated into the one-dimensional model by taking into account variations in membrane surface area along the dendrite (Rall, 1962b; Jack et al., 1988). The problem with this approach, however, is that it assumes in advance that changes in membrane surface area are the only feature of the dendrite that must be incorporated into the one-dimensional model to account for taper. This approach ignores other features such as variations in radial current that will be shown to be of the same order of magnitude as the effect that changes in membrane area have on the transmembrane potential. Starting with the full three-dimensional model avoids presupposing which features of the biophysics or morphology of the dendrite are necessary for an appropriate one-dimensional approximation.

The three-dimensional model of a dendritic section begins with a full description of the electrostatic potential in the intracellular and extracellular regions of the dendrite together with boundary conditions on the dendritic membrane connecting the gradients of these potentials with the transmembrane current. This procedure leads to a hierarchy of membrane equations, the first of which is the conventional cable equation. The second and higher order members of this hierarchy incorporate progressively more refined features of dendritic geometry. The representation of the transmembrane potential of a dendrite in terms of the solutions of a hierarchy of membrane equations provides guidance as to the magnitude of the errors incurred in the one-dimensional simplification, particularly with respect to the circumstances under which dendritic taper and applied currents can be accommodated successfully within the one-dimensional cable equation, and when they cannot.

11.2 Biophysical Preliminaries

One characteristic feature of dendrites is that they usually exhibit complex branching patterns. The basic unit of these patterns is the dendritic section, defined to be an unbranched element of dendrite connecting two branch points or connecting a branch point to either the soma or the dendritic terminal. Mathematical models of dendritic structure typically assume a mathematical model of the component sections together with joining conditions at dendritic branch points and boundary conditions at dendritic terminals or at the soma. Typical mathematical models for a dendritic section, for example, compartmental models in which a region of dendrite is described by a mathematical compartment, or continuum models in which a dendritic section is described by a partial differential equation such as the cable equation, presume a one-dimensional representation of a dendritic section on the basis of the observation that dendritic sections are usually long and thin. Data describing the morphology of a dendrite typically give dendritic diameter at prescribed distances from branch points and the soma. The nature of these data suggests that dendritic sections are best represented using cylindrical polar coordinates (r, θ, z) with the z-axis directed along the axis of the section and with dendritic membrane (separating the intracellular and extracellular regions) represented as a surface of revolution with equation $r = R(z)$.

11.2.1 Geometrical Considerations

Let $\boldsymbol{e}_r$ and $\boldsymbol{e}_z$ be the (orthogonal) unit base vectors in the radial and axial directions, respectively, for a system of cylindrical polar coordinates (r, θ, z). The membrane separates the intracellular dendritic region $0 \leq r < R(z)$ from the extracellular dendritic region $r > R(z)$. The outward unit normal to the dendritic membrane and the elemental surface area of the membrane at the point with cylindrical polar coordinates $(R(z), \theta, z)$ are, respectively

$$\boldsymbol{n} = \frac{\boldsymbol{e}_r - R_z \boldsymbol{e}_z}{\sqrt{1 + R_z^2}}, \qquad dS = R\sqrt{1 + R_z^2}\, d\theta\, dz\,, \tag{11.1}$$

where R_z is the usual derivative of R with respect to z. Let $\mathcal{A}(z)$ denote the interior of the circle $r = R(z)$ formed by the intersection of the dendrite with a plane of fixed axial coordinate z, and let $\partial\mathcal{A}(z)$ denote the boundary of $\mathcal{A}(z)$, that is, the circle $r = R(z)$. The intracellular volume of dendrite between the planes $z = a$ and $z = b$ is formed by sweeping the area $\mathcal{A}(z)$ from $z = a$ to $z = b$ and will be denoted symbolically by $\mathcal{A} \times (a, b)$. Similarly, the dendritic membrane is the surface formed by sweeping the circle $\partial\mathcal{A}(z)$ from $z = a$ to $z = b$ and will be denoted symbolically by $\partial\mathcal{A} \times (a, b)$. The extracellular region is by definition the region exterior to the surface $\partial\mathcal{A} \times (a, b)$.

11.2.2 Bioelectrical Considerations

In neuronal modeling, the current flow in the intracellular and extracellular regions is assumed to obey Ohms Laws. If $\psi(r, z, t)$ is an electrostatic potential and σ is a conductivity then the current density $\boldsymbol{J}$ is given by

$$\boldsymbol{J} = -\sigma\left(\frac{\partial\psi}{\partial r}, 0, \frac{\partial\psi}{\partial z}\right), \qquad \frac{1}{r}\frac{\partial}{\partial r}\left(r\frac{\partial\psi}{\partial r}\right) + \frac{\partial^2\psi}{\partial z^2} = 0, \tag{11.2}$$

where $\psi = \phi(r, z, t)$ and $\sigma = \sigma_{\mathrm{A}}$ (assumed constant) in the intracellular region, and $\psi = \Phi(r, z, t)$ and $\sigma = \sigma_{\mathrm{E}}$ (assumed constant) in the extracellular region. The presence of the membrane introduces a discontinuity in potential between the intracellular and extracellular regions. This discontinuity is called the *membrane potential* and is denoted here by

$$V_{\mathrm{M}}(z, t) = \phi(R(z), z, t) - \Phi(R(z), z, t) = \phi_{\mathrm{A}} - \Phi_{\mathrm{E}}, \tag{11.3}$$

where $\phi_{\mathrm{A}}(z, t) = \phi(R(z), z, t)$ and $\Phi_{\mathrm{E}}(z, t) = \Phi(R(z), z, t)$.

11.2.3 Specification of the Mathematical Problem

In the absence of intracellular injected current, the specification of the mathematical problem for a dendritic section begins by observing that the current density $\boldsymbol{J}$ satisfies $\operatorname{div} \boldsymbol{J} = 0$ since ϕ and Φ are both solutions of Laplace's equation. Gauss's theorem applied to the volume $\mathcal{A}(z) \times (a, b)$ gives

$$\begin{aligned}\int_{\mathcal{A}(z)\times(a,b)} (\operatorname{div} \boldsymbol{J}) r\, \mathrm{d}r\, \mathrm{d}\theta\, \mathrm{d}z = &-\int_{\mathcal{A}(a)} J_z(r, a, t) r\, \mathrm{d}r\, \mathrm{d}\theta + \int_{\mathcal{A}(b)} J_z(r, b, t) r\, \mathrm{d}r\, \mathrm{d}\theta \\ &+ \int_{\partial\mathcal{A}(z)\times(a,b)} \boldsymbol{J}\cdot\boldsymbol{n} R\sqrt{1 + R_z^2}\, \mathrm{d}\theta\, \mathrm{d}z = 0.\end{aligned} \tag{11.4}$$

On dividing Equation (11.4) by $2h$ with $a = z - h$ and $b = z + h$, and then taking the limit as $h \to 0^+$, it follows that

$$\frac{\partial}{\partial z}\left(\int_0^{R(z)} J_z(r, z, t) r\, \mathrm{d}r\right) + R\sqrt{1 + R_z^2}\, \boldsymbol{J}\cdot\boldsymbol{n} = 0, \tag{11.5}$$

where the integrations with respect to θ have been carried out. This identity will play a fundamental role in the development of the hierarchy of membrane equations described in the introduction, as well as in defining the relationship between the membrane potential and that derived directly from the conventional cable equation.

Further progress with the analysis of Equation (11.5) requires the biophysical properties of the dendritic membrane to be introduced through a constitutive law for the transmembrane current density $J_{\mathrm{M}} = \boldsymbol{J}\cdot\boldsymbol{n}$ (μA/cm^2). In general, there are three contributions to J_{M}, namely, the density of synaptic current, J_{SYN} (μA/cm^2), the density of intrinsic voltage-dependent current, J_{IVDC} (μA/cm^2), arising from ionic channels, and the density of capacitative current due to polarization of the membrane whose lipid bilayer structure causes it to behave locally like a parallel plate capacitor with plates raised to the potential difference of the transmembrane potential, V_{M}. The transmembrane current density due to these processes is

$$J_{\mathrm{M}}(V_{\mathrm{M}}) = c_{\mathrm{M}}\frac{\partial V_{\mathrm{M}}}{\partial t} + J_{\mathrm{SYN}}(V_{\mathrm{M}}) + J_{\mathrm{IVDC}}(V_{\mathrm{M}}), \tag{11.6}$$

where c_{M} (μF/cm^2) is the specific capacitance of the membrane and V_{M} is the transmembrane potential defined in Equation (11.3). In the absence of sources of free charge within the membrane, the current flow normal to the membrane predicted by the intracellular and extracellular potential

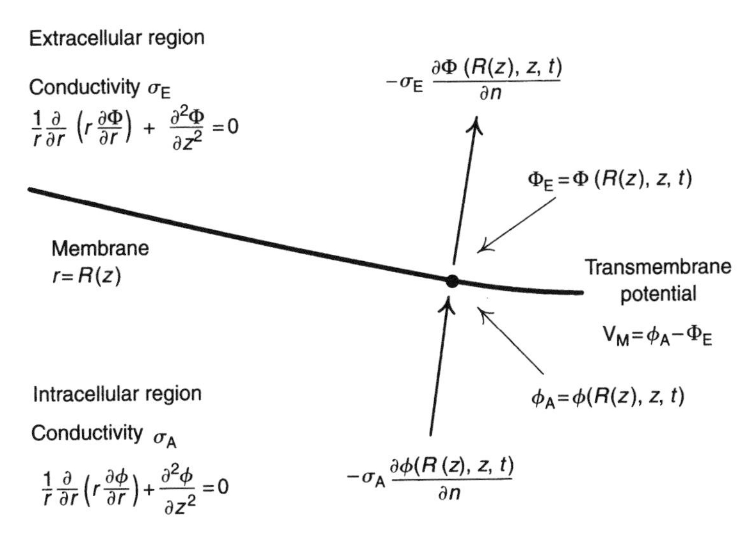

FIGURE 11.1 A diagram of a dendritic membrane bounded by intracellular and extracellular regions. The electrostatic potentials are solutions of Laplace's equation. The transmembrane potential $V_M = \phi_A - \Phi_E$ is the difference in these solutions on the membrane.

gradients at the membrane must equal J_M, the membrane current determined by the transmembrane potential. Conservation of current crossing the membrane gives rise to the boundary conditions

$$J_M(z,t) = -\sigma_A \frac{\partial \phi(R(z),z,t)}{\partial n} = -\sigma_E \frac{\partial \Phi(R(z),z,t)}{\partial n}, \tag{11.7}$$

where $\boldsymbol{n}$ is the unit normal to the membrane directed from the intracellular region to the extracellular region and partial differentiation with respect to $\boldsymbol{n}$ is the usual derivative in the direction of the normal to the membrane. The task is to construct a mathematical description of the transmembrane potential in space and time, bearing in mind that the intracellular and extracellular potentials individually satisfy Laplace's equation. The mathematical problem is presented schematically in Figure 11.1.

The problem posed in dendritic modeling is different from the classical boundary value problems involving Laplace's equation. In a classical application of Laplace's equation, either the potential or the current or a mixture of both is prescribed everywhere on the boundary of the region in which the solution is sought. In neurophysiological modeling, neither the intracellular potential, the extracellular potential, nor the transmembrane current are known on the dendritic membrane. Instead, these functions are determined by requiring continuity of transmembrane current as expressed in Equation (11.7) where the functional form of J_M is prescribed by constitutive formula (11.6).

11.3 IDENTIFICATION OF A ONE-DIMENSIONAL MEMBRANE POTENTIAL

The one-dimensional representation of the three-dimensional dendritic segment given by Equation (11.5) may be further simplified by recalling that $\boldsymbol{J} \cdot \boldsymbol{n} = J_M$, and by replacing J_z with its definition in terms of the axial gradient of the intracellular potential $\phi(r,z,t)$. The result of these substitutions is

$$-\frac{\partial}{\partial z}\left(\int_0^{R(z)} \sigma_A \frac{\partial \phi(r,z,t)}{\partial z} r\,\mathrm{d}r\right) + J_M(V_M)R(z)\sqrt{1+R_z^2} = 0, \tag{11.8}$$

where $V_M = \phi_A(z,t) - \Phi_E(z,t)$ is the transmembrane potential. The problem now is to reconcile Equation (11.8) with the one-dimensional cable equation (see Jack et al., 1988)

$$-\frac{\partial}{\partial z}\left(\sigma_A \frac{R^2(z)}{2}\frac{\partial \mathcal{V}}{\partial z}\right) + J_M(\mathcal{V})R(z)\sqrt{1+R_z^2} = 0, \tag{11.9}$$

where $\mathcal{V}(z,t)$ is the *membrane potential* derived from the one-dimensional treatment of a dendritic segment and $J_M(\mathcal{V})$ is the transmembrane current density at membrane potential $\mathcal{V}$ as defined in Equation (11.6). A comparison of the transmembrane currents in Equations (11.8) and (11.9) requires $\mathcal{V} = V_M$. With this identification, the compatibility of the terms representing the diffusion of axial current in Equations (11.8) and (11.9) must be investigated. To facilitate this comparison, it is convenient to consider the mathematical identity

$$\int_0^{R(z)} \sigma_A \frac{\partial \phi}{\partial z} r\,\mathrm{d}r = \frac{\sigma_A R^2(z)}{2}\frac{\partial V_M}{\partial z} + \int_0^{R(z)} \sigma_A \frac{\partial(\phi - \phi_A)}{\partial z} r\,\mathrm{d}r + \frac{\sigma_A R^2(z)}{2}\frac{\partial \Phi_E}{\partial z}, \tag{11.10}$$

where it is recognized that V_M and Φ_E are defined on the membrane and are therefore independent of r. Compatibility between Equations (11.8) and (11.9) requires the association

$$\frac{\partial}{\partial z}\left(\frac{\sigma_A R^2(z)}{2}\frac{\partial \mathcal{V}}{\partial z}\right) = \frac{\partial}{\partial z}\left(\frac{\sigma_A R^2(z)}{2}\frac{\partial V_M}{\partial z} + \int_0^{R(z)} \sigma_A \frac{\partial(\phi - \phi_A)}{\partial z} r\,\mathrm{d}r + \frac{\sigma_A R^2(z)}{2}\frac{\partial \Phi_E}{\partial z}\right). \tag{11.11}$$

But since $\mathcal{V} = V_M$, Equation (11.11) is only satisfied provided

$$\frac{\partial}{\partial z}\left(\int_0^{R(z)} \sigma_A \frac{\partial(\phi - \phi_A)}{\partial z} r\,\mathrm{d}r + \frac{\sigma_A R^2(z)}{2}\frac{\partial \Phi_E}{\partial z}\right) = 0. \tag{11.12}$$

The first component of the left-hand side of Equation (11.12) describes the discrepancy in intracellular axial current when the true intracellular potential is represented by its value at the inner boundary of the membrane, and the second component on the left-hand side of Equation (11.12) is a correction to the intracellular axial current arising from axial variation of the extracellular potential on the outer boundary of the membrane. In the development of the conventional cable equation, it is implicitly assumed that these terms are negligible. In fact, these terms represent irreducible components in the description of the membrane potential that have no representation in the traditional one-dimensional cable equation. In particular, the first term in expression (11.12) is of the same order of magnitude as the term in the conventional cable equation describing the axial diffusion of intracellular current.

11.4 Development of a Hierarchy of Membrane Equations

One important objective of cable theory in dendritic modeling is to capture the significant features of three-dimensional dendrites in a one-dimensional representation, and to use this representation to explore dendritic behavior. The aim of this section is to show that identity (11.8), in combination with suitable expressions for the intracellular potential ϕ and extracellular potential Φ, leads to an ordered set of one-dimensional equations in which successive members of the hierarchy embed a progressively more refined representation of the membrane potential of the dendrite within a one-dimensional mathematical model (see Section 11.8).

First, recall that the potentials $\phi(r,z,t)$ and $\Phi(r,z,t)$ must be finite-valued solutions of the Laplace equations

$$\frac{\partial^2\phi}{\partial r^2}+\frac{1}{r}\frac{\partial\phi}{\partial r}+\frac{\partial^2\phi}{\partial z^2}=0, \qquad \frac{\partial^2\Phi}{\partial r^2}+\frac{1}{r}\frac{\partial\Phi}{\partial r}+\frac{\partial^2\Phi}{\partial z^2}=0, \tag{11.13}$$

subject to the boundary conditions expressed Equations (11.6) and (11.7), namely that

$$\sigma_{\mathrm{A}}\frac{\partial\phi(R(z),z,t)}{\partial n}=\sigma_{\mathrm{E}}\frac{\partial\Phi(R(z),z,t)}{\partial n}=-J_{\mathrm{M}}(V_{\mathrm{M}}), \tag{11.14}$$

where V_{M} is the transmembrane potential defined in Equation (11.3) and partial differentiation with respect to n is the derivative taken in the direction of the normal to the membrane. Note that even when the conductivity of the extracellular region is very large by comparison with that of the intracellular region (so the potential of the extracellular region is approximately zero), the fact that the intracellular potential ϕ must satisfy Laplace's equation means that the membrane potential $\mathcal{V}(z,t)$ from a one-dimensional model cannot be interpreted as ϕ, because an intracellular potential field that is a function of z and t alone cannot support the radial potential gradients necessary to pass current across the dendritic membrane.

In dendritic geometries, typical radial distances are almost always significantly shorter than typical axial distances. Therefore, a comparison between radial and axial variations in potential must take account of the different distances over which these variations occur. This disparity can best be exploited by scaling the radial coordinate r so that it varies over an interval of length comparable to that of the axial coordinate z. If $\tilde{R}$ is a typical dendritic radius and $L \gg \tilde{R}$ is a typical axial distance[1] along a dendrite, it is beneficial to introduce a scaled radial coordinate $\hat{r}=(L/\tilde{R})r$ with respect to which the membrane has equation $\hat{r}=\hat{R}(z)$. The first of Equations (11.13) then becomes

$$\frac{\partial^2\phi}{\partial \hat{r}^2}+\frac{1}{\hat{r}}\frac{\partial\phi}{\partial \hat{r}}+\varepsilon\frac{\partial^2\phi}{\partial z^2}=0, \qquad \varepsilon=\left(\frac{\tilde{R}}{L}\right)^2, \tag{11.15}$$

in which $\varepsilon \approx 0.01$ for a typical dendrite. Since ε is a small parameter, Equation (11.15) suggests a representation of the intracellular potential field ϕ by the regular perturbation expansion[2]

$$\phi=\phi_0(\hat{r},z,t)+\varepsilon\,\phi_1(\hat{r},z,t)+\varepsilon^2\phi_2(\hat{r},z,t)+\cdots. \tag{11.16}$$

The requirement that ϕ must satisfy Equation (11.15) (i.e., Laplace's equation in cylindrical polar coordinates) for arbitrary ε is replaced by the requirement that the functions $\phi_0, \phi_1, \ldots,$ satisfy

$$\begin{aligned}&\frac{\partial^2\phi_0}{\partial \hat{r}^2}+\frac{1}{\hat{r}}\frac{\partial\phi_0}{\partial \hat{r}}=0,\\ &\frac{\partial^2\phi_k}{\partial \hat{r}^2}+\frac{1}{\hat{r}}\frac{\partial\phi_k}{\partial \hat{r}}=-\frac{\partial^2\phi_{k-1}}{\partial z^2}, \qquad k\geq 1.\end{aligned} \tag{11.17}$$

The choice of scaling factor in the extracellular region is motivated by the need to construct the transmembrane potential as a regular perturbation expansion of the difference between the limiting values of the expansions for the intracellular and extracellular potentials as the membrane is approached from the intracellular and extracellular regions respectively. The validity of this procedure requires that z have the same meaning for the intracellular and extracellular expansions, and that both expansions use the same small parameter ε. Consequently, under the change of radial coordinate $\hat{r} = (L/\tilde{R})r$, the extracellular potential $\Phi(\hat{r}, z)$ satisfies

$$\frac{\partial^2 \Phi}{\partial \hat{r}^2} + \frac{1}{\hat{r}} \frac{\partial \Phi}{\partial \hat{r}} + \varepsilon \frac{\partial^2 \Phi}{\partial z^2} = 0, \tag{11.18}$$

which in turn leads to the regular perturbation expansion

$$\Phi = \Phi_0(\hat{r}, z, t) + \varepsilon \Phi_1(\hat{r}, z, t) + \varepsilon^2 \Phi_2(\hat{r}, z, t) + \cdots \tag{11.19}$$

for the potential in the extracellular region where $\Phi_0, \Phi_1, \ldots,$ satisfy

$$\begin{aligned} &\frac{\partial^2 \Phi_0}{\partial \hat{r}^2} + \frac{1}{\hat{r}} \frac{\partial \Phi_0}{\partial \hat{r}} = 0, \\ &\frac{\partial^2 \Phi_k}{\partial \hat{r}^2} + \frac{1}{\hat{r}} \frac{\partial \Phi_k}{\partial \hat{r}} = -\frac{\partial^2 \Phi_{k-1}}{\partial z^2}, \quad k \geq 1. \end{aligned} \tag{11.20}$$

Furthermore, this rescaling of the radial coordinate imposes a scale length on the infinite extracellular region, a consequence of which is that important variations in the extracellular potential field lie close to the membrane itself. The regular perturbation expansions for the intracellular and extracellular potential fields in turn lead to the regular perturbation expansion

$$\begin{aligned} V_{\text{M}}(z,t) &= V_{\text{M}}^{(0)}(z,t) + \varepsilon V_{\text{M}}^{(1)}(z,t) + \varepsilon^2 V_{\text{M}}^{(2)}(z,t) + \cdots \\ &= \mathcal{V}^{(0)}(z,t) + \mathcal{V}^{(1)}(z,t) + \mathcal{V}^{(2)}(z,t) + \cdots \end{aligned} \tag{11.21}$$

for the transmembrane potential, where

$$\mathcal{V}^{(k)}(z,t) = \varepsilon^k V_{\text{M}}^{(k)}(z,t) = \varepsilon^k (\phi_k(\hat{R}(z), z, t) - \Phi_k(\hat{R}(z), z, t)). \tag{11.22}$$

The membrane potential is determined by solving the families of partial differential Equations (11.17) and (11.20) in their respective domains of definition. The potentials $\phi_0, \phi_1, \ldots,$ and $\Phi_0, \Phi_1, \ldots,$ must be finite-valued and must satisfy the membrane boundary conditions (11.14).

11.4.1 The Membrane Boundary Conditions

To appreciate the role played by the membrane boundary conditions (11.14) in the construction of the intracellular and extracellular potential fields, we start by noting that the normal derivative of the intracellular potential field on the membrane is

$$\frac{\partial \phi}{\partial n} = \frac{\partial \phi / \partial r - R_z (\partial \phi / \partial z)}{\sqrt{1 + R_z^2}} = \frac{1}{\sqrt{\varepsilon}} \frac{\partial \phi / \partial \hat{r} - \varepsilon \hat{R}_z (\partial \phi / \partial z)}{\sqrt{1 + \varepsilon \hat{R}_z^2}}, \tag{11.23}$$

where the partial derivatives on the right-hand side of Equation (11.23) are assumed to be evaluated on $\hat{r} = \hat{R}(z)$. An identical expression exists for $\partial \Phi / \partial n$ with ϕ replaced by Φ. Equation (11.23) may be reorganized immediately to give

$$\begin{aligned} \sigma_{\text{A}} \frac{\partial \phi}{\partial \hat{r}} &= \sqrt{\varepsilon} \sqrt{1 + \varepsilon \hat{R}_z^2}\, \sigma_{\text{A}} \frac{\partial \phi}{\partial n} + \varepsilon \hat{R}_z \sigma_{\text{A}} \frac{\partial \phi}{\partial z} \\ &= -\sqrt{\varepsilon} \sqrt{1 + \varepsilon \hat{R}_z^2}\, J_{\text{M}}(V_{\text{M}}) + \varepsilon \hat{R}_z \sigma_{\text{A}} \frac{\partial \phi}{\partial z}, \end{aligned} \tag{11.24}$$

where $J_M(V_M)$ is the transmembrane current density. Recall that the motivation for rescaling the radial coordinate was to facilitate the comparison of gradients of potential in the axial and radial directions where axial and radial distances are vastly different. A similar type of problem arises in the comparison of axial and membrane current densities. In this context, the magnitude of a typical transmembrane current density induced by a transmembrane potential V_M is compared with that of the axial current density induced by the same potential difference applied over the typical dendritic scale length L. In the case of a cylindrical dendritic segment of radius $\tilde{R}$ with membrane of constant conductance g_M and with intracellular fluid of conductance σ_A, the ratio of transmembrane current density to axial current density is

$$\frac{g_M V_M}{\sigma_A (V_M/L)} = \frac{g_M L}{\sigma_A} = \frac{\tilde{R}}{L}\left(\frac{g_M L^2}{\sigma_A \tilde{R}}\right) = \sqrt{\varepsilon}\left(\frac{g_M L^2}{\sigma_A \tilde{R}}\right). \tag{11.25}$$

Recall that $L = (\tilde{R}\sigma_A/g_M)^{1/2}$ is a typical axial scale length, and thus

$$\frac{\text{Transmembrane current density}}{\text{Axial current density}} = O(\sqrt{\varepsilon}).$$

This fundamental scaling, which must be reflected in the specification of transmembrane current, is expressed through the rescaled constitutive law

$$\begin{aligned} J_M(V_M) &= \sqrt{\varepsilon}\hat{J}_M(V_M) = \sqrt{\varepsilon}\left[\hat{J}_M^{(0)} + \varepsilon \hat{J}_M^{(1)} + \varepsilon^2 \hat{J}_M^{(2)} + \cdots\right] \\ &= \mathcal{J}_M^{(0)} + \mathcal{J}_M^{(1)} + \mathcal{J}_M^{(2)} + \cdots, \end{aligned} \tag{11.26}$$

where $\mathcal{J}_M^{(k)} = \varepsilon^{k+1/2}\hat{J}_M^{(k)}$. In discussing the role of the membrane boundary condition, it is sufficient to work with the specification (11.26), and not necessary to state explicitly how the coefficients in Equation (11.26) are computed from the constitutive formulae for intrinsic voltage-dependent and synaptic current density. Note that the computation of the coefficients of expansions (11.26) may involve a complex calculation because the constitutive law for J_{IVDC} (and perhaps also J_{SYN}) often contains auxiliary variables which in turn have their individual regular perturbation expansions, and that the coefficient $\hat{J}_M^{(k)}$ will not contain elements of the regular perturbation expansion of V_M of higher order than k.

The final specification of the membrane boundary condition for the determination of ϕ is constructed from Equation (11.24) by replacing ϕ and $J_M(V_M)$ by their regular perturbation expansions (11.16) and (11.26) to get

$$\begin{aligned} \sigma_A\left[\frac{\partial\phi_0}{\partial\hat{r}} + \varepsilon\frac{\partial\phi_1}{\partial\hat{r}} + \cdots\right] &= -\varepsilon\sqrt{1+\varepsilon\hat{R}_z^2}\left[\hat{J}_M^{(0)} + \varepsilon\hat{J}_M^{(1)} + \cdots\right] \\ &\quad + \varepsilon\hat{R}_z\sigma_A\left[\frac{\partial\phi_0}{\partial z} + \varepsilon\frac{\partial\phi_1}{\partial z} + \cdots\right]. \end{aligned} \tag{11.27}$$

A coefficient-by-coefficient comparison of the regular perturbation expansions of both sides of Equation (11.27) gives a sequence of conditions, the first two of which are

$$\frac{\partial\phi_0}{\partial\hat{r}} = 0, \qquad \frac{\partial\phi_1}{\partial\hat{r}} = -\frac{\hat{J}_M^{(0)}}{\sigma_A} + \hat{R}_z\frac{\partial\phi_0}{\partial z}. \tag{11.28}$$

In conclusion, the coefficients $\phi_0, \phi_1, \ldots$ are finite-valued solutions of the two-dimensional Poisson Equations (11.17) satisfying gradient boundary conditions, the first two of which are given in

Equations (11.28). A similar treatment applies to the boundary condition satisfied by Φ. By contrast with the determination of $\phi_0, \phi_1, \ldots$ which requires the solution of the hierarchy of membrane equations, the behavior of $\Phi_0, \Phi_1, \ldots$ is entirely determined by the configuration of stimulating currents applied in the extracellular region of the dendrite.

11.4.2 Consistency of Boundary Conditions

It is important to note that the gradient boundary condition, stated for ϕ in Equation (11.24) or on the coefficients of the regular perturbation expansion of ϕ in Equation (11.27), inherently impose restrictions on the behavior of $\phi_0, \phi_1, \ldots$. Consider, for example, the problem of finding a finite solution $\Psi = \Psi(r,t)$ of the Poisson problem

$$\frac{\partial^2 \Psi}{\partial r^2} + \frac{1}{r}\frac{\partial \Psi}{\partial r} = \rho(r), \quad r < a, \qquad \frac{\partial \Psi(a,t)}{\partial r} = Q, \tag{11.29}$$

where a is constant, $\rho(r)$ is a user-supplied function of position, and Q is constant. By multiplying Equation (11.29) by r and integrating the resulting equation with respect to r over $[0, a]$, it follows immediately that $\rho(r)$ and Q satisfy

$$\int_0^a \rho(r) r \, \mathrm{d}r = aQ. \tag{11.30}$$

Equation (11.30) is a simple example of a consistency condition in the sense that it is a condition which must be satisfied by $\rho(r)$ and Q for problem (11.29) to have a solution; the user-supplied choices for $\rho(r)$ and Q are not entirely arbitrary. When Equation (11.28) is considered in the light of this simple example, it can be seen that the sequence of boundary value problems arising in the treatment of the membrane boundary condition require the identification of a sequence of consistency conditions and the subsequent demonstration that these conditions are satisfied.

The task is now to identify the consistency conditions associated with the hierarchy of membrane equations and to verify that they are satisfied. The construction of the consistency condition begins by noting that the differential of arc length along the curve $\partial\hat{A}(z)$ is $\mathrm{d}s = \hat{R}\,\mathrm{d}\theta$. Therefore,

$$\sigma_{\mathrm{A}} \int_{\partial\hat{A}(z)} \frac{\partial \phi}{\partial \hat{r}} \mathrm{d}s = 2\pi\varepsilon \left[R_z \sigma_{\mathrm{A}} \frac{\partial \phi}{\partial z} - \sqrt{1 + \varepsilon R_z^2}\, \hat{J}_{\mathrm{M}}(V_{\mathrm{M}}) \right] \hat{R}, \tag{11.31}$$

where $\partial\phi/\partial\hat{r}$ has been replaced by the right-hand side of Equation (11.24). All the consistency conditions for the hierarchy of membrane equations are embedded in identity (11.31). Gauss's theorem applied to the left-hand side of this identity gives

$$\int_{\partial\hat{A}(z)} \frac{\partial \phi}{\partial \hat{r}} \mathrm{d}s = 2\pi \int_0^{\hat{R}(\theta,z)} \left(\frac{\partial^2 \phi}{\partial \hat{r}^2} + \frac{1}{\hat{r}}\frac{\partial \phi}{\partial \hat{r}} \right) \hat{r}\,\mathrm{d}\hat{r}, \tag{11.32}$$

which in turn reduces the consistency condition (11.31) to the identity

$$\sigma_{\mathrm{A}} \int_0^{\hat{R}(z)} \left(\frac{\partial^2 \phi}{\partial \hat{r}^2} + \frac{1}{\hat{r}}\frac{\partial \phi}{\partial \hat{r}} \right) \hat{r}\,\mathrm{d}\hat{r} = \varepsilon\sigma_{\mathrm{A}} \hat{R}\hat{R}_z \frac{\partial \phi}{\partial z} - \varepsilon \int_{-\pi}^{\pi} \hat{R}\sqrt{1 + \varepsilon \hat{R}_z^2}\, \hat{J}_{\mathrm{M}}(V_{\mathrm{M}})\,\mathrm{d}\theta. \tag{11.33}$$

However,

$$\frac{\partial^2 \phi}{\partial \hat{r}^2} + \frac{1}{\hat{r}}\frac{\partial \phi}{\partial \hat{r}} + \varepsilon \frac{\partial^2 \phi}{\partial z^2} = 0,$$

and so it follows immediately from Equation (11.33) that

$$-\varepsilon\sigma_{\mathrm{A}}\int_0^{\hat{R}(z)}\frac{\partial^2\phi}{\partial z^2}\hat{r}\,\mathrm{d}\hat{r}-\varepsilon\sigma_{\mathrm{A}}\hat{R}\hat{R}_z\frac{\partial\phi}{\partial z}+\varepsilon\hat{R}\sqrt{1+\varepsilon\hat{R}_z^2}\,\hat{J}_{\mathrm{M}}(V_{\mathrm{M}})=0, \tag{11.34}$$

or in terms of the unscaled variables that

$$-\sigma_{\mathrm{A}}\int_0^{R(z)}\frac{\partial^2\phi}{\partial z^2}r\,\mathrm{d}r-\sigma_{\mathrm{A}}RR_z\frac{\partial\phi}{\partial z}+R(z)\sqrt{1+R_z^2}\,J_{\mathrm{M}}(V_{\mathrm{M}})=0. \tag{11.35}$$

The fundamental theorem of calculus asserts that

$$\frac{\partial}{\partial z}\left(\int_0^{R(z)}\frac{\partial\phi}{\partial z}r\,\mathrm{d}r\right)=\int_0^{R(z)}\frac{\partial^2\phi}{\partial z^2}r\,\mathrm{d}r+RR_z\frac{\partial\phi}{\partial z},$$

and therefore the consistency condition (11.35) finally simplifies to

$$-\frac{\partial}{\partial z}\left(\int_0^{R(z)}\sigma_{\mathrm{A}}\frac{\partial\phi}{\partial z}r\,\mathrm{d}r\right)+R(z)\sqrt{1+R_z^2}\,J_{\mathrm{M}}(V_{\mathrm{M}})=0. \tag{11.36}$$

But this equation is known a priori to be satisfied — it is Equation (11.8) expressing conservation of charge — and so the boundary value problems arising from the membrane boundary condition are consistent. The importance of this result is that the coefficient functions $\phi_0, \phi_1, \ldots$ determined by solving Equations (11.17) will automatically satisfy the required consistency condition provided the regular perturbation expansion of Equation (11.36) is satisfied to a suitable order of ε.

11.5 The Membrane Equations

The hierarchy of one-dimensional membrane equations is derived from identity (11.5), namely

$$\frac{\partial}{\partial z}\left(\int_0^{R(z)}J_z(r,z,t)r\,\mathrm{d}r\right)+\boldsymbol{J}\cdot\boldsymbol{n}R(z)\sqrt{1+R_z^2}=0, \tag{11.37}$$

by replacing $\boldsymbol{J}\cdot\boldsymbol{n}$ by J_{M} and by taking advantage of identity (11.10) to obtain

$$-\frac{\partial}{\partial z}\left(\sigma_{\mathrm{A}}\frac{\partial V_{\mathrm{M}}}{\partial z}\frac{R^2(z)}{2}\right)-\frac{\partial}{\partial z}\left(\int_0^{R(z)}\sigma_{\mathrm{A}}\frac{\partial(\phi-\phi_{\mathrm{A}})}{\partial z}r\,\mathrm{d}r\right)-\frac{\partial}{\partial z}\left(\sigma_{\mathrm{A}}\frac{\partial\Phi_{\mathrm{E}}}{\partial z}\frac{R^2(z)}{2}\right)$$
$$+J_{\mathrm{M}}(V_{\mathrm{M}})R(z)\sqrt{1+R_z^2}=0. \tag{11.38}$$

This equation is now rescaled using the change of variable $\hat{r}=Lr/\tilde{R}$ to give

$$-\varepsilon\frac{\partial}{\partial z}\left(\sigma_{\mathrm{A}}\frac{\partial V_{\mathrm{M}}}{\partial z}\frac{\hat{R}^2(z)}{2}\right)-\varepsilon\frac{\partial}{\partial z}\left(\int_0^{\hat{R}(z)}\sigma_{\mathrm{A}}\frac{\partial(\phi-\phi_{\mathrm{A}})}{\partial z}\hat{r}\,\mathrm{d}\hat{r}\right)-\varepsilon\frac{\partial}{\partial z}\left(\sigma_{\mathrm{A}}\frac{\partial\Phi_{\mathrm{E}}}{\partial z}\frac{\hat{R}^2(z)}{2}\right)$$
$$+\varepsilon\hat{J}_{\mathrm{M}}(V_{\mathrm{M}})\hat{R}\sqrt{1+\varepsilon\hat{R}_z^2}=0. \tag{11.39}$$

The hierarchy of one-dimensional equations is constructed from identity (11.39) by replacing ϕ, Φ, V_M, and $\hat{J}_M$ with their regular perturbation expansions. The nth member of the hierarchy, called the nth membrane equation, is obtained by setting the coefficient of ε^n to zero in the regular perturbation expansion of (11.39). For example, the first membrane equation is

$$-\frac{\partial}{\partial z}\left(\sigma_A \frac{\partial V_M^{(0)}}{\partial z}\frac{\hat{R}^2(z)}{2}\right) - \frac{\partial}{\partial z}\left(\int_0^{\hat{R}(z)} \sigma_A \frac{\partial(\phi_0 - \phi_A^{(0)})}{\partial z}\hat{r}\,d\hat{r}\right) - \frac{\partial}{\partial z}\left(\sigma_A \frac{\partial \Phi_E^{(0)}}{\partial z}\frac{\hat{R}^2(z)}{2}\right) + \hat{J}_M^{(0)}\hat{R} = 0, \tag{11.40}$$

where $\phi_A = \phi_A^{(0)} + \varepsilon\phi_A^{(1)} + \cdots$ is the regular perturbation expansion of the intracellular potential on the boundary of the membrane with similar definitions for $\Phi_E^{(0)}, \Phi_E^{(1)} \ldots$. The second membrane equation is

$$-\frac{\partial}{\partial z}\left(\sigma_A \frac{\partial V_M^{(1)}}{\partial z}\frac{\hat{R}^2(z)}{2}\right) - \frac{\partial}{\partial z}\left(\int_0^{\hat{R}(z)} \sigma_A \frac{\partial(\phi_1 - \phi_A^{(1)})}{\partial z}\hat{r}\,d\hat{r}\right) - \frac{\partial}{\partial z}\left(\sigma_A \frac{\partial \Phi_E^{(1)}}{\partial z}\frac{\hat{R}^2(z)}{2}\right) + \left[\hat{J}_M^{(1)} + \frac{R_z^2}{2}\hat{J}_M^{(0)}\right]\hat{R} = 0. \tag{11.41}$$

To proceed further requires the construction of the regular perturbation expansions of ϕ and Φ by sequential integration of Equations (11.17) and (11.20) with appropriate boundary conditions. The process is now demonstrated for the first and second membrane equations.

11.5.1 First Membrane Equation

To complete the formulation of the first membrane equation requires ϕ_0 and Φ_0 to be determined. The function $\phi_A^{(0)}$ is simply the value of ϕ_0 on $\hat{r} = \hat{R}(z)$. Recall that ϕ_0 is the finite solution of the boundary value problem

$$\frac{\partial^2\phi_0}{\partial\hat{r}^2} + \frac{1}{\hat{r}}\frac{\partial\phi_0}{\partial\hat{r}} = 0, \quad \hat{r} \in [0, \hat{R}(z)), \qquad \frac{\partial\phi_0(\hat{R}(z), z, t)}{\partial\hat{r}} = 0. \tag{11.42}$$

Equation (11.42) has solution $\phi_0 = \phi_0(z, t)$, and therefore $\phi_0 - \phi_A^{(0)} = 0$ in the first membrane Equation (11.40) which subsequently becomes

$$-\frac{\partial}{\partial z}\left(\sigma_A \frac{\partial V_M^{(0)}}{\partial z}\frac{\hat{R}^2(z)}{2}\right) + \hat{J}_M^{(0)}\hat{R} - \frac{\partial}{\partial z}\left(\sigma_A \frac{\partial \Phi_E^{(0)}}{\partial z}\frac{\hat{R}^2(z)}{2}\right) = 0 \tag{11.43}$$

in the scaled coordinate system or

$$-\frac{\partial}{\partial z}\left(\sigma_A \frac{\partial \mathcal{V}^{(0)}}{\partial z}\frac{R^2(z)}{2}\right) + \mathcal{J}_M^{(0)}R(z) - \frac{\partial}{\partial z}\left(\sigma_A \frac{\partial \Phi_E^{(0)}}{\partial z}\frac{R^2(z)}{2}\right) = 0 \tag{11.44}$$

in the original unscaled coordinate system where it is understood that $\mathcal{J}_M^{(0)}$ is expressed in terms of $\mathcal{V}_M^{(0)}$ through the constitutive formula

$$\mathcal{J}_M^{(0)} = c_M \frac{\partial \mathcal{V}^{(0)}}{\partial t} + J_{IVDC}(\mathcal{V}^{(0)}) + J_{SYN}(\mathcal{V}^{(0)}). \tag{11.45}$$

The function $\Phi_E^{(0)}$ appearing in Equations (11.43) and (11.44) is the solution of a boundary value problem for the extracellular region in the presence or absence of external stimulation.

The representation of dendritic taper enters the first membrane Equation (11.44) implicitly through the formula for $R(z)$, but not explicitly as would occur, for example, by distinguishing axial length from membrane arc length (see Rall, 1962b; Jack et al., 1988). This distinction can be fully assessed only by considering higher order membrane equations.

11.5.2 Second Membrane Equation

To complete the formulation of the second membrane equation requires ϕ_1 and $\Phi_E^{(1)}$ to be determined. In order to retain mathematical simplicity, the subsequent development of the membrane equations will assume that θ can be ignored in the specification of the intracellular and extracellular potential fields and that the dendritic membrane is a surface of revolution with equations $r = R(z)$ in the unscaled coordinates and $\hat{r} = \hat{R}(z)$ in the scaled coordinates, respectively. Under these circumstances, the second membrane Equation (11.41) becomes

$$-\frac{\partial}{\partial z}\left(\sigma_A \pi \hat{R}^2(z)\frac{\partial V_M^{(1)}}{\partial z}\right) + 2\pi\hat{R}(z)\left[\hat{J}_M^{(1)} + \frac{1}{2}\left(\frac{d\hat{R}}{dz}\right)^2 \hat{J}_M^{(0)}\right] - \frac{\partial}{\partial z}\left(\sigma_A \pi \hat{R}^2(z)\frac{\partial \Phi_E^{(1)}}{\partial z}\right)$$
$$-\frac{\partial}{\partial z}\left(2\pi\int_0^{\hat{R}(z)} \sigma_A \frac{\partial(\phi_1 - \phi_A^{(1)})}{\partial z}\hat{r}\,d\hat{r}\right) = 0, \qquad (11.46)$$

where the functions ϕ_0, $\Phi_E^{(0)}$, $V_M^{(0)}$, and $\hat{J}_M^{(0)}$ are known from the solution of the first membrane equation. Since θ is an ignorable coordinate, the potential ϕ_1 is the finite solution of the gradient boundary value problem

$$\begin{aligned}\frac{\partial^2\phi_1}{\partial\hat{r}^2} + \frac{1}{\hat{r}}\frac{\partial\phi_1}{\partial\hat{r}} &= -\frac{\partial^2\phi_0}{\partial z^2}, \qquad \hat{r}\in[0,\hat{R}(z)),\\ \frac{\partial\phi_1(\hat{R}(z),z,t)}{\partial\hat{r}} &= -\frac{\hat{J}_M^{(0)}}{\sigma_A} + \frac{d\hat{R}}{dz}\frac{\partial\phi_0}{\partial z}.\end{aligned} \qquad (11.47)$$

This equation has general solution

$$\phi_1(\hat{r},z,t) = -\frac{\hat{r}^2}{4}\frac{\partial^2\phi_0}{\partial z^2} + \xi_1(z,t)\log\hat{r} + \eta_1(z,t)$$

in which $\xi_1(z,t)$ and $\eta_1(z,t)$ are unknown functions of z and t. Since ϕ_1 is required to be finite in the intracellular region, then $\xi_1(z,t) = 0$ and ϕ_1 becomes

$$\phi_1(\hat{r},z,t) = -\frac{\hat{r}^2}{4}\frac{\partial^2\phi_0}{\partial z^2} + \eta_1(z,t). \qquad (11.48)$$

Recall that the consistency condition demonstrated in Section 11.4.2 guarantees that ϕ_1 satisfies the boundary condition on $\hat{r} = \hat{R}(z)$ provided ϕ_0 is a solution of the first membrane equation. Since ϕ_0 is known, ϕ_1 is now known, and the focus of the argument shifts to the computation of the last term in Equation (11.46). Ignoring multiplicative constants, the steps in the computation of this term are

$$\int_0^{\hat{R}(z)} \frac{\partial(\phi_1 - \phi_A^{(1)})}{\partial z} \hat{r}\,\mathrm{d}\hat{r} = \frac{1}{4}\int_0^{\hat{R}(z)} \frac{\partial}{\partial z}\left[(\hat{R}^2(z) - \hat{r}^2)\frac{\partial^2 \phi_0}{\partial z^2}\right]\hat{r}\,\mathrm{d}\hat{r}$$
$$= \frac{1}{4}\int_0^{\hat{R}(z)} \left[(\hat{R}^2(z) - \hat{r}^2)\frac{\partial^3 \phi_0}{\partial z^3} + 2\hat{R}(z)\frac{\mathrm{d}\hat{R}}{\mathrm{d}z}\frac{\partial^2 \phi_0}{\partial z^2}\right]\hat{r}\,\mathrm{d}\hat{r}$$
$$= \frac{1}{4}\left[\frac{\hat{R}^4(z)}{4}\frac{\partial^3 \phi_0}{\partial z^3} + \hat{R}^3(z)\frac{\mathrm{d}\hat{R}}{\mathrm{d}z}\frac{\partial^2 \phi_0}{\partial z^2}\right]$$
$$= \frac{1}{16}\frac{\partial}{\partial z}\left[\hat{R}^4(z)\frac{\partial^2 \phi_0}{\partial z^2}\right]. \tag{11.49}$$

The result of this calculation is now used to replace the fourth term on the left-hand side of Equation (11.46) to get

$$-\frac{\partial}{\partial z}\left(\sigma_A \pi \hat{R}^2(z)\frac{\partial V_M^{(1)}}{\partial z}\right) + 2\pi\hat{R}(z)\left[\hat{J}_M^{(1)} + \frac{1}{2}\left(\frac{\mathrm{d}\hat{R}}{\mathrm{d}z}\right)^2 \hat{J}_M^{(0)}\right] - \frac{\partial}{\partial z}\left(\sigma_A \pi \hat{R}^2(z)\frac{\partial \Phi_E^{(1)}}{\partial z}\right)$$
$$-\frac{\pi\sigma_A}{8}\frac{\partial^2}{\partial z^2}\left[\hat{R}^4(z)\frac{\partial^2 \phi_0}{\partial z^2}\right] = 0. \tag{11.50}$$

The final step in the simplification of Equation (11.50) begins by noting that the definition of transmembrane potential ensures that $V_M^{(0)} + \Phi_E^{(0)} = \phi_0$, from which it follows that Equation (11.44) can be rewritten as

$$2\pi\hat{R}(z)\hat{J}_M^{(0)} = \frac{\partial}{\partial z}\left(\sigma_A \pi \hat{R}^2(z)\frac{\partial \phi_0}{\partial z}\right).$$

This identity is now used to replace $\hat{J}_M^{(0)}$ in Equation (11.50) to give the final version

$$-\frac{\partial}{\partial z}\left(\sigma_A \pi \hat{R}^2(z)\frac{\partial V_M^{(1)}}{\partial z}\right) + 2\pi\hat{R}(z)\hat{J}_M^{(1)} - \frac{\partial}{\partial z}\left(\sigma_A \pi \hat{R}^2(z)\frac{\partial \Phi_E^{(1)}}{\partial z}\right)$$
$$+\frac{\pi\sigma_A}{2}\left[\left(\frac{\mathrm{d}\hat{R}(z)}{\mathrm{d}z}\right)^2 \frac{\partial}{\partial z}\left(\hat{R}^2(z)\frac{\partial \phi_0}{\partial z}\right) - \frac{1}{4}\frac{\partial^2}{\partial z^2}\left(\hat{R}^4(z)\frac{\partial^2 \phi_0}{\partial z^2}\right)\right] = 0 \tag{11.51}$$

of the second membrane equation when expressed in the scaled coordinate system or

$$-\frac{\partial}{\partial z}\left(\sigma_A \pi R^2(z)\frac{\partial \mathcal{V}^{(1)}}{\partial z}\right) + 2\pi R(z)\mathcal{J}_M^{(1)} - \varepsilon\frac{\partial}{\partial z}\left(\sigma_A \pi R^2(z)\frac{\partial \Phi_E^{(1)}}{\partial z}\right)$$
$$+\frac{\pi\sigma_A}{2}\left[\left(\frac{\mathrm{d}R(z)}{\mathrm{d}z}\right)^2 \frac{\partial}{\partial z}\left(R^2(z)\frac{\partial \phi_0}{\partial z}\right) - \frac{1}{4}\frac{\partial^2}{\partial z^2}\left(R^4(z)\frac{\partial^2 \phi_0}{\partial z^2}\right)\right] = 0 \tag{11.52}$$

in terms of the unscaled coordinates. Clearly the solution of Equation (11.52) requires the solution of the first membrane equation. Furthermore, the first and second terms of Equation (11.52) take the same forms as the first and second terms of the first membrane equation. The last term in

Equation (11.52), namely

$$\frac{\pi\sigma_A}{2}\left(\frac{dR(z)}{dz}\right)^2\frac{\partial}{\partial z}\left(R^2(z)\frac{\partial\phi_0}{\partial z}\right)-\frac{\pi\sigma_A}{8}\frac{\partial^2}{\partial z^2}\left(R^4(z)\frac{\partial^2\phi_0}{\partial z^2}\right), \tag{11.53}$$

merits special comment. The first component of expression (11.53) incorporates explicitly the effect of dendritic taper (as opposed to implicitly as in the first membrane equation), and is exactly what might be anticipated as a consequence of changes in membrane area due to taper (see Rall, 1969b; Jack et al., 1988). However, what cannot be anticipated is the presence and importance of the variation of the intracellular potential with radial coordinate. The second component of expression (11.53) takes account of this effect, but unlike the contribution from dendritic taper, there is no obvious physical basis in traditional one-dimensional cable theory from which this term can be constructed. Moreover, it is clear from expression (11.53) that the correction for dendritic taper and that for the variation in the intracellular potential with radial coordinate have the same order of magnitude. Consequently, generalizations of the conventional cable equation based solely on the former (see Rall, 1969b; Jack et al., 1988) are incomplete. It is important to recognize that even in the absence of taper, the effect of a changing intracellular potential with radial coordinate remains, and the addition of a contribution from taper simply adds a term of the same order of magnitude.

Finally, in order to facilitate the numerical treatment of Equation (11.52), it is convenient to introduce the auxiliary function

$$\psi(z,t)=\frac{\partial}{\partial z}\left(\pi R^2(z)\frac{\partial\phi_0(z,t)}{\partial z}\right) \tag{11.54}$$

with respect to which the second membrane Equation (11.52) takes the modified form

$$\begin{aligned}&-\frac{\partial}{\partial z}\left(\sigma_A\pi R^2(z)\frac{\partial\mathcal{V}^{(1)}(z,t)}{\partial z}\right)+2\pi R(z)\mathcal{J}_M^{(1)}-\varepsilon\frac{\partial}{\partial z}\left(\sigma_A\pi R^2(z)\frac{\partial\Phi_E^{(1)}}{\partial z}\right)\\&+\frac{\sigma_A}{4}\left[2\left(\frac{dR}{dz}\right)^2\psi(z)-\frac{1}{2}\frac{\partial}{\partial z}\left(R^2\frac{\partial\psi}{\partial z}\right)+\frac{\partial}{\partial z}\left(\pi R^2(z)\frac{d}{dz}\left(R\frac{dR}{dz}\right)\frac{\partial\phi_0}{\partial z}\right)\right]=0.\end{aligned} \tag{11.55}$$

11.5.3 Computation of Axial Current

The construction of a mathematical model of a dendrite from the hierarchy of membrane equations requires regular perturbation expansions for the total intracellular axial current at segment end points. In the presence of strong symmetry for which the intracellular and extracellular potential fields are functions of $\hat{r}$ and z, the total axial current in the intracellular region is

$$I(z,t)=-2\sigma_A\pi\int_0^{R(z)}\frac{\partial\phi}{\partial z}r\,dr=-2\varepsilon\sigma_A\pi\int_0^{\hat{R}(z)}\frac{\partial\phi}{\partial z}\hat{r}\,d\hat{r}.$$

In the light of identity (11.10), this current becomes

$$I(z,t)=-\varepsilon\sigma_A\pi\hat{R}^2(z)\frac{\partial V_M}{\partial z}-2\varepsilon\sigma_A\pi\int_0^{\hat{R}(z)}\frac{\partial(\phi-\phi_A)}{\partial z}\hat{r}\,d\hat{r}-\varepsilon\sigma_A\pi\hat{R}^2(z)\frac{\partial\Phi_E}{\partial z}. \tag{11.56}$$

The regular perturbation expansion of $I(z,t)$ in the scaled coordinate system is derived from Equation (11.56) by replacing each term with its regular perturbation expansion in ε to get

$$I(z,t)=\mathcal{I}^{(0)}(z,t)+\mathcal{I}^{(1)}(z,t)+\mathcal{I}^{(2)}(z,t)+\cdots,\qquad \mathcal{I}^{(k)}(z,t)=O(\varepsilon^{k+1}). \tag{11.57}$$

The first term in expression (11.57) is

$$\mathcal{I}^{(0)}(z,t) = -\sigma_{\mathrm{A}}\pi R^2(z)\frac{\partial \mathcal{V}^{(0)}}{\partial z} - \sigma_{\mathrm{A}}\pi R^2(z)\frac{\partial \Phi_{\mathrm{E}}^{(0)}}{\partial z}, \tag{11.58}$$

while the second term of expansion (11.57) takes the form

$$\mathcal{I}^{(1)}(z,t) = -\sigma_{\mathrm{A}}\pi R^2(z)\frac{\partial \mathcal{V}^{(1)}}{\partial z} - \frac{\sigma_{\mathrm{A}}\pi}{8}\frac{\partial}{\partial z}\left(R^4(z)\frac{\partial^2 \phi_0}{\partial z^2}\right) - \varepsilon\sigma_{\mathrm{A}}\pi R^2(z)\frac{\partial \Phi_{\mathrm{E}}^{(1)}}{\partial z} \tag{11.59}$$

when expressed in terms of ϕ_0, and the alternative form

$$\mathcal{I}^{(1)}(z,t) = -\sigma_{\mathrm{A}}\pi R^2(z)\left[\frac{\partial \mathcal{V}^{(1)}}{\partial z} + \frac{1}{8\pi}\frac{\partial \psi}{\partial z} - \frac{1}{4}\frac{\mathrm{d}}{\mathrm{d}z}\left(R\frac{\mathrm{d}R}{\mathrm{d}z}\right)\frac{\partial \mathcal{V}^{(0)}}{\partial z} + \varepsilon\frac{\partial \Phi_{\mathrm{E}}^{(1)}}{\partial z}\right] \tag{11.60}$$

when expressed in term of the auxiliary function ψ.

11.6 Constitutive Equations for Transmembrane Current

The previous development of the membrane equations treated the transmembrane current as a single entity. To proceed with the implementation of a particular neuronal model requires the capacitative, intrinsic voltage-dependent, and synaptic components of the transmembrane current $J_{\mathrm{M}}(V_{\mathrm{M}})$ to be specified. The capacitative current density is usually taken to be $c_{\mathrm{M}}\partial V_{\mathrm{M}}/\partial t$, where c_{M} (μF/cm^2) is constant and refers to the specific capacitance of the dendritic membrane. The associated regular perturbation expansion of the capacitative current density is therefore

$$c_{\mathrm{M}}\frac{\partial \mathcal{V}^{(0)}}{\partial t} + c_{\mathrm{M}}\frac{\partial \mathcal{V}^{(1)}}{\partial t} + c_{\mathrm{M}}\frac{\partial \mathcal{V}^{(2)}}{\partial t} + \cdots .$$

The remaining contributions to transmembrane current density are now considered separately.

11.6.1 Intrinsic Voltage-Dependent Current

Intrinsic voltage-dependent current density is commonly modeled as the product of a voltage-dependent conductance and a potential difference. The simplest case occurs when the voltage-dependent conductance takes the constant value g_{M} (mS/cm^2) in which case the intrinsic voltage-dependent current density is $g_{\mathrm{M}}(V_{\mathrm{M}} - E_{\mathrm{M}})$ where E_{M} is a constant voltage. For this membrane, the regular perturbation expansion of the intrinsic voltage-dependent current density is

$$g_{\mathrm{M}}(\mathcal{V}^{(0)}(z,t) - E_{\mathrm{M}}) + g_{\mathrm{M}}\mathcal{V}^{(1)}(z,t) + g_{\mathrm{M}}\mathcal{V}^{(2)}(z,t) + \cdots . \tag{11.61}$$

In the more complex case of a nontrivial voltage-dependent conductance, the conductance g_{M} appearing in Equation (11.61) would be replaced by its regular perturbation expansion, which in turn may require the construction of regular perturbation expansions for a number of auxiliary functions arising in the specification of g_{M}.

11.6.2 Synaptic Current

Suppose a dendritic segment receives synaptic input at locations $0 < \eta_1 < \cdots < \eta_m < L$. The point-like nature of synapses requires that their input be expressed as a current density per unit length of segment. Therefore, the constitutive law for synaptic input is expressed in terms of its contribution to $2\pi R(z) J_{\rm M}$ and takes the value

$$\sum_{j=1}^{m} g_j(t)(V_{\rm M}(\eta_j, t) - E_j)\delta(z - \eta_j), \tag{11.62}$$

where E_j is the reversal potential for the ionic species acting at the jth synapse and $g_j(t)$ is the conductance profile at that synapse generated by the synaptic activity. As previously, the membrane potential $V_{\rm M}$ is now replaced by its regular perturbation expansion to obtain

$$\sum_{j=1}^{m} g_j(t)(\mathcal{V}^{(0)}(\eta_j, t) - E_j)\delta(z - \eta_j) + \sum_{j=1}^{m} g_j(t)\mathcal{V}^{(1)}(\eta_j, t)\delta(z - \eta_j) + \cdots \tag{11.63}$$

as the regular perturbation expansion of the contribution made by synaptic input to $2\pi R(z) J_{\rm M}$.

11.7 Membrane Equations for Segments of Constant Conductance

The regular perturbation expansion of the transmembrane current for a dendritic segment of constant conductance $g_{\rm M}$ is

$$J_{\rm M} = \mathcal{J}_{\rm M}^{(0)} + \mathcal{J}_{\rm M}^{(1)} + \mathcal{J}_{\rm M}^{(2)} + \cdots, \tag{11.64}$$

where the components of this expansion are

$$\begin{aligned} 2\pi R(z)\mathcal{J}_{\rm M}^{(0)} &= 2\pi R(z) c_{\rm M} \frac{\partial \mathcal{V}^{(0)}}{\partial t} + 2\pi R(z) g_{\rm M}(\mathcal{V}^{(0)}(z, t) - E_{\rm M}) \\ &\quad + \sum_{j=1}^{m} g_j(t)(\mathcal{V}^{(0)}(\eta_j, t) - E_j)\delta(z - \eta_j), \\ 2\pi R(z)\mathcal{J}_{\rm M}^{(k)} &= 2\pi R(z) c_{\rm M} \frac{\partial \mathcal{V}^{(k)}}{\partial t} + 2\pi R(z) g_{\rm M}\mathcal{V}^{(k)}(z, t) \\ &\quad + \sum_{j=1}^{m} g_j(t)\mathcal{V}^{(k)}(\eta_j, t)\delta(z - \eta_j), \quad k \geq 1. \end{aligned} \tag{11.65}$$

The first and second membrane equations for this segment are constructed from Equations (11.44) and (11.52) by choosing the appropriate expression in Equation (11.65) to represent the transmembrane current density. The potential $\mathcal{V}^{(0)}(z, t)$ for the first membrane equation satisfies

$$\begin{aligned} &-\frac{\partial}{\partial z}\left(\sigma_{\rm A}\pi R^2(z)\frac{\partial \mathcal{V}^{(0)}(z, t)}{\partial z}\right) + 2\pi R(z)\left[c_{\rm M}\frac{\partial \mathcal{V}^{(0)}(z, t)}{\partial t} + g_{\rm M}(\mathcal{V}^{(0)}(z, t) - E_{\rm M})\right] \\ &\quad + \sum_{k=1}^{m} g_k(t)(\mathcal{V}^{(0)}(\eta_k, t) - E_k)\delta(z - \eta_k) - \frac{\partial}{\partial z}\left(\sigma_{\rm A}\pi R^2(z)\frac{\partial \Phi_{\rm E}^{(0)}}{\partial z}\right) = 0, \end{aligned} \tag{11.66}$$

while the potential $\mathcal{V}^{(1)}(z,t)$ for the second membrane equation satisfies

$$-\frac{\partial}{\partial z}\left(\sigma_A \pi R^2(z)\frac{\partial \mathcal{V}^{(1)}(z,t)}{\partial z}\right) + 2\pi R(z)\left[c_M \frac{\partial \mathcal{V}^{(1)}(z,t)}{\partial t} + g_M \mathcal{V}^{(1)}(z,t)\right]$$
$$+\sum_{k=1}^{m} g_k(t)\mathcal{V}^{(1)}(\eta_k,t)\delta(z-\eta_k) - \varepsilon\frac{\partial}{\partial z}\left(\sigma_A \pi R^2(z)\frac{\partial \Phi_E^{(1)}}{\partial z}\right)$$
$$+\frac{\sigma_A}{4}\left[2\left(\frac{dR}{dz}\right)^2\psi(z) - \frac{1}{2}\frac{\partial}{\partial z}\left(R^2(z)\frac{\partial \psi}{\partial z}\right)\right.$$
$$\left.+\frac{\partial}{\partial z}\left(\pi R^2(z)\frac{d}{dz}\left(R(z)\frac{dR}{dz}\right)\frac{\partial \phi_0}{\partial z}\right)\right] = 0. \tag{11.67}$$

In the solution of Equations (11.66) and (11.67), the gradients of the extracellular potentials on the membrane describe exogenous current input. Further simplifications in Equations (11.66) and (11.67) can be achieved by redefining V_M to be the deviation of the transmembrane potential from E_M (rather than the membrane potential itself) and by introducing the differential operator

$$\mathcal{L}(V) = 2\pi R(z)\left[c_M\frac{\partial V}{\partial t} + g_M V\right] - \frac{\partial}{\partial z}\left[\sigma_A \pi R^2(z)\frac{\partial V}{\partial z}\right]. \tag{11.68}$$

With these changes, the first membrane equation becomes

$$\mathcal{L}(\mathcal{V}^{(0)}) + \sum_{k=1}^{m} g_k(t)(\mathcal{V}^{(0)}(\eta_k,t) - E_k)\delta(z-\eta_k) - \frac{\partial}{\partial z}\left(\sigma_A \pi R^2(z)\frac{\partial \Phi_E^{(0)}}{\partial z}\right) = 0, \tag{11.69}$$

while the second membrane equation takes the form

$$\mathcal{L}(\mathcal{V}^{(1)}) + \sum_{k=1}^{m} g_k(t)\mathcal{V}^{(1)}(\eta_k,t)\delta(z-\eta_k) - \varepsilon\frac{\partial}{\partial z}\left(\sigma_A \pi R^2(z)\frac{\partial \Phi_E^{(1)}}{\partial z}\right)$$
$$+\frac{\sigma_A}{8\pi}\left[4\pi\left(\frac{dR}{dz}\right)^2\psi(z) - \frac{\partial}{\partial z}\left(\pi R^2(z)\frac{\partial \psi}{\partial z}\right) + \frac{\partial}{\partial z}\left(\pi R^2(z)\frac{d}{dz}\left(2\pi R(z)\frac{dR}{dz}\right)\frac{\partial \phi_0}{\partial z}\right)\right] = 0, \tag{11.70}$$

where it is assumed in Equations (11.69) and (11.70) that the reversal potentials $E_0, \ldots, E_m$ are suitably adjusted by E_M.

11.7.1 Constant Conductance Dendritic Model

The first- and second-order linear dendritic models of a neuron are constructed from the core Equations (11.69) and (11.70) by imposing boundary conditions at terminal segments and the somato–dendritic junction, and by applying joining conditions at dendritic branch points. For example, a terminal segment of length L leaks no current provided $\mathcal{I}^{(0)}(L,t) = \mathcal{I}^{(1)}(L,t) = 0$. Segments meeting at dendritic branch points are connected by requiring continuity of membrane potential and conservation of axial current for each member of the hierarchy of membrane equations.

The somal-dendritic joining condition on a soma with membrane of constant conductance g_S (mS/cm^2) and constant specific capacitance c_S (μF/cm^2) is

$$A_S\left(c_S\frac{dV_S}{dt}+g_S V_S\right)=-\sum\mathcal{I}(0,t), \tag{11.71}$$

where V_S is the somal potential, A_S is the area of the somal membrane, and the summation on the right-hand side of Equation (11.71) is taken over all dendritic segments meeting at the soma. The potential V_M and the axial currents $\mathcal{I}$ are now assigned their regular perturbation expansions in terms of ε and terms of similar powers in ε are collected together to give the somal-dendritic joining conditions

$$\begin{aligned}A_S\left(c_S\frac{d\mathcal{V}^{(0)}(0,t)}{dt}+g_S\mathcal{V}^{(0)}(0,t)\right)&=-\sum\mathcal{I}^{(0)}(0,t),\\A_S\left(c_S\frac{d\mathcal{V}^{(k)}(0,t)}{dt}+g_S\mathcal{V}^{(k)}(0,t)\right)&=-\sum\mathcal{I}^{(k)}(0,t),\quad k\geq 1.\end{aligned} \tag{11.72}$$

The final form of the somal-dendritic joining condition requires continuity of transmembrane potential at the somal ends of segments meeting at the soma for each member of the hierarchy of membrane equations, and that conditions (11.72) should be satisfied by these potentials.

11.8 Summary of the Mathematical Model

The potential fields in the intracellular and extracellular regions are solutions of the three-dimensional Laplace equation with appropriate boundary conditions wherever current is injected. The intracellular and extracellular potential fields are discontinuous at the dendritic membrane, and it is the magnitude of this discontinuity that defines the transmembrane potential. The transmembrane potential in turn controls the current flowing across the membrane via a constitutive model for its electrical properties. The behavior of the transmembrane potential is such that the transmembrane current matches the current leaving the intracellular region and the current entering the extracellular region where the intracellular and extracellular currents on the dendritic membrane are defined from the gradients of the intracellular and extracellular potential fields.

The development of the one-dimensional model for the transmembrane potential begins with the equation $\operatorname{div}\boldsymbol{J}=0$ expressing point-wise conservation of current, and results in a general one-dimensional equation involving the transmembrane potential and dendritic geometry. A term-by-term comparison of this one-dimensional equation with the conventional cable equation for a dendritic membrane characterized by a surface of revolution indicates that, in this case, the transmembrane potential in the conventional cable equation and the true transmembrane potential are identical. However, the equation describing the true transmembrane potential contains additional terms due to the three-dimensional description of the dendrite; these terms have no representation in the conventional cable equation.

The development of the hierarchy of membrane equations exploits the fact that dendritic segments are long and thin. This property leads to the introduction of a small nondimensional parameter in terms of which the intracellular and extracellular potentials are represented as regular perturbation expansions. In turn, the membrane potential and all functions dependent on the membrane potential have regular perturbation expansions with coefficients that are functions of time and axial coordinate alone. The regular perturbation expansion of the membrane boundary condition gives rise to a series of two-dimensional Poisson equations describing the behavior of the intracellular and extracellular potentials in a plane perpendicular to the axis of the dendritic segment. To solve the kth boundary value problem in the intracellular region requires a consistency condition to be satisfied. The family of such conditions defines the hierarchy of membrane equations, and is shown to be equivalent to the generalized one-dimensional representation of a three-dimensional dendritic segment.

The first membrane equation contains a generalized one-dimensional cable equation and an additional term incorporating the effect of the extracellular potential gradient. In the presence of strong symmetry, this equation reduces to the conventional cable equation whenever the effect of the extracellular potential gradient is replaced by exogenous current input. However, second- and higher-order membrane equations have no analogy in conventional cable theory. The role of these equations is to provide an increasingly refined description of the membrane potential. For example, the first membrane equation includes the effect of taper implicitly, while the second membrane equation includes the effect of taper explicitly. In the first membrane equation, membrane area is approximated by the integral of the membrane perimeter with respect to axial coordinate, whereas in the second membrane equation the integral is taken with respect to membrane arc length — the Pythagorean contribution to membrane area is not present in the former but is present in the latter.

11.9 Application to an Extracellular Region at Constant Potential

The hierarchy of membrane equations is now applied to dendrites enclosed by an extracellular region at constant potential. The first and second membrane equations are solved numerically using a finite-element procedure. One advantage of this procedure is that it moves directly from the morphological data to the mathematical model thereby avoiding unnecessary discretization errors (see Altenberger et al., 2001). The development of the finite-element equations is given for an arbitrarily branched dendrite with nonuniform segments and membrane of constant conductance.

11.9.1 Finite-Element Representation of Functions

The fundamental idea in the finite-element procedure is to express all functions, both known and unknown, in terms of a user-supplied set of basis functions, denoted $u_0(z), \ldots, u_n(z)$ in this discussion. Let $z_0, \ldots, z_n$ be user-specified nodes along a dendritic segment of length L with $z_0 = 0$ at the somal end of the segment and $z_n = L$ at its distal end. The finite-element procedure in one dimension represents a function $f(z)$ by

$$\hat{f}(z) = \sum_{k=0}^{n} f_k u_k(z). \tag{11.73}$$

Of course, different functions f will have different sets of coefficients $f_0, \ldots, f_n$ but each function will be expressed in terms of the basis $u_0(z), \ldots, u_n(z)$. In numerical work involving the hierarchy of membrane equations, the set of triangular (or tent-like) basis functions

$$u_k(z) = \begin{bmatrix} \dfrac{z - z_{k-1}}{z_k - z_{k-1}}, & z \in [z_{k-1}, z_k], \\ \dfrac{z_{k+1} - z}{z_{k+1} - z_k}, & z \in (z_k, z_{k+1}], \\ 0, & \text{otherwise}, \end{bmatrix} \tag{11.74}$$

is the preferred choice. The function $u_k(z)$ is illustrated in Figure 11.2 as the shaded triangle, and the elements are the intervals $[z_{k-1}, z_k]$, where k takes all integer values from $k = 1$ to $k = n$.

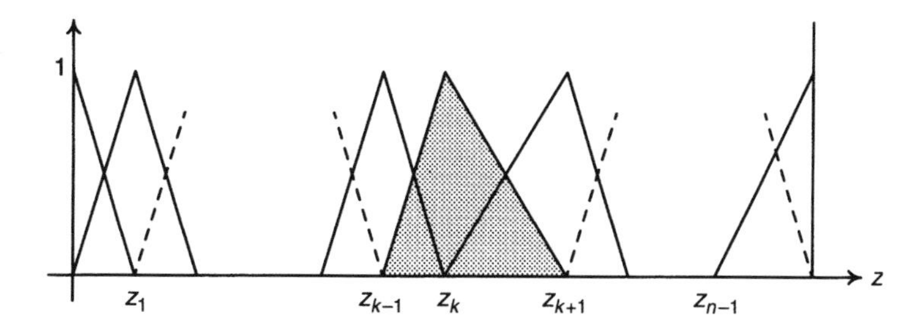

FIGURE 11.2 Basis functions $u_0(z), u_1(z), \ldots, u_n(z)$ with member $u_k(z)$ shaded.

Given a particular function $f(z)$, an immediate problem is how to choose the coefficients $f_0, \ldots, f_n$ so that expression (11.73) is an optimal approximation of $f(z)$. The *least-squares* criterion is used frequently for this purpose. The objective of this criterion is to minimize

$$E(f_0,f_1,\ldots,f_n) = \int_0^L [f(z) - \hat{f}(z)]^2 \,\mathrm{d}z = \int_0^L \left[f(z) - \sum_{k=0}^{n} f_k u_k(z) \right]^2 \mathrm{d}z \tag{11.75}$$

by an appropriate choice of the coefficients $f_0, \ldots, f_n$. The minimum of $E(f_0,f_1,\ldots,f_n)$ is attained when $f_0, \ldots, f_n$ are solutions of the equations

$$\frac{\partial E}{\partial f_j} = -2 \int_0^L \left[f(x) - \sum_{k=0}^{n} f_k u_k(z) \right] u_j(z) \,\mathrm{d}z = 0, \quad j = 0, \ldots, n, \tag{11.76}$$

which in turn asserts that $f_0, \ldots, f_n$ satisfy the $(n+1)$ simultaneous equations

$$\sum_{k=0}^{n} f_k \int_0^L u_j(z) u_k(z) \,\mathrm{d}z = \int_0^L f(z) u_j(z) \,\mathrm{d}z, \quad j = 0, \ldots, n. \tag{11.77}$$

Numerical work involving the finite-element procedure necessarily requires the evaluation of various integrals with integrands formed from products of triangular basis functions and their derivatives. All the necessary computations for this work are provided in the Appendix. Furthermore, the elementary nature of the triangular basis functions ensures that the coefficients of the finite-element expansion of $f(z)$ form a system of linear equations with a matrix that is symmetric, tridiagonal, and also diagonally dominated. A solution procedure based on the *LU* factorization algorithm is therefore guaranteed to succeed.

11.9.2 First Membrane Equation

With respect to the basis functions $u_0(z), \ldots, u_n(z)$, the potential $\mathcal{V}^{(0)}(z,t)$ and the dendritic cross-sectional area and perimeter are represented by the finite-element expansions

$$\mathcal{V}^{(0)}(z,t) = \sum_{k=0}^{n} V_k^{(0)}(t) u_k(z), \qquad \pi R^2(z) = \sum_{k=0}^{n} A_k u_k(z), \qquad 2\pi R(z) = \sum_{k=0}^{n} P_k u_k(z), \tag{11.78}$$

where $z \in [0, L]$. The objective of the finite-element procedure is to determine the time course of the coefficients $V_0^{(0)}(t), \ldots, V_n^{(0)}(t)$ appearing in the specification of the potential $\mathcal{V}^{(0)}$. In order to achieve this objective, the finite-element representation for $\mathcal{V}^{(0)}$ in (11.78) must be fitted to the first membrane equation

$$\mathcal{L}(\mathcal{V}^{(0)}) + \sum_{k=1}^{m} g_k(t)(\mathcal{V}^{(0)}(\eta_k, t) - E_k)\delta(z - \eta_k) = 0, \tag{11.79}$$

where it has now been assumed that the extracellular region is at constant potential. The fit is obtained using the least-squares procedure described in Section 11.9.1 for the basis functions $u_0(z), \ldots, u_n(z)$ and a generic function $f(z)$. The functions $V_0^{(0)}(t), \ldots, V_n^{(0)}(t)$ are chosen to ensure that each coefficient in the finite-element expansion of the left-hand side of Equation (11.79) is zero. It follows immediately from Equation (11.77) that this objective is achieved if the functions $V_0^{(0)}(t), \ldots, V_n^{(0)}(t)$ satisfy

$$\int_0^L \left[\mathcal{L}(\mathcal{V}^{(0)}) + \sum_{k=1}^{m} g_k(t)(\mathcal{V}^{(0)}(\eta_k, t) - E_k)\delta(z - \eta_k) \right] u_j(z)\, \mathrm{d}z = 0, \tag{11.80}$$

where j takes all integer values from $j = 0$ to $j = n$ and the reversal potential E_k is measured with respect to the potential E_{M}. Equation (11.80) yields immediately

$$\int_0^L \mathcal{L}(\mathcal{V}^{(0)}) u_j(z)\, \mathrm{d}z + \sum_{k=1}^{m} g_k(t) \left(\sum_{i=0}^{n} V_i^{(0)}(t) u_i(\eta_k) - E_k \right) u_j(\eta_k) = 0 \tag{11.81}$$

in which j takes all integer values from $j = 0$ to $j = n$. To complete the analysis of Equation (11.81), the operator $\mathcal{L}$ is now replaced by its definition (11.68) to give

$$\begin{aligned} \int_0^L 2\pi R(z) \left(c_{\mathrm{M}} \frac{\partial \mathcal{V}^{(0)}}{\partial t} + g_{\mathrm{M}} \mathcal{V}^{(0)} \right) u_j(z)\, \mathrm{d}z + \sum_{k=1}^{m} g_k(t) \left(\sum_{i=0}^{n} V_i^{(0)}(t) u_i(\eta_k) - E_k \right) u_j(\eta_k) \\ - \sigma_{\mathrm{A}} \int_0^L \frac{\partial}{\partial z} \left(\pi R^2(z) \frac{\partial \mathcal{V}^{(0)}}{\partial z} \right) u_j(z)\, \mathrm{d}z = 0. \end{aligned} \tag{11.82}$$

The last term on the right-hand side of Equation (11.82) is first integrated by parts and thereafter $\mathcal{V}^{(0)}$ is replaced by its finite-element expansion (11.78). The result of these operations is

$$\begin{aligned} &\sum_{i=k=0}^{n} P_i \left(c_{\mathrm{M}} \frac{\mathrm{d}V_k^{(0)}(t)}{\mathrm{d}t} + g_{\mathrm{M}} V_k^{(0)}(t) \right) \int_0^L u_i(z) u_k(z) u_j(z)\, \mathrm{d}z - \mathcal{I}^{(0)}(0, t) u_j(0) \\ &+ \mathcal{I}^{(0)}(L, t) u_j(L) + \sum_{k=1}^{m} g_k(t) \left(\sum_{i=0}^{n} V_i^{(0)}(t) u_i(\eta_k) - E_k \right) u_j(\eta_k) \\ &+ \sigma_{\mathrm{A}} \sum_{i=k=0}^{n} A_i V_k^{(0)}(t) \int_0^L u_i(z) \frac{\mathrm{d}u_k}{\mathrm{d}z} \frac{\mathrm{d}u_j}{\mathrm{d}z}\, \mathrm{d}z = 0. \end{aligned} \tag{11.83}$$

Each integer j in the range $j = 0$ to $j = n$ provides a differential equation to be satisfied by the functions $V_0^{(0)}(t), \ldots, V_n^{(0)}(t)$ giving a total of $(n + 1)$ equations. The case $j = 0$ yields

$$\begin{aligned}
&\sum_{i=k=0}^{n} P_i \left(c_{\mathrm{M}} \frac{\mathrm{d}V_k^{(0)}(t)}{\mathrm{d}t} + g_{\mathrm{M}} V_k^{(0)}(t) \right) \int_0^L u_i(z) u_k(z) u_0(z)\, \mathrm{d}z - \mathcal{I}^{(0)}(0, t) \\
&\quad + \sum_{k=1}^{m} g_k(t) \left(\sum_{i=0}^{n} V_i^{(0)}(t) u_i(\eta_k) - E_k \right) u_0(\eta_k) \\
&\quad + \sigma_{\mathrm{A}} \sum_{i=k=0}^{n} A_i V_k^{(0)}(t) \int_0^L u_i(z) \frac{\mathrm{d}u_k}{\mathrm{d}z} \frac{\mathrm{d}u_0}{\mathrm{d}z}\, \mathrm{d}z = 0.
\end{aligned} \tag{11.84}$$

The cases $0 < j < n$ yield

$$\begin{aligned}
&\sum_{i=k=0}^{n} P_i \left(c_{\mathrm{M}} \frac{\mathrm{d}V_k^{(0)}(t)}{\mathrm{d}t} + g_{\mathrm{M}} V_k^{(0)}(t) \right) \int_0^L u_i(z) u_k(z) u_j(z)\, \mathrm{d}z \\
&\quad + \sum_{k=1}^{m} g_k(t) \left(\sum_{i=0}^{n} V_i^{(0)}(t) u_i(\eta_k) - E_k \right) u_j(\eta_k) \\
&\quad + \sigma_{\mathrm{A}} \sum_{i=k=0}^{n} A_i V_k^{(0)}(t) \int_0^L u_i(z) \frac{\mathrm{d}u_k}{\mathrm{d}z} \frac{\mathrm{d}u_j}{\mathrm{d}z}\, \mathrm{d}z = 0.
\end{aligned} \tag{11.85}$$

Finally, the case $j = n$ yields

$$\begin{aligned}
&\sum_{i=k=0}^{n} P_i \left(c_{\mathrm{M}} \frac{\mathrm{d}V_k^{(0)}(t)}{\mathrm{d}t} + g_{\mathrm{M}} V_k^{(0)}(t) \right) \int_0^L u_i(z) u_k(z) u_n(z)\, \mathrm{d}z \\
&\quad + \mathcal{I}^{(0)}(L, t) + \sum_{k=1}^{m} g_k(t) \left(\sum_{i=0}^{n} V_i^{(0)}(t) u_i(\eta_k) - E_k \right) u_n(\eta_k) \\
&\quad + \sigma_{\mathrm{A}} \sum_{i=k=0}^{n} A_i V_k^{(0)}(t) \int_0^L u_i(z) \frac{\mathrm{d}u_k}{\mathrm{d}z} \frac{\mathrm{d}u_n}{\mathrm{d}z}\, \mathrm{d}z = 0.
\end{aligned} \tag{11.86}$$

To complete the specification of the finite-element equations requires the calculation of the various integrals appearing in Equations (11.84) to (11.86). These computations are included in the Appendix. By enforcing boundary conditions at dendritic terminals and conservation of axial current and continuity of potential at points where segments connect, Equations (11.84) and (11.86) may be used to terminate segments, connect segments together at dendritic branch points, or connect dendrites to the soma.

11.9.3 The Second Membrane Equation

The finite-element analysis of Equation (11.70) for the potential $\mathcal{V}^{(1)}$ mirrors that for the potential $\mathcal{V}^{(0)}$. Let $\mathcal{V}^{(1)}(z, t)$ and $\psi(z, t)$ have respective finite-element expansions

$$\mathcal{V}^{(1)}(z, t) = \sum_{k=0}^{n} V_k^{(1)}(t) u_k(z), \qquad \psi(z, t) = \sum_{k=0}^{n} \psi_k(t) u_k(z) \tag{11.87}$$

in terms of the basis functions $u_0(z), \ldots, u_n(z)$. By analogy with the treatment of $\mathcal{V}^{(0)}$, the functions $V_0^{(1)}, \ldots, V_n^{(1)}$ are required to satisfy the equations

$$\int_0^L \mathcal{L}(\mathcal{V}^{(1)})u_j(z)\,\mathrm{d}z + \sum_{k=1}^{m} g_k(t) \sum_{i=0}^{n} V_i^{(1)}(t)u_i(\eta_k)u_j(\eta_k) + \frac{\sigma_A}{8\pi}\int_0^L \left[4\pi\left(\frac{\mathrm{d}R}{\mathrm{d}z}\right)^2 \psi(z) - \frac{\partial}{\partial z}\left(\pi R^2(z)\frac{\partial\psi}{\partial z}\right) + \frac{\partial}{\partial z}\left(\pi R^2(z)\frac{\mathrm{d}}{\mathrm{d}z}\left(2\pi R(z)\frac{\mathrm{d}R}{\mathrm{d}z}\right)\frac{\partial\phi_0}{\partial z}\right)\right]u_j(z)\mathrm{d}z = 0, \tag{11.88}$$

where j takes all integer values from $j = 0$ to $j = n$. The analysis of this equation follows exactly that of the first membrane equation. The term $\mathcal{L}(\mathcal{V}^{(1)})$ is replaced by its definition and various terms in the resulting equation are integrated by parts to obtain the equation

$$\begin{aligned}
&\sum_{i=k=0}^{n} P_i\left(c_M\frac{\mathrm{d}V_k^{(1)}(t)}{\mathrm{d}t} + g_M V_k^{(1)}(t)\right)\int_0^L u_i(z)u_k(z)u_j(z)\,\mathrm{d}z \\
&\quad + \sigma_A\sum_{i=k=0}^{n} A_i V_k^{(1)}(t)\int_0^L u_i(z)\frac{\mathrm{d}u_k}{\mathrm{d}z}\frac{\mathrm{d}u_j}{\mathrm{d}z}\,\mathrm{d}z + \sum_{k=1}^{m} g_k(t)\sum_{i=0}^{n} V_i^{(1)}(t)u_i(\eta_k)u_j(\eta_k) \\
&\quad - \left[\sigma_A\pi R^2(z)\frac{\partial\mathcal{V}^{(1)}(z,t)}{\partial z}u_j(z)\right]_{z=0}^{z=L} + \frac{\sigma_A}{8\pi}\int_0^L 4\pi\left(\frac{\mathrm{d}R}{\mathrm{d}z}\right)^2\psi(z)u_j(z)\,\mathrm{d}z \\
&\quad - \frac{\sigma_A}{8\pi}\left[\pi R^2(z)\left(\frac{\partial\psi}{\partial z} - \frac{\mathrm{d}}{\mathrm{d}z}\left(2\pi R(z)\frac{\mathrm{d}R}{\mathrm{d}z}\right)\frac{\partial\phi_0}{\partial z}\right)u_j(z)\right]_{z=0}^{z=L} \\
&\quad + \frac{\sigma_A}{8\pi}\int_0^L \pi R^2(z)\left(\frac{\partial\psi}{\partial z} - \frac{\mathrm{d}}{\mathrm{d}z}\left(2\pi R(z)\frac{\mathrm{d}R}{\mathrm{d}z}\right)\frac{\partial\phi_0}{\partial z}\right)\frac{\mathrm{d}u_j}{\mathrm{d}z}\,\mathrm{d}z = 0.
\end{aligned} \tag{11.89}$$

Equation (11.89) is analogous to Equation (11.83) developed for $\mathcal{V}^{(0)}$. On taking account of the definition of $\mathcal{I}^{(1)}(z,t)$, namely

$$\mathcal{I}^{(1)}(z,t) = -\sigma_A\pi R^2(z)\left[\frac{\partial\mathcal{V}^{(1)}}{\partial z} + \frac{1}{8\pi}\frac{\partial\psi}{\partial z} - \frac{1}{8\pi}\frac{\mathrm{d}}{\mathrm{d}z}\left(2\pi R(z)\frac{\mathrm{d}R}{\mathrm{d}z}\right)\frac{\partial\phi_0}{\partial z}\right],$$

the finite-element representation of the second membrane equation for a dendritic segment becomes

$$\begin{aligned}
&\sum_{i=k=0}^{n} P_i\left(c_M\frac{\mathrm{d}V_k^{(1)}(t)}{\mathrm{d}t} + g_M V_k^{(1)}(t)\right)\int_0^L u_i(z)u_k(z)u_j(z)\,\mathrm{d}z - \mathcal{I}^{(1)}(0,t)u_j(0) \\
&\quad + \sigma_A\sum_{i=k=0}^{n} A_i V_k^{(1)}(t)\int_0^L u_i(z)\frac{\mathrm{d}u_k}{\mathrm{d}z}\frac{\mathrm{d}u_j}{\mathrm{d}z}\,\mathrm{d}z + \sum_{k=1}^{m} g_k(t)\sum_{i=0}^{n} V_i^{(1)}(t)u_i(\eta_k)u_j(\eta_k) \\
&\quad + \mathcal{I}^{(1)}(L,t)u_j(L) + \frac{\sigma_A}{8\pi}\int_0^L 4\pi\psi(z)\left(\frac{\mathrm{d}R}{\mathrm{d}z}\right)^2 u_j(z)\,\mathrm{d}z \\
&\quad + \frac{\sigma_A}{8\pi}\int_0^L \pi R^2(z)\left(\frac{\partial\psi}{\partial z} - \frac{\mathrm{d}}{\mathrm{d}z}\left(2\pi R(z)\frac{\mathrm{d}R}{\mathrm{d}z}\right)\frac{\partial\phi_0}{\partial z}\right)\frac{\mathrm{d}u_j}{\mathrm{d}z}\,\mathrm{d}z = 0.
\end{aligned} \tag{11.90}$$

Let the auxiliary functions $\xi(z)$ and $\omega(z)$, together with their finite-element expansions, be defined by the formulae

$$
\begin{aligned}
4\pi\left(\frac{\mathrm{d}R}{\mathrm{d}z}\right)^2 &= \xi(z) = \sum_{k=0}^{n} \xi_k u_k(z), \\
\pi R^2(z)\frac{\mathrm{d}}{\mathrm{d}z}\left(2\pi R(z)\frac{\mathrm{d}R}{\mathrm{d}z}\right) &= \omega(z) = \sum_{k=1}^{n} \omega_k u_k(z).
\end{aligned}
\tag{11.91}
$$

Then the finite-element representation of the second membrane equation has final form

$$
\begin{aligned}
&\sum_{i=k=0}^{n} P_i\left(c_\mathrm{M}\frac{\mathrm{d}V_k^{(1)}(t)}{\mathrm{d}t} + g_\mathrm{M}V_k^{(1)}(t)\right)\int_0^L u_i(z)u_k(z)u_j(z)\,\mathrm{d}z - \mathcal{I}^{(1)}(0,t)u_j(0) \\
&+ \sigma_\mathrm{A}\sum_{i=k=0}^{n} A_i V_k^{(1)}(t)\int_0^L u_i(z)\frac{\mathrm{d}u_k}{\mathrm{d}z}\frac{\mathrm{d}u_j}{\mathrm{d}z}\mathrm{d}z + \sum_{k=1}^{m} g_k(t)\sum_{i=0}^{n} V_i^{(1)}(t)u_i(\eta_k)u_j(\eta_k) \\
&+ \mathcal{I}^{(1)}(L,t)u_j(L) + \frac{\sigma_\mathrm{A}}{8\pi}\sum_{i=k=0}^{n} \xi_i\psi_k(t)\int_0^L u_i(z)u_k(z)u_j(z)\,\mathrm{d}z \\
&+ \frac{\sigma_\mathrm{A}}{8\pi}\sum_{i=k=0}^{n} (A_i\psi_k(t) - \omega_i V_k^{(0)}(t))\int_0^L u_i(z)\frac{\mathrm{d}u_k}{\mathrm{d}z}\frac{\mathrm{d}u_j}{\mathrm{d}z}\mathrm{d}z = 0,
\end{aligned}
\tag{11.92}
$$

where j again takes all integer values from $j = 0$ to $j = n$. The values of the various integrals appearing in Equations (11.92) are given in the Appendix. As in the case of the first membrane equation, segments may be terminated or connected to other segments or the soma by enforcing conservation of the order-ε^2 contribution to axial current at points of connection. To deal with these possibilities, Equation (11.92) is separated into the cases $j = 0$ (proximal/somal node of a dendritic section), $0 < j < n$ (internal nodes of a dendritic section), and $j = n$ (distal/terminal node of a dendritic section). The case $j = 0$ yields

$$
\begin{aligned}
&\sum_{i=k=0}^{n} P_i\left(c_\mathrm{M}\frac{\mathrm{d}V_k^{(1)}(t)}{\mathrm{d}t} + g_\mathrm{M}V_k^{(1)}(t)\right)\int_0^L u_i(z)u_k(z)u_0(z)\,\mathrm{d}z - \mathcal{I}^{(1)}(0,t) \\
&+ \sigma_\mathrm{A}\sum_{i=k=0}^{n} A_i V_k^{(1)}(t)\int_0^L u_i(z)\frac{\mathrm{d}u_k}{\mathrm{d}z}\frac{\mathrm{d}u_0}{\mathrm{d}z}\,\mathrm{d}z + \sum_{k=1}^{m} g_k(t)\sum_{i=0}^{n} V_i^{(1)}(t)u_i(\eta_k)u_0(\eta_k) \\
&+ \frac{\sigma_\mathrm{A}}{8\pi}\sum_{i=k=0}^{n} \xi_i\psi_k(t)\int_0^L u_i(z)u_k(z)u_0(z)\,\mathrm{d}z \\
&+ \frac{\sigma_\mathrm{A}}{8\pi}\sum_{i=k=0}^{n} (A_i\psi_k(t) - \omega_i V_k^{(0)}(t))\int_0^L u_i(z)\frac{\mathrm{d}u_k}{\mathrm{d}z}\frac{\mathrm{d}u_0}{\mathrm{d}z}\,\mathrm{d}z = 0,
\end{aligned}
\tag{11.93}
$$

the cases $0 < j < n$ yield

$$\sum_{i=k=0}^{n} P_i \left(c_{\mathrm{M}} \frac{\mathrm{d}V_k^{(1)}(t)}{\mathrm{d}t} + g_{\mathrm{M}} V_k^{(1)}(t) \right) \int_0^L u_i(z) u_k(z) u_j(z)\,\mathrm{d}z$$
$$+ \sigma_{\mathrm{A}} \sum_{i=k=0}^{n} A_i V_k^{(1)}(t) \int_0^L u_i(z) \frac{\mathrm{d}u_k}{\mathrm{d}z} \frac{\mathrm{d}u_j}{\mathrm{d}z}\,\mathrm{d}z + \sum_{k=1}^{m} g_k(t) \sum_{i=0}^{n} V_i^{(1)}(t) u_i(\eta_k) u_j(\eta_k)$$
$$+ \frac{\sigma_{\mathrm{A}}}{8\pi} \sum_{i=k=0}^{n} \xi_i \psi_k(t) \int_0^L u_i(z) u_k(z) u_j(z)\,\mathrm{d}z$$
$$+ \frac{\sigma_{\mathrm{A}}}{8\pi} \sum_{i=k=0}^{n} (A_i \psi_k(t) - \omega_i V_k^{(0)}(t)) \int_0^L u_i(z) \frac{\mathrm{d}u_k}{\mathrm{d}z} \frac{\mathrm{d}u_j}{\mathrm{d}z}\,\mathrm{d}z = 0, \tag{11.94}$$

and finally the case $j = n$ yields

$$\sum_{i=k=0}^{n} P_i \left(c_{\mathrm{M}} \frac{\mathrm{d}V_k^{(1)}(t)}{\mathrm{d}t} + g_{\mathrm{M}} V_k^{(1)}(t) \right) \int_0^L u_i(z) u_k(z) u_n(z)\,\mathrm{d}z + \mathcal{I}^{(1)}(L,t)$$
$$+ \sigma_{\mathrm{A}} \sum_{i=k=0}^{n} A_i V_k^{(1)}(t) \int_0^L u_i(z) \frac{\mathrm{d}u_k}{\mathrm{d}z} \frac{\mathrm{d}u_n}{\mathrm{d}z}\,\mathrm{d}z + \sum_{k=1}^{m} g_k(t) \sum_{i=0}^{n} V_i^{(1)}(t) u_i(\eta_k) u_n(\eta_k)$$
$$+ \frac{\sigma_{\mathrm{A}}}{8\pi} \sum_{i=k=0}^{n} \xi_i \psi_k(t) \int_0^L u_i(z) u_k(z) u_n(z)\,\mathrm{d}z$$
$$+ \frac{\sigma_{\mathrm{A}}}{8\pi} \sum_{i=k=0}^{n} (A_i \psi_k(t) - \omega_i V_k^{(0)}(t)) \int_0^L u_i(z) \frac{\mathrm{d}u_k}{\mathrm{d}z} \frac{\mathrm{d}u_n}{\mathrm{d}z}\,\mathrm{d}z = 0. \tag{11.95}$$

In practice, Equations (11.93) to (11.95) define the numerical problem for an arbitrarily branched dendrite with nonuniform segments and linear membrane in an extracellular region at zero potential.

11.9.4 The Finite-Element Expansion of ψ

The function $\psi(z)$ was introduced to assist in the treatment of the fourth-order derivative of $\phi_0(z,t)$ that arises in the formulation of the second membrane equation. The coefficients $\psi_0, \ldots, \psi_n$ in the finite-element expansion of $\psi(z)$ may be computed from the finite-element coefficients of $\mathcal{V}^{(0)}(z,t)$ by noting that $\phi_0(z,t) = \mathcal{V}^{(0)}(z,t) + \Phi_{\mathrm{E}}^{(0)}(z,t)$. When the extracellular region is at zero potential, it follows from the definition of $\psi(z)$ in Equation (11.54) that

$$\sigma_{\mathrm{A}} \int_0^L \psi(z) u_j(z)\,\mathrm{d}z = \sigma_{\mathrm{A}} \int_0^L \frac{\partial}{\partial z} \left(\pi R^2(z) \frac{\partial \mathcal{V}^{(0)}(z,t)}{\partial z} \right) u_j(z), \tag{11.96}$$

where j takes all integer values from $j = 0$ to $j = n$. The function $\psi(z)$ is replaced by its finite-element expansion on the left-hand side of this equation, while the right-hand side is integrated by parts to obtain

$$\sigma_{\mathrm{A}} \sum_{k=0}^{n} \psi_k \int_0^L u_k(z) u_j(z)\,\mathrm{d}z = \mathcal{I}^{(0)}(0,t)u_j(0) - \mathcal{I}^{(0)}(L,t)u_j(L) - \sigma_{\mathrm{A}} \sum_{i=k=0}^{n} A_i V_k^{(0)}(t) \int_0^L u_i(z) \frac{\mathrm{d}u_k}{\mathrm{d}z} \frac{\mathrm{d}u_j}{\mathrm{d}z}\,\mathrm{d}z, \tag{11.97}$$

where $\mathcal{V}^{(0)}(z,t)$ has been replaced by its respective finite-element expansion. Equation (11.97) is regarded as a system of $(n+1)$ simultaneous equations for $\psi_0, \ldots, \psi_n$ with right-hand side determined from the solution of the first membrane equation.

11.9.5 Time Integration

The finite-element procedure developed in the preceding section reduces the exact mathematical problem for a dendritic segment to one requiring the numerical solution of one or more systems of ordinary differential equations. These equations have generic form

$$\frac{\mathrm{d}V}{\mathrm{d}t} + A(t)V = B(t) \tag{11.98}$$

in which $A(t)$ is a known matrix and $B(t)$ is a known column vector. The task is to derive the time course of the solution vector $V(t)$ starting with a given initial condition, typically $V(0) = 0$. The interval over which the simulation is to be conducted is normally subdivided into N intervals of size Δt by the mesh points $t_0, t_1, \ldots, t_N$, where $t_k = k\Delta t$. Integration proceeds in a step-by-step manner. The solution is obtained as the series of vectors $V^0, V^1, \ldots, V^N$, where V^k is the estimated value of $V(t_k)$. Integration is affected by applying the trapezoidal quadrature

$$\int_{t_j}^{t_{j+1}} f(t)\,\mathrm{d}t = \frac{\Delta t}{2}\left[f(t_j) + f(t_{j+1})\right] + O(\Delta t)^3 \tag{11.99}$$

to the intervals $[t_0, t_1]$, $[t_1, t_2], \ldots, [t_{N-1}, t_N]$ in sequence. For example, the result of applying the quadrature to the interval $[t_j, t_{j+1}]$ is

$$V^{j+1} - V^j + \frac{\Delta t}{2}\left[A(t_{j+1})V^{j+1} + A(t_j)V^j\right] = \frac{\Delta t}{2}\left[B(t_{j+1}) + B(t_j)\right] + O(\Delta t)^3, \tag{11.100}$$

which may be rearranged into the form

$$\left[I + \frac{\Delta t}{2}A(t_{j+1})\right]V^{j+1} = \left[I - \frac{\Delta t}{2}A(t_j)\right]V^j + \frac{\Delta t}{2}[B(t_{j+1}) + B(t_j)] + O(\Delta t)^3. \tag{11.101}$$

Therefore, the integration procedure based on the trapezoidal rule generates at each stage a system of linear equations of form

$$A_{\mathrm{L}} V^{j+1} = A_{\mathrm{R}} V^j + B \tag{11.102}$$

in which the matrices A_{L}, A_{R}, and the vectors B and V^j are known. In dendritic work based on finite elements, the matrices A_{L} and A_{R} are largely tridiagonal, but will have off-tridiagonal entries in the presence of dendritic branching. The value of V^{j+1} is determined by solving the linear system (11.102). In dendritic work the structure of A_{L} is such that the solution to Equation (11.102) can be obtained by Thomas's algorithm or its generalization in the case of branched dendrites (e.g., see Lindsay et al., 2001c). Moreover, the eigenvalues of the matrix $A_L^{-1}A_R$ are guaranteed to lie within the unit circle, and so the procedure is stable with respect to rounding error due to finite-precision arithmetic. Thus the solution of the system (11.98) is obtained in a robust way by an algorithm which is locally $O(\Delta t)^3$ and globally $O(\Delta t)^2$ accurate.

In general, the procedure is to solve the membrane equations in ascending order inasmuch as the solution of the kth membrane equation will require the solution of the previous $(k-1)$ membrane equations. However, the solution of each membrane equation will require nothing more than the solution of a complex system of linear Equations (11.98) in which $A(t)$ and $B(t)$ are known.

11.10 Application to a Uniform Dendrite

Assuming an extracellular region at zero potential, the first membrane equation is the conventional cable equation. The aim of this application is to compare the predictions of the neuronal model based on the conventional cable equation with that based on the first and second membrane equations by examining the spike train output from both models for a uniform dendrite of length 1000 μm and radius R_D connected to a spherical soma of radius 15 μm. The soma generates spike train output using an integrate-to-threshold-and-fire condition with threshold set at 10 mV above the equilibrium transmembrane potential. A refractory period of 2 msec is enforced immediately after each spike by discharging the soma through a large leakage conductance with reversal potential 12 mV below the equilibrium transmembrane potential. The somal and dendritic membranes have specific capacitances $c_S = c_M = 1\ \mu\text{F/cm}^2$ and specific conductances $g_S = g_M = 0.091\ \text{mS/cm}^2$. Spike activity is driven by 100 synapses placed randomly along the length of the dendrite. Each synapse has reversal potential $E_k = 115$ mV and conductance profile

$$g(t) = \frac{g_{syn}t}{\tau}e^{1-t/\tau},$$

where t (msec) measures the time that has elapsed since the activation of the synapse, τ (0.25 msec in this application) is the time constant of the synapse, and g_{syn} (mS) is the maximum conductance of the synapse. In this experiment, each synapse is activated by an independent Poisson processes with mean rate 30 Hz.

11.10.1 Results

For a given 20 sec of synaptic activity, each experiment consisted of generating spike train output for a dendritic cylinder of fixed radius R_D using the conventional cable equation with potential $\mathcal{V}^{(0)}$, and using the neuronal model with potential $\mathcal{V}^{(0)} + \mathcal{V}^{(1)}$, where $\mathcal{V}^{(0)}$ and $\mathcal{V}^{(1)}$ are the potentials generated by the first and second membrane equations. The choice of dendritic radius ranged from $R_D = 1$ μm to $R_D = 10$ μm, and for each choice of radius, the synaptic conductance g_{syn} was adjusted so that each experiment involved the generation of approximately the same number of spikes. The modification of g_{syn} is necessary in order to compensate for the significant changes taking place in the value of the dendritic radius.

One immediate observation from this series of experiments is that the neuronal model based on the first- and second-order membrane equations consistently generated fewer spikes over a given interval of time than the model based on the first membrane equation (the conventional cable equation) alone. For example, 8% fewer spikes were generated over a 20 sec period of synaptic activity rising to 10% fewer spikes over a 100 sec period of synaptic activity. The local variations in spike rate are greater over shorter time intervals as illustrated in Figure 11.3. For a cylindrical dendrite of radius 5 μm and random synaptic activity, the neuronal model based on the first membrane equation alone generates 18 spikes in 500 msec and that based on the first and second membrane equations generates

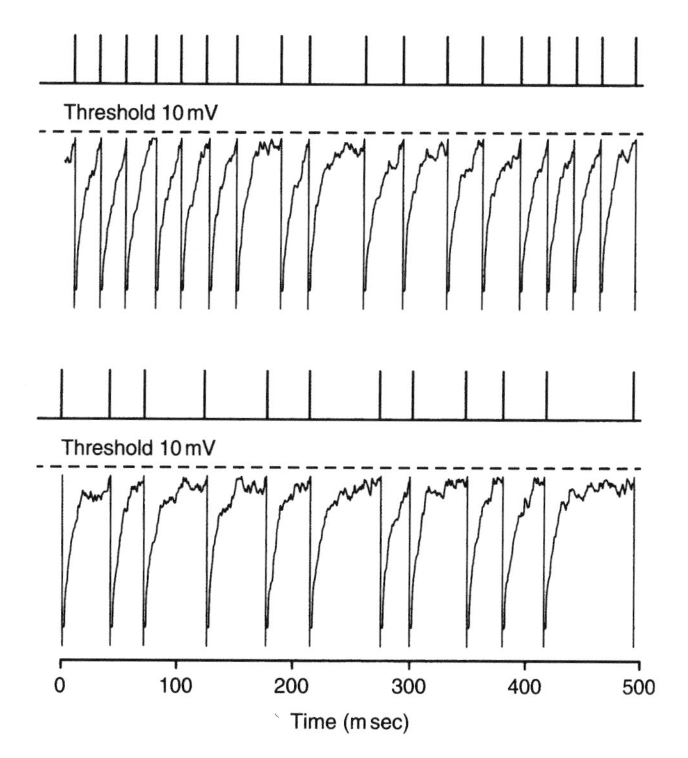

FIGURE 11.3 A sample of 500 msec of the spike train output generated by the neuronal model based on the first membrane equation (top panel) and generated for the same time period by the neuronal model based on the first and second membrane equations (lower panel). Each panel shows the trajectory of the membrane potential and the spikes which occur when it crosses threshold.

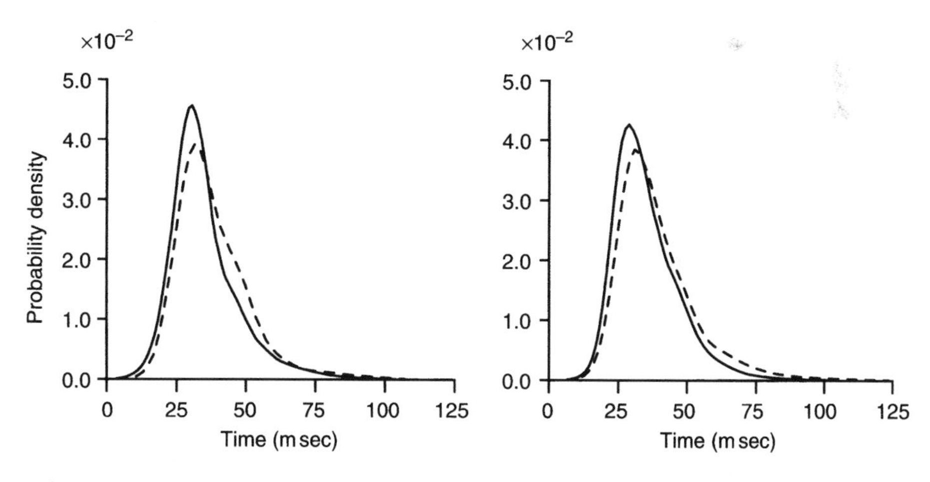

FIGURE 11.4 Kernel density estimates of the distribution of interspike intervals for a 20 sec run (left panel) and a 100 sec run (right panel). The solid line refers to estimates based on the first membrane equation alone and the dashed line refers to estimates based on the first and second membrane equations.

only 12 spikes in the same time interval. Moreover, the pattern of spike activity is different in the two cases.

Figure 11.4 shows kernel-density estimates of the distribution of interspike intervals for spike trains generated over 20 and 100 sec for the two models. A two-sample t-test of the interspike intervals found that for each sample duration, the mean interspike intervals generated by the two neuronal models were different ($p < 0.001$).

These results suggest that in this series of experiments the traditional one-dimensional cable equation (first membrane equation alone) not only gives spike train activity with higher mean rate than that generated by the first and second membrane equations, but also that the local patterns of spike activity differ in the two models.

11.11 CONCLUDING REMARKS

By starting with a three-dimensional model for the electrical behavior of a dendrite and reducing this model to a one-dimensional form, an exact prescription of how to incorporate three-dimensional properties of a dendrite into a one-dimensional representation is developed. This procedure generates a sequence of membrane equations, the first of which is the conventional cable equation. Higher-order membrane equations incorporate progressively the influence of three-dimensional properties of dendritic geometry on transmembrane potential. This method contrasts sharply with that in which an intuitive interpretation of the individual terms of the conventional cable equation is used as a guide to generalizing these terms to three dimensions. The latter is limited, for example, in the respect that radial variations in intracellular potential are not represented in the traditional one-dimensional cable equation, nor does the equation have an obvious component which can be used as a basis to generalize it to include such variations. Although these variations may be small, they are not necessarily negligible since they occur over small distances and induce gradients in intracellular potential that are of the same order of magnitude as those produced by changes in membrane surface area with axial distance.

The comparison of spike trains generated by the first membrane equation alone and the first and second membrane equations shows that they differ both in mean rate and detailed pattern of spike timing. To the extent that mean rate and pattern of spike activity are important features of neuronal activity, it is necessary to realize that spike generating models based on the cable equation alone can be misleading.

ACKNOWLEDGMENTS

Gayle Tucker is funded by the Future and Emerging Technologies programme (IST-FET) of the European Community, under grant IST-2000-28027 (POETIC). The information provided is the sole responsibility of the authors and does not reflect the Community's opinion. The Community is not responsible for any use that might be made of data appearing in this publication.

PROBLEMS

1. Show how the integrate-to-threshold-and-fire spike generation mechanism in the example given in the chapter can be replaced by:
   (a) Hodgkin–Huxley membrane kinetics.
   (b) Connors–Stevens membrane kinetics.
   (c) Morris–Lecar membrane kinetics.
2. Discuss the influence that the choice of kinetic model has on the generated spike train for the first membrane equation and for the first and second membrane equations for the model dendrite used in the example in the chapter.
3. Generate the third member of the hierarchy of cable equations for the constant conductance membrane model.
4. Derive the expression for the axial current for the third member of the hierarchy of cable equations.

5. In the case of an unmyelinated axon, how does the inclusion of the second membrane equation in the model for the membrane potential alter the predicted speed of the propagated action potential?
6. Develop the hierarchy of membrane equations in the case of intracellularly injected current, for example, when the current is injected via a wire oriented along the axis of an axon.
7. Extend the hierarchy of membrane equations given in the chapter to include the case of a dendritic section contained within a finite extracellular space.

APPENDIX

11.A1 Integrating Products of Basis Functions

The least-squares procedure described in Section 11.9.1 and the subsequent construction of the finite-element representation of the solutions to the first and second membrane equations require the evaluation of integrals with integrands formed from various products of the basis functions and their derivatives. The calculation of these integrals is illustrated for the contributions made by the element occupying $[z_j, z_{j+1}]$, and takes advantage of the fact that $u_0(z), \ldots, u_n(z)$ satisfy the algebraic properties

$$\begin{aligned} u_{j-1}(z) + u_j(z) &= 1, \quad z \in [z_{j-1}, z_j], \\ u_j(z) + u_{j+1}(z) &= 1, \quad z \in [z_j, z_{j+1}], \end{aligned} \tag{11.A1}$$

whenever both basis functions are defined.

11.A1.1 Integration of Products of Two Basis Functions

There are three products of two basis functions that contribute to integrals over the element occupying $[z_j, z_{j+1}]$, namely, $u_j^2(z)$, $u_j(z)u_{j+1}(z)$, and $u_{j+1}^2(z)$. Under the change of variable $z = z_j + (z_{j+1} - z_j)\zeta$, it is clear that $u_j(z) = (1-\zeta)$ and $u_{j+1}(z) = \zeta$. Consequently,

$$\begin{aligned} \int_{z_j}^{z_{j+1}} u_j^2(z)\,dz &= (z_{j+1} - z_j)\int_0^1 (1-\zeta)^2\,d\zeta = \frac{z_{j+1} - z_j}{3}, \\ \int_{z_j}^{z_{j+1}} u_j(z)u_{j+1}(z)\,dz &= (z_{j+1} - z_j)\int_0^1 (1-\zeta)\zeta\,d\zeta = \frac{z_{j+1} - z_j}{6}, \\ \int_{z_j}^{z_{j+1}} u_{j+1}^2(z)\,dz &= (z_{j+1} - z_j)\int_0^1 \zeta^2\,d\zeta = \frac{z_{j+1} - z_j}{3}. \end{aligned} \tag{11.A2}$$

Because basis functions interior to a segment span two elements and those at the ends of a segment span only one element (see Figure 11.2), it follows immediately that

$$\int_0^L u_k(z)u_j(z)\,dz = \int_{z_{j-1}}^{z_j} u_k(z)u_j(z)\,dz + \int_{z_j}^{z_{j+1}} u_k(z)u_j(z)\,dz, \tag{11.A3}$$

where each integral on the right-hand side is present only if the underlying element exists.

11.A1.2 Integration of Products of Three Basis Functions

There are four possible products of three basis functions that contribute to integrals over the element occupying $[z_j, z_{j+1}]$, namely, $u_j^3(z)$, $u_j^2(z)u_{j+1}(z)$, $u_j(z)u_{j+1}^2(z)$, and $u_{j+1}^3(z)$. The change of variable $z = z_j + (z_{j+1} - z_j)\zeta$ now yields

$$\begin{aligned}
&\int_{z_j}^{z_{j+1}} u_j^3(z)\,\mathrm{d}z = z_{j+1} - z_j) \int_0^1 (1-\zeta)^3\,\mathrm{d}\zeta = \frac{z_{j+1} - z_j}{4}, \\
&\int_{z_j}^{z_{j+1}} u_j^2(z)u_{j+1}(z)\,\mathrm{d}z = (z_{j+1} - z_j) \int_0^1 (1-\zeta)^2\zeta\,\mathrm{d}\zeta = \frac{z_{j+1} - z_j}{12}, \\
&\int_{z_j}^{z_{j+1}} u_j(z)u_{j+1}^2(z)\,\mathrm{d}z = (z_{j+1} - z_j) \int_0^1 (1-\zeta)\zeta^2\,\mathrm{d}\zeta = \frac{z_{j+1} - z_j}{12}, \\
&\int_{z_j}^{z_{j+1}} u_{j+1}^3(z)\,\mathrm{d}z = (z_{j+1} - z_j) \int_0^1 \zeta^3\,\mathrm{d}\zeta = \frac{z_{j+1} - z_j}{4}.
\end{aligned} \tag{11.A4}$$

Because basis functions interior to a segment span two elements and those at the ends of a segment span only one element, it again follows immediately that

$$\int_0^L u_i(z)u_k(z)u_j(z)\,\mathrm{d}z = \int_{z_{j-1}}^{z_j} u_i(z)u_k(z)u_j(z)\,\mathrm{d}z + \int_{z_j}^{z_{j+1}} u_i(z)u_k(z)u_j(z)\,\mathrm{d}z, \tag{11.A5}$$

where each integral on the right-hand side is present only if the underlying element exists.

11.A1.3 Integrals of Basis Functions and their Derivatives

In the construction of the finite-element representation of the first membrane equation it is necessary to compute

$$\int_0^L u_i(z)\frac{\mathrm{d}u_k(z)}{\mathrm{d}z}\frac{\mathrm{d}u_j(z)}{\mathrm{d}z}\mathrm{d}z = \int_{z_{j-1}}^{z_j} u_i\frac{\mathrm{d}u_k}{\mathrm{d}z}\frac{\mathrm{d}u_j}{\mathrm{d}z}\mathrm{d}z + \int_{z_j}^{z_{j+1}} u_i\frac{\mathrm{d}u_k}{\mathrm{d}z}\frac{\mathrm{d}u_j}{\mathrm{d}z}\mathrm{d}z, \tag{11.A6}$$

where it is again understood that each integral on the right-hand side of this identity is present only if the underlying element exists. In this calculation there are six possible products of basis functions that contribute to integrals over the element occupying $[z_j, z_{j+1}]$, namely,

$$\begin{aligned}
&\int_{z_j}^{z_{j+1}} u_j\left(\frac{\mathrm{d}u_j}{\mathrm{d}z}\right)^2\mathrm{d}z, \quad \int_{z_j}^{z_{j+1}} u_j\frac{\mathrm{d}u_j}{\mathrm{d}z}\frac{\mathrm{d}u_{j+1}}{\mathrm{d}z}\mathrm{d}z, \quad \int_{z_j}^{z_{j+1}} u_j\left(\frac{\mathrm{d}u_{j+1}}{\mathrm{d}z}\right)^2\mathrm{d}z, \\
&\int_{z_j}^{z_{j+1}} u_{j+1}\left(\frac{\mathrm{d}u_j}{\mathrm{d}z}\right)^2\mathrm{d}z, \quad \int_{z_j}^{z_{j+1}} u_{j+1}\frac{\mathrm{d}u_j}{\mathrm{d}z}\frac{\mathrm{d}u_{j+1}}{\mathrm{d}z}\mathrm{d}z, \quad \int_{z_j}^{z_{j+1}} u_{j+1}\left(\frac{\mathrm{d}u_{j+1}}{\mathrm{d}z}\right)^2\mathrm{d}z.
\end{aligned} \tag{11.A7}$$

The computation of the integrals in Equation (11.A7) is immediate once it is recognized that

$$\begin{aligned}
&\frac{\mathrm{d}u_j}{\mathrm{d}z} = \frac{-1}{z_{j+1} - z_j}, \qquad \int_{z_j}^{z_{j+1}} u_j(z)\mathrm{d}z = \frac{z_{j+1} - z_j}{2}, \\
&\frac{\mathrm{d}u_{j+1}}{\mathrm{d}z} = \frac{1}{z_{j+1} - z_j}, \qquad \int_{z_j}^{z_{j+1}} u_{j+1}(z)\mathrm{d}z = \frac{z_{j+1} - z_j}{2}.
\end{aligned}$$

The result of these computations is

$$\left.\begin{aligned}\int_{z_j}^{z_{j+1}} u_j \left(\frac{du_j}{dz}\right)^2 dz &= \int_{z_j}^{z_{j+1}} u_{j+1} \left(\frac{du_j}{dz}\right)^2 dz \\ \int_{z_j}^{z_{j+1}} u_j \left(\frac{du_{j+1}}{dz}\right)^2 dz &= \int_{z_j}^{z_{j+1}} u_{j+1} \left(\frac{du_{j+1}}{dz}\right)^2 dz\end{aligned}\right] = \frac{1}{2}\frac{1}{z_{j+1}-z_j},$$

$$\int_{z_j}^{z_{j+1}} u_j \frac{du_j}{dz}\frac{du_{j+1}}{dz} dz = \int_{z_j}^{z_{j+1}} u_{j+1} \frac{du_j}{dz}\frac{du_{j+1}}{dz} dz = -\frac{1}{2}\frac{1}{z_{j+1}-z_j}. \tag{11.A8}$$

11.A2 Notation and Definitions

(Vectors are written in bold face italic as, for example, $\boldsymbol{x}, \boldsymbol{r}$)

$\mathcal{A}(z)$	Area formed by the intersection of a dendritic segment with a plane of fixed axial coordinate z lying interior to the curve $r = R(z)$
$\partial\mathcal{A}$	Boundary of $\mathcal{A}(z)$ given by the curve $r = R(z)$
$\mathcal{A} \times (a, b)$	Intracellular volume of a section between the planes $z = a$ and $z = b$
$\partial\mathcal{A} \times (a, b)$	Dendritic membrane between the planes $z = a$ and $z = b$
c_M/c_S	Specific membrane capacitance of a dendritic section/soma (μF/cm^2)
$\delta(z - a)$	Dirac delta function $\delta(z-a) = 0$ if $z \neq a$, and $\int_{-\infty}^{\infty} \delta(z-a)\,dz = 1$, $\int_{-\infty}^{\infty} \delta(z-a) f(z)\,dz = f(a)$
$\operatorname{div} \boldsymbol{A}$	The divergence $\operatorname{div}\boldsymbol{A}$ of the vector $\boldsymbol{A} = A_r\boldsymbol{e}_r + A_z\boldsymbol{e}_z$ in cylindrical polar coordinates is the scalar $(1/r)(\partial/\partial r)(rA_r) + \partial A_z/\partial z$
$\boldsymbol{e}_r, \boldsymbol{e}_z$	The radial (increasing r) and axial (increasing z) base vectors, respectively, for cylindrical polar coordinates
ε	Scaling factor $(\tilde{R}/L)^2$, where $\tilde{R}$ is the section radius (cm) and L is the unit of axial length (cm)
η_m	Location of the mth synaptic input on a dendritic section
E_M	Resting membrane potential (mV) in the absence of synaptic activity
E_k	Reversal potential for the ionic species at the kth synapse (and may be biased by E_M) (mV)
g_M/g_S	Constant membrane conductance of a dendritic section/soma (mS/cm^2)
$g_k(t)$	Conductance profile at the kth synapse constructed from a sum of alpha-function representations $g_{max}(t/\tau)e^{1-t/\tau}$
$\boldsymbol{J}$	Current density (A/m^2)
$J_M(V_M)$	Total transmembrane current density (μA/cm^2)
$J_{SYN}(V_M)$	Synaptic current density (μA/cm^2)
$J_{IVDC}(V_M)$	Intrinsic voltage-dependent current density (μA/cm^2)
$\mathcal{J}^{(k)}$	The kth component in the regular perturbation expansion of $J_M(V_M)$ with order $\varepsilon^{k+1/2}$
$\mathcal{L}(\mathcal{V})$	Differential operator defined as $\mathcal{L}(\mathcal{V}) = 2\pi R(z)[c_M(\partial\mathcal{V}/\partial t) + g_M\mathcal{V}] - (\partial/\partial z)[\sigma_A \pi R^2(z)(\partial\mathcal{V}/\partial z)]$

$\boldsymbol{n}$	Unit outward normal vector to the segment membrane
$\partial\psi/\partial n$	Differential operator defined as $\partial\psi/\partial n = \boldsymbol{n} \cdot (\partial\psi/\partial r, 0, \partial\psi/\partial z)$
ϕ, Φ	Intracellular and extracellular potential fields respectively (mV)
$\phi_{\mathrm{A}}, \Phi_{\mathrm{E}}$	Potentials on the interior and exterior surfaces of the segment membrane, respectively (mV)
$R(z)$	Value of the segment radius at the point z
$\hat{r}$	Radial coordinate defined as $\hat{r} = (L/\tilde{R})r$, where L is the unit of axial length, $\tilde{R}$ a typical segment radius, and r is the radial coordinate
R_z	Derivative of $R(z)$ with respect to z
$\sigma_{\mathrm{A}}/\sigma_{\mathrm{E}}$	Conductivities of intracellular/extracellular fluids, respectively
$u_0(z), \ldots, u_n(z)$	Triangular basis functions
V_{M}	True membrane potential (mV)
$\mathcal{V}$	Membrane potential (mV) for the conventionally derived one-dimensional cable equation
$\mathcal{V}^{(0)}$	Membrane potential (mV) arising in the first membrane equation derived from the three-dimensional model
$\mathcal{V}^{(1)}, \mathcal{V}^{(2)}, \ldots$	Perturbations to $\mathcal{V}^{(0)}$ where $\mathcal{V}^{(k)}$ is the kth order perturbation to $\mathcal{V}^{(0)}$, and $\mathcal{V}^{(k)} = \varepsilon^k V_{\mathrm{M}}^{(k)}$ where $V_{\mathrm{M}}^{(k)}$ is the kth coefficient in the expansion of the membrane potential V_{M} in terms of ε
$z_0, \ldots, z_n$	Location of nodes along a dendritic segment

NOTES

1. If σ_{A} and g_{M} are typical axial and membrane conductances respectively, the distance $(\tilde{R}\sigma_{\mathrm{A}}/g_{\mathrm{M}})^{1/2}$ is often regarded as a realistic axial length scale and would therefore be a suitable choice for L.
2. Near segment endpoints, the assumptions underlying the regular perturbation expansion (11.16) are invalid and one recovers full balance through Equation (11.15).

12 Simulation Analyses of Retinal Cell Responses

Yoshimi Kamiyama, Akito Ishihara, Toshihiro Aoyama, and Shiro Usui

CONTENTS

12.1 Introduction

The vertebrate retina has been thought to be a window to the brain because of its accessibility and suitability for neuroscientific investigation. It is thought to be one of the few parts of the vertebrate brain where we can reasonably explain its purpose and how it works. Therefore, the retina is an ideal model of an information processing neural network (Dowling, 1987).

Physiological studies of the retina have uncovered a number of cellular and subcellular mechanisms such as phototransduction and the particular characteristics of the ion channels found in retinal neurons. Membrane ionic currents of retinal cells have been studied quantitatively using voltage-clamp techniques (Kaneko and Tachibana, 1986; Lasater, 1991). On the other hand, neuroanatomical studies have revealed the fundamental morphological principles governing the structure of the vertebrate retina (Masland, 2001). The former data provide information about the functional role of ionic currents in generating and shaping the light response of retinal neurons. Nevertheless, the detailed experimental data alone are not enough to understand how the retinal neurons operate.

A combination of experimental work and mathematical modeling is necessary to clarify the complicated retinal functions. Over the last decade, physiologically based mathematical models have been developed which have generated insight into cellular functions such as the role of interactions among ionic currents (Usui et al., 1991, 1996a, 1996b, 1996c; Kamiyama et al., 1996; Aoyama et al., 2000). For example, models based on ionic mechanisms are employed to understand individual ionic current dynamics under current-clamp conditions and light responses, which cannot be directly measured by conventional experimental techniques. These models can also easily reveal what happens if a particular current is active or inactive. Since ionic processes are fundamental for generating the electrical responses of neurons, simulation studies based on these processes make it possible to analyze nonlinear dynamical response properties and to understand the roles of particular ionic channels, and thus provide key insights for elucidating the underlying mechanisms in retinal processing. We believe that mathematical modeling and simulations using powerful computers play a necessary role in contributing to an understanding of the relationships between the microscopic characteristics of photoreceptors and associated neurons and their integrative functions.

In the present study, mathematical descriptions of membrane ionic currents in photoreceptors, horizontal cells, bipolar cells, and ganglion cells are given. The electrical properties of the photoreceptor and each type of neuron are described by parallel conductance circuits. The voltage- and time-dependent characteristics of ionic conductance are modeled by Hodgkin–Huxley membrane kinetics. The full model is able to reproduce accurately the voltage- and current-clamp responses of retinal neurons. Any electrical response, including light response, which depends on the dynamic balance of different ionic currents is quantitatively accounted for by the model. Therefore, hypotheses on how the retina processes visual information are understood at the cellular and subcellular level. These models can be used to summarize what we know and what we need to find out about the retina.

12.2 Description of Ionic Currents

12.2.1 The Parallel Conductance Model

The electrical properties of cellular membranes can be modeled by electrical circuits, that is, the membrane capacitance as a capacitor, membrane permeability as conductances, and the difference between the membrane potential and the reversal potential of each ionic species as the electrical driving force for that species (Johnston and Wu, 1995). The electrical properties of the rod photoreceptor, the horizontal cell, the bipolar cell, and the ganglion cell were described by parallel conductance models as shown in Figure 12.1 (Kamiyama et al., 1996; Usui et al., 1996a; Aoyama et al., 2000; Kamiyama and Usui, 2003). The total membrane current in a parallel conductance model can be written as

$$I_{\text{total}} = C\frac{\mathrm{d}V}{\mathrm{d}t} + \sum_{\text{ion}} I_{\text{ion}}(V,t) = C\frac{\mathrm{d}V}{\mathrm{d}t} + \sum_{\text{ion}} g_{\text{ion}}(V,t)(V - E_{\text{ion}}),$$

where C is the membrane capacitance (nF), V is the membrane potential (mV), I_{ion} is the ionic current (pA), g_{ion} is the conductance of the ionic current (nS), and E_{ion} is the reversal potential (mV) of the current. Since a particular ionic channel is selectively blocked by a chemical drug

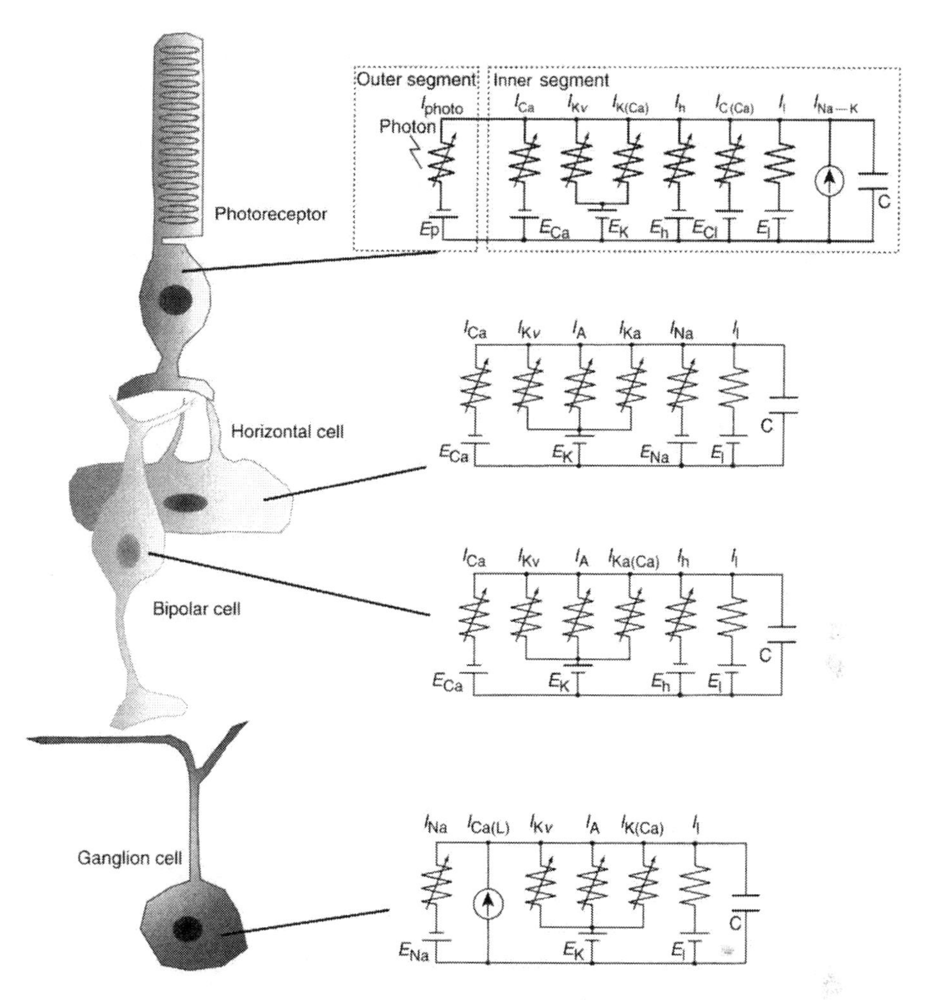

FIGURE 12.1 Parallel conductance models of retinal cells.

(channel blocker) in electrophysiological experiments, the equivalent circuit can be simplified to a single circuit, that is, each ionic current is analyzed separately. A space clamp of the membrane potential allowed Hodgkin and Huxley to fix this potential at particular values in order to determine the underlying macroscopic ionic currents (Hodgkin and Huxley, 1952b; Johnston and Wu, 1995). Such currents are macroscopic because they reflect an ensemble average of many localized channels. They resemble the results seen in intracellular and whole-cell patch recordings. In the Hodgkin–Huxley formulation, each current is described by an equation of the form

$$
\begin{aligned}
I_{\text{ion}}(V,t) &= \bar{g}_{\text{ion}} m^M h^N (V - E_{\text{ion}}), \\
\frac{\mathrm{d}m}{\mathrm{d}t} &= \alpha_m(V)(1-m) - \beta_m(V)m, \\
\frac{\mathrm{d}h}{\mathrm{d}t} &= \alpha_h(V)(1-h) - \beta_h(V)h,
\end{aligned}
\tag{12.1}
$$

where $\bar{g}_{\text{ion}}$ is the maximum conductance, m is an activation variable, h is an inactivation variable, and M, N are gating exponents. Gating variables m and h depend on both the membrane voltage and time.

Model parameters in the specification of the αs and βs in Equations (12.1) are estimated from the results of voltage-clamp experiments. If the membrane potential is stepped from a holding voltage V_h to a command voltage V_c, and the instantaneous transmembrane current density per unit membrane surface is represented through a linear relationship (i.e., Equation [12.1]), then the gating variables $x(= m, h)$ can be readily obtained by solving (Hodgkin and Huxley, 1952b)

$$x(V_c, t) = x_\infty(V_c) - (x_\infty(V_c) - x_\infty(V_h)) \exp\left(-\frac{t}{\tau_x(V_c)}\right), \tag{12.2}$$

where $x_\infty(V_h) = \alpha_x(V_h)/(\alpha_x(V_h) + \beta_x(V_h))$ is the initial value of $x(V_c, t)$, $x_\infty(V_c, t) = \alpha_x(V_c)/(\alpha_x(V_c) + \beta_x(V_c))$ is the steady-state value of $x(V_c, t)$, $\tau_x(V_c) = (\alpha_x(V_c) + \beta_x(V_c))^{-1}$ is the time-constant (sec). Steady-state activation curves $x_\infty(V)$ $(x = m, h)$ and the voltage dependence of the time constants $(\tau_x(V))$ are measured from a whole family of voltage-clamp records at different potentials. The rate coefficients, $\alpha_x(V)$ and $\beta_x(V)$, are then calculated from the relationships

$$\alpha_x(V) = \frac{x_\infty(V)}{\tau_x(V)}, \qquad \beta_x(V) = \frac{1 - x_\infty(V)}{\tau_x(V)}. \tag{12.3}$$

The functions $\alpha_x(V), \beta_x(V)$ are fitted by the empirical expression

$$\alpha_x(V), \beta_x(V) = \frac{C_1 \exp(C_2(V + C_3) + C_4(V + C_5))}{\exp(C_6(V + C_3) + C_7)}, \tag{12.4}$$

where the parameters C_1 to C_7 take suitable values for the determination of $\alpha_x(V)$ and $\beta_x(V)$. Values of measurable parameters, such as the maximum conductances $\bar{g}_{ion}$ and reversal potentials E_{ion} are directly introduced into the model. These procedures are the basic steps for modeling ionic conductance. Tables 12.1 to 12.9 show full descriptions of the proposed model of retinal neurons (Kamiyama et al., 1996; Usui et al., 1996c; Aoyama et al., 2000; Kamiyama and Usui, 2003).

12.2.2 Intracellular Calcium Concentration

Several processes affect the intracellular concentration of calcium ions (Yamada et al., 1989). After calcium ions enter the cell through calcium channels, they diffuse into the cell and bind to various buffers. Specialized channels act as transporters to extrude calcium ions from the cell. The intracellular mechanisms that control calcium ions in retinal cells are not well understood. We thus modeled the calcium system using the fewest mechanisms necessary to reproduce the calcium-dependent currents of the cells. Figure 12.2 shows a schematic illustration of the compartmental calcium model (Kamiyama et al., 1996; Usui et al., 1996a). The mathematical description and parameters for the photoreceptor model and the bipolar cell model are summarized in Tables 12.4 and 12.5, and Tables 12.6 and 12.7, respectively (Kamiyama et al., 1996; Usui et al., 1996a).

This model is based on the earlier work by DiFrancesco and Noble (1985) in cardiac cells and the work on rod outer segments by Forti et al. (1989). Intracellular calcium is controlled by the presence of the calcium current I_{Ca}, by extrusion through the transporters, and by binding and unbinding to high- and low-affinity internal buffers. Because our primary aim is to model the action of calcium on the calcium-dependent current, we divided the cell space into two compartments: the space just below the cellular membrane and the central space. The calcium concentration $[Ca^{2+}]_s$ just below the membrane was used for the calcium-dependent currents.

TABLE 12.1
A Mathematical Formulation of Each Ionic Current of a Photoreceptor

Delayed rectifying K current

$$\alpha_{mKv} = \frac{5(100 - V)}{\exp((100 - V)/42) - 1}$$

$$\beta_{mKv} = 9\exp\left(-\frac{V - 20}{40}\right)$$

$$\frac{dm_{Kv}}{dt} = \alpha_{mKv}(1 - m_{Kv}) - \beta_{mKv}m_{Kv}$$

$$\alpha_{hKv} = 0.15\exp\left(-\frac{V}{22}\right)$$

$$\beta_{hKv} = \frac{0.4125}{\exp((10 - V)/7) + 1}$$

$$\frac{dh_{Kv}}{dt} = \alpha_{hKv}(1 - h_{Kv}) - \beta_{hKv}h_{Kv}$$

$$I_{Kv} = \bar{g}_{Kv}(m_{Kv})^3 h_{Kv}(V - E_K)$$

$$\bar{g}_{Kv} = 2.0\ [\text{nS}]$$

$$E_K = -80\ [\text{mV}]$$

Ca-dependent Cl current

$$m_{Cl} = \frac{1}{1 + \exp((\text{Cl}_h - [\text{Ca}^{2+}]_s)/0.09)}$$

$$I_{Cl(Ca)} = \bar{g}_{Cl}m_{Cl}(V - E_{Cl})$$

$$\bar{g}_{Cl} = 6.5\ [\text{nS}]$$

$$E_{Cl} = -45\ [\text{mV}]$$

$$\text{Cl}_h = 0.37\ [\mu\text{M}]$$

Leakage current

$$I_L = gl(V - E_l)$$

$$gl = 0.5\ [\text{nS}]$$

$$E_l = -55\ [\text{mV}]$$

Na–K exchanger current

$$I_{Na\text{-}K} = 9.6\ [\text{pA}]$$

Hyperpolarization-activated current

$$\alpha_h = \frac{8}{\exp((V + 78)/14) + 1}$$

$$\beta_h = \frac{18}{\exp(-(V + 8)/19) + 1}$$

$$M = [\ C_1 \quad C_2 \quad O_1 \quad O_2 \quad O_3\ ]^t$$

Ca current

$$\alpha_{Ca} = \frac{300(80 - V)}{\exp((80 - V)/25) - 1}$$

$$\beta_{Ca} = \frac{1000}{1 + \exp((V + 38)/7)}$$

$$\frac{dm_{Ca}}{dt} = \alpha_{Ca}(1 - m_{Ca}) - \beta_{Ca}m_{Ca}$$

$$E_{Ca} = 12.5\log\left(\frac{[\text{Ca}^{2+}]_o}{[\text{Ca}^{2+}]_s}\right)$$

$$m_{Cah} = \frac{\exp((40 - V)/18)}{1 + \exp((40 - V)/18)}$$

$$I_{Ca} = \bar{g}_{Ca}(m_{Ca})^4 m_{Cah}(V - E_{Ca})$$

$$\bar{g}_{Ca} = 0.7\ [\text{nS}]$$

Ca-dependent K current

$$\alpha_{Kc} = \frac{15(80 - V)}{\exp((80 - V)/40) - 1}$$

$$\beta_{Kc} = 20\exp\left(-\frac{V}{35}\right)$$

$$\frac{dm_{Kc}}{dt} = \alpha_{Kc}(1 - m_{Kc}) - \beta_{Kc}m_{Kc}$$

$$m_{Kc1} = \frac{[\text{Ca}^{2+}]_s}{[\text{Ca}^{2+}]_s + 0.3}$$

$$I_{K(Ca)} = \bar{g}_{Kc}(m_{Kc})^2 m_{Kc1}(V - E_K)$$

$$\bar{g}_{Kc} = 0.5\ [\text{nS}]$$

Capacitance

$$C = 10\ [\text{pF}]$$

$$K = \begin{bmatrix} -4\alpha_h & \beta_h & 0 & 0 & 0 \\ 4\alpha_h & -(\beta_h + 3\alpha_h) & 2\beta_h & 0 & 0 \\ 0 & 3\alpha_h & -(2\beta_h + 2\alpha_h) & 3\beta_h & 0 \\ 0 & 0 & 2\alpha_h & -(3\beta_h + \alpha_h) & 4\beta_h \\ 0 & 0 & 0 & \alpha_h & -4\beta_h \end{bmatrix}$$

$$\frac{d}{dt}M = KM$$

$$m_h = O_1 + O_2 + O_3$$

$$I_h = \bar{g}_h m_h(V - E_h)$$

$$\bar{g}_h = 3\ [\text{nS}]$$

$$E_h = -32\ [\text{mV}]$$

TABLE 12.2
A Mathematical Formulation of Ionic Currents of a Horizontal Cell

Transient outward K current

$$\alpha_{m_A} = \frac{2400}{1 + \exp(-(V - 50)/28)}$$

$$\beta_{m_A} = 80 \exp\left(\frac{-V}{36}\right)$$

$$\frac{dm_A}{dt} = \alpha_{m_A}(1 - m_A) - \beta_{m_A} m_A$$

$$\alpha_{h_A} = 1 \exp\left(-\frac{V}{60}\right)$$

$$\beta_{h_A} = \frac{20}{\exp(-(V + 40)/5) + 1}$$

$$\frac{dh_A}{dt} = \alpha_{h_A}(1 - h_A) - \beta_{h_A} h_A$$

$$I_A = \bar{g}_A m_A^3 h_A (V - E_K)$$

$$\bar{g}_A = 15.0 \text{ [nS]}$$

$$E_K = -80 \text{ [mV]}$$

Anomalous rectifying K current

$$m_{Ka} = \frac{1}{1 + \exp((V + 60)/12)}$$

$$I_{Ka} = \bar{g}_{Ka} m_{Ka}^5 (V - E_K)$$

$$\bar{g}_{Ka} = 4.5 \text{ [nS]}$$

Na current

$$\alpha_{m_{Na}} = \frac{200(38 - V)}{\exp((38 - V)/25) - 1}$$

$$\beta_{m_{Na}} = 2000 \exp\left(-\frac{55 + V}{18}\right)$$

$$\frac{dm_{Na}}{dt} = \alpha_{m_{Na}}(1 - m_{Na}) - \beta_{m_{Na}} m_{Na}$$

$$\alpha_{h_{Na}} = 1000 \exp\left(-\frac{80 + V}{8}\right)$$

$$\beta_{h_{Na}} = \frac{800}{\exp((80 - V)/75) + 1}$$

$$\frac{dh_{Na}}{dt} = \alpha_{h_{Na}}(1 - h_{Na}) - \beta_{h_{Na}} h_{Na}$$

$$I_{Na} = \bar{g}_{Na} m_{Na}^3 h_{Na} (V - E_{Na})$$

$$\bar{g}_{Na} = 2.4 \text{ [nS]}$$

$$E_{Na} = 55 \text{ [mV]}$$

Delayed rectifying K current

$$\alpha_{m_{Kv}} = \frac{0.4(65 - V)}{\exp((65 - V)/50) - 1}$$

$$\beta_{m_{Kv}} = 4.8 \exp\left(\frac{45 - V}{85}\right)$$

$$\frac{dm_{Kv}}{dt} = \alpha_{m_{Kv}}(1 - m_{Kv}) - \beta_{m_{Kv}} m_{Kv}$$

$$\alpha_{h_{Kv}} = \frac{1500}{\exp((V + 92)/7) + 1}$$

$$\beta_{h_{Kv}} = \frac{80}{\exp((V + 100)/15) + 1}$$

$$\frac{dh_{Kv}}{dt} = \alpha_{h_{Kv}}(1 - h_{Kv}) - \beta_{h_{Kv}} h_{Kv}$$

$$I_{Kv} = \bar{g}_{Kv} m_{Kv}^4 h_{Kv} (V - E_K)$$

$$\bar{g}_{Kv} = 4.5 \text{ [nS]}$$

Ca current

$$\alpha_{m_{Ca}} = \frac{240(68 - V)}{\exp((68 - V)/21) - 1}$$

$$\beta_{m_{Ca}} = \frac{800}{\exp((55 + V)/55) + 1}$$

$$\frac{dm_{Ca}}{dt} = \alpha_{m_{Ca}}(1 - m_{Ca}) - \beta_{m_{Ca}} m_{Ca}$$

$$E_{Ca} = 12.9 \log \frac{[Ca]_o}{[Ca]_d}$$

$$I_{Ca} = \bar{g}_{Ca} m_{Ca}^4 (V - E_{Ca})$$

$$\bar{g}_{Ca} = 9.0 \text{ [nS]}$$

$$[Ca]_o = 2.0 \text{ [mM]}$$

$$[Ca]_d = 30 \text{ [}\mu\text{M]}$$

Leakage current

$$I_l = \bar{g}_l (V - E_l)$$

$$\bar{g}_l = 0.5 \text{ [nS]}$$

$$E_l = -80 \text{ [mV]}$$

Capacitance

$$C = 0.106 \text{ [nF]}$$

TABLE 12.3
A Mathematical Formulation of Ionic Currents of a Bipolar Cell

Delayed rectifying K current

$$\alpha_{mKv} = \frac{400}{\exp(-(V-15)/36)+1}$$

$$\beta_{mKv} = \exp\left(-\frac{V}{13}\right)$$

$$\frac{dm_{Kv}}{dt} = \alpha_{mKv}(1-m_{Kv}) - \beta_{mKv}m_{Kv}$$

$$\alpha_{hKv} = 0.0003\exp\left(-\frac{V}{7}\right)$$

$$\beta_{hKv} = \frac{80}{\exp((V+115)/15)+1} + 0.02$$

$$\frac{dh_{Kv}}{dt} = \alpha_{hKv}(1-h_{Kv}) - \beta_{hKv}h_{Kv}$$

$$g_{Kv} = \bar{g}_{Kv}m_{Kv}^3 h_{Kv}$$

$$I_{Kv} = g_{Kv}(V-E_K)$$

$$\bar{g}_{Kv} = 2.0\ [\text{nS}]$$

$$E_K = -58\ [\text{mV}]$$

Transient outward K current

$$\alpha_{mA} = \frac{1200}{\exp(-(V-50)/28)+1}$$

$$\beta_{mA} = 6\exp\left(-\frac{V}{10}\right)$$

$$\frac{dm_A}{dt} = \alpha_{mA}(1-m_A) - \beta_{mA}m_A$$

$$\alpha_{hA} = 0.045\exp\left(-\frac{V}{13}\right)$$

$$\beta_{hA} = \frac{75}{\exp(-(V+30)/15)+1}$$

$$\frac{dh_A}{dt} = \alpha_{hA}(1-h_A) - \beta_{hA}h_A$$

$$g_A = \bar{g}_A m_A^3 h_A$$

$$I_A = g_A(V-E_K)$$

$$\bar{g}_A = 35\ [\text{nS}]$$

Capacitance

$$C = 10\ [\text{pF}]$$

Hyperpolarization-activated current

$$\alpha_h = \frac{3}{\exp((V+110)/15)+1}$$

$$\beta_h = \frac{1.5}{\exp(-(V+115)/15)+1}$$

$$M = [C_1\quad C_2\quad O_1\quad O_2\quad O_3]^t$$

$$K = \begin{bmatrix} -4\alpha_h & \beta_h & 0 & 0 & 0 \\ 4\alpha_h & -(3\alpha_h+\beta_h) & 2\beta_h & 0 & 0 \\ 0 & 3\alpha_h & -(2\alpha_h+2\beta_h) & 3\beta_h & 0 \\ 0 & 0 & 2\alpha_h & -(\alpha_h+3\beta_h) & 4\beta_h \\ 0 & 0 & 0 & \alpha_h & -4\beta_h \end{bmatrix}$$

$$\frac{d}{dt}M = KM$$

$$m_h = O_1 + O_2 + O_3$$

$$g_h = \bar{g}_h m_h$$

$$I_h = g_h(V-E_h)$$

$$\bar{g}_h = 0.975\ [\text{nS}]$$

$$E_h = -17.7\ [\text{mV}]$$

Ca current

$$\alpha_{mCa} = \frac{12000(120-V)}{\exp(-(V-120)/25)-1}$$

$$\beta_{mCa} = \frac{40000}{\exp((V+68)/25)+1}$$

$$\frac{dm_{Ca}}{dt} = \alpha_{mCa}(1-m_{Ca}) - \beta_{mCa}m_{Ca}$$

$$h_{Ca} = \frac{\exp(-(V-50)/11)}{\exp(-(V-50)/11)+1}$$

$$E_{Ca} = 12.9\log\left(\frac{[\text{Ca}^{2+}]_o}{[\text{Ca}^{2+}]_s}\right)$$

$$g_{Ca} = \bar{g}_{Ca}m_{Ca}^4 h_{Ca}$$

$$I_{Ca} = g_{Ca}(V-E_{Ca})$$

$$\bar{g}_{Ca} = 1.1\ [\text{nS}]$$

$$[\text{Ca}^{2+}]_o = 2500\ [\mu\text{M}]$$

Ca-dependent K current

$$\alpha_{mKc} = \frac{100(230-V)}{\exp((230-V)/52)-1}$$

$$\beta_{mKc} = 120\exp\left(-\frac{V}{95}\right)$$

$$\frac{dm_{Kc}}{dt} = \alpha_{mKc}(1-m_{Kc}) - \beta_{mKc}m_{Kc}$$

$$m_{Kc1} = \frac{[\text{Ca}^{2+}]_s}{[\text{Ca}^{2+}]_s + 0.2}$$

$$g_{Kc} = \bar{g}_{Kc}m_{Kc}^2 m_{Kc1}$$

$$I_{K(Ca)} = g_{Kc}(V-E_K)$$

$$\bar{g}_{Kc} = 8.5\ [\text{nS}]$$

Leakage current

$$I_l = gl(V-E_l)$$

$$gl = 0.23\ [\text{nS}]$$

$$E_l = -21\ [\text{mV}]$$

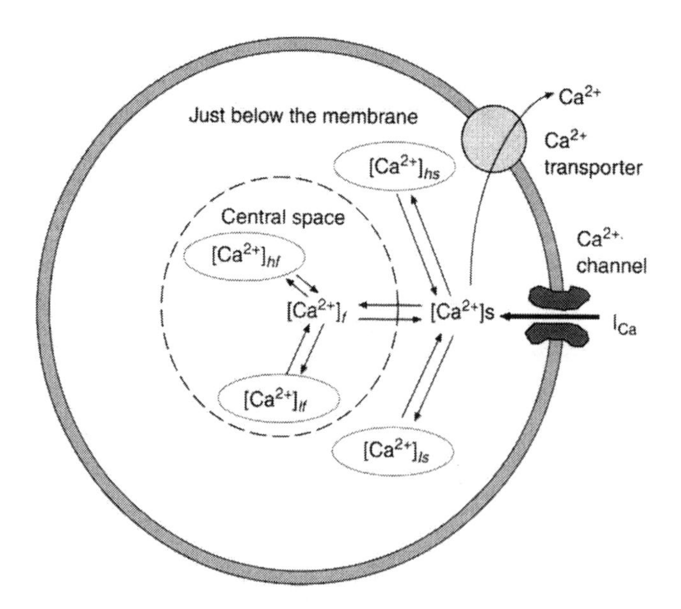

FIGURE 12.2 Compartmental model of intracellular calcium system (Kamiyama et al., 1996; Usui et al., 1996a). The rate of change of internal calcium concentration $[Ca^{2+}]_s$ below the membrane is determined by influx via I_{Ca}, by extrusion through the transporter, by diffusion into the central space ($[Ca^{2+}]_f$) and by binding and unbinding to high $[Ca^{2+}]_{hs}$ and low $[Ca^{2+}]_{ls}$ affinity buffers below the membrane. The calcium ions $[Ca^{2+}]_f$ in the central space also bind and unbind to the high $[Ca^{2+}]_{hf}$ and low $[Ca^{2+}]_{lf}$ affinity internal buffers.

TABLE 12.4
A Mathematical Description of the Calcium System of the Photoreceptor

Calcium concentration

$$\begin{aligned}\frac{d}{dt}[Ca^{2+}]_s &= -\frac{I_{Ca}}{2FV_1} - \frac{I_{ex}+I_{ex2}}{2FV_1} - D_{Ca}\frac{S_1}{V_1\delta}([Ca^{2+}]_s - [Ca^{2+}]_f) \\ &\quad + Lb_2[Ca^{2+}]_{ls} - Lb_1[Ca^{2+}]_s(BL - [Ca^{2+}]_{ls}) - Hb_1[Ca^{2+}]_s(BH - [Ca^{2+}]_{hs}) \\ &\quad + Hb_2[Ca^{2+}]_{hs}\end{aligned}$$

$$\begin{aligned}\frac{d}{dt}[Ca^{2+}]_f &= D_{Ca}\frac{S_1}{V_2\delta}([Ca^{2+}]_s - [Ca^{2+}]_f) + Lb_2[Ca^{2+}]_{lf} - Lb_1[Ca^{2+}]_f(BL - [Ca^{2+}]_{lf}) \\ &\quad - Hb_1[Ca^{2+}]_f(BH - [Ca^{2+}]_{hf}) + Hb_2[Ca^{2+}]_{hf}\end{aligned}$$

Calcium buffer concentration

$$\frac{d}{dt}[Ca^{2+}]_{ls} = Lb_1[Ca^{2+}]_s(BL - [Ca^{2+}]_{ls}) - Lb_2[Ca^{2+}]_{ls}$$

$$\frac{d}{dt}[Ca^{2+}]_{hs} = Hb_1[Ca^{2+}]_s(BH - [Ca^{2+}]_{hs}) - Hb_2[Ca^{2+}]_{hs}$$

$$\frac{d}{dt}[Ca^{2+}]_{lf} = Lb_1[Ca^{2+}]_f(BL - [Ca^{2+}]_{lf}) - Lb_2[Ca^{2+}]_{lf}$$

$$\frac{d}{dt}[Ca^{2+}]_{hf} = Hb_1[Ca^{2+}]_f(BH - [Ca^{2+}]_{hf}) - Hb_2[Ca^{2+}]_{hf}$$

Calcium transporters

$$I_{ex} = I_{max}\exp\left(-\frac{V+14}{70}\right)\frac{([Ca^{2+}]_s - Ca_e)}{([Ca^{2+}]_s - Ca_e) + 2.3}$$

$$I_{ex2} = I_{max2}\frac{([Ca^{2+}]_s - Ca_e)}{([Ca^{2+}]_s - Ca_e) + 0.5}$$

TABLE 12.5
Parameters of the Calcium System of the Photoreceptor

F	Faraday constant, 9.649×10^4 [C/mol]
V_1	volume of submembrane area, 3.812×10^{-13} [ℓ]
V_2	volume of deep intracellular area, 5.236×10^{-13} [ℓ]
δ	diffusion distance, 3×10^{-4} [cm]
S_1	submembrane area, 3.14×10^{-6} [cm^2]
D_{Ca}	Ca diffusion coefficient, 6.000×10^{-6} [$cm^2\ sec^{-1}$]
I_{max}	maximum Na^+–K^+ exchanger current, 20 [pA]
I_{max2}	maximum Ca-ATPase pump current, 20 [pA]
Ca_e	minimum intracellular Ca concentration, 0.01 [μM]
BH	maximum concentration of high-affinity buffer, 300 [μM]
BL	maximum concentration of low-affinity buffer, 500 [μM]
Lb_1	on rate constant to low-affinity buffer, 0.4 [$sec^{-1}\ \mu M^{-1}$]
Lb_2	off rate constant to low-affinity buffer, 0.2 [sec^{-1}]
Hb_1	on rate constant to high affinity buffer, 100 [$sec^{-1}\ \mu M^{-1}$]
Hb_2	off rate constant to high affinity buffer, 90 [sec^{-1}]

TABLE 12.6
A Mathematical Description of the Calcium System of the Bipolar Cell

Calcium concentration

$$\begin{aligned}\frac{d[Ca^{2+}]_s}{dt} &= -\frac{I_{Ca}}{2FV_s} - \frac{D_{Ca}S_{sd}}{V_s d_{sd}}([Ca^{2+}]_s - [Ca^{2+}]_d) - \frac{(I_{ex} + I_{ex2})}{2FV_s}\\ &\quad + \beta_{bl}[Ca^{2+}]_{bls} - \alpha_{bl}[Ca^{2+}]_s([Ca^{2+}]_{blmax} - [Ca^{2+}]_{bls})\\ &\quad + \beta_{bh}[Ca^{2+}]_{bhs} - \alpha_{bh}[Ca^{2+}]_s([Ca^{2+}]_{bhmax} - [Ca^{2+}]_{bhs})\end{aligned}$$

$$\begin{aligned}\frac{d[Ca^{2+}]_d}{dt} &= \frac{D_{Ca}S_{sd}}{V_d d_{sd}}([Ca^{2+}]_s - [Ca^{2+}]_d)\\ &\quad + \beta_{bl}[Ca^{2+}]_{bld} - \alpha_{bl}[Ca^{2+}]_d([Ca^{2+}]_{blmax} - [Ca^{2+}]_{bld})\\ &\quad + \beta_{bh}[Ca^{2+}]_{bhd} - \alpha_{bh}[Ca^{2+}]_d([Ca^{2+}]_{bhmax} - [Ca^{2+}]_{bhd})\end{aligned}$$

Calcium buffer concentration

$$\frac{d[Ca^{2+}]_{bls}}{dt} = \alpha_{bl}[Ca^{2+}]_s([Ca^{2+}]_{blmax} - [Ca^{2+}]_{bls}) + \beta_{bl}[Ca^{2+}]_{bls}$$

$$\frac{d[Ca^{2+}]_{bhs}}{dt} = \alpha_{bh}[Ca^{2+}]_s([Ca^{2+}]_{bhmax} - [Ca^{2+}]_{bhs}) + \beta_{bh}[Ca^{2+}]_{bhs}$$

$$\frac{d[Ca^{2+}]_{bld}}{dt} = \alpha_{bl}[Ca^{2+}]_d([Ca^{2+}]_{blmax} - [Ca^{2+}]_{bld}) + \beta_{bl}[Ca^{2+}]_{bld}$$

$$\frac{d[Ca^{2+}]_{bhd}}{dt} = \alpha_{bh}[Ca^{2+}]_d([Ca^{2+}]_{bhmax} - [Ca^{2+}]_{bhd}) + \beta_{bh}[Ca^{2+}]_{bhd}$$

Calcium pump and exchanger

$$I_{ex} = \frac{J_{ex}([Ca^{2+}]_s - [Ca^{2+}]_{min})}{[Ca^{2+}]_s - [Ca^{2+}]_{min} + 2.3}\exp\left(-\frac{V + 14}{70}\right)$$

$$I_{ex2} = \frac{J_{ex2}([Ca^{2+}]_s - [Ca^{2+}]_{min})}{[Ca^{2+}]_s - [Ca^{2+}]_{min} + 0.5}$$

TABLE 12.7
Parameters of the Calcium System of the Bipolar Cell

F	Faraday constant, 9.649×10^4 C mol^{-1}
D_{Ca}	Ca diffusion coefficient, 6.000×10^{-8} dm^2 sec^{-1}
V_s	volume of the submembrane area, 1.692×10^{-13} dm^3
V_d	volume of the deep intracellular area, 7.356×10^{-13} dm^3
S_{sd}	surface area of the submembrane and the deep intracellular area spherical boundary, 4.000×10^{-8} dm^2
d_{sd}	distance between submembrane area and the deep intracellular area, 5.800×10^{-5} dm
$[Ca^{2+}]_{blmax}$	total low-affinity buffer concentration, 400 μM
$[Ca^{2+}]_{bhmax}$	total high-affinity buffer concentration, 300 μM
α_{bl}, β_{bl}	on and off rate constants for the binding of Ca to low-affinity buffer, 0.4 sec^{-1} μM^{-1}, 0.2 sec^{-1}
α_{bh}, β_{bl}	on and off rate constants for the binding of Ca to high-affinity buffer, 100 sec^{-1} μM^{-1}, 90 sec^{-1}
J_{ex}	maximum Na–Ca exchanger current, 360 pA
J_{ex2}	maximum Ca-ATPase exchanger current, 380 pA
$[Ca^{2+}]_{min}$	minimum intracellular Ca concentration for Ca extrusion, 0.01 μM

12.3 MODEL OF RETINAL CELLS

12.3.1 Photoreceptor

12.3.1.1 Ionic Currents in Photoreceptors

There are at least three voltage-dependent ionic currents and two calcium-dependent ionic currents in the photoreceptor inner segments, namely, hyperpolarization-activated current I_h, delayed rectifying potassium current I_{Kv}, calcium current I_{Ca}, calcium-dependent potassium current $I_{K(Ca)}$, and calcium-dependent chloride current $I_{Cl(Ca)}$. These currents have been observed in a variety of rods, salamander cones, and lizard cones (Bader et al., 1982; Maricq and Korenbrot, 1988, 1990a, 1990b; Barnes and Hille, 1989). Since both rod and cone inner segments have similar ionic currents (Bader et al., 1982; Barnes and Hille, 1989), we made no distinction between rods and cones when referring to these ionic currents.

12.3.1.2 Voltage-Clamp Simulation

Voltage-clamp simulations were made under conditions identical to electrophysiological experiments by Maricq and Korenbrot (1988). Figure 12.3A shows the simulated total ionic current under three different pharmacological conditions corresponding to control, TEA-containing, and TEA- and Co^{2+}-containing solutions (Kamiyama et al., 1996). In these conditions, external solutions always contain Cs^+, and thus we blocked I_h completely. Figure 12.3B shows the current–voltage relation measured from the TEA responses. The currents were measured at the peak, 5 msec after the onset of the depolarizing pulse, and 135 msec after the onset of the command pulse. The experimental observations are well reproduced by our model.

Each ionic current simulated under control conditions is shown in Figure 12.3C. From these responses we can explain how the complicated current responses in Figure 12.3A are generated. An early inward current elicited by the voltage step to −20 mV is I_{Ca}, and the following outward component is mostly $I_{Cl(Ca)}$ due to the increase of the calcium concentration $[Ca^{2+}]_s$. The voltage step to 0 and +20 mV activates the outward currents, I_{Kv} and $I_{Cl(Ca)}$. The voltage step to +40 mV activates I_{Ca} to a lesser extent than the voltage step to 0 or +20 mV, thus the calcium concentration increases slowly and is reduced compared to these voltage-clamp levels. As a result, $I_{Cl(Ca)}$ is smaller and slower compared to the current elicited by 0 and +20 mV voltage steps.

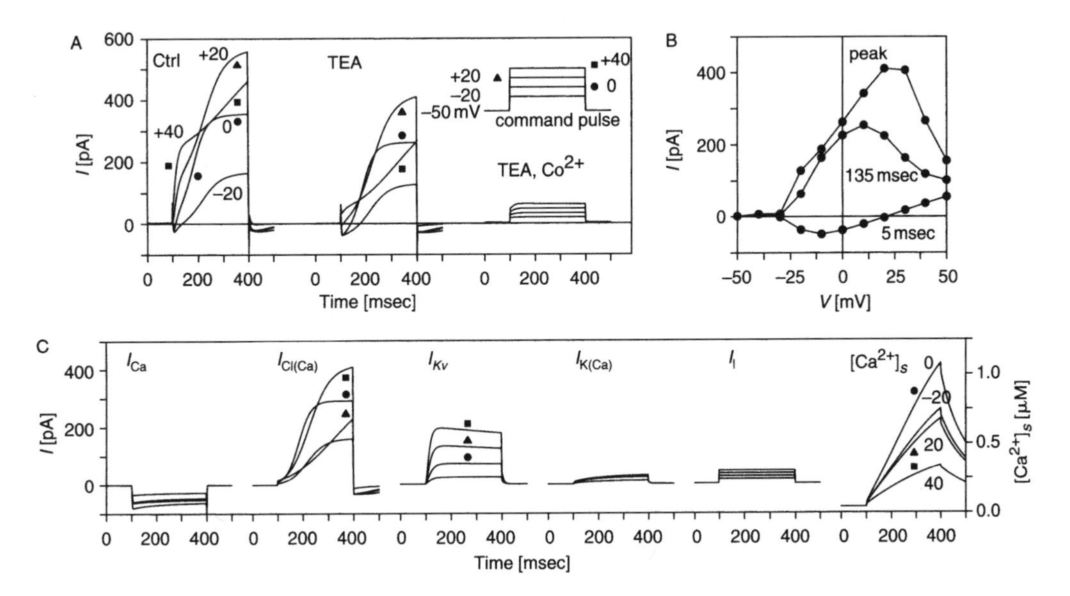

FIGURE 12.3 Simulated responses of the inner-segment model under voltage-clamp conditions (Kamiyama et al., 1996). A: Membrane currents under three different conditions corresponding to control, TEA, and TEA–Co solutions. The TEA responses were simulated by reducing $\bar{g}_{Kv}$ from 2.0 to 0.16 nS. B: Current–voltage relation at the peak, 5 msec, 135 msec, measured from the TEA responses. C: Ionic current components and calcium concentration under control conditions.

12.3.1.3 Current-Clamp Simulation

The ionic current models are derived from the data obtained from voltage-clamp experiments, and thus there are no interactions among ionic currents. Here, we simulate current-clamp responses to validate the proposed model using an independent set of data. The current-clamp simulation is a critical test of the ionic current model, since there is no guarantee that the model can reproduce the current-clamp responses, in which the simulation has to be made by use of all the ionic components.

Maricq and Korenbrot (1988) showed that, in isolated lizard cones, action potentials can be generated by depolarizing current steps. The corresponding spike responses simulated by our model are shown in Figure 12.4. The responses elicited by the depolarizing current under control and TEA conditions are reproduced well. Dotted traces in Figure 12.4A,B are results simulated by blocking $I_{Cl(Ca)}$ in the model. Current dynamics during spike responses are shown in Figure 12.4C,D. From these results, we can understand which ionic currents participate in the different phases of the spike responses. The membrane depolarization initially elicited by the applied current activates I_{Ca}, and I_{Ca} further depolarizes the cell. I_{Kv} and $I_{Cl(Ca)}$ participate in repolarization of the action potential in phases 1 and 2, respectively. As clearly shown by the dotted traces in Figure 12.4A,B, the membrane potentials remain depolarized without $I_{Cl(Ca)}$, that is, $I_{Cl(Ca)}$ terminates the action potential.

12.3.1.4 Simulated Light Responses of the Rod Photoreceptor

In general, ionic current models are typically used in the context of voltage- and current-clamp experiments. They can also be used in studies investigating several sorts of dynamical behaviors. Here, we simulate the light response of rod photoreceptors.

The biophysical processes underlying phototransduction are now well understood (Yau and Baylor, 1989; McNaughton, 1990; Pugh and Lamb, 2000; Ebrey and Koutalos, 2001), and a quantitative model of phototransduction has been proposed by Forti et al. (1989). In order to

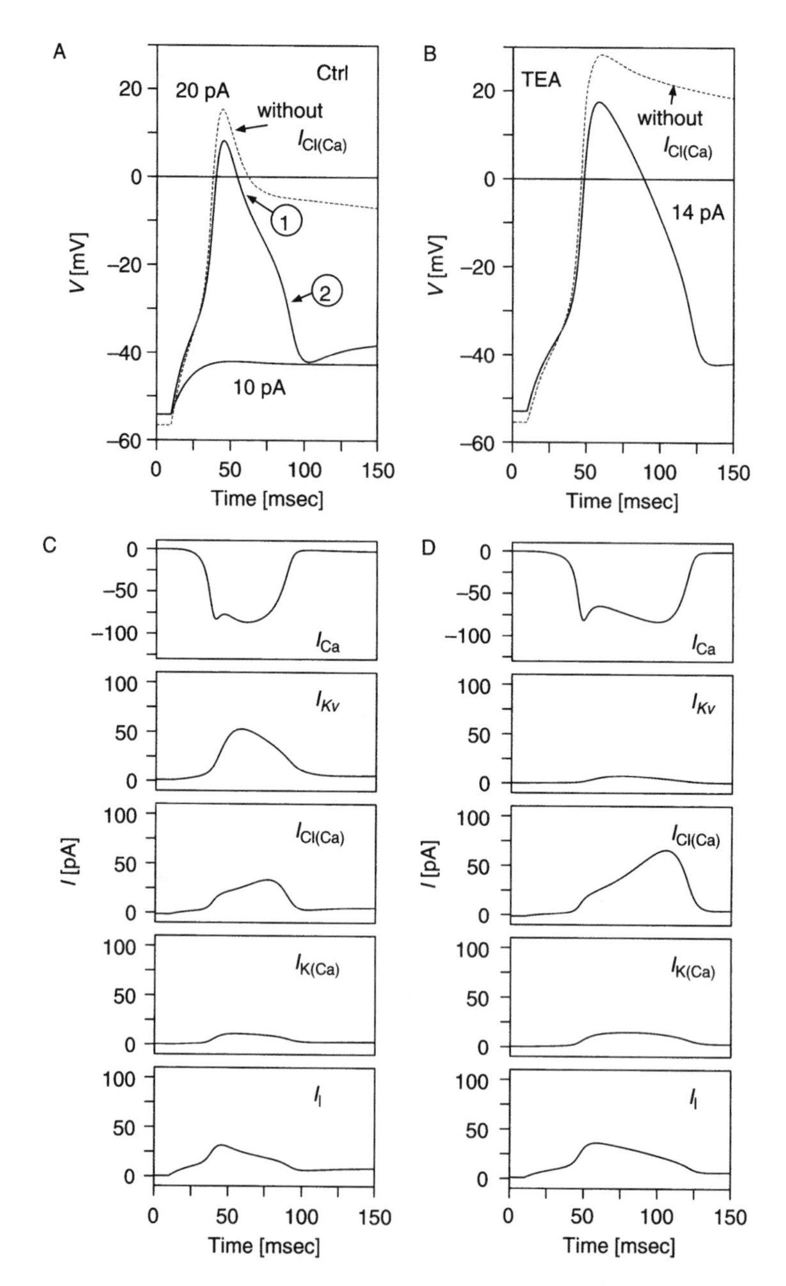

FIGURE 12.4 Simulations of the current-clamp experiments using the inner segment model (Kamiyama et al., 1996). A: Voltage responses of the model to depolarizing current steps of 10 and 20 pA under control condition. The response without $I_{Cl(Ca)}$ is plotted by the denoted line. B: Voltage response of the model to a depolarizing current step of 14 pA under the TEA condition. $\bar{g}_{Kv} = 0.16$ nS for the TEA condition. Dotted trace was simulated by blocking $I_{Cl(Ca)}$. C, D: Ionic-current components under control and TEA conditions, respectively.

simulate the light response of photoreceptors, we introduced a flow of photocurrent into the inner segment as shown in Figure 12.1. In the model of phototransduction (Forti et al., 1989), the cyclic GMP-activated current J is expressed as a function of the concentration of cyclic GMP g. The cyclic GMP-activated current can be measured in an excised-patch condition, and it is sufficient to use the current, J, for reproducing the current response in the outer segment. The current–voltage relation

TABLE 12.8
Model of Phototransduction in a Rod Photoreceptor

Change of intracellular Ca^{2+}

$$\frac{dc}{dt} = bJ - \gamma_{Ca}(c - c_0) - k_1(e_T - c_b)c + k_2 c_b$$

$$\frac{dc_b}{dt} = k_1(e_T - c_b)c - k_2 c_b$$

$b = 0.25$ [μM sec^{-1} pA^{-1}], $\gamma_{Ca} = 50$ [sec^{-1}],

$k_1 = 0.2$ [sec^{-1} μM], $k_2 = 0.8$ [sec^{-1}], $e_T = 500$ [μM]

Change of intracellular cGMP

$$\frac{dg}{dt} = A - g(\bar{V} + \sigma \mathrm{PDE}^*)$$

$$A = \frac{A_{max}}{1 + (c/K_c)^4}$$

$\sigma = 1$ [sec^{-1} μM^{-1}], $A_{max} = 65.6$ [μM sec^{-1}]

$\bar{V} = 0.4$ [sec^{-1}], $K_c = 0.1$ [μM]

Mechanism of phototransduction

$$\frac{dRh}{dt} = J_{h\nu}(t) - \alpha_1 Rh + \alpha_2 Rh_i$$

$$\frac{dRh_i}{dt} = \alpha_1 Rh - (\alpha_2 + \alpha_3)Rh_i$$

$$\frac{dT^*}{dt} = \varepsilon Rh(T_{Tot} - T^*) - \beta_1 T^* + \tau_2 \mathrm{PDE}^* - \tau_1 T^*(\mathrm{PDE}_{Tot} - \mathrm{PDE}^*)$$

$$\frac{d\mathrm{PDE}^*}{dt} = \tau_1 T^*(\mathrm{PDE}_{Tot} - \mathrm{PDE}^*) - \tau_2 \mathrm{PDE}^*$$

$\alpha_1 = 50$ [sec^{-1}], $\alpha_2 = 0.0003$ [sec^{-1}], $\alpha_3 = 0.03$ [sec^{-1}],

$T_{Tot} = 1000$ [μM], $\tau_1 = 0.2$ [sec^{-1} μM^{-1}], $\tau_2 = 5$ [sec^{-1}]

$\varepsilon = 0.5$ [sec^{-1} μM^{-1}], $\mathrm{PDE}_{Tot} = 100$ [μM]

Photosensitive current

$$J = J_{max}\frac{g^3}{g^3 + K^3}$$

$$I_{photo} = -J(1 - \exp(V - 8.5)/17)$$

$J_{max} = 5040$ [pA], $K^3 = 1000$ [μM]

of the cyclic GMP-activated current shows an outward rectification reversing near +5 mV, and thus the photocurrent flow into the inner segment depends on the membrane potential (Baylor and Nunn, 1986). We expressed the voltage dependence of the photocurrent I_{photo} by multiplying the cyclic GMP-activated current J by the electrochemical driving force term as shown in Table 12.8.

Figure 12.5 shows the responses of photocurrent (12.5A) and photovoltage (12.5B) to a series of light flashes of increasing intensity. The model produces responses highly similar to those measured experimentally (Baylor et al., 1984). In order to assess the importance of each ionic current in shaping the voltage response to light, we estimated the ionic current flow during the light responses as shown in Figure 12.5C. From these responses, we can explain why the voltage response of a rod to a bright flash is not the mirror image of the photocurrent. In normal conditions, I_h activates slowly as the membrane hyperpolarizes, and thus I_h shapes the initial transient in the voltage response. To confirm the effects of I_h, we simulated the responses by blocking I_h in the model as shown in Figure 12.5D. The voltage responses changed dramatically. The initial transient for bright stimuli disappeared, the time to peak of the voltage response was delayed, the amplitude was increased, and the duration of the response was prolonged. These results are consistent with the experimental results by Baylor et al. (1984), who showed that the initial spike characteristics of rod responses was abolished by the I_h blocker, Cs.

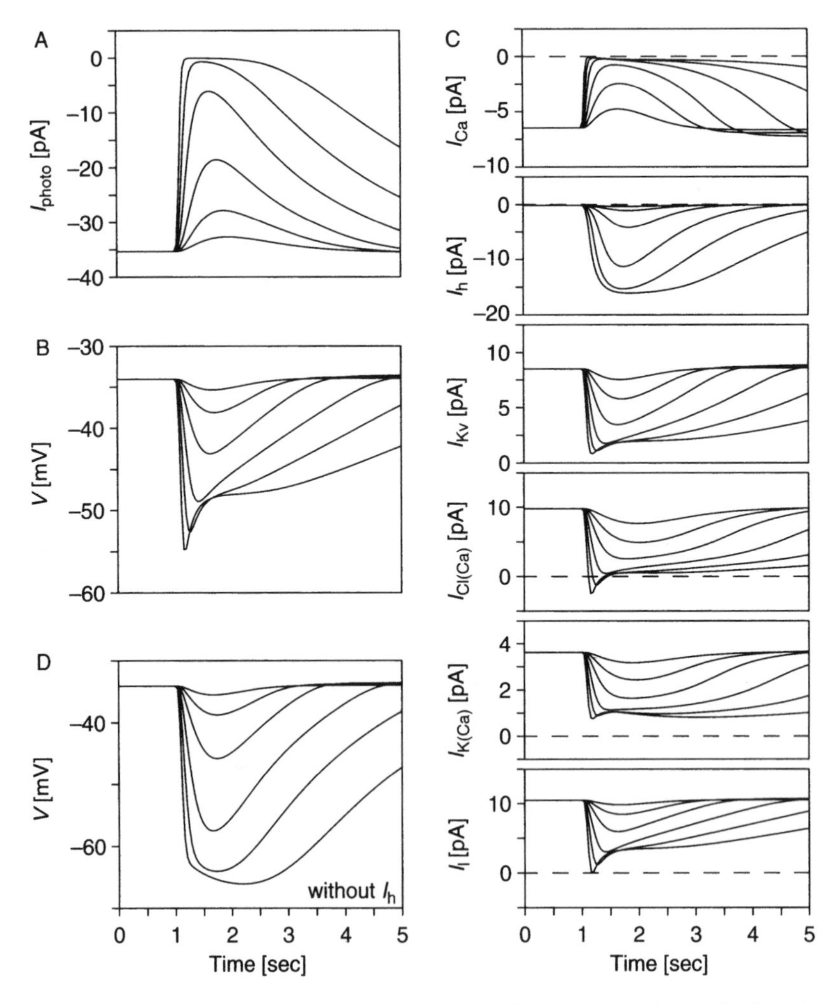

FIGURE 12.5 Simulated light responses of the photoreceptor (Kamiyama et al., 1996). Stimulus intensities $J_{h\nu}(t)$ were 0.03, 0.1, 0.3, 1.0, 3.0, 10.0 $\text{Rh}_t^* \text{ sec}^{-1}$. The duration was 50 msec. A: Photocurrents. B: Voltage responses. C: Each ionic current of the inner segment. D: Voltage responses without I_h. $\text{Cl}_h = 0.45\ \mu\text{M}$, $A_{max} = 131.2\ \mu\text{M sec}^{-1}$, $\bar{V} = 0.8\ \text{sec}^{-1}$ for simulating the rod responses.

12.3.2 Horizontal Cell

12.3.2.1 Ionic Currents in Horizontal Cells

Ionic currents in horizontal cells have been measured in a variety of animals including goldfish (Tachibana, 1981, 1983), catfish (Shingai and Christensen, 1983), white perch (Lasater, 1986), white bass (Sullivan and Lasater, 1990), cat (Ueda et al., 1992), and rabbit (Löhrke and Hofmann, 1994). Five types of voltage-dependent ionic currents have been found in the cell body: a sodium current I_{Na}, a calcium current I_{Ca}, a delayed rectifying potassium current $I_{\mathrm{K}v}$, a transient outward potassium current I_{A}, and an anomalous rectifying potassium current $I_{\mathrm{K}a}$. Glutamate, which is the neurotransmitter of photoreceptors, activates I_{glu} and suppresses $I_{\mathrm{K}a}$ by decreasing the number of active channels (Ishida et al., 1984; Kaneko and Tachibana, 1985b; Tachibana, 1985; O'Dell and Christensen, 1989; Massey, 1990; Lu et al., 1998; Shen et al., 1999). One of the differences in the ionic currents between lower-vertebrate and mammalian horizontal cells is the time course of the calcium current. I_{Ca} is inactivated by the accumulation of intracellular calcium ions in goldfish horizontal cells (Tachibana, 1983); however, this inactivation has never been observed in cat (Ueda et al., 1992) and rabbit (Löhrke and Hofmann, 1994) horizontal cells. The calcium-dependent potassium current $I_{\mathrm{K}c}$ is found in rabbit B-type horizontal cells (Löhrke and Hofmann, 1994), but has never been observed in lower vertebrates and cat. Figure 12.6 shows the simulated ionic-current responses elicited by voltage steps from the holding potential to various command voltages. The experimental properties of each current (Ueda et al., 1992; Löhrke and Hofmann, 1994) are reproduced well.

12.3.2.2 Simulated Action Potentials

Repetitive action potentials were observed in dissociated rabbit A-type horizontal cells (Blanco et al., 1996). In response to a large-amplitude current injection, an action potential or repetitive action potentials were observed in 7 of 17 horizontal cells. Long-lasting depolarization was observed in the rest of the cells. The underlying mechanism of spike generation was not clarified at that time. Although the horizontal cells normally respond to light with a graded potential, it is of interest to elucidate how the cells generate the repetitive spikes.

Figure 12.7A shows simulated voltage responses to current injection under control conditions. The membrane potential is kept at -80 mV, and currents of 10, 25, 40, and 70 pA are injected. Figure 12.7B shows the ionic current responses during current injection. In control conditions, repetitive action potentials are not observed, and only a long-lasting depolarization is observed. The membrane depolarization induced by current pulse over 25 pA activates the calcium current I_{Ca}, and the activated I_{Ca} further depolarizes the membrane. The delayed rectifier $I_{\mathrm{K}v}$ is activated slower than I_{Ca} as is clearly seen in Figure 12.7B. When the current injection is terminated, the potential stabilizes at a depolarized level in which I_{Ca} is balanced by $I_{\mathrm{K}v}$. This simulated result corresponds well with the long-lasting Co^{2+}-sensitive action potentials observed in the horizontal cell of the cat (Ueda et al., 1992) and goldfish (Tachibana, 1981).

Repetitive spikes require not only a calcium current but also ionic currents which hyperpolarize the membrane. In order to determine which current is involved in repetitive action potentials, we analyzed individual ionic currents. The current hyperpolarizing the membrane should flow outward and should activate slower than the calcium current. The delayed rectifying potassium current $I_{\mathrm{K}v}$ is the most likely candidate since this current is an outward current activated by depolarization. We simulated the response to current injection in models with different values of the maximum conductance of $I_{\mathrm{K}v}$. When the conductance is about eight times larger than the control condition, repetitive action potentials are generated. Figure 12.7C,D shows the voltage responses and the ionic current responses to current injections in the model with a large conductance of $I_{\mathrm{K}v}$. When a current pulse with amplitude over 25 pA was applied, an action potential or repetitive action potentials were produced by the model. These response are highly similar to those measured experimentally (Blanco et al., 1996).

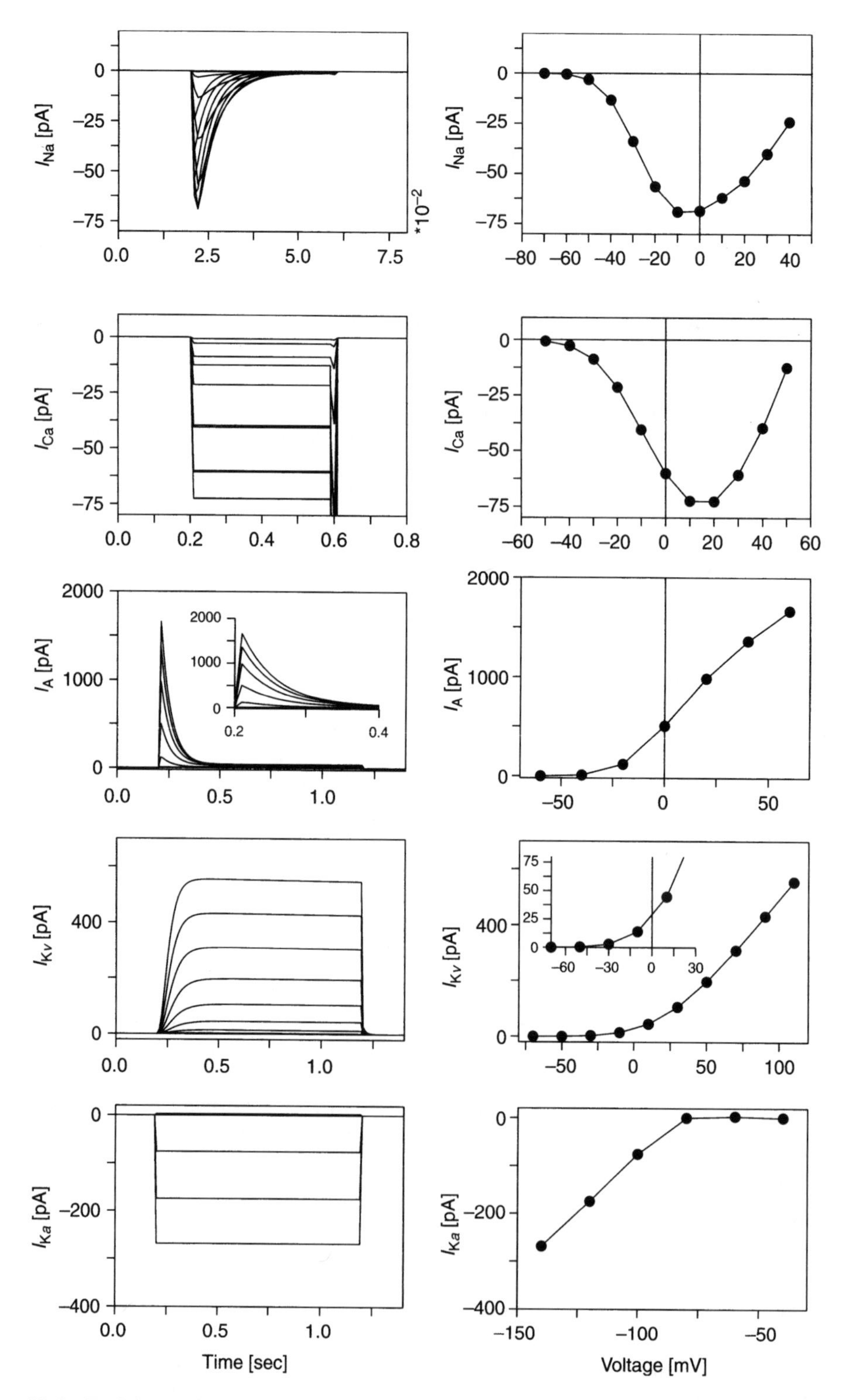

FIGURE 12.6 Each ionic current response of the horizontal cell model under voltage-clamp conditions. Each panel shows the time courses (left) and the I–V characteristics (right) for each ionic current. I_{Na}:V_h is −80 mV and V_c is −70 to 40 mV (10 mV steps). I_{Ca}:V_h is −80 mV and V_c is −50 to 50 mV (10 mV steps). I_A:V_h is −80 mV and V_c is −60 to 60 mV (20 mV steps). I_{Kv}:V_h is −80 mV and V_c is −70 to 110 mV (20 mV steps). I_{Ka}:V_h is −40 mV and V_c is −140 to 20 mV (20 mV steps).

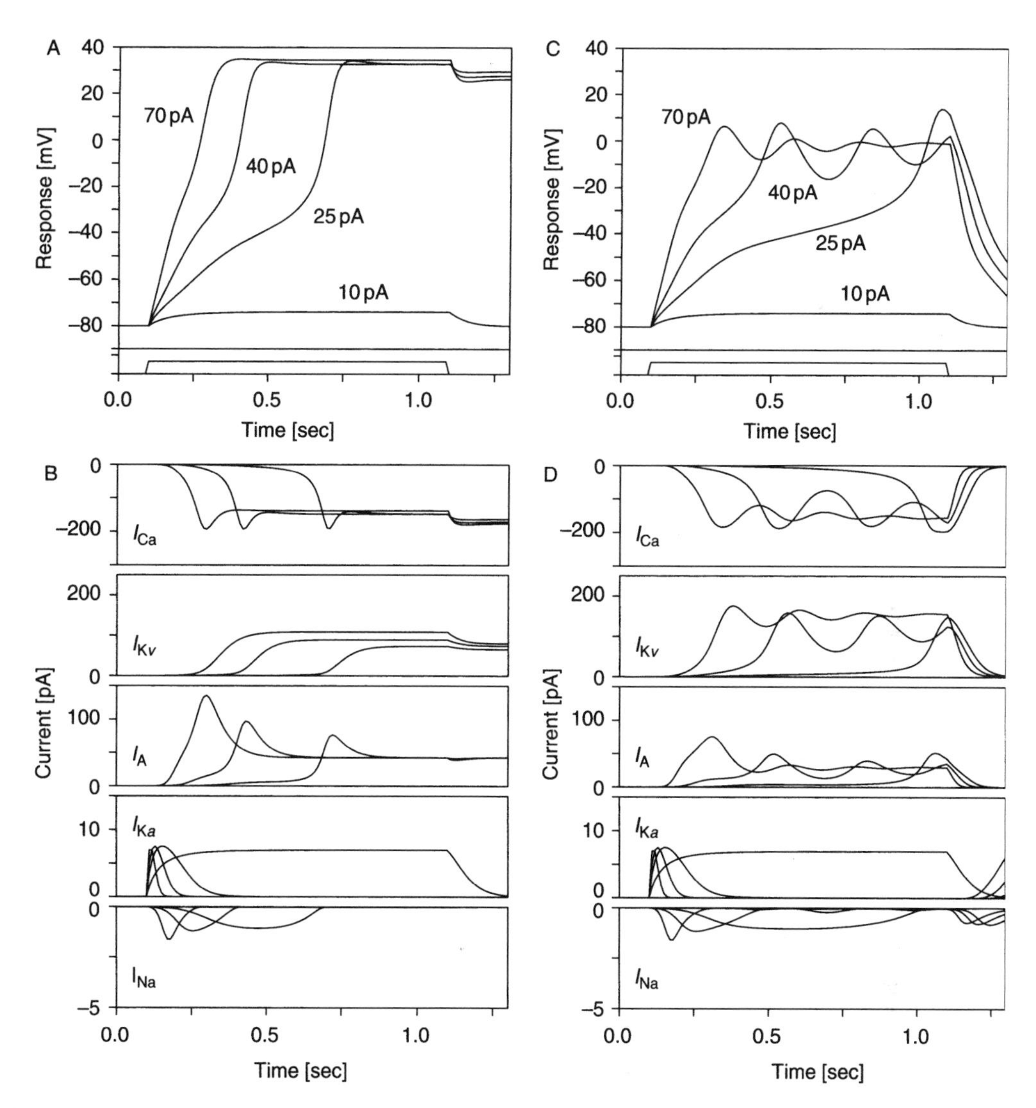

FIGURE 12.7 Simulated voltage responses to current injections. The membrane potential is kept at −80 mV, and currents of 10, 25, 40, and 70 pA are injected. A: The current injection induced a long-lasting depolarization in the control condition. B: The long-lasting depolarization is caused by the activation of the calcium current. C: Repetitive action potentials were generated in the model with a large conductance of I_{Kv} ($\bar{g}_{Kv} = 34$ nS). D: I_{Kv} and I_{Ca} are responsible for the repetitive action potentials.

12.3.3 Bipolar Cell

12.3.3.1 Ionic Currents in Bipolar Cells

The bipolar cell is morphologically divided into three compartments; the dendrite, the cell body, and the axon terminal. Each compartment contains unique ionic channels. Channels activated by the transmitter released from photoreceptors are distributed over the dendritic compartment (Nawy and Jahr, 1990, 1991). There are at least five membrane currents in the bipolar cell body, namely, hyperpolarization activated current I_h, delayed rectifying potassium current I_{Kv}, transient outward potassium current I_A, calcium current I_{Ca}, and voltage- and calcium-dependent potassium current $I_{K(Ca)}$ (Kaneko and Tachibana, 1985a; Lasater, 1988). Calcium channels are distributed not only over the cell body but are also present in high density over the axon terminal (Heidelberger and Matthews, 1992; Tachibana et al., 1993). The axon terminal has four ionic currents (Kaneko and Tachibana, 1985a; Lasater, 1988), two calcium-dependent currents (Kobayashi and Tachibana, 1995; Okada et al., 1995; Sakaba et al., 1997), electrogenic Ca^{2+} exchanger and pump (Kobayashi and Tachibana, 1995), and GABA- or glycine-induced currents (Attwell et al., 1987; Tachibana and Kaneko, 1988; Lukasiewicz et al., 1994; Matthews et al., 1994). There are two types of glutamate induced currents I_{glu} in the bipolar cells, namely, ionotropic and metabotropic types (Attwell et al., 1987; Nawy and Copenhagen, 1990; Nawy and Jahr, 1990; Shiells and Falk, 1994; Euler et al., 1996).

12.3.3.2 Voltage-Clamp Simulation

Figure 12.8A shows the simulated total ionic current in response to depolarizing command pulses to -10, $+10$, and $+30$ mV from a holding potential of -30 mV. An outward current develops with further depolarization and the amplitude peaks near a membrane potential of $+50$ mV (Figure 12.8C). The outward current is decreased by depolarization beyond $+50$ mV and shows little time dependency. Individual ionic currents and calcium concentrations change in response to the depolarizing pulses shown in Figure 12.8D. One may expect that since the cytosolic concentration of calcium ions changes slowly, a tail current would appear at the cessation of the command pulse. However, the calcium concentration $[Ca^{2+}]_s$ near the ionic channels returns within a few milliseconds after the command pulse, and therefore I_{Ca} and $I_{\mathrm{K(Ca)}}$ show no tail currents. These results explain why no tail current has been observed with the depolarizing pulse (Kaneko and Tachibana, 1985a; Lasater, 1988).

The total current induced by the hyperpolarizing pulses to -50, -80, and -100 mV is shown in Figure 12.8B. These inward currents show a strong time dependence, and tail components are observed as in the experimental results of Kaneko and Tachibana (1985a). Figure 12.8E provides an interpretation of these current responses to voltage pulses. The slowly activating inward current results from the time- and voltage-dependent properties of I_{h}. The tail currents following the hyperpolarizing steps are caused by I_{h} and $I_{\mathrm{K}\nu}$, which are activated by the depolarizing pulse of -30 mV from the hyperpolarizing potentials.

12.3.3.3 Current-Clamp Simulation

Kaneko and Tachibana (1985a) demonstrated that bipolar cells showed sustained depolarizations to depolarizing current pulses and, in response to hyperpolarizing pulses, showed complicated responses with an initial transient hyperpolarization followed by a smaller plateau hyperpolarized level. These response properties are well reproduced by our model as shown in Figure 12.9 and can be explained as follows.

Figure 12.9A shows that membrane depolarization gradually reaches a saturated value of around $+10$ mV as the injecting current is increased. This feature is determined by the magnitude of the currents, I_{Ca} and $I_{\mathrm{K(Ca)}}$. The initial transient and the subsequent plateau in response to the hyperpolarizing currents in Figure 12.9B is caused by activation of I_{h}, that is, activation of I_{h} slowly reduces the level of hyperpolarization. The spike-like rebound shown in Figure 12.9B at the cessation of the hyperpolarizing current pulses and after-hyperpolarization results from I_{h} and $I_{\mathrm{K}\nu}$.

12.3.4 Ganglion Cell

12.3.4.1 Stochastic Model of Spike Generation

Voltage-clamp studies of retinal ganglion cells have identified voltage- or ion-gated currents, which appear to play a role in generating spikes (Ishida, 1995). In the previous studies, the

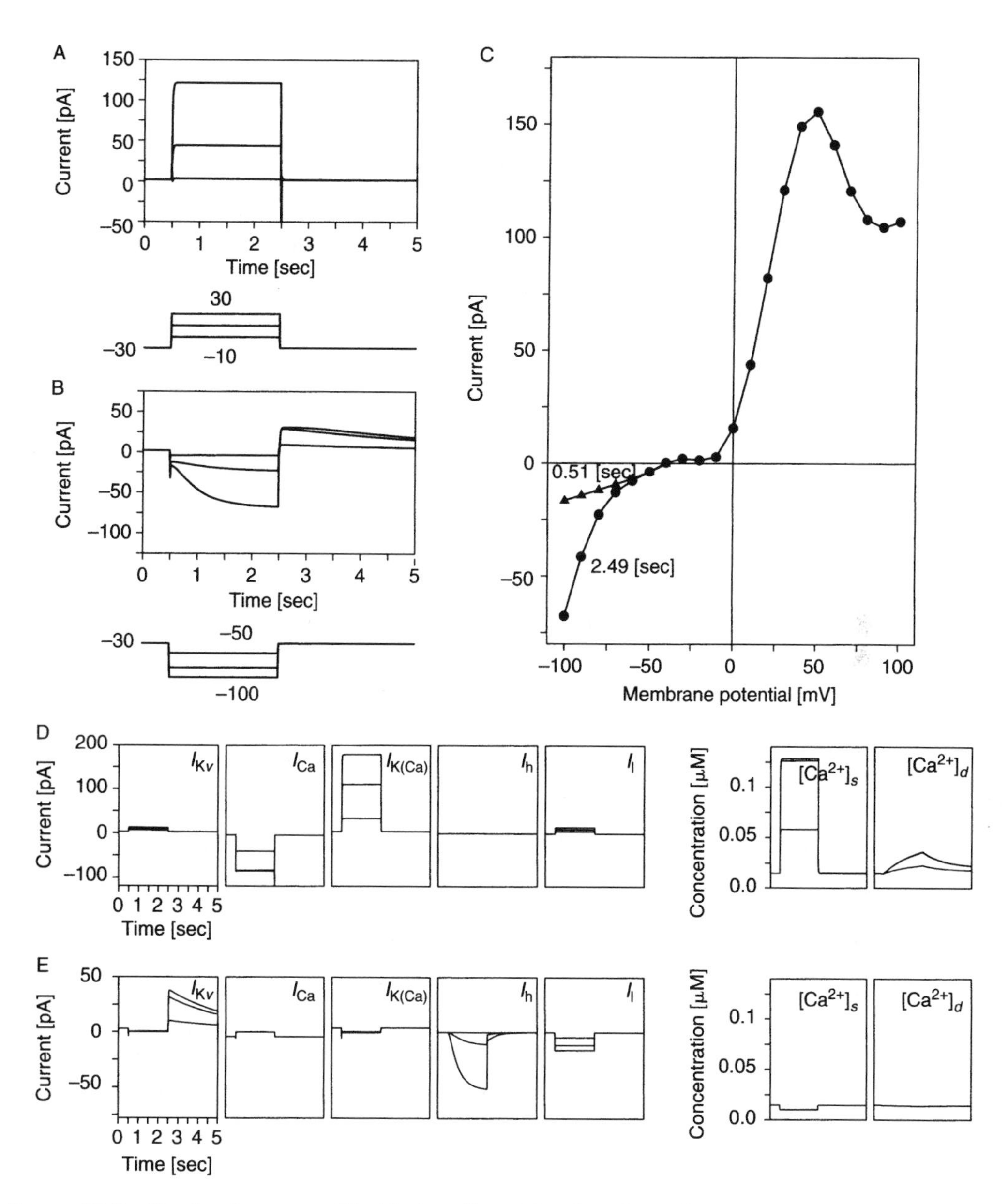

FIGURE 12.8 Simulated response of the bipolar cell model under voltage-clamp conditions (Usui et al.,1996a). The cell was held at −30 mV and 2 sec command pulses were applied in 10 mV steps. A: Simulated total ionic current to depolarizing command pulses to −10, +10, and +30 mV. B: Simulated total ionic current to hyperpolarizing command pulses to −50, −80, and −100 mV. C: *I–V* relationship under voltage clamp conditions. The *I–V* curves were obtained from the current data measured at 0.51 sec (triangle) and 2.49 sec (filled circle). D: Ionic currents and calcium concentrations under voltage-clamp conditions in response to the depolarizing pulses. E: Ionic currents and calcium concentrations under voltage-clamp conditions in response to the hyperpolarizing pulses.

ionic conductances have been modeled by means of deterministic differential equations similar to the Hodgkin–Huxley formulation (Hodgkin and Huxley, 1952b; Fohlmeister and Miller, 1997a). Recently, however, it was suggested that the stochastic properties of ionic channels are critical in determining the reliability and accuracy of the neuron firing (Schneidman et al., 1998). It is important,

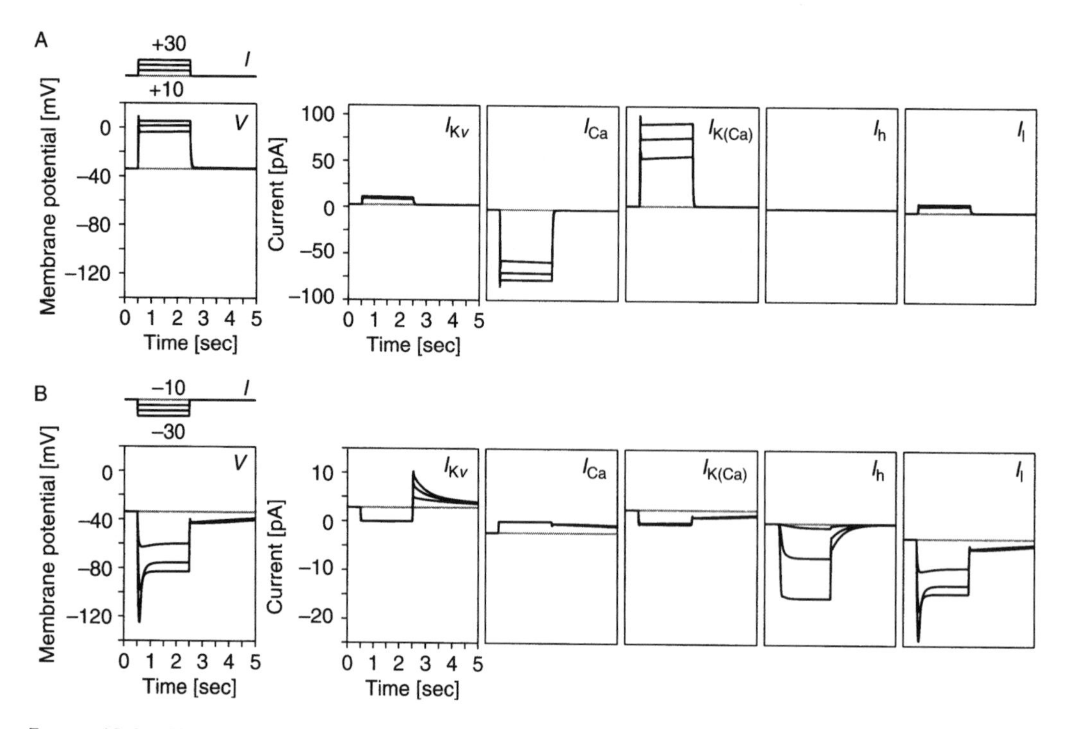

FIGURE 12.9 Simulated response of the bipolar cell model under current clamp conditions (Usui et al., 1996a). A: Voltage response and each ionic current to depolarizing current pulses of 10, 20, and 30 pA. B: Voltage response and each ionic current to hyperpolarizing current pulses of 10, 20, and 30 pA.

therefore, to clarify the relationship between membrane excitability and channel stochasticity in retinal ganglion cells.

As summarized in Table 12.9, the macroscopic membrane dynamics are modeled by five ionic currents: voltage-gated sodium current I_{Na}, calcium current I_{Ca}, transient outward current I_A, calcium-activated current $I_{K(Ca)}$, and the delayed rectifying potassium current I_K (Fohlmeister and Miller, 1997a). We developed a stochastic version of the retinal ganglion cell model, based on discrete stochastic ion channels represented by n-state Markov processes (Clay and DeFelice, 1983) as shown in Figure 12.10. Each channel is assumed to fluctuate randomly between stable states. Transition probabilities between these states are assumed to depend on the present state and the present membrane potential. The voltage-dependent rate constants from the deterministic model (Fohlmeister and Miller, 1997a) were used to estimate the transition probabilities. To simulate the channel behavior, we set random numbers for the initial state that the channel occupies, the duration of the state, and the subsequent state in each time step. We assumed that K(Ca) and leakage channels do not exhibit stochastic characteristics in the present model.

12.3.4.2 Simulated Results

Figure 12.11 shows the voltage-clamp step response of the membrane conductances, g_{Na}, g_K, g_A, and g_{Ca}. The membrane potential is held at −65 mV and then depolarized in a step to 20 mV. The responses of five different size membrane patches were simulated. As the membrane size is increased, the continuous deterministic behavior of each conductance emerges from the stochastic single-channel behavior.

Figure 12.12 shows the simulated firing patterns to DC and Gaussian current injections in the deterministic and stochastic models. The deterministic model shows a train of regular firing.

TABLE 12.9
A Mathematical Description of Ionic Channels of the Ganglion Cell

Na channel	$\alpha_m = \dfrac{-0.6(V+30)}{\exp(-0.1(V+30))-1}$	$\beta_m = 20\exp\left(-\dfrac{V+55}{18}\right)$
	$\alpha_h = 0.4\exp\left(-\dfrac{V+50}{20}\right)$	$\beta_h = \dfrac{6}{\exp(-0.1(V+20))+1}$
	$\dfrac{dm}{dt} = \alpha_m \cdot (1-m) - \beta_m \cdot m$	$\dfrac{dh}{dt} = \alpha_h \cdot (1-h) - \beta_h \cdot h$
Deterministic	$g_{Na} = \bar{g}_{Na} \cdot m^3 \cdot h$	$\bar{g}_{Na} = 50.0$ [mS cm^{-2}]
Stochastic	$g_{Na} = D_{Na} \cdot \gamma_{Na} \cdot [m_3h_1]$	$D_{Na} = 25.0$ [μm^{-2}] $\gamma_{Na} = 0.02$ [nS]
A channel	$\alpha_a = \dfrac{-0.006(V+90)}{\exp(-0.1(V+90))-1}$	$\beta_a = 0.1\exp\left(-\dfrac{V+30}{10}\right)$
	$\alpha_b = 0.04\exp\left(-\dfrac{V+70}{20}\right)$	$\beta_b = \dfrac{0.6}{\exp(-0.1(V+40))+1}$
	$\dfrac{da}{dt} = \alpha_a \cdot (1-a) - \beta_a \cdot a$	$\dfrac{db}{dt} = \alpha_b \cdot (1-b) - \beta_b \cdot b$
Deterministic	$g_A = \bar{g}_A \cdot a^3 \cdot b$	$\bar{g}_A = 36.0$ [mS cm^{-2}]
Stochastic	$g_A = D_A \cdot \gamma_A \cdot [a_3b_1]$	$D_A = 18.0$ [μm^{-2}] $\gamma_A = 0.02$ [nS]
K channel	$\alpha_n = \dfrac{-0.02(V+40)}{\exp(-0.1(V+40))-1}$	$\beta_n = 0.4\exp\left(-\dfrac{V+50}{80}\right)$
	$\dfrac{dn}{dt} = \alpha_n \cdot (1-n) - \beta_n \cdot n$	
Deterministic	$g_K = \bar{g}_K \cdot n^4$	$\bar{g}_K = 12.0$ mS cm^{-2}
Stochastic	$g_K = D_K \cdot \gamma_A \cdot [n_4]$	$D_K = 6.0$ [μm^{-2}] $\gamma_K = 0.02$ [nS]
Ca channel	$\alpha_c = \dfrac{-0.3(V+13)}{\exp(-0.1(V+13))-1}$	$\beta_c = 10\exp\left(-\dfrac{V+38}{18}\right)$
	$\dfrac{dc}{dt} = \alpha_c \cdot (1-c) - \beta_c \cdot c$	
Deterministic	$g_{Ca} = \bar{g}_{Ca} \cdot c^3$	$\bar{g}_{Ca} = 2.2$ [mS cm^{-2}]
Stochastic	$g_{Ca} = D_{Ca} \cdot \gamma_{Ca} \cdot [c_3]$	$D_{Ca} = 1.1$ [μm^{-2}] $\gamma_{Ca} = 0.02$ [nS]

Ca reversal potential

$$V_{Ca} = \frac{RT}{2F}\ln\left(\frac{[Ca^{2+}]_e}{[Ca^{2+}]_i}\right) \qquad [Ca^{2+}]_e = 1800\ [\mu M]$$

Ca concentration

$$\frac{d[Ca^{2+}]_i}{dt} = -\frac{3I_{Ca}}{2Fr} - \frac{[Ca^{2+}]_i - [Ca^{2+}]_{res}}{\tau_{Ca}}$$

$\tau_{Ca} = 50$ [msec] $[Ca^{2+}]_{res} = 0.1$ [μM] $r = 12.5$ [μm]

Ca dependent K conductance

$$g_{K(Ca)} = \bar{g}_{K(Ca)} \frac{([Ca^{2+}]_i/[Ca^{2+}]_{diss})^2}{1 + ([Ca^{2+}]_i/[Ca^{2+}]_{diss})^2}$$

$\bar{g}_{K(Ca)} = 0.05$ [mS cm^{-2}] $[Ca^{2+}]_{diss} = 1.0$ [μM]

Na channel

$$[m_0h_0] \underset{\beta_m}{\overset{3\alpha_m}{\rightleftharpoons}} [m_1h_0] \underset{2\beta_m}{\overset{2\alpha_m}{\rightleftharpoons}} [m_2h_0] \underset{3\beta_m}{\overset{\alpha_m}{\rightleftharpoons}} [m_3h_0]$$

$$\beta_h \uparrow\downarrow \alpha_h$$

$$[m_0h_1] \underset{\beta_m}{\overset{3\alpha_m}{\rightleftharpoons}} [m_1h_1] \underset{2\beta_m}{\overset{2\alpha_m}{\rightleftharpoons}} [m_2h_1] \underset{3\beta_m}{\overset{\alpha_m}{\rightleftharpoons}} [m_3h_1]$$

A channel

$$[a_0b_0] \underset{\beta_a}{\overset{3\alpha_a}{\rightleftharpoons}} [a_1b_0] \underset{2\beta_a}{\overset{2\alpha_a}{\rightleftharpoons}} [a_2b_0] \underset{3\beta_a}{\overset{\alpha_a}{\rightleftharpoons}} [a_3b_0]$$

$$\beta_b \uparrow\downarrow \alpha_b$$

$$[a_0b_1] \underset{\beta_a}{\overset{3\alpha_a}{\rightleftharpoons}} [a_1b_1] \underset{2\beta_a}{\overset{2\alpha_a}{\rightleftharpoons}} [a_2b_1] \underset{3\beta_a}{\overset{\alpha_a}{\rightleftharpoons}} [a_3b_1]$$

K channel

$$[n_0] \underset{\beta_n}{\overset{4\alpha_n}{\rightleftharpoons}} [n_1] \underset{2\beta_n}{\overset{3\alpha_n}{\rightleftharpoons}} [n_2] \underset{3\beta_n}{\overset{2\alpha_n}{\rightleftharpoons}} [n_3] \underset{4\beta_n}{\overset{\alpha_n}{\rightleftharpoons}} [n_4]$$

Ca channel

$$[c_0] \underset{\beta_c}{\overset{3\alpha_c}{\rightleftharpoons}} [c_1] \underset{2\beta_c}{\overset{2\alpha_c}{\rightleftharpoons}} [c_2] \underset{3\beta_c}{\overset{\alpha_c}{\rightleftharpoons}} [c_3]$$

FIGURE 12.10 Markov kinetic scheme of each ionic channel. Each Na channel is represented by eight different states. $[m_ih_i]$ refers to the number of channels in the state m_ih_i. $[m_3h_1]$ is the open state of the Na channel. The delayed rectifying K channel can exist in five different states. $[n_4]$ labels the open state of the K channel. The A channel can exist in eight different states. $[a_3b_1]$ is the open state of the A channel. The Ca channel can exist in four states. $[c_3]$ labels the open states of the Ca channel. The voltage-dependent rate functions were used from the deterministic model as summarized in Table 12.9.

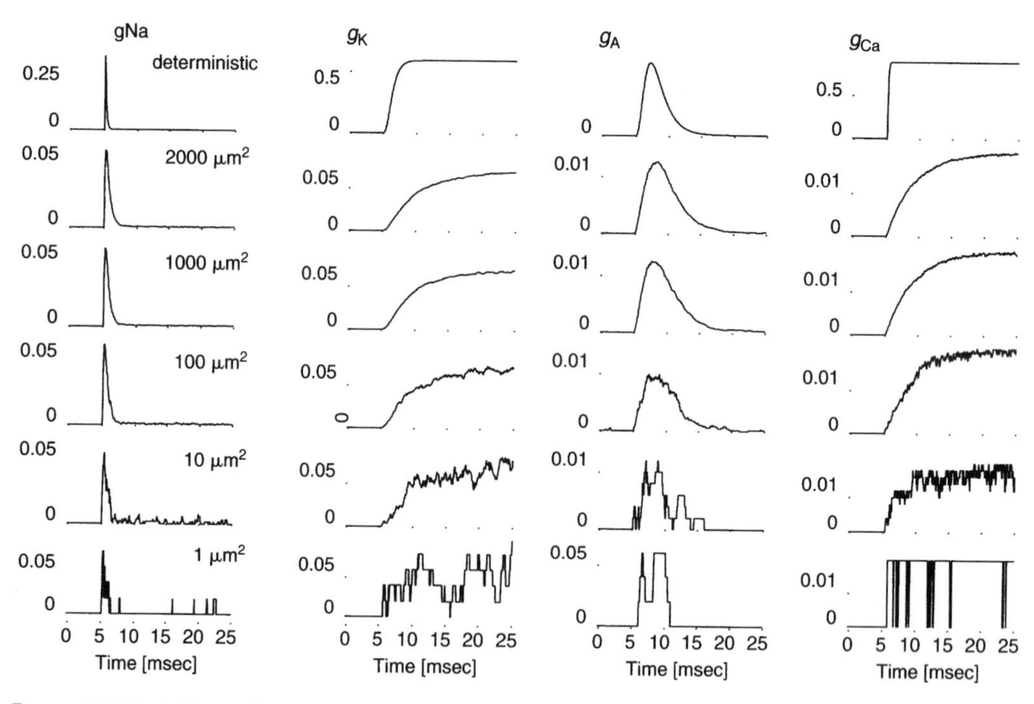

FIGURE 12.11 Voltage-clamp response of the membrane conductances, g_{Na}, g_K, g_A, and g_{Ca}. The membrane potential is held at -65 mV and then depolarized to 20 mV. The top trace represents the deterministic responses. The second and the rest of the traces are the responses by the stochastic model with variation of the membrane size to 2000, 1000, 100, 10, and 1 μm^2. Na, K, A, and Ca channel densities are assumed to be 25, 6, 18, and 1.1 μm^{-2}, respectively. All single channel conductances are assumed to be 0.02 nS.

The response of the stochastic model varies from trial to trial, and thus 20 superimposed responses to the same input current are shown. By contrast with the case of responses to DC pulses, when the stimulus is randomly fluctuating, the stochastic model shows relatively precise and stable spike timing as clearly seen in the raster plots.

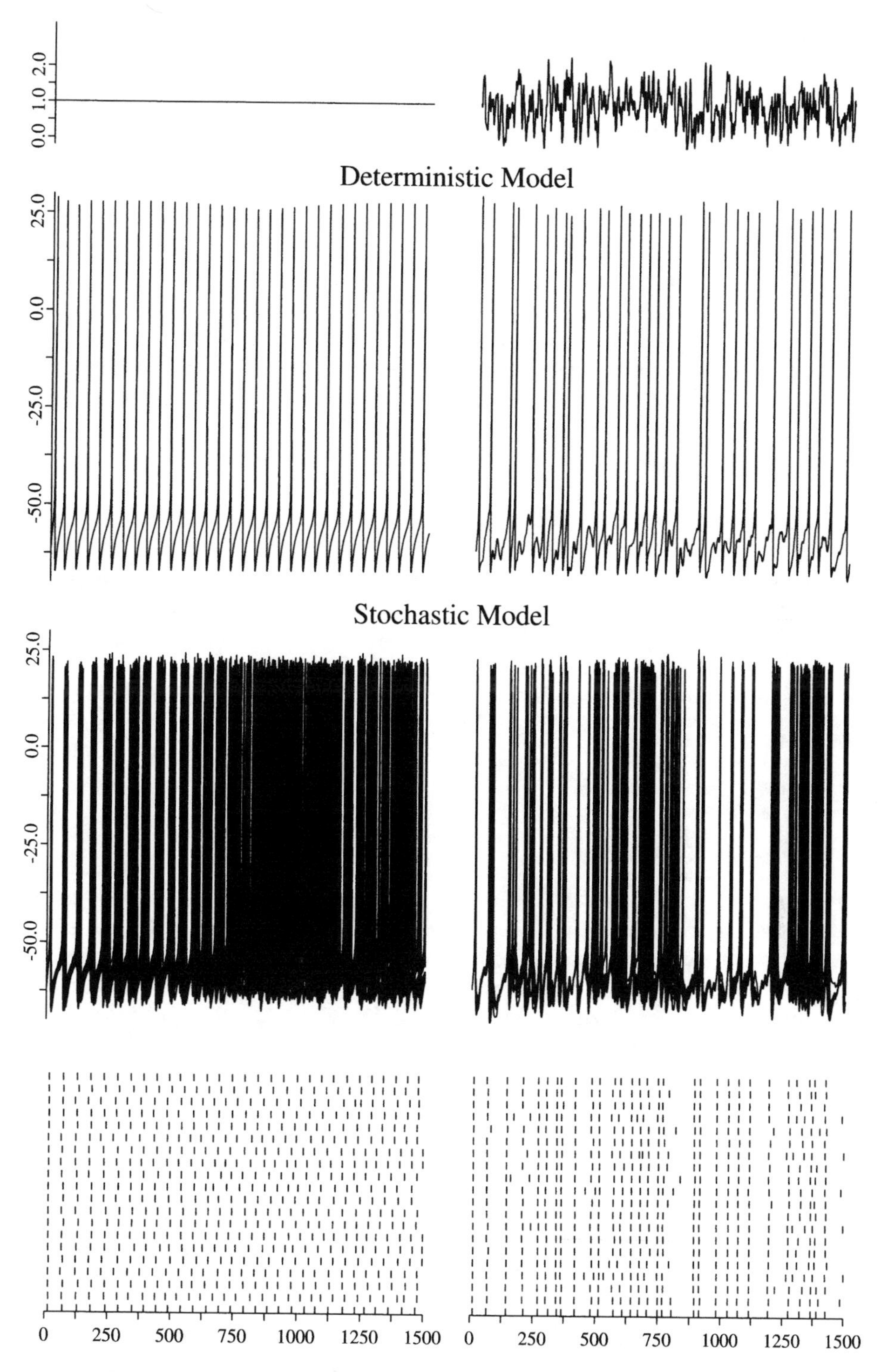

FIGURE 12.12 Firing patterns of the ganglion cell model in response to both DC and Gaussian current stimuli. Left: Twenty superimposed response to repeated DC current injections. A train of regular firing was evoked in the deterministic model. A jitter in the firing was observed in the stochastic model. The bottom frame shows the raster plot of the firing. Right: The model was stimulated 20 times with a Gaussian current. As can be clearly seen, the jitter in spike timing in the stochastic model is smaller than that of the DC stimulus case. The spike timing is reliable as shown in the raster plot (bottom frame). The membrane patch area used was 500 $\mu m^2 (r = 6.25\,[\mu m])$.

12.4 Discussion

12.4.1 Photoreceptor Model

We have obtained satisfactory agreement between simulated and experimental responses of rod photoreceptors. The model reproduces the voltage- and current-clamp responses reasonably well and the light response properties observed in the photoreceptors.

Attwell and Wilson (1980) also modeled the time- and voltage-dependent current in rods, and analyzed the electrical behavior of rods using model simulations. They found that the current[1] termed I_A plays a major role in shaping the light responses. Since their I_A is an inward current activated by membrane hyperpolarization, the current partially corresponds to I_h in our model. They did not separate the membrane current into each ionic current because the components were not fully identified in 1980. Our model, however, takes into account many of the ionic current components. This level of detail in the model makes it possible to predict the responses under various conditions. The predicted effects of I_h on the light responses suggest that I_h makes the response more transient as the membrane potential hyperpolarizes, which is consistent with the experimental finding by Baylor et al. (1984). We also found that zatebradine, a drug with a specific bradycardiac action, might induce visual symptoms as a side effect by blocking I_h channels in rods (Usui et al., 1996c).

The present model can account for many features of the experimental observations in photoreceptors. However, ionic currents identified in synaptic terminals (Kaneko and Tachibana, 1986) of cone photoreceptors were not included in the present model. Moreover, the detailed mechanism of phototransduction in cones is not well understood. We still have to develop a quantitative model of phototransduction in cones to simulate the light responses. Recently, it was found that mammalian rod photoreceptors express voltage-gated Na^+ channels and can generate sodium action potentials (Kawai et al., 2001), while cold-blooded vertebrates generate the Ca^{2+} action potentials shown in the present study. These species differences can also be modeled by introducing a new channel and estimating appropriate parameter sets.

12.4.2 Horizontal Cell Model

We constructed a model of rabbit horizontal cells based on ionic current mechanisms that can reproduce voltage- and current-clamp responses, and repetitive spikes. Using traditional biophysical techniques, it is not technically feasible to analyze how individual ionic currents participate in generating membrane responses. The mathematical model proposed here allows us to analyze the mechanism underlying complicated membrane responses, such as repetitive action potentials.

Simulated responses suggest that repetitive spikes are caused by I_{Ca} and a large conductance of $I_{K\nu}$ channels. We simulated the spikes with the conductance $\bar{g}_{K\nu}$ set at eight times its control value of 4.5 nS. There is room for argument concerning this choice of maximum conductance. One possibility is a difference in membrane properties among the cells. We have insufficient data to make a quantitative comparison of the maximum conductance in voltage-clamp experiments, but several studies suggest that the ionic current varies from cell to cell. For example, $I_{K\nu} = 0.06 \pm 0.04$ pA/μm^2 (Malchow et al., 1990), $I_A = 2777 \pm 1441$ pA/μm^2 (Sullivan and Lasater, 1990), and $I_{Ca} = 0.57 \pm 0.23$ pA/μm^2 (Kaneda and Kaneko, 1991). Differences in maximum conductance will result in differences in current magnitudes. These observations suggest that there may be several-fold differences in the maximum conductance from cell to cell.

Horizontal cells are depolarized (around −35 mV) in the dark by the glutamate release from photoreceptors (reviewed in Massey, 1990). Horizontal cells normally respond with graded hyperpolarization to light. Therefore, it is unlikely that an action potential is generated during the light response. In excitable cells, it is known that an action potential can be observed at the offset of a hyperpolarizing current step, called anode break (Johnston and Wu, 1995). Since the light stimulus behaves as a hyperpolarizing current to the horizontal cells, a similar phenomenon to the anode break may occur at the light offset. In the future, in order to evaluate this possibility we have to add a model of the glutamate receptor into the present model and simulate the responses to light.

12.4.3 Bipolar Cell Model

A quantitative reconstruction of the membrane ionic currents in the retinal bipolar cell soma has been realized in the present study. Our model is capable of accurately reproducing the voltage- and current-clamp responses of bipolar cells. The present model also has the ability to represent the subtype specificities found in bipolar cells. The membrane ionic currents of bipolar cells have been examined not only in lower vertebrates but also in mammals (Kaneko et al., 1989; Karschin and Wässle, 1990). These results show that most of the ionic currents in mammalian bipolar cells are of the same kind as those in lower vertebrates, although there is a difference in the dynamic properties of these currents. Since our model sufficiently reproduced the different dynamic characteristics of white bass bipolar cells, it should be possible to simulate mammalian bipolar cell responses as well by re-estimating the model parameters.

It was found that several ionic channels and receptors on bipolar cells are localized on the dendrite, cell body, or axon terminal. Nawy and Jahr (1990) found that cGMP-gated channels closed by glutamate in bipolar cells are responsible for the on-light response. Since the mechanism of the cGMP-mediated pathway is similar to that of phototransduction, it is possible to reconstruct a bipolar cell model which produces the light response based on the present model. Tachibana et al. (1993) found that most of the calcium channels on bipolar cells are localized at the axon terminal and that calcium ions diffuse towards the cell body through the axon. Although the present model can reproduce the calcium current in bipolar cells, it is important to incorporate the spatial characteristics of the calcium system into our model for the full reconstruction of the behavior of the bipolar cell. It has been reported that bipolar cells can produce calcium action potentials (Zenisek and Matthews, 1998; Protti et al., 2000) as well as sodium action potentials (Pan and Hu, 2000). To reproduce these characteristics in the model, we must introduce into the model more accurate morphological characteristics of the bipolar cell.

12.4.4 Ganglion Cell Model

The stochastic model of the retinal ganglion cell based on discrete ionic channels has been compared with a deterministic model based on the conventional Hodgkin–Huxley type equations. Under current-clamp conditions, the firing patterns simulated by the stochastic model showed a dramatic difference from those generated by the deterministic model. The predicted spiking characteristics should be demonstrated experimentally.

In the present study, the ganglion cell was modeled by five ionic currents. However, more than a dozen voltage-, ion-, and ligand-gated ionic channels have been identified in retinal ganglion cells (Ishida, 1995). A more realistic model of the spike encoding mechanisms which takes into account various ionic channels, as well as dendritic morphologies, should be developed.

12.5 Conclusions and Future Perspectives

A quantitative reconstruction of the membrane ionic currents in photoreceptors, horizontal cells, bipolar cells, and ganglion cells has been presented in this study. Our models accurately reproduce the voltage- and current-clamp responses of these cells. We believe that our models, which replicate the detailed physiological structures and functions of the retina, clearly illustrate what we know and what we still need to know about the retina. The structure of our model is such that it can easily incorporate new experimental results on the behavior of the retina, and can be used to assess how these results improve our understanding of retinal function. Therefore, the models we have developed provide not only a unified framework for understanding the functional organization of the retina, but also may be used to evaluate the clinical effects and side effects of specific drugs (e.g., see Usui et al. [1996b] for side effects induced by the bradycardic agent zatebradine).

One often hears the criticism that mathematical models with many parameters can fit almost anything. This is not a serious problem in our models for which the parameters were estimated from the results of voltage-clamp experiment assuming Hodgkin–Huxley membrane kinetics for ionic conductances in the cell membrane (Hille, 2001; Johnston and Wu, 1995). Parameter values of each ionic current were estimated independently current-by-current, that is, we did not require ourselves to estimate the model parameters simultaneously. The estimated values of measurable parameters such as the maximum conductances and the reversal potentials are well represented by the experimental data. However, there is no guarantee that the models can reproduce the behavior of the cells under other experimental conditions, since the modeling of each ionic current assumed pharmacologically isolated conditions. We examined the acceptability of the models by current-clamp experiments, and the results were extremely similar to the experimental data. These results clearly demonstrate the reality of our models.

ACKNOWLEDGMENTS

This work was supported in part by the Hori Information Science Promotion Foundation, the Grant-in-Aid for Scientific Research (No. 14550424), the Project on Neuroinformatics Research in Vision through Special Coordination Funds for Promoting Science and Technology from the Ministry of Education, Culture, Sports, Science and Technology of the Japanese Government.

PROBLEMS

1. The present photoreceptor model consists of two functional parts, the inner and outer segments. Extend the model to include the membrane properties of the synaptic terminal.
2. In darkness, glutamate, an excitatory neurotransmitter, is tonically released from the synaptic terminals of photoreceptors. Light closes the channels in the outer segment, hyperpolarizes the photoreceptor membrane, and lowers the rate of transmitter release. Reduction of transmitter release causes a graded light response in second-order neurons, bipolar and horizontal cells. Discuss the underlying biophysical mechanisms of the signal transmission among photoreceptors and second-order neurons.
3. The transient potassium current I_A is known to play an important role in repetitive firing of neurons. Discuss how I_A contributes to the repetitive firing of horizontal cells.
4. Bipolar cells are the key interneurons forming the signal pathway from the outer retina to the inner retina. Earlier models of chemically interacting retinal networks include Tate and Woolfson (1971), Abernethy (1974), Hara and Takabayashi (1975), Poznanski and Umino (1997), and Yang et al. (2003b). Discuss how visual signals are coded as neural signals in bipolar cells and how the ionic channels in the outer retina contribute to signal processing.
5. Retinal ganglion cell morphology is thought to control impulse generation and propagation effects. Extend the present single compartment model to include more realistic representations of dendritic trees, the soma, and the axon hillock.
6. Amacrine cells of the vertebrate retina are interneurons that interact at the synaptic pathway from bipolar cells to ganglion cells (Masland, 2001). Recent voltage-clamp analyses on amacrine cells have shown that rat GABAergic amacrine cells have two types of sodium currents and three types of calcium currents (Koizumi et al., 2001). Develop an ionic-current model of the amacrine cell and discuss the functional role of each current in the voltage responses.

NOTE

1. In their paper, the term I_A does not mean the A-type K current.

13 Modeling Intracellular Calcium: Diffusion, Dynamics, and Domains

Gregory D. Smith

CONTENTS

13.1 INTRODUCTION

Calcium signaling is an important aspect of cellular neurophysiology and biophysics (Ghosh and Greenberg, 1995; Berridge 1998). To give an obvious example, action-potential-dependent chemical neurotransmission is a process that begins with Ca^{2+} influx through presynaptic Ca^{2+} channels. As Ca^{2+} ions diffuse away from the mouth of voltage-gated plasma membrane Ca^{2+} channels into the cytosol, microdomains of elevated intracellular $[Ca^{2+}]$ activate effector proteins that are associated with neurotransmitter release (Neher, 1998; Augustine et al., 2003). These Ca^{2+} domains are formed in the presence of the ubiquitous Ca^{2+}-binding proteins of the presynaptic terminal (Stanley, 1995). The association of free Ca^{2+} with endogenous Ca^{2+}-binding proteins and other "Ca^{2+} buffers" (Celio et al., 1996) determines the range of action of Ca^{2+} ions, influences the time course of their effect, and facilitates Ca^{2+} clearance. Figure 13.1 shows a schematic representation of the Ca^{2+} fluxes that contribute to neuronal Ca^{2+} dynamics including Ca^{2+} influx across the plasma membrane via

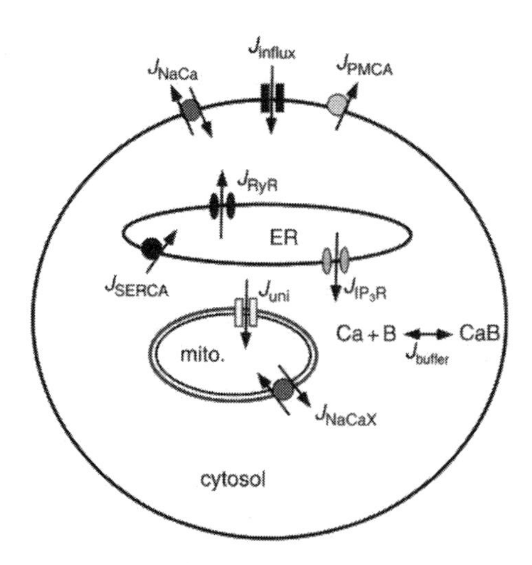

FIGURE 13.1 Schematic representation of Ca^{2+} fluxes that contribute to neuronal Ca^{2+} dynamics, including Ca^{2+} influx across the plasma membrane via voltage- and ligand-gated plasma membrane Ca^{2+} channels (J_{influx}), Ca^{2+} pumps (J_{PMCA}), and exchangers (J_{NaCa}), release from intracellular stores via either ryanodine receptors (J_{RyR}) or IP_3 receptors (J_{IP_3R}), sequestrations of Ca^{2+} into the ER by calcium pumps (J_{SERCA}), mitochondrial Ca^{2+} handling by the uniporter (J_{uni}) and Na^+–Ca^{2+} exchange (J_{NaCaX}), and association of Ca^{2+} with endogenous and exogenous buffers (J_{buffer}). Reproduced with permission from Jafri and Yang (2004).

voltage- and ligand-gated plasma membrane Ca^{2+} channels (J_{influx}), and the interaction of Ca^{2+} with Ca^{2+}-binding proteins (J_{buffer}).

While Ca^{2+} influx via voltage-gated Ca^{2+} channels of the plasma membrane is a major source of cytosolic Ca^{2+} in neurons, another source is the endoplasmic reticulum (ER), a continuous membrane-delimited intracellular compartment with integrative and regenerative properties analogous to those of the membranes of electrically excitable cells (Berridge, 1993, 1997, 1998; Clapham, 1995; Ghosh and Greenberg, 1995). In neurons and other cell types, activated metabotropic receptors of the plasma membrane stimulate the production of the intracellular messenger inositol 1,4,5-trisphosphate (IP_3) (Berridge, 1993). IP_3 in turn promotes Ca^{2+} release from intracellular stores by binding and activating IP_3R receptor Ca^{2+} channels (IP_3Rs) located on the ER membrane (Bezprozvanny and Ehrlich, 1995; Figure 13.1, J_{IP_3R}, and J_{RyR}). In addition to this IP_3-mediated pathway, a second parallel mechanism for Ca^{2+} release is mediated by ryanodine receptors (RyRs)—intracellular Ca^{2+} channels on the ER that are activated by cyclic ADP ribose (Ehrlich, 1995). Importantly, both IP_3Rs and RyRs can be activated or inactivated by intracellular Ca^{2+} leading to ER "Ca^{2+} excitability" (Keizer et al., 1995; Li et al., 1995a). This Ca^{2+} excitability is the physiological basis for IP_3- and Ca^{2+}-induced Ca^{2+} release (IICR and CICR) in excitable and nonexcitable cells. Other Ca^{2+} fluxes that contribute to neuronal Ca^{2+} dynamics include Ca^{2+} efflux via plasma membrane Ca^{2+}-ATPases (J_{PMCA}) and exchangers (J_{NaCa}), sequestration of Ca^{2+} into the ER by sarco-endoplasmic reticulum Ca^{2+}-ATPases (J_{SERCA}), and mitochondrial Ca^{2+} handling via the uniporter (J_{uni}) and Na^+–Ca^{2+} exchange (J_{NaCaX}).

We begin with a summary of the mathematical aspects of intracellular Ca^{2+} signaling emphasizing the dynamics of global Ca^{2+} responses, spatial phenomena such as buffered diffusion of Ca^{2+} and propagating Ca^{2+} waves, and biophysical theory and modeling of localized Ca^{2+} elevations. Some introductory material is adapted from previous work of Smith and coworkers (e.g., Smith, 2000a, 2000b; Smith et al., 2002). For a review of intracellular Ca^{2+} signaling from an empirical perspective, see Clapham (1995), Ghosh and Greenberg (1995), Denk et al. (1996), Dolphin (1996), Simpson et al. (1995), Berridge (1997), Stuart et al. (1999a), Sabatini et al. (2001), Schuster et al. (2002), and Augustine et al. (2003).

13.2 ODEs for Association of Ca^{2+} with Buffer

Experimental observations of intracellular Ca^{2+} dynamics are often made using microfluorimetry. The experimental technique involves the loading of Ca^{2+} indicator dyes into cells and the use of instruments to optically excite the indicators and measure their emissions. These Ca^{2+} indicator dyes are themselves Ca^{2+} buffers that can potentially affect intracellular Ca^{2+} signaling (Jafri and Keizer, 1995; Smith et al., 1998). Although a measured fluorescence signal is only indirectly related to the dynamics of intracellular Ca^{2+}, it is relatively straightforward to determine the free $[Ca^{2+}]$ during a cell-wide Ca^{2+} response using the time course of measured fluorescence (Grynkiewicz et al., 1985; Maeda et al., 1999).

The relationship between free $[Ca^{2+}]$ and indicator fluorescence can be derived by assuming a bimolecular association reaction between Ca^{2+} and indicator dye,

$$\mathrm{Ca}^{2+} + \mathrm{B} \underset{k_-}{\overset{k_+}{\rightleftharpoons}} \mathrm{CaB}, \tag{13.1}$$

where B and CaB are free and bound buffer, respectively. Assuming mass-action kinetics, we write the following system of ODEs for the time evolution of $[Ca^{2+}]$, [B], and [CaB] (Wagner and Keizer, 1994),

$$\frac{\mathrm{d}[\mathrm{Ca}^{2+}]}{\mathrm{d}t} = R + J, \tag{13.2}$$

$$\frac{\mathrm{d}[\mathrm{B}]}{\mathrm{d}t} = R, \tag{13.3}$$

$$\frac{\mathrm{d}[\mathrm{CaB}]}{\mathrm{d}t} = -R, \tag{13.4}$$

with reaction terms given by

$$R = -k_+[\mathrm{B}][\mathrm{Ca}^{2+}] + k_-[\mathrm{CaB}]. \tag{13.5}$$

In these equations, k_+ and k_- are the association and dissociation rate constants for Ca^{2+} buffer with units micromolar per second and $\sec^{-1}$, respectively. The J in Equation (13.2) represents increases ($J > 0$) or decreases ($J < 0$) in cytosolic $[Ca^{2+}]$ due to Ca^{2+} influx or efflux (J_{influx}, J_{PMCA}, and J_{NaCa} in Figure 13.1). Associated with Equations (13.2) to (13.4) are the initial conditions $[Ca^{2+}](0) = [Ca^{2+}]_0$, $[B](0) = [B]_0$, and $[CaB](0) = [CaB]_0$.

Note that by choosing to model the dynamics of Ca^{2+} and indicator with a system of non-linear ODEs (Equations [13.2] to [13.4]), we have implicitly assumed that Ca^{2+} and indicator dye concentrations are spatially homogeneous throughout a small, continuously stirred compartment (e.g., a neuronal cell body). This assumption of spatial homogeneity will be relaxed in subsequent sections. For now, note that adding Equation (13.4) to Equations (13.2) and (13.3) eliminates R and gives,

$$\frac{\mathrm{d}[\mathrm{Ca}^{2+}]_\mathrm{T}}{\mathrm{d}t} = J, \tag{13.6}$$

$$\frac{\mathrm{d}[\mathrm{B}]_\mathrm{T}}{\mathrm{d}t} = 0, \tag{13.7}$$

where the total cell calcium concentration $[Ca^{2+}]_T$ and the total buffer concentration $[B]_T$ are defined by

$$[\mathrm{Ca}^{2+}]_\mathrm{T} = [\mathrm{Ca}^{2+}] + [\mathrm{CaB}], \tag{13.8}$$

$$[\mathrm{B}]_\mathrm{T} = [\mathrm{B}] + [\mathrm{CaB}]. \tag{13.9}$$

Because we have assumed that there is no pathway for the buffer to leave the small cell, the total buffer concentration is constant and fixed by the initial conditions $[B]_T = [B]_0 + [CaB]_0$. Similarly, because we have assumed a balance between Ca^{2+} influx and efflux, total cell Ca^{2+} is fixed and given by $[Ca^{2+}]_T = [Ca^{2+}]_0 + [CaB]_0$.

When Ca^{2+} influx and efflux is balanced ($J = 0$), the steady state of Equations (13.2) to (13.4) can be found by setting the derivatives $d[Ca^{2+}]/dt$, $d[B]/dt$, and $d[CaB]/dt$ to zero. Thus, at steady state $R = 0$ and the association and dissociation rates are balanced, that is,

$$k_+[B]_\infty[Ca^{2+}]_\infty = k_-[CaB]_\infty,$$

where $[Ca^{2+}]_\infty$, $[B]_\infty$, and $[CaB]_\infty$ are the respective equilibrium concentrations of Ca^{2+} and buffer (free and bound). Using Equation (13.9), we can substitute $[B]_\infty = [B]_T - [CaB]_\infty$ in this equation to obtain the equilibrium relation

$$[CaB]_\infty = \frac{[Ca^{2+}]_\infty[B]_T}{K + [Ca^{2+}]_\infty}, \tag{13.10}$$

where the dissociation constant $K = k_-/k_+$ has units of concentration (μM).

13.3 Microfluorimetry and Back-Calculating $[Ca^{2+}]$

When Equation (13.10) accurately describes the relationship between bound indicator dye (CaB) and Ca^{2+}, one can "back-calculate" the $[Ca^{2+}]$ from a fluorescent signal. For a single excitation wavelength measurement (e.g., using a nonratiometric dye such as fluo-3), we can idealize indicator fluorescence as the sum of two components,

$$F = \eta_B[B] + \eta_{CaB}[CaB], \tag{13.11}$$

where η_B and η_{CaB} are proportionality constants for free and bound dye, respectively. When $\eta_B < \eta_{CaB}$, the maximum and minimum observable fluorescence are $F_{min} = \lim_{[Ca^{2+}]\to 0} F = \eta_B[B]_T$ and $F_{max} = \lim_{[Ca^{2+}]\to\infty} F = \eta_{CaB}[B]_T$. Using Equation (13.10) and the equilibrium relation for free buffer,

$$[B]_\infty = \frac{K[B]_T}{K + [Ca^{2+}]_\infty}, \tag{13.12}$$

it can be shown that (Grynkiewicz et al., 1985; Maeda et al., 1999)

$$[Ca^{2+}]_\infty = K\frac{[CaB]_\infty}{[B]_\infty} = K\frac{F_\infty - F_{min}}{F_{max} - F_\infty}. \tag{13.13}$$

The validity of this formula relies on the stability of instrument sensitivity, optical path length, and dye concentration between measurements of F, F_{min}, and F_{max}. Because determining F_{min} and F_{max} usually involves titrating the indicator released from lysed cells, this is difficult to achieve in practice.

Fluorescent intensities can also be obtained at two excitation wavelengths (λ and λ') using indicator dyes such as fura-2 (Grynkiewicz et al., 1985). Such *ratiometric* measurements can be related to the underlying free Ca^{2+} signal by supplementing Equation (13.11) with

$$F' = \eta'_B[B] + \eta'_{CaB}[CaB],$$

where the primes indicate the second excitation wavelength. Using the first equality of Equation (13.13), the fluorescence ratio, $R = F/F'$, as a function of $[Ca^{2+}]$ can be inverted to give

$$[Ca^{2+}]_{\infty} = K \frac{\eta_B - \eta'_B R_{\infty}}{\eta'_{CaB} R_{\infty} - \eta_{CaB}} = K \left(\frac{R_{\infty} - R_{min}}{R_{max} - R_{\infty}} \right) \left(\frac{\eta'_B}{\eta'_{CaB}} \right), \tag{13.14}$$

where $R_{min} = \lim_{[Ca^{2+}]\to 0} F/F' = \eta_B/\eta'_B$, and $R_{max} = \lim_{[Ca^{2+}]\to\infty} F/F' = \eta_{CaB}/\eta'_{CaB}$. If λ' is chosen to be a wavelength at which the calibration spectra at different Ca^{2+} concentrations cross one another, then $\eta'_B \approx \eta'_{CaB}$ and the last factor in Equation (13.14) is eliminated. An advantage of the ratiometric method is its insensitivity to changes in dye concentration and instrument sensitivity between measurements.

13.4 The Buffer Equilibration Time

How long does it take for the equilibrium relationships Equations (13.10) and (13.12) to become approximately valid? The answer depends on the "buffer equilibration time," a quantity easily derived under the assumption of a balance between Ca^{2+} influx and efflux ($J = 0$). First, use the fact that total buffer is conserved (Equation [13.9]) to eliminate Equation (13.3),

$$\frac{d[Ca^{2+}]}{dt} = R + J, \tag{13.15}$$

$$\frac{d[CaB]}{dt} = -R, \tag{13.16}$$

$$R = -k_{+} ([B]_T - [CaB]) [Ca^{2+}] + k_{-}[CaB]. \tag{13.17}$$

Next, use the fact that $J = 0$ and assume that both $[Ca^{2+}]$ and $[CaB]$ are slightly perturbed from their equilibrium values, that is,

$$[Ca^{2+}] = \delta[Ca^{2+}] + [Ca^{2+}]_{\infty}, \tag{13.18}$$

$$[CaB] = \delta[CaB] + [CaB]_{\infty}, \tag{13.19}$$

where $\delta[Ca^{2+}]$ and $\delta[CaB]$ are the magnitudes of the perturbations, and $[Ca^{2+}]_{\infty}$ and $[CaB]_{\infty}$ satisfy Equation (13.10). Because $[Ca^{2+}]_{\infty}$ and $[CaB]_{\infty}$ are constants, it is straightforward to substitute Equations (13.18) and (13.19) into Equations (13.15) and (13.16). Dropping terms involving the quadratic product $\delta[CaB]\delta[Ca^{2+}]$ gives the linearized system

$$\frac{d}{dt} \begin{pmatrix} \delta[Ca^{2+}] \\ \delta[CaB] \end{pmatrix} = \begin{pmatrix} -\eta & \xi \\ \eta & -\xi \end{pmatrix} \begin{pmatrix} \delta[Ca^{2+}] \\ \delta[CaB] \end{pmatrix}, \tag{13.20}$$

where $\eta = k_{+}[B]_{\infty}$ and $\xi = k_{+}[Ca^{2+}]_{\infty} + k_{-}$. The unique steady state of Equation (13.20) is $(0, 0)^T$, and the perturbation $(\delta[Ca^{2+}], \delta[CaB])^T$ will decay as Ca^{2+} and buffer equilibrate because the constant coefficient matrix in Equation (13.20) has eigenvalues $\lambda_1 = 0$ and $\lambda_2 = -\eta - \xi$. The zero eigenvalue is related to the conserved quantity

$$\delta[Ca^{2+}] + \delta[CaB] = 0$$

TABLE 13.1
Example Ca^{2+} Buffers and Parameters

Buffer	D_B $\mu m^2\ sec^{-1}$	k_+ $(\mu M^{-1}\ sec^{-1})$	k_- (s^{-1})	K (μM)	Reference
BAPTA	95	600	100	0.17	Tsien, 1980
EGTA	113	1.5	0.3	0.2^{c}	Pethig et al., 1989
Fura-2	$31–95^{a,d}$	600	80–100	0.13–0.60	Pethig et al., 1989
Fluo-3	$20–90^{d}$	80	90	1.13	Harkins et al., 1993
Calmodulin	32	500, 100^{a}	470, 37	0.9–2.0, 0.2–0.4	Falke et al., 1994
Calbindin-D28K	27	20	8.6	0.4–1.0	Koster et al., 1995
Parvalbumin	36	6^{b}	1	0.00037	Falke et al., 1994

[a] A range indicates different experimental conditions. Comma indicates distinct binding sites.

[b] Slower physiologically because Mg^{2+} must dissociate ($\tau \approx 1$ sec) before Ca^{2+} can bind (Falke et al., 1994).

[c] Strongly dependent on pH variation near 7.0 (Kao, 1994).

[d] Partially immobilized *in vivo* (Blatter and Wier, 1990; Harkins et al., 1993). Buffer parameter set B1: $D_B = 75\ \mu m^2\ sec^{-1}$, $k_+ = 100\ \mu M\ sec^{-1}$, $k_- = 100\ sec^{-1}$, $K = 1\ \mu M$, and $[B]_T = 50\ \mu M$. Set B2 is identical except $k_- = 1000\ sec^{-1}$ and $K = 10\ \mu M$.

Source: Adapted from Smith, G. et al., *Biophys. J.* 70(6), 2527–2539, 1996b. With permission.

For reviews see Baimbridge et al. (1992), Heizmann and Hunziker (1991), Kao (1994), and Celio et al. (1996).

that follows from Equations (13.6) and (13.8) when $J = 0$. The magnitude of the negative eigenvalue allows us to identify an exponential time constant for the relaxation of buffer,

$$\tau_B = \frac{1}{|\lambda_2|} = \frac{1}{\eta + \xi} = \frac{1}{k_+([B]_\infty + [Ca^{2+}]_\infty) + k_-}. \tag{13.21}$$

Assuming a background concentration of $[Ca^{2+}]_\infty = 0.1\ \mu M$ and the buffer parameters for fura-2 (see Table 13.1), Equation (13.21) gives a buffer equilibration time that is less than a millisecond. Because global Ca^{2+} responses such as Ca^{2+} oscillations with a period measured in seconds or tens of seconds are slow compared to τ_B, we expect the back-calculation formulas (Equations [13.11] and [13.14]) to be valid. Conversely, the back-calculation formula and, indeed, Equations (13.1) to (13.5) are not adequate descriptions for the association of Ca^{2+} with buffers such as parvalbumin that has mixed Ca^{2+}/Mg^{2+} binding sites characterized by slow kinetics due to rate-limiting dissociation of Mg^{2+} (Celio et al., 1996).

13.5 CONCENTRATION FLUCTUATIONS AND Ca^{2+} BUFFERS

In addition to assuming a small, continuously stirred compartment, the above calculations assume that Ca^{2+} and buffer are continuous variables. This is a good approximation for a compartment of the size of a neuron soma (say $10^3\ \mu m^3$) where a resting $[Ca^{2+}]$ of 0.1 μM corresponds to 6×10^4 free ions. However, subcellular compartments such as dendritic spine heads have volumes in the range of 0.001 to 1 μm^3 (Nimchinsky et al., 2002), the larger corresponding to only 60 free ions, and the smaller corresponding to 600 total Ca^{2+} ions (Ca^{2+} plus CaB) for $[B]_T = 120\ \mu M$ and $K = 1\ \mu M$. This suggests that it is not always appropriate to think of $[Ca^{2+}]$ and $[CaB]$ as continuous variables.

Another possibility is to explicitly model the elementary processes of Ca^{2+} binding and unbinding with buffer (Keizer, 1987). For example, if a small, continuously stirred compartment has volume V and contains N_C Ca^{2+} ions and N_B buffer binding sites, the arrows in Equation (13.1) can be reinterpreted as transition probabilities defining a continuous-time Markov process for $n_C(t)$, a random variable for the number of free Ca^{2+} ions (Smith, 2002b). When $J = 0$, Equations (13.8) and (13.9) indicate that the number of bound Ca^{2+} ions is $n_{CaB} = N_C - n_C$ and the number of free buffer binding sites is $n_B = N_B - n_{CaB} = N_B - N_C + n_C$. To ensure n_C, n_B, and n_{CaB} are nonnegative, n_C must be within the range

$$M_C = \max(0, N_C - N_B) < n_C < N_C,$$

where the first equality defines M_C. Writing $P(n,t)$ for the probability that $n_C = n$ (i.e., precisely n Ca^{2+} ions are free), we see that conservation of probability implies

$$\sum_{n=M_C}^{N_C} P(n,t) = 1.$$

Next, assume that the binding reactions represented by Equation (13.1) are independent and consider a time interval $[t, t + \Delta t]$ short enough so that only one Ca^{2+} ion has appreciable probability of either binding or unbinding. During this interval, four events can influence $P(n,t)$. For example, it is possible that there are currently n free Ca^{2+} ions, and that during the time interval $[t, t + \Delta t]$ one of these ions undergoes an association reaction with buffer. This probability is given by

$$\text{loss}_- = k_+ V^{-1} n\,(N_B - N_C + n)\,P(n,t)\Delta t,$$

where the bracketed term is the number of free-buffer binding sites, n_B. The reciprocal of the volume of the compartment scales the probability of this bimolecular association reaction and balances units. Similar reasoning leads to the expression

$$P(n, t + \Delta t) = P(n,t) + \text{gain}_+ - \text{loss}_+ + \text{gain}_- - \text{loss}_-, \tag{13.22}$$

where

$$\begin{aligned}
\text{gain}_- &= k_+ V^{-1}\,(n+1)\,(N_B - N_C + n + 1)\,P(n+1,t)\Delta t,\\
\text{loss}_+ &= k_-\,(N_C - n)\,P(n,t)\Delta t,\\
\text{gain}_+ &= k_-\,(N_C - n + 1)\,P(n-1,t)\Delta t.
\end{aligned}$$

Taking limit $\Delta t \to 0$ of Equation (13.22) gives the *master equation* (Keizer, 1987; Smith, 2002b)

$$\begin{aligned}
\frac{d}{dt}P(n,t) = &+ k_-\,(N_C - n + 1)\,P(n-1,t)\\
&- k_-\,(N_C - n)\,P(n,t)\\
&+ k_+ V^{-1}\,(n+1)\,(N_B - N_C + n + 1)\,P(n+1,t)\\
&- k_+ V^{-1} n\,(N_B - N_C + n)\,P(n,t).
\end{aligned} \tag{13.23}$$

This equation represents $N_C - M_C + 1$ coupled ODEs, one for each $P(n,t)$, where $M_C \le n \le N_C$.

The equilibrium solution to the master equation is $N_C - M_C + 1$ time-independent probabilities, $P_\infty(n)$, satisfying Equations (13.23) with the left-hand side equal to zero. Figure 13.2 shows four

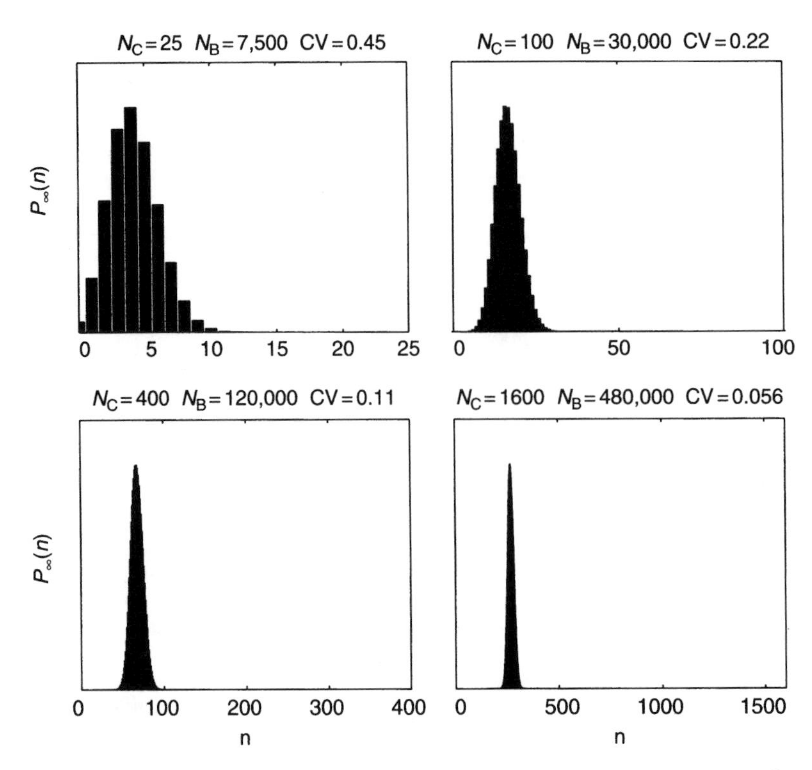

FIGURE 13.2 The equilibrium probability distribution, $P_{\infty}(n)$, for the number of free Ca^{2+} ions (n_C) in a restricted compartment calculated as the steady state of Equations (13.23). The total number of Ca^{2+} ions (N_C) and buffer binding sites (N_B) are increased as the volume V is increased so that the total Ca^{2+} and buffer concentrations are unchanged. As V is increased the mean and variance of n_C grow proportionately, but the coefficient of variation (CV) scales as $1/\sqrt{V}$ (see Equation [13.24]). Parameters as in set B1 of Table 13.1 with $[Ca^{2+}]_T = 0.17$ μM giving $[Ca^{2+}]_{\infty} = 0.028$ μM; $V = 0.25$, 1, 4, and 16 μm^3; $N_C = 25, 100, 400, 1600$; $N_B = 7.5 \times 10^3, 3 \times 10^4, 1.2 \times 10^5, 4.8 \times 10^5$.

such distributions for dendritic spine-head volumes of 0.25, 1, 4, and 16 μm^3. The total number of Ca^{2+} ions (N_C) and buffer binding sites (N_B) are chosen so that $[Ca^{2+}]_T = 0.17$ μM and $[B]_T = 50$ μM do not vary. Note that the expected value of the number of free Ca^{2+} ions,

$$\mathrm{E}[n_C] = \sum_{n=M_C}^{N_C} nP(n,t)$$

scales with system size (see x-axis). Conversely, a relative measure of variance known as the *coefficient of variation* and defined by

$$\mathrm{CV} = \frac{\sqrt{\mathrm{Var}[n_C]}}{\mathrm{E}[n_C]}, \qquad \mathrm{Var}[n_C] = \mathrm{E}[(n_C - \mathrm{E}[n_C])^2] = \sum_{n=M_C}^{N_C} (n - \mathrm{E}[n_C])^2 \, P(n,t), \tag{13.24}$$

indicates that the distributions in Figure 13.2 are narrow relative to their mean as spine-head volume increases. Figure 13.2 also shows that as volume is increased from $V = 0.25$ to 1, 4, and 16 μm^3, the distribution mean, $\mathrm{E}[n_C]$, grows proportionately (4.19, 16.75, 67.02, and 268.05, respectively). Conversely, the coefficient of variation scales as $1/\sqrt{V}$ (CV = 0.45, 0.22, 0.11, 0.056), the inverse of the square root of the system size. When modeled as a discrete random variable, $[Ca^{2+}]$ fluctuates over a range of 0.016 to 0.050 μM in the 1 μm^3 compartment (Figure 13.2), suggesting that it may be difficult to justify modeling Ca^{2+} dynamics in dendritic spine heads (even relatively large ones) using continuum equations.

13.6 PDEs for Reaction–Diffusion Problems

The dynamics of Ca^{2+} signaling in dendritic arborizations of neurons can be imaged via confocal microfluorimetry and two-photon laser scanning microscopy (Denk et al., 1996; Stuart et al., 1999a; Sabatini et al., 2001; Diaspro, 2002; Nimchinsky et al., 2002). Figure 13.3 shows the spatial-temporal pattern of the rise in $[Ca^{2+}]$ in a layer-5 pyramidal cell evoked by focal iontophoresis of glutamate on an apical dendrite along with the simultaneously recorded somatic membrane potential. Such experiments suggest that Ca^{2+}-dependent regenerative potentials are initiated by Ca^{2+} influx via voltage-gated Ca^{2+} channels. Regenerative Ca^{2+}-dependent potentials can be evoked at any sufficiently depolarized location on the main apical or basal trunk of these pyramidal cells and they may play a role in synaptic integration (Oakley et al., 2001).

Assuming concentration fluctuations can be ignored, the dynamics of intracellular Ca^{2+} influx and diffusion in a cylindrical dendrite with an aspect ratio (length/diameter) of 10 to 100 can be modeled using PDEs appropriate for reaction and diffusion in one dimension (Denk et al., 1996; Stuart et al., 1999b). Because diffusion times scale as (Carslaw and Jaeger, 1959; Crank, 1975; Berg, 1983)

$$\tau_D = L^2/D, \tag{13.25}$$

where L is a characteristic length and D is the diffusion coefficient, we can expect radial concentration gradients to dissipate 10^2 to 10^4 times faster than axial gradients. For this reason, we assume one-dimensional diffusion, that is, no radial variation in $[Ca^{2+}]$, [B], or [CaB]. With this geometry, the concentration $u(x,t)$ and flux $\phi(x,t)$ of a chemical species such as Ca^{2+}, B, or CaB produced at rate $R(x,t)$ satisfies the conservation law (Edelstein-Keshet, 1988; Keener and Sneyd, 1998; Keener, 2002),

$$\frac{\partial u}{\partial t} + \frac{\partial \phi}{\partial x} = R. \tag{13.26}$$

Combining Equation (13.26) with Fick's law of diffusion, namely, that

$$\phi(x,t) = -D\frac{\partial u}{\partial x},$$

gives the familiar reaction–diffusion equation

$$\frac{\partial u}{\partial t} = \frac{\partial}{\partial x}\left(D\frac{\partial u}{\partial x}\right) + R, \tag{13.27}$$

which simplifies to

$$\frac{\partial u}{\partial t} = D\frac{\partial^2 u}{\partial x^2} + R, \tag{13.28}$$

when D is not a function of x. Equations (13.27) and (13.28) are generalized to many dimensions by writing $\phi = -D\nabla u$ and replacing the axial gradient of ϕ in Equation (13.26) by its divergence. The result of this procedure is

$$\frac{\partial u}{\partial t} = \nabla \cdot [D\nabla u] + R \quad \text{and} \quad \frac{\partial u}{\partial t} = D\nabla^2 u + R,$$

where the latter equation assumes that D exhibits no spatial variation, and ∇^2 is the usual Laplacian operator.

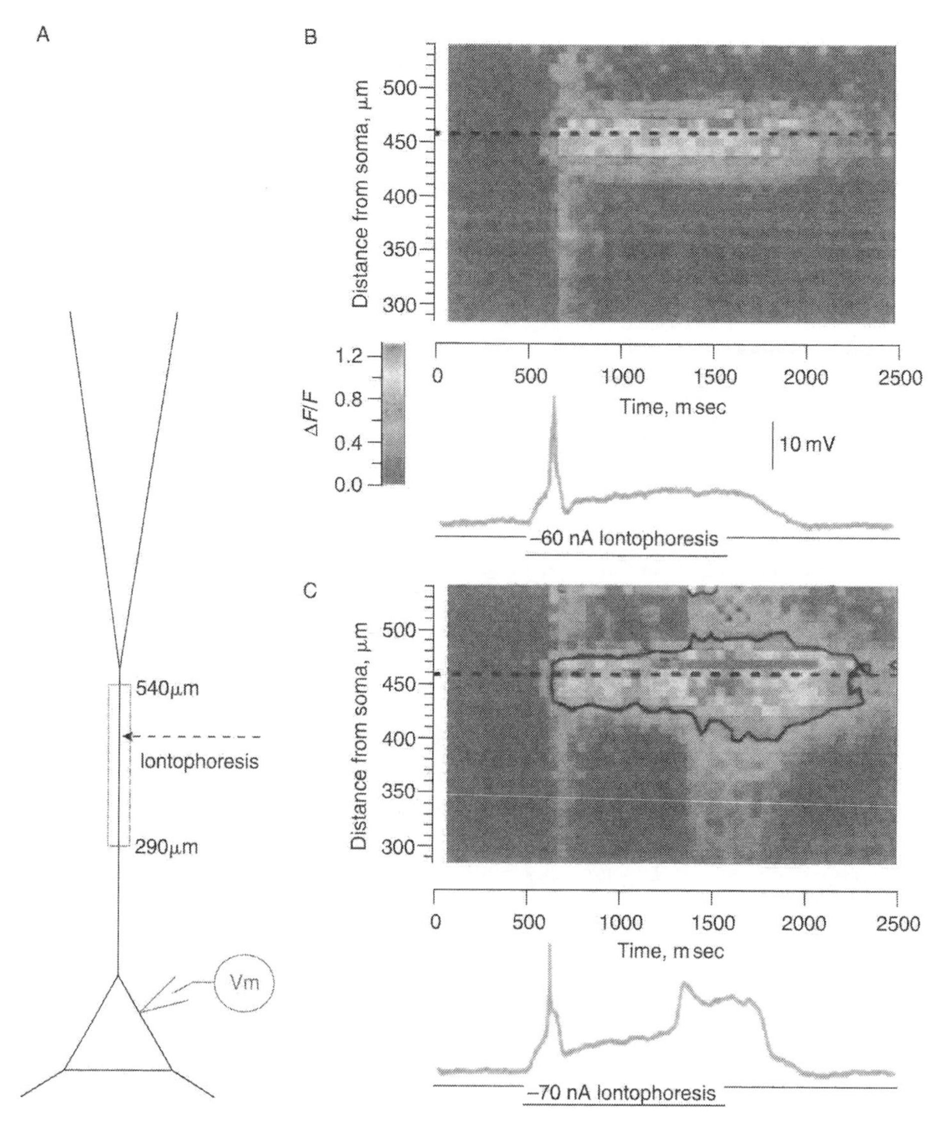

FIGURE 13.3 **(See color insert following page 366)** Spatial-temporal pattern of rise in [Ca^{2+}] in a layer-5 pyramidal cell evoked by focal iontophoresis of glutamate on an apical dendrite imaged using infrared-differential interference contrast microscopy. A: Schematic drawing showing extent of imaged area (*box* [red in color insert]), distance from soma, and position of iontophoretic electrode (*arrow*) during recording of somatic membrane potential (Vm). B, *top*: relative change in fluorescence ($\Delta F/F$ [pseudocolor in color insert]), an indicator of [Ca^{2+}], as a function of distance along the apical dendrite 290 to 540 μm from the soma (ordinate) and time (abscissa). Shown *below* is the corresponding somatic membrane potential (thick line [red in color insert], spike truncated) evoked by glutamate iontophoresis (−60 nA, thin line [black in color insert]). C: As *B* except for −70 nA iontophoresis. Black contour line indicates $\Delta F/F \geq 50\%$ of the peak $\Delta F/F$ evoked during the plateau. Figure reproduced and legend adapted with permission from Oakley, J. et al., *J. Neurophysiol.*, **86**(1), 503–513, 2001.

13.7 BUFFERED AXIAL DIFFUSION OF Ca^{2+}

Using Equation (13.28), the equations for the association of Ca^{2+} and buffer (Equations [13.15] to [13.17]) can be extended to model the buffered diffusion of intracellular Ca^{2+}. The axial diffusion of Ca^{2+} within a dendrite is described by the equations (Wagner and Keizer, 1994;

De Schutter and Smolen, 1998),

$$\frac{\partial[\mathrm{Ca}^{2+}]}{\partial t} = D_{\mathrm{C}}\frac{\partial^2[\mathrm{Ca}^{2+}]}{\partial x^2} + R + J, \tag{13.29}$$

$$\frac{\partial[\mathrm{B}]}{\partial t} = D_{\mathrm{B}}\frac{\partial^2[\mathrm{B}]}{\partial x^2} + R, \tag{13.30}$$

$$\frac{\partial[\mathrm{CaB}]}{\partial t} = D_{\mathrm{CaB}}\frac{\partial^2[\mathrm{CaB}]}{\partial x^2} - R. \tag{13.31}$$

Here D_{C}, D_{B}, and D_{CaB} are diffusion constants for free Ca^{2+}, free, and bound buffer, respectively. The function R is given by Equation (13.5) and $J(x,t)$ is the influx of free Ca^{2+} into the dendrite, for example, due to voltage-gated Ca^{2+} entry. To complete the description, we must associate boundary conditions with Equations (13.30) and (13.31), for example, assuming a finite domain ($0 \leq x \leq L$); where L is the length of the dendrite, we may assume "no flux" or von Neumann boundary conditions for each chemical species, so that

$$-D_{\mathrm{C}}\frac{\partial[\mathrm{Ca}^{2+}]}{\partial x}(0,t) = 0 \qquad -D_{\mathrm{C}}\frac{\partial[\mathrm{Ca}^{2+}]}{\partial x}(L,t) = 0, \tag{13.32}$$

and similarly for [B] and [CaB].

Note that when the buffer diffusion coefficient does not change upon binding of Ca^{2+} ($D_{\mathrm{B}} = D_{\mathrm{CaB}}$), Equations (13.30) and (13.31) can be summed to give

$$\frac{\partial[\mathrm{B}]_{\mathrm{T}}}{\partial t} = D_{\mathrm{B}}\frac{\partial^2[\mathrm{B}]_{\mathrm{T}}}{\partial x^2}.$$

This equation implies that if the total buffer concentration profile is initially uniform (i.e., $[\mathrm{B}]_{\mathrm{T}}(0)$ is not a function of x), then it will remain uniform all the time. We can thus eliminate Equation (13.30) and obtain a reduced system analogous to Equations (13.15) and (13.16),

$$\frac{\partial[\mathrm{Ca}^{2+}]}{\partial t} = D_{\mathrm{C}}\frac{\partial^2[\mathrm{Ca}^{2+}]}{\partial x^2} + R + J, \tag{13.33}$$

$$\frac{\partial[\mathrm{CaB}]}{\partial t} = D_{\mathrm{B}}\frac{\partial^2[\mathrm{CaB}]}{\partial x^2} - R, \tag{13.34}$$

with R defined by Equation (13.17). Alternatively, one can eliminate Equation (13.31) to arrive at

$$\frac{\partial[\mathrm{Ca}^{2+}]}{\partial t} = D_{\mathrm{C}}\frac{\partial^2[\mathrm{Ca}^{2+}]}{\partial x^2} + R + J, \tag{13.35}$$

$$\frac{\partial[\mathrm{B}]}{\partial t} = D_{\mathrm{B}}\frac{\partial^2[\mathrm{B}]}{\partial x^2} + R, \tag{13.36}$$

and use $R = -k_+[\mathrm{B}][\mathrm{Ca}^{2+}] + k_-\left([\mathrm{B}]_{\mathrm{T}} - [\mathrm{B}]\right)$.

13.8 An Explicit Numerical Scheme for Buffered Ca^{2+} Diffusion

The most intuitive approach to simulating the equations for the buffered diffusion of intracellular Ca^{2+} is to use the "method of lines" (Morton and Mayers, 1994; De Schutter and Smolen, 1998).

This technique reduces a PDE such as Equation (13.35) or Equation (13.36) to a system of ODEs that can be numerically integrated using, for example, the forward Euler or Runge–Kutta methods (Mascagni and Sherman, 1998). Assuming a finite domain ($0 \leq x \leq L$), we discretize space so that $C_j(t)$ is an approximation to $[\mathrm{Ca}^{2+}]$ evaluated at the mesh points $x_j = j\Delta x$ where $0 \leq j \leq J$ and $\Delta x = L/J$, that is, $C_j \approx [\mathrm{Ca}^{2+}](x_j, t)$. If we use a centered second difference for the Laplacian,

$$\frac{C_{j+1} - 2C_j + C_{j-1}}{(\Delta x)^2} \approx \frac{\partial^2 [\mathrm{Ca}^{2+}]}{\partial x^2}(x_j, t)$$

and similarly for $B_j \approx [\mathrm{B}](x_j, t)$, we can write

$$\frac{\mathrm{d}C_j}{\mathrm{d}t} = \nu_{\mathrm{C}}(C_{j+1} - 2C_j + C_{j-1}) + R_j + J_j, \tag{13.37}$$

$$\frac{\mathrm{d}B_j}{\mathrm{d}t} = \nu_{\mathrm{B}}(B_{j+1} - 2B_j + B_{j-1}) + R_j, \tag{13.38}$$

where $\nu_i = D_i/(\Delta x)^2$, $i \in \{C, B\}$, $1 \leq j \leq J - 1$, and $R_j = -k_+ B_j C_j + k_-([\mathrm{B}]_{\mathrm{T}} - B_j)$. To derive analogous equations for $j = 0$ or J, we apply our chosen boundary conditions. For example, the "no flux" boundary condition for Ca^{2+} at $x = 0$ (Equation [13.32]) implies $\partial[\mathrm{Ca}^{2+}]/\partial x|_{x=0} = 0$. To discretize this expression, note that a centered first difference for the spatial derivative

$$\frac{C_{j+(1/2)} - C_{j-(1/2)}}{\Delta x} \approx \frac{\partial [\mathrm{Ca}^{2+}]}{\partial x}(x_j, t)$$

is equivalent to

$$\frac{C_{j+1} - C_{j-1}}{2\Delta x} \approx \frac{\partial [\mathrm{Ca}^{2+}]}{\partial x}(x_j, t)$$

because $x_{j+(1/2)} = (x_j + x_{j+1})/2$ and $x_{j-(1/2)} = (x_{j-1} + x_j)/2$. Thus, $\partial[\mathrm{Ca}^{2+}]/\partial x|_{x=0} = 0$ can be discretized as

$$\frac{C_1 - C_{-1}}{2\Delta x} = 0,$$

an expression that can be interpreted as the definition of $[\mathrm{Ca}^{2+}]$ for a "ghost mesh point" at $j = -1$. Setting $j = 0$ and $C_{-1} = C_1$ in Equation (13.37) gives

$$\frac{\mathrm{d}C_0}{\mathrm{d}t} = 2\nu_{\mathrm{C}}\,(C_1 - C_0) + R_0 + J_0. \tag{13.39}$$

Similar arguments imply that a no-flux boundary condition for Ca^{2+} at $x = L$ can be enforced using $j = J$ and $C_{J+1} = C_{J-1}$ in Equation (13.37), that is,

$$\frac{\mathrm{d}C_J}{\mathrm{d}t} = 2\nu_{\mathrm{C}}\,(C_{J-1} - C_J) + R_J + J_J. \tag{13.40}$$

The buffered diffusion of intracellular Ca^{2+} can be simulated by integrating the nonlinear system of ODEs given by Equations (13.37) to (13.40), and similar equations for B_0 and B_J derived assuming no flux boundary conditions for [B].

13.9 An Implicit Numerical Scheme for Buffered Ca^{2+} Diffusion

Numerical stability of the method of lines can be a severe restriction when the spatial mesh size (Δx) is small. When the forward Euler method is used to integrate Equation (13.28), discretized using the method of lines with $R = 0$ (i.e., Ca^{2+} diffusion in the absence of buffers), a von Neumann stability analysis shows that the time step must satisfy

$$\Delta t < (\Delta x)^2/2D \tag{13.41}$$

for numerical stability (see Problem 3). This requirement is undesirable because if the number of mesh points is increased by a factor of two (so that Δx is halved), the time step Δt must decrease by a factor of four to maintain the numerical stability of the scheme.

An implicit finite difference scheme known as the Crank–Nicholson method is often preferred over the explicit method described above due to improved stability properties (Morton and Mayers, 1994). To describe this scheme, write C_j^n as an approximation to $[Ca^{2+}](x_j, t_n)$ where $t_n = n\Delta t$ and $n = 0, 1, 2, \ldots$. Using a forward difference for the time derivative,

$$\frac{C_j^{n+1} - C_j^n}{\Delta t} \approx \frac{\partial [Ca^{2+}]}{\partial t}(x_j, t_n)$$

and averaging the Laplacian evaluated at the current and future time steps,

$$\frac{1}{2}\left[\frac{C_{j+1}^{n+1} - 2C_j^{n+1} + C_{j-1}^{n+1}}{(\Delta x)^2} + \frac{C_{j+1}^n - 2C_j^n + C_{j-1}^n}{(\Delta x)^2}\right] \approx \frac{\partial^2 [Ca^{2+}]}{\partial x^2}(x_j, t_n),$$

we can discretize Equations (13.35) and (13.36) to obtain

$$U_j^{n+1} - \tfrac{1}{2}\eta_U\left(U_{j+1}^{n+1} - 2U_j^{n+1} + U_{j-1}^{n+1}\right) = U_j^n + \tfrac{1}{2}\eta_U\left(U_{j+1}^n - 2U_j^n + U_{j-1}^n\right) + \Delta t R_j^n, \tag{13.42}$$

where $\eta_U = D_U \Delta t/(\Delta x)^2$ and U stands for both C and B in turn. Thus U_j^{n+1} is obtained from U_j^n and R_j^n by solving a system of linear equations.

Writing the problem in matrix form gives two tridiagonal matrices of the form

$$A = \begin{pmatrix} b_0 & c_0 & & & \\ a_1 & b_1 & c_1 & & \\ & & \ddots & & \\ & & a_{J-1} & b_{J-1} & c_{J-1} \\ & & & a_J & b_J \end{pmatrix}, \tag{13.43}$$

where $a_j = -\eta_U/2$, $b_j = 1 + \eta_U$, $c_j = -\eta_U/2$ for $1 \le j \le J-1$. If no flux boundary conditions are used then $U_{-1} = U_1$ and $U_{J+1} = U_J$ for both n and $n+1$. Thus, for $j = 0$ and $j = J$ Equation (13.42) becomes

$$U_0^{n+1} - \eta_U\left(U_1^{n+1} - U_0^{n+1}\right) = U_0^n + \eta_U\left(U_1^n - U_0^n\right) + \Delta t R_0^n$$
$$U_J^{n+1} - \eta_U\left(U_{J-1}^{n+1} - U_J^{n+1}\right) = U_J^n + \eta_U\left(U_{J-1}^n - U_J^n\right) + \Delta t R_J^n$$

and $b_0 = 1 + \eta_U$, $c_0 = -\eta_U$, $a_J = -\eta_U$, and $b_J = 1 + \eta_U$ in Equation (13.43). The concentration profiles for Ca^{2+} and buffer are updated in each time step ($U_j^n \rightarrow U_j^{n+1}$) by solving the tridiagonal system

$$\begin{pmatrix} b_0 & c_0 & & & \\ a_1 & b_1 & c_1 & & \\ & & \ddots & & \\ & & a_{J-1} & b_{J-1} & c_{J-1} \\ & & & a_J & b_J \end{pmatrix} \begin{pmatrix} U_0^{n+1} \\ U_1^{n+1} \\ \vdots \\ U_{J-1}^{n+1} \\ U_J^{n+1} \end{pmatrix} = \begin{pmatrix} X_0^n \\ X_1^n \\ \vdots \\ X_{J-1}^n \\ X_J^n \end{pmatrix},$$

where the X_j are given by

$$X_0^n = U_0^n + \eta_U\left(U_1^n - U_0^n\right) + \Delta t R_0^n,$$
$$X_j^n = U_j^n + \tfrac{1}{2}\eta_U\left(U_{j+1}^n - 2U_j^n + U_{j-1}^n\right) + \Delta t R_j^n \quad (1 \le j \le J-1),$$
$$X_J^n = U_J^n + \eta_U\left(U_{J-1}^n - U_J^n\right) + \Delta t R_J^n.$$

For simplicity, we evaluate the nonlinear reaction terms at the current time step. In Problem 4, the reader can use von Neumann stability analysis to show that in the absence of buffers the Crank–Nicholson method is unconditionally stable.

13.10 Calculations of the Buffered Diffusion of Intracellular Ca^{2+}

Simulations of the buffered diffusion of intracellular Ca^{2+} using the Crank–Nicholson-like scheme outlined in the previous section are illustrated in Figure 13.4. Solid lines in Figure 13.4A demonstrate how the Crank–Nicholson-like scheme can be used to solve a simple diffusion problem with

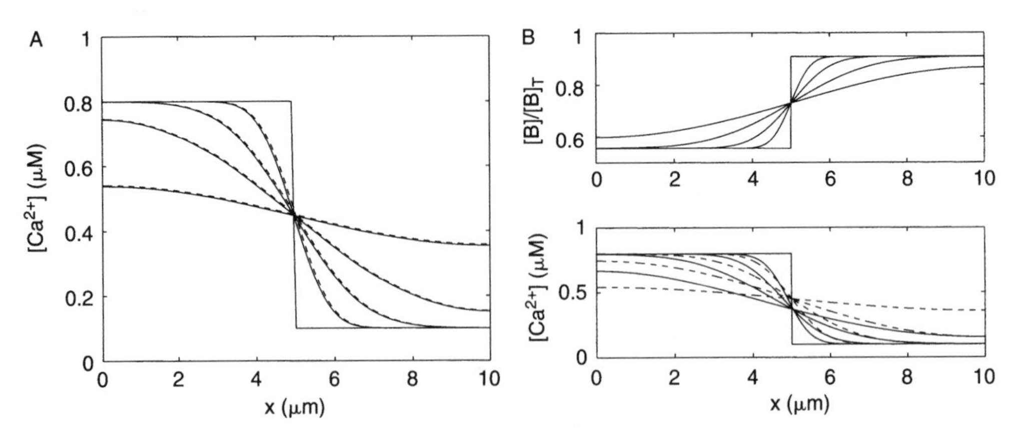

Figure 13.4 Comparison between the analytical and Crank–Nicholson-like numerical solutions for diffusion of Ca^{2+}. A: Simulations that do not include Ca^{2+} buffers ($[B]_T = 0$). The discrepancy between the numerical (solid lines) and analytical (dashed lines) calculations is eliminated when Δx is reduced from 0.1 to 0.01 μm and Δt is reduced from 0.1 to 0.01 μsec (not shown, results are indistinguishable from dashed lines). B: When 50 μM of mobile Ca^{2+} buffer is included ($D_B = 75$ μm^2 sec^{-1}), Ca^{2+} diffusion is slowed, and the dissipation of the initial Ca^{2+} gradient is delayed. Dashed lines re-plot the result from A where $[B]_T = 0$.

initial condition

$$[\mathrm{Ca}^{2+}](x,0) = \begin{cases} c_L & (x < h) \\ c_R & (x \geq h) \end{cases},$$

where $0 \leq x \leq 10\ \mu\mathrm{m}$, $h = 5\ \mu\mathrm{m}$, $c_L = 0.8\ \mu\mathrm{M}$, and $c_R = 0.1\ \mu\mathrm{M}$. Assuming $[\mathrm{B}]_\mathrm{T} = 0$ and $D_\mathrm{C} = 250\ \mu\mathrm{m}^2\ \mathrm{sec}^{-1}$, the Ca^{2+} profile is shown for $t = 0, 1, 16, 64$, and 256 msec. For comparison, the dotted lines in Figure 13.4A show the analytical result (Crank, 1975, p. 16)

$$[\mathrm{Ca}^{2+}] = c_R + \frac{1}{2}(c_L - c_R) \sum_{n=-\infty}^{\infty} \left[\mathrm{erf}\left(\frac{h + 2nL - x}{2\sqrt{D_\mathrm{C} t}} \right) + \mathrm{erf}\left(\frac{h - 2nL + x}{2\sqrt{D_\mathrm{C} t}} \right) \right],$$

for $h = L/2$. Note that at $t = 256$ msec the Ca^{2+} gradient is nearly dissipated in agreement with Equation (13.25) where $\tau_D = 400$ msec. The discrepancy between the numerical *solid lines* and analytical *dashed lines* is due to discretization error for $\Delta x = 0.1\ \mu\mathrm{m}$ and $\Delta t = 0.1\ \mu\mathrm{sec}$. When Δx and Δt are reduced by a factor of 10, the numerical calculation is indistinguishable by eye from the analytical calculation.

Figure 13.4B repeats the calculations of Figure 13.4A with $[\mathrm{B}]_\mathrm{T} = 50\ \mu\mathrm{M}$ (see Table 13.1 set B1). The buffer is initially at equilibrium with free Ca^{2+} so that $[\mathrm{B}](x,0) = 27.8\ \mu\mathrm{M}$ ($x < 5\ \mu\mathrm{m}$) or $45.5\ \mu\mathrm{M}$ ($x \geq 5\ \mu\mathrm{m}$). This mobile buffer diffuses more slowly than Ca^{2+} ($D_\mathrm{B} = 75\ \mu\mathrm{m}^2\ \mathrm{sec}^{-1}$) and impedes the diffusion of Ca^{2+} on this one-dimensional finite domain (compare profiles at $t =$ 256 msec, the most horizontal), that is, the initial Ca^{2+} gradient dissipates more slowly in the presence of Ca^{2+} buffer. Note that although the Ca^{2+} buffer is mobile, there is no facilitation of Ca^{2+} diffusion with these initial conditions.

13.11 The Rapid Buffer Approximation and Effective Diffusion Coefficient

If the buffer equilibration time constant (τ_B in Equation [13.21]) is small compared with the time constant (τ_D in Equation [13.25]) of diffusion, buffer reactions are rapid by comparison with diffusion. Under these conditions, Equations (13.15) and (13.16) can be simplified using the so-called rapid buffer approximation (Wagner and Keizer, 1994) leading to analytical formulas which give insight into the effect that buffers have on Ca^{2+} diffusion.

Assuming chemical equilibrium at every point in space, the equilibrium relations of Equations (13.10) and (13.12) will be satisfied approximately, and we may write

$$[\mathrm{Ca}^{2+}]_\mathrm{T} = [\mathrm{Ca}^{2+}] + [\mathrm{CaB}] = [\mathrm{Ca}^{2+}] + \frac{[\mathrm{Ca}^{2+}][\mathrm{B}]_\mathrm{T}}{[\mathrm{Ca}^{2+}] + K}.$$

We then eliminate reaction terms (and the fast time scale) by adding Equations (13.33) and (13.34) to get

$$\frac{\partial[\mathrm{Ca}^{2+}]_\mathrm{T}}{\partial t} = \frac{\partial[\mathrm{Ca}^{2+}]}{\partial t} + \frac{\partial[\mathrm{CaB}]}{\partial t} = D_\mathrm{C}\frac{\partial^2[\mathrm{Ca}^{2+}]}{\partial x^2} + D_\mathrm{B}\frac{\partial^2[\mathrm{CaB}]}{\partial x^2} + J. \tag{13.44}$$

The reader can confirm that substituting $[\mathrm{Ca}^{2+}]_\mathrm{T}$ and $[\mathrm{CaB}]$ in terms of $[\mathrm{Ca}^{2+}]$ and taking the partial derivatives $\partial[\mathrm{Ca}^{2+}]_\mathrm{T}/\partial t$ and $\partial^2[\mathrm{CaB}]/\partial x^2$ gives

$$\frac{\partial[\mathrm{Ca}^{2+}]}{\partial t} = \beta\left[(D_\mathrm{C} + D_\mathrm{B}\gamma)\frac{\partial^2[\mathrm{Ca}^{2+}]}{\partial x^2} - \frac{2\gamma D_\mathrm{B}}{K + [\mathrm{Ca}^{2+}]}\left(\frac{\partial[\mathrm{Ca}^{2+}]}{\partial x}\right)^2 + J \right], \tag{13.45}$$

where $\gamma = K[\mathrm{B}]_\mathrm{T}/(K+[\mathrm{Ca}^{2+}])^2$ and $\beta = 1/(1+\gamma)$. Although this equation may appear ominous, note that the rapid buffer approximation has reduced the buffered Ca^{2+} diffusion problem to a single transport equation.

The rapid buffer approximation provides insight into the effect of Ca^{2+} buffers on dendritic Ca^{2+} diffusion. For example, inspecting the first term of Equation (13.45) allows us to identify a Ca^{2+}-dependent effective diffusion coefficient

$$D_\mathrm{eff} = \beta\,(D_\mathrm{C} + \gamma D_\mathrm{B}) \tag{13.46}$$

and observe that $D_\mathrm{eff} < D_\mathrm{C}$ for physiological parameter values where $D_\mathrm{B} < D_\mathrm{C}$. If the buffer is not only rapid but also stationary (e.g., an immobilized Ca^{2+}-binding protein or a high molecular weight dextran-conjugated indicator), then $D_\mathrm{B} = 0$ and $D_\mathrm{eff} = \beta D_\mathrm{C}$. On the other hand, if a buffer is mobile and has low affinity, then $[\mathrm{Ca}^{2+}] \ll K$, $\gamma \approx [\mathrm{B}]_\mathrm{T}/K$, and the diffusion coefficient is approximately constant with value

$$D_\mathrm{eff} \approx \frac{K}{K+[\mathrm{B}]_\mathrm{T}}\left(D_\mathrm{C} + \frac{[\mathrm{B}]_\mathrm{T}}{K} D_\mathrm{B}\right).$$

There is a formal equivalence between the rapid buffer approximation in this low-affinity limit and the cable equation for the behavior of the membrane potential in dendritic trees. For example, if the J in Equation (13.45) includes contributions from constant Ca^{2+} influx (J_in) and a linear Ca^{2+} pump ($J = J_\mathrm{in} - k_\mathrm{out}[\mathrm{Ca}^{2+}]$), then in the presence of a rapid, stationary, low-affinity buffer,

$$\frac{\partial[\mathrm{Ca}^{2+}]}{\partial t} = \frac{K}{K+[\mathrm{B}]_\mathrm{T}}\left[D_\mathrm{C}\frac{\partial^2[\mathrm{Ca}^{2+}]}{\partial x^2} + J_\mathrm{in} - k_\mathrm{out}[\mathrm{Ca}^{2+}]\right].$$

This equation can be put into the form of the cable equation $\tau\,\partial u/\partial t = \lambda^2\partial^2 u/\partial x^2 - u$,

$$\frac{K+[\mathrm{B}]_\mathrm{T}}{k_\mathrm{out}K}\frac{\partial[\mathrm{Ca}^{2+}]}{\partial t} = \frac{D_\mathrm{C}}{k_\mathrm{out}}\frac{\partial^2[\mathrm{Ca}^{2+}]}{\partial x^2} - ([\mathrm{Ca}^{2+}] - [\mathrm{Ca}^{2+}]_{ss})$$

with time constant $\tau = (K+[\mathrm{B}]_\mathrm{T})/k_\mathrm{out}K$ and length constant $\lambda^2 = D_\mathrm{C}/k_\mathrm{out}$ where $[\mathrm{Ca}^{2+}]_{ss} = J_\mathrm{in}/k_\mathrm{out}$, and $u = [\mathrm{Ca}^{2+}] - [\mathrm{Ca}^{2+}]_{ss}$ (Zador and Koch, 1994; Koch, 1999).

13.12 The Validity of the Effective Diffusion Coefficient

The validity of the analytically derived effective diffusion coefficient (Equation [13.46]) can be demonstrated numerically by using a Gaussian initial condition for Ca^{2+} and observing the rate at which Ca^{2+} spreads. In the absence of buffers, the Gaussian function

$$u(x,t) = \frac{1}{\sqrt{4\pi Dt}}\mathrm{e}^{-x^2/4Dt} \tag{13.47}$$

is a solution of the one-dimensional diffusion problem on an infinite domain satisfying Equation (13.28) with $R = 0$. The square of the full width at half maximum (FWHM) of this evolving Gaussian function is given by

$$(\mathrm{FWHM})^2 = (16\ln 2)Dt, \tag{13.48}$$

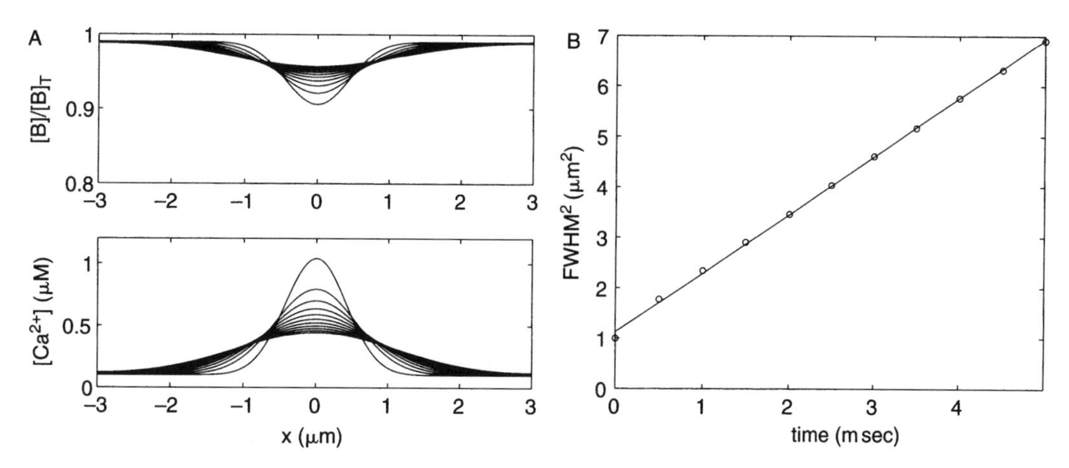

FIGURE 13.5 A: Numerical simulation of the diffusion of an initially Gaussian pulse of free Ca^{2+} and equilibrated Ca^{2+} buffer. B: The observed diffusion coefficient ($D_{obs} = 104\ \mu m^2\ sec^{-1}$) calculated from the square of the full width at half maximum of the evolving and approximately Gaussian free Ca^{2+} profile agrees with the analytical estimate ($D_{eff} = 105\ \mu m^2\ sec^{-1}$ in Equation [13.46]). Parameters: $D_C = 250\ \mu m^2\ sec^{-1}$, $[Ca^{2+}]_\infty = 0.1\ \mu M$, and set B2 as in Table 13.1.

that is, the FWHM grows at a rate proportional to the diffusion coefficient (see Problem 6). Thus, the slope of the square of the FWHM of an initially Gaussian pulse plotted as a function of time can be used to estimate an observed diffusion coefficient D_{obs}.

Figure 13.5A shows a numerical simulation of the diffusion of an initially Gaussian pulse of free Ca^{2+} in the presence of initially equilibrated buffer. In Figure 13.5B, the observed diffusion coefficient D_{obs} is calculated from the $[Ca^{2+}]$ profile using Equation (13.48). Note that the observed diffusion coefficient of $D_{obs} = 104\ \mu m^2\ sec^{-1}$ is in agreement with Equation (13.46) where $D_{eff} = 105\ \mu m^2\ sec^{-1}$ because the assumption of rapid equilibrium is valid. The buffer parameters used (Table 13.1, set B2) lead to a buffer equilibration time of $\tau_B = 0.17$ ms, much smaller than the diffusional time scale that is initially $\tau_D = 4$ ms as calculated from Equation (13.25) using $D_C = 250\ \mu m^2\ sec^{-1}$ and a characteristic length of 1 μm (the initial FWHM of the Gaussian).

When buffer parameters are such that reaction is *not* fast compared to diffusion, the assumption of rapid equilibrium used in deriving Equation (13.46) may not be satisfied and D_{obs} may disagree with D_{eff} (Equation [13.46]). Figure 13.6 shows a parameter study confirming that $D_{obs} \approx D_C$ for slow buffers (left, $k_\pm$ small) while $D_{obs} \approx D_{eff}$ for rapid buffers (right, $k_\pm$ large). In these calculations, the buffer dissociation constant $K = k_-/k_+ = 10\ \mu M$ is fixed and each data point corresponds to a calculation similar to Figure 13.5A and Figure 13.5B.

Note that in Figure 13.5 and Figure 13.6, the FWHM of the free Ca^{2+} profile is used to estimate D_{obs}. However, in the experimental application of this technique, the free $[Ca^{2+}]$ profile is not available and an effective diffusion coefficient of the *fluorescence signal* is often measured (Gabso et al., 1997). This quantity can be directly related to the effective diffusion coefficient for Ca^{2+} only when the indicator dye is sufficiently rapid.

13.13 DYNAMICS OF Ca^{2+} RELEASE AND RESEQUESTRATION BY INTERNAL STORES

As mentioned in Section 13.1, both Ca^{2+} influx via voltage-gated Ca^{2+} channels and Ca^{2+} release by intracellular Ca^{2+} channels (IP_3Rs and RyRs) of the ER are sources for cytosolic $[Ca^{2+}]$ (Berridge, 1998). While models of global Ca^{2+} responses often neglect Ca^{2+} diffusion, these models have played a major role in understanding Ca^{2+} dynamics in many cell types (Dupont et al., 1991;

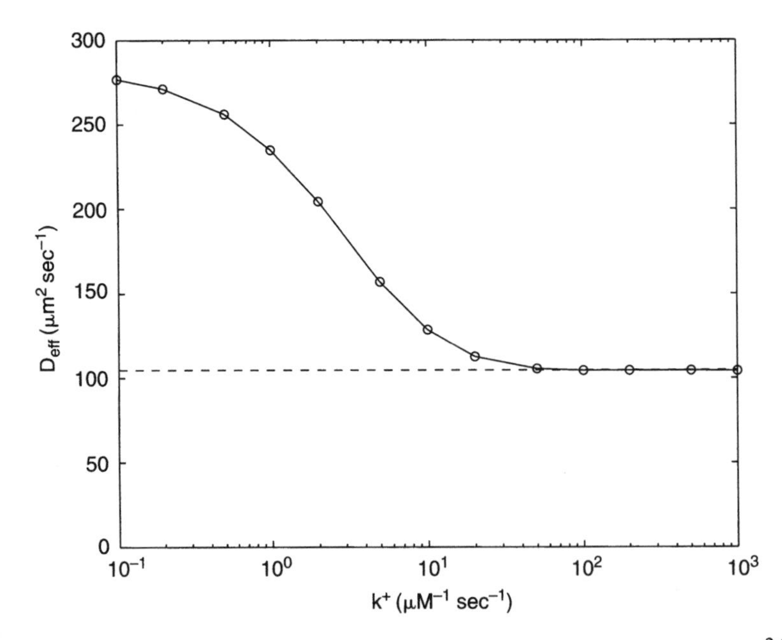

FIGURE 13.6 Parameter study showing that the observed diffusion coefficient (D_{obs}) for Ca^{2+} approaches the analytically calculated value (D_{eff} in Equation [13.46]) as the association (k_+) and dissociation (k_-) rate constants increase with constant ratio $K = k_-/k_+ = 10\ \mu M$.

Dupont and Goldbeter, 1992; Keizer et al., 1995; Li et al., 1995a; Sherman et al., 2002). Such "whole cell models" of Ca^{2+} handling are systems of nonlinear ODEs. For example,

$$\frac{d[Ca^{2+}]}{dt} = \underbrace{J_{rel} - J_{up}}_{J_{er}} + \underbrace{J_{in} - J_{out}}_{J_{pm}} \tag{13.49}$$

$$\frac{d[Ca^{2+}]_{ER}}{dt} = -J_{er}/\alpha_{er} \tag{13.50}$$

$$\frac{dw}{dt} = \frac{w_\infty - w}{\tau_w} \tag{13.51}$$

is a common form for whole cell models of IP_3-mediated Ca^{2+} responses with ER Ca^{2+} release and re-uptake fluxes (Figure 13.1) given by

$$J_{rel} = (v_l + v_{ip} f_o)([Ca^{2+}]_{ER} - [Ca^{2+}]), \tag{13.52}$$

$$J_{up} = \frac{v_p[Ca^{2+}]^2}{[Ca^{2+}]^2 + k_p^2}. \tag{13.53}$$

Such models are often very complex and involve many additional compartments including mitochondrial Ca^{2+} stores (Magnus and Keizer, 1998a, 1998b; Fall and Keizer, 2001) and restricted cytosolic compartments (Jafri et al., 1998). Whole cell models of Ca^{2+} handling also often include Hodgkin–Huxley-like membrane kinetics that couples with intracellular Ca^{2+} handling via voltage-gated Ca^{2+} entry, Ca^{2+}-activated K^+ conductances, and other mechanisms (Li et al., 1995b, 1997).

This relatively simple whole cell model includes an equation for the time evolution of w, a Hodgkin–Huxley-like gating variable representing the fraction of IP_3Rs *not* inactivated. The open fraction (or open probability) of IP_3Rs (f_o) is a function of w, $[Ca^{2+}]$, and $[IP_3]$ (see Section 13.14). Fluxes influencing the $[Ca^{2+}]$ in the cytosol include: Ca^{2+} release via the IP_3R and a passive leak, J_{rel}; Ca^{2+} reuptake via SERCA-type Ca^{2+}-ATPases, J_{up}; and influx (J_{in}) and efflux (J_{out}) across the cell plasma membrane (see Figure 13.1). By numerical integration of Equations (13.49) to (13.51), Ca^{2+} release and re-uptake by IP_3-sensitive intracellular Ca^{2+} stores can be simulated.

Whole cell models of Ca^{2+} handling are biophysically realistic to the extent that they include details of molecular mechanism. For example, sigmoidal kinetics of the SERCA-type Ca^{2+}-ATPase (Lytton et al., 1992) have been used in Equations (13.49) to (13.53) and the parameter $\alpha_{er} \approx 1/6$ accounts for the ER/cytosol volume ratio. As is common in whole cell models, the effect of Ca^{2+} buffers is not modeled explicitly; instead the rates v_l, v_{ip}, and v_p are adjusted to account for "constant fraction buffering" (Wagner and Keizer, 1994; Sherman et al., 2002) that reduces Ca^{2+} fluxes 100- to 1000-fold (Sala and Hernandez-Cruz, 1990; Allbritton et al., 1992; Gabso et al., 1997). Although this approximation is done for convenience, it can be justified by setting $\partial[Ca^{2+}]/\partial x = 0$ (or in more complicated geometries, $\nabla[Ca^{2+}] = 0$) in the rapid buffer approximation (Equation [13.45]) to give

$$\frac{d[Ca^{2+}]}{dt} = \beta J_T = \left(1 + \frac{K[B]_T}{(K + [Ca^{2+}])^2}\right)^{-1} J_T,$$

where $J_T = J_{er} + J_{pm}$. A similar equation gives the reduced dynamics of $[Ca^{2+}]_{ER}$. Constant fraction buffering further assumes low-affinity Ca^{2+} buffers so that $\beta^{-1} = 1 + \gamma \approx 1 + [B]_T/K$. Thus any difference in ER and cytosol buffer capacity is absorbed into the parameter α_{er} (Sherman et al., 2002).

13.14 A Minimal Intracellular Ca^{2+} Channel Model

The key to biophysical realism of whole cell models of intracellular Ca^{2+} responses is the functional form for the open probability of the intracellular Ca^{2+} channels and the equation for the kinetics of activation and inactivation of these channels. For example, Equation (13.51) represents the Li–Rinzel reduction of the De Young–Keizer model in which the type 1 IP_3R is viewed as a collection of n independent subunits, each of which has one binding site for IP_3 and two binding sites for Ca^{2+} (De Young and Keizer, 1992; Li and Rinzel, 1994). Thus, three processes (IP_3-potentiation, Ca^{2+}-activation, and Ca^{2+}-inactivation) produce eight possible states for each IP_3R subunit (see Figure 13.7 and Table 13.2).

After assuming certain symmetries in the rate constants, for example, the association rate constant (a_5) for Ca^{2+} binding to the activating site is the same whether the subunit is in state S_1, S_2, S_3, or S_4 where $d_1 d_2 = d_3 d_4$, DeYoung and Keizer chose parameters to fit three experimental observations:

1. The Ca^{2+}-dependent dissociation constant of IP_3 binding to the IP_3R (Joseph et al., 1989).
2. The bell-shaped equilibrium open probability curve of the type 1 IP_3R as a function of $[Ca^{2+}]$ (Bezprozvanny et al., 1991).
3. The rapidity of Ca^{2+}-activation compared to Ca^{2+}-inactivation (Parker and Ivorra, 1990).

When the De Young–Keizer IP_3R model is combined with a SERCA-type Ca^{2+}-ATPase and a passive leak as in Equations (13.49) to (13.53), the resulting whole cell model of Ca^{2+} handling can exhibit bistability, excitability, and oscillations in $[Ca^{2+}]$.

Li and Rinzel derived a simplified version of the De Young–Keizer model IP_3R by noticing that the fast processes of IP_3-potentiation and Ca^{2+}-activation are essentially at equilibrium with the slower process of Ca^{2+}-inactivation (Li and Rinzel, 1994). Retaining rate-constant symmetries

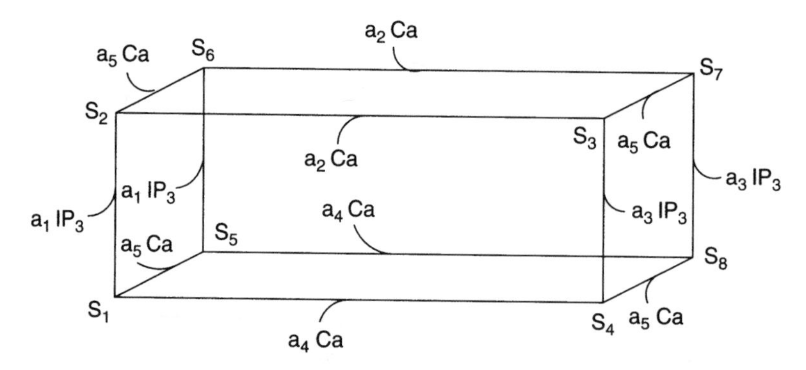

FIGURE 13.7 Transition and state diagram of one subunit of the De Young–Keizer IP_3R model (De Young and Keizer, 1992). The processes of Ca^{2+}-activation, Ca^{2+}-inactivation, and IP_3-potentiation result in eight possible states (S_1–S_8) for each IP_3R subunit. The channel is open when all n subunits of the model IP_3R are in the permissive state, S_6. For clarity, dissociation constants (d_i) and rates ($b_i = a_i d_i$) are not shown. Reproduced with permission from Smith, G., In C.A. Condat and A. Baruzzi, (eds), *Recent Research Developments in Biophysical Chemistry*, Research Signpost, Kerala, 2002a.

TABLE 13.2
Parameters Leading to Bistability, Excitability, and Oscillations in a Whole-Cell Model of Ca^{2+} Handling (Figure 13.9 and Figure 13.11)

Parameter	Value	Units
a_2	0.2	$\mu M^{-1}\ sec^{-1}$
a_4	0.2	$\mu M^{-1}\ sec^{-1}$
d_1	0.13	μM
d_2	1.049 or 5	μM
d_3	0.9434	μM
d_4	$d_1 d_2/d_3$	μM
d_5	0.08234	μM
v_{ip}	2	sec^{-1}
v_l	0.01 or 0.002 or 0.005	sec^{-1}
v_p	0.9	$\mu M^{-1}\ sec^{-1}$
k_p	0.1	μM
c_0	2	μM
α	0.185	—
$[IP_3]$	0.5	μM
D	30	$\mu m^2\ sec^{-1}$
n	4	—

With the exception of v_{ip}, v_l, d_5, and n, parameters used largely follow De Young and Keizer (1992). In Figure 13.9: $v_l = 0.01\ s^{-1}$, $d_2 = 1.049\ \mu M$; Figure 13.11A: $v_l = 0.002\ s^{-1}$, $d_2 = 5\ \mu M$; Figure 13.11B: $v_l = 0.005\ s^{-1}$, $d_2 = 1.049\ \mu M$.

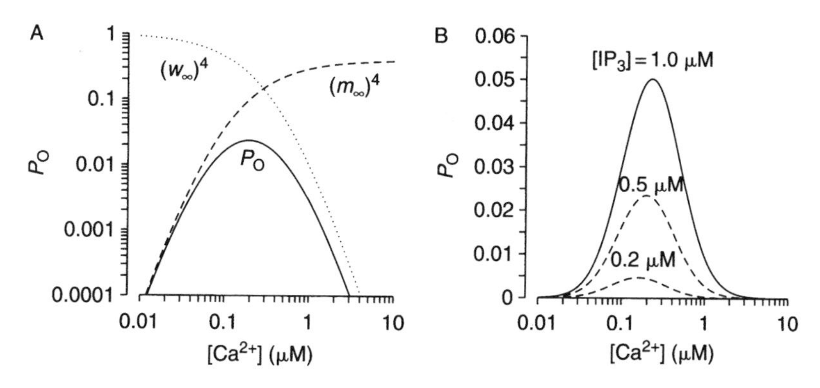

FIGURE 13.8 A: The equilibrium open probability, P_o^{eq}, of the Li–Rinzel reduction of the De Young–Keizer IP_3R model (De Young and Keizer, 1992; Li and Rinzel, 1994) is a bell-shaped function of $[Ca^{2+}]$ given by the product of activation (m_∞) and inactivation (w_∞) gates for four independent subunits. Higher $[IP_3]$ leads to greater equilibrium activity.

(e.g., $a_2 = a_4$) of the original De Young–Keizer model, the functional forms of w_∞ and τ_w in Equation (13.51) are (Li and Rinzel, 1994)

$$w_\infty = \frac{Q_2}{Q_2 + [Ca^{2+}]}, \qquad \tau_w^{-1} = a_2(Q_2 + [Ca^{2+}]),$$

where

$$Q_2 = d_2 \frac{[IP_3] + d_1}{[IP_3] + d_3}. \tag{13.54}$$

In this quasi-static limit, the fraction of open IP_3Rs is given by

$$f_o = m_\infty^n w^n, \tag{13.55}$$

where m_∞ is an instantaneously equilibrating activation gating variable given by

$$m_\infty = \left(\frac{[IP_3]}{[IP_3] + d_1}\right)\left(\frac{[Ca^{2+}]}{[Ca^{2+}] + d_5}\right). \tag{13.56}$$

The equilibrium open probability of the Li–Rinzel reduction is thus $P_o^{eq} = f_o^\infty = m_\infty^n w_\infty^n$. Figure 13.8 shows the bell-shaped dependence of P_o^{eq} on cytosolic $[Ca^{2+}]$ and the requirement for potentiating IP_3.

Figure 13.9A shows Ca^{2+} oscillations exhibited by the whole cell model given by Equations (13.49) to (13.53) for $[IP_3] = 0.5\ \mu M$. Also shown are oscillations in the fraction of IP_3R subunits not inactivated by Ca^{2+} (i.e., the gating variable w). The $(w, [Ca^{2+}])$ phase plane exhibits a single unstable equilibrium point (open circle) surrounded by a stable limit-cycle oscillation (solid line). Figure 13.9C shows a bifurcation diagram for $0 \leq [IP_3] \leq 3\ \mu M$ that reveals stable equilibria (solid line) becoming unstable (dot-dashed line) between two subcritical Hopf points at $[IP_3] = 0.5$ and $1.4\ \mu M$. Filled and open circles are stable and unstable periodic orbits, respectively. For further discussion of the dynamics of the whole-cell model of Ca^{2+} handling that includes the effect of Ca^{2+} influx on Ca^{2+} oscillations subserved by both the IP_3R and RyR, see Sherman et al. (2002), Li et al. (1995a), Keizer et al. (1995), Keizer and Levine (1996), and Smith et al. (1996a).

The IP_3R exists in multiple isoforms and splice variants (Joseph, 1996) with significant sequence homology (Sudhof et al., 1991; Blondel et al., 1993) and similar ion permeation properties

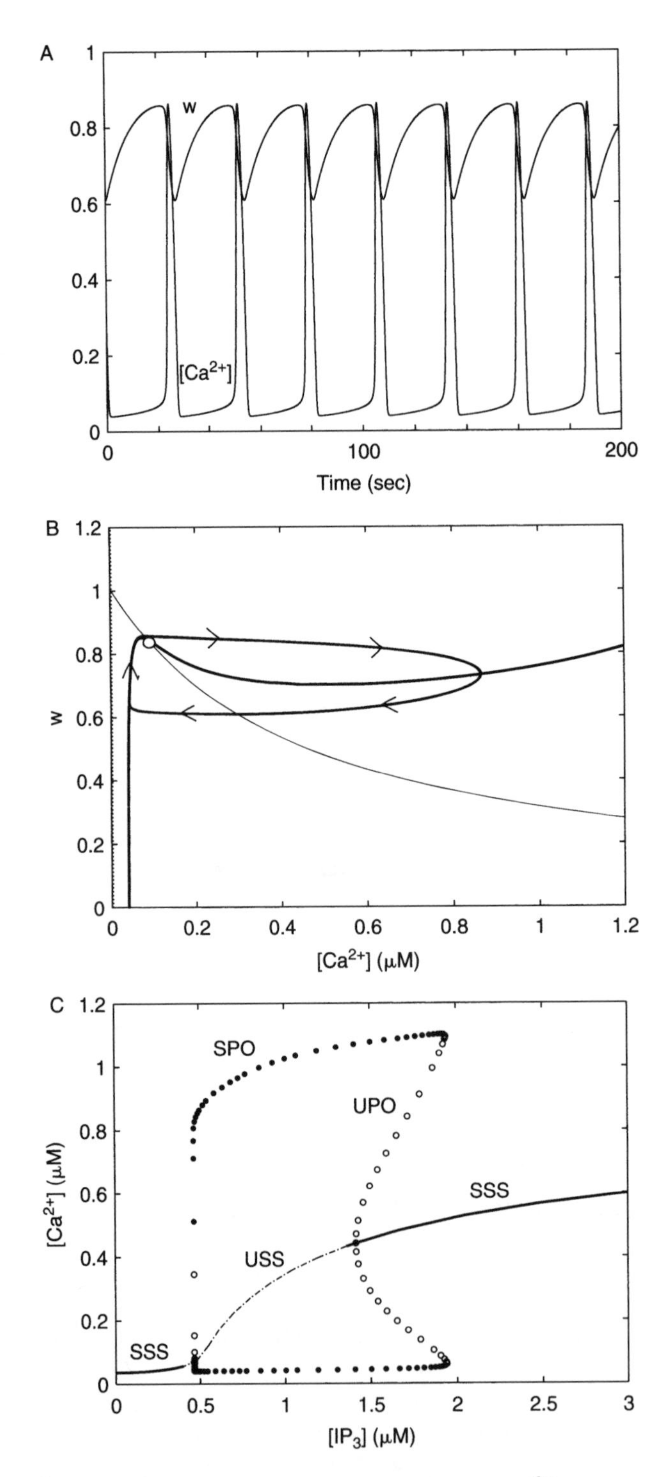

FIGURE 13.9 A: $[Ca^{2+}]$ oscillations and corresponding oscillations in Ca^{2+}-inactivation of IP_3Rs (i.e., the gating variable w) for the whole cell model of Ca^{2+} handling given by Equations (13.49) to (13.53). B: The (w, $[Ca^{2+}]$) phase plane exhibits a single unstable equilibrium (open circle) and a limit cycle oscillation (arrows). C: Bifurcation diagram for $0 \leq [IP_3] \leq 3$ μM shows stable steady states (SSS, solid line) becoming unstable (USS, dot-dashed line) between two subcritical Hopf points at $[IP_3] \approx 0.5$ and 1.4 μM. Filled and open circles are stable (SPO) and unstable (UPO) periodic orbits, respectively. Model parameters are given in Table 13.2. Calculations were performed using XPPAUT (http://www.pitt.edu/~phase/).

(Mak et al., 2000). While there are comparatively few IP_3R-2 and -3 models in the literature (Le Beau et al., 1999; Sneyd and Dufour, 2002), the De Young–Keizer model is one of several developed to reproduce single channel kinetics of the type 1 IP_3R (De Young and Keizer, 1992; Atri et al., 1993; Bezprozvanny and Ehrlich, 1994; Tang et al., 1995; Kaftan et al., 1997; Swillens et al., 1998; Moraru et al., 1999). Recent findings suggest that the "bell-shaped" curve relating $[Ca^{2+}]$ to IP_3R open probability may not apply to all IP_3R isoforms (Hagar et al., 1998; Ramos-Franco et al., 1998; Bootman and Lipp, 1999). For example, in addition to being more active than the type 1 IP_3R at low $[Ca^{2+}]$, the type 3 IP_3R may or may not inactivate at high $[Ca^{2+}]$ (Hagar et al., 1998; Mak et al., 2000). Note also that type 1 IP_3R can lose the bell-shaped equilibrium open probability in the presence of high $[IP_3]$ (Kaftan et al., 1997). The consequences of the differential Ca^{2+} regulation of the various IP_3R isoforms on global Ca^{2+} responses can be analyzed using whole-cell models of Ca^{2+} handling Equations (13.49) to (13.53) (LeBeau et al., 1999).

13.15 Propagating Waves of Ca^{2+} Release

Beginning with Equations (13.49) to (13.53), we can write a system of nonlinear PDEs to model propagating Ca^{2+} waves. The equations are

$$\frac{\partial[Ca^{2+}]}{\partial t} = D\frac{\partial^2[Ca^{2+}]}{\partial x^2} + \underbrace{J_{rel} - J_{up}}_{J_{er}} + \underbrace{J_{in} - J_{out}}_{J_{pm}}, \tag{13.57}$$

$$\frac{\partial[Ca^{2+}]_{ER}}{\partial t} = D_{ER}\frac{\partial^2[Ca^{2+}]_{ER}}{\partial x^2} - J_{er}/\alpha_{er}, \tag{13.58}$$

$$\frac{\partial w}{\partial t} = \frac{w_\infty - w}{\tau_w}, \tag{13.59}$$

where J_{rel} and J_{up} are given by Equations (13.52) and (13.53). We assume constant effective diffusion coefficients, D and D_{ER}, and constant fraction buffering that scales v_{ip}, v_l, and v_p in Equations (13.52) and (13.53). Note that Equation (13.59) does not include a Laplacian term since IP_3Rs (and thus the gating variable representing their inactivation) do not diffuse. If $D = D_{ER}$, a weighted sum of Equations (13.57) and (13.58) leads to

$$\frac{\partial[Ca^{2+}]_T}{\partial t} = \frac{\partial^2[Ca^{2+}]_T}{\partial x^2} + J_{pm},$$

where $[Ca^{2+}]_T = [Ca^{2+}] + \alpha_{er}[Ca^{2+}]_{ER}$. In the absence of plasma membrane fluxes, $J_{pm} = 0$ and we can write

$$[Ca^{2+}]_{ER} = \frac{1}{\alpha_{er}}([Ca^{2+}]_T - [Ca^{2+}]), \tag{13.60}$$

where, providing the total cell Ca^{2+} profile is initially spatially homogeneous, $[Ca^{2+}]_T$ is a constant determined by initial conditions. Similar to the equations for the buffered diffusion of intracellular Ca^{2+} just discussed Equations (13.57) and (13.59) can be integrated numerically using the method of lines or a Crank–Nicholson-like scheme. For example, when $[IP_3]$ is spatially uniform, the extension of Equation (13.37) to include the dynamics of ER Ca^{2+} release are

$$\frac{dC_j}{dt} = \nu_C(C_{j+1} - 2C_j + C_{j-1}) + J_{er}(C_j, W_j), \quad (0 \le j \le J), \tag{13.61}$$

$$\frac{dW_j}{dt} = \frac{w - w_\infty(C_j)}{\tau_w(C_j)}, \quad (0 \le j \le J), \tag{13.62}$$

where $\nu_C = D/(\Delta x)^2$, and zero flux boundary conditions for $[Ca^{2+}]$ are used to assert that $C_{-1} = C_1$ and $C_{J+1} = C_{J-1}$.

13.16 Traveling Fronts and Pulses

Figure 13.10A shows a simulation of a Ca^{2+} wave calculated using Equations (13.57) to (13.62). The horizontal axis is space, the vertical axis is time, and dark regions correspond to higher values of $[Ca^{2+}]$. Initial conditions are chosen so that elevated $[Ca^{2+}]$ in a small region triggers a rightward traveling Ca^{2+} wave. The $[Ca^{2+}]$ before and after the front passes is 0.2 and 0.8 μM, respectively. Because $[Ca^{2+}]$ does not return to basal values after the wave passes, this type of traveling wave is referred to as a *propagating front*.

The propagating front of Figure 13.10A arises because in the absence of diffusion (i.e., when $D = 0$ in Equation [13.61]) the Ca^{2+} wave model exhibits a dynamic phenomenon known as *bistability* (see Figure 13.11A). Imagine removing a small region of the spatial model and investigating its dynamics. If this region is sufficiently small, the time scale for diffusion in the isolated region ($T = L^2/D$) will be much faster than that associated with the reaction terms J_{er} and $(w - w_\infty)/\tau_w$. Thus the Ca^{2+} profile will be approximately uniform ($\partial[Ca^{2+}]/\partial x = 0$), and the region will behave as a small, continuously stirred compartment with phase plane given by Figure 13.10A. Furthermore, if the reaction terms of Equations (13.57) to (13.62) were not bistable, both the basal and elevated $[Ca^{2+}]$ on and behind the wave front could not persist.

In general, Ca^{2+} wave phenomena can be categorized by the qualitative dynamics of the ODEs obtained by assuming that all profiles are uniform (Murray, 1989). To give another example, Figure 13.10B shows a propagating Ca^{2+} wave that results when parameters are chosen so that the underlying dynamics of Ca^{2+} release and resequestration are *excitable* (see Figure 13.11B for phase plane). Although initial conditions similar to those used to generate Figure 13.10A are used, $[Ca^{2+}]$ returns to its basal level as the Ca^{2+} transient propagates past a given point, thus exhibiting a

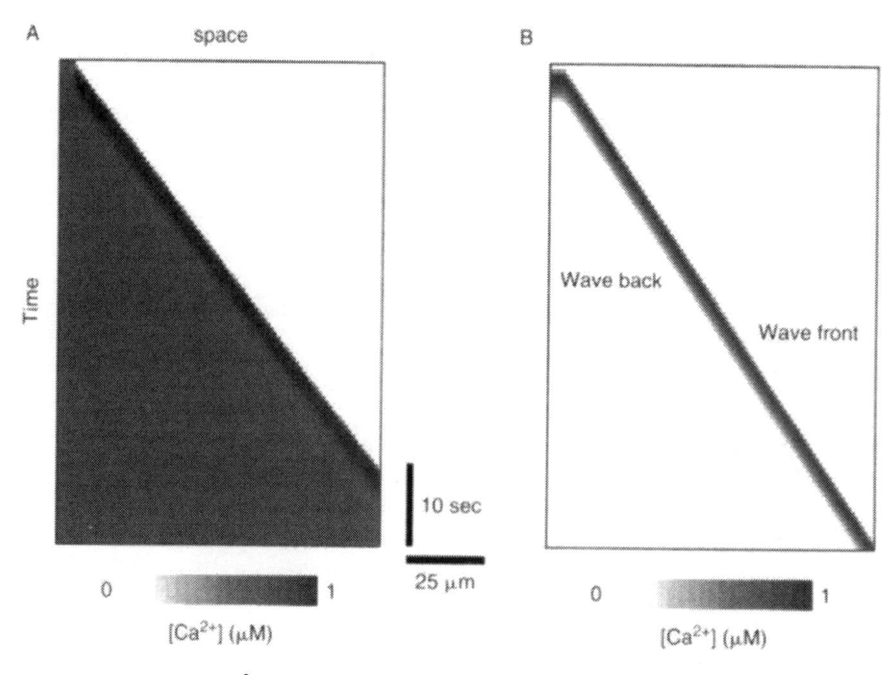

Figure 13.10 Simulation of Ca^{2+} wave phenomenon utilizing Equations (13.57) to (13.62). Both the traveling front (A) and the traveling pulse (B) can be understood as a consequence of the $(w, [Ca^{2+}])$ phase portrait for the homogeneous system (see Figure 13.11). Adapted with permission from Smith, G. et al., In C. Fall et al., *Computational Cell Biology*, Springer-Verlag, 2002, pp. 200–232.

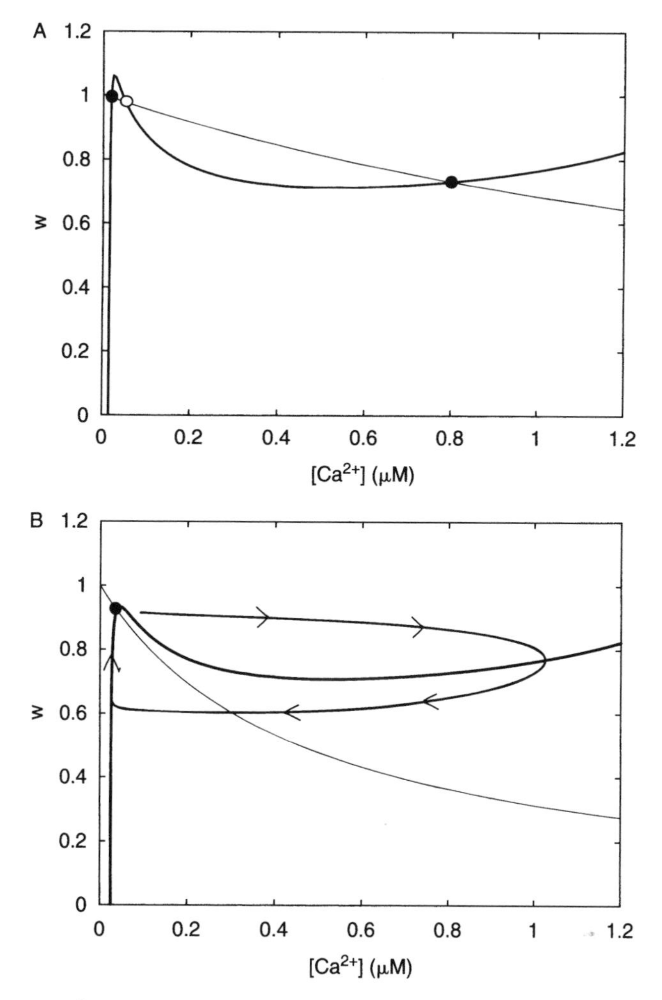

FIGURE 13.11 The (w,[Ca^{2+}]) phase plane for the reaction terms of the Ca^{2+} wave model, Equations (13.57) to (13.62), can exhibit bistability (A) or excitability (B) as well as oscillation (Figure 13.9). *Solid* and *open circles* are stable and unstable equilibria, respectively. Parameters as in Table 13.2.

wave phenomenon known as a *traveling pulse*. Traveling pulses produced by excitable dynamics are often called "trigger waves" because a perturbation at one end of the excitable medium triggers a signal that may propagate with undiminished amplitude. Repeated stimulation of an excitable medium often generates a *train* of traveling pulses. Oscillatory media (see Figure 13.9 for phase plane) support *kinematic waves* (also called *phase waves*) that are distinct from traveling fronts and pulses in that they can occur with arbitrary shape and do not require diffusion (Murray, 1989; Smith et al., 2002).

The dynamical phenomena just described have been studied in the context of IP_3-mediated Ca^{2+} wave propagation in neuroblastoma cells (Fink et al., 2000), spiral waves of Ca^{2+} release observed in immature *Xenopus* oocytes (Girard et al., 1992; Atri et al., 1993; Dupont and Goldbeter, 1994; Jafri, 1995), and the fertilization Ca^{2+} wave of mature oocytes (Wagner et al., 1998). The spatial-modeling approach has contributed to our understanding of the effect of endogenous Ca^{2+} buffers and indicator dyes on the dynamics of propagating Ca^{2+} waves (Jafri and Keizer, 1994, 1995; Sneyd et al., 1998). Similar calculations suggest that propagating Ca^{2+} waves in neurons could lead to graded persistent activity, influence temporal integration of synaptic inputs (Loewenstein and Sompolinsky, 2003), and introduce long time scales into neural dynamics (Berridge, 1998; Wang, 1998).

The dynamics of saltatory Ca^{2+} waves that arise when Ca^{2+} release fluxes (J_{IP_3R} and J_{RyR}) are spatially heterogeneous has also been studied using the whole-cell modeling approach (Keizer et al., 1998; Smith and Keizer, 1998; Dawson et al., 1999; Strier et al., 2003; Lemon, 2004). A related saltatory wave phenomenon occurs in models of active dendritic spines coupled with passive dendritic cables (Coombes and Bressloff, 2003). Although analysis of physically separated excitable units is beyond the scope of this chapter, we note that the manner in which wave-front velocity scales with model parameters is different for saltatory and continuous traveling waves (see Problem 7).

13.17 Modeling Localized Ca^{2+} Elevations

As discussed, intracellular Ca^{2+}-release events in neurons mediated by IP_3Rs and RyRs can be spatially localized, forming localized Ca^{2+} elevations analogous to the Ca^{2+} domains formed by Ca^{2+} influx during chemical neurotransmission. For example, repetitive activation of the synapse between parallel fibers and Purkinje cells evokes synaptic activity that produces IP_3 in discrete locations. In these neurons, subsequent IP_3-mediated Ca^{2+} release from intracellular stores is often localized to individual postsynaptic spines, or to multiple spines and adjacent dendritic shafts (Finch and Augustine, 1998). Localized Ca^{2+} elevations are also observed in neural preparations such as the presynaptic boutons of neuromuscular junction and hippocampal neurons in cell culture (Melamed-Book et al., 1999).

Spatially localized Ca^{2+} release also occurs in nonneuronal cell types. In fact, localized Ca^{2+} release events called Ca^{2+} "puffs" were first observed in the immature *Xenopus laevis* oocyte, where Ca^{2+} puffs mediated by IP_3Rs can be evoked in response to both flash photolysis of caged IP_3 and microinjection of nonmetabolizable IP_3 analog (Yao et al., 1995; Parker et al., 1996). In the immature *Xenopus* oocyte IP_3Rs occurs in clusters of 10 to 100 with inter-cluster spacing of the order of a few microns (Berridge, 1997). Under some conditions this organization of Ca^{2+} release sites reveals itself in cauliflower-like wave fronts (Bugrim et al., 1997). Similarly, localized Ca^{2+}-release events known as Ca^{2+} "sparks" are observed in cardiac myocytes (Cheng et al., 1993a, 1993b). Ca^{2+} sparks are mediated by RyRs located on the sarcoplasmic reticulum, the intracellular Ca^{2+} store of muscle cells. During cardiac excitation–contraction coupling, Ca^{2+} sparks activated by Ca^{2+} influx through sarcolemmal Ca^{2+} channels are the building blocks for the global Ca^{2+} responses that cause muscle contraction (Fabiato, 1985).

The interpretation of microfluorometric measurements of localized Ca^{2+} responses is complicated by the diffusion of Ca^{2+}, endogenous buffers, and indicator, all of which contribute to the dynamics of a fluorescent signal. Although the equilibrium relation (Equations [13.10] to [13.12]) is likely to hold between Ca^{2+} and indicator during global Ca^{2+} responses and Ca^{2+} waves, this is often not the case for localized Ca^{2+} elevations (Smith et al., 1996b). The interpretation of confocal microfluorometric measurements is further complicated by the optical blurring which occurs as a result of limited spatial resolution (Pratusevich and Balke, 1996; Smith et al., 1998).

The equations for the buffered diffusion of intracellular Ca^{2+} (Equations [13.35] and [13.36]) are a good starting point for a theoretical understanding of the dynamics of localized Ca^{2+} elevations. In the simplest scenario, a Ca^{2+} puff or spark is due to Ca^{2+} release through one channel (i.e., a "blip" or a "quark") or a tight cluster of channels (a "puff" or a "spark"). If Ca^{2+} is released from intracellular Ca^{2+} stores deep within a large cell (so that the plasma membrane is far away and does not influence the time course of the event), and the intracellular milieu is homogeneous and isotropic, then we have spherical symmetry. In this case, the evolving profiles of Ca^{2+} and buffer (though a function of time and distance from the source) are independent of the polar and azimuthal angles. In this case, the Laplacian becomes

$$\nabla^2 u = \frac{1}{r^2}\frac{\partial}{\partial r}\left(r^2\frac{\partial u}{\partial r}\right) = \frac{1}{r}\frac{\partial^2}{\partial r^2}(ru) = \frac{\partial^2 u}{\partial r^2} + \frac{2}{r}\frac{\partial u}{\partial r}.$$

Reasonable boundary conditions (Smith et al., 1996b, 1998) for the generalization of Equations (13.35) and (13.36) to spherically symmetric regions are

$$\lim_{r\to 0}\left\{-4\pi r^2 D_C \frac{\partial[\mathrm{Ca}^{2+}]}{\partial r}\right\} = \sigma(t), \qquad \lim_{r\to\infty}[\mathrm{Ca}^{2+}] = [\mathrm{Ca}^{2+}]_\infty,$$
$$\lim_{r\to 0}\left\{-4\pi r^2 D_B \frac{\partial[\mathrm{B}]}{\partial r}\right\} = 0, \qquad \lim_{r\to\infty}[\mathrm{B}] = [\mathrm{B}]_\infty = \frac{K[\mathrm{B}]_T}{K + [\mathrm{Ca}^{2+}]_\infty}. \tag{13.63}$$

Here $\sigma(t)$ denotes the rate of Ca^{2+} release in units of micromoles per second, and is related to the Ca^{2+} current i_{Ca} by

$$\sigma = \frac{i_{\mathrm{Ca}}}{zF}, \tag{13.64}$$

where $z = 2$ and $F = 9.648 \times 10^4$ C mol^{-1} is Faraday's constant.

Both the explicit and implicit numerical methods described in the context of one-dimensional buffered Ca^{2+} diffusion in a dendrite (Sections 13.8 and 13.9) can be adapted to this spherically symmetric problem. To use the method of lines, for example, we discretize the radial coordinate $r_j = j\Delta r$, $0 < j < J$, and choose $R_{max} = J\Delta r$ which is large enough that its chosen value does not influence calculations. The discretized Laplacian is

$$\frac{1}{r_i^2(\Delta r)^2}\left[r_{i+(1/2)}^2(U_{i+1} - U_i) - r_{i-(1/2)}^2(U_i - U_{i-1})\right] \approx \frac{1}{r^2}\frac{\partial}{\partial r}\left(r^2\frac{\partial u}{\partial r}\right)(r_j, t),$$

where $1 \le j \le J - 1$ and the equation for $r = J$ can be found by assuming an absorbing boundary conditions far from the source ($U_{J+1} = [\mathrm{Ca}^{2+}]_\infty$ or $[\mathrm{B}]_\infty$). To obtain a discretized Laplacian that can be evaluated at $r_0 = 0$, we consider $\nabla^2 u = \partial^2 u/\partial r^2 + 2r^{-1}\partial u/\partial r$ and use L'Hospital's rule to show that (Crank, 1975)

$$\lim_{r\to 0}\nabla^2 u = 3\frac{\partial^2 u}{\partial r^2}.$$

Thus the discretized representation of the Laplacian at the origin is

$$\frac{3}{(\Delta r)^2}[U_1 - 2U_0 + U_{-1}] \approx 3\frac{\partial^2 u}{\partial r^2}(0, t).$$

By using the discretized zero-flux boundary condition at $r = 0$, $U_{-1} = U_1$, we obtain

$$\frac{6}{(\Delta r)^2}[U_1 - U_0] \approx 3\frac{\partial^2 u}{\partial r^2}(0, t).$$

All that remains is to inject the correct amount of Ca^{2+} at the origin of the simulation. Given the spherical symmetry, C_0 represents $[\mathrm{Ca}^{2+}]$ for a sphere with radius $\Delta r/2$ and volume $\pi(\Delta r)^3/6$. Thus the rate of change in C_0 due to the source is $6\sigma/\pi(\Delta r)^3$ and

$$\frac{dC_0}{dt} = \frac{6}{(\Delta r)^2}[C_1 - C_0] + \frac{6\sigma}{(\pi\Delta r)^3}.$$

The computational efficiency of this numerical scheme can be significantly increased by using an expanding mesh, where $r_0 = 0$ and $r_1 = \Delta r_1$ and $r_j = (1 + \epsilon)\, r_{j-1}$ $(2 \leq j \leq J)$ with $0 < \epsilon \ll 1$. Defining $\Delta r_j = r_j - r_{j-1}$ $(1 \leq j \leq J)$ the finite difference expression for the first derivative becomes

$$\frac{U_{j+(1/2)} - U_{j-(1/2)}}{\Delta r_{j+(1/2)}} \approx \frac{\partial u}{\partial r}(r_j, t)$$

and the finite difference representation of the Laplacian of u is

$$\frac{1}{r_{j+(1/2)}^2 \Delta r_{j+(1/2)}} \left[r_{j+(1/2)}^2 \frac{U_{j+1} - U_j}{\Delta r_{j+1}} - r_{j-(1/2)}^2 \frac{U_j - U_{j-1}}{\Delta r_j} \right] \approx \frac{1}{r^2} \frac{\partial}{\partial r} \left(r^2 \frac{\partial u}{\partial r} \right) (r_j, t),$$

where $r_{j+(1/2)} = (r_j + r_{j+1})/2$ and $r_{j-(1/2)} = (r_{j-1} + r_j)/2$.

Figure 13.12 shows a Crank–Nicholson-like calculation (Section 13.9) of a localized Ca^{2+} elevation using a radial version of Equations (13.35) and (13.36) differenced as above. For simplicity

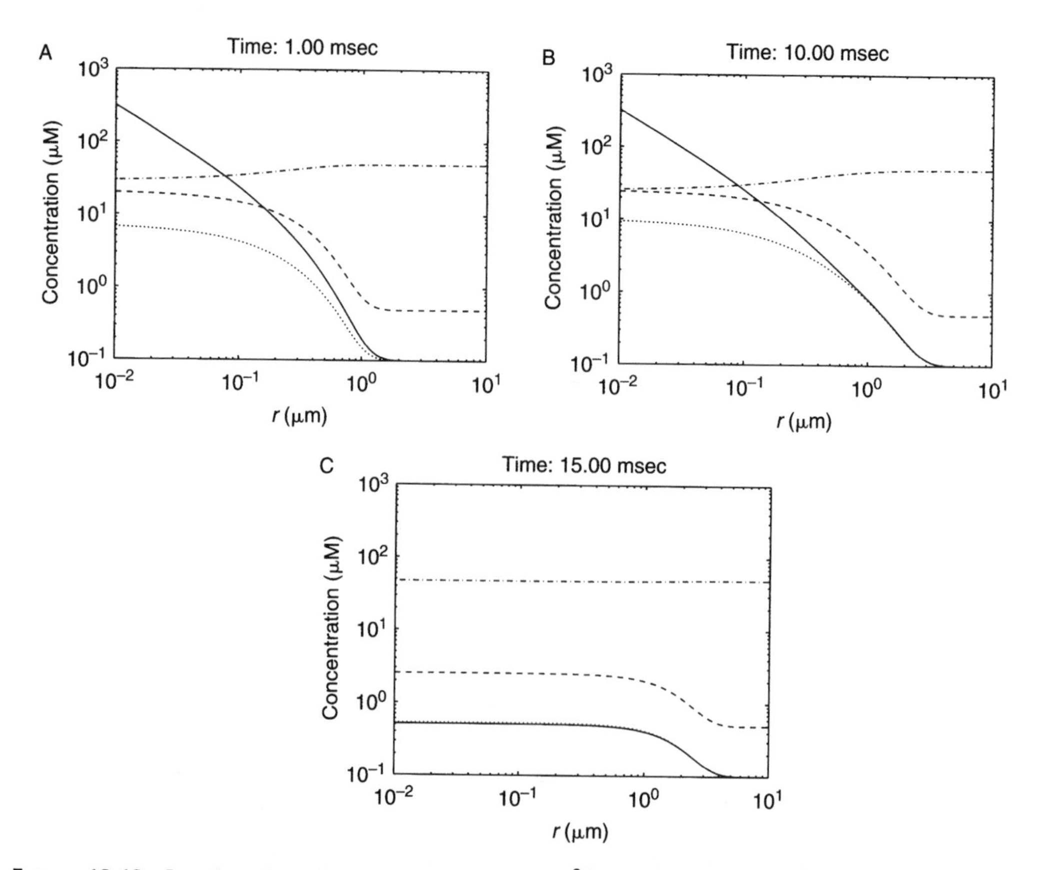

FIGURE 13.12 Log–log plots of a simulated localized Ca^{2+} elevation (i.e., a Ca^{2+} puff or spark) show $[Ca^{2+}]$ (solid lines), [B] (dot-dashed lines), and [CaB] (dashed lines) profiles at $t = 1$, 10, and 15 msec. The source amplitude is 2 pA for a duration of 10 msec. The dotted lines show "back-calculated" $[Ca^{2+}]$ estimated from [CaB].

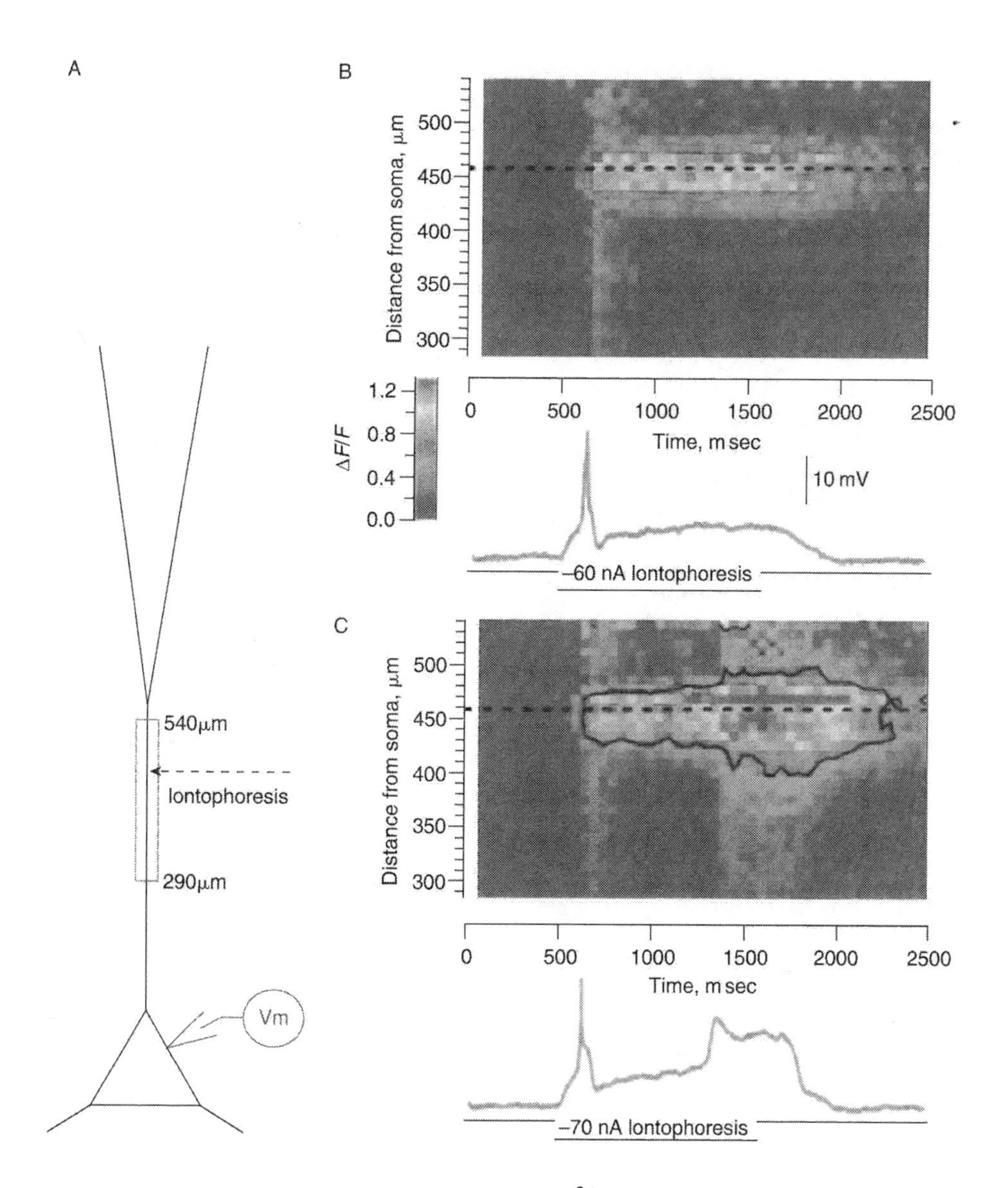

Color FIGURE 13.3 Spatial-temporal pattern of rise in [Ca^{2+}] in a layer-5 pyramidal cell evoked by focal iontophoresis of glutamate on an apical dendrite imaged using infrared-differential interference contrast microscopy. A: Schematic drawing showing extent of imaged area (*red box*), distance from soma, and position of iontophoretic electrode (*arrow*) during recording of somatic membrane potential (Vm). B, *top*: pseudocolor plot of the relative change in fluorescence ($\Delta F/F$), an indicator of [Ca^{2+}], as a function of distance along the apical dendrite 290 to 540 μm from the soma (ordinate) and time (abscissa). Shown *below* is the corresponding somatic membrane potential (*red trace*, spike truncated) evoked by glutamate iontophoresis (−60 nA, *black trace*). C: As *B* except for −70 nA iontophoresis. Black contour line indicates $\Delta F/F \geq 50\%$ of the peak $\Delta F/F$ evoked during the plateau. Figure reproduced and legend adapted with permission from Oakley, J. et al., *J. Neurophysiol.*, **86**(1), 503–513, 2001

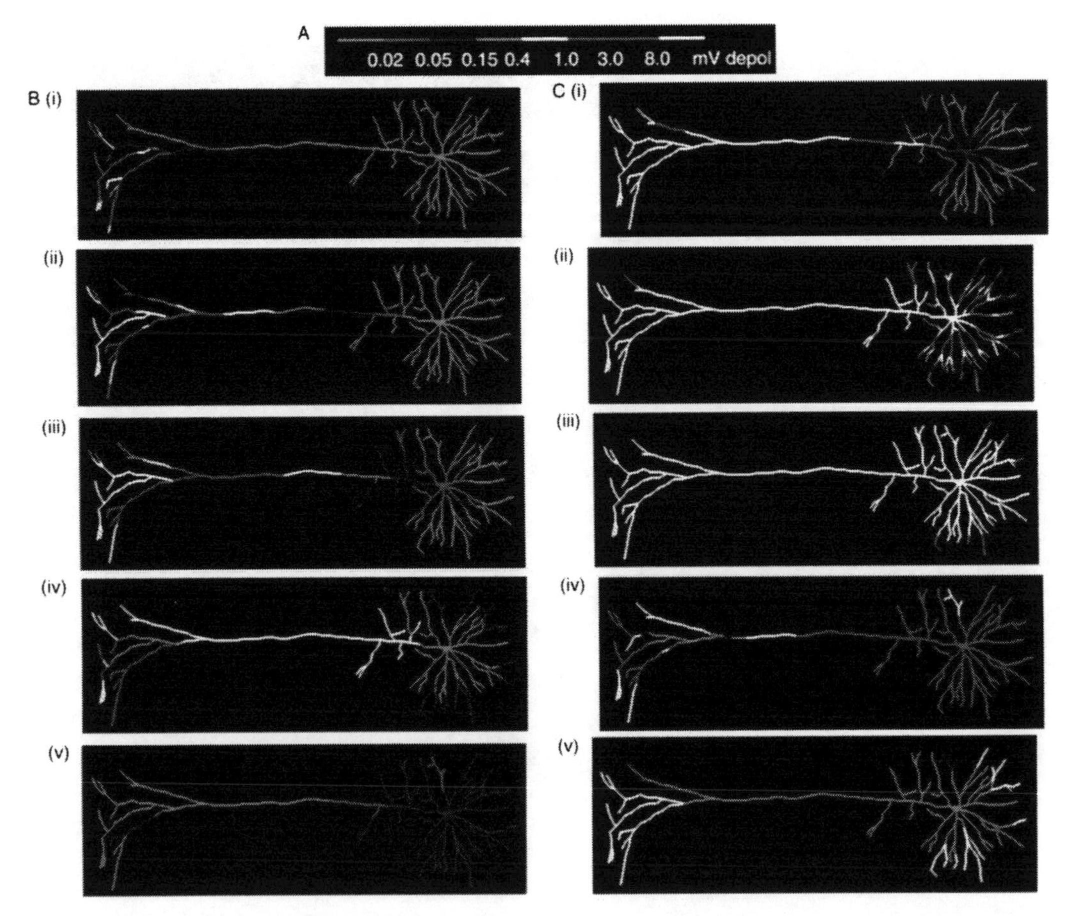

Color FIGURE 15.4 (A) The distribution of depolarization over the pyramidal cell as a function of time. The colors show the distribution using the log scale illustrated in the calibration. (B) The distribution following eight inputs to the apical dendrites and only passive conduction (i.e., depolarization of apical nexus not sufficient to initiate action potentials there): (i) 0.4 msec; (ii) 1.4 msec; (iii) 2.5 msec; (iv) 5.0 msec; (v) 14.0 msec. (C) The distribution following eight inputs to the apical dendrites leading to active conduction along the climbing apical dendrite: (i) 1.6 msec; (ii) 2.2 msec; (iii) 2.7 msec; (iv) 5.0 msec; (v) 16.0 msec

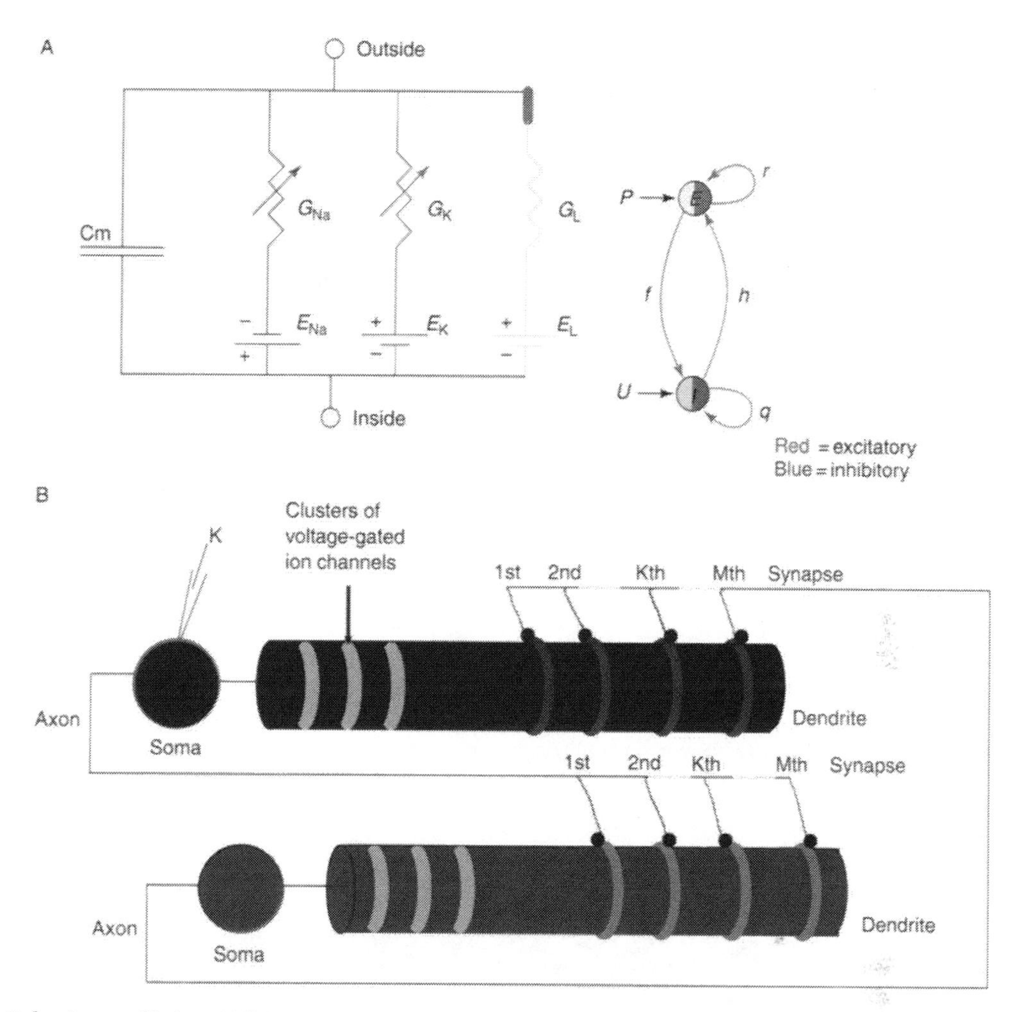

Color FIGURE 19.1 (A) Schematic illustration of a recurrent network comprising two spiking neurons coupled through synaptic weights. The parameters $\boldsymbol{h}, \boldsymbol{r}, \boldsymbol{f}, \boldsymbol{q}$, represent the strengths of the interactions between the units and $\boldsymbol{P}$, $\boldsymbol{U}$ represent excitatory and inhibitory input respectively. A circuit representation of one conductance-based point neuron is shown. [Kindly provided by Dr. Andrew Gillies, Edinburgh University.] (B) Schematic illustration of a biophysically realistic recurrent network comprising two synaptically coupled neurons. Each neuron is spatially distributed, with a single equivalent ionic cable representing the dendrites and coupled to a lumped isopotential soma. The ionic cable model assumes the cable to be "leaky" (i.e., a passive RC cable) with densities of voltage-dependent ionic channels, referred to as "hot spots," discretely imposed at specific locations along the cable. Nonlinearity is assumed to occur only at these hot spots rather than continuously. The first neuron, "0," is activated by a subthreshold current injection, I, at the soma (represented by the schematic pipette electrode), and the resultant graded electrotonic potential is boosted by persistent Na^+ ionic channels, resulting in neural transmitter release at M axodendritic synapses (chosen at arbitrary locations), which generate synaptic potentials in the second neuron, "1." The same process is repeated from neuron "1" back to neuron "0." (Reproduced from Poznanski, R.R., *J. Integr. Neurosci.*, 1, 69–99, 2002. With permission)

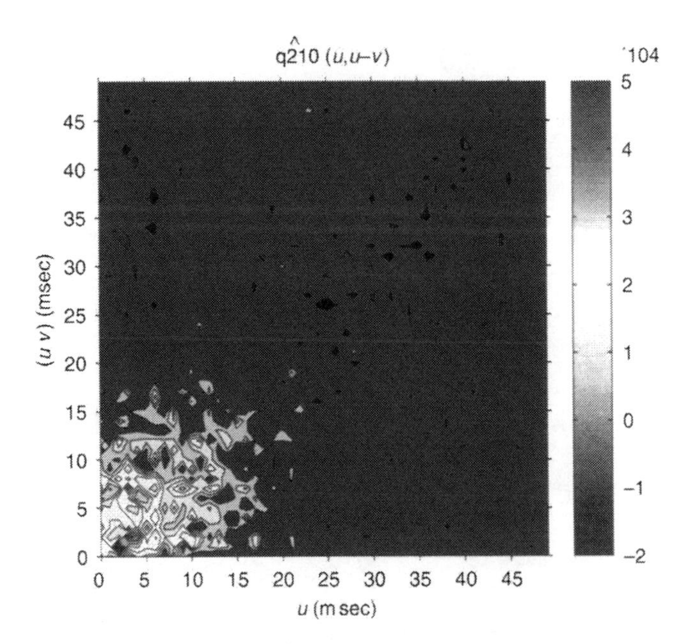

Color FIGURE 20.14 Contour plot of the estimated third-order cumulant-density function, $\hat{q}_{210}(u, u - v)$. The estimated upper and lower 95% confidence limits for this estimate are $\pm 9.9 \times 10^{-5}$. Values lying inside this region, which can be interpreted as an indication of no significant third-order interaction, are indicated by a single grey level in the plot. This level covers the majority of the plot. Significant values of the cumulant are located at lag values less than 15 msec, with a maximum around lag values of 2 to 3 msec. All these significant values are positive

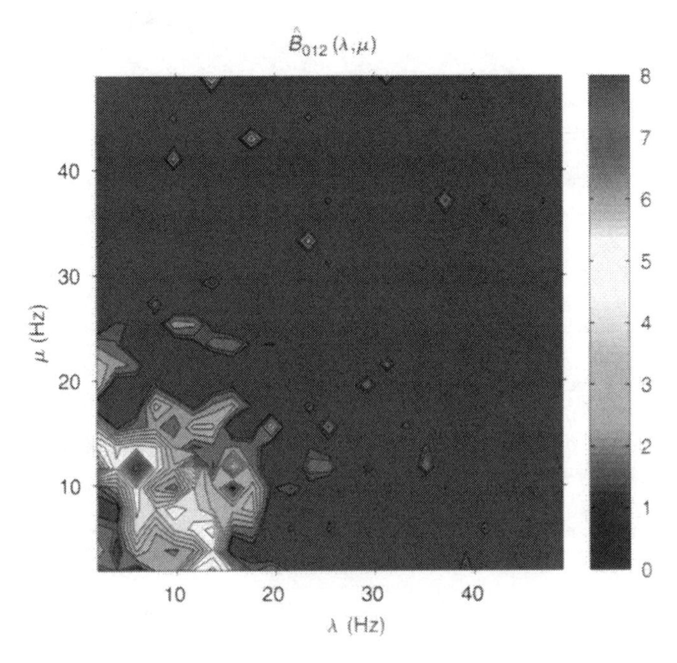

Color FIGURE 20.16 Contour plot of the estimated third-order bispectral coefficient, $\hat{B}_{012}(\lambda, \mu)$. From Equation (20.49), the estimated 95% confidence limit is 1.25. Values less than this can be interpreted as an indication of no significant third-order interaction, and are indicated by a single grey level in the plot. This level covers the majority of the plot. Significant values are located at frequencies below 20 Hz. Both frequency axes start at 2 Hz

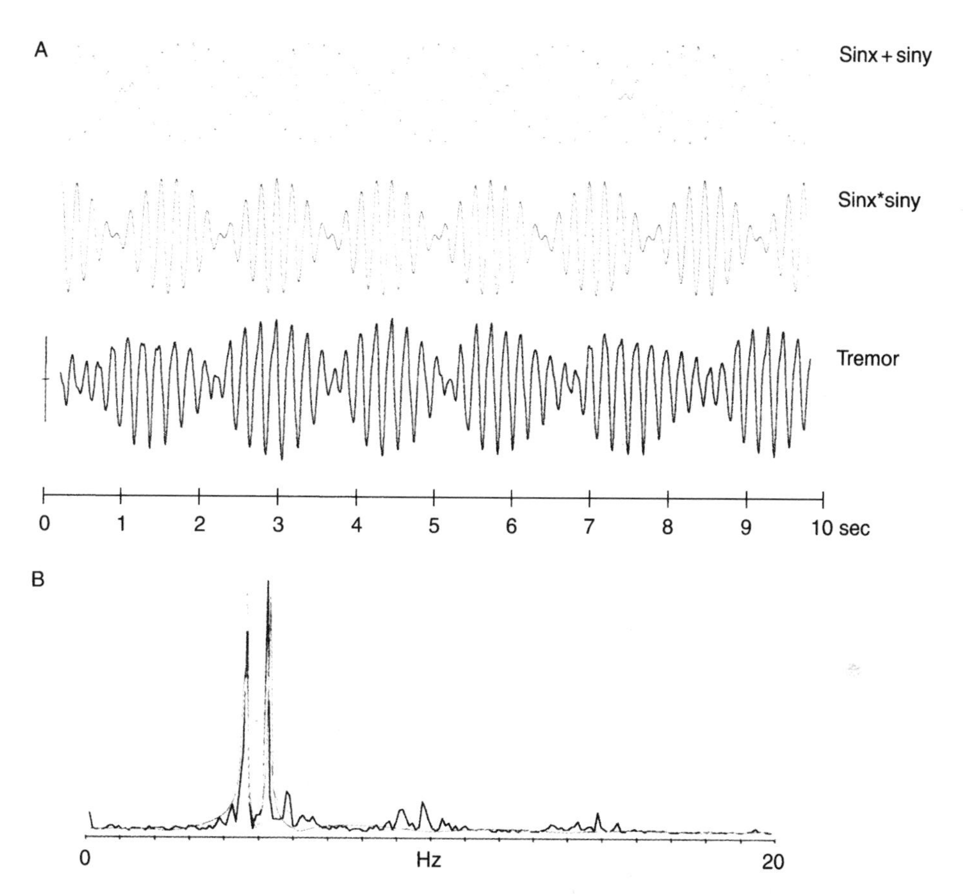

Color FIGURE 21.1 (A) Three 10 sec records. The bottom trace (black) is taken from a Parkinson patient with a resting tremor of the left hand. Sensors responding to angular velocity were oriented to record rotation about the pronation–supination axis of the wrist (pronation positive; scale: ±100°/sec). Note the quasiperiodic modulation of tremor amplitude. The two records above were computer generated. In the top record (blue), sinusoids of frequencies 5.2 and 4.6 Hz were added; their amplitudes were adjusted to compare with the real tremor. The middle record (red) was also generated by computer simulation, this time by multiplying two sinusoids. Amplitude modulation of the carrier sinusoid (at 4.9 Hz) by a sinusoid at 0.35 Hz creates two sideband peaks in the spectrum at 4.9 + 0.35 Hz and 4.9 – 0.35 Hz. (B) Spectra for the three records in (A), each in its own color. Note that in the spectrum of the real tremor there are smaller peaks at harmonic frequencies (near 10 and 15 Hz)

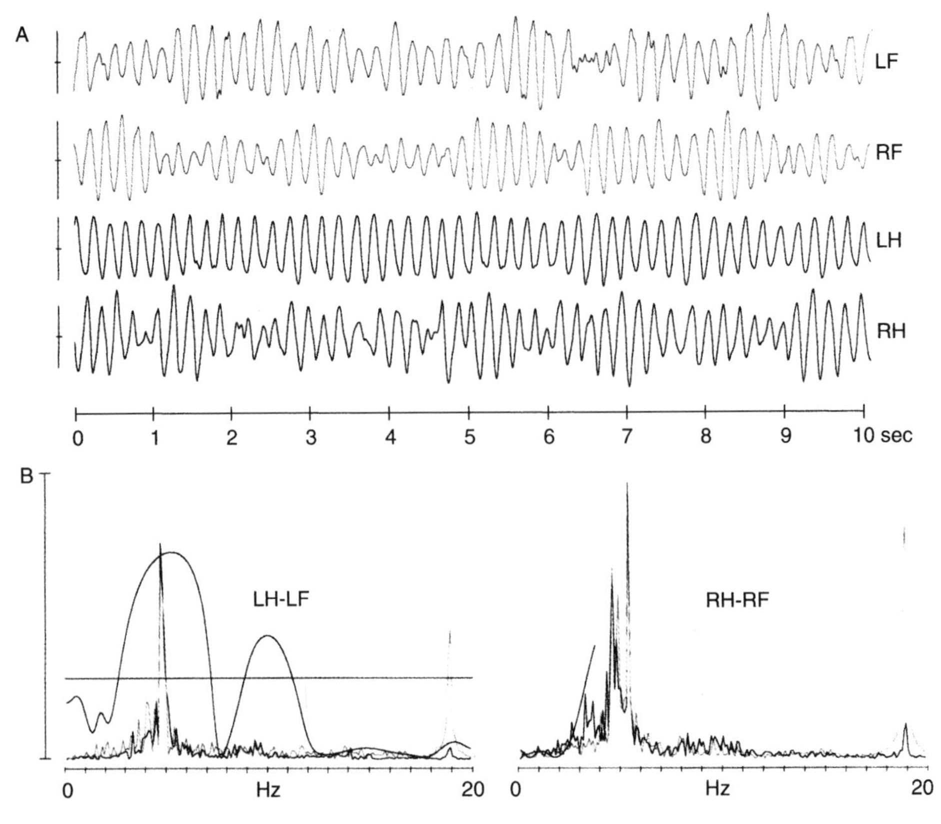

Color FIGURE 21.3 (A) Four 10 sec records of resting tremor concurrently measured in a patient with Parkinson's disease. Top: Flexion–extension rotation of the left foot about the ankle (extension positive; scale: ±20°/sec). Second from top: Flexion–extension rotation about the right ankle (extension positive; same scale). Bottom trace: Pronation–supination hand tremor about the right wrist (supination positive; scale: ±100°/sec). Second from bottom: Pronation–supination hand tremor about the left wrist (pronation positive; scale ±200°/sec). (B) Left: Power spectra for the left hand (black) and left foot (red) tremors are superimposed (using different scaling factors). They share a peak near 4.7 Hz. Right: The spectra for the right hand (black) and right foot (red) tremors are superimposed (also using different scaling factors), with large peaks shared near 5.4 Hz. The smooth curve superimposed on each pair is the coherence function which, in the range of greatest power, is considerably above the horizontal line, which represents the 95% confidence limit. Sinusoidal calibration signals of 19 Hz were added to each tremor record prior to calculating spectra and removed before calculating coherences. Their peaks appear toward the right in each spectrum and indicate how the data processing methods treat a pure sine wave. The sinusoids in each case had peak-to-peak amplitudes of 20°/sec. The different scale factors used to plot the spectra result in different heights for the four calibration peaks at 19 Hz

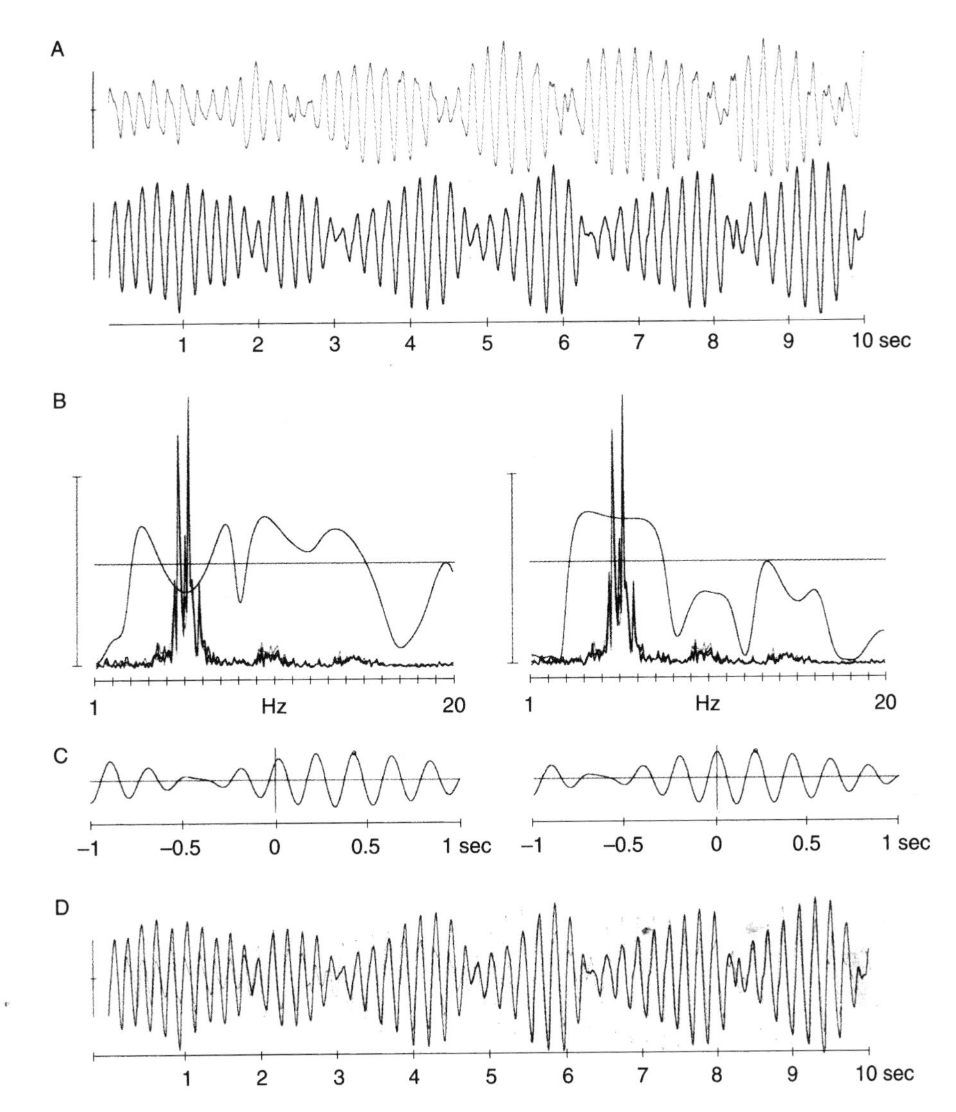

Color FIGURE 21.4 (A) Top trace (red): angular velocity of flexion–extension about the right wrist in a Parkinson patient (extension positive). Bottom trace (black): concurrently measured tremor along the pronation axis of the right wrist (supination positive; scales: ±100°/sec). (B) The spectra of the two records are superimposed and seem indistinguishable (colors as in A). The coherence function lies below the level of statistical significance in the important band of power (around 5 Hz). This indicates that the spectral frequencies are not shared concurrently. (C) Left: The cross-correlation function computed from the two tremors. The peak value (circled) is at +430 msec (flexion–extension leading pronation–supination). Right: The coherence function (and spectra) recomputed for the data, time shifted as in D, is significant in the 5 Hz band of shared power. (D) The two tremors from A are superimposed, with the flexion–extension record delayed by 430 msec (colors as in A)

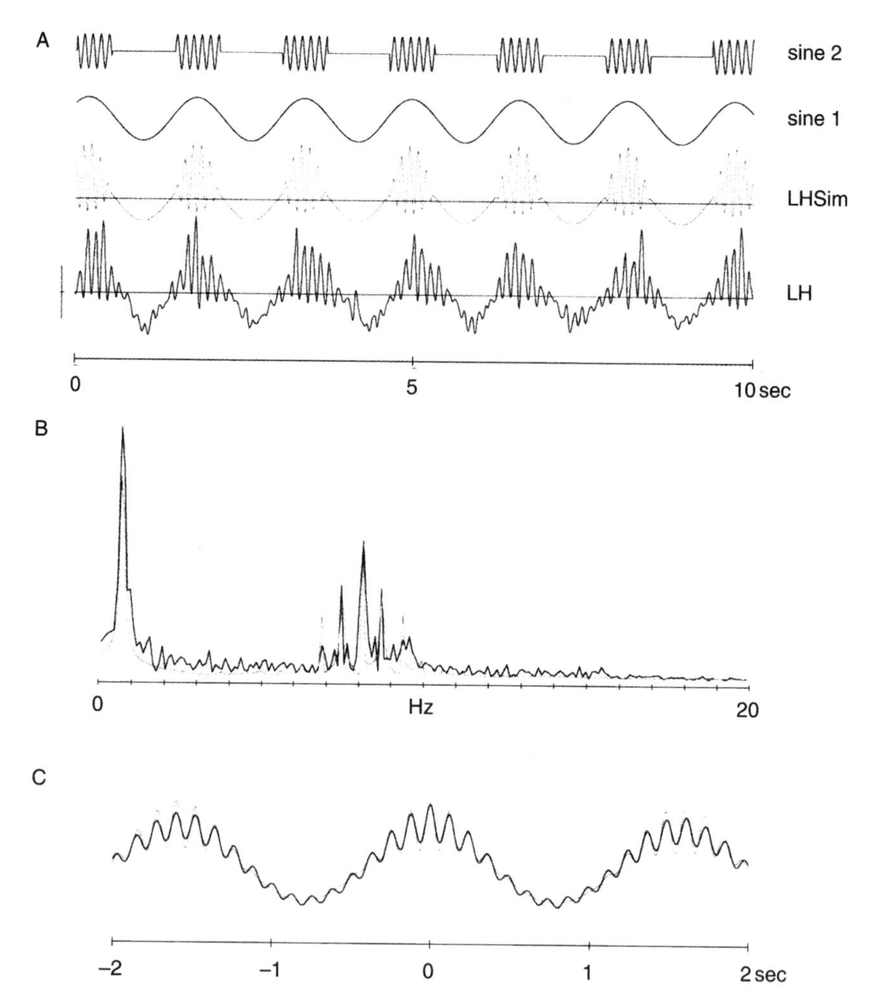

Color FIGURE 21.8 (A) Bottom trace (black): ten seconds of data from a Parkinson patient executing voluntary alternating pronation and supination movements of the left hand (pronation positive; scale ±200°/sec). Approximately seven complete cycles are shown. Superimposed on this slow oscillation can be seen larger- and higher-frequency (involuntary) oscillations constituting a so-called "action tremor." Above this trace is a simulation of the tremor (red). Two sinusoids were used: a higher-frequency sinusoid at 8.0 Hz. and a lower-frequency sinusoid at 0.6 Hz. The high-frequency signal is gated when the low-frequency signal exceeds zero, and this gated sinusoid (top trace) is added to the low-frequency sinusoid (next trace below) to produce the approximation shown directly above the actual tremor. (B) Spectra for the two lowest records in (A) (colors as in A). Both consist of a sharp peak near 0.6 Hz derived from the voluntary movement and a group of peaks from 7 to 10 Hz that are derived from the basic frequency of the action tremor. (C) The autocorrelation functions of the real and simulated records in (A) (colors as in A)

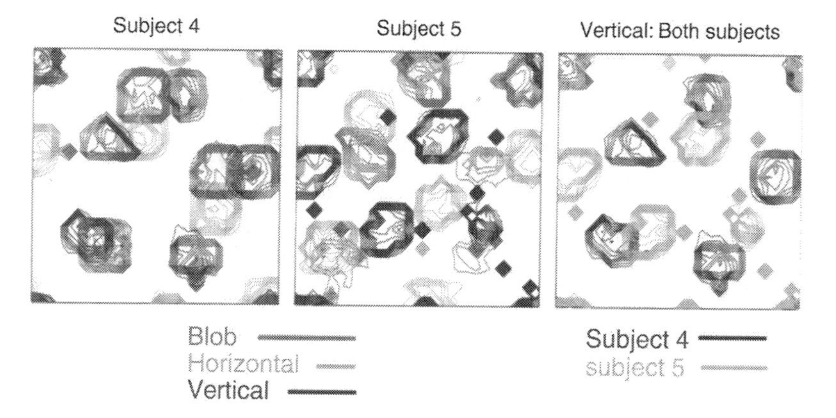

Color FIGURE 23.6 Comparison of patterns of activity in area IT for the three classes of visual stimuli across different Darwin VII subjects. The left and center contour plots show, for two representative subjects, the borders of neuronal group activity in response to blobs (red lines), horizontal stripes (green lines), and vertical stripes (blue lines). The contour plot on the right shows the IT activity for two different subjects in response to the same class of stimuli (vertical stripes): subject 4 (dark blue); subject 5 (light blue). Across all subjects tested ($n = 7$), the mean overlap of activity within each subject but between stimuli was 26.6% ($\sigma = 0.10$). The mean overlap of activity between different subjects was 22.1% ($\sigma = 0.09$) in response to blob patterns, 26.9% ($\sigma = 0.11$) in response to horizontal striped patterns, and 20.0% ($\sigma = 0.10$) in response to vertical striped patterns. (Adapted from Krichmar and Edelman, 2002)

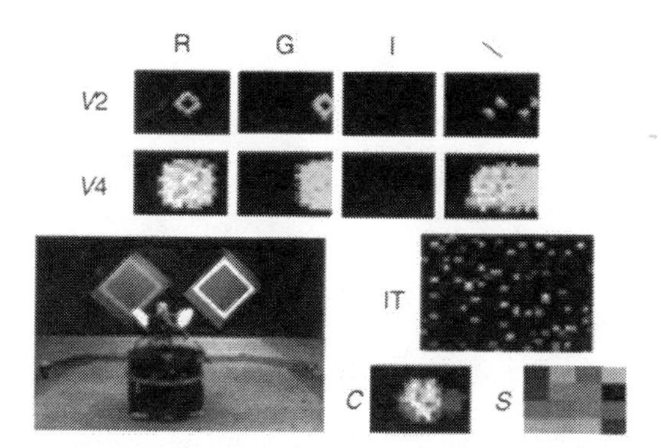

Color FIGURE 23.10 Snapshot of Darwin VIII's neuronal unit activity during a behavioral experiment. Darwin VIII is approaching (left) a red diamond target and (right) a green diamond distractor toward the end of a training session. Darwin VIII has not yet broken the beam that triggers the sound from the speakers located on the left side of the floor. The panels next to Darwin VIII show the activity and phase of selected neural areas (top row; *V*2-red, *V*2-green, *V*2-vertical, *V*2-diagonal, second row; *V*4-red, *V*4-green, *V*4-vertical, *V*4-diagonal, third row [to the right of Darwin VIII]; IT, fourth row [to the right of Darwin VIII]; *C* and *S*). Each pixel in the selected neural area represents a neuronal unit; the activity is normalized from no activity (black) to maximum activity (bright colors), and the phase is indicated by the color of the pixel (colors were chosen from an arbitrary pseudocolor map, there is no connection between the colors of the stimulus objects and the colors representing the phases of neuronal responses). The neuronal units responding to the attributes of the red diamond share a common phase (red–orange color), whereas the neuronal units responding to the green diamond share a different phase (blue–green color)

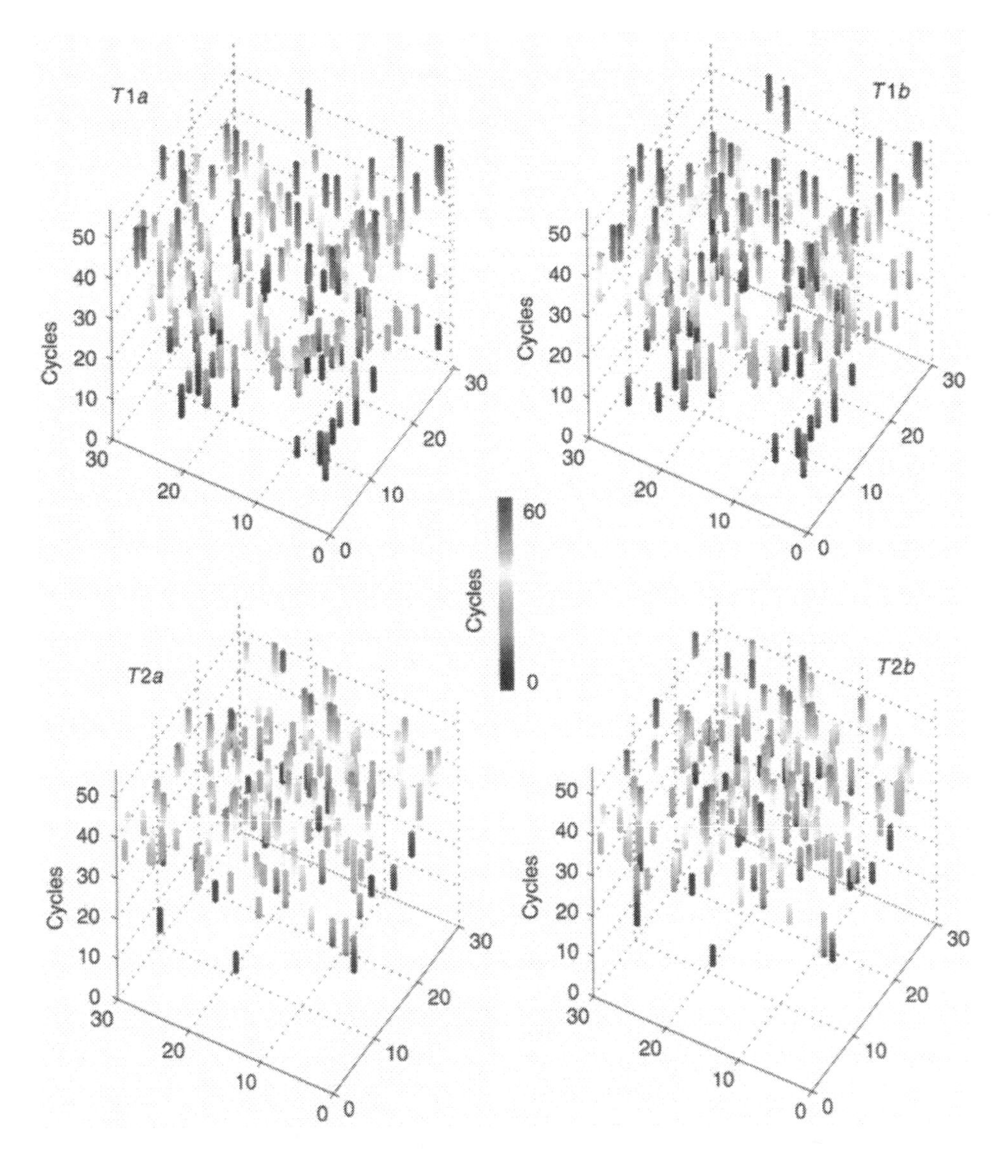

Color FIGURE 23.14 Spatiotemporal response properties of neuronal units in $S2$ in Darwin IX. The top panels show $S2$ activity during two separate encounters (a, b) with texture $T1$ by Darwin IX's left whisker column. The bottom panels show encounters of the same whiskers with texture $T2$. Each panel shows the activity of the 30×30 matrix of $S2$ neuronal units (x and y axes) over 60 poststimulus cycles (indicated by the z-axis as well as by the color scale). Neuronal units are shown if their activity exceeded a threshold (0.25)

only a single buffer has been included (B2 in Table 13.1). Figure 13.12A to Figure 13.12C are numerically calculated snapshots of concentration profiles at $t = 1$, 10, and 15 msec. A source of amplitude 1 pA was used for a duration of 10 msec. As expected, $[Ca^{2+}]$ (solid lines) is significantly elevated and the Ca^{2+} profile quickly reaches its steady-state value near the source. Because released free Ca^{2+} reacts with buffer, [CaB] (dashed lines) is also elevated near the source. Conversely, [B] (dot-dashed lines) decreases near the source. The dotted lines in Figure 13.12 show that $[Ca^{2+}]$ estimated from $[CaB](r,t)$ using the equilibrium relation (i.e., back-calculated $[Ca^{2+}]$) deviates significantly from that calculated directly until after the Ca^{2+} release event is nearly complete. The discrepancy between the solid and dotted lines is largest near the active source for free Ca^{2+} and indicates disequilibrium between Ca^{2+} and buffer during the formation of the localized Ca^{2+} elevation.

Simulations such as these have played a role in understanding the dynamics of puffs and sparks. Ca^{2+}-spark simulations performed in an axisymmetric environment using a physiologically realistic portfolio of endogenous Ca^{2+} buffers such as those present in cardiac myocytes (troponin C, calmodulin, and others) suggest that the time course of experimentally observed fluo-3 fluorescence can be explained by a 2 pA, 15 msec Ca^{2+}-release event from a loose cluster of RyRs located on the sarcoplasmic reticulum membrane (Pratusevich and Balke, 1996; Smith et al., 1998). Interestingly, the model indicates that Ca^{2+}-spark properties (such as brightness, FWHM, and decay time) are sensitive to indicator dye parameters (such as association rate constant, concentration, and diffusion coefficient). These relationships are not always intuitive. For example, an increase in indicator-dye concentration may decrease the brightness of a simulated Ca^{2+} spark (Smith et al., 1998). This is partly due to the fact that spark brightness is a normalized measure (peak/basal fluorescence) and partly due to the indicator perturbing the underlying free Ca^{2+} signal.

13.18 Estimates of Domain Ca^{2+} Near Open Ca^{2+} Channels

Numerical simulations like those in Figure 13.12 confirm that localized Ca^{2+} elevations achieve steady-state values very rapidly (within microseconds) near point sources. Steady-state solutions to the full radial equations for the buffered diffusion of Ca^{2+} are of interest since they provide estimates of the domain $[Ca^{2+}]$ near open Ca^{2+} channels. In the case of one mobile buffer, steady-state solutions to Equations (13.35) and (13.36) will satisfy the boundary-value problem (Roberts, 1994)

$$0 = D_C \nabla^2 [Ca^{2+}] - k_+ [B][Ca^{2+}] + k_- ([B]_T - [B]), \tag{13.65}$$

$$0 = D_B \nabla^2 [B] - k_+ [B][Ca^{2+}] + k_- ([B]_T - [B]), \tag{13.66}$$

where $\nabla^2 u = r^{-1} d^2 (ru)/dr^2$ and associated boundary conditions are similar to Equation (13.63). Note that while stationary buffers are important for the time-dependent evolution of localized Ca^{2+} elevations, they have no influence on the steady-state Ca^{2+} microdomain. This can be seen by setting $D_B = 0$ in Equation (13.66), in which case the balance of reaction terms, $k_+[B][Ca^{2+}] = k_- ([B]_T - [B])$, causes Equation (13.65) to reduce to $0 = D_C \nabla^2 [Ca^{2+}]$, the steady-state diffusion equation for free Ca^{2+} in the absence of buffers.

Equations (13.65) and (13.66) are nonlinear and the analysis of these equations usually involves linearization of reaction terms (Neher, 1986, 1998; Stern, 1992; Pape et al., 1995; Naraghi and Neher, 1997) or asymptotic methods that begin with nondimensionalization (Smith et al., 2001). The conclusion of this more recent study is that the steady-state "excess buffer approximation" (EBA) and the steady-state "rapid buffer approximation" (RBA) are complementary and can be viewed as "weak source" and "strong source" limits.

In dimensional form, the steady-state EBA for a single buffer is given by (Neher, 1986)

$$[\mathrm{Ca}^{2+}] = \frac{\sigma}{4\pi D_{\mathrm{C}} r} \mathrm{e}^{-r/\lambda} + [\mathrm{Ca}^{2+}]_{\infty}, \tag{13.67}$$

$$[\mathrm{B}] = [\mathrm{B}]_{\infty}, \tag{13.68}$$

where λ is the characteristic length constant for the mobile buffer given by $\lambda = \sqrt{D_{\mathrm{C}}/k_{+}[\mathrm{B}]_{\infty}}$. The EBA is valid when mobile buffer is at high concentration and when the source amplitude is small (Neher, 1986). Note that λ is a decreasing function of the association rate constant k_{+} and the free buffer concentration $[\mathrm{B}]_{\infty}$. When a buffer is in excess, increasing the free buffer concentration far from the source (e.g., by increasing $[\mathrm{B}]_{\mathrm{T}}$ or decreasing K) leads to smaller λ and a spatially restricted localized Ca^{2+} elevation. In addition, buffers with fast reaction kinetics (e.g., BAPTA, see Table 13.1) are expected to restrict localized Ca^{2+} elevations more than slow buffers (e.g., EGTA).

The steady-state RBA near a point source for Ca^{2+} results in the following implicit expression for $[\mathrm{Ca}^{2+}]$ (Smith, 1996; Smith et al., 1996b),

$$D_{\mathrm{C}}[\mathrm{Ca}^{2+}] + D_{\mathrm{B}}\frac{[\mathrm{Ca}^{2+}][\mathrm{B}]_{\mathrm{T}}}{[\mathrm{Ca}^{2+}] + K} \approx \frac{\sigma}{4\pi r} + D_{\mathrm{C}}[\mathrm{Ca}^{2+}]_{\infty} + D_{\mathrm{B}}\frac{[\mathrm{Ca}^{2+}]_{\infty}[\mathrm{B}]_{\mathrm{T}}}{[\mathrm{Ca}^{2+}]_{\infty} + K}, \tag{13.69}$$

where $[\mathrm{B}] = K[\mathrm{B}]_{\mathrm{T}}/(K + [\mathrm{Ca}^{2+}])$. The corresponding explicit expression for $[\mathrm{Ca}^{2+}]$ is

$$[\mathrm{Ca}^{2+}] = \frac{1}{2D_{\mathrm{C}}}\left(-D_{\mathrm{C}}K + \frac{\sigma}{4\pi r} + D_{\mathrm{C}}[\mathrm{Ca}^{2+}]_{\infty} - D_{\mathrm{B}}[\mathrm{B}]_{\infty} + \sqrt{\left(D_{\mathrm{C}}K + \frac{\sigma}{4\pi r} + D_{\mathrm{C}}[\mathrm{Ca}^{2+}]_{\infty} - D_{\mathrm{B}}[\mathrm{B}]_{\infty}\right)^2 + 4D_{\mathrm{C}}D_{\mathrm{B}}[\mathrm{B}]_{\mathrm{T}}K}\right). \tag{13.70}$$

This approximation tends to be valid when buffers are rapid, but interestingly, a sufficiently large source amplitude, σ, can compensate for modest binding rates, suggesting that the RBA can be thought of as a "strong source" approximation (Smith et al., 2001).

The fundamental assumptions used in deriving the steady-state excess and rapid buffer approximations are different. In the case of the RBA, it is assumed that buffer and Ca^{2+} are in local equilibrium (Equations [13.10] and [13.12]). Because $[\mathrm{Ca}^{2+}] \to \infty$ as $r \to 0$ in Equation (13.70), we see that according to the steady-state RBA

$$\lim_{r\to 0} [\mathrm{B}] \approx 0 \quad (\mathrm{RBA}), \tag{13.71}$$

that is, the steady-state RBA cannot be valid unless the source is strong enough to saturate the buffer. On the other hand, the derivation of the EBA assumes that buffer is not perturbed from its equilibrium value, even near the source, so that

$$\lim_{r\to 0} [\mathrm{B}] \approx [\mathrm{B}]_{\infty} \quad (\mathrm{EBA}). \tag{13.72}$$

Thus we expect the EBA and RBA approximations to be complementary in the sense that the steady-state solution to the full equations for the buffered diffusion of Ca^{2+} near a point source, Equations (13.65) and (13.66), cannot simultaneously be EBA-like, as Equation (13.72), and RBA-like, as Equation (13.71).

In the process of extending both the EBA and RBA to higher order, this expectation has been confirmed (Smith et al., 2001). The second-order EBA (not shown) is similar to the solution of a linearized version of Equations (13.65) and (13.66) given by Naraghi and Neher (1997), Pape et al. (1995), and Stern (1992).

$$[Ca^{2+}] = [Ca^{2+}]_\infty + \frac{\sigma}{4\pi r\,(D_C + \gamma_\infty D_B)}\left[1 + \frac{\gamma_\infty D_B}{D_C}e^{-r/\Lambda}\right] \tag{13.73}$$

$$[B] = [B]_\infty + \frac{\sigma\gamma_\infty}{4\pi r\,(D_C + \gamma_\infty D_B)}[e^{-r/\Lambda} - 1] \tag{13.74}$$

with

$$\frac{1}{\Lambda^2} = \frac{1}{\tau}\left(\frac{1}{D_B} + \frac{\gamma_\infty}{D_C}\right) \quad \frac{1}{\tau} = k_+[Ca^{2+}]_\infty + k_- \quad \gamma_\infty = \frac{K[B]_T}{(K + [Ca^{2+}]_\infty)^2}. \tag{13.75}$$

Importantly, the derivation of Equations (13.73) to (13.75) does not assume that the buffer (Equation [13.68]) cannot saturate. Similarly, the second-order RBA (not shown) does not assume complete saturation near the source (Smith et al., 2001).

13.19 Domain Ca^{2+}-Mediated Ca^{2+} Channel Inactivation

Experimental and theoretical evidence indicates that Ca^{2+}-inactivation of open IP_3R receptor Ca^{2+} channels may be mediated by elevated domain Ca^{2+}. For example, high concentrations of intra-luminal Ca^{2+} can reduce IP_3R activity as measured in planar lipid bilayer experiments (Bezprozvanny and Ehrlich, 1994). Further evidence is provided by a model of agonist-induced Ca^{2+} oscillations in pituitary gonadotrophs based on the IP_3R model discussed in Section 13.14 (Li and Rinzel, 1994; Li et al., 1995a). In a detailed comparison of experimental and theoretical Ca^{2+} responses, enhanced IP_3R inactivation (mediated by elevated ER Ca^{2+}) was required to account for the persistence of Ca^{2+} oscillations in gonadotrophs at low ER Ca^{2+} concentrations, that is, increased ER Ca^{2+} concentration was assumed to promote IP_3R inactivation by increasing the domain $[Ca^{2+}]$ near open IP_3Rs. As the ER concentration decreases (e.g., during agonist-stimulated Ca^{2+} oscillations in Ca^{2+} free extracellular media), domain $[Ca^{2+}]$ is reduced leading to the disinactivation of IP_3R receptor Ca^{2+} channels. This hypothesized disinhibitory effect of low ER Ca^{2+} was incorporated *post hoc* into the Li–Rinzel reduction of the De Young–Keizer IP_3R model by modifying Equation (13.55) so that $f_o = d_\infty^n m_\infty^n w^n$, where d_∞ represents indirect IP_3R activation due to falling lumenal Ca^{2+} (Li et al., 1994).

We now describe an extension of the De Young–Keizer IP_3R model that accounts for domain $[Ca^{2+}]$-mediated inactivation (first presented in Smith, 2002b). The effort is complicated by the fact that the IP_3R receptor experiences elevated domain $[Ca^{2+}]$ (abbreviated below as $[Ca^{2+}]_d$) only when it is *open* (all subunits in the permissive state, S_6 in Figure 13.7). This means that the n IP_3R subunits are no longer independent, that is, the $[Ca^{2+}]$ experienced by each IP_3R subunit depends on the state of the other $n-1$ subunits in the model IP_3R (thus invalidating both Equations [13.51] and [13.55]). In order to account for this subunit interdependence caused by $[Ca^{2+}]_d$-mediated inactivation, the De Young–Keizer transition-state diagram must be expanded to represent IP_3-potentiation, Ca^{2+}-activation, and Ca^{2+}-inactivation for an entire channel, as opposed to the one subunit in Figure 13.7. Because the IP_3R model has $3n$ binding sites (two for Ca^{2+}, one for IP_3) this expanded transition-state diagram has $2^{3n} = 4096$ states, 330 of which are distinguishable if the subunits are identical.

Fortunately, the expanded transition-state diagram that accounts for $[Ca^{2+}]_d$-mediated IP_3R inactivation can be greatly simplified by performing a quasistatic approximation (Li and Rinzel, 1994). The assumption that the fast processes of IP_3-potentiation and Ca^{2+}-activation are essentially at equilibrium with the slower process of Ca^{2+}-inactivation leads to the following reduced transition-state diagram (Smith, 2002a)

$$W_0 \underset{v_0^-}{\overset{4v_0^+}{\rightleftharpoons}} W_1 \underset{2v_1^-}{\overset{3v_1^+}{\rightleftharpoons}} W_2 \underset{3v_2^-}{\overset{2v_2^+}{\rightleftharpoons}} W_3 \underset{4v_3^-}{\overset{v_3^+}{\rightleftharpoons}} W_4, \tag{13.76}$$

where we have assumed $n = 4$ subunits and the state W_i corresponds to the channel having i Ca^{2+} ions bound to the four binding sites that lead to inactivation. The v_i^+ and v_i^- are inactivation and disinactivation transition rates for a single subunit.

Assuming the rate-constant symmetry of the original model, for example, $a_2 = a_4$ (De Young and Keizer, 1992), the inactivation and disinactivation rates become

$$v_3^- = v_2^- = v_1^- = v_0^- = a_2 Q_2,$$

$$v_3^+ = v_2^+ = v_1^+ = a_2[Ca^{2+}],$$

$$v_0^+ = a_2[Ca^{2+}]_*,$$

$$[Ca^{2+}]_* = (1 - m_\infty^4)[Ca^{2+}] + m_\infty^4 [Ca^{2+}]_d,$$

where Q_2 and m_∞ are given by Equations (13.54) and (13.56), respectively. Note that the expression for the inactivation rate

$$4v_0^+ = 4a_2\{(1 - m_\infty^4)[Ca^{2+}] + m_\infty^4[Ca^{2+}]_d\}$$

of the fully disinactivated IP_3R (state W_0) is a weighted average of $4a_2[Ca^{2+}]$ and $4a_2[Ca^{2+}]_d$. The value of the inactivation rate depends on the value of m_∞^4, the conditional probability that the channel is open given that it is disinactivated.

It can be shown that the equilibrium open probability, P_o^{eq}, of this extended De Young–Keizer IP_3R model is

$$P_o^{eq} = m_\infty^4 \left(\frac{v^3 v_*}{v^3 v_* + 4v^3 + 6v^2 + 4v + 1} \right), \tag{13.77}$$

where $v = Q_2/[Ca^{2+}]$ and $v_* = Q_2/[Ca^{2+}]_*$. The dependence on domain Ca^{2+} occurs via the v_* factor that is a function of $[Ca^{2+}]_d$.

At this point we could choose a value for elevated $[Ca^{2+}]_d$ — perhaps using the EBA (Equation [13.67]) or RBA (Equation [13.70]) — and use this value in the v_* term in Equation (13.77) to calculate the P_o^{eq} for the extended De Young–Keizer-like IP_3R model. However, because the source amplitude of an open IP_3R receptor Ca^{2+} channel is not fixed but time varying, and likely proportional to $[Ca^{2+}]_{ER}$ (Sherman et al., 1990), we assume instead that the dependence of $[Ca^{2+}]_d$ on $[Ca^{2+}]_{ER}$ is (Smith, 1996, 2002a)

$$[Ca^{2+}]_d = \frac{D[Ca^{2+}] + D_{ER}[Ca^{2+}]_{ER}}{D + D_{ER}}, \tag{13.78}$$

where D and D_{ER} are the effective diffusion coefficients for free Ca^{2+} in the cytosol and ER, respectively.

Figure 13.13A shows P_o^{eq} of the augmented IP_3R model assuming no Ca^{2+} fluxes cross the plasma membrane, that is, Equation (13.60) is satisfied. Interestingly, the dashed and dot-dashed lines of Figure 13.13B show that P_o^{eq} decreases as $[Ca^{2+}]_T$ (and $[Ca^{2+}]_{ER}$ and $[Ca^{2+}]_d$) increase. Similarly, when $[Ca^{2+}]_T$ is reduced from 2 to 1 μM, the activity of the IP_3R receptor Ca^{2+} channels increases

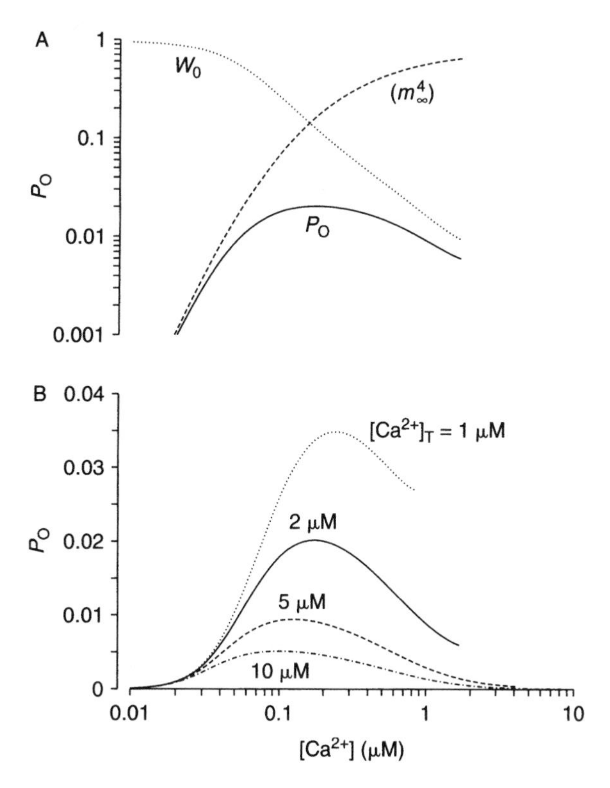

FIGURE 13.13 A: The equilibrium P_O (solid line) of the De Young–Keizer-like IP_3R model with $[Ca^{2+}]_d$-mediated inactivation where $[Ca^{2+}]_d$ is given by Equation (13.78) with $[Ca^{2+}]_T$= 2 μM. Dotted and dashed lines show the fraction of IP_3R receptor Ca^{2+} channels not inactivated (W_O in Equation [13.76]) and the proportion of the disinactivated channels that are activated by Ca^{2+} (m^4_∞). B: The dependence of P^{eq}_o on $[Ca^{2+}]_d$ is demonstrated using $[Ca^{2+}]_T$ = 1, 2, 5, and 10 μM. Reproduced with permission from Smith, G.D. in C.A. Condat and A. Baruzzi, eds., *Recent Research Developments in Biophysical Chemistry*, volume II. Research Signpost, Kerala, pp. 37–55.

(compare solid and dotted lines). This dependence of P^{eq}_o on $[Ca^{2+}]_d$ provides further justification for the assumption of increased activity of the IP_3R receptor Ca^{2+} channels at low lumenal Ca^{2+} concentrations (Li et al., 1994; 1995a). Note here that we have not hypothesized an indirect effect of intraluminal Ca^{2+}, but rather we have begun with an IP_3R transition-state diagram that includes $[Ca^{2+}]_d$-mediated inactivation and derived an IP_3R model that accounts for this phenomenon.

When this augmented IP_3R model is used in whole-cell models of Ca^{2+} handling that relate $[Ca^{2+}]_d$ to cytosolic and ER $[Ca^{2+}]$ according to Equation (13.78), $[Ca^{2+}]_d$-mediated inactivation does indeed lead to indirect lumenal disinhibition (Smith, 2002a). These observations do not rule out the possibility of lumenal disinhibition of the IP_3R occurring through other mechanisms, either directly or indirectly in gonadotrophs or other cell types, but may indicate that it is important to explicitly include $[Ca^{2+}]_d$-mediated inactivation in biophysically realistic whole-cell models based in part on Ca^{2+}-regulatory mechanisms of intracellular Ca^{2+} channels.

13.20 CONCLUSION

Computational and mathematical modeling of the buffered diffusion of intracellular Ca^{2+}, the dynamics of global Ca^{2+} responses, spatial modeling of localized Ca^{2+} elevations, and propagating Ca^{2+} waves, has made a significant contribution to our understanding of Ca^{2+} signaling in neurons and other cell types. This approach has helped us to interpret the reduced effective diffusion coefficient of intracellular Ca^{2+} (Allbritton et al., 1992; Wagner and Keizer, 1994) and has clarified the influence of Ca^{2+}-indicator dyes and exogenous Ca^{2+} chelators on spatial dynamics and Ca^{2+} microdomains (Jafri and Keizer, 1994, 1995; Neher, 1998; Smith et al., 2001). In addition to providing a theoretical foundation for cellular examples of dynamical phenomena such as excitability and bistability and their spatial analogues (propagating pulses and fronts), computational mathematical modeling is providing interesting lines of research associated with novel dynamics of cellular phenomena such as saltatory Ca^{2+} waves (Bugrim et al., 1997; Keizer et al., 1998; Smith and Keizer, 1998; Dawson et al., 1999; Falcke et al., 2000b; Coombes, 2001) and the assessment of the validity of the RBA in this context (Strier et al., 2003).

While considerable insight has been obtained through the analogy of plasma-membrane electrical excitability and ER Ca^{2+} excitability (Keizer et al., 1995; Li et al., 1995a), the disparity between electrical length scales (100 to 200 μm) and the range of action of intracellular Ca^{2+} (i.e., a chemical length scale of 1 to 2 μm) suggests that some aspects of the analogy are strained (Koch, 1999). For example, the Hodgkin–Huxley-like gating variable equations such as the Li–Rinzel reduction of the De Young–Keizer model (Equation [13.51]) are only strictly valid for a large number of intracellular channels globally coupled by the same cytosolic $[Ca^{2+}]$ (Smith, 2002b). While it is true that plasma-membrane ion channels in a small cell experience essentially the same time course of membrane voltage, intracellular Ca^{2+} channels experience radically different local $[Ca^{2+}]$ even during global Ca^{2+} responses, and are in fact only locally coupled via the buffered diffusion of intracellular Ca^{2+}. The relationship between single-channel kinetics and the dynamics of Ca^{2+} release is complicated by the fact that intracellular Ca^{2+} channels are often clustered and activated and inactivated in a concerted fashion. The next generation of compartmental and spatial models of Ca^{2+} handling will require realistic and minimal representations of the dynamics of Ca^{2+} regulation at Ca^{2+}-release sites. This will include stochastic activation and inactivation of channels consistent with our increasing knowledge of IP_3R and RyR diversity as well as a realistic account of the buffered diffusion of intracellular Ca^{2+} leading to concerted activity of intracellular Ca^{2+} channels.

ACKNOWLEDGMENTS

This work was supported in part by National Science Foundation Molecular and Cell Biology CAREER award #0133132, NSF Integrative Biology and Neuroscience grant #0228273, and the Thomas F. and Kate Miller Jeffress Memorial Trust. The work was performed in part using computational facilities at the College of William and Mary enabled by grants from the NSF and Sun Microsystems.

PROBLEMS

1. Show that the assumed form of the indicator-dye fluorescence, Equation (13.11), and the equilibrium relations, Equations (13.10) and (13.12), imply the "back-calculation" formulas, Equations (13.13) and (13.14), for a nonratiometric and ratiometric indicators respectively.
2. Show that the background or equilibrium concentration, $[Ca^{2+}]_\infty$, can be scaled out of the ODEs for the association of Ca^{2+} with buffer, Equations (13.2) to (13.5). Defining the shifted Ca^{2+} concentration $[Ca^{2+}]' = [Ca^{2+}] - [Ca^{2+}]_\infty$, write Equation (13.5), as

$$R = -k_+[B][Ca^{2+}] + k_- ([B]_T - [B]) \tag{13.79}$$

and rearrange these reaction terms to obtain

$$R = -k_+[B][Ca^{2+}]' + (k_+[Ca^{2+}]_\infty + k_-)([B]_\infty - [B]).$$

Note that this expression has the same form as Equation (13.79) if you make the correspondence

$$\begin{aligned} {k_+}' &\leftarrow k_+, \\ {k_-}' &\leftarrow k_+[\mathrm{Ca}^{2+}]_\infty + k_-, \\ K' &\leftarrow [\mathrm{Ca}^{2+}]_\infty + K, \\ [\mathrm{B}]_\mathrm{T}' &\leftarrow [\mathrm{B}]_\infty + K. \end{aligned}$$

3. Derive the stability condition for Equation (13.37) given by Equation (13.41) using von Neumann stability analysis when there is no influx or buffer ($R = 0$ or $J = 0$). Start by noting that the numerical integration of Equation (13.37) by Euler's algorithm gives (Morton and Mayers, 1994; Fitzpatrick, 2003)

$$C_j^{n+1} = C_j^n + \Lambda(C_{j+1}^n - 2C_j^n + C_{j-1}^n),$$

where $\Lambda = D_\mathrm{C}\Delta t/(\Delta x)^2$, and consider a Fourier mode of wave number k, that is, write

$$C_j^n = \bar{C}^n e^{ik\,x_j} = \bar{C}^n e^{ikj\Delta x}.$$

For stability, the complex amplification factor ξ defined by

$$\bar{C}^{n+1} = \xi \bar{C}^n$$

must be within the unit circle centered on the origin for all values of k. Show that this implies $\Lambda < \frac{1}{2}$ and thus Equation (13.41).

4. Following Problem 3, assume no influx or buffer ($R = 0$ or $J = 0$) in Equation (13.42) and show that the implicit Crank–Nicholson method

$$C_j^{n+1} - \frac{\Lambda}{2}(C_{j+1}^n - 2C_j^n + C_{j-1}^n) = C_j^n + \frac{\Lambda}{2}(C_{j+1}^n - 2C_j^n + C_{j-1}^n),$$

is unconditionally stable since $|\xi| < 1$ for all values of k.

5. Beginning with Equation (13.44), finish the derivation of the rapid buffer approximation by substituting equilibrium values for $[\mathrm{Ca}^{2+}]_\mathrm{T}$ and [CaB] to obtain

$$\frac{\partial}{\partial t}\left([\mathrm{Ca}^{2+}] + \frac{[\mathrm{Ca}^{2+}][\mathrm{B}]_\mathrm{T}}{[\mathrm{Ca}^{2+}] + K}\right) = \frac{\partial^2}{\partial x^2}\left(D_\mathrm{C}[\mathrm{Ca}^{2+}] + D_\mathrm{B}\frac{[\mathrm{Ca}^{2+}][\mathrm{B}]_\mathrm{T}}{[\mathrm{Ca}^{2+}]_i + K}\right) + J.$$

Now take spatial and temporal derivatives to get

$$\frac{1}{\beta}\frac{\partial[\mathrm{Ca}^{2+}]}{\partial t} = (D_\mathrm{C} + D_\mathrm{B}\gamma)\frac{\partial^2}{\partial x^2}[\mathrm{Ca}^{2+}] - \frac{2\gamma D_\mathrm{B}}{K + [\mathrm{Ca}^{2+}]}\left(\frac{\partial[\mathrm{Ca}^{2+}]}{\partial x}\right)^2 + J,$$

where $\gamma = K[\mathrm{B}]_\mathrm{T}/(K + [\mathrm{Ca}^{2+}])^2$ and $\beta = 1/(1+\gamma)$ (Wagner and Keizer, 1994).

6. Show that the Gaussian function given by Equation (13.47) solves Equation (13.28) with $R = 0$. Show that the square of the full width at half maximum (FWHM) of this evolving Gaussian is given by Equation (13.48).
7. Repeat the simulations of Figure 13.10 using a range of values of D and confirm numerically that the speed of the traveling front is proportional to the square root of the diffusion coefficient. You should aim to plot the speed of each observed traveling pulse or front against $\sqrt{D}$.

Change the governing equations (Equations [13.57] to [13.62]) so that the Ca^{2+} release flux J_{rel} is spatially inhomogeneous, for example, by setting $v_{ip}(x)$ to be alternately zero and nonzero without changing the spatially averaged value. Find a parameter regime where propagation is saltatory and the speed of the traveling front is proportional to D rather than $\sqrt{D}$ (Keizer et al., 1998; Smith and Keizer, 1998; Dawson et al., 1999).

8. Show that Equation (13.69) implies that the steady-state flux of total Ca^{2+} (diffusing in both free and bound forms) across any spherical surface centered on the source is equal to the source amplitude σ by evaluating $J_{total} = J_{free} + J_{bound}$ where

$$J_{free} = -4\pi r^2 D_C \frac{d[Ca^{2+}]}{dr} \qquad J_{bound} = -4\pi r^2 D_B \frac{d[CaB]}{dr}.$$

9. To derive rigorously the steady-state excess (EBA) and rapid (RBA) buffer approximations, asymptotic methods are required (Smith et al., 2001). Another analysis of the steady-state equations for the buffered diffusion of Ca^{2+} involves linearizing the equations around the equilibrium concentrations of Ca^{2+} and buffer, $[Ca^{2+}]_\infty$ and $[B]_\infty$ (Stern, 1992; Pape et al., 1995; Naraghi and Neher, 1997; Neher, 1998). Define $\delta[Ca^{2+}] = [Ca^{2+}] - [Ca^{2+}]_\infty$ and $\delta[B] = [B] - [B]_\infty$, substitute these expressions into Equations (13.65) and (13.66), and drop the quadratic terms $\delta[Ca^{2+}]\,\delta[B]$ to find

$$0 = D_C \nabla^2 \delta[Ca^{2+}] + R, \qquad (13.80)$$

$$0 = D_B \nabla^2 \delta[B] + R, \qquad (13.81)$$

where

$$R = -k_+[B]_\infty \delta[Ca^{2+}] - (k_+[Ca^{2+}]_\infty + k_-)\delta[B].$$

By writing the associated boundary conditions Equation (13.63) for $\delta[Ca^{2+}]$ and $\delta[B]$ as

$$\lim_{r\to 0}\left\{-4\pi r^2 \frac{d\delta[Ca^{2+}]}{dr}\right\} = \sigma, \qquad \lim_{r\to\infty} \delta[Ca^{2+}] = 0,$$

$$\lim_{r\to 0}\left\{-4\pi r^2 \frac{d\delta[B]}{dr}\right\} = 0, \qquad \lim_{r\to\infty} \delta[B] = 0,$$

derive Equations (13.73) to (13.75), the linearized steady-state solution of the radial equations for the buffered diffusion of Ca^{2+} near a point source. Show that when γ_∞ is large (e.g., lots of buffer), Equation (13.73) reduces to Equation (13.67).

14 Ephaptic Interactions Between Neurons

Robert Costalat and Bruno Delord

CONTENTS

14.1 INTRODUCTION

It has long been recognized that chemical synaptic transmission is the main way of information transfer between neurons (for a historical perspective, e.g., see Shepherd, 1991). However, other mechanisms can contribute to interaction between neurons, and possibly to information transfer, such as: (a) electrotonic coupling via gap junctions, (b) changes in ionic concentrations, and (c) electrical field effects (or ephaptic transmission, or extracellular field interaction) (e.g., Dudek and Traub, 1989).

The first documented example of an electrical field effect was the alteration of the excitability of axons by adjacent, stimulated axons within a peripheral nerve (Jasper and Monnier, 1938; Katz and Schmitt, 1940; Arvanitaki, 1942). These experimental studies were soon extended by thorough theoretical work (Lorente de No, 1947). Although the experimental conditions in most of these studies were somewhat artificial, it was strongly suggested that ephaptic transmission is a prominent mechanism of interaction between neurons, even in the central nervous system (CNS). This hypothesis fell into considerable disfavor when chemical synaptic transmission was experimentally demonstrated in the CNS in the early 1950s (see Eccles, 1964; Shepherd, 1991), and few articles on ephaptic transmission were published for some time. Recently, there has been a renewal of interest in the possible involvement of electrical field effects in the physiology and pathophysiology of nervous tissue, both in the peripheral nerves and the CNS. Electrical field coupling in peripheral nerves has been the subject of several experimental (Kocsis et al., 1982) and theoretical studies (Markin, 1970; Clark and Plonsey, 1970, 1971). Moreover, examples of physiological field effects in the CNS were documented in the early 1960s (Furukawa and Furshpan, 1963). Further experimental work has provided good evidence that electrical field effects can play a role (a) in synchronizing neuronal activity in the CNS, especially during epilepsy, (b) in cell growth, and (c) in membrane differentiation (Faber and Korn, 1989). Some of these experimental studies were closely correlated with precise modeling of the electrical field within nervous tissue, for example, in the hippocampus (Dudek and Traub, 1989).

This chapter reviews the main biophysical and mathematical methods used to model electrical field effects. Section 14.2 is devoted to the derivation of the biophysical equations used to describe electrical field interaction in a nervous tissue, whereas a simplified model is produced in Section 14.3. Section 14.4 examines the analytical and numerical methods available for solving these equations. A related approach, recently implemented by Reutskiy et al. (2003), is described in Section 14.5. The methods are discussed and examples of their application to real nervous tissues are given in Section 14.6.

14.2 Biophysical Basis

14.2.1 Definitions

Stimulated excitable cells generate electrical fields that can have detectable effects on neighboring cells; this effect is known as electrical field effect, electrical interaction, electrical field interaction, or electrical field coupling between neurons. This effect is sometimes referred to as extracellular field interaction, in order to emphasize the importance of the extracellular milieu. However, as shown below, the properties of both the intra- and extracellular milieux must be taken into account in any study of electrical field interaction. The term ephaptic transmission was first used in the 1930s to designate electrical interaction between two fibers that have been artificially brought together, so as "to touch" (Arvanitaki, 1942); in Greek, εφαψις is the action of touching. Nowadays, ephaptic transmission is generally used as a synonym for electrical field effects.

14.2.2 Electromagnetic Analysis of Nervous Tissue

The biophysics of excitable cells is primarily based upon the Maxwell equations:

$$\vec{\nabla} \cdot \vec{D} = \rho, \tag{14.1}$$

$$\vec{\nabla} \times \vec{E} = -\frac{\partial \vec{B}}{\partial t}, \tag{14.2}$$

$$\vec{\nabla} \cdot \vec{B} = 0, \tag{14.3}$$

$$\vec{\nabla} \times \vec{H} = \vec{j} + \frac{\partial \vec{D}}{\partial t}, \tag{14.4}$$

where $\vec{E}$ is the electric field, $\vec{D}$ is the electric induction (or electric displacement, or electric flux density), $\vec{B}$ is the magnetic induction or magnetic flux density, $\vec{H}$ is the magnetic field, ρ is the spatial charge density, and $\vec{j}$ is the (conduction) current density; $\vec{\nabla}$ is the del operator (e.g., Shadowitz, 1975). It was unambiguously shown by Lorente de No (1947) and Rosenfalck (1969) that the time derivative $\partial\vec{B}/\partial t$ can be neglected in excitable cell tissues (note that this does not preclude the possibility of a nonnegligible $\vec{B}$). Hence the curl of the vector field $\vec{E}$ is zero, so that

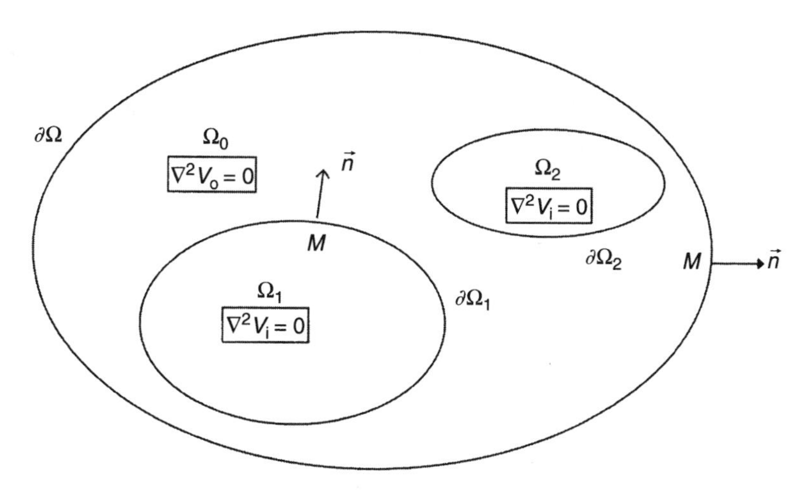

FIGURE 14.1 Intracellular and extracellular domains and their boundaries. Point M is a generic point at the exterior boundary.

$\vec{E}$ derives from a scalar potential V:

$$\vec{E} = -\vec{\nabla}V. \tag{14.5}$$

Let us consider the nervous tissue as a domain Ω of R^3 (Figure 14.1) and assume that Ω can be subdivided into (a) intracellular domains Ω_j (cells), $j = 1 - m$, with boundaries $\partial\Omega_j$ (infinitely thin cell membranes), and (b) an extracellular domain Ω_0, with boundaries $\partial\Omega$ (exterior boundary) and $\partial\Omega_j$, $j = 1 - m$. We can write $\vec{E}_\mathrm{i}$, electric field at a point of intracellular space Ω_j, and $\vec{E}_\mathrm{o}$, electric field at a point of extracellular space Ω_0, as:

$$\vec{E}_\mathrm{i} = -\vec{\nabla}V_\mathrm{i}, \tag{14.6}$$

$$\vec{E}_\mathrm{o} = -\vec{\nabla}V_\mathrm{o}. \tag{14.7}$$

It is generally assumed that capacitive effects are negligible at each point in the intracellular and extracellular domains (although transmembrane capacitive currents do exist): for a discussion based upon experimental results, see Nunez (1981). This means that in the charge conservation equation, which stems from Equations (14.1) and (14.4) (e.g., Shadowitz, 1975):

$$\vec{\nabla} \cdot \vec{j} + \frac{\partial \rho}{\partial t} = 0, \tag{14.8}$$

the term $\partial\rho/\partial t$ is zero, so that:

$$\vec{\nabla} \cdot \vec{j}_\mathrm{i} = 0, \tag{14.9}$$

$$\vec{\nabla} \cdot \vec{j}_\mathrm{o} = 0, \tag{14.10}$$

where $\vec{j}_\mathrm{i}$ and $\vec{j}_\mathrm{o}$ are the (conduction) current densities in the intracellular and extracellular domains, respectively. It is also generally assumed that the intracellular and extracellular milieux behave as linear conductors, so that Ohm's local law is valid (Jack et al., 1983):

$$\vec{j}_\mathrm{i} = \sigma_\mathrm{i}\vec{E}_\mathrm{i} = -\sigma_\mathrm{i}\vec{\nabla}V_\mathrm{i}, \tag{14.11}$$

$$\vec{j}_\mathrm{o} = \sigma_\mathrm{o}\vec{E}_\mathrm{o} = -\sigma_\mathrm{o}\vec{\nabla}V_\mathrm{o}, \tag{14.12}$$

σ_i and σ_o being the intracellular and extracellular conductivities; $\sigma_i = 1/\rho_i$ and $\sigma_o = 1/\rho_o$, where ρ_i and ρ_o are the corresponding resistivities. Note that Equations (14.11) and (14.12) may be wrong under some circumstances, especially if the diffusion fluxes of ionic species within the intracellular or extracellular spaces are not negligible. The high-frequency discharge of excitable cells may result in nonuniform spatial distributions of, for instance, the potassium or sodium ion concentrations within a confined extracellular space (Bergman, 1970), so that diffusion of these ions could occur. Hence, ephaptic effects can sometimes be strongly linked to changes in ionic concentrations. This possibility will not be examined in the present study; a discussion of the possible effects of ion distribution on membrane excitability behavior can be found in Genet and Cohen (1996).

Now we assume that σ_i and σ_o are constant scalars. Although this latter assumption is generally made (Plonsey, 1964; Clark and Plonsey, 1966, 1968), we must emphasize that it is not trivial, because precise measurements of conductivity are rather difficult. Combining Equations (14.9) through (14.12) we can write the Laplace equations:

$$\nabla^2 V_i = 0, \tag{14.13}$$

$$\nabla^2 V_o = 0, \tag{14.14}$$

which we assume to be valid at every point in the domains Ω_j, $j = 1 - m$, and Ω_0. To fully characterize the mathematical problem, we must now write the boundary conditions at boundaries $\partial\Omega_j$ and $\partial\Omega$.

14.2.3 The Boundary Conditions

The boundary conditions that are generally encountered in excitable cell modeling can be either Dirichlet or Neumann boundary conditions for potential. One says that a Dirichlet boundary condition occurs on a boundary (surface) ∂v of a domain (volume) v if the value of the potential V is given at every point of ∂v:

$$\forall M \in \partial V \;\; \forall t \quad V(M,t) = f_d(M,t), \tag{14.15}$$

where $f_d(M,t)$ is a given function of point $M \in \partial v$ and time t (note that in many problems of applied mathematics, the Dirichlet boundary condition does *not* depend on time). Furthermore, we shall say that a Neumann boundary condition occurs on a boundary ∂v of a domain v if the value of the normal derivative of potential $(\partial V/\partial n)$ is given at every point of ∂v:

$$\forall M \in \partial v \;\; \forall t \quad \frac{\partial V}{\partial n}(M,t) = f_n(M,t), \tag{14.16}$$

where $f_n(M,t)$ is a given function of point M and time.

The choice of the boundary condition for the domain Ω_0 at $\partial\Omega$ (Figure 14.1) depends upon experimental conditions, and can sometimes be nontrivial. Let us consider, for instance, an *in vitro* slice of neural tissue. If a portion of the boundary of the slice is in contact with an electrolyte solution of great enough conductivity and volume, we can assume that the electric potential is the same at every point of this solution and of the boundary surface. If we further assume that the potential of the solution is constant, we can take this potential as zero potential, and the following Dirichlet boundary condition will be fulfilled at every point of the surface:

$$\forall M \;\; \forall t \quad V_o(M,t) = 0. \tag{14.17}$$

Let us assume now that a portion of the boundary surface is in close contact with an insulator (for instance, a glass surface). Obviously there can be no current orthogonal to the surface (if we disregard any capacitive current), so the following Neumann boundary condition can be deduced at every point of the surface:

$$\forall M \ \forall t \quad \frac{\partial V_o}{\partial n}(M,t) = 0. \tag{14.18}$$

We now consider *in vivo* measurements. The choice of the boundary condition for the domain Ω_0 is more subtle. First, where is boundary $\partial\Omega$ in the body? Second, what kind of condition can be fulfilled? In most experimental devices, the position of the reference electrode is crucial. If the reference electrode is far enough (from an electrical point of view) from the excitable tissue being studied, it can be assumed that boundary $\partial\Omega$ is at an infinite distance and that the potential V_o at infinite distance is zero; this simplifies the mathematical study (see below). Finally, unless the problem of the boundary condition at $\partial\Omega$ is very carefully studied, both theoretically and experimentally, for every experimental device, the solution of potential equations may lead to nonsense; for further discussion, see Nunez (1981).

The conditions at boundaries Ω_j, $j = 1-m$, that is, at the cell membranes, are generally Neumann boundary conditions. The cell membranes are assumed to be infinitely thin, so that their detailed molecular structure is not taken into account. Current crossing the cell membrane or *transmembrane current* is orthogonal to the membrane (e.g., Johnston and Miao-Sin Wu, 1995). Owing to the continuity of the current density, we can write:

$$\forall M \in \partial\Omega_j \ \forall t \quad I_m(M,t) = j_{in}(M,t) = j_{on}(M,t), \tag{14.19}$$

where $I_m(M,t)$ is the transmembrane current per unit area, $j_{in}(M,t)$ and $j_{on}(M,t)$ are the projections of current density vectors $\vec{j}_i(M,t)$ and $\vec{j}_o(M,t)$ on the normal vector $\vec{n}$. Using Equations (14.11) and (14.12), we can write:

$$I_m(M,t) = -\sigma_i \frac{\partial V_i}{\partial n}(M,t) = -\sigma_o \frac{\partial V_o}{\partial n}(M,t), \tag{14.20}$$

which is a Neumann boundary condition for V_i at boundary $\partial\Omega_j$ and for V_o at boundary $\partial\Omega_j$.

Finally, the transmembrane current $I_m(M,t)$ must have an explicit form. Equations for membrane current may either be linear ("passive" membrane properties), or nonlinear ("active"); for a full discussion, see Tuckwell (1988a, 1988b). These terms "passive" and "active" must not be confused with active and passive transport through the membrane.

Linear membrane properties can be observed in the dendrites and in the soma or the axon below the firing threshold. Transmembrane current can be written in this case as:

$$I_m(M,t) = C_m \frac{\partial V}{\partial t}(M,t) + G_m V(M,t) = C_m \frac{\partial V}{\partial t}(M,t) + \frac{V(M,t)}{R_m}, \tag{14.21}$$

where

$$V(M,t) = V_i(M,t) - V_o(M,t) - V_{m,R}, \tag{14.22}$$

C_m and G_m are the membrane capacitance and conductance per unit area, so that $R_m = 1/G_m$ is the resistance of a unit area of membrane. $V_{m,R}$ is the membrane resting potential, which is assumed to be constant. $V(M,t)$ is often referred to as the *membrane depolarization*. Equation (14.22) is quite classical in theoretical neurobiology; however, in most neuron models (e.g., in dendrite modeling, see Jack et al., 1983), $V_o(M,t)$ is assumed to be constant and equal to zero. It is obviously necessary to disregard this latter assumption in order to properly model ephaptic interactions; this results in an important modification of the mathematical problem, as will be shown below.

The membrane properties in the axon, soma (especially during firing), and some regions of dendrites are nonlinear. Equations for transmembrane current are far more complex in this case, and several paradigms have emerged, for example, the Hodgkin–Huxley equations (Hodgkin and Huxley, 1952b) and the FitzHugh–Nagumo equations (FitzHugh, 1961; Nagumo et al., 1962). The (space-clamped) Hodgkin–Huxley equations include at least four dependent variables, while the space-clamped version of the FitzHugh–Nagumo model is two-dimensional. The Hodgkin–Huxley equations can be written in their simplest form:

$$I_{\mathrm{m}}(M,t) = C_{\mathrm{m}}\frac{\partial V}{\partial t}(M,t) + \bar{g}_{\mathrm{Na}}[m(M,t)]^3 h(M,t)[V(M,t) - V_{\mathrm{Na}}] + \bar{g}_{\mathrm{K}}[n(M,t)]^4[V(M,t) - V_{\mathrm{K}}] + \bar{g}_{\mathrm{l}}[V(M,t) - V_{\mathrm{l}}], \quad (14.23)$$

$$\frac{\partial m}{\partial t}(M,t) = \alpha_{\mathrm{m}}(V(M,t)) \cdot [1 - m(M,t)] - \beta_{\mathrm{m}}(V(M,t)) \cdot m(M,t), \quad (14.24)$$

$$\frac{\partial h}{\partial t}(M,t) = \alpha_h(V(M,t)) \cdot [1 - h(M,t)] - \beta_h(V(M,t)) \cdot h(M,t), \quad (14.25)$$

$$\frac{\partial n}{\partial t}(M,t) = \alpha_n(V(M,t)) \cdot [1 - n(M,t)] - \beta_n(V(M,t)) \cdot n(M,t), \quad (14.26)$$

where $\bar{g}_{\mathrm{Na}}$ and $\bar{g}_{\mathrm{K}}$ are the maximum conductances for sodium and potassium ions, and $\bar{g}_{\mathrm{l}}$ is the leakage conductance. m, h, and n are variables that were introduced by Hodgkin and Huxley to describe the activation and inactivation of ionic channels. V_{Na}, V_{K}, and V_{l} are the respective Nernst potentials, minus the membrane resting potential; for instance:

$$V_{\mathrm{Na}} = \frac{RT}{F}\ln\frac{[\mathrm{Na}^+]_{\mathrm{o}}}{[\mathrm{Na}^+]_{\mathrm{i}}} - V_{\mathrm{m,R}}, \quad (14.27)$$

where $[\mathrm{Na}^+]_{\mathrm{o}}$ and $[\mathrm{Na}^+]_{\mathrm{i}}$ are the concentrations (more rigorously, the activities) of the sodium ion at the outer and inner face of the membrane. R is the gas constant, T the absolute temperature, and F the Faraday constant. α_y and β_y, $y = m, h, n$, are functions of membrane depolarization; they can depend upon T, the absolute temperature, as a parameter. Hodgkin–Huxley equations have undergone many further developments in order to take into account other types of current, for example, calcium, persistent sodium, or slowly inactivating potassium currents (Hille, 2001; Johnston and Wu, 1995). Equations (14.23) through (14.26) have been written here with all independent variables, namely M and t. It should be noted that any change of outer or inner potential alters both the transmembrane driving forces $V(M,t) - V_\eta$, $\eta = \mathrm{Na, K, l}$, and the dynamics of the Hodgkin–Huxley m, h, n variables.

Finally, synaptic currents can be added to either Equation (14.21) or (14.23). For instance, in the case of a simple, AMPA-like synapse, Equation (14.21) can be replaced by the following (e.g., Johnston and Wu, 1995):

$$I_{\mathrm{m}}(M,t) = C_{\mathrm{m}}\frac{\partial V}{\partial t}(M,t) + G_{\mathrm{m}}V(M,t) + g_{\mathrm{syn}}(M,t)\cdot[V(M,t) - V_{\mathrm{rev}}], \quad (14.28)$$

where the synaptic conductance $g_{\mathrm{syn}}(M,t)$ is a given function of space and time that describes the distribution and dynamics of synaptic channels and V_{rev} is the synaptic reversal potential (minus membrane resting potential $V_{\mathrm{m,R}}$).

14.3 The Mathematical Model and its Simplification

14.3.1 The Mathematical Model

We can now summarize the mathematical problem as follows: we are given m intracellular domains $\Omega_j, j = 1 - m$, with boundaries $\partial\Omega_j$, and an envelope domain Ω_0, with boundaries $\Omega_j, j = 1 - m$, and $\partial\Omega$, as shown in Figure 14.1. We are looking for:

1. A function $V_o(x, y, z, t)$, where (x, y, z) is a generic point of the closed domain $\bar{\Omega}_0 = \Omega_0 \cup \partial\Omega \cup \partial\Omega_1 \cup \cdots \cup \partial\Omega_m$. We assume that the function V_o is continuous on $\bar{\Omega}_0$, and we can also assume that this function is smooth. Function V_o satisfies the Laplace equation (Equation [14.14]):

$$\nabla^2 V_o = \frac{\partial^2 V_o}{\partial x^2} + \frac{\partial^2 V_o}{\partial y^2} + \frac{\partial^2 V_o}{\partial z^2} = 0, \tag{14.29}$$

with the following boundary conditions:

$$\forall (x, y, z) \in \partial\Omega_j \;\; \forall t \quad \frac{\partial V_o}{\partial n}(x, y, z, t) = -\frac{1}{\sigma_o} I_m(x, y, z, t), \tag{14.30}$$

$$\forall (x, y, z) \in \partial\Omega \quad \left(\forall t \quad \frac{\partial V_o}{\partial n}(x, y, z, t) = 0 \right) \quad \text{or} \quad (\forall t \quad V_o(x, y, z, t) = 0). \tag{14.31}$$

2. For each $j = 1 - m$: a function $V_i(x, y, z, t)$, where (x, y, z) is now a generic point of the closed domain $\bar{\Omega}_j = \Omega_j \cup \partial\Omega_j$. We assume that function V_i is continuous and smooth in $\bar{\Omega}_j$. Function V_i satisfies the Laplace equation (Equation [14.14]):

$$\nabla^2 V_i = \frac{\partial^2 V_i}{\partial x^2} + \frac{\partial^2 V_i}{\partial y^2} + \frac{\partial^2 V_i}{\partial z^2} = 0, \tag{14.32}$$

with the following boundary condition:

$$\forall (x, y, z) \in \partial\Omega_j \;\; \forall t \quad \frac{\partial V_i}{\partial n}(x, y, z, t) = -\frac{1}{\sigma_i} I_m(x, y, z, t). \tag{14.33}$$

The initial conditions for V_o and V_i (as well as for the Hodgkin–Huxley variables m, h, and n) must be specified. As noted above, the equations for $I_m(x, y, z, t)$ can be, for example, Equations (14.21), (14.23), or (14.28). Obviously, the Laplace Equations (14.29) and (14.32) are linked through the boundary conditions, more precisely, through $I_m(x, y, z, t)$. Conversely, changes in V_o or V_i alter $I_m(x, y, z, t)$.

We must note that $I_m(x, y, z, t)$ must satisfy a zero flux condition for every cell $j = 1 - m$. Taking into account Equation (14.9) and the Gauss divergence theorem or Green's theorem (e.g., Kellogg, 1954; Shadowitz, 1975):

$$\iiint_{\Omega_j} \vec{\nabla} \cdot \vec{j}_i \, dV = \iint_{\partial\Omega_j} \vec{j}_i \cdot \vec{n} \, dS, \tag{14.34}$$

we can write:

$$\iint_{\partial\Omega_j} \vec{j}_i \cdot \vec{n} \, dS = 0 \tag{14.35}$$

or, according to Equation (14.19):

$$\iint_{\partial\Omega_j} I_m(x, y, z, t) dS = 0. \tag{14.36}$$

Therefore every choice of $I_m(x, y, z, t)$ must fulfill Equation (14.36).

To our knowledge, the problem of the existence and uniqueness of the solution of the system of Equations (14.29) through (14.33) and (14.21), (14.23) through (14.26), or (14.28) has not been thoroughly studied. However, based on the above hypothesis, we can deduce that: (a) at each time t, there is a unique solution for V_{i}, plus or minus an additive arbitrary constant; and (b) at each time t, there is a unique solution for V_{o}, provided that the Dirichlet boundary condition (14.17) is fulfilled at least over a part of boundary $\partial\Omega$.

The proof can be found in Courant and Hilbert (1962, p. 329), or in Kellogg (1929, p. 314). The problem of the additive arbitrary constant can generally be solved using biophysical considerations, as will be shown in Section 14.4.

14.3.2 A Simplification of the Model

The model can be simplified if we consider one-dimensional neurons in a three-dimensional extracellular space, as shown in Figure 14.2 (note that branching and/or tapering of the neurons can be taken into account). This approximation is based upon (a) the fact that the typical diameter of a neuron cross section (e.g., in the dendritic trees) is generally much smaller than the neuron length; for instance, hippocampal pyramidal cells are about 1 mm long, while the diameter of their dendritic trunk is about 1 μm; and (b) calculations taking into account electrophysiological data, for example, intracellular and membrane resistivities. A theoretical study of the one-dimensional neuron in a three-dimensional space was completed by Pickard (1968, 1969, 1971), who showed that the intracellular potential is almost uniform across a neuron's cross section.

Using these assumptions, Ohm's local law for intracellular fluid (Equation [14.11]) can be rewritten (we assume that the axis of the neuronal segment is the x-axis, see Figure 14.2):

$$j_{\mathrm{i}} = -\sigma_{\mathrm{i}}\frac{\partial V_{\mathrm{i}}}{\partial x}, \tag{14.37}$$

or, if the neuron cross section is circular:

$$i_a = \pi a^2 j_{\mathrm{i}} = -\pi a^2 \sigma_{\mathrm{i}}\frac{\partial V_{\mathrm{i}}}{\partial x}, \tag{14.38}$$

where a is the fiber radius and $i_a = \pi a^2 j_{\mathrm{i}}$ the intracellular axial current of the neuron segment. Note that $r_a = \rho_{\mathrm{i}}/(\pi a^2) = 1/(\sigma_{\mathrm{i}}\pi a^2)$ is the intracellular resistance per unit length. Kirchhoff's law of

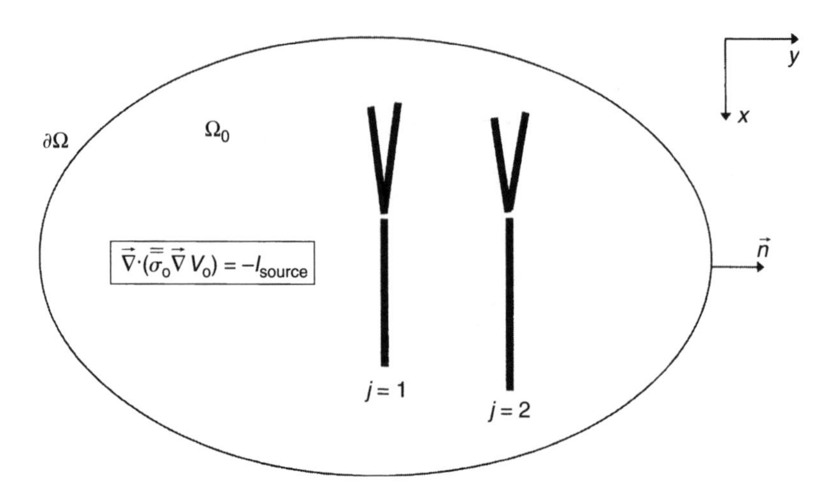

FIGURE 14.2 Simplified model of unidimensional neurons within an extracellular volume conductor.

total current conservation leads to:

$$\frac{\partial i_a}{\partial x} = -i_m = -2\pi a I_m, \tag{14.39}$$

where $i_m = 2\pi a I_m$ is the transmembrane current per unit length of fiber. Combining Equations (14.38) and (14.39) gives:

$$\frac{\partial}{\partial x}\left(a^2 \frac{\partial V_i}{\partial x}\right) = \frac{2a}{\sigma_i} I_m. \tag{14.40}$$

If the fiber radius is constant, that is, there is no tapering, then Equation (14.40) can be rewritten:

$$\frac{\partial^2 V_i}{\partial x^2} = r_a i_m = \frac{2I_m}{a\sigma_i}. \tag{14.41}$$

Since neurons are now considered to be one-dimensional, we assume that the extracellular potential is defined over the whole domain Ω, as well as on the boundary. The domain Ω contains a current source distribution due to neuron membrane currents. More precisely, the extracellular current density can be written (see Mitzdorf, 1985):

$$\vec{j}_o = \overline{\overline{\sigma}}_o \vec{E}_o = -\overline{\overline{\sigma}}_o \vec{\nabla} V_o, \tag{14.42}$$

where $\overline{\overline{\sigma}}_o$ is the conductivity tensor, which is a tensor of order two. It must be emphasized that tensor $\overline{\overline{\sigma}}_o$ is a macroscopic, "tissue," parameter, which includes neuronal tissue anisotropy (Nicholson, 1973). Since neuron spatial domains $\Omega_j, j = 1 - m$, are no longer taken into account explicitly, the macroscopic tensor $\overline{\overline{\sigma}}_o$ depends on both the extracellular "local" conductivity σ_o and on the local geometry of the neurons. For instance, for parallel neurons in a laminar structure (e.g., the hippocampus), the tensor $\overline{\overline{\sigma}}_o$ has different components along the axis of the neurons (x-axis) and along the orthogonal y- and z-axes; using this system of coordinates, the tensor reduces to its three main components, that is:

$$\overline{\overline{\sigma}}_o = \begin{bmatrix} \sigma_{xx} & 0 & 0 \\ 0 & \sigma_{yy} & 0 \\ 0 & 0 & \sigma_{zz} \end{bmatrix}, \tag{14.43}$$

with $\sigma_{yy} = \sigma_{zz}$ in the present case.

An infinitesimal volume element $\mathrm{d}V$ in Ω contains a volume current source density I_{source} due to the transmembrane current of the neurons present in $\mathrm{d}V$. Therefore the divergence of $\vec{j}_o$ is generally different from zero, and can be written in its most general form:

$$\vec{\nabla} \cdot \vec{j}_o = I_{\text{source}}. \tag{14.44}$$

For instance, if the volume element $\mathrm{d}V$ contains the membrane area element $\mathrm{d}S$ with a transmembrane current in the volume element $I_m\,\mathrm{d}S$, then the current source-density will be written as:

$$I_{\text{source}} = \frac{I_m\,\mathrm{d}S}{\mathrm{d}v}. \tag{14.45}$$

If several neurons and/or neuron segments are present in the same volume element, their respective current sources are simply added. Combining Equations (14.42) and (14.44) leads to the

Poisson equation:

$$\vec{\nabla} \cdot \left(\overline{\overline{\sigma}}_o \vec{\nabla} V_o\right) = -I_{\text{source}}. \tag{14.46}$$

Equations (14.40), (14.45), and (14.46) together with the equations for membrane currents (e.g., Equations [14.21], [14.23], or [14.28]) and boundary conditions at boundary $\partial\Omega$ will be referred to as the "simplified model." This model is strongly correlated with current source density analysis (CSDA) (Freeman and Stone, 1969; Nicholson and Llinás, 1971; Mitzdorf, 1985; Kossev et al., 1988). The goal of CSDA is to determine the current source distribution (as a function of space and time) from measurements of extracellular potentials, so that it includes Equations (14.45) and (14.46). CSDA has been applied, for instance, to the olfactory bulb (Rall and Shepherd, 1968), to the anuran cerebellum (Freeman and Nicholson, 1975; Nicholson and Freeman, 1975), and to cortically arranged neurons (Klee and Rall, 1977). The present model differs from CSDA in that it includes explicit equations for intracellular potential and transmembrane current.

To be rigorous, the boundary conditions for intracellular potential, that is, for Equation (14.40) or (14.41), must not be forgotten, and the corresponding current should be included in the volume current source density I_{source}. A thorough discussion of boundary conditions for intracellular potential can be found in Tuckwell (1988a). Here we can simply assume that a Neumann boundary condition with zero current holds at each neuron extremity, so that no term needs to be added to I_{source}.

If the tissue is isotropic and homogeneous, the tensor $\overline{\overline{\sigma}}_o$ reduces to a constant scalar, so that:

$$\nabla^2 V_o = -\frac{I_{\text{source}}}{\sigma_o}. \tag{14.47}$$

If we further assume that a Dirichlet condition (14.17) holds at boundary $\partial\Omega$, and that the boundary $\partial\Omega$ is at infinite distance from the neural tissue (remote reference electrode), then we can simply deduce from classical electrodynamics (Shadowitz, 1975):

$$V_o(N,t) = \frac{1}{4\pi\sigma_o} \iiint_{\Omega} \frac{I_{\text{source}}(P,t)}{r} \mathrm{d}V, \tag{14.48}$$

where r is the distance between the generic point $P \in \Omega$ and the observation point N; the integral is computed over the whole neural tissue that has been studied.

Finally, we should like to emphasize the fact that in the CNS, the "macroscopic" conductivity σ_o can be very low, for example, $\sigma_o = 0.5\ \text{mS cm}^{-1}$ or even lower (e.g., Faber and Korn, 1989, p. 838). These values strongly differ from the conductivity of an electrolyte Ringer-like solution (about $15\ \text{mS cm}^{-1}$).

14.4 Mathematical Methods for Solving the Resulting Partial Differential Equations (PDEs)

Both analytical and numerical methods have been used to solve the equations of the models described in Section 14.3. As in most domains of neuron modeling, analytical methods can be used only for linear membranes and relatively simple geometry: the example of two one-dimensional neurons is given in Section 14.4.1. A method of solution for models including nonlinear equations and a three-dimensional lattice is presented in Section 14.4.3.

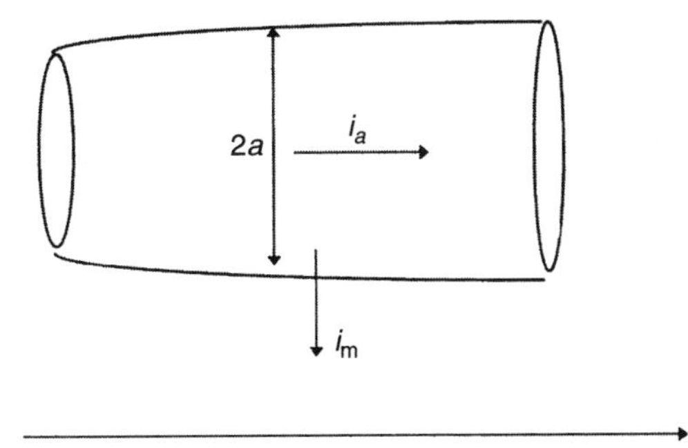

FIGURE 14.3 Segment of a one-dimensional neuron.

14.4.1 Analytical Methods in One Dimension: Field Effects in Nerve Trunks and Parallel Dendrites

Let us consider two parallel cylindrical neuron segments immersed in an extracellular space of constant section (Figure 14.3). If we neglect radial fields not only in the intracellular space but also in the extracellular milieu, we can write:

$$\frac{\partial V_o}{\partial x} = -r_o i_o \qquad \frac{\partial V_i^{(j)}}{\partial x} = -r_a^{(j)} i_a^{(j)} \quad \text{for } j = 1, 2, \tag{14.49}$$

where r_o is the extracellular resistance per unit length, i_o is the extracellular axial current, $r_a^{(j)}$ is the intracellular resistance per unit length, and $i_a^{(j)}$ the intracellular axial current of neuron j, $j = 1, 2$. $V_i^{(j)}$ is the intracellular potential of neuron j. Note that $r_a^{(j)} = 1/(\sigma_i \pi (a^{(j)})^2)$, where $a^{(j)}$ is the radius of the neuron j (see Equation [14.38]).

Kirchhoff's law of conservation of total current gives:

$$\frac{\partial i_o}{\partial x} = i_m^{(1)} + i_m^{(2)} \qquad \frac{\partial i_a^{(j)}}{\partial x} = -i_m^{(j)} \quad \text{for } j = 1, 2, \tag{14.50}$$

where $i_m^{(j)}$ is the transmembrane current per unit length of neuron j. Combining Equations (14.22), (14.49), and (14.50) and differentiating twice with respect to x, we obtain:

$$\frac{\partial^2 V^{(1)}}{\partial x^2} = (r_a^{(1)} + r_o) i_m^{(1)} + r_o i_m^{(2)}, \tag{14.51}$$

$$\frac{\partial^2 V^{(2)}}{\partial x^2} = r_o i_m^{(1)} + (r_a^{(2)} + r_o) i_m^{(2)}. \tag{14.52}$$

Equations (14.51) and (14.52) can be used for linear and nonlinear membranes; in particular, they can be valid for axons as well as for dendrites. These equations are closely related to the approach of Markin (1970, 1973), Scott and Luzader (1979), and Bell (1981) — see, for instance, equations (14.1) through (14.3) in Bell's paper (1981). These authors have studied the ephaptic interactions between parallel, unmyelinated axons, some of which are stimulated to carry propagated action potentials. Markin (1970, 1973) and Scott and Luzader (1979) use piecewise linear relations for $i_{\text{ion}}^{(j)} = f(V^{(j)})$, where $i_{\text{ion}}^{(j)}$ is the part of $i_m^{(j)}$ which is due to current through ionic channels. For instance:

$$i_{\text{ion}}^{(j)} = G(V^{(j)} - H(V^{(j)} - a)),$$

where $H(\cdot)$ is the Heaviside step function, a and G are constants, and $V^{(j)}$ is properly scaled. Note that this piecewise linear approximation is a further simplification of the FitzHugh–Nagumo model (FitzHugh, 1961; Nagumo et al., 1962). Markin (1970, 1973) then constructed traveling wave solutions, and showed in particular that (a) in the case of two axons having the same geometry and electrophysiological properties, the "active" (stimulated) axon cannot induce an action potential on the "inactive" (unstimulated) axon; (b) however, an unstimulated axon within a fiber bundle can become excited if it is surrounded by a sufficient number of stimulated axons. Scott and Luzader used a structural perturbation technique to study impulse condensation on bundles of parallel fibers. Bell (1981) used Hodgkin–Huxley equations and a perturbation theory approach to model pulse "trapping" due to ephaptic transmission. A somewhat different approach, based on fundamental electromagnetic equations, was developed by Clark and Plonsey (1970) and applied to fiber bundles (1971); this approach confirmed some of the conclusions of Markin's approach, especially the necessity for the presence of several stimulated axons in order to induce an action potential on an unstimulated one; furthermore, Clark and Plonsey pointed out the importance of several parameters, for example, extracellular resistivity.

The next section is devoted to the case in which the neuron membrane properties are linear (e.g., in dendrites).

14.4.2 Example: Electric Field Effects from Synaptic Potentials

An analytical solution has been derived for semiinfinite neurons with linear membrane properties (Costalat, 1991), in the stationary state and in response to a unit impulse. We shall briefly summarize the response to a unit impulse. The passive membrane properties imply that the membrane current per fiber unit length is:

$$i_{\mathrm{m}}^{(j)} = c_{\mathrm{m}}^{(j)} \frac{\partial V^{(j)}}{\partial t} + g_{\mathrm{m}}^{(j)} V^{(j)} = c_{\mathrm{m}}^{(j)} \frac{\partial V^{(j)}}{\partial t} + \frac{V^{(j)}}{r_{\mathrm{m}}^{(j)}} \quad j = 1, 2, \tag{14.53}$$

where $c_{\mathrm{m}}^{(j)} = 2\pi a^{(j)} C_{\mathrm{m}}$ is the capacitance per unit length of fiber and $g_{\mathrm{m}}^{(j)} = 1/r_{\mathrm{m}}^{(j)} = R_{\mathrm{m}}/(2\pi a^{(j)})$ is the corresponding conductance.

Simple transformations then lead to the following system:

$$\frac{\partial^2 V^{(1)}}{\partial X^2} = \frac{\partial V^{(1)}}{\partial T} + V^{(1)} + \beta k \left(\frac{\partial V^{(2)}}{\partial T} + V^{(2)} \right), \tag{14.54}$$

$$\frac{\partial^2 V^{(2)}}{\partial X^2} = k \left(\frac{\partial V^{(1)}}{\partial T} + V^{(1)} \right) + \alpha \left(\frac{\partial V^{(2)}}{\partial T} + V^{(2)} \right), \tag{14.55}$$

where

$$X = \frac{x}{\lambda} \qquad T = \frac{t}{\tau_{\mathrm{m}}} \qquad k = \frac{r_{\mathrm{o}}}{r_{a}^{(1)} + r_{\mathrm{o}}},$$
$$\alpha = \frac{1/(\sigma_{\mathrm{i}} \pi a^{(2)}) + r_{\mathrm{o}} a^{(2)}}{1/(\sigma_{\mathrm{i}} \pi a^{(1)}) + r_{\mathrm{o}} a^{(1)}} \qquad \beta = \frac{a^{(2)}}{a^{(1)}}, \tag{14.56}$$

$\tau_{\mathrm{m}} = R_{\mathrm{m}} C_{\mathrm{m}}$ is the membrane time constant and λ is defined as:

$$\lambda = \sqrt{\frac{r_{\mathrm{m}}^{(1)}}{r_{\mathrm{a}}^{(1)} + r_{\mathrm{o}}}}. \tag{14.57}$$

If the extracellular space resistance is zero, then λ is the first fiber space constant or characteristic length, so that X is simply the space coordinate expressed in units of this space constant. If the radii of the two fibers are equal, obviously $\alpha = \beta = 1$.

Equations (14.53) and (14.54) have been solved using various assumptions (see Costalat, 1991). First it is assumed that the system is closed, so that:

$$\forall X \quad i_a^{(1)} + i_a^{(2)} + i_o = 0. \tag{14.58}$$

Let us assume that (a) neurons 1 and 2 are semiinfinite, that is, $0 \leq X < +\infty$; (b) the fiber radii are equal, that is, $a^{(1)} = a^{(2)}$; (c) depolarizations at infinity are zero, that is:

$$\lim_{X \to +\infty} V^{(1)}(X, T) = \lim_{X \to +\infty} V^{(2)}(X, T) = 0, \tag{14.59}$$

and (d) fiber depolarization is initially zero, that is:

$$\forall X \quad V^{(1)}(X, 0) = V^{(2)}(X, 0) = 0. \tag{14.60}$$

We further assume that (e) neuron 1 and/or neuron 2 are stimulated at $T = 0$ and $X = 0$ by an impulse current, so that:

$$i_a^{(1)}(0, T) = Q_0^{(1)} \frac{\delta(T)}{\tau_m} \qquad i_a^{(2)}(0, T) = Q_0^{(2)} \frac{\delta(T)}{\tau_m}, \tag{14.61}$$

where $\delta(T)$ is the unit impulse (the so-called Dirac delta function) and $Q_0^{(j)}$ is the total amount of charge that enters the jth neuron; the presence of τ_m in the denominator is simply a consequence of using T (dimensionless) instead of t (msec). $Q_0^{(j)}$ will be either 0 (unstimulated neuron) or a fixed amount Q_0 (stimulated neuron). Equation (14.61) may be interpreted in the following way: $i_a^{(j)}(0, T)$ can be viewed as a brief synaptic current of fixed magnitude at $X = 0$ and $T = 0$, followed by a Neumann boundary condition (zero flux) at $X = 0$ and $T > 0$, which means that there is a sealed end. The response of the system to a more complex stimulus can be derived using convolution (see Jack et al., 1983). With the above assumptions it can be proved using Laplace transforms (Costalat, 1991) that:

$$V^{(j)} = V^{*}(X, T; Q_0^{(j)}, k, r_a, \lambda, \tau_m) + V^{**}(X, T; Q_0^{(l)}, k, r_a, \lambda, \tau_m) \quad j, l = 1, 2 \quad j \neq l \tag{14.62}$$

with

$$\begin{aligned} V^{*}(X, T; Q_0^{(j)}, k, r_a, \lambda, \tau_m) &= \frac{Q_0^{(j)} \lambda r_a}{2\tau_m} \left(\frac{\sqrt{1+k}}{1-k} \frac{\exp[(-X^2(1+k) - 4T^2)/(4T)]}{\sqrt{\pi T}} \right. \\ &\quad \left. + \frac{1}{\sqrt{1-k}} \frac{\exp[(-X^2(1-k) - 4T^2)/(4T)]}{\sqrt{\pi T}} \right) \\ V^{**}(X, T; Q_0^{(l)}, k, r_a, \lambda, \tau_m) &= \frac{Q_0^{(l)} \lambda r_a}{2\tau_m} \left(\frac{\sqrt{1+k}}{1-k} \frac{\exp[(-X^2(1+k) - 4T^2)/(4T)]}{\sqrt{\pi T}} \right. \\ &\quad \left. - \frac{1}{\sqrt{1-k}} \frac{\exp[(-X^2(1-k) - 4T^2)/(4T)]}{\sqrt{\pi T}} \right). \end{aligned} \tag{14.63}$$

An example of modeled fiber depolarizations is shown in Figure 14.4, where neuron 1 is stimulated and neuron 2 is unstimulated. The results are compared with the depolarization of one stimulated

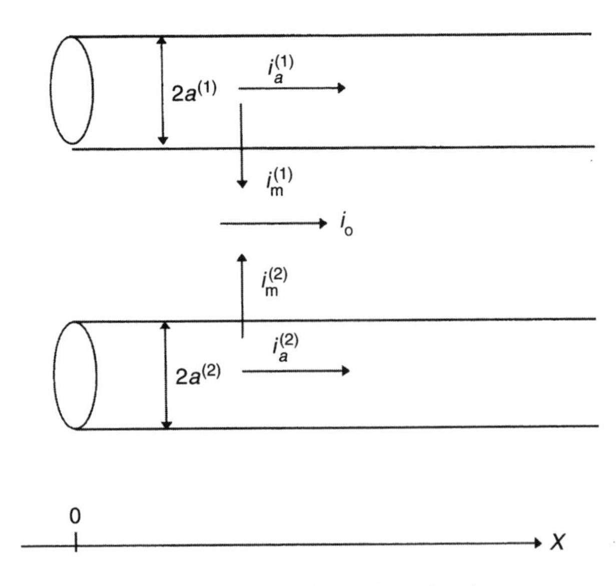

FIGURE 14.4 Two parallel one-dimensional neurons.

neuron if there is no ephaptic interaction, that is, if $k = 0$, so that Equation (14.62) obviously reduces to the classical Fatt and Katz equation (Fatt and Katz, 1951):

$$V(X,T) = \frac{Q_0 \lambda r_a}{\tau_m} \frac{\exp[(-X^2 - 4T^2)/(4T)]}{\sqrt{\pi T}}. \tag{14.64}$$

In this example, ephaptic interactions result in enhanced depolarization of the stimulated neuron and a nonmonotonic effect on the unstimulated neuron depolarization. More precisely, the unstimulated neuron is depolarized near the "synaptic region" of the stimulated neuron, and hyperpolarized in the more distal region. The depolarization front, that is, the point X where $V^{(2)}$ is zero, propagates with a velocity of:

$$v_0 = \frac{1}{2}\sqrt{\frac{\ln((1+k)/(1-k))}{kT}}. \tag{14.65}$$

As expected, k is a prominent parameter in the determination of the strength of the coupling between the neurons. An increase in k results in (a) increased stimulated fiber depolarization, (b) increased absolute values of the depolarization and hyperpolarization of the unstimulated fiber, and (c) increased v_0 velocity.

The parameter k depends directly on the extracellular volume fraction:

$$k = \left(1 + \frac{S_o}{\pi a^2}\frac{\sigma_o}{\sigma_i}\right)^{-1}, \tag{14.66}$$

where S_o is the cross-section of the extracellular milieu and σ_o is the extracellular conductivity. Lowering the $S_o/(\pi a^2)$ ratio results in increased ephaptic transmission. These results have been extended to branched dendrites (Costalat, 1991) and an arbitrary number of fibers, some of them stimulated.

14.4.3 Numerical Methods in Three Dimensions: Field Effects in Populations of Active Neurons

We now consider the numerical solution of Equations (14.40) or (14.41), (14.45), and (14.46), and (14.21), (14.23) through (14.26) or (14.28). We shall solve the elliptic PDE (14.46) using a finite difference method (see for instance Press et al., 1992). The time evolution will be described using a method we adapted from MacGregor (1987).

The extracellular milieu is first discretized using a three-dimensional grid (Figure 14.5). Points in the extracellular milieu will be referred to using their integer coordinates k, l, m, so that:

$$x = k\Delta x \quad y = l\Delta y \quad z = m\Delta z \qquad k = 0 - k_{\max} \quad l = 0 - l_{\max} \quad m = 0 - m_{\max}, \tag{14.67}$$

where Δx, Δy, Δz are the discretization steps along the x-, y-, and z-axes. Assuming that the conductivity tensor does not depend on the space coordinates and can be written as in Equation (14.43), we have:

$$\bar{\bar{\sigma}}_{\mathrm{o}} \vec{\nabla} V_{\mathrm{o}} = \begin{bmatrix} \sigma_{xx} \dfrac{\partial V_{\mathrm{o}}}{\partial x} \\ \sigma_{yy} \dfrac{\partial V_{\mathrm{o}}}{\partial y} \\ \sigma_{zz} \dfrac{\partial V_{\mathrm{o}}}{\partial z} \end{bmatrix} \tag{14.68}$$

and Equation (14.46) can be written as:

$$\sigma_{xx} \frac{\partial^2 V_{\mathrm{o}}}{\partial x^2} + \sigma_{yy} \frac{\partial^2 V_{\mathrm{o}}}{\partial y^2} + \sigma_{zz} \frac{\partial^2 V_{\mathrm{o}}}{\partial z^2} = -I_{\mathrm{source}}. \tag{14.69}$$

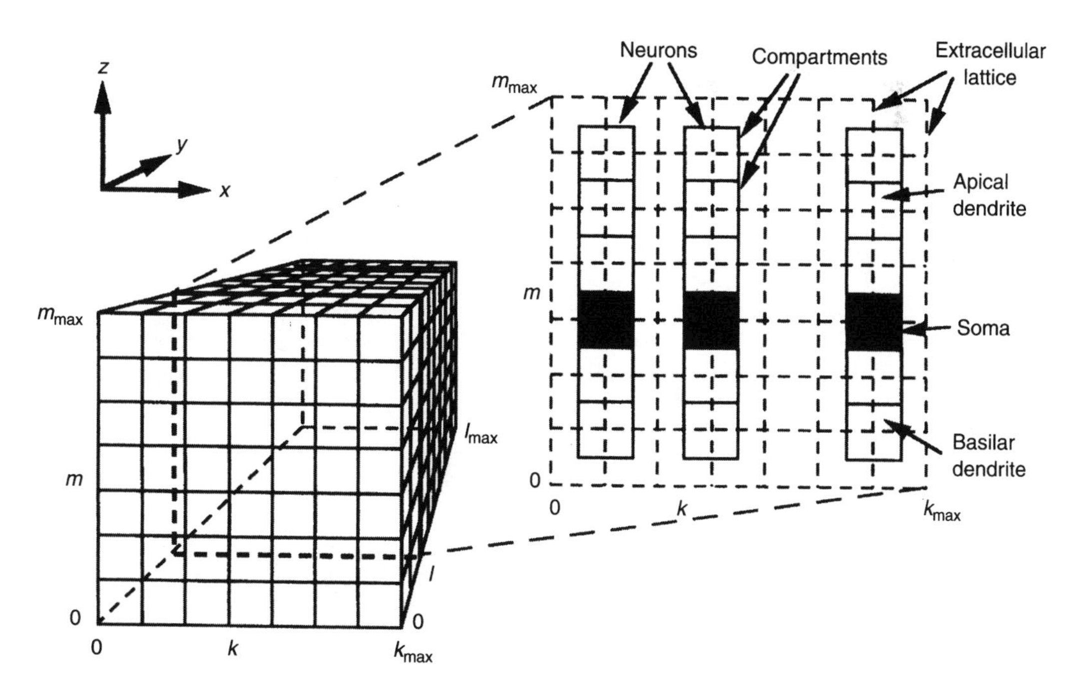

FIGURE 14.5 Schematic representation of the model structure. The extracellular milieu is modeled as a three-dimensional (x, y, z) lattice. The model neurons incorporate unbranched apical and basilar dendrites in addition to their somata.

In the finite-difference approximation, this latter equation can be replaced at each point by:

$$\sigma_{xx}\frac{V_{o;k+1,l,m} + V_{o;k-1,l,m} - 2V_{o;k,l,m}}{\Delta x^2} + \sigma_{yy}\frac{V_{o;k,l+1,m} + V_{o;k,l-1,m} - 2V_{o;k,l,m}}{\Delta y^2}$$
$$+ \sigma_{zz}\frac{V_{o;k,l,m+1} + V_{o;k,l,m-1} - 2V_{o;k,l,m}}{\Delta z^2} = -I_{source;k,l,m}, \tag{14.70}$$

where $V_{o;k,l,m}$ and $I_{source;k,l,m}$ are the extracellular potential and current source term at the point with integer coordinates (k, l, m). Note that both $V_{o;k,l,m}$ and $I_{source;k,l,m}$ depend on time. The current source term $I_{source;k,l,m}$ can be written, as a consequence of Equation (14.45):

$$I_{source;k,l,m} = \frac{I_{m;k,l,m}\Delta S_{k,l,m}}{\Delta x \Delta y \Delta z}, \tag{14.71}$$

where $\Delta S_{k,l,m}$ is the area of the membrane surface contained in the elementary volume $\Delta x \Delta y \Delta z$; for instance, in the case of the one cylindrical neuron placed along the z-axis, we have $\Delta S_{k,l,m} = 2\pi a \Delta z$. $I_{m;k,l,m}$ is the current that crosses this latter membrane surface (per unit area). Of course $I_{m;k,l,m}$ can be expressed using, for instance, Equations (14.21), (14.23) through (14.26) or (14.28).

Ad hoc boundary conditions must be chosen at the boundaries of the extracellular milieu (see Section 14.2.3). For instance, if a Dirichlet boundary condition (Equation [14.17]) is fulfilled at every point of $\partial\Omega$, then we can write:

$$(k = 0 \quad \text{or} \quad k = k_{max} \quad \text{or} \quad l = 0 \quad \text{or} \quad l = l_{max} \quad \text{or} \quad m = 0 \quad \text{or} \quad m = m_{max}) \Rightarrow V_{o;k,l,m} = 0 \tag{14.72}$$

Similarly, a Neumann boundary condition can be written at some parts of boundary $\partial\Omega$. For instance, let us assume that boundary condition (14.18) is fulfilled at every point of the surface $m = 0$; then we can write using a finite-difference approximation for the normal derivative:

$$\forall k, \ \forall l \quad \frac{V_{o;k,l,0} - V_{o;k,l,1}}{\Delta z} = 0. \tag{14.73}$$

More elaborate formulae for the derivative in Equation (14.73) can be used; see for example, Press et al. (1989).

We turn now to the discretization of the intracellular milieu. Let us consider a neuron segment, as shown in Figure 14.6. Here we assume that the one-dimensional neuron is placed along the z-axis. Using the finite-difference scheme once again, Equation (14.41) can now be written as:

$$\frac{V_{i;k,l,m+1} + V_{i;k,l,m-1} - 2V_{i;k,l,m}}{\Delta z^2} = \frac{2I_{m;k,l,m}}{a\sigma_i}. \tag{14.74}$$

Equation (14.74) is clearly derived from Ohm's law and the total current conservation law (Kirchhoff's law). Equations similar to (14.74) can readily be written using Ohm's and Kirchhoff's laws for particular points, such as branching points in a dendritic tree.

As previously mentioned, the ends of the "fibers," that is, of one-dimensional neurons, may deserve a careful discussion of the boundary conditions encountered there (e.g., see Tuckwell, 1988a, 130 ff.).

Within the described method, using the formula for transmembrane current (Equations [14.21], [14.23] through [14.26] or [14.28]) is straightforward. For instance, Equation (14.21) can simply be

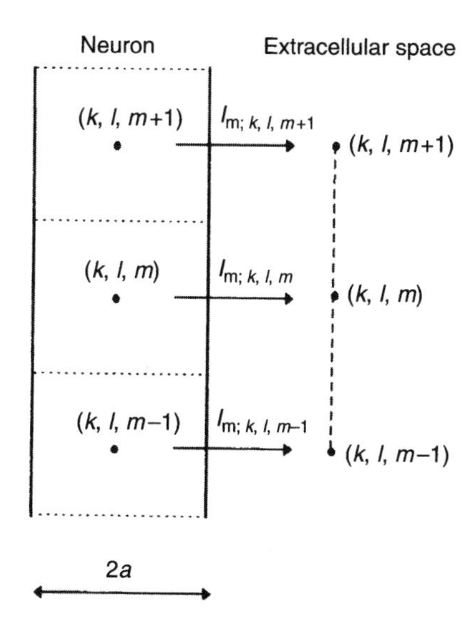

FIGURE 14.6 Discretization of the intracellular space and connection to extracellular space.

written as:

$$I_{m;k,l,m} = C_m \frac{dV_{k,l,m}}{dt} + G_m V_{k,l,m}, \tag{14.75}$$

where $V_{k,l,m} = V_{i;k,l,m} - V_{o;k,l,m} - V_{m,R}$ depends on time.

Similarly, Equation (14.23) would become:

$$\begin{aligned} I_{m;k,l,m} = C_m \frac{dV_{k,l,m}}{dt} &+ \bar{g}_{Na}(m_{k,l,m})^3 h_{k,l,m}[V_{k,l,m} - V_{Na}] \\ &+ \bar{g}_K(n_{k,l,m})^4[V_{k,l,m} - V_K] + \bar{g}_l[V_{k,l,m} - V_l] \end{aligned} \tag{14.76}$$

and Equations (14.24) through (14.26) would be replaced by the following:

$$\frac{d\phi_{k,l,m}}{dt} = \alpha_\phi(V_{k,l,m}) \cdot [1 - \phi_{k,l,m}] - \beta_\phi(V_{k,l,m}) \cdot \phi_{k,l,m} \tag{14.77}$$

with $\phi \in \{m, h, n\}$; $\phi_{k,l,m}$ depends on time.

Finally, one has to solve the system consisting of: extracellular milieu Poisson equation (Equation [14.70]); intracellular milieu Poisson equation (Equation [14.74]); Equation (14.71), which links the two latter equations; equations such as (14.75) or (14.76) and (14.77), which give explicit form to the transmembrane current; and *ad hoc* boundary conditions, such as Equation (14.73). This system is somewhat difficult to handle, due to the fact that the Poisson equation for intracellular space (Equation [14.74]) is written in terms of intracellular potential, while the equations for membrane current (Equations [14.75] through [14.77]) are written in terms of membrane depolarization. However, simple algebraic calculation leads to:

$$\begin{aligned} C_m \frac{dV_{k,l,m}}{dt} &= -\left[G_m + \frac{a\sigma_i}{\Delta z^2}\right].V_{k,l,m} \\ &+ \left[\frac{a\sigma_i}{2\Delta z^2}(V_{k,l,m-1} + V_{k,l,m+1} + V_{o;k,l,m-1} + V_{o;k,l,m+1} - 2V_{o;k,l,m})\right] \end{aligned} \tag{14.78}$$

in the linear case (Equation [14.75]), and to:

$$C_m \frac{dV_{k,l,m}}{dt} = -\left[\bar{g}_{Na}(m_{k,l,m})^3 h_{k,l,m} + \bar{g}_K(n_{k,l,m})^4 + \bar{g}_l + \frac{a\sigma_i}{\Delta z^2}\right].V_{k,l,m}$$
$$+ \left[\bar{g}_{Na}(m_{k,l,m})^3 h_{k,l,m} V_{Na} + \bar{g}_K(n_{k,l,m})^4 V_K + \bar{g}_l V_l\right.$$
$$\left. + \frac{a\sigma_i}{2\Delta z^2}(V_{k,l,m-1} + V_{k,l,m+1} + V_{o;k,l,m-1} + V_{o;k,l,m+1} - 2V_{o;k,l,m})\right] \quad (14.79)$$

in the nonlinear case (Equation [14.76]).

Equations (14.78) and (14.79) can simply be written:

$$\frac{dV_{k,l,m}}{dt} = aV_{k,l,m} + b, \quad (14.80)$$

where coefficients a and b do not depend on $V_{k,l,m}$, while they do depend on: (a) depolarization of adjacent points ("compartments") in the intracellular grid, namely, $V_{k,l,m+1}$ and $V_{k,l,m-1}$, (b) extracellular potential $V_{o;k,l,m}$, $V_{o;k,l,m+1}$, and $V_{o;k,l,m-1}$, and (c) possibly other variables such as the Hodgkin–Huxley m, h, n variables. At a given time Equation (14.80) can roughly be considered as a linear ordinary differential equation with constant coefficients, the integration of which leads to:

$$V_{k,l,m}(t + \Delta t) = V_{k,l,m}(t) \cdot e^{a\Delta t} - \frac{b}{a}(1 - e^{a\Delta t}). \quad (14.81)$$

This method of resolution in time is an extension of the method proposed and extensively used by MacGregor (1987) in the case of a neuron in an isopotential extracellular milieu. An alternative way would consist in a discretization of the expression of the time derivative $dV_{k,l,m}/dt$ using one of the classical finite-difference methods for solution of parabolic PDEs, namely, explicit, implicit, or Crank–Nicholson methods.

Finally, we propose an algorithm to solve the resulting system of equations (Costalat, 1991), which can be summarized as follows:

1. Initial values are assigned to variables, that is, to the depolarization $V_{k,l,m}$, the extracellular potential $V_{o;k,l,m}$, and maybe to other variables such as the m, h, n variables.
2. Transmembrane currents are computed using Equations (14.75) or (14.76) (if synaptic current is present, a time-dependent term $g_{syn;k,l,m}(V_{k,l,m} - V_{rev})$ can simply be added).
3. Poisson equation (Equation [14.74]) is solved using, for instance, a simultaneous over-relaxation technique (Press et al., 1989), so that we obtain new values for the extracellular potential.
4. New values for the membrane depolarizations $V_{k,l,m}$ are obtained from Equation (14.78) or (14.79), solved along the lines of Equation (14.81).
5. New values of the m, h, n variables are then computed using Equation (14.77).
6. Finally, new values for the transmembrane currents can be obtained; return to (2).

Each iteration corresponds to a time step Δt. In our experience, the numerical method appears to be stable if the time step is about 1 to 2 μsec, using classical values for the Hodgkin–Huxley theory parameters (resulting in an action-potential duration of about 1 msec).

This method was applied to study the influence of one stimulated neuron upon one unstimulated neuron, assuming active conductances in the soma and axon (Costalat and Chauvet, 1993). From a qualitative point of view, the depolarization profiles were similar to those obtained with the analytical model of Section 14.4.1. However, stimulation of the extremity of the dendrite of one neuron could result in a decrease of the excitability of the other neuron, due to the prolonged hyperpolarization of its soma. Hence this model suggests the possibility of complex changes, not only in the depolarization, but also in the excitability of unstimulated neurons within monolayered structures.

In the case where the neural tissue is homogeneous and isotropic, and a Dirichlet condition (14.17) holds at boundary $\partial\Omega$ situated at an infinite distance, Equations (14.47) and (14.48) are valid, so that the extracellular potential can simply be computed using Equation (14.48) (in step 3 of the algorithm). This assumption, which may not be bad *in vivo* (see Nunez, 1981), greatly lowers the calculation time. An example of the solutions obtained using this latter method is displayed in Figures 14.7, 14.8, and 14.9. In these figures, the stimulated and unstimulated neurons have a soma carrying Hodgkin–Huxley conductances and one dendrite with passive, linear conductance. The stimulated neuron is depolarized by a synaptic current (analogous to an AMPA current). When the stimulation effect is below the action-potential threshold (Figure 14.7), one can observe an excitatory postsynaptic potential (EPSP) at the stimulated neuron, together with a nonmonotonic variation of the unstimulated neuron depolarization; these results can be compared to those obtained using an analytical method in Figure 14.4 (however, note that in Figure 14.4 the first neuron is stimulated by a Dirac delta function). When the stimulation is increased, for instance if more synapses are stimulated, so that an action

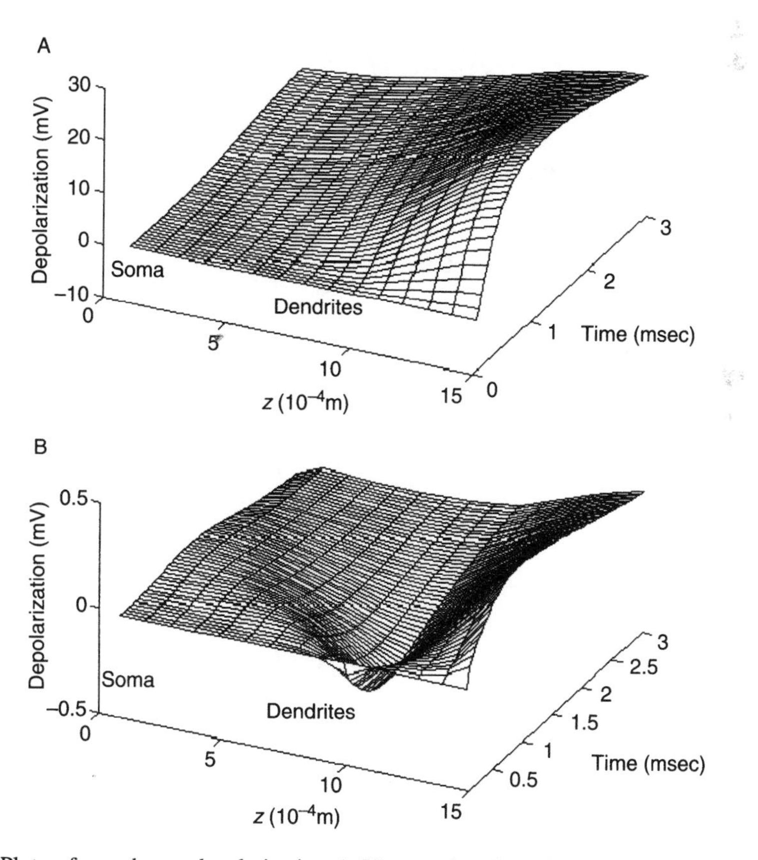

FIGURE 14.7 Plots of membrane depolarization (mV) as a function of space (in units of 10^{-4} m) and time (msec). (A) Stimulated neuron, (B) unstimulated neuron. Both neurons are composed of the soma (compartment 1) embedded with Hodgkin–Huxley conductances and dendrites (compartments 2 to 15) with linear conductance. At the tip of the stimulated neuron (five last compartments), an excitatory synaptic current is modeled using the formula $\bar{g}_{\text{syn}}(\exp(-t/\tau_1) - \exp(-t/\tau_2))(V - V_{\text{rev}})$, with $\tau_1 = 1$ msec, $\tau_2 = 0.1$ msec, $V_{\text{rev}} = 70$ mV. $\bar{g}_{\text{syn}}$ (maximal synaptic conductance) increases from compartment 11 ($0.28\ \text{mS cm}^{-2}$) to compartment 15 ($1.4\ \text{mS cm}^{-2}$).

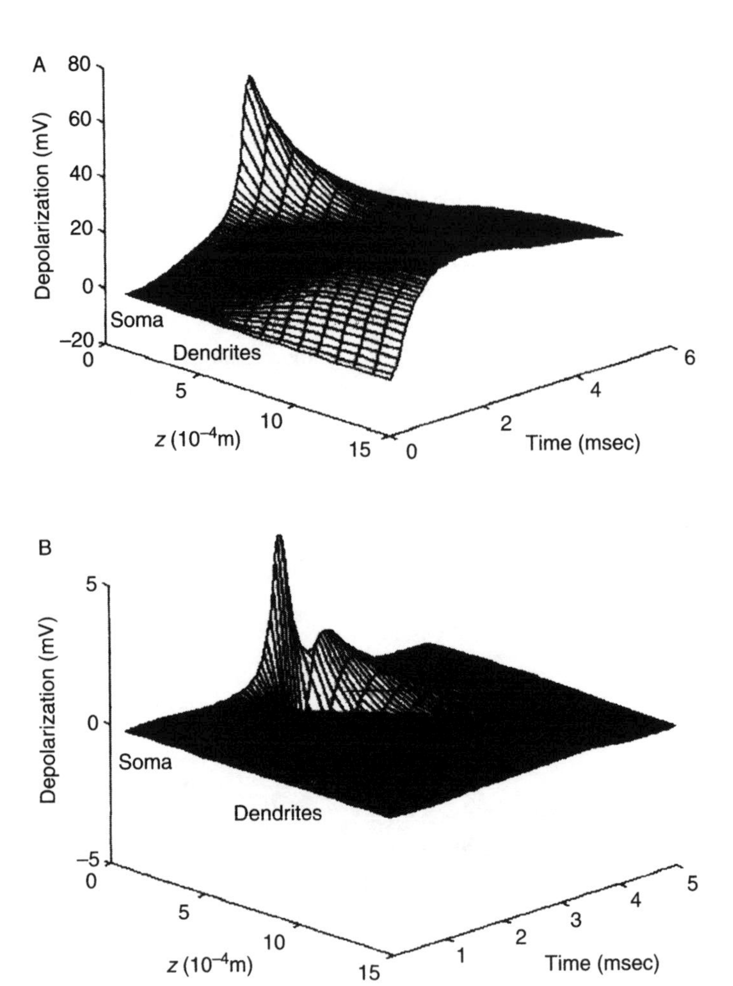

FIGURE 14.8 Plots of membrane depolarization (mV) as a function of space (in units of 10^{-4} m) and time (msec). (A) Stimulated neuron, (B) unstimulated neuron. The parameter values are the same as in Figure 14.7, with the exception of $\overline{g}_{syn}$ (maximal synaptic conductance) which now increases from compartment 6 (0.28 mS cm^{-2}) to compartment 15 (1.4 mS cm^{-2}). Note the action potential at the stimulated neuron.

potential is elicited at the stimulated neuron soma, the observed effects in the unstimulated neuron are more complex (Figure 14.8 and Figure 14.9): as in Figure 14.7, a small depolarization is observed at the distal part of the dendrite (Figure 14.9D), while a weak hyperpolarization appears at the dendritic compartments close to the soma (Figure 14.9C); then an action potential in the first neuron induces a strong depolarization of the second neuron soma (Figure 14.9A). In this example, the behavior of the soma potential of the unstimulated neuron is polyphasic, due to the superposition of (a) the dendritic phenomena, which can induce a hyperpolarization at the soma, and (b) the depolarization and subsequent hyperpolarization induced by the action potential in the first neuron. This results in nonmonotonic changes in the second neuron soma Na conductance, as can be seen on a plot of m^3h versus time (Figure 14.9B). In this example, m^3h is first increased, then decreased.

It should be emphasized that Figure 14.7 and Figure 14.8 were obtained with a relatively low value of the "macroscopic" extracellular conductivity ($\sigma_o = 0.1$ mS cm^{-1}); quantitatively similar results can be obtained with $\sigma_o = 0.5$ mS cm^{-1} and about five stimulated neurons around the unstimulated one. It is probable that such values of conductivity can be observed in the CNS (see the end of

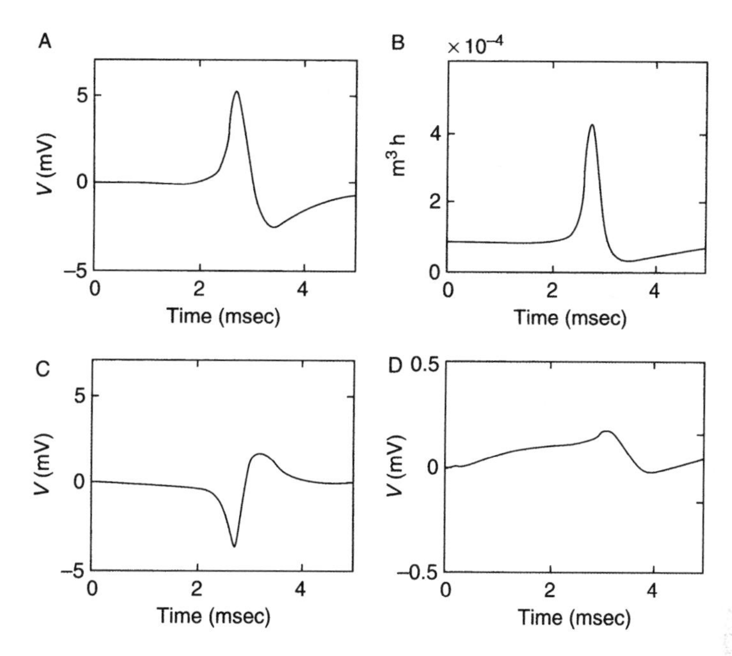

FIGURE 14.9 Unstimulated neuron. Plots of (A) somatic membrane depolarization (mV), (B) soma m^3h product (proportional to Na^+ conductance), (C) second compartment (dendrite) depolarization, (D) 12th compartment (dendrite) depolarization. Note that the second compartment is close to the soma, while the 12th is more remote. Parameter and initial condition values are the same as in Figure 14.8.

Section 14.3.2). This strongly suggests that ephaptic interactions may be important in the CNS, even at the local level.

14.5 CONDUCTION IN BUNDLES OF MYELINATED NERVE FIBERS

A related approach was recently proposed by Reutskiy et al. (2003), who modeled ephaptic interactions in bundles of myelinated nerve fibers in order to study the consequences of demyelination in multiple sclerosis. This paper extends a previous computational study of synchronized firing in the olfactory nerve by Bokil and colleagues (2001), and it is based on the work of Markin (1970). Here we shall not fully analyze the results contained in these works, but rather show how the basic biophysical equations (as described in the work by Reutskiy et al., 2003) are related to our previously described approach. The model of Reutskiy et al. is based on a set of assumptions, namely (modifying only slightly the authors' presentation):

(a) the axons are axis symmetric,
(b) the inner cross section of the fibers is constant,
(c) the Ranvier nodes are equally spaced (and aligned),
(d) active ionic currents are concentrated in the Ranvier nodes,
(e) the internodal membrane is totally passive,
(f) axon cross sections are isopotential,
(g) ion concentrations are constant,
(h) the fibers are parallel,
(i) all the fibers are identical,
(j) the nerve boundary is perfectly insulating, and
(k) each transverse section of the extracellular space is isopotential.

It can be noted that assumptions (a), (b), (f), (g), (h), (j), and (k) are equivalent to the assumptions we used for parallel dendrites (see Section 14.4). However, assumption (i) is more restrictive than the assumption we used, namely, we assume fibers with different radii. Finally, assumptions (c), (d), and (e) are specific to the model of Reutskiy, Rossoni, and Tirozzi.

In order to make the study of the article by Reutskiy, Rossoni, and Tirozzi easier, we summarize in the following table the correspondence between our notations and those adopted in their paper (Reutskiy et al., 2003).

	Our Equations (14.22), (14.49), (14.50), (14.56), and (14.66)	**Reutskiy, Rossoni, and Tirozzi (Reutskiy et al., 2003)**
Extracellular potential	V_o	ϕ_0
Intracellular potential	$V_i^{(j)} j = 1, 2$	$\phi_k, k = 1, \dots, N$
Extracellular axial current	i_o	I_0
Intracellular axial current	$i_o^{(j)} j = 1, 2$	$I_k, k = 1, \dots, N$
Extracellular resistance per unit length	r_o	r_0
Intracellular resistance per unit length	$r_a^{(j)} j = 1, 2$	r_f (*does not depend on which fiber is considered*)
Membrane potential	$V^{(j)} + V_{m,R}\, j = 1, 2$	$V_k, k = 1, \dots, N$
Cross section of the extracellular milieu	S_o	S_{free}
Cross section of one fiber	$\pi(a^{(j)})^2$ or πa^2	s

Hence, Reutskiy, Rossoni, and Tirozzi write:

$$I_0(x, t) + \sum_{k=1}^{N} I_k(x, t) = 0, \tag{14.82}$$

which is obviously equivalent to Equation (14.50), and:

$$\frac{\partial \phi_0}{\partial x} = -r_0 I_0 \qquad \frac{\partial \phi_k}{\partial x} = -r_f I_k \quad k = 1, \dots, N, \tag{14.83}$$

which is identical to Equation (14.49). It can easily be derived that:

$$\frac{\partial V_k}{\partial x} = -r_f I_k + r_0 I_0 \quad k = 1, \dots, N. \tag{14.84}$$

Then, using Equation (14.82), straightforward algebraic computations lead to:

$$I_k = -\frac{1}{r_f} \frac{\partial V_k}{\partial x} + \frac{\alpha_N}{r_f} \sum_{n=1}^{N} \frac{\partial V_n}{\partial x} \quad k = 1, \dots, N, \tag{14.85}$$

where the parameter α_N (the so-called coupling strength) is defined by Reutskiy et al. as:

$$\alpha_N = \frac{\alpha_0}{1 + N\alpha_0}, \tag{14.86}$$

the parameter α_0 being defined as:

$$\alpha_0 = \frac{r_0}{r_f}. \tag{14.87}$$

Obviously, α_0 is related to our parameter k in Equation (14.56) by the relation:

$$\frac{1}{k} = 1 + \frac{1}{\alpha_0} \tag{14.88}$$

in the special case where all fibers are identical. The parameter α_0 can be written as:

$$\alpha_0 = \frac{\rho_0}{\rho_{ax}} \frac{s}{S_{\text{free}}}, \tag{14.89}$$

ρ_0 and ρ_{ax} being the resistivity of the extracellular milieu and the fiber, respectively. This latter equation, together with Equation (14.88), is equivalent to Equation (14.66), taking into account the fact that:

$$\rho_0 = \frac{1}{\sigma_o} \qquad \rho_{ax} = \frac{1}{\sigma_i}, \tag{14.90}$$

σ_o and σ_i being the extracellular and intracellular conductivities, respectively, as defined in Section 14.4.

Furthermore, Reutskiy, Rossoni, and Tirozzi define the ephaptic current, caused by the electrical interaction between the fibers, as:

$$I_{\text{eph}} = -\frac{\alpha_N}{r_f} \sum_{n=1}^{N} \frac{\partial V_n}{\partial x}. \tag{14.91}$$

Using this definition, the following simple formula can be used for each fiber:

$$I_k = -\frac{1}{r_f} \frac{\partial V_k}{\partial x} - I_{\text{eph}} \quad k = 1, \ldots, N, \tag{14.92}$$

where I_{eph} is zero when no ephaptic interaction occurs. Finally, using current conservation laws as was described in Section 14.4 (see Equation [14.53]), Reutskiy et al., find a modified cable equation:

$$c_{m,k} \frac{\partial V_k}{\partial t} = \frac{1}{r_f} \frac{\partial^2 V_k}{\partial x^2} + \frac{\partial I_{\text{eph}}}{\partial x} - j \quad k = 1, \ldots, N, \tag{14.93}$$

j being the ionic membrane current per unit length (possibly, j also includes the stimulating current per unit length). In the case of internodal regions, $c_{m,k}$ depends on the abscissa x, according to the formula proposed by Goldman and Albus (1968), and, assuming passive membrane properties, j must be an affine function of V_k, namely:

$$j = g_{m,k}(x)(V_k - V_{\text{m,R}}), \tag{14.94}$$

where $V_{\text{m,R}}$ is the resting potential of the fibers. It can be noted that in the paper of Reutskiy, Rossoni, and Tirozzi, Equation (14.94) is written as (see equation 9 in Reutskiy et al., 2003):

$$j = g_{m,k}(x)V_k. \tag{14.95}$$

This seems erroneous, since it would imply, in the limiting case where $I_{\text{eph}} = 0$, that we have a stationary state with $V_k \equiv 0$ for $k = 1, \ldots, N$, which contradicts classical membrane biophysics.

In the case of the membranes at the nodes of Ranvier, Reutskiy, Rossoni, and Tirozzi use the nonlinear equations of Frankenhaeuser and Huxley (1964), which include the sodium and potassium "active" currents, a leakage current, and a nonspecific, mainly sodium, current. Numerical solution of these equations allows these authors to point out several interesting phenomena, for instance, ephaptic interactions can lead to synchronized firing, as was already suggested by the computational model of olfactory nerve developed by Bokil and colleagues (2001). Moreover, Reutskiy, Rossoni, and Tirozzi show for the first time that strong coupling can lead to the generation of spurious spikes, which can have clinical implications. Finally, they suggest that the damaged axons could overcome conduction block by means of ephaptic interactions with undamaged axons.

14.6 Discussion

14.6.1 Biophysical and Mathematical Methods

The biophysical basis of ephaptic interaction modeling was established approximately between 1945 and 1970. Two main assumptions were made that still appear to be valuable: (a) the time derivative of the magnetic induction (Rosenfalck, 1969), and (b) the variation of the intracellular potential inside a neuron cross section (Pickard, 1971), are negligible. These hypotheses allow one to derive a three-dimensional Poisson equation for the extracellular potential and a one-dimensional Poisson equation for the intracellular potential. In both cases, the source term is proportional to the transmembrane current, which is generally nonlinear. Analytical solutions are available when the membranes are passive and the extracellular space is assumed to be one-dimensional. Although these assumptions are rather strong, this simple case makes it possible to show some general properties of ephaptic interactions, such as the nonmonotonic spatial influence of stimulated neurons on the membrane potential of unstimulated neurons. For neurons bearing active conductances, or when the extracellular space is considered as three-dimensional, numerical methods have to be employed. To our knowledge, no attempt has been made to use finite-element instead of the more classical finite-difference methods. Finite-difference discretization of extracellular space and neurons has been used extensively by Traub et al. (1985a, 1985b, 1987a, 1987b, 1987c). Here we propose a relatively simple approach, which is also based on finite-difference discretization in space; however, this method differs from Traub's approach by (a) using a time-solution procedure derived from the MacGregor method (1987); and (b) an explicit link to the Poisson equation and CSDA. Obviously, the stability of this method, as well as others, should be tested in every concrete case using at least varying time steps.

Ideally, the three-dimensional and nonlinear PDEs obtained in these studies deserve a more complete mathematical study, including analysis of the existence and uniqueness of solutions, and methods of solution using, for example, finite-element methods, together with a study of stability and convergence. However, the solution would still remain rather cumbersome and time-consuming, especially if one tries to include detailed aspects of neuronal geometry (e.g., dendritic tree branching). Besides, it would be highly desirable to get simplified models of ephaptic transmission. Some models of neural networks include an additional term for diffusion of activity through the extracellular space (see Parodi et al., 1996), which may result from ephaptic transmission. Although interesting, these models are based on *a priori* simplifications and do not include, for instance, nonmonotonic variation of depolarization in the unstimulated neuron, and, more generally, spatial effects observed along neurons. In our view, an alternative approach would consist in a careful reduction of the biophysical equations from complete three-dimensional models, in accordance with specific experiments.

14.6.2 Electrical Field Effects in the Hippocampus and the Cerebral Cortex

Ephaptic interactions have been studied in several nervous tissues (spinal cord, cerebellar cortex, hippocampus), but most of the attention has been paid to the hippocampus, for historical and experimental reasons. This central nervous structure has long been a major center of interest because of its functional involvement in learning and memory (Eichenbaum et al., 1992). It is now relatively well-known in terms of its connectivity (Rolls, 1989), electrophysiology (Wong and Prince, 1981; Brown and Griffith, 1983), and synaptic plasticity (Bliss and Lynch, 1988).

Clear evidences of ephaptic interactions in the hippocampus were found in the early 1980s (Taylor and Dudek, 1982), when spontaneous or evoked synchronized bursts of population spikes were observed in hippocampal slices in which chemical synaptic transmission had been blocked in low-[Ca^{2+}] solution. Differential recordings of intracellular and extracellular potentials clearly showed that field effects could be observed in the slice during the population burst spikes (Taylor and Dudek, 1982). The extracellular potential around the recorded somata was found to decrease transiently during the successive spikes of the bursts, depolarizing the membrane potential by the same amount. The excitability of the unstimulated pyramidal neurons was thus increased since the gating mechanisms of the action-potential fast sodium channels (as many other types of channels) are sensitive to depolarization of the membrane potential. The instantaneous firing probability of these neurons was consequently increased, eventually leading to their recruitment as additional bursting neurons. These effects were found to remain substantial when normal chemical synaptic transmission was functional (Snow and Dudek, 1986). Ephaptic interactions can thus be a major mechanism of synchronized firing, not only in epileptiform activity but also under normal physiological conditions.

The modeling effort of Traub et al. (1985a, 1985b, 1987a, 1987b, 1987c; review in Dudek and Traub, 1989) has provided a biophysical theoretical framework for understanding ephaptic interactions in the hippocampus. A model of an array of 2000 pyramidal neurons and their surrounding extracellular space was developed. Each of the model neurons possessed purely passive dendritic processes, including an eight-compartment basilar dendrite, and a ten-compartment apical dendrite. The model neurons also included a somatic compartment, which carried standard Hodgkin–Huxley Na^+ and K^+ conductances. The extracellular space was modeled as a resistive lattice of 19,000 points (with the notations of Section 14.4 and Figure 14.5, $k_{max} = 19$, $l_{max} = 49$, $m_{max} = 18$). Although these authors do not give all the details of the method employed, it clearly appears that it is based on a finite-difference scheme (see Section 14.4). The simulations allowed the authors to reproduce various experimental results, including synchronization properties of the CA3 hippocampal network in the normal condition (excitatory and inhibitory synaptic transmission), low-[Ca^{2+}] condition (blockade of the synaptic transmission), and epileptogenic condition (no inhibitory synaptic transmission).

A hypothetical mechanism of ephaptic interactions in the hippocampus was described (Dudek and Traub, 1989). It took into account several basic anatomical features of this structure: the hippocampus CA1–CA3 contains radially oriented pyramidal neurons, with dendritic trees parallel to each other and perpendicular to the surface; furthermore, the CA1–CA3 can be seen as composed of three principal layers, one layer containing the cell bodies and two layers containing the apical dendritic processes and the basal dendritic processes, respectively. Briefly, the hypothetical scheme can be summarized as follows: the synchronized firing of numerous neurons implies large transient inward (sodium, calcium) ionic currents from the extracellular space into the cell bodies of the discharging cells. A "current sink" ($I_m < 0$) appears at the cell-bodies layer, so that the extracellular potential decreases locally. The unstimulated neurons are consequently locally depolarized at their cell bodies. As a consequence, a "current source" ($I_m > 0$) appears in the dendrite layer (see Equation [14.36]). It locally hyperpolarizes unstimulated neurons at their dendritic tips (these findings can be compared with those obtained from the simple model of Section 14.4.1: see Figures 14.4, 14.8, and 14.9). This dual effect has great potential functional implications: it predicts at the same time an increased excitability of neurons and a more selective dendritic integration of their synaptic inputs. In a neural network, such a tendency may favor a "winners take all" mode of computation, that is, a state of increased competition between neurons.

Most of the cited studies were concerned with antidromically evoked electrical fields. What happens when a neuron or a group of neurons are stimulated by a postsynaptic potential, for example, an EPSP, which can be followed (or not) by an action potential? Yim et al. (1986) proposed a simple model of the unstimulated neuron, consisting of a lumped resistance and capacitance for the neuron membrane and an internal longitudinal cell resistance; the extracellular field was modeled using for example, a half sine wave. They combined this one-neuron model with formulae proposed by Patlak (1955) to calculate the extracellular field generated by the synchronous activity of *n* fibers. Yim et al. concluded that if the maximum amplitude of the stimulated neurons' EPSPs is about 30 mV, then the depolarization generated in unstimulated neurons can be about 0.3 to 1 mV in the hippocampus. They also noted that the action potentials of the stimulated neurons can depolarize the unstimulated neurons' somata by 5 mV or even more. Our results (Figures 14.4, 14.7 to 14.9), although obtained using a different method, are consistent with these latter conclusions; however, they do not preclude the possibility of an inhibition of the unstimulated neuron. Finally, the possible ephaptic effects due to EPSPs in the hippocampus might have amplitudes ranging from 0.1 to 10 mV, depending on the specific situations encountered. Such influences may contribute to integration of synaptic inputs and to synaptic plasticity phenomena.

The question arises whether some functional aspects of other central structures could be influenced by ephaptic interactions. Of special interest is the cerebral cortex, which has been poorly studied in terms of ephaptic interaction. This is surprising for several reasons. First, the cerebral cortex is one of the major loci of epileptiform discharges, and ephaptic interaction has been demonstrated to be an important factor in epileptiform discharge in the hippocampus. Second, synchronized discharges of large populations of pyramidal cortical neurons under physiological conditions are now considered to underlie perceptual processes (Gray and Singer, 1989). Third, several features indicate that ephaptic interactions may be substantial in the cerebral cortex. The anatomical arrangement of the neocortex is relatively similar to that of the hippocampus. It is two-dimensional and contains a dense array of radially oriented cells and dendritic trees. The presence of dendrite bundles implies regions of small extracellular volume and therefore high extracellular resistance (Petsche et al., 1975). The sizes and shapes of cortical pyramidal neurons are very similar to those of hippocampal CA1–CA3 pyramidal neurons, which support substantial field effects. Furthermore, cortical dendritic trees are especially long (0.5 to 2 mm), favoring spatial segregation between current sinks and sources, and the subsequent impact of field effects. One of the major anatomical features that distinguishes the neocortex from the hippocampus is that the neocortex is constituted of several superimposed layers of neurons whereas the hippocampus is monolayered. The rather simple ephaptic interactions observed in unstimulated hippocampal neurons (depolarized soma, hyperpolarized dendrites) might not be good models for the intricate vertical organization of the cortical tissue. For example, the soma of a neuron in a superficial layer could be both subject to the depolarizing field effect of neurons firing in a near vicinity (of the same layer), while being at the same time under the hyperpolarizing influence of a bundle of dendrites projecting from a group of stimulated deep layer neurons. This hypothetical example shows how the interplay between dendritic and somatic components of field interactions might be much more complex in the cortex than in the hippocampus. Investigations of field effects in the cortical structure remain to be done and should provide an invaluable contribution to the understanding of information processing and epileptic states of the cortex.

14.7 Conclusions and Future Perspectives

Electric field effects offer several challenges to modelers. The basic biophysical laws appear to be well established, so that writing down the corresponding PDEs can be achieved — at least in a general form — for virtually every concrete system. However, studying concrete systems and/or finding general properties of ephaptic transmission remains a rather difficult issue, due to the relative lack of experimental data and the complexity of the mathematical aspects. Our impression is that future work in this domain should follow two distinct, albeit complementary, directions. First, the effects of the specific geometry of the nervous tissues studied should be taken into account in a more detailed fashion. In this context, three-dimensional models of real membrane systems could be highly valuable; see for instance the three-dimensional model of human cochlea cells developed by Tylstedt and Rask-Andersen (2001). This would allow for a deeper understanding of ephaptic interactions for particular neuronal arrangements. One of the major interests of this approach would be to evaluate the contribution of field effects to information transfer between computational units at various scales (neurons, dendrites, spines). Second, carefully simplified models of ephaptic interactions should be developed and included in neuronal networks models. These models should significantly contribute to the understanding of the dynamics of neuronal populations and to the computational processes they subserve. Finally, here as well as in many other areas of neural modeling, we think that the development of both analytical and numerical methods will enable a better understanding of experimental data.

14.8 Summary

Although chemical synaptic transmission is the main pathway of information transfer between neurons, other mechanisms can contribute to interactions between neurons (and other excitable cells): (a) electrotonic coupling via gap junctions, (b) changes in ionic concentrations, and (c) electrical field interaction (ephaptic transmission). Recently, much attention has been paid to the possible role of ephaptic transmission in the physiology of invertebrate as well as vertebrate neurons, and in their pathophysiology (e.g., epilepsy). The aim of this study is to review the basic biophysical and mathematical methods of electrical-field interaction modeling. We start from biophysical equations used to describe electrical fields in nervous tissue, viz. Maxwell's equations, and their simplification due to Lorente de No, and specific equations for membranes, either linear ("passive") or nonlinear ("active"). We then derive general PDEs for electric fields and ephaptic transmission in a neural tissue, together with suitable *ad hoc* boundary conditions. We show that this approach is strongly correlated to the so-called "Current Source Density Analysis" used in the study of extracellular fields. In the second part, we study the main methods for solving these PDEs. On the one hand, analytical methods have been used in the case of very simplified models (e.g., one-dimensional with passive membranes). On the other hand, numerical methods are based on finite-difference schemes (sometimes called "compartmental analysis"), and have been applied to large sets of neurons. Finally, we review some examples of the application of modeling methods to real neural tissue, for example, recent computational approaches to ephaptic interactions in bundles of demyelinated fibers.

ACKNOWLEDGMENT

We thank O. Parkes for editing the English text.

15 Cortical Pyramidal Cells

Roger D. Orpwood

CONTENTS

15.1 INTRODUCTION

The main input/output cells of the cortex are the pyramidal cells (Shepherd, 1979) and these cells are the subject of the work described in this chapter. The pyramidal neuron is primarily involved in receiving information from other cells and making decisions on the basis of that information and its past experience about whether to provide an output or not. It is subject to developmental processes like all cells and the mature configuration of the cell is one to which it adapts following its early experiences, possibly in a Darwinian fashion (Edelman, 1987). However, even when the cell has achieved its mature status it is in a constant dynamic state of adapting to its experiences. Its responses to incoming information are not rigid but constantly changing. Any model of the mature pyramidal cell must include all the main factors that enable it to behave in this way.

Neurons, and pyramidal cells in particular, have a unique structure and there must be an important link between their cellular geometry and the way they behave. The input side of the cell in particular, with its vast arborizations of dendrites providing a very large surface area for interactions with other cells, must be emulated in any model of the cell. In addition many pyramidal cells have a dendritic structure that is divided into two parts. Proximal and basal dendrites are present close to the cell body, echoing the kind of dendritic structure seen in most other neurons, but there are also apical dendrites separated from the cell body by a long climbing fiber. The cell appears to be divided into two separate dendritic trees; the proximal dendrites around the cell body, and the distal dendrites, with the apical climbing fiber linking the two. This geometry is a highly distinctive feature of the pyramidal cell and must be included in any model that simulates its behavior.

The adaptive capabilities of pyramidal cells result from the properties of the regions of the cell that receive incoming signals. All facilitatory inputs to the cell occur at dendritic spines. The spine has a very typical structure with a spine-head receptor region connected to the dendrite via the spine shaft. Individual receptor sites on dendritic spines have been shown to have the capability to change their sensitivity to inputs, and these sensitivity changes can remain over a long period (Baudry and Davis, 1991; Madison et al., 1991). It is widely suspected that these long-term changes underlie information storage and retrieval in the cortex, and constitute the biological basis of memory (Bliss and Collingridge, 1993).

Facilitatory signals are transferred from incoming axons to the pyramidal cell by means of the excitatory transmitter glutamate. The glutamate receptors are through-membrane ion channels in the dendritic spines. They can be generally classified into two main types. The first (the AMPA receptor) is simply opened by the glutamate transmitter, allowing Na^+ ions through the membrane at the receptor site and causing a local depolarization. The second kind of receptor (the NMDA receptor) has a more complex response. It is also a through-membrane ion channel but at resting membrane potential the channel is blocked by Mg^{++} ions. Consequently at resting membrane potential, glutamate causes no response from the receptor. Only when the local membrane has been depolarized by around 30 mV is the Mg^{++} block removed and the ion channel can then be opened up by afferent glutamate (Mayer et al., 1984; Nowak et al., 1984). The activated channel then allows an influx of both Na^+ and Ca^{2+} ions (Regehr and Tank, 1990), again leading to a local membrane depolarization. The Ca^{2+} influx has another effect. Calcium acts as a second messenger within the cell and provides the starting point for a series of intracellular reactions. The end point of this chain of reactions is a change in the sensitivity of the AMPA receptors on (mostly) the same spine (Bekkers and Stevens, 1989).

The receptor mechanisms described above allow the cell to associate one activation with another. Depolarization of the local membrane potential by more than 30 mV implies that the cell is receiving a large significant input elsewhere. Any local input received at the same time will ultimately lead to the AMPA receptors in the spine increasing their sensitivity. Any subsequent input to the same spine will then lead to a larger depolarization from its AMPA receptors. The ability to associate information in this way is a key aspect of the studies discussed in this chapter.

Therefore it appears that there are three important properties of individual pyramidal cells that should be included in any realistic model:

1. The model must include a realistic representation of the cell's geometry.
2. The model must include the mechanisms that exist at individual receptor sites to enable the transformation of the input chemical signal to an electrical one, and must include the ability to change the sensitivity of the receptor regions depending on their experiences.
3. The model must include the capabilities of the cell to communicate intracellularly. The communication must enable the activities at a given receptor site to be transmitted to other receptor regions and it must enable this activity to be transmitted to the soma of the cell.

This chapter explores all these core aspects of individual pyramidal cells.

Detailed studies have been carried out on the behavior of pyramidal cells, both in terms of their single-cell behavior and in terms of the more detailed functioning of their component parts (see Creutzfeldt, 1995). The problem for any modeling study is that not all of the myriad properties demonstrated can be included and there is always the danger that some key factor will be left out. Nevertheless the models do provide an extension of our cognitive capabilities to mentally model the behavior of pyramidal cells, given the measured capabilities of their component parts. As long as care is taken over the choice of parameters included in the model the results should provide some insight into properties of the system being modeled and enable experimental work to be more closely focused.

Pyramidal cells are the main information processing cells in the cerebral cortex and the way in which they interconnect is clearly of importance to cortical function. The basic structure of the cortex is a hierarchy, or series of hierarchies, of networks of pyramidal cells. The pyramidal cells have a very distinctive connectivity within these networks, and also between the network layers. Connections from network layers lower down in the hierarchy tend to synapse with dendrites near the cell body. Connections from layers higher up tend to synapse with dendrites in the apical tuft. The interaction between these inputs is likely to have a major impact on the pyramidal cell's ability to carry out learning. This chapter illustrates the importance of sophisticated neuron models by showing how they can be used to explore these interactions between different parts of the distinctive pyramidal cell geometry.

In this chapter there is a progression from a sophisticated model using simulations of physiological processes to a much simpler model that uses empirical equations to describe pyramidal cell behavior. Other researchers have followed the same process (Traub et al., 1987b; Bush and Sejnowski, 1993; Mainen and Sejnowski, 1996; Poirazi et al., 2003). The more sophisticated models enable the influence of larger numbers of variables to be explored. Once the behavior of these models has been examined, and the key variables affecting this behavior have been identified, the models can be expressed in an empirical form. The models can be simplified and checked to make sure the same key properties can still be demonstrated. The simpler models can then be used in studies of larger-scale biological systems. It is felt that this process of building up from models of detailed intracellular biological processes to models of complex cellular systems is a useful one in that it enables the key properties of each level of complexity to be elucidated and ensures that the larger-scale models are based on realistic assumptions about the properties of their component elements. As far as the single cell is concerned then, the model evolves from a physiological model of the cellular system to a much simpler empirical model that demonstrates the same key properties. This chapter will describe the development of the physiological model and discuss the simplifications embodied in the empirical one. It will compare the performance of the two models to demonstrate that the simpler one is a useful basis on which to explore larger-scale cortical simulations.

15.2 Compartmentalization of Dendrites and Dendritic Spines

15.2.1 Cable Model of Dendrites

Information is communicated at a high speed throughout the cell by means of changes in the membrane potential. This communication is both a passive transfer of potential and an active process, depending on the region of the cell involved (Llinás, 1988). There have been many studies of the way that transient membrane potentials are transmitted around neurons, mostly building on the pioneering studies of Rall (1977).

The potential distribution over a dendrite can be described by the one-dimensional cable equation with synaptic reversal potentials:

$$\frac{a}{2R_{\mathrm{i}}}\frac{\partial^2 V}{\partial x^2} = I_{\mathrm{m}} = C_{\mathrm{m}}\frac{\partial V}{\partial t} + I_{\mathrm{i}} + I_{\mathrm{syn}}, \tag{15.1}$$

where V = potential difference across the membrane, in terms of the departure from its resting value, a = radius of dendrite, R_i = specific resistance of the cytoplasm, x = distance along the dendrite, C_m = membrane capacitance, I_m = membrane current density, I_i = current carried by the movement of ions through the membrane, and I_syn = junctional (synaptic) current through receptors.

Analytical solutions to Equation (15.1) have been derived for many simple cellular geometries (Rall, 1977), and simplifications have provided analytical solutions to more complex but still basic geometries (Rall and Rinzel, 1973; Tuckwell, 1985; Stratford et al., 1989; see also Chapter 8). It is very important to note that in Equation (15.1) both the ionic and synaptic conductances are assumed to be continuous functions of position. In reality the conductivity of the membrane is of discrete, not continuous character and therefore to allow for the discrete (noncontinuous) nature of biological membrane conductance, the last two terms on the right-hand side (r.h.s) of Equation (15.1) each need to be multiplied by a series of one-dimensional Dirac delta functions in space and separated by a specific ansatz uniformly or nonuniformly separating ionic channels and synaptic receptors (see Mozrzymas and Bartoszkiewicz, 1993).

15.2.2 Compartmental Model of Dendrites

This section describes the basis for the powerful numerical technique of compartmental modeling, which enables the membrane potential distribution to be calculated over any complex neuron geometry. Several software packages are available, which enable this modeling technique to be applied without involving an understanding of its mathematical basis, such as GENESIS (Bower and Beeman, 1994), Surf-Hippo (Borg-Graham, 1995), and NEURON (Hines and Carnevale, 1997). NEURON in particular has a good graphical user interface aimed at enabling neuroscientists to apply these modeling techniques using the terminology and methodologies they are familiar with.

The work described in this chapter uses a compartmental model to simulate the transient potential changes throughout the cell (see Rall, 1964; Dodge and Cooley, 1973; Carnevale and Lebeda, 1987; De Schutter and Steuber, 2000a, for discussion). The compartmental approach divides the cell into a large number of small compartments that are all assumed to be spatially isopotential. So for the jth dendritic compartment,

$$C_\mathrm{m}\frac{\mathrm{d}V_j}{\mathrm{d}t} = I_{\mathrm{m}_j} - I_{\mathrm{i}_j} - I_{\mathrm{syn}_j}. \tag{15.2}$$

The potential difference between compartments, and between the compartment and the extracellular space, can be used to calculate the movements of charge between the compartments and out of the cell. From Figure 15.1 the membrane current density I_m for the jth compartment is given by the difference between the current entering from compartment $j-1$ and that leaving to compartment $j+1$, divided by the surface area of the compartment.

$$\begin{aligned}
I_{\mathrm{m}_j} &= \frac{(I_\mathrm{in} - I_\mathrm{out})}{2\pi a\Delta x} \\
I_\mathrm{in} &= \frac{(V_{j-1} - V_j)}{\Delta x R_\mathrm{i}/\pi a^2} \quad \text{and} \quad I_\mathrm{out} = \frac{(V_j - V_{j+1})}{\Delta x R_\mathrm{i}/\pi a^2} \\
I_{\mathrm{m}_j} &= \frac{a}{2R_\mathrm{i}}\frac{(V_{j+1} - 2V_j + V_{j-1})}{(\Delta x)^2}.
\end{aligned} \tag{15.3}$$

The membrane ionic current I_i comprises two main components: I_leak, the passive leakage current, and I_act, the voltage-gated current through various membrane ion channels.

$$\begin{aligned}
I_\mathrm{i} &= I_\mathrm{leak} + I_\mathrm{act} \\
&= G_\mathrm{leak}(V - E_\mathrm{leak}) + G_\mathrm{act}(V - E_\mathrm{act}).
\end{aligned} \tag{15.4a}$$

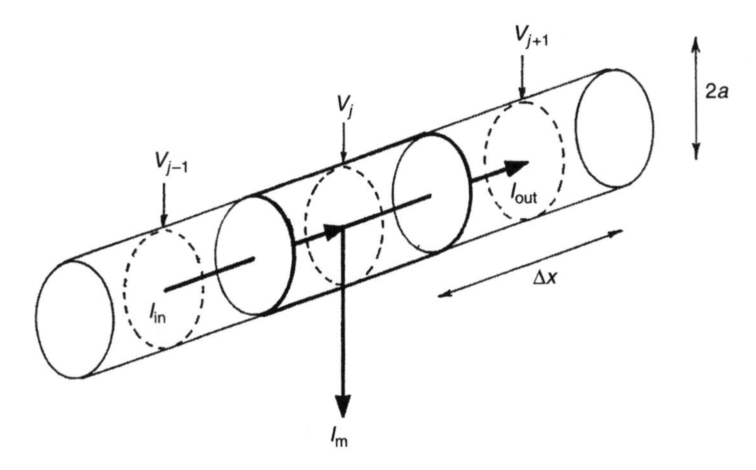

FIGURE 15.1 Diagram of a short length of dendrite illustrating the main variables used in the compartmental model.

The ligand-gated synaptic current, I_{syn}, is based on simple first-order kinetics (e.g., Destexhe et al., 1994).

$$I_{syn} = G_{syn}S(V - E_{syn})$$
$$\frac{dS}{dt} = \alpha[T](1 - S) - \beta S, \tag{15.4b}$$

where G is the transmembrane conductance and E is the reversal potential for the receptors and ionic species involved. $\alpha = 1.1 \times 10^6\,M^{-1}$ (sec^{-1}), $\beta = 190$ (sec^{-1}), $[T] = 1$ mM, and S is the synaptic activation variable. It is important to remember that many receptor systems generate second messengers that are sensitive to the local membrane potential (e.g., see Orpwood, 1990).

In the first instance just consider the passive leakage component I_{leak} of the membrane ionic current. The reversal potential for leakage current is zero, so

$$I_{leak} = G_{leak}V = \frac{V}{R_m}, \tag{15.5}$$

where R_m is the membrane resistance.

The rate of change of membrane potential can be expressed as a simple difference approximation of the first derivative with respect to time:

$$\frac{dV_j}{dt} \approx \frac{V_j^{t+1} - V_j^t}{\Delta t}, \tag{15.6}$$

where V_j^t is the potential difference at the jth compartment at time t and V_j^{t+1} is the potential difference at time $t + \Delta t$.

The beauty of using the simple approximation in Equation (15.6) is that it is very easy to implement and enables an explicit solution to be provided for V^{t+1} in terms of V^t. Unfortunately the solution is unstable unless the following condition applies (Mascagni and Sherman, 1998):

$$\Delta t \leq \frac{R_i C_m(\Delta x^2)}{a}. \tag{15.7}$$

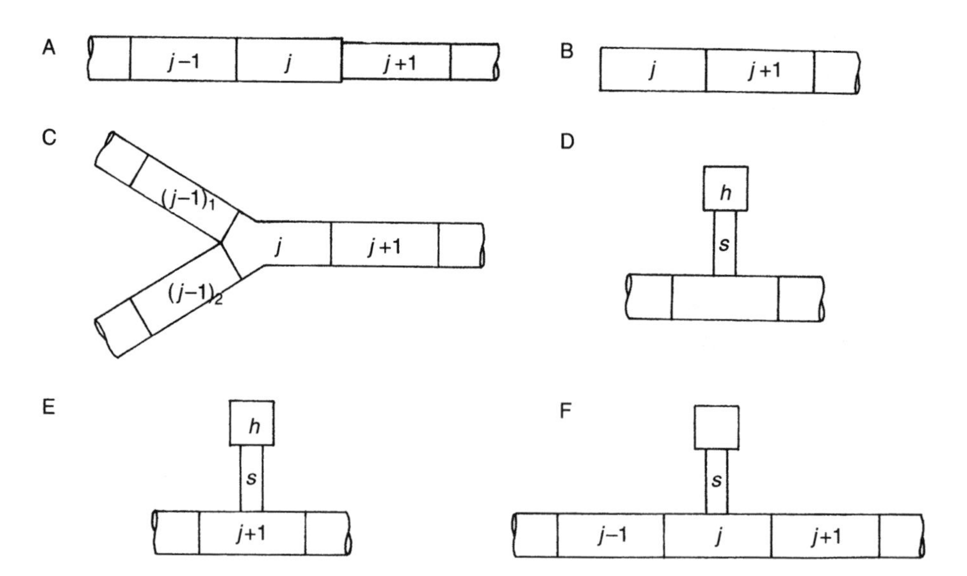

FIGURE 15.2 The main dendritic geometries for which difference equations are derived for the compartmental model.

This stability criterion (Equation [15.7]) inevitably leads to a necessity to use small time steps and consequently makes the numerical model computationally rather expensive.

The difference approximation (Equation [15.6]) and the expressions for membrane current (Equation [15.3]) and leakage current (Equation [15.5]) can be applied to Equation (15.2) to yield a simple expression for the membrane potential of the jth compartment at time $t + \Delta t$:

$$V_j^{t+1} = \left(\frac{\lambda}{\Delta x}\right)_j^2 \frac{\Delta t}{\tau}\left[V_{j+1}^t + V_{j-1}^t - V_j^t\left(2 + \left(\frac{\Delta x}{\lambda}\right)_j^2\right)\right] + V_j^t, \tag{15.8}$$

where $\tau = R_m C_m$ and $\lambda^2 = aR_m/2R_i$.

Equation (15.8) enables transient potential changes to be plotted along a single cable. The geometry of the neuron may be complex with many bifurcations and changes in diameter of dendrites and so similar expressions need to be derived for compartments in different regions of the cell. Figure 15.2 illustrates the main geometries needed for this study and the difference equations that result are provided below.

Figure 15.2A, different diameters

$$V_j^{t+1} = \left(\frac{\lambda}{\Delta x}\right)_j^2 \frac{\Delta t}{\tau}\left[\alpha V_{j+1}^t + \beta V_{j-1}^t - V_j^t\left(\alpha + \beta + \left(\frac{\Delta x}{\lambda}\right)_j^2\right)\right] + V_j^t, \tag{15.9}$$

where $\alpha = 2/(1 + (a_j/a_{j+1})^2)$ and $\beta = 2/(1 + (a_j/a_{j-1})^2)$.

Figure 15.2B, dendrite ends

$$V_j^{t+1} = \left(\frac{\lambda}{\Delta x}\right)_j^2 \frac{\Delta t}{\tau}\left[\alpha V_{j+1}^t - V_j^t\left(\alpha + \left(\frac{\Delta x}{\lambda}\right)_j^2\right)\right] + V_j^t. \tag{15.10}$$

Figure 15.2C, bifurcations

$$V_j^{t+1} = \left(\frac{\lambda}{\Delta x}\right)_j^2 \frac{\Delta t}{\tau}\left[\alpha V_{j+1}^t + \gamma_1 V_{(j-1)_1}^t + \gamma_2 V_{(j-1)_2}^t - V_j^t\left(\alpha + \gamma_1 + \gamma_2 + \left(\frac{\Delta x}{\lambda}\right)_j^2\right)\right] + V_j^t, \tag{15.11}$$

where $\gamma_1 = 2/(1 + (a_j/a_{(j-1)_1})^2)$ and $\gamma_2 = 2/(1 + (a_j/a_{(j-1)_2})^2)$.

Equations (15.8) through (15.11) provide the basis for a numerical model of the potential distribution over a pyramidal cell that takes into account its complex geometry.

15.2.3 Model of Dendritic Spines

Several studies have explored the significance of dendritic spinal structure for the functioning of the spinal receptors and their communication with the dendrite (Brown et al., 1988; Zador et al., 1990). When the spine receives an input, a large potential difference can be set up between the spine-head membrane and the local dendritic membrane. This potential is important for the way that currents are injected into the dendrite and for the way the receptor region is able to change its sensitivity depending on prior activity, as will be described below. Consequently, each spine that receives an input in the neuron model must be modeled explicitly. In the model used here, each spine is modeled as two compartments, one representing the spine head and one the spine shaft. These compartments are illustrated in Figure 15.2 and the additional difference equations are as follows.

Figure 15.2D, spine heads

$$V_h^{t+1} = \left(\frac{\lambda}{\Delta x}\right)_h^2 \frac{\Delta t}{\tau}\left[\left(E - V_h^t\right)\rho G + \eta_A V_s^t - V_h^t\left(\rho G + \eta_A + \left(\frac{\Delta x}{\lambda}\right)_h^2\right)\right] + V_h^t, \tag{15.12}$$

where $\eta_A = 2/(1 + (a_h/a_s)^2(\Delta x_s/\Delta x_h))$ and $\rho = R_i \Delta x_h/\pi a_h^2$. G is the conductance of the spine head and E the reversal potential, as discussed below.

Figure 15.2E, spine shafts

$$V_s^{t+1} = \left(\frac{\lambda}{\Delta x}\right)_s^2 \frac{\Delta t}{\tau}\left[\eta_B V_{j+1}^t + \eta_C V_h^t - V_j^t\left(\eta_B + \eta_C + \left(\frac{\Delta x}{\lambda}\right)_s^2\right)\right] + V_s^t, \tag{15.13}$$

where $\eta_B = 2/(1 + (a_s/a_h)^2(\Delta x_h/\Delta x_s))$ and $\eta_C = 2/(1 + (a_s/a_{j+1})^2(\Delta x_{j+1}/\Delta x_s))$.

Figure 15.2F, parent dendrite

$$V_j^{t+1} = \left(\frac{\lambda}{\Delta x}\right)_j^2 \frac{\Delta t}{\tau}\left[\alpha V_{j+1}^t + \beta V_{j-1}^t + \eta_D V_s^t - V_j^t\left(\alpha + \beta + \eta_D + \left(\frac{\Delta x}{\lambda}\right)_j^2\right)\right] + V_j^t, \tag{15.14}$$

where $\eta_D = 2/(1 + (a_j/a_s)^2(\Delta x_s/\Delta x_j))$.

If all the spines present on the neuron had to be modeled explicitly it would complicate the model enormously as each cortical pyramidal cell has around 10,000 spines arranged over its dendrites. Fortunately, when current is passing from the dendrite to the spine the situation is simplified by the fact that only a tiny potential difference exists between the dendrite and the spine head (Jack et al., 1975; Segev and Rall, 1988). Consequently, the presence of the spines can be modeled implicitly through appropriate changes in the properties of the compartments to which the spines are linked. The method used in this work used changes in the specific membrane properties so as to effectively include the area of the spine in the model (Holmes, 1989):

$$R_{\mathrm{m}}^{*} = \frac{R_{\mathrm{m}}}{F}, \qquad C_{\mathrm{m}}^{*} = C_{\mathrm{m}}F, \qquad F = \frac{A_{\mathrm{d}} + A_{\mathrm{s}}}{A_{\mathrm{d}}},$$

where A_{d} is the area of the dendrite and A_{s} the area of the spine.

The transformed values of R_{m} and C_{m} can then be used in the dendritic compartment to which the spines are attached. The value of F depends on the density of spines existing in the dendritic compartment. The values used in this study took into account the fact that the density varies over the pyramidal cell (Feldman, 1984) and they are indicated in Figure 15.3. The transformation only

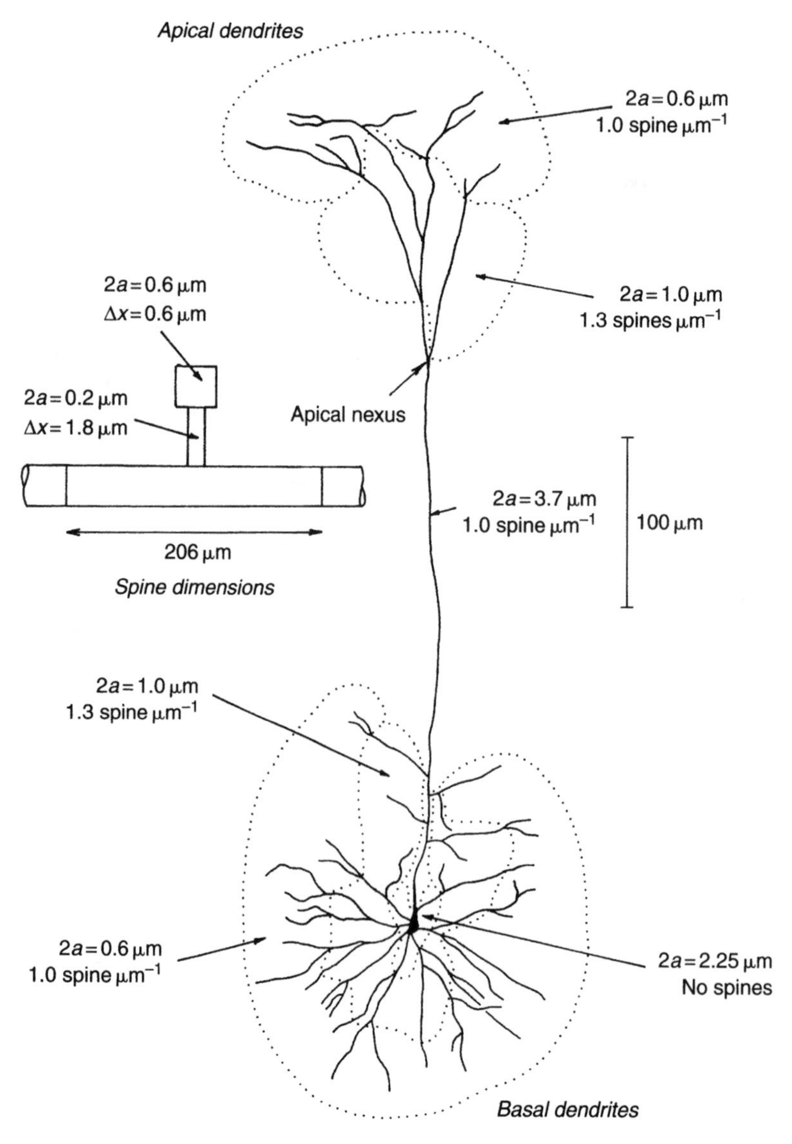

FIGURE 15.3 Diagram of the pyramidal neuron that was modeled, showing the main geometric variables and their distribution over the cell.

works, of course, if the membrane properties of the dendrite and the spine are the same. Alternative methods for incorporating the spines into the dendritic compartment have been discussed by Stratford et al. (1989).

The values of R_m, R_i, and C_m are crucial to the performance of the numerical model. The membrane capacitance has been widely accepted as having a value around 0.01 F m^{-2} (Cole, 1972) and that value is used in this study. Unfortunately, physiological measurements cannot provide unique values for the two resistances. The measured data can be matched to a wide range of pairs of values for R_m and R_i. Stratford et al. (1989) and Major et al. (1994) injected a brief current pulse into a pyramidal cell and then compared the measured voltage response to that predicted by a simplified model of the cell. All the major parameters were allowed to vary within an optimization routine until an optimal least-squares fit was obtained. Cauller and Connors (1992) used a fixed range of R_i values and adjusted the other parameters to fit the observed input resistance and time constant. Their values for R_m and R_i pairs varied from 0.8 Ω m^2 and 0.69 Ω m, respectively, to 1.73 Ω m^2 and 4.0 Ω m. The authors mentioned above showed that the time courses of potential changes in the soma predicted by compartmental models were not unduly affected by the range of resistance values used, except following current injections to the more distal dendrites. Unfortunately inputs to the distal regions are of great interest. Stuart and Spruston (1998) compared patch-clamp recordings at the soma and dendrites with a morphologically realistic compartmental model to predict a low R_i coupled to multiple resting conductances to increase attenuation in the dendrites.

As the cell used in this study was the one published by Cauller and Connors (1992) the values for R_m and R_i finally chosen were a pair from their study, with $R_m = 0.8159$ Ω m^2 and $R_i = 1.81$ Ω m. Until further work is completed on exploring more definitive values of R_i for cortical pyramidal cells, the uncertainty over these two resistances will remain a major problem for compartmental modeling of these neurons.

15.3 Receptor Dynamics and Active Conductances

15.3.1 Modeling Adaptive Receptors

The main reactions taking place following activation of the glutamate receptor are explored in detail in Orpwood (1988, 1990), where simulations demonstrate the association properties discussed in Section 15.2.2. For the current work it was felt that the principles demonstrated by these more comprehensive biochemical simulations matched the measured response of receptors well enough for simplifications to be introduced. The receptor dynamics used in these whole cell simulations therefore were empirical in nature as shown below.

As indicated in Equation (15.4b) the synaptic current that passes into the cell through the receptors is as follows (Jack et al., 1983):

$$I_{syn}(t) = G(t)[V(t) - E],$$

where $I_{syn}(t)$ is the synaptic current at the spine head, $G(t)$ the membrane conductance caused by the activation of the glutamate receptor, $V(t)$ the local membrane potential, and E the reversal potential for the channel given the ion species being admitted.

The actual value for the reversal potential will depend on the Nernst potential for the ion species involved. Most measurements of the reversal potentials for NMDA and AMPA channels indicate values close to zero (Hestrin et al., 1990; Keller et al., 1991) and the values used here are those adopted by Jensen and Shepherd (1993) of 70 mV relative to resting potential for the AMPA channel and 75 mV for the NMDA.

The channel conductance change, $G(t)$, encompasses the synaptic activation variable, S, expressed in Equation (15.4b), and uses an empirically derived time course that has been used by several authors (e.g., Rall, 1967; Jack et al., 1975):

$$G(t) = \varepsilon t \exp\left(-\frac{t}{t_{\mathrm{p}}}\right), \tag{15.15}$$

where t_{p} is the time to the peak conductance change. A value of 0.5 msec is used for the AMPA channel and 2.5 msec for the NMDA. The constant, ε, is set at 5.437×10^{-6} to provide a peak conductance change of 1 nS for the AMPA channel.

If the local membrane depolarization is less than 30 mV then only the AMPA channel is activated and the local depolarization provides an input into the compartmental model and is conducted elsewhere. However, if the local membrane is depolarized by more than the crucial 30 mV, the NMDA channel is also activated and this results in a sensitivity change in the AMPA channel. The feedback link between the activation of the NMDA channel and the AMPA sensitivity change was modeled physiologically in Orpwood (1990) but an empirical algorithm was incorporated here. The main ion species admitted by the NMDA receptor are Na^+ and Ca^{2+} although the crucial process is the admittance of Ca^{2+} as this ion is the second messenger that initiates the AMPA sensitivity changes. It is assumed that conductance time courses for the two ions are the same so that the empirical expression for the conductance change (Equation [15.15]) can be assumed to reflect the level of activation of the glutamate receptors. While the local membrane was depolarized by more than 30 mV, the NMDA receptor membrane current, therefore, modeled the rate of admittance of Ca^{2+}. The local buildup of Ca^{2+} is derived in the model by summing the membrane current at each computational time step:

$$[\mathrm{Ca}^{2+}]_{\mathrm{local}} \propto \sum I_{\mathrm{NMDA}} \quad \text{while} \quad V_h > 30\ \mathrm{mV},$$

where $[\mathrm{Ca}^{2+}]_{\mathrm{local}}$ is the local calcium ion concentration.

This is a gross simplification of the mechanisms involved, as there are many processes taking place in the cell to remove the free Ca^{2+} from the intracellular space (McBurney and Neering, 1987). However, it is felt that the proportionality is a valid one. In the model the increase in sensitivity of the AMPA receptor is made proportional to the change in Ca^{2+} concentration.

15.3.2 Modeling Active Conduction

The other key membrane current presented in Equation (15.4) is the active current passing through voltage-gated membrane ion channels. Useful summaries of different channel properties are contained in Hille (2001), Brown et al. (1990), Reyes (2001), and Migliore and Shepherd (2002). Magee (1999a) contains a good review of voltage-gated ion channels, and Bekkers (2000) contains excellent detail of K^+ channels in layer 5 pyramidal neurons. Many authors have shown the existence of actively conducting regions within pyramidal cells (Llinás, 1988; Huguenard et al., 1989; Regehr et al., 1993; Hirsch et al., 1995). Microelectrode recordings (Pockberger, 1991) together with patch clamping (Stuart and Sakmann, 1994) and fluorescence imaging (Yuste et al., 1994) all indicate that the dendrites of neocortical pyramidal cells are capable of supporting action potentials (see Johnston et al., 1996, 2003 for useful reviews). These studies demonstrate that the dendrites contain one type of Na^+ channel but several types of Ca^{2+} channels, with the distribution of the Ca^{2+} channels varying over the dendritic tree.

Stuart and Sakmann (1994) quite clearly showed the role of Na^+ channels in facilitating active conduction from the soma up to the apical dendritic tuft. In this last study cell firing in the axon hillock was necessary before antidromic Na^+ spikes could be generated to carry the depolarization up into the apical tuft. Rapp et al. (1996) were able to simulate some aspects of these results by stipulating a particular distribution of Na^+ channels.

There are many types of voltage-gated Ca^{2+} channels in the dendrites of pyramidal cells, according to studies exploring their responses to pharmacological agents (Johnston et al., 1996). However, they can be grouped into two main categories. High-voltage activated (HVA) channels require a depolarization from resting of around 20 mV before they will open, whereas low-voltage activated (LVA) channels have a much lower threshold and will open with depolarizations of only a few millivolts from resting potential. HVA channels can be activated by the Na^+ action potentials backpropagating into the dendrites. The LVA channels can be activated by EPSPs and appear to be more involved in the forward propagation of activity initiated in the dendrites by synaptic activity.

However, several recent studies have shown the capability of the neuron to establish active conduction throughout the dendritic tree, both in the proximal and apical regions (Stuart et al., 1997a; Larkum et al., 2001). As was discussed by Cauller and Conners (1992) it seems reasonable to expect that such conductances should allow inputs to the apical dendrites to have a major impact on cell firing rather than just being modulatory. It has been indicated that the active properties of pyramidal neurons are mostly facilitated by Na^+ channels in young and juvenile animals, with Ca^{2+} channels only really coming into their own in the adult cell (Zhu, 2000). During the early stages of the cell's existence, while its receptor regions are still undergoing many changes as the cell learns to respond to significant patterns, it is the Na^+ channels that underlie the active conductance properties of the cell. The main interest of the present work was to explore the capability of the cell to learn new patterns, so it was assumed that a young neuron was being modeled and that the active conductances were generated via Na^+ channels. The cell model needs to include active conductances, particularly those underpinned by Na^+ channels.

The current passing through these active channels is a function of both time and membrane potential. The ionic current passing through a voltage-gated channel, I_{act}, can be expressed by the Hodgkin–Huxley formulation as follows (Hodgkin and Huxley, 1952b):

$$I_{\text{act}_k} = \overline{G}_k m^{p_k} h^{q_k} (V - E_k),$$

where $\overline{G}_k$ is the maximum channel conductance for ion species k, E_k is the reversal potential for this ion species, and m and h are the activation and inactivation variables, respectively. The exponents p_k and q_k are parameters dependent on the particular ion species, for example, for fast sodium channels $p_{\text{Na}} = 3$ and $q_{\text{Na}} = 1$. The values of the activation and inactivation variables were expressed by Hodgkin and Huxley in terms of differential equations:

$$\frac{\mathrm{d}m}{\mathrm{d}t} = \alpha_m(V)(1 - m) - \beta_m(V)m \quad \text{and} \quad \frac{\mathrm{d}h}{\mathrm{d}t} = \alpha_h(V)(1 - h) - \beta_h(V)h,$$

where the values for α and β are empirical functions of the membrane potential adjusted to fit experimental data (see Appendix 15.A3 for further details).

The incorporation of Hodgkin–Huxley kinetics in models of this kind can add considerably to the computational cost of implementation. It has been estimated by Izhikevich (2004) that 1200 floating-point operations are needed per compartment for 1 msec of simulation even when considering only two ion species. In addition, it was felt that much of the channel data needed for a full Hodgkin–Huxley model, and importantly data on channel distributions, were not easily available for the various channel types in pyramidal cells.

The main interest with this study was the need to include regenerative behavior based on Na^+ channels. Consequently an empirical function describing regenerative Na^+ channel conductance changes was used, similar to the approach taken for spine-head conductance changes following the reception of an input. The function's constants were adjusted to match published data on potential changes during action potentials in pyramidal cell dendrites (Stuart et al., 1997a).

We present a simple method for the realistic description of conductance changes referred to as the composite approach; it combines in a one-compartment model elements of both the leaky integrator cell and the conductance-based formalism of Hodgkin and Huxley (1952). Composite models treat the cell membrane as an equivalent circuit that contains ligand-gated synaptic, voltage-gated, and voltage- and concentration-dependent conductances. The time dependences of these various conductances are assumed to correlate with their spatial locations in the real cell. Thus, when viewed from the soma, ligand-gated synaptic and other dendritically located conductances can be modeled as either single alpha or double exponential functions of time (Coop and Reeke, 2001).

It was assumed that every compartment in the cell model had regenerative properties, and the conductance changes that underlay these properties were assumed to follow the same time course throughout the cell, and were approximated by the following empirical relationship:

$$G_{\mathrm{act}}(t) = \alpha_1 t \exp(1 - \beta_1 t) - \alpha_2 t \exp(1 - \beta_2 t). \tag{15.16}$$

Typical values of these parameters for the base of the climbing apical dendrite are:

$$\alpha_1 = 14.0 \times 10^{-5}, \qquad \alpha_2 = 1.2 \times 10^{-5}, \qquad \beta_1 = 5 \times 10^3, \qquad \beta_2 = 2 \times 10^3.$$

Data on the time course of potential changes published by Stuart et al. (1997a) and data on the conductance density in different parts of pyramidal cells enabled values for the constants in Equation (15.16) to be derived for different regions of the dendritic tree. For Na^+, Rhodes and Llinás (2001) estimated a value of 12 mS cm^{-2} for the conductance density at the base of the climbing apical dendrite from the measurements of the rate of rise of the backpropagating action potentials published by Stuart et al. (1997a). Several researchers (Jung et al., 1997; Stuart et al., 1997a; Larkum et al., 1999a, 2001) have reported a reduction in the amplitude of the backpropagating action potential as it travels up the apical dendrite which Rhodes and Llinás (2001) interpret as a reduction in Na^+ conductance density to 8 mS cm^{-2} at the most distal end of the climbing fiber. These authors made estimates of 6 mS cm^{-2} for more proximal dendrites and 2 mS cm^{-2} for the dendrites in the apical tuft.

For each compartment the conductance change described by Equation (15.16) was used to calculate the regenerative current occurring there:

$$I_{\mathrm{act}}(t) \approx G_{\mathrm{act}}(t)[V(t) - E],$$

where t is the time starting from the time when the compartment membrane potential is first depolarized by more than 18 mV. The reversal potential used to calculate active currents in and out of the cell was set to 140 mV relative to resting. When the active current was incorporated into the derivation of the difference equation, the following equation was obtained for the standard compartment:

$$V_j^{t+1} = \left(\frac{\lambda}{\Delta x}\right)_j^2 \frac{\Delta t}{\tau} \left[V_{j+1}^t + V_{j-1}^t - V_j^t \left(2 + \sigma + \left(\frac{\Delta x}{\lambda}\right)_j^2 \right) + E\sigma \right] + V_j^t, \tag{15.17}$$

where $\sigma = G_{\mathrm{act}} \Delta x R_{\mathrm{i}} / \pi a^2$.

15.4 Predicted Response to Simple Input Patterns

15.4.1 Structure of the Model

It was considered important that the neuron model be based on a real cell. The neuron used was illustrated by Cauller and Connors (1992) and was a layer 5 rat cortical pyramidal cell. The simplified geometry of the cell illustrated by these authors was digitized as a two-dimensional image and divided into 280 compartments. Each compartment was 20 μm in length except for the cell body, which was pyramidal in shape, 50 μm in length, and 8 μm across the base. The dendrites varied in diameter depending on how far away they were from the soma. Figure 15.3 summarizes the dimensions used. As discussed above, the membrane properties for each compartment were adjusted to take account of the extra membrane area afforded by the synaptic spines. For external inputs into the spines a marked potential difference developed between the spine head and the dendrite, and consequently extra compartments were added to the model for each input. A total of 28 inputs was used during the explorations of the model behavior, 16 on the distal dendrites and 12 on the proximal ones. The dimensions of the spinal compartments were held fixed and are shown in Figure 15.3.

The compartmental model was implemented in "C" language on a fast PC with two forms of output display. The main display consisted of a to-scale representation of the digitized neuron. Each compartment of the model was separately represented and was given a false color to represent its level of depolarization. The colors were arranged on a logarithmic scale and the values used for the color bands are shown on the calibration photo of Figure 15.4. Thus, as the simulation was being run the spatial changes in membrane potential were illustrated in real time on the monitor. In this way the

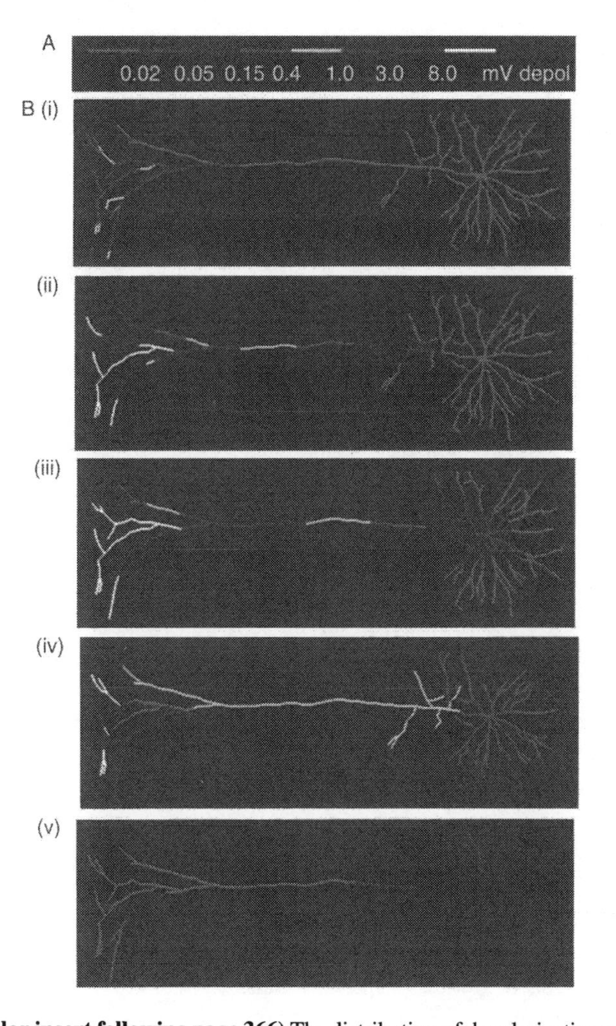

FIGURE 15.4 (See color insert following page 366) The distribution of depolarization over the pyramidal cell as a function of time. (A) Calibration of activity levels on a log scale (see color insert). (B) The distribution following eight inputs to the apical dendrites and only passive conduction (i.e., depolarization of apical nexus not sufficient to initiate action potentials there): (i) 0.4 msec; (ii) 1.4 msec; (iii) 2.5 msec; (iv) 5.0 msec; (v) 14.0 msec. (C) The distribution following eight inputs to the apical dendrites leading to active conduction along the climbing apical dendrite: (i) 1.6 msec; (ii) 2.2 msec; (iii) 2.7 msec; (iv) 5.0 msec; (v) 16.0 msec.

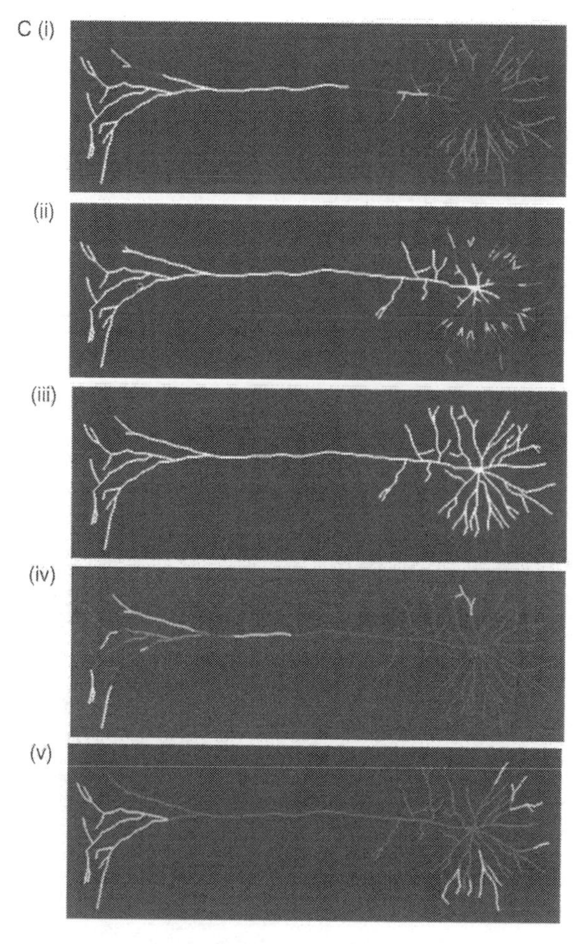

FIGURE 15.4 *Continued*

distribution of membrane potential could be monitored from the initial injection of current until the cell had settled back to its resting state. Using the computer system, 20 msec of cellular activity took about 1 min of display time. This rate was slow enough to see the details of the changes taking place but quick enough not to be frustrating. In addition to the display of potential distribution, the user could call up the time course of potential changes at any point in the cell and the peak values of potential change at that point. These two display techniques gave the user the bulk of the information needed to explore the response of the cell to inputs.

15.4.2 Behavior Following Simple Input Patterns

A simple spatial pattern of eight single inputs was presented to the receptor spines, each with a peak conductance of 1 nS, and the temporal changes of membrane potential were observed. Activity initiated in the distal dendrites quickly spread around the dendritic tuft. Depolarizations of the soma were only a few microvolts because the inputs were small enough not to cause firing in the climbing apical dendrite. Figure 15.4B shows a series of photos of the potential distribution following such a single input pattern to the apical dendrites. Figure 15.5 illustrates the time course of the depolarization achieved, both at one of the spines and at the soma. Similarly, activation of the proximal dendrites caused extensive local activation and depolarized the soma by up to 1 mV, but caused little depolarization up the climbing apical dendrite if the soma was not triggered to

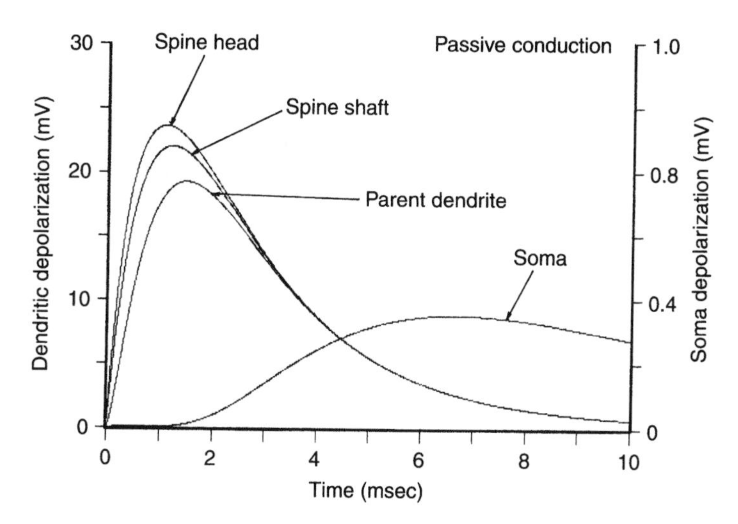

FIGURE 15.5 Time course of depolarization at several points on the cell following eight inputs to the apical dendrites and assuming passive conduction only (note different scales for somatic and dendritic depolarizations).

generate an action potential. These results agree with previous simulations run on simpler cell models (Orpwood, 1994) and with the simulations run by Cauller and Connors (1992), and follow the results of Stratford et al. (1989), Holmes and Woody (1989), and Rhodes and Llinás (2001). The peak values and time course obtained for soma depolarizations following activation of the proximal dendrites are close to those measured in hippocampal pyramidal cells (Barrionuevo et al., 1986; Anderson, 1990) and a time-to-peak similar to that measured in neocortical pyramidal cells (Artola et al., 1990; Larkum et al., 2001). All these results instil confidence as to the validity of the model and enable the intended work on pattern recognition to proceed.

15.4.3 Response to Active Conduction

By involving active conduction, there was improved communication between the soma and basal dendrites and the apical dendrites. The behavior of the cell following eight inputs to apical dendritic receptors, each with a peak conductance of 3.5 nS, is illustrated in Figure 15.6. Inputs to the distal apical dendrites caused EPSPs that were always quite large because of the small diameter of the dendrites in this region. The initial region of the climbing apical dendrite, just proximal to the apical nexus, was much influenced by these potential changes. As the depolarizations in the distal apical dendrite followed their time course, the climbing dendrite progressively depolarized until threshold was reached and a Na^+ action potential occurred. Once the action potential was initiated, it progressively activated subsequent sections of the climbing fiber and rapidly propagated to the soma where it fired the cell body. The steady depolarization of the initial section of the climbing apical fiber that led to its triggering an action potential was also reported by Rhodes and Llinás (2001). With the regenerative conductance algorithm in place, it was relatively easy to generate action potentials in the climbing fiber from inputs to the distal dendrites. This result was also reported by Rhodes (1999). If the soma of the cell fired an action potential it was backpropagated up the climbing fiber and into the distal dendrites. Cell firing spread efficiently throughout the cell, ensuring that this significant cell event was communicated to all receptor regions.

If inputs to the distal dendrites initiated action potentials down the climbing fiber, they would cause soma firing, and this in turn could lead to oscillatory behavior. The action potential that was backpropagated up the climbing fiber could initiate another large depolarization of the distal dendrites. This depolarization could generate another action potential down the climbing fiber leading to further soma firing. This oscillatory behavior was very reminiscent of the burst firing of some

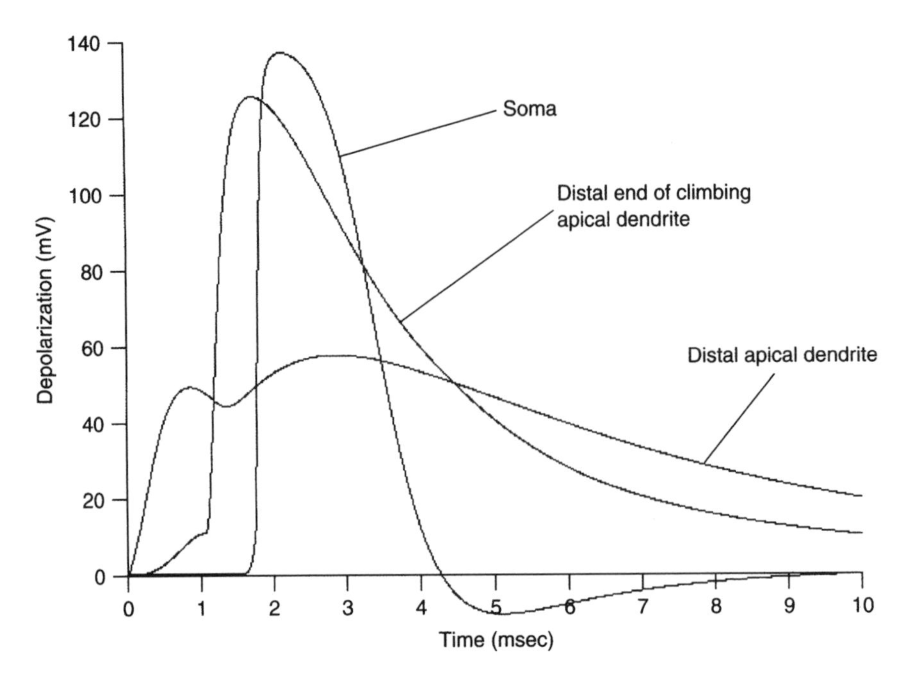

FIGURE 15.6 Time course of depolarizations at different parts of the cell following eight inputs to the apical dendrites sufficient to cause an action potential at the apical nexus.

layer 5 pyramidal neurons in the neocortex. The layer 5 bursting cells tend to be ones with elaborate apical dendritic tufts, as modeled here (Larkman and Mason, 1990). The oscillations were also noted by Rhodes and Llinás (2001) in their simulations. Several authors have shown that dendritic current injection can lead to this bursting behavior (Schwindt and Crill, 1999; Williams and Stuart, 1999). Larkum et al. (1999b) clearly demonstrated how simultaneous inputs to apical and proximal regions can interact in a facilitatory manner to generate burst firing.

Figure 15.4C shows the distribution of depolarization at several times following the same input pattern as used in Figure 15.4B, but with apical inputs sufficient to cause active firing down the climbing apical dendrite. As can be seen, the depolarization soon spread throughout the basal dendritic field. There is no doubt that the incorporation of active conduction provides a much improved communication from the distal apical dendrites to the soma.

15.4.4 Control of Apical Influence

The climbing apical dendrite is obviously an important channel of communication between the apical dendritic field and the proximal dendrites and the soma. It is also interestingly the site where much of the inhibitory input is received by the cell (Shepherd, 1979; Feldman, 1984). This inhibition seems to be ideally located to have a maximal effect on the communication between the two regions of the neuron. If conduction is blocked at the climbing fiber, then the fact that input patterns have been recognized in the apical dendrites cannot be communicated to the soma.

The ability of inhibition of the climbing fiber to block this communication was explored with the pyramidal cell model. Inhibition was modeled by providing hyperpolarizing currents in the compartments of the climbing fiber. It was assumed the inhibition was being provided by Cl^- channels and the reversal potential for these inputs was set at 20 mV relative to resting potential with a peak conductance change of 1 nS. The inhibitory inputs were provided at the soma end of the climbing apical dendrite and the cell was provided with active conduction in this fiber.

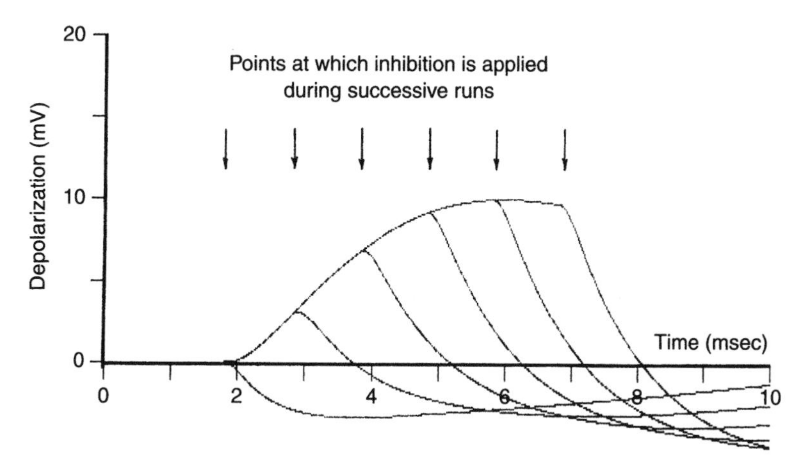

FIGURE 15.7 The effect on somatic depolarization of an inhibitory input to the climbing apical dendrite. The graph shows six successive runs in which the inhibitory input is applied at different times after the arrival of an active spike from the apical dendrites.

The results are shown in Figure 15.7, where the somatic depolarization following activity triggered at the apical nexus is plotted against time. The inhibition was very effective at reducing the depolarization of the soma. The figure shows the combined response of several tests in which the inhibitory input was provided at different delays relative to the arrival of the depolarizing transient. Even with very large delays, the activity of the soma is severely reduced once the inhibitory input is provided.

15.5 Predicted Response of the Model Neuron to Paired Patterns: Learning in the Dendrites

It was hoped that numerical models of pyramidal cells could provide an indication of the pattern-associating capabilities of these neurons. It is felt in this work that even the higher manifestations of cortical function ultimately depend on the ability of cortical networks to process patterns (Orpwood, 1994). However, it is also felt that the key to the ability of cortical networks to carry out their information processing lies with the pattern-processing capabilities of their individual cells. A few studies have been published on the capabilities of single cortical cells to deal with patterns of information (Mel, 1992; Orpwood, 1992).

For one pattern of inputs to be associated with another, the first pattern must cause a large depolarization in the vicinity of the second. The large local depolarization can be generated in two ways:

1. The depolarizations could be caused by activity in spines adjacent to the one in question. If the dendrites received a spatial pattern of inputs which activated a number of adjacent receptors, then the local membrane depolarization would be quite large and would spread to spines in the vicinity.
2. The large depolarization could also arise because the cell has fired. The action potential generated would propagate down the cell's axon but the large depolarization generated in the soma would also transmit antidromically back into the dendrites, causing depolarizations there.

Investigations were carried out to explore the impact on the pyramidal model of both sources of large depolarization described above. The initial study explored the impact of one pattern of strong

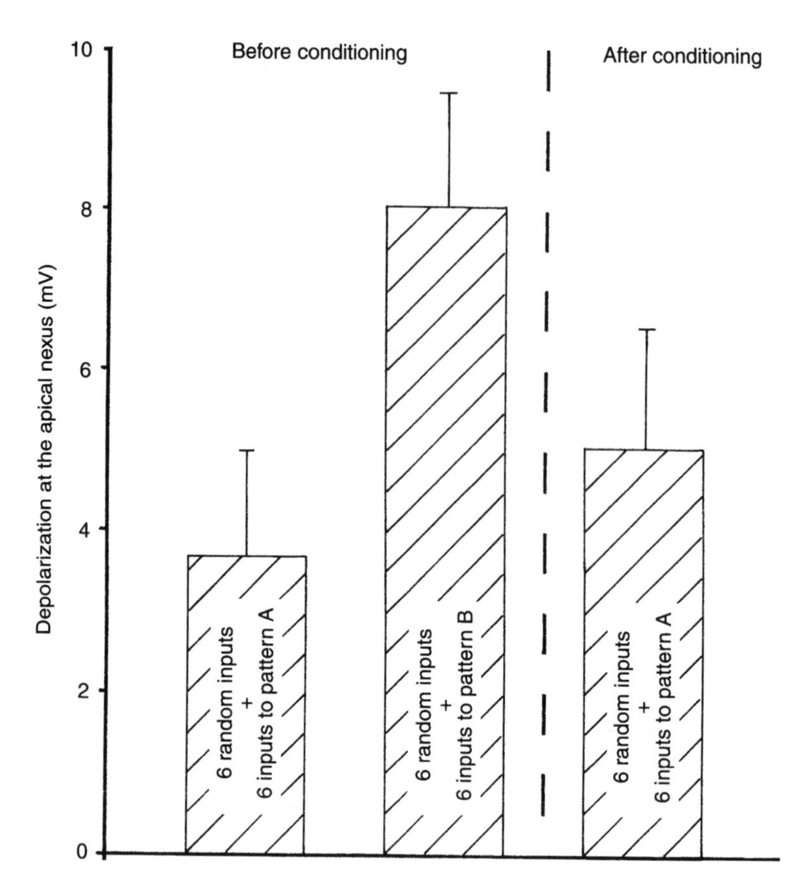

FIGURE 15.8 The results of the conditioning studies. The mean depolarization achieved at the apical nexus is shown before and after conditioning. The conditioning involved simultaneous exposure to pairs of input patterns, one to a set of weak receptors (pattern A) and one including strong receptors (pattern B), as described in the text.

inputs on another pattern of weaker ones, and just focused on patterns presented to the apical tuft. The weaker pattern (pattern "A") consisted of six inputs presented to the same receptors each time. The weaker pattern was always presented together with another six weak inputs chosen at random from the potential 72 receptors in the apical dendrites. Thus, the initial ten runs used inputs to pattern "A" together with another six random inputs to find a control response for the weak input. The mean response at the apical nexus and its standard deviation are shown in Figure 15.8.

The study was then repeated using the pattern of six strong inputs (pattern "B") instead of pattern "A." Each of the strong inputs was presented to a receptor with a peak conductance of 3 nS. The strong inputs were always presented together with another six weak inputs chosen at random. Another series of ten runs were carried out and the response at the apical nexus monitored. Figure 15.8 again shows the mean response and standard deviation. As expected, the pattern of inputs that included the six strong inputs generated a much larger response.

Following these control studies a process of association was carried out in which weak pattern "A" was presented together with strong pattern "B." Following the mechanisms underlying changes in the receptor sensitivity described above, each coincident presentation of the two patterns led to an increase in the sensitivity of the weaker receptors. The coincident presentations were repeated seven times, after which the impact of the association on the cell's response was explored. The initial series of runs was repeated, with pattern "A" presented together with six random inputs. Because of the change in the sensitivity of the receptors activated by pattern "A," the response at the apical nexus was increased. Figure 15.8 again illustrates the mean response and standard deviation.

There is no doubt that the apical dendritic region is able to associate two separate patterns provided at the same time. The depolarization of the apical nexus following only seven presentations of the two patterns shows a very significant increase. The number of receptors activated in these tests is very small and so the resulting depolarizations of the nexus are not large. However, even with only the 12 inputs to the apical tuft used in the above tests a peak receptor conductance of only 2.2 nS for all of them was sufficient to cause an action potential in the climbing apical fiber, which rapidly led to cell firing.

15.5.1 Interactions Between Different Regions of the Cell During Learning

As was mentioned in the Introduction, pyramidal cells have a very distinctive geometry and the way that they are connected to other pyramidal neurons in both the previous network of the cortical hierarchy and the subsequent one is very interesting. Feedforward inputs from networks at a lower level in the hierarchy tend to synapse with dendrites in the more proximal areas near the cell body. Feedback inputs, however, tend to synapse with the distal dendrites (Cauller et al., 1998). From the pyramidal cell geometry it might be thought that the feedback inputs would therefore have less effect than the feedforward ones, and there has long been an assumption that the feedback connections are therefore mostly modulatory. Modeling work such as that presented here shows that inputs to the apical tuft, and therefore feedback connections, are in fact very effective at firing the cell when active dendritic properties are employed.

Given these active properties and the nature of the connections to pyramidal cells, it was considered important to explore how paired inputs to different parts of the cell could interact with each other. A series of runs was carried out where inputs to one part of the cell were used to condition learning among inputs to another region. We asked, for example, whether feedforward inputs to the proximal dendrites could condition learning among feedback inputs to the distal dendrites. Timing is quite important in these interactions, so each pattern of the pair was separated from the other by the delay that was found to provide the greatest change in receptor sensitivity.

The pairs of patterns that were used were as follows:

1. Conditioning from cell firing, and test inputs to the proximal dendrites.
2. Conditioning inputs to the apical dendrites that led to firing in the climbing fiber, and test inputs to the proximal dendrites.
3. Conditioning from cell firing, and test inputs to the apical dendrites.
4. Conditioning and test inputs to the apical dendrites (as discussed above, and at input levels that did not cause action potentials to be generated in the climbing apical fiber).
5. Conditioning and test inputs to the proximal dendrites (again at levels that did not lead to cell firing).

Figure 15.9 shows the results of these series of paired inputs, where the percentage change in input sensitivity is plotted against the number of paired events. As can be seen, the results show that in all cases significant levels of learning could take place. The learning was particularly marked following cell firing or action potential propagation down the climbing fiber. The results show that feedforward input patterns to the proximal dendrites sufficient to fire the cell can very effectively condition the learning of patterns presented by other feedforward inputs to the proximal dendrites. They also show that these feedforward patterns that fire the cell can condition the learning of patterns presented by feedback inputs to the apical dendrites. In addition, the results also show that feedback input patterns to the apical dendrites that lead to firing in the climbing fiber are very effective at conditioning learning in feedforward inputs presented to the proximal dendrites. It is felt that these results, particularly the ability of feedback inputs to condition feedforward ones, has much significance for the way that the cortex operates when learning novel patterns. Several authors have previously conjectured

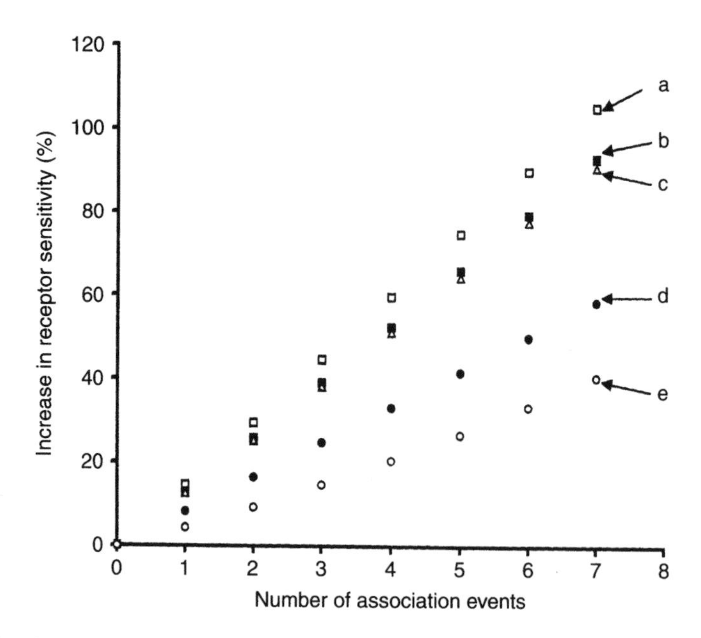

FIGURE 15.9 Graph showing the mean change in receptor sensitivity for six receptors following different numbers of association events. (a) Conditioning from cell firing with test inputs to the proximal dendrites. (b) Conditioning inputs to the apical dendrites (leading to firing in the climbing fiber) with test inputs to the proximal dendrites. (c) Conditioning from cell firing with test inputs to the apical dendrites. (d) Conditioning and test inputs to the apical dendrites. (e) Conditioning and test inputs to the proximal dendrites (no cell firing).

that top–down activity (i.e., feedback to distal dendrites) could be the stimulus that ensures that pyramidal cells can learn patterns presented from bottom up (i.e., feedforward) activity (e.g., see Rolls and Treves, 1998). The fact that the influence of the feedback activity can be very effectively modulated by the levels of inhibition in the climbing fiber provides a possible pointer to mechanisms of attentional control of the learning process.

15.6 A Reduced Pyramidal Cell Model

15.6.1 The Need for a Simple Model

The behavior described in Section 15.5.1 has many implications for the pattern-processing capabilities of networks of pyramidal cells. Feedforward patterns can be recognized by the proximal dendrites and that fact passed on to subsequent layers of cells. Feedback patterns can be recognized by the apical dendrites, and that fact passed on to the cell body but in a conditional way dependent on the level of inhibition in the climbing apical dendrite. Unfortunately, large-scale compartmental models such as the one described are fairly computationally intensive, and if simple networks are to be explored using such models they become unacceptably slow if large numbers of cells are involved. In order for such network studies to be carried out using reasonably representative neuron models, a simplification of the cell model is desired. The subsequent part of this work was therefore to generate such a model with the aim of emulating the behavior of the full-scale compartmental system.

15.6.2 Key Features Indicated by a Comprehensive Model

The studies described above illustrate the key features that enable the pyramidal neuron to process information. The cell appears to have two separate integrating regions; the apical dendrites at a long networks (Amit and Brunel, 1997) and with careful choice of parameters can model experimental data quite closely. For example, Rauch et al. (2003) showed such a model that closely matched *in vivo* data as far as the mean spike frequency-dependent adaptation to soma injected currents was concerned. It is important that these reduced models are able to include all the key features demonstrated by more comprehensive multicompartmental models. Izhikevich (2003) reviewed a number of spiking neuron models to compare their abilities to reproduce observed pyramidal cell behavior and to compare their computational cost. Simplifying neuron models to enable large-scale network simulations is motivating many studies. Poirazi et al. (2003) showed that the firing behavior of a comprehensive neuron model to a wide range of input stimulation could be closely replicated by much simpler models. If the cell was just firing according to stimulus input power then a single-point conductance-based neuron model was good. If the geometry of the cell was important for dealing with patterns of input information, then a simple two-layer network model with a sigmoidal link performed quite well. However, neuron geometry is clearly important when it comes to learning, as are the interactions between different inputs and the interactions between cell firing and these inputs, as shown by the work presented here. Schaefer et al. (2003) showed that the detection of the coincidence between cell firing and dendritic inputs is tuned by their proximal dendritic branching pattern. Indeed in another modeling study, Sakatani and Hirose (2003) showed that aspects such as dendrite/soma communication can be influenced by even the shape of the apical dendrite hillock. Clearly if features such as learning are to be included, then any reduced model has to carefully consider the basis for interactions between inputs as well as the interactions between inputs and cell firing. As Poirazi et al. (2003) concluded, different simplifying abstractions may apply in different neural contexts.

15.8 Conclusions and Future Perspectives

This chapter has described some modeling work that was applied to the most important information processing cell in the cerebral cortex, the pyramidal cell. It has used a straightforward compartmental technique to explore the potential distributions generated over the surface of the cell following input patterns. It has used a hybrid construction in which both receptor dynamics and learning and active conduction within the cell are simply modeled in an empirical fashion. The model is able to demonstrate pattern recognition on the part of the cell. It is also able to demonstrate the ability of the cell to increase its repertoire of learnt patterns through pattern association. Pattern association was shown to be dependent on the good communication between the apical dendrites and the cell body provided by active conduction throughout the cell. The climbing apical dendrite was shown to be ideally placed to allow inhibitory inputs to provide effective control over the communication between the apical dendrites and the soma. The work also demonstrated the ability of inputs to one part of the cell to condition learning of inputs to other parts. Finally, a much simplified version of the cell model was described that could demonstrate the same basic behavior but in a form that increased the speed of computing to a point where large networks of such cells could be efficiently modeled.

Future developments of pyramidal cell models depend on a number of gaps being filled in neurobiological data. Values for membrane resistance are still far from being settled. The lack of information about channel dynamics and distribution inhibits the development of accurate physiologically based models, although as mentioned in the discussion, this problem is rapidly diminishing. And, of course, any properties that are uncovered by cell models need to be verified by experiment before they can be assumed in larger-scale models. It would be useful, for example, to have a more direct demonstration of the nature of the communication between the apical dendritic tuft and the soma, and its impact on learning. The results of Stuart et al. (1997a) and Larkum et al. (2001) described in the discussion are useful indications of the kind of detail about this communication that should be forthcoming. Further modeling refinements probably would involve longer timescale events such as the neuromodulatory effects of some transmitters, and the link between second messenger generation and the laying down of changes in synapse sensitivity.

the climbing apical dendrite. The receptors on each of the dendritic structures were arranged over a two-dimensional grid folded to provide a continuous surface in the form of a toroid. Each of the receptor nodes on the grid provided a local depolarization when activated, modeled as a step change in local potential followed by an exponential decay:

$$V = V_0 \exp\left(\frac{t}{\Delta t}\ln(\phi)\right), \tag{15.18}$$

where V_0 is the step change in local potential at $t = 0$ following an input and ϕ the fractional change at each Δt time step.

The exponential decay (Equation [15.18]) was modeled computationally using shift operators. For example, if the fractional change at each time step was 0.5, then a right shift of one bit divided the operand, V, by two at each computational step and gave a very effective model of the local potential decay that was computationally very fast. Different fractional changes can easily be modeled in a similar fashion (see Appendix 15.A1 for a discussion of the use of shift operators). The computational time steps were chosen to allow the time course of the decrement achieved in this way to match that obtained in the full compartmental model. Steps of 1 msec were chosen.

The potentials at each receptor point were communicated to the other receptor points in their vicinity. Several schemes were explored but the one finally chosen is as below:

$$V_{i,j}^{t+1} = \frac{\left(4V_{i,j}^t + V_{i,j+1}^t + V_{i,j-1}^t + V_{i+1,j}^t + V_{i-1,j}^t\right)}{8},$$

or using shift operators

$$V_{i,j}^{t+1} = \left(V_{i,j}^t \ll 2 + V_{i,j+1}^t + V_{i,j-1}^t + V_{i+1,j}^t + V_{i-1,j}^t\right) \gg 3,$$

where $\ll 2$ signifies a left shift of the operand by two bits (i.e., multiplication by 2^2) and $\gg 3$ signifies a right shift of the operand by three bits (i.e., division by 2^3). See Appendix 15.A1 for more details.

15.6.4 Basic Behavior of the Simple Model

An input event received at any of the receptor positions led to a local depolarization that spread across the sheet of dendritic receptors and ultimately decayed away to zero. Each of the receptor points also incorporated a learning algorithm which ensured that if an input was received when it was locally depolarized below a threshold, then the sensitivity of that receptor point was increased and any subsequent inputs led to larger local depolarizations.

The model also assumed each of the receptor points was at the same electrotonic distance from a summing node. For the proximal dendritic toroid, the summing node represented the soma, or more accurately the axon hillock region. For the apical dendritic toroid, the summing node represented the apical nexus. As with the physiological neuron model, the summing nodes received the activations generated by the receptors and then leaked the summed activation in an exponential fashion. The decay of depolarization at the summing nodes was achieved by using a fractional change of depolarization at each computational time step, in a manner similar to the decay of the receptor activation. The time interval had already been chosen to provide a realistic decay at the receptor sites. Given this time step, the relationship found to provide a realistic decay at the summing node and to maintain stability is given below. Note the use of the shift operator again:

$$V_{\text{sum}}^{t+1} = \left(V_{\text{sum}}^t + \sum V_{i,j}^t\right) \gg 2.$$

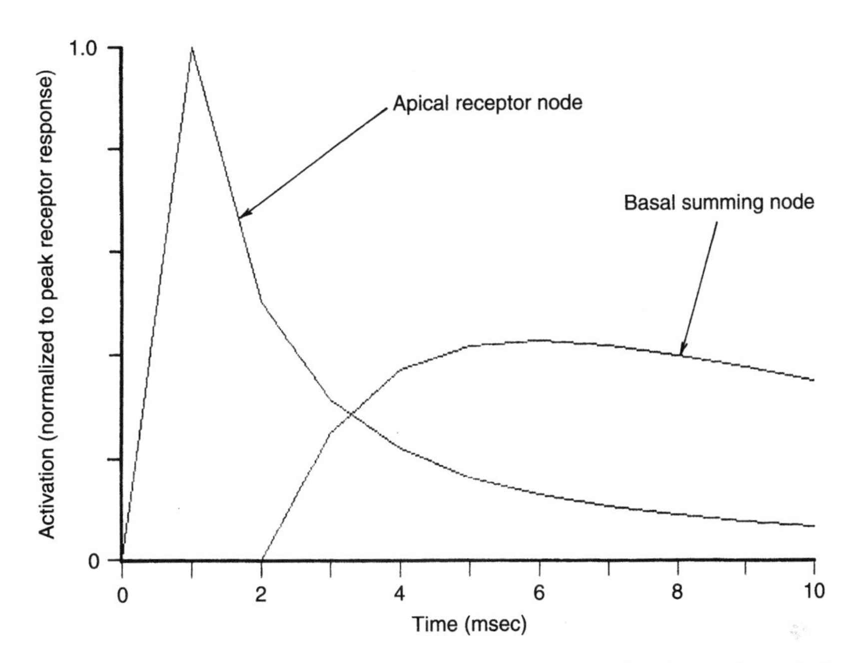

FIGURE 15.11 Time course of activity for the apical receptor node and the basal summing node (soma) using the simple neuron model.

Figure 15.11 shows the time course of depolarization at a receptor node in the apical toroid and at the basal summing node (representing the soma). A very good match is achieved between this much simplified model and the depolarizations produced by the full compartmental model (compare with Figure 15.5).

15.6.5 Adaptive Response

A series of three inputs was provided to low-sensitivity receptor nodes in the apical toroidal net. At the same time, another set of three inputs was provided to receptor nodes that had a sensitivity three times as large. Figure 15.12 shows the activity generated at one receptor following different numbers of these pairings, and as can be seen the response was rapidly enhanced.

The simplified pyramidal cell was exposed to procedures similar to those applied to the compartmental model. A series of six low-sensitivity inputs was provided to the apical toroid, three random and three to a set pattern. Ten sets of inputs were used and the response of the summing node was monitored. The three random inputs were replaced by three inputs to a set of receptors with three times the sensitivity. These paired patterns of strong and weak inputs were repeated three times. The three strong inputs were then replaced by three random ones again and another ten sets of responses were monitored at the summing node. Figure 15.13 compares the responses, and as can be seen the association has led to an enhanced response.

The two-dimensional sheet of receptors therefore formed a very effective pattern-recognizing and learning system. As with the compartmental model, the sensitivity of the receptor regions increased and after conditioning the same patterns of inputs, generated a much larger response at the apical nexus.

15.6.6 Computation Speeds

It is felt that the simplified pyramidal cell model provides a behavior that emulates that of the full pyramidal cell model quite effectively. It demonstrates the key properties demonstrated by the full

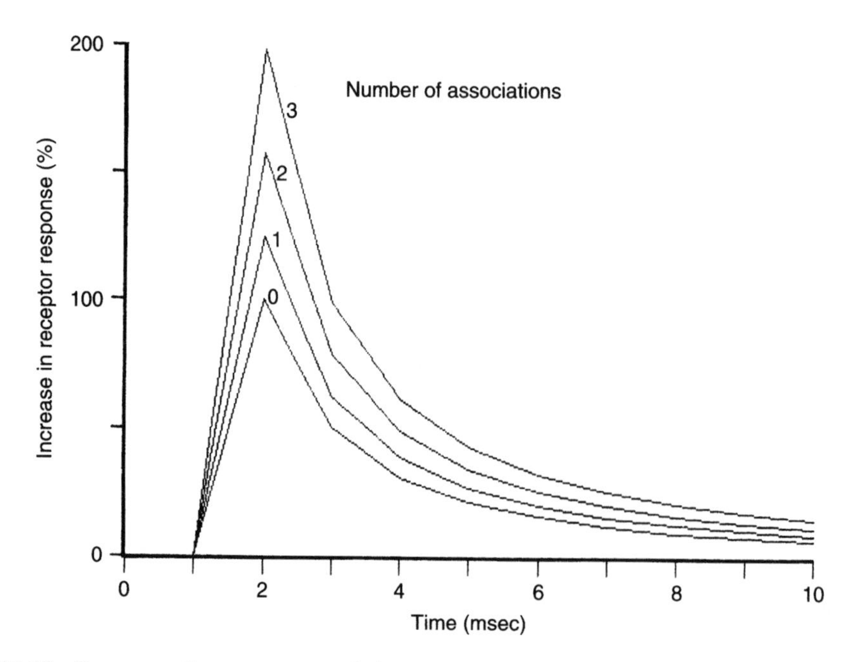

FIGURE 15.12 Response of one receptor node in the simple neuron model following differing numbers of pattern associations.

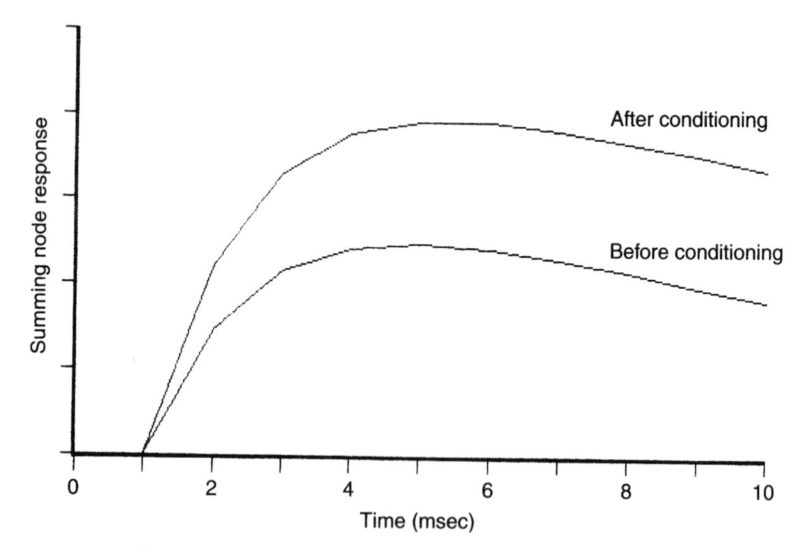

FIGURE 15.13 The mean response at the apical summing node for ten sets of eight inputs. Activity is shown before conditioning and after three pattern associations.

model but at a much reduced cost in terms of computing time. For 10 msec of cell activity in the full model, the simulation took 25 sec using the computer system. With the same computer system and for a cell with the same number of receptor channels, the 10 msec of activity took only 2.5 msec with the simplified model, that is, simulation was four times faster than a real neuron of the same level of complexity. The speeds achieved enable exploration of large networks of such cells to be carried out in the secure knowledge that the individual pattern-processing elements used are reasonable representations of the pyramidal cells involved in biological networks.

15.6.7 Integrate-and-Fire Neuron Models

The described reduced model has many features that are similar to a two-compartment integrate-and-fire neuron model. (For a recent review of such models see Gerstner and Kistler, 2002.) The important feature of the model described in this chapter is the ability of input regions to interact with one another in the same fashion as was found to occur in the full large-scale compartment model. This interaction has been shown to be very important during learning activity. The model described also uses a very simple numerical technique for modeling the time course of changes in membrane potential that makes it very fast. See Appendix 15.A2 for a brief description of integrate-and-fire neuron models.

15.7 Discussion

The aim of the work described in this chapter was to provide a building block that would allow more accurate predictions of cortical network behavior. Much information about cortical structure and properties is now available, but the links between this knowledge and the higher-level capabilities of the cortex are difficult to predict. Simulation studies have an important role to play in exploring these links but the network models must include realistic simulations of neuron behavior. As is so obviously indicated by neurobiological studies, the behavior of cortical neurons is very sophisticated (e.g., Hausser et al., 2000). If the predictions of network models are to be believed, it is essential that their component neuron models reflect the known properties of neuron behavior. Comprehensive neuron models such as that described in this chapter would require too much computing time to form the basis of a large network model and for this reason the reduced model was developed. Therefore, the ultimate aim of the work described is to enable the exploration of cortical networks, but for this to be done in a meaningful way the comprehensive model of the pyramidal cell has to be shown to reflect biological behavior, and the reduced model has to be shown to reflect the core properties of the more comprehensive one. Only in this way can large-scale network models be generated with confidence.

A major theme of this chapter concerned interactions of activity in different parts of the neuron to enable associative memories to be established. Such learning behavior is clearly a major role of pyramidal cell activity. The work predicted the ability of the cell to associate one pattern of input activity with another, and to associate patterns of activity with cell firing. Recent neurobiological work has very clearly demonstrated an association between patterns of inputs and cell firing. Markram et al. (1997) showed that backpropagating action potentials can cause a long-term potentiation (LTP) of the response of neocortical cells to a physiological synaptic input. The synaptic input and the action potential both needed to occur within 10 msec of each other for the association to occur and an LTP to take place, and the mechanism involved the NMDA glutamate receptor. This overall mechanism is very similar indeed to that demonstrated by the neuron model described in this chapter. Magee and Johnston (1997) demonstrated a similar role for backpropagating action potentials in causing LTP in hippocampal pyramidal cells. A major prediction of the modeling work has been confirmed by these experiments.

A further key conclusion of the modeling work presented here is the likelihood of local interactions between different inputs to the cell. Experimental evidence for interactions of this kind has been published recently. Golding et al. (2002) have shown that dendritic spikes allow synaptic potentiation between inputs, both in distal and proximal dendrites. Schiller et al. (2000) also demonstrated such interactions in basal dendrites where coactivated neighboring synapses could interact both directly and through the presence of prebound glutamate on NMDA receptors. Again, these experimental studies provide confirmation of predictions made by the models.

For efficient association between apical dendritic inputs and cell firing, it would be necessary for backpropagating action potentials to penetrate deeply into the apical dendritic tuft, as predicted by the model incorporating regenerative Na^+ spikes that is presented here. In another simulation study, Rhodes (1999) also showed that efficient spread into the apical dendrites requires active

Na^+ channels. Originally, experimental work reported a poor spread of backpropagating action potentials into the dendritic tuft (Stuart and Sakmann, 1994). However, later work from Sakmann's group ultimately confirmed these predictions and showed much better penetration (Stuart et al., 1997a; Larkum et al., 1999a). Larkum et al. (2001) showed the penetration to be much greater for the first action potential of a train compared to later spikes. In the hippocampal pyramidal cell Hoffman et al. (1997) also showed a controlled spread of activity into the dendrites. Antic (2003) has more recently demonstrated very strong backpropagation of somatic firing into basal and oblique dendrites, to the extent that this author feels that, as far as backpropagation is concerned, these dendrites should be considered an integral part of the axosomatic compartment. A further observation is that trains of backpropagating action potentials have been shown to have a decreasing spike amplitude (Callaway and Ross, 1995). This decrease was shown by Jung et al. (1997) and Colbert et al. (1997) to be due to an inactivation of the sparse Na^+ channels in the dendrites.

It is clear that active propagation within the pyramidal cell has a profound impact on internal communication. Information arising from activity in the dendritic fields can be communicated actively to the soma. Information about the firing of the cell can be communicated back to dendritic areas through backpropagating action potentials. These properties are underpinned by the properties of the voltage-gated channels in the cell membrane. The study in this chapter explored the behavior of young cells where regenerative properties are governed by voltage-gated Na^+ channels. The work used an empirical description of the regenerative behavior. However, an increasing amount of interesting data on channel properties and distribution is becoming available that should enable full conductance-based models to be generated. Indeed Rhodes and Llinás (2001) have made an excellent attempt to exploit published data on channel distribution and dynamics to generate a model that is more physiologically based.

Although not modeled in this study it is clear that Ca^{2+} channels have a major impact on cell behavior. The multipatch pipette recordings of Stuart et al. (1997a) have demonstrated active conduction involving Ca^{2+} spikes within the distal apical dendrites, but only passive conduction to the soma along the climbing fiber. Using Ca^{2+} fluorescence measurements, Schiller et al. (1997) confirmed these results, showing marked Ca^{2+} influx in the distal apical dendrites via high-threshold channels but little activity along the climbing fiber. Although in the young animal active propagation only requires Na^+ channels, in the more mature animal LVA Ca^{2+} channels are also activated in the apical dendrites (Zhu, 2000; Rhodes and Llinás, 2001).

It is also very clear from the data being generated that there is a complex interplay between voltage-gated channels. The interaction between Ca^{2+} transients and Na^+ channel activity was demonstrated very clearly *in vivo* by Svoboda et al. (1997). Hoffman et al. (1997) also provided much insight into the interaction between various channel types. Magee (1999b) also found that the time course of EPSPs as they propagated to the soma was helped to stay more constant through hyperpolarization-activated ion channels that increased in density the further they were from the soma.

New experimental data are becoming available all the time and show just how intricate and subtle is the behavior of pyramidal cells. In hippocampal pyramidal neurons Magee and Cook (2000) found that distal synapses had a higher strength compared to more proximal ones and thereby improved passive communication between the apical tuft and the soma. However, Williams and Stuart (2002) found that this was not the case in neocortical pyramidal neurons. In these cells, synaptic conductance stayed the same although distal EPSPs showed paired-pulse depression that was not shown by proximal inputs. Larkum et al. (2001) have shown active propagation to be very variable between cells. In the forward direction they found two categories of active propagation, with some activations being maintained in amplitude but others dropping to a low level by the time they reached the soma in the form of a boosted EPSP. A long-standing notion included in many neural network models, that the mean output of a cell remains constant despite changes in input sensitivity, has been shown by Turrigiano et al. (1998) to have a biological foundation, at least over 48 h timescales.

There is a lot of interest in the use of reduced cell models, such as the one described in this chapter. Simple integrate-and-fire models have shown much potential for the exploration of large-scale networks (Amit and Brunel, 1997) and with careful choice of parameters can model experimental data quite closely. For example, Rauch et al. (2003) showed such a model that closely matched *in vivo* data as far as the mean spike frequency-dependent adaptation to soma injected currents was concerned. It is important that these reduced models are able to include all the key features demonstrated by more comprehensive multicompartmental models. Izhikevich (2003) reviewed a number of spiking neuron models to compare their abilities to reproduce observed pyramidal cell behavior and to compare their computational cost. Simplifying neuron models to enable large-scale network simulations is motivating many studies. Poirazi et al. (2003) showed that the firing behavior of a comprehensive neuron model to a wide range of input stimulation could be closely replicated by much simpler models. If the cell was just firing according to stimulus input power then a single-point conductance-based neuron model was good. If the geometry of the cell was important for dealing with patterns of input information, then a simple two-layer network model with a sigmoidal link performed quite well. However, neuron geometry is clearly important when it comes to learning, as are the interactions between different inputs and the interactions between cell firing and these inputs, as shown by the work presented here. Schaefer et al. (2003) showed that the detection of the coincidence between cell firing and dendritic inputs is tuned by their proximal dendritic branching pattern. Indeed in another modeling study, Sakatani and Hirose (2003) showed that aspects such as dendrite/soma communication can be influenced by even the shape of the apical dendrite hillock. Clearly if features such as learning are to be included, then any reduced model has to carefully consider the basis for interactions between inputs as well as the interactions between inputs and cell firing. As Poirazi et al. (2003) concluded, different simplifying abstractions may apply in different neural contexts.

15.8 Conclusions and Future Perspectives

This chapter has described some modeling work that was applied to the most important information processing cell in the cerebral cortex, the pyramidal cell. It has used a straightforward compartmental technique to explore the potential distributions generated over the surface of the cell following input patterns. It has used a hybrid construction in which both receptor dynamics and learning and active conduction within the cell are simply modeled in an empirical fashion. The model is able to demonstrate pattern recognition on the part of the cell. It is also able to demonstrate the ability of the cell to increase its repertoire of learnt patterns through pattern association. Pattern association was shown to be dependent on the good communication between the apical dendrites and the cell body provided by active conduction throughout the cell. The climbing apical dendrite was shown to be ideally placed to allow inhibitory inputs to provide effective control over the communication between the apical dendrites and the soma. The work also demonstrated the ability of inputs to one part of the cell to condition learning of inputs to other parts. Finally, a much simplified version of the cell model was described that could demonstrate the same basic behavior but in a form that increased the speed of computing to a point where large networks of such cells could be efficiently modeled.

Future developments of pyramidal cell models depend on a number of gaps being filled in neurobiological data. Values for membrane resistance are still far from being settled. The lack of information about channel dynamics and distribution inhibits the development of accurate physiologically based models, although as mentioned in the discussion, this problem is rapidly diminishing. And, of course, any properties that are uncovered by cell models need to be verified by experiment before they can be assumed in larger-scale models. It would be useful, for example, to have a more direct demonstration of the nature of the communication between the apical dendritic tuft and the soma, and its impact on learning. The results of Stuart et al. (1997a) and Larkum et al. (2001) described in the discussion are useful indications of the kind of detail about this communication that should be forthcoming. Further modeling refinements probably would involve longer timescale events such as the neuromodulatory effects of some transmitters, and the link between second messenger generation and the laying down of changes in synapse sensitivity.

Thus, the basic accuracy of pyramidal cell models is probably close enough for their incorporation into larger scale models. Preliminary work using a cortical network of 400 of the reduced cell models described in this chapter, together with 100 interneurons, has shown interesting attractor behavior and the ability to generalize categories of input patterns (Orpwood, 1996). It is hoped that the basic pattern-processing capabilities of individual pyramidal cells can be used to explore the more sophisticated pattern-processing capabilities of networks of such cells, and of hierarchies of such networks. The ultimate aim of such simulations is to attempt to shed some light on the profound high-level capabilities of the most challenging of all biological organs, the cerebral cortex.

PROBLEMS

1. The neuron model discussed in this chapter was a young cell in which active properties have been shown to be dominated by Na^+ channel activity. For more mature cells, Ca^{2+} channels are also involved in regenerative activity. What impact might this be expected to have on the behavior of pyramidal cells?
2. The model demonstrated an oscillatory firing behavior reminiscent of the bursting behavior of some layer 5 pyramidal neurons. What factors not included in the model might limit the number of spikes that take place within a burst?
3. How could the observations of Turrigiano et al. (1998) about the total synaptic strength of a neuron be incorporated in the reduced cell model described in the chapter? Develop appropriate algorithms that would enable the model neuron to behave in this manner.
4. Using the principles outlined in Section 15.6, construct a model of an inhibitory interneuron that could be used alongside pyramidal cells in a model of a cortical neural network.
5. Use the numerical equations provided in the chapter and the data on channels published by Hoffman et al. (1997) to construct a limited compartmental model of the climbing apical dendrite, and incorporate Hodgkin–Huxley dynamics for the Na^+ and K^+ channels. What distribution of K^+ channels is necessary to emulate the decrement in backpropagating action potentials measured by Stuart and Sakmann (1994)? Can the reducing amplitude of a train of spikes, as shown by Callaway and Ross (1995), be simulated using the Na^+ channel inactivation that is discussed in Jung et al. (1997)?
6. *Experimental suggestion.* Using a pattern of inputs to the distal apical dendrites of a layer 5 pyramidal neuron via layer I/II stimulation, sufficient to cause firing in the soma, explore whether concurrent patterns of input to the proximal dendrites are conditioned by this activity and undergo LTP.

APPENDIX

15.A1 USE OF SHIFT OPERATORS

This appendix briefly describes the use of shift operators to simulate exponential decays in the simplified neuron model. The value of the local depolarization is represented in the computer as a binary number. If this number is shifted one bit to the right and the end bit is lost the effect is to divide the binary number by two. Similarly, a shift of one bit to the left multiplies the number by two. So if a right shift is used at each computer time step the local depolarization would decrease by one-half each time and an exponential decay can be simulated, with a time course dependent on the time step used. Such operations are computationally very much faster than carrying out floating point divisions or using exponential functions.

In many computer languages the shift operator is signified as follows:

$$x \gg n \text{ is equivalent to } x \text{ shifted right } n \text{ times, that is, } x/2^n,$$

$$x \ll n \text{ is equivalent to } x \text{ shifted left } n \text{ times, that is, } x \times 2^n.$$

It is easy to represent different rational changes at each time step. For example,

$$((x \ll 1) + x) \gg 2 = x \times \tfrac{3}{4}$$

15.A2 Leaky Integrate-and-Fire Neuron Models

The basic integrate-and-fire model assumes a single-compartment neuron where the cell membrane acts as a simple parallel resistance, R, and capacitance, C. Any driving current across this membrane, $I(t)$, can be expressed as

$$I(t) = \frac{V_{\mathrm{m}}(t)}{R} + C\frac{\mathrm{d}V_{\mathrm{m}}}{\mathrm{d}t},$$

that is,

$$\tau_{\mathrm{m}}\frac{\mathrm{d}V_{\mathrm{m}}}{\mathrm{d}t} = -V_{\mathrm{m}}(t) + RI(t),$$

where τ_{m} is the membrane time constant, RC.

Consequently, if there is no driving current, then for any starting membrane potential, V_0, the neuron "leaks" charge and the membrane potential decays as follows (compare with Equation [15.15] describing the reduced model used in the work presented):

$$V_{\mathrm{m}}(t) = V_0 \exp\left(\frac{-t}{\tau}\right).$$

For a simple constant current input, $I(t) = I_0$, and for a starting condition of $V_{\mathrm{m}} = 0$, the differential expression for the membrane potential integrates to

$$V_{\mathrm{m}}(t) = RI_0\left[1 - \exp\left(\frac{-t}{\tau}\right)\right].$$

When this exponential rise in membrane potential reaches a preset threshold the neuron is said to fire and the membrane potential is reset to its starting value. The driving current is normally a complex function of the various inputs being received by the neuron. The model basically integrates these inputs against a background of leaky charge loss and when the membrane potential reaches the threshold, the cell provides an output to the downstream neurons connected to it and the membrane potential is reset. The models can use input currents that are functions of time such as alpha functions, and can incorporate various degrees of extra detail such as refractory periods after firing or use models of action potential changes rather than just signaling a firing event with no information content.

15.A3 Conductance-Based Integrate-and-Fire Neuron Models

The integrate-and-fire models can be extended to reproduce more complex firing behavior but their empirical nature makes it difficult to relate them to measured neurobiological information such as voltage-clamp data. However, conductance-based models include representations of channel behavior, usually based on the Hodgkin–Huxley formulation introduced in Section 15.3.2.

As shown in Section 15.3.2 the current, I_k, through the channel conducting ion species, k, can be expressed as

$$I_k = \hat{G}_k m^p h(V - E_k),$$

where m and h are the activation and inactivation variables respectively. For the activation variable, m,

$$\frac{dm}{dt} = \frac{m_\infty(V) - m}{\tau_m(V)} \qquad \text{where } m_\infty(V) = \frac{\alpha_m(V)}{\alpha_m(V) + \beta_m(V)} \quad \text{and} \quad \tau_m(V) = \frac{1}{\alpha_m(V) + \beta_m(V)}$$

and for the inactivation variable, h,

$$\frac{dh}{dt} = \frac{h_\infty(V) - h}{\tau_h(V)} \qquad \text{where } h_\infty(V) = \frac{\alpha_h(V)}{\alpha_h(V) + \beta_h(V)} \quad \text{and} \quad \tau_h(V) = \frac{1}{\alpha_h(V) + \beta_h(V)}.$$

The Hodgkin–Huxley model was derived for an axon, the giant squid axon, and needs to be applied carefully if it is used with neuron models. Mainen et al. (1995) and Mainen and Sejnowski (1996) have used this approach to develop a two-compartment model of a neocortical pyramidal cell, with one compartment representing the axosomatic region and the other the dendrites, with an axial resistance connecting them varying between 1 and 10 MΩ. Four voltage-dependent currents were modeled: fast Na^+ current, I_{Na}, fast K^+ current, I_{Kv}, slow noninactivating K^+ current, I_{Km}, and high-voltage activated Ca^{2+} current, I_{Ca}. One Ca^{2+}-dependent K^+ current, I_{KCa}, was also used. For the axosomatic compartment they used conductance densities (in pS μm^{-2}) of $\hat{G}_{Na} = 30{,}000$, $\hat{G}_{Kv} = 1{,}500$, and for the dendritic compartment $\hat{G}_{Na} = 15$, $\hat{G}_{Ca} = 0.3$, $\hat{G}_{KCa} = 3$, and $\hat{G}_{Km} = 0.1$. The values used by these authors for reversal potentials were: $E_{Na} = 50$ mV, $E_K = -90$ mV, and $E_{Ca} = 140$ mV, and the expressions they derived for the α and β variables are described below for the different ion currents.

For the fast Na^+ current, I_{Na},

$$p = 3$$

$$\alpha_m = \frac{0.182(V + 25)}{1 - e^{-(V+25)/9}} \qquad \beta_m = \frac{-0.124(V + 25)}{1 - e^{-(V+25)/9}}$$

$$\alpha_h = \frac{0.024(V + 40)}{1 - e^{-(V+40)/5}} \qquad \beta_h = \frac{-0.0091(V + 65)}{1 - e^{-(V+65)/5}}$$

For the high-voltage activated Ca^{2+} current, I_{Ca},

$$p = 2$$

$$\alpha_m = \frac{0.055(V + 27)}{1 - e^{-(V+27)/3.8}} \qquad \beta_m = 0.94e^{-(V+75)/17}$$

$$\alpha_h = 4.57 \times 10^{-4} e^{-(V+13)/50} \qquad \beta_h = \frac{0.0065}{1 + e^{-(V+15)/28}}$$

For the fast K^+ current, I_{Kv},

$$p = 1$$

$$\alpha_m = \frac{0.02(V - 25)}{1 - e^{-(V-25)/9}} \qquad \beta_m = \frac{-0.002(V - 25)}{1 - e^{(V-25)/9}}$$

For the slow noninactivating K^+ current, I_{Km},

$$p = 1$$

$$\alpha_m = \frac{1 \times 10^{-4}(V + 30)}{1 - e^{-(V+30)/9}} \qquad \beta_m = \frac{-1 \times 10^{-4}(V + 30)}{1 - e^{(V+30)/9}}$$

For the Ca^{2+}-dependent K^+ current, I_{KCa}

$$p = 1$$

$$\alpha_m([Ca^{2+}]_i) = 0.01 \times [Ca^{2+}]_i \qquad \beta_m = 0.02$$

Internal calcium concentration, $[Ca^{2+}]_i$, is computed using Ca^{2+} entry via I_{Ca} and removal by a first-order calcium pump

$$\frac{d[Ca^{2+}]_i}{dt} = \frac{-1 \times 10^{-5} \times I_{Ca}}{2F} - \frac{[Ca^{2+}]_i - [Ca^{2+}]_\infty}{\tau_R},$$

where $[Ca^{2+}]_\infty = 0.1\ \mu M$, $\tau_R = 200$ msec, and F is Faraday's constant ($9.649 \times 10^{13}\ C\,mM^{-1}$).

16 Semi-Quantitative Theory of Bistable Dendrites with Potential-Dependent Facilitation of Inward Current

Aron Gutman, Armantas Baginskas, Jorn Hounsgaard, Natasha Svirskiene, and Gytis Svirskis

CONTENTS

16.1 INTRODUCTION

In many central neurons, including motoneurons, accommodation of the spike threshold is slow, partial, or absent (Bradley and Somjen, 1961; Koike et al., 1968; Schlue et al., 1974). This means that these cells possess a subthreshold slowly- or noninactivating inward current (Gutman, 1971). Noninactivating sodium currents (Magee and Johnston, 1995; Lipowsky et al., 1996; Schwindt and Crill, 1996) and calcium currents (Hounsgaard and Kiehn, 1993; Avery and Johnston, 1996; Magee et al., 1996; Hsiao et al., 1998; Morisset and Nagy, 1999) have been found in many nerve cells. These currents mediate slow dendritic spikes in many CNS neurons (Hounsgaard and Kiehn, 1993; Kim and Connors, 1993; Andreasen and Lambert, 1995; Schwindt and Crill, 1999), and plateau potentials in motoneurons (Hounsgaard and Kiehn, 1989; Hsiao et al., 1998; Lee and Heckman, 1998), Purkinje cells (Llinás and Sugimori, 1980a, 1980b), nociceptive deep dorsal horn neurons (Russo and Hounsgaard, 1994; Morisset and Nagy, 1998), neocortex pyramidal neurons (Schwindt and Crill, 1999), and striatal neurons (Dunia et al., 1996). In neurons *bistability* means that there is a net slow inward current that can cause a prolonged plateau potential without stimulation or active synaptic input. In bistable neurons the input current–voltage characteristic (I–V_{in}) is N-shaped and has two stable points of zero current.

Direct measurements revealed that slow inward currents are distributed throughout soma–dendritic membrane (Llinás and Sugimori, 1980a, 1980b; Hounsgaard and Kiehn, 1993; Magee and Johnston, 1995; Schwindt and Crill, 1995; Lipowsky et al., 1996; Magee et al., 1996). This spatial distribution of slow inward current creates a possibility for *dendritic bistability*. Earlier theoretical work (Gutman, 1971; Butrimas and Gutman, 1978, 1981) found that a dendrite with an N-shaped stationary membrane current–voltage relation can be bistable. When voltage at one end of a bistable dendrite is clamped, the other end can have one of two different potential values. However, the dendrite has to be sufficiently long electrotonically and the dendritic inward current has to be sufficiently strong (see following sections). The input current–voltage and frequency–stimulus current (f–I) relations of neurons with such dendrites have hysteresis (Butrimas and Gutman, 1981, 1983; Gutman, 1984, 1991). Bistability of the distal dendritic membrane strongly influences the input current, I_{in}, needed to clamp to potential, V_{in}. The input current–voltage relation (I–V_{in}) becomes not only N-shaped but may even split into two branches (Figure 16.3): there may be two I_{in} values at the given value of V_{in}. The larger value corresponds to the distal membrane potential near resting potential, RP, and the lower value — near to steady depolarization, SD. Whether the distal membrane potential is near RP or steady depolarization depends on the previous activity of the cell and previous V_{in} values. In other words, hysteresis occurs. Such hysteresis was found in mathematical models (Gutman, 1971; Butrimas and Gutman, 1978) and observed experimentally in cat motoneurons (Schwindt and Crill, 1980b; Lee and Heckman, 1996, 1998; Svirskis and Hounsgaard, 1997), suggesting dendritic bistability. Dendritic steady depolarization was observed following direct dendritic excitation in mammalian Purkinje cells (Llinás and Sugimori, 1980b) and in turtle motoneurons (Hounsgaard and Kiehn, 1993). In Purkinje cells the dendritic slow inward current is mediated by P-type Ca-channels (Llinás et al., 1992) and in turtle spinal neurons by L-type Ca-channels (Hounsgaard and Mintz, 1988; Hounsgaard and Kiehn, 1989, 1993).

The theory of dendritic bistability was developed for the simplest case of a time-independent I–V relation (Gutman, 1984, 1991). Later, experiments on interneurons and motoneurons in slices of turtle spinal cord (Russo and Hounsgaard, 1994, 1999; Svirskis and Hounsgaard, 1997) revealed that the inward current that underlies neuronal bistability develops slowly during depolarization. In other words, the inward current undergoes depolarization-dependent facilitation. Wind-up of the spiking response during a series of afferent stimuli (Mendell, 1966; Sivilotti et al., 1993) could, at least partly, be explained by this facilitation. Thus, the depolarization-dependent facilitation of an inward current is termed wind-up of the slow inward current or simply wind-up. This wind-up phenomenon is in accordance with known features of the short-time plasticity of L-type calcium channels (Bourinet et al., 1994; Sculptoreanu et al., 1995; Dolphin, 1996; Kavalali and Plummer, 1996; Parri and Lansman, 1996; Perrier et al., 2000).

In this chapter, the theory of bistable dendrites is enhanced by including wind-up of the inward current. In particular, we analyzed a cylindrical bistable cable with wind-up of the slow inward current. The wind-up by itself causes hysteresis of the I–V_{in} relation because the membrane I–V relation becomes deeper with depolarization. The wind-up hysteresis interferes with the bistability hysteresis. We demonstrated that bistability hysteresis is an all-or-none event while wind-up hysteresis is not. For processes that are much slower or much faster than the rate of wind-up, the theory of dendritic bistability for the time-independent I–V relation can still provide satisfactory understanding. Most important, potential-dependent facilitation of the inward current enables much shorter dendritic cables to be bistable, which can lead to multistability of neurons with many dendrites.

16.2 Bistable Dendrites with Time-Independent Membrane Current–Voltage Relations

We will start with a review of the simplest approximation of a cable with time-independent I–V relation. The corresponding cable equation differs from the usual Ohmic cable equation by only one

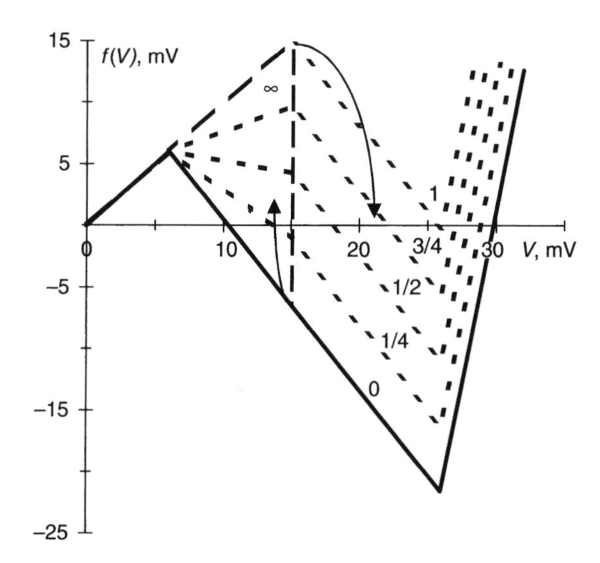

FIGURE 16.1 Model piecewise-linear current–voltage characteristics of bistable dendritic membranes, Equations (16.2), (16.9), and (16.10). Lowest solid line: fully developed wind-up, $f(V,N = 0)$, or time-independent $f(V)$. Stippled lines: incompletely developed wind-up, $f(V,N > 0)$, values of N are marked at the curves. Bold dashed line: stationary membrane current–voltage characteristic in the case of wind-up, $f_\infty(V)$. For $V < V_S = 15\,\text{mV}$, $f_\infty(V)$ coincides with the membrane current–voltage characteristic without wind-up, $f(V,N = 1)$, then jumps down to the current–voltage characteristic of fully developed wind-up, $f(V,N = 0)$, and coincides with this function for $V \geq V_S$. Arrows indicate the change of the I–V relation during moderately fast voltage ramps when N is decreasing during voltage increase and increasing during voltage decrease.

term that describes the I–V relation. The equation in electrotonic units is:

$$\frac{\partial^2 V}{\partial x^2} = \frac{\partial V}{\partial t} + f(V), \tag{16.1}$$

where V is the change of transmembrane potential relative to RP; x is the electrotonic axial coordinate, expressed in units of the electrotonic length constant (λ); t is electrotonic time, expressed in units of the membrane time constant (τ_m); $f(V)$ is the I–V relation multiplied by the membrane resistance near the RP, $f(V) \to V$ as $V \to 0$. Let us analyze the case of a membrane I–V relation with a net inward current. Such a membrane (or its I–V relation) is bistable in the following sense: the function $f(V)$ has three points of zero membrane current (Figure 16.1, solid curve): $V = 0$ (stable RP), H (SD), and h (unstable depolarization, $0 < h < H$). This implies that steady depolarization, and thus spiking activity of the neuron, may persist without stimulation or synaptic current. The steady depolarization of the model is equivalent to the plateau depolarization observed experimentally (see Introduction). The simplest representation of this function is a piecewise linear curve (Figure 16.1, solid curve) (Butrimas and Gutman, 1981; Gutman, 1984, 1991), which we will use for illustration:

$$f(V) = \begin{cases} V, & V \leq V_1; \\ k(h - V), & V_1 < V \leq V_2; \\ K(V - H), & V > V_2. \end{cases} \tag{16.2}$$

We use the following values of the parameters: $V_1 = 6\,\text{mV}$, $k = 1.38$, $h = 10.3\,\text{mV}$, $V_2 = 26\,\text{mV}$, $K = 5.7$, $H = 30\,\text{mV}$. These values were chosen (Butrimas and Gutman, 1983) to imitate the I–V_{in} relation and the relation between discharge frequency and constant stimulus current for the cat motoneuron (Granit et al., 1966; Schwindt and Calvin, 1973; Gustafsson, 1974; Schwindt and Crill, 1977, 1980a).

In order to understand the solutions of Equation (16.1), let us exclude the membrane capacitance and consider the stationary version of the equation:

$$\frac{d^2V}{dx^2} = f(V), \tag{16.3a}$$

which is similar to the dynamic problem of a classical particle. Here, the particle moves in a potential well (bottom at $V = h$) between two finite barriers (tops at 0 and H). The phase plane of trajectories, $(dV/dx)(V)$, for the general solution is presented in Figure 16.2A. The phase trajectories reflect the first integral of the stationary equation, the "energy" conservation,

$$\frac{1}{2}\left(\frac{dV}{dx}\right)^2 - \int_0^V f(y)\,dy = E.$$

There are no crossings between phase trajectories because the solutions are unique when the "potential" and its gradient are defined. Let us consider the boundary value problem meaningful for the dendritic cable:

$$\begin{aligned} \frac{d^2V}{dx^2} &= f(V), \qquad V = V_{in}, \quad x = 0; \\ \frac{dV}{dx} &= 0, x = X. \end{aligned} \tag{16.3b}$$

Equation (16.3b) describes potential fixation at the "impaled" proximal end and the sealed distant end. The second condition implies that the phase trajectory terminates on the abscissa, $dV/dx = 0$. Thus, there are two types of solutions, that is, potential distributions along a bistable dendritic cable: (1) monotonic solutions that do not cross the abscissa, and (2) solutions that may cross the abscissa and oscillate around h.

The boundary value problem (16.3b) may have nonunique solutions since the potential V_{in} at the beginning of the cable may correspond to more than one trajectory ending on the abscissa. We are going to define conditions for the solution of this boundary value problem to be nonunique by analyzing solutions near the unstable equilibrium point h. The nonuniqueness of the solution implies that the I–V_{in} relation of the cable, which is equivalent to the dependence of $(dV/dx)|_{x=0}$ on V_{in}, is also nonunique (Figure 16.3B). Analysis has shown that the uniqueness or nonuniqueness of the input current–voltage relation depends on the electrotonic length of the cable. For electrotonically very short cables, the I–V_{in} relation is proportional to the membrane I–V relation and is unique. In this case the input resistance, $R_{in} = dV_{in}/dI_{in}$, is negative near the point $V_{in} = h$, $I_{in} = 0$. For electrotonically long cables the I–V_{in} relation deforms and may lose its uniqueness. The stationary solution becomes nonunique when and only when a range of positive R_{in} appears between ranges of negative R_{in} in the I–V_{in} relation as the cable length increases (Figure 16.3) (Alaburda et al., 2001). The sufficient condition for the nonuniqueness is: $R_{in} > 0$ at the point $V_{in} = h$, $I_{in} = 0$.

The stationary solutions for potentials near the nonstable equilibrium point (Figure 16.2A), $[V(x) - h] \approx 0$, satisfy the following equation:

$$\frac{d^2V}{dx^2} = k(h - V) = \frac{1}{L^2}(h - V), \tag{16.4}$$

where $L = 1/\sqrt{k}$ is the critical length parameter for the oscillatory solution around h in units of the length constant, λ, and $k = -(df(V)/dV)|_h$. Here $f(V)$ is for bistable functions in general, not only

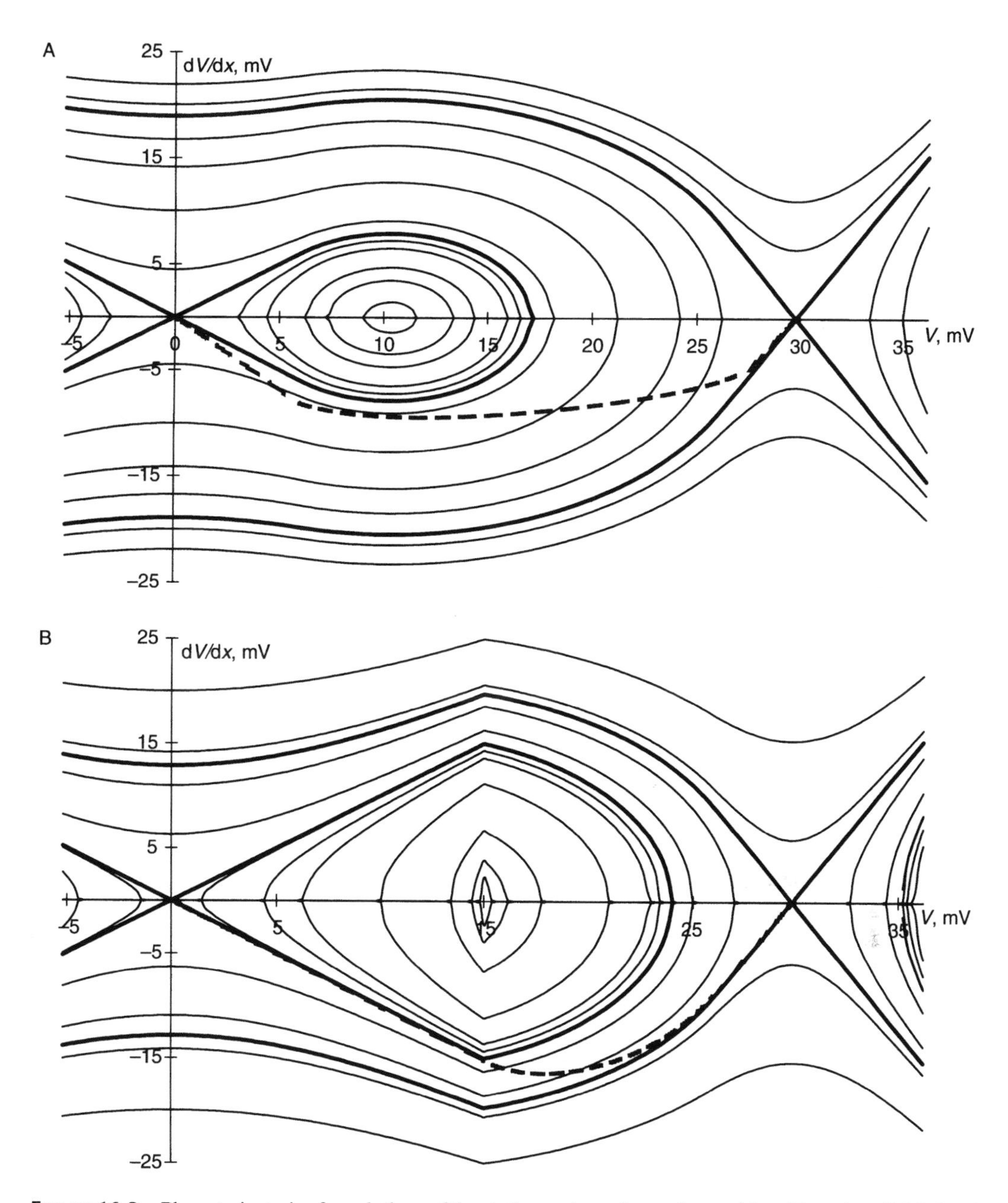

FIGURE 16.2 Phase trajectories for solutions of the stationary boundary value problem (Equations [16.3a] and [16.3b] solid curves) and for the wave of a cable recharging (Equation [16.8], dashed curves). Thick lines: separatrices; their asymptotes, $V \rightarrow H + A\exp(\pm x\sqrt{K})$ or $A\exp(\pm x)$, $x \rightarrow \mp\infty$. The separatrices segregate the phase plane into areas of qualitatively different solutions. (A) Time-independent membrane I–V relation (Equation [16.2]). (B) Stationary membrane I–V_∞ relation, when the wind-up of the inward current is present (Equation [16.10]). Note the breaks and the absence of the smooth maxima of dV/dx at V_S in (B).

for the piecewise-linear I–V relation of Equation (16.2). The corresponding harmonic solution of the boundary value problem Equation (16.3b) is:

$$V = h + (V_{\text{in}} - h)\frac{\cos[(x - X)/L]}{\cos(X/L)}. \tag{16.4a}$$

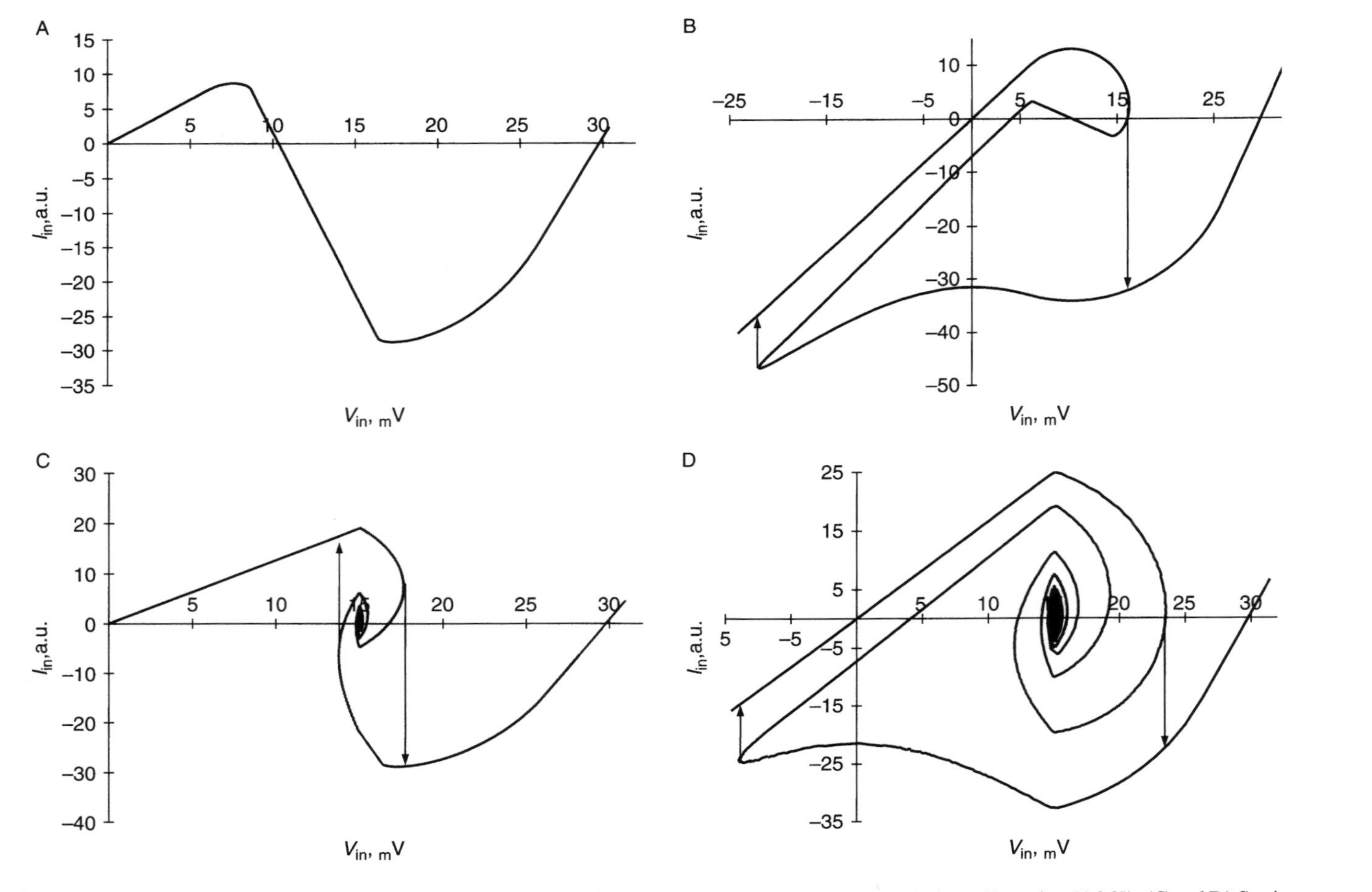

FIGURE 16.3 Input current–voltage relations of cylindrical cables. (A and B) Time-independent membrane I–V relations (Equation [16.2]). (C and D) Stationary membrane I–V_∞ relations, when the wind-up of the inward current is present (Equation [16.10]). In these cases, an infinite number of oscillating solutions merge in a patch. The electrotonic lengths, X, of the cables are 1 (A and C) and 3 (B and D). Input current is given in arbitrary units. All stable and unstable solutions are presented.

If $X < \pi L/2$, then R_{in} is negative at $V_{\text{in}} = h$, $I_{\text{in}} = 0$ (Figure 16.3A). If $\pi L/2 < X < \pi L$, R_{in} becomes positive at this point and I–V_{in} surely loses uniqueness. We can introduce a critical length X_{cr} for bistability of a cable:

$$X_{\text{cr}} = \frac{\pi L}{2} = \frac{\pi}{2\sqrt{k}}. \tag{16.5}$$

In the case of the membrane I–V relation described by Equation (16.2), $X_{\text{cr}} = 1.34$. Thus, when the electrotonic length X of a cable exceeds the critical length X_{cr}, some range of fixed V_{in} appears in the I–V_{in} relation where two stable values of I_{in} correspond to a given value of V_{in}. The cable is bistable in this range of fixed V_{in} and the input current–voltage relation has hysteresis. When the electrotonic length of a cable is less than the critical length, the input current–voltage relation is unique (only one value of I_{in} corresponds to a given value of V_{in}), and there is no hysteresis.

Now we give some results (Gutman, 1984) relating the number and stability of stationary solutions to the boundary value problem, Equation (16.3b). These results (Figure 16.3A and B) are the following:

1. There are at least $(2n + 1)$ qualitatively different solutions when $\pi L(n - 1/2) < X < \pi L(n + 1/2)$.
2. The solution that includes the largest positive value I_{in} monotonically changes from V_{in} to $V(X)$, which is close to the RP. This RP solution is stable. The RP solution can occur if the input voltage values are in the range $-\infty < V_{\text{in}} < V_{\text{RP}}$. At V_{RP}, which is the largest possible V_{in} of RP-solutions, the input resistance, $(\mathrm{d}V_{\text{in}}/\mathrm{d}I_{\text{in}})|_{V_{\text{RP}}} = 0$, and $V_{\text{RP}} > h$.
3. The solution which includes the largest negative value I_{in} monotonically approaches steady depolarization at the distal end. This SD solution is also stable. The SD solution exists for all $V_{\text{SD}} < V_{\text{in}} < \infty$. At V_{SD}, which is the smallest possible V_{in} of the SD solutions, the input resistance, $(\mathrm{d}V_{\text{in}}/\mathrm{d}I_{\text{in}})|_{V_{\text{SD}}} = 0$, and $V_{\text{SD}} < h$.
4. All the other $(2n - 1)$ solutions with intermediate values of I_{in} are unstable. Almost all of them oscillate along the cable.
5. If the cable length, X, is increased, the value of V_{RP} is shifted rightward and the value of V_{SD} leftward.
6. The monotonic stable solutions are qualitatively similar. Their quantitative difference depends on a parameter of the I–V relation, $u = \int_0^H f(V)\,\mathrm{d}V$. With u increasing, V_{RP} and V_{SD} also increase. In the case of a semi-infinite cable and positive u, the RP solution is absolutely stable, $V_{\text{RP}} = \infty$, while V_{SD} is determined by the equation $\int_{V_{\text{SD}}}^H f(V)\,\mathrm{d}V = 0$. For $X = \infty$ and negative u, the SD solution is absolutely stable, $V_{\text{SD}} = -\infty$, while V_{RP} is determined by $\int_0^{V_{\text{RP}}} f(V)\,\mathrm{d}V = 0$. The I–V relation of Equation (16.2) gives an example of an $f(V)$ with negative u. In the case of the semi-infinite cable, $I_{\text{in}}(V_{\text{SD}}) = I_{\text{in}}(V_{\text{RP}}) = 0$.

Thus, there may be two stable values of I_{in} at the same fixed V_{in}. In other words, the cable may be bistable even in voltage-clamp conditions. We shall use the notion of cable bistability only in this sense. There exists a hysteresis of the cable's I–V_{in} relation for $V_{\text{SD}} < V_{\text{in}} < V_{\text{RP}}$ (Figure 16.3B). When V_{in} exceeds V_{RP} or decreases below V_{SD}, I_{in} must jump. This reflects changes from the RP solutions to SD solutions, in other words, the distal potential shifts from values close to RP to values close to steady depolarization, and vice versa. Experimental observations suggest that the same could happen in spinal motoneurons. A voltage clamp to RP after a strong depolarization reveals an inward current which does not decay for at least hundreds of milliseconds (Schwindt and Crill, 1980b; Lee and Heckman, 1996). The I–V_{in} relation of a branching dendrite or a cell with several dendrites has several jumps reflecting different V_{RP} and V_{SD} and abrupt changes of potential distribution in individual dendrites or even their branches (Figure 16.4) (Baginskas and Gutman,

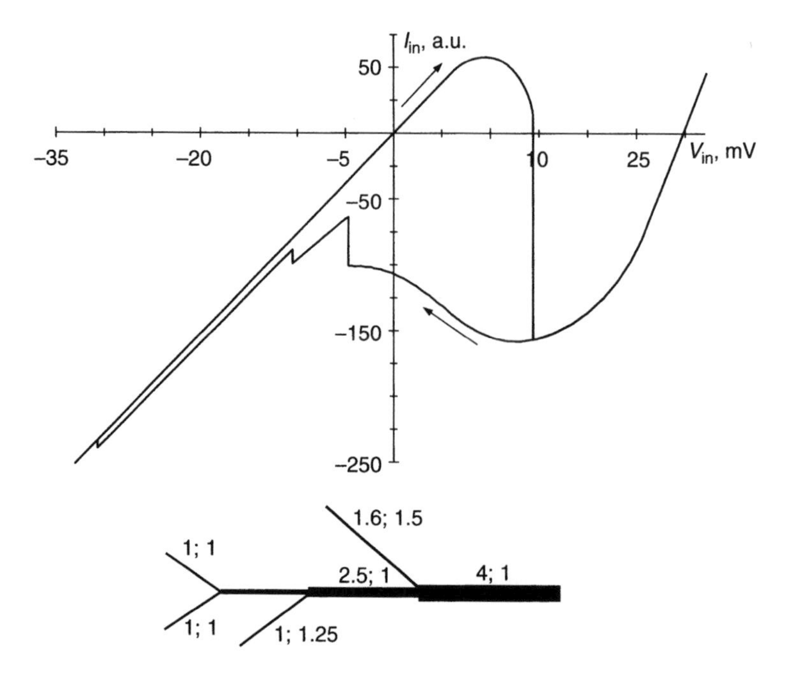

FIGURE 16.4 Input current–voltage relation of a branching dendrite in the case of a time-independent membrane I–V relation. Only the stable solutions are shown. The model dendrite is presented below. Diameters and electrotonic lengths, X, of the branches are marked in the following way: diameter; X. There is a single jump in I_{in} at depolarization $V_{RP} \approx 14$ mV. This occurs because the jump to the SD solution of the proximal branch induces the jumps of the more distal branches. There are three independent jumps at hyperpolarization $V_{SD} \approx -5, -10$, and -30 mV, for the proximal, intermediate, and distal branches, respectively. This difference between jumps from RP to steady depolarization and from steady depolarization to RP reflects considerable negativity of the parameter u: negative current at $h < V < H$ is much larger than positive current at $0 < V < h$. Thus, the SD solutions are more stable than the RP solutions.

1990). Figure 16.4 illustrates that individual branches of dendrites can maintain a state of steady depolarization independently from each other (Svirskis et al., 1998; Alaburda et al., 2000). This means that a neuron with an elaborate dendritic tree and a bistable membrane can be multistable, that is, there are multiple nonzero stable stationary values of soma potential, each corresponding to a different history of stimulation or synaptic activity. For instance, after synaptic excitation ends, the somatic potential can stabilize to different nonzero stationary values, depending on the strength and pattern of previous synaptic activity (Baginskas and Gutman, 1990; Svirskis et al., 1998; Alaburda et al., 2000).

The next problem is to analyze the process of switching between RP and SD solutions (Nagumo et al., 1965). In an infinite cable the switch proceeds as a wave, $V(x, t) = V(x - ct)$ (Gutman, 1984). Thus, Equation (16.1) simplifies to:

$$\frac{d^2V}{dx^2} = -c\frac{dV}{dx} + f(V). \tag{16.6}$$

In the case of absolute stability of the SD solution, that is, when $u < 0$, the boundary conditions for the wave traveling from $-\infty$ to ∞ are:

$$V|_{x=\infty} = 0; \quad V|_{x=-\infty} = H. \tag{16.7}$$

In the phase plane Equations (16.6) and (16.7) transform to:

$$\frac{1}{2}\frac{\mathrm{d}}{\mathrm{d}V}\left(\frac{\mathrm{d}V}{\mathrm{d}x}\right)^2 + c\frac{\mathrm{d}V}{\mathrm{d}x} = f(V); \qquad \left.\frac{\mathrm{d}V}{\mathrm{d}x}\right|_{V=0 \text{ or } H} = 0. \tag{16.8}$$

In the equivalent dynamic problem, the wave speed c corresponds to the coefficient of viscous friction. One must adjust its value numerically in order to satisfy the boundary conditions of Equation (16.8) (Appendix 16.A2). The mechanical analogy of the adjustment means that the particle falls from rest at the top of the high barrier, $V = H$, and stops exactly at the top of the low barrier, $V = 0$ (Figure 16.2A). In case of the piecewise-linear I–V relation, Equation (16.2), the wave has the following numerical parameter values (Appendix 16.A2): $c = 0.63$; length of the wave front, defined as, $\Lambda = H/(\mathrm{d}V/\mathrm{d}x)|_{V=h}$, equals 2.7λ; duration of the wave front, $T = \Lambda/c = 4.3\tau_m$.

Thus, in addition to the electrotonic parameters λ and τ_m (both equal 1 in the equations used here), the bistable cable with stationary I–V relation is characterized by the additional parameters L, c, Λ, and T. If $X < \pi L/2$, the cable is short in the sense of bistability and there is no hysteresis of its I–V_{in} relation (Figure 16.3A). If $X > \Lambda$, the cable is long. The wave of switching from RP to steady depolarization may develop fully in such long cables. If the potential, V_{in}, at the proximal end of the long cable is fixed at the steady depolarization, at least for a time $\approx T$, then the cable recharges to steady depolarization (there is sufficient time for the wave to develop).

The I–V_{in} relation of a cell determines its discharge qualities. In the model (Butrimas and Gutman, 1983), the trigger zone confined to the soma was described by Hodgkin–Huxley equations with the addition of a post-spike hyperpolarization. There are two possibilities in bistable dendrites (Figure 16.5). In long dendrites, $X \gg L$, the potential of the distal membrane jumps to steady depolarization and remains there in spite of the post-spike hyperpolarization. The discharge frequency–stimulus current relation (F–C) has a jump-and-hysteresis, reflecting the jump-and-hysteresis of the dendritic I–V_{in} relation. This pattern has been observed in cat and turtle motoneurons (Schwindt and Crill, 1977, 1980a; Hounsgaard et al., 1984, 1988). The frequency–current relation hysteresis ends at small depolarizing or even hyperpolarizing currents.

In dendrites that are shorter, but still longer than the critical length (which is equal to $\pi L/2$), the steady depolarization of the distal membrane cannot resist maximal values of the postspike hyperpolarization and there is a steady depolarization for part of the time during the interspike interval and RP during the remaining time. In some range of the current strength, the F–C relation rises more steeply than for Ohmic dendrites but jump-and-hysteresis is absent (see Figure 16.5). This range of steeper rise could explain the phenomenon of two linear ranges in the F–C relation (Granit et al., 1966; Schwindt and Calvin, 1973; Gustafsson, 1974). The primary range for small suprathreshold currents is linear and ends with a break point that is followed by a steeper secondary linear range for larger stimulus currents. In the model, the primary range corresponds to the Ohmic range of the membrane I–V relation and the secondary, steeper range is determined by the part-time steady depolarization and part-time RP during the inter-spike interval.

The secondary steeper range in the F–C relation could also be explained by the prolonged process of recharging the dendritic cable to steady depolarization. Modeling has shown that the secondary steeper range is "observed" with a quasi-stationary F–C relation, obtained using stimulus current pulses of shorter duration than the time T needed to recharge the dendritic cable from RP to a state of steady depolarization. For a long, bistable dendritic cable, and with the values for the membrane I–V relation parameters used in the study, the recharging time $T \approx 0.5$ sec. As we will see below, the recharging process of a bistable cable with wind-up of the inward current is much slower; the time T exceeds 1 sec. The experimentally recorded F–C relations are, probably, quasi-stationary since stimulus current pulses of duration about 1 sec often are used in the corresponding experiments. Thus, the secondary steeper range in the F–C relation could be attributed to the prolonged recharging of dendrites to steady depolarization.

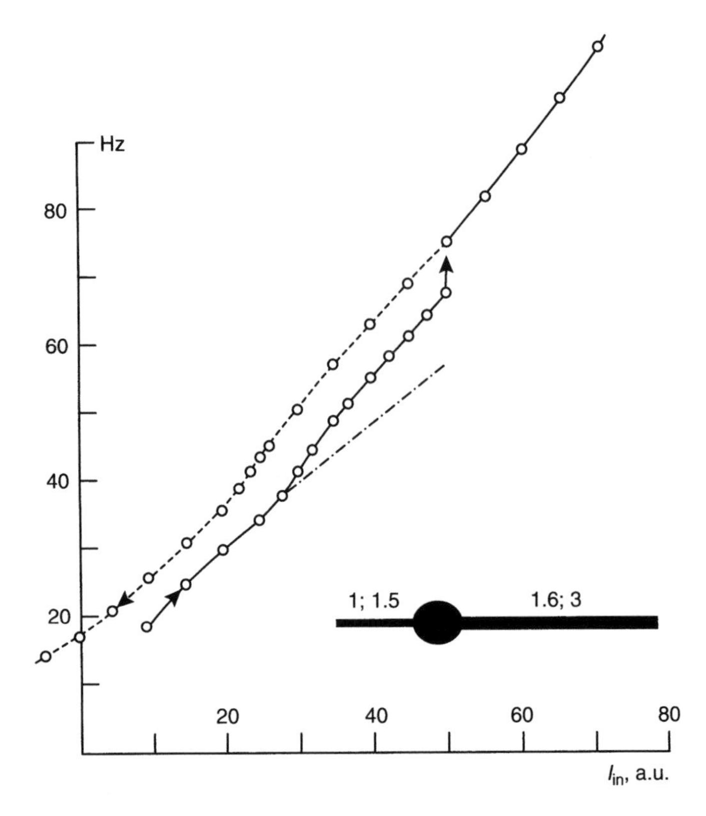

FIGURE 16.5 Discharge frequency–stimulus current relation for a model neuron (inset) with a time-independent bistable membrane I–V relation. Two dendrites of electrotonic length $X = 1.5$ and 3 are attached to the electrotonically compact soma. The diameter of the long dendrite is 1.6 times larger than the diameter of the short dendrite. The soma sodium and potassium channels are described by Hodgkin–Huxley equations. The jump-and-hysteresis in the discharge frequency–stimulus current relation reflects the steady depolarization of the long dendrite. The smooth increase in frequency is due to the steady depolarization of the short dendrite during part of the interspike interval (see text).

Thus, the theory of bistable dendrites with time-independent I–V relation explains the experimental observations on I versus V_{in} and the discharge properties of the neurons.

16.3 POTENTIAL-DEPENDENT FACILITATION (WIND-UP) OF THE SLOW INWARD CURRENT

Experiments on interneurons and motoneurons in slices of turtle spinal cord (Russo and Hounsgaard, 1994, 1999; Svirskis and Hounsgaard, 1997) have demonstrated that the inward current that underlies bistability develops slowly with a characteristic time of hundreds of milliseconds or even seconds (Figure 16.6A). Such a phenomenon was observed earlier in cat motoneurons (Schwindt and Crill, 1980b; Hounsgaard et al., 1988), but these early observations were inconclusive since slow development of the observed I_{in} might have been explained as a result of a slow propagation of the steady depolarization along the dendritic cable (Gutman, 1984, 1991). Only experiments with repeated application of current or potential pulses or ramps (Svirskis and Hounsgaard, 1997; Russo and Hounsgaard, 1999) revealed that there is an additional process of slow facilitation of the slow inward wind-up current. The slow inward current develops during a train of repeated pulses, becoming larger with each successive pulse (see Figure 16.6, and also Russo and Hounsgaard, 1999). Under some conditions, wind-up results in a steady depolarization that persists beyond the train of current pulses

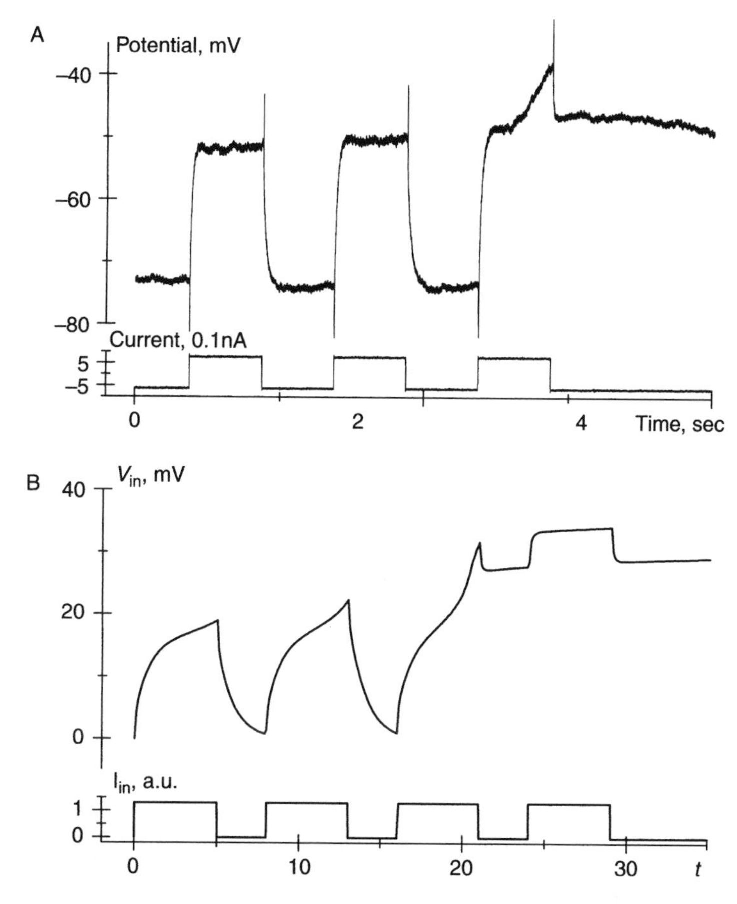

FIGURE 16.6 Wind-up of the inward current induced by current pulses. (A) Experimental curve, motoneuron from turtle spinal cord slice. (B) Model, cable electrotonic length $X = 1$. *Note:* the wind-up time constant, θ, is too small for detailed imitation of the experimental curves.

(see Figure 16.6, response to third pulse). This observation means that the I–V relation has a net inward current, as in Figure 16.1 and Equation (16.2), after development of the wind-up.

Formally, wind-up may be described as reverse channel accommodation, that is, slow inward current inactivation at RP or during hyperpolarization and deinactivation during significant depolarization. We have modeled the wind-up of a slow inward current as a slow shift of the membrane I–V relation from an initial "shallow" to the final "deep" shape (Figure 16.1). In other words, the voltage sensitivity of the inward current increases (i.e., the threshold voltage of the current activation decreases) during the wind-up. This assumption is in agreement with tail-current development in parallel with wind-up (Russo and Hounsgaard, 1994, 1999; Svirskis and Hounsgaard, 1997). The transition from a "shallow" to a "deep" N-shaped membrane I–V relation is expressed in the model as a continuous shift of the conductance in the range of voltage $V_1 < V_S$ from the initial Ohmic to the negative final value of Equation (16.2) (Figure 16.1): $df/dV = -k(1 - N) + N$. Here $N(t)$ is the variable of the wind-up process. There is no wind-up when $N = 1$, wind-up is completely developed and the membrane I–V relation corresponds to Equation (16.2) when $N = 0$. Obviously, the membrane current–voltage relation becomes deeper at $V > V_S$ (Figure 16.1) during the wind-up process. For illustration, we choose the membrane I–V relation corresponding to the completely developed wind-up to be identical to the stationary function of Equation (16.2) (Figure 16.1, solid curve). Thus, we choose the simplest description of the wind-up as follows

(Figure 16.1, stippled curves):

$$f(V,N) = \begin{cases} V, & V \le V_1; \\ V - (1+k)(1-N)(V - V_1), & V_1 \le V \le V_S; \\ V - (1+k)(1-N)(V - V_1) - (1+k)N(V - V_S), & V_S \le V \le V_2; \\ V - (1+k)(1-N)(V - V_1) - (1+k)N(V - V_S) + (k+K)(V - V_2), & V \ge V_2. \end{cases} \tag{16.9}$$

The variable of the wind-up is described by first-order kinetics:

$$\begin{aligned} \frac{dN}{dt} &= \frac{1-N}{\theta}; \quad V < V_S; \\ \frac{dN}{dt} &= -\frac{N}{\theta}; \quad V \ge V_S; \\ N(t) &= 1, \quad \text{if } V(a) < V_S \text{ at all } a \le t. \end{aligned} \tag{16.10}$$

The process has a threshold voltage V_S. The return of the membrane I–V relation to a more shallow shape takes place at potential values $V < V_S$ as the potential is decreasing from large depolarizations, $V > V_S$, that had caused wind-up (Figure 16.1, upward arrow). The time constant of the wind-up, θ, in our models equals $10\tau_m$, and the threshold, $V_S = 15$ mV, coincides with the maximum of the I–V relation before the wind-up starts.

Let us consider a stationary potential distribution along the cable. In this case, $t = \infty$, thus, $N = 0$ or 1 and the I–$V(I$–$V_\infty)$ relation corresponds to the following function (Figure 16.1, dashed bold curve):

$$f_\infty(V) = \begin{cases} V, & V < V_S; \\ k(h - V), & V_2 \ge V \ge V_S; \\ K(V - H), & V_2 \le V. \end{cases} \tag{16.11}$$

Equation (16.11) differs from the Equation (16.2) by the Ohmic segment extending to V_S and a jump of current at this threshold voltage of the wind-up (Figure 16.1). In our model $f_\infty(V_S + 0) = f(V_S, N = 0) < 0$. The unstable zero-current point is now V_S, not h, and $df_\infty/dV|_{V_S} = -\infty$. Thus, the critical length for a cable to be bistable [Equations (16.4) and (16.4a)], L_∞, is equal to zero. This means that even the shortest cable is bistable. Of course, in more realistic and complicated models the dependence of the wind-up of the inward current on the membrane potential should be continuous in some range of potentials, not thresholded, as in our model. The thresholding of the wind-up is an oversimplification that, nevertheless, facilitates understanding. Correspondingly, in such a model there will be a smooth transition of N from 1 to 0, that is, a smooth transition from Ohmic to negative current in f_∞. But still the parameter $k = (-df/dV)|_h$ [see Equation (16.4)] must be larger than in the case of stationary $f(V) \equiv f(V, N = 0)$, which means that the critical length of the cable [Equation (16.9)] should be less. The theory of stationary bistable cables is useful as a marginal approximation of the theory of bistable cables with time- and potential-dependent wind-up of the slow inward current. Cable bistability with very slow ramps of V_{in} is determined by the I–V_∞ relation [Equation (16.11)], while during the relatively fast ramps the wind-up parameter N remains approximately constant and cable bistability is determined by df/dV at the medial zero point of $f(V,N)$ (Figure 16.1, stippled curves). The corresponding critical length parameter, $L(N)$, is greater than or equal to the L of a stationary bistable cable [Equation (16.4)]: $L(N) \ge L$. In general, bistability is more probable for slow potential ramps. We may conclude that even in more realistic models of the wind-up the critical length of the cable [Equation (16.5)] is different for extremely slow and relatively fast potential ramps.

The bistability of infinitely short cables may be seen in the phase plane of the stationary equation, $d^2V/dx^2 = f_\infty(V)$ (Figure 16.2B). d^2V/dx^2 has a jump at V_S due to the jump of $f_\infty(V)$. This jump is reflected by breaks in the phase trajectories (Figure 16.2B). The jump (Figure 16.1, dashed solid curve) at the point of unstable equilibrium, V_S, means that $d^2V/dx^2 \neq 0$ also for trajectories that are very close to the point, $V = V_S, I = 0$, in contrast to trajectories with continuous $f(V)$ (Figure 16.2A). Thus, oscillating solutions (the phase trajectory crosses the abscissa one or more times before ending on this axis) with infinitely small $|V - V_S|$ and $|dV/dx|$ are possible for infinitely small cable length X. The oscillating solutions are unstable and hysteresis around V_S exists for infinitely short cables. The physical meaning of this hysteresis may be explained by using a simple two-compartment approximation (Gutman, 1984; Booth and Rinzel, 1995) of a short cable. When the input potential is rising and $V_{in} = V_S + 0$, wind-up develops in the proximal compartment while still absent in the distal compartment. At the same V_{in}, wind-up is developed in both compartments when returning from larger depolarizations. Thus, I_{in} is different at the same V_{in}. A similar situation exists for $V_{in} = V_S - 0$. In a short cable, the stationary hysteresis reflecting the bistability is approximately as narrow as the potential difference, $[V_{in} - V(X)]$, between the ends of the cable. In short cables with sealed ends this difference rises proportionally to X^2, thus, the width of the hysteresis region increases considerably faster than X. An example of I versus V_{in} hysteresis for a short cable in case of $f_\infty(V)$ is presented at Figure 16.3C. There is no hysteresis for this cable in the case of stationary $f(V)$ or $f(V, N = 0)$, Figure 16.3A.

16.4 Discussion

Our model imitates the basic experimental phenomena of wind-up of the inward current when the cell is stimulated with a repeated current pulse (Figure 16.6B). The electrotonic response is followed by a delayed rise of V_{in} which increases in amplitude and begins earlier with each pulse in the series. Finally, this increased potential may persist as a steady depolarization even after the stimulus current has been switched off. The wind-up is a prerequisite for bistability. The apparent difference between the experimental and model curves is due to a large difference in timescale (compare Figure 16.6A and Figure 16.6B). The time constant for wind-up in the model, θ, is too small and thus the electrotonic component of the response is overexpressed. The much smaller value of θ in the model was chosen for convenience in the calculations.

Now, we may advance to the more complex problem of defining the I–V_{in} relation in the case of wind-up. The basic observation comes from the experiment with application of a train of voltage-clamp ramps (Figure 16.7A). The I–V_{in} relation becomes deeper and diverges downwards from the straight Ohmic line at a lower level of depolarization during the second application of the voltage ramp. There is a hysteresis loop in the I–V_{in} relation, but in contradiction with the theory of cables with time-independent membrane, the I–V relation it is not accompanied by a potential jump. This observation deserves a little more analysis.

While V_{in} is moving from small depolarizations and becomes larger than V_S, the membrane I–V relation corresponds to $f(V, N \approx 1)$ (Figure 16.1, upper curves). During the return from large depolarization, $V \approx$ SD or more, to values below V_S, the membrane I–V relation corresponds to $f(V, N \approx 0)$ (Figure 16.1, lower curves). This means that a relatively fast ramp of fixed V_{in} gives two types of hysteresis curves. One type is for and only for long cables, $X > X_{cr}$, where X_{cr} is defined according to Equation (16.5) for $f(V, N \approx 0)$. This hysteresis is essentially similar to the hysteresis described for cables with time-independent bistable I–V relation (Figure 16.7C). Indeed, the potential distribution at the minimum of the lower hysteresis loop of Figure 16.7C corresponds to the potential at the distal end of the cable being close to steady depolarization. We can call this phenomenon

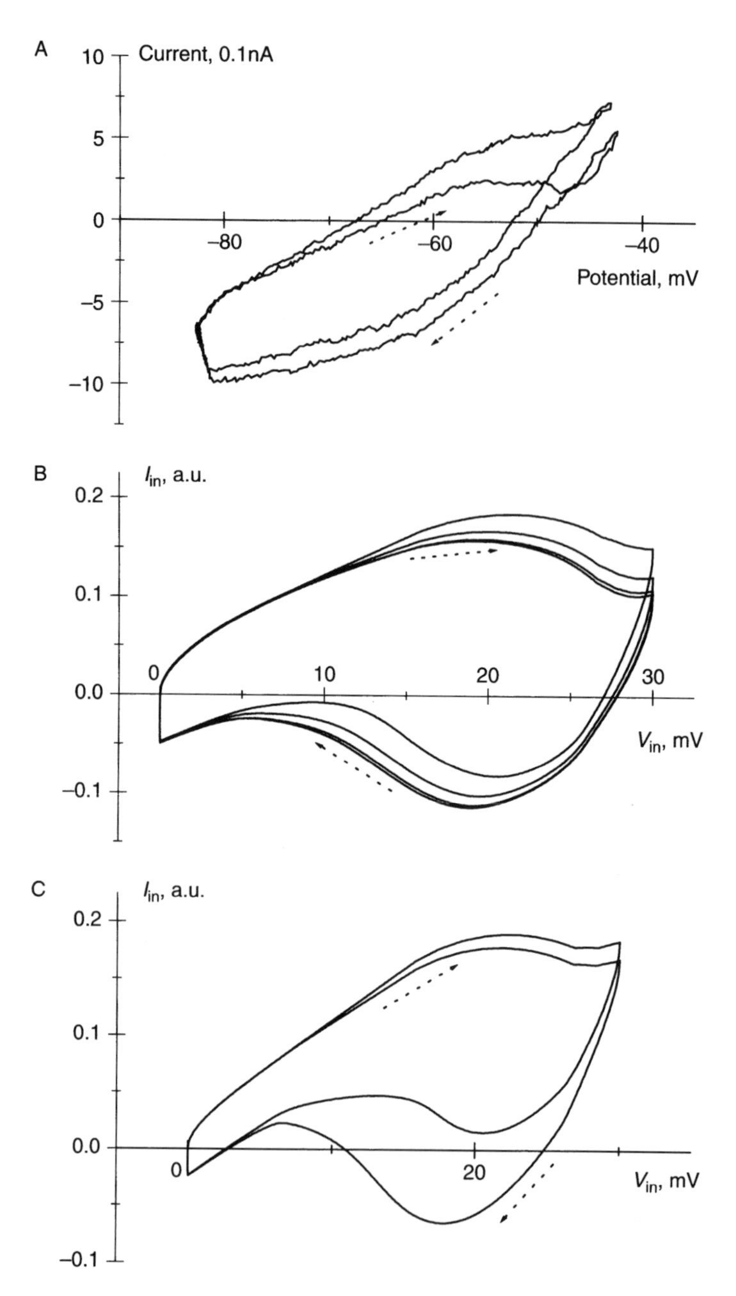

FIGURE 16.7 Wind-up expression when triangular voltage-ramp series are used for stimulation. (A) Wind-up observed in turtle motoneuron from spinal cord slices; duration of the ramp 0.43 sec, time between two ramps 0.57 sec. (B) Model, $X = 1$; duration of ramp $7.5\tau_m$, time between ramps $3\tau_m$. Note saturation of the model wind-up at the 3rd ramp. (C) Model, $X = 1.5$, duration of ramp $15\tau_m$, time between ramps $7\tau_m$. The duration of the ramp was made rather short in order to see the development of the wind-up. This causes capacitive hysteresis to be prominent and clearly seen near the RP (Butrimas and Gutman, 1980; Gutman, 1984).

"bistability hysteresis." The second type of hysteresis, with fast voltage ramps, has no jumps and is not characterized by the marginal potentials V_{RP} or V_{SD} (see Section 16.2). This hysteresis may be caused by the change of N during the V_{in} ramp (Figure 16.7B). This type of hysteresis can be expressed even in the proximal membrane. We call this phenomenon "wind-up hysteresis."

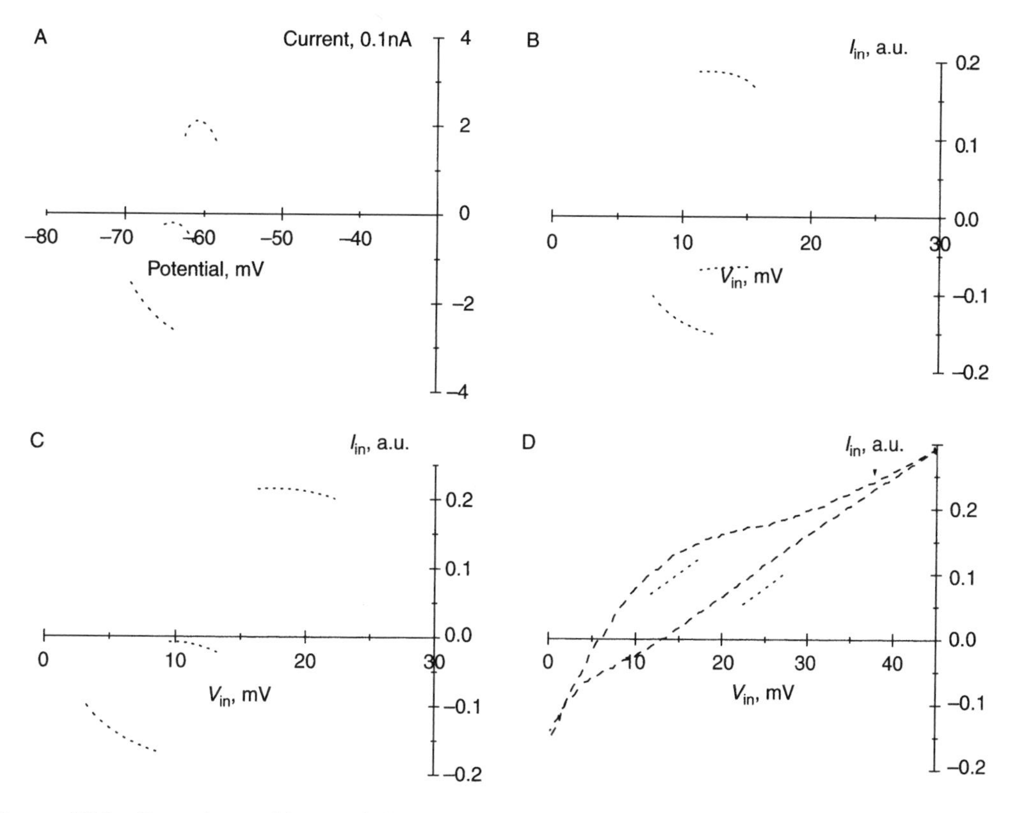

FIGURE 16.8 Dependence of hysteresis loop in the input current–voltage relation on maximal V_{in} of triangular voltage ramp. (A) Experimental hysteresis loops with two values of maximal clamped potential, turtle motoneuron, duration of the ramp, 4 sec. (B, C, and D) Model. (B) $X = 1.5$, duration of the ramp $100\tau_m$. (C) $X = 1$, duration of the ramp $100\tau_m$. (D) $X = 1$, duration of the ramp $390\tau_m$.

In contrast, very slow ramps that pass through the negative resistance region compared with the characteristic time of the wind-up process θ, always give rise to bistability hysteresis in our model (Figure 16.3C). For more realistic models of the wind-up of the inward current, the bistability hysteresis is also more probable with slow ramps of V_{in} in relatively short cables. This should be due to a steeper range of negative resistance in the stationary membrane I–V_∞ relation. Both types of hysteresis are illustrated in Figure 16.8. As the maximal V_{in} gradually shifts to more depolarized values with every ramp in the train, there are narrow hysteresis loops, at first due to the change of N (wind-up hysteresis, Figure 16.8B, three upper loops; Figure 16.8C, two upper loops; Figure 16.8D, two upper loops). When V_{in} exceeds some threshold value there is a dramatic increase in the width of the loop (Figure 16.8B, two lower loops; Figure 16.8D, three lower loops). Thus, two families of hysteresis are clearly distinguished in the case of long cables or very slow ramps. It is difficult to distinguish them in the case of short cables and moderately slow ramps (Figure 16.8C). The family of narrow loops corresponds to the wind-up hysteresis, caused by the wind-up of the inward current, while the family of broad loops corresponds to the bistability hysteresis, caused by the bistability of dendrites according to the shape of the I–V_∞ relation (Figure 16.8D) or $f(V, N \approx 0)$ (Figure 16.8B). This conclusion is confirmed by examining the distribution of the potential along the cable (not shown).

There is one additional criterion for discrimination between wind-up and bistability hysteresis. When hysteresis is caused only by the wind-up of the inward current, all loops return to the straight Ohmic lines at the same point (Figure 16.8B, three upper loops; Figure 16.8C and D, two upper loops). According to Equation (16.10), $N(t)$ decays independently of its maximal value. The loops of bistability hysteresis return to an Ohmic I–V_{in} relation at distinctly different points (Figure 16.8B,

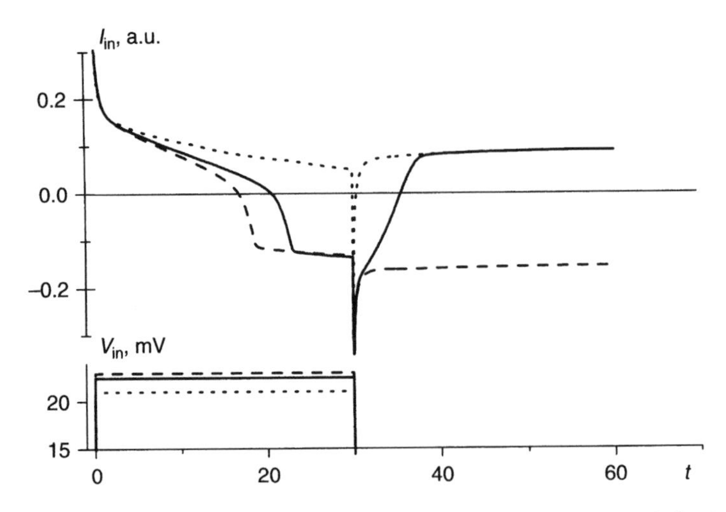

FIGURE 16.9 Model tail currents at clamped potential of 10 mV following large depolarization. Dotted curve, almost Ohmic tail current; solid curve, slow decaying tail current of the wind-up of the inward current; dashed curve, the nondecaying tail current caused by the dendritic bistability, $X = 1.5$.

two lower loops; Figure 16.8D, three lower loops). These points may change slightly if steady depolarization is not completely established at the distal end of the cable (Figure 16.8B, compare two lower loops). Thus, if the experimental loops merge with the Ohmic I–V_{in} relation at the same point, then the hysteresis is predominantly caused by the wind-up. If these loops end at clearly different points, then they may be caused by dendritic bistability.

A different manifestation of the two types of hysteresis may be the tail current observed by Schwindt and Crill (1980b). The experimental paradigm was the following: first, the somatic potential was clamped at RP, then at high depolarization, then again at RP, or at low depolarization. Two types of tail currents were observed on return to RP or low depolarization: a nondecaying inward current or a slowly decaying inward current. The latter decayed an order of magnitude slower than simple electrotonic decay. We have modeled this observation. The nondecaying current corresponds to a depolarization that is strong enough to induce steady depolarization at the distal end of the cable (Figure 16.9, dashed curve). A slightly lower depolarization, which is insufficient to induce the steady depolarization, is followed by a tail current that decays in accordance with the time-constant of wind-up (Figure 16.9, solid curve).

Finally, let us consider the wave of steady depolarization in a long cable in case of wind-up (Figure 16.10). The analytical solution of the wave equation in a semi-infinite cable is presented in Appendix 16.A3. The important result is the very slow speed of the wave, $c = 0.023$. This wave proceeds 27 times slower than in a cable with a stationary membrane I–V relation. The length of the wave front, $\Lambda = H/(\mathrm{d}V/\mathrm{d}x)|_{V_S} \approx 2\lambda$, calculated at the point of unstable equilibrium, is a little shorter than the corresponding length in a cable with time-independent membrane I–V relation. Correspondingly, the duration of the development of the wave in the case of wind-up becomes almost 20 times longer, $T \approx 80\tau_m$. Taking into account that wind-up in our model is too fast compared with wind-up in experiments (Figure 16.6, see also Russo and Hounsgaard, 1994, 1999; Svirskis and Hounsgaard, 1997), we suppose that the speed of the real wave could be much slower and its duration much longer. The wave development may be tested using the same paradigm as in the experiments of Schwindt and Crill (1980b). If the wave has not developed completely in the long cable, then the return of V_{in} to RP from moderately long-lasting fixation of steady depolarization may block the advance of the wave, and the whole cable repolarizes (Figure 16.10B). The critical duration of steady depolarization fixation is shorter for shorter cables (Figure 16.10B). In very long cables and for moderate durations of the steady depolarization fixation, the wave occupies only the proximal

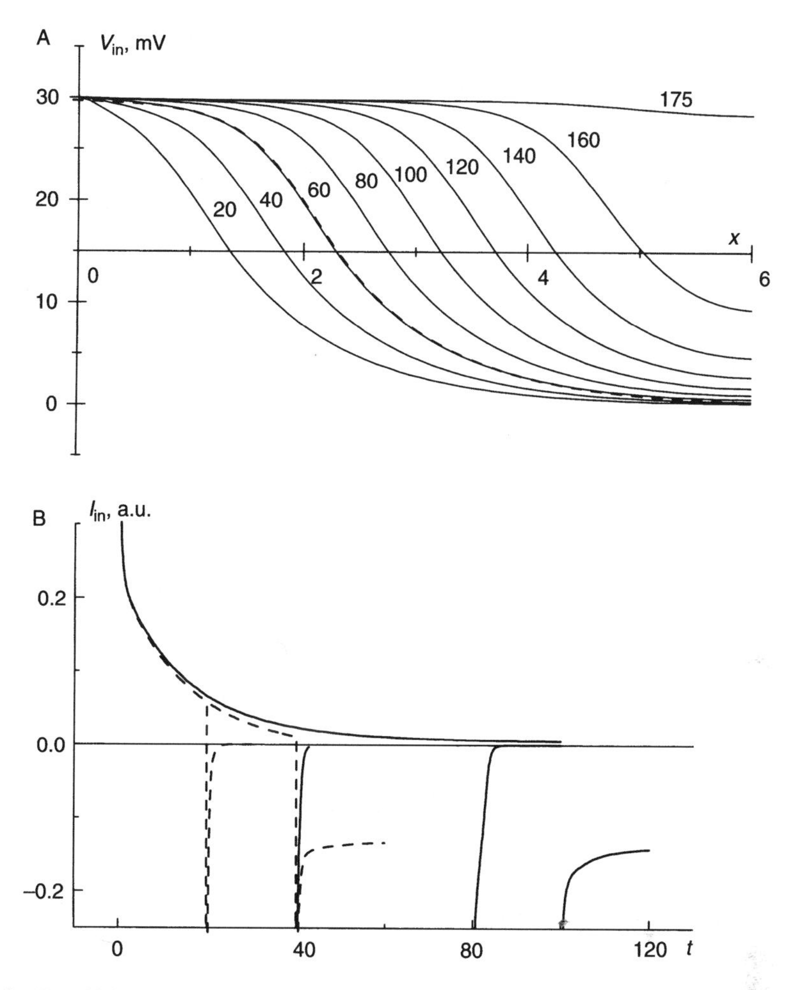

FIGURE 16.10 Dendritic cable recharging from the resting state to the steady depolarization state. (A) Development of the recharging wave front in a long cable ($X = 6$) with wind-up. The time points (in units of membrane time constant τ_m) are indicated at the curves. Dashed curve, analytical solution for the wave front propagated in a semi-infinite cable (Appendix 16.A3). (B) Current response of dendritic cables of electrotonic lengths $X = 6$ (solid curve) and $X = 3$ (dashed curve) to depolarizing voltage steps of 30 mV. The durations of the steps were 20 and 40 msec for the cable with $X = 3$ and 40, 80, and 100 msec for the cable with $X = 6$. The marginal duration for the recharging wave of a cable to develop completely (i.e., the marginal duration for the nondecaying tail current) is equal to $\sim 40\tau_m$ for the electrotonicaly shorter ($X = 3$) and $\sim 80\tau_m$ for the electrotonicaly longer ($X = 3$) cable, respectively.

part of the cable. Thus, following the clamp of the proximal end at RP, the wave occurs loaded (or shunted) from both ends of the cable and may collapse. In shorter cables and with the same duration of the steady depolarization clamp, the wave can reach the distal membrane of the cable and develop fully. Thus, following the fixation of RP, the wave occurs loaded only from the proximal end and may survive. The slow development of steady depolarization in the distal end of a cable, particularly prominent in the long cables, explains why the bistability hysteresis is not accompanied by a current jump in recordings during moderately slow ramps of fixed potential. The jump simply requires a longer time to develop (compare Figure 16.8B and C).

Therefore, cables with wind-up differ from cables with time-independent bistable membrane I–V relation in displaying much slower establishment of the bistability and a shorter critical length for cable bistability. In neurons with many dendrites, for example, motoneurons, the shorter critical

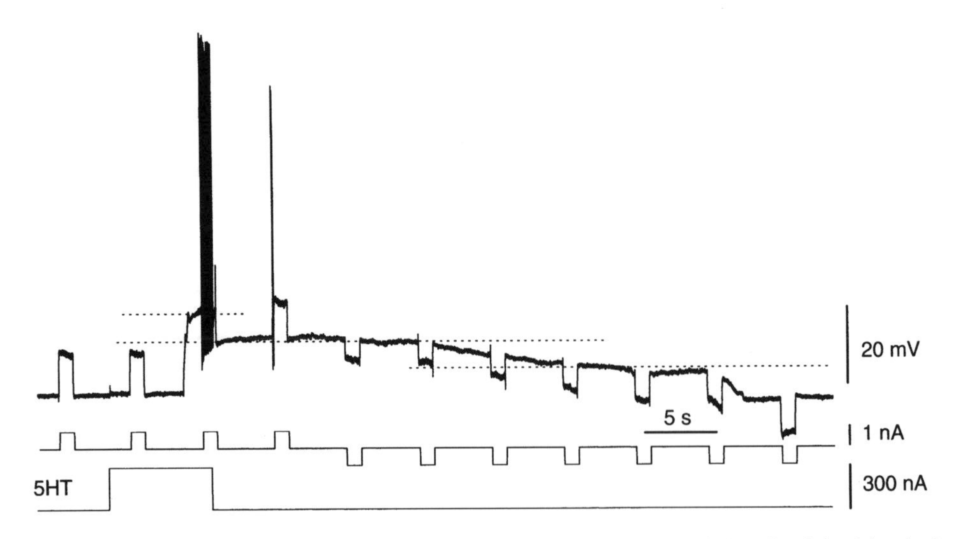

FIGURE 16.11 Multistability of turtle spinal motoneurons. Serotonin was applied to distal dendrites by iontophoresis. This evoked a plateau potential which decayed through multiple levels. Note that the input resistance is decreased and different at each plateau level.

length increases the probability of multistability, which could be functionally important in fine tuning of a prolonged firing rate. In agreement with this notion, recent experimental observations show multilevel decay of plateau potentials evoked by the iontophoretic application of serotonin to distal dendrites (Figure 16.11). Very slow switching between different levels (Figure 16.11) also indicates very slow recharging of the dendritic cables.

16.5 CONCLUSIONS and FUTURE PERSPECTIVES

The theory of bistable dendrites with time-independent membrane I–V relations explains important electrophysiological features of some central neurons including bistable behavior and the discharge pattern evoked by depolarizing current. However, it fails to account for the slow development, or wind-up, of the inward current responsible for bistability.

We have introduced the wind-up of the membrane I–V relation in the model by means of a threshold-like dependence of a slow deinactivation of the inward current on transmembrane voltage. There exist two types of hysteresis in the cable input current–voltage relation during input-voltage ramps in the bistable cable with wind-up of the inward current. A smaller hysteresis loop corresponds to the deinactivation–inactivation sequence during the ramp. It develops if the ramp duration is at least comparable to the time constant for the wind-up of the inward current but vanishes in the case of rather long ramps. A larger hysteresis loop corresponds to bistability of the distal membrane of the cable. The latter bistability should be considered in two contexts, namely, in a cable with an approximately constant level of inactivation and in a cable with completely developed deinactivation of the inward current. In the latter case, the bistability develops much more slowly and may be established in shorter cables.

Dendritic bistability brings about neuronal multistability or at least bistability. There are a few trivial implications for the modeling of neural networks with multistable cells (Gutman, 1984). First, long dendritic branches may accomplish all logical functions (Gutman, 1984, 1991). In this way, the neuron becomes an integrating circuit. We could consider such a neuron as an equivalent of a small integrating chip in an electronic device. Second, a sheet of noninteracting multistable cells may serve as a memory screen. Such a screen imitates the iconic memory of human beings.

More interesting is a nontrivial relation of multi- or bistable cells (Gutman, 1994; Gutman and Svirskis, 1997) to a neurodynamical principle postulated by Gelfand and Tsetlin (1969) for the CNS in accordance with games studied in automata. They stated that afference (in a general sense) is minimized, that is, afference is meant to be punishment. Neural networks with bistable cells and balanced excitation and inhibition may minimize their afferent input (Gutman, 1994; Gutman and Svirskis, 1997) by changing their state from the depolarization plateau (SD) to the RP and vice versa under transient afferent command, ultimately achieving a stable state when the afference becomes small enough, that is, when the network with bistable neurons functions properly.

The Gelfand–Tsetlin principle of minimal afference probably reflects a real mechanism of CNS functioning. Macefield et al. (1991) recorded afferent activity in humans during long and strong contraction. In contradiction to the servomotor theory, they found that 1a-afference declines with time in spite of fatigue. If the same contraction is repeated several times, afferent activity at the start of the contraction becomes smaller and smaller. The last observation seems to be a general phenomenon: the afferent firing is significant during learning and declines during memory consolidation, at least in networks involved in motor control (Kandel et al., 2000).

The Gelfand–Tsetlin principle of minimal afference and dendritic bistability can illuminate two existing theories of the motor control — the Feldman λ-model (also known as the equilibrium-point hypothesis) (Latash and Gutman, 1993; Feldman and Levin, 1995) and the Hogan–Flash principle of minimal jerk or smooth movement (Hogan and Flash, 1987; Viviani and Flash, 1995). The main gist of the equilibrium-point hypothesis is that the stable equilibrium length of a loaded muscle is determined by a neural command. In accordance with the Gelfand–Tsetlin principle, a network with bistable or multistable neurons reaches a stable equilibrium state when the afference becomes minimal. The Hogan–Flash principle of minimal jerk also follows from the Gelfand–Tsetlin principle of minimal afference. A smooth movement between fixed starting and ending points during time T corresponds to min H, where $H \sim \int_0^T (\mathrm{d}F/\mathrm{d}t)^2\,\mathrm{d}t$ is a mathematical expression for jerk; here F is a muscle force. F is approximately proportional to the number of active motoneurons, n, if agonists and antagonists are not active at the same time, thus: $H \sim \int_0^T (\mathrm{d}n/\mathrm{d}t)^2\,\mathrm{d}t$. For bistable motoneurons, $\mathrm{d}n/\mathrm{d}t$ is a measure of afference; then, $H \sim \int_0^T (\text{afferentation})^2\,\mathrm{d}t$, and minimization of H presents an example of the Gelfand–Tsetlin principle.

Finally, bistable neurons can be introduced in the concept of the classical quasineuron. If a classical quasineuron has an excitatory suprathreshold autapse then the cell becomes functionally bistable: transient excitation evokes and transient inhibition stops steady firing. We hope that the Gelfand–Tsetlin principle together with the concept of neuronal multi- or bistability may be exploited in the modeling of neural networks.

16.6 Summary

In many central neurons slow inward currents are distributed throughout the soma–dendritic membrane. This spatial distribution of slow inward current creates a possibility for dendritic bistability: when the membrane potential at one end of a bistable dendrite is clamped the membrane potential at the other end can have one or two different stable values. Here, we review the theory of bistable dendrites with an instantaneously activating inward current. It is known that in spinal motoneurons slowly activating inward currents have potential-dependent facilitation. We included the facilitation of the membrane current–voltage relation in the model by means of a threshold-like dependence of a slow deinactivation of the inward current on transmembrane voltage. The stationary properties of such dendritic cables were explored analytically. Numerical results from simulations were compared with experimental observations in spinal motoneurons. The analysis shows that potential-dependent facilitation of the inward current brings about dendritic bistability for electrotonically short cables, which leads to multistability and increased computational complexity in neurons with branched dendrites.

ACKNOWLEDGMENT

We would like to thank Jean-Francois Perrier for providing experimental data for Figure 16.11.

PROBLEMS

1. Long-lasting inward currents have been observed in many cell types. Show how the presence of dendritic bistability in these cell types can be determined from the I–V_{in} relation obtained with voltage-clamp ramps.
2. This work employs a very simplistic model of wind-up of the inward current. Several mechanisms could account for wind-up: the transmembrane potential may influence (a) the voltage threshold of channel activation; (b) channel opening probability; (c) channel-open state duration (lifetime); (d) channel conductivity; or (e) it may induce a transition of a subpopulation of channels, residing in an inactivated state, to a deinactivated state. The mechanism by which depolarization induces wind-up is unknown but may involve G-proteins, channel phosphorylation, or changes in intracellular calcium concentration. Create a more elaborated model that simulates some of these possibilities. This could allow a quantitative comparison between models and experimental data.
3. We modeled simple cables in order to show qualitative features of dendritic bistability. Complex interactions between the dendritic branches may take place in real neurons with nonlinear membrane properties. Investigate this interaction in realistic models of morphologically reconstructed neurons.
4. Cat motoneurons and some other nerve cells have characteristic discharge frequency–stimulus current relations with two linear ranges with different slopes. This property may be explained as a result of the recharging of bistable dendrites from the resting potential state to the steady depolarization state. Investigate in the model of the dendritic cable how the potential-dependent facilitation (wind-up) of the inward current influences the discharge frequency–stimulus current relation.
5. Dendritic bistability allows basic logic functions to be performed on the separate dendrites or even dendritic branches. Investigate the information processing capabilities of neural networks with bi- or multi-stable neurons.

APPENDIX

16.A1 List of Symbols and Abbreviations

I–V — stationary membrane current–voltage relation

I–V_{in} — cable input current–voltage relation

I–V_{∞} — stationary membrane current–voltage relation in case of wind-up of the inward current

F–C — discharge frequency–stimulus current relation

RP — resting membrane potential

RP solution — solution approaching RP at the distal end of the cable

SD — steady depolarization

SD solution — solution approaching SD at the distal end of the cable

c — wave velocity in an infinite bistable cable in electrotonic units (λ/τ_m)

$f(V)$ — membrane I–V multiplied by membrane resistance, mV

$f(V,N)$ — $f(V)$ in the case of wind-up

$f_\infty(V)$ — stationary $f(V,N)$ in the case of wind-up

h — potential of unstable equilibrium depolarization, medial zero-point of $f(V)$

H — potential of steady depolarization

I_{in} — input current

$k = -df/dV|_h$, or negative membrane conductance at $V = h$, normalized to conductance at RP

$K = df/dV|_H$, membrane conductance at SD normalized to conductance at RP

$L = 1/\sqrt{k}$, critical length parameter for oscillatory solution around h, in length constant units

$N(t)$ — variable of wind-up, dimensionless

$R_{in} = dV_{in}/dI_{in}$, differential input resistance of the cable

t — electrotonic time, in membrane time constant units

$u = \int_0^H f(V)\,dV$, I–V parameter

V — transmembrane potential relative to the resting potential

$V_{in} = V(x = 0)$, input potential at the proximal end of the cable

V_{RP} — maximal possible V_{in} for RP-solution

V_S — threshold of wind-up

V_{SD} — minimal possible V_{in} for SD-solution

V_1 — potential at current maximum of linear-broken membrane I–V

V_2 — potential at current minimum of linear-broken membrane I–V

x — cable axial electrotonic coordinate, in length constant units

X — cable electrotonic length

X_{cr} — cable critical electrotonic length

λ — electrotonic length constant

Λ — electrotonic length of wave front in infinite bistable cable

τ_m — membrane time constant

θ — wind-up time constant

T — duration of wave front in infinite bistable cable in membrane time constant units.

16.A2 Wave Propagation in Infinite Bistable Dendrites with Time-Independent *I–V* Relation

The numerical solution of wave propagation in a bistable dendrite with time-independent I–V relation was given previously (Butrimas and Gutman, 1981; Gutman, 1984, 1991). Here we give an analytical solution. The wave solution of Equations (16.6) and (16.7) for the piecewise-linear I–V relation [Equation (16.2)] is presented as an ensemble of solutions for the three linear segments of $f(V)$: for the Ohmic segment near the RP, the negative resistance segment near h, and the Ohmic segment near steady depolarization:

$$\begin{aligned}
&\frac{\mathrm{d}^2V}{\mathrm{d}x^2} + c\frac{\mathrm{d}V}{\mathrm{d}x} - V = 0, \quad V \le V_1, \quad V \to 0 \text{ as } x \to \infty;\\
&\text{let } V(0) = V_1; \quad \text{thus, } V = V_1 \exp\left[-x\left(c/2 + \sqrt{c^2/4+1}\right)\right];\\
&\frac{\mathrm{d}^2V}{\mathrm{d}x^2} + c\frac{\mathrm{d}V}{\mathrm{d}x} + k(V-h) = 0, \quad V_1 \le V \le V_2;\\
&\qquad \text{thus, } V = h + \frac{V_1 - h}{\cos\varphi}\exp(-cx/2)\cos\left(x\sqrt{k - c^2/4} + \varphi\right);\\
&\frac{\mathrm{d}^2V}{\mathrm{d}x^2} + c\frac{\mathrm{d}V}{\mathrm{d}x} - K(V-H) = 0, \quad V \ge V_2, \quad V \to H \text{ as } x \to -\infty;\\
&\text{let } V(-l) = V_2; \quad \text{thus, } V = H + (V_2 - H)\exp\left[(x+l)\left(\sqrt{K + c^2/4} - c/2\right)\right].
\end{aligned} \tag{16.A2.1}$$

The potential continuity at V_1 was ensured by boundary conditions and taken into account in expressions for membrane potential in Equation (16.A2.1). Continuity of the axial current along the wave front at potentials V_1 and V_2 and the potential continuity at V_2 give the required equations for establishing of the parameters l, φ, and c:

$$\begin{aligned}
&-V_1\left(\sqrt{1 + c^2/4} + c/2\right) = (h - V_1)\left(\tan\varphi\sqrt{k - c^2/4} + c/2\right);\\
&V_2 = h + \frac{V_1 - h}{\cos\varphi}\exp(cl/2)\cos\left(l\sqrt{k - c^2/4} + \varphi\right);\\
&\frac{h - V_1}{\cos\varphi}\exp(cl/2)\left[\sqrt{k - c^2/4}\sin\left(l\sqrt{k - c^2/4} + \varphi\right) + \frac{c}{2}\cos\left(l\sqrt{k - c^2/4} + \varphi\right)\right]\\
&\qquad = (V_2 - H)\left(\sqrt{K + c^2/4} - c/2\right).
\end{aligned} \tag{16.A2.2}$$

The obvious relations from Equation (16.2) were used for (16.A2.2): $h = (1 + 1/k)V_1$ and $H = V_2(1 + k/K) - V_1(1 + k)/K$.

16.A3 WAVE PROPAGATION IN INFINITE DENDRITES WITH WIND-UP OF A BISTABLE *I–V* RELATION

Analytical solution of the wave propagation is possible when wind-up is considered as a threshold phenomenon. Wind-up develops along the wave front at points behind the point with the potential $V = V_S$. Thus, for the wave-front segment ahead of this point, that is, for $V < V_S$, the Ohmic solution and corresponding continuity conditions of Equations (16.A2.1) and (16.A2.2) are valid:

$$V = V_S \exp\left[-x\left(\sqrt{1+c^2/4}+c/2\right)\right], \qquad V < V_S. \tag{16.A3.1}$$

Here, $V(0) = V_S$ and thus Equation (16.A3.1) is valid for $x \geq 0$. The wind-up Equation (16.10) holds for all $V \geq V_S$, that is, for points $x \leq 0$:

$$N = \exp(-t/\theta) = \exp(x/c\theta), \qquad V \geq V_S. \tag{16.A3.2}$$

Thus, according to Equation (16.9):

$$\begin{aligned}
&\frac{d^2V}{dx^2} + c\frac{dV}{dx} + k(V-h) - [V_S(1+k) - kh]\exp(x/c\theta) = 0;\\
&V(0) = V_S, \quad \left.\frac{dV}{dx}\right|_{V_S} = -V_S\left(\sqrt{1+c^2/4}+c/2\right); \quad V_S \leq V \leq V_2;\\
&\frac{d^2V}{dx^2} + c\frac{dV}{dx} - K(V-H) - [V_S(1+k) - kh]\exp(x/c\theta) = 0;\\
&V(-\infty) = H; \quad V \geq V_2.
\end{aligned} \tag{16.A3.3}$$

Continuity of potential and its derivative at $V = V_2$ for both solutions of Equation (16.A3.3) is assumed. These solutions may be expressed analytically. For $V_S \leq V \leq V_2$ we get:

$$\begin{aligned}
V &= h + \frac{V_S(1+k)-kh}{1/(c\theta)^2+1/\theta+k}\exp\left(\frac{x}{c\theta}\right)\\
&\quad + \left[V_S - h - \frac{V_S(1+k)-kh}{1/(c\theta)^2+1/\theta+k}\right]\frac{\exp(-xc/2)}{\cos\varphi}\cos\left(x\sqrt{k-c^2/4}+\varphi\right);\\
\tan\varphi &= \frac{V_S(\sqrt{1+c^2/4}+c/2) + (V_S(1+k)-kh)/(1/c\theta+c+c\theta k)}{[V_S - h - (V_S(1+k)-kh)/(1/(c\theta)^2+1/\theta+k)]\sqrt{k-c^2/4}} - \frac{c}{2\sqrt{k-c^2/4}}
\end{aligned} \tag{16.A3.4}$$

For $V \geq V_2$:

$$V = H - \frac{V_S(1+k)-kh}{1/(c\theta)^2+1/\theta-K}\exp\left(\frac{x}{c\theta}\right) + u\exp\left[x\left(\sqrt{K+c^2/4}-c/2\right)\right]. \tag{16.A3.5}$$

Now we have three equations for establishing parameters c, u, and l. At the point $x = l$, solutions of both Equations (16.A3.4) and (16.A3.5) equal V_2 and their derivatives are equal.

17 Bifurcation Analysis of the Hodgkin–Huxley Equations

Shunsuke Sato, Hidekazu Fukai, Taishin Nomura, and Shinji Doi

CONTENTS

17.1 Introduction

Single neurons are the basic elements for information processing in the central nervous system. Neuronal information is embedded in the spike trains generated by neurons (Perkel and Bullock, 1968; Rieke et al., 1997). In the 1950s the discharge of single neurons was enthusiastically studied using electrophysiological methods and it was elucidated that the generation of action potentials crucially involves the electrical properties of the membrane and its ionic conductances. Hodgkin and Huxley (1952a) showed by their famous experiments with the giant axon of *Loligo* that the electrical excitation of a nerve axon is propagated by changes in the concentration differences of several ions inside and outside the axon. The membrane permeabilities for Na^+ and K^+ are responsible for these changes, and these permeabilities vary as a function of the membrane potential. Hodgkin and Huxley derived a set of differential equations describing the changes in the selective ion-channel conductances and the associated membrane-potential changes. These equations (Hodgkin and Huxley, 1952b) were later called the Hodgkin–Huxley equations, hereinafter the H–H equations for the sake of simplicity. The H–H equations are now considered as the prototypical phenomenological model of electrical excitation in single nerve cells. They consist of a set of nonlinear differential equations in four variables. The equations include several adjustable parameters and show qualitatively different behaviors depending on the values of those parameters. In the present chapter, we analyze the behavior of the H–H equations from the point of view of bifurcation theory in nonlinear dynamical systems theory.

17.2 The Hodgkin–Huxley Equations

Hodgkin and Huxley (1952b) proposed that the membrane potential of the squid giant axon can be described by the differential equation

$$C\frac{\mathrm{d}V}{\mathrm{d}t} = I_{\text{ext}} - g_{\text{Na}}m^3h(V - V_{\text{Na}}) - g_{\text{K}}n^4(V - V_{\text{K}}) - g_{\text{L}}(V - V_{\text{L}}), \tag{17.1}$$

where $V(t)$ (mV) is the membrane potential, t is time (msec), C (usually set to $1\,\mu\text{F/cm}^2$) is the specific membrane capacitance, V_{Na} (115 mV) is the sodium equilibrium potential, V_{K} (-12 mV) is the potassium equilibrium potential, V_{L} (10.613 mV) is the leakage equilibrium potential, g_{Na} (120 mS/cm^2) is the maximum sodium conductance, g_{K} (36 mS/cm^2) is the maximum potassium conductance, g_{L} (0.3 mS/cm^2) is the leak conductance, and I_{ext} ($\mu\text{A/cm}^2$) is an externally applied current. The auxiliary variables m, h, and n (which may be interpreted as probabilities) satisfy

$$\frac{\mathrm{d}m}{\mathrm{d}t} = \phi(\theta)[\alpha_m(1-m) - \beta_m m], \tag{17.2}$$

$$\frac{\mathrm{d}h}{\mathrm{d}t} = \phi(\theta)[\alpha_h(1-h) - \beta_h h], \tag{17.3}$$

$$\frac{\mathrm{d}n}{\mathrm{d}t} = \phi(\theta)[\alpha_n(1-n) - \beta_n n], \tag{17.4}$$

where $\phi(\theta) = 3^{(\theta-6.3)/10}$ expresses their dependence on temperature θ (°C). By fitting experimental data, Hodgkin and Huxley (1952b) represented the rate functions in Equations (17.2) to (17.4) by the formulae

$$\begin{aligned}
\alpha_m &= 0.1\frac{-V+25}{\exp((-V+25)/10)-1}, & \beta_m &= 4\exp(-V/18),\\
\alpha_h &= 0.07\exp(-V/20), & \beta_h &= \frac{1}{\exp((-V+30)/10)+1},\\
\alpha_h &= 0.01\frac{-V+10}{\exp((-V+10)/10)-1}, & \beta_m &= 0.125\exp(-V/80).
\end{aligned} \tag{17.5}$$

Equation (17.1) together with Equations (17.2) through (17.4) will subsequently be referred to as the Hodgkin–Huxley equations in this work. These equations have no known closed-form solutions but are easily solved numerically, for example, by the Runge–Kutta method. For a steady current I_{ext}, Equations (17.1) to (17.4) have a steady-state solution, V, satisfying

$$I_{\text{ext}} = g_{\text{Na}}m^3h(V - V_{\text{Na}}) + g_{\text{K}}n^4(V - V_{\text{K}}) + g_{\text{L}}(V - V_{\text{L}}),$$

where the functions m, h, and n take on steady-state values

$$m_\infty(V) = \frac{\alpha_m(V)}{\alpha_m(V)+\beta_m(V)}, \qquad h_\infty(V) = \frac{\alpha_h(V)}{\alpha_h(V)+\beta_h(V)}, \qquad n_\infty(V) = \frac{\alpha_n(V)}{\alpha_n(V)+\beta_n(V)}. \tag{17.6}$$

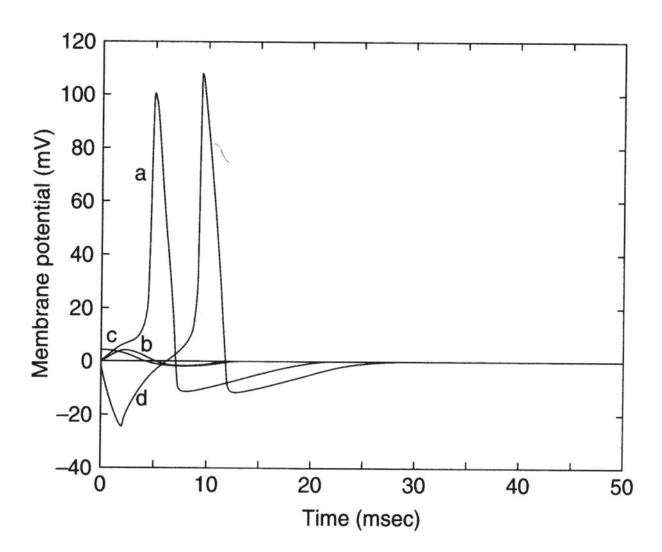

FIGURE 17.1 Membrane potential responses to four current stimuli. (a) Stimulus intensity $A = 4\ \mu A/cm^2$, duration $w = 2$ msec. (b) $A = 3\ \mu A/cm^2$, $w = 2$ msec. (c) $A = 50\ \mu A/cm^2$, $w = 0.1$ msec. (d) $A = -20\ \mu A/cm^2$, $w = 2$ msec.

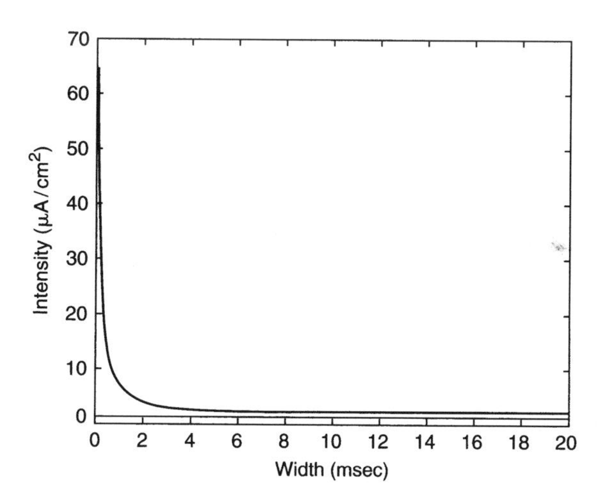

FIGURE 17.2 The Hodgkin–Huxley neuron at equilibrium discharges an action potential in response to a single current stimulus with a set of parameters located at the upper region of the hyperbolic curve. Abscissa: width or duration w (msec) and ordinate: intensity $A(\mu A/cm^2)$.

Figure 17.1 illustrates the time course of the membrane potential generated by the H–H equations in response to current pulses of different magnitudes, $A\ \mu A/cm^2$, and durations, w msec. The figure illustrates combinations of A and w that do or do not lead to the generation of an action potential. Curves (b) and (c) show, respectively, the time courses of the membrane potential for a small current pulse of prolonged duration and for a large current pulse of short duration, neither of which elicits a membrane action potential. Anode-break excitation is illustrated in curve (d). Figure 17.2 shows the strength–duration curve for the generation of membrane action potentials with single square-wave pulses. For the Hodgkin–Huxley membrane, this threshold curve is well described by the relation $(A - A_0)(w - w_0) \geq S_0$, where $A_0 \approx 0.89\ \mu A/cm^2$, $w_0 \approx 0.009$ msec, and $S_0 \approx 6.0\ \mu A \cdot msec/cm^2$.

17.2.1 Linearization of the Hodgkin–Huxley Equations

For a fixed externally applied current I_{ext}, the H–H equations have a corresponding stationary solution, referred to as an equilibrium (the alternative terms "equilibrium solution" and "equilibrium point" will also be used here), in which the membrane potential and the auxiliary variables m, n, and h are constant. The nature of this equilibrium solution can be investigated by calculating the eigenvalues of the H–H equations when linearized about an equilibrium solution $(\overline{V}, \overline{m}, \overline{h}, \overline{n})$. The linearized equations can be represented as

$$\frac{\mathrm{d}}{\mathrm{d}t}\begin{pmatrix} V \\ m \\ h \\ n \end{pmatrix} = J\left(\overline{V}, \overline{m}, \overline{h}, \overline{n}\right)\begin{pmatrix} V - \overline{V} \\ m - \overline{m} \\ h - \overline{h} \\ n - \overline{n} \end{pmatrix}.$$

The Jacobian matrix J with respect to an equilibrium solution, ignoring the temperature dependence of the Hodgkin–Huxley variables, is

$$J = \begin{pmatrix} -(g_{\text{Na}}m^3h + g_{\text{K}}n^4 + g_{\text{L}}) & -3g_{\text{Na}}m^2h(V - V_{\text{Na}}) & -g_{\text{Na}}m^3(V - V_{\text{Na}}) & -4g_{\text{K}}n^3(V - V_{\text{K}}) \\ \dfrac{\mathrm{d}\alpha_m}{\mathrm{d}V} - m\left(\dfrac{\mathrm{d}\alpha_m}{\mathrm{d}V} + \dfrac{\mathrm{d}\beta_m}{\mathrm{d}V}\right) & -(\alpha_m + \beta_m) & 0 & 0 \\ \dfrac{\mathrm{d}\alpha_h}{\mathrm{d}V} - h\left(\dfrac{\mathrm{d}\alpha_h}{\mathrm{d}V} + \dfrac{\mathrm{d}\beta_h}{\mathrm{d}V}\right) & 0 & -(\alpha_h + \beta_h) & 0 \\ \dfrac{\mathrm{d}\alpha_n}{\mathrm{d}V} - n\left(\dfrac{\mathrm{d}\alpha_n}{\mathrm{d}V} + \dfrac{\mathrm{d}\beta_n}{\mathrm{d}V}\right) & 0 & 0 & -(\alpha_n + \beta_n) \end{pmatrix}.$$

Figure 17.3 shows the eigenvalues of the Jacobian matrix J when plotted against the externally applied current I_{ext}. For small I_{ext}, the eigenvalues of the linearized problem are real and negative, and therefore the equilibrium solution is stable. As I_{ext} is increased, a pair of complex-conjugate eigenvalues with negative real part arise, and, as I_{ext} increases beyond $I_{\text{ext}} \approx 9.769$, the real part of these complex eigenvalues becomes positive, indicating that the equilibrium solution is now unstable, and the model's solutions start to oscillate. The real part of the complex eigenvalues remains positive until $I_{\text{ext}} > 154.25$, at which point it becomes negative and the equilibrium solution of the H–H equations recovers stability (see Sabah and Spangler, 1970; Evans, 1972; Evans and Feroe, 1977).

17.2.2 Reduction of the Hodgkin–Huxley Equations

A number of procedures have been proposed to reduce the dimension of the H–H equations without losing their essential properties (e.g., FitzHugh, 1960; Morris and Lecar, 1981; Kepler et al., 1992). For example, FitzHugh (1960) observed that m responds to changes in membrane potential on a submillisecond timescale, and on this basis proposed that occurrences of m in the H–H equations be replaced by its equilibrium value

$$\overline{m}(V) = \frac{\alpha_m(V)}{\alpha_m(V) + \beta_m(V)}.$$

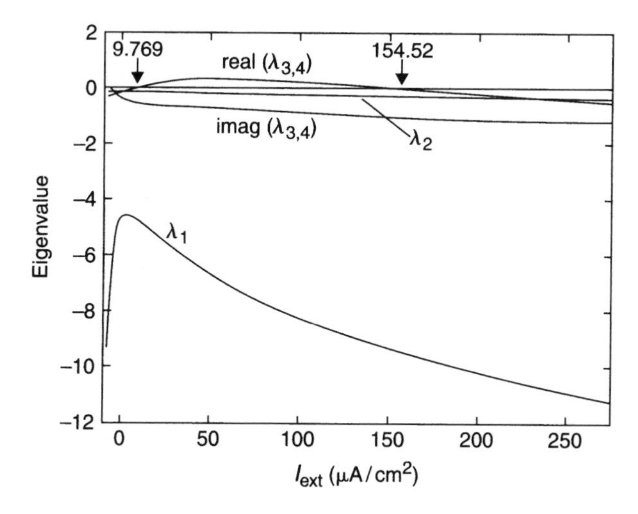

FIGURE 17.3 Eigenvalues of the Jacobian matrix.

FitzHugh also observed from a phase–space analysis of the behavior of n and h that $h + n \approx 0.85$. Taken together, these two observations lead to the simplified the H–H equations

$$C\frac{dV}{dt} = I_{ext} - g_{Na}\overline{m}^3(V - V_{Na})(0.85 - n) - g_K(V - V_K)n^4 - g_L(V - V_L), \tag{17.7}$$

$$\frac{dn}{dt} = \alpha_n(1 - n) - \beta_n n. \tag{17.8}$$

Figure 17.4A and B shows the lines dV/dt and $dn/dt = 0$ for Equations (17.7) and (17.8). These lines are called nullclines. Their crossover point(s) give the equilibrium solution(s) of the reduced H–H equations. In Figure 17.4A, the stationary solution is stable, while in Figure 17.4B it is unstable and one sees a trajectory of a limit–cycle oscillation. Figure 17.4C and D illustrates that the set of equations behave as a model of an excitatory membrane or of an oscillatory membrane (see Krinskii and Kokoz, 1973; Abbott and Kepler, 1990; Kepler et al., 1992; Meunier, 1992; Kistler et al., 1997).

17.3 Global Structure of Bifurcation in the Hodgkin–Huxley Equations

In Section 17.2.1 we showed that the linear stability of the equilibrium solution of the H–H equations changes as the constant current I_{ext} increases. In this section, we explore behaviors of the H–H equations for various values of the parameters involved in the equations. When we look at the temporal dynamics of the model's membrane potential for a given set of parameter values, we fix I_{ext} at some value, providing a corresponding equilibrium solution, and consider the behavior of the H–H equations in response to small perturbations from the equilibrium solution. These perturbations may be generated by instantaneously depolarizing the membrane, or alternatively, by applying a current stimulus with intensity A μA/cm^2 and duration w msec in addition to the fixed external current I_{ext}.

17.3.1 Bifurcation due to I_{ext}

When $I_{ext} = 0$, the H–H equations have one stable equilibrium solution, which means that in response to a small perturbation, the model's state point returns to the equilibrium solution without generating

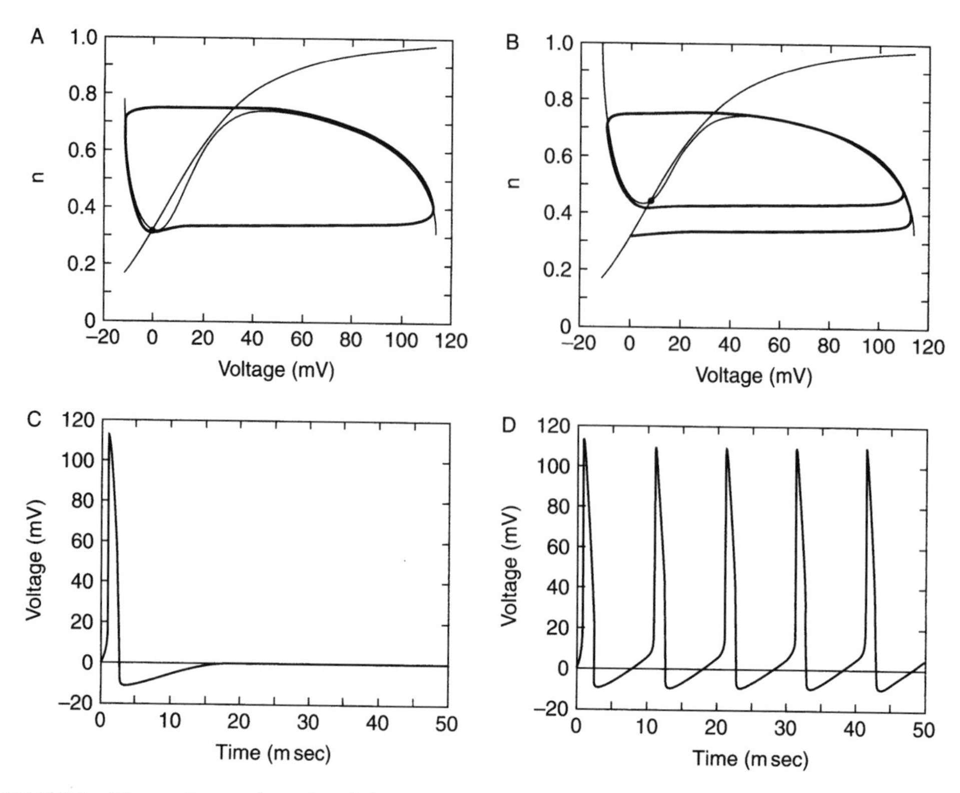

FIGURE 17.4 Phase–plane trajectories of the two-dimensional reduced H–H equations. (A) Thin lines indicate V and n-nullclines. Thick lines show the trajectories (V, n). No constant current was applied ($I_{ext} = 0$). The stimulus condition was $A = 10\ \mu A/cm^2$, $w = 2$ msec. (C) Corresponding membrane potential. (B) and (D) DC current of $I_{ext} = 15\ \mu A/cm^2$ was applied. The membrane potential oscillates in this case.

an action potential. Figure 17.5C shows the time course of the membrane potential in response to an instantaneous depolarization of the membrane potential by 10 mV at 10 msec after the beginning of the simulation.

When the externally applied constant current I_{ext} increases slightly from zero, the H–H equations exhibit multiple action potentials. This is a transient behavior and the membrane potential converges eventually to the equilibrium point corresponding to the value of I_{ext}. Figure 17.5D illustrates the time course of the membrane potential for such a case with $I_{ext} = 6.2$ in response to a perturbation of 5 mV. The amplitude of the transient action potentials decreases as I_{ext} is close to 6.264, and eventually the transient oscillations become subthreshold. For $6.264 < I_{ext} < 9.769$, the membrane potential waveform can exhibit both a subthreshold oscillation that asymptotes at the equilibrium solution and stable periodic waveforms depending on the amplitude of the small perturbation delivered. Figure 17.5E illustrates such an example with two different waveforms, viz., periodic discharges and an action potential converging to the stable equilibrium point for $I_{ext} = 6.3$, depending on the amplitude of the perturbation (6 mV for the former and 3 mV for the latter).

At $I_{ext} = 9.769$, the equilibrium solution of the H–H equations becomes unstable and only the periodic discharges are stable (Figure 17.5F, $I_{ext} = 10$). Further increases in the value of I_{ext} lead to small amplitude, but sustained, oscillations (Figure 17.5G, $I_{ext} = 100$) that disappear altogether at $I_{ext} = 154.25$ (the point at which stability is reestablished), and only damped oscillations of small amplitude remain for a range of I_{ext} slightly larger than 154.25 (Figure 17.5H, $I_{ext} = 160$).

In Figure 17.5A and B, the equilibrium potential is plotted as a function of I_{ext}, where it appears as a monotone increasing curve (Figure 17.5B is an enlargement of Figure 17.5A for

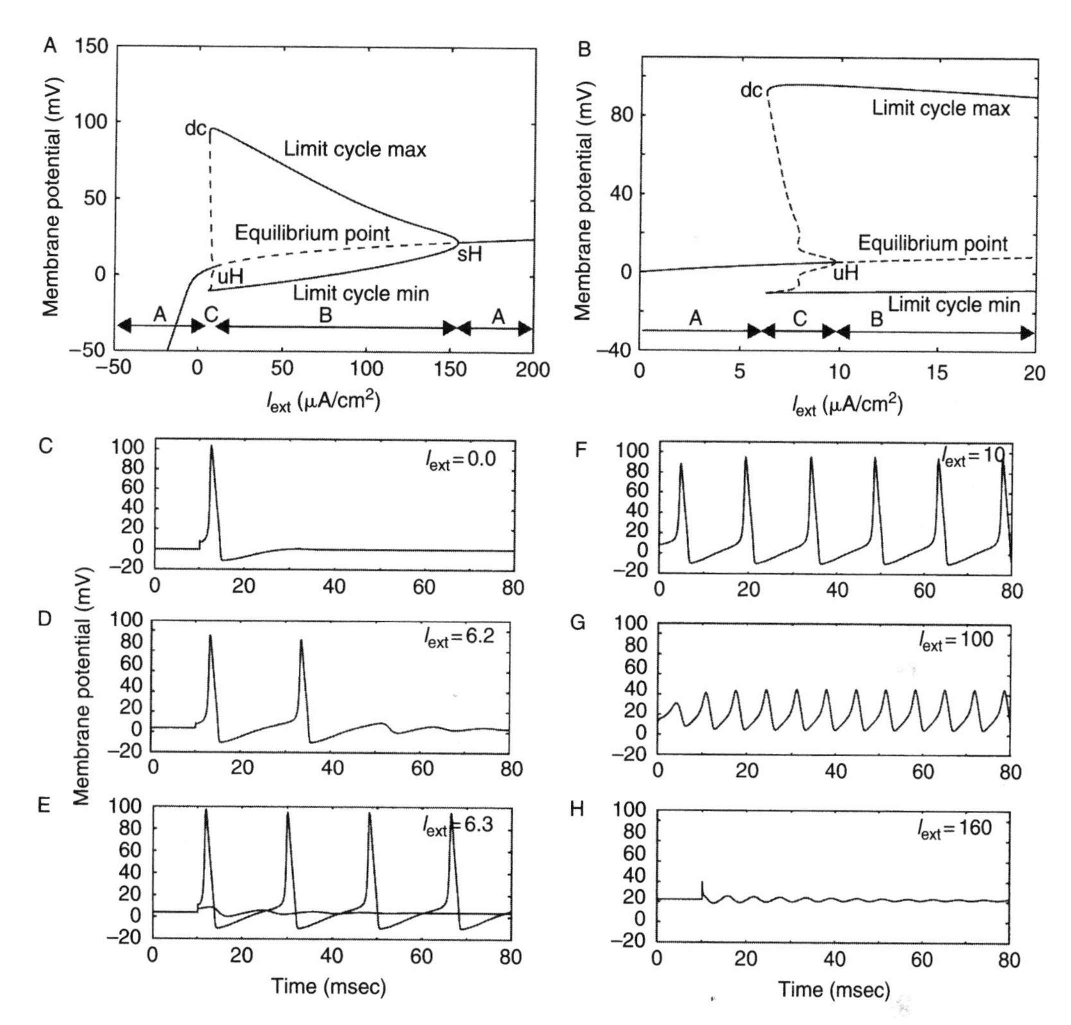

FIGURE 17.5 (A) One-parameter bifurcation diagram of the H–H equations with the original parameter values. Abscissa is the constant applied current, I_{ext}, and the ordinate is the membrane potential. (B) Enlargement of the region C in (A). (C) to (H) Time courses of the membrane potential for $I_{\text{ext}} = 0, 6.2, 6.3, 10, 100, 160\ \mu\text{A/cm}^2$, respectively. In (C), (D), (E), and (H), the membrane potential is perturbed instantaneously from the equilibrium potential at $t = 10$ msec. In (E), the system is bistable so that, depending on the intensity of perturbation, the membrane potential tends to either periodic firing or equilibrium potential.

the region $0 < I_{\text{ext}} < 12$). At the points $I_{\text{ext}} = 9.769$ and $I_{\text{ext}} = 154.25$ marked by the symbols uH and sH, the stability of the equilibrium point of the H–H equations changes as described above, and the equilibrium curve is indicated by the dotted line between these two points, implying that the equilibrium point is unstable. Curves started from these two points represent the branches of the periodic solutions. The curves departing for upper and lower sides of the equilibrium curve represent, respectively, the maximum and the minimum values of the periodic solution, specifying the amplitude of the periodic oscillations. As in the equilibrium curve, the periodic solution is stable for the solid branch and unstable for the dotted branch.

The horizontal axis of Figure 17.5A and B is divided into regions A, B, and C where the H–H equations behave in qualitatively different manners. The H–H equations possess one stable equilibrium point for any point in region A. In region C, the H–H equations can have one stable equilibrium point and a stable and an unstable limit cycle depending on the initial position. In region B, an unstable equilibrium point and a stable limit cycle are obtained. Figures 17.5C to H show time courses of membrane potential in the regions A, A, C, B, B, and A respectively. Rigorously speaking, in region C a period-doubling bifurcation takes place midway along the branch of the unstable periodic solution generated at the Hopf bifurcation point (defined below), viz., at $I_{ext} = 9.769$ (Rinzel and Miller, 1980).

Thus, at $I_{ext} = 6.264, 9.769, 154.25$, the behavior of the H–H equations changes drastically. The parameter points where the behavior of the equations changes qualitatively are called bifurcation points. A bifurcation point is defined as a critical point, ν_0, of a controlled parameter, ν, where the geometry of the basin of attractors in the phase plane changes (Thompson and Stewart, 1993).

Note that Figures 17.5C and D apply to region A. The difference between the waveforms in these figures is that the former shows damped oscillations of the membrane potential not seen in the latter. These oscillations are due to the presence in the former of complex-conjugate eigenvalues (with negative real part) of the Jacobian matrix at the equilibrium solution.

We now consider the bifurcation points at the edges of each region. The points $I_{ext} = 9.769$ and $I_{ext} = 154.25$, where the stability of the equilibrium point changes, are called Hopf bifurcation points. Hopf bifurcation points are of two types depending on the direction of the generation of the periodic solution. If a stable equilibrium point becomes unstable and a stable periodic solution shows up as the value of the bifurcation parameter increases, the point is a supercritical Hopf bifurcation point (sH in Figure 17.5A). If an unstable equilibrium point becomes stable and an unstable periodic solution takes place, the point is called a subcritical Hopf bifurcation point (uH in Figure 17.5A and B) (Thompson and Stewart, 1993). At $I_{ext} = 6.264$, a stable and an unstable periodic solution collide and disappear.

At the Hopf bifurcation point $\nu = \bar{\nu}$, the Jacobian matrix has a pair of purely imaginary eigenvalues. Since the H–H equations have dimension four, the characteristic equation of the Jacobian has the form

$$\lambda^4 + C_3\lambda^3 + C_2\lambda^2 + C_1\lambda + C_0 = 0, \tag{17.9}$$

where $C_0, \ldots, C_3$ are real. Since this matrix has eigenvalues $\pm i\omega$, Equation (17.9) takes the form

$$(\lambda - \alpha)(\lambda - \beta)(\lambda^2 + \omega^2) = 0. \tag{17.10}$$

By comparing the coefficients of both equations, it may be demonstrated that this Hopf bifurcation requires that

$$C_1^2 + C_0C_3^2 - C_1C_2C_3 = 0, \quad C_1C_3 > 0, \tag{17.11}$$

where C_0, C_1, C_2, and C_3 are functions of the parameter ν. The condition for a saddle-node bifurcation point is obtained by putting $C_0 = 0$, because in that case the characteristic equation has at least one zero eigenvalue.

17.3.2 Why Study the Global Organization of Bifurcations?

The remainder of this chapter will be devoted to the bifurcation analysis of the H–H equations. Nonlinear dynamical system models described by multiple differential equations may involve several or many parameters. I_{ext}, V_K, and V_{Na} are such parameters for the H–H equations. We can say that such a dynamical model is embedded in the multidimensional parameter space. Every point in the parameter space corresponds to a single realization of the model.

As we have observed above for the H–H equations, a nonlinear dynamical system parameterized in a suitable manner shows bifurcations, that is, the models associated with different points in the parameter space may behave differently. It is informative to know the global organization of the bifurcations that can take place in the model when the values of many or possibly all the parameters involved in the model change, even if those changes in some parameter values or ranges of investigation in the parameter space are physically implausible.

Some biophysicists might find fault with varying a parameter of a model in a range that could not be physically realized. However, we should be aware of the fact that the construction of the H–H equations based on the experimental data from the squid giant axon was achieved with just a single set of physical parameter values chosen such that the model could reproduce the behavior of the axon with those parameter values. This means that there is no guarantee for the constructed model to be capable of reproducing well a behavior of the axonal membrane for another set of parameter values apart from the one used for the model construction.

A good mathematical model describing a physical process should be generic, that is, the model's behavior should be robust and insensitive to small perturbations in its parameter values. This is not the case, however, if the set of parameter values used to construct the model is located in the model's parameter space near the bifurcation set, where the model is structurally unstable and nearby parameter values can exhibit qualitatively different behaviors. In the following sections, we look at behaviors of the H–H equations for cases, for example, with $V_K = 10$ mV, and even with V_K near -100 mV, which depart largely from the case of the squid giant axon ($V_K = -12$ mV). Nevertheless, we will observe behaviors and transitions from one behavior to another that are qualitatively similar to those encountered in some other neuronal membranes with certain sets of parameter values quantitatively different from the ones used for the H–H equations. This approach should help us get a better insight into the generic dynamics of excitable membranes from the H–H equations. In other words, the H–H equation, which is, after all, a rather simple model of the excitable membrane, might exhibit dynamics that are common to membranes other than the squid axon, and thus be able to play a core role in our understanding of the general nature of action potential generation.

17.3.3 Bifurcation Diagram for I_{ext} at $V_K = 10$ mV

In this subsection, we fix $V_K = 10$ mV and consider the behavior of the H–H equations as I_{ext} changes. Figure 17.6A shows a one-parameter bifurcation diagram for I_{ext} when $V_K = 10$ mV. In contrast to Figure 17.5A, the equilibrium curve is not a simple monotonic increasing curve, but is S-shaped. Hence in the region $-13.4 < I_{ext} < -6.8$, there are three equilibrium points. When $I_{ext} = -13.4$ and -6.8, saddle-node bifurcation takes places where two of the equilibrium points collide and disappear (two sn's in Figure 17.6A). The periodic solution generated at the Hopf bifurcation point at $I_{ext} = 29.8$ collides with the saddle near $I_{ext} = -11.9$ and disappears in the region of negative I_{ext}. In this case the unstable manifold for the saddle and the stable manifold coincide with each other. Such a solution trajectory is called a homoclinic trajectory, and the bifurcation point from which a homoclinic orbit originates and the periodic orbit disappears, is called a homoclinic bifurcation point (see Figure 17.6A, sl). As the trajectory of a periodic solution approaches the homoclinic trajectory, the period increases toward infinity. Figure 17.6B is a conceptual phase diagram showing the homoclinic trajectory. Figure 17.6C illustrates the time course of the membrane potential near the homoclinic bifurcation point when $V_K = 10$ mV and $I_{ext} = -11.5$.

17.3.4 Bifurcation Diagram of I_{ext} and V_K

In this section, we consider I_{ext} as a primary bifurcation parameter and V_K a secondary one to obtain a two-parameter bifurcation diagram. In the two-parameter bifurcation diagram, we pay attention to bifurcation points of the primary parameter and investigate how these points move in the two-parameter space, thus obtaining bifurcation curves in the two-parameter space. The two-dimensional parameter space is divided into several regions by the bifurcation curves. In this

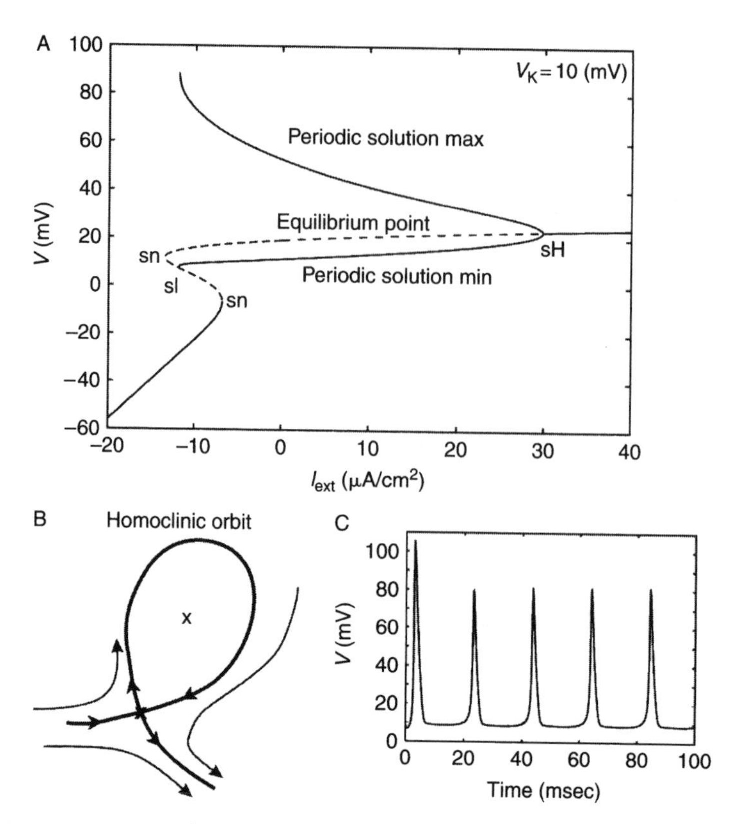

FIGURE 17.6 (A) One-parameter bifurcation diagram for $V_K = 10$ mV. (B) Schematic phase portrait of a homoclinic orbit. (C) An example of the time course of the membrane potential near the homoclinic bifurcation, $V_K = 10$ mV, $I_{ext} = -11.5\ \mu A/cm^2$.

section, we qualitatively classify different behaviors of the H–H equations on regions in the space (see Figure 17.7).

We analyze the bifurcation structure of the H–H equations when I_{ext} and V_K are changed over a wide range. Figure 17.8A illustrates a bifurcation diagram of the H–H equations, where the abscissa is I_{ext} and the ordinate the equilibrium potential V_K of K^+. The original value of V_K is -12 mV and we set $V_{Na} = 115$ mV. In Figure 17.8A, there are three bifurcation curves that separate the parameter space into several regions. The V-shaped dotted curve (sn) on the upper left part of the diagram consists of saddle-node bifurcation points. Inside the V-shaped saddle-node bifurcation curve, the equation possesses three equilibrium points. The U-shaped curve (uH, sH) at the center of the diagram is a Hopf bifurcation curve. Inside the Hopf bifurcation curve, the equilibrium point of the equation is unstable and there exists at least one stable periodic solution. Figures 17.8B and E are one-parameter bifurcation diagrams of I_{ext} for $V_K = -60, -80, -120, -140$ mV, respectively, where $V_{Na} = 115$ mV.

The system's phase diagrams are briefly illustrated on each of the parameter regions separated by the bifurcation curves, where a dot (•) indicates a stable equilibrium point and a cross (×) an unstable one. A solid circle around (×) implies a stable periodic solution, while a dotted one around (•) is an unstable periodic solution. A thin solid line with arrow shows a typical one-dimensional manifold or a solution trajectory.

The equilibrium point is stable in the region A in Figure 17.8A, where the H–H equations show a monophasic discharge to a stimulus of appropriate intensity with convergence to the equilibrium potential. Needless to say, we are not interested in exactly how the membrane potential decays to

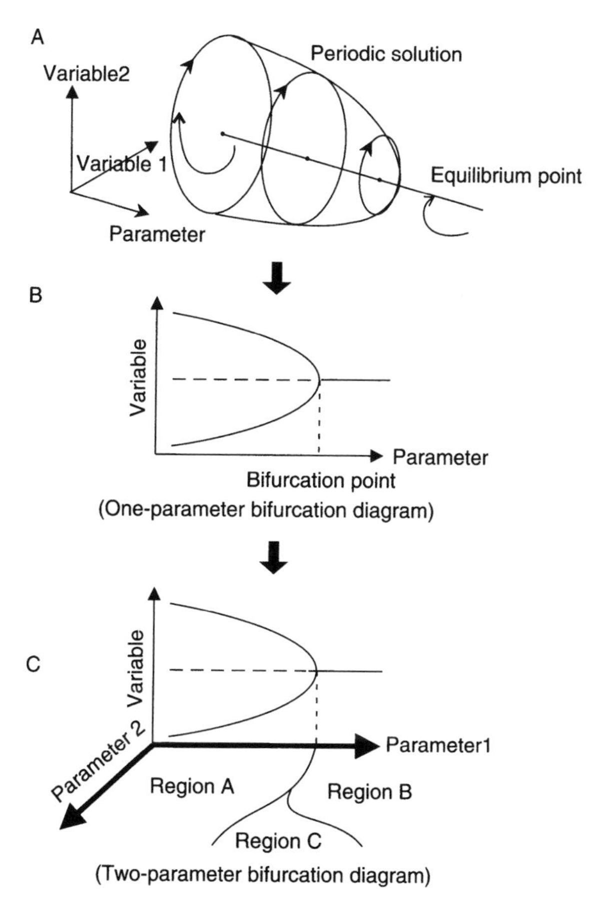

FIGURE 17.7 (A) Schematic diagram of the bifurcation. As the parameter changes, the attractor changes between the equilibrium point and the periodic solution. (B) One-parameter bifurcation diagram. The center line represents the equilibrium point and the upper and lower curves represent the maximum and the minimum values of the periodic solution, respectively. (C) An example of a two-parameter bifurcation diagram.

the equilibrium potential (see Figure 17.5C and D). The region A includes a point $((I_{\text{ext}}, V_{\text{K}}) = (0, -12.0))$, which is utilized by the Hodgkin–Huxley neuron.

A periodic solution is only stable in region B in Figure 17.8A, where the solution approaches asymptotically to a sustained oscillation (see Figure 17.5F and G). As I_{ext} increases, both the period of oscillation and the amplitude become smaller. As the value of V_{K} decreases, the period of oscillation becomes smaller but the amplitude becomes larger.

A stable equilibrium point and stable and unstable periodic solutions coexist in the region C. The H–H equations approach either a stable periodic solution or an equilibrium point, depending on the initial conditions. Figure 17.5E shows the membrane waveform in region C in Figure 17.8A, where we have set $(I_{\text{ext}}, V_{\text{K}}) = (6.3, -12.0)$. Figure 17.9A also illustrates an example of the waveform of the membrane potential for $(I_{\text{ext}}, V_{\text{K}}) = (9.0, -12.0)$ in region C, but switching takes place between a stable state and a stable oscillation by perturbation to the membrane potential at times $t = 50$ msec and 180 msec, respectively. Guttman et al. (1980) showed in their physiological experiment with the squid giant axon, that bistability between an equilibrium and an oscillatory state appears by providing an appropriate stimulus to the giant axon immersed in a solution of low Ca^{2+} concentration.

In region D are one equilibrium point and three periodic solutions, two of which are stable. Hence, the H–H equations can take either one of two oscillatory solutions with different amplitudes

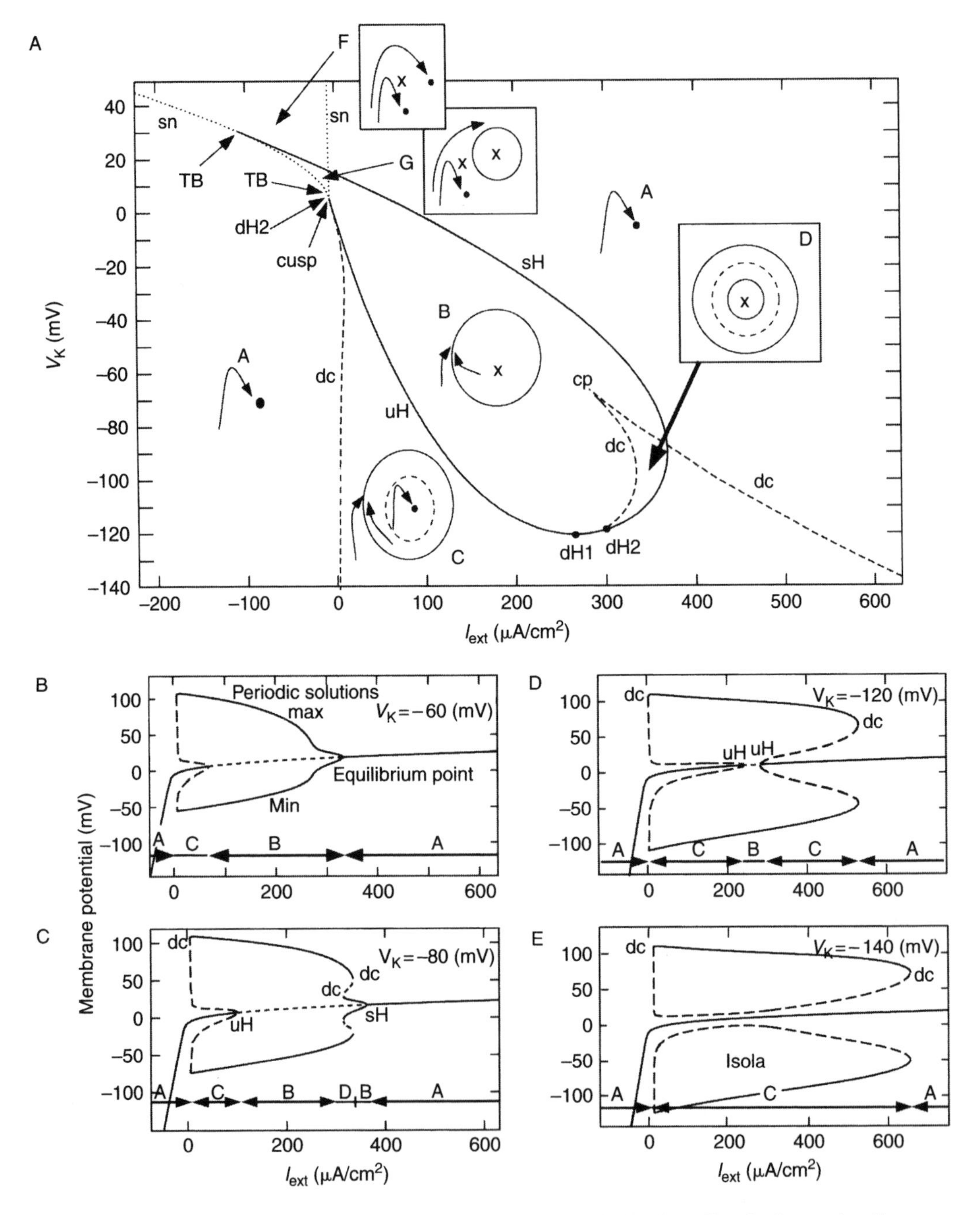

FIGURE 17.8 (A) 2BD of I_{ext}–V_K for $V_{Na} = 115$ mV, the original value. The abscissa and ordinate are I_{ext} and V_K, respectively. The supercritical Hopf, subcritical Hopf, double-cycle, and saddle-node bifurcation curves are labeled as "sH," "uH," "dc," and "sn," respectively. (B) to (E) 1BDs for $V_K = -60, -80, -120$, and -140 mV, respectively.

depending on the initial conditions (bistability of two periodic solutions). Figure 17.9D illustrates an example of the membrane waveforms observed in region D. At first the H–H equations give a stable periodic solution with a small amplitude; after perturbation to the membrane voltage, the solution transforms to the one with a bigger amplitude.

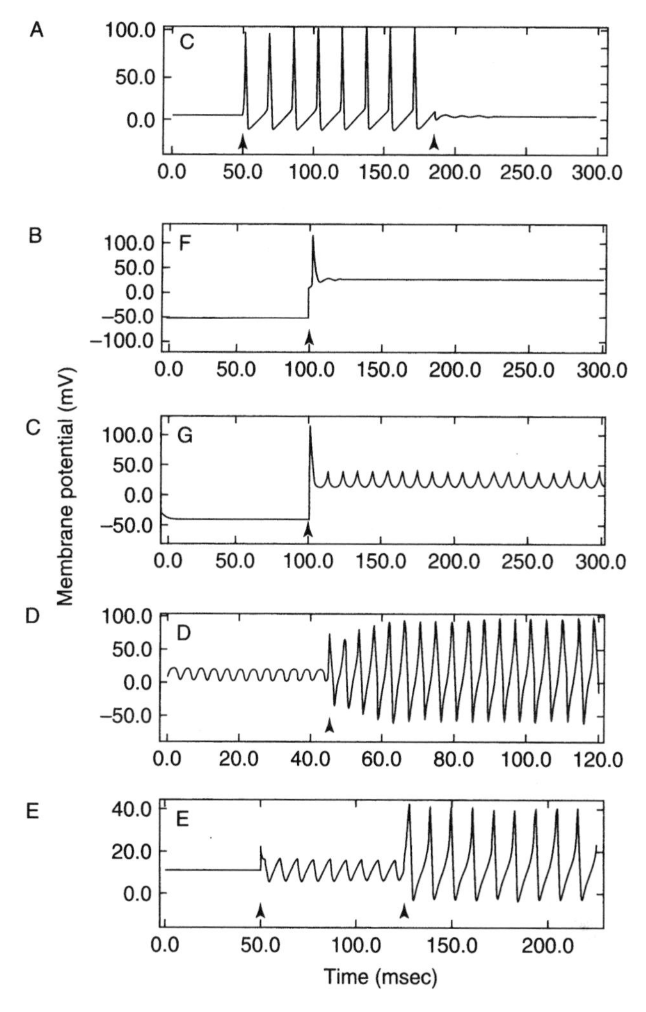

FIGURE 17.9 Examples of membrane-potential waveforms in regions C to G in Figure 17.8A. The abscissa is time (msec) and the ordinate is the membrane potential (mV). In each plot, all the parameters of the H–H equations are fixed, and appropriate impulse perturbations delivered to the membrane potential, V, lead to transitions from one steady state to the other. See text for details. From Fukai et al. (2000b), reproduced with permission.

In region F inside the V-shaped saddle-node bifurcation curve (sn) at the left upper part of Figure 17.8A, the H–H equations possess three equilibrium points, two of which are stable. The solution approaches one of two equilibrium points depending on the initial state (bistability of equilibrium points). Figure 17.9B illustrates an example of the membrane potential waveform observed in region F. The membrane potential began at the equilibrium potential with lower value but transferred to the other stable point when the perturbation was added at time 100 msec. The H–H equations show complicated bifurcations near the cusp where two sn curves contact tangentially with each other and vanish.

By changing the parameters V_{Na} and V_K, the equilibrium potentials of Na^+ and K^+, respectively, of the H–H equations in the region including physiologically unrealistic cases, we could investigate the bifurcation structure in a two-dimensional plane of the primary parameter I_{ext} and either V_{Na} or V_K. Results are shown in the two-dimensional bifurcation diagrams. U-shaped Hopf bifurcation curves and V-shaped saddle-node bifurcation curves are seen in both spaces (I_{ext}, V_K) and (I_{ext}, V_{Na}).

17.3.5 Codimension and Organizing Center

We saw in Figure 17.8 that two bifurcation curves cross or contact tangentially on a two-parameter bifurcation diagram such as the I_{ext}–V_K plane. This point satisfies two bifurcation conditions at the same time. The codimension of a bifurcation is defined as the minimum dimension of a parameter space in which the bifurcation arises (Guckenheimer and Holmes, 1983). A bifurcation of codimension 1 appears as a critical point of a bifurcation parameter in a one-dimensional bifurcation diagram and as a curve in a two-dimensional bifurcation diagram. A bifurcation of codimension 2 appears as a point in a two-dimensional bifurcation diagram as seen in Figure 7.17C. Bifurcations of codimensions 1, 2, and 3 appear respectively as a plane, a curve, and a point in the three-parameter space. A bifurcation of codimension 2 in a two-dimensional bifurcation diagram appears because two bifurcation conditions are satisfied simultaneously.

In this way, one sees a singular point (a highly degenerate bifurcation point) where multiple bifurcation conditions are satisfied simultaneously by changing several parameters of a given nonlinear system. Several less degenerate bifurcation sets meet together at the highly degenerate bifurcation point. Hence one may observe several qualitatively different behaviors near a highly degenerate bifurcation point. Such a point is often called an organizing center (Guckenheimer and Holmes, 1983; Golubitsky and Schaeffer, 1985). It is a particular set of parameter values around which all or many of the qualitatively different behaviors of a nonlinear system appear (Golubitsky and Schaeffer, 1985).

17.3.6 Bifurcation in the I_{ext}–V_K–V_{Na} Parameter Space

Fukai et al. (2000a) and Fukai (2000) showed that a highly degenerate point, that is, an organizing center, exists in the three-dimensional parameter space I_{ext}–V_K–V_{Na}. Figure 17.10 illustrates schematically a global organization of the bifurcation of the H–H equations in the I_{ext} –V_K –V_{Na} parameter space. In Figure 17.10, cross section (a), the bifurcation structure in the two-dimensional parameter space I_{ext}–V_K is shown where $V_{Na} = 115$ mV (the sodium equilibrium potential used by Hodgkin and Huxley [1952b]) and both I_{ext} and V_K were varied (see also Figure 17.8A). The cross sections (b), (c), and (d) are two-dimensional bifurcation diagrams in (I_{ext}–V_K) parameter space corresponding to $V_{Na} = 110, 90, 85$ mV, respectively. In each I_{ext}–V_K parameter space, a set of bifurcation points of codimension 1 appears as a curve. The arc-shaped solid lines appearing at the center in (a) and drawn at the upper left side in each of panels (b), (c), and (d), indicate the Hopf bifurcation curves (sH and uH). Dotted lines depict the left and right bottoms of the Hopf bifurcation curves in each panel and show the double-cycle bifurcation (constant). The V-shaped curve at the upper left corner of each panel is the saddle-node bifurcation (sn) of the equilibrium points. These are bifurcations of codimension 1 and constitute a surface in the three-dimensional parameter space. This surface divides the parameter space into several regions in each of which the H–H equations show a qualitatively different behavior.

Each of the bifurcations dH1 and dH2 of codimension 2 appears as a Hopf bifurcation point in a two-parameter bifurcation diagram and forms a curve in the three-parameter space. The points labeled P and Q on the curves dH1 and dH2, the existence of which was confirmed by Fukai et al. (2000a) and Fukai (2000), are the doubly degenerate Hopf bifurcation points of codimension 3. See also Labouriau (1985, 1989) for the degenerate bifurcation point.

The doubly degenerate bifurcation point shown by the H–H equations can also be analyzed in the framework of singularity theory (see, Fukai et al. 2000a; Fukai, 2000).

17.4 Discussion

Let us briefly discuss a possible contribution of the bifurcation analysis to understanding the dynamics of neuronal cell membranes. The bifurcation analysis of the membrane cell model could provide

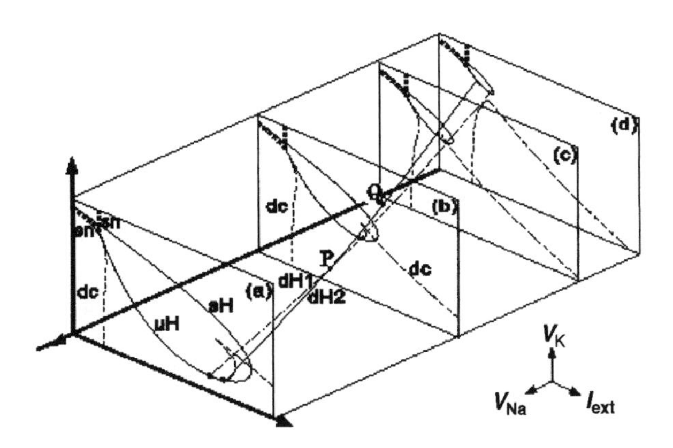

FIGURE 17.10 Global view of the bifurcation structure in the I_{ext}–V_K–V_{Na} three-dimensional parameter space of the H–H equations. The two-dimensional planes (a) to (d) are the I_{ext}–V_K 2BDs for V_{Na} = 115, 100, 90, and 85 mV, respectively. In each plane, solid, dashed, and dotted curves represent the Hopf ("uH," "sH"), the double-cycle ("dc"), and the saddle-node bifurcation ("sn"), respectively. "dH1" and "dH2" (dotted and solid lines) represent the loci of the degenerate Hopf bifurcations. "P" and "Q" represent doubly degenerate Hopf bifurcation points. From Fukai et al. (2000b), reproduced with permission.

a way to evaluate the plausibility of the mathematical model. This could be done by comparing the bifurcation structure of the model with the experimentally observed behaviors in the corresponding physiological system (membrane cell) under various conditions.

Tasaki (1959) showed that when the outward current is injected under a $[K^+]_o$-rich condition, the squid giant axon exhibits the bistability of two stable equilibrium potentials. Aihara and Matsumoto (1983) observed the same behavior in the Hodgkin–Huxley model when some parameters of the equations were tuned appropriately to match the experimental environment.

In our study, this condition corresponds to the parameter region F inside the V-shaped curve (saddle node) in the two-parameter (I_{ext}–V_K) bifurcation diagram (Figure 17.8). Guttman et al. (1980) showed that the squid giant axon under Ca^{2+}-rich conditions has a bistable equilibrium point and a periodic oscillation when the appropriate external applied current is injected. This corresponds to region C in Figure 17.8. These are examples in which the behavior of the model and experiment coincide. The results described in this chapter show wide regions of V_{Na} and V_K parameter space in which the H–H equations exhibit the bistability of periodic solutions.

Qualitative agreement between the bifurcation diagrams of the model and the physical system can be used as a new methodology for nonlinear system (including physiological system) identification. Conventionally, a system identification for given dynamic behaviors has been done by constructing a model that can reproduce the behaviors of the system with conditions in which most of the parameters of the target system are fixed. This was basically the case also for the derivation of the H–H equations as we described in the earlier part of this chapter. In such a case, we can hardly expect the model to reproduce the behavior of the system when the parameter values are changed. This means there is no guarantee that the H–H equations will behave similarly to the actual physiological membrane for a wider parameter range. However, at least qualitatively, some experimental results are in good agreement with the Hodgkin–Huxley behavior even for parameter values far from those used to derive the original H–H equations. These facts likely imply that the Hodgkin–Huxley model captures the essential mechanism of action potential generation of the physical system. To clarify the parameter regions where the model shows bistability of periodic solutions as we have shown is a good starting point to achieve the suggested study along this line.

17.5 Conclusion

In this chapter, we illustrated the properties of a neuron based on the H–H equations. The bifurcation theory was used to classify qualitatively different behaviors of the equations in appropriate parameter spaces when values of the primary parameter I_{ext} and other controllable parameters are changed. The U-shaped Hopf bifurcation curves and V-shaped saddle-node bifurcation curves are commonly seen in the two-parameter spaces of (I_{ext}, V_K) and (I_{ext}, V_{Na}). A similar global bifurcation structure is seen for the other Hodgkin–Huxley-type equations, which show apparently different behaviors from their original parameter settings. This implies that the H–H equations not only provide a simple model reproducing excitability of the squid giant axon but also a base model among those proposed to explain complicated behaviors shown by other neurons. The system has three equilibrium points inside the V-shaped saddle-node bifurcation curves. The membrane of the squid giant axon does not possess multiple equivalent potentials in its natural state. Tasaki (1959) reported that the axon can possess multiple equilibrium potentials by adjusting the ionic concentrations in the intracellular and extracellular regions of the axon. This suggests that the H–H equations give a plausible model for a wide range in the parameter space.

It was confirmed that the H–H equations possess multiple stable periodic solutions in a couple of two-dimensional spaces of parameters different from those studied in this chapter. These multiple stable periodic solutions originate from the highly degenerate Hopf bifurcation. The highly degenerate bifurcation point is called an organizing center, and multiple less degenerate bifurcation surfaces join at the point, around which all of or many of the qualitatively different behaviors exhibited by the system are observed.

It is important to compare the types of the highly degenerate bifurcation points when we investigate the possible equivalency of several systems. It is possible to say whether systems are equivalent by comparing their highly degenerate bifurcation points even though the two systems have different global bifurcation structures (Fukai, 2000).

PROBLEMS

1. Write a computer program to study the response of the membrane voltage of the H–H equations to a current stimulus with intensity I $\mu A/cm^2$ and width w msec. Plot the trajectories starting from various initial values of (V, m, h, n) on the V–m space. Confirm that the H–H equations generate action potentials both for depolarizing ($I > 0$) and for hyperpolarizing ($I < 0$) current stimuli with fixed w.
2. Show that, if you fix $I_{ext} = 10\ \mu/cm^2$, the membrane potential V oscillates with time.
3. To investigate the stability of an equilibrium point (e.g., see Figure 17.3), say $(\overline{V}, \overline{m}, \overline{h}, \overline{n})$, in the H–H equations, one needs a 4×4 Jacobian matrix, J, of the equations evaluated at the equilibrium point. Notice a relatively simple form of the matrix due to a special formalism of the H–H equations, where each gate variable x ($= m, h,$ or n) is a function of only V and x itself. Show that rank $(J) \geq 3$ and $\mathrm{tr}(J) < 0$ for any choice of the equilibrium point. The former may restrict the types of bifurcation that can take place in the H–H equations, and the latter means that the dynamics of the H–H equations are volume contractive.
4. The bifurcation analyses shown in this chapter mostly used a one-parameter bifurcation diagram in which the membrane potential of the steady state, either the equilibrium (resting) point or the oscillatory state, was plotted as a function of the intensity of the external constant current I_{ext}. For example, in Figure 17.6A, one can see an S-shaped curve representing the bifurcation of the equilibrium. In order to produce this curve, one usually solves the nonlinear algebraic equations by setting $dV/dt = dm/dt = dh/dt = dn/dt = 0$ to find an equilibrium point $(\overline{V}, \overline{m}, \overline{h}, \overline{n})$ for a given set of paramters in the H–H equations, perhaps by using Newton's method among others. However, in the case of H–H equations, one does not necessarily do that. Once again, this is

due to the special formalism of the H–H equations as mentioned in Problem 3. Look at Equation (17.1) in the text. At the equilibrium point, $dV/dt = 0$ so that one can obtain the following equation:

$$I_{\mathrm{ext}} = g_{\mathrm{Na}}(\overline{V} - V_{\mathrm{Na}}) + g_{\mathrm{K}}(\overline{V} - V_{\mathrm{K}}) + g_{\mathrm{L}}(\overline{V} - V_{\mathrm{L}}).$$

Plot the value of I_{ext} as a function of $\overline{V}$ so that the above equation holds. Then, note that the result is the one-parameter bifurcation diagram if you see it as a graph of $\overline{V}$ as a function of I_{ext}.

5. This problem is a combination of Problems 3 and 4. Define the steady state I–V relationship for the H–H equations as in Problem 4:

$$i_{\mathrm{ss}}(\overline{V}) = g_{\mathrm{Na}}(\overline{V} - V_{\mathrm{Na}}) + g_{\mathrm{K}}(\overline{V} - V_{\mathrm{K}}) + g_{\mathrm{L}}(\overline{V} - V_{\mathrm{L}}).$$

We want to investigate whether $i_{\mathrm{ss}}(\overline{V})$ is a monotonic or nonmonotonic function of $\overline{V}$. Show that

$$\frac{\mathrm{d}i_{\mathrm{ss}}(\overline{V})}{\mathrm{d}\overline{V}} = \det(J).$$

Note that the sign of $\det(J)$ is a function of $\overline{V}$ as well as of the values of parameters such as V_{K} and V_{Na}. If $i_{\mathrm{ss}}(\overline{V})$ is monotonic, that is, if $\det(J)$ is always positive (it may not always be negative in the H–H equations) for any value of $\overline{V}$, saddle-node bifurcations will not arise in the one-parameter bifurcation diagram of the equilibrium point as I_{ext} changes. If not, they will arise at critical values of I_{ext} such as:

$$I_{\mathrm{ext}}^{*} = i_{\mathrm{ss}}(\overline{V}^{*}),$$

where $\overline{V}^{*}$ satisfies the following equation

$$\frac{\mathrm{d}i_{\mathrm{ss}}(\overline{V}^{*})}{\mathrm{d}\overline{V}} = 0.$$

APPENDIX

17.A1 Analysis and Tools

In the present chapter, the externally applied current, I_{ext}, was always set constant and was considered as a primary bifurcation parameter. The H–H equations were analyzed roughly in two ways: (1) Phase–space analysis. System dynamics were analyzed for fixed parameters; and (2) parameter–space analysis. Qualitative changes of system behavior were analyzed by changing parameters. Here we explain the methods of the analyses in these two spaces and we also describe several tools for the analysis of dynamical systems used in this chapter together with their advantages and efficiencies.

We first consider the phase–space analysis. Analysis in phase space is important for describing the system's behavior in detail for a given set of parameter values and also for understanding the bifurcation structure found in parameter space. It is hard to understand the bifurcation structure just by tracing bifurcation curves analytically in parameter space. It can be very helpful in this regard to perform state–space analyses at typical points of each parameter region.

The fourth-order Runge–Kutta method with preassigned time step or a modified Runge–Kutta method with variable time step was used to obtain the trajectories in phase space and/or the time courses of variables of interest. Many of the neuron models represented by Hodgkin–Huxley-type equations are stiff systems with fast dynamics representing the generation of the action potential and slow dynamics at the refractory period. The Runge–Kutta method with variable step size is effective in this case. By taking a negative time step in the method, it is also possible to catch an unstable limit–cycle oscillation for a system described by a set of two-dimensional differential equations.

DsTool (DynamicalSystemToolkit) is a tool for analyzing dynamical systems and trajectories, finding equilibrium points, and tracking one-dimensional stable and unstable manifolds of diffeomorphic maps for sets of ordinary differential equations. It is useful for the analysis of dynamics in the phase plane for fixed parameters (Guckenheimer et al., 1992). DsTool can compute the eigenvalues and corresponding eigenvectors of the Jacobian matrix at equilibrium points and can classify the types of equilibrium points. However, it cannot trace the equilibrium points or the periodic solutions for continuous changes of parameter values and cannot classify the types of bifurcation. If one makes use of it together with a graphical user interface, DsTool is highly useful for finding the features of the phase space inasmuch as the initial conditions as well as the parameter values can be easily changed.

The shooting method (Parker and Chua, 1989) was used to obtain changes in periodic solutions dependent on the parameter values and also to study unstable periodic solutions in detail. The shooting method applies the Newton–Raphson method to a Poincaré map on the Poincaré section to obtain a periodic solution. However, for the method to be applied, it is necessary to choose the initial point appropriately, given the properties of the Newton–Raphson method. In the present chapter, we have used DsTool to search for appropriate initial values. Note also that the shooting method is effective for finding stable or unstable periodic solutions far from an equilibrium point, but because of the relatively slow speed of convergence, it is not so efficient for obtaining periodic solutions near an equilibrium point or near a double-cycle bifurcation point, around which multiple solutions of the H–H equations exist.

The Net Method is well-known as a powerful method for obtaining periodic solutions (Rinzel and Miller, 1980). In the Net Method, periodic solutions are discretized at several points, and the problem of finding them is reduced to a two-point boundary-value problem with periodic boundary conditions. Hence in the Net Method there is no problem with errors caused by numerical computation along the trajectory, and the solution does not deviate from the real solution trajectory. The periodicity is considered as a boundary condition. Therefore, the error at one of the two points is corrected by the true value of the other point, and a net of coarse mesh enables one to obtain an approximate solution from a correct initial point. Hence it is irrelevant whether the periodic solution is stable or not.

In this chapter, XPP-Aut (X-Windows Phase Plane plus Auto) was used to obtain double-cycle bifurcations or a precise form of the branch of a periodic solution near the Hopf bifurcation point. XPP-Aut was developed by combining the AUTO software for continuation and bifurcation problems in ordinary differential equations (Doedel and Kernevez, 1986) with XPP (X-Windows Phase Plane), a tool for phase–plane analysis. The user obtains an equilibrium point or periodic solution of the system with a given set of initial parameter values, and then from data on the parameters, he or she is able to analyze the bifurcation when the parameter values are changed by AUTO. Therefore, using XPP-Aut, it is possible to perform a phase–plane analysis and to obtain a one-parameter bifurcation diagram (BD) or even a two-parameter BD if it is simple. In particular, it is possible to trace branches of the periodic solution generated at Hopf bifurcation points in a one-parameter BD. In a two-dimensional diagram, Hopf bifurcation curves, saddle nodes, and double-cycle bifurcation curves can be traced by this method.

We obtained double-cycle bifurcation points by utilizing AUTO or by computing the regions of stable or unstable periodic solutions by varying the parameter value systematically. Homoclinic bifurcations were obtained by computing the regions of existence of the periodic solutions. To this end, the parameter value was systematically varied while referring to the shape of the one-dimensional BD. AUTO includes a package to compute a homoclinic bifurcation, but we did not use it.

17.A2 The Hopf Bifurcation

Let us explain the Hopf bifurcation briefly. Consider an n-dimensional dynamical system with $k+1$ parameters, $\boldsymbol{\alpha} = (\nu, \alpha_1, \alpha_2, \ldots, \alpha_k) \in R^{k+1}$.

$$\frac{\mathrm{d}\mathbf{y}}{\mathrm{d}t} = F(\mathbf{y}, \boldsymbol{\alpha}), \quad R^n \times R^{k+1} \to R^n, \tag{17.A1}$$

($\mathbf{y} \in R^n$). We fix the values of k parameters, $\alpha_1, \alpha_2, \ldots, \alpha_k$, and vary the parameter, ν. We assume that $F(\mathbf{y}, \boldsymbol{\alpha})$ is continuously differentiable with respect to $\mathbf{y}, \boldsymbol{\alpha}$ for as many times as necessary. Let $\mathbf{y} = \mathbf{y}_0$ be the equilibrium point of Equation (17.A1) for a fixed parameter value $\boldsymbol{\alpha}$ ($F(\mathbf{y}_0, \boldsymbol{\alpha} = 0)$. The stability of the equilibrium point is determined from the eigenvalues of the Jacobian matrix $A(\boldsymbol{\alpha}) = (\mathrm{d}F)|_{\mathbf{y}_0, \boldsymbol{\alpha}}$ of the linearized equation of the system (17.A1) around the equilibrium point $\mathbf{y}_0$. Continuous change of the value of the parameter ν causes continuous change of the eigenvalues of the matrix A. Suppose that the matrix A possesses a pair of complex eigenvalues $\lambda(\nu) = \sigma(\nu) \pm i\omega(\nu)$ at ν. If the following conditions H1 and H2 are met at $\nu = \nu_0$ when ν is varied, a standard Hopf bifurcation appears at $\nu = \nu_0$:

(H1) The real part $\lambda(\nu)$ of the complex eigenvalue changes its sign at $\nu = \nu_0$.

$$\sigma(\nu_0) = 0, \quad \omega(\nu_0) \neq 0.$$

(H2) The change of sign takes place at a finite speed, that is,

$$\left.\frac{\partial \sigma(\nu)}{\partial \nu}\right|_{\nu=\nu_0} \neq 0.$$

In this case, the stability of the equilibrium point $\mathbf{y} = \mathbf{y}_0$ changes, and there exists only one periodic solution with amplitude x satisfying $\nu = \mu_2 x^2 + \mu_4 x^4 + \cdots$

(H3) $\mu_2 \neq 0$.

The stability of the periodic solution depends on the sign of μ_2. If $\mu_2 < 0$, the periodic solution originating from the Hopf bifurcation is stable and the Hopf bifurcation is called supercritical, while if $\mu_2 > 0$, the periodic solution originating from the Hopf bifurcation is unstable and the Hopf bifurcation is called subcritical.

17.A3 The Degenerate Hopf Bifurcation

So far, we have varied the parameter ν and fixed the rest. However, if the other parameters are changed, either one or both of (H2) and (H3) may break down. In this case, degenerate Hopf bifurcation takes place. It is not very interesting to view the one-parameter bifurcation diagram of ν when the remaining parameters $\alpha_1, \alpha_2, \ldots, \alpha_k$ are set such that the Hopf bifurcation is degenerate. Rather, an interesting bifurcation diagram is obtained when the parameter values of $\alpha_1, \alpha_2, \ldots, \alpha_k$ are slightly changed from the point of degeneration. First, let us consider the case where (H2) is broken, that is,

$$\left.\frac{\partial \sigma(\nu)}{\partial \nu}\right|_{\nu=\nu_0} = 0.$$

Put in words, the trajectory of the complex eigenvalue of the matrix A contacts the imaginary axis tangentially as ν changes. Holding the parameter ν at ν_0, we change the values of $\alpha_1, \alpha_2, \ldots, \alpha_k$

slightly. Then two Hopf bifurcation points emerge.[1] By going back through this process, we see that the condition (H2) is broken by changing the values of $\alpha_1, \alpha_2, \ldots, \alpha_k$ appropriately and two Hopf bifurcation points in the one-parameter bifurcation diagram with respect to the parameter ν collide with each other and disappear. If the equation is written in an appropriate coordinate system near the bifurcation point (in other words, if the equation is written in the normal form of the bifurcation and if it is embedded in the parameter family as a universal unfolding), then the degenerate Hopf bifurcation can be obtained by changing only one parameter, but not ν. (This is the bifurcation problem with codimension 1, bifurcation of codimension 2.)

When condition (H3) breaks down, that is, when $\mu_2 = 0$, the periodic solution emerging from the Hopf bifurcation changes its stability. At the same time, a double-cycle bifurcation (a saddle-node bifurcation of the periodic solution) appears, where a (generically) stable periodic solution and an unstable periodic solution collide and disappear. This degenerate Hopf bifurcation can also be obtained by changing the value of only one parameter if the equation is written in the appropriate coordinate system near the bifurcation point. (This is again the bifurcation problem with codimension 1, bifurcation of codimension 2.)

A difficulty in finding the degenerate Hopf bifurcation point is to choose which one among the parameters $\alpha_1, \alpha_2, \ldots, \alpha_k$ to change. Generally speaking, one cannot necessarily find the degenerate Hopf bifurcation point by changing all the parameters in the equation one-by-one, because the parameters in the default coordinate system are actually functions of multiple parameters in the universal unfolding.

NOTE

1. The number of newly emerging Hopf bifurcation points depends on the smoothness of the trajectory of a complex eigenvalue at the contact point in the complex plane. n Hopf bifurcations emerge if all the derivatives of the trajectory up to the n-th order vanish at the contact point.

18 Highly Efficient Propagation of Random Impulse Trains Across Unmyelinated Axonal Branch Points: Modifications by Periaxonal K^+ Accumulation and Sodium Channel Kinetics

Mel D. Goldfinger

CONTENTS

18.1 INTRODUCTION

This chapter addresses the divergence of axonal information at branch points, specifically comparing the extent to which impulse-based information[1] incident to a branch point differs from that which is emitted from that branch point. In the CNS, extensive axonal arborization provides part of the substrate for the divergence of action potential-coded information within and between neuronal circuits. (By contrast, the axonal arborization of peripheral afferent units — in which a group of mechanoreceptors is innervated by branches of the same primary afferent axon — is a site for information convergence [Eagles and Purple, 1974; Goldfinger and Amassian, 1980; Goldfinger, 1990; Banks et al., 1997].) In general, an axonal branch point represents a site where divergent information may be either recapitulated on each connected branch, lost through the failure of propagation, or modified such that different branches carry different information (Swadlow et al., 1980; Stoney, 1994).

Impulse propagation past a branch point can be reliable (e.g., Dityatev and Clamann, 1996; MacKenzie et al., 1996; MacKenzie and Murphy, 1998; Cox et al., 2000; Koester and Sakmann, 2000; Raastad and Shepherd, 2003). However, when impulse propagation fails at a branch point (e.g., Krnjevic and Miledi, 1959; Baccus et al., 2000) or when parent and daughter branches carry different information (Swadlow et al., 1980), some mechanism(s) must either extinguish a propagating action potential or distinguish between the different branch paths, respectively. Possible examples include: asymmetric branch geometry (Stockbridge, 1988; Korogod et al., 1994), a synaptically mediated shunt (Wall, 1995), the accumulation of extracellular K^+ (Grossman et al., 1973; Parnas et al., 1976; Smith and Hatt, 1976; cf. Graham, 1996), a difference in channel densities (Mellon and Kaars, 1974; Vandenberg and Bezanilla, 1991a; Johnson et al., 1996), activity-dependent changes in intracellular Ca^{++} (Stuart and Sakmann, 1994; Markram et al., 1995; Spruston et al., 1995; Helmchen et al., 1996; Lüscher et al., 1996; MacKenzie et al., 1996; cf. Miralles and Solsona, 1996; Schwindt and Crill, 1997), local K^+ currents (Obaid and Salzberg, 1996; Debanne et al., 1997), impulse extinction by collisions (e.g., Tauc and Hughes, 1963; Eagles and Purple, 1974; Goldfinger and Amassian, 1980; Fromherz and Gaede, 1993; Jansen et al., 1996; Zecevic, 1996), and channel noise (Horikawa, 1993). Assuming only bifurcations and normal temperature (cf. Westerfield et al., 1978), each mechanism could manifest itself upon all branches or a subset thereof, the latter case causing impulse propagation to become differential subsequent to the branch point(s).

If information is to diverge through an axonal arborization without modification, the cable properties of the parent axon, its branch points, and the daughter branches must provide a sufficiently high safety factor for impulse propagation. The safety factor for impulse propagation through a branch point has been characterized as low, primarily due to the theoretical work of Goldstein and Rall (1974) and others (Khodorov et al., 1969; Markin and Pashtushenko, 1969; Khodorov and Timin, 1975; Joyner et al., 1978, 1980; Pauwelussen, 1982; Carnevale et al., 1990; Lüscher and Shiner, 1990; Rinzel, 1990; Manor et al., 1991b; Bove et al., 1994; Zhou and Bell, 1994; Graham, 1996). However, a recent computational study showed that the propagation of a single impulse at a branch point has a high safety factor in a Hodgkin–Huxley-type axon (Goldfinger, 2000). The difference in results was due in part to the boundary conditions applied at the branch point (see Section 18.1.1).

In the study described here, the propagation of random trains of impulses across single or serially arranged symmetrical branch points in unmyelinated axons was simulated. Single-impulse propagation illustrated the axial modifications of the point-to-point conduction velocity (hereafter: "CV(x)") through primary and secondary serial branch points. Poisson-like random stimulation elicited wide-band impulse trains to assess possible frequency-modulated distortion during impulse propagation within the arborization. Possible modifications due to periaxonal K^+ accumulation or the kinetic description of Na^+ conductance were also considered.

18.1.1 Computational Considerations

18.1.1.1 Reconstructing Propagating Action Potentials

Computations were performed in FORTRAN on x86-based PC systems under Windows NT or XP, unless otherwise indicated. The computations utilized compartmental models (Rall, 1964; Cooley and Dodge, 1966). The flow of ionic currents within axons was given by the nonlinear, parabolic partial differential cable equation (Kelvin, 1855; Davis and Lorenté de Nó, 1946; Hodgkin and Rushton, 1946; Rall, 1977; Scott, 2002; cf. Sangrey et al., 2004):

$$I_{\text{m}} + I_{\text{stim}} = I_{\text{capac}} + I_{\text{ionic}} \tag{18.1a}$$

$$(1000 \cdot a/2\rho) \cdot \partial^2 V/\partial x^2 + I_{\text{stim}} = C_{\text{m}} \cdot \partial V/\partial t + \sum I_i, \tag{18.1b}$$

where a = diameter (cm), ρ = axial resistivity ($\Omega \cdot$ cm), V = membrane potential (mV), t = time (msec), x = axial distance (cm), C_{m} = membrane capacitance (=1 μF/cm^2), I_{m} = membrane current density (μA/cm^2), I_{capac} = capacitive current (μA/cm^2), I_{ionic} = ionic current (μA/cm^2), and I_{stim} is any extrinsic stimulus current (μA/cm^2). The single spatial dimension reflects the conventional assumption that there were no potential differences over the width of the fiber at any point ($\partial V(a) \equiv 0$). Extracellular isopotentiality was another conventional assumption (Fatt and Katz, 1951).

The cable equation was discretized with finite-difference expressions. For loci with uniform diameter and using Equation (18.1b), I_{m} at compartment j is:

$$I_{\text{m},j} = 1000 \cdot (a/2\rho) \cdot [(V_{j+\Delta x} - V_j) - (V_j - V_{j-\Delta x})]/\Delta x^2, \tag{18.1c}$$

where Δx is the length of each compartment, which defines the limit of spatial resolution.

Excitation was represented by the Hodgkin–Huxley equations and constants ("H–H," Hodgkin and Huxley, 1952b; Cronin, 1987), which provided the $\sum I_i$ term in Equation (18.1b) and included expressions for driving forces and voltage- and time-dependent (Na^+ and K^+) and Ohmic (leakage; cf. Lamas et al., 2002) macroscopic conductances. The H–H temperature (constant in these studies) was 10°C. The excitation formalism did not include an inactivation of $g_{\text{K}}(V,t)$ (Clay, 1989; Inoue et al., 1997) or any slow components of Na^+ conductance (Crill, 1996; Rakowski et al., 2002). However, later sections of this chapter present two modifications of the H–H equations contributed by Clay (1998). The axial point-to-point action potential conduction velocity CV(x) was determined from the time of occurrence T of a propagating action potential's positive peak in a given compartment at $x = i$:

$$\text{CV}(x) = \Delta x/(T_i - T_{i-\Delta x}), \quad i \geq 2 \tag{18.2}$$

(Ramon et al., 1975; Goldfinger et al., 1992).

In some cases, small single-compartmental departures from the apparent underlying CV(x)($\pm$0.11 m/sec) occurred. These could be attributed to slight inaccuracies in determining the exact time of the action potential peak due to the finite value of Δt (the temporal integration step), since the amplitude of the departure was spatially isolated and was diminished or eliminated with smaller Δt without affecting the apparent underlying CV(x) (Goldfinger, unpublished).

In parent and daughter fibers, respective channel densities (in H–H terms: maximal macroscopic conductances for Na^+, 120 mS/cm^2; K^+, 36 mS/cm^2; leakage, 0.3 mS/cm^2) and biophysical properties (cf. Colbert and Pan, 2002) were assumed to be equal and evenly distributed (cf. Johnson et al., 1996; Clay and Kuzirian, 2000; Savtchenko et al., 2001), and remained activity independent (cf. Siegel et al., 1994; Booth et al., 1997; Golowasch et al., 1999a; Stemmler and Koch, 1999; Falcke et al., 2000a). Other assumptions were the constancy over all fibers and branches of intracellular ionic concentrations (cf. Qian and Sejnowski, 1989; Kager et al., 2000), and the lack of noise in any conductance function (White et al., 2000; cf. Kuriscak et al., 2002) or individual stimulus waveform (Pakdaman, 2002). All fibers and branches had a 1 μm diameter unless otherwise indicated. In all fibers, axial resistivity was 108.5 $\Omega \cdot$ cm (Rall, 1977), the resting transverse membrane resistance was 1453.2 $\Omega \cdot \text{cm}^2$, and τ_m was 1.45 msec.

For the combined system of cable and excitation equations, numerical solution instabilities associated with insufficiently small values for the spatial (Δx) and/or temporal (Δt) integration steps are well-known (e.g., FitzHugh, 1962; Goldman and Albus, 1968; Parnas and Segev, 1979; Moore, 1987; Mascagni and Sherman, 1998). For a given computation, the values of Δt and Δx remained unchanged. By using $\Delta x = 3\,\mu\text{m}$ ($= 0.0164 \cdot \lambda_{d=1\,\mu\text{m}}$) and $\Delta t = 100\,\text{nsec}$ ($= 6.99 \cdot 10^{-5}\tau_m$), unless otherwise indicated, oscillatory instabilities were avoided. Stability was also promoted through implicit trapezoidal integration of all differential equations discretized as finite-difference expressions (Cooley and Dodge, 1966; Bhalla et al., 1992; Goldfinger et al., 1992; Press et al., 1992).

For the H–H axon under isopotential conditions, a variety of numerical methods applied to the integration of the H–H ordinary differential equations could generate similar reconstructions of the action potential, as shown in Figure 18.1. This result suggested that the H–H equations' stiffness may not be excessive (Mascagni and Sherman, 1998) and was handled similarly by each numerical method. The major difference between the solutions generated with the various methods was in the immediate poststimulus $V(t)$ trajectory, leading — at varied delay — to the rapidly rising phase of the action potential. The complex parametric dynamics of this period between single stimulus perturbation and action potential initiation have been described for the isopotential H–H regime (FitzHugh, 1960; Cronin, 1987).

Representation of the spatial dimension was the predominant source of instability in the combined cable/H–H system of partial differential equations, where stiffness increased with increasing compartmentalization (Mascagni and Sherman, 1998). To maintain consistent spatial and temporal resolution at all loci at all times, no predictor/corrector methods were used. The relatively small Δt value allowed the representation of the fastest significant components associated with distributed cable properties for which $\tau \ll \tau_m$, including the rapid spread of axial currents (cf. France et al., 1992). In separate testing, using $\Delta t = 10\,\text{nsec}$ yielded solutions for the propagating action potential which differed negligibly from those in which Δt was 100 nsec (as per Sherman and Rinzel, 1991; also see Appendix. For a discussion of stiffness and solution convergence, see Mascagni and Sherman, 1998).

Several tests against standard results under nonisopotential conditions have been successfully performed to validate the present computational methods for constant-diameter fibers. These tests included:

(1) reconstruction of the experimental (Pumphrey and Young, 1938) and theoretical (Hodgkin, 1954) linear relationship between the square of the fiber diameter and the conduction velocity (Goldfinger, 1978);
(2) increasing conduction velocity as the impulse approached a sealed end (Goldstein and Rall, 1974);
(3) the extinction of a propagating impulse as it approached an axial wire inserted from the other end of a uniform-diameter axon, as shown in Figure 18.2 (top);
(4) a step change in the axial conduction velocity at a diameter expansion for which the electrotonic step ($\Delta x/\lambda_d$) is kept unchanged by appropriately adjusting the value of Δx, as shown in Figure 18.2 (bottom), such that the impedance per compartment remained constant; and
(5) the analytic solutions for charging transients in passive fibers of infinite or finite length (Jack et al., 1975) were reconstructed with negligible error (Goldfinger, 2000).

To establish the initial conditions at all x, it was assumed that the initial solution could be suitably estimated by allowing the unperturbed system "at rest" to relax into temporal stability in the steady

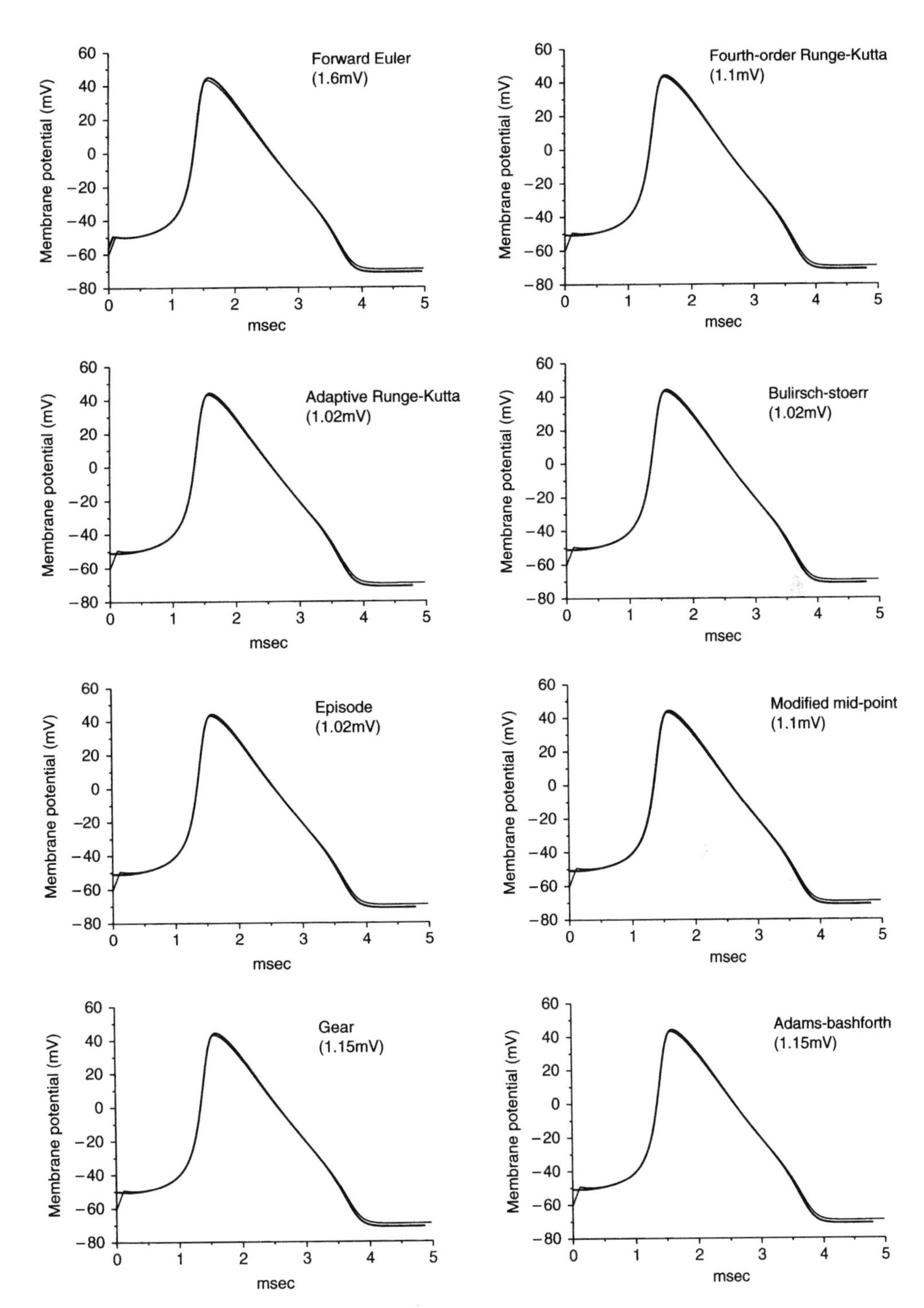

FIGURE 18.1 Isopotential membrane action potentials reconstructed with various numerical methods. Each solution (thicker traces) is superimposed upon solution using trapezoidal integration (thin traces). Each pair of responses is aligned with respect to the point of maximum dV/dt on the rising limb. Differences in action potential peak height (mV) are given in parentheses. Gear (via IMSL Library) and trapezoidal solutions were performed on a DEC 4000/610; other solutions shown were generated using a commercial ODE solver ("The Scientist," previously "DIFFEQ," MicroMath Scientific Software, Salt Lake City, Utah) on a 486-PC. In all runs, stimulus was $100\,\mu s \times 100\,\mu A/cm^2$, and $\Delta t = 10\,\mu s$.

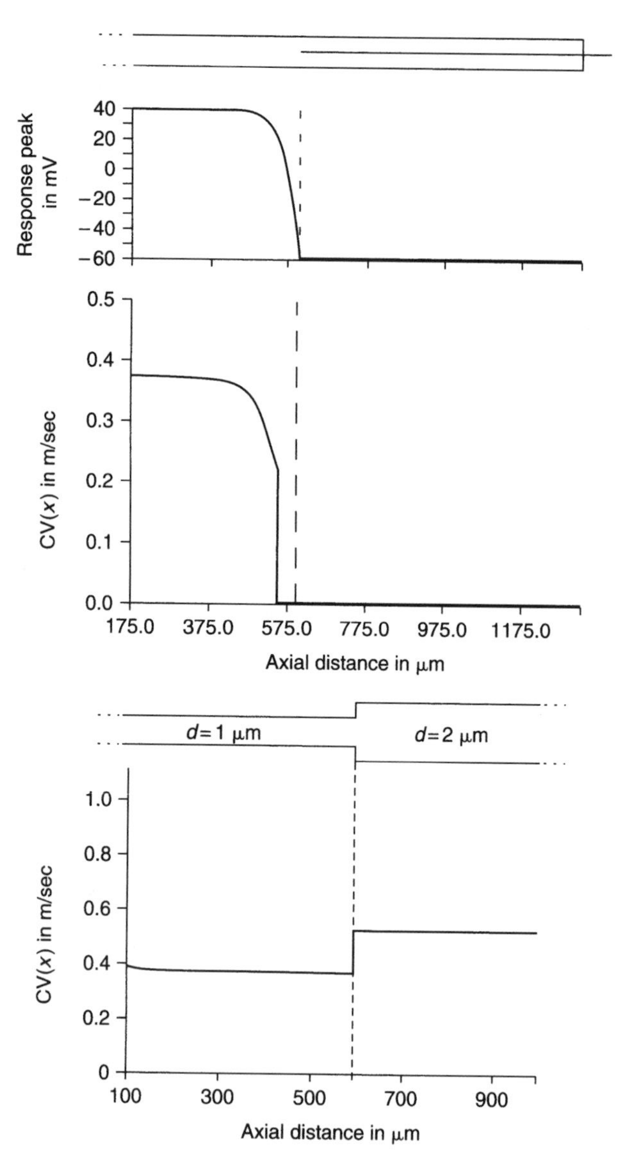

FIGURE 18.2 Tests of computed impulse propagation with structural inhomogeneities. (Top) Effect of an axial wire inserted into a 1.0 μm-diameter axon. (Bottom) Δx a function of diameter such that the electrotonic step is constant ($\Delta z = \Delta x_d / \lambda_d \equiv 0.0164$); $\Delta x_{d=1\,\mu m} = 3\,\mu m$, $\Delta x_{d=2\,\mu m} = 4.24\,\mu m$.

state, wherein any parameter changed only negligibly with respect to time. For the nonisopotential H–H regime with $d = 1\,\mu m$, the system relaxed spontaneously from an initial membrane potential value of -60 mV to a new value of -60.13 mV within approximately 0.05 msec and remained at this value indefinitely thereafter; voltage-dependent parameters were also determined in the new stable steady state. Then, the collection of fully relaxed new steady-state values defined the initial conditions used thereafter.

18.1.1.2 Cable Equation Representation of a Branch Point and its Branches

A branch point and its branches have been represented by literal cable models, in which the daughter branches are explicitly designated by serially arranged compartments constituting separate cables

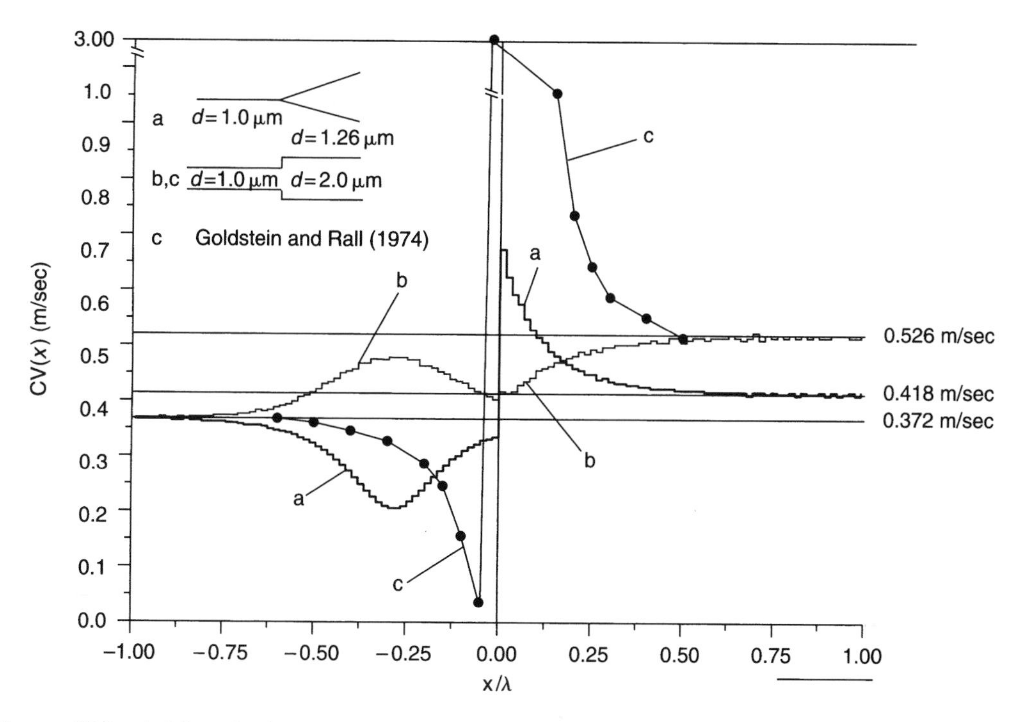

FIGURE 18.3 Axial conduction velocity $\mathrm{CV}(x)$ as a function of electrotonic distance (x/λ), where the geometrical inhomogeneity is at $x/\lambda = 0$. (a) A 1 μm-diameter parent axon which bifurcates into two 1.26 μm-diameter daughter branches each having electrotonic length z_d. (b) Same as (a) except the branches are represented by an "equivalent" cable having electrotonic length z_d. The respective $\mathrm{CV}(x)$ values averaged between $x/\lambda = 0.75$ to 1.00 are given at right. (c) Goldstein and Rall's (1974) values for $\mathrm{CV}(x)$ for the same "equivalent" cable.

(Goldfinger, 2000), or by "equivalent" cables in which the surface areas of all the components of a given order of branching are summed linearly and represented collectively as a single larger-diameter cylinder (Rall, 1962a, 1962b; Manor et al., 1991a, 1991b). The terms "completely branched," "fully-branched" (Rall et al., 1992), and "real" (Fohlmeister and Miller, 1997b) have been applied to literal representations, to contrast with "equivalent" cables and cylinders.

The use of an "equivalent" cable to represent a branched axon is not appropriate for the following reasons:

1. Given the symmetrical bifurcations used in the present study, the diameter of the equivalent cable representation of two equal-diameter sister branches, d_{eq}, would be given by $d_{\mathrm{eq}} = [2 \cdot (d_{\mathrm{br}}^{3/2})]^{2/3}$ (Manor et al., 1991a). Thus, the diameter of the equivalent cable always exceeds that of either individual branch (d_{br}). Further: $\theta_{\mathrm{eq}} = [\theta_{\mathrm{br}} \cdot (d_{\mathrm{eq}}/d_{\mathrm{br}})^{0.5}]$ (Goldstein and Rall, 1974, equation 9), where θ is conduction velocity. Then: $\theta_{\mathrm{eq}} > \theta_{\mathrm{br}}$, where θ_{eq} and θ_{br} are the conduction velocity in meters per second in the equivalent cable and literal branch, respectively, as shown in Figure 18.3. Therefore, using such an "equivalent" cable representation would overestimate the propagation velocity and underestimate the arrival times of the information propagating over electrotonic loci corresponding to the branches in real (vs. electrotonic) space.

2. The pattern of axial current dissipation at a literal branch point is different from that of an "equivalent" expansion: at a symmetrically bifurcating branch point where parent and branch diameters are equal, downstream axial current divides (Koch, 1999), as shown in passive cable (Goldfinger, 2000). By contrast, at a diameter expansion, downstream axial current is enhanced by distributing into a cable with lower axial resistance. Thus, as shown in Figure 18.3 for H–H axons, the conduction velocity of an impulse approaching a uniform-diameter symmetrical bifurcation is slowed,

while the conduction velocity of an impulse approaching an expansion is enhanced (Goldfinger et al., 1992; Goldfinger, 2000). Therefore, using an "equivalent" cable representation could inaccurately characterize the timing of information propagation incident to and emitted from a branch point.

3. For testing the possibility of differential propagation in axonal arborizations, it is necessary to be able to localize a modification (such as periaxonal K^+ accumulation) to one branch but not to the other(s) within a given branching order. For this purpose, a literal representation is essential.

18.1.1.3 Boundary Condition at a Branch Point

Cable solutions at points of geometrical inhomogeneities such as diameter changes, branches, and varicosities require specific boundary conditions. At such points, current conservation — appropriate to analytical regimes — has been applied to compartmental representations (e.g., Goldstein and Rall, 1974; Mascagni, 1989a; Bressloff and Taylor, 1993; Mascagni and Sherman, 1998; also see Lindsay et al., 2001c, 2002). For example, in the diameter transition from d_1 to d_2:

$$[(1/r_{i(d1)}) \cdot \partial V/\partial x]_1 = [(1/r_{i(d2)}) \cdot \partial V/\partial x]_2, \tag{18.3a}$$

where d is diameter, $r_{i(d1)} = (1000 \cdot a_1/2\rho)^{-1}$, and constant intracellular resistivity (ρ, geometry-independent) is assumed (Goldstein and Rall, 1974, equation 27). However, for a compartmental discretization, the terms of Equation (18.3a) necessarily become equivalent to

$$[I_{\text{axial}}]_1 = [I_{\text{axial}}]_2, \tag{18.3b}$$

where at any point k, I_{axial} is axially flowing current and $[I_{\text{axial}}]_k \propto \partial V/\partial x_k$. Then, at any geometrical inhomogeneity, I_{m} would become zero since $I_{\text{m}} \propto \{\partial/\partial x(\partial V/\partial x)\}$ Equation (18.1b).

Application of Equation (18.3) at a geometrical inhomogeneity in a compartmental representation is untenable because: (a) the diameter change does not occur within the confines of a single compartment, but rather is formed in this example by two adjacent compartments with diameters d_1 and d_2, respectively; and (b) within a given compartment, V is not a continuous variable. Thus, when Equation (18.3) is imposed as a boundary condition, current cannot flow axially across the inhomogeneity since by definition *there is no net axial current*. In the case of numerical simulation of impulse propagation in an axon whose diameter changed from 1 to 2 μm, it was found that applying the Equation (18.3) boundary condition caused the cessation of propagation. Also, when approximating the Equation (18.3) boundary at the diameter transition by imposing the condition:

$$I_{\text{m}} \rightarrow I_{\text{m}}/1000,$$

results similar to those of Goldstein and Rall (1974) were obtained, as shown in Figure 18.4 (cf. Goldfinger, 2000; cf. Figure 18.3). This approximation introduced a pre-widening decline in the action potential peak and a distorted $\text{CV}(x)$ distribution.

The correct boundary condition at a symmetrical literal branch point has been validated directly by analog computation, as shown in Figure 18.5 (Goldfinger, 2000). An electrical circuit was constructed with the same specifications as those of the compartmental representation of a passive (i.e., all membrane conductances voltage independent) symmetrical branch point consisting of a parent fiber and two equal daughter branches. This analog permitted a direct determination of the axial current in the vicinity of the branch point. The demonstration of a trifurcation of axial

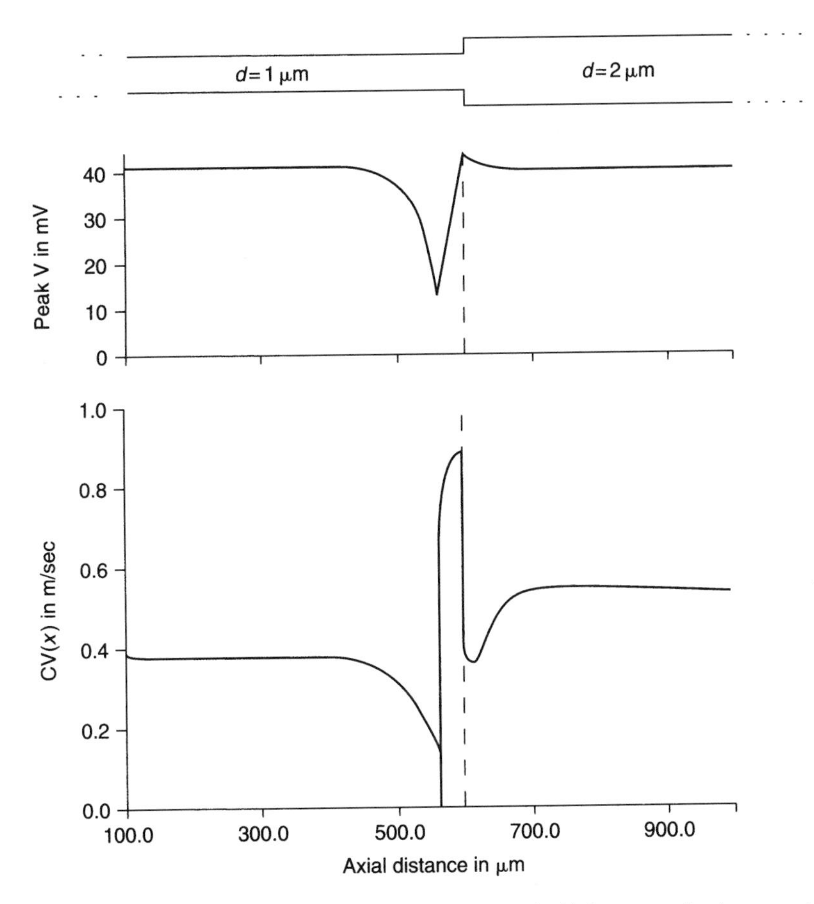

FIGURE 18.4 For an axon whose diameter doubled, at the geometrical inhomogeneity the current conservation principle of an analytic solution regime was approximated in the numerical (compartmental) solution regime as: $I_m \rightarrow I_m/1000$, yielding results similar to those of Goldstein and Rall (1974), where the impulse action potential peak voltage dropped drastically prior to the widening and $CV(x)$ declined to nearly 0.

current clarified the discretizations of the cable equation at the branch point. A passive numerical compartmental model (i.e., an exact numerical version of the circuit) was tested for various boundary expressions; the expression which reconstructed the circuit solution for net current at the branch point (compartment k) was:

$$I_k = 1000 \cdot (a/2\rho_i) \cdot [(V_{k-\Delta x} - V_k) - \{(V_k - V_q) + (V_k - V_r)\}]/\Delta x^2 \tag{18.4a}$$

$$= 1000 \cdot (a/2\rho_i) \cdot (V_{k-\Delta x} + V_q + V_r - 3V_k)/\Delta x^2, \tag{18.4b}$$

where the branch point at compartment k contacted compartments $k - \Delta x$ (the last compartment of the parent axon), and q and r (first compartments of each daughter branch, respectively). Using this boundary condition permitted the numerical reconstruction of the responses recorded from the analog circuit, including transients of membrane potential and axial current at any point [cf. Equation (18.1c)].

18.1.1.4 Statistical Analysis

All trains of events (propagating action potentials, stimulus pulses) were assessed as stationary point processes (Hagiwara, 1954; Cox and Lewis, 1966; Perkel et al., 1967a) and contained at least 1001

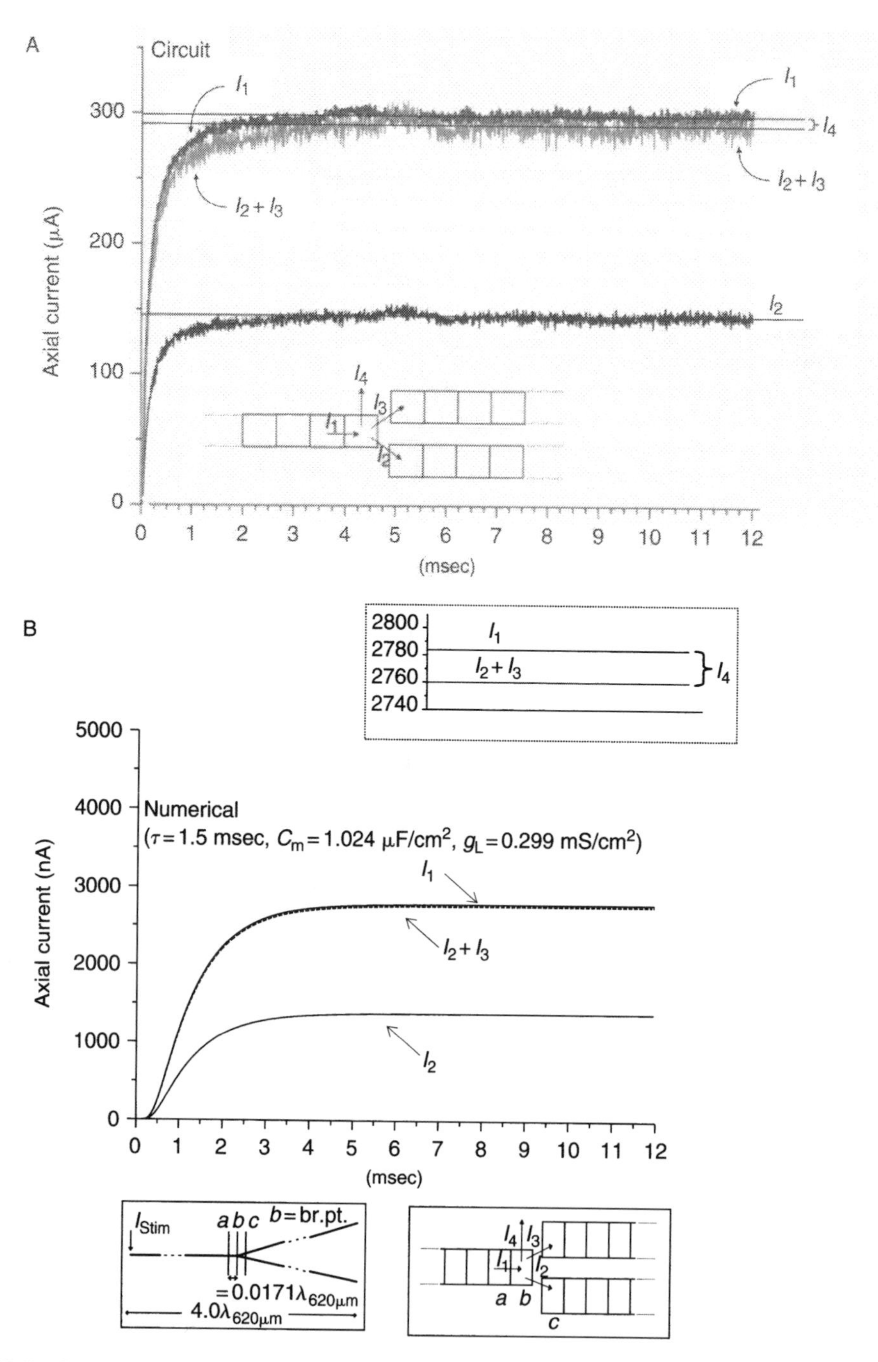

FIGURE 18.5 Components of axial current at a branch point. Axial currents (I_a) were measured using the recorded voltage differences between pairs of adjacent compartments:

$$I_a = (r_i)^{-1} \cdot \partial V_{(x,t)} / \partial x,$$

where $r_i = \rho/(\pi \cdot a_x^2)$. I_1 = axial current entering branch point; I_2, I_3 = axial current entering branches; I_4 = axial current leaking out at branch point, that is, difference between I_1 and $I_2 + I_3$. (A) Circuit recordings (single sweeps, 1 μsec sampling; each point is the average of 10 sequential 1 μsec values). Light trace: $I_2 + I_3$ represented by $2 \times I_2$ for clarity. Dashed lines: steady-state values averaged between t = 10 to 12 msec. (B) Simulations using numerical model with τ = 1.5 msec, C_m = 1.024 μF/cm^2, and g_L = 0.299 mS/cm^2. All d = 620 μm; stimulus current differed in A & B. (Reprinted with permission: ©*NeuroReport*/Lippincott Williams & Wilkins.)

events. The small changes in event train parameters over successive thirds of the run (viz., less than 1%, Moradmand and Goldfinger, 1995a) indicated that the trains were statistically stable over the duration of the run. Such stability was sufficient for the present analytical goals and no further tests for stationarity were performed. For each train, the Interevent Interval Distribution (IID, i.e., the sampled observation of the underlying probability density function, pdf) and Expectation Density (ED, i.e., the postevent probability of subsequent events, Poggio and Viernstein, 1964) were determined with the binwidth equal to 0.5 msec. The shortest event-train interval constituted a dead time (DT), which for spike trains mainly reflected absolute refractoriness. The IID shape parameters were mean (weighted as per Moradmand and Goldfinger, 1995a), standard deviation (σ), and coefficient of variation: cvar $= \sigma/(\text{mean} - \text{DT})$, given that the sampled pdf was limited to interevent intervals equal to or greater than the duration of DT (Tuckwell, 1989). The ED algorithm computed the postevent probability of subsequent events using sequential bins of Δt duration (here $\Delta t = 0.5$ msec). Thus, given an event (action potential or stimulus pulse), the ED showed the incidence of events at times $n_i \cdot \delta t$ thereafter, where n_i is the ith bin and $i \geq 1$. ED amplitude variability was mainly confined to $\pm 1\sigma$ of the average found in the steady-state plateau ($t \geq 25$ msec); this belt of confined variability is shown in each ED plot. The ED steady-state plateau average predicted the IID mean to within 4% (Poggio and Viernstein, 1964; Moradmand and Goldfinger, 1995a). When normalized to the steady-state average incidence, the amplitude of ED(t) is independent of the mean rate and allows direct comparison of ED trajectories under different conditions (Goldfinger and Amassian, 1980); in the present study, all EDs were so normalized.

18.1.1.5 Stochastic Stimulation

For stochastic stimulation studies, a Poisson-like train of equal-amplitude/equal-duration (0.2 msec) stimulus pulses was applied to fibers at $x = 0$ (Goldfinger, 1986; Moradmand and Goldfinger, 1995a). A Poisson process is a sequence of nonsuperimposable point events occurring randomly at a statistically stable mean rate R with a coefficient of variation equal to 1. Interevent intervals are distributed as a single-exponential falloff; on a log-ordinate plot, the slope equals $-R$ (Griffith, 1971). The ED is a constant proportional to the mean rate (Cox and Lewis, 1966).

As applied, the stimulus process (derived from nuclear disintegration[2]) departed from the Poisson ideal because of the deadtime, finite duration of events, and low-level ED noise. This Poisson-like stimulus process had mean rate = 95.13/sec, mean interval = 10.51 msec, minimum interval (deadtime) = 0.2 msec, and coefficient of variation = 0.912. The correlation coefficient of the linear regression single-exponential fit to the entire log-ordinate IID was −0.969. The postdeadtime ED was a constant with overlying variability equal to ±8.97% of the average probability computed between 50 and 100 msec (with 1 msec binwidth); this average probability predicted the IID mean within ±4.4%. All impulse trains elicited by such Poisson-like stimulation consisted of at least 1001 successively generated action potentials which propagated to the distal end(s) of the fiber.

The transformation of such Poisson-like stimuli into membrane potential transients was a spatially modulated process (Goldfinger, 1986). First, at the stimulus site (the sealed end of the fiber) or electrotonically close ($\leq 0.5\lambda$) to this site, voltage trajectories elicited by sequential stimuli were highly variable in amplitude and waveform (due to ongoing changes in membrane resistance). Second, by about 1λ downstream from the stimulus site, the response was considerably simplified, as some of the voltage transients occurring at or close to the stimulus site did not develop into corresponding propagating impulses and were dissipated electrotonically. Here as well as further downstream, variability in both impulse amplitude and conduction velocity were evident, due to impulse initiations in the temporal wake of preceding conditioning short-interval impulses (Miller and Rinzel, 1981; Moradmand and Goldfinger, 1995a); the propagation of relatively low-amplitude impulses was first described by Huxley (1959).

18.2 Impulse Propagation through Branch Points

18.2.1 Single Action Potential Propagation through Serial Branch Points

The axon consisted of a parent fiber with two orders of branching, that is, an initial (primary) branch point and a secondary branch point downstream on each primary daughter branch, where each daughter branch had the same length and diameter. A single action potential was initiated at the origin of the parent axon and propagated toward the primary branch point with constant $CV(x)$. For the simplest case, the distance between primary and secondary branch points was large, as shown in Figure 18.6A. In the vicinity of the primary branch point, $CV(x)$ changed biphasically, with a depression and return to the initial value prior to the branch point, an elevation at the branch point, and a rapid relaxation back to the initial value just beyond the branch point (Goldfinger, 2000). Goldstein and Rall (1974) computed a somewhat similar $CV(x)$ pattern for a single branch point represented by an "equivalent" cable (cf. Figure 18.3 and Figure 18.4). There were no impulse reflections from primary or secondary branch points.

This same $CV(x)$ pattern also occurred at the secondary branch points if separated from the primary branch point by a sufficiently large distance, as shown in Figure 18.6A. With closer spacing between primary and secondary branch points (e.g., 0.84λ, Figure 18.6B), the end phase of the primary branch point $CV(x)$ pattern merged with the beginning phase of the secondary branch point pattern.

With further reduction of the electrotonic spacing between primary and secondary branch points, the overall $CV(x)$ pattern became more complex, as shown in Figure 18.6C. For a spacing of 0.4λ, elimination of one of the secondary branches slightly lessened the depth of the return limb of the $CV(x)$ depression prior to the primary branch point but otherwise caused little alteration to the subsequent $CV(x)$ pattern. Extreme reduction in the distance between primary and secondary branch points (to 0.03λ) further depressed $CV(x)$ prior to the primary branch point and enhanced the elevation of $CV(x)$ after the primary branch point, as shown in Figure 18.6D. Separate testing showed that the overall $CV(x)$ patterns with relatively close primary-to-secondary spacing (Figure 18.6C and D) were not exact spatial superpositions of two $CV(x)$ patterns from isolated branch points (as in Figure 18.6A).

The depression in $CV(x)$ prior to the primary branch point added a delay to the impulse transmission time of approximately 90 μsec for a single isolated branch point. With more closely spaced branching (Figure 18.6C and D), the more complex $CV(x)$ depression prior to the branch point(s) added further to the delay (Figure 18.6D, inset). The increments in delay were spatially linked to the positions of the serial branch points. Between and beyond branch points, where the conduction velocity was spatially constant, action potential arrival time increased linearly with time. As the impulse approached the sealed-end fiber terminal, $CV(x)$ grew progressively (Goldstein and Rall, 1974). As a result, beginning approximately 0.6λ upstream from the terminal, the rate of increase of the impulse arrival time decreased progressively to zero at the terminal. Thus, the delay added previously by the branch point(s) was present at the axon terminal.

18.2.2 Propagation of Random Action Potential Trains through Serial Branch Points

Do serial branch points alter a propagating neuronal code? This question was addressed using random impulse trains, since: (a) axonal firing due to information convergence in a neuron's somatodendritic tree is likely to yield random firing in the axon (Goldfinger, 1986), and (b) a long random sequence provides a wide range of condition-test intervals where the conditioning itself contains diverse permutations and combinations of impulse sequences, thus addressing the question more thoroughly than would be possible by testing selected individual sequences. To determine whether the stochastic

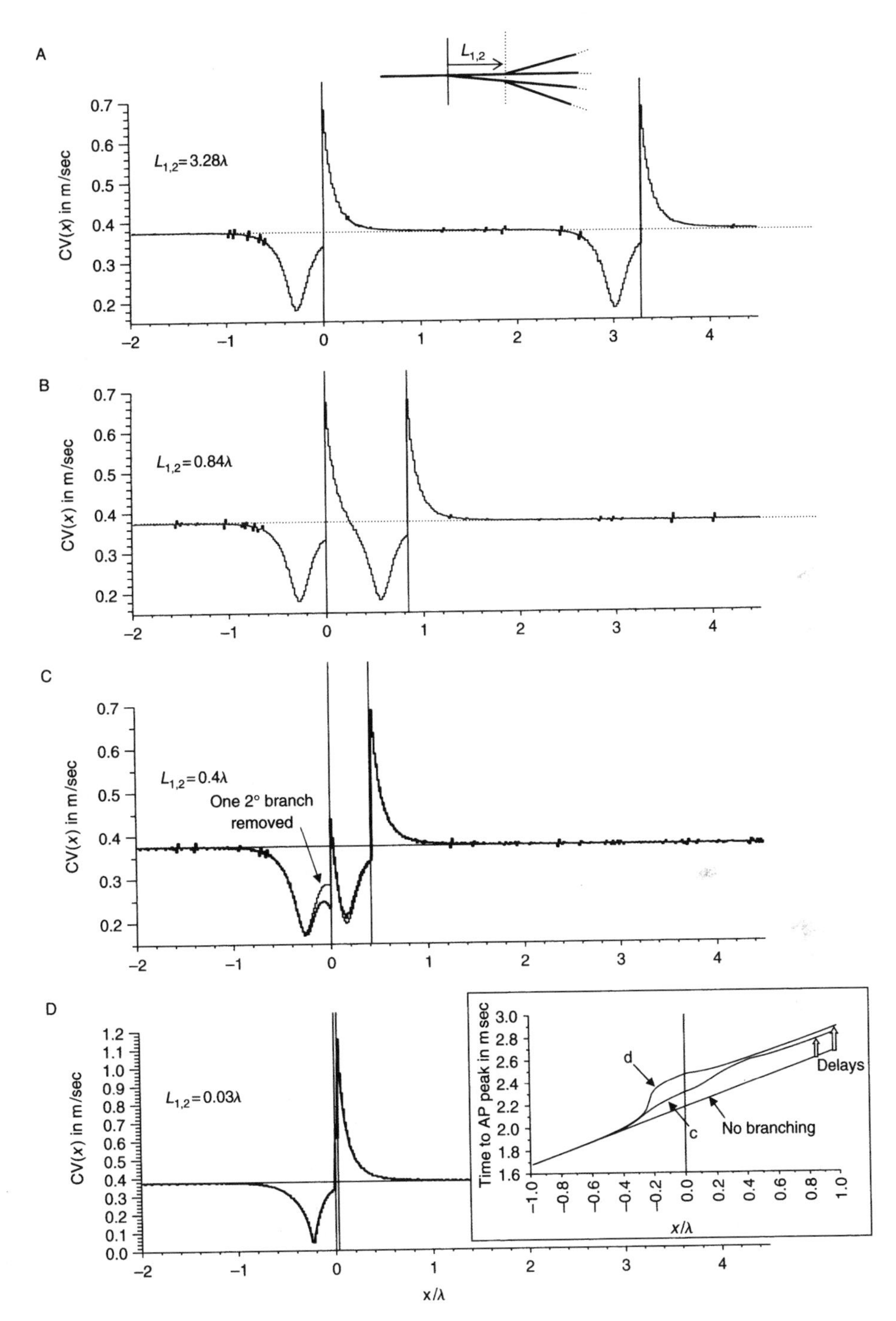

FIGURE 18.6 Changes in CV(x) with various primary-to-secondary branch point spacings. A single impulse propagated from a parent axon into four arborization configurations (A–D) differing in the distance between primary and secondary branch points ($L_{1,2}$); all diameters were 1 μm ($\lambda = 183$ μm) and each path length was 9.83λ. Abscissae, electrotonic distance ($z = x/\lambda$). Data shown are from one of the four possible symmetrical propagation paths for each case. (C) CV(x) shown with (thick trace) and without (thin trace) one of the secondary branches. (D, inset) Ordinate is the arrival time of the impulse at each z; comparisons are to the unbranched equal-path-length case, where increased arrival times indicate a delay. The primary branch point was at $z = 0$ and the secondary branch points were at (C) $z = 0.4$ or (D) 0.03.

properties of propagating information are altered by serial branch points, random impulse trains were elicited by Poisson-like stimulation of the parent axon origin and propagated into the arborization. The trains of propagated impulses were recorded at several loci incident to and within the arborization. The results for a single-branch point arborization are given in Figure 18.7.

Unbranched fibers served as the control for fibers with a single branch point. At either of the two recording locations, diameter reductions (37 or 68%) did not change the mean rate by more than 0.14%. The unbranched fibers' ED had a characteristic form with three successive components: (a) a deadtime (i.e., minimum interspike interval, 5.26–5.88 msec), due mainly to absolute refractoriness; (b) an early peak (6–8 msec), where the lack of preceding excitation during the deadtime briefly enhanced the subsequent firing probability (Goldfinger, 1986; Moradmand and Goldfinger, 1995a); another contributor to this higher probability phase was the dispersal between the stimulation and recording sites of the conduction velocity of the second of a short-interval impulse pair (Miller and Rinzel 1981; Moradmand and Goldfinger, 1995a); and (c) a maintained period of noisy probability centered around a mean (100% in the ED) with variability mainly distributed between $\pm 1\sigma$; in this Poisson-like plateau, the variability reflected the diversity of conditioning for $t < 0$ (Goldfinger, 1986). These stochastic properties characterized the event process incident to the primary branch points.

For a single-branch-point axon, the response patterns were recorded at the branch point as well as at a site 900 μm downstream from the branch point (sites *a*, *b*, respectively, in Figure 18.7). Representatives of three classes of single-branch-point axons were classified from the diameters of parent and branch fibers using Rall's (1959) "geometric ratio," $\text{GR} = \sum_i d_i^{3/2}/d_{\text{p}}^{3/2}$, where d_i and d_{p} are the diameters of the *i*th branch and the parent axon, respectively. Thus, given symmetrical bifurcations, for GR $= 2$, all diameters were equal. For GR < 2, branch diameters were less than that of the parent fiber. At a given axial position, branched and unbranched control fibers had the same diameter.

At each recorded position, the responses of branched fibers were similar to those of their unbranched controls. In addition to nearly superimposable normalized EDs, the shape parameters of the IIDs (mean interval and coefficient of variation) were also nearly equivalent. At site *a*, the impulse trains' stochastic properties were unaffected by the addition of the branch point. Further downstream at site *b*, the same result obtained. That is, the information incident to, transmitted through, and emitted from the branch point's position was largely unaffected by the branch point. However, branching introduced a small but consistent increase in the deadtime. For each of the three cases, this effect was larger at the branch point and smaller at the downstream site. The largest deadtime prolongation was 0.55 msec (Figure 18.7, GR $= 0.26$, site *a*); nevertheless, the mean interval only increased by 0.27%. These results showed that (a) the primary branch point introduced no significant modifications to the coding of the random propagating action potential trains, and (b) impulse train propagation through the primary branch point was highly efficient in that most incident frequencies were emitted without omission.

To assess the role of secondary branching, we tested axons including symmetrical primary and secondary branching with primary and secondary branch points separated by 3.3λ, 0.4λ, or 0.03λ, as shown in Figure 18.8. Impulse trains were assessed at and beyond the secondary branch point; separate fibers with only primary branching served as the controls. The similarity of response properties at and beyond the secondary branch points to those of the singly branched fiber controls at the same axial positions showed minimal coding alterations due to the secondary branch points. However, secondary branch points introduced some modifications in the impulse train patterns. First, in several cases the EDs' early probability peak was shifted by 0.5 msec and enhanced in amplitude by as much as 31%. Second, the deadtime duration was increased by as much as 12.84%. However, in all cases the mean interval did not change by more than 0.54%.

These results showed that (a) the secondary branch point introduced no significant modification to the coding of the random propagating action potential trains, (b) impulse train propagation through the secondary branch point was highly efficient in that most incident

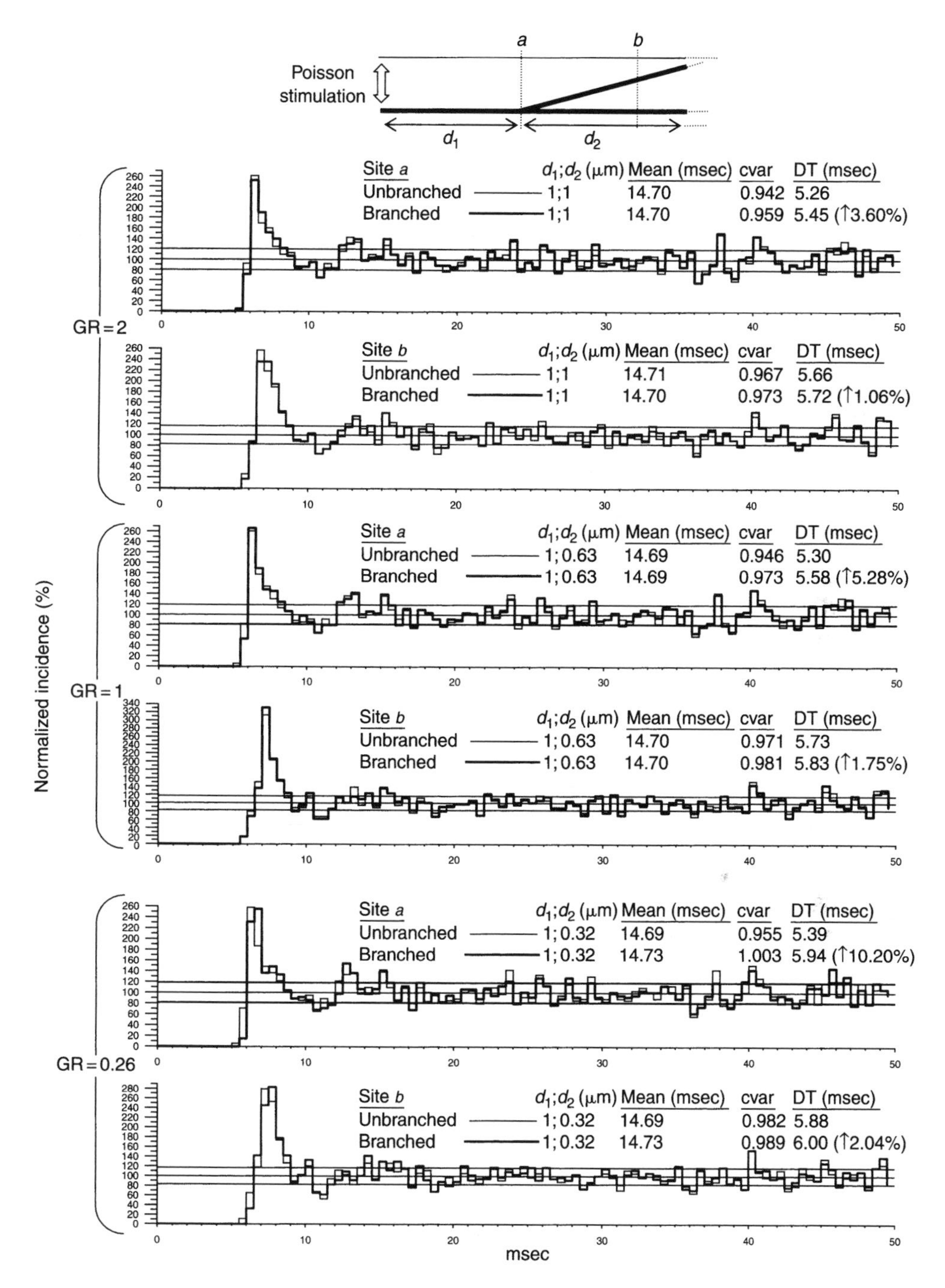

FIGURE 18.7 Responses to Poisson-like stimulation in three axons with a single symmetrical branch point. Stimulation process applied at parent axonal origin ($x = 0$). Recording sites were at the branch point ($x = 600\,\mu$m: site *a*) and 900 μm downstream from the branch point (site *b*). Ordinates: Normalized Expectation Densities (ED, 0.5 msec binwidth), with average probability (100%) determined in the steady state between $t = 25$ to 50 msec; dashed lines, average $\pm 1\sigma$. Control data from unbranched equal-path-length axons are shown with thin traces; branched axon data are shown with thicker traces. Interspike Interval Distributions (IID) parameters: mean, coefficient of variation (cvar, deadtime-corrected), and deadtime (DT, minimal interval). In the fiber configuration diagram (top), d_1 and d_2 are the parent and branch diameters, respectively. GR, geometric ratio (see text). $N_{\text{spikes}} = 1001$.

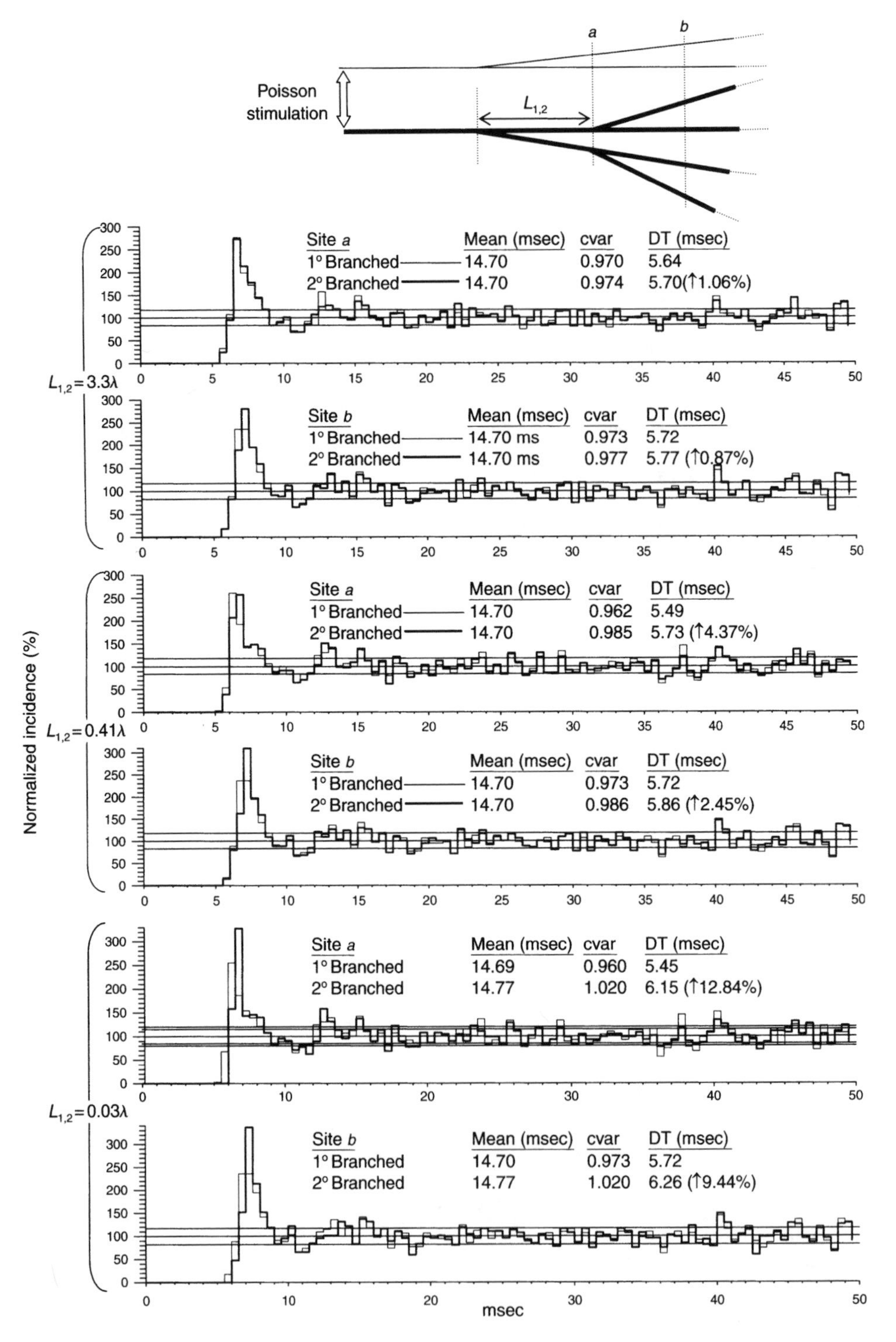

FIGURE 18.8 Responses to Poisson-like stimulation applied at $x = 0$ in three axons with primary and secondary branching. Normalized EDs and IID statistics as in Figure 18.7. In each case all fibers were symmetrical; all diameters were 1 μm; the primary branch point was at $x = 600\,\mu\text{m}$; the proximal recording site a was at one of the secondary branch points; and the distal recording site b was 900 μm downstream from the primary branch point. The fibers differed in the distance between primary and secondary branch points ($L_{1,2}$). Control fibers (thin traces) had only the primary branch point. $N_{\text{spikes}} = 1001$.

frequencies were emitted without omission, and (c) impulse train stochastic properties were sustained independently of the short electrotonic distance between primary and secondary branch points.

Thus, for each type of branching fiber, impulse-based information incident to a primary or secondary branch point was transmitted with negligible distortion. Neither the primary nor the secondary arborizations significantly altered the rates or patterns of impulse propagation over a wide range of incident frequencies. Therefore, impulse propagation through primary and secondary branch points was highly efficient.

18.2.3 Discussion

The computations in this section have shown that the propagation of random impulse-based wide-band information through symmetrical unmyelinated axonal branch points under normal conditions can occur virtually without distortion. Altenberger et al. (2001) computed an analogous result for impulse train propagation through an axonal flare (a region of gradually increasing diameter) using a finite-element method. This general result was consistent with earlier demonstrations of the robust propagation of single impulses at branch points despite differences between parent and daughter diameters or a widening or narrowing of the parent fiber (Goldfinger, 2000).

Each serial branch point introduces an increment in an impulse's downstream arrival time at any location downstream from the branch point(s), including the axon terminals. When serial branch points are closely spaced, propagation is additionally delayed but not compromised. However, the delays themselves are believed to be of significance with respect to interneuronal synchronization and other timing-dependent functions within or between neural networks (e.g., Bi and Poo, 1999; Diesmann et al., 1999; Golomb and Ermentrout, 1999; Engelhardt and Chase, 2001; Fox et al., 2001; Kempter et al., 2001; Panzeri et al., 2001; Petersen et al., 2002).

The presence of a literal branch point introduces three modifications to impulse propagation. First, for a single impulse, the local alterations of $CV(x)$ yield a delay in downstream arrival time. Second, during random activity, there is a slight lengthening in the shortest interspike interval, a type of highest-frequency elimination (low-pass filtering). Third, also during random activity, there is a slight shift in the early probability peak of the autocorrelation as represented by the normalized ED.

Each of these results is attributable to the effect of a branch point on the axial current flowing downstream ahead of a propagating action potential. An aliquot of such current is outward and stimulates the immediately downstream fiber segments; the larger this stimulus is, the sooner the downstream action potential is initiated, hence a larger $CV(x)$. If such axial current is reduced, action-potential initiation in the invaded region is necessarily either delayed or precluded. With a literal branch point, as an impulse approaches the branch-point site of axial current division (Figure 18.5), axial current invading a given branch is increasingly attenuated such that $CV(x)$ is necessarily reduced. This effect reverses as stimulating axial currents activate membrane in the branch(es) just downstream from the branch point. Such spatial resurgence in axial current is due to the constancy and magnitude of Na^+ channel density in parent and branch fibers assumed in the model.

The opposite effect occurs at an axonal widening. As the impulse approaches the widening, the flow of downstream axial current is enhanced due to the lowered axial resistance associated with the larger diameter. This enhanced flow of axial current causes a more rapid stimulation of the invaded fiber, such that $CV(x)$ necessarily increases on approaching a widening (Goldfinger et al., 1992; Goldfinger, 2000).

During random activity, the branch point also affects the flow of downstream axial currents. When two closely spaced impulses — forming a short interspike interval — approach a branch point, the first impulse conditions the invaded membrane. Without any branch point, the axial currents generated by the second impulse are sufficient to overcome the relative refractoriness and elicit a second impulse, thus allowing the short interspike interval to propagate. However, at a branch point, axial currents are attenuated. For the shortest interspike intervals, the attenuated version of the axial current from the second upstream impulse is too small to stimulate the second spike at the expected time; thus the second spike fails to develop, and therefore the smallest interspike interval entering the branch point region is not emitted. This accounts for the small increase in deadtime observed at the branch point. When two branch points are closely spaced (see Figure 18.8, $L_{1,2} = 0.03\lambda$, site a), this axial current attenuation effect is necessarily enhanced, such that the minimum interval is larger than if just one branch point were present locally.

These reductions in the probability for generating the shortest interspike intervals near branch points are indicated in the ED, whose initial peak of elevated probability is shifted slightly to a larger value. Since the shortest intervals do not occur, following a given impulse the system has sufficient time to recover from relative refractoriness such that stimuli at these intervals are on average more likely to be suprathreshold, hence the shift of the ED peak to slightly longer conditional times. However, these local branch-point-related effects are relatively infrequent and effectively minor in that they do not appreciably affect the content of the code conveyed. For the wide-band spike trains elicited by Poisson-like stimulation, both the distribution of interspike intervals as well as the patterns of impulse occurrences are nearly the same as those occurring with propagation over the same distances without traversing a branch point. With primary branching, the presence of diameter reductions does not alter the impulse code.

The early ED peak is not a property of the stimulus process and is associated with propagation in unbranched fibers, specifically related mainly to the effects of refractoriness (Miller and Rinzel, 1981; Goldfinger, 1986; Moradmand and Goldfinger, 1995a). The introduction of branching sustains this initial transient elevation of conditional probability. Such response patterns have been observed in experimental axons driven by a Poisson-like stimulus process (Goldfinger, 1990; Moradmand and Goldfinger, 1995b). These experimental axons were myelinated primary afferent fibers associated with cat cutaneous $G1$ (velocity-detector) hair receptors; they were stimulated electrically in the periphery (forepaw skin) and the elicited trains of propagated impulses were recorded centrally near their terminations in the ipsilateral cuneate nucleus. Interestingly, given the anatomy of these particularly long and highly branched axons (Weinberg et al., 1990), the recorded impulse trains may have traversed as many as 26 sites of collateral bifurcational branch points with interbranch point distances between 50 and 1350 μm.

18.3 Role of Periaxonal K^+ Accumulation

18.3.1 Periaxonal K^+ Accumulation Model

During a propagating impulse, K^+ efflux could cause a temporary accumulation of K^+ in the periaxonal space (Frankenhaeuser and Hodgkin, 1956). Given the effects of extracellular K^+ on macroscopic Na^+ conductance in the isopotential squid axon (Adelman and Palti, 1969) and the suggestions of conduction block by extracellular K^+ variation in other preparations (Parnas et al., 1976; Spira et al., 1976), the possibility that periaxonal K^+ accumulation could cause a modification of information propagating through branch points was addressed.

The dynamics of periaxonal K^+ accumulation with activity were represented by the first-order differential equation given by Clay (1998, equation 4). In this model, K^+ accumulates in a virtual periaxonal space (versus a literal anatomical space, Inoue et al., 1997) due to the K^+ efflux during each action potential, and K^+ is cleared from the periaxonal space by two mechanisms: (a) passive diffusion and (b) K^+ uptake by glia (conceptually including Schwann cells, Araque et al., 1999):

$$dK_s/dt = R_s - R_d - R_g, \tag{18.5}$$

where K_s is the concentration of K^+ in the periaxonal space, R_s is the rate of K^+ accumulation in the periaxonal space due to voltage-dependent K^+ efflux during axonal activity, R_d is the rate of K^+ clearance from the periaxonal space due to diffusion into the bulk extracellular solution, and R_g is the rate of K^+ clearance from the periaxonal space due to glial cells (see Clay, 1998, equation A4; cf. Inoue et al., 1997; Zhou and Chiu, 2001). This model's K^+ transport by glial cells is described entirely by (see Clay, 1998, equation 4):

$$R_g = (K_s - K_o)/\{\tau_g \cdot [1 + (K_s - K_o)/K_{diss}]^3\}, \tag{18.6}$$

where K_o is the extracellular (i.e., unstirred bulk) K^+ concentration and K_{diss} is a dissociation constant (see subsequent sections). The vectorial direction of R_g is positive for $K_s > K_o$ and negative when $K_s < K_o$ and $[(K_s - K_o)/K_{diss}] < 1$.

The removal of K^+ from the periaxonal space by a glial transport mechanism is represented as a dual-state kinetic process where K_{diss} is the dissociation constant for glial K_s removal; its dimensions are:

$$K_{diss} = (k_{off}/k_{on}) = [t^{-1}/((mM) \cdot t)^{-1}] = mM,$$

such that K^+ removal from the periaxonal space is enhanced by larger values of K_{diss}.

Figure 18.9 shows this model's periaxonal K^+ accumulation and removal during a propagating action potential. In the present study, parameter values were: extracellular K^+ concentration, $K_o = 22$ mM; intracellular K^+ concentration = 410 mM; $K_{diss} = 2$ mM; glial K^+-removal time constant, $\tau_g = 0.23$ msec; diffusion time constant = 12 msec; virtual periaxonal space width = 2.4 nm. These values provided for significantly high levels of K_s accumulation so as to clearly delineate the role of periaxonal K^+ accumulation in impulse coding. Normal conditions (i.e., where K^+ is cleared rapidly from the space and the action potential waveform is normal) obtained for $K_{diss} = 20$ mM and all other constants unchanged. With the smaller K_{diss}, the rate of K^+ clearance from the space was attenuated considerably (Figure 18.9, bottom trace), causing the accumulation of periaxonal K^+ and a lengthening of the action potential falling phase. The reduction in the net electrochemical driving force on K^+ diminished I_K and therefore slowed the trajectory of $V_x(t)$.

With K_{diss} equal to 2 mM, the rate of K^+ removal by glia was highest when K_s concentration was in the 22 to 30 mM range, and monotonically approached saturation at higher values (Figure 18.9). Subsequently, as K_s returned to its initial value, the glial transport rate rose again transiently; this did not occur with $K_{diss} = 20$ mM since the small maximum (22.47 mM, an elevation of 2.1%) did not reoccur during the normal action potential. The single-impulse conduction velocity (0.375 M/sec at $x > 1\lambda$ from sealed ends) was unaffected by the value of K_{diss}.

Clay's model of periaxonal K^+ accumulation/clearance is essentially generic. The present application did not attempt to reconstruct any specific experimental data. This model did not incorporate other explicit mechanisms of K_o regulation such as those required for glia-mediated K^+ buffering (Orkand et al., 1966; Reichenbach, 1991; cf. Brown and Kukita, 1996; Ceelen et al., 2001). The rapid removal of periaxonal K^+ by glia has been described for optic nerve axons (Ransom et al., 2000).

Intracellular and nonperiaxonal extracellular concentrations of K^+ and Na^+ were assumed to be constant. In some experiments, this periaxonal K^+ accumulation model was only applied to a portion of the cable. The excitation formalism did not include a persistent K_o- and V-dependent Na^+ conductance (Somjen and Müller, 2000) nor empirical relations between α_h and β_h with K_o (Adelman and Palti, 1969).

18.3.2 Effects of Periaxonal K^+ Accumulation

The accumulation of periaxonal K^+ (Frankenhauser and Hodgkin, 1956) has been implicated as a factor which may influence impulse propagation at branch points (Parnas et al., 1976;

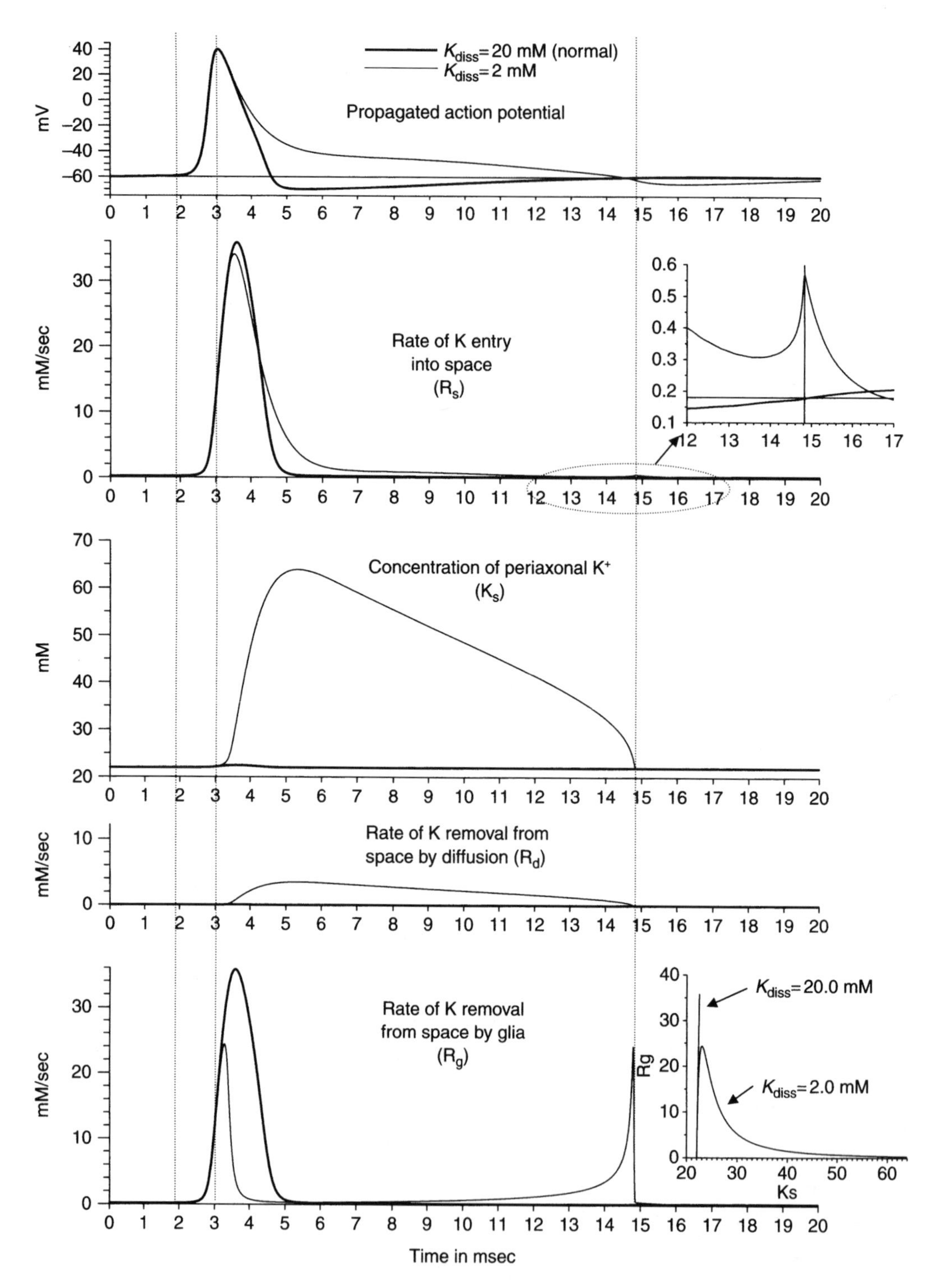

FIGURE 18.9 Dynamics associated with K^+ accumulation and removal in the periaxonal space, based on Clay's (1998) model. Recordings from the center of a 9.9λ-length, 1 μm-diameter unbranched axon with uniform K_{diss}. A single action potential was initiated at $x = 0$ and propagated at 0.375 m/sec, with K_{diss} of 20.0 mM (thick traces, normal) or 2.0 mM (thin traces). Lower trace inset: K_S–R_g phase plane during respective action potentials.

Smith and Hatt, 1976). This hypothesis was tested for action potentials propagating within axonal arborizations using Clay's model for periaxonal K^+ accumulation (Clay, 1998). The effects of periaxonal K^+ accumulation during a propagating action potential in an unbranched axon are given in Figure 18.10. During the impulse, the value of the K^+ equilibrium potential in the periaxonal space ($V_{Ks} \propto \log[K_{space}(t)/K_{axoplasm}]$) rose over 1.77 msec to its peak, declined slowly over 10 msec, and finally returned rapidly to baseline. The K_s elevation (to a nearly three-fold peak, Figure 18.9) and prolonged V_{Ks} transient lowered the K^+ driving force; this decreased the magnitude of the ($I_K > I_{Na}$) inequality during repolarization and slowed the $V(t)$ slope.

After the action potential peak, the relaxation times of Na^+-channel population activation and inactivation, K^+-channel activation, and macroscopic K^+ conductance were all prolonged by at least 10 msec. These changes prolonged the excitation recovery cycle, as shown in Figure 18.11. The duration of the absolute refractory period was prolonged by approximately 5 msec. Also, there was an inflection in the normally monotonic initial falloff of relative refractoriness; this coincided with a small transient rise in the rate of K^+ entry into the periaxonal space due to a small transient increase in the driving force for K^+ as periaxonal K^+ concentration declined to its initial value (Figure 18.9, inset). The supernormal period (Adrian, 1920; Graham, 1934; Gasser, 1941; also known as "superexcitable," Raymond and Lettvin, 1978; i.e., a phase of the recovery cycle when excitability is higher than it was before conditioning; e.g., Swadlow and Waxman, 1978; Kocsis et al., 1979; McIntyre et al., 2002) was delayed by 8 msec and was slightly attenuated in amplitude but not in duration. The second phase of relative refractoriness (Moradmand and Goldfinger, 1995a) was similarly delayed. The cycle ended with a second phase of supernormality lasting 5 msec that was partially concurrent with a small delayed depolarization.

Inasmuch as the waveform of a propagating action potential is in part determined by axial voltage gradients (i.e., $\partial^2 V/\partial x^2$ in the cable equation, Equation [18.1b]), the effects of periaxonal K^+ accumulation on the action potential waveform were in part determined by the spatial distribution of periaxonal K^+ accumulation. In an unbranched axon, the amplitude and time course of the elevation of periaxonal K^+ were modulated by the length of fiber supporting periaxonal K^+ accumulation, as shown in Figure 18.12. While a single point (length $= 0.0164\lambda$) of periaxonal K^+ accumulation had little effect on the membrane potential trajectory at that same point, the extension of periaxonal K^+ accumulation into adjacent fiber areas caused a progressive enhancement of the amount of periaxonal K^+ accumulation and the duration of action potential recovery; the limit beyond which no further change occurred was $\pm 0.5\lambda$ surrounding the recording point.

This spatial effect also occurred with a single-branch point fiber where the distribution of periaxonal K^+ accumulation was increased symmetrically from the branch point, as shown in Figure 18.13. However, the spatial limit beyond which no further waveform prolongation occurred was greater than $\pm 0.5\lambda$, beyond which the duration — but only slightly the magnitude — of the periaxonal K^+ accumulation were enhanced. When a periaxonal K^+ accumulating branch was added to a periaxonal K^+ accumulation unbranched fiber (lower records in Figure 18.12 versus Figure 18.13), the addition of the branch prolonged the duration of the branch point action potential.

18.3.3 Differential Propagation Due to Periaxonal K^+ Accumulation

Although the conduction velocity (0.375 m/sec) of a single action potential was unaltered by periaxonal K^+ accumulation in the unbranched 1 μm-diameter axon, periaxonal K^+ accumulation caused significant alterations in postimpulse recovery, as shown in Figure 18.10 and Figure 18.11. Thus, it was of interest to consider whether periaxonal K^+ accumulation could cause differential propagation of subsequent action potentials at a branch point (Swadlow et al., 1980). This was studied in single-branched axons where the entire length of one of the branches supported periaxonal K^+ accumulation while the other branch — as well as the parent axon and the branch point — did not. A pair of action

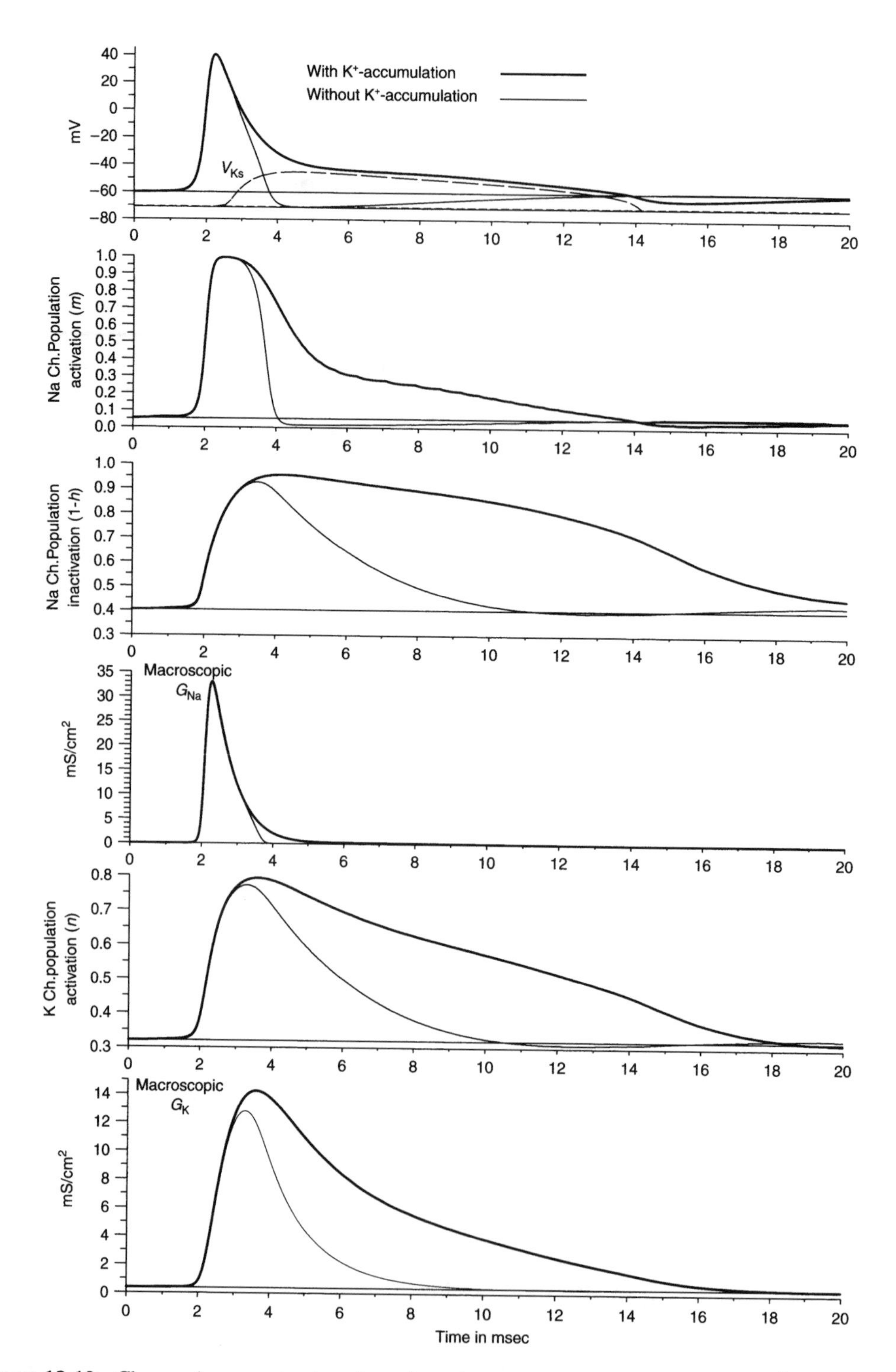

FIGURE 18.10 Changes in macroscopic voltage-dependent parameters due to periaxonal K^+ accumulation during a single propagating action potential (top). Unbranched 1 μm-diameter axon had total electrotonic length of 9.85λ with periaxonal K^+ accumulation (i.e., $K_{diss} = 2$ mM) over the entire length; recordings from $x = 3.28\lambda$. V_{Ks} (dashed line), K^+ equilibrium potential in the periaxonal space, where $V_{Ks} \propto \log[K_{space}(t)/K_{axoplasm}]$; G_{Na}, G_K, macroscopic conductances for Na^+ and K^+, respectively; $\Delta t = 10^{-4}$ msec; plot resolution $= 10^{-2}$ msec.

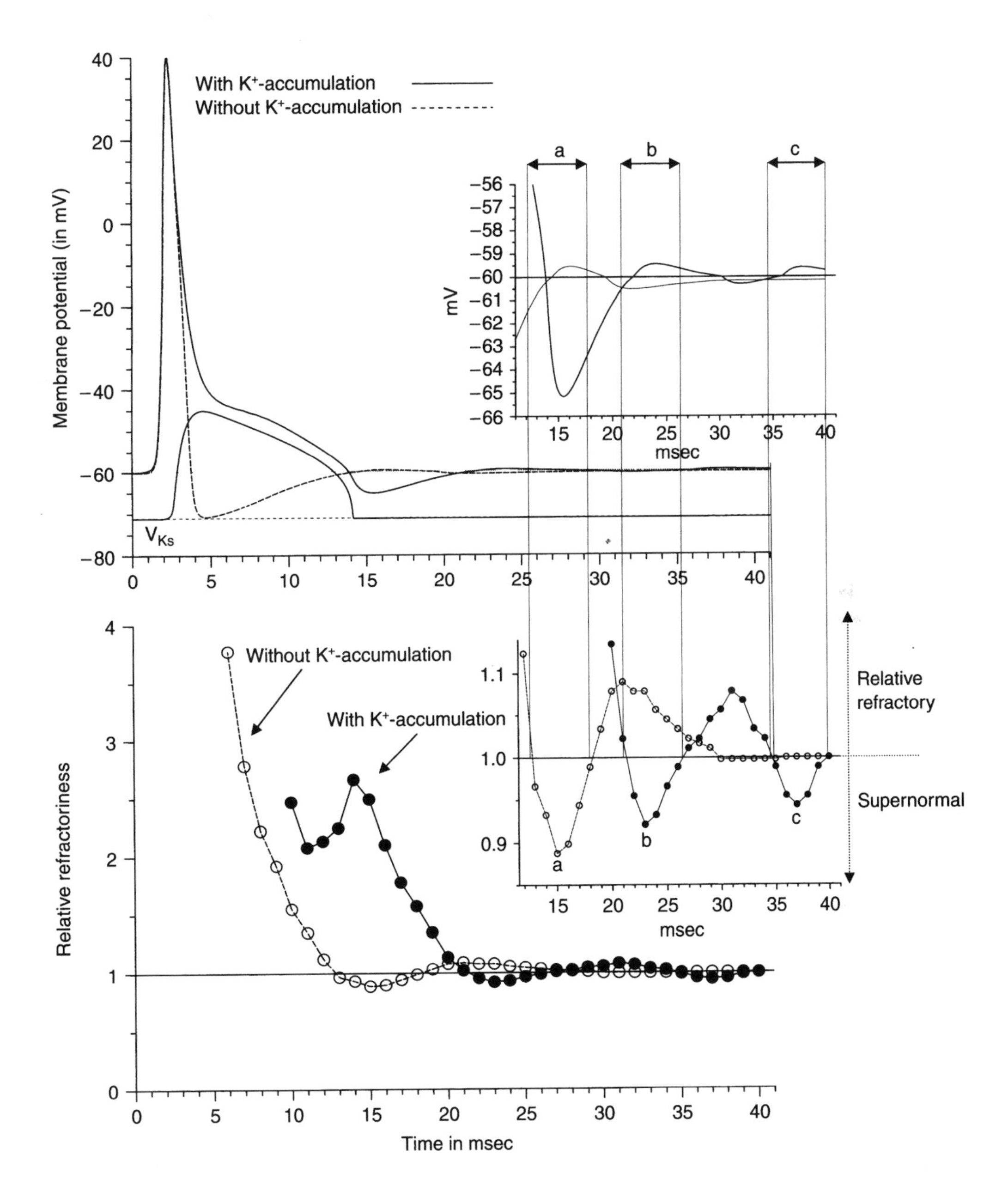

FIGURE 18.11 Refractoriness during the response of Figure 18.10. (Upper) Membrane potential trajectories with and without periaxonal K^+ accumulation. (Lower) Recovery cycle determined with conditioning spike initiated at $x = 0$ and test pulse (0.2 msec duration) to elicit a second impulse delivered at the recording site. Abscissa origin is the beginning of the propagating conditioning spike. Ordinate: relative refractoriness in terms of stimulus pulse amplitude, $(I_{\text{stim}(t)}/I_{\text{stim}(t=\infty)})$. Insets show periods of supernormality; vertical lines delineating supernormal periods (A: control; B, C: with periaxonal K^+ accumulation) allow comparison of timing with periods of concurrent depolarizing oscillations.

potentials at a given time interval was initiated at the parent fiber's end and propagated into the daughter branches; recordings were made at several locations along the branches, where the responses of the daughter branch without periaxonal K^+ accumulation served as a control which included the cable effects of the branch point *per se*. Single interspike intervals were tested separately to determine

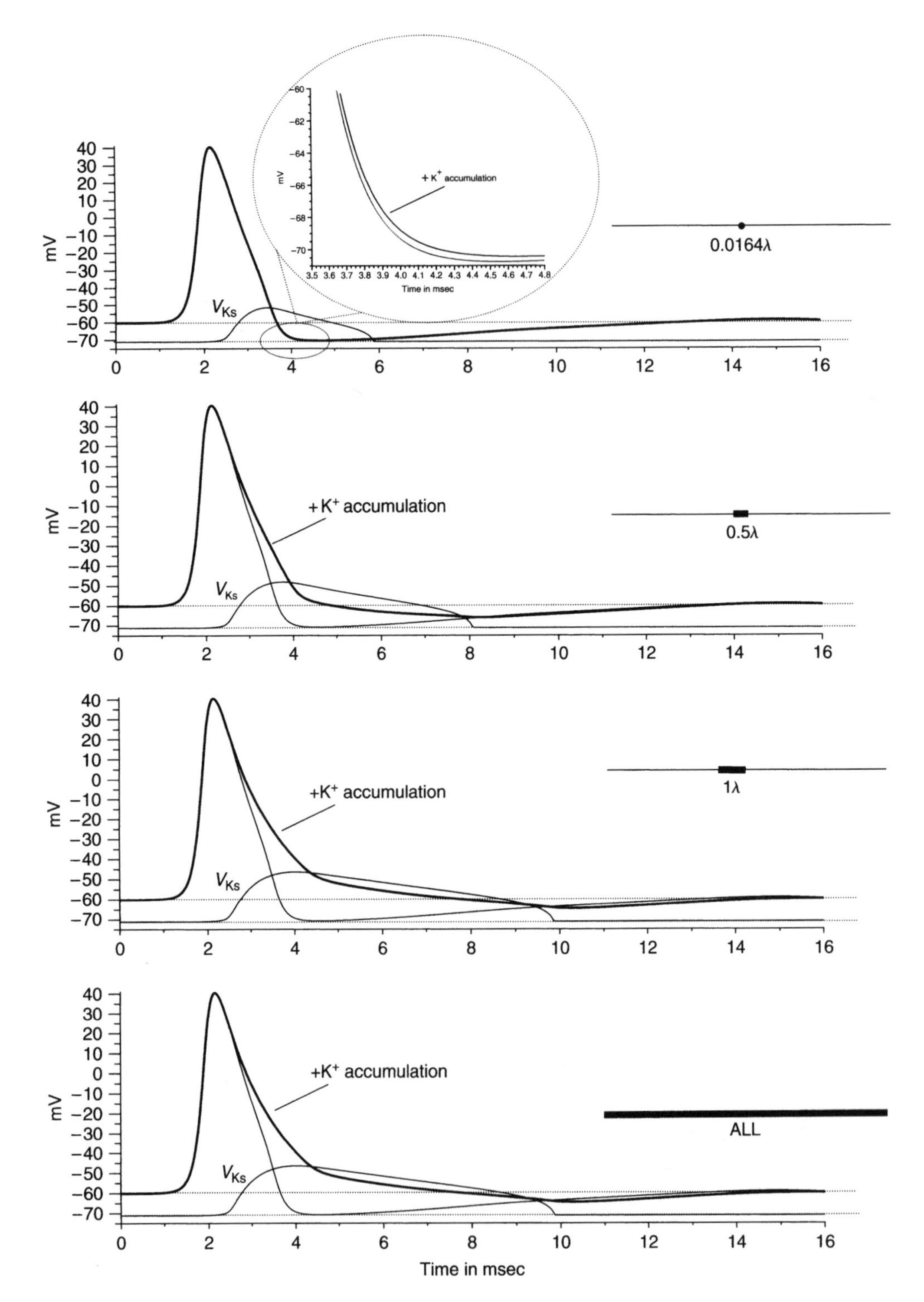

FIGURE 18.12 Effects of the spatial distribution of periaxonal K^+ accumulation on a single propagating action potential. A single impulse was initiated at the left end of the fiber and propagated into region(s) of periaxonal K^+ accumulation. V_{Ks} and membrane potential recorded with (heavier traces) or without (lighter traces) periaxonal K^+ accumulation, which was distributed symmetrically from the recording site at the center of the fiber (heavier line in diagrams at right). $\Delta t = 10^{-4}$ msec; plot resolution $= 10^{-2}$ msec.

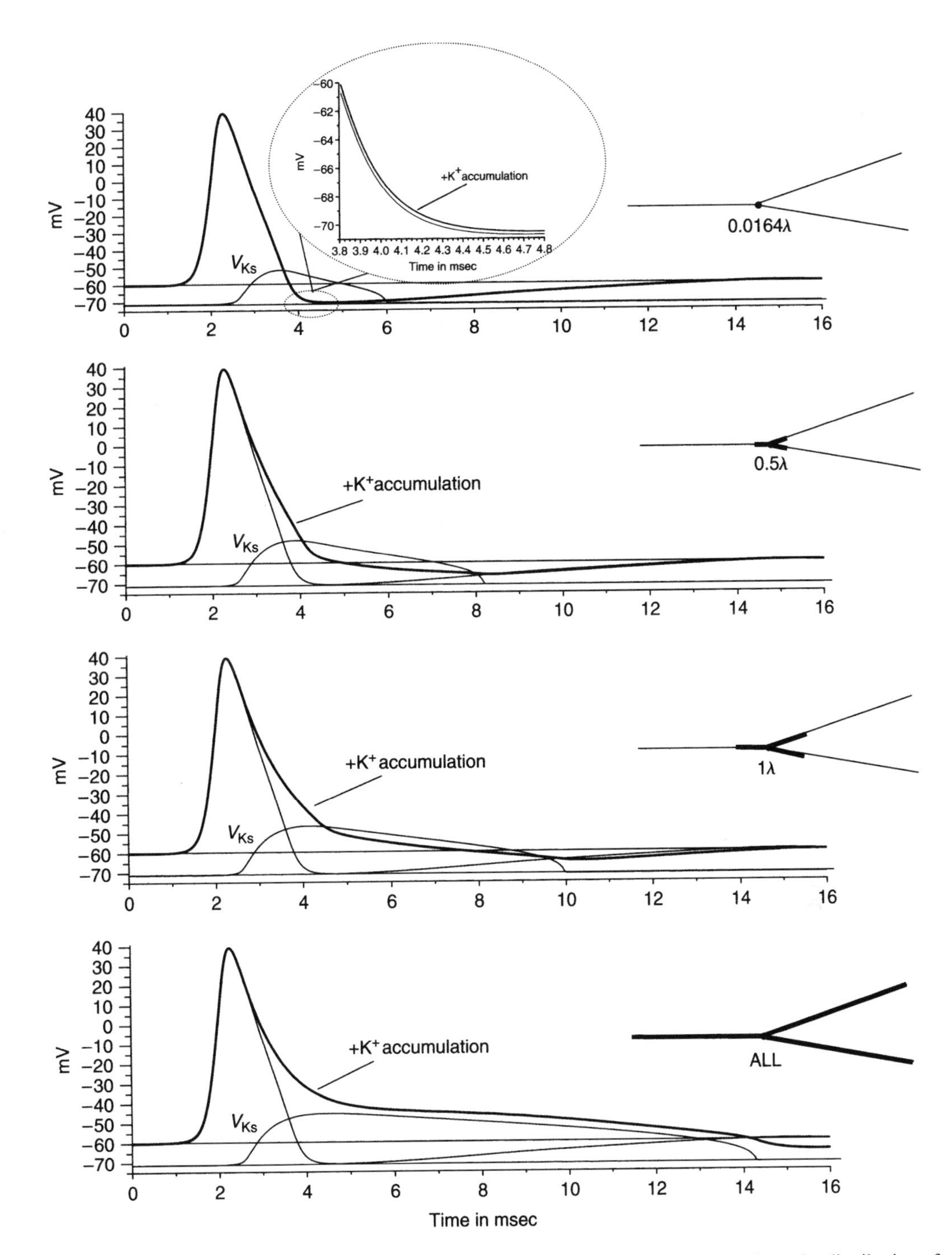

FIGURE 18.13 Same analysis as in Figure 18.12 but using a singly branched axon, where the distribution of periaxonal K^+ accumulation was centered around the recording site at the branch point.

the minimum value possible for propagation through and past the branch point to the end of each of the two branches. The results are given in Table 18.1.

When one or both of the two branches had periaxonal K^+ accumulation, the minimum interspike interval increased from 4.3 to 7.0 msec, independent of the other branch. Therefore, periaxonal

TABLE 18.1
Minimum Interspike Interval Propagated (msec)

Fiber type	Branch 1	Branch 2
1	N: 4.3	N: 4.3
2	N: 4.3	K: 7.0
3	K: 7.0	K: 7.0

Note: Branch Property: K = periaxonal K^+ accumulation; N = normal.

K^+ accumulation could cause a type of differential impulse propagation, where branches with periaxonal K^+ accumulation did not support the shorter interspike intervals.

The parameters underlying such differential impulse propagation were assessed for the example of a 6-msec interval at several axial locations in single-branched fibers where only one of the branches had periaxonal K^+ accumulation; for comparison, the same parameters also were assessed for the same conditions in a fiber with no periaxonal K^+ accumulation in either branch, as shown in Figure 18.14. First, the amplitudes of both sequential action potentials changed with axial distance (Figure 18.14A). The decrease in amplitude just prior to the branch point for all impulses corresponded to the decrease in conduction velocity at the same loci (Figure 18.6); second-spike amplitudes were smaller than those of the first due to the residual refractoriness from the first impulse (Miller and Rinzel, 1981; Moradmand and Goldfinger, 1995a). Differential propagation was evident for the second spike in the branch with periaxonal K^+ accumulation; the second spike amplitude fell progressively beyond the branch point whereas in controls this amplitude regained its initial value subsequent to the branch point. Within 3λ past the branch point, the second spike failed to propagate in the branch with periaxonal K^+ accumulation.

The failure of the second spike was due to axially diminishing net currents, given as $(C_m \cdot \partial V/\partial t)_x$, which is the immediate vectorial determinant of the peak height in propagating action potentials (Hunter et al., 1975; Moradmand and Goldfinger, 1995a; Figure 18.14B). This parameter normally decreased progressively within 1λ before the branch point and regained its initial value thereafter; however, for the branch with periaxonal K^+ accumulation, $\{C_m \cdot \partial V/\partial t\}_x$ declined to almost zero over 3λ downstream from the branch point. Thus, the second impulse eventually failed to develop.

A propagating action potential is stimulated by an initial transient of net outward extrinsic current I_m from upstream loci, where $I_m \propto \partial^2 V/\partial x^2$ (Hodgkin and Rushton, 1946; Cooley and Dodge, 1966; Equation [18.1b]). Figure 18.14C shows that while this parameter normally diminished just prior to the branch point and regained its initial value within 1λ beyond the branch point, in the branch with periaxonal K^+ accumulation it diminished monotonically from 0.5λ beyond the branch point and reached zero by 3λ beyond the branch point. Thus, the failure of propagation was due to the attenuation of downstream-flowing outward stimulation currents.

The diminution of I_m with distance in the branch with periaxonal K^+ accumulation (Figure 18.14C) reflected the diminution of upstream ionic currents. The prolonged recovery of action potentials elicited with periaxonal K^+ accumulation altered the values of voltage-dependent conductance parameters by the time of the second impulse's expected arrival, including an ~81% greater degree of inactivation of macroscopic Na^+ conductance (i.e., parameter $1 - h$) and ~400% greater macroscopic K^+ conductance (i.e., $G_K \cdot n^4$, where G_K is the maximal macroscopic K^+ conductance, 36 mS/cm^2, and n is the macroscopic K^+ conductance activation; Figure 18.10). In order to demonstrate directly that such conditioning caused effective refractoriness during propagation, the following test was performed: subsequent to a conditioning action potential, the major components of the I_m waveform associated with a normal propagating action potential were used as an artificial test stimulus applied to a site of propagation failure in the branch with periaxonal K^+ accumulation. This test stimulus was applied at the time when the second impulse of a 6-msec duration interspike interval arrived at the same site on the non-K^+-accumulating branch. The results

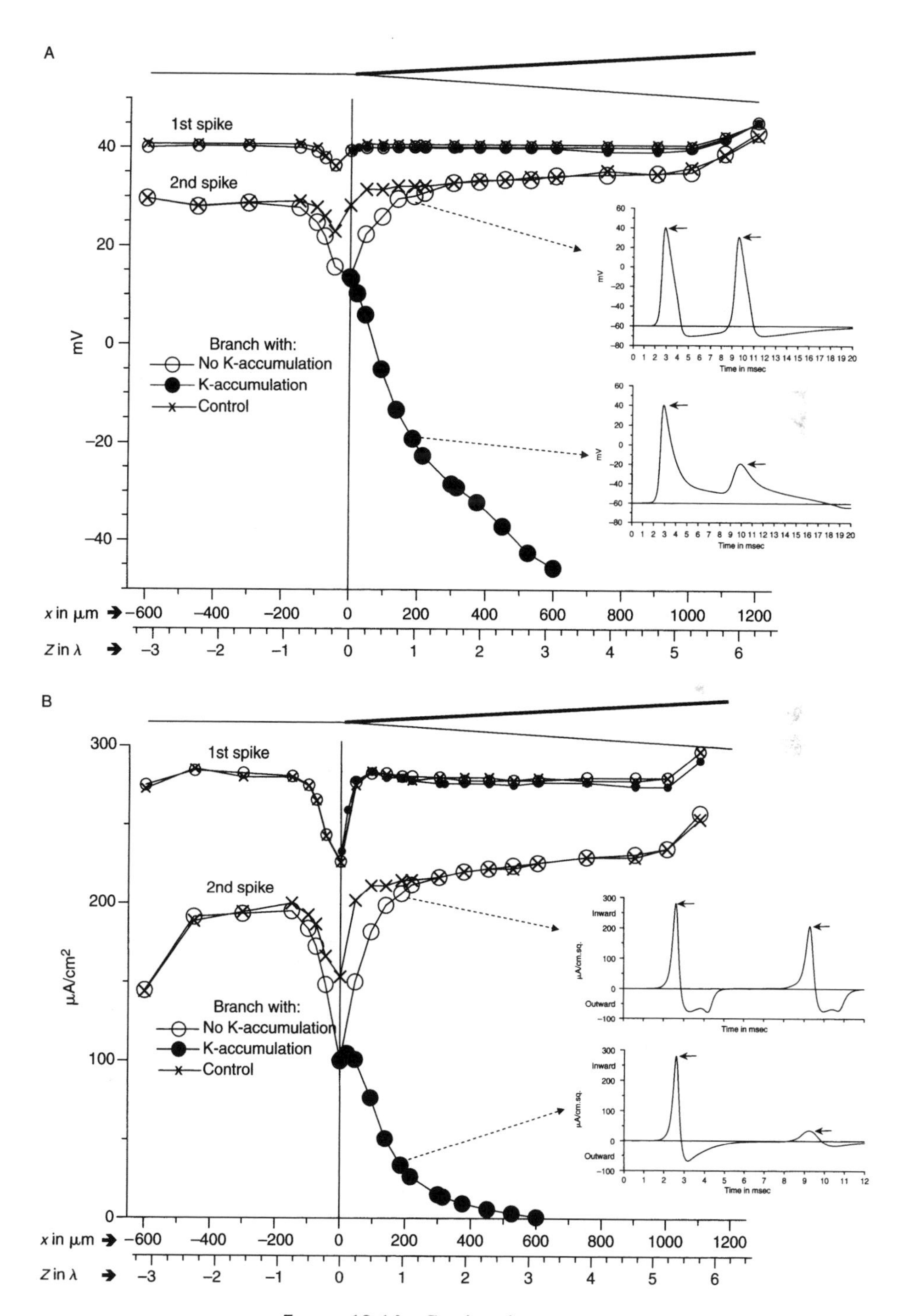

FIGURE 18.14 Continued on next page

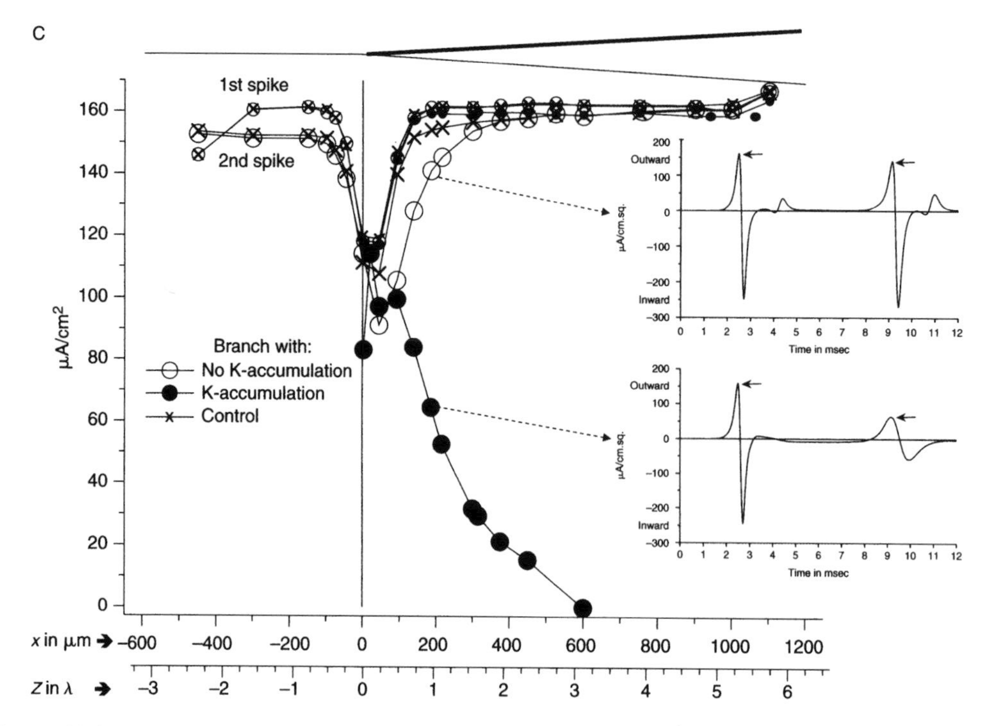

FIGURE 18.14 Parameters of differential propagation due to periaxonal K^+ accumulation restricted to a single branch. Parameters for each action potential of the 6-msec interval plotted with respect to axial position (literal, x, or electrotonic, λ). In the fiber diagrams (top), the darker branch had periaxonal K^+ accumulation over the entire length (except at the branch point). (A) Action potential peak amplitudes. (B) Action potential rising slope peak values; that is, peak values of first component of current $\propto (\partial V/\partial t)_{x,t}$. (C) Stimulating outward current peak values; that is, peak values of first component of current $\propto (\partial^2 V/\partial x^2)_{x,t}$. Symbols: open circles, from the branch without periaxonal K^+ accumulation; closed circles, from the branch with periaxonal K^+ accumulation; $\times$ (control), from a single-branched fiber with no periaxonal K^+ accumulation in either branch. Data are shown for first and second action potentials of the interval pair; symbols' relative size refers to first (smaller) or second (larger) action potential. Insets show sample waveforms recorded from either branch at $x = 186\,\mu\text{m}$ ($= 1.02\,\lambda$); parameter amplitudes were measured from respective baseline (A: -60 mV; B, C: $0\,\mu\text{A/cm}^2$) to the initial peak.

are shown in Figure 18.15. Thus, following the conditioning impulse, the artificial I_m stimulus did not elicit a second propagating impulse for 6-msec timing, but did elicit a second action propagating potential in a separate test if applied 2 msec later. That is, the effect of periaxonal K^+ accumulation elicited by the first spike was to transiently depress the excitability at the test site such that even normal physiological stimulation did not elicit a second impulse. This result corresponded to the observation that in the branch with periaxonal K^+ accumulation, while the second impulse of a 6-msec interval did not propagate, both impulses of an 8-msec interval did propagate.

18.3.4 Modifications of Random Impulse Trains by Periaxonal K^+ Accumulation

These experiments considered whether periaxonal K^+ accumulation modifies the stochastic properties of propagating impulse trains elicited by Poisson-like stimulation. Preliminarily, it was necessary to consider unbranched fibers, as shown in Figure 18.16. With periaxonal K^+ accumulation, the IID mean interval was longer, corresponding to a 17.65% decrease in the mean firing rate. Periaxonal

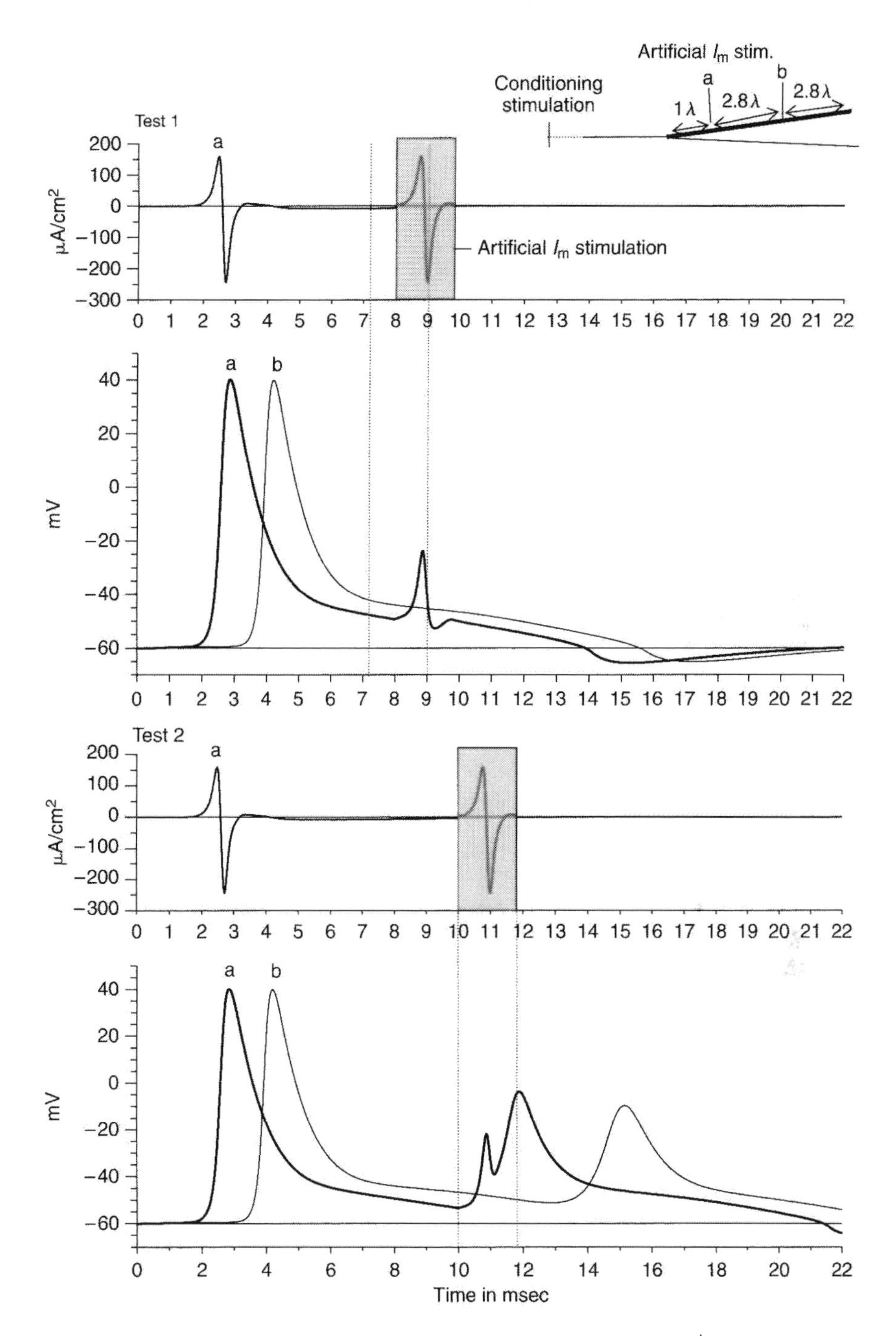

FIGURE 18.15 Testing short-interval excitability in the branch with periaxonal K^+ accumulation. The major waveform of I_m for an unconditioned propagating impulse was used as a test stimulus (artificial I_m stimulation) following a conditioning action potential. For Tests 1 and 2: upper traces, I_m at test stimulation site *a*; lower traces, responses at *a* (1λ downstream from the branch point, thicker lines) and at site *b* (2.8λ downstream from a, thinner lines); conditioning action potential elicited at $x = 0$. For a 6-msec interval set up at $x = 0$, the second action potential arrived at 1λ downstream from the branch point in the branch without periaxonal K^+ accumulation at $t = 8$ msec. Test 1: Applying the test stimulation at $t = 8$ msec at *a* elicited neither a second impulse at *a* nor any response at *b*. Test 2: Application of the test stimulation 2 msec later ($t = 10$ msec) yielded an action potential at *a* which propagated to *b*. $\Delta t = 10^{-4}$ msec; plot resolution $= 10^{-2}$ msec.

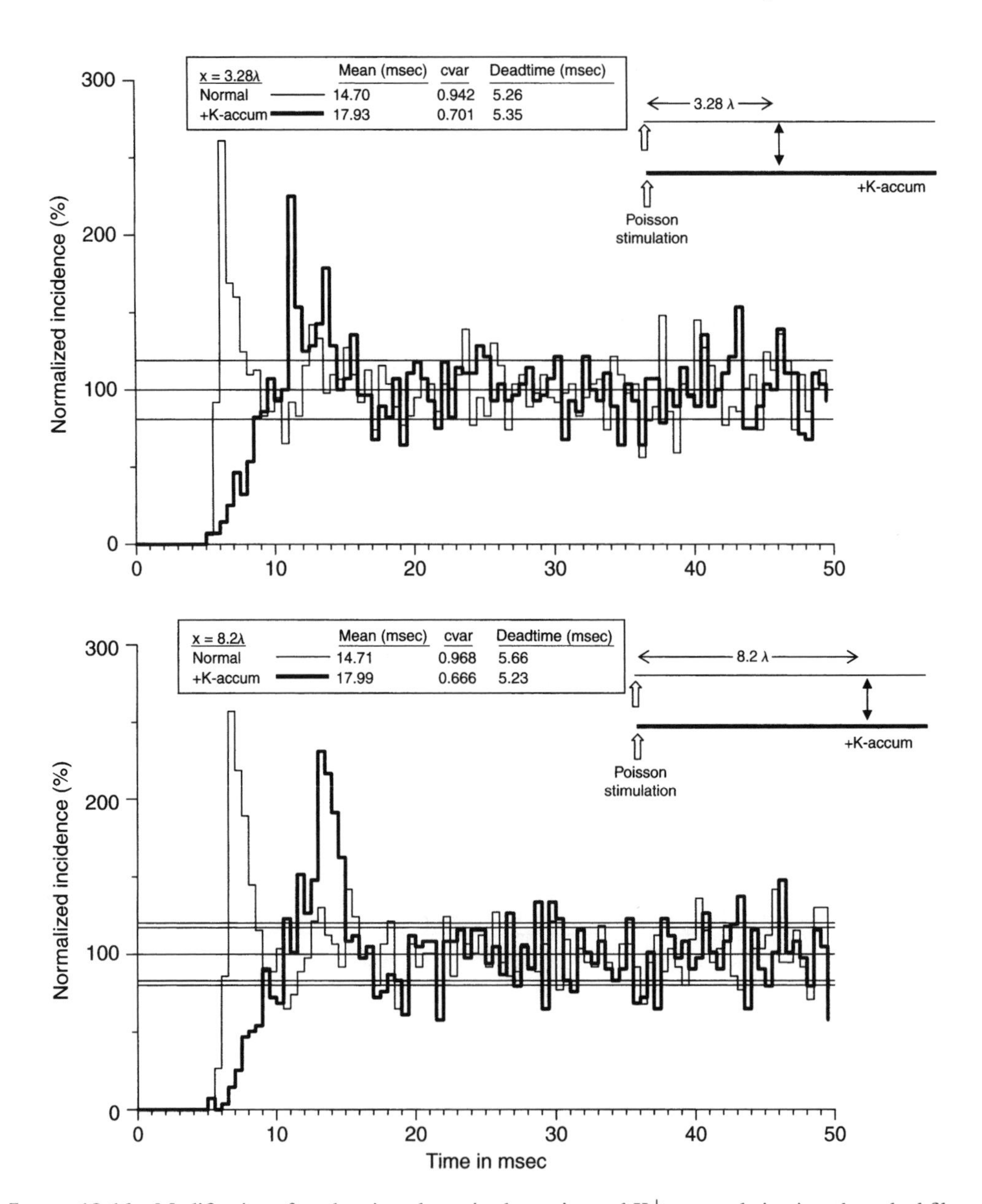

FIGURE 18.16 Modification of random impulse trains by periaxonal K^+ accumulation in unbranched fibers. Recordings from two points of a 9.9λ length, 1 μm-diameter axon. Normalized EDs (0.5 msec binwidth) for fibers with (darker traces) and without (lighter traces) periaxonal K^+ accumulation over the entire fiber, together with respective IID mean interval, coefficient of variation (cvar $= \sigma/(\text{mean} - \text{deadtime})$), and directly measured deadtime (i.e., shortest interspike interval). Poisson-like stimulation was applied in the parent axon at $x = 0$; $N_{\text{spikes}} \geq 1001$. Dashed lines: normalized ED mean $\pm 1\sigma$.

K^+ accumulation also diminished the conditional probability of spike train frequencies between 200 and 125 per sec (i.e., 5 to 8 msec incidence in the ED) and enhanced the occurrence of frequencies between 91 and 71 per sec (i.e., 11 to 14 msec incidence in the ED). Periaxonal K^+ accumulation slowed the normally rapid rise of ED(t) and shifted the initial ED peak to longer times. In the steady state ($t > 20$ msec), conditional probability was similar to that occurring without periaxonal K^+ accumulation. Interestingly, the duration of absolute refractoriness was not significantly altered by periaxonal K^+ accumulation (see later). The same testing was performed for single-branched fibers where periaxonal K^+ accumulation was restricted to one of the daughter branches. At those loci without periaxonal K^+ accumulation, firing patterns were unaffected by periaxonal K^+ accumulation elsewhere in the arbor, as shown in Figure 18.17 (top; bottom). However, in the branch with periaxonal K^+ accumulation, coding was altered significantly. The mean interval was prolonged, decreasing the mean firing rate by 10.3%. The conditional probabilities for the highest impulse train frequencies (200 to 125 per sec; 5 to 8 msec in the ED) were markedly reduced. The early-time ED trajectory rose to a new peak between 10 and 11 msec; the steady-state average level was unaffected. Also, as in the unbranched case, the duration of absolute refractoriness was not increased by periaxonal K^+ accumulation. Thus, periaxonal K^+ accumulation in one branch caused a lower-rate altered-pattern impulse train in that branch, that is, a form of differential propagation.

Examination of the $V(t)$ trajectories of the impulse trains elicited by Poisson-like stimulation explained some of the changes in the stochastic properties due to periaxonal K^+ accumulation. Figure 18.18 shows an excerpt from the impulse trains recorded at the branch point and 4.92λ downstream in the branch with periaxonal K^+ accumulation. First, propagation was delayed in the branch with periaxonal K^+ accumulation; this delay was in addition to that introduced by the branch point without periaxonal K^+ accumulation (Figure 18.6). The time between spikes $1 \rightarrow 2$ (i.e., time for the propagation of the first spike from the branch point to the distal recording site) was less than that between spikes $3 \rightarrow 4$; this delay lengthened the first interspike interval from 8.76 msec at the branch point (spikes $1 \rightarrow 3$) to 10.88 msec (spikes $2 \rightarrow 4$) downstream in the branch with periaxonal K^+ accumulation. The 6.11 msec interval at the branch point (spikes $3 \rightarrow 5$) was elongated to 22.12 msec (spikes $4 \rightarrow 7$) due to the propagation failure of branch point spike 5. These data showed how the initial ED peak was shifted to longer times (from $\sim$7 to 11 msec in Figure 18.17, middle) and the mean rate was decreased as a result of periaxonal K^+ accumulation.

During random firing in the branch with periaxonal K^+ accumulation, there was considerable variation among impulse waveforms. Variability in action potential peak value (+40 to -10 mV) with respect to interspike interval duration was expected as a property of impulse propagation at a relatively high rate (Moradmand and Goldfinger, 1995a). Impulses grew in amplitude and $CV(x)$ on approaching the branch terminal (Goldstein and Rall, 1964). The shortest interspike intervals with periaxonal K^+ accumulation — though relatively rare — occurred because the first impulse of the pair had a briefer duration from the absence of the delayed afterdepolarization, as shown in Figure 18.19. Such action potentials had faster recovery and therefore shorter relative refractoriness, such that a subsequent impulse could be initiated at a short interval (here, 5.48 msec). These briefer-duration propagating impulses also had lower amplitude.

Periaxonal K^+ accumulation was absent in association with such impulses because of the substantial conditioning of K_s regulation by the immediately preceding impulses; the relevant parameters during a 4-impulse sequence which included such a brief-duration action potential are shown in Figure 18.19. Spike 1 (unconditioned as per Figure 18.9) conditions spike 2, whose I_K transient was reduced such that K^+ accumulation in the periaxonal space was reduced in amplitude and duration. With spike 3, the I_K transient was further reduced to the extent that K_s filling was entirely compensated by immediate glial and diffusional removal of K^+ from the space, despite the relatively low value of K_{diss}. Spike 3 had no afterdepolarization because there was almost no periaxonal K^+ accumulation. With sufficient recovery time, the I_K transient during the subsequent spike 4 was large enough to saturate glial and diffusional K^+ clearance, thus eliciting periaxonal K^+ accumulation.

18.3.5 Discussion

The accumulation of periaxonal K^+ during activity — demonstrated first by Frankenhauser and Hodgkin (1956) — is well-known for experimental axons (e.g., Smith, 1983), where the

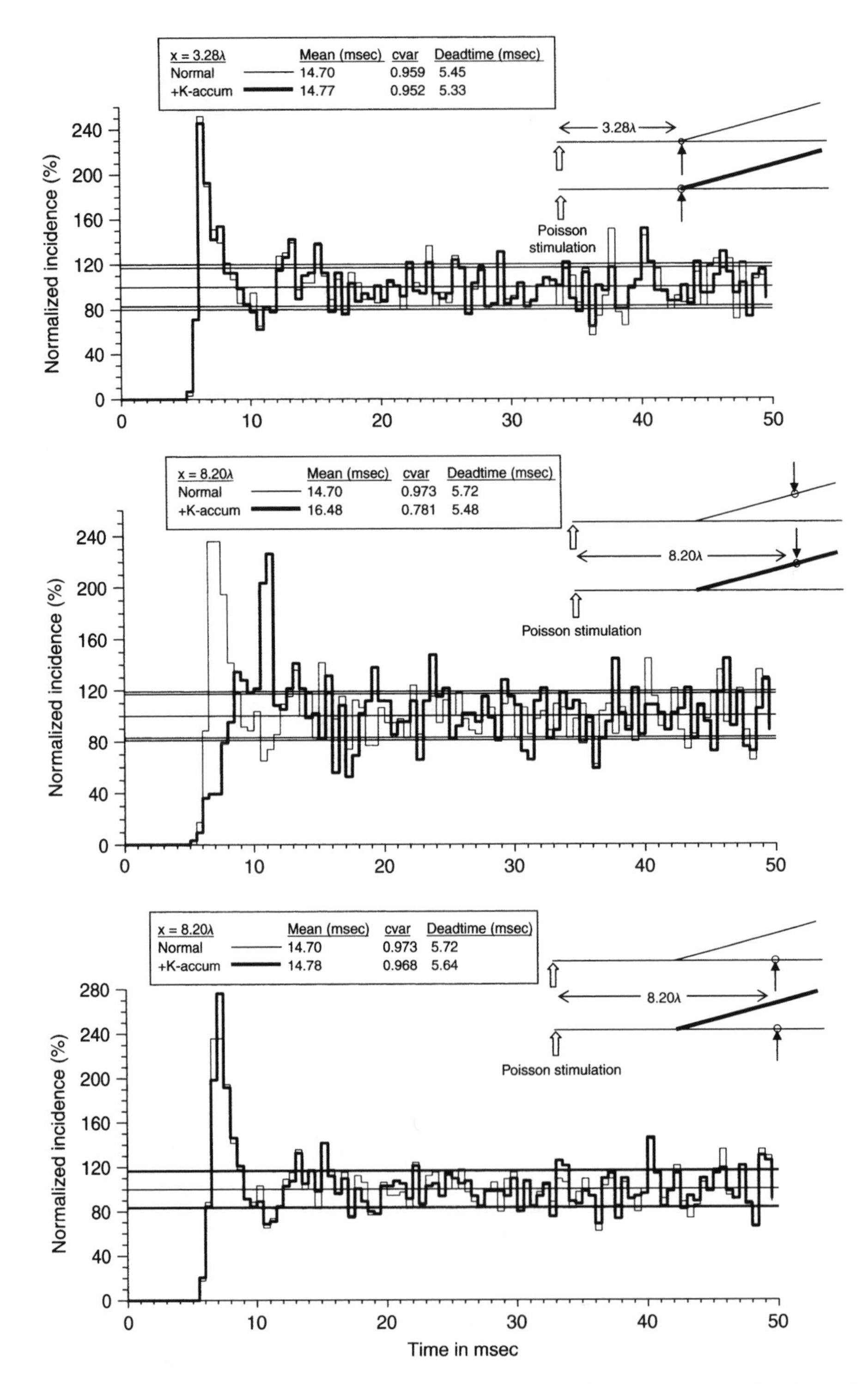

FIGURE 18.17 Modification of a random impulse train by periaxonal K^+ accumulation restricted to one branch. Recordings from two points of a 9.9λ-path length 1 μm-diameter axon. Normalized EDs (0.5 msec binwidth) from branches with (darker traces) and without (lighter traces) periaxonal K^+ accumulation, together with respective IID mean interval, coefficient of variation, and directly measured deadtime. Poisson-like stimulation was applied in the parent axon at $x = 0$; arrows point to the recording sites. Dashed lines: normalized ED mean $\pm 1\sigma$. $N_{\text{spikes}} \geq 1001$.

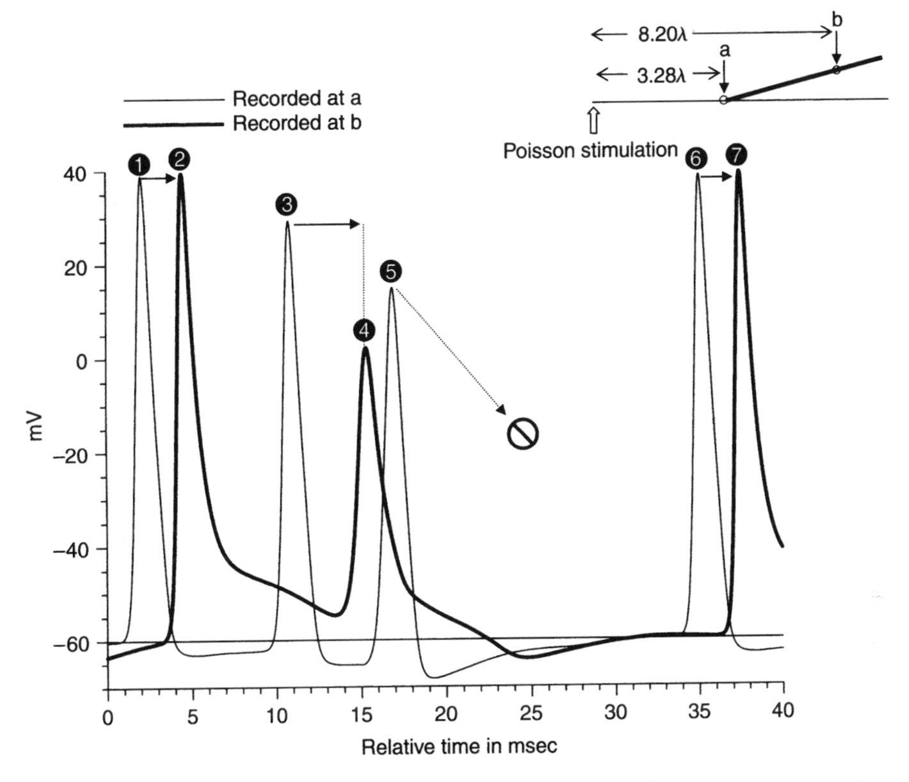

FIGURE 18.18 Modification of interspike interval values by periaxonal K^+ accumulation. Data from the run given in Figure 18.17 (middle). Sequential spikes labeled 1, 3, 5, and 6 recorded from the branch point (site *a*); spikes 2, 4, and 7 recorded 4.92λ downstream from the branch point (site *b*) in the branch with periaxonal K^+ accumulation. Spikes 1, 3, and 6 at site *a* propagated to become spikes 2, 4, and 7, respectively, at recording site *b*. Branch point spike 5 did not propagate. The input interval 1 → 3 is lengthened (interval 2 → 4) in the branch with periaxonal K^+ accumulation. $\Delta t = 10^{-4}$ msec; plot resolution $= 10^{-2}$ msec.

transiently higher periaxonal K^+ slows the action potential falling phase (i.e., a delayed after-depolarization) and limits the degree of afterhyperpolarization. Previous theoretical studies and the present results for random impulse trains are in agreement that the propagation of high impulse rates may be compromised by periaxonal K^+ accumulation. The possibility of a depression of higher-frequency firing due to periaxonal K^+ accumulation was raised by Jack et al. (1983). In a study of impulse propagation through branch points of computed myelinated fibers, Zhou and Chiu (2001) reported a low-pass filtering effect due in part to local periaxonal K^+ accumulation.

In the present computations of propagating action potentials, it was shown that these effects of periaxonal K^+ accumulation were spatially modulated; the degree to which the propagating action potential waveform was modified by periaxonal K^+ accumulation was proportional over a limited range to the axial distance over which periaxonal K^+ accumulation was distributed in unbranched as well as branched fibers (Figure 18.12 and Figure 18.13). This occurred because the propagating action potential waveform was determined by extrinsic ($\propto \partial^2 V/\partial x^2$) as well as intrinsic currents. In the case in which only a small length (0.0164λ) of fiber supported periaxonal K^+ accumulation, while K^+ concentration in the periaxonal space over that short length of fiber increased transiently by approximately 11 mM and changed the K^+ equilibrium potential from −72 to −54 mV, the propagating action potential waveform changed by not more than 0.5 mV during recovery (Figure 18.12, top). The same results were obtained when periaxonal K^+ accumulation

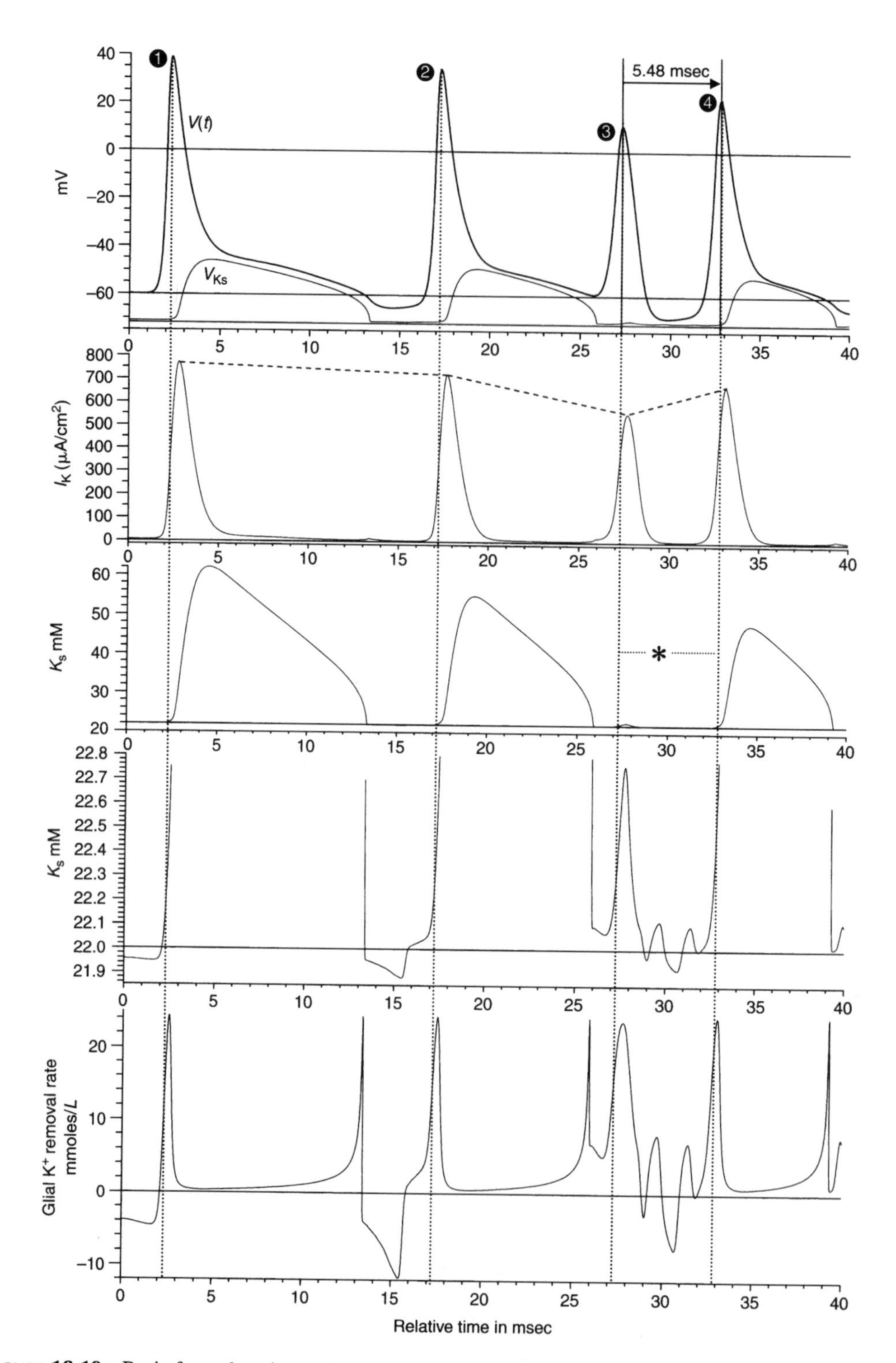

FIGURE 18.19 Basis for a short interspike interval via temporary absence of periaxonal K^+ accumulation. Data from the run given in Figure 18.17 (middle), recorded 4.92λ downstream from the branch point in the branch with periaxonal K^+ accumulation. Vertical dotted lines mark the peak of each action potential. The decrease of I_K during impulses 2 and 3 allows nearly complete clearance of periaxonal K^+ accumulation during spike 3 (. . . ✱ . . .) such that spike 4 can be generated at the short 5.48-msec interval. $\Delta t = 10^{-4}$ msec; plot resolution $= 10^{-2}$ msec.

was restricted to a branch point (Figure 18.13, top), showing that periaxonal K^+ accumulation restricted only to the branch point that did not compromise impulse propagation. That is, there was intrinsic protection against modification of the impulse code by periaxonal K^+ accumulation if such accumulation were spatially restricted.[3] In experimental tissue, periaxonal K^+ accumulation may be spatially widespread since the efficiency of periaxonal K^+ clearance can be greatly decreased due to trauma from dissection (Brown, 1993).

Dual-pulse stimulation showed that for impulse propagation in a fiber where all loci had periaxonal K^+ accumulation, absolute refractoriness was prolonged. This result predicted correctly that shorter interspike intervals could not be supported where periaxonal K^+ accumulation had a sufficiently large spatial distribution. For fibers in which only one of the daughter branches had periaxonal K^+ accumulation, the propagation failure of the second of a short-interval impulse pair was spatially progressive. The conditioning impulse initiated the K^+ accumulation which — primarily through the depolarization-induced temporary enhancement of g_K, Figure 18.10 — depressed the ability of one site to generate sufficiently large net inward current to spread outward at adjacent downstream loci and successfully stimulate a second impulse to complete the short interspike interval. Thus, under these conditions, differential propagation was strictly interval-dependent.

By contrast, with Poisson-like stimulation, while periaxonal K^+ accumulation did not lengthen the minimum interval value found without periaxonal K^+ accumulation, the incidence of such intervals was lowered. In general, periaxonal K^+ accumulation suppressed the highest firing frequencies (Figure 18.16 and Figure 18.17), due to the complex conditioning inherent in Poisson-elicited trains. The underlying mechanisms included the prolonged delay of the second impulse of an interval pair (Figure 18.18) and the elimination of a delayed afterdepolarization in the first impulse of an interval pair (Figure 18.19). Any effects of supernormality due to periaxonal K^+ accumulation were not appreciable in that no elevations in conditional probability above the noise were evident in the normalized EDs at the times corresponding to either supernormal period (Figure 18.16 and Figure 18.17). Thus, with random impulse trains, differential propagation due to periaxonal K^+ accumulation took the form of lower mean rate and depressed incidence of short intervals between 5 to 10 msec. But in no case did a significantly high degree of periaxonal K^+ accumulation prevent other components of these complex random impulse trains from propagating successfully through the branch point and to the end of the branch.

Periaxonal K^+ accumulation is of interest due to the relatively large amounts of K^+ being injected into relatively small periaxonal volumes by each action potential, a design feature requiring rapid and highly efficient K^+ clearance. Brown's results (1993) suggested that periaxonal K^+ accumulation — at least in the squid axon preparation — is an imposed condition which does not normally occur, and more generally raised the possibility that dissected neuronal tissue is vulnerable to the development of periaxonal K^+ accumulation. Under isopotential conditions, using the unmodified H–H equations as a control and assuming the only difference with experimental data is the effect of periaxonal K^+ accumulation, comparison with Clay's experimental results (1998) suggested an example of the extent to which periaxonal K^+ accumulation could occur, as shown in Figure 18.20. In terms of the present model for periaxonal K^+ accumulation, this extent of action potential distortion could be simulated for propagating impulses either by a significant periaxonal K^+ accumulation distributed over a relatively small membrane area (cf. Figure 18.12), a reduced value of K_{diss}, or some combination of the two. In squid axons carefully dissected (Astion et al., 1988; Abbott et al., 1988) and at less than 4°C (Pichon et al., 1987), periaxonal K^+ accumulation was minimized. Thus, the inclusion of periaxonal K^+ accumulation in models of nonisopotential axonal activity must consider the spatial distribution of periaxonal K^+ accumulation as well as the degree to which normal periaxonal K^+ clearance is modified.

In the space-clamped squid axon, response reconstructions by Clay (1998) required neither the incorporation of a direct effect of K^+ on Na^+ channel inactivation — suggested for the same preparation by Adelman and Palti (1969) — nor the inclusion of a depolarization-induced K^+ efflux from glia (MacVicar et al., 1988). From a modeling perspective, the modifications of

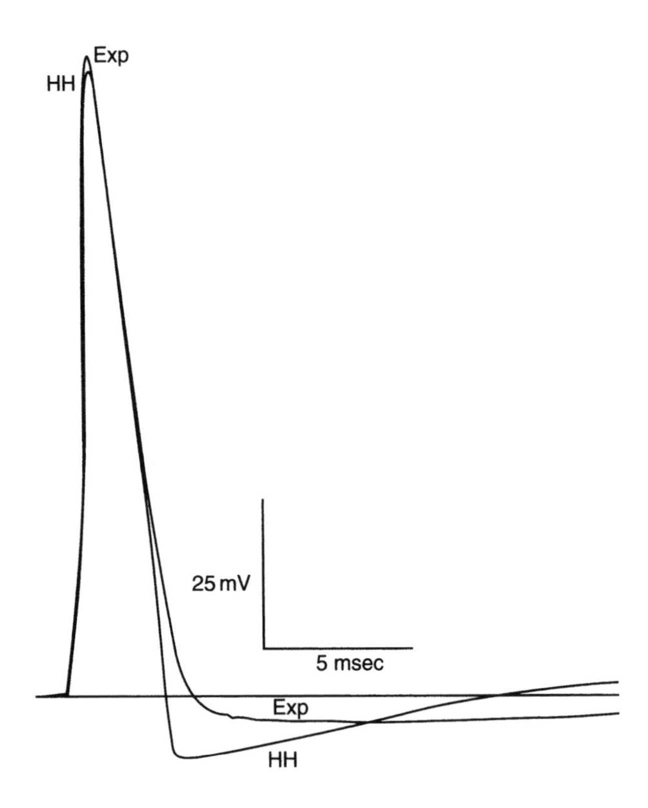

FIGURE 18.20 Experimental squid axon action potential ("Exp") compared with that of the unmodified H–H equations ("HH") under space-clamped conditions (isopotential). All data are from Clay (1998), Figure 18.5A and B. Departure of the experimental falling limb from that of the H–H action potential is presumed to be due to periaxonal K^+ accumulation in the experimental data. (Reproduced with permission of Dr. Clay and the American Physiological Society ©)

impulse patterns — specifically, the lowered probability for higher frequencies due to periaxonal K^+ accumulation — may be of particular significance at axon terminals, where transmitter release is subject to frequency-modulated modification (e.g., Dobrunz and Stevens, 1999; Migliore and Lansky, 1999; Dittman et al., 2000; Kreitzer and Regehr, 2000; Bi and Poo, 2001; Stewart and Froehring, 2001).

18.4 ROLE OF A MORE COMPLETE AND ACCURATE DESCRIPTION OF SODIUM CHANNEL POPULATION KINETICS

18.4.1 The "PC" Model

In the H–H 1952 formalism, Na^+ channel activation and inactivation are independent kinetic processes. However, subsequent studies have shown that Na^+ channel activation and inactivation are kinetically coupled (Goldman, 1976; Bezanilla and Armstrong, 1977; Patlak, 1991). Vandenberg and Bezanilla (1991b) have provided a quantitative description of individual and population Na^+ channel kinetics which incorporates the coupling between closed, activated, and inactivated states, with a total of nine possible substates. Clay (1998) has applied this model to a reconsideration of macroscopic excitation in the squid axon under isopotential conditions.

In this section, this more complete and accurate model of squid axon Na^+ channel kinetics was applied to considerations of impulse propagation. The formalism used was essentially the

Vandenberg–Bezanilla model (1991b) as represented by Clay (1998). The Markov kinetic scheme for individual Na channels was as follows:

$$
\begin{array}{ccccccccccc}
 & & & & & & I_4 & \underset{b}{\overset{a}{\rightleftharpoons}} & I_5 & \underset{d}{\overset{c}{\rightleftharpoons}} & I \\
 & & & & & & f\updownarrow e & & & & f\updownarrow e \\
C_1 & \underset{z}{\overset{y}{\rightleftharpoons}} & C_2 & \underset{z}{\overset{y}{\rightleftharpoons}} & C_3 & \underset{z}{\overset{y}{\rightleftharpoons}} & C_4 & \underset{b}{\overset{a}{\rightleftharpoons}} & C_5 & \underset{d}{\overset{c}{\rightleftharpoons}} & O
\end{array}
$$

The constituent states are labeled as either closed [**C**], open [**O**], or inactive [**I**], with **C** and **I** having respective substates. The transitional forward and backward rate constants were:

$$a = 7.55 \times \exp(0.017 \times v); \quad b = 5.6 \times \exp(-0.00017 \times v); \quad c = 21.0 \times \exp(0.06 \times v)$$

$$d = 1.8 \times \exp(-0.02 \times v); \quad e = 0.0052 \times \exp(-0.038 \times v); \quad f = 1.12 \times \exp(0.00004 \times v)$$

$$y = 22.0 \times \exp(0.014 \times v); \quad z = 1.26 \times \exp(-0.048 \times v)$$

Clay's voltage shift was used ($v = V - 10$; Clay, 1998). Mass action statements are written as derivatives with respect to time directly from the kinetic diagram shown above; thus:

$$
\begin{aligned}
\mathrm{d}C_2/\mathrm{d}t &= ((y \times C_1) + (z \times C_3) - (z \times C_2) - (y \times C_2)) \times \Phi \\
\mathrm{d}C_3/\mathrm{d}t &= ((y \times C_2) + (z \times C_4) - (y \times C_3) - (z \times C_3)) \times \Phi \\
\mathrm{d}C_4/\mathrm{d}t &= ((y \times C_3) + (b \times C_5) + (e \times I_4) - (z \times C_4) - (a \times C_4) - (f \times C_4)) \times \Phi \\
\mathrm{d}C_5/\mathrm{d}t &= ((a \times C_4) + (\mathrm{d} \times \mathbf{O}) - (b \times C_5) - (c \times C_5)) \times \Phi \\
\mathrm{d}\mathbf{I}/\mathrm{d}t &= ((f \times \mathbf{O}) + (c \times I_5) - (\mathrm{d} \times \mathbf{I}) - (e \times \mathbf{I})) \times \Phi \\
\mathrm{d}I_4/\mathrm{d}t &= ((f \times C_4) + (b \times I_5) - (e \times I_4) - (a \times I_4)) \times \Phi \\
\mathrm{d}I_5/\mathrm{d}t &= ((\mathrm{d} \times \mathbf{I}) + (a \times I_4) - (b \times I_5) - (c \times I_5)) \times \Phi \\
\mathrm{d}\mathbf{O}/\mathrm{d}t &= ((c \times C_5) + (e \times \mathbf{I}) - (\mathrm{d} \times \mathbf{O}) - (f \times \mathbf{O})) \times \Phi
\end{aligned}
$$

where Φ is the H–H temperature factor ($Q_{10} = 3$), that is, $\Phi = 3^{[(T-6.3)/10]}$ with $T = 10°\mathrm{C}$, and

$$C_1 = 1 - (C_2 + C_3 + C_4 + C_5 + \mathbf{I} + I_4 + I_5 + \mathbf{O})$$

The values of C_i, I_i, and **O** are population probabilities varying between 0 and 1; at any t, all such probabilities sum to 1. The Na^+ current at any x, t is given by

$$I_{\mathrm{Na}(x,V,t)} = G_{\mathrm{Na\ max}} \cdot \mathbf{O}_{(x,V,t)} \cdot (V_{x,t} - V_{\mathrm{Na}}),$$

where $G_{\mathrm{Na\,max}}$ is the maximal macroscopic Na^+ conductance in any compartment ($= 120\,\mathrm{mS/cm^2}$; Hodgkin and Huxley, 1952b), and V_{Na} is the Na^+ equilibrium potential.

This scheme differs slightly from Clay's version (1998) in that: (a) the "*g*," "*j*," and "*i*" rate constants are replaced by *e*/*f* rate constants so as to maintain symmetrical rate constant products within the closed kinetic loop (Onsager, 1931; see Hille, 2001, Chapter 18; Varghese and Boland, 2002); and (b) the values of *c* and *f* are doubled, so as to achieve a best fit of the H–H isopotential action potential. For convenience to distinguish from the unmodified H–H equations, this model of excitation will be referred to as a "PC" (partially-coupled) model. Initial conditions at all *x* were determined by a 100 msec relaxation method (as described above in computational considerations, Section 18.1.1), where $V_{\text{Istim}=0,t=100\,\text{msec}}$ was stably equal to -59.7 mV at all x. Other aspects of the H–H cable equation system (including g_L and $g_K(V,t)$) were unaltered. Periaxonal K^+ accumulation was not included unless otherwise indicated.

18.4.2 PC Model's Basic Properties

PC and H–H predictions in a typical voltage clamp paradigm revealed two immediate differences between the two models, as shown in Figure 18.21. First, with regard to the time course of rising inactivation in the Na^+ channel population, this transient rose early in H–H but with sigmoidal delay in PC (N.B.: while this PC response fit experimental observation, the H–H response did not; see Bezanilla and Armstrong, 1977; Fain, 1999, Chapter 6). Second, macroscopic (i.e., population) Na^+ conductance reached a higher maximum in PC; as a result, PC generated larger macroscopic Na^+ current throughout the clamp period.

Under nonisopotential conditions, both PC and H–H models generated a propagating action potential, as shown in Figure 18.22. Due to a larger initial macroscopic net inward current, the PC spike had a faster maximal rate of rise and thereby ascended to a greater peak height (Hunter et al., 1975; Moradmand and Goldfinger, 1995a). This net inward current contributed to the axial current flowing into immediately downstream loci and initiating propagated excitation. On this basis, for a given diameter axon, the PC spike should have a somewhat larger conduction velocity than the H–H spike, as was in fact found and is shown in Figure 18.23.

The PC model showed the *statu variabilis* of the Na^+ channel population during activity, as is illustrated in Figure 18.24A and Figure 18.24B. At a given recording point prior to the arrival of the propagating action potential, the Na^+ channel population distributed into a variety of (sub)states (Figure 18.24A). Initially, while some Na^+ channels are in the I_4 or I_5 substates, most are in one of the closed substates, the greatest number being in C_1. As the recording site becomes invaded by stimulating axial currents from upstream excitation, there is a transitional departure from C_1 into the kinetic scheme. There occur sequentially relative peaks in $C_2 - C_5$, together with a departure from I_4 and I_5. These transitions lead to a rapid rise of the **O** population, followed shortly thereafter by an even larger rise in the **I** population and a related falloff of the **O** population. With the peak and falloff in the **I** population, there is a considerable rise in I_4 and I_5 as well as the $C_1 - C_3$ substate populations. Transitions out of the I_4, I_5 substates and into the C_1 and C_2 substates occur relatively slowly with an overall time constant of approximately 15 msec (Figure 18.24B), as the C_1 population regains predominance and the other substates relax to their initial values.

The recovery processes in the PC and H–H models of a propagated action potential were compared, as is shown in Figure 18.25. While both models had similarly timed absolute and relative refractory periods, the cycles differed mainly in that the PC model did not generate a supernormal period. That is, the PC and H–H recovery functions had the same inflections but the PC cycle did not drop into supernormality, yielding greater refractoriness during the initial relative refractory period. When periaxonal K^+ accumulation (as described above) was added to the PC axon properties: (a) the action potential waveform changed in a manner similar to that of the H–H axon, as given in Figure 18.26 (cf. Figure 18.11); (b) relative refractoriness was prolonged, with the inflection shifted to more delayed times; and (c) the recovery cycle ended with a slight supernormal period.

18.4.3 Propagation of Random Action Potential Trains in the PC Model

Responses of the PC axon to Poisson-like stimulation (as described above) were recorded and compared to those of the H–H axon. In unbranched axons, the responses of PC and H–H axons were

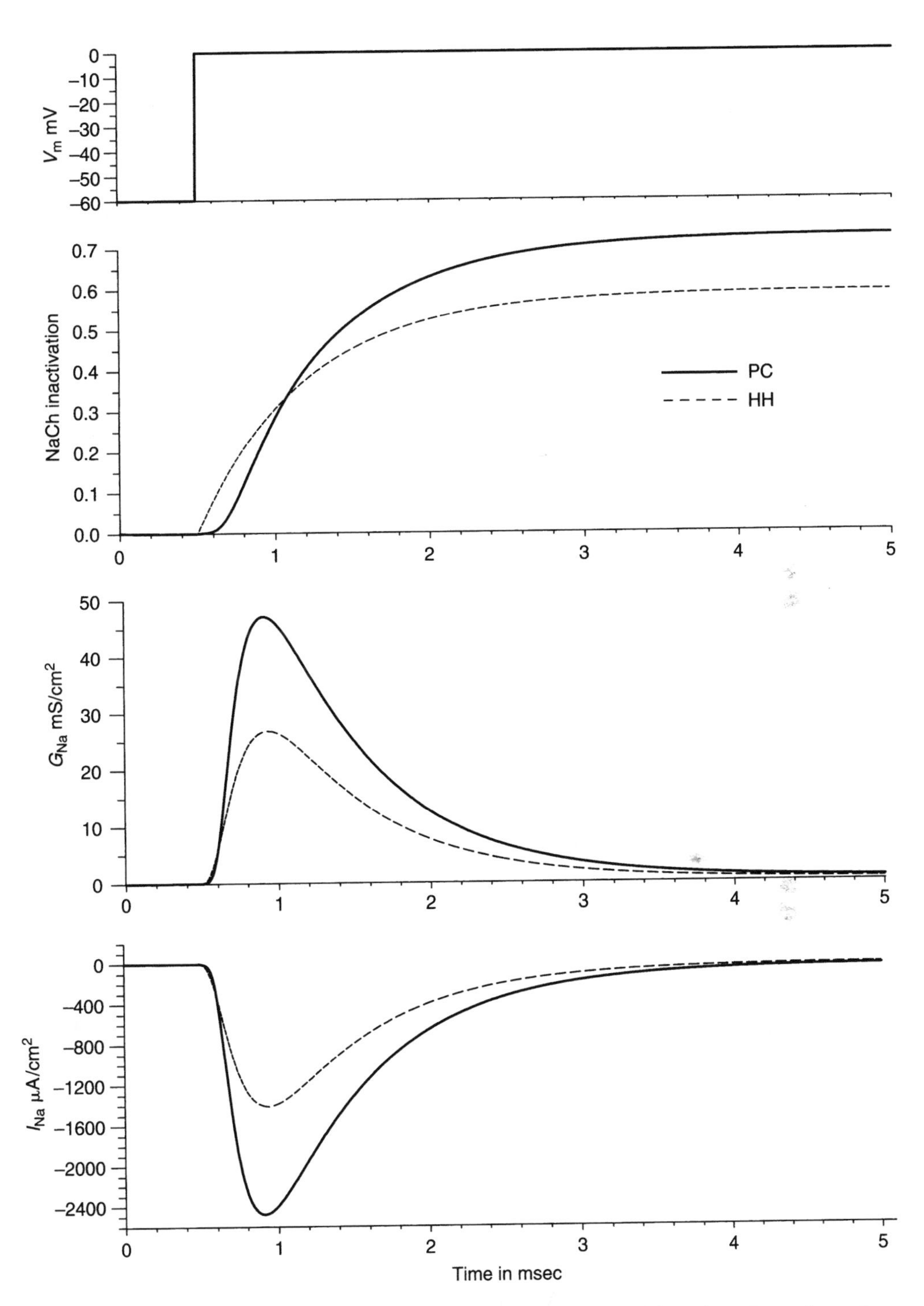

FIGURE 18.21 Comparison of responses of H–H and PC excitation models under voltage-clamp (isopotential) conditions. Membrane potential (V_m) was stepped from −60.0 to 0 mV. Variables plotted: Na^+ channel population inactivation (H–H: $1 - h_{(V,t)}$; PC: kinetic parameter I_t), macroscopic Na^+ conductance (G_{Na}), and macroscopic Na^+ current (I_{Na}). For both models, maximum possible Na^+ conductance was 120 mS/cm^2.

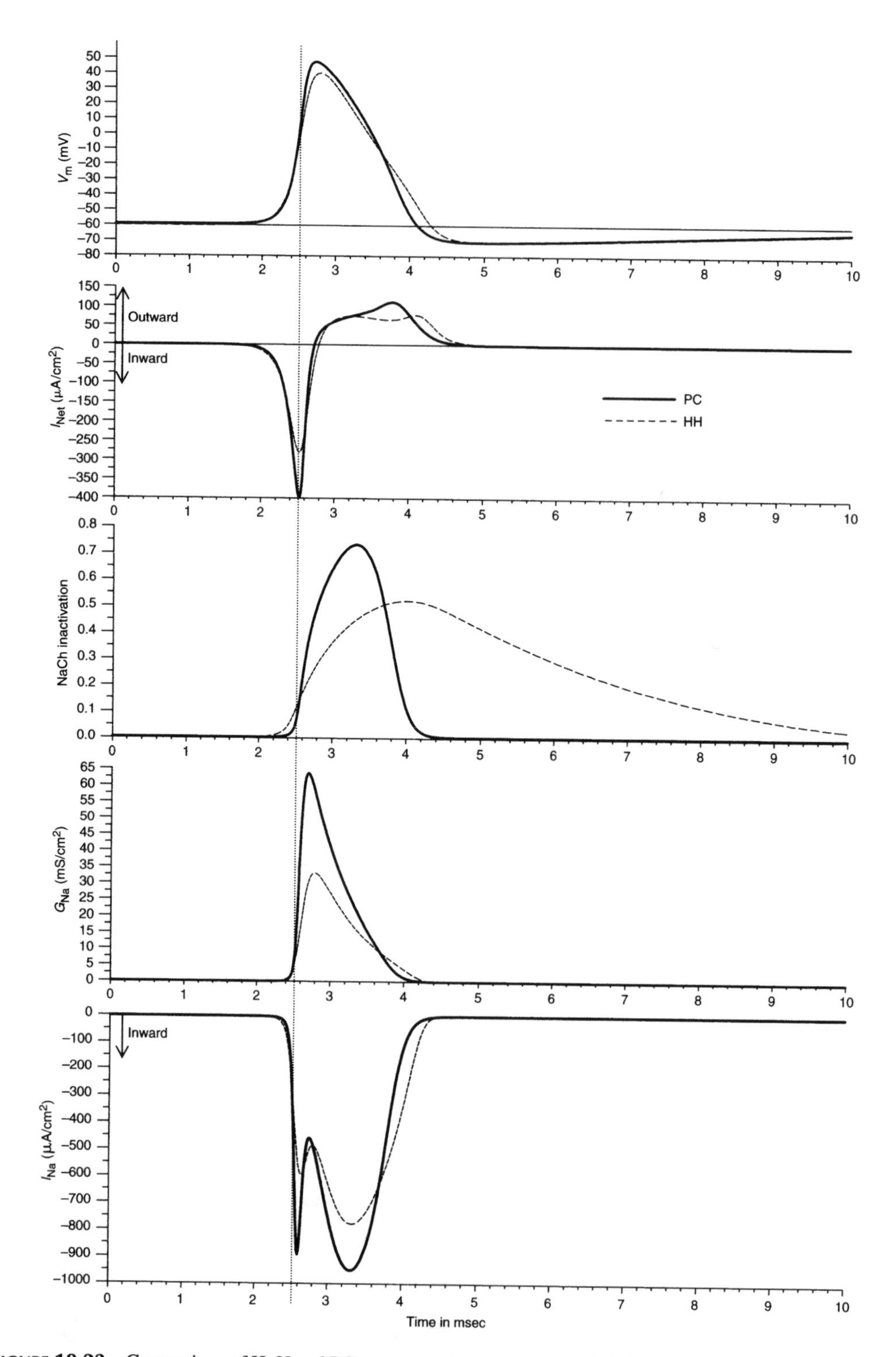

FIGURE 18.22 Comparison of H–H and PC macroscopic parameters underlying a single propagating impulse initiated at $x = 0$. Each axon had $1\,\mu$m diameter and total electrotonic length 9.9λ. Recordings were made at the fiber's center. For H–H, Na^+ channel population inactivation was plotted as $\{h_{\text{initial}} - h_{(V,t)}\}$. I_{net} was $\{I_{\text{ionic}} - I_{\text{m}}\}$ (Equation [18.1]). Each plot was aligned with respect to the time of maximal $\partial V/\partial t$ at the recording site (dotted vertical line). $\Delta t = 10$ nsec; $\Delta x = 3\,\mu$m.

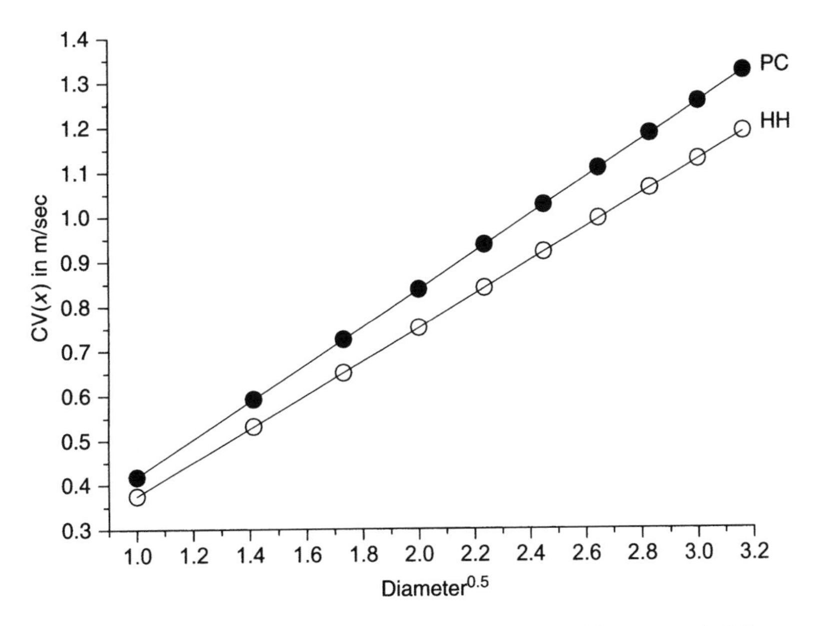

FIGURE 18.23 Comparison of H–H and PC axial conduction velocity, CV(x), recorded electrotonically distant from either sealed end. Each axon had total electrotonic length $10\lambda_d$; diameters tested were between 1 to 10 μm, inclusive. For each diameter, Δx_d was adjusted such that $\{\Delta x_d/\lambda_d\} = 0.0164$. In all runs, $\Delta t = 100$ nsec.

somewhat similar, as shown in Figure 18.27. The mean firing rate of the H–H axon was larger than that of the PC fiber by 4.6% at either recording site (each of which was sufficiently distant from either sealed end). The PC deadtime exceeded that of the H–H axon by about 0.42 to 0.54 msec, an increase of 8 to 10%. The initial peak in ED probability for the PC axon was slightly reduced and shifted by an amount equal to the deadtime prolongation. There was no representation of the presence or absence of supernormality (Figure 18.25) in either ED trajectory. The steady-state plateaus had similar variability restricted mainly to within $\pm 1\sigma$.

In singly branched axons, H–H and PC responses were also mainly similar, but slight differences were evident, as shown in Figure 18.28. As for the unbranched fibers, the H–H mean rate exceeded that of the PC fiber by 4.6% at either recording site. At the branch point, the respective deadtimes were equal, while further downstream the PC deadtime exceeded that of H–H by 0.54 msec (9.44%). The ED trajectories were similar in shape but the PC's ED peak at the branch point was reduced and shifted by an amount similar to the PC deadtime prolongation (0.54 msec). In singly branched axons with periaxonal K^+ accumulation throughout, the mean firing rates of the two fiber types were again similar, as shown in Figure 18.29. In both models, widespread periaxonal K^+ accumulation caused the initial ED trajectory to be slowed with a peak delayed by approximately 7 msec. The PC model had a consistently longer deadtime.

18.4.4 Discussion

The PC model incorporates an experimentally validated kinetic scheme for voltage- and time-dependent macroscopic Na^+ conductance, which differs significantly from the empirical H–H m^3h scheme. The PC model's characteristics were clearly essential for voltage-clamp considerations (Figure 18.21). For coding considerations in nonisopotential cables, the PC and H–H models' predictions of random firing properties did differ, notably in the higher mean firing rates for the H–H axon. While the two recovery cycles differed, no correlate of this difference was observed in the ED plateaus. Although the recovery cycle is a classical experimental method for axonal classification (Gasser, 1941), its test paradigm (i.e., short-duration current pulses constituting conditioning and test stimuli) has no correlate for coding *in situ*, where the stimulus

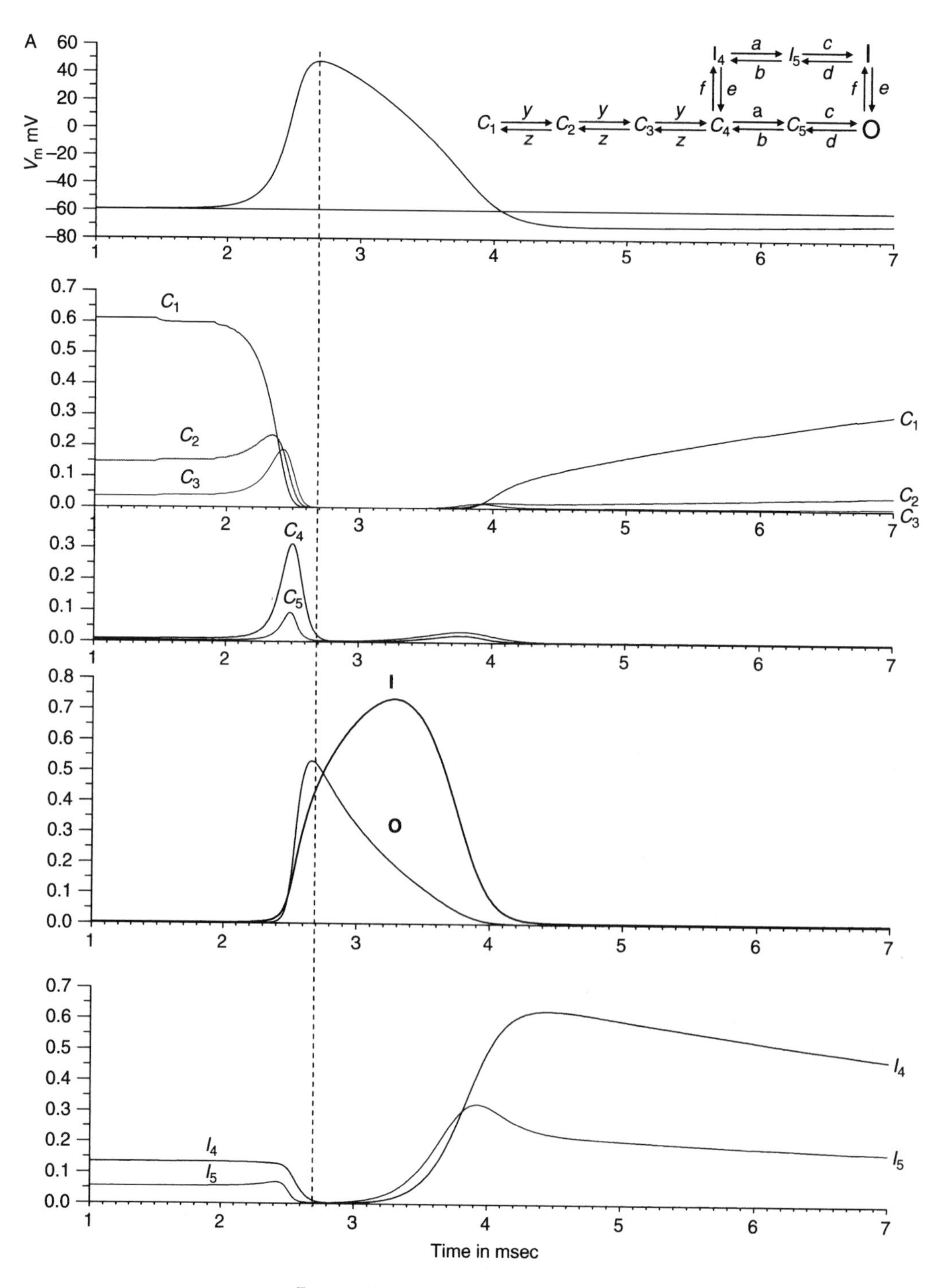

FIGURE 18.24 Continued on next page

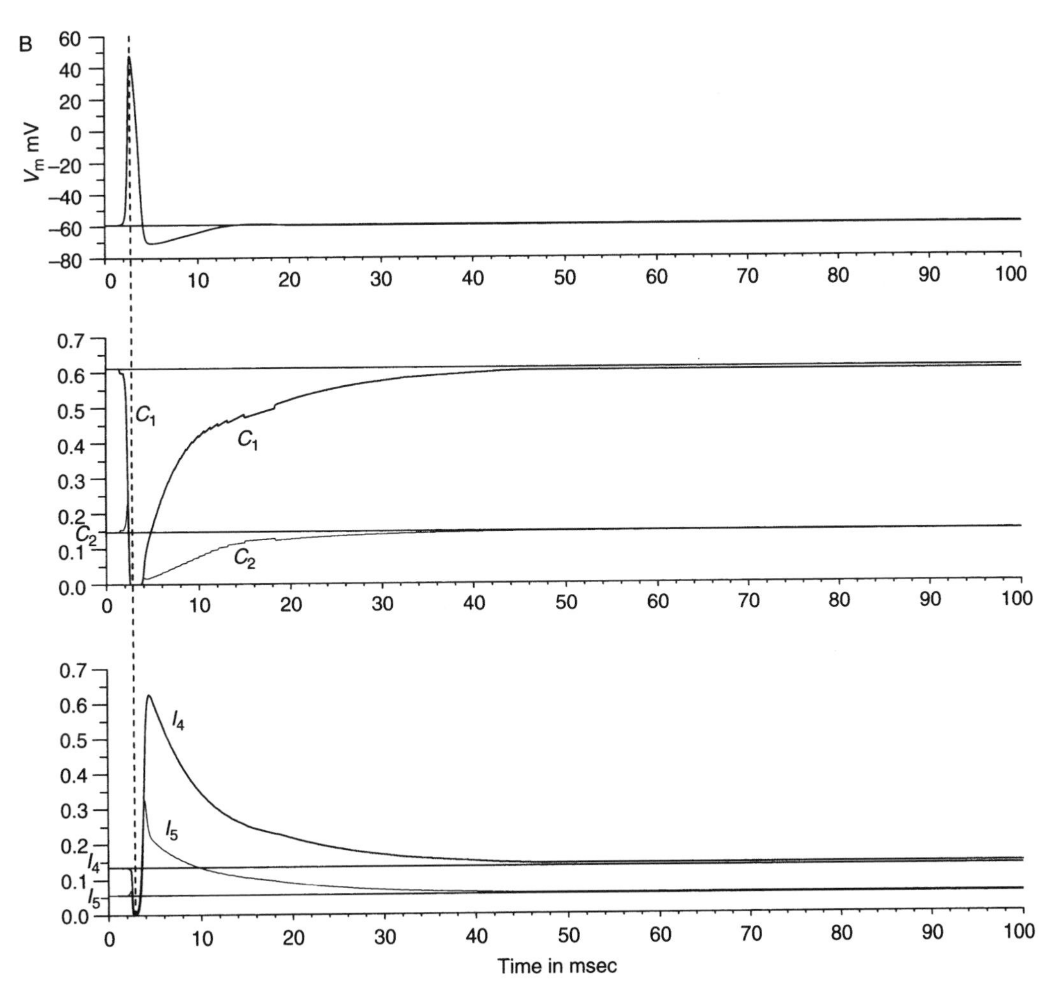

FIGURE 18.24 Macroscopic conductance parameters in the PC model of a propagating action potential. Axon diameter was 1.0 μm and total electrotonic length was $9.9\lambda_d$. Action potential initiated at $x = 0$ and propagated over the entire fiber length. Recordings were made at the fiber center ($=4.92\lambda_d$). Populations of states **O** (open), **I** (inactivated), and **C** (closed) expressed as probability (i.e., percent of Na^+ channel population). **I** and **C** are subdivided into substates defined by kinetic scheme (see text). $\Delta t = 100$ nsec; plot resolution $= 20$ μsec. (A) Events up to 7 msec total delay. (B) Longer-term events up to 100 msec.

for impulse initiation in axons has a complex and conditionable waveform (e.g., Figure 18.14C and Figure 18.15).

Should the PC-type formulation become a preferred or even necessary substitute for the H–H-type G_{Na} component in models of excitable cables? As shown here, both models predict a highly efficient propagation of wide-band random impulse trains across unmyelinated branch points. The PC model parameters can be modified (e.g., rate constant values; other kinetic schemes as in Patlak, 1991; cf. Miro et al., 2002) so as to provide response properties closer to those of the H–H model, including the H–H recovery cycle. However, in models which include inactivating conductance regimes, the kinetic scheme needs to address the explicit coupling between activation and inactivation. The inclusion of activation–inactivation coupling may be particularly important for such problems as the generation of functionally explicit codes (i.e., motor, sensory, affective), repetitive firing, spontaneous activity, and the role of noise. Also, the PC model's relatively slow removal of inactivation (Figure 18.24B)

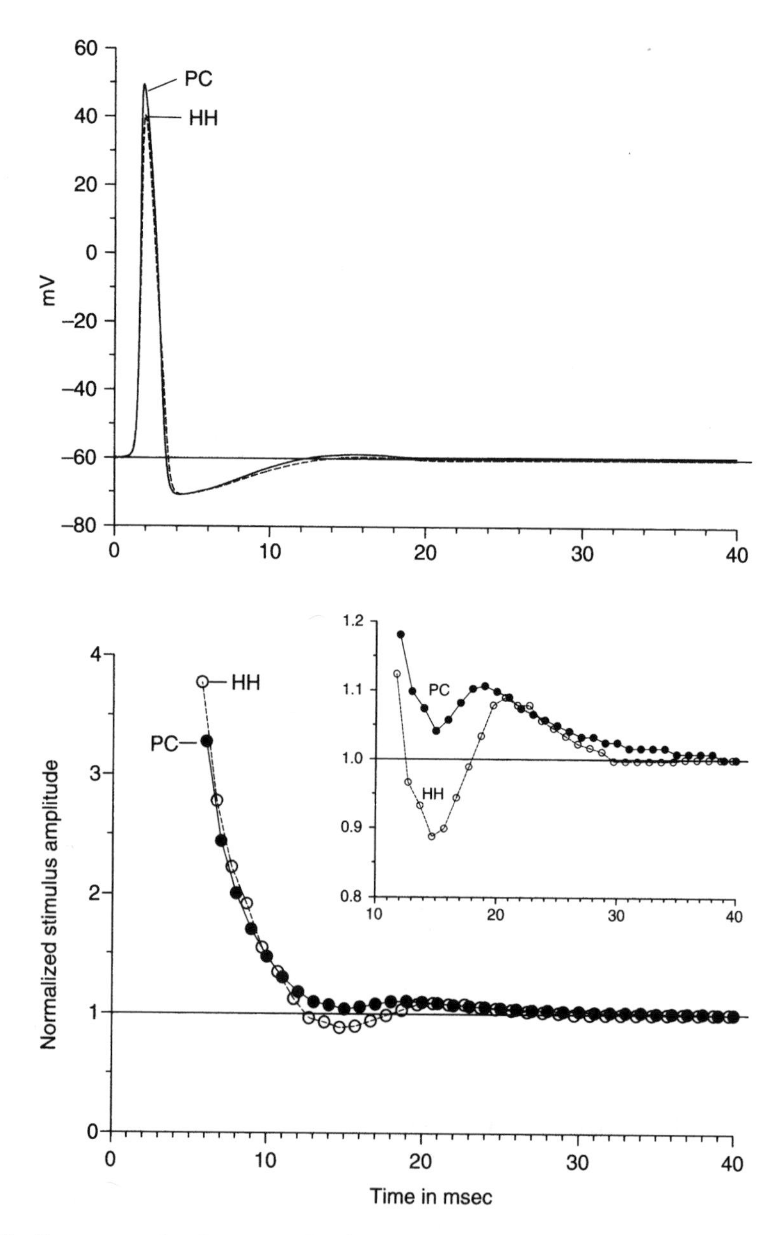

FIGURE 18.25 Recovery cycle of propagated action potential in H–H versus PC axons. Axon had a uniform diameter of 1.0 μm and total electrotonic length of $9.9\lambda_d$. Conditioning action potential initiated at $x = 0$ and propagated over the entire fiber length. Recording and testing performed at $x = 3.28 \cdot \lambda_d$.

may contribute uniquely to limiting any type of higher-frequency firing (cf. Tsutsui and Oka, 2002). The importance of a PC-type model for the reconstruction of experimental responses of isopotential squid axon has been demonstrated (Clay, 1998).

18.5 DIRECTIONS FOR FUTURE RESEARCH

The computed results in this chapter have shown that, in theory, wide-band impulse trains propagate with little distortion over unmyelinated axonal branch points. Epiphenomena — including

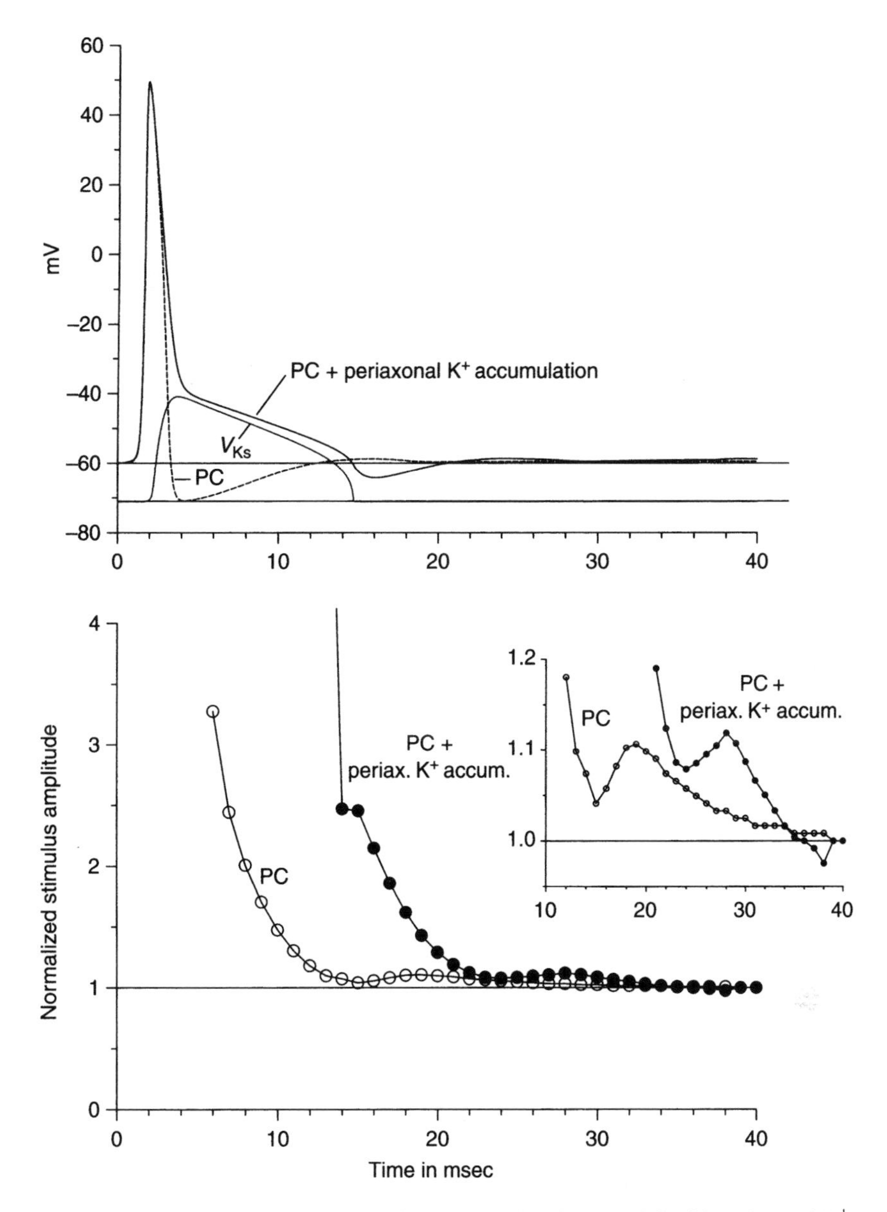

FIGURE 18.26 Recovery cycle of propagated action potential in PC versus PC with periaxonal K^+ accumulation axons. Same fiber and experimental configuration as in Figure 18.25. Periaxonal K^+ accumulation model was as described above, using $K_{diss} = 2$ mM. V_{Ks}, K^+ equilibrium potential in the periaxonal space.

a less-than-normal effectiveness in the regulation of periaxonal K^+ accumulation — could compromise this efficiency. Impulse train propagation over arborized axons appears to be highly reliable through such features as an optimized density of Na^+ channels (as per Hodgkin, 1954), whose kinetics provide for an efficient extraction of macroscopic Na^+ conductance and current (as per Figure 18.22). Thus, information transmitted on branching axons will be faithfully delivered to their many axonal terminals, where wide-band activity can elicit frequency-modulated modifications of transmitter release; such interactions may be highly complex. But if it is assumed that, analogous to axonal branch points, the larger design property of frequency-modulated chemical synaptic transmission is to promote — with perhaps some degree of filtering — the intercellular

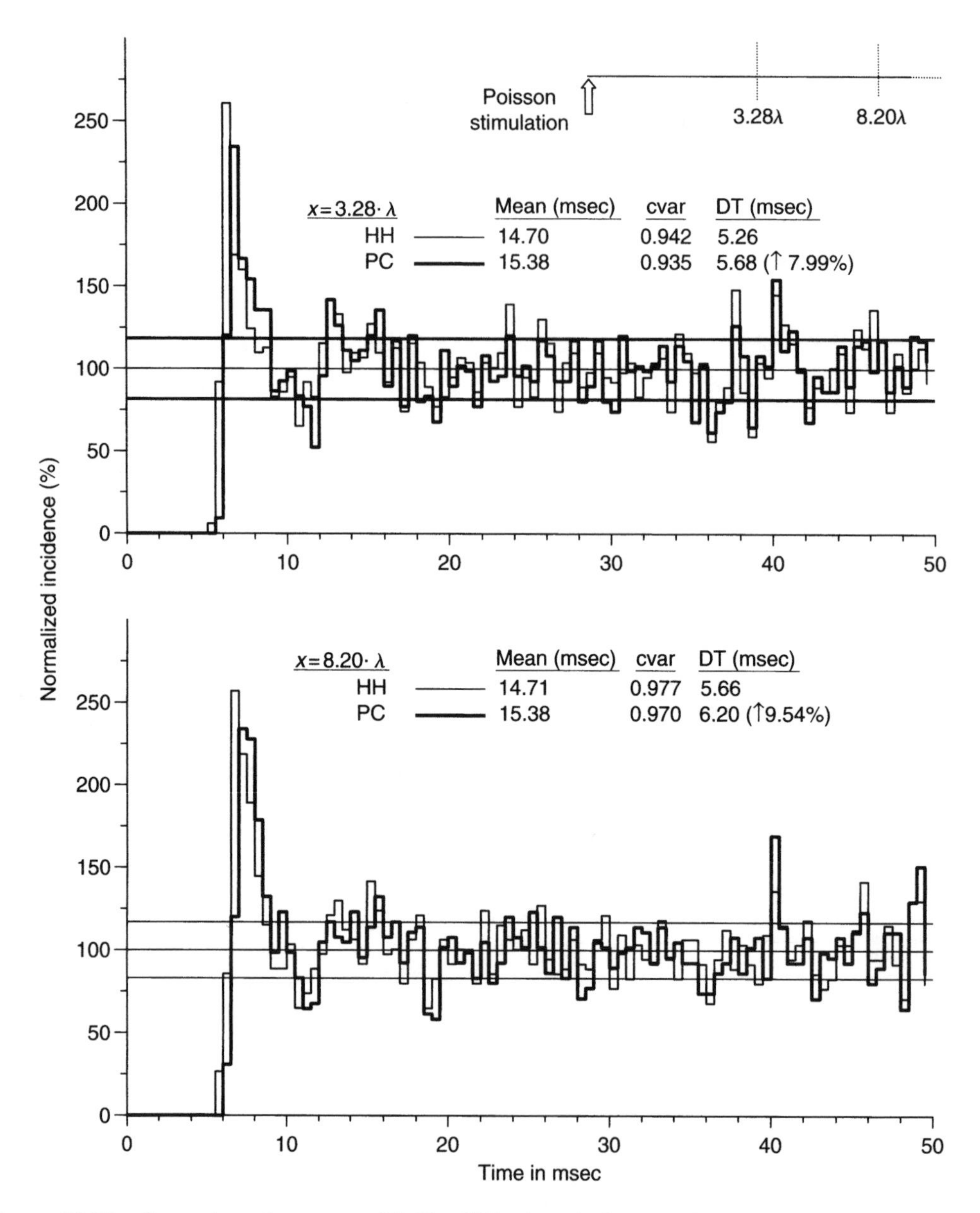

FIGURE 18.27 Comparison of responses of H–H and PC unbranched axons to Poisson-like stimulation. Propagated impulse trains were recorded at two locations on the 1 μm-diameter axon. Normalized ED trajectories with IID shape parameters were computed at 0.5 msec binwidth. Plateau noise average and $\pm 1\sigma$ (fine lines) were computed between $t = 25$ and 50 msec. Fiber configuration is shown at upper right.

communication of minimally distorted information (e.g., Kistler and Gerstner, 2002), then the net pathway of propagating information may be considered to be itself highly efficient.

The vertebrate CNS is organized to a significant degree as multicomponent closed loops (e.g., the reciprocal connectivities between cerebral cortex and cerebellum; Eccles, 1973). Given the efficiency of wide-band impulse propagation over axons and their branch points (encompassing such constructs as an optimized rate code: Levy and Baxter, 2002), and assuming a matching efficiency for the transmission of such information at chemical synapses, then propagating information — once initiated and further assuming all else remains equal — is destined to propagate continuously

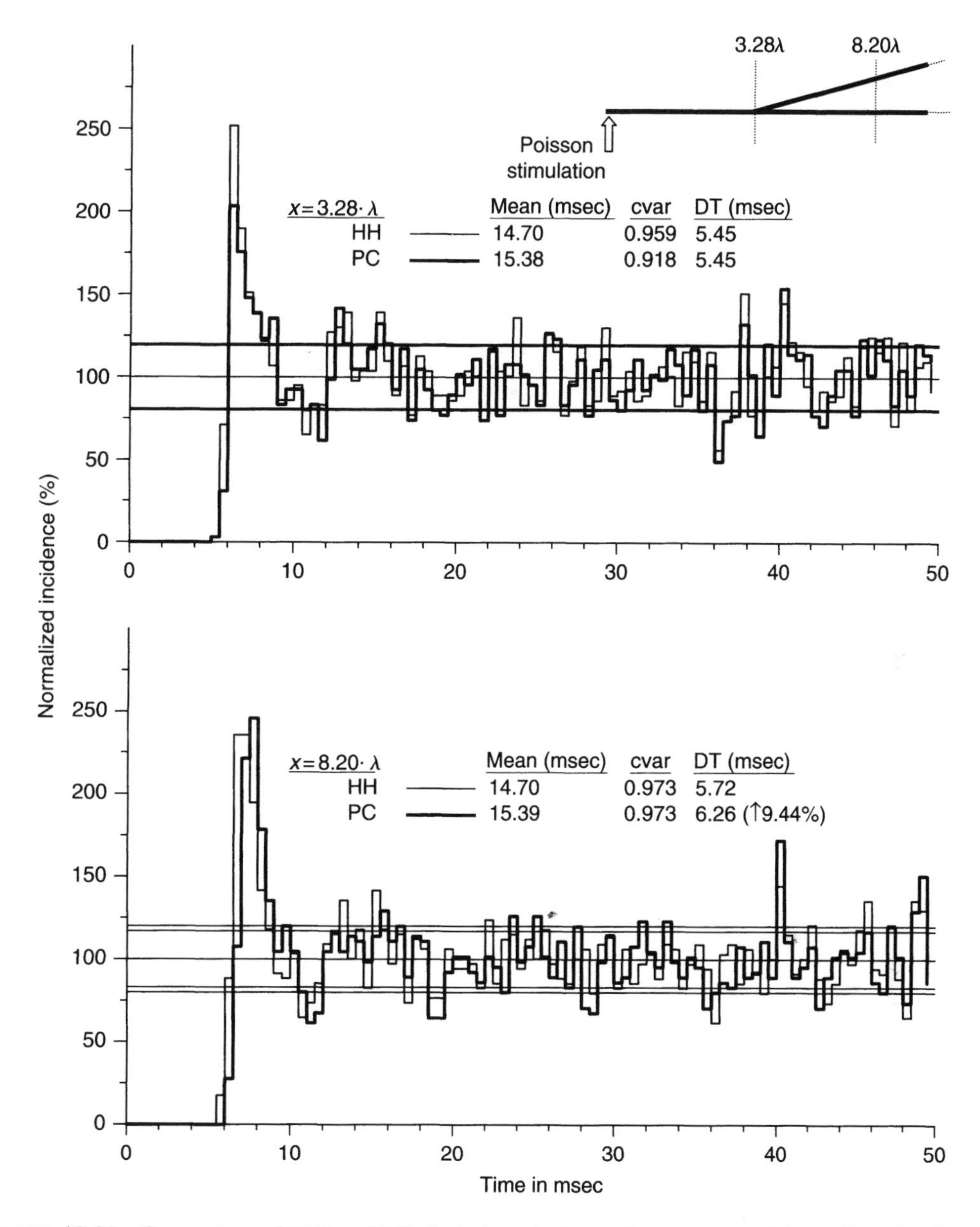

FIGURE 18.28 Comparison of H–H and PC singly branched axons' responses to Poisson-like stimulation. Normalized ED trajectories with IID shape parameters were computed at 0.5 msec binwidth. Plateau noise average and $\pm 1\sigma$ (fine lines) were computed between $t = 25$ and 50 msec. Fiber configuration is shown at upper right; all diameters were 1 μm.

over these variously sized circular pathways, repeating indefinitely, with the only modifications being perhaps a limited focusing of mean firing rates and patterns (e.g., Moradmand and Goldfinger, 1995a). Does impulse-based information reverberating in such significantly efficient loops constitute a form of information processing and/or storage (e.g., Martin, 2002; Whittington and Traub, 2003)? Does the repetition of an impulse sequence define a clock specific to a frame of reference? Given that each finite-duration impulse sequence occupies a finite length of axonal path, is the total axonal path length a limit to the number of unique sequences propagating within the loop?

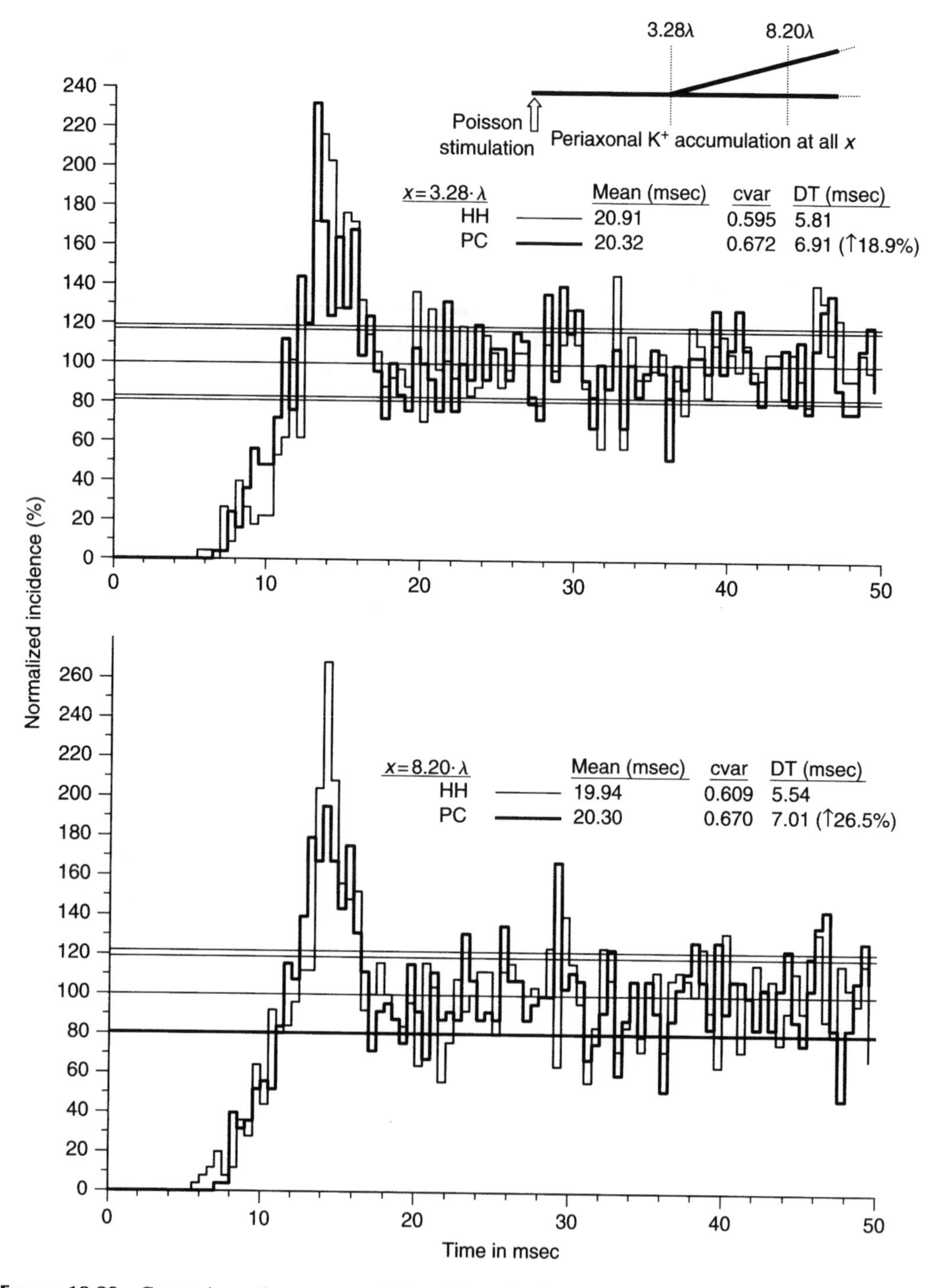

FIGURE 18.29 Comparison of responses to Poisson-like stimulation by H–H and PC singly branched axons with periaxonal K^+ accumulation ($K_{diss} = 2\,mM$) at all spatial locations. Normalized ED trajectories with IID shape parameters were computed at 0.5 msec binwidth. Plateau noise average and $\pm 1\sigma$ (fine lines) were computed between $t = 25$ and 50 msec. Fiber configuration is shown at upper right; all diameters were 1 μm.

ACKNOWLEDGMENTS

This work was supported in part by the U.S. National Science Foundation (B.C.S. 9315856) and by the College of Science and Mathematics and the School of Medicine, Wright State University. The author thanks Dr. V.E. Amassian (S.U.N.Y.-Downstate Medical Center) for the nuclear disintegration data, and Dr. John Clay (NINDS/NIH) for helpful correspondence. The author also thanks *NeuroReport*/Lippincott, Williams & Wilkins and the *Journal of Neurophysiology*/American Physiological Society for permissions to use copyright material in Figure 18.5 and Figure 18.20, respectively.

PROBLEMS

1. Add a multistate kinetic scheme for $g_K(V, t)$ to the PC model, with the criteria for adequacy being the reconstruction of the H–H isopotential and propagating action potentials. Does this modification compromise or support the axon's ability to propagate wide-band Poisson-like impulse trains?
2. What modifications of the nonisopotential PC model's g_{Na} kinetics are needed to make its recovery cycle more closely equal to that of the nonisopotential H–H model?
3. Regarding periaxonal K^+ accumulation: add to the present model more glial details relating to K^+ buffering, such as K^+ release from the transporter into the glia intracellular space; K^+ diffusion intercellularly through connexin channels in gap-junction membranes; the extrusion of excess K^+ into a venous or other compartment of extracellular space.
4. Does the PC model differ from H–H in any response optimizations due to noise, as in stochastic resonance (Yu et al., 2001)?

APPENDIX

In H–H axons with total length of 9.9λ whose diameter doubled in its distal half from 1 to 2 μm, fourth-order Runge–Kutta integration yielded several different stable solutions for the propagation of a single action potential, depending on the value of the temporal integration step, as shown in Figure 18.30A. In some stable solutions, the computed conduction velocity in the expanded diameter region differed significantly from the value expected given the average CV(x) in the 1-μm region distal from the geometrical inhomogeneity (Goldfinger, 2000). At the extreme tested ($\Delta t = 250$ nsec), the action potential peak declined from 34 mV (e.g., 1λ upstream from the widening) to 6 mV (e.g., 1λ downstream from the widening) and propagated at a low conduction velocity (0.11 m/sec; Huxley, 1959; Moradmand and Goldfinger, 1995a). Below a critical Δt size (12 nsec), fourth-order Runge–Kutta-generated solutions converged, with a 40 mV action potential peak changing less than 2 mV during the increase in CV(x) prior to the diameter expansion. With trapezoidal integration, stable solutions obtained only below a critical Δt value; all such solutions converged.

For a single literal branch point, CV(x) solutions were also sensitive to the Δt value, as shown in Figure 18.30B. With trapezoidal integration, sufficiently small values of Δt yielded convergent stable solutions; above a critical value (173 nsec), solutions were unstable. By contrast, fourth-order Runge–Kutta solutions were stable for Δt at and below 700 nsec. However, in this stable range, CV(x) solutions were not convergent until Δt values decreased to 20 nsec. For solutions obtained with Δt values exceeding this convergent value and below the stability limit, the entire CV(x) function shifted upward on the abscissa with shorter Δt.

Numerical solution dependency upon Δt size has been described by Mascagni and Sherman (1998) and by Yee et al. for another system of differential equations (1991; Stewart, 1992).

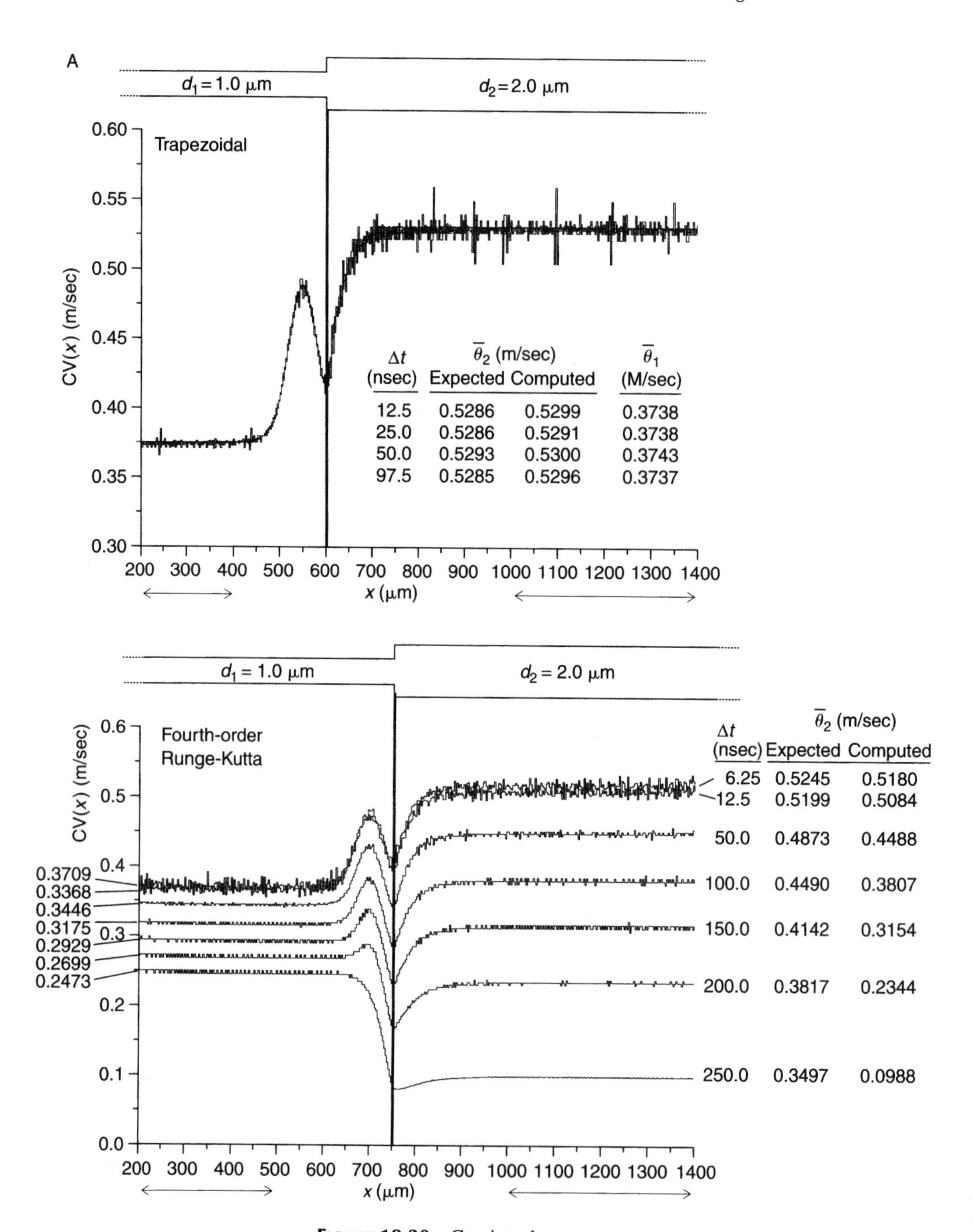

FIGURE 18.30 Continued on next page

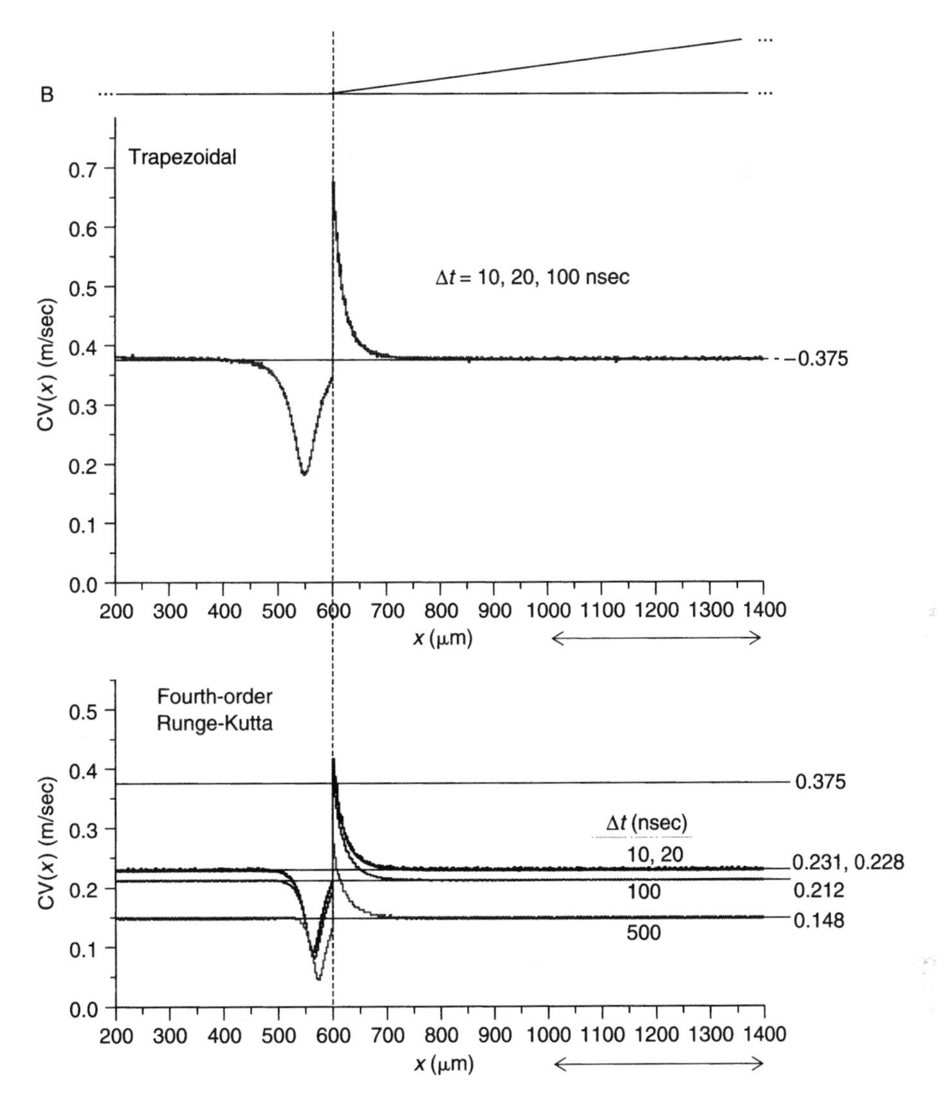

FIGURE 18.30 Sensitivity of trapezoidal versus fourth-order Runge–Kutta solutions to temporal integration step (Δt). (Top) Trapezoidal or (bottom) fourth-order Runge–Kutta integration method used to compute axial conduction velocity [CV(x)] for a single propagating action potential. The spatial integration step, Δx (=3 μm), was constant in all compartments. Average conduction velocity was determined between 1000 and 1400 μm. (A) Fiber diameter doubled in the center. Stable Runge–Kutta solutions varied with values of the temporal integration step, Δt. Expected conduction velocity values in the 2 μm-diameter region were obtained using:

$$\theta_{\text{expected}} = \theta_{d=1\,\mu\text{m}} \cdot (d_2/d_1)^{0.5} \text{ (Goldfinger, 2000).}$$

(B) Computations of CV(x) for a single action potential propagating in an H–H axon with a single literal branch point

NOTES

1. In Poznanski's conception of "Integrative Neuroscience" (Poznanski, 2001c), axonal information is a contributor to the larger functional connectivity of synaptic transmission. Axonal information may be defined in different ways, by use of explicit impulse timing (e.g., Singer, 1993; Mainen and Sejnowski, 1995), a mean rate-based code (e.g., Moradmand and Goldfinger, 1995a), Shannon entropy (Shannon and Weaver, 1949; e.g., Rieke et al., 1997), or a characterization of the frequency content and its resultant effect(s) at

the presynaptic terminals of the axon carrying the code (e.g., Migliore and Lansky, 1999; Usrey, 2002). The present chapter does not address this issue of functional definition.

2. A sequence of delta function-like events with Poisson-like statistical properties can be produced in at least two other ways: (a) as a superposition of appropriately selected periodic event trains (Goldfinger, 1984); and (b) by generation with a modified multivibrator circuit (Moradmand and Goldfinger, 1995b).
3. An analogous spatial requirement was observed with respect to repetitive spontaneous activity elicited by higher-than-normal extracellular K^+. In 1-μm-diameter fibers (total length $= 9.9\lambda$) without periaxonal K^+ accumulation, twice-normal extracellular K^+ (localized to one axonal end) elicited bistability (Hahn and Durand, 2001), causing repetitive propagating impulses. A minimum fiber length of 0.656λ was required to elicit such action potentials at a constant firing rate, which increased with greater fiber length exposed. If smaller lengths (l_{ex}) of fiber were exposed to twice-normal extracellular K^+, only a single ($0.492\lambda < l_{ex} < 0.656\lambda$) or no ($0.0164\lambda < l_{ex} < 0.492\lambda$) impulse was initiated.

19 Dendritic Integration in a Two-Neuron Recurrent Excitatory Network Model

Roman R. Poznanski

CONTENTS

19.1 INTRODUCTION

The most advanced neural network models include "spiking" neuron models (also referred to as conductance-based point neuron models or integrate-and-fire models) as single isopotential compartments (i.e., without spatial structure) that fire impulses whenever a voltage threshold is exceeded. Although the field is slowly undergoing reform, these "spiking" neurons are still being used in many network models (e.g., see Gerstner and Kistler, 2002). However, if dendritic structure is taken into consideration, then axodendritic synaptic connections may not reliably evoke impulses when stimulated, due to the cable properties of dendrites and the sparse distribution of voltage-dependent ionic channels as shown by *ionic cable theory* (Poznanski, 2001b). Therefore, to adequately represent dendritic integration, it is vital to construct neuronal networks as biologically constrained models, focusing on the inclusion of the relevant physicochemical processes. Such an approach to modeling in the neurosciences urgently needs to be developed.

In the past, the status quo for modeling active dendrites was to assume fully active Hodgkin–Huxley-like channel kinetics in a network of multicompartmental models (Gardner, 1993; Lansner and Fransen, 1995; Bibbig et al., 2002; Nenadic et al., 2003). The problem with such models is their inability to link hierarchical neuronal structures (at the subcellular, cellular, and network level) in a practical manner. For instance, in the words of Traub and Jefferys (1994, p. 116): "others have modeled individual neurons with many thousands of compartments, but such models do not lend themselves to network simulations." Most often multicompartmental models of single neurons are reduced ranging from a single to a dozen compartments for practical purposes leaving the validity of the approach in question. Invariably a multicompartmental model representation of a single neuron is inept at advancing to the network level in an efficient way because of the sheer complexity in terms of the thousands of parameters it has required to adequately represent a single neuron, as for example, exemplified by Spruston and Kath (2004):

> One strategy for developing more sophisticated neural networks is to replace simple elements with sophisticated computational models of neurons, including branching dendritic trees, thousands of synapses, and dozens of voltage-gated conductances. The problem with this approach is that it is not manageable to use such computationally expensive neuronal models in large-scale networks.

The need to construct integrative as opposed to computational neural models of networks is therefore fundamental to new advances in our understanding of dendritic integration.

All multicompartmental models suffer from the nonuniqueness problem (Rall, 1990, 1995a). Therefore the lack of robustness of such models to parameter changes could in principle be the most significant cause of variability in response characteristics. This often overlooked anomaly adds emphasis to the conclusion found by Krichmar et al. (2002), that altered morphology per se (keeping all other factors constant) drastically modified the spiking activity of their simulated multicompartment hippocampal neurons. Our aim is to show that changes in the distribution of ionic channels and numerous other nonmorphological parameters can also bring about major changes in response characteristics. Unequivocally, compartmentalization of the neuronal membrane, by which the continuous nature of the neuronal membrane is sliced to pieces, introduces discontinuities that are not intrinsic to real neuronal circuitry. For instance, Calabrese et al. (2000) offer a basic modeling strategy for small-scale neural networks which "boils down" to solving the following current balance equation for any whole single-compartment neuron model or any single compartment within a multicompartment neuron model:

$$C\,\mathrm{d}V/\mathrm{d}t = -\sum(I_{\mathrm{ion}}) - \sum(I_{\mathrm{syn}}) + \sum(I_{\mathrm{inject}}),$$

where I_{ion} represents membrane currents (e.g., voltage-gated and leak currents), I_{syn}, synaptic currents, and I_{inject}, injected currents (or the current from connected compartments). This equation is used without comment in almost all software packages. The work presented here, however, suggests that it will be important in new models of neurons in networks to incorporate wherever possible a continuously distributed spatial structure.

A significant challenge in the construction of biophysically realistic neurons in networks is to identify and incorporate the cellular behavior in the mathematical description of the macroscopic population behavior. We have developed a model for a recurrent excitatory network of two "weakly" active neurons from the result of a *subthreshold* current injection into a point close to the fictitious soma of one neuron. Such modeling is new because recurrent network models with a distributed dendritic structure (see Bressloff, 1995, 1996, 1999; Bressloff and Coombes, 1997a, 1997b; Coombes and Lord, 1997; Bressloff and De Souza, 1998) were developed without any voltage-dependent ionic channels. We propose to show, using a two-neuron recurrent excitatory network model, that "weakly" excitable dendrites enable the dendritic spike to propagate passively along its arbor, without too much decrement in its amplitude if the spatial distribution between clusters (or hot spots) of voltage-dependent ionic channels is small, approximating an almost uniform-density distribution of voltage-gated ion channels. We also aim to show how feedback in a system of two neurons can maintain peak amplitude in the postsynaptic neuron at a distally placed location along the dendrite. This is important because experimental studies showing that somatic feedback to the dendrites becomes stronger with increasing network activity (Waters and Helmchen, 2004) are incapable of showing the effect on distal fine dendrites.

19.2 Ionic Cable Theory for Conduction of Dendritic Potentials

19.2.1 Ionic Cable Equation for a Single Neuron

Let V be the depolarization (i.e., membrane potential less than the resting potential) in mV, and I_{ion} be the inward ionic-current density through the membrane enclosing a unit length of cable (i.e., per unit membrane surface of cable [mA/cm]), as defined in Section 19.2.2. The voltage response or depolarization as a result of synaptic input, as defined in Section 19.2.4, in a leaky cable representation of a cylindrical passive dendritic cable with I_{ion} occurring at discrete points, that is, $x = x_{\text{p}}$, can be shown to satisfy:

$$C_{\text{m}} V_t = \left(\frac{d}{4R_{\text{i}}}\right) V_{xx} - \frac{V}{R_{\text{m}}} - \sum_{j=1}^{N} I_{\text{ion}}(x,t;V)\,\delta(x-x_j) - \sum_{k=1}^{M} g(t)[V - V^{\text{rev}}]\delta(x-x_k)$$
$$-\sum_{z=1}^{N} P\delta(x-x_z), \quad t > 0, \tag{19.1}$$

where x is the distance (cm), t is time (sec), d is the diameter of the cable (cm), C_{m} is the membrane capacitance (F/cm^2), R_{m} is the membrane resistivity ($\Omega\,\text{cm}^2$), R_{i} is the intracellular resistivity ($\Omega\,\text{cm}$), N is the number of hot spots of voltage-dependent ionic channels, M synapses control the ionic channels at specified positions along the dendritic cable (i.e., $x = x_k$), with time-dependent conductance changes $g(t)$ (S/cm) and reversal potentials V^{rev} (mV). P is the sodium-pump current density representing ion fluxes due to active transport (mA/cm), and δ is the Dirac delta function reflecting the position on the membrane where the ionic current is represented along the circumference of the cable per cm. Subscripts x and t indicate partial derivatives with respect to these variables, and by convention, inward current is negative and outward current is positive.

19.2.2 Representation of Noninactivating Voltage-Dependent Ionic Currents

The integration of synaptic signals in dendrites has been shown experimentally to be affected by persistent Na^+ (Na^+P) channels, resulting in amplified synaptic potentials (see MacVicar, 1985; Schwindt and Crill, 1995; Stuart and Sakmann, 1995; Lipowsky et al., 1996; Urban et al., 1998; Fricker and Miles, 2000; Berman et al., 2001; Gonzalez-Burgos and Barrionuevo, 2001; Koizumi et al., 2001) as well as enhancement of backpropagating action potentials (bAPs) (Pan and Colbert, 2001; Doiron et al., 2003). The spatial location and distribution of Na^+ channels in dendrites remains unknown, yet several studies have shown a high differential in the densities of sodium channels along axonal and dendritic membranes (Safronov, 1999). Furthermore, it is safe to assume that Na^+P channels are present in low density (about 2% of the Na^+ channels) and make the membrane "weakly" excitable to support a nonregenerative dendritic spike in the subthreshold voltage range.

Dendritic spikes are mediated predominately by Na^+ channels, although K^+ channels have been found in some dendrites (see Hoffman et al., 1997). Thus, the experimental observation of peak amplitude reduction of backpropagating dendritic spikes could be due to a sparse density of Na^+ channels or to a high concentration of K^+ ions in the extracellular medium pointing indirectly to the presence of K^+ channels on dendrites (cf. Johnston et al., 1999). It is well known that Na^+ channels may be induced to lose their capability for inactivation as a result of chemical treatments or exposure to proteolytic enzymes (e.g., protease and papaine), yet these treatments have a relatively minor or no effect on activation in these channels. When Na^+ channels lose inactivation, Na^+ currents show no inactivation; persistent currents without transient components are a common occurrence for freshly dissociated cells (cf. Poznanski, 2001d).

The inward ionic-current density per unit membrane surface of the cable (mA/cm) may be represented as

$$I_{\mathrm{ion}}(x,t;V) = g_{\mathrm{ion}}m(V)[V - V_{\mathrm{ion}}], \quad t > 0, \tag{19.2}$$

where $V_{\mathrm{ion}} = E_{\mathrm{ion}} - E_{\mathrm{r}}$, E_{r} is the resting membrane potential, E_{ion} is the equilibrium potential (mV), and g_{ion} (S/cm) is the conductance per unit membrane length of the cable obtained by multiplying the maximum attainable conductance of a single ionic channel by the number of channels contained in a unit length of cable. Note that the inward current in Equation (19.2) by convention is negative, which should not be confused with a current applied or injected via a pipette, which is positive because it is an intracellular current flowing outward. The dimensionless activation variable $m(V)$ is governed by a first-order reaction equation (Hodgkin and Huxley, 1952b):

$$\partial m/\partial t = \alpha_m(1 - m) - \beta_m m, \tag{19.3}$$

where α_m and β_m are the rate constants (in $\sec^{-1}$) of the ionic channels independent of t. Equations (19.2) and (19.3) correspond to various *nonsynaptic* voltage- and time-dependent ionic currents.

19.2.3 Representation of Axodendritic Chemical Synapses

Modeling of synaptic transfer in artificial neural networks usually entails the inclusion of synaptic weights (see Kurogi, 1987; Seung et al., 2000), but real neuronal networks form axodendritic synapses that are activated by action potentials, resulting in a discrete or noncontinuous release of neural transmitters that leads to a change in conductance and subsequent generation of synaptic current in the postsynaptic neurons (see Miftakhov and Wingate, 1995). Modeling of this process is based on the following assumptions:

1. The rate of neural transmitter release is proportional to the activity of individual spikes arriving at the presynaptic terminals. Hence, to avoid the continuous representation of activity in terms of a mean firing rate, discrete spiking activity that releases neurotransmitter into the synaptic cleft can be included and used to investigate the consequences of the nonuniform firing of spikes. The rate of neural transmitter release depends upon the timing of the action potentials arriving at the presynaptic terminals, which can be represented by a series of Dirac delta functions:

$$E(V^{\mathrm{pre}}) = \sum_{k=1}^{N_{\mathrm{sp}}} \delta(t - t_k - \Delta t)\mathrm{H}[V^{\mathrm{pre}}(0,t) - \theta], \tag{19.4}$$

where N_{sp} is the total number of spikes, $t_k + \Delta t$ is the time of arrival of the kth spike at the synapse, where t_k is the time of generation of the spike at the soma or axon hillock and Δt is the axonal delay (sec), θ is the presynaptic threshold voltage, and $\mathrm{H}[\cdot]$ is the Heaviside step function.

Equation (19.4) records a spike only if the presynaptic voltage is above the threshold. Hence, a spike may not be recorded for every measured time, especially for recurrent networks as discussed in Section 19.3.

2. The amount of neural transmitter release into the synaptic cleft is driven by the individual spike activity, which provides the source term in the first-order kinetic equation governing the amount of transmitter in the synaptic cleft. This process is often neglected and replaced by converting the presynaptic signal into postsynaptic activation via a conductance change based on first-order kinetics, as has been carried out by Wang et al. (see Golomb et al., 1994; Rinzel et al., 1998). The amount of neural transmitter in the synaptic cleft is given by:

$$\tau_c \, dC/dt = -C(t) + \tau_c C_0 E(V^{pre}), \tag{19.5}$$

where E is the input (presynaptic) spike-train activity (with dimension sec^{-1}) expressed through Equation (19.4), C_0 represents the maximum amount of transmitter (mM), and τ_c is the time constant of transmitter release in response to presynaptic depolarization (sec). It should be noted that in artificial neural networks, $C(t)$ is often ignored and replaced in Equation (19.6) with a source term represented as a series of Dirac delta functions for the conductance change (e.g., see Treves et al., 1996) because it is assumed that the duration of the neural transmitter release is sufficiently short (i.e., <1 msec) or the amount of neural transmitters in the synaptic cleft is sufficiently small, that $C(t)$ does not need to be temporally resolved (Chapeau-Blondeau and Chambert, 1995).

3. The neural transmitter–receptor channel kinetics obey a first-order process in which the synaptic conductance is assumed to be proportional to the total number of channels occupied by neural transmitter (Bernard et al., 1994), so the opening of the synaptic conductance is the product of the free neural transmitter and the number of unoccupied receptors, while the closing is determined by a simple exponential decay process (Leibovic and Sabah, 1969; Chapeau-Blondeau and Chambert, 1995). These kinetics lead to a complicated nonlinear equation for the time course of the synaptic conductance, which under most circumstances can be simplified to a damped harmonic oscillator equation (see Holmes and Levy, 1990; Bernard et al., 1994; Kalitzin et al., 1997; Rolls and Treves, 1998). More complex dynamics are necessary when neural transmitter–receptor kinetics are mediated by second messengers, such as a calcium influx (see Byrne and Gingrich, 1989). The following equation is based on first-order kinetics (Leibovic and Sabah, 1969; Chapeau-Blondeau and Chambert, 1995):

$$\tau_g \, dg/dt = -g(t) + \omega C(t)[g_{sat} - g(t)], \tag{19.6}$$

where τ_g is the time constant of the decay of $g(t)$ (sec), ω is the equilibrium constant for the neural transmitter–receptor interaction in mM^{-1}, g_{sat} $(=g_{max}M^*/\pi d)$ represents the saturating conductance associated with a noncontinuous release of neural transmitter per unit membrane surface of cable (S/cm), with g_{max} being the maximum conductance of an ionic channel (S), and $M^*/\pi d$ being the number of postsynaptic channels per unit membrane surface of cable. $C(t)$ is the amount of neural transmitter in the synaptic cleft (mM). The neural transmitters are released by the arrival of action potentials from the presynaptic neurons (cf. Hoppensteadt, 1986; Kurogi, 1987).

19.2.4 Network Equations without Synaptic Weights

Consider a recurrent excitatory network of two chemically coupled neurons with and without synaptic weights, as depicted in Figure 19.1. The first neuron, whose membrane potential is $V^0(x,t)$, provides

M synaptic inputs to the postsynaptic neuron, whose membrane potential is $V^1(x,t)$. Note that $U^0(x,t)$ reflects a quiescent neuron in the absence of synaptic activity and $V^0(x,t)$ reflects on the feedback from the synaptically coupled first neuron. The formulation of the system can be expressed in terms of the following equations:

$$C_m U_t^0 = \left(\frac{d_0}{4R_i}\right) U_{xx}^0 - \frac{U^0}{R_{m0}} - \sum_{j=1}^{N} I_{ion}^0(U^0)\delta(x-x_j) + I(t)\delta(x-x_0) - \sum_{z=1}^{N} P^0\delta(x-x_z), \quad t>0, \tag{19.7}$$

$$C_m V_t^1 = \left(\frac{d_1}{4R_i}\right) V_{xx}^1 - \frac{V^1}{R_{m1}} - \sum_{j=1}^{N} I_{ion}^1(V^1)\delta(x-x_j) - \sum_{j=1}^{\infty} \mathrm{H}[t-(2j-1)\Delta t] I_{syn}^1(V^1) - \sum_{z=1}^{N} P^1\delta(x-x_z), \quad t>0, \tag{19.8}$$

$$C_m V_t^0 = \left(\frac{d_0}{4R_i}\right) V_{xx}^0 - \frac{V^0}{R_{m0}} - \sum_{j=1}^{N} I_{ion}^0(V^0)\delta(x-x_j) - \sum_{j=1}^{\infty} \mathrm{H}[t-2j\Delta t] I_{syn}^0(V^0) - \sum_{z=1}^{N} P^0\delta(x-x_z), \quad t>0, \tag{19.9}$$

where $I_{ion}^i (i=0,1)$ is the current density of various ionic channels embedded in the membrane at $x=x_j$, I (mA/cm) is the input current density of the "0" neuron (only) at $x=x_0$ (see Figure 19.1), and Δt is the axonal delay associated with spike propagation from pre- to postsynaptic neurons (sec). Note that the neurons may have different diameters.

For a network of two chemically coupled neurons, each providing M synaptic inputs to the postsynaptic neurons, the synaptic current is

$$I_{syn}^i(x,t;V^i) = \sum_{k=1}^{M} \delta(x-x_k) g^i(t)[V^i(x,t) - V_{rev}^i], \quad i=0,1, \tag{19.10}$$

where $g^i(t)$ represents the synaptic conductance (S/cm) of the ith neuron at a location governed by the kth synapse from the voltage in the presynaptic neuron whose membrane potential varies with space and time and is usually measured at the soma, and V_{rev}^i is the reversal potential of synaptic receptors on the ith dendritic cable.

Equation (19.10) assumes that all the synapses in the model can be located at arbitrary distances from the cell bodies and can generate conductance changes of different values in the postsynaptic cells due to their varying presynaptic voltages. These assumptions are more realistic than those of previous models. Equation (19.10) accounts for various transmitter-gated ionic currents that are activated at synapses. The voltage of the presynaptic cell is always measured at the soma location (cf. Equations [19.4] and [19.5]), and the dimension of $g^i(t)$ reflects a somewhat unrealistic synaptic knob of infinitesimal width encompassing the entire circumference of the dendrite.

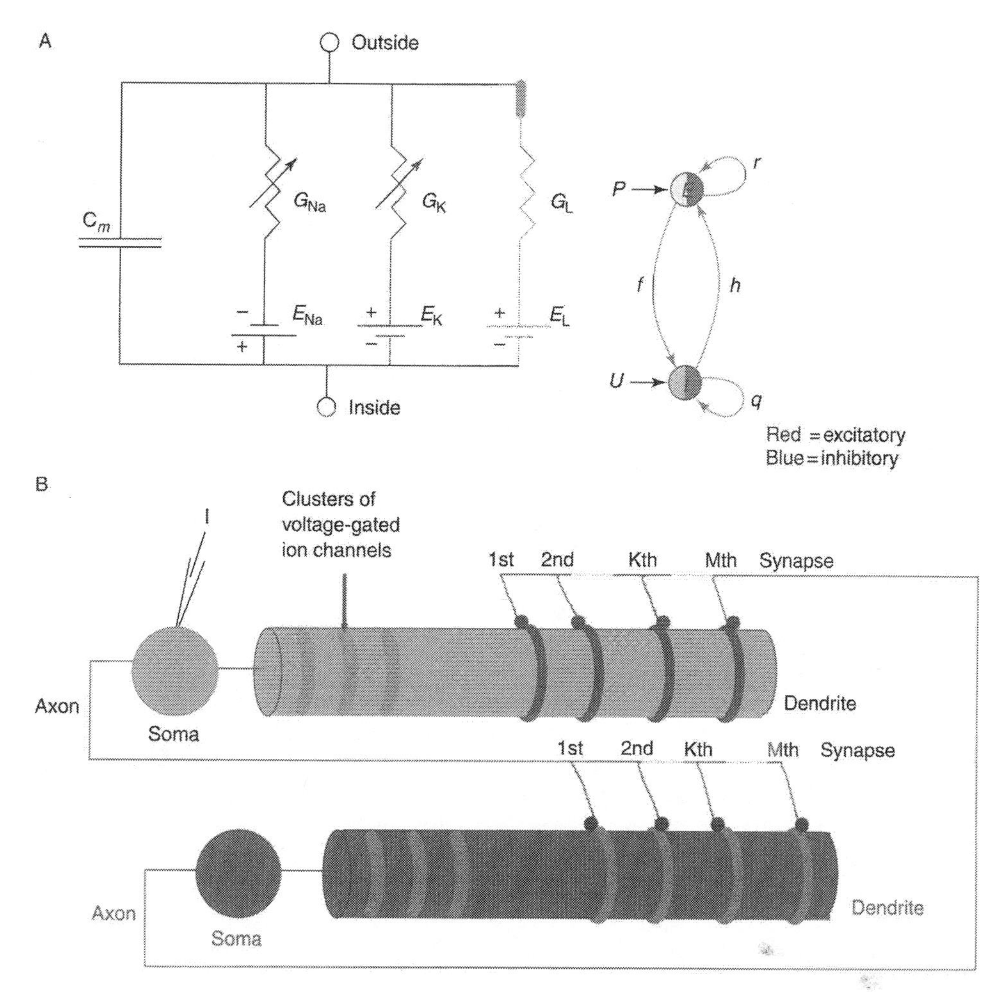

FIGURE 19.1 **(See color insert following page 366)** (A) Schematic illustration of a recurrent network comprising two spiking neurons coupled through synaptic weights. The parameters $\boldsymbol{h}, \boldsymbol{r}, \boldsymbol{f}, \boldsymbol{q}$, represent the strengths of the interactions between the units and $\boldsymbol{P}$, $\boldsymbol{U}$ represent excitatory and inhibitory input respectively. A circuit representation of one conductance-based point neuron is shown. [Kindly provided by Dr. Andrew Gillies, Edinburgh University.] (B) Schematic illustration of a biophysically realistic recurrent network comprising two synaptically coupled neurons. Each neuron is spatially distributed, with a single equivalent ionic cable representing the dendrites and coupled to a lumped isopotential soma. The ionic cable model assumes the cable to be "leaky" (i.e., a passive RC cable) with densities of voltage-dependent ionic channels, referred to as "hot spots," discretely imposed at specific locations along the cable. Nonlinearity is assumed to occur only at these hot spots rather than continuously. The first neuron, "0," is activated by a subthreshold current injection, I, at the soma (represented by the schematic pipette electrode), and the resultant graded electrotonic potential is boosted by persistent Na^+ ionic channels, resulting in neural transmitter release at M axodendritic synapses (chosen at arbitrary locations), which generate synaptic potentials in the second neuron, "1." The same process is repeated from neuron "1" back to neuron "0." (Reproduced from Poznanski, R.R., *J. Integr. Neurosci.*, 1, 69–99, 2002. With permission.)

19.3 Recurrent Dynamics of Two Synaptically Coupled Neurons

19.3.1 Linearization of the Persistent Sodium Current

One way of analytically obtaining the response due to synaptic interactions on ionic cable structures with a finite number of spatially localized ion channels is to linearize the voltage-dependent

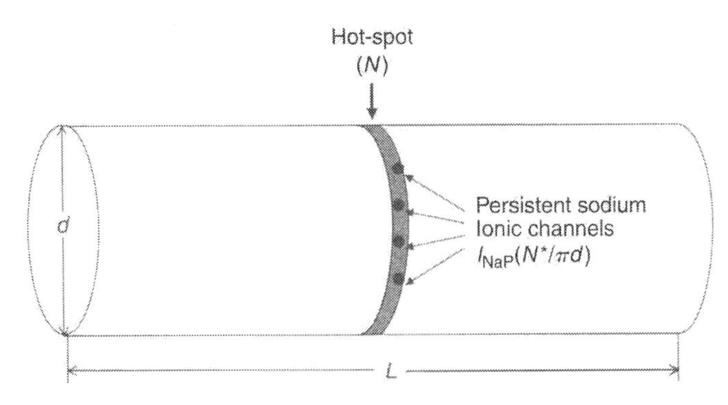

FIGURE 19.2 A schematic illustration of a dendrite of diameter d (cm) studded with clusters or "hot spots" of persistent sodium ionic channels. The arrow above the hot spot reflects the notion of I_{NaP} representing a point source of current applied to an infinitesimal area on the cable. Symbol N denotes the number of hot spots, and N^* denotes the number of persistent sodium channels, represented schematically as dark spots, in each hot spot per unit membrane surface of cable (Modified from Poznanski and Bell, 2000a, 2000b.) (*Source:* From Poznanski, R.R., *J. Integr. Neurosci.*, 1, 69–99, 2002a. With permission.)

ionic currents. This approach was first advocated by Hodgkin and Huxley (1952b) and was further developed by Sabah and Leibovic (1969), who obtained solutions for so-called "graded pulses" or g-pulses in infinite cables by applying linearization theory to the voltage-dependent ionic currents (see Leibovic, 1972 for a review). The linearization of voltage-dependent currents introduces an inductive component that can generate oscillatory responses, but the inclusion of inductive components at discrete locations along the cable, as shown schematically in Figure 19.2, does not alter the cable equation from a parabolic to a wave or a hyperbolic equation. The wave or hyperbolic equation governs oscillatory propagation (cf. Sirovich and Knight, 1977), but in the nerve membrane inductance is negligibly small (see comment made by Rall [1977]).

We assume that the ionic current can be represented solely by a persistent Na^+ inward current density per unit length of cable (mA/cm), described by

$$I^i_{NaP}(x,t;V^i) = g^i_{NaP}m(V^i)[V^i - V^i_{NaP}], \quad i = 0,1, \tag{19.11}$$

with $V^i_{NaP} = E^i_{NaP} - E_r$, $E_r = -70\,\text{mV}$ is the resting membrane potential, $E^i_{NaP} = 65\,\text{mV}$ is the persistent sodium equilibrium potential (mV), and the strength (conductance) of the persistent sodium ion channel densities is given by (see Hodgkin, 1975) $g^i_{NaP} = g^*_{NaP}N^*$, where $N^*/\pi d_i$ is the number of persistent sodium channels per unit membrane surface of cable in cm^{-1}, and g^*_{NaP} is the maximum attainable conductance of a single sodium channel (=18 pS as measured by Sigworth and Neher [1980] and Stuhmer et al. [1987]). However, other voltage-dependent ionic currents could easily be included in the theory.

Linearization of Equation (19.11) yields

$$\begin{aligned} I^i_{NaP} = {} & g^i_{NaP}m_r(V^i - V^i_{NaP}) + \tau_{mr}g^i_{NaP}V^i_{NaP}\exp(-t/\tau_{mr}) \\ & \times \{m_r[(d\alpha_m/dV^i)_r + (d\beta_m/dV^i)_r] - (d\alpha_m/dV^i)_r\}V^i, \qquad i = 0,\ 1, \quad t > 0, \end{aligned} \tag{19.12}$$

where $m_r = \alpha_{mr}/(\alpha_{mr}+\beta_{mr})$, and $\tau_{mr} = 1/(\alpha_{mr}+\beta_{mr})$; α_{mr} and β_{mr} are given by (French et al., 1990):

$$\alpha_{mr} = -1.74(E_r - 11)/\{\exp[-(E_r - 11)/12.94] - 1\}, \tag{19.13}$$

$$\beta_{mr} = 0.06(E_r - 5.9)/\{\exp[(E_r - 5.9)/4.47] - 1\}. \tag{19.14}$$

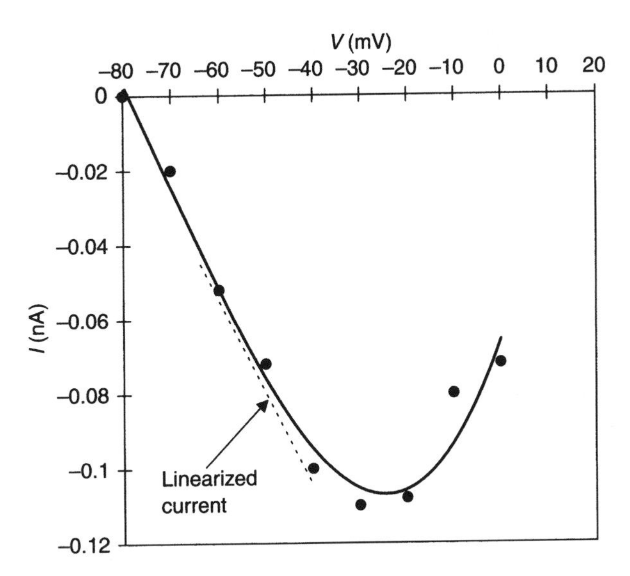

FIGURE 19.3 Input I–V relationship for persistent sodium current in cultured hippocampal pyramidal neurons. Data taken from French et al. (1990). Dashed curve represents a linearized approximation governed by Equation (19.12). (*Source:* From Poznanski, R.R., *J. Integr. Neurosci.*, 1, 69–99, 2002a. With permission.)

Equation (19.12) represents a good approximation before the voltage–current relationship bends upward. Thus, linearization is appropriate for membrane potentials typically under 30 mV, as depicted in Figure 19.3.

In the present situation, the maintenance of zero ionic current at rest (i.e., when $I = 0$) can be accomplished with a sodium pump. The inclusion of an outward sodium-pump current density would effectively remove the $g^{i}_{\mathrm{NaP}} m_{\mathrm{r}} V^{i}_{\mathrm{NaP}}$ term in Equation (19.12), which corresponds to the nondecaying (persistent) level of sodium current, and thus cause the plateau potential to decay with time until it reaches a resting state. Consequently, the current balance equation at the site of the hot spot becomes $I^{i}_{\mathrm{NaP}} - P^{i} = 0$, where $P^{i} = g^{i}_{\mathrm{NaP}} m_{\mathrm{r}} V^{i}_{\mathrm{NaP}}$ is the outward sodium-pump current density.

19.3.2 Electrotonic Potential in the Stimulated Neuron

If the outward sodium-pump current density is positive (by convention), then $P^{0} = g_{\mathrm{NaP}} m_{\mathrm{r}} V^{0}_{\mathrm{NaP}}$. If we assume that the hot-spot points "i" and sodium-pump points "z" are juxtaposed on the cable, and take advantage of linearity and use Green's function methods to solve Equation (19.7) (see Tuckwell, 1988a, p. 191), we can readily obtain a Neumann–Volterra series expansion for the voltage in response to a current injection at the soma in the presence of persistent sodium ion channels at localized points on the cable:

$$U^{0}(x,t) = \int_{0}^{t} \left[I(s) G^{*0}(x,0;t-s) - g^{0}_{\mathrm{NaP}} \sum_{j=1}^{N} \rho^{u0}(s) U^{0}(x_j,s) G^{0}(x,x_j;t-s) \right] \mathrm{d}s, \quad t > 0, \tag{19.15}$$

where

$$\rho^{u0}(t) = m_{\mathrm{r}} - \tau_{m\mathrm{r}} V^{0}_{\mathrm{NaP}} \{ (\mathrm{d}\alpha_m/\mathrm{d}U^{0})_{\mathrm{r}} - m_{\mathrm{r}} [(\mathrm{d}\alpha_m/\mathrm{d}U^{0})_{\mathrm{r}} + (\mathrm{d}\beta_m/\mathrm{d}U^{0})_{\mathrm{r}}] \} \exp(-t/\tau_{m\mathrm{r}})$$

is dimensionless, and the so-called alpha function (Rinzel and Rall, 1974):

$$I(t) = \beta\alpha t \exp(1 - \alpha t)$$

is the applied subthreshold current (mA) with α and β taken as constants. Note that β' is the input charge corresponding to the area found by integrating I, and has a dimension of β per unit time, that is, $\beta' = \beta/\text{sec}$, $\rho^{0\prime}$ has the dimension of ρ^0 per unit time, that is, $\rho^{0\prime} = \rho^0/\text{sec}$ found by integrating ρ^0, and G^{*0} and G^0 are Green's functions corresponding to the solution of the standard linear cable equation in dimensions of ohms and ohms centimeters, respectively (see Appendix).

Equation (19.15) can be solved numerically or analytically using techniques found in Poznanski (1990). Adopting this latter approach, with notation $U^0(x_p, t) \equiv U_p^0(t)$ and $G(x_p, x_0, t - s) \equiv G_{p0}(t - s)$, Equation (19.15) can be rewritten as

$$U_p^0(t) = f_p^0(t) - g_{\text{NaP}}^0 \int_0^t \sum_{j=1}^{N} \aleph_{0pj}(t, s) U_j^0(s)\, \mathrm{d}s, \quad t > 0, \tag{19.16}$$

where

$$f_p^0(t) = \int_0^t \lambda_0^{-1} I(s) G_{p0}^0(t - s)\, \mathrm{d}s \quad \text{and} \quad \aleph_{0pj}(t, s) = G_{pj}^0(t - s)\rho^{u0}(s).$$

The system of Volterra integral equations becomes amenable to the following analytical solution corresponding to the voltage response at $x = x_p$ in response to a current injection at $x = 0$, and hot spots at $x = x_j$:

$$U_p^0(t) = f_p^0(t) - g_{\text{NaP}}^0 \int_0^t \sum_{j=1}^{N} \Re_{pj}(t, s) f_j^0(s)\, \mathrm{d}s, \quad t > 0, \tag{19.17}$$

where the kernels, $\Re_{pj}$, are the sums of the uniformly convergent series

$$\Re_{pj}(t, s) = \aleph_{0pj}(t, s) + \sum_{\nu=1}^{\infty} (g_{\text{NaP}}^0)^\nu (-1)^\nu \aleph_{0pj}^{(\nu)}(t, s) \tag{19.18}$$

and the ν-fold iterated kernels, $\aleph_{0pj}^{(\nu)}$, are defined inductively by the relation

$$\begin{aligned}\aleph_{0pj}^{(\nu)}(t, s) &= \sum_{\gamma=1}^{N} \int_s^t \aleph_{0p\gamma}(t, \xi) \aleph_{0\gamma j}^{(\nu-1)}(\xi, s)\, \mathrm{d}\xi \\ &= \sum_{\gamma=1}^{N} \int_s^t \aleph_{0p\gamma}(t, \xi_1) \aleph_{0\gamma j}(\xi_1, s) \prod_{h=2}^{\nu} \left\{ \int_s^{\xi_{h-1}} \aleph_{0p\gamma}(\xi_{h-1}, \xi_h)\, \mathrm{d}\xi_h \right\} \mathrm{d}\xi_1\end{aligned} \tag{19.19}$$

with

$$\aleph_{0\gamma j}^{(0)} \equiv \aleph_{0\gamma j} \quad \text{and} \quad \aleph_{0pj}^{(1)}(t, s) = \sum_{\gamma=1}^{N} \int_s^t \aleph_{0p\gamma}(t, \xi_1) \aleph_{0\gamma j}(\xi_1, s)\, \mathrm{d}\xi_1.$$

Finally, after some simplification, the following solution is obtained for the electrotonic potential in current-injected neuron "0":

$$U_p^0(t) \approx \int_0^t \lambda_0^{-1} I(s) G_{p0}^0(t-s)\,\mathrm{d}s - g_{\mathrm{NaP}}^0 \int_0^t \sum_{j=1}^{N} \left\{ \int_0^s \lambda_0^{-1} I(\xi_0) G_{j0}^0(s-\xi_0)\,\mathrm{d}\xi_0 \right\}$$
$$\times \left[G_{pj}^0(t-s)\rho^0(s) - g_{\mathrm{NaP}}^0 \sum_{\gamma=1}^{N} \int_s^t G_{p\gamma}^0(t-\xi_1) G_{\gamma j}^0(\xi_1 - s)\rho^{u0}(s)\rho^{u0}(\xi_1)\,\mathrm{d}\xi_1 \right] \mathrm{d}s, \quad t>0. \tag{19.20}$$

19.3.3 Dendritic Potentials in the Stimulated Neuron with Synaptic Feedback

If outward sodium-pump current density is positive by convention, then $P^i = g_{\mathrm{NaP}}^i m_{\mathrm{r}} V_{\mathrm{NaP}}^i$. By taking advantage of linearity and applying the Green's function method of solution (see Tuckwell, 1988b, p. 191) to Equations (19.8) and (19.9), a Volterra series expansion for the voltage in response to synaptic inputs treated as conductance changes impinging at localized points on the cable with localized synaptic conductance changes for specific ions can be readily obtained by linearizing Equation (19.10), that is, $V^i(x_k, t) \ll V_{\mathrm{rev}}^i$, which is a valid approximation when considering the NMDA and AMPA receptor channels. The linearization of synaptic current has been taken into consideration by Amit and Tsodyks (1992), Tuckwell et al. (1996), and Chapeau-Blondeau and Chambert (1995).

Let the membrane potential of the neuron activated by a current injection to the soma be $V^0(x,t)$ and the membrane potential of the synaptically coupled neuron as $V^1(x,t)$. As the first neuron has synaptic feedback from the coupled neuron a similar expression to that for the electrotonic potential can be obtained for the synaptically coupled neurons, as depicted in Figure 19.1, viz.

$$V^0(x,t) = U^0(x,t) - \int_0^t \left[\sum_{y=1}^{\infty} \mathrm{H}[s - 2y\Delta t] \sum_{k=1}^{M} g^0(s) V_{\mathrm{rev}}^0 G^0(x, x_k; t-s) \right.$$
$$\left. + g_{\mathrm{NaP}}^0 \sum_{j=1}^{N} \rho^{v0}(s) V^0(x_j, s) G^0(x, x_j; t-s) \right] \mathrm{d}s, \quad t>0, \tag{19.21}$$

where

$$\rho^{v0}(t) = m_{\mathrm{r}} - \tau_{m\mathrm{r}} V_{\mathrm{NaP}}^0 \{(\mathrm{d}\alpha_m/\mathrm{d}V^0)_{\mathrm{r}} - m_{\mathrm{r}}[(\mathrm{d}\alpha_m/\mathrm{d}V^0)_{\mathrm{r}} + (\mathrm{d}\beta_m/\mathrm{d}V^0)_{\mathrm{r}}]\} \exp(-t/\tau_{m\mathrm{r}})$$

and $g^{0\prime}$ has the dimension of g^0 per unit time, that is, $g^{0\prime} = g^0/\mathrm{sec}$ found from integrating g^0.

The system of Volterra integral equations governed by Equation (19.21) may be rewritten as

$$V_p^0(t) = f_p^0(t) - g_{\mathrm{NaP}}^0 \int_0^t \sum_{j=1}^{N} \aleph_{1pj}(t,s) V_j^0(s)\,\mathrm{d}s, \quad t>0, \tag{19.22}$$

where

$$f_p^0(t) = U_p^0(t) - V_{\mathrm{rev}}^0 \int_0^t \sum_{y=1}^{\infty} \mathrm{H}[s - 2y\Delta t] \sum_{k=1}^{M} g^0(s) G_{pk}^0(t-s)\,\mathrm{d}s$$

and $\aleph_{1pj}(t,s) = G^0_{pj}(t-s)\rho^{v0}(s)$. The above system of Volterra integral equations is amenable to the following analytical solution corresponding to the voltage response at $x = x_p$ in response to current injection at the soma, a composite synaptic input at location $x = x_k$, and hot spots at $x = x_j$:

$$V^0_p(t) = f^0_p(t) - g^0_{\text{NaP}} \int_0^t \sum_{j=1}^{N} \Re_{pj}(t,s) f^0_j(s)\,\mathrm{d}s, \quad t > 0, \tag{19.23}$$

where the resolvent kernels, $\Re_{pj}$, are the sums of the uniformly convergent series given by Equation (19.18) with $\aleph_0$ replaced by $\aleph_i$, and the v-fold iterated kernels, $\aleph^{(v)}_{1pj}$, are defined inductively by the relation given by Equation (19.19) with $\aleph_0$ replaced with $\aleph_1$. After some simplification, the following solution is obtained for the synaptic potentials:

$$\begin{aligned}
V^0_p(t) \approx U^0_p(t) &- V^0_{\text{rev}} \int_0^t \sum_{y=1}^{\infty} \mathrm{H}[s - 2y\Delta t] \sum_{k=1}^{M} g^0(s) G^0_{pk}(t-s)\,\mathrm{d}s \\
&- g^0_{\text{NaP}} \int_0^t \left(\sum_{j=1}^{N} \left\{ U^0_p(s) - V^0_{\text{rev}} \int_0^s \sum_{y=1}^{\infty} \mathrm{H}[\xi_0 - 2y\Delta t] \right.\right. \\
&\left. \times \sum_{k=1}^{M} g^0(\xi_0) G^0_{jk}(s-\xi_0)\,\mathrm{d}\xi_0 \right\} \\
&\times \left[G^0_{pj}(t-s)\rho^{v0}(s) - g^0_{\text{NaP}} \sum_{\gamma=1}^{N} \int_s^t G^0_{p\gamma}(t-\xi_1) G^0_{\gamma j}(\xi_1 - s)\rho^{v0}(s) \right. \\
&\left.\left. \times \rho^{v0}(\xi_1)\,\mathrm{d}\xi_1 \right] \right) \mathrm{d}s, \quad t > 0.
\end{aligned} \tag{19.24}$$

The conductance change $g^0(t)$ at the kth synapse on the stimulated neuron emanating from the coupled neuron was found by solving Equation (19.6) subject to initial condition $g(2y\Delta t) = 0$ (Chapeau-Blondeau and Chambert, 1995):

$$\begin{aligned}
g^0(t) = (g^0_{\text{sat}}\omega/\tau_g) &\int_{2y\Delta t}^{t} C^0(t'') \exp\left[\frac{t'' - 2y\Delta t}{\tau_g} + \left(\frac{\omega}{\tau_g}\right) \int_{2y\Delta t}^{t''} C^0(t')\,\mathrm{d}t' \right] \mathrm{d}t'' \\
&\times \exp\left[\frac{t - 2y\Delta t}{\tau_g} - \left(\frac{\omega}{\tau_g}\right) \int_{2y\Delta t}^{t} C^0(t')\,\mathrm{d}t' \right],
\end{aligned} \tag{19.25}$$

where $C^{0\prime}$ has the dimension of C^0 per unit time, that is, $C^{0\prime} = C^0/\text{sec}$ found by integrating C^0. The concentration of neural transmitter in the synaptic cleft, as a result of spiking activity at the kth synapse on the stimulated neuron emanating from the coupled neuron, was found by substituting Equation (19.4) into Equation (19.5) with initial condition $C^0(2y\Delta t) = C_0$ to yield:

$$C^0(t) = C_0 \exp\left[-\left(\frac{t - 2y\Delta t}{\tau_c} \right) \right] \sum_{k=1}^{N_{\text{sp}}} \mathrm{H}[V^1(t_k + (2y-1)\Delta t) - \theta], \tag{19.26}$$

where t_k is measured from $t = (2y-1)\Delta t$, with $y = 1, 2, \ldots$.

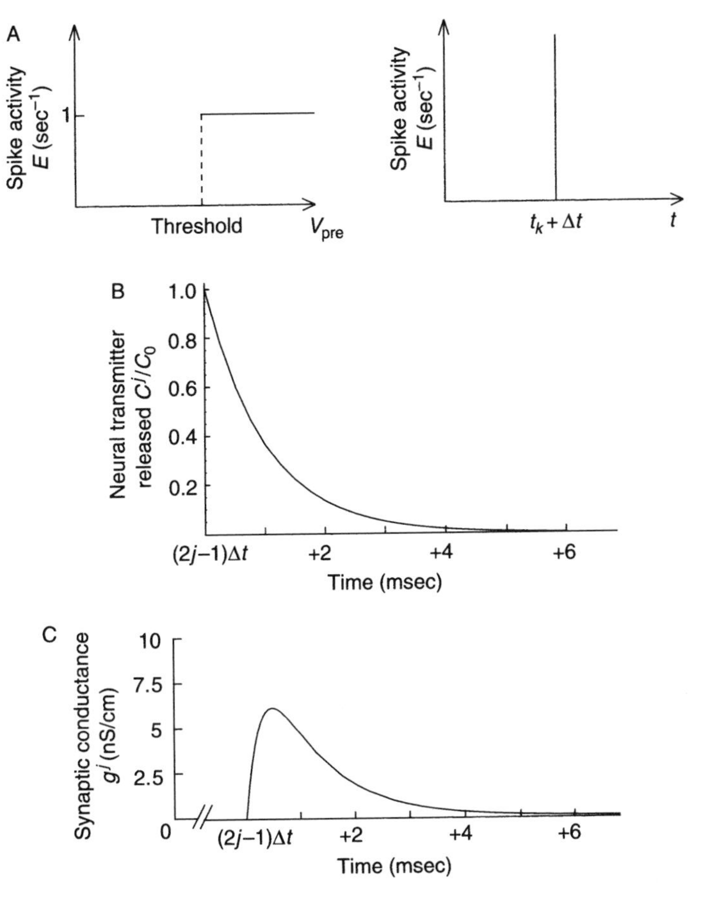

FIGURE 19.4 (A) Spike-train activity is recorded at the synapse every t_k sec if the presynaptic voltage (at a point near the fictitious soma or axon hillock) is above threshold, measured at time intervals of $t_k + \Delta t$ where Δt is the axonal delay. (B) The concentration of neural transmitter in the synaptic cleft as a function of time. (C) The conductance change (nS/cm) resulting at each synapse as a function of time. See Table 19.1 for parameter values used. (*Source*: From Poznanski, R.R., *J. Integr. Neurosci.*, 1, 69–99, 2002a. With permission.)

Figure 19.4 shows typical concentrations of a neural transmitter in the synaptic clefts of all synapses with a slightly more rapid decay for the synapses presynaptic to the second neuron, and the conductance change as a result of the neural transmitter attaching to the receptors in the postsynaptic membrane.

19.3.4 Dendritic Potentials in the Nonstimulated Neuron via Feedforward Synapses

A similar expression results for the second synaptically coupled neuron as outlined in Section 19.3.3:

$$V^1(x,t) = -\int_0^t \sum_{y=1}^{\infty} \mathrm{H}[s-(2y-1)\Delta t]\left[\sum_{k=1}^{M} g^1(s)V_{\mathrm{rev}}^1 G^1(x,x_k;t-s)\right.$$

$$\left. + g_{\mathrm{NaP}}^1 \sum_{j=1}^{N} \rho^1(s)V^1(x_j,s)G^1(x,x_j;t-s)\right] \mathrm{d}s, \qquad (19.27)$$

where

$$\rho^1(t) = m_r - \tau_{mr} V^1_{\text{NaP}}\{(\mathrm{d}\alpha_m/\mathrm{d}V^1)_r - m_r[(\mathrm{d}\alpha_m/\mathrm{d}V^1)_r + (\mathrm{d}\beta_m/\mathrm{d}V^1)_r]\} \exp(-t/\tau_{mr}).$$

Note that $g^{1\prime}$ and $\rho^{1\prime}$ are of dimension g^1 and ρ^1 per unit time, respectively, that is, $g^{1\prime} = g^1/\text{sec}$ and $\rho^{1\prime} = \rho^1/\text{sec}$, found by integrating g^1 and ρ^1, respectively. The systems of Volterra integral equations governed by Equation (19.27) may be rewritten as:

$$V^1_p(t) = f^1_p(t) - g^1_{\text{NaP}} \int_0^t \sum_{j=1}^{N} \aleph_{2pj}(t,s) V^1_j(s)\,\mathrm{d}s, \tag{19.28}$$

where

$$f^1_p(t) = -V^1_{\text{rev}} \int_0^t \sum_{y=1}^{\infty} \mathrm{H}[s - (2y-1)\Delta t] \sum_{k=1}^{M} g^1(s) G^1_{pk}(t-s)\,\mathrm{d}s$$

and

$$\aleph_{2pj}(t,s) = G^1{}_{pj}(t-s)\rho^1(s).$$

The systems of Volterra integral equations are amenable to the following analytical solution corresponding to the voltage response at location $x = x_p$ in response to a synaptic input at location $x = x_k$, and hot spot locations at $x = x_j$

$$V^1_p(t) = f^1_p(t) - g^1_{\text{NaP}} \int_0^t \sum_{j=1}^{N} \Re_{pj}(t,s) f^1_j(s)\,\mathrm{d}s, \quad t > 0. \tag{19.29}$$

where the resolvent kernels, $\Re_{pj}$, are the sums of the uniformly convergent series given by Equation (19.18) with $\aleph_0$ replaced by $\aleph_2$, and the ν-fold iterated kernels $\aleph^{(\mu)}_{2pj}$ are defined inductively by the relation given by Equation (19.19) with $\aleph_0$ replaced with $\aleph_2$. After some simplification the following solution is obtained for the synaptic potentials:

$$\begin{aligned} V^1_p(t) \approx V^1_{\text{rev}}\Bigg(&- \int_0^t \sum_{y=1}^{\infty} \mathrm{H}[s - (2y-1)\Delta t] \sum_{k=1}^{M} g^1(s) G^1_{pk}(t-s)\,\mathrm{d}s \\ &+ g^1_{\text{NaP}} \int_0^t \sum_{j=1}^{N} \left\{ \int_0^s \sum_{y=1}^{\infty} \mathrm{H}[\xi_0 - (2y-1)\Delta t] \sum_{k=1}^{M} g^1(\xi_0) G^1_{jk}(s-\xi_0)\,\mathrm{d}\xi_0 \right\} \\ &\times \Bigg[G^1_{pj}(t-s)\rho^1(s) + g^1_{\text{NaP}} \sum_{\gamma=1}^{N} \int_s^t G^1_{p\gamma}(t-\xi_1) \\ &\times G^1_{\gamma j}(\xi_1 - s)\rho^1(s)\rho^1(\xi_1)\,\mathrm{d}\xi_1 \Bigg] \mathrm{d}s \Bigg), \quad t > 0. \end{aligned} \tag{19.30}$$

Conductance change $g^1(t)$ at the kth synapse on the postsynaptic neuron emanating from the presynaptic neuron was found by solving Equation (19.6) subject to initial condition $g[(2y-1)\Delta t] = 0$

(Chapeau-Blondeau and Chambert, 1995):

$$g^1(t) = (g^1_{\text{sat}}\omega/\tau_g) \int_{(2y-1)\Delta t}^{t} C^1(t'') \exp\left[\frac{t'' - (2y-1)\Delta t}{\tau_g} + \left(\frac{\omega}{\tau_g}\right) \int_{(2y-1)\Delta t}^{t''} C^1(t')\,\mathrm{d}t'\right] \mathrm{d}t''$$
$$\times \exp\left[\frac{t - (2y-1)\Delta t}{\tau_g} - \left(\frac{\omega}{\tau_g}\right) \int_{(2y-1)\Delta t}^{t} C^1(t')\,\mathrm{d}t'\right], \qquad (19.31)$$

where $C^{1\prime}$ has the dimension of C^1 per unit time, that is, $C^{1\prime} = C^1/\text{sec}$ found by integrating C^1.

The amount of neural transmitter in the synaptic cleft at the kth synapse on the postsynaptic neuron emanating from the presynaptic neuron was found by substituting Equation (19.4) into Equation (19.5) with initial condition $C^1[(2y-1)\Delta t] = C_0$ to yield:

$$C^1(t) = C_0 \exp\left[-\left(\frac{t - (2y-1)\Delta t}{\tau_c}\right)\right] \sum_{k=1}^{N_{\text{sp}}} \mathrm{H}[V^0(t_k + 2(y-1)\Delta t) - \theta], \qquad (19.32)$$

where $U^0 \equiv V^0$ for $y = 1$ and t_k is measured from $t = 2(y-1)\Delta t$, with $y = 1, 2, \ldots$.

19.3.5 Illustrative Simulation without Repetitive Firing (i.e., $y = 1$)

In this section we will illustrate the spread of membrane potential in the neural network of Figure 19.1 for both sparse and dense synaptic connectivities between pre- and postsynaptic neurons in the network.

A second-order approximation to Equation (19.20) for the voltage response at $x = x_p$ is

$$U_p^0(t) \approx \int_0^t \lambda_0^{-1} I(s) G_{p0}^0(t-s)\,\mathrm{d}s - g_{\text{NaP}} \int_0^t \left\{ \sum_{j=1}^{N} \rho^{u0}(s) G_{pj}^0(t-s) \int_0^s \lambda_0^{-1} I(\xi_0) G_{j0}^0(s-\xi_0)\,\mathrm{d}\xi_0 \right\}. \qquad (19.33)$$

A second-order approximation to Equation (19.24) for the voltage response at $x = x_p$ with $y = 1$ is

$$V_p^0(t) \approx U_p^0(t) - V_{\text{rev}}^0 \int_{2\Delta t}^{t} \sum_{k=1}^{M} g^0(s) G_{pk}^0(t-s)\,\mathrm{d}s$$
$$+ g_{\text{NaP}}^0 \int_{2\Delta t}^{t} \sum_{j=1}^{N} G_{pj}^0(t-s) \rho^{v0}(s) \int_{2\Delta t}^{s} V_{\text{rev}}^0 \sum_{k=1}^{M} g^0(\xi_0) G_{jk}^0(s-\xi_0)\,\mathrm{d}\xi_0\,\mathrm{d}s$$
$$- g_{\text{NaP}}^0 \int_0^t \sum_{j=1}^{N} \rho^{v0}(s) G_{pj}^0(t-s) U_p^0(s)\,\mathrm{d}s, \quad t > 0, \qquad (19.34)$$

where

$$g^0(t) = (g_{\text{sat}}^0 \omega/\tau_g) \int_{2\Delta t}^{t} C^0(t'') \exp\left[\frac{t'' - 2\Delta t}{\tau_g} + \left(\frac{\omega}{\tau_g}\right) \int_{2\Delta t}^{t''} C^0(t')\,\mathrm{d}t'\right] \mathrm{d}t''$$
$$\times \exp\left[\frac{t - 2\Delta t}{\tau_g} - \left(\frac{\omega}{\tau_g}\right) \int_{2\Delta t}^{t} C^0(t')\,\mathrm{d}t'\right] \qquad (19.35)$$

and

$$C^0(t) = C_0 \exp\left[-\left(\frac{t-2\Delta t}{\tau_c}\right)\right]\sum_{k=1}^{N_{sp}} \mathrm{H}[V^1(t_k+\Delta t)-\theta]. \tag{19.36}$$

A second-order approximation to Equation (19.30) with $y = 1$ is

$$\begin{aligned} V_p^1(t) \approx & V_{rev}^1\Bigg(-\int_{\Delta t}^{t}\sum_{k=1}^{M} g^1(s)G_{pk}^1(t-s)\,\mathrm{d}s \\ & + g_{NaP}^1\int_{\Delta t}^{t}\left\{\sum_{j=1}^{N}\rho^1(s)G_{pj}^1(t-s)\int_{\Delta t}^{s}\sum_{k=1}^{M}G_{jk}^1(s-\xi_0)g^1(\xi_0)\,\mathrm{d}\xi_0\right\}\mathrm{d}s\Bigg), \\ & t > \Delta t, \end{aligned} \tag{19.37}$$

where

$$\begin{aligned} g^1(t) = & (g_{sat}^1\omega/\tau_g)\int_{\Delta t}^{t} C^1(t'')\exp\left[\frac{t''-\Delta t}{\tau_g}+\left(\frac{\omega}{\tau_g}\right)\int_{\Delta t}^{t''}C^1(t')\,\mathrm{d}t'\right]\mathrm{d}t'' \\ & \times \exp\left[\frac{t-\Delta t}{\tau_g}-\left(\frac{\omega}{\tau_g}\right)\int_{\Delta t}^{t}C^1(t')\,\mathrm{d}t'\right] \end{aligned} \tag{19.38}$$

and

$$C^1(t) = C_0 \exp\left[-\left(\frac{t-\Delta t}{\tau_c}\right)\right]\sum_{k=1}^{N_{sp}} H[U^0(t_k)-\theta]. \tag{19.39}$$

Equations (19.33), (19.34), and (19.37) were evaluated using the *Mathematica*™ (Version 3.0) software. The computation required under 60 min on a HITACHI Flora™ 370 workstation with a 400 MHz CPU and 384 MB memory provided the first three terms in the Green's function are used. The first term of the Green's function dominates; the other terms decrease exponentially with time, therefore no significant difference in results occurred by truncating the Green's function. However, at small times ($t < 0.01$) an alternative Green's function should be used that converges more rapidly (see Section 19.3.5). Simulations were done using the parameter values given in Table 19.1 for a nonuniform distribution of hot spots and in the presence of no hot spots (Figure 19.5 and Figure 19.6).

In Figure 19.5 we have illustrated the passive and weakly regenerative backpropagation of electrotonic potential in neuron "0" as a consequence of stimulating the soma by current injection and without synaptic feedback from neuron "1." It is clear that the backpropagating potential falls in amplitude below threshold (i.e., under 10 mV). The distribution of voltage-dependent ion channels in close proximity to the soma yields only a minor increase in the peak amplitude as measured at the terminal end of the cable, suggesting that the axosomatic distribution of voltage-dependent ion channels plays a minor role in dendritic integration.

In Figure 19.6 we have illustrated the formation of dendritic spike-like potentials or excitatory postsynaptic-like potentials (EPSLPs) at a distally placed location along the dendrite of stimulated neuron "0" as a consequence of synaptic feedback from neuron "1" and both passive and weakly regenerative backpropagation of electrotonic potentials. Some important results that have

TABLE 19.1
Parameters for Synaptic Transmission

Parameter	Value	Reference
θ	7–15 mV	Hodgkin and Huxley (1952b)
Δt	6 msec	Kawaguchi and Fukunishi (1998)
τ_k	20 msec	Kawaguchi and Fukunishi (1998)
τ_c	1 msec	Kurogi (1987)
τ_g	1 msec	Chapeau-Blondeau and Chambert (1995)
ω	2 mM^{-1}	Chapeau-Blondeau and Chambert (1995)
C_0	1 mM	Chapeau-Blondeau and Chambert (1995)
g_{max}	20 pS	Hille (2001)
M^*	10–1000	Korn and Faber (1991)
V_{rev}^i	100 mV	Tuckwell et al. (1996)
g_{sat}^0	25.46 μS/cm	Based on $d_0 = 2.5$ μm and $M^* = 1000$
g_{sat}^1	17.2 μS/cm	Based on $d_1 = 3.7$ μm and $M^* = 1000$
N_{sp}	1	A single spike activates transmitter

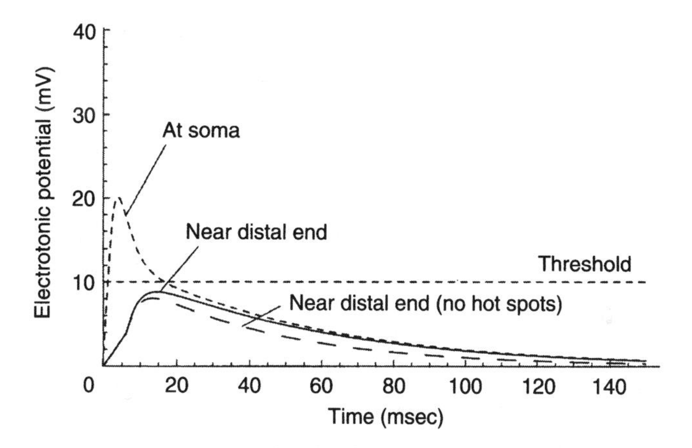

FIGURE 19.5 Electrotonic potentials in neuron "0" measured at $x_p = 0.0$ (dotted line) and $x_p = 0.9L$ (solid line) along the dendritic cable in the presence of persistent sodium ionic channels and at $x_p = 0.9L$ (dashed line) in the absence of persistent sodium hot spots. A current step of magnitude I_0 is injected at the soma (i.e., at $x_p = 0$) with the resultant amplification of electrotonic potentials mediated by $N = 16$ persistent sodium hot spots distributed close to the soma with a constant density of $N^* = 10$. The key parameters were $R_a = 200\,\Omega$ cm, $R_{m0} = 40{,}000\,\Omega\,\text{cm}^2$, $L_0 = 800$ μm, $d_0 = 2.5$ μm, $\lambda_0 = 0.1118$ cm, $g_{NaP}^0 = 0.229$ μS/cm, $\beta = 0.3$ mA, and $\alpha = 0.25$/msec. The values of β and α were selected arbitrarily to yield a response at the soma of approximately 9.0 mV in the absence of persistent sodium hot spots. (*Source:* From Poznanski, R.R., *J. Integr. Neurosci.*, 1, 69–99, 2002a. With permission.)

been observed experimentally clearly indicate the data shown in Figure 19.6. In Figure 19.6A it can be seen that, unlike the results presented in Figure 19.5 where the distal electrotonic potential falls below threshold, the peak amplitude of the synaptic potential as a result of feedback from neuron "1" is well above threshold and produces a "dendritic spike." The effect of back-propagation is to further amplify the dendritic spike from about 28 to 37 mV. As is the case with the data shown in Figure 19.5, the results in Figure 19.6A reveal that hot spots concentrated in close proximity to the soma are of negligible consequence for the membrane potential recorded at the distal tip of neuron "0." However, in Figure 19.6B the voltage-dependent ionic channels are

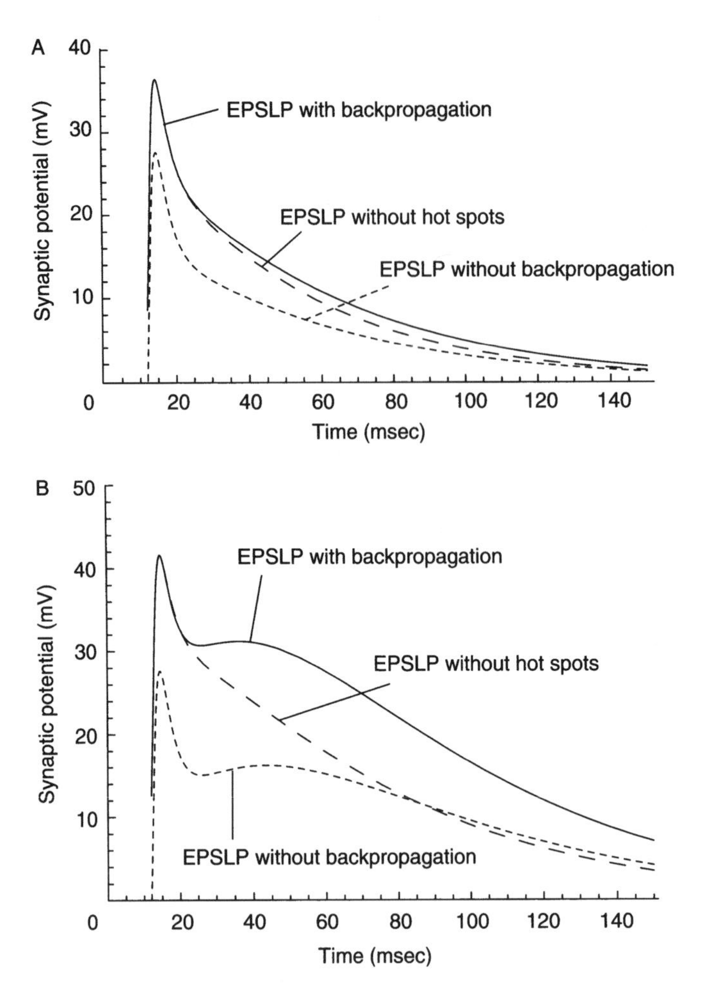

FIGURE 19.6 The effect of boosting on the shape of the dendritic spike-like potentials mediated by persistent sodium hot spots distributed (A) close to the soma, and (B) uniformly along the simulated neuron as a result of synaptic feedback from the nonstimulated neuron "0" with inputs distributed peripherally along the dendritic cable of the stimulated neuron ($M = 24$). The curves are for dendritic spikes at $x_p = 0.9L$ (solid line) and without hot spots (dashed line), and without backpropagating electrotonic potentials (dotted line) for a subthreshold current input at $x_0 = 0$ (soma). The key parameters were $R_a = 200\,\Omega\,\text{cm}$, $R_{m0} = 40{,}000\,\Omega\,\text{cm}^2$, $R_{m1} = 50{,}000\,\Omega\,\text{cm}^2$, $L_0 = 800\,\mu\text{m}$, $L_1 = 1{,}000\,\mu\text{m}$, $d_0 = 2.5\,\mu\text{m}$, $d_1 = 3.7\,\mu\text{m}$, $\lambda_0 = 0.1118\,\text{cm}$, $\lambda_1 = 0.152\,\text{cm}$, $g^0_{\text{NaP}} = 0.229\,\mu\text{S/cm}$, $g^1_{\text{NaP}} = 0.155\,\mu\text{S/cm}$, $\beta = 0.3\,\text{mA}$, and $\alpha = 0.25/\text{msec}$. The values of β and α were selected arbitrarily to yield a response at the soma of approximately 9.0 mV in the absence of sodium persistent hot spots. (*Source:* From Poznanski, R.R., *J. Integr. Neurosci.*, 1, 69–99, 2002a. With permission.)

distributed uniformly throughout the dendritic cable of neuron "0" and neuron "1." The results in this case reveal a further amplification in the peak amplitude of the dendritic spike in comparison with the synaptic potentials seen in Figure 19.6A. In particular, the effect of spatially distributing the persistent sodium channels is to boost the peak amplitude of the dendritic spike to about 42 mV (Figure 19.6B), in comparison to about 37 mV when the persistent sodium channels are located proximal to the soma (Figure 19.6A). It is particularly interesting that the peak amplitude of the synaptic potential measured at the distal end of neuron "0" is left relatively unchanged by changes in the spatial distribution of persistent sodium channels in the case when backpropagation of the signal is not taken into consideration. As shown in both Figure 19.6A and Figure 19.6B the peak amplitude of the dendritic spike remains constant at 28 mV. This result suggests that the spatially distributed voltage-dependent ion channels boost the peak amplitude of the dendritic spikes during their propagation. Another important result that can be seen by comparing the results in Figure 19.6A and Figure 19.6B is that the peak amplitude of the EPSLP without hot spots reaches about 42 mV in Figure 19.6B but only 37 mV for the same EPSLP in Figure 19.6A. What causes this discrepancy in peak amplitude when the only difference in the models generating the results in the two figures is the spatial distribution of voltage-dependent ionic channels? The answer is that the synaptic input driving the EPSLP in Figure 19.6B is marginally stronger than that driving the EPSLP in Figure 19.6A. In other words, the presynaptic voltage drives the kinetics of the synaptic conductance in the postsynaptic neuron. Such effects on synaptic potentiation as a result of network properties is a testament to the power of biophysically realistic neural networks, such as the ones presented here, and suggests that such networks may be useful for the investigation of associative learning principles.

A final observation that is also evident by comparing Figure 19.6A with Figure 19.6B is the fact that the spatial distribution of ionic channels produces a broadening in the time course of the EPSLP both with and without backpropagation. Such broadening is of course absent in EPSLPs without hot spots (dashed line in Figure 19.6B). The broadening in the time course of the dendritic spikes may also play a role in dendritic integration and associative learning.

19.4 Discussion

The need for a spatially continuous dendritic structure as opposed to discretized caricatures can be exemplified through a recent paper by Polsky et al. (2004) showing nonlinear (or superlinear) summation of subthreshold synaptic inputs when located on the same dendritic branch, but linear summation when synaptic inputs are located on more distal dendrites. Although these conclusions can be easily grasped by solving the ionic cable equation without linearization of the synaptic current as used in present simulations, it is the first experimental observation showing nonlinear summation of EPSLPs in distal dendrites of cortical pyramidal neurons. However, the authors argue haphazardly in their paper that such experimental results support a two-layer neural network model of pyramidal neurons as postulated by Poirazi et al. (2003). This claim is fortuitous as it gives pretense to the existence of the "integration zones" or "subunits" commonly discussed in the computational neuroscience literature. Experimental evidence provided by Polsky et al. (2004) does not support the notion of single neurons having any segregating mechanisms allowing for compartmentalization of dendrites to form computational subunits, but simply advocates the importance of the spatial distribution of synaptic inputs in dendritic integration, which is impossible to examine with point neuron models. In order to circumvent stringent criteria for generating computational subunits in dendrites of cortical pyramidal neurons corresponding to a one-to-one mapping between dendrite and subunit, Polsky et al. (2004) advocate a less stringent notion of "sliding window" or "sliding" subunits of less than 100 μm as a basic integrative compartment. However, the experimental results of Polsky et al. (2004) pertaining to a "sliding" subunit are not functionally equivalent to a two-layer model as suggested by Spruston and Kath (2004). Two-layer models assume a rigid subunit structure with a clearly defined "integration zone," yet in the Polsky et al. (2004) paper a more fluid conception with a "sliding" subunit organization is advocated. The latter can be modeled only when spatially continuous dendrites with hot spots of synaptic activity manifest into an artificially transient or "sliding" subunit aurora in the membrane potential. This differs drastically from the rigid "computational" subunits explored in two-layer models, yet is compatible with the methodology pioneered by Poznanski (2002a).

Recently, Izhikevich et al. (2004) have presented a neuronal network model based on an elaborated kind of integrate-and-fire "point" neuron that incorporates receptor kinetics, axonal delays, and short-term as well as spike-timing-dependent long-term synaptic plasticity. Although such an approach is convenient for large-scale simulations of neuronal networks while going well beyond the level of detail in other integrate-and-fire neuron models, it still does not capture the fine detail possible with biophysically realistic continuous-membrane neuronal models. Here are some essential characteristics which we consider that biological networks should possess to be dynamically suited to model the processing of neural information:

- The network has the capacity for adaptability.
- Firing rate or transfer function at the axon hillock is dependent on ionic channels that are modulated by various neurotransmitters, neuromodulators, and second messengers.
- Synaptic weights pertaining to synaptic efficacy of the networks are chemically tuned.
- Functional connectivity is dynamic, that is, changes "fluidly."

The above characteristics can be interpreted as being essential for modeling biophysically realistic neural networks. To obtain them, ionic channels should be included at discrete locations and synaptic weights should be replaced with a conductance changes occurring on the postsynaptic neurons. The notion of "fluidity" of connectivity can be mimicked with the introduction of spatial dimensionality and the movement of synaptic connections along the neuronal structure in response to appropriate electric and chemical signals. Neuromorphic models of networks with each neuron represented by a reduced ionic cable provide three hierarchic levels (subcellular, cellular, and network) of neural organization. Such multilevel models should reproduce neurophysiologic data involving spatiotemporal processing with greater dynamical and functional prowess.

Biophysical mechanisms governing signal processing at the individual neuron level do not succeed in explaining brain function, yet coupled with a view of the brain as a collection of interacting distributed systems of biophysical neural networks; a more powerful approach to the understanding of brain function is possible. Research on how brain function arises from biophysical mechanisms modeled at the neural network level is vital to understanding brain function from an integrative viewpoint. Network models with neuromorphic components are likely to possess greater information processing capabilities than the models studied at present. This leads to the question: what minimal properties of real neurons would be required to construct such networks? If ionic channels, neural transmitters, neural regulators, and neuronal geometry are all crucial factors governing electrical signaling, then the key point is to show how the connectivity among individual neurons brings about a coherent functional response. In other words, a way must be sought to bridge the gap between the various levels of the neural hierarchy, especially between the cellular, network, and systems levels.

19.5 Conclusions and Future Perspectives

In this chapter we have presented an analytical solution for a system of equations describing a model of a recurrent neuronal network. Further work is needed to extend the modeling to large-scale neuronal networks. Hasselmo and Kapur (2001) in their attempt at modeling large-scale neural networks wrote: "realistic models must take advantage of single cell properties beyond what is already simulated in an integrate-and-fire model." In this vein, the methodology presented herein goes beyond the older approaches such as the use of binary neurons to investigate dendritic integration in recurrent networks (Ascoli, 2003) or the population density utilizing "average neurons" (integrate-and-fire model neurons) to simulate large-scale neuronal networks (e.g., Haskell et al., 2001; Alvarez and Vibert, 2002; Toth and Crunelli, 2002; Renart et al., 2004).

Many examples of biological neural processing have been highly optimized through the action of evolution. Appreciation of the underlying algorithms is considered to be a valuable step towards implementing artificial systems of equivalent or better functionality. The opportunities for exploiting biological design in man-made systems have increased greatly with the development of very large-scale integrated (VLSI) circuits in neurochips and other automata design. The possibility of uncovering new design principles can emerge from a program involving realistic modeling. The long-term focus of this research will be the foundation of biologically plausible large-scale neural networks focusing on the inclusion of the relevant biophysical processes, enabling the neurobiology of semantics to capture the essence of what will constitute "neuromimetic robotics" as the final frontier of modeling in the neurosciences.

ACKNOWLEDGMENTS

The author would like to thank G.N. Reeke, R.J. MacGregor S. Hidaka, and G.A. Chauvet for valuable discussions. I am indebted to Imperial College Press for permission to reproduce earlier work published in the *Journal of Integrative Neuroscience*.

PROBLEMS

1. On hippocampal CA1 pyramidal cells, synaptic plasticity (i.e., the strengthening and weakening of synaptic connections) has been shown to involve the movement of ligand-gated α-amino-3-hydroxy-5-methyl-4-isoxazoleproprioniate (AMPA) receptors (Baringa, 1999). Long-term potentiation (LTP) involves a modification to a synapse that depends on a relationship between the level of presynaptic activity and postsynaptic activity. This phenomenon can be accomplished by modifying Equation (2.6) with M^* now a function of presynaptic voltage, that is, $M^*[V\mathrm{pre}(t)]$. Investigate whether LTP is observed in the model.
2. Extend the simulations to include repetitive firing (i.e., $j > 1$) of the presynaptic signals.
3. Consider a large-scale neuronal network consisting of realistic neurons represented by spatially distributed units (or dendritic cables without ionic channels), functionally connected in groups of neural assemblies (see Figure 19.7). Investigate the functional connectivity of such a system by extending the scalar analysis to matrix analysis. Derive an analytical solution in terms of Green's function matrices for a large-scale simulation of approximately 10,000 neurons arranged as a two-dimensional array of cortical columns. Obtain the integrated membrane potentials arising from spiking activity in the dendrites of such simulated neuronal networks. (Hint: Compute $e^{\mathbf{A}t}$ by finding the eigenvalues of $\mathbf{A}t$ where $\mathbf{A}$ is an $n \times n$ diagonal matrix and utilize the corollary given in Tuckwell [1988c]).
4. An extension to a large-scale recurrent neuronal network model will require an extensive use of matrix algebra and Green's function matrices. This type of research is still in its infancy and would represent an enormous leap forward over the models employed today. Determine Green's function matrices for a system of neuronal cables with ionic channels by taking Laplace transforms and obtaining a linear system of coupled ordinary differential equations.

APPENDIX: THE GREEN'S FUNCTION FOR A PASSIVE CABLE

The Green's function for a passive cable, $G^i(x, x_0; t), i = 0, 1$ corresponds to the response at time t and location x to a unit impulse at location x_0 at time $t = 0$ and is given by the solution to the following initial-value problem:

$$C_m G_t^i(x, x_0; t) = (d_i/4R_a) G_{xx}^i(x, x_0; t) - G^i(x, x_0; t)/R_{mi}, \tag{19A.1a}$$

$$G^i(x, x_0; 0) = \delta(x - x_0). \tag{19A.1b}$$

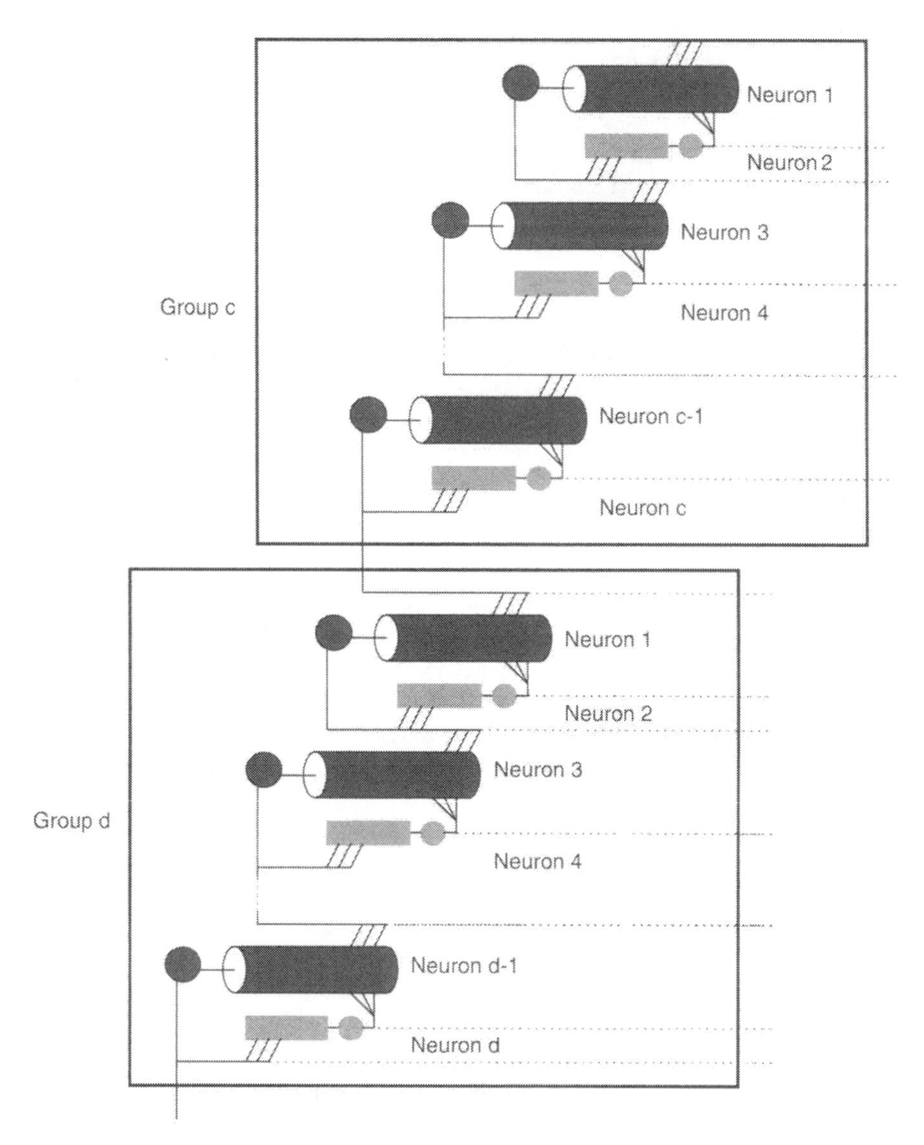

FIGURE 19.7 A schematic illustration showing a large-scale recurrent excitatory network of two neuronal groups located in different cortical areas (cf. Izhikevich et al., 2004). (*Source:* From R.R. Poznanski, ed, *Biophysical Neural Networks*, 2001b. With permission.)

For a single finite cable of length L_i with both ends sealed (i.e., $G_x^i(0, x_0; t) = G_x^i(L_i, x_0; t) = 0$), a representation that converges quickly for large t values was given by Tuckwell (1988a):

$$G^i(x, x_0; t) = (R_{mi}/L_i)\Bigg[\exp(-t/\tau_{mi}) + 2\sum_{n=1}^{\infty}\cos(n\pi x/L_i)\cos(n\pi x_0/L_i) \times \exp[-\{1 + (n\lambda_i\pi/L_i)^2\}(t/\tau_{mi})]\Bigg], \quad t > 0, \tag{19A.2}$$

where $\tau_{mi} = R_{mi}C_m$ and $\lambda_i = (R_{mi}d_i/4R_a)^{1/2} = (r_{mi}/r_a)^{1/2}$. An alternative representation that converges more quickly at small t values was also given by Tuckwell (1988a):

$$G^i(x, x_0; t) = (R_{mi}/\lambda_i)\exp(-t/\tau_{mi})\sqrt{(\tau_{mi}/4\pi t)}$$
$$\times \Bigg\{ \exp[-\tau_{mi}(x - x_0)^2/4\lambda_i^2 t] + \exp[-\tau_{mi}(x + x_0)^2/4\lambda_i^2 t]$$
$$+ \sum_{n=1}^{\infty}[\exp[-\tau_{mi}(x - 2nL_i - x_0)^2/4\lambda_i^2 t] + \exp[-\tau_{mi}(x - 2nL_i + x_0)^2/4\lambda_i^2 t]$$
$$+ \exp[-\tau_{mi}(x + 2nL_i - x_0)^2/4\lambda_i^2 t] + \exp[-\tau_{mi}(x + 2nL_i + x_0)^2/4\lambda_i^2 t]] \Bigg\}. \tag{19A.3}$$

For a single finite cable of length L_i with a lumped soma attached to the end at $x = 0$ (i.e., $G^i(0, x_0, t) + G_t^i(0, x_0, t) - \rho G_x^i(0, x_0, t) = 0$, where ρ is the dendritic-to-somatic conductance ratio), and with a sealed end at $x = L_i$ (i.e., $G_x^i(L_i, x_0, t) = 0$), a representation that converges quickly for large t values was also given by Tuckwell (1988a):

$$G^i(x, x_0; t) = (R_{mi}/\lambda_i)\exp(-t/\tau_{mi})\sum_{n=0}^{\infty}\varphi_n^i(x)A_n^i(x_0)\exp[-(\psi_n^{i2}t/\tau_{mi})], \quad t > 0, \tag{19A.4}$$

where $\varphi_n^i(x) = \cos(\psi_n^i x/\lambda_i) - (\psi_n^i/\rho)\sin(\psi_n^i x/\lambda_i)$. The eigenvalues, ψ_n^i, are the roots of $\rho\tan(\psi_n^i L_i/\lambda_i) + \psi_n^i = 0$, and the coefficients are $A_n^i(x_0) = 2\rho\cos\theta_n/[\sin\psi_n^i L_i/\lambda_i\{(\rho/\psi_n^i - \psi_n^i L_i/\lambda_i) + (\rho/\lambda_i)(x_0 - L_i)\tan\theta_n\} + \cos\psi_n^i L_i/\lambda_i\{(2 + \rho L_i/\lambda_i) + \theta_n\tan\theta_n\}]$ and $A_0^i(x_0) = \rho/(1 + \rho L_i/\lambda_i)$, where $\theta_n^i = (x_0 - L_i)/\lambda^i\psi_n^i$ and $n = 1, 2, \ldots$. A representation that converges quickly for small t values can be found in Tuckwell (1988a).

For current injection at the end of the cable or at the soma, as opposed to injection along the cable, which reflects current density per unit length, it can be shown that the Green's function (G^{*i}) is simply $\lambda_i^{-1}G^i$, where G^i corresponds to the Green's function defined in (19A.2)–(19A.4).

20 Spike-Train Analysis for Neural Systems

David M. Halliday

CONTENTS

20.1 Introduction

The human central nervous system processes vast amounts of sensory information, most of which is conveyed in the form of neural spike trains. For example, as we walk along an uneven surface we are sensitive to the undulations that we experience as changes of tension in the muscles and tendons of our lower limbs. The bioengineer would describe this situation using the language of forces, torques, and moments. The neuroscientist who wishes to study how the brain and spinal cord represent and process this ever-changing sensory input must use the language of spikes. For our purposes, spikes may be considered to be stereotyped all-or-none electrical events generated by sensory receptors and neurons. In most cases, spikes will propagate down an axon, using a regenerative active process, and exert an influence on a postsynaptic neuron. Spikes are very much part of the language of computational neuroscience, particularly studies of neural coding. One of the challenges in computational neuroscience is how information can be extracted from sequences of spike timings (e.g., see Rieke et al., 1997, chapter 1).

This chapter is about the analysis of spike-train data, in particular how to build up a picture of the relationship between several simultaneously observed spike trains. The approach used is that of linear and nonlinear correlation analysis, undertaken in both the time and frequency domains. Our hypothesis is that the presence of correlated activity between the discharges of two or more neurons provides an indicator of some common functional role for the cells.

The reference points and antecedents of the work in this chapter are the early pioneering studies of Gerstein, Moore, Perkel, and coworkers (Perkel et al., 1967a, 1967b; Moore et al., 1970) on spike-train analysis, and the formal development of a multivariate point-process framework (Brillinger, 1972, 1975; Rosenberg et al., 1989). The framework presented in Section 20.2 makes extensive use of the Fourier transform for the definition and estimation of parameters. An important aspect of the framework is the ability to set confidence limits on all parameter estimates, as these are an essential part of any rigorous statistical analysis. In this context, a key result is that the large-sample statistical properties of the finite Fourier transform of a stochastic point process are simpler than those of the underlying process that generates the signal (Brillinger, 1974).

The chapter is arranged in six sections. Section 20.2 describes linear and nonlinear time-domain analysis of spike-train data; Section 20.3 describes linear and nonlinear frequency-domain analysis. The methods outlined in Sections 20.2 and 20.3 are illustrated through their application to data generated by a simulation of a 2D sheet of cortical neurons. The cortical neural network simulation is described in Section 20.4. Section 20.5 is the results section, which describes and illustrates the application of the methods in Sections 20.2 and 20.3 to the data generated by the simulation described in Section 20.4. Section 20.6 contains some concluding remarks.

20.2 Linear and Nonlinear Time-Domain Point-Process Analysis

In this section we consider the definition and estimation of linear and nonlinear point-process parameters that can be used to characterize interactions between neuronal spike trains. This section deals solely with time-domain measures; frequency-domain techniques and the relationship between the time and frequency domains are considered in the next section.

The quantities normally available for analysis of spike-train data are the firing times from one or more neurons. The duration of action potentials is short compared with the spacing between them, thus the sequences of spike times can be considered as realizations of stochastic point processes. A stochastic point process is defined formally as a random nonnegative integer-valued measure (Brillinger, 1978). This point process represents the ordered times of occurrence of spikes (or events) as integer multiples of the sampling interval, $\mathrm{d}t$. In practice, $\mathrm{d}t$ should be chosen sufficiently small that at most one event occurs in any interval. A point process that satisfies this condition is described as *orderly*.

The present analysis assumes that the point-process data are weakly stationary, that is, the parameters that characterize the data do not change with time, and widely spaced differential increments are effectively independent. This latter is known as a mixing condition. Discussion of these assumptions can be found in Cox and Isham (1980), Cox and Lewis (1972), and Daley and Vere-Jones (1988), and in relation to neuronal spike trains in Conway et al. (1993). The assumption of orderliness is important, since it allows certain point-process parameters to be interpreted as probabilities (Cox and Lewis, 1972; Srinivasan, 1974; Brillinger, 1975).

For a point process, denoted by N_1, the counting variate $N_1(t)$ counts the number of events in the interval $(0, t]$. An important elementary function of point processes is the differential increment. The differential increment for process N_1 is denoted by $\mathrm{d}N_1(t)$, and defined as $\mathrm{d}N_1(t) = N_1(t+\mathrm{d}t)-N_1(t)$. It can be considered as a counting variate that counts the number of events in a small interval of duration $\mathrm{d}t$ starting at time t. For an orderly point process, $\mathrm{d}N_1(t)$ will take on the value 0 or 1 depending on the occurrence of a spike in the sampling interval $\mathrm{d}t$. Other point-process parameters may be defined in terms of the differential increments (Brillinger, 1975, 1976; Rosenberg et al., 1982, 1989; Conway et al., 1993).

First-order point-process statistics include the mean intensity, P_1, of the point process N_1 defined as

$$E\{\mathrm{d}N_1(t)\} = P_1\mathrm{d}t, \tag{20.1}$$

where E{ } denotes the mathematical expectation of a random variable. The assumption of orderliness allows expression (20.1) to be interpreted in a probabilistic manner as

$$\mathrm{Prob}\{N_1 \text{ event in}(t, t+\mathrm{d}t]\}. \tag{20.2}$$

The mean intensity, P_1, for a sample of duration R from the point process N_1 can be estimated as

$$\hat{P}_1 = \frac{N_1(R)}{R}, \tag{20.3}$$

where R is the number of sample points in the record. Throughout this chapter, we use the hat symbol to indicate an estimate.

Second-order point-process statistics include product- and cumulant-density functions. The product density between the two point processes N_0 and N_1, $P_{10}(u)$, at lag u is defined as

$$E\{\mathrm{d}N_1(t+u)\,\mathrm{d}N_0(t)\} = P_{10}(u)\,\mathrm{d}u\,\mathrm{d}t. \tag{20.4}$$

This expression can be interpreted as

$$\mathrm{Prob}\{N_1 \text{event in} (t+u, t+u+\mathrm{d}u] \text{ and } N_0 \text{ event in} (t, t+\mathrm{d}t]\}. \tag{20.5}$$

The second-order product-density functions, $P_{00}(u)$ and $P_{11}(u)$, are defined as in Equation (20.4) by equating N_0 and N_1. From the mixing condition, the differential increments $\mathrm{d}N_1(t+u)$ and $\mathrm{d}N_0(t)$ become independent as u becomes large. Therefore we can write

$$\lim_{|u|\to\infty} P_{10}(u) = P_1 P_0. \tag{20.6}$$

The cumulant (or cross-covariance) function, $q_{10}(u)$, is defined as

$$q_{10}(u)\,\mathrm{d}u\,\mathrm{d}t = E\{(\mathrm{d}N_1(t+u) - P_1\,\mathrm{d}u)(\mathrm{d}N_0(t) - P_0\,\mathrm{d}t)\}. \tag{20.7}$$

The cumulant-density functions, $q_{00}(u)$ and $q_{11}(u)$, are defined similarly by using the appropriate differential increments and mean rates. The cumulant density is related to the product density according to the relation

$$q_{10}(u) = P_{10}(u) - P_1 P_0. \tag{20.8}$$

Equation (20.6) shows that the cumulant has an expected value of zero for independent differential increments, that is, $\lim_{|u|\to\infty} q_{10}(u) = 0$.

Estimates of the second-order product densities defined above can be constructed by the following procedure. If we denote the set of spike times for N_0 as $\{r_i; i = 1, \ldots, N_0(R)\}$ and the set of spike times for N_1 as $\{s_j; j = 1, \ldots, N_1(R)\}$, we can construct a counting variate $J_{10}^R(u)$ (Griffith and Horn, 1963; Cox, 1965)

$$J_{10}^R(u) = \#\left\{(r_i, s_j), \quad \text{such that} \quad \left(u - \frac{b}{2}\right) < (s_j - r_i) < \left(u + \frac{b}{2}\right)\right\}, \tag{20.9}$$

where $\#\{A\}$ indicates the number of events in set A. The variate $J_{10}^R(u)$ counts the number of occurrences of N_1 events falling in a bin of width b, whose midpoint is u time units away from an N_0 event. The parameter b is the bin width of the estimate, $b \geq 1.0$; this is normally 1.0. In the neurophysiological literature, the variate $J_{10}^R(u)$ is often called the cross-correlation histogram. The expected value of this variate is (Cox, 1965; Cox and Lewis, 1972)

$$E\{J_{10}^R(u)\} \approx bRP_{10}(u). \tag{20.10}$$

Equation 20.10 makes precise the relationship between the cross-correlation histogram and the point-process formalism we are using. The product density can be estimated as

$$\hat{P}_{10}(u) = \frac{J_{10}^R(u)}{bR}. \tag{20.11}$$

The cumulant density, $q_{10}(u)$, can be estimated from Equation (20.8) as

$$\hat{q}_{10}(u) = \hat{P}_{10}(u) - \hat{P}_1\hat{P}_0 = J_{10}^R(u)/bR - \hat{P}_1\hat{P}_0. \tag{20.12}$$

The large-sample properties of the above estimate of $P_{10}(u)$ are considered in Brillinger (1976), where it is shown that the estimate of Equation (20.11) is approximately Poissonian, which implies it can be approximated by a normal distribution $N\{P_{10}(u), P_{10}(u)/bR\}$. Since the variance of the estimate depends on the value of the parameter, Brillinger (1976) proposes a square-root variance-stabilizing transformation

$$\hat{P}_{10}(u)^{1/2} \approx N\{P_{10}(u)^{1/2}, (4bR)^{-1}\}. \tag{20.13}$$

The asymptotic distribution and 95% confidence limits for $(\hat{P}_{10}(u))^{1/2}$ are then

$$(\hat{P}_1\hat{P}_0)^{1/2} \pm 1.96(4bR)^{-1/2}. \tag{20.14}$$

Estimated values lying inside the upper and lower confidence limits can be interpreted as evidence of uncorrelated spike trains.

The asymptotic distribution of the estimated cumulant density, $\hat{q}_{10}(u)$, is (Rigas, 1983)

$$\mathrm{var}\{\hat{q}_{10}(u)\} \approx \frac{2\pi}{R}\int_{-\pi/b}^{\pi/b} f_{11}(\lambda)f_{00}(\lambda)\,\mathrm{d}\lambda + O(R^{-1}). \tag{20.15}$$

The quantities $f_{11}(\lambda)$ and $f_{00}(\lambda)$ are the auto spectra of processes N_1 and N_0, respectively, considered in Section 20.3. Here we propose a simplified expression for the estimation of confidence limits. Under the assumption of Poisson spike trains, the variance in Equation (20.15) can be approximated by (P_1P_0/Rb) (for a more detailed discussion of this simplified method, see Section 20.3). Therefore, for two independent spike trains, an asymptotic value and upper and lower confidence limits for the estimate, Equation (20.12), of $\hat{q}_{10}(u)$ is

$$0 \pm 1.96(\hat{P}_1\hat{P}_0/Rb)^{1/2}. \tag{20.16}$$

Once the cross-correlation histogram, $J_{10}^R(u)$, has been obtained, the product density and cumulant density can be readily estimated, and confidence limits set, using simple expressions that depend only on the quantities b, R, $N_0(R)$, and $N_1(R)$.

Third-order point-process parameters allow nonlinear dependencies between spike trains to be characterized within the point-process formulation. These are defined and estimated by extension of the equivalent second-order measures. In the following equation we assume we have three point processes, N_0, N_1, and N_2 with mean rates, P_0, P_1, and P_2, respectively. The third-order product density is defined as

$$E\{\mathrm{d}N_2(t+u)\,\mathrm{d}N_1(t+v)\,\mathrm{d}N_0(t)\} = P_{210}(u,v)\,\mathrm{d}u\,\mathrm{d}v\,\mathrm{d}t. \tag{20.17}$$

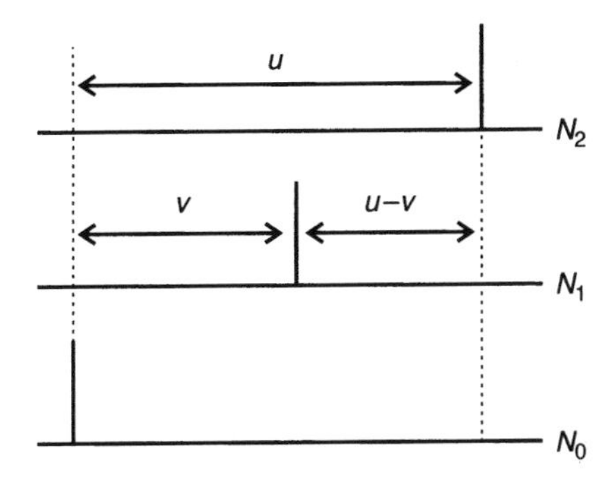

FIGURE 20.1 Timing convention used in third-order analysis between three point processes, N_0, N_1, and N_2. The thick vertical lines indicate the positions of spikes in each point process.

This may be interpreted as

$$\begin{aligned}&\text{Prob}\{N_2 \text{ event in}(t+u, t+u+\mathrm{d}u] \quad \text{and} \quad N_1 \text{ event in}(t+v, t+v+\mathrm{d}v]\\ &\quad \text{and} \quad N_0 \text{ event in}(t, t+\mathrm{d}t]\},\end{aligned} \tag{20.18}$$

providing a probabilistic measure related to the occurrence of an N_2 event u time units after an N_0 event and an N_1 event v time units after an N_0 event. The third-order cumulant density between the three processes is defined as

$$q_{210}(u)\,\mathrm{d}u\,\mathrm{d}v\,\mathrm{d}t = E\{(\mathrm{d}N_2(t+u) - P_2\,\mathrm{d}u)(\mathrm{d}N_1(t+v) - P_1\,\mathrm{d}v)\,(\mathrm{d}N_0(t) - P_0\,\mathrm{d}t)\}. \tag{20.19}$$

By applying the rules for expectation as above, this can be expanded to give the expression

$$q_{210}(u,v) = P_{210}(u,v) - P_{21}(u-v)P_0 - P_{20}(u)P_1 - P_{10}(v)P_2 + 2P_2P_1P_0. \tag{20.20}$$

The timing convention represented by this cumulant is illustrated in Figure 20.1. In some situations it is convenient to represent the cumulant as a function of time lags u and $(u - v)$. This function is defined as

$$q_{210}(u,u-v) = P_{210}(u,u-v) - P_{21}(u-v)P_0 - P_{20}(u)P_1 - P_{10}(v)P_2 + 2P_2P_1P_0. \tag{20.21}$$

The lower-order contributions are the same in each case. The interpretation of product densities as probability descriptions allows us to provide the following heuristic interpretation of the third-order cumulant densities as defined in Equations (20.20) and (20.21): All the terms on the RHS are concerned with triplets of spikes with the timing convention illustrated in Figure 20.1. The first term, $P_{210}(u, v)$, represents the joint probability of occurrence of spikes with this particular timing. The following three terms subtract the contributions to this joint probability that result when the timings of spikes from two of the three processes are correlated, but are independent of the third process. For example, the product $P_{21}(u - v)P_0$ subtracts the contribution that would result from correlated N_2 and N_1 spikes with an N_0 spike occurring by chance, with timing as illustrated in Figure 20.1. The final term on the RHS represents the situation that the three spike trains consist of independent increments. Since the contribution of three independent processes is subtracted three times by the three terms with the second-order product densities, the final term required is a corrective addition.

Estimates of the third-order product densities can be constructed in a similar manner to those for second order from the two counting variates

$$J^R_{210}(u,v) = \#\left\{(r_i, s_j, t_k), \quad \text{such that} \quad \left(u - \frac{b}{2}\right) < (t_k - r_i) < \left(u + \frac{b}{2}\right), \right.$$
$$\left. \text{and} \quad \left(v - \frac{b}{2}\right) < (s_j - r_i) < \left(v + \frac{b}{2}\right)\right\}, \tag{20.22}$$

$$J^R_{210}(u,u-v) = \#\left\{(r_i, s_j, t_k), \quad \text{such that} \quad \left(u - \frac{b}{2}\right) < (t_k - r_i) < \left(u + \frac{b}{2}\right), \right.$$
$$\left. \text{and} \quad \left(u - v - \frac{b}{2}\right) < (t_k - s_j) < \left(u - v + \frac{b}{2}\right)\right\}, \tag{20.23}$$

where the convention for the N_0 and N_1 spikes is as above, and the set of spike times for N_2 is defined as $\{t_k; k = 1, \ldots, N_2(R)\}$. The variate $J^R_{210}(u,v)$ counts the number of triplets of spikes where N_2 and N_0 events are separated by lag u, and N_1 and N_0 events are separated by lag v. In this variate the N_0 event is common to both lags. The variate $J^R_{210}(u,u-v)$ counts the number of triplets of spikes where N_2 and N_0 events are separated by lag u, and N_2 and N_1 events are separated by lag $u-v$; in this case the N_2 event is common to both lags. The third-order product densities, $P_{210}(u,v)$ and $P_{210}(u,u-v)$ can be estimated as

$$\hat{P}_{210}(u,v) = \frac{J^R_{210}(u,v)}{b^2 R}, \tag{20.24}$$

$$\hat{P}_{210}(u,u-v) = \frac{J^R_{210}(u,u-v)}{b^2 R}. \tag{20.25}$$

The third-order cumulants, $q_{210}(u,v)$ and $q_{210}(u,u-v)$, can be estimated by substituting the appropriate first-, second-, and third-order product density estimates, Equations (20.3), (20.11), (20.22), and (20.23) into Equations (20.20) and (20.21).

The variance of estimates of the third-order cumulant, $q_{210}(u,v)$, under the assumption of three independent processes, is given by Conway et al. (1993)

$$\text{var}\{\hat{q}_{210}(u,v)\} \approx \frac{2\pi}{R} \int_{-\pi/b}^{\pi/b} \int_{-\pi/b}^{\pi/b} f_{22}(\lambda) f_{11}(\mu) \overline{f_{00}(\lambda+\mu)}\, d\lambda\, d\mu + O(R^{-1}), \tag{20.26}$$

where R is the record length, $f_{22}(\lambda)$, $f_{11}(\mu)$, and $f_{00}(\lambda+\mu)$ are the auto spectra of the component processes, and b is the bin width of the estimate, $b \geq 1.0$. Direct substitution of the appropriate spectral estimates can be used to estimate this. In the case of $b = 1.0$ this gives an expression of the form

$$\text{var}\{\hat{q}_{210}(u,v)\} \approx \left(\frac{2\pi}{R}\right)\left(\frac{2\pi}{T}\right)^2 \sum_{i=1}^{T-1} \sum_{j=1}^{T-1} \hat{f}_{22}(\lambda_i) \hat{f}_{11}(\lambda_j) \hat{f}_{00}(\lambda_k), \tag{20.27}$$

where $\lambda_i = 2\pi i/T, \lambda_j = 2\pi j/T, \lambda_k = 2\pi k/T, (k = i + j, \text{mod} T)$, R is the record length, and T the segment length used to form the spectral estimates (see Section 20.3). As in the second-order case a simplified procedure can be adopted by substituting the expected value for the spectrum of a point process into the above equations. This leads to the expression $\text{var}\{\hat{q}_{210}(u,v)\} \approx (P_2 P_1 P_0 / R b^2)$.

Thus the asymptotic value and confidence limits for an estimate of the third-order cumulant can be set as

$$0 \pm 1.96(\hat{P}_2\hat{P}_1\hat{P}_0/Rb^2)^{1/2}. \tag{20.28}$$

The above expressions for the definition and estimation of third-order cumulants are valid for other combination of point-process data. For example, the cumulant $q_{211}(u, u-v)$ can be defined as

$$q_{211}(u,u-v) = P_{211}(u,u-v) - P_{21}(u-v)P_1 - P_{21}(u)P_1 - P_{11}(v)P_2 + 2P_2P_1^2; \qquad v \neq 0. \tag{20.29}$$

The condition imposed on the v lag, $v \neq 0$, results from the singularity in $P_{11}(v)$ at zero lag. This will need to be removed in any estimate of $q_{211}(u, u-v)$. Singularities may also be present at lags $u = 0$, and $(u-v) = 0$ depending on the particular case being studied. For example, both $q_{111}(u, u-v)$ and $q_{111}(u, v)$ would require the conditions $u \neq 0$; $v \neq 0$; $(u-v) \neq 0$ to be imposed on the definition, and the singularities would need to be removed from any estimate.

20.3 Linear and Nonlinear Frequency-Domain Point-Process Analysis

In this section we describe a complementary frequency-domain approach to the analysis of neural spike-train data. Where appropriate, we highlight the relationship with the time-domain parameters defined in the previous section. In all cases the equivalence is through the use of a Fourier transform. The Fourier transform is central to the frequency-domain approach; all the parameters defined in this section can be estimated by averaging across the finite Fourier transforms of short segments of point-process data. In calculating the finite Fourier transform, we assume that our record R consists of L nonoverlapping segments of length T, $R = LT$. The finite Fourier transform of a segment of length T from point process N_1 is defined as

$$\mathrm{d}_{N_1}^T(\lambda_j) \approx \sum_{t=0}^{T-1} \mathrm{e}^{-i\lambda_j t}\,\mathrm{d}N_1(t). \tag{20.30}$$

The λ_j are the Fourier frequencies, $\lambda_j = 2\pi j/T$, $j = 0, \ldots T/2$. This defines a range for λ_j of $(0, \pi)$, where π is the Nyquist frequency. In hertz the Fourier frequencies are $(j/T\,\mathrm{d}t)$, where T is the number of points in the segment, and $\mathrm{d}t$ is the sampling interval (s). In practice, $\mathrm{d}_{N_1}^T(\lambda_j)$ can be readily calculated using a fast-Fourier transform (FFT) algorithm to transform the sequence of differential increments. As for time-series analysis, it is usual to subtract the mean of the sequence prior to applying the transform. We will use the notation $\mathrm{d}_{N_1}^T(\lambda, l)$ to represent the finite Fourier transform of segment number l, $l = 1, \ldots, L$.

The second-order spectrum between point processes N_0 and N_1 can be estimated as

$$\hat{f}_{10}(\lambda) = \frac{1}{2\pi LT}\sum_{l=1}^{L} \mathrm{d}_{N_1}^T(\lambda, l)\,\overline{\mathrm{d}_{N_0}^T(\lambda, l)}. \tag{20.31}$$

This estimation procedure follows from the definition of the spectrum in terms of the expectation $f_{10}(\lambda) = \lim_{T\to\infty}(1/2\pi T)E\{\mathrm{d}_{N_1}^T(\lambda)\,\overline{\mathrm{d}_{N_0}^T(\lambda)}\}$. The auto spectrum of the process N_1, $f_{11}(\lambda)$, is defined and estimated by replacing N_0 with N_1 in the above equations. The auto spectrum of a point process is a useful quantity since it provides an indication of any departures from a Poisson process and can indicate the presence of distinct rhythmic components or rhythmic modulation of a spike-train discharge. The definition of the point-process spectrum in terms of the Fourier transform of the cumulant density is (Bartlett, 1963)

$$f_{11}(\lambda) = \frac{P_1}{2\pi} + \frac{1}{2\pi}\int_{-\infty}^{\infty} q_{11}(u)\,\mathrm{e}^{-i\lambda u}\,\mathrm{d}u, \tag{20.32}$$

where the additional term, $P_1/2\pi$, arises from the inclusion of the Dirac delta function at zero lag in the cumulant for a point process. For u large, $q_{11}(u)$ will tend to zero, which gives the asymptotic distribution of $f_{11}(\lambda)$ as $\lim_{\lambda\to\infty} f_{11}(\lambda) = P_1/2\pi$. For a Poisson spike train, $q_{11}(u)$ will be zero, thus the hypothesis that a spike train represents a sample from a Poisson process can be tested by looking for departures from the asymptotic value. This approximation underlies the simplified expressions derived for the confidence limits of cumulant-density estimates in Equations (20.16) and (20.28).

The variance of the auto spectrum can be approximated by $\mathrm{var}\{\hat{f}_{11}(\lambda)\} \approx L^{-1}(f_{11}(\lambda))^2$ (Brillinger, 1972, 1981; Bloomfield, 2000), where L is the number of disjoint sections in the spectral estimate. An appropriate variance-stabilizing transform is the natural log, giving $\mathrm{var}\{\ln(\hat{f}_{11}(\lambda))\} \approx L^{-1}$. It is customary practice to plot spectra on a $\log_{10}$ scale, giving

$$\mathrm{var}\{\log_{10}(\hat{f}_{11}(\lambda))\} = (\log_{10}(e))^2 L^{-1}. \tag{20.33}$$

Therefore in log plots of the point-process auto spectrum the following can be used as estimates of the asymptotic and upper and lower 95% confidence limits, based on the assumption of a Poisson distribution of intervals

$$\log_{10}\left(\frac{\hat{P}_1}{2\pi}\right), \qquad \log_{10}\left(\frac{\hat{P}_1}{2\pi}\right) \pm 0.851L^{-1/2}. \tag{20.34}$$

The above values provide a guide to interpret features in the estimate $\hat{f}_{11}(\lambda)$.

The dependence between two point-process signals can be characterized by the magnitude squared of the correlation between the Fourier transforms of the two signals under consideration. For a bivariate point process (N_0, N_1) this is written as (Brillinger, 1975; Rosenberg et al., 1989)

$$\lim_{T\to\infty} |\mathrm{corr}\{\mathrm{d}_{N_1}^T(\lambda), \mathrm{d}_{N_0}^T(\lambda)\}|^2 \tag{20.35}$$

and is called the coherence function after Wiener (1930). Coherence functions provide a measure of the strength of correlation between N_0 and N_1 as a function of frequency. The correlation $\mathrm{corr}\{\mathrm{d}_{N_1}^T(\lambda), \mathrm{d}_{N_0}^T(\lambda)\}$ between the Fourier transforms of the two point processes N_0 and N_1 can be defined in terms of variance and covariance as:

$$\mathrm{corr}\{\mathrm{d}_{N_1}^T(\lambda), \mathrm{d}_{N_0}^T(\lambda)\} = \mathrm{cov}\{\mathrm{d}_{N_1}^T(\lambda), \mathrm{d}_{N_0}^T(\lambda)\}/\sqrt{\mathrm{var}\{\mathrm{d}_{N_1}^T(\lambda)\}\mathrm{var}\{\mathrm{d}_{N_0}^T(\lambda)\}}.$$

This leads to an alternative definition of the coherence as

$$|R_{10}(\lambda)|^2 = \frac{|f_{10}(\lambda)|^2}{f_{11}(\lambda) f_{00}(\lambda)}. \tag{20.36}$$

Coherence functions provide a normative measure of linear association between two processes on a scale from 0 to 1, with 0 occurring in the case of independent processes (Brillinger, 1975; Rosenberg et al., 1989). An estimate can be constructed by direct substitution of the spectral estimates, $\hat{f}_{00}(\lambda)$, $\hat{f}_{11}(\lambda)$, and $\hat{f}_{10}(\lambda)$ into Equation (20.36). A confidence interval at the $100\alpha\%$ point, which is based on the assumption of independence, that is, $|R_{10}(\lambda)|^2 = 0$, is given by the value $1 - (1-\alpha)^{1/(L-1)}$, where L is the number of disjoint sections used to estimate the second-order spectra (Brillinger, 1981). An upper 95% confidence limit can be set at the constant level

$$1 - (0.05)^{1/(L-1)}. \tag{20.37}$$

Values of coherence falling below this level can be taken as evidence for a lack of any linear correlation between point processes N_0 and N_1 at a particular Fourier frequency. When significant values of coherence are present, a method for setting pointwise confidence limits about the estimated value of coherence is discussed in Halliday et al. (1995).

The point-process cross spectrum is a complex-valued function. Information relating to timing can be obtained by examining the phase estimate. This is defined as the argument of the cross spectrum

$$\Phi_{10}(\lambda) = \arg\{f_{10}(\lambda)\} \tag{20.38}$$

and is directly estimated using this equation. The coherence estimate $|\hat{R}_{10}(\lambda)|^2$ can be used to indicate regions where $\hat{\Phi}_{10}(\lambda)$ has a valid interpretation; in regions where there is no significant coherence between N_0 and N_1, the phase will be uniformly distributed over its range. The arctan function can be used to obtain the argument of the cross spectrum, resulting in a phase estimate over the range $[-\pi/2, \pi/2]$ radians. The signs of the real and imaginary parts of $\hat{f}_{10}(\lambda)$ can be used to determine in which quadrant the arctangent falls, as implemented by the atan2 function in C and MATLAB, thereby extending the range to $[-\pi, \pi]$ radians.

Phase estimates can often be interpreted according to different theoretical models. One important model is that for a fixed time delay, where the phase is a straight line passing through the origin with slope equal to the delay (see Jenkins and Watts, 1968). In situations where there is significant correlation over a wide range of frequencies and a pure delay is a valid model to describe the correlation structure between N_0 and N_1, it is reasonable to extend the phase estimate outside the range $[-\pi, \pi]$ radians, which avoids discontinuities in phase estimates. Such a phase estimate is often referred to as an unconstrained phase estimate. The representation of phase estimates is discussed in Brillinger (1981). Different theoretical phase curves for other forms of correlation structure are discussed in Jenkins and Watts (1968). Details for the construction of confidence limits about estimated phase values can be found in Halliday et al. (1995). A weighted least-squares regression approach to estimate the delay from the phase is described in Rosenberg et al. (1989; appendix). This method also provides an estimate of the standard error for the estimated delay.

Equation (20.39) makes explicit the relationship between time- and frequency-domain measures of correlation. This equation defines the cumulant as the inverse Fourier transform of the cross spectrum

$$q_{10}(u) = \int_{-\pi}^{\pi} f_{10}(\lambda)\, e^{i\lambda u}\, d\lambda. \tag{20.39}$$

This quantity can be estimated as

$$\hat{q}_{10}(u) = \frac{2\pi}{T} \sum_{|j| \le T/2} \hat{f}_{10}(\lambda_j)\, e^{i\lambda_j u}, \tag{20.40}$$

where $\lambda_j = 2\pi j/T$. Estimation of the cumulant density via the frequency domain has a number of advantages. It allows a unified approach to the implementation, and it allows hybrid time-series and point-process data to be analyzed using the same algorithm (Halliday et al., 1995). In addition, estimation of cumulant densities via the frequency domain allows confidence limits to be estimated according to Equation (20.15), without the assumption of Poisson processes.

A frequency-domain approach allows a multivariate approach, through the use of partial and multiple spectra to describe the relationship between several simultaneously recorded spike trains. Full details of this approach are described in Rosenberg et al. (1989). Here we reproduce the essential elements of a multivariate frequency-domain approach to analysis of spike-train data. A multivariate frequency-domain approach is based largely on regression methods. This framework is described in detail in Brillinger (1981) for time-series data; the extension to the point-process case is set out in Rosenberg et al. (1989). When three simultaneously recorded spike trains, N_0, N_1, and N_2 are available, a first-order partial coherence estimate can be used to test the hypothesis that any significant correlation between N_0 and N_1 can be accounted for by the common influence of N_2. This estimate is defined as

$$|R_{10/2}(\lambda)|^2 = \lim_{T\to\infty} \left| \text{corr} \left\{ d_{N_1}^T(\lambda) - \left(\frac{f_{12}(\lambda)}{f_{22}(\lambda)}\right) d_{N_2}^T(\lambda), d_{N_0}^T(\lambda) - \left(\frac{f_{02}(\lambda)}{f_{22}(\lambda)}\right) d_{N_2}^T(\lambda) \right\} \right|^2. \quad (20.41)$$

This equation defines the partial coherence as the magnitude squared of the correlation between the finite Fourier transforms of N_0 and N_1 after removing any linear contribution that process N_2 makes to each of these processes. The two terms $(f_{12}(\lambda)/f_{22}(\lambda))$ and $(f_{02}(\lambda)/f_{22}(\lambda))$ represent the regression coefficients that give the optimum linear predictions of $d_{N_1}^T(\lambda)$ and $d_{N_0}^T(\lambda)$, respectively, in terms of $d_{N_2}^T(\lambda)$. Equation (20.41) can be expanded in terms of ordinary coherence functions as

$$|R_{10/2}(\lambda)|^2 = \frac{|R_{10}(\lambda) - R_{12}(\lambda)R_{20}(\lambda)|^2}{(1 - |R_{12}(\lambda)|^2)(1 - |R_{20}(\lambda)|^2)}. \quad (20.42)$$

The terms in the numerator are complex-valued coherency functions. For two point processes N_0 and N_1, these are defined, for example, as $R_{10}(\lambda) = f_{10}(\lambda)/\sqrt{f_{00}(\lambda)f_{11}(\lambda)}$. The partial cross spectrum between N_0 and N_1 taking into account process $N_2, f_{10/2}(\lambda)$, can be defined as

$$f_{10/2}(\lambda) = f_{10}(\lambda) - \frac{f_{12}(\lambda)f_{20}(\lambda)}{f_{22}(\lambda)}. \quad (20.43)$$

The first-order partial coherence can also be defined in terms of partial spectra as

$$|R_{10/2}(\lambda)|^2 = \frac{|f_{10/2}(\lambda)|^2}{f_{11/2}(\lambda)f_{00/2}(\lambda)}, \quad (20.44)$$

leading directly to an estimation procedure by substitution of partial spectral estimates.

Adopting a vector- and matrix-valued notation allows extension of the multivariate parameters to order greater than 1. Here we adopt the notation of Brillinger (1981) and Rosenberg et al. (1989). Thus, $\boldsymbol{M}(t)$ represents an r-vector-valued stationary point process, and $\boldsymbol{N}(t)$ an s-vector-valued stationary point process, where the component processes, $M_1, \ldots, M_r$ and $N_1, \ldots, N_s$, are point processes. Note we are using boldface to represent vector-valued and matrix-valued processes. We will assume that the vector $\boldsymbol{M}(t)$ represents the r inputs, and the vector $\boldsymbol{N}(t)$ represents the s outputs to a point-process system. $F_{MM}(\lambda)$ is an $r \times r$ matrix of spectral densities whose principal diagonal consists of the auto spectra of the components of $\boldsymbol{M}(t)$, and the off-diagonal elements the cross spectra between these components. Similarly $F_{NN}(\lambda)$ is an $s \times s$ spectral density matrix. These matrices are formed by direct substitution of estimates, Equation (20.31), of the appropriate spectra at each frequency.

The elements of the $s \times s$ matrix

$$F_{NN}(\lambda) - F_{NM}(\lambda)F_{MM}^{-1}(\lambda)F_{MN}(\lambda) \quad (20.45)$$

are defined as partial spectra between components of $\boldsymbol{N}(t)$ after removing the linear contribution of $\boldsymbol{M}(t)$. The -1 superscript indicates a complex-valued matrix inversion operation. For $r = s = 1$, Equation (20.45) can be seen to reduce to the same form as Equation (20.43). Evaluation of Equation (20.45) involves filling in the appropriate elements of each matrix and solving by standard complex-matrix arithmetic at each frequency. The elements of Equation (20.45) may be denoted as $f_{N_iN_j/M}(\lambda)$ where $i = j$ gives the partial auto spectra, and $i \neq j$ the partial cross spectra of the components N_i and N_j of $\boldsymbol{N}(t)$ after removing the linear contribution of the r point processes in $\boldsymbol{M}(t)$. Higher-order partial coherence estimates can be defined and estimated using an equation similar to Equation (20.44).

The distribution of partial coherence estimates can be evaluated in terms of the incomplete Beta function with parameters 1 and $(L - r - 1)$. The upper 95% confidence limit for nonzero partial coherence estimates is then given at the $100\alpha\%$ point by $1 - (1 - \alpha)^{1/(L-r-1)}$ where r is the number of predictors and L the number of disjoint sections (Brillinger, 1981; Amjad, 1989). An estimate of the upper 95% confidence limit is then provided by

$$1 - (0.05)^{1/(L-r-1)}. \tag{20.46}$$

Values of estimated partial coherence lying below this line can be interpreted as evidence for a lack of a linear association between the two processes after removal of the effects of the r predictors. The extension of partial coherence estimates to order greater than 1 is discussed in Rosenberg et al. (1989) and Halliday et al. (1995), as are details of setting confidence limits on partial coherence estimates.

The third-order spectrum between three point processes, N_0, N_1, and N_2 can be estimated as

$$\hat{f}_{012}(\lambda, \mu) = \frac{1}{(2\pi)^2 LT} \sum_{l=1}^{L} \mathrm{d}_{N_0}^{T}(\lambda, l)\, \mathrm{d}_{N_1}^{T}(\mu, l)\, \overline{\mathrm{d}_{N_2}^{T}(\lambda + \mu, l)}. \tag{20.47}$$

This is a direct extension of the second-order spectrum in Equation (20.31). This third-order spectrum provides an estimate of the dependency between components at frequency λ in process N_0 and components at frequency μ in process N_1 with components at frequency $(\lambda + \mu)$ in process N_2.

A number of bispectral coefficients have been proposed to investigate higher-order interactions based on specific models (Brillinger, 1965; Godfrey, 1965; Brillinger and Rosenblatt, 1967a, 1967b; Subba Rao and Gabr, 1984). We will consider the frequency-beating model, which investigates to what extent components in one point process can be accounted for by the product of components at different frequencies in the other two point processes. This results in frequencies that add or "beat" together, which would occur in a nonlinear system where input signals are multiplied together. The frequency-beating coefficient between the three point processes, $B_{012}(\lambda, \mu)$, is defined as (Brillinger, 1965)

$$B_{012}(\lambda, \mu) = \frac{|f_{012}(\lambda, \mu)|^2}{f_{00}(\lambda) f_{11}(\mu) f_{22}(\lambda + \mu)}. \tag{20.48}$$

We estimate this coefficient by direct substitution of the appropriate second- and third-order spectral estimates. This coefficient is the modulus squared of the standardized third-order bispectrum (Brillinger and Rosenblatt, 1967b), which has the advantage that it should not contain features that are due to power in the second-order spectra (Godfrey, 1965). Some authors refer to Equation (20.48) as bicoherence, although this term has also been used for other bispectral coefficients. For uncorrelated signals, estimates of this bispectral coefficient have an exponential distribution (Huber et al., 1971). Thus a 95% confidence limit can be set as

$$\frac{3T}{2\pi L}, \tag{20.49}$$

under the assumption of three independent point processes. Estimated values that exceed this level may indicate the presence of a nonlinear interaction between the three point processes.

20.4 Cortical Neural Network Simulation

In this section, we illustrate the application of the above techniques to simulated neural network data. The data were obtained from a network of simulated cortical neurons, based on the point-neuron model of Troyer and Miller (1997). The model is based on a conductance formulation commonly used in biophysical neural model descriptions (Orpwood, Chapter 15, this volume). In conductance models, transmembrane ionic currents are assumed to flow through channels with linear instantaneous current–voltage relationships obeying Ohm's law (Hille, 1984). For the point cortical-neuron model, the intracellular membrane potential for each cell is governed by Equation (20.50) in conjunction with a partial reset mechanism

$$C_{\mathrm{m}}\frac{\mathrm{d}V_{\mathrm{m}}}{\mathrm{d}t} = -I_{\mathrm{leak}}(V_{\mathrm{m}}) - \sum_{j=1}^{n} I_{\mathrm{syn}}^{j}(V_{\mathrm{m}}, t). \tag{20.50}$$

In this equation, V_{m} represents the membrane potential at time t, C_{m} the cell capacitance, and $I_{\mathrm{leak}}(V_{\mathrm{m}})$ the passive leakage current. $I_{\mathrm{syn}}^{j}(V_{\mathrm{m}}, t)$ is the current due to the jth presynaptic spike, with the summation over the total number of presynaptic spikes, denoted by n. The cell leakage current is calculated as $I_{\mathrm{leak}}(V_{\mathrm{m}}) = (V_{\mathrm{m}} - V_{\mathrm{r}})/R_{\mathrm{m}}$, where R_{m} is the cell input resistance and V_{r} is the cell resting potential. In the model the passive membrane parameters were set according to Troyer and Miller (1997), with $R_{\mathrm{m}} = 40\,\mathrm{M}\Omega$, $C_{\mathrm{m}} = 0.5\,\mathrm{pF}$ ($\tau_{\mathrm{m}} = 20\,\mathrm{msec}$), and $V_{\mathrm{r}} = -74\,\mathrm{mV}$. The partial reset mechanism is specified by two parameters, a fixed firing threshold, V_{thresh}, and a fixed reset value, V_{reset}. When the cell membrane potential exceeds V_{thresh}, the cell is deemed to have fired an output spike; following this the membrane potential, V_{m}, is reset instantaneously to V_{reset}. The values of these parameters are $V_{\mathrm{thresh}} = -54\,\mathrm{mV}$ and $V_{\mathrm{reset}} = -60\,\mathrm{mV}$. The partial reset mechanism allows point cortical-neuron models to mimic the variability seen in the discharge of *in vivo* recordings from cortical neurons (Softky and Koch, 1993; Bugmann et al., 1997; Troyer and Miller, 1997).

The synaptic current due to a single presynaptic spike at time $t = 0$ is estimated as $I_{\mathrm{syn}}(V_{\mathrm{m}}, t) = g_{\mathrm{syn}}(t)(V_{\mathrm{m}} - V_{\mathrm{syn}})$, where $g_{\mathrm{syn}}(t)$ is a time-dependent conductance change associated with the opening of ionic channels following neurotransmitter release, and V_{syn} is the equilibrium potential for this ionic current. Each presynaptic input spike activates one term in the summation over n in Equation (20.50), which lasts for the duration of the $g_{\mathrm{syn}}(t)$ for that input. Each cell in our network receives input from a population of 100 excitatory inputs and 25 inhibitory inputs as well as synaptic inputs from other cells in the network. Thus the cells operate in a balanced input regime (Shadlen and Newsome, 1998). Each input is triggered by a separate random spike train firing at 40 spikes/sec, generated from an exponential interspike-interval distribution. These two populations provide a background synaptic activation level for each cell of 4000 EPSPs/sec and 1000 IPSPs/sec. EPSP and IPSP conductances are modeled by an alpha function (Rall, 1967)

$$g_{\mathrm{syn}}(t) = G_{\mathrm{syn}}/\tau_{\mathrm{syn}} \exp(-t/\tau_{\mathrm{syn}}), \tag{20.51}$$

−	+	+	+	+	+	−	+	+	+
+	+	+	−	+	+	+	+	+	−
+	+	−	−	−	−	+	−	−	−
+	+	+	+	+	+	+	−	+	+
−	+	+	+	+ *	+p	+	+	−	−
+	+	+	−2	+1	+	+	+	+	+
+	+	+	+	+3	+	−	+	+	+
+	+	+	+	+4	+	+	+	+	+
+	+	+	−	+5	−	−	+	+	−
−	+	−	+	+6	−	+	−	+	+

FIGURE 20.2 Relative positions of excitatory and inhibitory cortical neuron models in a 10×10 2D sheet of neurons. The positions of excitatory neurons are indicated by "+", the positions of inhibitory neurons by "−". The superscripts indicate the positions of neurons in the network, the spiking activity of which is analyzed in Section 20.5.

requiring two parameters to be specified, the conductance scaling factor, G_s, and the time constant τ_s. For excitatory background inputs $G_{\text{syn}} = 4.14 \times 10^{-9}$, $\tau_{\text{syn}} = 1.0$ msec, and $V_{\text{syn}} = 0$ mV. For inhibitory background inputs, $G_{\text{syn}} = 2.28 \times 10^{-9}$, $\tau_{\text{syn}} = 10.0$ msec, and $V_{\text{syn}} = -74$ mV. Thus the peak synaptic conductance for excitatory inputs is 1.52 nS at 1 msec, and for inhibitory inputs it is 0.839 nS at 10 msec. Since V_{syn} for the inhibitory inputs is identical to the resting potential, V_r, the inhibition acts as silent, or shunting inhibition. We therefore characterized the EPSP and IPSP magnitudes at V_{thresh}. In addition, since this is the upper end of the normal working range for the cell, it gives a better indication of the typical magnitude of EPSPs and IPSPs just prior to firing an output spike. The EPSP magnitude, rise time, and half width at V_{thresh} were 366 μV, 0.575 msec and 5.75 msec. The corresponding values for the IPSP inputs were −366 μV, 2.75 msec, and 11.75 msec; the negative magnitude indicates a hyperpolarization. In our balanced input model we have chosen to use identical EPSP and IPSP magnitudes at V_{thresh}. The EPSP magnitude when activated from the resting potential, V_r, is 500 μV. The IPSP activated from rest is silent.

The simulated network consists of 100 neurons arranged in a 10×10 2D sheet. The network contains 75 excitatory neurons and 25 inhibitory neurons. This ratio was chosen to match the ratio of excitatory to inhibitory cells found in the cortex (Douglas and Martin, 1998). Excitatory and inhibitory neurons are identical except for the pattern of connectivity and their action on postsynaptic cells. Inhibitory neurons were randomly placed in the sheet, as illustrated in Figure 20.2.

Connectivity within the sheet adopts a center-surround pattern, as illustrated in Figure 20.3. Center-surround connectivity is thought to play an important role in visual processing, for example, the spatial organization of some retinal cell receptive fields exhibits center-surround characteristics (e.g., see Kuffler et al., 1984, chapter 2). By promoting local excitation and inhibition of more distant regions, this pattern of connectivity can result in highly variable firing rates and promote local network oscillations (Usher et al., 1994).

Each excitatory cell in the network adopts the pattern of connectivity illustrated in Figure 20.3A; each inhibitory cell adopts the pattern of connectivity illustrated in Figure 20.3B. The EPSP magnitude for the synaptic connections in the network was increased to 500 μV, that for the IPSP was increased to −500 μV. This was achieved by increasing G_s to 5.67×10^{-9} for EPSPs and 3.13×10^{-9} for IPSPs. Other synaptic conductance parameters are as for the background inputs. The rise times and half widths of these larger synaptic inputs are the same as for the background inputs. No wrapping of connections is used at the edge of the network, resulting in a single planar sheet of neurons. This allows us to make statements about the sphere of influence of a single neuron in influencing the firing times of other cells in the network, without contamination from wrapping effects.

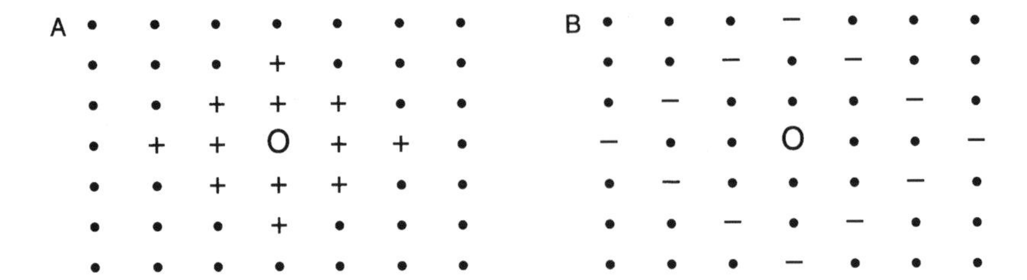

Figure 20.3 Center-surround connectivity used in 2D network simulations. Each character indicates the position of one cell in the network. The position of the presynaptic cell is indicated by the letter 'O'. (A) Connections from excitatory neurons: postsynaptic cells receiving excitatory connections are indicated by "+". (B) Connections from inhibitory neurons: postsynaptic cells receiving inhibitory connections are indicated by "−". Cells with no synaptic connection from the presynaptic cell are indicated by "•". Each presynaptic cell makes only excitatory (A), or inhibitory (B), connections, depending on whether it is designated as an excitatory or inhibitory neuron in Figure 20.2.

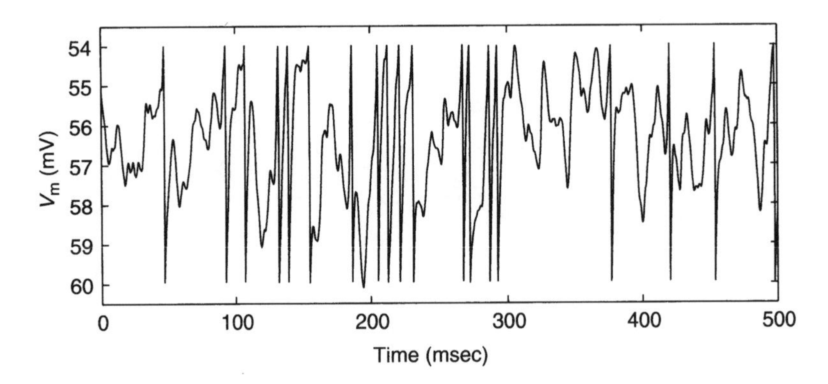

Figure 20.4 A 500 msec segment of the membrane potential in one neuron in the 2D sheet. The partial rest mechanism can be seen following each output spike. The firing threshold is −54 mV, the partial reset potential is −60 mV.

The simulation was used to generate 100 sec of data, and the output spike times for all neurons were logged with a 1 msec sampling interval. This data set, which contains 100 spike trains, was subject to the analyses described in Sections 20.2 and 20.3.

20.5 Results

Figure 20.4 illustrates a 500 msec segment of the membrane potential from one cell in the network during the simulation. The partial reset mechanism constrains the voltage to lie mostly within the region −54 mV to −60 mV, although a concentration of inhibitory input can pull the membrane potential below −60 mV on occasion. This figure also illustrates the partial reset mechanism following every output spike; in the segment illustrated there are 19 output spikes.

Additionally, the flexibility of simulations allows us selectively to adjust model parameters away from their usual values. Figure 20.5 illustrates the membrane potential fluctuations in the same cell with an identical input sequence to that used in Figure 20.4, but with the cell threshold removed so that no output spikes are generated. This illustrates the range of potential values in the cell that result purely from the balanced input regime, without the partial reset mechanism superimposed. Figure 20.5 may provide a more indicative plot for characterizing the membrane-potential fluctuations

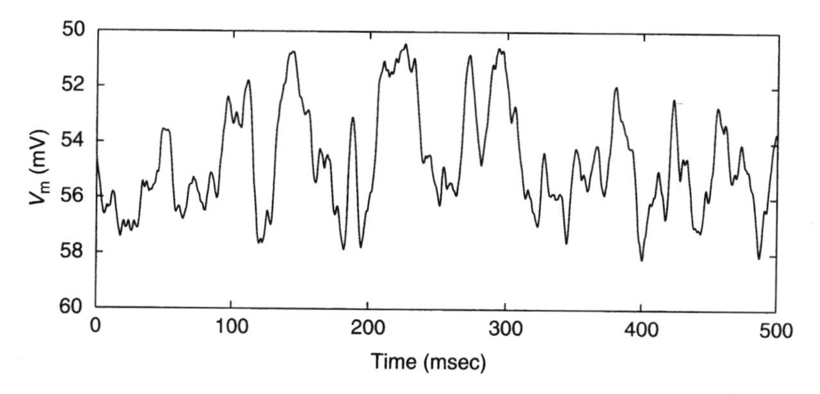

FIGURE 20.5 A 500 msec segment of the membrane potential from the same neuron as in Figure 20.3, subject to an identical input sequence, but with the firing threshold increased so that no output spikes are generated.

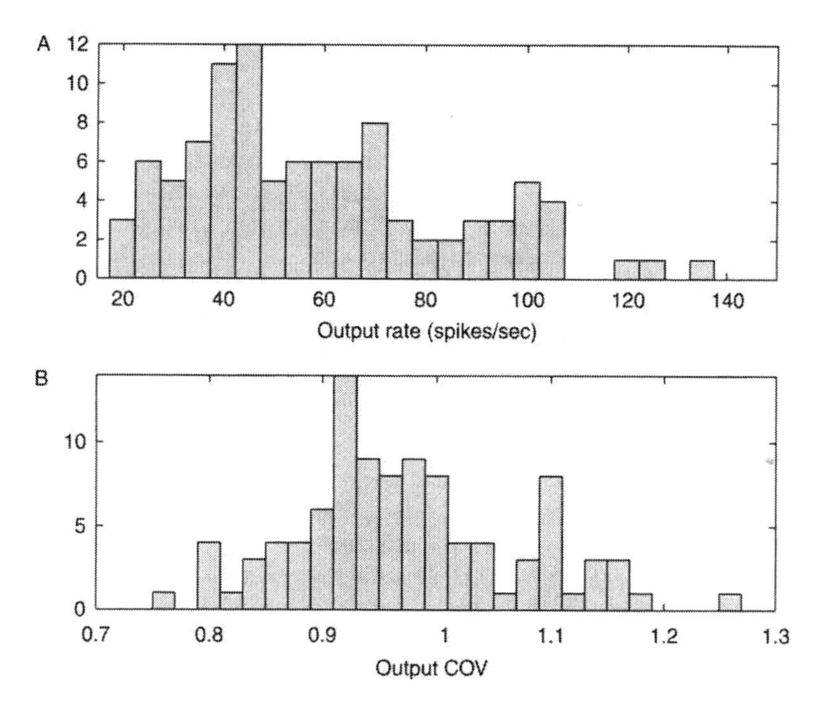

FIGURE 20.6 Histogram plots of (A) the output firing rates, and (B) output COV for the 100 cells in the simulation.

resulting from a balanced input regime, however, a more detailed statistical analysis of these data is outside the scope of this chapter. All further analysis is based on the spike firing time data.

The average value of $\hat{P}$ across the network was 0.059, equivalent to a mean rate of 59 spikes/sec (range 19 to 135 spikes/sec). The mean coefficient of variation was 0.97, with a range of 0.76 to 1.26. This is slightly higher than the range of 0.5 to 1.0 for the coefficient of variation obtained from recordings of nonbursting neurons in V1 and MT cells *in vivo* (Softky and Koch, 1993). Thirty two of the 100 cells in the present simulation had an output coefficient of variation (COV) greater than 1.0. Figure 20.6 shows histograms of the mean rate (in spikes/sec) and COV for all 100 cells. The COV is a dimensionless measure widely used to assess the variability of neural spike trains. It is defined and calculated from the interspike-interval (ISI) histogram as the ratio $\sigma_{\text{ISI}}/\mu_{\text{ISI}}$, where σ_{ISI} is the standard deviation of the ISI histogram, and μ_{ISI} is the mean interval.

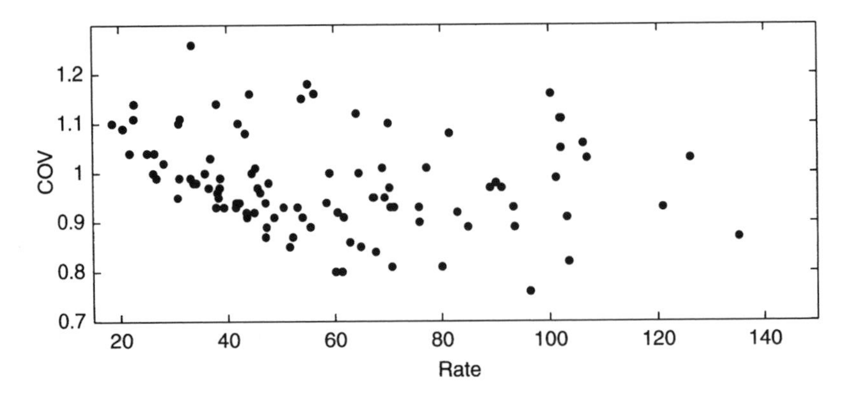

FIGURE 20.7 Scatter plot of firing rate (spikes/sec) vs COV for the 100 cells in the network.

While the above histograms give a good visual impression of the distribution of the firing rate and COV values in our data set, they do not provide any indication of whether a systematic relationship exists between the two variables. To complete our analysis of the first-order statistics describing our 100 cell network, Figure 20.7 shows a scatter plot of the firing rate and COV data for the 100 cells. The visual impression of the data in Figure 20.7 is one of little dependence between firing rate and COV. This conclusion is backed up by a regression analysis. The best-fit straight line to the above data has a slope of -0.0008 and an intercept of 1.03; the r^2 value is 0.049. Thus, for this particular network, firing rate and COV would appear to be unrelated.

For a random, or Poisson, spike train, which has an exponential ISI, the expected value for the COV is 1.0, that is, the standard deviation of the distribution is the same as the mean. Many of the spike trains generated in the simulation have a COV that is very close to 1. Figure 20.8A and Figure 20.8B illustrate the point-process spectral estimates for two output spike trains that have a COV of 1.0. For comparison, Figure 20.8C shows the point-process spectral estimate for one of the background input spike trains in the network, which also has a COV of 1.0.

The spectra in Figure 20.8 were estimated according to Equation (20.31), with $L = 97$. The sampling rate for the spike-train data is 1 msec. This rate defines the Nyquist frequency as 500 Hz; the plots in Figure 20.8 show the spectral estimates over the entire frequency range. The segment length in Equation (20.31) was $T = 1024$, and this value was used for all the linear parameters described below. The confidence limits for each estimate are defined according to Equation (20.34); since all estimates use the same number of segments, the spacing between the upper and lower confidence limits is the same in each estimate. Despite the fact that all three spike trains have a COV of 1 there are clear differences in the spectra illustrated in Figure 20.8. The spectra for the two output discharges have a reduced magnitude below 150 Hz; in addition that for the cell with the higher output rate (Figure 20.8B) has increased power below 20 Hz. In contrast, the estimated spectrum for the input spike train (Figure 20.8C), which was artificially generated from an exponential ISI distribution, does not deviate from the expected value for a Poisson spike train. Thus the second-order point-process spectrum of a spike train provides a further means of distinguishing the characteristics of spike trains with similar first-order statistics, in this case a COV of 1.

Next we consider a second-order (linear) correlation analysis between pairs of output spike trains from the network. The cells analyzed are indicated in Figure 20.2. The common (or input) cell is indicated by "$*$", and the correlation of the discharge with cells numbered 1 to 6 is illustrated in Figure 20.9 and Figure 20.10. Figure 20.9 illustrates the six coherence estimates, $|\hat{R}_{10}(\lambda)|^2$, with point process N_0 obtained from the output firing times of the common cell and point process N_1 obtained from the output firing times of the cells labeled 1 to 6, respectively.

As might be expected, the magnitudes of the coherence estimates drop off with increasing distance of each cell from the reference cell, indicating a reduced strength of correlation in the network.

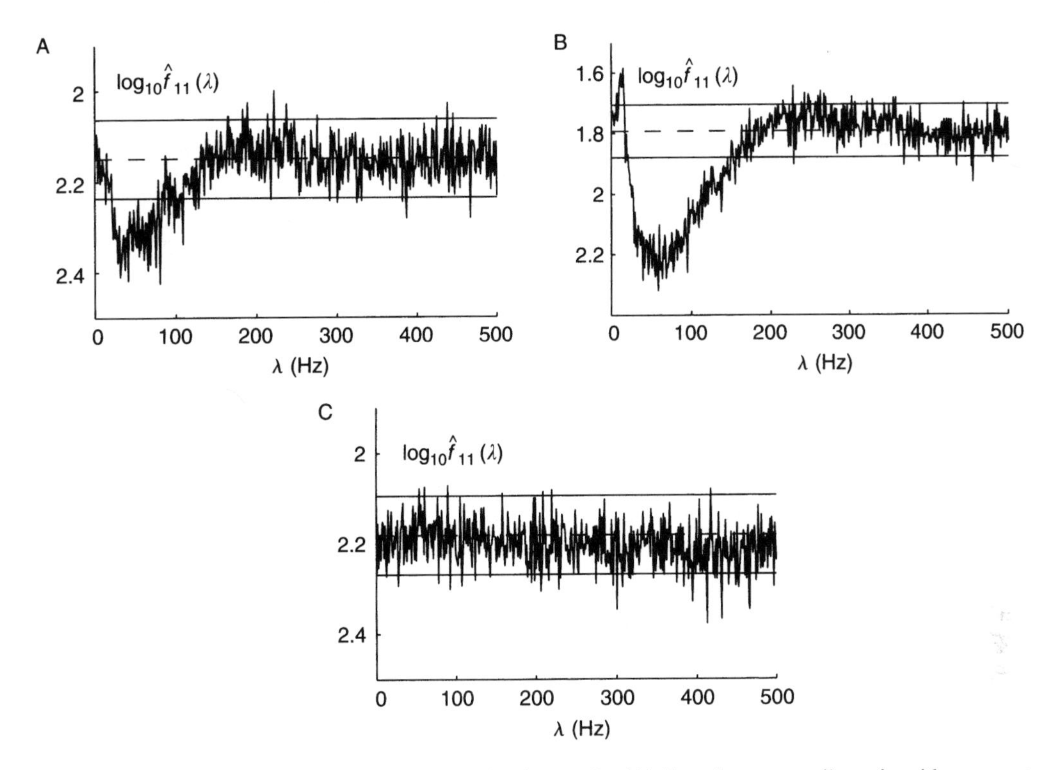

FIGURE 20.8 Log plots of point-process spectral estimates for (A) Sample output spike train with mean rate 44.5 spikes/sec and COV 1.0; (B) Sample output spike train with mean rate 101.2 spikes/sec and COV 0.99; and (C) Sample background input spike train with mean rate 41.4 spikes/sec and COV 0.98. The horizontal lines are the asymptotic values of the point-process spectra (dashed lines) and the upper and lower 95% confidence limits for each estimate (solid lines), see Equation (20.34).

Adjacent cells (e.g., cell 1 and cell 2) are strongly correlated, with a peak coherence of around 0.6 at frequencies around 20 Hz. We interpret this as an indication that the variability in the firing times of the common cell can account for around 60% of the variability in the discharge of the two adjacent cells at around 20 Hz. This large value of coherence may well reflect the recurrent nature of the connections in the present network. Figure 20.10 illustrates the corresponding cumulant-density estimates, $\hat{q}_{10}(u)$ for the same data as analyzed in Figure 20.9.

The cumulant estimates in Figure 20.10 are suggestive of a similar relationship between the spike trains in the network to that indicated by Figure 20.9 — the strength of correlation between pairs of neurons decreases with increasing separation between cells. It is also interesting to note that the general form of correlation structure between the reference cell and cell 1, and between the reference cell and cell 2 is similar, despite the fact that cell 2 is an inhibitory neuron.

Use of the partial parameters described above allows us to explore the interactions between the spike trains in the network in more detail. We next consider first-order partial parameters. Figure 20.11 shows the first-order partial coherence estimates, $|\hat{R}_{10/2}(\lambda)|^2$, between the reference cell and cells 1 to 6 using the spike times from an adjacent cell as predictor. The letter "p" in Figure 20.2 indicates the cell used as predictor in this case. This figure has the same layout as Figure 20.9; for comparison the ordinary coherence estimates from Figure 20.9 are shown superimposed on each panel in Figure 20.11.

As described above, a frequency-domain approach to the definition and estimation of partial parameters allows us to construct estimates of partial time-domain parameters. Figure 20.12 shows the estimates of the six first-order partial cumulant-density estimates, $\hat{q}_{10/2}(u)$, corresponding to the frequency-domain estimates in Figure 20.11.

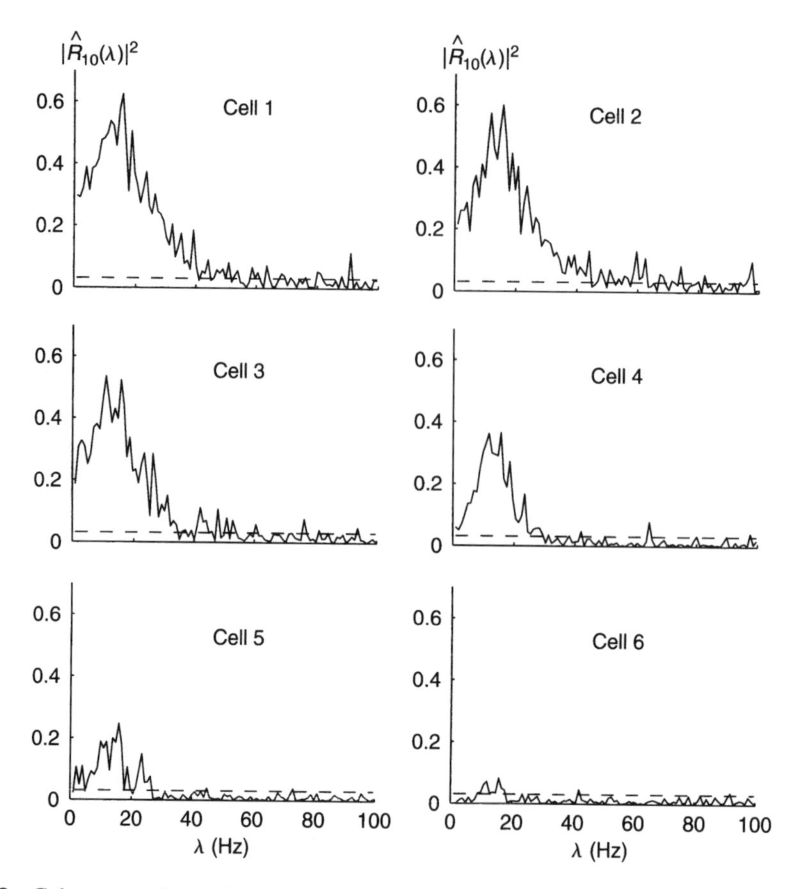

FIGURE 20.9 Coherence estimates between the output discharge of the common cell, labeled "∗", and the cells labeled 1 to 6 in Figure 20.2. The horizontal dashed lines are an estimate of the upper 95% confidence limit based on the assumption of independence, Equation (20.37).

The partial coherence estimates between the reference cell and cells 1 to 6 (Figure 20.11, solid lines) all show a marked reduction in magnitude when compared with the ordinary coherence estimates (Figure 20.11, dotted lines). This supports the idea that there is some redundancy in the moment-to-moment fluctuations in interspike intervals from nearby cells in the network. The predictor cell used for the partial coherence estimates (cell "p" in Figure 20.2) only has direct synaptic connections to the reference cell (cell "∗") and cell 1. Despite this, there is a marked reduction in the partial coherence compared with the ordinary coherence in all the pairwise combinations examined in Figure 20.11. This is likely to result from the large degree of interconnectivity and the large number of reciprocal connections present in the network. The reduction in the magnitude of the partial coherence estimates when compared with the ordinary coherence estimates is mirrored by the time-domain results. The partial cumulant-density estimates (Figure 20.12) exhibit a reduction in both the magnitude and lag range of significant features when compared with the ordinary cumulant-density estimates (Figure 20.10).

The concept of local interactions can be investigated further using higher-order partial parameters. Figure 20.13 shows seventh-order partial coherence and partial cumulant-density estimates between the reference cell (cell "∗") and cell 1. The seven predictors are the seven nearest neighbors to the reference cell, not including cell 1 (it is not valid to include the same spike train more than once in a multivariate analysis). The seventh-order partial coherence in Figure 20.13 has almost no significant features when compared with the ordinary coherence (Figure 20.9, panel: Cell 1), suggesting that the concept of local interactions discussed above may well be valid in this network. There is a suggestion

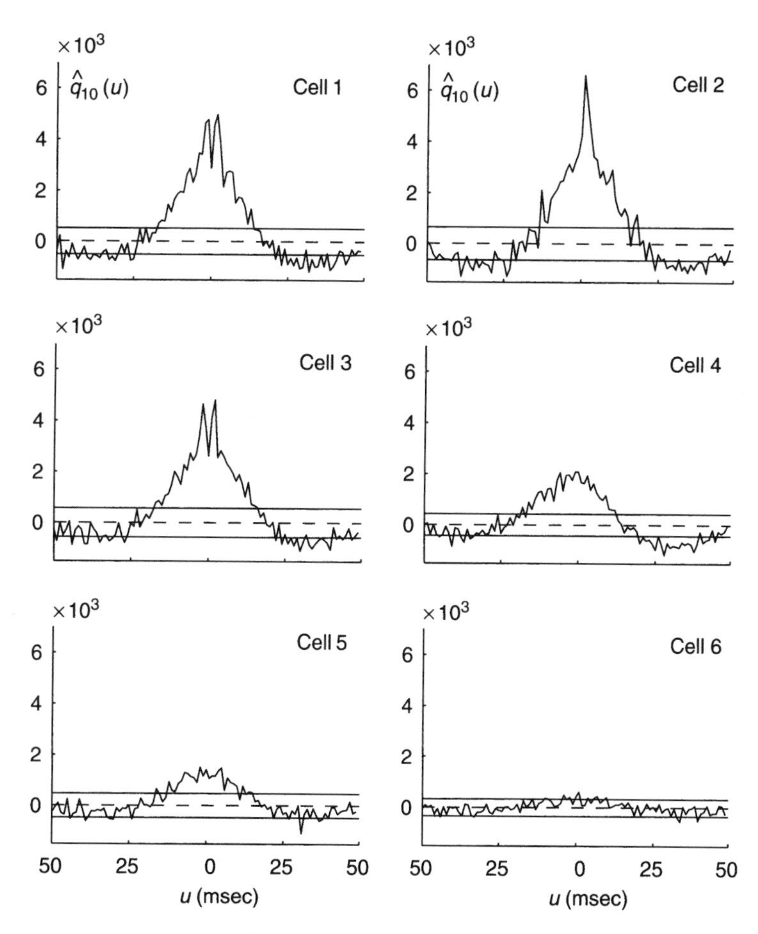

FIGURE 20.10 Cumulant-density estimates between the output discharge of the common cell, labeled "*", and cells labeled 1 to 6 in Figure 20.2. The horizontal lines indicate the asymptotic values (dashed lines at zero) and the estimated upper and lower 95% confidence limits (solid lines), based on the assumption of independence. Values obtained from an evaluation of Equation (20.15).

of some significant features still remaining in the partial coherence estimate, although these are weak and variable. In addition, the partial cumulant-density estimate, although greatly reduced in magnitude when compared with the ordinary cumulant (Figure 20.10, panel: Cell 1) still exhibits some residual features around zero lag. Further analyses using different multivariate models may be appropriate here. The residual features in Figure 20.13 may also result from nonlinear interactions between the cells in the network; such nonlinear interactions cannot be accounted for by the linear analysis used so far.

We will now consider nonlinear analysis of the 2D network, starting with a time-domain analysis using third-order cumulant-density functions. Figure 20.14 shows a contour plot of the third-order cumulant density, $\hat{q}_{210}(u, u - v)$. In this case point process N_2 is the discharge of the reference cell in the network (cell "*" in Figure 20.2), process N_1 is the discharge of cell 1, and process N_0 is the discharge of cell "p". This cumulant can be used to determine whether there is any nonlinear interaction between different synaptic inputs to the reference cell. It will characterize the nonlinear dependency of the timing of output (N_2) spikes which are preceded by an N_1 spike at an interval of $(u - v)$, and an N_0 spike at an interval of u. This timing convention is illustrated in Figure 20.1.

The cumulant in Figure 20.14 provides strong evidence in favor of nonlinear interactions between the cells in the network. The maximum values are located at similar lag values around 2 to 3 msec,

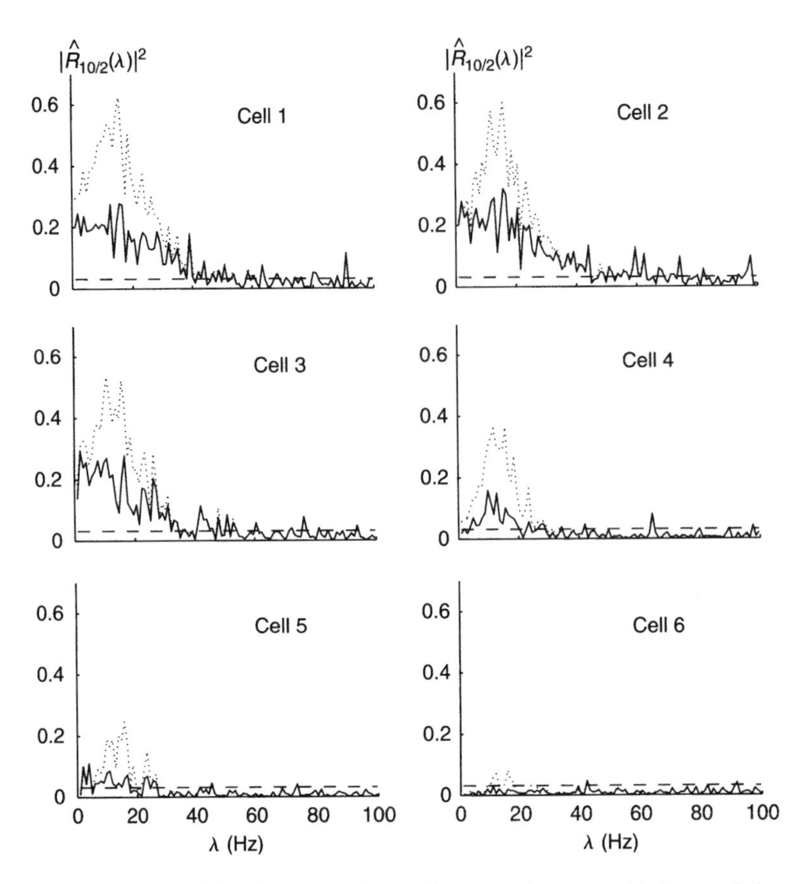

FIGURE 20.11 First-order partial coherence estimates between the output discharge of the common cell, labeled "∗", and those of the cells labeled 1 to 6, using the spike timings from the cell labeled "p" as predictor in Figure 20.2. The horizontal dashed lines are estimates of the upper 95% confidence limits, based on the assumption of independence, Equation (20.46). The ordinary partial coherence estimates (copied from Figure 20.9) are shown as dotted lines for comparison.

and are positive. This indicates that there is an increased likelihood of the reference cell firing an output spike around 2 to 3 msec following a pair of closely grouped input spikes, one from each input. This may well reflect a process of nonlinear EPSP summation in the cortical neuron model. Sections through the third-order cumulant allow the characteristics of the nonlinear interaction in a specific region to be represented in a more informative manner. Figure 20.15 shows two sections through the third-order cumulant at fixed lags $u = 2$ msec and $(u - v) = 1$ msec. The sections through the estimate $\hat{q}_{210}(u, u-v)$ indicate that a nonlinear interaction is present for presynaptic spikes occurring up to 20 msec prior to an output spike, and that the effect of the nonlinear interaction is a facilitation of the output discharge, that is, an increased tendency for output spikes to occur.

Third-order spectral coefficients allow the nonlinear interaction to be studied in the frequency domain. Figure 20.16 shows an estimate of the third-order bispectral coefficient, $\hat{B}_{012}(\lambda, \mu)$, for the same three spike trains analyzed in Figure 20.14. This estimate was constructed using a segment length of $T = 512$, giving $L = 195$. The envelope of the bispectral coefficient is similar to those seen in the linear case (Figure 20.9), with a distinct peak which decays monotonically. The envelope of the bispectral coefficient can be seen more clearly in the section illustrated in Figure 20.17, which shows a section through $\hat{B}_{012}(\lambda, \mu)$ at a fixed lag of $\lambda = 6$ Hz.

The maximum value of the section shown in Figure 20.17 occurs at 12 Hz. This higher-order frequency-domain analysis suggests that the two different inputs to the reference cell

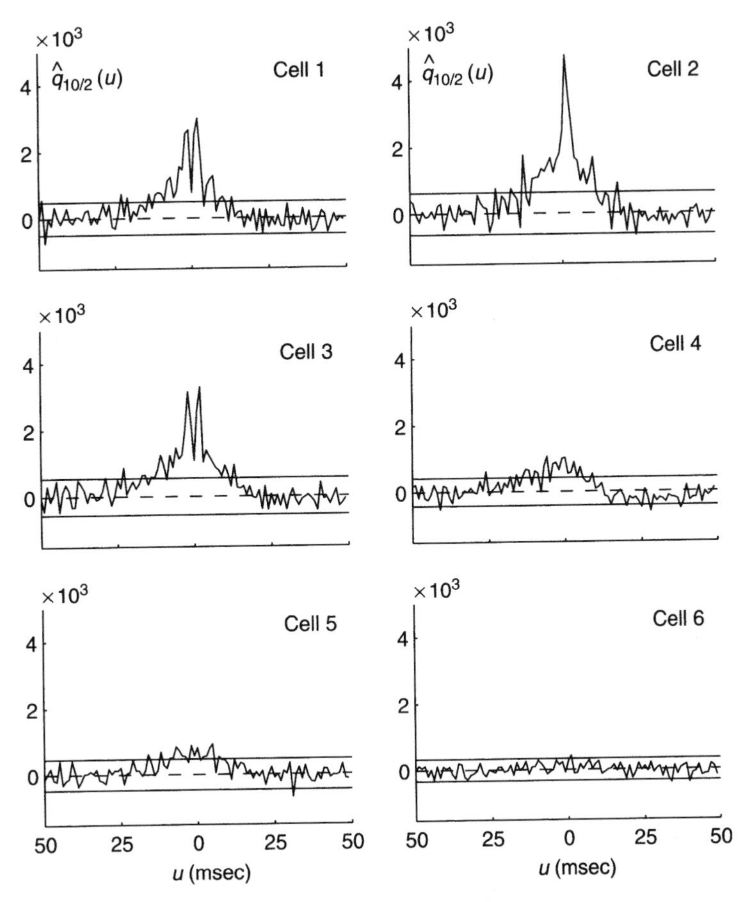

FIGURE 20.12 First-order partial cumulant-density estimates between the output discharge of the common cell, labeled "∗", and the cells labeled 1 to 6, with the discharge of cell "p" as predictor in Figure 20.2. The horizontal lines indicate the asymptotic values (dashed lines) and the estimated upper and lower 95% confidence limits (solid lines) based on the assumption of independence.

(cell "∗" in Figure 20.2) interact at different frequencies. The bispectral coefficient is based on a frequency-beating model (Brillinger, 1965), characterizing a nonlinear interaction between frequency components λ, μ, and $\lambda + \mu$. In conjunction with the section in Figure 20.17, this model suggests a nonlinear interaction between frequency components at 6 Hz in the discharge of cell "p" with components at 12 Hz in the discharge of cell 1 and components at 18 Hz in the discharge of the reference cell "∗".

20.6 Concluding Remarks

We have described linear and nonlinear methods for the analysis of multivariate spike-train data within a stochastic point-process framework. The advantages of our approach are that the methods are suitable for analysis of an arbitrary number of simultaneously recorded spike trains (e.g., Figure 20.13 presents a multivariate analysis of nine simultaneous spike trains). In addition the framework presented allows confidence limits to be set for all parameter estimates. In many cases the expressions depend only on the number of segments averaged in the spectral estimates and can be calculated independently of the data. Confidence limits enhance the statistical rigor and ease of interpretation. Our framework allows a unified approach to time- and frequency-domain analysis. Nonetheless,

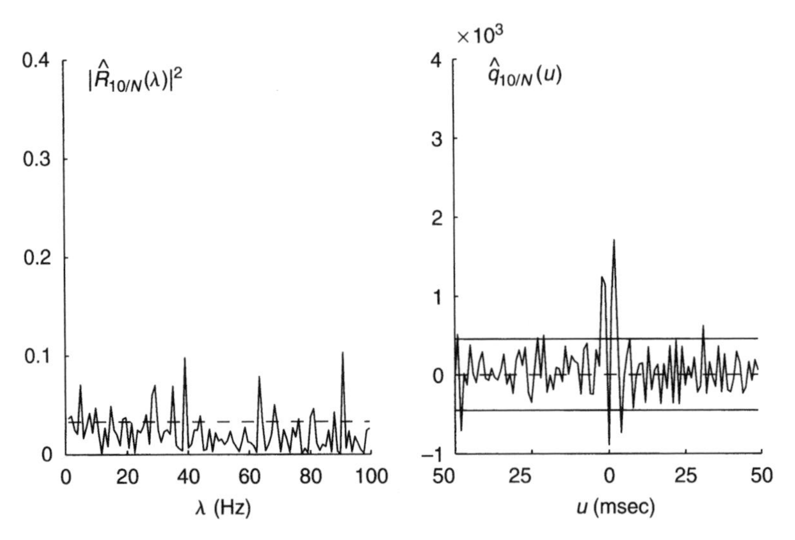

FIGURE 20.13 (A) Seventh-order partial coherence estimate, $|\hat{R}_{10/N}(\lambda)|^2$, between the output discharge of the reference cell, labeled "$*$", and cell 1, using the spike timings from the seven nearest neighbors to the reference cell as predictors. The activity of these seven cells is denoted by the matrix N. The horizontal dashed line is an estimate of the upper 95% confidence limit, based on the assumption of independence, Equation (20.46). (B) Corresponding seventh-order partial cumulant-density estimate, $\hat{q}_{10/N}(u)$. The horizontal lines indicate the asymptotic value (dashed line at zero) and the estimated upper and lower 95% confidence limits (solid lines), based on the assumption of independence.

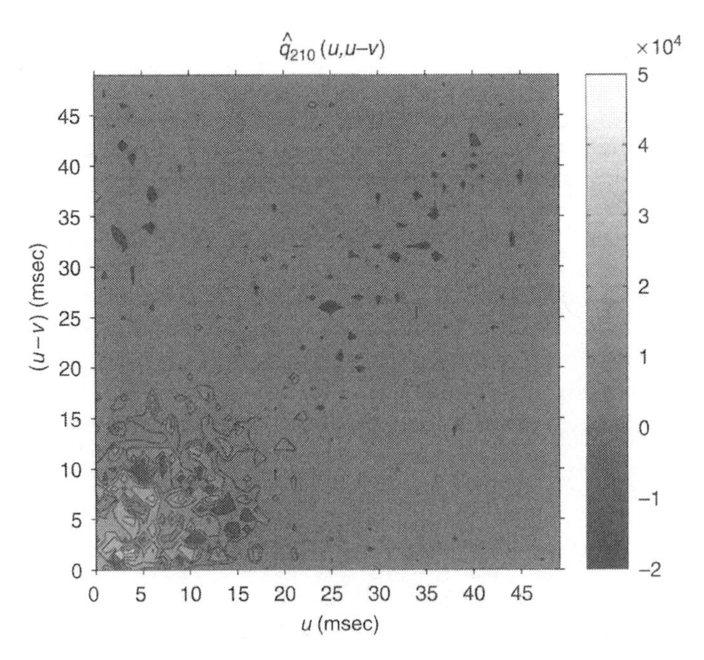

FIGURE 20.14 (See color insert following page 366) Contour plot of the estimated third-order cumulant-density function, $\hat{q}_{210}(u, u - v)$. The estimated upper and lower 95% confidence limits for this estimate are $\pm 9.9 \times 10^{-5}$. Values lying inside this region, which can be interpreted as an indication of no significant third-order interaction, are indicated by a single grey level in the plot. This level covers the majority of the plot. Significant values of the cumulant are located at lag values less than 15 msec, with a maximum around lag values of 2 to 3 msec. All these significant values are positive.

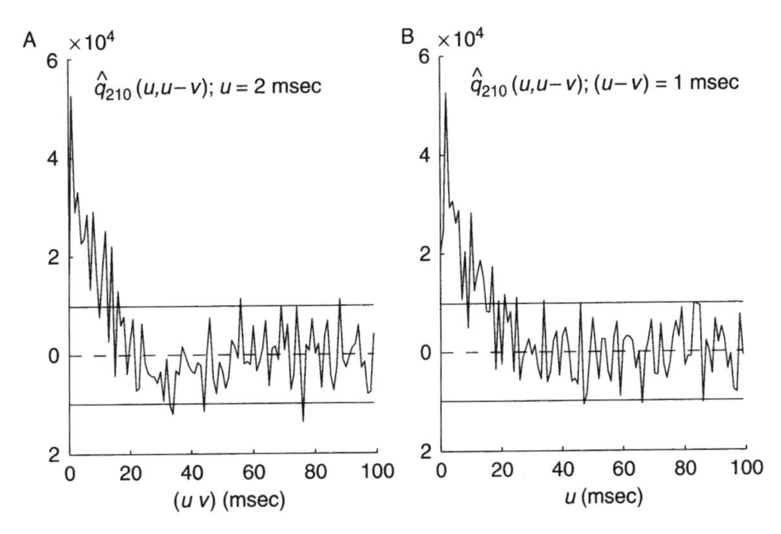

FIGURE 20.15 Sections through the estimate $\hat{q}_{210}(u, u-v)$ in Figure 20.14 at fixed lags (A) $u = 2$ msec and (B) $(u-v) = 1$ msec. The horizontal lines indicate the asymptotic value (dashed line at zero) and the estimated upper and lower 95% confidence limits (solid lines), based on the assumption of independence, Equation (20.27).

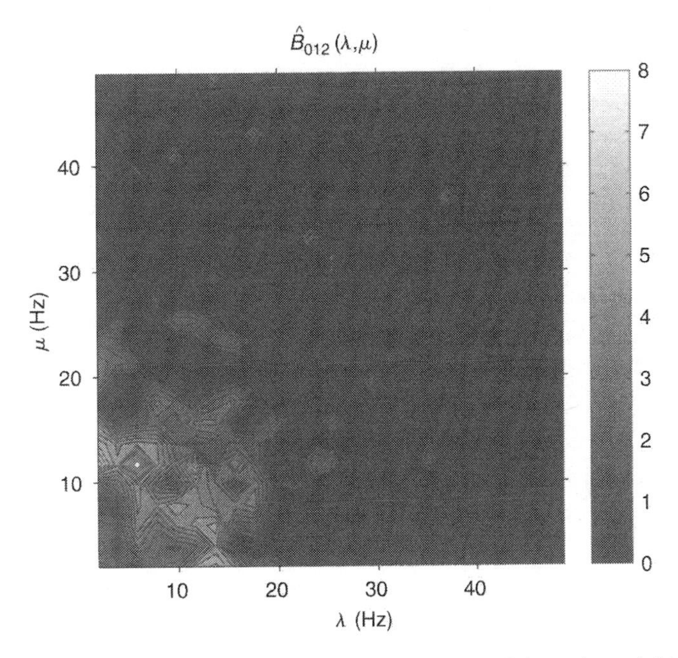

FIGURE 20.16 **(See color insert following page 366)** Contour plot of the estimated third-order bispectral coefficient, $\hat{B}_{012}(\lambda, \mu)$. From Equation (20.49), the estimated 95% confidence limit is 1.25. Values less than this can be interpreted as an indication of no significant third-order interaction, and are indicated by a single grey level in the plot. This level covers the majority of the plot. Significant values are located at frequencies below 20 Hz. Both frequency axes start at 2 Hz.

all the linear parameters presented above have been estimated via the frequency domain. This reflects the usefulness of Fourier-based methods in the analysis of multivariate data. Details of how the above approach can be applied to time-series data, and hybrid time-series/point-process data can be found in Halliday et al. (1995). Our hypothesis as stated in the Introduction is that correlated spike-train

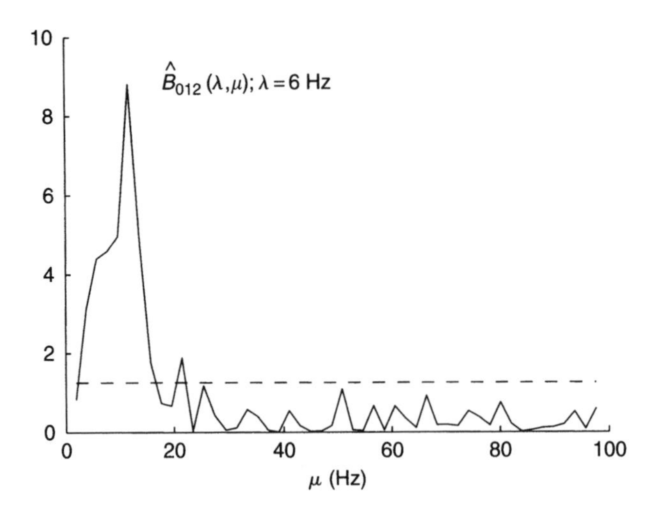

FIGURE 20.17 Section through the estimate $\hat{B}_{012}(\lambda, \mu)$ in Figure 20.16 at fixed frequency $\lambda = 6$ Hz. The horizontal dashed line indicates the estimated upper 95% confidence limit based on the assumption of independence, Equation (20.49).

activity may provide an indicator of a possible functional role for two or more neurons. In addition, these techniques are also suited to the study of neural pathways. In some situations partial parameters can be used to distinguish the effects of common inputs from those of direct connections (Rosenberg et al., 1998). With developments in multielectrode recording (Laubach et al., 2000; Nicolelis and Ribeiro, 2002), there is a requirement for analysis methods suited to large multichannel spike-train data sets. The continued development of the above methods, in particular partial parameters, is an area that might be usefully pursued.

ACKNOWLEDGMENTS

Supported in part by grants from the UK Engineering and Physical Sciences Research Council (EPSRC GR/R12350/01) and the European Community Future and Emerging Technologies program (IST-2000-28027; POETIC).

SOFTWARE ARCHIVE

MATLAB™ routines to implement some of the above analyses are available from the NeuroSpec archive at http://www.neurospec.org/.

PROBLEMS

1. As a precursor to application of the techniques described in this chapter to your own data, you may wish to become familiar with the use and interpretation of the above methods. The data set analyzed above is available from: www-users.york.ac.uk/~dh20. Much of the linear analyses described above can be implemented using the MATLAB routines available from the NeuroSpec archive at: www.neurospec.org. A key parameter in the analysis is the segment length T. A value of 1024 was used in the analysis illustrated in Figure 20.8 to Figure 20.12 (equivalent to a segment length of 1.024 sec). This parameter, in conjunction with the total record length, R, defines the number of segments, L. Try repeating the analysis with T increased and then decreased by a factor of two (the segment length, T, in NeuroSpec is specified as a power of 2). When you have done this,

two effects should become apparent. The first is the increased frequency resolution with longer segments, that is, there are more frequency points. However, this increased frequency resolution occurs at a cost of a reduced number of segments, which, importantly, results in an increased variability in parameter estimates (e.g., see Equations [20.34] and [20.37] above). Comparison of spectral and coherence estimates with different values of T should make apparent these two effects. Consideration of this trade-off between frequency resolution and statistical variability should guide the choice of parameters to use in the analysis of your own data.

2. How can we summarize the interactions within the network? Figure 20.9 illustrates only 6 of the possible 4950 coherence estimates describing the pairwise interactions within the network. Do these representative examples present an adequate summary of the interactions within the network, or do we need to look at a much larger subset of the 4950 possible coherence estimates to claim this? Is it reasonable to consider the average level of interaction between the cells in our network, and if so what does this tell us about the network? One way of approaching the problem of summarizing large data sets is through the use of pooled parameters (Amjad et al., 1997). This framework allows construction of time- and frequency-domain parameters which describe large populations of data, and includes a test to indicate whether all the individual components can be regarded as equivalent (in a statistical sense). However, this approach still requires all pairwise combinations to be evaluated. A frequency-domain approach to the description of stochastic data has much to learn from ordinary multivariate data analysis — are there additional tools and techniques which can be successfully mapped across and used to provide compact descriptions of large data sets, such as the one considered here?
3. Implications for neural coding. There is currently much discussion relating to the mechanisms of neural coding, that is, the representation of "information" by neuronal spike trains (e.g., Softky, 1994; Lestienne and Tuckwell, 1998; Shadlen and Newsome, 1998). One particular question of interest is the occurrence and role of precisely timed patterns of spikes (Lestienne and Tuckwell, 1998). Figure 20.14 and Figure 20.15 explore the interactions between specific triplets of spikes related to three cells in the network. This would suggest that the cells in this simulation are sensitive to specific temporal patterns in their inputs. Higher-order cumulant-density functions provide measures of statistical dependence between random processes (Rosenblatt, 1983; Mendel, 1991). Is there a role for higher-order statistics (both time- and frequency-domain) in studies of neural coding? It is known from both theoretical studies (Tuckwell, 2000; Rall, 1964) and modeling studies (Ianella et al., 2004) that the position of excitatory and inhibitory synapses on the dendritic tree can affect neural integration and neural coding. This extends the issue to that of the suitability of higher-order parameters to the study how spatial temporal integration contributes to neural coding mechanisms.

21 The Poetics of Tremor

G.P. Moore and Helen M. Bronte-Stewart

"...the exact meaning of poetics is the study of work to be done."
–Igor Stravinsky

CONTENTS

21.1 INTRODUCTION

Today, scores of clinics devoted to movement disorders routinely record the many manifestations of tremor seen in humans. In addition, research laboratories are investigating tremor and the processes that produce it in experimental animals and in normal humans. In parallel with this, it is now almost routine for neurosurgeons and neurologists to record electrical activity from single neurons in the human brain as their electrodes pass through regions believed to be either involved in the generation of tremor or at least suitable targets for therapeutic ablation or electrode implantation procedures.

Compared to the cognitive impairments that accompany some neurological disorders, tremor may be of far less importance. But tremor is one of the few immediate manifestations of central pathologies that can be easily measured. Though tremor may be but a neural spandrel, it proceeds on a timescale sufficiently fast and observable to be of great use and interest to those searching for evidence of central neuronal processes associated with it. Yet the Holy Grail — finding the structures from which tremors emanate — remains elusive. In principle, it should be easy for a clinician monitoring electrical activity in the brain to observe when that activity is correlated with a tremor simultaneously exhibited by the patient. However this has not yet occurred except in isolated instances. Perhaps it is because crucial areas have not been plumbed, or perhaps because signals from neurons intimately involved in tremor production are not so easily correlated with the casually observed features of tremor.

With a phenomenon as obvious as clinical tremor, often with highly periodic features and a sizable amplitude, why are there so few quantitative models of the tremors themselves, and no models whatever to account for their production, though the brain structures hypothesized to generate tremor are so often represented in block diagrams that imply a sound theoretical understanding[1]? More than 60 years after Norbert Wiener called attention to tremor as an obvious phenomenon suitable for engineering analysis (Wiener, 1948), we are no closer to realizing that objective.

And yet, we have ample opportunities to measure tremor with increasingly refined instrumentation, and the computational power to reveal subtleties in its temporal structure and its relation to other signals within the brain with which it might be associated. Sophisticated tremor analysis remains a most powerful tool for guiding the explorations of neurosurgeons and for testing theoretical models of tumor genesis. We can hope that the increasing sophistication of tremor analysis will be matched by corresponding developments in analyzing the expanding database of electrical activity recorded from living human brains and in creating more sophisticated protocols for intra-operative procedures. These in turn should stimulate the development of more useful and informative protocols for clinical testing that could supersede the gothic procedures now practiced.

21.2 Models of Tremor Data

To introduce the subject from the point of view of an investigator interested in developing a deeper theoretical understanding of how tremor might be generated and how it might be correlated with observable brain activity, we introduce the example shown in Figure 21.1A. This figure shows ten seconds of data from a patient diagnosed with Parkinson's disease. Plotted here is the angular velocity of a resting rotational hand tremor consisting of alternating pronation and supination about the left wrist. The periodicity is evident as is the waxing and waning of amplitude. The regularity of the amplitude modulation is an interesting feature, and though relatively uncommon, serves to make a point. The power spectrum of this tremor record is shown in Figure 21.1B. It has two major peaks, both in the narrow 4 to 6 Hz range common for the Parkinson tremor[2].

Also shown in Figure 21.1A are two other tremor-like signals. One (top) was obtained by adding two sinusoids of fixed, but equal, amplitude, each tuned to the frequency of one of the two peaks in the spectrum. This is an almost literal interpretation of the spectrum of the real tremor. In theory, the original tremor can be completely reconstructed by Fourier analysis from sinusoids of multiple frequencies whose amplitude and phase are determined from the spectrum. Here, for simplicity, we have constructed a pseudo-tremor from just two sinusoids. Its resemblance to the real tremor is obvious.

A second tremor-like signal (middle trace) is also shown, again generated from two sinusoids, but with different frequencies. The first was at a frequency midway between the peaks of the real tremor spectrum. The second was a sinusoid whose frequency was chosen to match the periodic waxing and waning of the tremor amplitude. To obtain the pseudo-tremor in the figure, the two sinusoids were multiplied together. The first model was obtained by adding two constant-amplitude sinusoids; the second by multiplying two constant-amplitude sinusoids with frequencies different from the added sinusoids. It should be remembered that these are models of the data. We note that the tremor records look very similar, and, not surprisingly, so do their spectra.

This example was chosen to illustrate a fundamental problem in the study of tremor: conventional data processing techniques, such as spectral computation, do not distinguish the different ways by which the observational data might have been generated; that is, between entirely different conceptual models of the tremor-generation process. Only carefully designed experiments and procedures can do that, and each should be chosen so as to rule out, if possible, one or more of the candidate hypotheses. But it is essential to remember that there may be several equally valid hypotheses (all of which could be wrong). But without further testing, none can be ruled out. And just as certainly, there are alternative models that might just as well (though probably not as simply) account for the observed data.

Our second point starts with the observation that at present there are surprisingly few unequivocal and convincing observations of single central neurons that could be claimed to be involved in tremor generation (e.g., see Lenz et al., 1994, 2002; Zirh et al., 1998; Hurtado et al., 1999; Levy et al., 2000). This may reflect the fact that clinicians and experimenters have not placed electrodes in brain structures whose activity is the progenitor of tremor. But it may also be that their expectation of what might be encountered in the presence of a specific tremor reflects only one particular model of the tremor and hence one model of its genesis.

Different models of tremor, such as the simple ones shown above, would lead to entirely different expectations concerning the central correlates of its genesis. In particular, as this example shows, central neurons involved in tremor genesis may not be active at frequencies that dominate the spectra

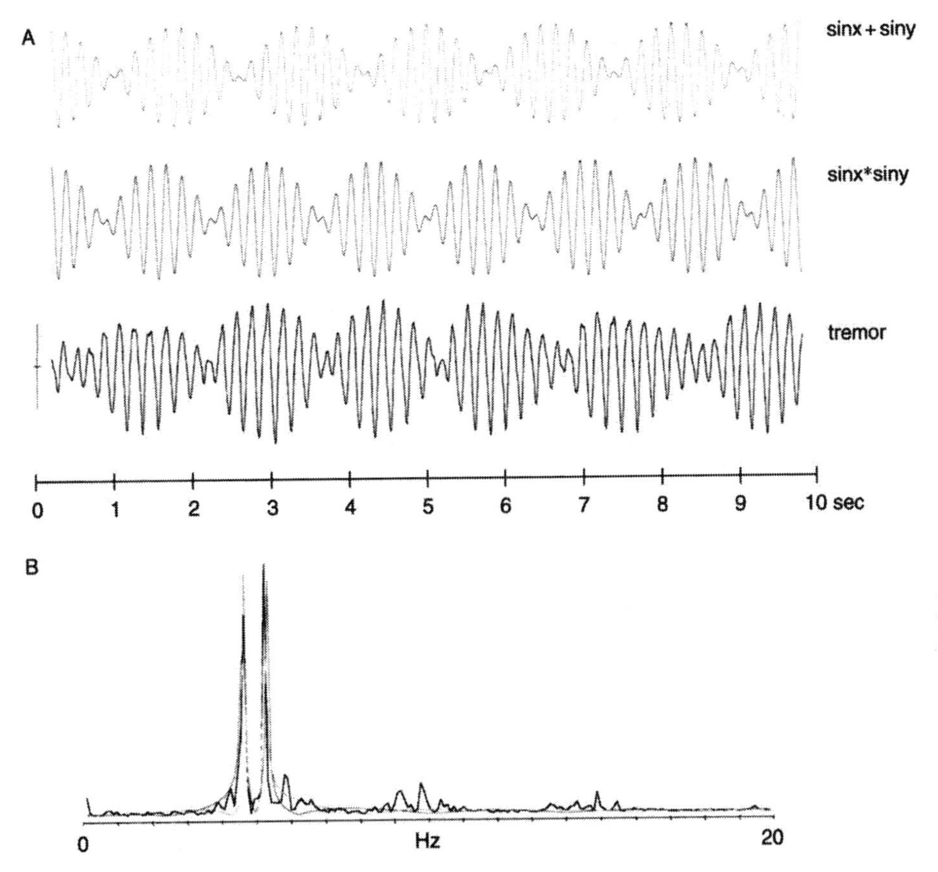

FIGURE 21.1 **(See color insert following page 366)** (A) Three 10 sec records. The bottom trace (black) is taken from a Parkinson patient with a resting tremor of the left hand. Sensors responding to angular velocity were oriented to record rotation about the pronation–supination axis of the wrist (pronation positive; scale: $\pm 100°$/sec). Note the quasiperiodic modulation of tremor amplitude. The two records above were computer generated. In the top record (blue), sinusoids of frequencies 5.2 and 4.6 Hz were added; their amplitudes were adjusted to compare with the real tremor. The middle record (red) was also generated by computer simulation, this time by multiplying two sinusoids. Amplitude modulation of the carrier sinusoid (at 4.9 Hz) by a sinusoid at 0.35 Hz creates two sideband peaks in the spectrum at $4.9 + 0.35$ Hz and $4.9 - 0.35$ Hz. (B) Spectra for the three records in (A), each in its own color. Note that in the spectrum of the real tremor there are smaller peaks at harmonic frequencies (near 10 and 15 Hz).

calculated from the peripheral tremor data, nor be correlated, in the conventional statistical sense, with that activity.

Different models create different expectations. And different styles in data processing implicitly reflect — or encourage — different assumptions (models) about how tremors are generated. In the example shown, the hypothesis that the tremor results from the superposition (addition) of two sinusoidal components at slightly different frequencies leads to the expectation that (at least) two central processes operating at those frequencies could be observed.

An alternative hypothesis for the same tremor — all candidate hypotheses must account for the observed data — leads to the expectation that there is a low-frequency/constant-amplitude central sinusoidal process that modulates the amplitude of another constant-frequency/constant-amplitude sinusoid with a higher frequency. Since neither of the frequencies of these two sinusoidal processes appear in the spectrum of the observed tremor, their central correlates might be assumed to be unrelated to the tremor when tested using standard methods of analysis.

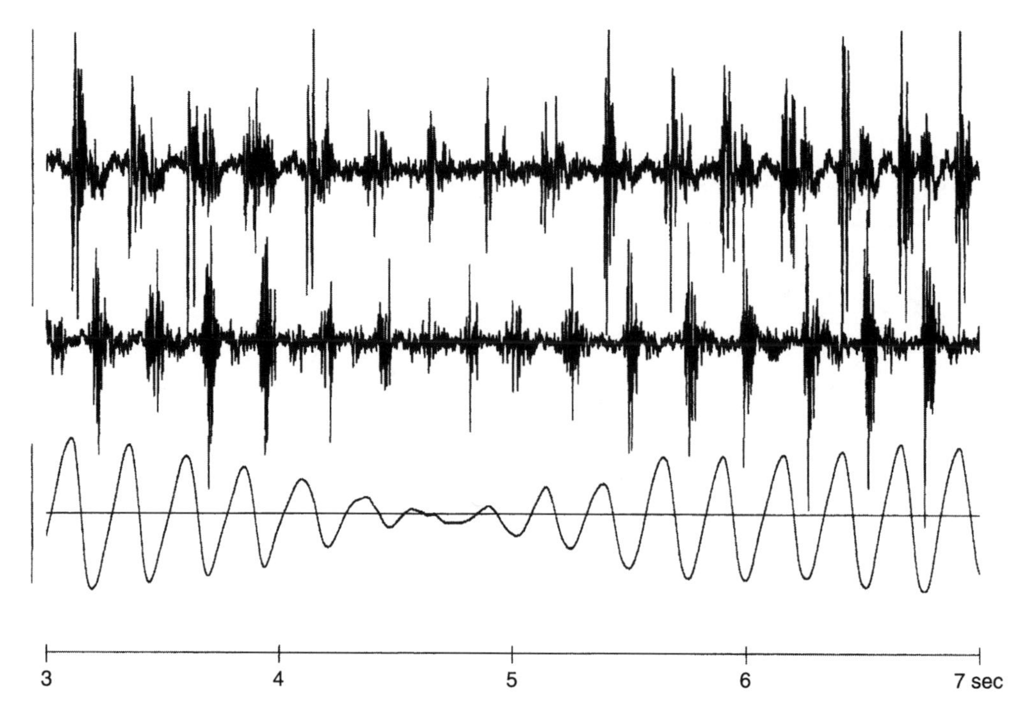

FIGURE 21.2 Four seconds of a resting flexion–extension tremor from a Parkinson patient (flexion positive). A null point is approached at around 4.5 sec. Shown above are concurrently recorded EMG bursts from the forearm extensors (top trace) and flexors (middle trace). At the beginning and the end of the record the flexor and extensor bursts alternate and have large amplitudes. As the null point approaches, the EMG phase relations are altered and cocontraction occurs; simultaneously, the burst amplitudes are diminished. Scales $\pm 200°$/sec; ± 0.1 mV.

Our third point is that when there are multiple hypotheses about the nature of the tremor data, there are opportunities for further clinical and experimental investigations of the tremor in the periphery. For example, for the tremor shown in Figure 21.1A, what is the pattern of muscle activity at the null points of the tremor? From the superposition hypothesis — the hypothesis that the tremor results from the addition of two sinusoidal components — one would expect to see alternating bursts of electromyographic activity in the agonist–antagonist muscles producing the tremor; the bursts would have fixed amplitudes, but a constantly changing phase, so that at times they would be out-of-phase and, in effect, additive and at other times in-phase (coactive) and canceling each other, creating null points in the tremor. The alternative hypothesis — amplitude modulation — leads to the prediction that electromyographic activity would consist of alternating agonist–antagonist bursts with a fixed phase relation but with a markedly diminished amplitude during the null points.

In Figure 21.2, we show one example of a tremor null point. It is of interest to note, although we make no claim it is a general finding, that as the null point is reached, the amplitude of the EMG signals is attenuated (a result anticipated by the amplitude-modulation hypothesis), but also that the alternating bursts of activity shift from out-of-phase to in-phase (a result anticipated by the superposition hypothesis). This example, therefore, is consistent with both models shown in Figure 21.1.

We include this example to bring home the point that, whereas tremor is basically a continuous and (in this example) single-channel process, its immediate antecedent generators are multichannel spike trains emanating from the spinal cord (themselves preceded by multichannel neuronal spike trains in the central nervous system), here combined in the final common pathways of muscle groups. Thus, the peripherally recorded tremor reflects central generation and transmission, peripheral effects contributed by normal neuromuscular biomechanics (the basic motor unit twitch response and its length–tension and force–velocity characteristics, all subject to potential pathological variations), physical constraints on movement, and effects generated by the sensors themselves (such as the gravitational components recorded by accelerometers[3]). Processing methods do not tell us which aspects of the tremor signal are peripheral or central, nor do they tell us which models accounting for the data may be correct.

21.3 Models of Multidimensional (Multiaxial) Tremor Data

Tremors are often spatially complex movements. For example, hand tremors are generally composed of concurrent rotations of more than one axis of the wrist: pronation–supination, for example, combined with wrist flexion and extension and/or ulnar flexion and extension. Tremors in more than one extremity may also occur concurrently. Central neuronal activity might be correlated with some but not all dimensions of these multidimensional tremors. If the tremor dimensions were independent, one might observe central neurons actually involved in tremor genesis but not be aware of it.

An example[4] is shown in Figure 21.3. Here, concurrently measured tremors from all four extremities are shown. At first, except for amplitude differences, they appear to be very similar (Figure 21.3A) and one would naturally be led to ask whether they were related in any way. In Figure 21.3B, their power spectra are shown. On the left side, spectra from the left hand and foot are superimposed; on the right, spectra from the right hand and right foot. Already, the spectral pairings show that the spectra from the right extremities, exhibiting multiple peaks, seem more related than the spectra from the left, which seem to share only a single peak (at a lower frequency than the peaks from the right-side spectra). This type of observation has led some authors to conclude that "multiple oscillators" are operating concurrently (Raethjen et al., 2000; Ben-Pazi et al., 2001).

Along with the spectra are the coherence functions for the pairs of spectra. The peaks of the coherence functions are well above the line of statistical significance[5], indicating that each of the two pairs share power (in effect, appear to have sinusoids in common, though perhaps of different amplitude) at the principal tremor frequencies. Indeed, the left-side tremors share power in the first harmonic band, even though the power levels there are not significant. This highlights the utility of the coherence function: since the power at the harmonic frequencies is hardly noticeable (and hence the cross-power spectrum would also be of low amplitude), the coherence function calls our attention to the shared power at those frequencies. The two right-side spectra also have comparably small amounts of power in the first-harmonic region, but they do not appear to be shared. All other combinations of tremor pairs in this example were found to be unrelated. It would not be possible to reach these conclusions by merely inspecting the original records. The left-hand tremor here would most simply be modeled as a constant-amplitude sinusoid, in contrast to the left foot tremor, which has noticeable amplitude variations. The right-side tremors share several peaks, implying that several periodic processes — at least — might be invoked in creating a model. These would be independent of, and in addition to, the single process on the left side.

A multiaxial (compound) tremor is shown in Figure 21.4, where two sensors have been placed on the right hand. One measures pronation–supination velocity (bottom trace), the other flexion–extension. The spectra for the two records are shown in Figure 21.4B (left). The spectra appear to be almost identical. But the coherence function in the region of greatest activity dips below the significance line. This can only mean that though the spectral components of the two records are almost identical, they are not present in the two records at the same time.

In Figure 21.4C (left) is shown the cross-correlation function for the two tremors. The largest peak in the plot is at +430 msec. Following this clue, the tremor records are replotted in Figure 21.4D with the flexion–extension record delayed by 430 msec. It can be seen that the two records now appear to be nearly identical, as if one tremor component were a copy of the other but delayed by over 400 msec, a neurological eternity. When the coherence function is recalculated using the delayed flexion–extension record, the coherence in the 5 Hz band becomes highly significant (Figure 21.4B, right)[6].

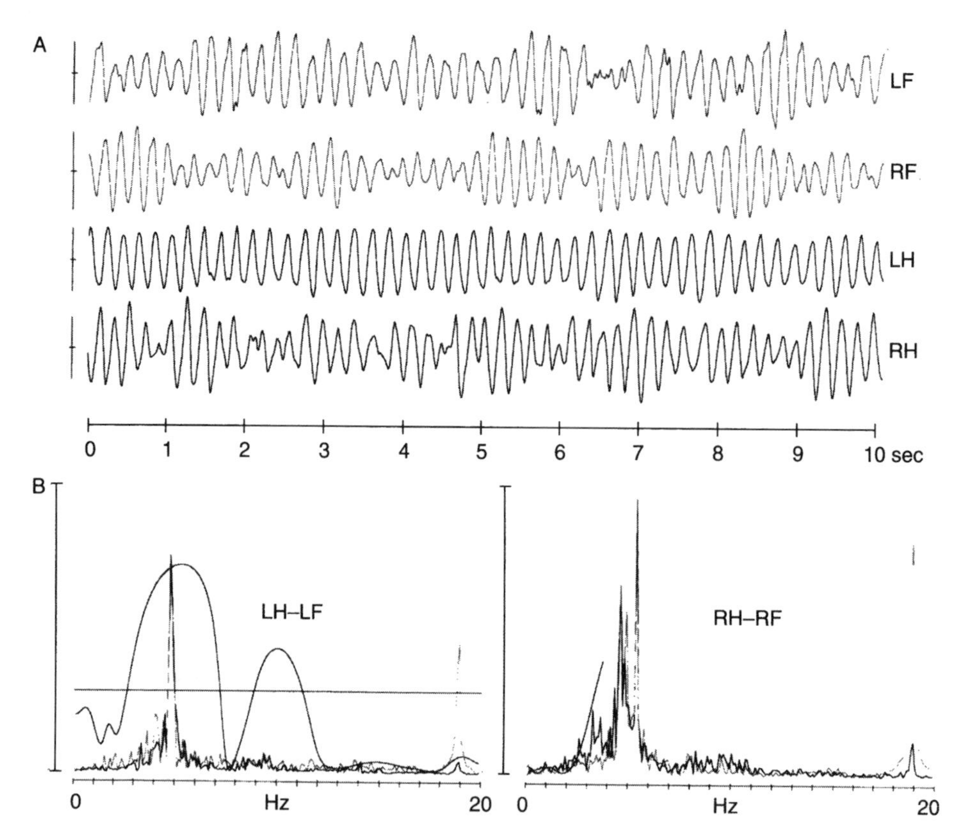

FIGURE 21.3 **(See color insert following page 366)** (A) Four 10 sec records of resting tremor concurrently measured in a patient with Parkinson's disease. Top: Flexion–extension rotation of the left foot about the ankle (extension positive; scale: ±20°/sec). Second from top: Flexion–extension rotation about the right ankle (extension positive; same scale). Bottom trace: Pronation–supination hand tremor about the right wrist (supination positive; scale: ±100°/sec). Second from bottom: Pronation–supination hand tremor about the left wrist (pronation positive; scale ±200°/sec). (B) Left: Power spectra for the left hand (black) and left foot (red; color designations here and in other figure legends refer to the color insert) tremors are superimposed (using different scaling factors). They share a peak near 4.7 Hz. Right: The spectra for the right hand (black) and right foot (red) tremors are superimposed (also using different scaling factors), with large peaks shared near 5.4 Hz. The smooth curve superimposed on each pair is the coherence function which, in the range of greatest power, is considerably above the horizontal line, which represents the 95% confidence limit. Sinusoidal calibration signals of 19 Hz were added to each tremor record prior to calculating spectra and removed before calculating coherences. Their peaks appear toward the right in each spectrum and indicate how the data processing methods treat a pure sine wave. The sinusoids in each case had peak-to-peak amplitudes of 20°/sec. The different scale factors used to plot the spectra result in different heights for the four calibration peaks at 19 Hz.

In this example, the coherence function showed that although the spectra of the two tremor components were nearly identical, the spectral components were not present and in-phase concurrently. The correlation function provided a clue as to how the tremors might be related.

21.4 HARMONICS AND PHASE

Another way of examining the internal temporal structure of tremor data is shown in Figure 21.5. At the top (Figure 21.5A) is shown ten seconds of a resting pronation–supination tremor of the right hand. This tremor clearly exhibits a prominent harmonic component during pronation (positive velocities). To the right of the record is shown an amplitude histogram of the data, indicating that most of the

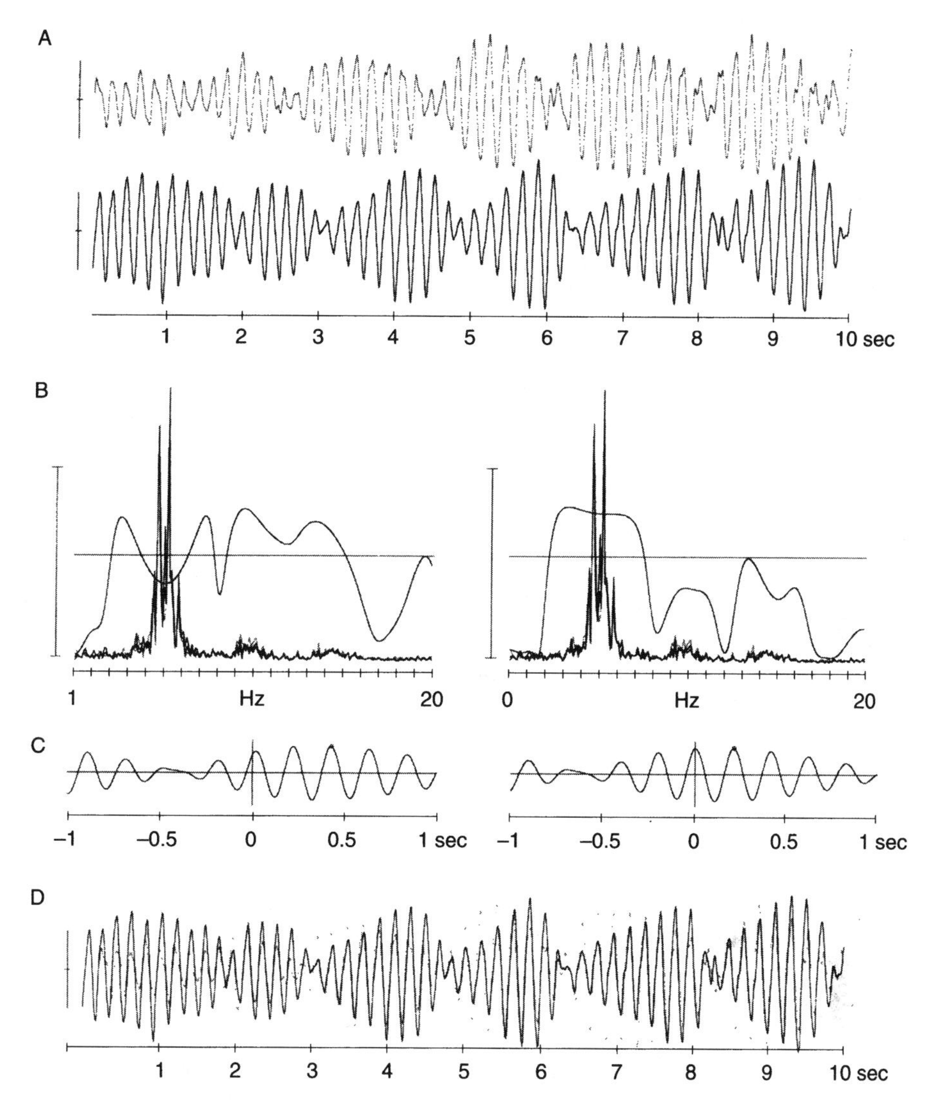

FIGURE 21.4 **(See color insert following page 366)** (A) Top trace (red): angular velocity of flexion–extension about the right wrist in a Parkinson patient (extension positive). Bottom trace (black): concurrently measured tremor along the pronation axis of the right wrist (supination positive; scales: ±100°/sec). (B) The spectra of the two records are superimposed and seem indistinguishable (colors as in A). The coherence function lies below the level of statistical significance in the important band of power (around 5 Hz). This indicates that the spectral frequencies are not shared concurrently. (C) Left: The cross-correlation function computed from the two tremors. The peak value (circled) is at +430 msec (flexion–extension leading pronation–supination). Right: The coherence function (and spectra) recomputed for the data, time shifted as in D, is significant in the 5 Hz band of shared power. (D) The two tremors from A are superimposed, with the flexion–extension record delayed by 430 msec (colors as in A).

time the hand is pronating, but with smaller velocities on average than during the supination phase. Harmonics are often prominent in tremor data, yet no studies have determined even the simplest aspects of their production, for example, the manner in which muscle groups responsible for tremor participate in the production of its harmonic components[7].

In Figure 21.5B are shown phase-loop plots of the tremor data. Here, the angular velocity of the tremor at various times (horizontal axis) is plotted against the angular velocity of the tremor *T* msec later. The value of T is shown near each loop. At zero delay is the identity loop, that is, a line at 45 degrees; the correlation coefficient for those points would be +1.0. At 110 msec delay, the loop has a figure-8 shape, a feature derived from the harmonics that create two cycles of pronation for each supination cycle. Since the duration of the tremor cycle is about 210 msec, as the delay T reaches that value, the loop for that delay closely resembles the loop for 0 delay. The values of the autocorrelation function at each of these delays are also shown in the center of the figure. The average relation between values of the tremor velocity at one moment relative to its later values are seen to be far more complex, in a geometric sense, than the scalar autocorrelation function alone would suggest.

Although the phase loops appear complex, it is possible to create a simple model for the data. Noting the period of the tremor cycle, which corresponds approximately to 4.6 Hz, and the presence in the record of what appears to be a first harmonic (which then ought to be at 9.2 Hz — twice the fundamental frequency) with a fairly stable phase relation, we construct a model that is the sum of two sinusoids whose amplitude and phase relation are chosen to match the experimental data. The resulting model and its phase loops are shown in Figure 21.5C and D.

21.5 Temporal Stability

The previous example illustrated a tremor which had a certain degree of stability in its temporal structure. But not all tremors show such stability. In Figure 21.6A we show a 10 sec excerpt from a

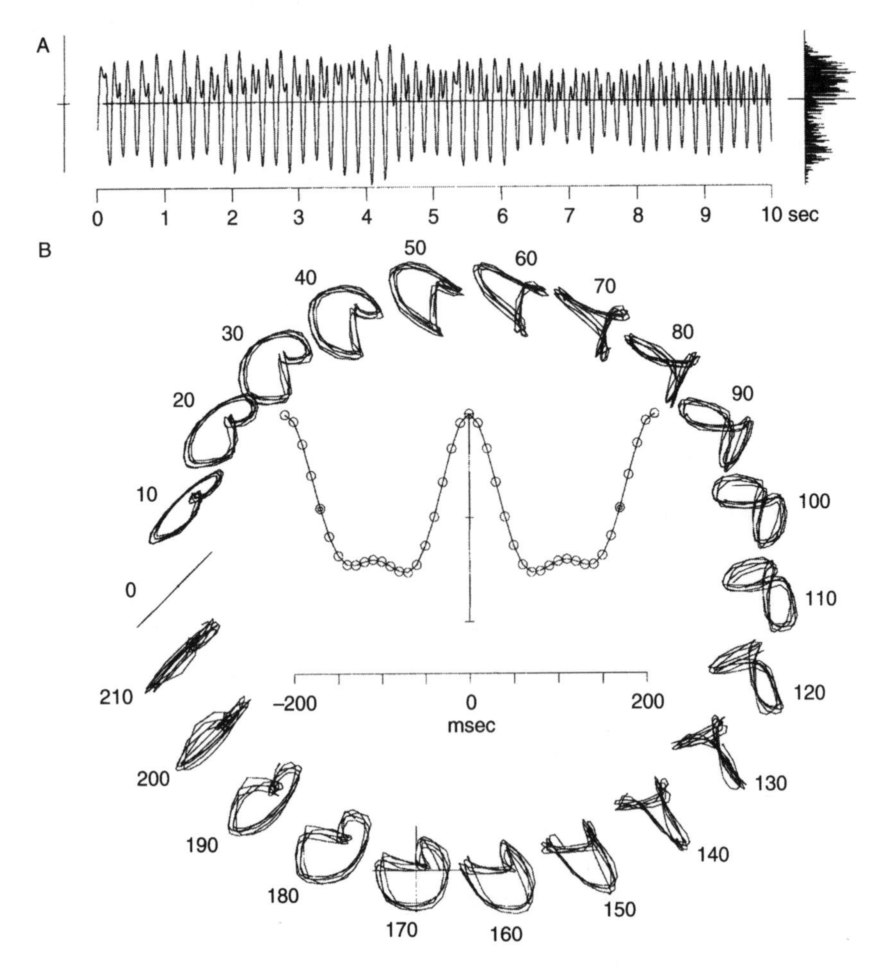

Figure 21.5 Continued on next page

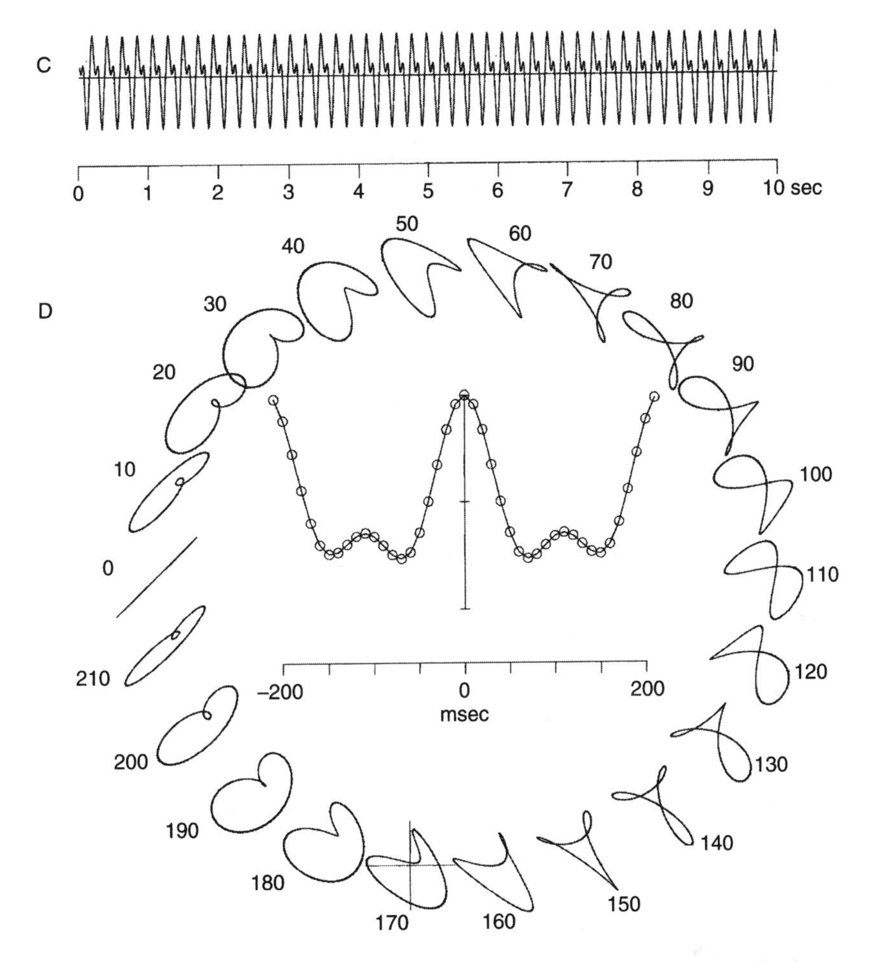

Figure 21.5 (A) Ten seconds of angular velocity data during a resting pronation–supination tremor at the right wrist of a Parkinson patient (pronation positive; calibration: ±100°/sec). To the right is shown the highly asymmetric amplitude histogram of the data. Note the strong presence of harmonics when the hand is pronating (positive velocity). (B) Lower portion shows phase loop plots in which the tremor velocity at time t (horizontal axis) is plotted against the tremor velocity at time $t + T$ (vertical axis), where T is the delay (in msec) indicated near each loop. Each loop plots only 100 points (0.1 sec) beginning at 0.5 sec. That is, only data from the first third of the entire record are used. The loop for 170 msec includes scale axes of ±100°/sec. Within the circle of loops is shown the autocorrelation function from 0 to 210 msec delay. The value of the function lies between −1.0 and +1.0, as indicated by the central axis. The value at zero delay is, by definition, 1.0. The maximum negative correlation is at a delay of 70 msec. The correlation is near zero at delays of 40 and 170 msec. Note that mirror image loops appear at symmetric parts of the cycle (e.g., 70 and 140 msec delays). (C) Model of the data in A simulated using two sinusoids of frequencies 4.6 Hz (fundamental) and 9.2 Hz (first harmonic). The amplitude of the lower-frequency sinusoid was ±180°/sec; the higher-frequency sinusoid had an amplitude of 130°/sec and a phase difference of 2.1 radians. (D) Phase loops for the simulated data, as in (B).

30 sec resting tremor of the left hand of a Parkinson patient. Sensors were oriented to record angular velocity of rotation along the pronation–supination and flexion–extension axes. In Figure 21.6B are the two power spectra and the coherence function. The cross-correlation function (not shown) had a peak at −20 msec, indicating that one of the rotations is delayed, on average, by 20 msec with respect to the other. These tremor records seem uncomplicated. Yet, in Figure 21.6C, where we plot extension velocity (horizontal axis) against pronation velocity in 1.5 sec subsections of the data, we see a more complex relationship. Though the tremor components seemed nearly identical

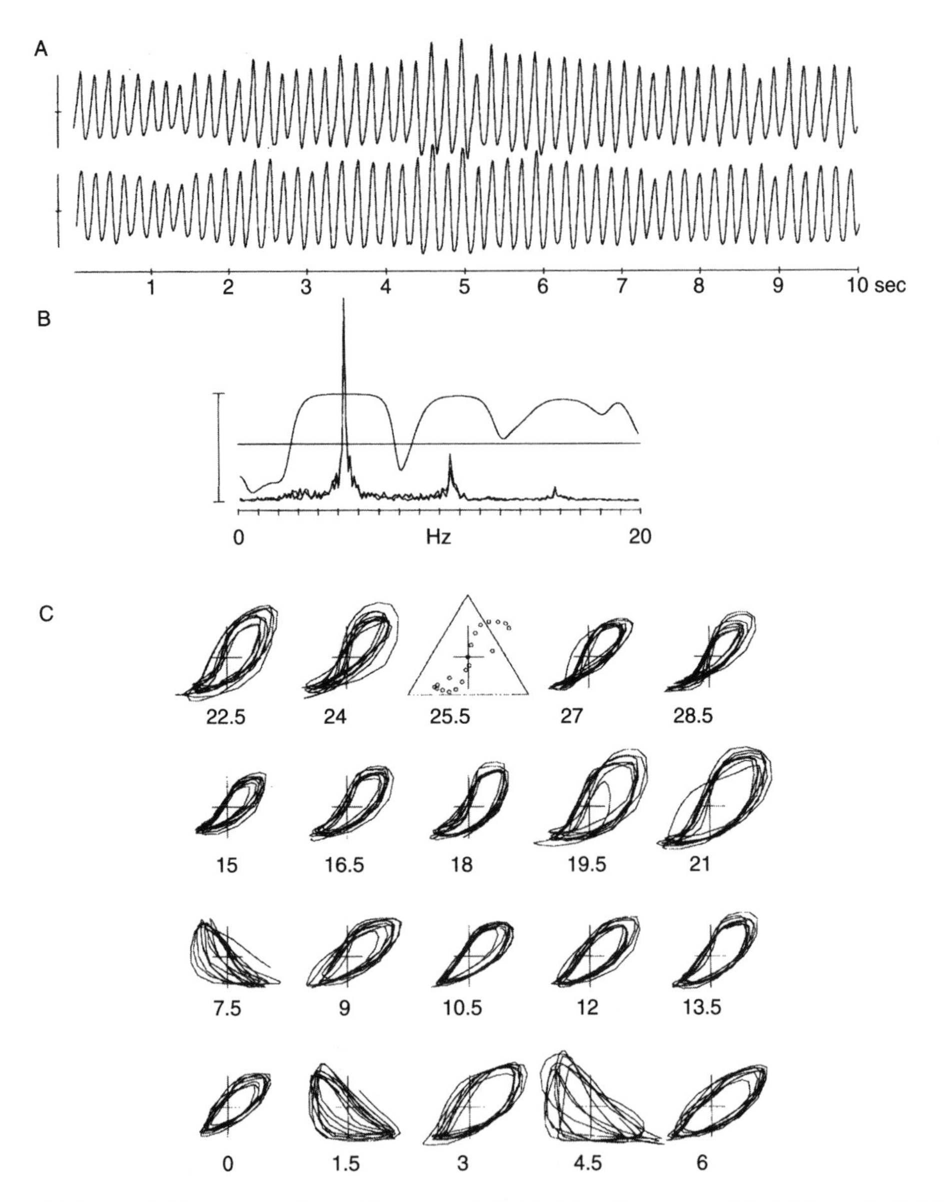

FIGURE 21.6 (A) A 10 sec excerpt from a 30 sec record of a biaxial resting tremor of the left hand of a Parkinson patient. Top trace: velocity about the flexion–extension axis (extension positive); bottom trace: rotation about the pronation–supination axis (pronation positive). Calibrations: ±100°/sec. (B) Spectra and the coherence function for the 10 sec excerpt shown in (A). (C) Successive phase-loop plots of extension velocity (horizontal axis) vs pronation velocity. The numbers below each loop show the time segment from which the loop was derived. Each point in the loop defines the instantaneous vector axis of rotation. The changing loop geometry indicates the time-varying orientation of the tremor axis in the plane. The first ten points from the segment beginning at 25.5 sec, plotted as circles, form a figure-8. Each loop plots 150 consecutive points and includes scaling axes of ±100°/sec.

we see now the more intimate and time-varying phase relation between them. The general shapes and amplitudes of the loops are variable, sometimes undergoing a radical shift quite suddenly, reflecting a change in the basic coupling between the two concurrent rotations (e.g., the abrupt shape changes occurring at 1.5 and 3.0 seconds). During the first ten seconds the phase loops alternate between two semistable configurations. There is nothing to suggest this in the raw data. At around

24 sec the tremor loop has become a figure-8, as shown in the insert for 25.5 sec. This reflects the presence of harmonics of significant amplitude at that time, though they are not obvious in the data records.

The locus of points in these loops describes the time-varying vector axis of rotation of the hand. Where the data points lie on the vertical axis, the hand tremor is — momentarily — purely pronation–supination; at moments when the data points cross the horizontal axis, the motion is purely flexion–extension. The vector axis rarely lies in the quadrant of concurrent pronation and flexion, and when it does, the velocities of rotation are quite small. Indeed, the radical shape changes involve large concurrent pronation–flexions.

In general, tremors are best approached as nonstationary processes[8], though again this depends on the length of time over which a tremor is observed. Records with a duration of ten seconds or less may appear to have stable spectra and correlation functions. But even on a timescale as short as 1 min, it can be seen that tremors are generally highly labile. Even if the tremor persists as one continuous record, its component frequencies shift, split into sidebands, exhibit harmonics, and have variable power.

An example is shown in Figure 21.7, where sequential spectra computed from overlapping 10 sec epochs are plotted. Over shorter subsections of the record, the spectra seem stable. Overall, however, the impression is one of ever-evolving and evanescent characteristics. In this example there is initially a single frequency in the tremor at around 5.5 Hz. After about 10 sec another spectral component appears near 4.2 Hz and begins to drift toward 4.5 Hz; the peak at 5.5 Hz begins to disintegrate, but later (around 30 sec) reappears before splitting into what may be sidebands that coexist with a very sharp, but apparently unrelated, peak at about 4.5 Hz. At around 40 sec the 4.5 Hz peak splits, possibly into sidebands, but reemerges as a single peak near the end of the record. At various points in the record, the tremor becomes briefly quiescent; as reflected in the absence of significant peaks in the corresponding parts of the spectral array. Further studies of the changes in muscle activity that accompany these changes in spectral composition are needed.

21.6 Models of Other Complex Tremor Data

In Figure 21.8 is shown a different Parkinson tremor, known as an "action" tremor because it accompanies voluntary movement. When this tremor was recorded, the patient was executing a series of large-amplitude voluntary pronation–supination rotations of his left hand. The voluntary cycles recurred at a rate of about 0.6 cycles per second; they are seen as the large slow positive and negative velocity waves in the record. This slower periodic movement was accompanied by a higher-frequency involuntary tremor at around 8/sec. The tremor is notable for its large amplitude when pronation occurs (when the overall curve is above the baseline) and its much reduced amplitude during supination. The spectrum of the tremor record shows a very strong peak near 0.6 Hz and a series of peaks from 7 to 9 Hz. The 0.6 Hz peak is generated by the pronation–supination movement, but what is the source of the many peaks at the higher frequencies? Are multiple generators operating here?

We note that the tremor record is definitely not a sum of two sinusoids, one at 0.6 Hz and the other near 9 Hz. We can rule that out because such a sum would produce a constant-amplitude high-frequency tremor superimposed on the lower-frequency pronation–supination movement. Moreover, the spectrum of such a record would have but a single peak near 9 Hz (as in one of the models in Figure 21.1). Since the amplitude of the high-frequency component is so different in the two phases of the cycle, we could imagine, in a simplified model of the tremor data, that the action tremor might be triggered by pronation and suppressed by supination. We know that such a model is not correct because the action tremor does not entirely disappear during supination, but we can explore this model and determine its consequences.

There are a number of ways the action tremor might be related to pronation. In one model, each pronation cycle might start a 9 Hz sine wave that begins at zero velocity; another model might consist of a continuously running sine wave that is gated by pronation and simply added to the

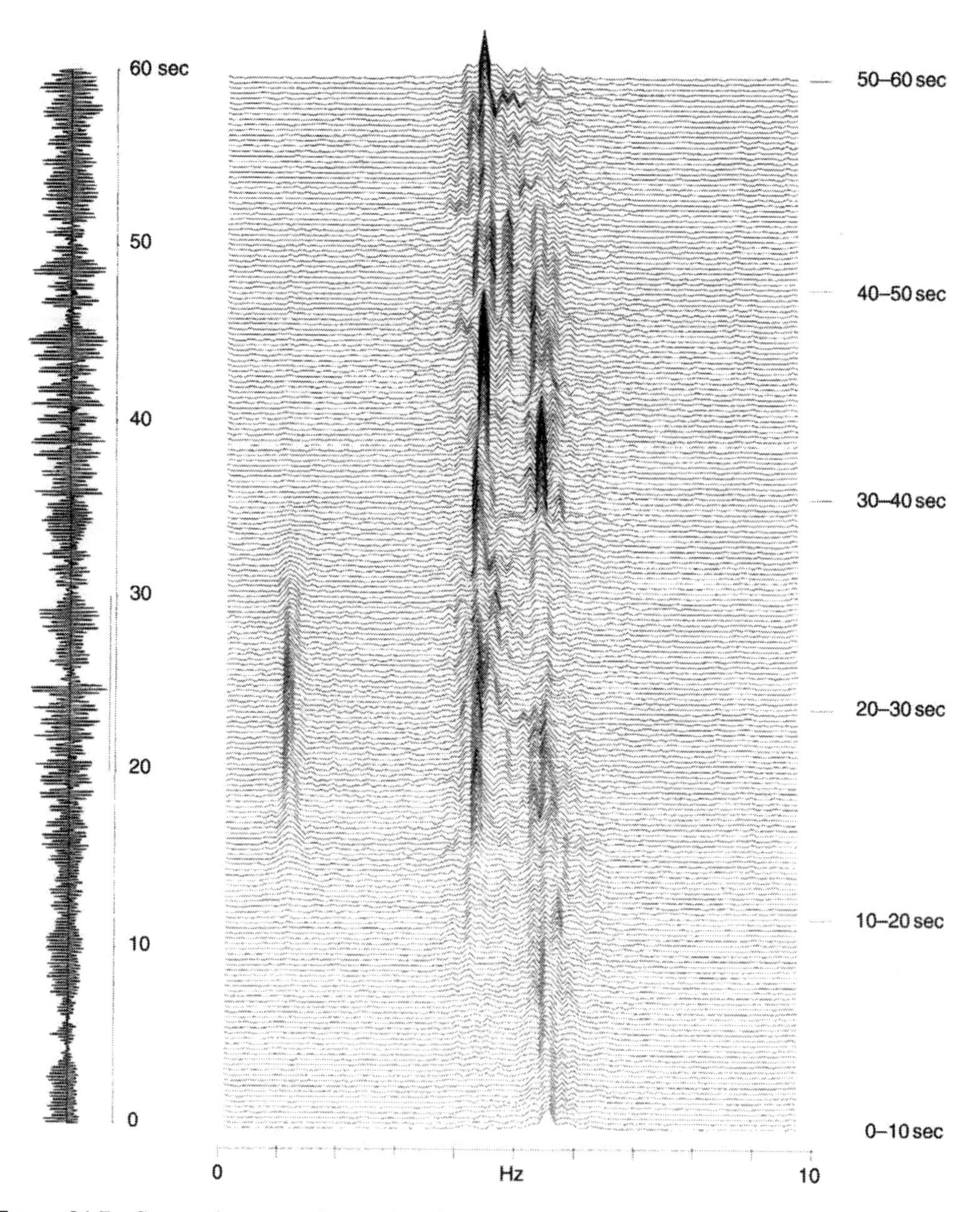

FIGURE 21.7 Consecutive spectra for a series of overlapping 10 sec segments of a 60 sec resting tremor are plotted sequentially beginning with (bottom trace) 0 to 10 sec and ending with (top) 50 to 60 sec. The entire tremor record is shown to the left. To clarify the effect on the spectra of the overlapping nature of the data segments, a 1.1 Hz constant-amplitude sinusoid was added to the tremor record during the time period 20 to 30 sec (indicated by the solid line on the tremor time base). It can be seen to emerge (at the 1.1 Hz frequency band) in the spectra at around 10 sec, and to decline after 20 sec. Though it is present only from 10 to 20 sec, the impression here is one of gradual emergence and disappearance. The appearance, disappearance, and shifting of other peaks should be interpreted accordingly. The width of the 1.1 Hz peak indicates how a single fixed frequency is represented.

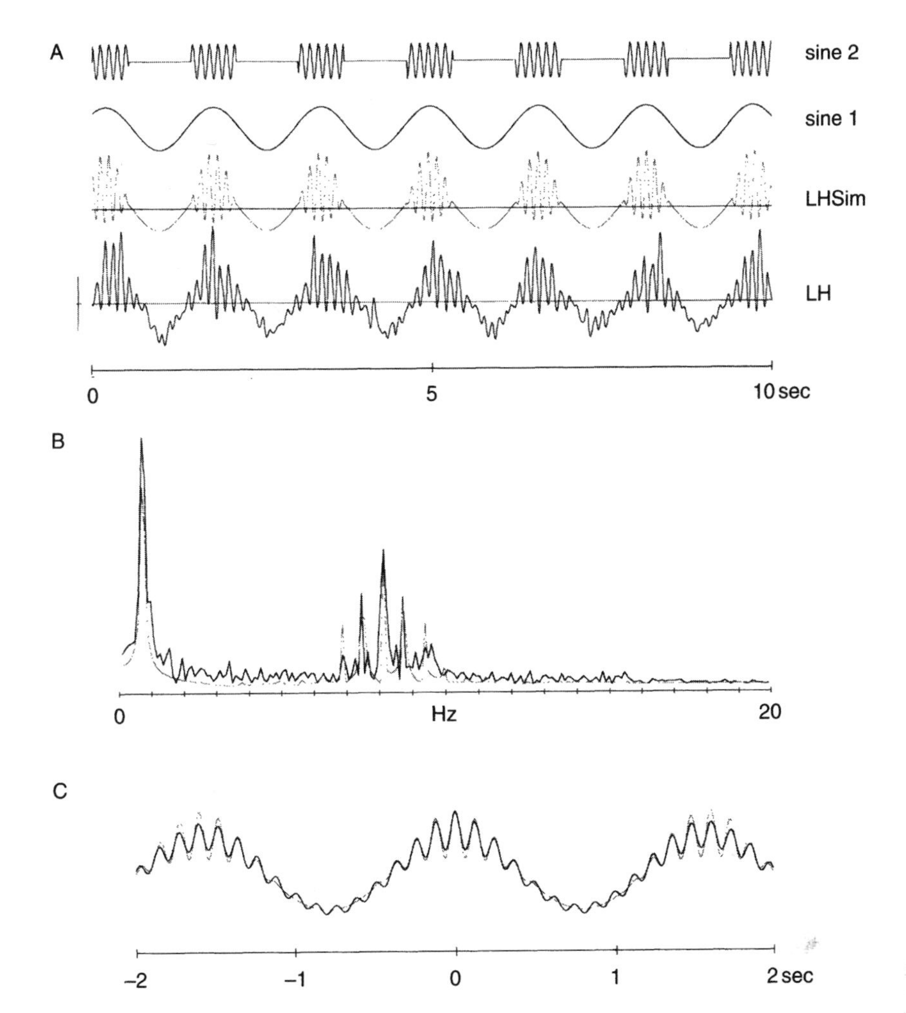

FIGURE 21.8 **(See color insert following page 366)** (A) Bottom trace (black): ten seconds of data from a Parkinson patient executing voluntary alternating pronation and supination movements of the left hand (pronation positive; scale ±200°/sec). Approximately seven complete cycles are shown. Superimposed on this slow oscillation can be seen larger- and higher-frequency (involuntary) oscillations constituting a so-called "action tremor." Above this trace is a simulation of the tremor (red). Two sinusoids were used: a higher-frequency sinusoid at 8.0 Hz. and a lower-frequency sinusoid at 0.6 Hz. The high-frequency signal is gated when the low-frequency signal exceeds zero, and this gated sinusoid (top trace) is added to the low-frequency sinusoid (next trace below) to produce the approximation shown directly above the actual tremor. (B) Spectra for the two lowest records in (A) (colors as in A). Both consist of a sharp peak near 0.6 Hz derived from the voluntary movement and a group of peaks from 7 to 10 Hz that are derived from the basic frequency of the action tremor. (C) The autocorrelation functions of the real and simulated records in (A) (colors as in A).

background velocity record. Various models might also assume some amplitude modulation of the sine wave during pronation. Whatever the choice of hypothesis, modeling allows us to explore its consequences.

In Figure 21.8A (top), we show how one simple, although nonlinear, model is created. In this model, two continuous sinusoids were used: one at 0.6 Hz, the frequency of the voluntary movement, the other at 9 Hz. When the lower-frequency sinusoid is positive it gates the 9 Hz sinusoid. The gated sine wave is then added to the low-frequency signal. This is an example of a nonlinear interaction between two sinusoids (as in one of the models in Figure 21.1). The spectrum for this model is superimposed on the spectrum of the real tremor in Figure 21.8B. What is immediately obvious is that the model generates multiple spectral peaks near 9 Hz. The central 9 Hz peak corresponds to the high-frequency sinusoid. The sidebands arise from the nonlinear interaction between the low- and high-frequency sinusoids. They are spaced on either side of the 9 Hz peak at intervals equal to 0.6 Hz, the frequency of the pronation cycle. The autocorrelation functions for the two records are superimposed in Figure 21.8C and are remarkably similar, suggesting that the model in which the underlying action tremor runs continuously is consistent with the data. Indeed, such a sinusoid can be superimposed on much of the pronation portion of the record in Figure 21.8A, although the same sinusoid is not a particularly good fit to the supination phase, suggesting that pronation and supination might each gate a different sinusoid. Of course, such hypotheses have to be challenged and tested further. Note, however, that our results do not distinguish whether the asymmetric amplitude of the tremor at different parts of the cycle is a peripheral or a central effect. The physiological basis of this tremor is an ideal target for additional clinical experimentation and testing.

21.7 Future Issues

In the course of this chapter we have pointed to a number of issues that need further work and clarification. Here, we wish to make some more general observations.

Models are a measure of the maturity of a field of investigation. They give an indication of the range of experimental or clinical phenomena for which hypotheses have been sufficiently specific and comprehensive to be put in mathematical form and subjected to comparison with observational data. When successful, they offer explanations for puzzling observations, and can be used to formulate future experiments and to predict their outcomes. Even when they are unsuccessful, they point to errors in the governing hypotheses, and suggest alternative hypotheses and new experiments.

We noted earlier the almost total absence of models — in this sense — in the study of tremors and the normal and clinical conditions in which they arise. This remark applies to models of the tremor data as well as of the central physiological processes by which tremors might be generated. There may be several explanations for this, the simplest being that it is too early, given our scanty grasp of basic phenomena and mechanisms, to expect serious models of the observations and results such as those discussed here. Or, conversely, some may feel that our empirical knowledge is already so advanced that we are beyond the need for theoretical models. Modern neurology has seen many examples of early simple models that, in an effort to conform to biological reality, eventually became so complex as to lose their usefulness, becoming as complex and inscrutable as the phenomena they attempted to describe. But even relatively simple models can confer order in a field, establish a useful terminology, pose key questions, and erect a conceptual framework within which further studies can evolve — even to the point where the original theoretical formulations have to be abandoned as incomplete or misguided. Whether we continue to restrict our attention to an empirical study of tremor remains to be seen. In the meantime, an attitude of humility in the face of the challenge would be appropriate.

Research Problems[9]

1. Harmonics are a common feature of tremors and their spectra. (See samples in the accompanying database.) Consider an angular velocity sensor placed on the dorsum of the hand and oriented to monitor pronation-supination rotations produced by the forearm at a single fundamental frequency. If the hand movement also had a component of flexion-extension at the wrist and/or ulnar flexion-extension, i.e., if the hand were rotating about several axes at the same fundamental frequency, could the sensor signal produce harmonics? It is important to determine whether harmonics result from neural signals at harmonic frequencies or from complex movements in multi-axial planes.

2. Action tremor, a tremor in Parkinson patients that is generated especially during repetitive voluntary movements, is easily seen in records taken from velocity sensors that record the voluntary movement as well as the tremor. An example was shown in Figure 21.8, whose data are included in the database. A prominent feature of the tremor is the phase dependence of its amplitude. Is it reasonable to postulate that this dependence could result from the known biomechanical properties of an agonist–antagonist pair of muscles, such as their length–tension relations? What additional experimental data or tests would be desirable?
3. Given the usual clinical practice of taking only short records of tremor, it is often claimed that concurrent tremors from different extremities are uncorrelated. Yet, a few instances where this is clearly not the case have also been reported. In all likelihood, sustained concurrent tremors are sometimes correlated and sometimes independent. What has been lacking are relatively long records of three or more tremors. Given the availability of such records, develop a useful way of displaying the time-varying relations between multiple tremors. Test data may be found in the database.
4. One of the enduring questions of tremor physiology is whether the peripheral tremor generates sensory afferent signals that affect the subsequent history of the tremor. Conversely, when recording tremor and the electrical activity of single central neurons simultaneously, how can one distinguish sensory from motor components of activity? Is the question meaningful? Is the issue resolvable? Would peripheral perturbations be useful? Some records of tremor and single neuron activity in the brain are included in the database.
5. In almost all cases, tremors are monitored by systems that record movement — either position, velocity, or acceleration. Would there be advantages instead in monitoring force in tremors that are generated isometrically, for example, by using force transducers associated with some sort of device that restricts movement in the periphery?

NOTES

1. When we consider normal movement and its pathological variations we must, on the basis of considerable clinical, anatomical, and physiological evidence, include several related and interconnected subcortical structures known, collectively, as the basal ganglia. At the present time, their interrelationships with each other and with other motor and sensory areas of the brain are often represented diagrammatically (e.g., see Albin et al., 1989; Chesselet and Delfs, 1996; Volkmann et al., 1996; Wichman and DeLong, 1996; Bergman et al., 1998a; Lang and Lozano, 1998; Lozano et al., 1998; Bar-Gad et al., 2001; Vitek, 2002). These diagrams, which borrow heavily from the technical terminology of electrical engineering ("networks," "circuits," "feedback," "information processing," etc.), use that terminology in a purely metaphorical sense, and should not be confused with engineering black-box diagrams, which they resemble. Rather, they are equivalent to a diagram that shows the tripartite structure of the American federal government, with lines connecting executive, legislative, and judicial branches. Such a diagram does not begin to account for the nature of each branch nor the complex legal and political interactions taking place within and between them.

   Brain-related diagrams attempt to sketch some basic anatomical and physiological relationships but are essentially mnemonic devices. Expectations that they might explain some aspects of normal and abnormal motor and sensory behavior have largely been disappointed. At the present time, they are but a preliminary sketch of a very complex reality, without either explanatory or predictive power. They await additional contributions from future research that, in filling out the picture with new or presently unappreciated details, may possibly help us to understand what, for now, are simply puzzling, complex, and relatively inexplicable phenomena. We can see how far they are from being models (in the engineering sense) by noting that one recent diagram had 11 compartments interconnected by 25 directed lines. Assuming that each compartment was at least second order, the system would require the determination of some 50 parameters in order to become computable. What is not clear is how, in principle, even a single parameter can be estimated.
2. It is important to point out that the data in Figure 21.1A are one of several equivalent representations of the tremor. They are in the form of a time series showing the velocity of rotation. This particular representation — or model — of the data captures but one aspect of the tremor, and it is but a sample of the tremor. The spectrum

of the time-series data is another representation, and when understood to include both its real and imaginary components (or, alternatively its amplitude and phase components), is entirely equivalent to the time-series representation, or model. But it calls our attention to the frequency aspects of the data. Either can be derived from the other by means of a Fourier Transform (e.g., see Bendat and Piersol, 1971, or any other textbook on signal analysis). The power spectra we show in these figures are obtained from both the real and imaginary components. A third representation (model) of the data, which is also related by means of a Fourier Transform, is the autocorrelation function, an example of which is shown in Figure 21.4C. This representation emphasizes the degree to which the tremor signal in Figure 21.1A is correlated with itself at past and future moments in time. These three models of the tremor record are only first-order descriptors of the data.

3. The traditional sensor for measuring tremor has been the accelerometer. These have the distinct disadvantage of reflecting changes in the orientation of the sensor with respect to gravity. Hence, if tremor motion does not lie entirely in the horizontal plane, a time-varying component of the transducer signal is contributed by the gravitational field. More recently, low-noise angular-velocity-sensitive transducers have been used to measure tremor. We used a gyro-based sensor to obtain all records shown here (Motus Bioengineering, Benicia California). These sensors do not respond to changes in their orientation with respect to gravity. Some caution needs to be used in interpreting their signals when the tremor motion is not planar.
4. This chapter is not intended to discuss data processing methods for tremor, and in most respects our methods are not different from the methods used by others except insofar as our questions are different. Methods used in this chapter have been described elsewhere (Moore et al., 2000) and can be summarized briefly: signals from the sensors were sampled at 100/sec. Spectra were computed using standard Fast Fourier Transform subroutines. Coherence spectra were averaged from 10 to 15 nonoverlapping 64-point subsections of the record.

   For conventional methods as developed by others, see Timmer et al., 1993, 1996, 1998; Halliday et al., 1995; Spieker et al., 1998; Suilleabhain and Matsumoto, 1998. Some authors have developed very different models from those employed here, in particular autoregressive models of the data (Randall, 1973; Speyers-Ashby et al., 1998). Gresty and Buckwell (1990) offer a useful introduction to the interpretation of tremor spectra; Arihara and Sakamoto (1999) introduce methods specific for testing a mechanical model of tremor.

   Several authors have developed techniques different from ours in an effort to classify differentially tremors arising in different clinical conditions: essential tremor, Parkinson tremor, physiological tremor, and others (e.g., see Gantert et al., 1992; Edwards and Beuter, 1999, 2000; Jakubowski et al., 2002). We do not deal with those issues here, nor with the thorny issues raised by the need to classify tremors in the absence of objective and definitive clinical criteria. For this, the reader can consult the Consensus Statement on Tremor (Deuschl et al., 1998) and other sources (Cleeves et al., 1994; Deuschl et al., 2000; McAuley and Marsden, 2000; Sullivan and Lees, 2000).
5. All of the statistical estimates utilized in the study of tremor — spectra, coherence, and correlation — depend on procedures that are perilous for the unwary, especially as they may be carelessly invoked by a simple keystroke in readily available data analysis software. As David Brillinger has pointed out, significance tests depend on assumptions about — models of — the data. At present we have no useful model for tremor data; our sensors have very low noise levels in comparison to the magnitude of the tremor signals, so the statistical problem of significance is not one of signal : noise ratio, for example, of detecting a signal buried in noise. Thus if we want to determine whether a spectral component is present simultaneously in two tremors we are ordinarily not looking for a spectral peak buried among other peaks. Here we have used a measure of significance given by Rosenberg et al. (1989) based on a different model than we believe is strictly applicable to our tremor data. In practical applications such as the ones discussed here, we rely on data simulations to determine the limits of detectability, discriminability, and/or resolvability in our estimates of coherence functions, and, to a lesser extent, in our spectral and correlation estimates.
6. There is some ambiguity about the time shift for one of the tremor records in Figure 21.4D. A shift of 430 msec was used here, but a shift of half that, which is one full tremor cycle less, seems equally good in terms of superposition fit and coherence in the spectra. The notion of a central delay of these magnitudes would pose a challenge to models of the central generators.
7. Harmonics — up to three or four multiples of the fundamental(s) — are a prominent feature of some tremors. They recall reports of single units in the brain of experimental animals that show activity at harmonics of their fundamental burst frequency (e.g., Bergman et al., 1998b), and reports of harmonics in EMG signals

at the periphery. Here again is an opportunity for additional investigation at the site of the tremor in which EMG and tremor activity would be recorded concurrently.

8. There is some debate about the taxonomic classification of tremor signals. The theoretical classification of tremors (e.g., as being stationary vs nonstationary, chaotic vs stochastic, linear vs nonlinear, or some combinations of these) seems not to have clear meaning or deep significance at this point. Some confusion arises simply from differences in methodology, especially in the choice of record length. Long records inevitably have more complex spectra, but a single spectrum hardly conveys the dynamic and evolving nature of a longer time series. Short records often appear stable.
9. Data sets related to this chapter and to the research problems given here may be found at two web sites: www.neurosciences.glasgow.ac.uk/~book and www.motusbioengineering.com

22 Principles and Methods in the Analysis of Brain Networks

Olaf Sporns

CONTENTS

22.1 Introduction

Throughout the history of neuroscience, there has been intensive philosophical and theoretical discussion over whether brain function should be conceptualized as being predominantly localized or distributed. In modern cognitive neuroscience, the problem of localization of brain function has remained a central issue in terms of methodology and brain mapping (Brett et al., 2002), as well as in terms of global theories of brain function (e.g., Edelman, 1987; Friston, 1997; Mesulam, 1998; Fuster, 2003). In recent years, the debate over localization of function in the brain has been transformed, with simplistic notions of modularity or holism giving way to a new view of the brain as a complex network, whose global functional properties are determined by the structure and dynamics of its constituent neuronal systems, especially those of the cerebral cortex and associated thalamocortical circuits. More than a decade of functional neuroimaging has provided overwhelming evidence that virtually all states of human perception and cognition are accompanied by widespread neural activation patterns within the corticothalamic system. The duality of localized and distributed aspects of brain function is naturally encompassed by a network perspective.

The advent of sophisticated methods for the structural and functional analysis of brain states demands the development of quantitative methods capable of capturing and untangling network aspects of brain function. In this chapter, I review several fundamental principles of brain networks and explore quantitative methods for the analysis of these networks (Sporns et al., 2004).

22.2 Anatomical, Functional, and Effective Connectivity

Because of the close relationship between neural connectivity and neural activity throughout the brain, it is important to consider the relationship between structural connection patterns and the patterns of dynamic ("functional" or "effective") interactions they support. While many kinds of natural or artificial networks have recently become objects of special scientific interest (e.g., Strogatz, 2001; Albert and Barabasi, 2002; Barabasi, 2002; Watts, 2003), the closeness and intricacy of the relationship between structure and function is perhaps a unique characteristic of brain networks. Before we begin our discussion of quantitative methods for analyzing brain networks, we present general definitions of anatomical (structural), functional, and effective connectivity in the brain. Most of these definitions have emerged from cognitive neuroscience and functional neuroimaging research, and the distinction of anatomical, functional, and effective connectivity is primarily applied here to large-scale networks involving macroscopic brain regions or populations of neurons and their long-range interregional pathways. We note that these definitions, especially the distinction between functional and effective connectivity, are still somewhat in flux (Horwitz, 2003).

22.2.1 Anatomical Connectivity

Anatomical connectivity is, at first glance, equivalent to the set of physical or structural (synaptic) connections linking neuronal units at a given time. Given the nature of brain anatomy, any structural analysis of neural connection patterns involves a choice about the level of the spatial scale at which the analysis is performed. Analyses carried out at the local circuit level would most likely focus on the pattern of synaptic connections between individual neurons. Examples are "canonical microcircuits," for example, those linking different types of cortical neurons arranged in specific layers (Douglas and Martin, 1991). Analyses of connections that link distant neurons within a given brain region ("intra-areal" connections) would involve "connection bundles" or "synaptic patches" linking local neuronal populations (such as neuronal groups or columns). Examples are the network of horizontal (also called "tangential" or "intrinsic") connections linking groups of neurons separated by several millimeters within primary visual cortex (Gilbert and Wiesel, 1989; Angelucci et al., 2002). Analyses of large-scale connection patterns would focus on interregional connection pathways linking segregated areas of the brain. Such pathways generally comprise many thousands or millions of individual fibers.

Irrespective of the level of the spatial scale, anatomical connection patterns are relatively static and unchanging at shorter timescales (seconds to minutes), but may be plastic and changing at longer timescales (hours to days). Learning and plasticity operate, at least in part, by changing the structural patterns of interconnections between neurons and brain regions. A key component of anatomical connectivity is synaptic strength or efficacy, which constitutes a "structural" parameter in most neural network models. The physical substrate of synaptic strength is multifaceted, with contributions from the morphology and physical shapes of synapses, as well as from the biophysics of their constituent receptors and neurotransmitter release machinery. Thus, anatomical connectivity is closely linked with the functioning of individual synapses and their plasticity.

Anatomical connectivity arises through developmental processes, that, at least in part, are dependent upon neural activity (Singer, 1995). Many brain networks remain plastic throughout the lifetime of the organism, exhibiting specific modifications of synaptic efficacy at multiple timescales, as well as continuous morphological change. Thus, the detailed structural organization of brain networks will to some extent reflect the developmental and experiential history of the individual organism (Edelman, 1987; Merzenich and deCharms, 1996; Sporns, 1998). This point deserves special emphasis. While it is possible (and perhaps desirable) to analyze brain networks as static entities, without reference to how they were generated, it is nonetheless essential to realize that their fine structure and morphology is the result of continuous interaction between neural structures, neuronal firing, and the behavioral activity of an individual organism within an environment. Theoretical neuroscience has only just begun to address this central issue.

22.2.2 Functional Connectivity

As originally defined (Friston, 1993, 1994), functional connectivity refers to the pattern of temporal correlations (or, more generally, deviations from statistical independence) that exists between distinct, distributed, and often spatially remote, neuronal units. In many cases, such temporal correlations can be the result of neuronal interactions along anatomical or structural connections. In other cases, observed correlations may be due to common input from an external neuronal or stimulus source. Deviations from statistical independence between neuronal elements are commonly captured in a covariance matrix (or a correlation matrix), which, under certain statistical assumptions, may be viewed as an approximate representation of the neuronal system's functional connectivity. While temporal cross correlations are often used to represent statistical patterns in neuronal networks, other measures such as spectral coherence (Srinivasan et al., 1999) or consistency in relative phase relationships (Lachaux et al., 1999) also serve as statistical indicators of functional connectivity.

According to this definition, functional connectivity is essentially a statistical measure of brain activity and is time dependent on a fast timescale (hundreds of milliseconds). Horwitz (2003) has pointed out the different methodologies for measuring brain activity, differences in the setting of baseline or null hypothesis conditions, or even different experimental designs (e.g., single versus blocked trials) will generally result in different estimates of functional connectivity. These multiple ways of estimating functional connectivity will not necessarily produce consistent findings or conclusions about neural interactions, even within a single experiment or task. Functional connectivity is essentially "model-free," that is, purely based on evaluating the degree to which two units or brain regions display statistical interdependence (mutual information) without reference to their causes.

22.2.3 Effective Connectivity

In the context of single-cell neurophysiology, Aertsen and Preissl (1991) originally defined effective connectivity as "the simplest possible circuit diagram" that could account for the observed temporal (and thus causal) relationships among a set of recorded neurons. In human neuroimaging, effective connectivity was originally defined as the set of causal effects or influences that one neural system exerts over another (Friston, 1994). As both these definitions show, effective connectivity emphasizes causality and attempts to isolate those regions or neurons within the experimental system that participate in generating a particular neural activity pattern or state. Unlike functional connectivity, effective connectivity aims to exclude statistical dependencies that are not mediated by some sort of a directed action of one neuron or system on another, usually as a result of a physical connection. In a sense, effective connectivity represents the union of functional connectivity with an underlying structural model. Experimentally, effective connectivity can be derived by measuring all functional or statistical interactions and then fitting this set of measurements to an explicit structural model, for example, an anatomical connection set. Another way of distinguishing between functional and effective connectivity is to point out that functional connectivity delivers a *description* of patterns of neural activity, while effective connectivity provides a possible *explanation* of their neural causes (Lee et al., 2003).

Effective connectivity, unlike functional connectivity, is not "model-free," but requires the specification of a causal model that is based on a combination of anatomical and physiological characteristics. In terms of experimental access, effective connectivity may be inferred through the use of perturbations, the effects of which can be directly measured throughout the brain and must be the result of causal interactions. Another strategy would be to analyze temporal data sets for signs of temporal ordering of events in the brain. Causal interactions must obey fundamental laws of temporal precedence. Signals arising in a neural unit or brain area can affect other components within the network only at a later time.

An important similarity between effective and functional connectivity is that both descriptions of brain networks are time dependent. Statistical interactions between brain regions change rapidly in the course of normal brain function and these changes involve the participation of varying subsets of brain regions and pathways. Effective connectivity patterns may also depend on the overall behavioral or attentional state of the system (Büchel and Friston, 1997). The extraction of effective connectivity patterns from brain data sets requires the mapping of time-dependent statistical patterns onto a (presumably relatively time-invariant) structural substrate of anatomical connections. The assumption of temporal invariance of structural connections may not always be valid. If structural substrates undergo learning-related changes in the course of an experiment, the resulting effective connectivity patterns may be found to change accordingly (Büchel et al., 1999; Büchel and Friston, 2000).

22.2.4 Relationships between Anatomical, Functional, and Effective Connectivity

The relationship between the structural and functional dimensions of brain connectivity is mutual and reciprocal. Clearly, structural connectivity is a major constraint on the kinds of patterns of functional connectivity that can be generated. For example, patterns of structural inputs and outputs of given cortical regions, its connectional fingerprint (Passingham et al., 2002), are major determinants of the localized functional properties of cortical regions and neurons. In the other direction, functional interactions can contribute to the shaping of the underlying anatomical substrate. This is accomplished either directly through activity (covariance)-dependent synaptic modification, or, over longer timescales, through effects of functional connectivity on an organism's perceptual, cognitive, or behavioral capabilities, which in turn affect adaptation and survival. The reciprocity between anatomical and functional/effective networks deserves emphasis as it captures some of the unique aspects of brain networks.

22.3 Segregation and Integration as Principles of Brain Organization

Experimental research in behavioral and cognitive neuroscience has revealed several organizing principles found throughout the mammalian cerebral cortex, the part of the brain with which we are mainly concerned in the remainder of this chapter. Landmark studies of the anatomy and physiology of the visual cortex have revealed parallel processing streams concerned with spatial and form-related visual attributes (Mishkin et al., 1983) and a strong tendency toward hierarchical processing (Van Essen et al., 1992). Throughout the cortex, interregional pathways exhibit a very high level of reciprocity, with a majority of pathways forming reentrant loops (Edelman, 1987), irrespective of the relative position of the constituent brain areas within the cortical hierarchy. These and other structural features of cortex provide significant clues as to the nature of cortical processing and function.

Here, we focus on two main principles of the structural and functional organization of the mammalian cerebral cortex, *segregation* and *integration* (Tononi et al., 1994, 1998; Friston, 2002). Segregation and integration are found throughout different systems of the cortex and represent general organizational principles. For now, we briefly examine these two principles in their anatomical and functional incarnations. The development of quantitative methods for their analysis within brain networks will then allow us to gain insights into their potential neurobiological relevance (see discussion).

22.3.1 Segregation

Anatomical and functional segregation refers to the existence of specialized neurons and brain areas, often organized into distinct neuronal populations (groups or columns) or cortical areas. The concept of anatomical segregation has its roots in the notion that specific brain processes or functions can be localized to specific anatomical regions of the human brain. Historically, most conceptualizations of localization revolved around neuropsychological studies of human patients with localized brain lesions, which often gave rise to specific functional defects. The anatomical investigations of Korbinian Brodmann yielded a cortical parcellation scheme based on cytoarchitectonics that has remained in use till the present day. By the middle of the 20th century, cerebral localization was so dominant an idea that even as ardent a critic as Karl Lashley remarked that "in the field of neurophysiology no fact is more firmly established than the functional differentiation of various parts of the cerebral cortex" (Lashley, 1931; quoted after Zeki, 1993).

With the development of single-cell neurophysiology it became evident that many of these anatomically distinct brain regions contained neurons with distinct physiological properties. Ungerleider and Mishkin (1982), van Essen and Maunsell (1983), Zeki and Shipp (1988), and Felleman and van Essen (1991) developed increasingly refined network diagrams of multiple anatomically and functionally distinct areas of the primate visual cortex. These specialized and segregated brain regions contain neurons that selectively respond to specific input features (such as orientation, spatial frequency, or wavelength), or conjunctions of features (such as faces). Even within single cortical regions such as V1 or V2, functionally distinct populations of neurons remain spatially segregated, forming substructures such as blobs and interblobs (in V1), or thick stripes, thin stripes, and interstripes (in area V2).

22.3.2 Integration

When considering the brain as a network, it immediately becomes obvious that segregated and specialized neuronal units do not operate in isolation. When Brodmann, in 1909, discussed "mental faculties" (what we now might call cognitive functions), he presciently observed that "one cannot think of their taking place in any other way than through an infinitely complex and involved interaction and cooperation of numerous elementary activities. In each particular case [these] supposed elementary functional loci are active in differing numbers, in differing degrees and in differing combinations" (Brodmann, 1909; quoted after Cabeza and Kingstone, 2001).

Modern neuroscience has provided abundant evidence that coherent perceptual and cognitive states require the coordinated activation, that is, the *functional integration*, of very large numbers of neurons within the distributed system of the cerebral cortex (Bressler, 1995; Friston, 2002). Electrophysiological studies have shown that perceptual or cognitive states are associated with specific and highly dynamic (short-lasting) patterns of temporal correlations (functional connectivity) between different regions of the thalamocortical system (Varela et al., 2001). Human neuroimaging experiments have revealed that virtually all perceptual or cognitive tasks, for example, object recognition, memory encoding and retrieval, reading, working memory, attentional processing, motor planning, and awareness are the result of activity within large-scale and distributed brain networks (cf. McIntosh, 1999, 2000).

Integration across widely distributed brain regions requires neuronal interactions along interregional pathways. In the cortex, such interactions are mediated by the extensive and massive network of "extrinsic" corticocortical connections. When these structural substrates of integration are disabled or disrupted, resulting in the disconnection of neuronal populations, specific functional deficits are often observed. For example, the corpus callosum is a massive bundle of reentrant fibers linking the two cerebral hemispheres of the mammalian brain. Its disruption results in the failure of interhemispheric integration. Engel et al. (1991; see also Munk et al., 1995) reported that interhemispheric cross correlations in the primary visual cortex of cats disappeared when callosal fibers linking the two hemispheres were cut. Human patients whose corpus callosum is severed, resulting in a "split brain," exhibit highly characteristic cognitive deficits, most of which involve an inability to integrate neural states between the two disconnected hemispheres (Gazzaniga, 2000).

22.4 ANALYSIS OF ANATOMICAL CONNECTIVITY

Neurons in the cerebral cortex maintain thousands of input and output connections with other neurons, forming a dense network spanning the entire thalamocortical system. According to a detailed quantitative study (Murre and Sturdy, 1995), the human cerebral cortex contains approximately 8.3×10^9 neurons and 6.7×10^{13} connections. The length of all connection paths within a single human brain is estimated at between 100,000 and 10,000,000 km. Despite this massive connectivity, cortical networks are exceedingly sparse, with an overall connectivity factor (number of connections present out of all possible connections) of around 10^{-6}. Even a cursory examination of neuroanatomical data sets shows that brain networks are not random, but form highly specific patterns. A predominant feature of brain networks is that neurons tend to connect predominantly with other neurons in local groups. Thus, local connectivity ratios are likely to be significantly higher than those suggested by a random connection topology.

Neuronal networks consist of units connected by directed links (synapses). Such networks can be described using methods derived from the theory of directed graphs (Harary, 1969; Bang-Jensen and Gutin, 2001). The structure of a given graph (network) is captured by its connection matrix or adjacency matrix, with binary elements a_{ij} that represent the presence or absence of a directed edge between vertices j (source) and i (target). If such an edge exists, vertex j can directly communicate signals (spikes) to vertex i. However, in brain networks, such direct connections are not the only way in which one neuronal element can influence another. Indirect interactions can proceed along paths, defined as ordered sequences of distinct vertices and edges. It is important to note that if there is neither a direct connection nor a path between vertices j and i, then j cannot functionally influence i in any way.

The analysis of structural connections and paths within networks allow the quantification of a broad range of network characteristics. Recent reviews by Hilgetag et al. (2002) and Sporns et al. (2004) provide surveys of most of the relevant measures, including analytical and numerical methods for their computation (see also www.indiana.edu/~cortex/CCNL.html). For example, the adjacency matrix of a network allows the derivation of the reachability matrix and the distance matrix. The reachability matrix indicates, for each ordered pair of vertices j and i, whether a path (of any length) exists from j to i. If all entries of the reachability matrix are ones, the network consists of only one component and is strongly connected. Partitioning of the reachability matrix into nonoverlapping subsets of vertices with no paths between them indicates the existence of multiple components. The entries of the distance matrix give the length of the shortest (directed) path between the two vertices j and i. The global maximum of the distance matrix is also called the diameter.

The adjacency matrix, reachability matrix, and distance matrix of a graph allow the subsequent derivation of a broad range of measures of network connectivity. For example, the adjacency matrix allows the examination of the degree distribution of a given network. The *degree* of a vertex is the number of incident edges, sometimes sorted into *indegree* and *outdegree*, for incoming and outgoing edges, respectively. The degree distribution of a network provides important insights into whether the network contains vertices with approximately equal degrees (i.e., conforming to a Gaussian distribution) or whether the network's vertices show an exponential degree distribution. These two kinds of networks are called "one scale" or "scale free," respectively (Amaral et al., 2000), and result in very different dynamical behavior. Scale-free networks are found in many biological systems, including metabolic and genetic regulatory networks (see Chapter 2, this volume). It is less clear whether anatomical brain networks are scale free (discussed in Sporns and Zwi, 2004), while functional brain networks may indeed exhibit scale-free organization (Eguiluz et al., 2004). For any given vertex (brain region), its indegree and outdegree have straightforward functional interpretations. A high indegree indicates that the vertex is influenced by a large number of other vertices, while a high outdegree indicates a large number of potential functional targets. The pattern of incident edges (connections) on brain regions has been called the "connectional fingerprint" (Passingham et al., 2002), found to be unique for a broad range of brain regions and thought to confer functional specificity.

Much interest in the recent literature on network statistics and dynamics has focused on a specific class of networks exhibiting so-called "small-world" properties (Watts, 1999, 2003). A small-world network combines a high degree of local clustering with the ability to communicate between all its constituent vertices along short paths. Two measures capture these two essential properties (Watts and Strogatz, 1998), the characteristic path length and the clustering coefficient (or cluster index). The average of all the entries of the distance matrix constitutes the characteristic path length of a graph, $\lambda(G)$. The cluster index of a vertex $\gamma(v)$ indicates how many connections are maintained between this vertex's neighbors. For digraphs (graphs containing only directed edges), neighbors are all those vertices that are connected, either through an incoming or an outgoing connection, to the central vertex v. The number of neighbors per vertex is b_v. The vertex's cluster index is defined as the ratio of actually existing connections between the b_v neighbors and the maximal number of such connections possible $(b_v^2 - b_v)$. If $b_v = 0$, $\gamma(v) = 0$. The average of the cluster indices for all the individual vertices is the cluster index of the graph $\gamma(G)$. The cluster index for a network expresses the extent to which the units within the network share common neighbors that "talk" among each other, an attribute that has been called the "cliquishness" of the network. In the brain, a high cluster index $\gamma(G)$ points to a global organizational pattern consisting of groups of units that mutually share structural connections and can thus be surmised to be closely related functionally.

Recently, Milo et al. (2002) have introduced a novel method for the structural analysis of large biological networks. The approach involves the quantification of subgraphs (called "motifs") within larger networks and the detection of motifs that are statistically increased in biological networks as compared with equivalent random networks. Milo et al. have found a specific set of motifs that are increased in genetic regulatory networks and some that occur in high abundance in the neuronal network of the nematode *Caenorhabditis elegans*. The identification of network motifs represents an important extension of graph-theoretical methods for network analysis (Sporns and Kötter, 2004).

In addition to these methods of network analysis that are based on the theory of graphs and digraphs, there is a broad range of methods from multivariate statistics that have been applied to neuronal connection patterns. The aim of these methods is to provide a global overview of network structure by visualizing consistent patterns or systematic relationships between network elements. For example, such methods utilize cluster and similarity analysis, dimension reduction methods such as multidimensional scaling (MDS) or principal components analysis (PCA), or multiple correspondence analysis (MCA). In the analysis of brain networks, nonmetric multidimensional scaling was originally introduced by Young (1992) and applied to the macaque cortex. The analysis involved the derivation of distance measures between brain regions from the mutual similarity or dissimilarity of their connectional relationships with each other and with other brain regions. The resulting distance matrix was then mapped into two dimensions, thus allowing visualization of the entire network. This analysis led to important insights into the large-scale organization of the visual cortex, for example, confirming the existence of distinct ventral and dorsal processing streams (Young, 1992) and allowing the identification of distinct sensory systems and their associated multimodal areas in the entire macaque cortex (Young, 1993; Young et al., 1995). Other multivariate techniques include hierarchical analysis (Hilgetag et al., 1996) and structural cluster analysis (Hilgetag et al., 2000).

These methods and measures for characterizing anatomical connection patterns have been applied to large-scale connectivity matrices of the cerebral cortex, which have been assembled from hundreds of neuroanatomical studies conducted in a variety of species, including cat (Scannell et al., 1999) and nonhuman primates (Felleman and van Essen, 1991; Young, 1993). Results indicate that the cerebral cortex comprises clusters of densely and reciprocally coupled cortical areas that are globally interconnected (Sporns et al., 2000a, 2000b). Regarding this clustered architecture, there is strong agreement between different clustering methods (Hilgetag et al., 2000; Sporns et al., 2000a). Importantly, large-scale cortical networks share some attributes of small-world networks, including high values for cluster indices and short characteristic path lengths (Hilgetag et al., 2000; Sporns et al., 2000a). A recent detailed analysis revealed that these properties are shared by a variety of large-scale connection matrices across a variety of species and cortical systems, and are also found in connection matrices generated by making empirically based probabilistic assumptions about local connection densities and arborizations of corticocortical connections (Sporns and Zwi, 2004).

What is the reason for the existence of small-world connectivity patterns in the cerebral cortex? For an answer, we have to consider the network of functional and effective interactions that such anatomical patterns generate.

22.5 Analysis of Functional Connectivity

Measures of statistical dependence are the elementary basis of functional connectivity. We will focus here on methods for analyzing patterns of functional connectivity that are based on estimates of mutual information between two neuronal units or brain regions. The measures reviewed in this section are valid for all systems, whether they are linear or nonlinear, neural or nonneural. It is essential, however, to derive valid estimates for entropy and mutual information for any given application.

Mutual information between A and B is defined as the difference between the sum of their individual entropies and their joint entropy:

$$\mathrm{MI}(A, B) = H(A) + H(B) - H(AB). \tag{22.1}$$

Note that $\mathrm{MI}(A, B) \geq 0$ and $\mathrm{MI}(A, B) = \mathrm{MI}(B, A)$. The mutual information $\mathrm{MI}(A, B)$ is zero if no statistical relationship exists between A and B, that is, if A and B behave statistically independent of each other, such that $H(AB) = H(A) + H(B)$. Any reduction of their joint entropy indicates a degree of statistical dependence and results in a positive value for the mutual information. Note also that $\mathrm{MI}(A, B)$ does not necessarily express causal influences from A on B or vice versa. In any real or simulated system, the estimation of mutual information ultimately depends on correct estimates for the individual and joint entropies. These estimates can be difficult to derive from small or sparse data sets and, in many cases, additional statistical assumptions have to be made (Cover and Thomas, 1991; Papoulis, 1991; Paninski, 2003).

A and B may be individual units or brain regions, or they may be subsets of populations of elements that are considered jointly. For example, let us consider a system X composed of n elements that is partitioned into two complementary subsets. One subset consists of k elements and is denoted as X^k, while its complement contains the remaining $n - k$ elements and is denoted as $X - X^k$. The mutual information between these two subsets is

$$\mathrm{MI}(X^k, X - X^k) = H(X^k) + H(X - X^k) - H(X). \tag{22.2}$$

While mutual information captures the degree of statistical dependence between two elements (or subsets), a measure of the total amount of statistical dependence among an arbitrarily large set of elements, called integration, was first introduced by Tononi et al. (1994). Let us consider a system X composed of a set of elements $\{x_i\}$. The integration $I(X)$ is defined as the difference between the sum of the entropies of the individual elements and their joint entropy:

$$I(X) = \sum_i H(x_i) - H(X). \tag{22.3}$$

Note that $I(X) \geq 0$. Any amount of statistical dependence between elements will express itself in a reduction of those elements' joint entropy and thus in a positive value for $I(X)$. If all elements are statistically independent, their joint entropy is the sum of their individual entropies and $I(X) = 0$. Thus, integration quantifies the total amount of statistical structure or statistical dependencies present within the system. In comparing Equations (22.1) and (22.3), it becomes clear that integration is, in a sense, the multivariate generalization of mutual information.

Intuitively, as a system X is divided into smaller and smaller subsets, the total amount of statistical dependencies (i.e., the integration of these subsets) should decrease. We denote the average integration of subsets of size k as $\langle I(X^k)\rangle$, noting that the average is taken over all k-out-of-n subsets. Thus, $\langle I(X^n)\rangle = I(X)$ and $\langle I(X^1)\rangle = 0$. It can be proven that the spectrum of average integration for all values of $k(1 \leq k \leq n)$ must increase monotonically, that is, $\langle I(X^{k+1})\rangle \geq \langle I(X^k)\rangle$ (Tononi et al., 1994).

The characterization of the spectrum of average integration across all levels of scale (subset size k) within a given system allows us to examine how and where statistical structure within the system is distributed. How is this possible? Let us say we find a system that has a certain amount of statistical structure, measured by its integration, $I(X) > 0$. This means that some statistical dependencies exist somewhere within the system X. But the global estimate of $I(X)$ does not provide information as to whether these dependencies are homogeneously distributed throughout the system, or are localized and concentrated among specific units or subsets. In the first case, the system would be, in terms of its functional connectivity, totally undifferentiated, essentially presenting the same view to an observer zooming in on different levels of scale. We might say that such a system lacks functional segregation. In the second case, there would be parts of the system that are more integrated than others and these integrated subsets would represent local structure. Such a system does express functional segregation in addition to the global functional integration expressed by $I(X)$.

To distinguish between these possibilities, Tononi et al. (1994) introduced a measure of complexity that captures the extent to which a system is both functionally segregated (small subsets of the system tend to behave independently) and functionally integrated (large subsets tend to behave coherently). A statistical measure of neural complexity, $C_N(X)$, takes into account the full spectrum of subsets. $C_N(X)$ can be derived either from the ensemble average of integration for all subset sizes from 1 to n, or (equivalently) from the ensemble average of the mutual information between subsets of a given size (ranging from 1 to $n/2$) and their complements. $C_N(X)$ is thus defined as:

$$\begin{aligned} C_N(X) &= \sum_k (k/n)I(X) - \langle I(X^k)\rangle \\ &= \sum_k \langle \mathrm{MI}(X^k, X - X^k)\rangle. \end{aligned} \tag{22.4}$$

According to the second expression for $C_N(X)$, the complexity of a system is high when, on average, the mutual information between any subset of the system and its complement is high.

Another, closely related, measure of complexity expresses the portion of the entropy that is accounted for by the interactions among all the components of a system (Tononi et al., 1998). There are three mathematically equivalent expressions for this measure, called $C(X)$:

$$\begin{aligned} C(X) &= H(X) - \sum_i H(x_i \mid X - x_i) \\ &= \sum_i \mathrm{MI}(x_i, X - x_i) - I(X) \\ &= (n-1)I(X) - n\langle I(X - x_i)\rangle. \end{aligned} \tag{22.5}$$

These three expressions for complexity are mathematically equivalent. Note that neither $C_N(X)$ nor $C(X)$ can take on negative values. Here $H(x_i \mid X - x_i)$ denotes the conditional entropy of each element x_i, given the entropy of the rest of the system $X - x_i$.

To illustrate how $C(X)$ can be conceptualized, the second formulation of $C(X)$ is perhaps most useful. $C(X)$ is obtained as the difference of two terms: the sum of the mutual information between each individual element and the rest of the system, minus the total amount of integration. Thus, $C(X)$ takes on large values if single elements are highly informative about the system to which they belong, while not being overly alike (as they would be if their total integration, or total shared information, is high). $C_N(X)$ and $C(X)$ are closely related, but not mathematically equivalent. In graphical representations (see Tononi et al., 1998), $C(X)$ corresponds to a specific component at level $n - 1$ of the full spectrum of ensemble averages of integration.

Within the context of applications of functional connectivity to brain networks, it is essential to underscore that complexity captures the degree to which a neural system combines functional segregation and functional integration. Extensive computational explorations (Tononi et al., 1994, 1998; Sporns and Tononi, 2002) have shown that complexity is high for systems that contain specialized elements capable of global (system-wide) interactions. On the other hand, complexity is low for random systems, or for systems that are highly uniform (or, in other words, systems that lack either global integration or local specialization).

While this consideration of random and highly, uniformly integrated systems might suggest that integration and complexity are largely incompatible, it can easily be shown that these two measures can associate, specifically that there are network architectures that give rise to high complexity accompanied by high levels of global integration (Tononi et al., 1994). In general, random networks show broad ranges of both integration and complexity, irrespective of whether linear or nonlinear local dynamics is employed. Which are the crucial structural features of such networks that result in high complexity and/or integration?

To examine this theoretical question, one approach is to use complexity (and other measures of functional connectivity, such as entropy or integration) as cost functions in simulations designed to optimize network architectures by modifying their connection matrices. Networks that are optimized for high complexity develop structural motifs that are very similar to those observed in real cortical connection matrices (Sporns et al., 2000a, 2000b; Sporns and Tononi, 2002). Specifically, such networks exhibit an abundance of reciprocal (reentrant) connections, a strong tendency to form clusters, and they have short characteristic path lengths. Other measures (entropy or integration) produce networks with strikingly different structural characteristics. While nonlinear dynamics cannot readily be studied in the context of optimizations (due to extensive computational requirements), a closer examination of specific connection topologies (sparse, uniform, and cortex-like) that are simulated as nonlinear systems has revealed that the association of small-world attributes and complex functional dynamics persists for more realistic models of cortical architectures as well (Sporns, 2004). Thus, high complexity, a measure of global statistical features and of functional connectivity, appears to be strongly and uniquely associated with the emergence of small-world networks.

Other quantitative measures of functional connectivity or interregional interactivity not discussed in detail in this chapter have been used to analyze various empirical brain data sets. These include cross correlation between neuronal spike trains, local field potential recordings, or fMRI/PET voxel time series. For example, Bressler and colleagues have carried out numerous studies examining task-dependent large-scale networks of phase synchronization in primate and human cortex (Liang et al., 2000; Bressler and Kelso, 2001). Patterns of interregional cross correlations have been found to accompany the performance of specific cognitive tasks in cats (e.g., Roelfsema et al., 1997), primates (Bressler, 1995), and humans (e.g., Srinivasan et al., 1999; von Stein et al., 1999; Varela et al., 2001; Munk et al., 2002). McIntosh has documented changes in brain functional connectivity related to awareness (McIntosh et al., 1999), and most recently through recording differential interactivity of the human medial temporal lobe with other regions of the neocortex (McIntosh et al., 2003).

Patterns of functional connectivity are statistical signatures of causal processes occurring within and among a specific and time-varying subset of neurons and brain regions. The identification of which subset is currently engaged in a given task requires the adoption of a structural model to access effective interactions.

22.6 Analysis of Effective Connectivity

Effective connectivity attempts to reconstruct or "explain" recorded time-varying activity patterns in terms of underlying causal influences of one brain region upon another. This involves the combination of (essentially covariance-based) functional connectivity patterns with a structural system-level model of interconnectivity. A technique called "covariance structural equation modeling" is used to assign effective connection strengths to anatomical pathways that best match observed covariances in a given task (McIntosh and Gonzalez-Lima, 1994; Horwitz et al., 1999). Striking differences in effective connectivity between given sets of brain regions are found for different cognitive tasks, illustrating the time- and task-dependent nature of these patterns.

Another approach to identifying highly interactive brain regions and their causal interactions involves the use of "effective information," a novel measure of the degree to which two brain regions or systems causally influence each other (Tononi and Sporns, 2003). Given a neural system that is partitioned into two complementary subsets, A and B, we obtain the effective information from A to B by imposing maximal entropy on all outputs of A. Under these conditions the amount of entropy that is shared between A and B is due to causal effects of A on B, mediated by connections linking A and B. These connections can either be direct connections or indirect links via a surrounding neural context. Effective information may be formulated as

$$\mathrm{EI}(A \rightarrow B) = \mathrm{MI}(A^{H\max}, B). \tag{22.6}$$

Note that unlike $\mathrm{MI}(A, B)$, effective information may be nonsymmetric, that is, $\mathrm{EI}(A \rightarrow B) \neq \mathrm{EI}(B \rightarrow A)$, owing to nonsymmetric connection patterns. Furthermore, effective information requires perturbations of units or connections.

As discussed earlier, the integration of information is considered essential for the functioning of large-scale brain networks (e.g., Tononi and Edelman, 1998; Tononi et al., 1998). In considering information integration the notion of causality, or effectiveness, is crucial. A system that integrates information effectively must do so via actual causal interactions occurring within it. Mere statistical coincidences are insufficient to characterize the participating entities as truly integrated. Tononi and Sporns (2003) defined a measure for information integration based on effective information that captures the maximal amount of information that can be integrated within the system. This measure allows the simultaneous quantification of information integration and the identification of all those system elements that participate in it. It can thus be used to delineate integrated functional clusters or networks of effective connectivity within larger sets of brain regions.

Currently, this measure of information integration has only been tested in computer simulations of small model systems with varying anatomical architectures. The results indicate that information integration is maximized by two main attributes of the anatomical connection pattern. First, each element maintains a different connection pattern, or connectional "fingerprint," a property that strongly promotes regional functional specialization. Second, the pattern maintains global connectedness and ensures that a large amount of information can be exchanged across any bipartition of the network, which in turn promotes global functional integration. Simple models of the connectional organization of specific neural architectures, such as the thalamocortical system, are found to be well suited to information integration, while others, such as the cerebellum, are not.

Other methods for analyzing causal influences in the brain have been proposed, many of which utilize the temporal dynamics of the observed neural system to provide information about effective interactions. Several methods are based on interpretations or adaptations of the concept of Granger causality, involving estimates of how much information a set of variables provides for the prediction of another. For example, Kaminski et al. (2001) develop an approach based on exploiting the directed transfer function between two neural signals. Additional causality measures can discriminate between direct causality and effects mediated through extraneous system components (see also Liang et al., 2000). Bernasconi and König (1999) develop statistical measures that allow the detection of directed dependencies within temporal brain data sets. Schreiber (2000) describes a measure called transfer entropy, which is able to detect directed exchange of information between two systems by considering the effects of the state of one element on the state transition probabilities of the other element. This yields a nonsymmetric measure of the effects of one element on the other, exploiting the entire system's temporal dynamics.

The experimental application of measures of effective connectivity presents a number of difficult problems. For example, effective information, as defined above, may be difficult to estimate in real neuronal system, due to the need for invasive perturbations of restricted sets of neuronal elements. Some promising avenues have recently been pursued. The combination of transcranial magnetic stimulation with functional neuroimaging for the first time allows the quantification of effects of localized perturbations on extended brain networks engaged in the performance of specific tasks (Paus, 1999; Pascual-Leone et al., 2000).

22.7 Discussion: Brain Networks, Complexity, and Cognition

Brain networks represent a description of macroscopic brain activity of thousands or millions of neurons integrated through "synaptic connectivity," "effective connectivity," and "functional connectivity." To model such integration, one might use spatially coarse-grained modeling approaches in which the dynamics of local populations of neurons are treated as statistical ensembles or neural masses in a field description of nervous tissue (e.g., Wilson and Cowan, 1973). Similar modeling approaches have been used in the past to study the dynamics of cortical networks, including piriform cortex, hippocampus, and somatosensory cortex (for a review see Hasselmo and Kapur, 2001). Although the macroscopic field theories of the brain are derivable from the mesoscopic neural dynamics of Wilson and Cowan (1972) (see e.g., Jirsa and Haken, 1997), how to relate such theories to brain networks which can be classified not in terms of excitatory and inhibitory synaptic connectivity patterns per se, but in terms of effective and functional connectivity patterns, is still an open research problem. One approach to elucidating the connectivity patterns between and within brain regions is to represent brain networks not as systems of differential equations (see, for example, Wilson, 1999a), but to utilize principles and methods from graph theory in order to gain insight into the integrative dynamics of the brain. Indeed, there have been some attempts at such multiscale simulation of brain networks addressing data at multiple levels (e.g., Aertsen et al., 1989; Johnsen et al., 2002), but progress has been slow.

This chapter has provided a brief overview of current research into principles and methods of studying brain connectivity. Two main principles of brain organization were identified, namely, functional segregation and integration. From an anatomical and an informational perspective, these two principles appear, in some sense, antagonistic if not mutually exclusive. Functional segregation is consistent with the information-theoretical idea that neurons attempt to extract specialized information from their inputs, eliminating redundancy and maximizing information transfer. Segregation tends to favor the analysis of inputs into (orthogonal and independent) principal components, ultimately represented in the activation of dedicated sets of neurons. Functional integration, on the other hand, establishes statistical relationships (temporal correlations) between different and distant cell populations and cortical areas, leading to the generation of mutual information between brain regions. By creating these mutual dependencies, local neuronal specialization may be degraded — thus the apparent antagonism between segregation and integration.

However, both segregation and integration are needed for the global functioning of the brain. From a computational and information-theoretical perspective, two of the major problems that brain has to solve are the extraction of information (statistical regularities) from their inputs and the generation of coherent states that allow coordinated perception and action in real time. A major *leitmotif* of our review of brain connectivity is the idea that the solutions to these problems are reflected in the dual organizational principles of functional segregation and functional integration, found throughout the cerebral cortex. The requirement to simultaneously achieve segregation and integration imposes severe constraints on the set of possible cortical connection patterns. Not surprisingly, the structural characteristics of real cortical connection matrices are strongly associated with the functional capacity of the cortex to combine functional segregation and integration. As discussed earlier, complexity (as a measure of functional connectivity) captures the extent to which a system combines segregation and integration. Complexity meets the challenges of generating and integrating information at the same time.

There is much we do not understand yet about the structure and function of brain networks. But it seems certain that as we gain more and more insight into the integrative dynamics of the brain, the nature of its function will be increasingly illuminated.

ACKNOWLEDGMENT

Portions of this work were supported by the U.S. government contract NMA201-01-C-0034. The views, opinions, and findings contained in this report are those of the author and should not be construed as an opinion, policy, or decision of the U.S. government.

PROBLEMS

1. Generate random graphs (connection matrices) composed of $N = 128$ vertices and $K = \{1, 16, 32, \ldots, 1280\}$ edges. For each setting of K generate 100 copies of these matrices and determine whether the graphs are strongly connected. Plot the average size of the largest component as a function of K. Then calculate the distance matrix (Sporns, 2002a) and the characteristic path length for each graph, omitting all entries for which $d_{ij} = \infty$. Relate the two plots to each other.
2. For linear systems with Gaussian stationary variables, the covariance matrix COV(X), which provides an estimate of the functional connectivity of the system, can be derived analytically from the connection matrix CON(X). Generate a random connection matrix CON(X) with $N = 30$ and $K = 311$ and assign a weight of $w = 0.05$ to each connection. Then calculate the covariance matrix, following the analytical method described in Tononi et al. (1994). Given the covariance matrix, calculate the entropy, the integration, and the complexity, using the expression $H(X) = -0.5 \log[(2\pi e)^n |\text{COV}(X)|]$, with $|\text{COV}(X)|$ denoting the determinant of the covariance matrix (Cover and Thomas, 1991; Papoulis, 1991). Consult www.indiana.edu/~cortex/complexity_toolbox.html and www.indiana.edu/~cortex/connectivity_toolbox.html for more information on computational methods and sample programs.
3. Now substitute a real neuroanatomical connection matrix, describing the network of interregional pathways in the macaque visual cortex (Felleman and van Essen, 1991). The matrix ($N = 30$, $K = 311$) is available for download from www.indiana.edu/~cortex/complexity_toolbox.html. What differences do you see in the values for entropy, integration, and complexity? Explain these differences in terms of the pattern of anatomical connections actually present in macaque visual cortex (discussion in Sporns et al., 2000).
4. Use the connection matrix of macaque visual cortex as a starting point and introduce different levels of random variations into the connection pattern, for example, by making and breaking a fraction of the connections (but not changing either N or K). What effect do these random variations have on average on the values for entropy, integration, and complexity?
5. In the networks with varying degrees of random variation generated as part of Problem 4, determine how these random variations affect the following structural connectivity measures: (a) characteristic path length, (b) cluster index. Consult www.indiana.edu/~cortex/connectivity_toolbox.html for sample programs. Discuss your results in the context of small-world models of networks.
6. Prove that $I(X) \geq I(X^k) + I(X - X^k)$.

23 The Darwin Brain-Based Automata: Synthetic Neural Models and Real-World Devices

Jeffrey L. Krichmar and George N. Reeke

CONTENTS

23.1 INTRODUCTION

In this chapter, we describe the "synthetic neural modeling" (Reeke et al., 1990a, 1990b) approach to the construction of large-scale models of neuronal systems and illustrate this approach with several examples from the "Darwin" series of automata. These models are designed to demonstrate how the theory of neuronal group selection (TNGS) proposed by Edelman (1978, 1987, 1989, 1992) leads to working models of perceptual and motor systems capable of carrying out behavioral tasks that are simplified, but still realistic. The earlier models in the series were entirely synthetic in nature, that is, their sensory and effector organs as well as their environments were simulated in a computer, whereas the later models comprise simulated nervous systems embodied in physical devices capable of interacting via real sensors and real effectors with a real-world environment.

The synthetic neural modeling approach per se is a general method of testing theories of brain function at the systems level. These models comprise multiple neural areas based on the neuro-anatomy and connection patterns of real brain regions and contain populations of different types of neurons that interact via simulated synapses. In principle it may be applied to any well defined and sufficiently detailed theory of brain function, and we hope the examples presented here will inspire the reader to consider whether synthetic modeling might be of use in addressing his or her own research problems. Nonetheless, the TNGS has been a critical ingredient in the models presented here, and so we begin with a brief summary of that theory. This is followed by a general description of the synthetic neural modeling methodology, which we illustrate with brief descriptions of four examples. Throughout, we emphasize how the principles of the TNGS lead one to consider only the most realistic possible simulations of neural systems in order to arrive at a correct understanding of their function. Naturally, the line defining the most realistic possible simulations constantly shifts in the direction of larger systems and more realistic neurons, synapses, and learning rules as the capabilities of available computer systems continue to increase at a rapid rate.

23.1.1 Summary of the Theory of Neuronal Group Selection

The TNGS is, at its heart, a selectionist theory of brain function. That is, it is based on the premise that neuronal circuits in the brain, formed with functional properties, the value of which is not knowable *a priori*, compete with one another to participate in the determination of behavior via strengthening of their connections with other brain circuits. Thus, selection in the brain operates in loose analogy with the way organisms, having individual variations, the adaptive value of which is not knowable *a priori*, compete with one another under natural selection to participate in the determination of the population distribution in future generations. (See Edelman [1987] for more details on all the material in this section.)

The basic units of selection in the TNGS are neuronal groups, collections of dozens, or hundreds of neurons that are strongly connected to each other via synapses (or possibly via gap junctions or other modes of intercellular communication) so as to form cooperative functional units. There is no implication in the theory that neuronal groups compete by multiplication and selective death as do organisms under natural selection; rather, neuronal groups sample sensory inputs from the environment and compete by forming and breaking connections with other neurons so as to stabilize their own boundaries and enhance their contribution, directly or indirectly, to the output motor tracts that ultimately generate the behavior of the organism. Some neuronal groups may have more or less permanent membership during the life of the organism; others may add or subtract members among neighboring neurons as environmental contingencies change during ongoing experience.

The theory postulates that neuronal groups are initially formed during development under the broad control of genetically specified processes. Through natural selection, these developmental processes provide appropriate neuronal cell types, their organization into brain regions, and tracts of connections between and among them, that are appropriate for the ecological requirements of their species. Competitive processes are already at work prior to the animal's exposure to the environment as individual neurons put forth axonal and dendritic outgrowths (neurites) that must reach appropriate target regions via a combination of chemical and sometimes electric signals in order to survive. Those neurites that do not reach their targets are withdrawn, and the cells that spawned them in many cases undergo apoptotic death. This entire set of developmental processes is referred to as developmental selection, and its product is a *primary repertoire* of neuronal groups, ready to undergo the next stage of selection after the animal is born into the world (see Figure 23.1A).

The young animal is equipped with a certain set of built-in responses, supplied by evolution via the structure of its nervous system, with which it must be able to feed itself and begin the exploration of its surroundings. The TNGS postulates that during this early exploration, and continuing indefinitely into adulthood, the spontaneous and sensory-driven activity of neuronal groups results in the production of behaviors. Some of these behaviors are of value to the animal, resulting in the satisfaction of basic needs such as hydration and nutrition; others are neutral; still others result in the sensation of pain. These consequences of behavior provide the substrate for further, experiential competition leading to selection among neuronal groups and the formation of *secondary repertoires* that more accurately and more rapidly generate the kinds of behavioral responses that have proven to be adaptive in the commonly encountered environmental situations (see Figure 23.1B).

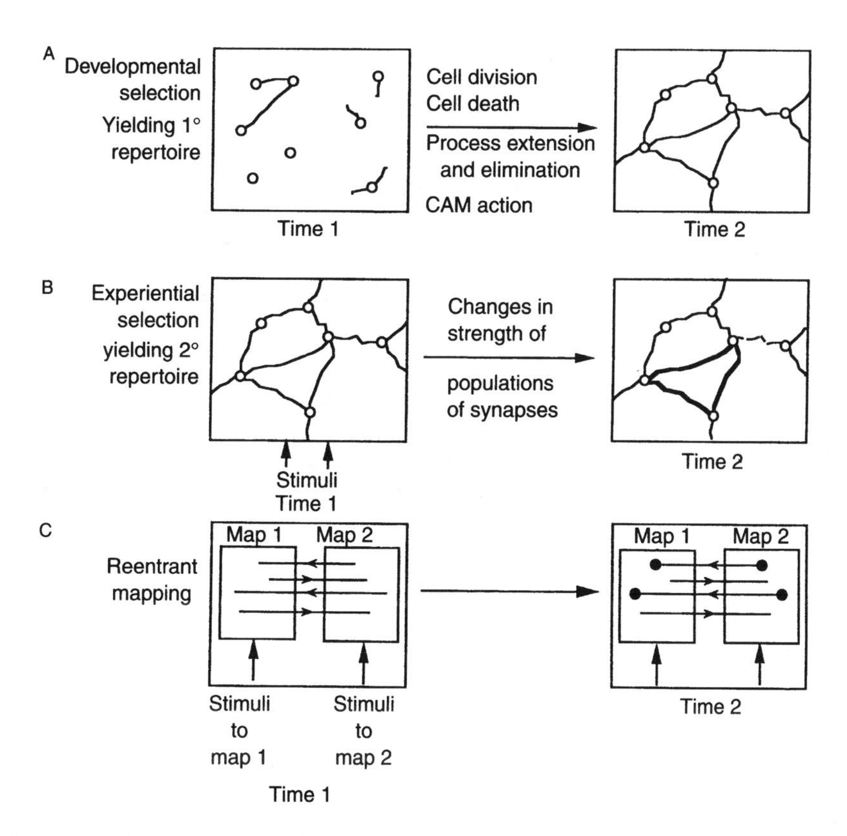

FIGURE 23.1 Schematic illustration of the main principles of the theory of neuronal group selection. (A) Developmental selection leading to formation of primary repertoires. CAM = cell adhesion molecule. (B) Experiential selection leading to formation of secondary repertoires. (C) Reentry leading to formation of mappings between neuronal areas that capture essential regularities and relationships in the world through the behavior–environmental change–sensation–neuronal response loop. Filled circles (right panel) = connections strengthened during map formation. (From Edelman [1989] *The Remembered Present.* Reproduced with permission of the Perseus Books Group.)

Experiential selection proceeds via a combination of local and global mechanisms. Locally, the strengths of the synaptic connections between neurons are constantly changing through processes such as long-term potentiation and long-term depression brought about by the combined activities of the presynaptic and postsynaptic neurons. These phenomena affect the probability and quantity of transmitter release, the magnitude of transmitter effects on postsynaptic receptors, or other details of synaptic transmission and cellular tone. These Hebbian (1949) changes, however, are sensitive only to levels of local neuronal activity and cannot by themselves sense the direction in which change is required to generate veridical cognition and behavior of adaptive value to the organism as a whole. The TNGS proposes that this essentially global function is provided through the operation of innate mechanisms, called *value systems*, that enable the nervous system to sense the consequences of the animal's behavior via its interoceptive or exteroceptive effects on the organism (Reeke et al., 1990a). Value systems provide nonspecific, modulatory signals (via e.g., dopaminergic, serotoninergic, cholinergic, histaminergic, and noradrenergic projections) to the rest of the brain that bias the outcome of local changes in synaptic efficacy in the direction needed to satisfy global needs. This bias is not computationally adjusted to provide just the right amount of change at each local synapse at each moment in time, nor could it ever be in the absence of an all-knowing supervisory system, if for no other reason than that of the estimation of value is itself imperfect and subject to error. Stated in the simplest possible terms, behavior that evokes positive responses in value systems biases synaptic change to make production of the same behavior more likely when the situation in the environment (and thus the local synaptic inputs) is similar; behavior that evokes negative value biases synaptic change in the opposite direction. The existence of simple innate value systems, based for the most part on circulation-based chemical senses (e.g., thirst, hunger) and pain and their opposites, permit the selective brain to "bootstrap" from infantile to mature function using its own experiences as its best teacher. Because value is an imperfect measure of behavioral success, and multiple behaviors in the past may contribute in different directions to current value signals, many iterations of the learning process are necessary to reach finely tuned adult performance levels. However, it should be noted that value systems, at least in higher animals, may themselves be subject to selection-based learning (Friston et al., 1994), and this feature allows learning rates to improve as the animal matures.

Value-based learning so far has been described as a mechanism of selection operating on individual neurons or neuronal groups. A further mechanism is needed to coordinate both ongoing responses and synaptic change within and across brain areas. Operating independently, no number of neuronal groups could provide the kind of coordinated, anticipatory responses to complex and rapidly changing environmental situations that are characteristics of animals with advanced nervous systems. Neuroanatomy tells us that collections of neuronal groups are organized hierarchically into strongly interconnected local circuits (e.g., the so-called minicolumns and columns in the mammalian visual cortex), and these local circuits are in turn assembled, via successively less dense connections, into brain areas or nuclei, gyri, and lobes. The TNGS holds that the purpose of these interconnections is to allow the local circuits to exchange signals in a process called *reentry*. Reentry includes both excitatory and inhibitory interactions. Reentry subsumes the functions of feedback and recurrence proposed in other models (Ashby, 1956; Wiener, 1961; Elman, 1990; Cleeremans, 1993) but its main function is to support the generation of *mappings* by which the responses of connected neuronal groups are brought into broad concordance with each other and with the correlations that hold between objects and events in the world outside the nervous system (see Figure 23.1C). Thus, if groups that are reentrantly connected with each other happen to respond in discordant ways to a particular stimulus, the value-based learning mechanism will act not only to weaken the connections between those groups, but also to weaken their contributions to the motor output as well. On the other hand, value-based learning operating on the richness of reentrant connections across the entire brain ensures the nuanced categorization of stimuli in correspondence with their natural hierarchies and joint probability distributions rather than the development of a single, nondiscriminatory pattern of response to any and all stimuli as might occur with purely excitatory recurrence.

Selective models can help us understand aspects of brain structure and function that may otherwise seem paradoxical. In a selective system, developmental mechanisms are not required to construct circuits with exquisite genetically encoded precision to carry out each of the tasks the brain is responsible for; instead, the developmentally inevitable variability of neuronal structure and connectivity at the finest level provides the population of units from which those that are most adapted to carry out a particular function can be recruited and then tuned. Some of these circuits may never be used; this appears at first glance to be terribly wasteful, but in fact these extra units provide backups that can be brought into service when other units fail. There are many ways by which any particular behavioral response can be generated by different neuronal pathways. If these pathways were identical, they would be simply redundant and that would be useful enough for dealing with neuronal aging and death; but in fact, the pathways are not identical and the behaviors they produce may be similar but never exactly identical. This condition is called *degeneracy* after the quantum-mechanical concept of states that are different but have the same energy. Degeneracy (Edelman and Gally, 2001) is critical to the ability of selection to work in the face of novelty because it provides preexisting repertoires of solutions to environmentally posed problems that may have already been partially selected and tuned in similar situations but which may work even better in some new situation for which they were not tuned. The existence of variant degenerate circuits, in other words, allows a kind of symmetry breaking to occur, analogous to speciation in natural selection, in which circuits that generate alternative responses to one kind of situation can become individually specialized by selection to deal with related, but different, environmental situations.

Clearly, more is needed than just model neuronal networks to simulate a system operating on selective principles. Because selection by its very nature depends on the interaction of an organism with its environment, using motor organs to perturb the environment, sensory organs to detect the resulting changes, and value systems to assess the results of those changes, model selective systems must include all of these elements. In addition, they must incorporate model neurons that accurately simulate the responses of real neurons to modulatory as well as synaptic inputs and the temporal evolution of those responses as parts of perceptual and gestural sequences that must match the rhythm of events in the outside world. Synthetic neural modeling is a methodology designed to meet these requirements.

23.1.2 Synthetic Neural Modeling and Brain-Based Devices

A synthetic neural model (SNM) comprises a real or simulated device with sensory and motor organs, an environment for the device to interact with and explore, and a neural system ("brain") simulation to implement memory and guide the device's behavior. Each of these components must necessarily be present for a model to be an SNM, because the fundamental idea behind the SNM approach is that nervous systems acquire the ability to generate adaptive behavior only through interaction with an environment, not through programming. Thus, all three components, brain, body, and environment, must exist and interact for successful performance to be evinced.

When the organismic component of an SNM is real, that is, a robotlike device, the system is referred to as a brain-based device (BBD). Unlike the actions of conventional robots, the behavior of BBDs is not programmed by computer instructions, although their nervous systems may be simulated by a computer program. Instead as the name suggests, BBDs operate according to the principles thought to be followed by the brain.

One essential property of SNMs, including BBDs, is that, like living organisms, they must organize the unlabeled signals they receive from the environment into categories. This organization of signals, which in general depends on a combination of sensory modalities (e.g., vision, sound, taste, or touch), is called *perceptual categorization*. Perceptual categorization in models (Edelman and Reeke, 1982) as well as living organisms makes object recognition possible based on experience, but without *a priori* knowledge or instruction. An SNM selects and generalizes the signals it receives with its sensors, puts these signals into categories without instruction, and learns the appropriate actions when confronted with objects under conditions that produce responses in value systems.

The behavior of SNMs and the activity of their simulated nervous systems must be recorded to allow comparisons with experimental data acquired from animals. This comparison serves two purposes: First, SNMs are powerful tools to test theories of brain function. The construction of a complete behaving model forces the designer to specify theoretical and implementational details that are easy to overlook in a purely verbal description and it forces those details to be consistent among themselves. The level of analysis permitted by having a recording of the activity of every neuron and synapse in the simulated nervous system during its behavior is just not possible with animal experiments. Second, by using the animal nervous system as a metric, designers can continually make their simulated nervous systems closer to that of the model animal. This in turn allows the eventual creation of practical devices that may approach the sophistication of living organisms.

The neural simulation in a SNM is modeled after the neuroanatomy (i.e., structure) and neurophysiology (i.e., dynamics) of a real (usually vertebrate) brain. It always contains these three key elements: (a) neuronal elements or units representing individual cells or groups of cells, (b) neural areas containing multiple such units with specific dynamic properties, and (c) connections between neuronal units both within areas and between areas. We now discuss these three components in turn.

23.1.3 The Single-Cell or Neuronal Group Model

In SNMs, neuronal groups are the processing elements of the simulated nervous system. If the model is to produce meaningful results, these neuronal units must be modeled after neurons found in real nervous systems (also see the discussion in Chapter 15, this volume). However, a vast range of such models with different levels of dynamical detail is possible in principle. The appropriate level of detail must be chosen according to the demands of the task being studied. Indeed, variation of cellular dynamics is one of the key means of testing hypotheses about the neural basis of brain function. Thus, for some studies, complete multicompartment single-cell models are appropriate.

For the models described in this chapter, neuronal units representing groups in the TNGS (see Figure 23.2) were simulated with mean firing-rate models. The activity of each such unit is represented by a firing-rate variable, s, corresponding roughly to the average firing activity of a group of approximately 100 neurons averaged over a time period of 100 to 200 msec. These mean firing-rate values are constrained to range from 0 (i.e., quiescent) to 1 (i.e., maximal firing).

A neuronal unit gets its input either directly from sensory input (e.g., camera, microphone) or from other neuronal units in the form of simulated connections, each representing a synapse or collection of synapses. Changes in the strengths of the connections over time enable the model to form new memories and to learn. Specifically, each connection has an associated numerical variable (c_{ij} in Equation [23.1]) that represents its current relative strength, with positive values representing excitatory connections and negative values, inhibitory. Stronger connections have more influence on the activity of the postsynaptic neuronal unit than weaker connections. The total contribution of all synaptic inputs to unit i is given by,

$$A_i(t) = \sum_{l=1}^{M} \sum_{j=1}^{N_l} c_{ij} s_j(t), \tag{23.1}$$

where M is the number of different anatomically defined classes of connections and N_l is the number of connections to unit i from units of class l, s_j is the activity of neuronal unit j, and c_{ij} is the strength of the connection(s) from unit j to unit i. The activity of neuronal unit i is subjected to a sigmoidal

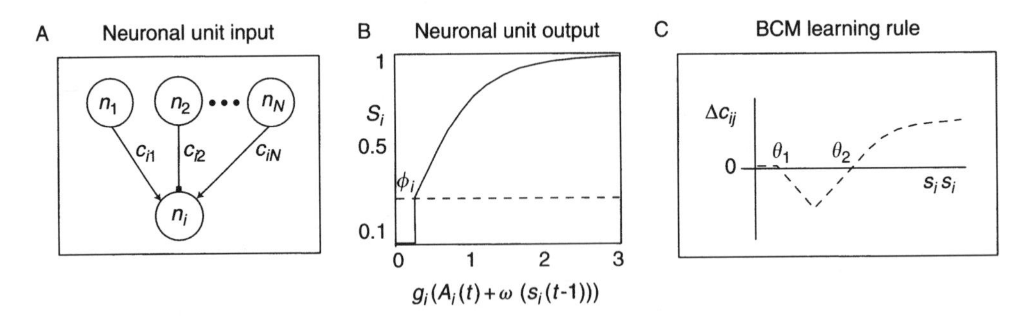

FIGURE 23.2 Typical neuronal unit models as used in Darwin II through Darwin IX. (A) Neuronal unit i receives input from N neuronal units via synaptic connections labeled $c_{ij}, j = 1, 2, \ldots, N$. Each connection has a relative strength that can be either excitatory (e.g., c_{i1} and c_{iN} are positive) or inhibitory (e.g., c_{i2} is negative). (B) The output of a neuron is subject to a function based on its current activity and the input from other neurons. Below a specific threshold ϕ_i, the output of the neuronal unit is 0. Above the threshold, the neuronal unit activity is a positive number that asymptotes to 1. (C) Learning is governed by a rule that changes the connection strength between two neuronal units. Here the so-called BCM rule is illustrated. When the activity of the two units is weakly correlated ($s_1 s_2$ product on the abscissa between θ_1 and θ_2), the connection is depressed, whereas when the activity of the two units is strongly correlated ($s_1 s_2$ product greater than θ_2), the connection is potentiated.

flattening function, for example,

$$s_i(t+1) = \max(\tanh(g_l(A_i(t) + N + \omega s_i(t))), 0), \tag{23.2}$$

where ω determines the persistence of unit activity from one cycle to the next, N represents a Gaussian noise term, and g_l is a scale factor dependent mainly on N_l (see Figure 23.2B for a diagram of this activity function). (In some simulations, a piecewise-linear approximation to the tanh function was used.) The persistence, scaling, and, in some models, firing threshold values allow the neural simulation designer to closely match the neuronal units in each part of a simulated brain to the dynamics of real neurons found in the corresponding parts of the nervous system (see table A1 in Krichmar and Edelman [2002] and table 1 in Seth et al. [2004c] for specific examples).

Some connections, for example, those implementing built-in or instinctive behaviors, have fixed connection strengths; others, those involved in implementing circuit learning and memory, are plastic. In general, the rules governing synaptic plasticity are another hypothesis-driven element of the theory being tested in a particular SNM. In the models described here, plasticity was based on the principles of value-based learning described earlier. Value-dependent modification of synaptic strength was implemented in the early SNMs as a variant of the well-known "Hebb rule" based on the proposal by Hebb (1949) that "When an axon of cell A excite[s] cell B and repeatedly or persistently takes part in firing it, some growth process or metabolic change takes place in one or both cells so that A's efficiency as one of the cells firing B is increased" (quoted in Kandel et al., 2000, p. 1260). It has been widely assumed that the phenomenon of long-term potentiation (LTP, Bliss and Lomo, 1973) is the biological manifestation of the Hebb rule. The rule is generally expressed mathematically by making the change in c_{ij} at each time step proportional to the product of the pre- and postsynaptic activities at connection ij. Value dependence is introduced by multiplying this product by a third factor proportional to the activity in the value system that modulates the class of connections, l, to which connection ij belongs. Considerations of stability further require that there be a saturation factor or other mechanism to keep c_{ij} values in a restricted range (here -1 to $+1$). A mechanism corresponding to long-term depression (LTD, Abraham and Goddard, 1983) in real neurons is also needed to weaken connections as well as to strengthen them; this is implemented in the value-dependent version of the Hebb rule by subtracting a fixed threshold from each of the three factors in the rule, thus allowing negative changes in connection strength to occur when one or all three of the inputs is below threshold. Thus the final rule is (Reeke et al., 1990b),

$$c_{ij}(t+1) = c_{ij}(t) + \delta_l \cdot \phi(c_{ij}) \cdot (s_i(t) - \Theta_l^I) \cdot (s_j(t) - \Theta_l^J) \cdot (v_l(t) - \Theta_l^V), \tag{23.3}$$

where δ_l is a modification (or "learning") rate parameter for connection type l, which is generally set to a value in the range 0.01 to 0.05; ϕ is the saturation function, a polynomial that evaluates to 1.0 when c_{ij} is 0 and to 0 when c_{ij} is ± 1; and Θ_l^I, Θ_l^J, and Θ_l^V are respectively the modification thresholds for postsynaptic activity, presynaptic activity, and value system activity for connection type l. Provision was made to set δ_l to 0 for certain combinations of signs of the three thresholded factors in the rule, for example, when all three were negative. As a practical matter, $s_i(t)$ may be replaced by $s_i(t-1)$ so that the change in each connection strength can be evaluated at the same time as its contribution to $s_i(t)$.

In the BBDs described here, the modified Hebb rule was replaced with a variant of the Bienenstock–Cooper–Munro (1982) rule, in which weakly correlated activity is depressed and strongly correlated activity is potentiated (see Figure 23.2C). This rule has some biological justification (Kirkwood, 1996). It takes the form

$$\mathrm{BCM}(x) = \begin{cases} 0 & x < \theta_1 \\ k_1(\theta_1 - x) & \theta_1 \leq x < \dfrac{\theta_1 + \theta_2}{2} \\ k_1(x - \theta_2) & \dfrac{\theta_1 + \theta_2}{2} \leq x < \theta_2 \\ \dfrac{k_2 \tanh(\rho(x - \theta_2))}{\rho} & \text{otherwise,} \end{cases} \tag{23.4}$$

where x is the postsynaptic activity, θ_1 and θ_2 are thresholds defining the regions of depression and potentiation, k_1 and k_2 are scaling constants, and ρ is a saturation constant.

23.1.4 Neural Areas and Network Architecture

The neural areas and their connectivities within the neural system of a SNM are based in a hypothesis-driven manner on real neuroanatomy and their details are critical for the model's performance and thus for the evaluation of the hypotheses on which it is based. Each neural area may have a multiplicity of cell types, representing, for example, cortical pyramidal cells, cerebellar Purkinje cells, inhibitory cells of various kinds, and so on. Each type of cell may have different dynamical parameters and different patterns of connectivity appropriate to its location and function. For example, area *IT* in Figure 23.3 corresponds to an area called the inferotemporal cortex that is known for visual object recognition in the primate. Some of these areas get their input directly from the environment (e.g., *R* in Figure 23.3 gets input directly from the CCD camera) whereas other areas directly affect actuators in the device (e.g., activity in area *C* in Figure 23.3 is converted into left and right wheel speeds). Neural areas between the primary sensory and motor areas get their input from the activity of other neuronal units and send their output to other neuronal units. The connections between these neuronal units are loosely given by the arrows shown in Figure 23.3, and more specifically, are given by constraints dictating the probability of connectivity and the initial strength of that connectivity (see table 1 in Reeke et al. [1990b], table A2 in Krichmar and Edelman [2002], and table 2 in Seth et al. [2004c] for specific examples). The sample wiring diagram in Figure 23.3 closely follows the synaptic connectivity found in vertebrate brains and nervous systems. Possibly more than any other factor, it is this wiring that dictates the specialization of different areas in the nervous system and gives them the specificity to guide behavior.

23.1.5 Embodying the Brain

SNMs must have sensory and effector organs and must interact with an environment. This bedrock principle of synthetic neural modeling is based on the obvious but often ignored fact that brains and bodies do not function separately but through mutual interactions. Brains are embedded in bodies and bodies are embedded in the real world. In purely synthetic models, sensory and effector organs and indeed the external environment as well are all simulated in a computer. In the case of BBDs, the sensory and effector organs are physically instantiated with a particular morphology that allows for active sensing and autonomous movement in a real-world environment. This environment may be a room in a laboratory, containing arrangements of stimuli for the BBD to sample and spaces for it to explore, or it may be the actual outdoor world. Senses may include vision, implemented via one or a pair of color CCD video cameras; touch, implemented via touch-sensitive resistive tapes or via mechanical switches connected at the bases of flexible "whiskers"; hearing, implemented via ordinary microphones; or even senses that humans do not have, such as infrared (IR) proximity detectors analogous perhaps to the echolocation senses of bats. In each case, input from the environment in the form of sensor data is converted into a neural signal having the characteristics of the corresponding

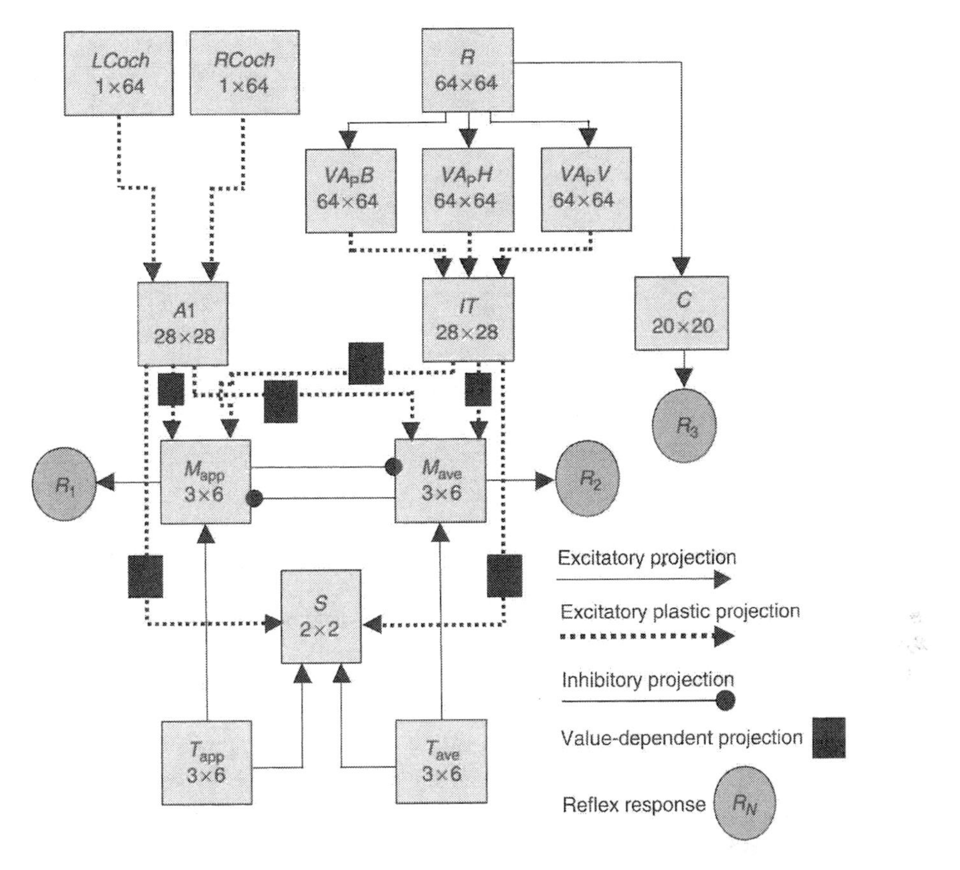

FIGURE 23.3 Schematic of the regional and functional neuroanatomy of Darwin VII. Boxes denote neural areas, the numbers in the boxes denote the numbers of rows and columns of neuronal units, and the arrows denote the projections connecting neuronal units from one area to another. Six major systems make up the simulated nervous system: an auditory system, a visual system, a taste system, sets of motor neurons capable of triggering behavior, a visual tracking system, and a value system. In the version described here, the simulated nervous system contains 18 neuronal areas, 19,556 neuronal units, and approximately 450,000 synaptic connections. The 64×64 gray-level pixel image captured by the CCD camera was relayed to a retinal area R and transmitted via topographic connections to a primary visual area, *V*Ap. *V*Ap had subareas selective for blob-like features (*V*Ap*B*), for short horizontal line segments (*V*Ap*H*), and for short vertical line segments (*V*Ap*V*). Responses within *V*Ap closely followed stimulus onset and projected nontopographically via activity-dependent plastic connections to a secondary visual area, analogous to the inferotemporal cortex (*IT*). The frequency and amplitude information captured by Darwin VII's microphones was relayed to right and left simulated cochlear areas (*LCoch* and *RCoch*, respectively) and transmitted via mapped tonotopic and activity-dependent plastic connections to a primary auditory area $A1$. The activity of each cochlear neuronal unit was broadly tuned to a preferred frequency and scaled according to the signal amplitude. $A1$ and *IT* contained local excitatory and inhibitory interactions producing firing patterns that were characterized by focal regions of excitation surrounded by inhibition. $A1$ and *IT* sent plastic projections to the value system S and to motor areas M_{app} and M_{ave}. These two neuronal areas were capable of triggering two distinct behaviors, appetitive and aversive, respectively, when the difference in instantaneous activity between motor areas M_{app} and M_{ave} exceeded a behavioral threshold ($\beta = 0.3$). The taste system (T_{app} and T_{ave}) consisted of two kinds of sensory units responsive to either the presence or absence of conductivity across the surface of stimulus objects as measured by sensors in Darwin VII's gripper. Picking up and sampling the conductivity of a block is innate to Darwin VII's behavior. In all the experiments, strongly conductive blocks activated T_{app} and weakly conductive blocks activated T_{ave}. The taste system sent information to the motor areas (M_{app} and M_{ave}) and to the value system (S). Area S projects diffusely with long-lasting value-dependent activity to the auditory, visual, and motor behavior neurons. Tracking commands were issued to Darwin VII's wheels based on activity in area C. The visual tracking system controlled navigational movements, in particular the approach to objects identified by brightness contrast with respect to the background. (From Krichmar and Edelman [2002]. Reprinted with permission of Cerebral Cortex, Oxford University Press.)

biological sense. For example, a pixel array from a CCD camera can be converted into an array of a neural activation that simulates retinal input into the visual system of the BBD. Similarly, outputs of motor areas in the neural simulation can be converted to commands to operate mechanical devices. 'Effector organs may be as simple as motor-powered wheels for locomotion on a flat floor, or as complex as multijointed robotic arms. Another way to generate motor output is to initiate a motor reflex when the neural activity in a motor area of the brain goes above some threshold. An example of such a reflex is seen when Darwin VII picks up a block with its gripper when neural activity in its motor area for sampling objects is above threshold (see Figure 23.3).

SNMs based on the TNGS have *value systems* as described above. When an important environmental event occurs, one or more value systems are triggered by salient environmental cues and their activity leads to large-scale changes in the model nervous system such that the SNM or BBD will "remember" this event the next time it occurs and modify its behavior accordingly. An example of a value system is the dopaminergic system, which has been implicated in the prediction of reward in primates (Schultz et al., 1997). The value system allows the BBD to learn what it needs to successfully explore an environment without explicit instruction.

A BBD needs to engage in a task that is initially constrained by a minimal set of innate behaviors or reflexes. The combination of the model's sensors, effectors, value systems, and the innate nervous-system structures that determine its reflexive behaviors are collectively referred to its *phenotype*. Similar to a newborn animal, a BBD has some minimal reflexive behaviors (e.g., do not crash into walls) that it can build upon to create more interesting behaviors. For example, the reflex for avoiding walls could initially be activated by IR proximity detectors. In turn, the IR proximity detectors could activate the value system causing the BBD's nervous system to remember visual attributes of the wall. The next time the BBD approaches the wall, the visual system might trigger the "avoid wall reflex" prior to the IR sensors being activated. This is an example of a *conditioned response*.

23.1.6 Software Simulator

A variety of software tools has been developed for the simulation of SNMs. One of these is "CNS," the cortical network simulator (Reeke and Edelman, 1987), which was used to generate the Darwin III and Darwin IV models (see Table 23.1). Similar software with additions required for the various special hardware was used for the other models described here. CNS permits the user to construct SNMs comprising multiple interconnected brain areas or nuclei, each of which may contain cells of many different types. Systems constructed with CNS can receive input from real or simulated senses and can provide output to real or simulated motor systems. CNS can be used to simulate purely perceptual as well as behaving systems and can be used as a neurally based robot controller. It is written in C language for portability and speed, and provides a general-purpose interface for connection to real or simulated sensors and motor output devices.

CNS can model a variety of rate-coded and integrate-and-fire cells. Cells in CNS can integrate synaptic inputs, some of which may be voltage-dependent, and respond to inputs above a specified threshold in a graded fashion or by emitting single spikes or spike trains. Production of a spike may induce a refractory period, possibly followed by synapse-specific paired pulse facilitation. Cell responses may be subject to graded depression (habituation). Probe currents or probabilistic noise may be injected into cells. Transmission delays between cells can be simulated. Time constants may be specified to control decay of cell activity and of postsynaptic potentials. The geometry and initial strengths of connections between cells can be specified according to a variety of built-in options, or they can be read from files. Several versions of the value-dependent Hebb rule for synaptic modification are implemented.

Output from CNS is available in the form of printable text, graphics, and binary data records that can be read into MATLAB™ (The Mathworks, Natick, MA) programs for further analysis. Text output includes a tabular summary of the cell group and network setup parameters, detailed information about individual cells as requested, and statistics. Graphical output is flexible, but

TABLE 23.1
The "Darwin" models

Darwin I	1981	Pattern recognition task Degeneracy, amplification, selection
Darwin II	1982	Categorization task Neuronal groups, reentrant connections
Darwin III	1990	Develops sensorimotor coordination Oculomotor system, reach system, tactile system Synthetic neural modeling with an organism interacting in a simulated environment
Darwin IV[a]	1992	Tracking and conditioning task Synthetic neural modeling in a real-world artifact
Darwin V[a]	1998	Translation invariance and pattern selectivity emerge due to the continuity of self-generated movement
Darwin VI[b]	2000	Sense-perceive-act concurrently in real time Role of history in perceptual categorization
Darwin VII[b]	2001	Multimodal sensing and conditioning
Darwin VIII[b]	2004	Binding through synchrony Visual cortex model with reentrant signaling
Darwin IX[b]	2004	Whisker barrel model and texture discrimination

[a] Physical devices interacting in the real world.

[b] Physical devices interacting in the real world in real time.

typically shows the activity of each cell in the system after each simulation cycle. Graphics can be recorded in a file for later viewing or for transmission to another computer. The state of the simulated system can be recorded at any time in a file that may be used to restart the run later, possibly with altered parameters.

23.2 THE DARWIN SERIES OF MODELS

A series of SNMs and BBDs have been constructed over the past 25 years at The Neurosciences Institute by Gerald Edelman, the authors, and their colleagues. These models have been named the "Darwin" series because all of them operate on selective principles as proposed by Edelman in the TNGS (Edelman, 1987, 1993). The characteristics of the entire series are summarized in Table 23.1. In the following sections, we briefly summarize the construction and performance of four of these models, Darwins III, VII, VIII, and IX. Darwin III (Reeke et al., 1990a, 1990b) was an entirely simulated model which demonstrated perceptual categorization, experience-based refinement of innate behaviors, and associative learning. Darwin VII (Krichmar and Edelman, 2002) is a device capable of perceptual categorization and operant conditioning. Darwin VIII (Seth et al., 2004c) binds visual objects and segments scenes. Darwin IX (Seth et al., 2004a, 2004b) performs texture discrimination by the use of artificial whiskers. Other models in the series are described in Edelman et al. (1992), Almassy et al. (1998), and Krichmar and Snook (2002).

23.2.1 A Totally Synthetic Model: Darwin III

Darwin III is a simulated, behaving automaton operating in a simulated environment with simulated senses and effectors (Reeke et al., 1990a, 1990b). As with the other models to be described later, the units in its nervous system represent small groups of neurons; their dynamical properties and the probability distributions of their connections are specified in advance by the experimenter, but no prior information about particular stimuli and no explicit algorithms for "neural computations" are specified. Training is carried out without any external specification of desired responses ("supervised learning") or detailed information about deviations from those desired responses ("back propagation of errors"). Instead, innate, internal value systems determine the adaptive value of responses by sensing their consequences as proposed in the TNGS.

Darwin III is a sessile creature with a single eye capable of scanning its immediate surroundings and an arm (two- or four-jointed in different experiments) capable of reaching toward objects in the environment and moving them. Its nervous system comprises some 50 reentrantly connected networks containing about 50,000 neural units and about 620,000 synaptic connections arranged in two major sensory subsystems for optical and tactile sensing of stimuli, two major motor subsytems for reaching to and palpating stimuli with the arm, and a perceptual categorization subsystem for recognizing stimuli and generating responses to them. The environment is a 32×32 pixel geometric array on which stimulus objects of roughly 10×12 pixels with various shapes (accessible to vision) and textures (accessible to touch) move about at random with occasional jumps. The "retina" of the eye samples a 16×16 area of the environment which can be moved about on the overall environmental array by an oculomotor subsystem.

Because detailed descriptions of Darwin III and its behavior have been available for some time (Reeke et al., 1990a, 1990b), we give only a brief summary here. In only a few hundred trials, visual tracking of moving objects develops by selection from among initially random movements of the eye generated by random connectivity in the oculomotor subsystem. Guided by a value system that simply responds positively to larger amounts of "light" from objects sensed on the retina, with a greater weighting toward the foveal center, connections between a neural area representing a superior colliculus and the oculomotor cortex are modified, resulting in an enhanced probability of successful tracking. This result is robust to changes in the initial numbers and strengths of the modifiable connections, as well as to lesioning of up to half of them after training.

The neural circuitry associated with the arm is more complicated, but a similar training principle is followed. A primary motor area controls movements of the arm's joints via intermediate relay areas representing brainstem nuclei. At the beginning of a run, arm movements take the form of random gestures engendered by spontaneous activity of the motor area units. Kinesthetic feedback from joints in the arm and visual input responsive to arm position enter an area modeled loosely on the granule cell layer of the cerebellum, the units of which respond to arbitrary combinations of these inputs. These cells are connected via connections subject to value-based modification with output units modeled on cerebellar Purkinje cells. As with real Purkinje cells, output from the model Purkinje cells is inhibitory and acts upon the brainstem motor relay areas that transmit gesture signals to the arm. These latter connections are also subject to value-dependent modification. The value system that modulates synaptic change in the model cerebellum responds simply to touch, giving positive value when the arm manages to contact a stimulus object. The overall result is inhibition of useless gestures and disinhibition of those gestures that lead to the arm's touching the stimulus. After a short period of training (300 simulation cycles), the arm usually manages to contact any stimulus object in reach from any starting configuration by repeating (with variation) some successful past reaching movement made when a similar constellation of visual and kinesthetic inputs happened to have occurred. Once contact has been established with a stimulus object, the arm proceeds to explore its surface with tactile sensors located at the end of the arm (the "hand"). For this function, the arm in the reported experiments assumed a stiff posture that did not require coordinated movement of its multiple joints.

Perceptual categorization in Darwin III is based upon reentrant interaction between responses in visual and tactile areas. Visual areas respond to visual features of stimuli, such as dark and light stripes. Tactile areas respond to textural features of stimuli, such as having a bumpy or smooth surface. A short-term memory system is required for these textural discriminations, inasmuch as the object surface is explored by sequential tactile probing, such that the input from the touch sensors from different parts of the stimulus object is not presented to the neural simulation all at the same time. The tactile probing system operates until no new features are discovered, then output is provided to the categorization system, where a "classification couple" responds to particular pairings of visual and tactile responses. In Darwin III, recognition of an object that is simultaneously bumpy and striped is signaled by a built-in motor response by which the arm "swats" the object away from the immediate environment. Conditioning of this response was deferred to studies with the later Darwin models.

23.2.2 Darwin VII — Multimodal Sensing and Conditioning

Darwin VII (Krichmar and Edelman, 2002) consists of a mobile base equipped with a movable head and a gripping manipulator having one degree of freedom (Figure 23.4). The head has a CCD

FIGURE 23.4 Darwin VII. The model consists of a mobile base equipped with several sensors and effectors and a neural simulation running on a remote computer workstation. The mobile unit contains a radio modem to transmit sensory and status information to the computer workstation carrying out the neural simulation and to receive motor commands from the simulation. This arrangement allows untethered exploration. A gripper (foreground) is able to pick up small objects. A color CCD camera (top), a pair of microphones on either side of the camera, and conductivity sensors embedded in the gripper provide sensory input to the neuronal simulation. Eight IR sensors mounted at 45-degree intervals around the mobile platform are used to trigger reflexes for obstacle avoidance. (From Krichmar and Edelman [2002]. Reprinted with permission of Cerebral Cortex, Oxford University Press.)

camera for vision and microphones for hearing; the gripper has conductivity sensors that work upon contact with gripped objects to provide a sense analogous to taste. Darwin VII's behavior is guided by a nervous system shown schematically in Figure 23.3. Its environment is an enclosed area with black walls and a black floor on which are distributed stimulus blocks (6 cm metallic cubes) in various arrangements. The sides of these blocks may be conductive or nonconductive, their tops are decorated with visual patterns, and they have speakers that emit sounds at different frequencies. Strongly conductive blocks with a striped pattern and a 3.9 kHz tone are assigned arbitrarily to be positive (appetitive) value exemplars, whereas weakly conductive blocks with a blob pattern and a 3.3 kHz tone represent negative (aversive) value exemplars. These assignments are initially unknown to Darwin VII; its task is to associate the visual and auditory properties of the blocks, which can be sensed at a distance, with their intrinsically positive- or negative-valued tastes, which can only be sampled on contact. Figure 23.4 shows Darwin VII interacting with some of these blocks.

Basic modes of behavior built into Darwin VII include IR sensor-dependent obstacle avoidance, visual exploration, visual approach and tracking, gripping, and "tasting." Its innate (unconditioned) response upon encountering a block is to pick up the block with its gripper, taste it, drop it, back away, and execute a counterclockwise turn. As a result of conditioning, distinctive responses develop to blocks having visual and auditory features that become associated during unconditioned responses with appetitive and aversive value. Respectively, these responses are: appetitive — pick up the block, taste it, drop it, and back away; aversive — do not pick up the block, back away. Both of these responses are marked by execution of a clockwise turn to make them distinguishable by an observer from an unconditioned response. With the exception of obstacle avoidance, selection among all these behaviors is under control of the simulated nervous system. Visual and auditory patterns are established and associated with each other via value-independent (BCM rule) modification of synaptic connection strengths in the pathways indicated by dotted lines in Figure 23.3. Gripping a block and measuring its conductivity with the gripper activates Darwin VII's value system and causes value-dependent modification of connection strengths in different pathways, those indicated by gray boxes in Figure 23.3. As a result, Darwin VII learns to associate the conductivity (i.e., taste) of a block with the visual patterns and sounds emitted by that block during the approach. Thus the visual patterns and sounds come to form two perceptual categories, to which the corresponding appetitive or aversive responses, respectively, become associated. Darwin VII thus learns to avoid bad-tasting blocks based on their visual and auditory configuration, and no longer approaches them when offered in the environment.

The development of perceptual categories through value-independent plasticity, and the development of conditioned behavior through value-dependent plasticity, are both essential for success in Darwin VII's tasks. Plasticity between primary sensing areas and associative areas (see $LCoch/RCoch \rightarrow A1$ and $VA_P \rightarrow IT$ in Figure 23.3) allows the formation of perceptual categories in response to auditory and visual signals without instruction. Darwin VII's value-dependent learning (see $A1/IT \rightarrow M_{\mathrm{app}}/M_{\mathrm{ave}}$ in Figure 23.3), allows it to associate the blob visual pattern and low-frequency beeping tone with negative value, and the striped visual pattern and high-frequency beeping tone with positive value. This learning requires only 5 to 10 samplings of each kind of block exemplar. After this association of visual and auditory patterns with taste, Darwin VII continues to pick up and "taste" stripe-patterned, high-tone blocks, but avoids blob-patterned, low-tone blocks over 90% of the time (see Figure 23.5).

While performance improves with training, it never reaches perfection and occasional "mistakes" are made. The unpredictability of behavioral responses in Darwin VII coupled with the variability of a complex environment does not, however, prevent the device from learning after mistakes or from generalizing over sensory inputs. Because a BBD does not reach a rigid, precise level of performance, it is able to adapt to novel situations.

Darwin VII never displays identical patterns of neural activity, even during repetitions of the same behavior (see Figure 23.6). The adaptive behaviors tend to remain similar, however, despite

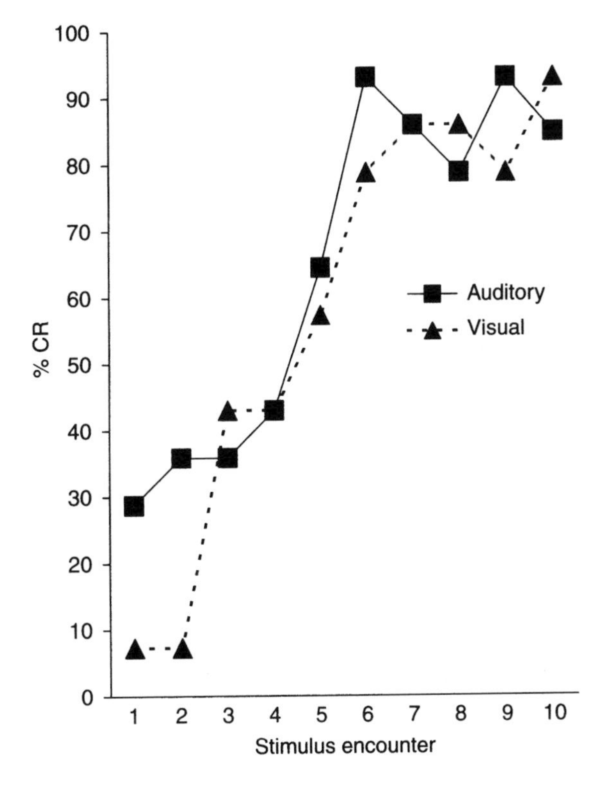

FIGURE 23.5 Percentage of conditioned responses (CR) during learning trials for both visual and auditory conditioning as a function of stimulus block encounters (seven subjects). The responses following aversive and appetitive stimulus encounters were pooled together.

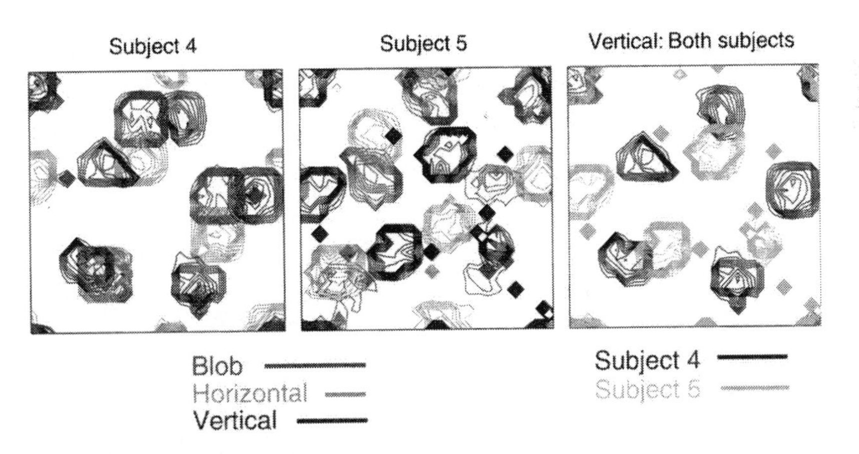

FIGURE 23.6 **(See color insert following page 366)** Comparison of patterns of activity in area IT for the three classes of visual stimuli across different Darwin VII subjects. The left and center contour plots show, for two representative subjects, the borders of neuronal group activity in response to blobs (red lines; colors mentioned here and in the later figure legends of this chapter refer to the figures in the color insert), horizontal stripes (green lines), and vertical stripes (blue lines). The contour plot on the right shows the IT activity for two different subjects in response to the same class of stimuli (vertical stripes): subject 4 (dark blue); subject 5 (light blue). Across all subjects tested ($n = 7$), the mean overlap of activity within each subject but between stimuli was 26.6% ($\sigma = 0.10$). The mean overlap of activity between different subjects was 22.1% ($\sigma = 0.09$) in response to blob patterns, 26.9% ($\sigma = 0.11$) in response to horizontal striped patterns, and 20.0% ($\sigma = 0.10$) in response to vertical striped patterns. (Adapted from Krichmar and Edelman [2002] with permission from Cerebral Cortex, Oxford University Press.)

variation in system properties resulting from different initial connectivities, multiple interactions across noisy circuitry, plastic synaptic connections, fluctuating value systems, and variable object encounters. In this respect, Darwin VII is an example of a degenerate system; different circuits and dynamics can yield similar behavior (Tononi et al., 1999; Edelman and Gally, 2001). Experiments comparing the development of responses to strongly biased samples of appetitive or aversive stimuli demonstrate that even with identical starting architectures, changes in experiential sequences can have profound effects on subsequent behavior. While this has been documented phenomenologically with living organisms, the experiments described here may suggest possible mechanisms underlying such epigenetic biases.

The behavior of Darwin VII successfully showed that a BBD operating on biological principles and without prespecified instructions can carry out perceptual categorization and conditioned responses.

23.2.3 Darwin VIII — Visual Cortex Model with Reentrant Signaling

Darwin VIII (Seth et al., 2004c) was designed to demonstrate visual categorization and selective conditioning in a rich visual environment. The mobile device is similar in configuration to Darwin VII. Figure 23.7 shows Darwin VIII's environment, which consists of an enclosed area with various shapes hung on the walls. Near the boundaries (the areas forbidden to the device by its built-in boundary avoidance response) of the walls containing visual shapes, an IR beam is set up that controls a speaker, such that when Darwin VIII's movement breaks the IR beam, a tone is emitted from the speaker (see Figure 23.7A). Darwin VIII innately orients toward the source of the sound and gradually comes to associate the sound with whatever object it sees near the sound source. After conditioning, the sound is no longer necessary; Darwin VIII approaches visual objects that have become associated with preferred sounds.

Internally, Darwin VIII's brain includes simulated cortical areas of the visual system that respond to shape and color, a motor system, an auditory system, and a value system (Figure 23.8). Similar to Darwin VII's neural simulation, neuronal units in Darwin VIII roughly correspond to the activity of 100 real neurons over 100 msec, Additionally, the neuronal units have a firing phase parameter, which specifies the relative timing of this activity within each simulation cycle (see Seth et al., 2004c, for details; Tononi et al., 1992). This modeling feature provides temporal specificity without incurring the computational costs associated with modeling spiking neurons in real time. Simulated

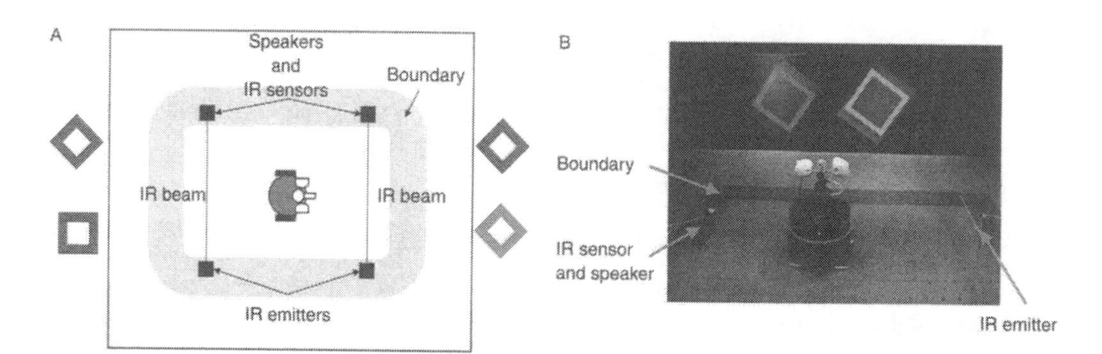

FIGURE 23.7 Darwin VIII's environment. (A) Darwin VIII views objects on two walls of an arena. The area Darwin VIII explores ($2.29 \times 1.68\,m^2$) is constrained by a boundary of reflective construction paper. Detection of this boundary by Darwin VIII's IR sensor triggers a reflexive turn (boundary avoidance reflex). When Darwin VIII breaks the beam from the IR emitter to the IR sensor, a tone is emitted is from the speaker. (B) Darwin VIII photographed in the experimental arena. (From Seth et al. [2004c]. Reprinted with permission of Cerebral Cortex, Oxford University Press.)

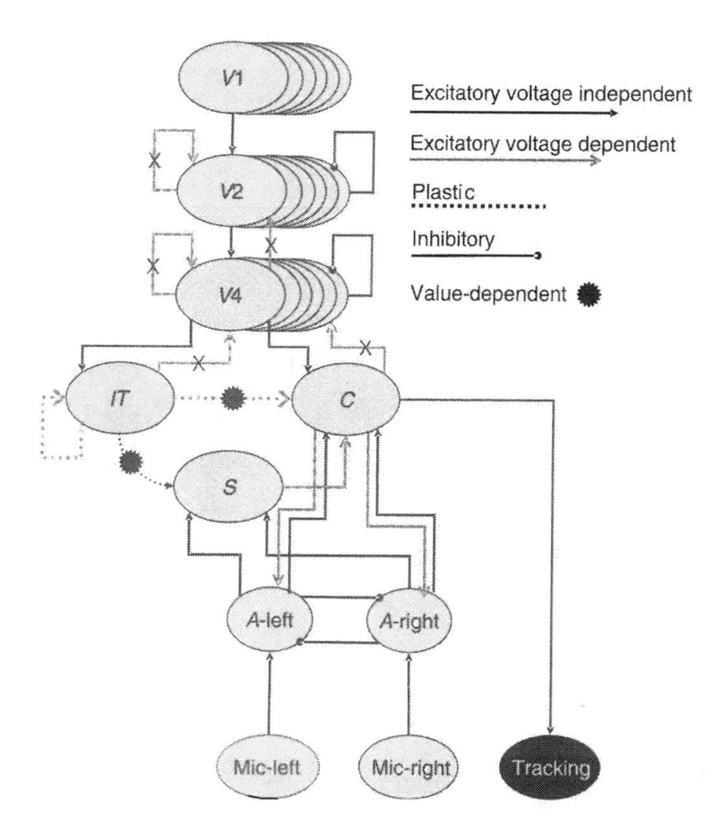

FIGURE 23.8 Schematic of the regional and functional neuroanatomy of Darwin VIII. In the version described here, the simulated nervous system contains 28 neuronal areas, 53,450 neuronal units, and approximately 1.7 million synaptic connections. The gray ellipses denote neural areas. Arrows between areas denote projections from one area to another. Projections marked with a sunburst are subject to value-dependent modification. Projections marked with an "X" are removed during lesion experiments. Tracking commands are issued to Darwin VIII's wheels based on activity in area *C*. The neural simulation consists of a visual stream $V1 \rightarrow V2 \rightarrow V4 \rightarrow IT$ in which receptive fields progressively widen from narrow to broad. The early visual areas ($V1$) contain neuronal units that respond preferentially to different colors and line orientations. The value system *S* is initially activated by salient environmental events, such as an auditory signal (A-left $\rightarrow S$, A-right $\rightarrow S$). After conditioning, value-dependent plasticity could enable a visual cue to trigger the value system ($IT \rightarrow S$) and influence tracking ($IT \rightarrow C$). See Seth et al. (2004c) for further details. (From Seth et al. [2004c]. Reprinted with permission of Cerebral Cortex, Oxford University Press.)

synaptic connections follow known vertebrate neuroanatomical projections and include extensive reentrant connectivity within and among neural areas. In Darwin VIII, reentrant connections among neuronal units encourage phase convergence and therefore lead to the emergence of neural synchrony.

When the BBD triggers a speaker as it approaches a visual object, the tone emitted by the speaker activates its value system. At this time all of the value-dependent connections between neural areas (pathways marked with a sunburst in Figure 23.8) are subject to value-dependent modification according to a rule similar to the BCM rule, with an added sensitivity to the relative phases of the activities. For detailed equations describing this plasticity rule, see Appendices C and D in Krichmar and Edelman (2002).

As a consequence of these anatomical and dynamical characteristics, Darwin VIII autonomously approaches and views multiple visual shapes containing overlapping features (e.g., red squares, red diamonds, green squares, and green diamonds) and can be trained to prefer one of these shapes by associating that shape with a positive-value tone (see Figure 23.7). It demonstrates this preference by orienting toward the preferred object. When confronted by a pair of these shapes, Darwin VIII learns successfully to track toward the preferred object, designated as the target, and to avoid

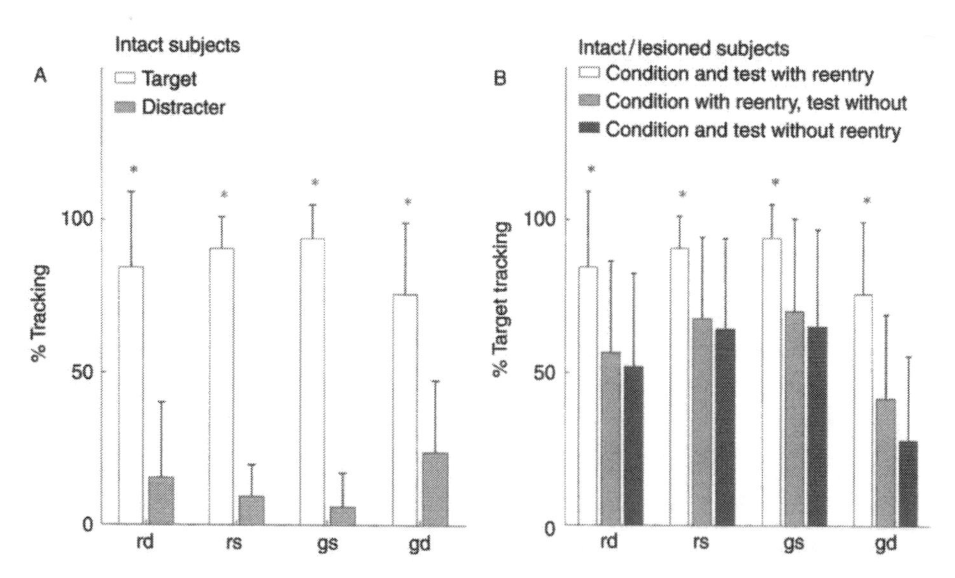

FIGURE 23.9 Darwin VIII's behavior following conditioning. (A) Three separate Darwin VIII subjects were conditioned to prefer one of four target shapes ("rd" = red diamond, "rs" = red square, "gs" = green square, "gd" = green diamond). Bars represent the mean percentage tracking time with error bars denoting the standard deviation. (B) Darwin VIII subjects with intact reentrant connections tracked the targets (white bars) significantly more than the distractors (gray bars) for each target shape, averaging over all approaches. Subjects with intact reentrant connections (white bars) tracked targets significantly better than subjects with lesions only during testing (light gray bars), and subjects with lesions during both training and testing (black bars). Asterisks denote $p < 0.01$ using a paired sample nonparametric sign test. (From Seth et al. [2004c]. Reprinted with permission of Cerebral Cortex, Oxford University Press.)

the other objects, designated as the distractors. At first, this orientation is in response to the tone, but after approximately 10 to 15 min of viewing pairs of objects, the visual pattern alone is enough to elicit this preference.

All subjects successfully track the four different targets over 80% of the time (Figure 23.9A). Successful performance on this task is not trivial. Targets and distractors appear in the visual field at many different scales and at many different positions as Darwin VIII explores its environment. Moreover, because of shared properties, targets cannot be reliably distinguished from distractors on the basis of color or shape alone. Thus, the behavior of Darwin VIII demonstrates visual categorization and selective conditioning in a rich visual environment. To investigate the importance of the presence of reentrant connections in the model, certain interareal reentrant connections were lesioned at different stages of the experimental paradigm (projections marked with an "X" in Figure 23.8). This had the effect of transforming the simulated nervous system into a pure "feedforward" model of visual processing. In one case, previously trained subjects were retested after lesioning. In a second, reentrant connections were lesioned in both training and testing stages. To compensate for the reduction in activity due to these lesions, neuronal unit outputs in areas $V2$ and $V4$ were amplified. Figure 23.9B shows that subjects with intact reentrant connections performed significantly better than either lesioned group (Wilcoxon Ranksum test; $p < 0.01$). The decrease in performance observed in the absence of reentry indicates that reentrant connections are essential for behavior above chance in the discrimination task.

During the behavior of an intact Darwin VIII subject, we observed circuits comprising synchronously active neuronal groups distributed throughout different areas in the simulated nervous system. Multiple objects were distinguishable by the differences in phase between the corresponding active circuits. A snapshot of Darwin VIII's neural responses is given in Figure 23.10, in which the device is approaching a red diamond target and a green diamond distractor toward the end of a training session.

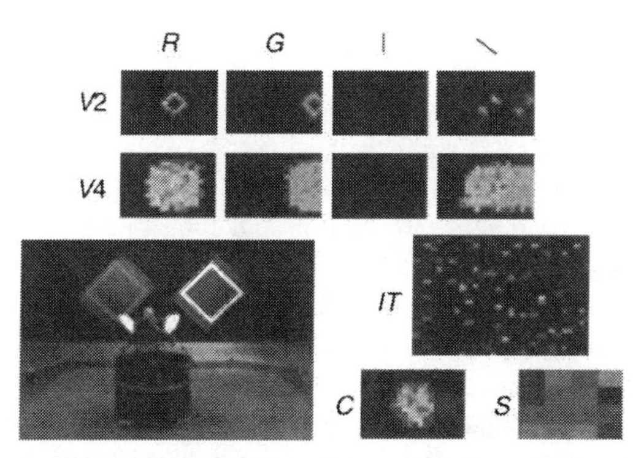

FIGURE 23.10 (See color insert following page 366) Snapshot of Darwin VIII's neuronal unit activity during a behavioral experiment. Darwin VIII is approaching (left) a red diamond target and (right) a green diamond distractor toward the end of a training session. Darwin VIII has not yet broken the beam that triggers the sound from the speakers located on the left side of the floor. The panels next to Darwin VIII show the activity and phase of selected neural areas (top row; *V*2-*red*, *V*2-*green*, *V*2-*vertical*, *V*2-*diagonal*, second row; *V*4-*red*, *V*4-*green*, *V*4-*vertical*, *V*4-*diagonal*, third row [to the right of Darwin VIII]; *IT*, fourth row [to the right of Darwin VIII]; *C* and *S*). Each pixel in the selected neural area represents a neuronal unit; the activity is normalized from no activity (black) to maximum activity (bright colors), and the phase is indicated by the color of the pixel (colors were chosen from an arbitrary pseudocolor map, there is no connection between the colors of the stimulus objects and the colors representing the phases of neuronal responses). The neuronal units responding to the attributes of the red diamond share a common phase (red–orange color), whereas the neuronal units responding to the green diamond share a different phase (blue–green color). (From Seth et al. [2004c]. Reprinted with permission of Cerebral Cortex, Oxford University Press.)

Each pixel in each neural area represents the activity (brightness) and phase (color) of a single neuronal unit. The figure shows two simultaneous dynamic neural responses differentiated by their distinct phases, which were elicited respectively by the red diamond and the green diamond. As shown in the figure, Darwin VIII had not yet reached the beam that triggers the speaker to emit a tone. The activity of area *S* was nonetheless in phase with the activity in areas *V*2 and *V*4 corresponding to the target, and is therefore predictive of the target's saliency or value. Area *IT* has two patterns of activity, indicated by the two different phase colors, which reflect two perceptual categories. The increased activity in area *C* on the side of the target is causing Darwin VIII to orient toward the target (i.e., the red diamond).

Reliable perceptual categorization is achieved by Darwin VIII despite continual changes in the scales and positions of stimuli in the visual field resulting from self-generated movement. Key mechanisms incorporated into Darwin VIII are reentrant connections within and among areas, neuronal units with both a mean firing rate and a relative firing phase, and a value system that modulates synaptic plasticity. The operation of these mechanisms, in conjunction with the sensorimotor correlations generated by self-motion, enable Darwin VIII to segment a scene, bind the features of visual objects, categorize visual objects, and demonstrate selective behavior in a rich real-world environment.

Darwin VIII solves the so-called binding problem, that is, it links responses in different brain areas and modalities to yield a coherent perceptual response in the absence of any superordinate control from a master or executive brain area (Treisman, 1998). The behavior of Darwin VIII exploits interaction between neural areas, and shows that reentrant activity (i.e., reciprocal excitatory activity brought about by reentrant connections between neuronal units in different neural areas) is sufficient for recognizing and selecting multiple objects in a scene. In contrast to previous models of target selection, which required external intervention or an artificial environment (Olshausen et al., 1993; Itti et al., 1998; Itti and Koch, 2000), Darwin VIII autonomously solved the binding problem in a rich environment even in the face of self-movement that generated changes in the sizes and locations of visual stimuli.

23.2.4 Darwin IX — Whisker Barrel Model and Texture Discrimination

Haptic sensory information provided by mystacial vibrissae (whiskers) allows the rat to discriminate among different textures in its environment (Harvey et al., 2001; Prigg et al., 2002). Although whisker-based perception lacks the fine resolution and long range of vision, whiskers have the advantage of allowing navigation and discrimination in the dark. Despite this advantage, whisker-based perception has received much less attention from roboticists than has vision (although see Russell, 1992; Jung and Zelinsky, 1996; Kaneko et al., 1998; Fend et al., 2003). To explore how haptic data may be integrated into perceptual categories by a BBD, we constructed Darwin IX, which is equipped with artificial whiskers and a simulated nervous system based on the neuroanatomy of the rat somatosensory system. We tested the hypothesis that neuronal units with time-lagged response properties, together with value-based modulation of synaptic plasticity, provide a plausible neural mechanism for the spatiotemporal transformations of whisker input needed for both texture discrimination and selective conditioning to textures.

In experiments with Darwin IX, the device autonomously explores a walled environment containing two distinct textures consisting of patterns of pegs embedded in the walls. It becomes conditioned to avoid one of the textures by associating this texture with an innately aversive simulated "foot shock," which is triggered by an IR sensor pointing downwards that detects reflective construction paper on the floor. Darwin IX demonstrates its conditioned behavior by freezing and then moving away from walls containing the texture corresponding to the innately aversive stimulus.

Darwin IX is based on the physical platform of Darwin VIII augmented by a whisker array on each side (Figure 23.11A and B). Each array consists of seven whiskers arranged in an inverted "T"

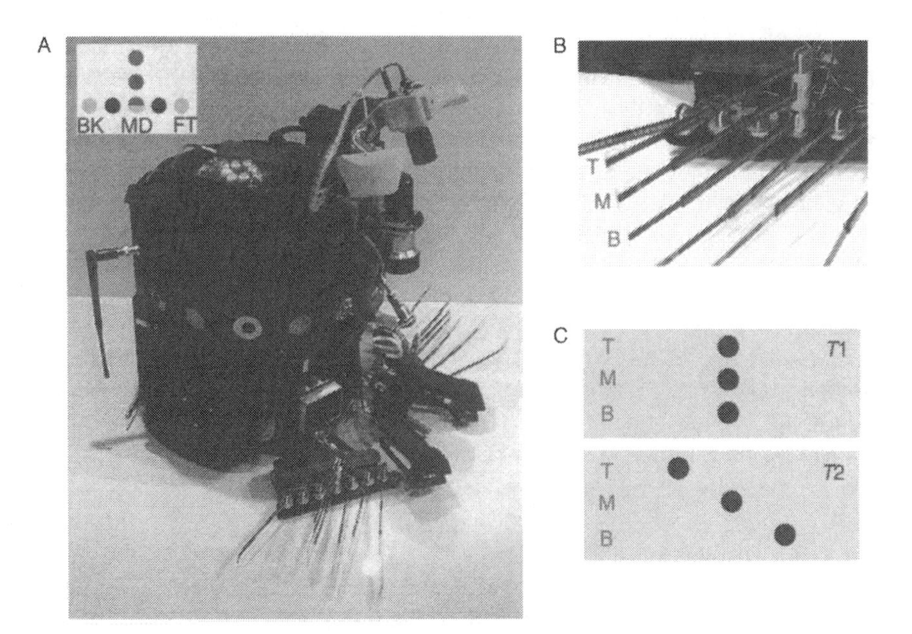

FIGURE 23.11 Darwin IX morphology. (A) Darwin IX with its left and right whisker arrays. The arrangement of a whisker array is shown in the inset. Each array has seven whiskers arranged in a row of five and a column of three. Whiskers used for wall following are marked in gray (FT, MD, BK). The column of MD whiskers provide input to the neural simulation. Note that one whisker is used for both purposes. (B) Detail of a whisker array: The top (T), middle (M), and bottom (B) whiskers in the column provide input to the neural simulation. (C) Schematic of textures $T1$ and $T2$. Each texture consists of pegs embedded in a wall; pegs are aligned in rows corresponding to the whiskers in a column. Pegs in the top row deflect the top whisker (T), and similarly for pegs in the middle row (M) and the bottom row (B).

comprising a column of three rising from a row of five. Whisker columns supply input to the simulated nervous system, while whiskers in the rows support innate avoidance and wall-following behaviors. Each whisker consists of two 4×0.63 cm polyamide strips, adhered back-to-back and held together with small polyamide sleeves that are responsive to bend (Abrams Gentile Entertainment, New York, NY). These strips are typically used as strain sensors in devices such as virtual reality gloves. Each strip has 20 resistive areas embedded regularly along its length, providing resistances of $\sim$10 KΩ when the strip is unbent and $\sim$50 KΩ when the strip is maximally bent. Each strip detects bending in only one direction, hence the back-to-back arrangement. Note that because of the regular arrangement of resistive areas, whisker signals can be generated by bending along any part of the whisker. Analog voltages from each whisker are converted into digital signals on a scale of 0 to 256. During each simulation cycle (100 msec), a "packet" of four values for each whisker is transmitted to Darwin IX's neural simulation.

As with previous Darwin automata, Darwin IX is equipped with innate behavioral responses. The default behavior of Darwin IX is to move at a constant speed. If Darwin IX nears a wall head-on, an avoidance response is triggered by either of two IR sensors, one facing front-left, and the other facing front-right: the device stops, backs up, and then turns away from the wall before resuming default behavior. Darwin IX has an innate freezing/escape response to the IR-triggered "foot shock." This response consists of freezing (stopping for $\sim$4 sec), then turning away from the nearest wall (as determined by the whisker array most recently deflected). Darwin IX also has an innate wall-following capability such that, on encountering a wall, the device moves parallel to the wall at a distance suitable for the detection of embedded textures. Wall following is based on signals from three whiskers on each side (Figure 23.11A); the rearmost (BK), the lowest of the vertical stack (MD), and the frontmost (FT). For further details regarding whisker parameters used for innate behaviors, see Seth et al. (2004a, 2004b).

Darwin IX's simulated nervous system comprises 17 areas, 1101 neuronal units, and $\sim$8400 synaptic connections (Figure 23.12). It contains areas analogous to the somatosensory pathway in the rat brain, specifically the ventromedial nuclei of the thalamus and primary and secondary somatosensory areas (in our model, $Th \rightarrow S1 \rightarrow S2$). Areas Th and $S1$ are subdivided into left (L) and right (R) regions and further into "top" (T), "middle" (M) and "bottom" (B) subregions, such that each subregion receives input from a single whisker in the column on the corresponding side. These subregions are analogous to so-called whisker "barrels" in rat $S1$ and "barreloids" in rat thalamus; as in Darwin IX, each whisker barrel/barreloid in the rat contains cells that respond preferentially to a specific whisker (Woolsey and Van der Loos, 1970; Jensen and Killackey, 1987).

Neuronal units in area Th respond to whisker input with unit-specific time delays. These units project topographically to the corresponding units in $S1$. Each barrel of $S1$ has local inhibitory connections which serve to increase the activity contrast among neuronal units. All barrels in $S1$ project to area $S2$ such that each neuronal unit in $S2$ takes input from three neuronal units, each of which is in a different barrel of either the left subarea or the right subarea of $S1$. This arrangement, which is similar to that used between $V4$ and IT in Darwin VIII, ensures that synaptic input to a neuronal unit in $S2$ is sparse and balanced. A deflection of a particular sequence of Darwin IX's whiskers leads to a spatiotemporal pattern of activity in $S2$. Such a dynamic sequence is comparable to that observed in the rat brain (Ghazanfar and Nicolelis, 1999). Darwin IX's nervous system also contains areas supporting the acquisition of conditioned aversion. Area FS is activated by detection of a "foot shock." It projects to areas S, Amy, and M_{ave}. Area S is a value system similar to the corresponding area in Darwin VIII. Area Amy is analogous to the amygdala, a neural area which has been widely implicated in the acquisition of conditioned fear (LeDoux, 1995; Maren and Fanselow, 1996). Area M_{ave} is analogous to a motor cortical area, activity in which elicits an innate aversive freezing/escape response.

The values from the whisker in each column provide input to the corresponding barreloids of area Th. Each barreloid contains 20 "lag" cells; neuronal units which have time-lagged response properties similar to those found in the lateral geniculate nucleus of the cat (Saul and Humphrey, 1992; Wolfe and Palmer, 1998). Because of differences among lag cells, cells in a barreloid fire with

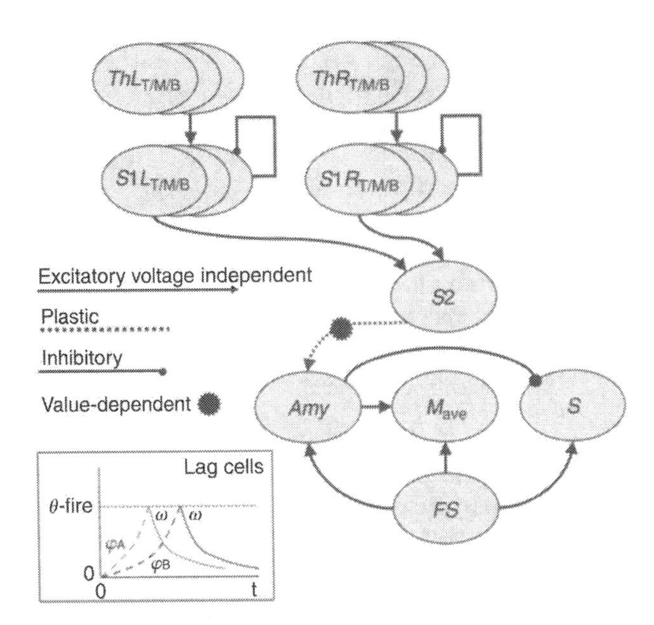

FIGURE 23.12 Global neuroanatomy of Darwin IX. Aversive freezing/escape responses are evoked by activity in area M_{ave}. Areas $ThL_{\text{T/M/B}}$ and $ThR_{\text{T/M/B}}$ receive input from the corresponding whiskers in the whisker columns on the appropriate side. These areas contain "lag cells" with temporal response properties. The operation of two idealized lag cells (A and B) is shown in the inset. The poststimulus internal state of cell A rises quickly (gray dashed line), at a rate determined by φ_A. The internal state of B rises more slowly (black dashed line, φ_B. When the internal state of each cell reaches a threshold, output is generated (solid lines) which decays at a rate determined by ω.

a range of delays, from 1 to 20 simulation cycles, following deflection of the corresponding whisker (see Figure 23.12, inset). This value is fed as input into neuronal units in the corresponding barrels of *S*1.

With the exception of the lag cells in area *Th*, neuronal units in Darwin IX use the same mean firing-rate model as in Darwin VII. Synaptic plasticity in Darwin IX acts to strengthen connections between simultaneously active neuronal units in areas *S*2 and *Amy*, according to the BCM rule. Synaptic plasticity supports conditioned aversion to texture as follows. Area *S* maintains a baseline level of activity (0.1) in the absence of input. Detection of a simulated foot shock causes neuronal units in area *FS* to produce a steady output of magnitude 1.0. This output pushes activity in area *S* above baseline, causing potentiation of synapses onto neuronal units in area *Amy* from units in *S*2 corresponding to the currently present texture. Freezing/escape responses are triggered by activity in M_{ave} exceeding a threshold (0.5) as a result of input from areas *FS* and/or *Amy*. This model also supports extinction of conditioned responses: If *Amy* is activated without any corresponding foot shock, area *S* is inhibited such that its firing rate drops below baseline and currently active synapses between *S*2 and *Amy* are weakened (for further details regarding the plasticity see Seth et al., 2004a, 2004b).

Experiments were divided into training and testing stages. During training, either *T*1 or *T*2 was paired with foot shock and Darwin IX autonomously explored its enclosure for 25,000 simulation cycles, corresponding to ~48 encounters with each wall and ~24 aversive responses to the simulated foot shock. During testing, foot shock was inactivated and Darwin IX was allowed to explore its enclosure for a further 15,000 simulation cycles. Training and testing was repeated using three different Darwin IX "subjects" initialized with different random seeds, and pairing both *T*1 and *T*2 with foot shock (six training/testing episodes in total).

Texture discrimination by Darwin IX subjects was assessed by monitoring trajectories during testing. Figure 23.13 shows the behavior of a single Darwin IX subject during training in which foot

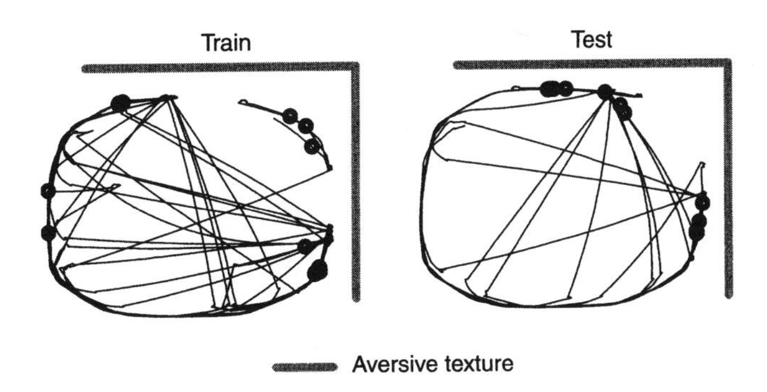

FIGURE 23.13 Conditioned aversion to textures by Darwin IX. The line tracings show the trajectory of a single Darwin IX subject (left) during aversive conditioning to texture $T1$, and (right) during testing. Large black dots show locations of aversive responses, gray shading shows locations of $T1$. During testing all freezing/escape responses occurred in proximity to the aversive texture T1.

shock was paired with $T1$, and during testing, in which foot shock was inactivated. In each case line tracings indicate the trajectory with locations of aversive responses marked by black circles. Most aversive responses, both to foot shock and to $T1$, were in regions associated with the aversive texture. This response pattern is similar to that observed during the conditioning of animals to aversive stimuli (Mackintosh, 1983).

Taking into account data from all subjects reveals near perfect conditioned avoidance of aversive textures. During testing, Darwin IX subjects which were trained to avoid $T1$ made aversive responses on 96.6% (SE = 0.18%) of encounters with $T1$. When trained to avoid $T2$, these subjects made aversive responses on 97.9% (SE = 0.14%) of encounters with $T2$ during testing. Only 3.2% of all aversive responses during testing occurred inappropriately, that is, in response to whisker deflections by walls or by the texture *not* associated with foot shock.

Darwin IX's ability to categorize textural stimuli is supported by spatiotemporal patterns of activity in $S2$. Each texture deflects whiskers in a column in a specific temporal order. The lag cells in area *Th* and neural units downstream in $S1$ present a pattern of activity with both a spatial component (i.e., the particular whisker) and a temporal component (i.e., the time since deflection). $S2$ responds to particular combinations of this $S1$ activity. The population response of $S2$ to a texture is specific and repeatable. Figure 23.14 shows representative $S2$ activity patterns during testing of a Darwin IX subject. The top panels show $S2$ activity during encounters of the left whisker column with $T1$. The bottom panels show encounters of the same whiskers with $T2$. While all panels show complex spatiotemporal patterns of activity, the top panels are highly similar to each other; the bottom panels are also highly similar to each other, but the top and bottom panels are dissimilar.

Darwin IX demonstrates the ability to categorize objects based on haptic sensing and to learn behavioral responses selective to specific textures. Observations of Darwin IX show that time-lagged neuronal responses to somatosensory input together with value-dependent synaptic plasticity provide a plausible mechanism for the spatiotemporal transformations of sensory input needed for texture discrimination, and can provide a basis for selective conditioned aversion to textures.

23.3 CONCLUSIONS

SNMs, such as Darwins III through IX, embed principles not seen previously in robots or intelligent systems. We have demonstrated that SNMs can address many difficult and unsolved problems, without instruction or intervention, such as object recognition, visual binding of objects in a scene, localization based on object cues, and conditioned behavior. Their performance provides a solid basis on which to develop autonomous, intelligent agents that follow neurobiological rather than computational principles in their construction. Unlike conventional robots, the behavior of these

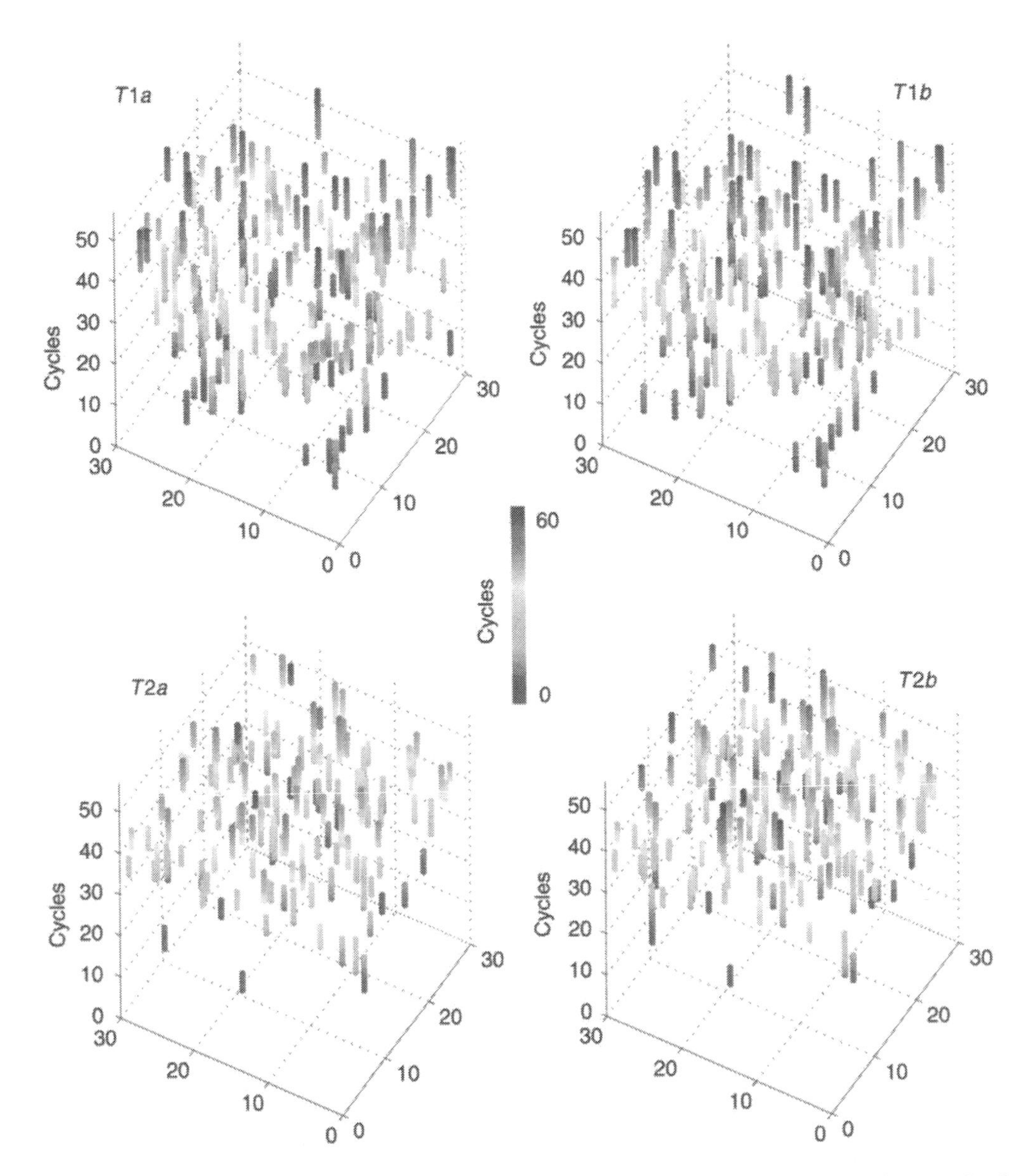

FIGURE 23.14 (See color insert following page 366) Spatiotemporal response properties of neuronal units in $S2$ in Darwin IX. The top panels show $S2$ activity during two separate encounters (a, b) with texture $T1$ by Darwin IX's left whisker column. The bottom panels show encounters of the same whiskers with texture $T2$. Each panel shows the activity of the 30×30 matrix of $S2$ neuronal units (x and y axes) over 60 poststimulus cycles (indicated by the z-axis as well as by the color scale). Neuronal units are shown if their activity exceeded a threshold (0.25).

devices is not programmed by computer instructions and is not purely reactive in nature. Instead, like living organisms, they operate according to the brain principles through which they form categorical memory, associate categories with innate values, and adapt to the environment.

ACKNOWLEDGMENT

Supported by Neurosciences Research Foundation and the W.M. Keck Foundation.

PROBLEMS

1. By estimating approximately the number of multiplications, additions, and exponentiations necessary to simulate rate-coded neural units, integrate-and-fire neurons, Hodgkin–Huxley neurons, and multicompartment neurons of 1000 dendritic compartments, calculate the relative amounts of time required to simulate models based on these classes of model neurons.
2. What is the most efficient way to allocate the work of a neural simulation among the separate processors of a shared-memory parallel processing computer system? A message-passing system? How does your answer change as the number of processors and the number of neural connections are scaled up?
3. Design a synthetic neural model along the lines of Darwin III that would operate in a three-dimensional, rather than a two-dimensional world.
4. In a simulated environment, how would you arrange for these properties of real-world objects that autonomously moving brain-based devices must contend with: (a) rotation of visual patterns, (b) size scaling of visual objects as a function of their distance from the BBD, (c) friction, (d) inexact execution of motor commands by real-world effector devices, (e) rotations between "retinal" (CCD camera), "head," and "body" coordinates as the camera is moved relative to the base?
5. Design a synthetic neural model to study a problem of interest to you.

24 Toward Neural Robotics: From Synthetic Models to Neuromimetic Implementations

Olaf Sporns

CONTENTS

24.1 INTRODUCTION

In 1948, the British neurophysiologist W. Grey Walter constructed a pair of "robot tortoises" (named Elmer and Elsie), simple goal-seeking mechanisms constructed from wartime surplus and scrap materials, which achieved international media attention as models of animal behavior. Grey Walter's motivation for constructing these creatures was to test a provocative hypothesis about brain function: that behavioral complexity could be generated by connecting relatively few components, or nerve cells. Later, he wrote that "... the elaboration of cerebral functions may possibly derive not so much from the number of units, as from the richness of their interconnection" (Walter, 1953, p. 77). This richness of interconnections between nerve cells was recognized as a possible generator of tremendous complexity and behavioral flexibility.

Remarkably, these prescient ideas have not been vigorously pursued until recently. Classical artificial intelligence (AI), with its reliance on the computer metaphor of brain and mind, stifled biologically based approaches for decades. But a new era of neural robotics has arrived, driven by dramatic new discoveries and technologies in neuroscience, cognition, and complex systems. Grey Walter's ideas about complex adaptive systems and autonomous robots resonate strongly in the work of his contemporary W. Ross Ashby (1960), in the more recent explorations of "vehicles" by Valentino Braitenberg (1984), and in Rodney Brooks' highly influential critique of classical AI (1991). What unites these different approaches is the fundamental insight that complex behavior can be generated through relatively simple (internal) mechanisms, especially if these mechanisms are dynamically coupled to an (external) environment.

Unlike AI, much of the new research in neural robotics is fundamentally grounded in the biology of brains and organisms and aims at capturing basic principles of biological organization, both structural (morphological) and functional. Thus, neural robotics quite explicitly serves a dual purpose. On the one hand is the creation of capable autonomous machines that represent new technological advances and, for example, might aid humans in various tasks and environments. On the other hand, robots might serve as research tools for exploring otherwise elusive system aspects of brain function and behavior.

This chapter is designed to provide a brief survey of some of the major scientific and intellectual issues facing neural robotics today. The selection of topics discussed here is by no means exhaustive, nor are these topics discussed at the level of depth and breadth that they deserve. Rather, this chapter is meant as a point of departure, an invitation to the reader to extend the path of integrative neural modeling out into the future, and into the realm of the whole nervous system and organism, real or artificial.

24.2 The Synthetic Approach

Grey Walter's ideas may have been influenced by those of a contemporary psychologist, Kenneth J. Craik. In the posthumously published draft of a book on *The Mechanism of Human Action* (Sherwood, 1966), Craik introduced the distinction between analytic and synthetic approaches to psychology. Analytic work included "the anatomical, psychological, and physiological examination of the actual structures and processes involved," while the synthetic approach focused on "the theoretical investigation of the basic principles" of the functioning of the brain and behavior as well as, supplementing this theoretical approach, "the construction of mechanical devices to indicate the possibilities and the shortcomings" of brain structures and mechanisms. Since Craik's definition of the "synthetic method" in the 1940s it has found only few followers within neuroscience, and much of the progress in understanding the cellular organization and neural processes of the brain has been due to "analytic" probing and experimentation.

Gerald Edelman and coworkers, in the 1980s, originated a modeling approach that explicitly addressed issues of cross-level integration, ranging from synapses all the way to behavior, called "synthetic neural modeling" (Edelman, 1987; Reeke et al., 1990a; Edelman et al., 1992; see also Chapter 23). This approach combines the design of anatomically and physiologically realistic neural models with the integration of such models into virtual (simulated) or real behaving artifacts or robots. The Darwin series of automata represents an attempt to design increasingly complex neural models that are capable of increasingly complex behavioral and cognitive functions. These automata are continually engaged in dynamic sensorimotor interactions with an environment and are capable of learning and synaptic plasticity as a result of their sensory experience. Unlike other autonomous robots designed around the same time (e.g., those incorporating Brooks' subsumption architecture) the emphasis of synthetic neural modeling is on designing sophisticated neural architectures and biological synaptic rules, rather than devising strategies that effectively deemphasize central mechanisms and control. In other words, rather than eliminating or marginalizing brain mechanisms in order to achieve complex behavior without central control, synthetic neural modeling embraces and harnesses the complexity of the central nervous system to deliver neural accounts of higher brain function in real behaving autonomous systems.

The synthetic approach has been developed in another direction by Rolf Pfeifer and coworkers, who distilled a set of principles that might underlie the operation of both artificial and biological autonomous systems (Pfeifer and Scheier, 1999). These principles intersect with the modern notion of the "embodied" mind (Varela et al., 1991; Sporns, 2002b), a set of ideas that views internal mental processes as intricately interwoven with and ultimately grounded in dynamic interactions between the organism and the environment.

These first explorations of the synthetic approach have revealed that the design and the construction of an artificial embodied system often offer unique insights into the relationship between brain and behavior that would be difficult if not impossible to obtain via analytic experimental study (Reeke and Sporns, 1993; Pfeifer and Scheier, 1999). The synthetic experimenter can manipulate and record a broad range of neural and behavioral variables, including anatomical pathways, biophysical characteristics of synapses, neuronal tuning parameters, physical characteristics of the robot body, the behavioral repertoire, and stimuli and events in the external world. In fact, as synthetic modeling has shown, even seemingly localized and discrete attributes of single neurons such as their receptive-field properties may be the result of concurrent and integrative processes occurring at multiple levels, from synaptic plasticity to experience-driven changes in neuromodulatory systems (e.g., Almassy et al., 1998; Krichmar and Edelman, 2002; Sporns and Alexander, 2002). As has been demonstrated in several working examples, such models can provide insights into normal brain and behavioral processes as well as those involved in brain disease and dysfunction.

Synthetic models quite naturally embrace the idea that cognition emerges from interactions between a developing and dynamic nervous system and an environment that the nervous system acts on and which serves as a source of sensory inputs (Figure 24.1). Many of the complex properties and functions of neurons develop as a result of experience and behavior and of encountering particular kinds of sensory stimuli and motor challenges. Synthetic models never begin structureless, as a *tabula rasa*. In fact, even at the initial stage of their construction they contain abundant structure, both physical and neural. The morphology of the body, the capabilities of sensors and effectors, the kinematics and dynamics of the motor repertoire all impose severe constraints on the kinds of interactions between the system and the environment that are possible. The structure of the nervous system, the topology of its connection patterns, the biophysical properties of synapses and cells, the capacity of connections to change or remain constant all define a specific set of structural parameters.

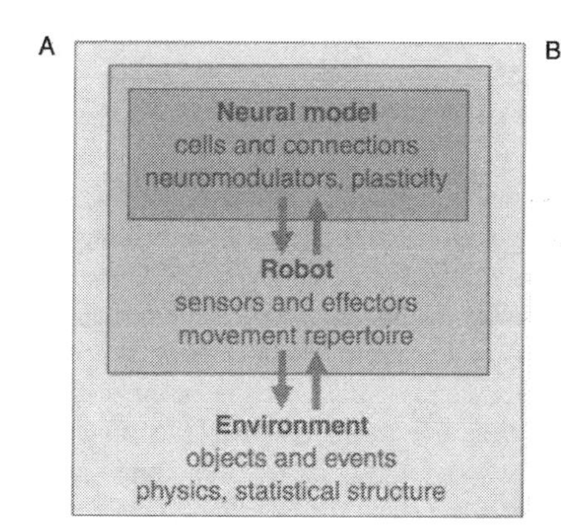

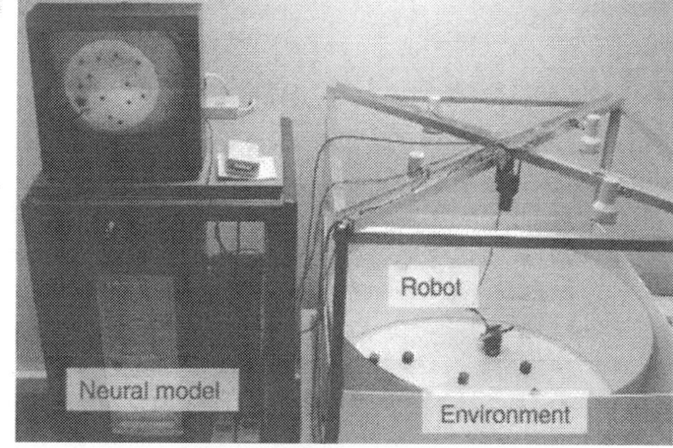

FIGURE 24.1 (A) Basic structure of a synthetic model incorporating a nervous system, a robot, and an environment. The emphasis is on the dynamic and integrative interactions between these components, in addition to the integrative nature of the neural model itself. (B) An example of an experimental neurorobotic setup as used in the work of Sporns and Alexander (2002). (C) A simple neural robot with sensors (camera, infrared sensors, and a gripper) and effectors (wheels and arm). The neural model is not contained within the robot itself, but is simulated on an off-base computer workstation.

The structure of any embodied system strongly deviates from randomness, and thus, in a statistical sense, incorporates a huge amount of information right from the start.

As synthetic models, neurorobotic systems consist of three main components (see also Chapter 23 and Chiel and Beer, 1997): (a) a neuronal model incorporating anatomical and physiological properties of real nervous systems, (b) an autonomous robot with defined body morphology and movement repertoire, and (c) the environment containing objects and events. It is important to realize that these three components are necessarily integrated and inseparable, with many opportunities for dynamical and reciprocal coupling. While brain and behavior are frequently studied in isolation or in reduced laboratory settings, neural robotics asks a fundamentally different set of questions that are centered on their mutual interactions. What is obvious to most observers is the simple fact that neural signals can cause movements of the body and thus produce action in the environment. It is perhaps less obvious, but equally important to note, that the effects of neural states on the environment can have an impact on the statistics and nature of sensory inputs reaching the nervous system. In other words, an embodied neurorobotic system influences what its future inputs will be and thus imposes structure on its own sensory input space.

The synthetic approach to studying brain and behavior is a natural adjunct to integrative modeling in the neurosciences. As the present volume demonstrates, integrative models attempt to bridge and unify processes that occur at multiple levels of scale, from genes and molecules to the coordinated action of large neural populations. Synthetic modeling specifically emphasizes cross-level interactions and the discovery of unifying principles, in particular regarding the emergence of integrated and coordinated states.

24.3 Complexity and Cognition

One of the most compelling lessons of the past decade or so of studies involving autonomous systems has been that complex (or seemingly complex) behavior can be achieved by using little (if any) internal control, modeling, or representation. Rodney Brooks' famous expression that "the world is its own best model" contrasts with the longstanding belief within classical AI that intelligent systems had to operate on a veridical and detailed internal model of their environment, and that any action was the end result of much internal computation involving such models.

However, the dramatic demonstrations of sophisticated behavior by simple creatures that, in some cases (e.g., Collins et al., 2001), lacked any neural control, cannot obscure the simple fact that complex organisms also have complex brains. The human cerebral cortex alone consists of approximately 10^{10} neurons and 10^{14} connections. Most of these neurons and connections are continually engaged in exchanging neural signals, responding to sensory inputs, and generating motor outputs. The complexity of the structural elaboration and of the dynamic configuration of large-scale brain networks represents one of the foremost challenges within neuroscience today. An explosion of research in cognitive neuroscience, driven by the development of new techniques that allow the functional imaging of the human brain has led to literally thousands of research studies and publications addressing human cognitive function in the context of brain processes and revealing in ever growing detail the intricate neural processing that accompanies even the most humble bits of human cognition. Even the most "high-level," most uniquely human aspects of our mental lives, the capability for abstract thought and language, the perception and production of music, the propensity for moral judgment, the generation of mental imagery, and consciousness itself, can be seen as reflections of the complex workings of a large integrated network of neurons and are now within the realm of scientific inquiry.

Network dynamics is a rapidly evolving and interdisciplinary field and its application to the structural and functional networks of the brain will likely contribute to our understanding of brain function (see Chapter 22, this volume). Another line of theoretical research involves the design of large-scale neuronal networks that incorporate anatomical and functional (physiological) principles and quantitatively capture the dynamical neural processes that underlie behavioral and cognitive function (Horwitz and Sporns, 1994). Such large-scale models have been designed to illuminate mechanisms of perceptual grouping and binding (Tononi et al., 1992), of categorization, object recognition, working memory (Tagamets and Horwitz, 1998), and auditory processing (Husain et al., 2004). In some cases, the neural signals obtained from such models accurately match signals recorded in human functional neuroimaging experiments and could be used to quantitatively predict the outcome of additional research studies. Neural robotics must strive to integrate such large-scale models of the brain into working and behaving autonomous systems. A successful integration of realistic large-scale models within a robot will allow synthetic experiments that would be difficult or entirely impossible to carry out inside the confines of an MRI scanner or with the spatial and temporal resolution offered by current experimental techniques. For example, the generation of synthetic fMRI signals from cellular models, using well developed mathematical techniques (Tagamets and Horwitz, 2000) and drawing on our growing insight into the biological bases of fMRI (Logothetis et al., 2001), will soon allow the recording of "robot fMRI" signals in the course of unrestrained behavior and over arbitrarily long periods of time.

Conceptualizing human cognition as a manifestation of a complex network will ultimately lead away from traditional and impoverished notions of cerebral localization (derisively called "neo-phrenology") and toward a network theory of cognition. While the shape of such a theory is still largely undetermined, we may surmise that it will have a strong mathematical and statistical underpinning and combine aspects of graph theory, biophysics, and information theory. A future generation of robots will have brains that implement a network theory of cognition to yield fully neuromimetic systems. Neuromimetic robots will not only utilize and embody this theory but will also offer a unique research tool to explore its ramifications and the many dynamic linkages between brain, body, and world.

24.4 Information and Morphology

Throughout most of the 20th century, nervous systems were viewed as "information processing devices." This view is still held by the large majority of researchers in neuroscience, cognitive science, and psychology. Although its original incarnation, perhaps most succinctly labeled as the "computer metaphor of the brain" (see Chapter 1), has now been abandoned by many (not all!), the perceived importance of neural information in its statistical context has grown tremendously in recent years. Influential theories state that nervous systems must use "efficient coding" strategies in representing their environments to optimally utilize the available neural computing resources. For example, neurons might maximize their mutual information to the external environment, capturing statistical regularities and eliminating excess (redundant) dimensions of inputs. Efficient coding theory makes a number of concrete and testable predictions about the nature of the probability distributions of neural firing found in single and between multiple neurons (Simoncelli and Olshausen, 2001).

Environmental (i.e., input) statistics have an important role to play in shaping neural processing. In this context it is important to note that the statistics of laboratory stimuli differ dramatically from those of natural environments. Monochromatically illuminated bars moving at constant velocity (a sclassical laboratory stimulus in visual neurophysiology) produce different patterns of neural activity than unconstrained visual scenery (as in a movie, or a color photograph; see Vinje and Gallant, 2000). Natural visual stimuli contain certain statistical regularities such as local correlations in intensity, orientation, and motion, which are readily picked up and preferentially "encoded" by biological nervous systems.

The theoretical investigation of information processing in the nervous systems and the role of environmental statistics present unique challenges and opportunities for neural robotics. One challenge is that neurorobotic systems should utilize biological informational strategies and capabilities as they sample the sensory contents of their environment. Their neural architecture needs to effectively embody such informational strategies. A unique opportunity for neural robotics arises from the seemingly trivial observation that autonomous systems are free to choose which sensory stimuli or parts of their physical environment they encounter. Their own sensorimotor activity structures the input space mapped onto their sensory surfaces, thus ensuring and indeed enforcing an active role of the system or organism in selecting its own inputs. This active participation of sensorimotor activity in shaping inputs has been studied in "active vision" systems, as well as in neurorobotic models of perceptual categorization (Scheier and Lambrinos, 1996) and coordinated motor activity (Fitzpatrick and Metta, 2003). A formal analysis of the informational consequences of embodied action is at its very beginning (Lungarella and Pfeifer, 2001; Sporns and Pegors, 2004). Initial results indicate that embodied action can have significant effects on the statistical structure of neural inputs. These effects may constitute a powerful rationale favoring embodied over disembodied implementations of nervous systems.

A fascinating corollary of the role played by the embodiment of robots or organisms in determining sensory inputs and thus shaping neural properties is that body morphology and the repertoire of movements available to a given creature play a causal role in how its internal control architecture (i.e., its brain) is configured. A series of simulated and physically instantiated models have addressed this issue of coevolution of body morphology and neural structures (Hara and Pfeifer, 2003), in some cases in the context of evolutionary robotics, in other cases explicitly modeling developmental changes in body and brain (Ayers et al., 2002). While much of this work is still in its infancy and is exploratory in nature, the message for integrative modeling approaches is clear: integration does not stop at the cellular boundaries of the nervous system but extends to and includes bodily action, developmental history, and the environment itself.

24.5 Hybrid Systems: Real Neurons Moving Robots

The cyborg is a hybrid artifact–organismic system, which "deliberately incorporates exogenous components extending [. . .] the organism in order to adapt it to new environments" (Clynes and Kline, 1960). Some neurorobotics research is about creating cyborgs, systems composed of robotic hardware that is connected to a biological nervous system, or in some cases, just a number of dissociated nerve cells grown *in vitro*. The latter approach, also called "neurally controlled animats" (DeMarse et al., 2001), attempts to address the problem of studying information processing in cultured neuronal networks by interfacing these networks with virtual or simulated robots. Interestingly, this hybrid system approach does not require neural modeling at all. Instead, the neurons constituting its "processors" are real living cells, capable of receiving and sending signals and naturally endowed with synaptic plasticity.

Interfacing entire biological nervous systems with mechanical or electrical prostheses is an active and growing branch of biomedical engineering. Cochlear implants are routinely used to aid severely hearing-impaired or deaf individuals and have high rates of success, especially in prelingually deaf children. The bionic eye, essentially an electronic camera interfaced with the surface of the visual cerebral cortex is much less common although limited success has been achieved with a few volunteers. Primate neurophysiologists have used their knowledge of the coding properties of neurons in the motor cortex to perform experiments in which remote robotic hardware is controlled in real time by neural activations generated by a monkey implanted with a multielectrode recording device. Such experiments have even been carried out in "closed-loop" settings in which the implanted animal can directly observe the movements generated by its own brain activity (Taylor et al., 2002; Carmena et al., 2003). These experiments raise tantalizing technological and biomedical possibilities, but perhaps most perplexingly, they illustrate the capacity of the brain to (literally) incorporate external objects into its internal operations, rendering a robotic arm an extension of the physical body that is controlled by its neurons. As the philosopher Andy Clark has argued, such extensions of the human mind into the surrounding world are fairly common and may represent a major driving force behind the expansion of the capacities of human cognition (Clark, 2003).

How far the design of hybrid systems or cyborgs may reach is uncertain. The incorporation of an external (mechanical or electrical) prosthetic device, be it sensory or motor, into a nervous system requires this system to exhibit a high degree of plasticity and adaptability. It will almost certainly be accompanied by a large-scale reorganization of the functional anatomy of the brain. Such reorganization can only occur within limits. However, working with hybrid neurorobotic systems may teach us a great deal about how brains learn to interpret sensory inputs and to control bodies, and how they adapt to continuing changes in sensory organs, body morphology, and dynamics over time. Certainly, the study of cyborgs is neurorobotics in its most literal sense and definition.

24.6 Humanoid Robots: The Search for Principles of Development

Humanoid robotics aims for the design of robots that are fully integrated within human communities, families, and societies. They are designed to perform human tasks alongside human beings as their partners and helpmates. Humanoid robots will need to have very high levels of cognitive abilities in order to flexibly adapt to changing circumstances and tasks and to communicate with their human counterparts. To this end, humanoid robots must integrate sensory information arriving from different sources and modalities, perform coordinated and planned actions, understand human speech and read facial expressions, and they must be sensitive to the success or failure of their operation in order to learn to improve.

While much effort has gone into designing humanoid robotic hardware and providing simple sensorimotor capabilities such as standing upright, walking, getting up, reaching, pointing, or simple imitation learning, there is currently relatively little integration between humanoid robotic design and sophisticated neural modeling. Recently, particular emphasis has been put on the role of development in shaping a cognitive architecture and adapting it to the varying demands of an environment. Autonomous mental development (Weng et al., 2001) may hold the key to creating truly intelligent artificial systems, through the gradual development of cognitive capabilities as the system interacts with its environment. The emphasis on autonomous development has important consequences for the design of future humanoid robots. Their initial design should endow them with general developmental mechanisms that allow autonomous growth and that are shared across different sensory, motor, and cognitive systems. If such common developmental mechanisms can be found and incorporated into robots, their capabilities may need extended periods of learning and adaptation to develop.

24.7 Outlook

With his emphasis on autonomy and self regulation, W. Grey Walter is now widely regarded as one of the pioneers of biologically inspired robotics (for a recent appraisal of Walter's role see Holland, 2003). The goals he laid out more than 50 years ago in his book "The Living Brain" remain in force today (Walter, 1953):

> "[...] If the performance of a model is to be demonstrably a fair imitation of cerebral activity, the conditions of stimulation and behavior must equally be comparable with those of the brain. *Not in looks, but in action, the model must resemble an animal.* Therefore it must have these or some measure of these attributes: exploration, curiosity, free-will in the sense of unpredictability, goal-seeking, self-regulation, avoidance of dilemmas, foresight, memory, learning, forgetting, association of ideas, form recognition, and the elements of social accommodation. Such is life." (pp. 78/79, emphasis added)

As we look toward the next five or ten years of neural robotics, there are daunting challenges ahead, but there is also great promise in the growing synergy of integrative modeling, neuroscience, systems biology, and engineering. While much of the future of neural robotics remains uncharted territory, at least some of the ingredients are now more sharply defined, more theoretically sound, more compelling than ever before. It seems the time for neural robotics has finally arrived.

ACKNOWLEDGMENT

O.S. was supported by NIH/NIDA grant R21DA15647.

Bibliography

Abbott, L.F. (1992). Simple diagrammatic rules for solving dendritic cable problems. *Physica A* **185**, 343–356.

Abbott, L.F. and Kepler, T.B. (1990). Model neurons: from Hodgkin–Huxley to Hopfield. In L. Garrido, ed., *Statistical Mechanics of Neural Networks*. Springer-Verlag, Berlin, pp. 5–18.

Abbott, L.F. and LeMasson, G. (1993). Analysis of neuron models with dynamically regulated conductances. *Neural Comput.* **5**, 823–842.

Abbott, L.F., Farhi, E., and Gutmann, S. (1991). The path integral for dendritic trees. *Biol. Cybern.* **66**, 49–60.

Abbott, N.J., Lieberman, E.M., Pichon, Y., and Larmet, Y. (1988). Periaxonal K^+ regulation in the small squid *Alloteuthis*; studies on isolated and *in situ* axons. *Biophys. J.* **53**, 275–279.

Abernethy, J.D. (1974). A dynamic model of a two-synapse feedback loop in the vertebrate retina. *Kybernetik* **14**, 187–200.

Abers, J., Davis, J.L., and Rudolph, A., eds. (2002). *Neurotechnology for Biomimetic Robots*. MIT Press, Cambridge, MA.

Abraham, W.C. and Goddard, G.V. (1983). Asymmetric relationships between homosynaptic long-term potentiation and heterosynaptic long-term depression. *Nature* **305**, 717–719.

Acebes, A. and Ferrus, A. (2000). Cellular and molecular features of axon collaterals and dendrites. *Trends Neurosci.* **23**, 557–565.

Adelman, W.J. and FitzHugh, R. (1975). Solutions of the Hodgkin–Huxley equations modified for potassium accumulation in a periaxonal space. *Fed. Proc.* **34**, 1322–1329.

Adelman, W.J. and Palti, Y. (1969). The influence of external potassium on the inactivation of sodium currents in the giant axon of the squid, *Loligo pealei*. *J. Gen. Physiol.* **53**, 685–703.

Adrian, E.D. (1920). The recovery process of excitable tissues. *J. Physiol. (Lond.)* **54**, 1–31.

Aertsen, A. and Preissl, H. (1991). Dynamics of activity and connectivity in physiological neuronal networks. In H.G. Schuster, ed., *Nonlinear Dynamics and Neuronal Networks*. VCH, New York, pp. 281–302.

Aertsen, A., Gerstein, G.L., Habib, M.K., and Palm, G. (1989). Dynamics of neuronal firing correlation: modulation of 'effective connectivity'. *J. Neurophysiol.* **61**, 900–917.

Aihara, K. and Matsumoto, G. (1983). Two stable steady states in the Hodgkin–Huxley axons. *Biophys. J.* **41**, 87–89.

Alaburda, A., Alaburda, M., Baginskas, A., Gutman, A., and Svirskis, G. (2001). Criteria of bistability of the cylindrical dendrite with variable negative slope of N-shaped current–voltage (I–V) membrane characteristic. *Biophysics-USSR* **46**, 337–340.

Alaburda, A., Baginskas, A., and Gutman, A. (2000). Realization of exclusive disjunction function by bistable dendrite. *Biophysics-USSR* **45**, 328–334.

Albert, R. and Barabási, A.-L. (2000). Topology of evolving networks: local events and universality. *Phys. Rev. Lett.* **85**, 5234–5237.

Albert, R. and Barabási, A.-L. (2002). Statistical mechanics of complex networks. *Rev. Mod. Phys.* **74**, 47–97.

Albert, R., Jeong, H., and Barabási, A.-L. (1999). The diameter of the World Wide Web. *Nature* **401**, 130–131.

Albin, R.L., Young, A.B., and Penney, J.B. (1989). The functional anatomy of basal ganglia disorders. *Trends Neurosci.* **12**, 366–375.

Aldrich, R.W., Corey, D.P., and Stevens, C.F. (1983). A reinterpretation of mammalian sodium channel gating based on single channel recording. *Nature* **306**, 436–441.

Allbritton, N.L., Meyer, T., and Streyer, L. (1992). Range of messenger action of calcium ion and inositol 1,4,5-trisphosphate. *Science* **258**, 1812–1815.

Almassy, N., Edelman, G.M., and Sporns, O. (1998). Behavioral constraints in the development of neuronal properties: a cortical model embedded in a real world device. *Cereb. Cortex* **8**, 346–361.

Alon, U. (2003). Biological networks: the tinkerer as an engineer. *Science* **301**, 1866–1867.

Altenberger, R., Lindsay, K.A., Ogden, J.A., and Rosenberg, J.R. (2001). The interaction between membrane kinetics and membrane geometry in the transmission of action potentials in non-uniform excitable fibres: a finite element approach. *J. Neurosci. Meth.* **112**, 101–117.

Alter, O., Brown, P.O., and Botstein, D. (2000). Singular value decomposition for genome-wide expression data processing and modeling. *Proc. Natl. Acad. Sci. USA* **97**, 10101–10106.

Alvarez, F.P. and Vibert, J.-F. (2002). Self-oscillatory dynamics in recurrent excitatory networks. *Neurocomputing* **44–46**, 257–262.

Amaral, L.A.N., Scala, A., Barthelemy, M., and Stanley, H.E. (2000). Classes of small-world networks. *Proc. Natl. Acad. Sci. USA* **97**, 11149–11152.

Amit, D.J. and Brunel, N. (1997). Model of global spontaneous activity and local structured activity during delay periods in the cerebral cortex. *Cereb. Cortex* **7**, 237–252.

Amit, D.J. and Tsodyks, M.V. (1992). Effective neurons and attractor neural networks in cortical environment. *Network: Comput. Neural Syst.* **3**, 121–137.

Amitai, Y., Gibson, J.R., Beierlein, M., Patrick, S.L., Ho, A.M., Connors, B.W., and Golomb, D. (2002). The spatial dimensions of electrically coupled networks of interneurons in the neocortex. *J. Neurosci.* **22**, 4142–4152.

Amjad, A.M. (1989). *Identification of Point Process Systems with Application to Complex Neuronal Networks.* Ph.D. Dissertation, University of Glasgow, UK.

Amjad, A.M., Halliday, D.M., Rosenberg, J.R., and Conway, B.A. (1997). An extended difference of coherence test for comparing and combining several independent coherence estimates — theory and application to the study of motor units and physiological tremor. *J. Neurosci. Meth.* **73**, 69–79.

Anderson, P. (1990). Synaptic integration in hippocampal CAl pyramids. In J. Storm-Mathisen, J. Zimmer, and O.P. Ottersen, eds., *Progress in Brain Research*, Volume 83. Elsevier, Amsterdam, pp. 215–222.

Andreasen, M. and Lambert, J.D.C. (1995). Regenerative properties of pyramidal cell dendrites in area CA1 of the rat hippocampus. *J. Physiol. (Lond.)* **483**, 421–441.

Angelucci, A., Lebitt, J.B., Walton, E.J.S., Hipe, J.-M., Bullier, J., and Lund, J.S. (2002). Circuits for local and global signal integration in primary visual cortex. *J. Neurosci.* **22**, 8633–8646.

Anger, G. (1990). *Inverse Problems in Differential Equations*. Plenum, New York.

Antic, S. (2003). Action potentials in basal and oblique dendrites of rat neocortical pyramidal neurons. *J. Physiol. (Lond.)* **550**, 35–50.

Aoyama, T., Kamiyama, Y., Usui, S., Blanco, R., Vaquero, C.F., and de la Villa, P. (2000). Ionic current model of rabbit retinal horizontal cell. *Neurosci. Res.* **37**, 141–151.

Araque, A., Parpura, V., Sanzgiri, R.P., and Haydon, P.G. (1999). Tripartite synapses: glia, the unacknowledged partner. *Trends Neurosci.* **22**, 208–214.

Arihara, M. and Sakamoto, K. (1999). Evaluation of spectral characteristics of physiological tremor of finger based on mechanical model. *Electromyo. Clin. Neur.* **39**, 289–304.

Arkin, A., Ross, J., and McAdams, H. (1998). Stochastic kinetic analysis of developmental pathway bifurcation in phage l infected *E. coli* cells. *Genetics* **149**, 1633–1648.

Armstrong, C.M. and Bezanilla, F. (1977). Inactivation of the sodium channel. II. Gating current experiments. *J. Gen. Physiol.* **70**, 567–590.

Artola, A., Brocher, S., and Singer, W. (1990). An in vitro model of plasticity in neocortical slices. In L.R. Squire and E. Lindenlaub, eds., *Biology of Memory*. Schattauer Verlag, Stuttgart, pp. 319–329.

Arvanitaki, A. (1942). Effects evoked in an axon by the activity of a contiguous one. *J. Neurophysiol.* **5**, 89–108.

Ascoli, G.A. (2002a). Neuroanatomical algorithms for dendritic modeling. *Network: Comput. Neural Syst.* **13**, 247–260.

Ascoli, G.A., ed. (2002b). *Computational Neuroanatomy — Principles and Methods*. Humana, Totowa, NJ.

Ascoli, G.A. (2003). Passive dendritic integration heavily affects spiking dynamics of recurrent networks. *Neural Netw.* **16**, 657–663.

Ascoli, G.A., Krichmar, J.L., Scorcioni, R., Nasuto, S.J., and Senft, S.L. (2001). Computer generation and quantitative morphometric analysis of virtual neurons. *Anat. Embryol.* **204**, 283–301.

Ashby, W.R. (1956). *An Introduction to Cybernetics*. Wiley, New York.

Ashby, W.R. (1960). *Design for a Brain*. Wiley, London.

Ashida, H. (1985). General solution of cable theory with both ends sealed. *J. Theor. Biol.* **112**, 727–740.

Astion, M.L., Coles, J.A., Orkand, R.K., and Abbott, N.J. (1988). K^+ accumulation in the space between giant axon and Schwann cell in the squid *Alloteuthis*. *Biophys. J.* **53**, 281–285.

Atri, A., Amundson, J., Clapham, D., and Sneyd, J. (1993). A single-pool model for intracellular Ca^{2+} oscillations and waves in the *Xenopus laevis* oocyte. *Biophys. J.* **65**, 1727–1739.

Attwell, D. and Jack, J.J.B. (1978). The interpretation of membrane current–voltage relations: a Nernst–Planck analysis. *Prog. Biophys. Mol. Bio.* **34**, 81–107.

Attwell, D. and Wilson, M. (1980). Behaviour of the rod network in the tiger salamander retina mediated by membrane properties of individual rods. *J. Physiol. (Lond.)* **309**, 287–315.

Attwell, D., Mobbs, P., and Tessier-Lavigne, M. (1987). Neurotransmitter-induced currents in retinal bipolar cells of the axolotl *Ambystoma mexicanum*. *J. Physiol. (Lond.)* **387**, 125–161.

Augustine, G., Santamaria, F., and Tanaka, K. (2003). Local calcium signaling in neurons. *Neuron* **40**, 331–346.

Avery, R.B. and Johnston, D. (1996). Multiple channel types contribute to the low-voltage-activated calcium current in hippocampal CA3 pyramidal neurons. *J. Neurosci.* **16**, 5567–5582.

Ayers, J., Davis, J.L., and Rudolph, A. (2002). *Neurotechnology for Biomimetic Robots*. MIT Press, Cambridge, MA.

Baccus, S.A., Burrell, B.D., Sahley, C.L., and Muller, K.J. (2000). Action potential reflection and failure at axon branch points cause stepwise changes in EPSPs in a neuron essential for learning. *J. Neurophysiol.* **83**, 1673–1700.

Bader, C.R., Bertrand, D., and Schwartz, E.A. (1982). Voltage-activated and calcium-activated currents studied in solitary rod inner segments from the salamander retina. *J. Physiol. (Lond.)* **331**, 253–284.

Baer, S.M. and Rinzel, J. (1991). Propagation of dendritic spikes mediated by excitable spines: a continuum theory. *J. Neurophysiol.* **65**, 874–890.

Bagley, R.J. and Glass, L. (1996). Counting and classifying attractors in high dimensional systems. *J. Theor. Biol.* **183**, 269–284.

Baginskas, A. and Gutman, A. (1990). Dependence of the excitatory synaptic currents on the clamped potential of the soma in a model of a neuron with non-linear dendrites. *Biophysics-USSR* **35**, 495–502.

Baimbridge, K.G., Celio, M.R., and Rogers, J.H. (1992). Calcium-binding proteins in the nervous system, *Trends Neurosci.* **15**, 303–308.

Balázsi, G., Kay, K.A., Barabási, A.-L., and Oltvai, Z. (2003). Spurious spatial periodicity of co-expression in microarray data due to printing design. *Nucleic Acids Res.* **31**, 4425–4433.

Bang-Jensen, J. and Gutin, G. (2001). *Digraphs: Theory, Algorithms and Applications*. Springer-Verlag, Berlin.

Banks, R.W., Hullinger, M., Scheepstra, K.A., and Otten, E. (1997). Pacemaker activity in a sensory ending with multiple encoding sites: the cat muscle spindle primary ending. *J. Physiol. (Lond.)* **498**, 177–199.

Barabási, A.-L. (2002). *Linked: The New Science of Networks*. Perseus, Cambridge, MA.

Barabási, A.-L. and Albert, R.A. (1999). Emergence of scaling in random networks. *Science* **286**, 509–512.

Bar-Gad, I. and Bergman, H. (2001). Stepping out of the box: information processing in the neural networks of the basal ganglia. *Curr. Opin. Neurobiol.* **11**, 689–695.

Barinaga, M. (1999). New clues to how neurons strengthen their connections. *Science* **284**, 1755–1757.

Barlow, H.B., Hill, R.M., and Levick, W.R. (1964). Rabbit retinal ganglion cells responding selectively to direction and speed of image motion in the rabbit. *J. Physiol. (Lond.)* **173**, 377–407.

Barnes, S. and Hille, B. (1989). Ionic channels of the inner segment of tiger salamander cone photoreceptors. *J. Gen. Physiol.* **94**, 719–743.

Barrett, J.N. and Crill, W. (1974). Influence of dendritic location and membrane properties on the effectiveness of synapses on cat motoneurons. *J. Physiol. (Lond.)* **239**, 325–345.

Barrionuevo, G., Kelso, S.R., Johnson, D., and Brown, T.H. (1986). Conductance mechanisms responsible for long-term potentiation in monosynaptic and isolated excitatory synaptic inputs to hippocampus. *J. Neurophysiol.* **55**, 540–550.

Bartlett, M.S. (1963). The spectral analysis of point processes. *J. R. Stat. Soc.* **25**, 264–281.

Bastian, J. and Nguyenkim, J. (2001). Dendritic modulation of burst-like firing in sensory neurons. *J. Neurophysiol.* **85**, 10–22.

Baudry, M. and Davis, J.L. (1991). *Long-Term Potentiation: A Debate of Current Issues*. MIT Press, Cambridge, MA.

Baylor, D.A. and Nunn, B.J. (1986). Electrical properties of the light-sensitive conductance of salamander rods. *J. Physiol. (Lond.)* **371**, 115–145.

Baylor, D.A., Matthews, G., and Nunn, B.J. (1984). Location and function of voltage-sensitive conductances in retinal rods of the salamander, *Ambystoma tigrinum*. *J. Physiol. (Lond.)* **354**, 203–223.

Beer, D.G., Kardia, S.L.R., Huang, C.C., Giordano, T.J., Levin, A.M., Misek, D.E., Lin, L., Chen, G., Gharib, T.G., Thomas, D.G., Lizyness, M.L., Kuick, R., Hayasaka, S., Taylor, J.M.G., Iannettoni, M.D., Orringer, M.B., and Hanash, S. (2002). Gene-expression profiles predict survival of patients with lung adenocarcinomas. *Nat. Med.* **8**, 816–824.

Bekkers, J.M. (2000). Distribution and activation of voltage-gated potassium channels in cell-attached and outside-out patches from large layer 5 cortical pyramidal neurons of the rat. *J. Physiol. (Lond.)* **525**, 611–620.

Bekkers, J.M. and Stevens, C.F. (1989). NMDA and non-NMDA receptors are co-localised at individual excitatory synapses in cultured rat hippocampus. *Nature* **341**, 230–233.

Bell, J. (1981). Modeling parallel, unmyelinated axons: pulse trapping and ephaptic transmission. *SIAM J. Appl. Math.* **41**, 168–180.

Bell, J. (1984). Behavior of some models of myelinated axons. *IMA J. Math. Appl. Med. Biol.* **1**, 149–167.

Bell, J. (1986). Parametric dependence of conduction speed for a diffusive model of a myelinated axon. *IMA J. Math. Appl. Med. Biol.* **3**, 289–300.

Bell, J. and Cook, L.M. (1978). On the solutions of a nerve conduction equation. *SIAM J. Appl. Math.* **35**, 678–688.

Bell, J. and Cook, L.M. (1979). A model of the nerve action potential. *Math. Biosci.* **46**, 11–36.

Bell, J. and Cosner, C. (1986). Threshold conditions for two diffusion models suggested by nerve impulse conditions. *SIAM J. Appl. Math.* **46**, 844–855.

Bell, J. and Cracium, G. (2003). A distributed parameter identification problem in neuronal cable theory models. *Technical Report No. 6*, Mathematical Bioscience Institute, Columbus, Ohio.

Bell, J. and Holmes, M.H. (1992). Model of the dynamics of receptor potential in a mechano-receptor. *Math. Biosci.* **110**, 139–174.

Bell, J., Bolanowski, S., and Holmes, M.H. (1994). Structure and function of Pacinian corpuscles: a review. *Prog. Neurobiol.* **42**, 79–128.

Bendat, J.S. and Piersol, A.G. (1971). *Random Data: Analysis and Measurement Procedures*. Wiley-Interscience, New York.

Bennett, M.R. and Kearns, J.L. (2000). Statistics of transmitter release at nerve terminals. *Prog. Neurobiol.* **60**, 545–606.

Ben-Pazi, H., Bergman, H., Goldberg, J.A., Giladi, N., Hansel, D., Reches, A., and Simon, E.S. (2001). Synchrony of rest tremor in multiple limbs in Parkinson's disease: evidence for multiple oscillators. *J. Neural Transm.* **108**, 287–296.

Berg, H. (1983). *Random Walks in Biology*. Princeton University Press, Princeton, NJ.

Bergman, C. (1970). Increase of sodium concentration near the inner surface of the nodal membrane. *Pflüg Arch. ges. Physiol.* **317**, 287–302.

Bergman, H., Feingold, A., Nini, A., Raz, A., Abeles, M., and Vaadia, E. (1998a). Physiological aspects of information processing in the basal ganglia of normal and parkinsonian primates. *Trends Neurosci.* **21**, 32–38.

Bergman, H., Raz, A., Feingold, A., Nini, A., Nelken, I., Hansel, D., Ben-Pazi, H., and Reches, A. (1998b). Physiology of MPTP tremor. *Mov. Disord.* **13** (Suppl. 3), 29–34.

Berkowitz, A. (1996). Networks of neurons, networks of genes. *Neuron* **17**, 199–202.

Berman, N., Dunn, R.J., and Maler, L. (2001). Function of NMDA receptors and persistent sodium channels in a feedback pathway of the electrosensory system. *J.Neurophysiol.* **86**, 1612–1621.

Bernander, O., Douglas, R., Martin, K., and Koch, C. (1991). Synaptic background activity influences spatio-temporal integration in single pyramidal cells. *Proc. Natl. Acad. Sci. USA* **88**, 11569–11573.

Bernard, C. and Johnston, D. (2003). Distance-dependent modifiable threshold for action potential back-propagation in hippocampal dendrites. *J. Neurophysiol.* **90**, 1807–1816.

Bernard, C., Ge, Y.C., Stockley, E., Willis, J.B., and Wheal, H.V. (1994). Synaptic integration of NMDA and non-NMDA receptors in large neuronal network models solved by means of differential equations. *Biol. Cybern.* **70**, 267–273.

Bernasconi, C. and König, P. (1999). On the directionality of cortical interactions studied by structural analysis of electrophysiological recordings. *Biol. Cybern.* **81**, 199–210.

Berridge, M. (1993). Inositol trisphosphate and Ca^{2+} signaling. *Nature* **361**, 315–325.

Berridge, M. (1997). Elementary and global aspects of Ca^{2+} signalling. *J. Physiol. (Lond.)* **499**, 291–306.

Berridge, M. (1998). Neuronal Ca^{2+} signaling. *Neuron* **21**, 13–26.

Berry, H. (2002). Monte Carlo simulations of enzyme reactions in two dimensions: fractal kinetics and spatial segregation. *Biophys. J.* **83**, 1891–1901.

Berry, M. and Bradley, P.M. (1976). The application of network analysis to the study of branching patterns of large dendritic fields. *Brain Res.* **109**, 111–132.

Besag, J. and Green, P.J. (1993). Spatial statistics and Bayesian computation. *J. R. Stat. Soc. B* **55**, 25–37.

Bezanilla, F. and Armstrong, C.M. (1977). Inactivation of the sodium channel. I. Sodium current experiments. *J. Gen. Physiol.* **70**, 549–566.

Bezprozvanny, I. and Ehrlich, B. (1994). Inositol (1,4,5)-trisphosphate Ip-gated Ca^{2+} channels from cerebellum: conduction properties for divalent cations and regulation by intraluminal Ca^{2+}. *J. Gen. Physiol.* **104**, 821–856.

Bezprozvanny, I. and Ehrlich, B. (1995). The inositol 1,4,5-trisphosphate receptor. *J. Membr. Biol.* **145**, 205–216.

Bezprozvanny, I., Watras, J., and Ehrlich, B. (1991). Bell-shaped Ca-response curves of Ins(1,4,5)P3- and Ca-gated channels from endoplasmic reticulum of cerebellum. *Nature* **351**, 751–754.

Bhalla, U.S. (2001). Modeling networks of signaling pathways. In E. De Schutter, ed., *Computational Neuroscience: Realistic Modeling for Experimentalists*. CRC Press, Boca Raton, FL, pp. 25–48.

Bhalla, U.S. (2002). Biochemical signaling networks decode temporal patterns of synaptic input. *J. Comput. Neurosci.* **13**, 49–62.

Bhalla, U.S. and Iyengar, R. (1999). Emergent properties of networks of biological signaling pathways. *Science* **283**, 381–387.

Bhalla, U.S., Bilitch, D.H., and Bower, J.M. (1992). Rallpacks: a set of benchmarks for neuronal simulators. *Trends Neurosci.* **15**, 453–458.

Bhan, A., Galas, D.J., and Dewey, T.G. (2002). A duplication growth model of gene expression networks. *Bioinformatics* **18**, 1486–1493.

Bi, G.Q. and Poo, M.M. (1998). Synaptic modifications in cultured hippocampal neurons: dependence on spike timing, synaptic strength, and postsynaptic cell type. *J. Neurosci.* **18**, 10464–10472.

Bi, G.Q. and Poo, M.M. (1999). Distributed synaptic modification in neural networks induced by patterned stimulation. *Nature* **401**, 792–796.

Bi, G.Q. and Poo, M.M. (2001). Synaptic modification by correlated activity: Hebb's postulate revisited. *Annu. Rev. Neurosci.* **24**, 139–166.

Bibbig, A., Traub, R.D., and Whittington, M.A. (2002). Long-range synchronization of gamma and beta oscillations and the plasticity of excitatory and inhibitory synapses: a network model. *J. Neurophysiol.* **88**, 1634–1654.

Bickhard, M.H. and Terveen, L. (1996). *Foundational Issues in Artificial Intelligence and Cognitive Science*. North-Holland, Amsterdam.

Bienenstock, E.L., Cooper, L.N., and Munro, P.W. (1982). Theory for the development of neuron selectivity: orientation specificity and binocular interaction in visual cortex. *J. Neurosci.* **2**, 32–48.

Black, M.M. (1994). Microtubule transport and assembly cooperate to generate the microtubule array of growing axons. In J. van Pelt, M.A. Corner, H.B.M. Uylings, and F.H. Lopes da Silva, eds., *The Self-Organizing Brain: From Growth Cones to Functional Networks. Progress in Brain Research*, Volume 102. Elsevier, Amsterdam, pp. 61–77.

Blake, W.J., Kaern, M., Cantor, C.R., and Collins, J.J. (2003). Noise in eukaryotic gene expression. *Nature* **422**, 633–637.

Blanco, R., Vaquero, C., and de la Villa, P. (1996). Action potentials in axonless horizontal cells isolated from the rabbit retina. *Neurosci. Lett.* **203**, 57–60.

Blatter, L. and Wier, W. (1990). Intracellular diffusion, binding, and compartmentalization of the fluorescent calcium indicators indo-1 and fura-2. *Biophys. J.* **58**, 1491–1499.

Bliss, T.V. and Collingridge, G.L. (1993). A synaptic model of memory: long-term potentiation in the hippocampus. *Nature* **361**, 31–39.

Bliss, T.V. and Lomo, T. (1973). Long-lasting potentiation of synaptic transmission in the dentate area of the anaesthetized rabbit following stimulation of the perforant path. *J. Physiol. (Lond.)* **232**, 331–356.

Bliss, T.V.P. and Lynch, M.A. (1988). Long-term potentiation of synaptic transmission in the hippocampus: properties and mechanisms. In P.W. Landfield and S. Deadwyler, eds., *Long-Term Potentiation: From Biophysics to Behavior*. Liss, New York, pp. 3–72.

Blondel, O., Takeda, J., Janssen, H., Seino, S., and Bell, G. (1993). Sequence and functional characterization of a third inositol trisphosphate receptor subtype, Ipr-3, expressed in pancreatic islets, kidney, gastrointestinal tract, and other tissues. *J. Biol. Chem.* **268**, 11356–11363.

Bloomfield, P. (2000). *Fourier Analysis of Time Series: An Introduction*, 2nd edition. Wiley, New York.

Bluman, G.W. and Tuckwell, H.C. (1987). Techniques for obtaining analytical solutions for Rall's model neuron. *J. Neurosci. Meth.* **20**, 151–166.

Boden, J. (1997). Programming the Drosophila embryo. *J. Theor. Biol.* **188**, 391–445.

Bokil, H., Laaris, N., Blinder, K., Ennis, M., and Keller, A. (2001). Ephaptic interactions in the mammalian olfactory system. *J. Neurosci.* **21**, RC173.

Bolshakov, V.Y. and Siegelbaum, S.A. (1995). Regulation of hippocampal transmitter release during development and long-term potentiation. *Science* **269**, 1730–1734.

Bonhoeffer, T. and Yuste, R. (2002). Spine motility: phenomenology, mechanisms, and function. *Neuron* **35**, 1019–1027.

Booth, V. and Rinzel, J. (1995). A minimal, compartmental model for a dendritic origin of bistability of motoneuron firing pattern. *J. Comput. Neurosci.* **2**, 299–312.

Booth, V., Rinzel, J., and Kiehn, O. (1997). Compartmental model of vertebrate motoneurons for Ca^{++}-dependent spiking and plateau potentials under pharmacological treatment. *J. Neurophysiol.* **78**, 3371–3385.

Bootman, M. and Lipp, P. (1999). Ringing changes to the 'bell-shaped curve'. *Curr. Biol.* **9**, R876–R878.

Borg-Graham, L. (1995). The Surf-Hippo neuron simulation program. http://www.cnrs-gif.fr./iaf/iaf9/surf-hippo.html, v.2.5.

Bota, M., Dong, H.-W., and Swan, L.W. (2003). From gene networks to brain networks. *Nat. Neurosci.* **6**, 795–799.

Bourinet, E., Charnet, P., Tomlinson, J.W., Stea, A., Snutch, T.P., and Nargeot, J. (1994). Voltage-dependent facilitation of a neuronal Ca^{2+} L-type calcium channel. *EMBO J.* **13**, 101–108.

Bove, M., Massobrio, G., Martinoia, S., and Grattarola, M. (1994). Realistic simulations of neurons by means of an ad hoc modified version of SPICE. *Biol. Cybern.* **71**, 137–145.

Bower, J.M. and Beeman, D. (1994). *The Book of GENESIS: Exploring Realistic Neural Models with the GEneral NEural SImulation System.* Springer-Verlag, New York.

Bower, J.M. and Beeman, D. (1998). *The Book of GENESIS: Exploring Realistic Neural Models with the GEneral NEural SImulation System*, 2nd edition. Springer-Verlag, New York.

Bradley, R. and Somjen, G.G. (1961). Accommodation of motoneurons of the rat and the cat. *J. Physiol. (Lond.)* **156**, 75–92.

Bradshaw, J.M., Kubota, Y., Meyer, T., and Schulman, H. (2003). An ultrasensitive Ca^{2+}/calmodulin-dependent protein kinase II–protein phosphatase 1 switch facilitates specificity in postsynaptic calcium signaling. *Proc. Natl. Acad. Sci.* **100**, 10512–10517.

Braitenberg, V. (1984). *Vehicles. Experiments in Synthetic Psychology.* MIT Press, Cambridge, MA.

Bras, H., Gogan, P., and Tyc-Dumont, S. (1987). The dendrites of single brain-stem motoneurons intracellulary labelled with horseradish peroxidase in the cat. Morphological and electrical differences. *Neuroscience* **22**, 947–970.

Braun, M., Coleman, C.S., Drew, D.A., and Lucas W.S., eds. (1983). *Differential Equation Models.* Springer-Verlag, New York.

Bray, D. (1998). Signaling complexes: biophysical constraints on intracellular communication. *Annu. Rev. Bioph. Biom.* **27**, 59–75.

Bray, D. (2003). Molecular networks: the top–down view. *Science* **301**, 1864–1865.

Bressler, S.L. (1995). Large-scale cortical networks and cognition. *Brain Res. Rev.* **20**, 288–304.

Bressler, S.L. and Kelso, J.A.S. (2001). Cortical coordination dynamics and cognition. *Trends Cogn. Sci.* **5**, 26–36.

Bressloff, P.C. (1995). Integro-differential equations and the stability of neural networks with dendritic structure. *Biol. Cybern.* **73**, 281–290.

Bressloff, P.C. (1996). New mechanisms for neural pattern formation. *Phys. Rev. Lett.* **76**, 4644–4647.

Bressloff, P.C. (1999). Resonant-like synchronization and bursting in a model of pulse-coupled neurons with active dendrites. *J. Comput. Neurosci.* **6**, 237–249.

Bressloff, P.C. and Coombes, S. (1997a). Synchrony in an array of integrate-and-fire neurons with dendritic structure. *Phys. Rev. Lett.* **78**, 4665–4668.

Bressloff, P.C. and Coombes, S. (1997b). Physics of the extended neuron. *Int. J. Mod. Phys. B* **11**, 2343–2392.

Bressloff, P.C. and De Souza, B. (1998). Neural pattern formation in networks with dendritic structure. *Physica D* **115**, 124–144.

Bressloff, P.C. and Taylor, J.G. (1993). Compartmental-model response function for dendritic trees. *Biol. Cybern.* **70**, 199–207.

Brett, M., Johnsrude, I., and Owen, A.M. (2002). The problem of functional localization in the human brain. *Nat. Rev. Neurosci.* **3**, 243–249.

Brillinger, D.R. (1965). An introduction to polyspectra. *Ann. Math. Stat.* **36**, 1351–1374.

Brillinger, D.R. (1972). The spectral analysis of stationary interval functions. In L.M. Le Cam, J. Neyman, and E.L. Scott, eds., *Proc. Sixth Berkeley Symposium Mathematical Statistics Probability.* University of California Press, Berkeley, CA, pp. 483–513.

Brillinger, D.R. (1974). Fourier analysis of stationary processes. *P. IEEE* **62**, 1628–1643.

Brillinger, D.R. (1975). Identification of point process systems. *Ann. Probab.* **3**, 909–929.

Brillinger, D.R. (1976). Estimation of second-order intensities of a bivariate stationary point process. *J. R. Stat. Soc. B* **38**, 60–66.

Brillinger, D.R. (1978). Comparative aspects of the study of ordinary time series and of point processes. In P.R. Krishnaiah, ed., *Developments in Statistics*, Volume 1. Academic, New York, pp. 33–133.

Brillinger, D.R. (1981). *Time Series — Data Analysis and Theory*, 2nd edition. Holden Day, San Francisco, CA.

Brillinger, D.R. and Rosenblatt, M. (1967a). Asymptotic theory of estimates of K-th order spectra. In B. Harris, ed., *Spectral Analysis of Time Series.* Wiley, New York, pp. 153–158.

Brillinger, D.R. and Rosenblatt, M. (1967b). Computation and interpretation of kth-order spectra. In B. Harris, ed., *Spectral Analysis of Time Series.* Wiley, New York, pp. 189–232.

Brismar, T. (1980). Potential clamp analysis of membrane currents in rat myelinated nerve fibres. *J. Physiol. (Lond.)* **298**, 171–184.

Britton, N.F. (1986). *Reaction-Diffusion Equations & Their Applications to Biology.* Academic, London.

Broder, A., Kumar, R., Maghoul, F., Raghavan, P., Rajalopagan, S., Stata, R., Tomkins, A., and Wiener J. (2000). Graph Structure in the Web. *Proceedings of the Ninth International World Wide Web Conference*, pp. 309–320.

Brodmann, K. (1909). *Vergleichende Lokalisationslehre der Grosshirnrinde in ihren Prinzipien dargestellt auf Grund des Zellenbaues.* J.A. Barth, Leipzig.

Brooks, R.A. (1991). New approaches to robotics. *Science* **253**, 1227–1232.

Brown, D.A. and Griffith, W.H. (1983). Calcium-activated outward current in voltage-clamped hippocampal neurons of the guinea-pig. *J. Physiol. (Lond.)* **337**, 287–301.

Brown, D.A., Gahwiler, B.H., Griffith, W.H., and Halliwell, J.V. (1990). Membrane currents in hippocampal neurons. In J. Storm-Mathisen, J. Zimmer, and O.P. Ottersen, eds., *Progress in Brain Research*, Volume 83. Elsevier, Amsterdam, pp. 141–160.

Brown, E.R. (1993). K^+ accumulation around the giant axon of the squid: comparison of electrical and morphological measurements. *Jpn. J. Physiol.* **43** (Suppl. 1), S279–S284.

Brown, E.R. and Kukita, F. (1996). Coupling between giant axon Schwann cells in the squid. *Proc. R. Soc. Lond.* **B263**, 667–672.

Brown, T.H., Chang, V.C., Ganong, A.H., Keenan, C.L., and Kelso, S.R. (1988). Biophysical properties of dendrites and spines that may control the induction and expression of long-term synaptic potentiation. In P.W. Landfield and S.A. Deadwyler, eds., *Long-term Potentiation: From Biophysics to Behaviour.* Liss, New York, pp. 201–264.

Brown, T.H., Friske, R.A., and Perkel, D.H. (1981). Passive electrical constants in three classes of hippocampal neurons. *J. Neurophysiol.* **46**, 812–827.

Büchel, C. and Friston, K.J. (1997). Modulation of connectivity in visual pathways by attention: cortical interactions evaluated with structural equation modeling and fMRI. *Cereb. Cortex* **7**, 768–778.

Büchel, C. and Friston, K.J. (2000). Assessing interactions among neuronal systems using functional neuroimaging. *Neural Netw.* **13**, 871–882.

Büchel, C., Coull, J.T., and Friston, K.J. (1999). The predictive value of changes in effective connectivity for human learning. *Science* **283**, 1538–1541.

Bugmann, G., Christodoulou, C., and Taylor, J.G. (1997). Role of temporal integration and fluctuation detection in the highly irregular firing of a leaky integrator neuron model with partial reset. *Neural Comput.* **9**, 985–1000.

Bugrim, A., Zhabotinsky, A., and Epstein, I. (1997). Ca^{2+} waves in a model with a random spatially discrete distribution of Ca^{2+} release sites. *Biophys. J.* **73**, 2897–2906.

Buhl, E.H., Halasy, K., and Somogyi, P. (1994). Diverse source of hippocampal unitary inhibitory postsynaptic potentials and the number of synaptic release sites. *Nature* **368**, 823–828.

Burke, R.E., Marks, W.B., and Ulfhake, B. (1992). A parsimonious description of motoneuron dendritic morphology using computer simulation. *J. Neurosci.* **12**, 2403–2416.

Bush, P.C. and Sejnowski, T.J. (1993). Reduced compartmental models of neocortical pyramidal cells. *J. Neurosci. Meth.* **46**, 159–166.

Butrimas, P. and Gutman, A. (1978). Theoretical analysis of an experiment with voltage clamping in the motoneuron. Proof of the N-shape pattern of the steady voltage-current characteristic of the dendrite membrane. *Biophysics-USSR* **23**, 897–904.

Butrimas, P. and Gutman, A. (1980). Theoretical current–voltage curve of neuron input during even change of clamped soma potential–RC-dendrites. *Biofizika* **25**, 1077–1080.

Butrimas, P. and Gutman, A. (1981). Theoretical input voltage–current characteristic of the neuron with NC-dendrites on uniform shift of the clamped potential of the soma. *Biophysics-USSR* **26**, 340–345.

Butrimas, P. and Gutman, A. (1983). Theoretical analysis of impulse activity of a neuron with stationary N-shaped current–voltage characteristic of dendritic membrane. Model considerations. *Neurofiziologija* **15**, 404–411 (in Russian).

Butz, E.G., and Cowan, J.D. (1974). Transient potentials in dendritic systems of arbitrary geometry. *Biophys. J.* **14**, 661–689.

Buzsaki, G., Penttonen, M., Nadasdy, Z., and Bragin, A. (1996). Pattern and inhibition-dependent invasion of pyramidal cell dendrites by fast spikes in the hippocampus in vivo. *Proc. Natl. Acad. Sci. USA* **93**, 9921–9925.

Byrne, J.H. and Gingrich, K.J. (1989). Mathematical model of cellular and molecular processes contributing to associative and nonassociative learning in Aplysia. In J.H. Byrne and W.O. Berry, eds., *Neural Models of Plasticity*. Academic, San Diego, CA, pp. 58–72.

Cabeza, R. and Kingstone, A. (2001). *Handbook of Functional Neuroimaging of Cognition*. MIT Press, Cambridge, MA.

Calabrese, R.L., Hill, A.A.V., and van Hooser, S.D. (2000). Realistic modeling of small neuronal circuits. In E. De Schutter, ed., *Computational Neuroscience: Realistic Modeling for Experimentalists*. CRC, Boca Raton, FL, pp. 259–288.

Caldwell, J.H., Schaller, K.L., Lasher, R.S., Peles, E., and Levinson, S.R. (2000). Sodium channel Na(v) 1.6 is localized at nodes of Ranvier, dendrites, and synapses. *Proc. Natl. Acad. Sci. USA* **97**, 5616–5620.

Caley, A. (1859). On the analytical forms called trees. *Philos. Mag.* **18**, 374–378.

Callaway, J.C. and Ross, W.N. (1995). Frequency dependent propagation of sodium action potentials in dendrites of hippocampal CAl pyramidal neurons. *J. Neurophysiol.* **74**, 1395–1403.

Calvin, W.H. (1989). *The Cerebral Symphony: Seashore Reflections on the Structure of Consciousnes.* Bantam Books, New York.

Cameron, W.E., Averill, D.B., and Berger, A.J. (1985). Quantitative analysis of the dendrites of cat phrenic motoneurons stained intracellularly with horseradish peroxidase. *J. Comp. Neurol.* **230**, 91–101.

Cameron, W.E., Binder, M.D., Botterman, B.R., Reinking, R.M., and Stuart, D.G. (1981). 'Sensory partitioning' of cat medial gastrocnemius muscle by its muscle spindles and tendon organs. *J. Neurophysiol.* **46**, 32–47.

Canon, J.R. (1984). *The One-Dimensional Heat Equation*, Addison-Wesley, Menlo Park, CA.

Cao, B.J and Abbott, L.F. (1993). A new computational method for cable theory problems. *Biophys. J.* **64**, 303–313.

Cao, G. and West, M. (1997). Bayesian analysis of mixtures of mixtures. *Biometrics* **52**, 221–227.

Carmena, J.M., Lebedev, M.A., Crist, R.E., O'Doherty, J.E., Santucci, D.M., Dimitrov, D.F., Patil, P.G., Henriquez, C.S., and Nicolelis, M.A.L. (2003). Learning to control a brain-machine interface for reaching and grasping by primates. *PLOS Biol.* **1**, 1–16.

Carnevale, N.T. and Lebeda, F.J. (1987). Numerical analysis of electrotonus in multicompartmental neuron models. *J. Neurosci. Meth.* **19**, 69–87.

Carnevale, N.T., Woolf, T.B., and Shepherd, G.M. (1990). Neuron simulations with SABER. *J. Neurosci. Meth.* **33**, 135–148.

Carriquiry, A.L., Ireland, W.P., Kliemann, W., and Uemura, E. (1991). Statistical evaluations of dendritic growth models. *B. Math. Biol.* **53**, 579–589.

Carslaw, H. and Jaeger, J. (1959). *Conduction of Heat in Solids*, 2nd edition. Clarendon, Oxford.

Casten, R.G., Cohen, H., and Lagerstrom, P.A. (1975). Perturbation analysis of an approximation to the Hodgkin–Huxley theory. *Q. Appl. Math.* **32**, 365–402.

Cauller, L.J. and Connors, B.W. (1992). Functions of very distal dendrites: experimental and computational studies of layer 1 synapses on neocortical pyramidal cells. In T. McKenna, J. Davis, and S.F. Zornetzer, eds., *Single Neuron Computation*. Academic, San Diego, CA, pp. 199–229.

Cauller, L.J., Clancy, B., and Connors, B.W. (1998). Backward cortical projections to primary somatosensory cortex in rats extend long horizontal axons in layer I. *J. Comp. Neurol.* **390**, 297–310.

Ceelen, P.W., Lockridge, A., and Newman, E. (2001). Electrical coupling between glial cells in the rat retina. *Glia* **35**, 1–13.

Celio, M., Pauls, T., and Schwaller, B., eds. (1996). *Introduction to EF-Hand Calcium-Binding Proteins.* Oxford University Press, New York.

Chan, S.F. and Sucher, N.J. (2001). An NMDA receptor signaling complex with protein phosphatase 2A. *J. Neurosci.* **21**, 7985–7992.

Chapeau-Blondeau, F. and Chambert, N. (1995). Synapse models for neural networks: from ion channel kinetics to multiplicative coefficient w_{ij}. *Neural Comput.* **7**, 713–734.

Cheng, H., Lederer, W., and Cannell, M. (1993a). Ca^{2+} sparks: elementary events underlying excitation–contraction coupling in heart muscle. *Science* **262**, 740–744.

Cheng, H., Lederer, M., Lederer, W., and Cannell, M. (1993b). Ca^{2+} sparks and Ca_i waves in cardiac myocytes. *Am. J. Physiol.-Cell* **270**, C148–C159.

Chesselet, M.-F. and Delfs, J.M. (1996). Basal ganglia and movement disorders: an update. *Trends Neurosci.* **19**, 417–422.

Chiang, C. (1978). On the nerve impulse equation: the dynamic responses of nerve impulse. *B. Math. Biol.* **40**, 247–255.

Chiel, H.D. and Beer, R.D. (1997). The brain has a body: adaptive behavior emerges from interactions of nervous system, body, and environment. *Trends Neurosci.* **20**, 553–557.

Chiu, S.Y., Ritchie, J.M, Rogart, R.B., and Stagg, D. (1979). A quantitative description of membrane currents in rabbit myelinated nerve. *J. Physiol. (Lond.)* **292**, 149–166.

Chomsky, N. (1988). *Language and Problems of Knowledge.* The Managua Lectures. MIT Press, Cambridge, MA.

Churchill, R.V. (1942). Expansions in series of non-orthogonal functions. *Am. Math. Soc. Bull.* **48**, 143–149.

Churchland, P.S. and Sejnowski, T.J. (1992). *The Computational Brain.* MIT Press, Cambridge, MA.

Clapham, D. (1995). Calcium signaling. *Cell* **80**, 259–268.

Clark, A. (2003). *Natural Born Cyborgs. Minds, Technologies, and the Future of Human Intelligence.* Oxford University Press, Oxford.

Clark, J. and Plonsey, R. (1966). A mathematical evaluation of the core conductor model. *Biophys. J.* **6**, 95–112.

Clark, J. and Plonsey, R. (1968). The extracellular potential field of a single active nerve fiber in a volume conductor. *Biophys. J.* **8**, 842–864.

Clark, J.W. and Plonsey, R. (1970). A mathematical study of nerve fiber interaction. *Biophys. J.* **10**, 937–957.

Clark, J.W. and Plonsey, R. (1971). Fiber interaction in a nerve trunk. *Biophys. J.* **11**, 281–294.

Clay, J.R. (1989). Slow inactivation and reactivation of the K^+ channel in squid axons. *Biophys. J.* **55**, 407–414.

Clay, J.R. (1998). Excitability of the squid axon revisited. *J. Neurophysiol.* **80**, 903–913.

Clay, J.R. (2003). On the persistent sodium current in squid giant axons. *J. Neurophysiol.* **89**, 640–644.

Clay, J.R. and DeFelice, L.J. (1983). Relationship between membrane excitability and single channel open-close kinetics. *Biophys. J.* **42**, 151–157.

Clay, J.R. and Kuzirian, A.M. (2000). Localization of voltage-gated K^+ channels in squid giant axons. *J. Neurobiol.* **45**, 172–184.

Cleeremans, A. (1993). *Mechanisms of Implicit Learning: Connectionist Models of Sequence Processing.* MIT Press, Cambridge, MA.

Cleeves, L., Findley, L.J., and Marsden, C.D. (1994). Odd tremors. In C.D. Marsden and S. Fahn, eds., *Movement Disorders 3.* Butterworth, Heinemann, Oxford, pp. 443–458.

Clements, J.D. (1984). *Synaptic Transmission and Integration in Spinal Motoneurons.* Ph.D. Dissertation, Expt. Neurology Unit, John Curtin School of Medical Research, Australian National University, Canberra.

Clements, J.D. and Redman, S.J. (1989). Cable properties of cat spinal motoneurons measured by combining voltage clamp, current clamp, and intracellular staining. *J. Physiol. (Lond.)* **409**, 63–87.

Cline H.T. (1999). Development of dendrites. In G. Stuart, N. Spruston, and M. Hausser, eds., *Dendrites.* Oxford University Press, Oxford, pp. 35–56.

Clynes, M. and Kline, N.S. (1960). Cyborgs and space. *Astronautics*, **Sept.**, 26–27, 74–75.

Coates, M., Hero III, A.O., Nowak, R., and Yu, B. (2002). Internet tomography. *IEEE Signal Proc. Mag.* **19**, 47–65.

Cohen, E.D. (2001). Voltage-gated calcium and sodium currents of starburst amacrine cells in the rabbit retina. *Visual Neurosci.* **18**, 799–809.

Colbert, M.C., Magee, I.C., Hoffman, D.A., and Johnston, D. (1997). Slow recovery from inactivation of Na^+ channels underlies the activity-dependent attenuation of dendritic action potentials in hippocampal CAl pyramidal neurons. *J. Neurosci.* **17**, 6512–6521.

Cole, K.S. (1972). *Membranes, Ions and Impulses.* University of California Press, Berkeley, CA.

Cole, K.S., Antosiewicz, H.A., and Rabinowitz, P. (1955). Automatic computation of nerve excitation. *J. Soc. Ind. Appl. Math.* **3**, 153–172.

Cole, K.S., Guttman, R., and Bezanilla, F. (1970). Nerve membrane excitation without threshold. *Proc. Natl. Acad. Sci. USA* **65**, 884–891.

Colbert, C., Magee, J.C., Hoffman, D.A., and Johnston, D. (1997). Slow recovery from inactivation of Na^+ channels underlies the activity-dependent attenuation of dendritic action potentials in hippocampal CA1 pyramidal neurons. *J. Neurosci.* **17**, 6512–6521.

Colbert, C.M. and Pan, E. (2002). Ion channel properties underlying axonal action potential initiation in pyramidal neurons. *Nat. Neurosci.* **5**, 533–538.

Colbran, R.J. (1993). Inactivation of Ca^{2+}/calmodulin-dependent protein kinase II by basal autophosphorylation. *J. Biol. Chem.* **268**, 7163–7170.

Collins, S.H., Wisse, M., and Ruina, A. (2001). A three-dimensional passive-dynamic walking robot with two legs and knees. *Int. J. Robot Res.* **20**, 607–615.

Conway, B.A., Halliday, D.M., and Rosenberg, J.R. (1993). Detection of weak synaptic interactions between single Ia-afferents and motor-unit spike trains in the decerebrate cat. *J. Physiol. (Lond.)* **471**, 379–409.

Cooley, J.W. and Dodge, F.A. (1966). Digital computer solutions for excitation and propagation of the nerve impulse. *Biophys. J.* **6**, 583–599.

Cooley, J.W., Dodge, F.A., and Cohen, H. (1965). Digital computer solutions for excitable membrane models. *J. Cell. Comp. Physiol.* **66**, 99–110.

Coomber, C. (1998a). Current theories of neuronal information processing performed by Ca^{2+}/calmodulin-dependent protein kinase II with support and insights from computer modeling and simulation. *Comput. Chem.* **22**, 251–263.

Coomber, C.J. (1998b). Site-selective autophosphorylation of Ca^{2+}/calmodulin dependent protein kinase II as a synaptic encoding mechanism. *Neural. Comput.* **10**, 1653–1678.

Coombes, S. (2001). The effect of ion pumps on the speed of travelling waves in the fire-diffuse-fire model of Ca^{2+} release. *B. Math. Biol.* **63**, 1–20.

Coombes, S. and Bressloff, P.C. (2000). Solitary waves in a model of dendritic cable with active spines. *SIAM J. Appl. Math.* **61**, 432–453.

Coombes, S. and Bressloff, P.C. (2003). Saltatory waves in the spike-diffuse-spike model of active dendritic spines. *Phys. Rev. Lett.* **91**, 028102.

Coombes, S. and Lord, G.J. (1997). Intrinsic modulation of pulse-coupled integrate-and-fire neurons. *Phys. Rev. E* **56**, 5809–5818.

Coop, A.D. and Reeke, G.N., Jr. (2001). The composite neuron: a realistic one-compartment purkinje cell model suitable for large-scale neuronal network simulations. *J. Comput. Neurosci.* **10**, 173–186.

Cooper, K., Jakobsson, E., and Wolynes, P. (1985). The theory of ion transport through membrane channels. *Prog. Biophys. Mol. Biol.* **46**, 51–96.

Costa, L.F., Manoel, E.T.M., Faucereau, F., Chelly, J., Van Pelt, J., and Ramakers, G. (2002). A shape analysis framework for neuromorphomery. *Network: Comput. Neural Syst.* **13**, 283–310.

Costalat, R. (1991). *Etude analytique et numérique des interactions neuronales par effet de champ électrique. Application à l'hippocampe.* Ph.D. Disseration, University of Angers, France.

Costalat, R. and Chauvet, G.(1993). Electrical interactions in biological neural networks. World Congress on Neural Networks '93, Portland, **II**, 614–617.

Coulter, D.A., Sombati, S., and DeLorenzo, R.J. (1992). Electrophysiology of glutamate neurotoxicity *in vitro*: induction of calcium-dependent extended neuronal depolarization. *J. Neurophysiol.* **68**, 362–373.

Courant, R. and Hilbert, D. (1962). *Methods of Mathematical Physics*, Volume 2. Wiley, New York.

Cover, T.M. and Thomas, J.A. (1991). *Elements of Information Theory.* Wiley, New York.

Cox, C.L., Denk, W., Tank, D.W., and Svoboda, K. (2000). Action potentials reliably invade axonal arbors of rat neocortical neurons. *Proc. Natl. Acad. Sci. USA* **97**, 9724–9728.

Cox, D.R. (1965). On the estimation of the intensity function of a stationary point process. *J. R. Stat. Soc.* **B27**, 332–337.

Cox, D.R. and Isham, V. (1980). *Point Processes*. Chapman & Hall, London.

Cox, D.R. and Lewis, P.A.W. (1966). *The Statistical Analysis of Series of Events*. Methuen, London.

Cox, D.R. and Lewis, P.A.W. (1972). Multivariate point processes. In L.M. LeCamm, J. Neyman, and E.L. Scott, eds., *Proceedings 6th Berkeley Symposium Mathematics Statistics Probability*, Volume 3. University of California Press, Berkeley, CA, pp. 401–488.

Cox, S.J. (1998). A new method for extracting cable parameters from input impedance data. *Math. Biosci.* **153**, 1–12.

Cox, S.J. (2004). Estimating the location and time course of synaptic input from multi-site potential recordings. *J. Comput. Neurosci.* **17**, 225–243.

Cox, S.J. and Griffith, B. (2001). Recovering quasi-active properties of dendrites from dual potential recordings. *J. Comput. Neurosci.* **11**, 95–110.

Cox, S.J. and Ji, L. (2000). Identification of the cable parameters in the somatic shunt model. *Biol. Cybern.* **83**, 151–159.

Cox, S.J. and Ji, L. (2001). Discerning ionic currents and their kinetics from input impedance data. *B. Math. Biol.* **63**, 909–932.

Cox, S.J. and Raol, J.H. (2004). Recovering the passive properties of tapered dendrites from single and dual potential recordings. *Math. Biosci.* **190**, 9–37.

Craig, A.M. and Boudin, H. (2001). Molecular heterogeneity of central synapses: afferent and target regulation. *Nat. Neurosci.* **4**, 569–578.

Crank, J. (1975). *The Mathematics of Diffusion*, 2nd edition. Clarendon, Oxford.

Creutzfeldt, O.D. (1995). *Cortex Cerebri: Performance, Structural and Functional Organization of the Cortex*. Oxford University Press, Oxford.

Crick, F. (1982). Do dendritic spines twitch? *Trends Neurosci.* **5**, 44–46.

Crick, F. (1999). The impact of molecular biology on neuroscience. *Philos. T. R. Soc. Lond.* **B354**, 2021–2025.

Crill, W.E. (1996). Persistent sodium current in mammalian central neurons. *Annu. Rev. Physiol.* **58**, 349–362.

Cronin, J. (1987). *Mathematical Aspects of Hodgkin–Huxley Neural Theory*. Cambridge University Press, Cambridge.

Cullheim, S., Fleshman, J.W., Glenn, L.L., and Burke, R.E. (1987a). Membrane area and dendritic structure in type-identified triceps surae alpha motoneurons. *J. Comp. Neurol.* **255**, 68–81.

Cullheim, S., Fleshman, J.W., Glenn, L.L., and Burke, R.E. (1987b). Three-dimensional architecture of dendritic trees in type-identified alpha motoneurons. *J. Comp. Neurol.* **255**, 82–96.

D'Aguanno, A., Bardakjian, B.J., and Carlen, P.L. (1986). Passive neuronal membrane parameters: comparison of optimization and peeling methods. *IEEE T. Bio-Med. Eng.* **33**, 1188–1196.

D'Alcantara, P., Schiffmann, S.N., and Swillens, S. (2003). Bidirectional plasticity as a consequence of interdependent Ca^{2+}-controlled phosphorylation and dephosphorylation pathways. *Eur. J. Neurosci.* **17**, 2521–2528.

Daley, D.J. and Vere-Jones, D. (1988). *An introduction to the Theory of Point Processes*. Springer-Verlag, New York.

Davidson E.H., Rast, J.P., Oliveri, P., Ransick, A., Calestanni, C., Yuh, C., Minokawa, T., Amore, G., Hinman, V., Arenas-Mena, C., Otmin, O., Brown, C.T., Livi, C.B., Lee, P.Y., Revilla, R., Rust, A.G., Pan, Z.J., Schilistra, M.J., Clarke, P.J.C., Arnone, M.I., Rowen, L., Cameron, R.A., McClay, D.R., Hood, L., and Bolouri, H. (2002). A genomic regulatory network for development. *Science* **295**, 1669–1678.

Davis, L., Jr. and Lorenté de Nó, R. (1946). Contribution to the mathematical theory of the electrotonus. *Stud. Rockefeller Inst. Med. Res.* **131**, 442–496.

Dawson, S., Keizer, J., and Pearson, J. (1999). Fire-diffuse-fire model of dynamics of intracellular calcium waves. *Proc. Natl. Acad. Sci. USA* **96**, 6060–6063.

Debanne, D., Guerineau, N.C., Gahwiler, B.H., and Thompson, S.M. (1997). Action-potential propagation gated by an axonal I_A-like K^+ conductance in hippocampus. *Nature* **389**, 286–289.

Debanne, D., Gahwiler, B.H., and Thompson, S.M. (1999). Heterogeneity of synaptic plasticity at unitary CA3-CA1 and CA3-CA3 connections in rat hippocampal slice cultures. *J. Neurosci.* **19**, 10664–10671.

Dehaene, S. and Changeux, J.P. (2000). Reward-dependent learning in neuronal networks for planning and decision making. In H.B.M. Uylings, C.G. Van Eden, J.P.C. De Bruin, M.G.P. Feenstra, and C.M.A. Pennartz, eds., *Cognition, Emotion and Autonomic Responses: The Integrative Role of the Prefrontal Cortex and Limbic Structures*. Elsevier, Amsterdam, pp. 217–229.

Del Castillo, J. and Katz, B. (1954). Quantal components of the end-plate potential. *J. Physiol. (Lond.)* **124**, 560–573.

Delorme, A. and Thorpe, S.J. (2003). SpikeNET: an event-driven simulation package for modelling large networks of spiking neurons. *Network: Comput. Neural Syst.*. **14**, 613–627.

DeMarse, T.B., Wagenaar, D.A., Blau, A.W., and Potter, S.M. (2001). The neurally controlled animat: biological brains acting with simulated bodies. *Auton. Robot.* **11**, 305–310.

Denk, W., Yuste, R., Svoboda, K., and Tank, D.W. (1996). Imaging calcium dynamics in dendritic spines. *Curr. Opin. Neurobiol.* **6**, 372–378.

De Ruiter, J.P. and Uylings, H.B.M. (1987). Morphometric and dendritic analysis of fascia dentata granule cells in human aging and senile dementia. *Brain Res.* **402**, 217–229.

Desai, N.S., Rutherford, L.C., and Turrigiano, G.G. (1999). Plasticity in the intrinsic excitability of cortical pyramidal neurons. *Nat. Neurosci.* **2**, 515–520.

De Schutter, E. and Smolen, P. (1998). Calcium dynamics in large neuronal models. In C. Koch and I. Segev, eds., *Methods in Neuronal Modeling: From Ions to Networks*, 2nd edition. MIT Press, Cambridge, MA, pp. 211–250.

De Schutter, E. and Steuber, V. (2000a). Modeling of neurons with active dendrites. In E. De Schutter, ed., *Computational Neuroscience: Realistic Modeling for Experimentalists*. CRC, Boca Raton, FL, pp. 233–257.

De Schutter, E. and Steuber, V. (2000b). Modeling simple and complex active neurons. In E. De Schutter, ed., *Computational Neuroscience: Realistic Modeling for Experimentalists*. CRC, Boca Raton, FL.

Desmond, N.L. and Levy, W.B. (1984). Dendritic caliber and the 3/2 power relationship of dentate granule cells. *J. Comp. Neurol.* **227**, 589–596.

Desmond, N.L. and Levy, W.B. (1985). Granule cell dendritic spine density in the rat hippocampus varies with spine shape and location. *Neurosci. Lett.* **54**, 219–224.

Destexhe, A., Mainen, Z., and Sejnowski, T.T. (1994). An efficient method for computing synaptic conductances based on a kinetic model of receptor binding. *Neural Comput.* **6**, 14–18.

Destexhe, A., Rudolph, M., and Pare, D. (2003). The high-conductance state of neocortical neurons in vivo. *Nat. Rev. Neurosci.* **4**, 739–751.

Deuschl, G., Bain, G.P., and Brin, M. (1998). Consensus statement of the Movement Disorders Society on Tremor. *Movement Disord.* **13** (Suppl. 3), 2–23.

Deuschl, G., Wenzelburger, R., and Raethjen, J. (2000). Tremor. *Curr. Opin. Neurol.* **13**, 437–443.

Dewey, T.G. and Galas, D.J. (2001). Dynamic models of gene expression and classification. *Funct. Integr. Genomics* **1**, 269–278.

De Young, G. and Keizer, J. (1992). A single-pool inositol 1,4,5-trisphosphate-receptor-based model for agonist-stimulated oscillations in Ca^{2+} concentration. *Proc. Natl. Acad. Sci. USA* **89**, 9895–9899.

D'haeseleer, P., Wen, X., Fuhrman, S., and Somogyi, R. (1999). Linear modeling of mRNA expression levels during CNS development and injury. *Pacific Symposium on Biocomputing* **4**, 41–52.

D'haeseleer, P., Liang, S., and Somogyi, R. (2000). Genetic network inference: from coexpression clustering to reverse engineering. *Bioinformatics* **16**, 707–726.

Diaspro, A. (2002). *Confocal and Two-Photon Microscopy: Foundations, Applications, and Advances*. Wiley-Liss, New York.

Diesmann, M., Gewaltig, M.-O., and Aertsen, A. (1999). Stable propagation of synchronous spiking in cortical neural networks. *Nature* **402**, 529–533.

DiFrancesco, D. and Noble, D. (1985). A model of cardiac electrical activity incorporating ionic pumps and concentration changes. *Philos. T. R. Soc. Lond. B* **307**, 353–398.

Dittman, J.S., Kreitzer, A.C., and Regehr, W.G. (2000). Interplay between facilitation, depression, and residual calcium at three presynaptic terminals. *J. Neurosci.* **20**, 1374–1385.

Dityatev, A.E. and Clamann, H.P. (1996). Reliability of spike propagation in arborizations of dorsal root fibers studied by analysis of postsynaptic potentials mediated by electrotonic coupling in frog spinal cord. *J. Neurophysiol.* **76**, 3451–3459.

Dityatev, A.E., Chmykhova, N.M., Studer, L., Karamian, O.A., Kozhanov, V.M., and Clamann, H.P. (1995). Comparison of the topology and growth rules of motoneuronal dendrites. *J. Comp. Neurol.* **363**, 505–516.

Dobrunz, L.E. and Stevens, C.F. (1999). Response of hippocampal synapses to natural stimulation patterns. *Neuron* **22**, 157–166.

Dodge, F.A. (1979). The nonuniform excitability of central neurons as exemplified by a model of the spinal motoneuron. In F.O. Schmidt and F.G. Worden, eds., *The Neurosciences Fourth Study Progam.* MIT Press, Cambridge, MA.

Dodge, F.A. and Cooley, J.W. (1973). Action potentials of the motorneuron. *IBM J. Res. Dev.* **17**, 219–229.

Dodge, F.A. and Frankenhaeuser, B. (1959). Sodium currents in the myelinated nerve fibre of *Xenopus laevis* investigated with the voltage clamp technique. *J. Physiol. (Lond.)* **148**, 188–200.

Doedel, E. and Kernevez, J.P. (1986). *AUTO: software for continuation and bifurcation problems in ordinary differential equations.* Applied Mathematics Report, California Institute of Technology, Pasadena, CA.

Doiron, B., Noonan, L., Lemon, N., and Turner, R.W. (2003). Persistent Na^+ current modifies burst discharge by regulating conditional backpropagation of dendritic spikes. *J. Neurophysiol.* **89**, 324–337.

Dolphin, A.C. (1996). Facilitation of Ca^{2+} current in excitable cells. *Trends Neurosci.* **19**, 35–43.

Dosemeci, A. and Albers, R.W. (1996). A mechanism for synaptic frequency detection through autophosphorylation of CaM kinase II. *Biophys. J.* **70**, 2493–2501.

Dosemeci, A., Reese, T.S., Petersen, J., and Tao-Cheng, J.H. (2000). A novel particulate form of Ca^{2+}/calmodulin-dependent protein kinase II in neurons. *J. Neurosci.* **20**, 3076–3084.

Douglas, R.J. and Martin, K.A.C. (1991). A functional microcircuit for cat visual cortex. *J. Physiol. (Lond.)* **440**, 735–769.

Douglas, R.J. and Martin K.A.C. (1992). Exploring cortical microcircuits: a combined anatomical, physiological, and computational approach. In T. McKenna, J. Davis, and S.F. Zornetzer, eds., *Single Neuron Computation.* Academic, San Diego, CA, pp. 381–412.

Douglas, R.J. and Martin, K.A.C. (1998). Neocortex. In G.M. Shepherd, ed., *The Synaptic Organization of the Brain*, 4th edition. Oxford University Press, Oxford.

Dowling, J. (1987). *The Retina: An Approachable Part of the Brain.* Harvard University Press, Cambridge, MA.

Doya, K. (2002). Metalearning and neuromodulation. *Neural Netw.* **15**, 495–506.

Dudek, F.E. and Traub, R.D. (1989). Local synaptic and electrical interactions in the hippocampus: experimental data and computer simulations. In J.H. Byrne and W.O. Berry, eds., *Neural Models of Plasticity. Experimental and Theoretical Approaches.* Academic, Orlando, FL, pp. 378–402.

Duijnhouwer, J., Remme, M.W.H., Van Ooyen, A., and Van Pelt, J. (2001). Influence of dendritic morphology on firing patterns in model neurons. *Neurocomputing* **38–40**, 183–189.

Dunaevsky, A., Tashiro, A., Majewska, A., Mason, C.A., and Yuste, R. (1999). Developmental regulation of spine motility in mammalian CNS. *Proc. Natl. Acad. Sci. USA* **96**, 13438–13443.

Dunia, R., Buckwalter, G., Defazio, T., Villar, F.D.S., Mcneill, T.H., and Walsh, J.P. (1996). Decreased duration of Ca^{2+}-mediated plateau potentials in striatal neurons from aged rats. *J. Neurophysiol.* **76**, 2353–2363.

Dupont, G. and Goldbeter, A. (1992). Oscillations and waves of cytosolic Ca^{2+}: insights from theoretical models. *BioEssays* **14**, 485–493.

Dupont, G. and Goldbeter, A. (1994). Properties of intracellular Ca^{2+} waves generated by a model based on Ca^{2+}-induced Ca^{2+} release. *Biophys. J.* **67**, 2191–2204.

Dupont, G., Berridge, M., and Goldbeter, A. (1991). Signal-induced Ca^{2+} oscillations: properties of a model based on Ca^{2+}-induced Ca^{2+} release. *Cell Calcium* **12**, 73–85.

Durand, D. (1984). The somatic shunt cable model for neurons. *Biophys. J.* **46**, 645–653.

Durand, D., Carlen, P.L., Gurevich, N., Ho, A., and Kunov, H. (1983). Electrotonic parameters of rat dentate granule cells measured using short current pulses and HRP staining. *J. Neurophysiol.* **50**, 1080–1097.

Durand, G.M., Kovalchuk, Y., and Konnerth, A. (1996). Long-term potentiation and functional synapse induction in developing hippocampus. *Nature* **381**, 71–75.

Eagles, J.P. and Purple, R.L. (1974). Afferent fibers with multiple encoding sites. *Brain Res.* **77**, 187–193.

Ebrey, T. and Koutalos, Y. (2001). Vertebrate photoreceptors. *Prog. Retin. Eye Res.* **20**, 49–94.

Eccles, J.C. (1964). *The Physiology of Synapses.* Springer-Verlag, Berlin.

Eccles, J.C. (1973). *The Understanding of the Brain.* McGraw-Hill, New York.

Edelman, G.M. (1978). Group selection and phasic reentrant signaling: A theory of higher brain function. In G.M. Edelman and V.B. Mountcastle, eds., *The Mindful Brain: Cortical Organization and the Group-Selective Theory of Higher Brain Function.* MIT Press, Cambridge, MA, pp. 51–100.

Edelman, G.M. (1987). *Neural Darwinism: The Theory of Neuronal Group Selection.* Basic Books, New York.

Edelman, G.M. (1989). *The Remembered Present: A Biological Theory of Consciousness.* Basic Books, New York.

Edelman, G.M. (1992). *Bright Air, Brilliant Fire. On the Matter of the Mind.* Basic Books, New York.

Edelman, G.M. (1993). Neural Darwinism: selection and reentrant signalling in higher brain function. *Neuron* **10**, 115–125.

Edelman, G.M. and Gally, J.A. (2001). Degeneracy and complexity in biological systems. *Proc. Natl. Acad. Sci. USA* **98**, 13763–13768.

Edelman, G.M. and Reeke, G.N., Jr. (1982). Selective networks capable of representative transformations, limited generalizations, and associative memory. *Proc. Natl. Acad. Sci. USA* **79**, 2091–2095.

Edelman, G.M., Reeke, G.N., Gall, W.E., Tononi, G., Williams, D., and Sporns, O. (1992). Synthetic neural modeling applied to a real-world artifact. *Proc. Natl. Acad. Sci. USA* **89**, 7267–7271.

Edelstein-Keshet, L. (1988). *Mathematical Models in Biology*. McGraw-Hill, Boston, MA.

Edwards, R. and Beuter, A. (1999). Indexes for identification of abnormal tremor using computer evaluation systems. *IEEE T. Bio-Med. Eng.* **46**, 895–898.

Edwards, R. and Beuter, A. (2000). Using time domain characteristics to discriminate physiologic and Parkinsonian tremors. *J. Clin. Neurophysiol.* **17**, 87–100.

Eguiluz, V.M., Cecchi, G., Chialvo, D.R., Baliki, M., and Apkarian, V. (2003). Scale-free structure of brain functional networks. arXiv:cond-mat/0309092.

Ehrlich, B. (1995). Functional properties of intracellular calcium-release channels. *Curr. Opin. Neurobiol.* **5**, 304–309.

Eichenbaum, H., Otto, T., and Cohen, N.J. (1992). The hippocampus: what does it do? *Behav. Neural Biol.* **57**, 2–36.

Eisen, M.B., Spellman, P.T., Brown, P.O., and Botstein, D. (1998). Cluster analysis and display of genome-wide expression patterns. *Proc. Natl. Acad. Sci. USA* **95**, 14863–14868.

Elman, J.L. (1990). Finding structure in time. *Cognitive Sci.* **14**, 179–211.

Elowitz, M.B., Levine, A.J., Siggia, E.D., and Swain, P.S. (2002). Stochastic gene expression in a single cell. *Science* **297**, 1183–1186.

Emptage, N.J., Bliss, T.V.P., and Fine, A. (1999). Single synaptic events evoke NMDA receptor-mediated release of calcium from internal stores in hippocampal dendritic spines. *Neuron* **22**, 115–124.

Emptage N.J., Reid C.A., Fine A., and Bliss, T.V.P. (2003). Optical quantal analysis reveals a presynaptic component of LTP at hippocampal Schaffer-associational synapses. *Neuron* **38**, 797–804.

Engel, A.K., König, P., Kreiter, A.K., and Singer, W. (1991). Interhemispheric synchronization of oscillatory neuronal responses in cat visual cortex. *Science* **252**, 1177–1179.

Engelbrecht, J. (1981). On theory of pulse transmission in a nerve fibre. *Proc. Roy. Soc. (Lond.)* **A375**, 195–209.

Engelhardt, J.K. and Chase, M.H. (2001). Neuronal network analysis based in arrival times of active-sleep specific inhibitory postsynaptic potentials in spinal cord motoneurons of the cat. *Brain Res.* **908**, 75–85.

Engl, H.W. and Rundell, W., eds. (1995). *Inverse Problems in Diffusion Processes*, SIAM, Philadelphia.

Erdos, P. and Renyi, A. (1959). On random graphs I. *Publ. Mat.* **6**, 290–297.

Escobar, D.M. and West, M. (1995). Bayesian density estimation and inference using mixtures. *J. Am. Stat. Assoc.* **90**, 577–588.

Euler, T. and Denk, W. (2001). Dendritic processing. *Curr. Opin. Neurobiol.* **11**, 415–422.

Euler, T., Detwiler, P.B., and Denk, W. (2002). Directionally selective calcium signals in dendrites of starburst amacrine cells. *Nature* **418**, 845–852.

Euler, T., Schneider, H., and Wässle, H. (1996). Glutamate responses of bipolar cells in a slice preparation of the rat retina. *J. Neurosci.* **16**, 2934–2944.

Evans, J.D. (2000). Analysis of a multiple equivalent cylinder model with generalized taper. *IMA J. Math. Appl. Med. Biol.* **17**, 347–377.

Evans, J.D. and Kember, G.C. (1994). Analytical solutions to the multi-cylinder somatic shunt cable model for passive neurons with differing dendritic electrical parameters. *Biol. Cybern.* **71**, 547–557.

Evans, J.D. and Kember, G.C. (1998). Analytical solutions to a tapering multi-cable somatic shunt cable model for passive neurons. *Math. Biosci.* **149**, 137–165.

Evans, J.D., Kember, G.C., and Major, G. (1992). Techniques for obtaining analytical solutions to the multi-cylinder somatic shunt cable model for passive neurons. *Biophys. J.* **63**, 350–365.

Evans, J.D., Kember, G.C., and Major, G. (1995). Techniques for the application of the analytical solutions to the multi-cylinder somatic shunt cable model for passive neurons. *Math. Biosci.* **125**, 1–50.

Evans, J.W. (1972). Nerve axon equations. I. Linear approximations. *Indiana U. Math. J.* **21**, 877–885.

Evans, J.W. and Feroe, J. (1977). Local stability theory of the nerve impulse. *Math. Biosci.* **37**, 23–50.

Evans, J.W. and Shenk, N. (1970). Solutions to axon equations. *Biophys. J.* **10**, 1090–1101.

Faber, D.S. and Korn, H. (1989). Electrical field effects: their relevance in central neural networks. *Physiol. Revs.* **69**, 821–863.

Faber D.S., Young W.S., Legendre P., and Korn, H. (1992). Intrinsic quantal variability due to stochastic properties of receptor-transmitter interactions. *Science* **258**, 1494–1498.

Fabiato, A. (1985). Simulated calcium current can both cause calcium loading in and trigger calcium release from the sarcoplasmic reticulum of a skinned canine cardiac purkinje cell. *J. Gen. Physiol.* **85**, 291–320.

Fain, G.L. (1999). *Molecular and Cellular Physiology of Neurons*. Harvard University Press, Cambridge, MA.

Falcke, M., Huerta, R., Rabinovich, M.I., Abarbanel, H.D.I., Elson, R.C., and Selverston, A.I. (2000a). Modeling observed chaotic oscillations in bursting neurons: the role of calcium dynamics and IP3. *Biol. Cybern.* **82**, 517–527.

Falcke, M., Tsimring, L., and Levine, H. (2000b). Stochastic spreading of intracellular Ca^{2+} release. *Phys. Rev. E* **62**, 2636–2643.

Falke, J., Drake, S., Hazard, A., and Peersen, O. (1994). Molecular tuning of ion binding to calcium signaling proteins. *Q. Rev. Biophys.* **27**, 219–290.

Fall, C. and Keizer, J. (2001). Mitochondrial modulation of intracellular Ca^{2+} signaling. *J. Theor. Biol.* **210**, 151–165.

Fall, C.P., Marland, E.S., Wagner, J.M., and Tyson, J.J. (2004). *Computational Cell Biology*. Springer-Verlag, New York.

Fatt, P. and Katz, B. (1951). An analysis of the end-plate potential recorded with an intracellular electrode. *J. Physiol. (Lond.)* **115**, 320–370.

Feldman, A.G. and Levin, M.F. (1995). The origin and use of positional frames of reference in motor control. *Behav. Brain Sci.* **18**, 723–806.

Feldman, M.L. (1984). Morphology of the neocortical pyramidal neuron. In A. Peters and E.G. Jones, eds., *Cerebral Cortex*, Volume 1. Plenum, New York, pp. 123–200.

Felleman, D.J. and Van Essen, D.C. (1991). Distributed hierarchical processing in the primate cerebral cortex. *Cereb. Cortex* **1**, 1–47.

Fend, M., Bovet, S., Yokoi, H., and Pfeifer, R. (2003). An active artificial whisker array for texture discrimination. In *IEEE/RSJ International Conference on Intelligent Robots and Systems (IROS)*. Las Vegas, NV, pp. 1044–1049.

Fewell, J.H. (2003). Social insect networks. *Science* **301**, 1867–1870.

Finch, E.A. and Augustine, G.J. (1998). Local calcium signalling by inositol-1,4,5-trisphosphate in Purkinje cell dendrites. *Nature* **396**, 753–756.

Fink, C., Slepchenko, B., Moraru, I., Watras, J., Schaff, J., and Loew, L. (2000). An image-based model of calcium waves in differentiated neuroblastoma cells. *Biophys. J.* **79**, 163–183.

Fischer, M., Kaech, S., Knutti, D., and Matus, A. (1998). Rapid actin-based plasticity in dendritic spines. *Neuron* **20**, 847–854.

FitzHugh, R. (1960). Thresholds and plateaus in the Hodgkin–Huxley nerve equations. *J. Gen. Physiol.* **43**, 867–896.

FitzHugh, R. (1961). Impulses and physiological states in theoretical models of nerve membrane. *Biophys. J.* **1**, 445–466.

FitzHugh, R. (1962). Computation of impulse initiation and saltatory conduction in a myelinated nerve fiber. *Biophys. J.* **2**, 11–21.

FitzHugh, R. (1966). Theoretical effect of temperature from threshold on the Hodgkin–Huxley nerve model. *J. Gen. Physiol.* **49**, 989–1005.

FitzHugh, R. (1973). Dimensional analysis of nerve models. *J. Theor. Biol.* **40**, 517–541.

Fitzpatrick, P. and Metta, G. (2003). Grounding vision through experimental manipulation. *Philos. T. R. Soc. Lond. A* **361**, 2165–2185.

Fitzpatrick, R. (2003). Course Notes for Introduction to Computational Physics. http://farside.ph.utexas.edu/teaching/329/329.html.

Fleshman, J.W., Segev, I., and Burke, R.E. (1988). Electrotonic architecture of type-identified α-motoneurons in the cat spinal cord. *J. Neurophysiol.* **60**, 60–85.

Fohlmeister, J.F. and Miller, R.F. (1997a). Impulse encoding mechanisms of ganglion cells in the tiger salamander retina. *J. Neurophysiol.* **78**, 1935–1947.

Fohlmeister, J.F. and Miller, R.F. (1997b). Mechanisms by which cell geometry controls repetitive impulse firing in retinal ganglion cells. *J. Neurophysiol.* **78**, 1948–1964.

Forti, S., Menini, A., Rispoli, G., and Torre, V. (1989). Kinetics of phototransduction in retinal rods of the newt *Triturus cristatus*. *J. Physiol. (Lond.)* **419**, 265–295.

Fox, J.J., Jayaprakesh, C., Wang, D.L., and Campbell, S.R. (2001). Synchronization in relaxation oscillator networks with conduction delays. *Neural Comput.* **13**, 1003–1021.

France, J., Thornley, J.H.M., Baldwin, R.L., and Crist, K.A. (1992). On solving stiff equations with reference to simulating ruminant metabolism. *J. Theoret. Biol.* **156**, 525–539.

Frankenhaeuser, B. and Hodgkin, A.L. (1956). The after-effects of impulses in the giant nerve fibres of *Loligo*. *J. Physiol. (Lond.)* **131**, 341–376.

Frankenhaeuser, B. and Huxley, A.F. (1964). The action potential in the myelinated nerve fibre of *Xenopus Laevis* as computed on the basis of the voltage clamp data. *J. Physiol. (Lond.)* **171**, 302–315.

Franks, K.M., Bartol, T.M., and Sejnowski, T.J. (2001). An MCell model of calcium dynamics and frequency response dependence of calmodulin activation in dendritic spines. *Neurocomputing* **38**, 9–16.

Freeman, J.A. and Nicholson, C. (1975). Experimental optimization of current source-density technique for anuran cerebellum. *J. Neurophysiol.* **38**, 369–382.

French, C.R., Sah, P., Buckett, K.J., and Gage, P.W. (1990). A voltage-dependent persistent sodium current in mammalian hippocampal neurons. *J. Gen. Physiol.* **95**, 1139–1157.

Frerking, M. and Wilson, M. (1999). Differences in uniquantal amplitude between sites reduce uniquantal variance when few release sites are active. *Synapse* **32**, 276–287.

Fricker, D. and Miles, R. (2000). EPSP amplification and the precision of spike timing in hippocampal neurons. *Neuron* **28**, 559–569.

Fried, S.I., Munch, T.A., and Werblin, F.S. (2002). Mechanisms and circuitry underlying directional selectivity in the retina. *Nature* **420**, 411–414.

Friston, K.J. (1993). Functional connectivity: the principal-component analysis of large (PET) data sets. *J. Cerebr. Blood F Met.* **13**, 5–14.

Friston, K.J. (1994). Functional and effective connectivity in neuroimaging: a synthesis. *Hum. Brain Mapp.* **2**, 56–78.

Friston, K.J. (1997). Imaging cognitive anatomy. *Trends Cogn. Sci.* **1**, 21–27.

Friston, K.J. (2002). Beyond phrenology: what can neuroimaging tell us about distributed circuitry? *Annu. Rev. Neurosci.* **25**, 221–250.

Friston, K.J., Tononi, G., Reeke, G.N., Jr., Sporns, O., and Edelman, G.M. (1994). Value-dependent selection in the brain: simulation in a synthetic neural model. *Neuroscience* **59**, 229–243.

Fromherz, P. and Gaede, V. (1993). Exclusive-OR function of single arborized neuron. *Biol. Cybern.* **69**, 337–344.

Fu, P., Bardakjian, B.J., D'Aguanno, A., and Carlen, P.L. (1989). Computation of the passive electrical parameters of neurons using a system model. *IEEE T. Bio-Med. Eng.* **36**, 55–64.

Fukai, H. (2000). *Study on Neural Excitation Based on the Bifurcation Theory in a Nonlinear Dynamical System*. Ph.D. Dissertation, Osaka University, Japan.

Fukai, H., Doi, S., Nomura, T., and Sato, S. (2000a). Hopf bifurcations in multiple parameter space of the Hodgkin–Huxley equations. I. Global organization of bistable periodic solutions. *Biol. Cybern.* **82**, 215–222.

Fukai, H., Nomura, T., Doi, S., and Sato, S. (2000b). Hopf bifurcations in multiple parameter space of the Hodgkin–Huxley equations. II. Singularity theoretic approach and highly degenerate bifurcations. *Biol. Cybern.* **82**, 223–229.

Fukunaga, K., Muller, D., and Miyamoto, E. (1996). CaM kinase II in long-term potentiation. *Neurochem. Int.* **28**, 343–358.

Furukawa, T. and Furshpan, E.J. (1963). Two inhibitory mechanisms in the Mauthner neurons of goldfish. *J. Neurophysiol.* **26**, 140–176.

Fuster, J. (2003). *Cortex and Mind*. Oxford University Press, New York.

Gabso, M., Neher, E., and Spira, M. (1997). Low mobility of the Ca^{2+} buffers in axons of cultured aplysia neurons. *Neuron* **18**, 473–481.

Gaertner, T.R., Putkey, J.A., and Waxham, M.N. (2004). RC3/neurogranin and Ca^{2+}/calmodulin-dependent protein kinase II produce opposing effects on the affinity of calmodulin for calcium. *J. Biol. Chem.* **279**, 39374–39382.

Gamble, E. and Koch, C. (1987). The dynamics of free calcium in dendritic spines in response to repetitive synaptic input. *Science* **236**, 1311–1315.

Gantert, C., Honerkamp, J., and Timmer, J. (1992). Analyzing the dynamics of hand tremor time series. *Biol. Cybern.* **66**, 479–484.

Gardoni, F., Caputi, A., Cimino, M., Pastorino, L., Cattabeni, F., and DiLuca M. (1998). Calcium/calmodulin-dependent protein kinase II is associated with NR2A/B subunits of NMDA receptor in postsynaptic densities. *J. Neurochem.* **71**, 1733–1741.

Gardoni, F., Schrama, L.H., van Dalen, J.J.W., Gispen, W.H., Cattabeni, F., Di Luca, M. (1999). αCaMKII binding to the C-terminal tail of NMDA receptor subunit NR2A and its modulation by autophosphorylation. *FEBS Lett.* **456**, 394–398.

Gardoni, F., Schrama, L.H., Kamal, A., Gispen, W.H., Cattabeni, F., and DiLuca, M. (2001). Hippocampal synaptic plasticity involves competition between Ca^{2+}/calmodulin-dependent protein kinase II and postsynaptic density 95 for binding to the NR2A subunit of the NMDA receptor. *J. Neurosci.* **21**, 1501–1509.

Gardner, D. ed. (1993). *The Neurobiology of Neural Networks*. MIT Press, Cambridge, MA.

Gardner, T.S., Cantor, C.R., and Collins, J.J. (2000). Construction of a genetic toggle switch in *Escherichia coli*. *Nature* **403**, 339–342.

Gasser, H.S. (1941). The classification of nerve fibers. *Ohio J. Sci.* **41**, 145–159.

Gaude, B.W. (2001). Solving nonlinear aeronautical problems using the Carleman linearization method. *Sand. Report No. SAND2001–3064*. Sandia National Laboratories, Albuquerque, NM.

Gavrikov, K.E., Dmitriev, A.V., Keyser, K.T., and Mangel, S.C. (2003). Cation-chloride contransporters mediate neural computation in the retina. *Proc. Natl. Acad. Sci. USA* **100**, 16047–16052.

Gazzaniga, M.S. (2000). Cerebral specialization and interhemispheric communication: does the corpus callosum enable the human condition? *Brain* **123**, 1293–1326.

Gelfand A.E. and Smith A.F.M. (1990). Sampling based approaches to calculating marginal densities. *J. Am. Stat. Assoc.* **85**, 398–409.

Genet, S. and Cohen, J. (1996). The cation distribution set by surface charges explains a paradoxical membrane excitability behavior. *C. R. Acad. Sci. Paris Sciences de la Vie/Life Sciences* **319**, 263–268.

Gerendasy, D.D. and Sutcliffe, J.G. (1997). RC3/neurogranin, a postsynaptic calpacitin for setting the response threshold to calcium influxes. *Mol. Neurobiol.* **15**, 131–163.

Gerstner, W. and Kistler, W.M. (2002). *Spiking Neuron Models: Single Neurons, Populations, Plasticity*. Cambridge University Press, New York.

Getting, P.A. (1974). Modification of neuron properties by electrotonic synapses. I. Input resistance, time constant, and integration. *J. Neurophysiol.* **37**, 846–857.

Ghazanfar, A.A. and Nicolelis, M.A. (1999). Spatiotemporal properties of layer V neurons of the rat primary somatosensory cortex. *Cereb. Cortex* **9**, 348–361.

Ghosh, A. and Greenberg, M. (1995). Calcium signaling in neurons: molecular mechanisms and cellular consequences. *Science* **268**, 239–247.

Gibson, J.J. (1966). *The Senses Considered as Perceptual Systems*. Houghton Mifflin, Boston, MA.

Gibson, J.J. (1979). *The Ecological Approach to Visual Perception*. Houghton-Mifflin, Boston, MA.

Gilbert, C.D. and Wiesel, T.N. (1989). Columnar specificity of intrinsic horizontal and corticocortical connections in cat visual cortex. *J. Neurosci.* **9**, 2432–2442.

Gillespie, D.T. (1976). A general method for numerically simulating the stochastic time evolution of coupled chemical reactions. *J. Comput. Phys.* **22**, 403–434.

Gillespie, D.T. (1977). Concerning the validity of the stochastic approach to chemical kinetics. *J. Stat. Phys.* **16**, 331–338.

Giordano, T.J., Shedden, K.A., Schwartz, D.R., Kuick, R., Taylor, J.M.G., Lee, N., Misek, D.E., Greenson, J.K., Kardia, S.L.R., Beer, D.G., Rennert, G., Cho, K.R., Gruber, S.B., Fearon, E.R., and Hanash, S. (2001). Organ-specific molecular classification of primary lung, colon, and ovarian adenocarcinomas using gene expression profiles. *Am. J. Pathol.* **159**, 1231–1238.

Giot, L., Bader, J.S., Brouwer, C., Chaudhuri, A., Kuang, B., Li, Y., Hao, Y.L., Ooi, C.E., Godwin, B., Vitols, E., Vijayadamodar, G., Pochart, P., Machineni, H., Welsh, M., Kong, Y., Zerhusen, B., Malcolm, R., Varrone, Z., Collis, A., Minto, M., Burgess, S., McDaniel, L., Stimpson, E., Spriggs, F., Williams, J., Neurath, K., Ioime, N., Agee, M., Voss, E., Furtak, K., Renzulli, R., Aanensen, N., Carrolla, S., Bickelhaupt, E., Lazovatsky, Y., DaSilva, A., Zhong, J., Stanyon, C.A.,

Finley, R.L., Jr., White, K.P., Braverman, M., Jarvie, T., Gold, S., Leach, M., Knight, J., Shimkets, R.A., McKenna, M.P., Chant, J., and Rothberg, J.M. (2003). A protein interaction map of *Drosophila melanogaster*. *Science* **302**, 1727–1736.

Girard, S., Luckhoff, A., Lechleiter, J., Sneyd, J., and Clapham, D. (1992). Two-dimensional model of Ca^{2+} waves reproduces the patterns observed in Xenopus oocytes. *Biophys. J.* **61**, 509–517.

Glass, L. and Kauffman, S.A. (1973). The logical analysis of continuous, nonlinear biochemical control networks. *J. Theor. Biol.* **39**, 103–129.

Glavinovic, M.I. and Rabie, H.R. (2001). Monte Carlo evaluation of quantal analysis in the light of Ca^{2+} dynamics and the geometry of secretion. *Pfluegers Arch.–Eur. J. Physiol.* **443**, 132–145.

Glenn, L.L. (1988). Overestimation of the electrical length of neuron dendrites and synaptic electrotonic attenuation. *Neurosci. Lett.* **91**, 112–119.

Glenn, L.L., Samojla, B.G., and Whitney, J.F. (1987). Electrotonic parameters of cat spinal alpha motorneurons evaluated with an equivalent cable model that incorporates non-uniform membrane resistivity. *Brain Res.* **435**, 398–402.

Glenn, L.L., Whitney, J.F., Rewitzer, J.S., Salamone, J.S., and Mariash, S.A. (1988). Method for stable intracellular recordings of alpha motoneurons during treadmill walking in awake, intact cats. *Brain Res.* **439**, 396–401.

Godfrey, M.D. (1965). An explanatory study of the bispectra of economic time series. *Appl. Stat-J. R. St. C* **14**, 48–69.

Gold, J.I. and Bear, M.F. (1994). A model of dendritic spine Ca^{2+} concentration exploring a possible basis for sliding synaptic modification. *Proc. Natl. Acad. Sci. USA* **91**, 3941–3945.

Goldberg, J.H., Tamas, G., and Yuste, R. (2003). Ca^{2+} imaging of mouse neocortical interneuron dendrites: la-type K^+ channels control action potential backpropagation. *J. Physiol. (Lond.)* **551**, 49–65.

Goldfinger, M.D. (1978). Propagated responses in nerve and muscle cable models. In A.I. Karshmer and M.A. Arbib, eds., *Brain Theory Newsletter 3*. Center for Systems Neuroscience, University of Massachusetts, Amherst, MA, pp. 63–65.

Goldfinger, M.D. (1984). Superposition of impulse activity in a rapidly-adapting afferent unit model. *Biol. Cybern.* **50**, 385–394.

Goldfinger, M.D. (1986). Poisson process stimulation of an excitable membrane cable model. *Biophys. J.* **50**, 27–40.

Goldfinger, M.D. (1990). Random-sequence stimulation of the G1 hair afferent unit. *Somatosens. Mot. Res.* **7**, 19–45.

Goldfinger, M.D. (2000). Computation of high safety-factor impulse propagation at axonal branch points. *NeuroReport* **11**, 449–456.

Goldfinger, M.D. and Amassian, V.E. (1980). Contributions by individual guard hairs and their interactions in response of forelimb guard hair afferent unit. *J. Neurophysiol.* **44**, 979–1001.

Goldfinger, M.D., Roettger, V.R., and Pearson, J.C. (1992). Theoretical studies of impulse propagation in serotonergic axons. *Biol. Cybern.* **66**, 399–406.

Golding, N.L. and Spruston, N. (1998). Dendritic sodium spikes are variable triggers of axonal action potentials in hippocampal CA1 pyramidal neurons. *Neuron* **21**, 1189–1200.

Golding, N.L., Kath, W.L., and Spruston, N. (2001). Dichotomy of action-potential backpropagation in CA1 pyramidal neuron dendrites. *J. Neurophysiol.* **86**, 2998–3010.

Golding, N.L., Staff, N.P., and Spruston, N. (2002). Dendritic spikes as a mechanism for cooperative long-term potentiation. *Nature* **418**, 326–331.

Goldman, L. (1975). Quantitative description of the sodium conductance of the giant axon of *Myxicola* in terms of a generalized second-order variable. *Biophys. J.* **15**, 119–136.

Goldman, L. (1976). Kinetics of channel gating in excitable membranes. *Q. Rev. Biophys.* **9**, 491–526.

Goldman, L. and Albus, J.S. (1968). Computation of impulse conduction in myelinated fibers: theoretical basis of the velocity diameter relation. *Biophys. J.* **8**, 596–607.

Goldman, L. and Schauf, C.L. (1972). Inactivation of the sodium current in Myxicola giant axons. Evidence for coupling to the activation process. *J. Gen. Physiol.* **59**, 659–675.

Goldman, L. and Schauf, C.L. (1973). Quantitative description of sodium and potassium currents and computed action potentials in Myxicola giant axons. *J. Gen. Physiol.* **61**, 361–384.

Goldstein, S.S. and Rall, W. (1974). Changes of action potential shape and velocity for changing core conductor geometry. *Biophys. J.* **14**, 731–757.

Golomb, D. and Ermentrout, G.B. (1999). Continuous and lurching traveling pulses in neuronal networks with delay and spatially decaying connectivity. *Proc. Natl. Acad. Sci. USA* **96**, 13480–13485.

Golomb, D., Wang, X.-J., and Rinzel, J. (1994). Synchronization properties of spindle oscillations in a thalamic reticular nucleus model. *J. Neurophysiol.* **72**, 1109–1126.

Golomb, D., Hansel, D., and Mato, G. (2001). Mechanisms of synchrony of neural activity in large networks. In F. Moss and S. Gielen, eds., *Handbook of Biological Physics, Volume 4: Neuro-Informatics and Neural Modelling*. Elsevier, Amsterdam, pp. 887–968.

Golowasch, J., Abbott, L.F., and Marder, E. (1999a). Activity-dependent regulation of potassium currents in an identified neuron of the stomatogastric ganglion of the crab *Cancer borealis*. *J. Neurosci.* **19**, RC33 (1–5).

Golowasch, J., Casey, M., Abbott, L.F., and Marder, E. (1999b). Network stability from activity-dependent regulation of neuronal conductances. *Neural Comput.* **11**, 1079–1096.

Golub, G.H. and Van Loan, C.F. (1989). *Matrix Computations*, 2nd edition. Johns Hopkins University Press, Baltimore.

Golubitsky, M. and Schaeffer, D.G. (1985). *Singularities and groups in bifurcation theory 1*. Springer-Verlag, Berlin.

Gonzalez-Burgos, G. and Barrionuevo, G. (2001). Voltage-gated sodium channels shape subthreshold EPSPs in layer 5 pyramidal neurons from rat prefrontal cortex. *J. Neurophysiol.* **86**, 1671–1684.

Gradshteyn, I.S. and Ryzhik, I.M. (1980). *Table of Integrals, Series, and Products*. Academic, New York.

Graham, B.P. (1996). Computer simulation of mechanisms for unblocking axons. In J.M. Bower, ed., *Computational Neuroscience*. Academic, New York.

Graham, H.T. (1934). Supernormality, a modification of the recovery process in nerve. *Am. J. Physiol.* **110**, 225–242.

Granato, A. and Van Pelt, J. (2003). Effects of early ethanol exposure on dendrite growth of cortical pyramidal neurons: inferences from a computational model. *Develop. Brain Res.* **142**, 223–227.

Granit, R., Kernell, D., and Lamarre, Y. (1966). Synaptic stimulation superimposed on motoneurons firing in the secondary range to injected current. *J. Physiol. (Lond.)* **187**, 401–415.

Grant, S.G.N. and O'Dell, T.J. (2001). Multiprotein complex signaling and the plasticity problem. *Curr. Opin. Neurobiol.* **11**, 363–368.

Gray, C.M. and Singer, W. (1989). Stimulus-specific neuronal oscillations in neuronal columns of cat visual cortex. *Proc. Natl. Acad. Sci. USA* **86**, 1698–1702.

Gresty, M. and Buckwell, D. (1990). Spectral analysis of tremor: understanding the results. *J. Neurol. Neurosur. & Ps.* **53**, 976–981.

Griffith, J.S. (1971). *Mathematical Neurobiology*. Academic, New York.

Griffith, J.S. and Horn, G. (1963). Functional coupling between cells in the visual cortex of the unrestrained cat. *Nature* **199**, 876, 893–895.

Grossman, Y., Spira, M.E., and Parnas, I. (1973). Differential flow of information into branches of a single axon. *Brain Res.* **64**, 379–386.

Grynkiewicz, G., Poenie, M., and Tsien, R. (1985). A new generation of Ca^{2+} indicators with greatly improved fluorescence properties. *J. Biol. Chem.* **260**, 3440–3450.

Guckenheimer, J. and Holmes, P. (1983). *Nonlinear Oscillations, Dynamical Systems, and Bifurcations of Vector Fields*. Springer-Verlag, Berlin.

Guckenheimer, A.J., Myers, M., Wicklin, F., and Worfolk, P. (1992). DsTool: computer assisted explanation of dynamical systems. *Not. Am. Math. Soc.* **39**, 303–309.

Guerini, D. and Carafoli, E. (1999). The calcium pumps. In E. Carafoli and C. Klee, eds., *Calcium as a Cellular Regulator.* Oxford University Press, New York, pp. 249–278.

Gustafsson, B. (1974). After hyperpolarization and the control of repetitive firing in spinal neurons of the cat. *Acta Physiol. Scand.* (Suppl.) **416**, 1–47.

Gutman, A. (1971). Further remarks on the effectiveness of the dendritic synapses. *Biophysics-USSR* **16**, 131–138.

Gutman, A. (1984). *Nerve Cell Dendrites. Theory, Electrophysiology, Function*. Mokslas, Vilnius, pp. 9–126 (in Russian).

Gutman, A. (1991). Bistability of dendrites. *Int. J. Neural Syst.* **1**, 291–304.

Gutman, A. (1994). Gelfand–Tsetlin principle of minimal afferentation and bistability of dendrites. *Int. J. Neural Syst.* **5**, 83–86.

Gutman, A. and Svirskis, G. (1997). Bistability of motoneurons and minimization of afferent flow in motor control. In V.S. Gurfinkel and Yu.S. Levik, eds., *Proceedings of International Symposium on Brain and Movement*. Institute of Information Transmission Problems, Russian Academy of Sciences, Moscow, pp. 88–89.

Guttman, R., Lewis, S., and Rinzel, J. (1980). Control of repetitive firing in squid axon membrane as a model for a neuron oscillator. *J. Physiol. (Lond.)* **305**, 377–395.

Hagar, R., Burgstahler, A., Nathanson, M., and Ehrlich, B. (1998). Type III Ip receptor channel stays open in the presence of increased Ca^{2+}. *Nature* **396**, 81–84.

Hagiwara, S. (1954). Analysis of interval fluctuation of the sensory nerve impulse. *J. Jpn. Physiol.* **4**, 234–240.

Hahn, P.J. and Durand, D. (2001). Bistability dynamics in simulations of neural activity in high-extracellular-potassium conditions. *J. Comput. Neurosci.* **11**, 5–18.

Halliday, D.M., Rosenberg, J.R., Amjad, A.M., Breeze, P., Conway, B.A. and Farmer, S.F. (1995). A framework for the analysis of mixed time series/point process data–theory and application to the study of physiological tremor, single motor unit discharges and electromyograms. *Prog. Biophys. Mol. Bio.* **64**, 237–278.

Halpain, S. (2000). Actin and the agile spine: how and why do dendritic spines dance? *Trends Neurosci.* **23**, 141–146.

Halter, J.A. and Clark, J.W. (1991). A distributed-parameter model of the myelinated nerve fiber. *J. Theor. Biol.* **148**, 345–382.

Hameroff, S., Nip, A., Porter, M., and Tuszynski, J. (2002). Conduction pathways in microtubules, biological quantum computation, and consciousness. *Biosystems* **64**, 149–168.

Hanse, E. and Gustafsson, B. (2001). Quantal variability at glutamatergic synapses in area CA1 of the rat neonatal hippocampus. *J. Physiol. (Lond.)* **531**, 467–480.

Hanson, J.E., Smith, Y., and Jaeger, D. (2004). Sodium channels and dendritic spike initiation at excitatory synapses in *Globus Pallidus* neurons. *J. Neurosci.* **24**, 329–340.

Hanson, P.I. and Schulman, H. (1992). Neuronal Ca^{2+}/calmodulin-dependent protein-kinases. *Annu. Rev. Biochem.* **61**, 559–601.

Hara, F. and Pfeifer, R. (2003). *Morpho-Functional Machines: The New Species*. Springer-Verlag, New York.

Hara, K. and Takabayashi, A. (1975). A dynamic model of retinal cells in the vertebrate retina. *Biol. Cybern.* **20**, 61–67.

Harary, F. (1969). *Graph Theory*. Addison-Wesley, Reading, MA.

Harding, E.F. (1971). The probabilities of rooted tree-shapes generated by random bifurcation. *J. Appl. Probab.* **3**, 44–77.

Harkins, A., Kurebayashi, N., and Baylor, S. (1993). Resting myoplasmic free calcium in frog skeletal muscle fibers estimated with fluo-3. *Biophys. J.* **65**, 865–881.

Harnad, S. (1990). The symbol grounding problem. In S. Forrest, ed., *Emergent Computation*. North-Holland, Amsterdam, pp. 335–346.

Harris, K.M. and Stevens, J.K. (1989). Dendritic spines of CA1 pyramidal cells in the rat hippocampus: serial electron microscopy with reference to their biophysical characteristics. *J. Neurosci.* **8**, 4455–4469.

Harris, K.M. and Sultan, P. (1995). Variation in the number, location and size of synaptic vesicles provides an anatomical basis for nonuniform probability of release at hippocampal CA1 synapses. *Neuropharmacology* **34**, 1387–1395.

Hartermink, A.J., Gifford, D.K., Jaakkola, T.S., and Young, R.A. (2001). Using graphical models and genomic expression data to statistically validate models of genetic regulatory networks. *Proc. Pac. Symp. Biocomput.* **6**, 422–433.

Harvey, M.A., Bermejo, R., and Zeigler, H.P. (2001). Discriminative whisking in the head-fixed rat: optoelectronic monitoring during tactile detection and discrimination tasks. *Somatosens. Mot. Res.* **18**, 211–222.

Haskell, E., Nykamp, D.Q., and Tranchina, D. (2001). Population density methods for large-scale modeling of neuronal networks with realistic synaptic kinetics: cutting the dimension down to size. *Network: Comput. Neural Syst.* **12**, 141–174.

Hasselmo, M.E. and Kapur, A. (2001). Modeling of large networks. In E. De Schutter, ed., *Computational Neuroscience: Realistic Modeling for Experimentalists*. CRC, Boca Raton, FL, pp. 289–316.

Hastings, S.P. (1976). On traveling wave solutions of the Hodgkin–Huxley equations. *Arch. Rat. Mech. Anal.* **60**, 229–257.

Hatzimanikatis, V. and Lee, K.H. (1999). Dynamical analysis of gene networks requires both mRNA and protein expression information. *Metab. Eng.* **1**, 275–281.

Hausser, M., Spruston, N., and Stuart, G.J. (2000). Diversity and dynamics of dendritic signaling. *Science* **290**, 739–744.

Hebb, D.O. (1949). *The Organization of Behavior: A Neuropsychological Theory*. Wiley, New York.

Heidelberger, R. and Matthews, G. (1992). Calcium influx and calcium current in single synaptic terminals of goldfish retinal bipolar neurons. *J. Physiol. (Lond.)* **447**, 235–256.

Heizmann, C.W. and Hunziker, W. (1991). Inracellular calcium-binding proteins: more sites than insights. *Trends Biochem. Sci.* **16**, 98–103.

Helmchen, F., Imoto, K., and Sakmann, B. (1996). Ca^{2+} buffering and action potential-evoked Ca^{2+} signaling in dendrites of pyramidal neurons. *Biophys. J.* **70**, 1069–1081.

Hestrin, S., Nicoll, R.A., Perkel, D.J., and Sah, P. (1990). Analysis of excitatory synaptic action in pyramidal cells using whole-cell recording from rat hippocampal slices. *J. Physiol. (Lond.)* **422**, 203–255.

Hilgetag, C.C., O'Neill, M.A., and Young, M.P. (1996). Indeterminate organization of the visual system. *Science* **271**, 776–777.

Hilgetag, C.C., Burns, G.A., O'Neill, M.A., Scannell, J.W., and Young, M.P. (2000). Anatomical connectivity defines the organization of clusters of cortical areas in the macaque monkey and the cat. *Philos. T. R. Soc. Lond.* **B355**, 91–110.

Hilgetag, C.C., Kötter, R., Stephan, K.E., and Sporns, O. (2002). Computational methods for the analysis of brain connectivity. In G. Ascoli, ed., *Computational Neuroanatomy: Principles and Methods*. Humana, Totowa, NJ, pp. 295–335.

Hille, B. (1984). *Ionic Channels of Excitable Membranes*. Sinauer, Sunderland MA.

Hille, B. (2001). *Ionic Channels of Excitable Membranes*, 3rd edition. Sinauer, Sunderland, MA.

Hillman, D.E. (1979). Neuronal shape parameters and substructures as a basis of neuronal form. In F.O. Schmitt and F.G. Worden, eds., *The Neurosciences, 4th Study Program*. MIT Press, Cambridge, MA, pp. 477–498.

Hillman, D.E. (1988). Parameters of dendritic shape and substructure: intrinsic and extrinsic determination? In R.J. Lasek and M.M. Black, eds., *Intrinsic Determinants of Neuronal Form and Function*. Liss, New York, pp. 83–113.

Hines, M.L. (1984). Efficient computation of branched nerve equations. *Int. J. Biomed. Comput.* **15**, 69–79.

Hines, M.L. (1989). A program for simulation of nerve equations with branching geometries. *Int. J. Biomed. Comput.* **24**, 55–68.

Hines, M.L. (1993). NEURON — a program for simulation of nerve equations. In F. Eeckman, ed., *Neural Systems: Analysis and Modeling*. Kluwer, Norwell, MA.

Hines, M.L. and Carnevale, N.T. (1997). The NEURON simulation environment. *Neural Comput.* **9**, 1179–1209.

Hirsch, J.A., Alonso, I., and Reid, C.R. (1995). Visually evoked calcium action potentials in cat striate cortex. *Nature* **378**, 612–616.

Hlavacek, W.S. and Savageau, M.A. (1996). Rules for coupled expression of regulator and effector genes in inducible circuits. *J. Mol. Biol.* **255**, 121–39.

Hodgkin, A.L. (1954). A note on conduction velocity. *J. Physiol. (Lond.)* **125**, 221–224.

Hodgkin, A.L. (1958). Ionic movements and electrical activity in giant nerve fibers. *P. Roy. Soc. Lond. B.* **148**, 1–37.

Hodgkin, A.L. (1975). The optimal density of sodium channels in an unmyelinated nerve. *Philos. T. Roy. Soc. Lond. B* **270**, 297–300.

Hodgkin, A.L. and Huxley, A.F. (1952a). Current carried by sodium and potassium ions through the membrane of the giant axon of *Loligo*. *J. Physiol. (Lond.)* **116**, 449–472.

Hodgkin, A.L. and Huxley, A.F. (1952b). A quantitative description of membrane current and its application to conduction and excitation in nerve. *J. Physiol. (Lond.)* **117**, 500–544.

Hodgkin, A.L. and Rushton, W.A.H. (1946). The electrical constants of a crustacean nerve fibre. *Proc. Roy. Soc. Lond. B* **133**, 444–479.

Hoffman, D.A., Magee, J.C., Colbert, C.M., and Johnston, D. (1997). K^+ channel regulation of signal propagation in dendrites of hippocampal pyramidal cells. *Nature* **387**, 869–875.

Hogan, N. and Flash, T. (1987). Moving gracefully: quantitative theories of motor coordination. *Trends Neurosci.* **10**, 170–174.

Holcman, D., Schuss, Z., and Korkotian, E. (2004). Calcium dynamics in dendritic spines and spine motility. *Biophys. J.* **87**, 81–91.

Holden, A.V. (1980). Autorhythmicity and entrainment in excitable membranes. *Biol. Cybern.* **38**, 1–8.

Holland, J.H. (1975). *Adaptation in Natural and Artificial Systems. An Introductory Analysis with Applications to Biology, Control, and Artificial Intelligence*. University of Michigan, Ann Arbor, MI.

Holland, O. (2003). Exploration and high adventure: the legacy of Grey Walter. *Philos. T. R. Soc. Lond. A* **361**, 2085–2121.

Holmes, W.R. (1986). A continuous cable method for determining the transient potential in passive dendritic trees of known geometry. *Biol. Cybern.* **55**, 115–124.

Holmes, W.R. (1989). The role of dendritic diameter in maximizing the effectiveness of synaptic inputs. *Brain Res.* **478**, 127–137.

Holmes, W.R. (1990). Is the function of dendritic spines to concentrate calcium? *Brain Res.* **519**, 338–342.

Holmes, W.R. (2000). Models of calmodulin trapping and CaM Kinase II activation in a dendritic spine. *J. Comput. Neurosci.* **8**, 65–85.

Holmes, W.R. and Aradi, I. (1998). Modeling the contributions of calcium channels and NMDA receptor channels to calcium current in dendritic spines. In J.M. Bower, ed., *Computational Neuroscience: Trends in Research.* Plenum, New York, pp. 191–196.

Holmes, W.R. and Levy, W.B. (1990). Insights into associative LTP from computational models of NMDA receptor-mediated calcium influx and intracellular calcium concentration changes. *J. Neurophysiol.* **63**, 1148–1168.

Holmes, W.R. and Li, Y. (2001). Modeling CaM Kinase II activation in dendritic spines of different shape during LTP induction. *Soc. Neurosci. Abstr. 27*, Program No. 611.12.

Holmes, W.R. and Rall, W. (1992a). Electrotonic length estimates in neurons with dendritic tapering or somatic shunt. *J. Neurophysiol.* **68**, 1421–1437.

Holmes, W.R. and Rall, W. (1992b). Estimating the electrotonic structure of neurons with compartment models. *J. Neurophysiol.* **68**, 1438–1452.

Holmes, W.R. and Woody, C.D. (1989). Effects of uniform and non-uniform synaptic activation distributions on the cable properties of modelled cortical pyramidal cells. *Brain Res.* **505**, 12–22.

Holmes, W.R., Segev, I., and Rall, W. (1992). Interpretation of time constant and electrotonic length estimates in multicylinder or branched neuronal structures. *J. Neurophysiol.* **68**, 1401–1420.

Holter, N.S., Maritian, A., Cieplak, M., Fedoroff, N.V., and Banavar, J.R. (2001). Dynamic Modeling of gene expression data. *Proc. Natl. Acad. Sci. USA* **98**, 1693–1698.

Holthoff, K., Tsay, D., and Yuste, R. (2002). Calcium dynamics of spines depend on their dendritic location. *Neuron* **33**, 425–437.

Hoppensteadt, F.C. (1986). *An Introduction to the Mathematics of Neurons*. Cambridge University Press, New York.

Horikawa, Y. (1993). Simulation study of the effects of channel noise on differential conduction at an axon branch. *Biophys. J.* **65**, 680–686.

Horikawa, Y. (1998). Bifurcations in the decremental propagation of a spike train in the Hodgkin–Huxley model of low excitability. *Biol. Cybern.* **79**, 251–261.

Horsfield, K., Woldenberg, M.J., and Bowes, C.L. (1987). Sequential and synchronous growth models related to vertex analysis and branching ratios. *B. Math. Biol.* **49**, 413–430.

Horwitz, B. (1983). Unequal diameters and their effects on time-varying voltages in branched neurons. *Biophys. J.* **41**, 51–66.

Horwitz, B. (2003). The elusive concept of brain connectivity. *Neuroimage* **19**, 466–470.

Horwitz, B. and Sporns, O. (1994). Neural modeling and functional neuroimaging. *Hum. Brain Mapp.* **1**, 269–283.

Horwitz, B., Tagamets, M.-A., and McIntosh, A.R. (1999). Neural modeling, functional brain imaging, and cognition. *Trends Cogn. Sci.* **3**, 91–98.

Hounsgaard, J. and Kiehn, O. (1989). Serotonin-induced bistability of turtle motoneurons caused by a nifedipine-sensitive calcium plateau potential. *J. Physiol. (Lond.)* **414**, 265–282.

Hounsgaard, J. and Kiehn, O. (1993). Calcium spikes and calcium plateaux evoked by differential polarization in dendrites of turtle motoneurons in vitro. *J. Physiol. (Lond.)* **468**, 245–259.

Hounsgaard, J. and Mintz, I. (1988). Calcium conductance and firing properties of spinal motoneurons in the turtle. *J. Physiol. (Lond.)* **398**, 591–603.

Hounsgaard, J., Hultborn, H., Jespersen, B., and Kiehn, O. (1984). Intrinsic membrane properties causing bistable behavior of α-motoneurons. *Exp. Brain Res.* **55**, 391–394.

Hounsgaard, J., Hultborn, H., Jespersen, B., and Kiehn, O. (1988). Bistability of α-motoneurons in the decerebrate cat and in the acute spinal cat after intravenous 5-hydroxytryptophan. *J. Physiol. (Lond.)* **405**, 345–367.

Hoyt, R.C. (1984). A model of the sodium channel. *Biophys. J.* **45**, 55–57.

Hsiao, C., Christopher, A.N., Trueblood, P.R., and Chandler, S.H. (1998). Ionic basis for serotonin-induced bistable membrane properties in guinea pig trigeminal motoneurons. *J. Neurophysiol.* **79**, 2847–2856.

Huber, P.J., Kleiner, B., Gasser, P. and Dumermuth, G. (1971). Statistical methods for investigating phase relations in stationary stochastic processes. *IEEE T. Acoust. Speech* **19**, 78–86.

Huberman, B.A. and Adamic, A.L. (1999). Internet: growth dynamics of the world-wide web. *Nature* **401**, 131–136.

Huguenard, D.R., Hamill, O.P., and Prince, D.A. (1989). Sodium channels in dendrites of rat cortical pyramidal neurons. *Proc. Natl. Acad. Sci. USA* **86**, 2473–2477.

Hull, D.L., Langman, R.E., and Glenn, S.S. (2001). A general account of selection: biology, immunology, and behavior. *Behav. Brain Sci.* **24**, 511–573.

Hunter, P.J., McNaughton, P.A., and Noble, D. (1975). Analytical models of propagation in excitable cells. *Prog. Biophys. Mol. Bio.* **30**, 99–144.

Hurtado, J.M., Gray, C.M., Tamas, L.B., and Sigvardt, K.A. (1999). Dynamics of tremor-related oscillations in the human globus pallidus: a single case study. *Proc. Natl. Acad. Sci. USA* **96**, 1674–1679.

Husain, F.T., Tagamets, M.-A., Fromm, S.J., Braun, A.R., and Horwitz, B. (2004). Relating neuronal dynamics for auditory object processing to neuroimaging activity: a computational modeling and an fMRI study. *NeuroImage* **21**, 1701–1720.

Husi, H. and Grant, S.G.N. (2001). Proteomics of the nervous system. *Trends Neurosci.* **24**, 259–265.

Huxley, A.F. (1959). Can a nerve propagate a subthreshold disturbance? *J. Physiol. (Lond.)* **148**, 80p–81p. (p — Proceedings of the Physiological Society)

Iannella, N., Tuckwell, H.C., and Tanaka, S. (2004). Firing properties of a stochastic PDE model of a rat sensory cortex layer 2/3 pyramidal cell. *Math. Biosci.* **188**, 117–132.

Ignaczak, J. (1998). Saint-Venant type decay estimate for transient heat conduction in a composite rigid semispace. *J. Therm. Stresses* **21**, 185–204.

Inoue, I., Tsutsui, I., and Brown, E.R. (1997). K^+ accumulation and K^+conductance inactivation during action potential trains in giant axons of the squid *Sepioteuthis*. *J. Physiol. (Lond.)* **500**, 355–366.

Ireland, W., Heidel, J., and Uemura, E. (1985). A mathematical model for the growth of dendritic trees. *Neurosci. Lett.* **54**, 243–249.

Isaac, J.T., Nicoll, R.A., and Malenka, R.C. (1995). Evidence for silent synapses: implications for the expression of LTP. *Neuron* **15**, 427–434.

Isakov, V. (1998). *Inverse Problems in Partial Differential Equations*. Springer-Verlag, New York.

Ishida, A.T. (1995). Ion channel components of retinal ganglion cells. *Prog. Retin. Eye Res.* **15**, 261–280.

Ishida, A.T., Kaneko, A., and Tachibana, M. (1984). Responses of solitary retinal horizontal cells from *Carassius auratus* to L-glutamate and related amino acids. *J. Physiol. (Lond.)* **348** 255–270.

Itti, L. and Koch, C. (2000). A saliency-based search mechanism for overt and covert shifts of visual attention. *Vision Res.* **40**, 1489–1506.

Itti, L., Koch, C., and Niebur, E. (1998). A model of saliency-based visual attention for rapid scene analysis. *IEEE T. Pattern Anal.* **20**, 1254–1259.

Izhikevich, E.M. (2001). Resonate-and-fire neurons. *Neural Networks* **14**, 883–894.

Izhikevich, E.M. (2004). Which model to use for cortical spiking neurons? *IEEE T. Neural Networks* **15**, 1063–1070.

Izhikevich, E.M., Gally, J.A., and Edelman, G.M. (2004). Spike-timing dynamics of neuronal groups. *Cereb. Cortex* **14**, 933–944.

Jack, J.J.B. (1979). An introduction to linear cable theory. In F.O. Schmidt and F.G. Worden, eds., *The Neurosciences Fourth Study Program*. MIT Press, Cambridge, MA, pp. 423–438.

Jack, J.J.B. and Redman, S.J. (1971). An electrical description of a motoneuron, and its application to the analysis of synaptic potentials. *J. Physiol. (Lond.)* **215**, 321–352.

Jack, J.J.B., Noble, D., and Tsien, R.W. (1983). *Electric Current Flow in Excitable Cells*. Clarendon, Oxford.

Jack, J.J.B., Noble, D., and Tsien, R.W. (1988). *Electric Current Flow in Excitable Cells*. Clarendon, Oxford. [This is a reprint of the 1983 edition.]

Jackson, M.B. (1993). Passive current flow and morphology in the terminal arborizations of the posterior pituitary. *J. Neurophysiol.* **69**, 692–702.

Jaeger, D. (2000). Accurate reconstruction of neuronal morphology. In E. De Schutter, ed., *Computational Neuroscience: Realistic Modeling for Experimentalists*. CRC, Boca Raton, FL, pp. 159–178.

Jafri, M. (1995). A theoretical study of cytosolic Ca^{2+} waves in Xenopus oocytes. *J. Theor. Biol.* **172**, 209–216.

Jafri, M. and Keizer, J. (1994). Diffusion of inositol 1,4,5-trisphosphate but not Ca^{2+} is necessary for a class of inositol 1,4,5-trisphosphate-induced Ca^{2+} waves. *Proc. Natl. Acad. Sci. USA* **91**, 9485–9489.

Jafri, M. and Keizer, J. (1995). On the roles of Ca^{2+} diffusion, Ca^{2+} buffers, and the endoplasmic reticulum in Ip-induced Ca^{2+} waves. *Biophys. J.* **69**, 2139–2153.

Jafri, M. and Yang, K.H. (2004). Modelling neuronal calcium dynamics. In J. Feng, ed., *Computational Neuroscience: A Comprehensive Approach*. Chapman and Hall/CRC, Boca Raton, FL, pp. 75–95.

Jafri, M., Rice, J., and Winslow, R. (1998). Cardiac Ca^{2+} dynamics: the roles of ryanodine receptor adaptation and sarcoplasmic reticulum load. *Biophys. J.* **74**, 1149–1168.

Jakubowski, J., Kwiatos, K., Chwaleba, A., and Osowski, S. (2002). Higher order statistics and neural network for tremor recognition. *IEEE T. Bio-Med. Eng.* **49**, 152–159.

Jankowska, E. (1992). Interneuronal relay in spinal pathways from propioceptors. *Prog. Neurobiol.* **38**, 335–378.

Jansen, R.F., Pieneman, A.W., and TerMaat, A. (1996). Spontaneous switching between ortho- and antidromic spiking as the normal mode of firing in the cerebral giant neurons of freely behaving *Lymnaea stagnalis*. *J. Neurophysiol.* **76**, 4206–4209.

Jasper, H.H. and Monnier, A.M. (1938). Transmission of excitation between excised non-myelinated nerves. An artificial synapse. *J. Cell. Comp. Physiol.* **11**, 259–277.

Jenkins, G.M. and Watts, D.G. (1968). *Spectral Analysis and Its Applications*. Holden-Day, San Francisco, CA.

Jensen, K.F. and Killackey, H.P. (1987). Terminal arbors of axons projecting to the somatosensory cortex of the adult rat. II. The altered morphology of thalamocortical afferents following neonatal infraorbital nerve cut. *J. Neurosci.* **7**, 3544–3553.

Jensen, R.V. and Shepherd, G.M. (1993). NMDA-activated conductances provide short-term memory for dendritic spine logic computations. In F.H. Eeckman and J.M. Bower, eds., *Computation and Neural Systems*. Kluwer, Norwell, MA, pp. 425–429.

Jeong, H., Tombor, B., Albert, R., Oltvai, Z.N., and Barabási, A.L. (2000). The large-scale organization of metabolic networks. *Nature* **407**, 651–654.

Jeong, H., Mason, S.P., Oltvai, Z.N., and Barabási, A.L. (2001). Lethality and centrality in protein networks. *Nature* **411**, 41–42.

Jeong, H., Néda, Z., and Barabási, A.-L. (2003). Measuring preferential attachment for evolving networks. *Europhys. Lett.* **61**, 567–572.

Jirsa, V.K. and Haken, H. (1997). A derivation of a macroscopic field theory of the brain from the quasi-microscopic neural dynamics. *Physica D* **99**, 503–526.

Johnsen, J.D., Nielsen, P., Luhmann, H.J., Northoff, G., and Kotter, R. (2002). Multi-level network modeling of cortical dynamics built on the GENESIS environment. *Neurocomputing* **44–46**, 863–868.

Johnson, M. (1987). *The Body in the Mind. The Bodily Basis of Meaning, Imagination, and Reason*. University of Chicago Press, Chicago, IL.

Johnson, W.L., Dyer, J.R., Castellucci, V.F., and Dunn, R.J. (1996). Clustered voltage-gated Na^{+} channels in *Aplysia* axons. *J. Neurosci.* **16**, 1730–1739.

Johnston, D. and Miao-Sin Wu, S. (1995). *Foundations of Cellular Neurophysiology*. MIT Press, Cambridge, MA.

Johnston, D., Magee, J.C., Colbert, C.M., and Christie, B.R. (1996). Active properties of neuronal dendrites. *Annu. Rev. Neurosci.* **19**, 165–186.

Johnston, D., Hoffman, D.A., Colbert, C.M., and Magee, J.C. (1999). Regulation of backpropagating action potentials in hippocampal neurons. *Curr. Opin. Neurobiol.* **9**, 288–292.

Johnston, D., Christie, B.R., Frick, A., Gray, R., Hoffman, D.A., Schexnayder, L.K., Watanabe, S., and Yuan, L.-L. (2003). Active dendrites, potassium channels and synaptic plasticity. *Philos.T. R. Soc. Lond. B* **358**, 667–674.

Joseph, S. (1996). The inositol triphosphate receptor family. *Cell. Signal.* **8**, 1–7.

Joseph, S., Rice, H., and Williamson, J. (1989). The effect of external Ca^{2+} and pH on inositol trisphosphate-mediated Ca^{2+} release from cerebellum microsomal fractions. *Biochem. J.* **258**, 261–265.

Joyner, R.W., Westerfield, M., Moore, J.W., and Stockbridge, N. (1978). A numerical method to model excitable cells. *Biophys. J.* **22**, 155–169.

Joyner, R.W., Westerfield, M., and Moore, J.W. (1980). Effects of cellular geometry on current flow during a propagated action potential. *Biophys. J.* **31**, 183–194.

Jung, D. and Zelinsky, A. (1996). Whisker-based mobile robot navigation. *Proceedings of IEEE/RSJ International Conference on Intelligent Robots and Systems*. 497–504.

Jung, H., Mickus, T., and Spruston, N. (1997). Prolonged sodium channel inactivation contributes to dendritic action potential attenuation in hippocampal pyramidal neurons. *J. Neurosci.* **17**, 6639–6646.

Kaftan, E., Ehrlich, B., and Watras, J. (1997). Inositol 1,4,5-trisphosphate (Ip) and Ca^{2+} interact to increase the dynamic range of Ip receptor-dependent Ca^{2+} signaling. *J. Gen. Physiol.* **110**, 529–538.

Kager, H., Wadman, W.J., and Somjen, G.G. (2000). Simulated seizures and spreading depression in neuron model incorporating interstitial space and ion concentrations. *J. Neurophysiol.* **84**, 495–512.

Kalitzin, S., van Dijk, B.W., Spekreijse, H., and van Leeuwen, W.A. (1997). Coherency and connectivity in oscillating neural networks: linear partialization analysis. *Biol. Cybern.* **76**, 73–82.

Kaminski, M., Ding, M., Truccolo, W.A., and Bressler, S.L. (2001). Evaluating causal relations in neural systems: granger causality, directed transfer function and statistical assessment of significance. *Biol. Cybern.* **85**, 145–157.

Kamiyama, Y. and Usui, S. (2003). A retinal ganglion cell model based on discrete stochatic ion channels. The Neural Basis of Early Vision. *Keio University International Symposia for Life Science and Medicine* **11**, 134–137.

Kamiyama, Y., Ogura, T., and Usui, S. (1996). Ionic current model of the vertebrate rod photoreceptor. *Vision Res.* **36**, 4059–4068.

Kandel, E.R., Schwartz, J.H., and Jessel, T.M. (2000). *Principles of Neuroscience*, 4th edition. McGraw Hill, Norwalk, Connecticut.

Kaneda, M. and Kaneko, A. (1991). Voltage-gated calcium currents in isolated retinal ganglion cells of the cat. *Jpn. J. Physiol.* **41**, 35–48.

Kaneko, A. and Tachibana, M. (1985a). A voltage-clamp analysis of membrane currents in solitary bipolar cell dissociated from *Carassius auratus*. *J. Physiol. (Lond.)* **358**, 131–152.

Kaneko, A. and Tachibana, M. (1985b). Effects of L-glutamate on the anomalous rectifier potassium current in horizontal cells of *Carassius auratus* retina. *J. Physiol. (Lond.)* **358**, 169–182.

Kaneko, A. and Tachibana, M. (1986). Membrane properties of solitary retinal cells. *Prog. Retin. Res.* **5**, 125–146.

Kaneko, M., Kanayama, N., and Tsuji, T. (1998). Active antenna for contact sensing. *IEEE T. Robotic. Autom.* **14**, 278–291.

Kaneko, A., Pinto, L.H., and Tachibana, M. (1989). Transient calcium current of retinal bipolar cells of the mouse. *J. Physiol. (Lond.)* **410**, 613–629.

Kao, J. (1994). Practical aspects of measuring [Ca] with fluorescent indicators. *Methods Cell. Biol.* **40**, 155–181.

Karschin, A. and Wässle, H. (1990). Voltage- and transmitter-gated currents in isolated bipolar cells of rat retina. *J. Neurophysiol.* **63**, 860–876.

Kashef, B. and Bellman, R. (1974). Solution of the partial differential equation of the Hodgkin–Huxley model using differential quadrature. *Math. Biosci.* **19**, 1–8.

Kater, S.B. and Rehder, V. (1995). The sensory-motor role of growth cone filopodia. *Curr. Opin. Neurobiol.* **5**, 68–74.

Kater, S.B., Davenport, R.W., and Guthrie, P.B. (1994). Filopodia as detectors of environmental cues: signal integration through changes in growth cone calcium levels. In J. van Pelt, M.A. Corner, H.B.M. Uylings, and F.H. Lopes da Silva, eds., *The Self-Organizing Brain: From Growth Cones to Functional Networks, Progress in Brain Research*, Volume 102. Elsevier, Amsterdam, pp. 49–60.

Katz, B. (1969). *The Release of Neural Transmitter Substances*. Liverpool University Press, Liverpool.

Katz, B. and Schmitt, O.H. (1940). Electrical interaction between two adjacent fibers. *J. Physiol. (Lond.)* **97**, 471–488.

Kauffman, S.A. (1969). Metabolic stability and epigenesis in randomly connected nets. *J. Theor. Biol.* **22**, 437–467.

Kauffman, S.A. (1974). The large-scale structure and dynamics of gene control circuits: an ensemble approach. *J. Theor. Biol.* **44**, 167–190.

Kauffman, S.A. (1984). Emergent properties in random complex automata. *Physica D* **10**, 145–156.

Kauffman, S.A. (1993). *Origins of Order*. Oxford University Press, Oxford, NY.

Kavalali, E.T. and Plummer, M.R. (1996). Multiple voltage-dependent mechanisms potentiate calcium channel activity in hippocampal neurons. *J. Neurosci.* **16**, 1072–1082.

Kawaguchi, H. and Fukunishi, K. (1998). Dendritic classification in rat hippocampal neurons according to signal propagation properties: observation by multichannel optical recording in cultured neuronal networks. *Exp. Brain Res.* **122**, 378–392.

Kawai, F., Horiguchi, M., Suzuki, H., and Miyachi, E.-I. (2001). Na^+ action potentials in human photoreceptors. *Neuron* **30**, 451–458.

Kawato, M. (1984). Cable properties of a neuron model with non-uniform resistivity. *J. Theor. Biol.* **111**, 149–169.

Keen, R.E. and Spain, J.D. (1992). *Computer Simulation in Biology.* Wiley-Liss, New York.

Keener, J. (2000). Propagation of waves in an excitable medium with discrete release sites. *SIAM J. Appl. Math.* **61**, 317–334.

Keener, J. (2002). Spatial modeling. In C. Fall, E. Marland, J. Wagner, and J. Tyson, eds, *Computational Cell Biology*. Springer-Verlag, Berlin, pp. 172–197.

Keener, J. and Sneyd, J. (1998). *Mathematical Physiology*. Springer-Verlag, New York.

Keizer, J. (1987). *Statistical Thermodynamics of Nonequilibrium Processes*. Springer-Verlag, Berlin.

Keizer, J. and Levine, L. (1996). Ryanodine receptor adaptation and calcium-induced calcium release-dependent calcium oscillations. *Biophys. J.* **71**, 3477–3487.

Keizer, J., Li, Y., Stojilkovic, S., and Rinzel, J. (1995). Ip-induced Ca^{2+} excitability of the endoplasmic reticulum. *Mol. Biol. Cell.* **6**, 945–951.

Keizer, J., Smith, G., Ponce-Dawson, S., and Pearson, J. (1998). Saltatory propagation of Ca^{2+} waves by Ca^{2+} sparks. *Biophys. J.* **75**, 595–600.

Keller, B.U., Konnerth, A., and Yaari, V. (1991). Patch clamp analysis of excitatory synaptic currents in Granule cells of rat hippocampus. *J. Physiol. (Lond.)* **435**, 275–293.

Kellogg, O.D. (1954). *Foundations of Potential Theory*. Dover, New York.

Kelly, J.J. and Ghausi, M.S. (1965). Tapered distributed RC networks with similar immittances. *IEEE T. Circuit Theory* **12**, 554–558.

Kelvin, W.T. (1855). On the theory of the electric telegraph. *Philos. R. Soc. Lond.* **7**, 382–399.

Kember, G.C. and Evans, J.D. (1995). Analytical solutions to a multicylinder somatic shunt cable model for passive neurons with spines. *IMA J. Math. Appl. Med. Biol.* **12**, 137–157.

Kempter, R., Leibold, C., Wagner, H., and van Hemmen, J.L. (2001). Formation of temporal feature maps by axonal propagation of synaptic learning. *Proc. Natl. Acad. Sci. USA* **98**, 4166–4171.

Kennedy, M.B. (1998). Signal transduction molecules at the glutamatergic postsynaptic membrane. *Brain Res. Rev.* **26**, 243–257.

Kennedy, M.B. (2000). Signal-processing machines at the postsynaptic density. *Science* **290**, 750–754.

Kepler, T.B., Abbott, L.F., and Marder, E. (1992). Reduction of conductance-based neuron models. *Biol. Cybern.* **66**, 381–387.

Kerszberg, M. and Changeux, J. (1998). A simple molecular model of neurulation. *Bioessays* **20**, 758–770.

Khodorov, B.I. and Timin, E.N. (1975). Nerve impulse propagation along nonuniform fibres. *Prog. Biophys. Mol. Bio.* **30**, 145–184.

Khodorov, B.I., Timin, Y.N., Vilenkin, S.Y., and Gul'ko, F.B. (1969). Theoretical analysis of the mechanisms of conduction of a nerve pulse over an inhomogeneous axon. I. Conduction through a portion with increased diameter. *Biophysics-USSR* **14**, 304–315.

Kim, H.G. and Connors, B.W. (1993). Apical dendrites of the neocortex: correlation between sodium- and calcium-dependent spiking and pyramidal cell morphology. *J. Neurosci.* **13**, 5301–5311.

Kirkwood, A.R., Rioult, M.G., and Bear, M.F. (1996). Experience-dependent modification of synaptic plasticity in visual cortex. *Nature* **381**, 526–528.

Kirsch, A. (1995). *An Introduction to the Mathematical Theory of Inverse Problems*. Springer-Verlag, New York.

Kistler, W.M. and Gerstner, W. (2002). Stable propagation of activity pulses in populations of spiking neurons. *Neural Comput.* **14**, 987–997.

Kistler, W.M., Gerstner, W., and van Hemmen, J.L. (1997). Reduction of the Hodgkin–Huxley equations to a single-variable threshold model. *Neural Comput.* **9**, 1015–1045.

Klee, M. and Rall, W. (1977). Computed potentials of cortically arranged populations of neurons. *J. Neurophysiol.* **40**, 647–666.

Kliemann, W.A. (1987). Stochastic dynamic model for the characterization of the geometrical structure of dendritic processes. *B. Math. Biol.* **49**, 135–152.

Knisley, J. (2003). Linear embedding of a dendritic electrotonic model with a Hodgkin–Huxley hot spot. *Paper presented at the 23rd Southeastern-Atlantic Regional Conference on Differential Equations*. Knoxville, TN, Oct. 11–12, 2003.

Kobayashi, K. and Tachibana, M. (1995). Ca^{2+} regulation in the presynaptic terminals of goldfish retinal bipolar cells. *J. Physiol. (Lond.)* **483**, 79–94.

Koch, C. (1999). *Biophysics of Computation: Information Processing in Single Neurons*. Oxford University Press, Oxford.

Koch, C. and Poggio, T. (1983). A theoretical analysis of electrical properties of spines. *Proc. R. Soc. Lond. B* **218**, 455–477.

Koch, C. and Segev, I., eds. (1989). *Methods in Neuronal Modeling: From Synapses to Networks*. MIT Press, Cambridge, MA.

Koch, C. and Segev, I., eds. (1998). *Methods in Neuronal Modeling: From Ions to Networks*, 2nd edition. MIT Press, Cambrige, MA.

Koch, C. and Zador, A. (1993). The function of dendritic spines: devices subserving biochemical rather than electrical compartmentalization. *J. Neurosci.* **13**, 413–422.

Koch, C., Poggio, T., and Torre, V. (1982). Retinal ganglion cells: a functional interpretation of dendritic morphology. *Philos. T. R. Soc. Lond. B* **298**, 227–264.

Koch, C., Poggio, T., and Torre, V. (1983). Nonlinear interactions in a dendritic tree: localization, timing, and role in information processing. *Proc. Natl. Acad. Sci. USA* **80**, 2799–2802.

Kocsis, J.D., Swadlow, H.A., Waxman, S.G., and Brill, M.H. (1979). Variation of conduction velocity during the relative refractory and supernormal periods: a mechanism for impulse entrainment in central axons. *Exp. Neurol.* **65**, 230–236.

Kocsis, J.D., Ruiz, J.A., and Cummins, K.L. (1982). Modulation of axonal excitability mediated by surround electric activity: an intra-axonal study. *Exp. Brain Res.* **47**, 151–153.

Koester, H.J. and Sakmann, B. (1998). Calcium dynamics in single spines during coincident pre- and post-synaptic activity depend on relative timing of back-propagating action potentials and subthreshold excitatory postsynaptic potentials. *Proc. Natl. Acad. Sci. USA* **95**, 9596–9601.

Koester, H.J. and Sakmann, B. (2000). Calcium dynamics associated with action potentials in single nerve terminals of pyramidal cells in layer 2/3 of the young rat neocortex. *J. Physiol. (Lond.)* **529**, 625–646.

Koike, H., Okada, Y., Oshima, T., and Takahashi, K. (1968). Accommodative behaviour of cat pyramidal tract cells investigated with intracellular injection of currents. *Exp. Brain Res.* **5**, 173–188.

Koizumi, A., Watanabe, S.-I., and Kaneko, A. (2001). Persistent Na^+ current and Ca^{2+} current boost graded depolarization of rat retinal amacrine cells in culture. *J. Neurophysiol.* **86**, 1006–1016.

Korn, H. and Faber, D.S. (1991). Quantal analysis and synaptic efficacy in the CNS. *Trends Neurosci.* **14**, 439–445.

Korogod, S.M., Bras, H., Sarana, V.N., Gogan, P., and Tyc-Dumont, S. (1994). Electrotonic clusters in the dendritic arborization of abducens motoneurons of the rat. *Eur. J. Neurosci.* **6**, 1517–1527.

Kossev, A., Gydikov, A., and Trayanova, N. (1988). Comparison of the different variants of the current source density analysis methods in neurophysiological studies. *Acta. Physiol. Pharm. B* **14**, 75–82.

Koster, H., Hartog, A., Os, C.-V., and Bindels, R. (1995). Calbindin-d28k facilitates cytosolic calcium diffusion without interfering with calcium signaling. *Cell Calcium* **18**, 187–196.

Kovalchuk, Y., Eilers, J., Lisman, J., and Konnerth, A. (2000). NMDA receptor-mediated subthreshold Ca^{2+} signals in spines of hippocampal neurons. *J. Neurosci.* **20**, 1791–1799.

Kowalski, K. and Steeb, W. (1991). *Nonlinear Dynamical Systems and Carleman Linearization*. World Scientific, Singapore.

Kraushaar, U. and Jonas, P. (2000). Efficacy and stability of quantal GABA release at a hippocampal interneuron-principal neuron synapse. *J. Neurosci.* **20**, 5594–5607.

Kreitzer, A.C. and Regehr, W.G. (2000). Modulation of transmission during trains at a cerebellar synapse. *J. Neurosci.* **20**, 1348–1357.

Krichmar, J.L. and Edelman, G.M. (2002). Machine psychology: Autonomous behavior, perceptual categorization, and conditioning in a brain-based device. *Cereb. Cortex* **12**, 818–830.

Krichmar, J.L. and Snook, J.A. (2002). A neural approach to adaptive behavior and multisensor action selection in a mobile device. In *IEEE Conference on Robotics and Automation*. Washington, D.C., pp. 3864–3869.

Krichmar, J.L., Nasuto, S.J., Scorcioni, R., Washington, S.D., and Ascoli, G.A. (2002). Effects of dendritic morphology on CA3 pyramidal cell electrophysiology: a simulation study. *Brain Res.* **941**, 11–28.

Krinskii, V.I. and Kokoz, Yu.M. (1973). Analysis of equations of excitable membranes. I. Reduction of the Hodgkin–Huxley equations to a second order system. *Biophysics-USSR* **18**, 533–539.

Krnjevic, K. and Miledi, R. (1959). Presynaptic failure of neuromuscular propagation in rats. *J. Physiol. (Lond.)* **149**, 1–22.

Krzyzanski, W. and Bell, J. (1999). Analysis of a model of membrane potential for a skin receptor. *Math. Biosci.* **158**, 1–45.

Krzyzanski, W., Bell, J., and Poznanski, R.R. (2002). Neuronal integrative analysis of the 'Dumbbell' model for passive neurons. *J. Integr. Neurosci.* **1**, 217–239.

Kubota, Y. and Bower, J.M. (2001). Transient versus asymptotic dynamics of CaM kinase II : possible roles of phosphatase. *J. Comput. Neurosci.* **11**, 263–279.

Kuffler, S.W., Nicholls, J.G., and Martin R.A. (1984). *From Neuron to Brain*, 2nd edition. Sinauer, Sunderland, MA.

Kumar, R., Raghavan P., Rajalopagan S., and Tomkins, A. (1999). Extracting large-scale knowledge bases from the web. *Proceedings of the 9th ACM Symposium on Principles of Database Systems*, pp. 639–650.

Kupfer, F.-S. and Sachs, E.W. (1992). Numerical solution of a nonlinear parabolic control problem by a reduced SQP method. *Comp. Opt. Appl.* **1**, 113–135.

Kuriscak, E., Trojan, S., and Wünsch, Z. (2002). Model of spike propagation reliability along the myelinated axon corrupted by axonal intrinsic noise sources. *Physiol. Res.* **51**, 202–215.

Kurogi, S. (1987). A model of neural network for spatiotemporal pattern recognition. *Biol. Cybern.* **57**, 103–114.

Labouriau, I.S. (1985). Degenerate Hopf bifurcation and nerve impulse. *SIAM J. Math. Anal.* **16**, 1121–1133.

Labouriau, I.S. (1989). Degenerate Hopf bifurcation and nerve impulse. Part II. *SIAM J. Math. Anal.* **20**, 1–12.

Lachaux, J.P., Rodriguez, E., Martinerie, J., and Varela, F. (1999). Measuring phase-synchrony in brain signals. *Hum. Brain Mapp.* **8**, 194–208.

Laird, J., Rosenbloom, P., and Newell, A. (1986). *Universal Subgoaling and Chunking. The Automatic Generation and Learning of Goal Hierarchies*. Kluwer, Boston, MA.

Lakoff, G. (1987). *Women, Fire, and Dangerous Things: What Categories Reveal About the Mind.* University of Chicago Press, Chicago, IL.

Lamas, J.A., Reboreda, A., and Codesido, V. (2002). Ionic basis of the resting membrane potential in cultured rat sympathetic neurons. *NeuroReport* **13**, 585–591.

Lang, A.E. and Lozano, A.M. (1998). Parkinson's disease. *New Engl. J. Med.* **339**, 1130–1143.

Lansner, A. and Fransen, E. (1995). Improving the realism of attractor models by using cortical columns as functional units. In J.M. Bower, ed., *The Neurobiology of Computation*. Kluwer Academic, Boston, MA.

Larkman, A.K., Major, G., Stratford, K.J., and Jack, J.J.B. (1992). Dendritic morphology of pyramidal neurons of the visual cortex of the rat. IV: Electrical geometry. *J. Comp. Neurol.* **323**, 137–152.

Larkman, A.U. (1991a). Dendritic morphology of pyramidal neurons of the visual cortex of the rat: I. Branching patterns. *J. Comp. Neurol.* **306**, 307–319.

Larkman, A.U. (1991b). Dendritic morphology of pyramidal neurons of the visual cortex of the rat: III. Spine distributions. *J. Comp. Neurol.* **306**, 332–343.

Larkman, A.U. and Mason, A.J.R. (1990). Correlations between morphology and electrophysiology of pyramidal neurons in slices of rat visual cortex. 1. Establishment of cell classes. *J. Neurosci.* **10**, 1407–1414.

Larkum, M.E., Kaiser, K.M.M., and Sakmann, B. (1999a). Calcium electrogenesis in distal apical dendrites of layer 5 pyramidal cells at a critical frequency of back-propagating action potentials. *Proc. Natl. Acad. Sci. USA* **96**, 14600–14604.

Larkum, M.E., Zhu, J.J., and Sakmann, B. (1999b). A new cellular mechanism for coupling inputs arriving at different cortical layers. *Nature* **398**, 338–341.

Larkum, M.E., Zhu, J.J., and Sakmann, B. (2001). Dendritic mechanisms underlying the coupling of the dendritic with the axonal action potential initiation zone of adult rat layer 5 pyramidal neurons. *J. Physiol. (Lond.)* **533**, 447–466.

Lasater, E.M. (1986). Ionic currents of cultured horizontal cells isolated from white perch retina. *J. Neurophysiol.* **55**, 499–513.

Lasater, E.M. (1988). Membrane currents of retinal bipolar cells in culture. *J. Neurophysiol.* **60**, 1460–1480.

Lasater, E.M. (1991). Membrane properties of distal retinal neurons. *Prog. Retin. Res.* **11**, 215–246.

Lashley, K.S. (1931). Mass action in cerebral function. *Science* **73**, 245–254.

Latash, M L. and Gutman, S.R. (1993). Variability of fast single-joint movements and the equilibrium-point hypothesis. In K.M. Newell and D.M. Corcos, eds., *Variability and motor control.* Human Kinetics, Champaign, IL, pp. 157–182.

Laubach, M., Wessberg, J., and Nicolelis, M.A.L. (2000). Cortical ensemble activity increasingly predicts behaviour outcomes during learning of a motor task. *Nature* **405**, 567–571.

Laughlin, S.B. and Sejnowski, T.J. (2003). Communication in Neuronal Networks. *Science* **301**, 1870–1874.

Lauk, M., Koster, B., Timmer, J., Guschlbauer, B., Deuschl, G., and Lucking, C.H. (1999). Side-to-side correlation of muscle activity in physiological and pathological human tremors. *Clin. Neurophysiol.* **110**, 1774–1783.

LeBeau, A., Yule, D., Groblewski, G., and Sneyd, J. (1999). Agonist-dependent phosphorylation of the inositol 1,4,5-trisphosphate receptor: a possible mechanism for agonist-specific calcium oscillations in pancreatic acinar cells. *J. Gen. Physiol.* **113**, 851–872.

LeDoux, J.E. (1995). Emotion: clues from the brain. *Annu. Rev. Psychol.* **46**, 209–235.

Lee, L., Harrison, L.M., and Mechelli, A. (2003). A report of the functional connectivity workshop, Dusseldorf 2002. *NeuroImage* **19**, 457–465.

Lee, R.H. and Heckman, C.J. (1996). Influence of voltage-sensitive dendritic conductances on bistable firing and effective synaptic current in cat spinal motoneurons in vivo. *J. Neurophysiol.* **76**, 1967–1970.

Lee, R.H. and Heckman, C.J. (1998). Bistability in spinal motoneurons in vivo: systematic variations in persistent inward currents. *J. Neurophysiol.* **80**, 583–593.

Lee, S.H., Rosenmund, C., Schwaller, B., and Neher, E. (2000). Differences in Ca^{2+} buffering properties between excitatory and inhibitory hippocampal neurons from the rat. *J. Physiol. (Lond.)* **525**, 405–418.

Lee, T.I., Rinaldi, N.J., Robert, F., Odom, D.T., Bar-Joseph, Z., Gerber, G.K., Hannett, N.M., Harbison, C.T., Thompson, C.M., Simon, I., Zeitlinger, J., Jennings, E.G., Murray, H.L., Gordon, D.B., Ren, B., Wyrick, J.J., Tagne, J.B., Volkert, T.L., Fraenkel, E., Gifford, D.K., and Young, R.A. (2002). Transcriptional regulatory networks in *Saccharomyces cerevisiae*. *Science* **298**, 799–804.

Leibovic, K.N. (1972). *Nervous System Theory: An Introductory Study*. Academic, New York.

Leibovic, K.N. and Sabah, N.H. (1969). On synaptic transmission, neural signals and psychophysiological phenomena. In K.N. Leibovic, ed., *Information Processing in the Nervous System*. Springer-Verlag, Berlin, pp. 273–292.

Leloup, J.C. and Goldberger, A. (1998). A model for circadian rhythms in Drosophila incorporating the formation of a complex between the PER and TIM proteins. *J. Biol. Rhythm* **13**, 70–87.

LeMasson, G., Marder, E., and Abbott, L.F. (1993). Activity-dependent regulation of conductances in model neurons. *Science* **259**, 1915–1917.

Lemon, G. (2004). Fire-diffuse-fire calcium waves in confined intracellular spaces. *B. Math. Biol.* **66**, 65–90.

Lemon, N. and Turner, R.W. (2000). Conditional spike backpropagation generates burst discharge in a sensory neuron. *J. Neurophysiol.* **84**, 1519–1530.

Lenz, F.A., Kwan, H.C., Martin, R.L., Tasker, R.R., Ostrovsky, J.O., and Lenz, Y.E. (1994). Single unit analysis of the human ventral thalamic nuclear group. Tremor-related activity in functionally identified cells. *Brain* **117**, 531–543.

Lenz, F.A., Jaeger, C.J., Seiko, M.S., Lin, Y.C., and Reich, S.G. (2002). Single-neuron analysis of human thalamus in patients with intention tremor and other clinical signs of cerebellar disease. *J. Neurophysiol.* **87**, 2084–2094.

Leonard, A.S., Lim, I.A., Hemsworth, D.E., Horne, M.C., and Hell, J.W. (1999). Calcium/calmodulin-dependent protein kinase II is associated with the N-methyl-D-aspartate receptor. *Proc. Natl. Acad. Sci. USA* **96**, 3239–3244.

Lestienne, R. and Tuckwell, H.C. (1998). The significance of precisely replicating patterns in mammalian CNS spike trains. *Neuroscience* **82**, 315–336.

Letourneau, P.C., Snow, D.M., and Gomez, T.M. (1994). Growth cone motility: substratum-bound molecules, cytoplasmic [Ca^{2+}] and -regulated proteins. In J. van Pelt, M.A. Corner, H.B.M. Uylings, and F.H. Lopes da Silva, eds., *The Self-Organizing Brain: From Growth Cones to Functional Networks, Progress in Brain Research*, Volume 102. Elsevier, Amsterdam, pp. 35–48.

Levy, R., Hutchison, W.D., Lozano, A.M., and Dostrovsky, J.O. (2000). High-frequency synchronization of neuronal activity in the subthalamic nucleus of parkinsonian patients with limb tremor. *J. Neurosci.* **20**, 7766–7775.

Levy, W.B. and Baxter, R.A. (2002). Energy-efficient neuronal computation via quantal synaptic failures. *J. Neurosci.* **22**, 4746–4755.

Levy, W.B. and Steward, O. (1983). Temporal contiguity requirements for long-term associative potentiation/depression in the hippocampus. *Neuroscience* **8**, 791–797.

Li, Y. and Holmes, W.R. (2000). Comparison of CaMKinase II activation in a dendritic spine computed with deterministic and stochastic models of the NMDA synaptic conductance. *Neurocomputing* **32–33**, 1–7.

Li, Y. and Rinzel, J. (1994). Equations for Ipr-mediated Ca_i oscillations derived from a detailed kinetic model: a Hodgkin–Huxley like formalism. *J. Theor. Biol.* **166**, 461–473.

Li, Y., Rinzel, J., Keizer, J., and Stojilkovic, S. (1994). Ca^{2+} oscillations in pituitary gonadotrophs: comparison of experiment and theory. *Proc. Natl. Acad. Sci. USA* **91**, 58–62.

Li, Y., Rinzel, J., Vergara, L., and Stojilkovic, S. (1995b). Spontaneous electrical and Ca^{2+} oscillations in unstimulated pituitary gonadotrophs. *Biophys. J.* **69**, 785–795.

Li, Y., Keizer, J., Stojilkovic, S., and Rinzel, J. (1995a). Ca^{2+} excitability of the ER membrane: an explanation for Ip-induced Ca^{2+} oscillations. *Am. J. Physiol.* **269**, C1079–C1092.

Li, Y., Stojilkovic, S., Keizer, J., and Rinzel, J. (1997). Sensing and refilling calcium stores in an excitable cell. *Biophys. J.* **72**, 1080–1091.

Liang, H., Ding, M., Nakamura, R., and Bressler, S.L. (2000). Causal influences in primate cerebral cortex during visual pattern discrimination. *NeuroReport* **11**, 2875–2880.

Liang, S., Fuhrman, S., and Somogyi, R. (1998). Reveal, a general reverse engineering algorithm for inference of genetic network architectures. *Proc. Pac. Symp. Biocomput.* **3**, 18–29.

Liebovitch, L.S. and Zochowski, M. (1997). Dynamics of neural networks relevant to properties of proteins. *Phys. Rev. E* **56**, 931–935.

Liebovitch, L.S., Arnold, N.D., and Selector, L.Y. (1994). Neural networks to compute molecular dynamics. *J. Biol. Syst.* **2**, 193–228.

Liebovitch, L.S., Todorov, A.T., Zochowski, M., Scheurle, D., Colgin, L., Wood, M.A., Ellenbogen, K.A., Herre, J.M., and Berstein, R.C. (1999). Nonlinear properties of cardiac rhythm abnormalities. *Phys. Rev. E* **59**, 3312–3319.

Lindsay, K.A., Ogden, J.M., Halliday, D.M., and Rosenberg, J.R. (1999). An introduction to the principles of neuronal modelling. In U. Windhorst and H. Johannson, eds., *Modern Techniques in Neuroscience Research*. Springer-Verlag, Berlin, pp. 213–306.

Lindsay, K.A., Ogden, J.M., and Rosenberg, J.R. (2001a). Dendritic subunits determined by dendritic morphology. *Neural Comput.* **13**, 2465–2476.

Lindsay, K.A., Ogden, J.M., and Rosenberg, J.R. (2001b). Equivalence transformations for dendritic Y-junctions: a new definition of dendritic sub-unit. *Math. Biosci.* **170**, 133–154.

Lindsay, K.A., Ogden, J.M., and Rosenberg, J.R. (2001c). Advanced numerical methods for modeling dendrites. In R.R. Poznanski, ed., *Biophysical Neural Networks: Foundations of Integrative Neuroscience*. Liebert, Larchmont, NY, pp. 411–493.

Lindsay, K.A., Ogden, J.M., and Rosenberg, J.R. (2002). An investigation into the influence of boundary condition specification in finite difference methods on the behaviour of passive and active neuronal models. *Prog. Biophys. Mol. Biol.* **78**, 3–43.

Lindsay, K.A., Rosenberg, J.R., and Tucker, G. (2003). Analytical and numerical construction of equivalent cables. *Math Biosci.* **184**, 137–164.

Lindsay, K.A., Rosenberg, J.R., and Tucker, G. (2004). From Maxwell's equations to the cable equation and beyond. *Prog. Biophys. Mol. Biol.* **85**, 71–116.

Lipowsky, R., Gillessen, T., and Alzheimer, C. (1996). Dendritic Na^+ channels amplify EPSPs in hippocampal pyramidal cells. *J. Neurophysiol.* **76**, 2181–21191.

Lisman, J. (1985). A mechanism for memory storage insensitive to molecular turnover: a bistable autophosphorylating kinase. *Proc. Natl. Acad. Sci. USA* **82**, 3055–3057.

Lisman, J. (1989). A mechanism for the Hebb and the anti-Hebb processes underlying learning and memory. *Proc. Natl. Acad. Sci. USA* **86**, 9574–9578.

Lisman, J. (1994). The CaM kinase II hypothesis for the storage of synaptic memory. *Trends Neurosci.* **17**, 406–412.

Lisman, J. (2003). Long-term potentiation: outstanding questions and attempted synthesis. *Philos. T. R. Soc. Lond. B* **358**, 829–842.

Lisman, J.E. and Goldring, M.A. (1988). Feasibility of long-term storage of graded information by the Ca^{2+}/calmodulin-dependent protein kinase molecules of the postsynaptic density. *Proc. Natl. Acad. Sci. USA* **85**, 5320–5324.

Lisman, J., Malenka, R.C., Nicoll, R.A., and Malinow, R. (1997). Learning mechanisms: the case for CaM-KII. *Science* **276**, 2001–2002.

Liu, G. and Tsien, R.W. (1995). Properties of synaptic transmission at single hippocampal synaptic boutons. *Nature* **375**, 404–408.

Liu, Z., Golowasch, J., Marder, E., and Abbott, L.F. (1998). A model neuron with activity-dependent conductances regulated by multiple calcium sensors. *J. Neurosci.* **18**, 2309–2320.

Llinás, R.R. (1968). Dendritic spikes and their inhibition in alligator purkinje cells. *Science* **160**, 1132–1135.

Llinás, R.R. (1988). The intrinsic electrophysiological properties of mammalian neurons: insights into central nervous system function. *Science* **242**, 1654–1664.

Llinás, R.R. and Nicholson, C. (1971). Electrophysiological properties of dendrites and somata in alligator Purkinje cells. *J. Neurophysiol.* **34**, 532–551.

Llinás, R.R. and Sugimori, M. (1980a). Electrophysiological properties of *in vitro* Purkinje cell somata in mammalian cerebellar slices. *J. Physiol. (Lond.)* **305**, 171–196.

Llinás, R.R. and Sugimori, M. (1980b). Electrophysiological properties of *in vitro* Purkinje cell dendrites in mammalian cerebellar slices. *J. Physiol. (Lond.)* **305**, 197–214.

Llinás, R.R., Nicholson, C., Freeman, J.A., and Hillman, D.E. (1968). Dendritic spikes and their inhibition in alligator Purkinje cells. *Science* **160**, 1132–1135.

Llinás, R.R., Nicholson, C., Freeman, J.A., and Hillman, D.E. (1969). Dendritic spikes versus cable properties. *Science* **163**, 97.

Llinás, R., Sugimori, M., Hillman, D.E., and Cherskey, B. (1992). Distribution and functional significance of the P-type, voltage-dependent Ca^{2+} channels in the mammalian central nervous system. *Trends Neurosci.* **15**, 251–355.

Loewenstein, Y. and Sompolinsky, H. (2003). Temporal integration by calcium dynamics in a model neuron. *Nat. Neurosci.* **6**, 961–967.

Logothetis, N.K., Pauls, J., Augath, M., Trinath, T., and Oeltermann, A. (2001). Neurophysiological investigation of the basis of the fMRI signal. *Nature* **412**, 150–157.

Löhrke, S. and Hofmann, H.D. (1994). Voltage-gated currents of rabbit a- and b-type horizontal cells in retinal monolayer cultures. *Visual Neurosci.* **11**, 369–378.

Lorente de No, R. (1947). *A Study of Nerve Physiology*. Rockefeller Institute, New York.

Lozano, A.M., Lang, A.E, Hutchinson, W.D., and Dostrovsky, J.O. (1998). New developments in understanding the etiology of Parkinson's disease and its treatment. *Curr. Opin. Neurobiol.* **8**, 783–790.

Lu, T., Shen, Y., and Yang, X.L. (1998). Desensitization of AMPA receptors on horizontal cells isolated from crucian carp retina. *Neurosci. Res.* **31**, 123–135.

Luby-Phelps, K., Hori, M., Phelps, J.M., and Won, D. (1995). Ca^{2+}-regulated dynamic compartmentalization of calmodulin in living smooth muscle cells. *J. Biol. Chem.* **270**, 21532–21538.

Ludwig, M. and Pittman, Q.J. (2003). Talking back: dendritic neurotransmitter release. *Trends Neurosci.* **26**, 256–261.

Lukasiewicz, P.D., Maple, B.R., and Werblin, F.S. (1994). A novel GABA receptor on bipolar cell terminals in the tiger salamander retina. *J. Neurosci.* **14**, 1202–1212.

Lungarella, M. and Pfeifer, R. (2001). Robots as cognitive tools: information-theoretic analysis of sensory-motor data. *Proceedings of the 2001 IEEE-RAS International Conference on Humanoid Robots*, pp. 245–252.

Lüscher, C., Lipp, P., Lüscher, H.-R., and Niggli, E. (1996). Control of action potential propagation by intracellular Ca^{2+} in cultured rat dorsal root ganglion cells. *J. Physiol. (Lond.)* **490**, 319–324.

Lüscher, H.-R. and Shiner, J.S. (1990). Computation of action potential propagation and presynaptic bouton activation in terminal arborizations of different geometries. *Biophys. J.* **58**, 1377–1388.

Lux, H.D., Schubert, P., and Kreutzberg, G.W. (1970). Direct matching of morphological and electrophysiological data in cat spinal motorneurons. In P. Anderson and J.K.S. Janse, eds., *Excitatory Synaptic Mechanisms*. Universitetsforlaget, Oslo, pp. 189–198.

Lynch, G., Larson, J., Kelso, S., Barrioneuvo, G., and Schottler, F. (1983). Intracellular injections of EGTA block induction of hippocampal long-term potentiation. *Nature* **305**, 719–721.

Lytton, J., Westlin, M., Burk, S., Shull, G., and MacLennan, D. (1992). Functional comparisons between isoforms of the sarcoplasmic or endoplasmic reticulum family of calcium pumps. *J. Biol. Chem.* **267**, 14483–14489.

Ma, M. and Koester, J. (1996). The role of K^+ currents in frequency-dependent spike broadening in Aplysia R20 neurons: a dynamic-clamp analysis. *J. Neurosci.* **16**, 4089–4101.

Macefield, G., Hagbarth, K.-E., Gorman, R., Gandevia, S.C., and Burke, D. (1991). Decline in spindle support to α-motoneurons during sustained voluntary contractions. *J. Physiol. (Lond.)* **440**, 497–512.

MacGregor, R.J. (1987). *Neural and Brain Modeling*. Academic, New York.

MacKenzie, P.J. and Murphy, T.H. (1998). High safety factor for action potential conduction along axons but not dendrites of cultured hippocampal and cortical cells. *J. Neurophysiol.* **80**, 2089–2101.

MacKenzie, P.J., Umemiya, M., and Murphy, T.H. (1996). Ca^{2+} imaging of CNS axons in culture indicates reliable coupling between single action potentials and distal functional release sites. *Neuron* **16**, 783–795.

Mackintosh, N.J. (1983). *Conditioning and Associative Learning*. Oxford University Press, Oxford.

MacVicar, B.A. (1985). Depolarizing prepotentials are Na^{+} dependent in CA1 pyramidal neurons. *Brain Res.* **333**, 378–381.

MacVicar, B.A. and Dudek, F.E. (1980a). Local synaptic circuits in rat hippocampus: interactions between pyramidal cells. *Brain Res.* **184**, 220–223.

MacVicar, B.A. and Dudek, F.E. (1980b). Dye-coupling between CA3 pyramidal cells in slices of rat hippocampus. *Brain Res.* **196**, 494–497.

MacVicar, B.A. and Dudek, F.E. (1981). Electrotonic coupling between pyramidal cells: a direct demonstration in rat hippocampal slices. *Science* **213**, 782–785.

MacVicar, B.A., Ropert, N., and Krnjevic, K. (1982). Dye-coupling between pyramidal cells of rat hippocampus in vivo. *Brain Res.* **238**, 239–244.

MacVicar, B.A., Baker, K., and Crichton, S.A. (1988). Kainic acid evokes a potassium efflux from astrocytes. *Neuroscience* **25**, 721–725.

Madison, D.V., Malenka, R.C., and Nicoll, R.A. (1991). Mechanisms underlying long-term potentiation of synaptic transmission. *Annu. Rev. Neurosci.* **14**, 379–397.

Maeda, H., Ellis-Davies, G.C.R., Ito, K., Miyashita, Y., and Kasai, H. (1999). Supralinear Ca^{2+} signaling by cooperative and mobile Ca^{2+} buffering in Purkinje neurons. *Neuron* **24**, 989–1002.

Magee, J.C. (1998). Dendritic hyperpolarization-activated currents modify the integrative properties of hippocampal CA1 pyramidal neurons *J. Neurosci.* **18**, 7613–7624.

Magee, J.C. (1999a). Voltage-gated ion channels. In G. Stuart, N. Spruston, and M. Hausser, eds., *Dendrites*. Oxford University Press, New York, pp. 139–160.

Magee, J.C. (1999b). Dendritic I_h normalises temporal summation in hippocampal CA1 neurons. *Nat. Neurosci.* **2**, 508–514.

Magee, J.C. and Cook, E.P. (2000). Somatic EPSP amplitude is independent of synapse location in hippocampal pyramidal neurons. *Nat. Neurosci.* **3**, 895–903.

Magee, J.C. and Johnston, D. (1995). Characterization of single voltage-gated Na^{+} and Ca^{2+} channels in apical dendrites of rat CA1 pyramidal neurons. *J. Physiol. (Lond.)* **487**, 67–90.

Magee, J.C. and Johnston, D. (1997). A synaptically controlled, associative signal for Hebbian plasticity in hippocampal neurons. *Science* **275**, 209–213.

Magee, J.C., Avery, R.B., Christie, B.R., and Johnston, D. (1996). Dihydropyridine-sensitive, voltage-gated Ca^{2+} channels contribute to the resting intracellular Ca^{2+} concentration of hippocampal CA1 pyramidal neurons. *J. Neurophysiol.* **76**, 3460–3470.

Magnus, G. and Keizer, J. (1998a). Model of beta-cell mitochondrial calcium handling and electrical activity. I. Cytoplasmic variables. *Am. J. Physiol.* **274**, C1158–C1173.

Magnus, G. and Keizer, J. (1998b). Model of beta-cell mitochondrial calcium handling and electrical activity. II. Mitochondrial variables. *Am. J. Physiol.* **274**, C1174–C1184.

Mahaffy, J.M., Jorgensen, D.A., and van der Heyden, R.L. (1992). Oscillations in a model of repression with external control. *J. Math. Biol.* **30**, 669–691.

Mainen, Z.F. and Sejnowski, T.J. (1995). Reliability of spike timing in neocortical neurons. *Science* **268**, 1503–1506.

Mainen, Z.F. and Sejnowski, T.J. (1996). Influence of dendritic structure on firing pattern in model neocortical neurons. *Nature* **382**, 363–366.

Mainen, Z.F., Joerges, I., Huguenard, J.R., and Sejnowski, T.J. (1995). A model of spike initiation in neocortical pyramidal neurons. *Neuron* **15**, 1427–1439.

Mainen, Z.F., Malinow, R., and Svoboda, K. (1999). Synaptic calcium transients in single spines indicate that NMDA receptors are not saturated. *Nature* **399**, 151–155.

Majewska, A., Brown, E., Ross, J., and Yuste, R. (2000a). Mechanisms of calcium decay kinetics in hippocampal spines: role of spine calcium pumps and calcium diffusion through the spine neck in biochemical compartmentalization. *J. Neurosci.* **20**, 1722–1734.

Majewska, A., Tashiro, A., and Yuste, R. (2000b). Regulation of spine calcium dynamics by rapid spine motility. *J. Neurosci.* **20**, 8262–8268.

Major, G. (1993). Solutions for transients in arbitrarily branching cables: III. Voltage clamp problems. *Biophys. J.* **65**, 469–491.

Major, G. and Evans, J.D. (1994). Solutions for transients in arbitrarily branching cables: IV. Nonuniform electrical parameters. *Biophys. J.* **66**, 615–634.

Major, G., Evans, J.D., and Jack, J.J. (1993a). Solutions for transients in arbitrarily branching cables: I. Voltage recording with a somatic shunt. *Biophys. J.* **65**, 423–449.

Major, G., Evans, J.D., and Jack, J.J. (1993b). Solutions for transients in arbitrarily branching cables: II. Voltage clamp theory. *Biophys. J.* **65**, 450–468.

Major, G., Larkman, A.U., Jonas, P., Sakmann, B., and Jack, J.J.B. (1994). Detailed passive cable models of whole-cell recorded CA3 pyramidal neurons in rat hippocampal slices. *J. Neurosci.* **14**, 4613–4638.

Mak, D., McBride, S., Raghuram, V., Yue, Y., Joseph, S., and Foskett, J. (2000). Single-channel properties in endoplasmic reticulum membrane of recombinant type 3 inositol trisphosphate receptor. *J. Gen. Physiol.* **115**, 241–256.

Makino, T. (2003). A discrete-event neural network simulator for general neuron models. *Neural Comput. Appl.* **11**, 210–223.

Malchow, R.P., Qian, H., Ripps, H., and Dowling, J.E. (1990). Structual and functional properties of two types of horizontal cell in the skate retina. *J. Gen. Physiol.* **95**, 177–198

Maltenfort, M.G. and Hamm, T.M. (2004). Estimation of the electrical parameters of spinal motoneurons using impedance measurements. *J. Neurophysiol.* **92**, 1433–1444.

Manor, Y., Gonczarowski, J., and Segev, I. (1991a). Propagation of action potentials along complex axonal trees. *Biophys. J.* **60**, 1411–1423.

Manor, Y., Koch, C., and Segev, I. (1991b). Effect of geometrical irregularities on propagation delay in axonal trees. *Biophys. J.* **60**, 1424–1437.

Mangan, S. and Alon, U. (2003). Structure and function of the feed-forward loop network motif. *Proc. Natl. Acad. Sci. USA* **100**, 11980–11985.

Mangan, S., Zaslaver, A., and Alon, U. (2003). The coherent feedforward loop serves as a sign-sensitive delay element in transcription networks. *J. Mol. Biol.* **334**, 197–204.

Manwani, A. and Koch, C. (2000). Detecting and estimating signals over noisy and unreliable synapses: information-theoretic analysis. *Neural Comput.* **13**, 1–33.

Marcus, R. (1974). Parabolic Ito equations. *T. Am. Math. Soc.* **198**, 177–190.

Marder, E. and Prinz, A.A. (2002). Modeling stability in neuron and network function: the role of activity in homeostasis. *BioEssays* **24**, 1145–1154.

Maren, S. and Fanselow, M.S. (1996). The amygdala and fear conditioning: has the nut been cracked? *Neuron* **16**, 237–240.

Maricq, A.V. and Korenbrot, J.I. (1988). Calcium and calcium-dependent chloride currents generate action potentials in solitaly cone photoreceptors. *Neuron* **1**, 503–515.

Maricq, A.V. and Korenbrot, J.I. (1990a). Inward rectification in the inner segment of single retinal cone photoreceptors. *J. Neurophysiol.* **64**, 1917–1928.

Maricq, A.V. and Korenbrot, J.I. (1990b). Potassium currents in the inner segment of single retinal cone photoreceptors. *J. Neurophysiol.* **64**, 1929–1940.

Markin, V.S. (1970). Electrical interaction of parallel non-myelinated nerve fibres. I. Change in excitability of the adjacent fibre. *Biophysics-USSR* **15**, 122–133.

Markin, V.S. (1973). Electrical interaction of parallel non-myelinated nerve fibres. III. Interaction in bundles. *Biophysics-USSR* **18**, 324–332.

Markin, V.S. and Pashtushenko, V.F. (1969). Spread of excitation in a model of inhomogeneous nerve fibre. I. Slight changes in dimensions of fibre. *Biophysics-USSR* **14**, 335–344.

Markram, H., Helm, P.J., and Sakmann, B. (1995). Dendritic calcium transients evoked by single back-propagating action potentials in rat neocortical pyramidal neurons. *J. Physiol. (Lond.)* **485**, 1–20.

Markram, H., Lubke, J., Frotscher, M., and Sakmann, B. (1997). Regulation of synaptic efficacy by coincidence of postsynaptic APs and EPSPs. *Science* **275**, 213–215.

Marom, S. and Abbott, L.F. (1994). Modeling state-dependent inactivation of membrane currents. *Biophys. J.* **67**, 575–520.

Marr, D. (1982). *Vision*. Freeman, San Francisco, CA.

Martin, K.A.C. (2002). Microcircuits in visual cortex. *Curr. Opin. Neurobiol.* **4**, 418–425.

Mascagni, M.V. (1989a). Numerical methods for neuronal modeling. In C. Koch and I. Segev, eds., *Methods in Neuronal Modeling: From Synapses to Networks*. MIT Press, Cambridge, MA.

Mascagni, M.V. (1989b). An initial-boundary value problem of physiological significance for equations of nerve conduction. *Commun. Pur. Appl. Math.* **42**, 213–227.

Mascagni, M.V. and Sherman, A.S. (1998). Numerical methods for neuronal modeling. In C. Koch and I. Segev, eds., *Methods in Neuronal Modeling: From Ions to Networks*, 2nd edition. MIT Press, Cambridge, MA, pp. 569–606.

Masland, R.H. (2001). The fundamental plan of the retina. *Nat. Neurosci.* **4**, 877–886.

Mason, A. and Larkman, A. (1990). Correlations between morphology and electrophysiology of pyramidal neurons in slices of rat visual cortex. II. Electrophysiology. *J. Neurosci.* **10**, 1415–1428.

Massey, S.C. (1990). Cell types using glutamate as a neurotransmitter in the vertebrate retina. *Prog. Retin. Res.* **9**, 399–425.

Matthews, G., Ayoub, G.S., and Heidelberger, R. (1994). Presynaptic inhibition by GABA is mediated via two distinct GABA receptors with novel pharmacology. *J. Neurosci.* **14**, 1079–1090.

Matus, A. (2000). Actin-based plasticity in dendritic spines. *Science* **290**, 754–758.

Mayer, M.L. and Westbrook, G.L. (1987). Permeation and block of N-methyl-D-aspartic acid receptor channels by divalent cations in mouse cultured central neurons. *J. Physiol. (Lond.)* **394**, 501–527.

Mayer, ML., Westbrook, G.E., and Guthrie, P.B. (1984). Voltage dependent block by Mg^{++} of NMDA receptors in spinal cord neurons. *Nature* **309**, 261–263.

McAdams, H. and Shapiro, L. (1995). Circuit simulation of gene networks. *Science* **269**, 650–656.

McAdams, H. and Shapiro, L. (2003). A bacterial cell-cycle regulatory network operating in time and space. *Science* **301**, 1874–1877.

McAllister, A.K. (2002). Conserved cues for axon and dendritic growth in the developing cortex. *Neuron* **33**, 2–4.

McAuley, J.H. and Marsden, C.D. (2000). Physiological and pathological tremors and rhythmic central motor control. *Brain* **123**, 1545–1567.

McBurney, R.N. and Neering, I.R. (1987). Neuronal calcium homeostasis. *Trends Neurosci.* **10**, 164–169.

McClelland, J.L., Rumelhart, D.E., and The PDP Research Group (1986). *Parallel Distributed Processing: Explorations in the Microstructure of Cognition, Volume 2: Psychological and Biological Models.* MIT Press, Cambridge, MA.

McGeer, P.L., Eccles, J.C., and McGeer, E.G. (1987). *Molecular Neurobiology of the Mammalian Brain*, 2nd edition. Plenum, New York.

McIntosh, A.R. (1999). Mapping cognition to the brain through neural interactions. *Memory* **7**, 523–548.

McIntosh, A.R. (2000). Towards a network theory of cognition. *Neural Netw.* **13**, 861–870.

McIntosh, A.R. and Gonzalez-Lima, F. (1994). Structural equation modeling and its application to network analysis in functional brain imaging. *Hum. Brain Mapp.* **2**, 2–22.

McIntosh, A.R., Rajah, M.N., and Lobaugh, N.J. (1999). Interactions of prefrontal cortex in relation to awareness in sensory learning. *Science* **284**, 1531–1533.

McIntosh A.R., Rajah M.N., and Lobaugh N.J. (2003). Functional connectivity of the medial temporal lobe relates to learning and awareness. *J. Neurosci.* **23**, 6520–6528.

McIntyre, C.C., Richardson, A.D., and Grill, W.M. (2002). Modeling the excitability of mammalian nerve fibers: influence of afterpotentials on the recovery cycle. *J. Neurophysiol.* **87**, 995–1006.

McNaughton, P.A. (1990). The light response of vertebrate photoreceptors. *Physiol. Rev.* **70**, 847–883.

Mel, B.W. (1992). NMDA-based pattern discrimination in a modelled cortical neuron. *Neural Comput.* **4**, 502–517.

Melamed-Book, N., Kachalsky, S., Kaiserman, I., and Rahamimoff, R. (1999). Neuronal calcium sparks and intracellular calcium noise. *Proc. Natl. Acad. Sci. USA* **96**, 15217–15221.

Mellon, D. and Kaars, C. (1974). Role of regional cellular geometry in conduction of excitation along a sensory neuron. *J. Neurophysiol.* **37**, 1228–1238.

Mendel, J. (1991). Tutorial on higher-order statistics (spectra) in signal processing and system theory: theoretical results and some applications. *P. IEEE* **79**, 278–305.

Mendell, L.M. (1966). Physiological properties of nonmyelinated fibre projection to the spinal cord. *Exp. Neurol.* **16**, 316–332.

Merzenich, M. and deCharms, C. (1996). Neural representations, experience, and change. In R. Llinás and P.S. Churchland, eds., *Mind Brain Continuum.* MIT Press, Cambridge, MA, pp. 61–81.

Mestl, T., Lemacy, C., and Glass, L. (1996). Chaos in high-dimensional neural and gene networks, *Physica D* **98**, 33–52.

Mesulam, M.M. (1998). From sensation to cognition. *Brain* **121**, 1013–1052.

Meunier, C. (1992). Two and three dimensional reductions of the Hodgkin–Huxley system: separation of time scales and bifurcation schemes. *Biol. Cybern.* **67**, 461–468.

Meunier, C. and Segev, I. (2002). Playing the Devil's advocate: is the Hodgkin–Huxley model useful? *Trends Neurosci.* **25**, 558–563.

Meyer, T., Hanson, P.I., Stryer, L, and Schulmann, H. (1992). Calmodulin trapping by calcium-calmodulin-dependent protein kinase. *Science* **256**, 1199–1202.

Michaels, G.S., Carr, D.B., Askenazi, M., Fuhrman, S., Wen, X., and Somogyi, R. (1998). Cluster analysis and data visualization of large-scale gene expression data. *Proc. Pac. Symp. Biocomput.* **3**, 42–53.

Michelson, S. and Schulman, H. (1994). CaM kinase: a model for its activation and dynamics. *J. Theor. Biol.* **171**, 281–290.

Mickus, T, Jung, H., and Spruston, N. (1999). Properties of slow, cumulative sodium channel inactivation in rat hippocampal CA1 pyramidal neurons. *Biophys. J.* **76**, 846–860.

Miftakhov, R.N. and Wingate, D.L. (1995). Mathematical modelling of the enteric nervous network: excitation propagation in a planar neural network. *Med. Eng. Phys.* **17**, 11–19.

Migliore, M. (1996). Modeling the attenuation and failure of action potentials in the dendrites of hippocampal neurons. *Biophys. J.* **71**, 2394–2403.

Migliore, M. and Lansky, P. (1999). Computational model of the effects of stochastic conditioning on the induction of long-term potentiation and depression. *Biol. Cybern.* **81**, 291–298.

Migliore, M. and Shepherd, G.M. (2002). Emerging rules for the distributions of active dendritic conductances. *Nat. Rev. Neurosci.* **3**, 362–370.

Migliore, M., Hoffman, D.A., Magee, J.C., and Johnston, D. (1999). Role of an A-type K^+ conductance in the back-propagation of action potentials in the dendrites of hippocampal pyramidal neurons. *J. Comput. Neurosci.* **7**, 5–15.

Milgram, S. (1967). The small-world problem. *Psychol. Today* **1**, 61–67.

Miller, J.P., Rall, W., and Rinzel, J. (1985). Synaptic amplification by active membrane in dendritic spines. *Brain Res.* **325**, 325–330.

Miller, L.D., Long, P.M., Wong, L., Mukherjee, S., McShane, L.M., and Liu, E.T. (2002). Optimal gene expression analysis by microarrays. *Cancer Cell* **2**, 353–361.

Miller, R.N. and Rinzel, J. (1981). The dependence of impulse propagation speed on firing frequency and dispersions for the Hodgkin–Huxley model. *Biophys. J.* **34**, 227–259.

Milo, R., Shen-Orr, S., Itzkovitz, S., Kashtan, N., Chklovskii, D., and Alon, U. (2002). Network motifs: simple building blocks of complex networks. *Science* **298**, 824–827.

Minsky, M.L. (1961). Steps toward artificial intelligence. *P. IRE* **49**, 8–30.

Miralles, F. and Solsona, C. (1996). Activity-dependent modulation of the presynaptic potassium current in the frog neuromuscular junction. *J. Physiol. (Lond.)* **495**, 717–732.

Miranker, W.L. (1981). *Numerical Methods for Stiff Equations*. Reidel, Boston, MA.

Miro, H. (2002). Comparison of algorithms for the simulation of action potentials with stochastic sodium channels. *Ann. Biomed. Eng.* **30**, 578–587.

Mishkin, M., Ungerleider, L.G., and Macko, K.A. (1983). Object vision and spatial vision: two cortical pathways. *Trends Neurosci.* **6**, 414–417.

Mitzdorf, U. (1985). Current source-density method and application in cat cerebral cortex: investigation of evoked potentials and EEG phenomena. *Physiol. Rev.* **65**, 37–100.

Montroll, E.W. and Helleman, R.H.G. (1976). On a nonlinear perturbation theory without secular terms. *AIP Conf. Proc.* **27**, 75–111.

Moore, G.P., Segundo, J.P., Perkel, D.H., and Levitan, H. (1970). Statistical signs of synaptic interaction in neurons. *Biophys. J.* **10**, 876–900.

Moore, G.P., Ding, L., and Bronte-Stewart H.M. (2000). Concurrent Parkinson tremors. *J. Physiol. (Lond.)* **529**, 273–281.

Moore, J.A. and Appenteng, K. (1991). The morphology and electrical geometry of rat jaw-elevator motoneurons. *J. Physiol. (Lond.)* **440**, 325–343.

Moore, J.W. (1987). Computer simulations as a component of a neuroscience research program. In M.A. Arbib and S.A. George, eds., *Computational Neuroscience*. Soc. Neuroscience, Washington, D.C., pp. 25–36.

Moore, J.W. and Adelman, W.J. (1961). Electronic measurements of the intracellular concentration and net flux of Na^+ in the squid axon. *J. Gen. Physiol.* **45**, 77–92.

Moore, J.W. and Ramon, F. (1974). On numerical integration of the Hodgkin–Huxley equations for a membrane action potential. *J. Theor. Biol.* **45**, 249–273.

Moore, J.W., Joyner, R.W., Brill, M.H., Waxman, S.D., and Najar-Joa, M. (1978). Simulations of conduction in uniform myelinated fibers. *Biophys. J.* **21**, 147–160.

Moradmand, K. and Goldfinger, M.D. (1995a). Computation of long-distance propagation of impulses elicited by Poisson process stimulation. *J. Neurophysiol.* **74**, 2415–2426.

Moradmand, K. and Goldfinger, M.D. (1995b). Poisson-process electrical stimulation: circuit and axonal responses. *J. Neurosci. Meth.* **63**, 113–120.

Moraru, I., Kaftan, E., Ehrlich, B., and Watras, J. (1999). Regulation of type 1 inositol 1,4,5-trisphosphate-gated Ca^{2+} channels by Ip and Ca^{2+}: simulation of single channel kinetics based on ligand binding and electrophysiological analysis. *J. Gen. Physiol.* **113**, 837–849.

Morisset, V. and Nagy, F. (1998). Nociceptive integration in the rat spinal cord: role of non-linear membrane properties of deep dorsal horn neurons. *Eur. J. Neurosci.* **10**, 3642–3652.

Morisset, V. and Nagy, F. (1999). Ionic basis for plateau potentials in deep dorsal horn neurons of the rat spinal cord. *J. Neurosci.* **19**, 7309–7316.

Morris, C. and Lecar, H. (1981). Voltage oscillations in the barnacle giant muscle fiber. *Biophys. J.* **35**, 193–213.

Morton, K. and Mayers, D. (1994). *Numerical Solution of Partial Differential Equations: An Introduction.* Cambridge University Press, Cambridge.

Mozhayeva, M.G., Yildirim, S., Liu, X., and Kavalali, E.T. (2002). Development of vesicle pools during maturation of hippocampal synapses. *J. Neurosci.* **22**, 654–665.

Mozrzymas, J.W. and Bartoszkiewicz, M. (1993). The discrete nature of biological membrane conductance, channel interaction through electrolyte layers and the cable equation. *J. Theor. Biol.* **162**, 371–380.

Mukherji, S. and Soderling, T.R. (1994). Regulation of Ca^{2+}/calmodulin-dependent protein kinase II by inter- and intrasubmit-catalyzed autophosphorylations. *J. Biol. Chem.* **269**, 13744–13747.

Muller, W. and Connor, J.A. (1991). Dendritic spines as individual neuronal compartments for synaptic Ca^{2+} responses. *Nature* **354**, 73–76.

Munk, M.H., Nowak, L.G., Nelson, J.I., and Bullier, J. (1995). Structural basis of cortical synchronization. II. Effects of cortical lesions. *J. Neurophysiol.* **74**, 2401–2414.

Munk, M.H., Linden, D.E., Muckli, L., Lanfermann, H., Zanella, F.E., Singer, W., and Goebel, R. (2002). Distributed cortical systems in visual short-term memory revealed by event-related functional magnetic resonance imaging. *Cereb. Cortex* **12**, 866–876.

Muratov, C.B. (2000). A quantitative approximation scheme for the traveling wave solutions in the Hodgkin–Huxley model. *Biophys. J.* **79**, 2893–2901.

Murray, J.D. (1989). *Mathematical Biology*. Springer-Verlag, Berlin.

Murre, J.M.J. and Sturdy, D.P.F. (1995). The connectivity of the brain: multi-level quantitative analysis. *Biol. Cybern.* **73**, 529–545.

Murthy, V.N., Sejnowski, T.J., and Stevens, C.F. (2000). Dynamics of dendritic calcium transients evoked by quantal release at excitatory hippocampal synapses. *Proc. Natl. Acad. Sci. USA* **97**, 901–906.

Nagumo, J.S., Arimoto, S., and Yoshizawa, S. (1962). An active pulse transmission line simulating nerve axon. *P. IRE.* **50**, 2061–2071.

Nagumo, J., Yoshizawa, S., and Arimoto, S. (1965). Bistable transmission lines. *IEEE T. Circuits Syst.* **12**, 400–412.

Nakamura, T., Lasser-Ross, N., Nakamura, K., and Ross, W.N. (2002). Spatial segregation and interaction of calcium signaling mechanisms in rat hippocampal CA1 pyramidal neurons. *J. Physiol. (Lond.)* **543**, 465–480.

Naraghi, M. and Neher, E. (1997). Linearized buffered Ca^{2+} diffusion in microdomains and its implications for calculation of [Ca] at the mouth of a Ca^{2+} channel. *J. Neurosci.* **17**, 6961–6973.

Nawy, S. and Copenhagen, D.R. (1990). Intracellular cesium separates two glutamate conductances in retinal bipolar cells of goldfish. *Vision Res.* **30**, 967–972.

Nawy, S. and Jahr, C.E. (1990). Suppression by glutamate of cGMP-activated conductance in retinal bipolar cells. *Nature* **346**, 269–271.

Nawy, S. and Jahr, C.E. (1991). cGMP-gated conductance in retinal bipolar cells is suppressed by the photoreceptor transmitter. *Neuron* **7**, 677–683.

Neher, E. (1986). Concentration profiles of intracellular Ca^{2+} in the presence of diffusible chelator. *Exp. Brain Res.* **14**, 80–96.

Neher, E. (1998). Vesicle pools and Ca^{2+} microdomains: new tools for understanding their roles in neurotransmitter release. *Neuron* **20**, 389–399.

Nenadic, Z., Ghosh, B.K., and Ulinski, P. (2003). Propagating waves in visual cortex: a large-scale model of turtle visual cortex. *J. Comput. Neurosci.* **14**, 161–184.

Newman, E.A. and Raymond, S.A. (1971). Activity dependent shifts in excitability of frog peripheral nerve axons. *Quart. Prog. Rep.-M.I.T.* **102**, 165–187.

Nicholson, C. (1973). Theoretical analysis of field potentials in anisotropic ensembles of neuronal elements. *IEEE T. Bio-Med. Eng.* **20**, 278–288.

Nicholson, C. and Freeman, J.A. (1975). Theory of current source-density analysis and determination of conductivity tensor for anuran cerebellum. *J. Neurophysiol.* **38**, 356–368.

Nicholson, C. and Llinás, R. (1971). Field potentials in the alligator cerebellum and theory of their relationship to Purkinje cell dendritic spikes. *J. Neurophysiol.* **34**, 509–531.

Nicolelis, M.A.L. and Ribeiro, S. (2002). Multielectrode recordings: the next steps. *Curr. Opin. Neurobiol.* **12**, 602–606.

Nimchinsky, E.A., Sabatini, B., and Svoboda, K. (2002). Structure and function of dendritic spines. *Annu. Rev. Physiol.* **64**, 313–353.

Nimchinsky, E.A., Yasuda, R., Oertner, T.G., and Svoboda, K. (2004). The number of glutamate receptors opened by synaptic stimulation in single hippocampal spines. *J. Neurosci.* **24**, 2054–2064.

Nitzan, R., Segev, I., and Yarom, Y. (1990). Voltage behavior along the irregular dendritic structure of morphologically and physiologically characterized vagal motoneurons in the guinea pig. *J. Neurophysiol.* **63**, 333–345.

Noldus, E. (1973). A perturbation method for the analysis of impulse propagation in a mathematical neuron model. *J. Theor. Biol.* **38**, 383–395.

Nowak, L., Bregestovski, P., Asher, P., Herbert, A., and Prochiantz, A. (1984). Magnesium gates glutamate activated channels in mouse central neurons. *Nature* **307**, 462–465.

Nowakowski, R.S., Hayes, N.L., and Egger, M.D. (1992). Competitive interactions during dendritic growth: a simple stochastic growth algorithm. *Brain Res.* **576**, 152–156.

Nunez, P.L. (1981). *Electrical Fields of the Brain. The Neurophysics of EEG.* Oxford University Press, New York.

Oakley, J., Schwindt, P., and Crill, W. (2001). Initiation and propagation of regenerative Ca-dependent potentials in dendrites of layer 5 pyramidal neurons. *J. Neurophysiol.* **86**, 503–513.

Obaid, A.L. and Salzberg, B.M. (1996). Micromolar 4-aminopyridine enhances invasion of a vertebrate neurosecretory terminal arborization. *J. Gen. Physiol.* **107**, 353–368.

O'Dell, T.J. and Christensen B.N. (1989). Horizontal cells isolated from catfish retina contain two types of excitatory amino acid receptors. *J. Neurophysiol.* **61**, 1097–1109.

Oertner T.G. (2002). Functional imaging of single synapses in brain slices. *Exp. Physiol.* **87**, 733–736.

Oertner T.G., Sabatine B.L., Nimchinsky E.A., and Svoboda K. (2002). Facilitation at single synapses probed with optical quantal analysis. *Nat. Neurosci.* **5**, 657–664.

Ohme, M. and Schierwagen, A. (1998). An equivalent cable model for neuronal trees with active membrane. *Biol. Cybern.* **78**, 227–243.

Okada, T., Horiguchi, H., and Tachibana, M. (1995). Ca^{2+}-dependent Cl^- current at the presynaptic terminals of goldfish retinal bipolar cells. *Neurosci. Res.* **23**, 297–303.

Okamoto, H. and Ichikawa, K. (2000a). Switching characteristics of a model for biochemical-reaction networks describing autophosphorylation versus dephosphorylation of Ca^{2+}/calmodulin-dependent protein kinase II. *Biol. Cybern.* **82**, 35–47.

Okamoto, H. and Ichikawa, K. (2000b). A model for molecular mechanisms of synaptic competition for a finite resource. *BioSystems* **55**, 65–71.

Olave, M.J., Puri, N., Kerr, R., and Maxwell, D.J. (2002). Myelinated and unmyelinated primary afferent axons form contacts with cholinergic interneurons in the spinal dorsal horn. *Exp. Brain Res.* **145**, 448–456.

Olshausen, B.A., Anderson, C.H., and VanEssen, D.C. (1993). A neurobiological model of visual attention and invariant pattern recognition based on dynamic routing of information. *J. Neurosci.* **13**, 4700–4719.

Onsager, L. (1931). Reciprocal relations in irreversible processes. Part 1. *Phys. Rev.* **37**, 405–426.

Orkand, R.K., Nicholls, J.G., and Kuffler, S.W. (1966). The effect of nerve impulses on the membrane potential of glial cells in thee central nervous system of amphibia. *J. Neurophysiol.* **29**, 788–806.

Orpwood, R.D. (1988). Basic module for an adaptive control system based on neuron information processing. *J. Biomed. Eng.* **10**, 201–205.

Orpwood, R.D. (1990). Mechanisms of association learning in the post-synaptic neuron. *J. Theor. Biol.* **143**, 145–162.

Orpwood, R.D. (1992). Computer simulation of neuron pattern processing. *J. Biomed. Eng.* **14**, 222–228.

Orpwood, R.D. (1994). A possible neural mechanism underlying consciousness based on the pattern processing capabilities of pyramidal neurons in the cerebral cortex. *J. Theor. Biol.* **169**, 403–418.

Orpwood, R.D. (1996). A demonstration of core features of perception using networks of cortical pyramidal cell models. *Brain Res. Abstr.* **13**, 78.

Overdijk, J., Uylings, H.B.M., Kuypers, K., and Kamstra, A.W. (1978). An economical, semi-automatic system for measuring cellular tree structures in three dimensions, with special emphasis on Golgi-impregnated neurons. *J. Microsci.* **114**, 271–284.

Ozbudak, E.M., Thattai, M., Kurtser, I., Grossman, A.D., and van Oudenaarden, A. (2002). Regulation of noise in the expression of a single gene. *Nat. Genet.* **31**, 69–73.

Pakdaman, K. (2002). White-noise stimulation of the Hodgkin–Huxley model. *Biol. Cybern.* **86**, 403–417.

Pan, E. and Colbert, C.M. (2001). Subthreshold inactivation of Na^{+} and K^{+} channels supports activity-dependent enhancement of backpropagating action potentials in hippocampal CA1. *J. Neurophysiol.* **85**, 1013–1016.

Pan, Z.H. and Hu, H.J. (2000). Voltage-dependent Na^{+} currents in mammalian retinal cone bipolar cells. *J. Neurophysiol.* **84**, 2564–2571.

Paninski, L. (2003). Estimation of entropy and mutual information. *Neural Comput.* **15**, 1191–1254.

Panzeri, S., Petersen, R.S., Schultz, S.R., Lebedev, M., and Diamond, M.E. (2001). The role of spike timing in the coding of stimulus location in rat somatosensory cortex. *Neuron* **29**, 769–777.

Pape, H.C. (1996). Queer current and pacemaker: the hyperpolarization-activated cation current in neurons. *Annu. Rev. Physiol.* **58**, 299–327.

Pape, P., Jong, D., and Chandler, W. (1995). Calcium release and its voltage dependence in frog cut muscle fibers equilibrated with 20 mM EGTA. *J. Gen. Physiol.* **106**, 259–336.

Papoulis, A. (1991). *Probability, Random Variables, and Stochastic Processes*. McGraw-Hill, New York.

Parker, I. and Ivorra, I. (1990). Inhibition by Ca^{2+} of inositol trisphosphate-mediated Ca^{2+} liberation: a possible mechanism for oscillatory release of Ca^{2+}. *Proc. Natl. Acad. Sci. USA* **87**, 260–264.

Parker, I., Choi, J., and Yao, Y. (1996). Elementary events of Ip-induced Ca^{2+} liberation in Xenopus oocytes: hot spots, puffs and blips. *Cell Calcium* **20**, 105–121.

Parker, T.S. and Chua, L.O. (1989). *Practical Numerical Algorithms for Chaotic Systems*. Springer-Verlag, Berlin.

Parnas, I. and Segev, I. (1979). A mathematical model for conduction of action potentials along bifurcating axons. *J. Physiol. (Lond.)* **295**, 323–343.

Parnas, I., Hochstein, S., and Parnas, H. (1976). Theoretical analysis of parameters leading to frequency modulation along an inhomogeneous axon. *J. Neurophysiol.* **39**, 909–923.

Parodi, O., Combe, P., and Ducom, J.-C. (1996). Temporal coding in vision: coding by the spike arrival times leads to oscillations in the case of moving targets. *Biol. Cybern.* **74**, 497–509.

Parri, H.R. and Lansman, J.B. (1996). Multiple components of Ca^{2+} channel facilitation in cerebellar granule cells: expression of facilitation during development in culture. *J. Neurosci.* **16**, 4890–4902.

Pascual-Leone A., Walsh V., and Rothwell J. (2000). Transcranial magnetic stimulation in cognitive neuroscience–virtual lesion, chronometry, and functional connectivity. *Curr. Opin. Neurobiol.* **10**, 232–237.

Passingham R.E., Stephan K.E., and Kotter R. (2002). The anatomical basis of functional localization in the cortex. *Nat. Rev. Neurosci.* **3**, 606–616.

Patlak, C.S. (1955). Potential and current distribution in nerve: the effect of the nerve sheath, the number of fibers, and the frequency of alternating current stimulation. *B. Math. Biophys.* **17**, 287–307.

Patlak, J. (1991). Molecular kinetics of voltage-dependent Na^{+} channels. *Physiol. Rev.* **71**, 1047–1080.

Paus T. (1999). Imaging the brain before, during, and after transcranial magnetic stimulation. *Neuropsychologia* **37**, 219–224.

Pauwelussen, J. (1982). One way traffic of pulses in a neuron. *J. Math. Biol.* **15**, 151–171.

Pavlidis, P. and Madison, D.V. (1999). Synaptic transmission in pair recordings from CA3 pyramidal cells in organotypic culture. *J. Neurophysiol.* **81**, 2787–2797.

Penney, M.S. and Britton, N.F. (2002). Modeling natural burst firing in Nigral dopamine neurons. *J. Theor. Biol.* **219**, 207–223.

Penrose, R. (1989). *The Emperor's New Mind: Concerning Computers, Minds, and the Laws of Physics.* Oxford University Press, New York.

Perkel, D.H. and Bullock, T.H. (1968). Neural coding. *Neurosci. Res. Prog.* B. **6**, 221–348.

Perkel, D.H. and Perkel, D.J. (1985). Dendritic spines: role of active membrane in modulating synaptic efficacy. *Brain Res.* **325**, 331–335.

Perkel, D.H., Gerstein, G.L., and Moore, G.P. (1967a). Neuronal spike trains and stochastic point processes. I. The single spike train. *Biophys. J.* **7**, 391–418.

Perkel, D.H., Gerstein, G.L., and Moore, G.P. (1967b). Neuronal spike trains and stochastic point processes. II. Simultaneous spike trains. *Biophys. J.* **7**, 419–440.

Perrier, J.F., Mejia-Gervacio, S., and Hounsgaard, J. (2000). Facilitation of plateau potentials in turtle motoneurons by a pathway depending on calcium and calmodulin. *J. Physiol. (Lond.)* **528**, 107–113.

Peters, A. and Kaiserman-Abramof, I.R. (1970). The small pyramidal neuron of the rat cerebral cortex. The perikaryon, dendrites and spines. *Am. J. Anat.* **127**, 321–356.

Petersen, R.S., Panzeri, S., and Diamond, M.E. (2002). Population coding in somatosensory cortex. *Curr. Opin. Neurobiol.* **12**, 441–447.

Pethig, R., Kuhn, M., Payne, R., Adler, E., Chen, T., and Jaffe, L. (1989). On the dissociation constants of bapta-type calcium buffers. *Cell Calcium* **10**, 491–498.

Petrozzino, J.J., Pozzo Miller, L.D., and Connor, J.A. (1995). Micromolar Ca^{2+} transients in dendritic spines of hippocampal pyramidal neurons in brain slice. *Neuron* **14**, 1223–1231.

Petsche, M., Promaska, O., Rappelsberger, P., and Vollmer, R. (1975). The possible role of dendrites in EEG synchronization. *Adv. Neurol.* **12**, 53–70.

Pfeifer, R. and Scheier, C. (1999). *Understanding Intelligence*. MIT Press, Cambridge, MA.

Pfeuty, B., Mato, G., Golomb, D., and Hansel, D. (2003). Electrical synapses and synchrony: the role of intrinsic currents. *J. Neurosci.* **23**, 6280–6294.

Philipson, K.D. (1999). Sodium–calcium exchange. In E. Carafoli and C. Klee, eds., *Calcium as a Cellular Regulator*. Oxford University Press, New York, pp. 279–294.

Pichon, Y., Abbott, N.J., Lieberman, E.M., and Larmet, Y. (1987). Potassium homeostasis in the nervous system of cephalopods and crustaceans. *J. Physiol. (Paris)* **82**, 346–356.

Pickard, W.F. (1968). A contribution to the electromagnetic theory of the unmyelinated axon. *Math. Biosci.* **2**, 111–121.

Pickard, W.F. (1969). The electromagnetic theory of electrotonus along an unmyelinated axon. *Math. Biosci.* **5**, 471–494.

Pickard, W.F. (1971). Electrotonus of a cell of finite dimensions. *Math. Biosci.* **10**, 201–213.

Pickard, W.F. (1974). Electrotonus on a nonlinear dendrite. *Math. Biosci.* **20**, 75–84.

Pietroban, D. (2002). Calcium channels and channelopathies of the central nervous system. *Mol. Neurobiol.* **25**, 31–50.

Pinker, S. (1994). *The Language Instinct*. Morrow, New York.

Plonsey, R. (1964). Volume conductor fields of action currents. *Biophys. J.* **4**, 317–328.

Pockberger, H. (1991). Electrophysiological and morphological properties of rat motor cortex neurons in vivo. *Brain Res.* **539**, 181–190.

Poggio, G.F. and Viernstein, L.J. (1964). Time series analysis of impulse sequences of thalamic somatic sensory neurons. *J. Neurophys.* **27**, 517–545.

Pogorzelski, W. (1966). *Integral Equations and their Applications*, Volume 1. Pergamon, Oxford.

Poirazi, P., Brannon, T., and Mel, B.W. (2003). Pyramidal neuron as two-layer neural network. *Neuron* **37**, 989–999.

Polsky, A., Mel, B.W., and Schiller, J. (2004). Computational subunits in thin dendrites of pyramidal cells. *Nat. Neurosci.* **7**, 621–627.

Pongracz, F., McClintock, T.S., Aches, B.W., and Shepherd, G.M. (1993). Signal transmission in lobster olfactory receptor cells: functional significance of electrotonic structure analysed by a compartmental model. *Neuroscience* **55**, 325–338.

Poolos, N.P. and Johnston, D. (1999). Calcium-activated potassium conductances contribute to action potential repolarization at the soma but not the dendrites of hippocampal CA1 pyramidal neurons. *J. Neurosci.* **19**, 5205–5212.

Poznanski, R.R. (1987). Transient response in a somatic shunt cable model for synaptic input activated at the terminal. *J. Theor. Biol.* **127**, 31–50.

Poznanski, R.R. (1988). Membrane voltage changes in passive dendritic trees: a tapering equivalent cylinder model. *IMA J. Math. Appl. Med. Biol.* **5**, 113–145.

Poznanski, R.R. (1990). Analysis of a postsynaptic scheme based on a tapering equivalent cable model. *IMA J. Math. Appl. Med. Biol.* **7**, 175–197.

Poznanski, R.R. (1991). A generalized tapering equivalent cylinder model for dendritic neurons. *B. Math. Biol.* **53**, 457–456.

Poznanski, R.R. (1992). Modelling the electrotonic structure of starburst amacrine cells in the rabbit retina: a functional interpretation of dendritic morphology. *B. Math. Biol.* **54**, 905–928.

Poznanski, R.R. (1994a). Electrotonic length estimates of CA3 hippocampal pyramidal neurons. *Neurosci. Res. Commun.* **14**, 93–100.

Poznanski, R.R. (1994b). Estimating the effective electrotonic length of dendritic neurons with reduced equivalent cable models. *Neurosci. Res. Commun.* **15**, 69–76.

Poznanski, R.R. (1996). Transient response in a tapering cable model with somatic shunt. *NeuroReport* **7**, 1700–1704.

Poznanski, R.R. (1998). Electrophysiology of a leaky cable model for coupled neurons. *J. Aust. Maths Soc. B* **40**, 59–71.

Poznanski, R.R. (2001a). On recent cable models in neurophysiology. *Math. Scientist* **26**, 74–86.

Poznanski, R.R., ed. (2001b). *Biophysical Neural Networks: Foundations of Integrative Neuroscience*. Liebert, New York.

Poznanski, R.R. (2001c). Introduction to integrative neuroscience. In R.R. Poznanski, ed., *Biophysical Neural Networks: Foundations of Integrative Neuroscience*. Liebert, Larchmont, NY, pp. 1–21.

Poznanski, R.R. (2001d). Conduction velocity of dendritic potentials in a cultured hippocampal neuron model. *Neurosci. Res. Commun.* **28**, 141–150.

Poznanski, R.R. (2002a). Dendritic integration in a recurrent network. *J. Integr. Neurosci.* **1**, 69–99.

Poznanski, R.R. (2002b). Towards an integrative theory of cognition. *J. Integr. Neurosci.* **1**, 145–156.

Poznanski, R.R. (2004). Analytical solutions of the Frankenhaeuser-Huxley equations. I. Minimal model for backpropagation of action potentials in sparsely excitable dendrites. *J. Integr. Neurosci.* **3**, 267–299.

Poznanski, R.R. and Bell, J. (2000a). A dendritic cable model for the amplification of synaptic potentials by an ensemble average of persistent sodium channels. *Math. Biosci.* **166**, 101–121.

Poznanski, R.R. and Bell, J. (2000b). Theoretical analysis of the amplification of synaptic potentials by small clusters of persistent sodium channels in dendrites. *Math. Biosci.* **166**, 123–147.

Poznanski, R.R. and Glenn, L.L. (1994). Estimating the effective electrotonic length of dendritic neurons with reduced equivalent cable models. *Neurosci. Res. Commun.* **15**, 69–76.

Poznanski, R.R. and Umino, O. (1997). Syncytial integration by a network of coupled bipolar cells in the retina. *Prog. Neurobiol.* **53**, 273–291.

Poznanski, R.R., Gibson, W.G., and Bennett, M.R. (1995). Electrotonic coupling between two CA3 hippocampal pyramidal neurons: a distributed cable model with somatic gap-junction. *B Math. Biol.* **57**, 865–881.

Prather, J.F., Powers, R.K., and Cope, T.C. (2001). Amplification and linear summation of synaptic effects on motoneuron firing rate. *J. Neurophysiol.* **85**, 43–53.

Pratusevich, V. and Balke, C. (1996). Factors shaping the confocal image of the Ca^{2+} spark in cardiac muscle cells. *Biophys. J.* **71**, 2942–2957.

Press, W.H., Teukolsky, S.A., Vetterling, W.T., and Flannery, B.P. (1992). *Numerical Recipes in Fortran: The Art of Scientific Computing*, 2nd edition. Cambridge University Press, New York.

Prigg, T., Goldreich, D., Carvell, G.E., and Simons, D.J. (2002). Texture discrimination and unit recordings in the rat whisker/barrel system. *Physiol. Behav.* **77**, 671–675.

Protti, D.A., Flores-Herr, N., and von Gersdorff, H. (2000). Light evokes Ca^{2+} spikes in the axon terminal of a retinal bipolar cell. *Neuron* **25**, 215–227.

Publicover, N.G. (1989). Mathematical models of intercellular communication. In N. Sperelakis and W.C. Cole, eds., *Cell Interactions and Gap Junctions*, Volume 1. CRC Press, Boca Raton, FL, pp. 183–201.

Pugh, E.N., Jr. and Lamb, T.D. (2000). Phototransduction in vertebrate rods and cones: molecular mechanisms of amplification, recovery and light adaptation. In D.G. Stavenga, W.J. de Grip, and E.N. Pugh, Jr., eds., *Handbook of Biological Physics*, Volume 3. Elsevier, Amsterdam, pp. 183–254.

Pumphrey, R.J. and Young, J.Z. (1938). The rates of conduction of nerve fibers of various diameters in cephalopods. *J. Exp. Biol.* **15**, 453–467.

Qian, N. and Sejnowski, T.J. (1989). An electro-diffusion model for computing membrane potentials and ionic concentrations in branching dendrites, spines, and axons. *Biol. Cybern.* **62**, 1–15.

Raastad, M. and Shepherd, G.M.G. (2003). Single-axon action potentials in the rat hippocampal cortex. *J. Physiol. (Lond.)* **548**, 745–752.

Raastad, M., Enriquez-Denton, M., and Kiehn, O. (1998). Synaptic signalling in an active central network only moderately changes passive membrane properties. *Proc. Natl. Acad. Sci. USA* **95**, 10251–10256.

Raethjen, J., Lindemann, M., Schmaljohan, H., Wenzelburger, R., Pfister, G., and Deuschl, G. (2000). Multiple oscillators are causing Parkinsonian and essential tremor. *Movement Disord.* **15**, 84–94.

Rakowski, R.F., Gadsby, D.C., and DeWeer, P. (2002). Single ion occupancy and steady-state gating of Na channels in squid giant axon. *J. Gen. Physiol.* **119**, 235–249.

Rall, W. (1959). Branching dendritic trees and motoneuron membrane resistivity. *Exp. Neurol.* **1**, 491–527.

Rall, W. (1960). Membrane potential transients and membrane time constants of motoneurons. *Exp. Neurol.* **2**, 503–532.

Rall, W. (1962a). Electrophysiology of a dendritic neuron model. *Biophys. J. (Suppl.)* **2**, 145–167.

Rall, W. (1962b). Theory of physiological properties of dendrites. *Ann. N.Y. Acad. Sci.* **96**, 1071–1092.

Rall, W. (1964). Theoretical significance of dendritic trees for neuronal input–output relations. In R.F. Reiss, ed., *Neural Theory and Modelling*. Stanford University Press, Stanford, CA, pp. 73–97.

Rall, W. (1967). Distinguishing theoretical synaptic potentials computed for different soma–dendritic distributions of synaptic inputs. *J. Neurophysiol.* **30**, 1138–1168.

Rall, W. (1969a). Time constants and electrotonic length of membrane cylinders and neurons. *Biophys. J.* **9**, 1483–1508.

Rall, W. (1969b). Distributions of potential in cylindrical coordinates for a membrane cylinder. *Biophys. J.* **9**, 1509–1541.

Rall, W. (1974). Dendritic spines, synaptic potency and neuronal plasticity. In C.D. Woody, K.A. Brown, T.J. Crow, and J.D. Knispel, eds., *Cellular Mechanisms Subserving Changes in Neuronal Activity*. Brain Information Services, Los Angeles, CA, pp. 13–21.

Rall, W. (1977). Core conductor theory and cable properties of neurons. In E.R. Kandel, ed., *Handbook of Physiology. Section 1: The Nervous System*, Volume 1. Am. Physiol. Soc., Bethesda, MD, pp. 39–97.

Rall, W. (1978). Dendritic spines and synaptic potency. In R. Porter, ed., *Studies in Neurophysiology.* Cambridge University Press, New York, pp. 203–209.

Rall, W. (1981). Functional aspects of neuronal geometry. In A. Roberts and B.M. Bush, eds., *Neurons Without Impulses*. Cambridge University Press, Cambridge.

Rall, W. (1989). Cable theory for dendritic neurons. In C. Koch and I. Segev, eds., *Methods in Neuronal Modeling: From Synapses to Networks*. MIT Press, Cambridge, MA, pp. 9–62.

Rall, W. (1990). Perspectives on neuron modeling. In M.D. Binder and L.M. Mendell, eds., *The Segmental Motor System*. Oxford University Press, Oxford, pp. 129–149.

Rall, W. (1995a). Perspective on neuron model complexity. In M.A. Arbib, ed., *The Handbook of Brain Theory and Neural Networks*. MIT Press, Cambridge, MA, pp. 728–732.

Rall, W. (1995b). The theoretical foundation of dendritic function: selected papers of Wilfrid Rall with commentaries. I. Segev, J. Rinzel, and G.M. Shepherd, eds. MIT Press, Cambridge, MA.

Rall, W. and Rinzel, J. (1971a). Dendritic spines and synaptic potency explored theoretically. *Proc. Int. Union Physiol. Sci. (XXV Int. Congr.) IX*, 466.

Rall, W. and Rinzel, J. (1971b). Dendritic spine function and synaptic attenuation calculations. *Prog. Abstr. Soc. Neurosci.* **1**, 64.

Rall, W. and Rinzel, J. (1973). Branch input resistance and steady attenuation for input to one branch of a dendritic neuron model. *Biophys. J.* **13**, 648–688.

Rall, W. and Shepherd, G.M. (1968). Theoretical reconstruction of field potentials and dendrodendritic synaptic interactions in olfactory bulb. *J. Neurophysiol.* **31**, 884–915.

Rall, W., Burke, R.E., Holmes, W.R., Jack, J.J.B., Redman, S.J., and Segev, I. (1992). Matching dendritic neuron models to experimental data. *Physiol. Rev.* **72**, S159–S186.

Ramakers, G.J.A. (2002). Rho proteins, mental retardation and the cellular basis of cognition. *Trends Neurosci.* **25**, 191–199.

Ramakers, G.J.A., Winter, J., Hoogland, T.M., Lequin, M.B., Van Hulten, P., Van Pelt, J., and Pool, C.W. (1998). Depolarization stimulates lamellipodia formation and axonal but not dendritic branching in cultured rat cerebral cortex neurons. *Dev. Brain Res.* **108**, 205–216.

Ramakers, G.J.A., Avci, B., Van Hulten, P., Van Ooyen, A., Van Pelt, J., Pool, C.W., and Lequin, M.B. (2001). The role of calcium signaling in early axonal and dendritic morphogenesis of rat cerebral cortex neurons under non-stimulated growth conditions. *Dev. Brain Res.* **126**, 163–172.

Ramon, F., Joyner, R.W., and Moore, J.W. (1975). Propagation of action potentials in inhomogeneous axon regions. *Fed. Proc.* **34**, 1357–1363.

Ramón y Cajal, S. (1952). *Histologie du Systèm Nerveux de l'Homme et des Vertébrés.* CSIC, Madrid.

Ramos-Franco, J., Fill, M., and Mignery, G. (1998). Isoform-specific function of single inositol 1,4,5-trisphosphate receptor channels. *Biophys. J.* **75**, 834–839.

Randall, J.E. (1973). A stochastic time series model for hand tremor. *J. Appl. Physiol.* **34**, 390–395.

Ransom, C.B., Ransom, B.R., and Sontheimer, H. (2000). Activity-dependent extracellular K+ accumulation in rat optic nerve: the role of glial and axonal Na^+ pumps. *J. Physiol. (Lond.)* **522**, 427–442.

Rapp, M., Yarom, Y., and Segev, I. (1992). The impact of parallel fiber background activity on the cable properties of cerebeller purkinje cells. *Neural Comput.* **4**, 518–533.

Rapp, M., Segev, I., and Yarom, Y. (1994). Physiology, morphology and detailed passive models of guinea-pig cerebellar Purkinje cells. *J. Physiol. (Lond.)* **474**, 101–118.

Rapp, M., Yarom, V., and Segev, I. (1996). Modeling back propagating action potentials in weakly excitable dendrites of neocortical pyramidal cells. *Proc. Natl. Acad. Sci. USA* **93**, 11985–11990.

Rauch, A., La Camera, G., Luescher, H.-R., Senn, W., and Fusi, S. (2003). Neocortical pyramidal cells respond as integrate-and-fire neurons to *in vivo-like* input currents. *J. Neurophysiol.* **90**, 1598–1612.

Ravasz, E. and Barabási, A.-L. (2003). Hierarchical organization in complex networks. *Phys. Rev. E* **67**, 26–112.

Raychaudhuri, S., Stuart, J.M., and Altman, R.B. (2000). Principal components analysis to summarize microarray experiments: application to sporulation time series. *Proc. Pac. Symp. Biocomput.* **5**, 452–463.

Raymond, S.A. and Lettvin, J.Y. (1978). After-effects of activity in peripheral axons as a clue to nervous coding. In S.G. Waxman, ed., *Physiology and Pathology of Axons.* Raven, New York.

Redman, S.J. (1973). The attenuation of passively propagating dendritic potentials in a motoneuron cable model. *J. Physiol. (Lond.)* **234**, 637–664.

Redman, S.J. (1990). Quantal analysis of synaptic potentials in neurons of the central nervous system. *Physiol. Rev.* **70**, 165–198.

Reeke, G.N., Jr. (2001). Replication in selective systems: multiplicity of carriers, variation of information, iteration of encounters. *Behav. Brain Sci.* **24**, 552–553.

Reeke, G.N., Jr. and Edelman, G.M. (1987). Selective neural networks and their implications for recognition automata. *Int. J. Supercomput. Ap.* **1**, 44–69.

Reeke, G.N., Jr. and Edelman, G.M. (1988). Real brains and artificial intelligence. *Daedalus* **117**, 143–173.

Reeke, G.N., Jr. and Sporns, O. (1993). Behaviorally based modeling and computational approaches to neuroscience. *Annu. Rev. Neurosci.* **16**, 597–623.

Reeke, G.N., Jr., Finkel, L.H., Sporns, O., and Edelman, G.M. (1990a). Synthetic neural modeling: a multilevel approach to the analysis of brain complexity. In G.M. Edelman, W.E. Gall, and W.M. Cowan, eds., *Signal and Sense: Local and Global Order in Perceptual Maps.* Wiley, New York, pp. 607–706.

Reeke, G.N., Jr., Sporns, O., and Edelman, G.M. (1990b). Synthetic neural modeling: the 'Darwin' series of automata. *P. IEEE* **78**, 1498–1530.

Regehr, W.G. and Tank, D.W. (1990). Postsynaptic NMDA receptor-mediated calcium accumulation in hippocampal CAl pyramidal cell dendrites. *Nature* **345**, 807–810.

Regehr, W.G., Kehoe, I., Ascher, P., and Armstrong, C. (1993). Synaptically triggered action potentials in dendrites. *Neuron* **11**, 145–151.

Reichenbach, A. (1991). Glial K^+ permeability and CNS K^+ clearance by diffusion and spatial buffering. *Ann. N.Y. Acad. Sci.* **633**, 272–286.

Renart, A., Brunel, N., and Wang, X.-J. (2004). Mean-field theory of irregularly spiking neuronal populations and working memory in recurrent cortical networks. In J. Feng, ed., *Computational Neuroscience: A Comprehensive Approach.* Chapman & Hall/CRC Press, Boca Raton, FL.

Reutskiy, S., Rossoni, E., and Tirozzi, B. (2003). Conduction in bundles of demyelinated nerve fibers: computer simulation. *Biol. Cybern.* **89**, 439–448.

Reyes, A. (2001). Influence of dendritic conductances on the input–output properties of neurons. *Annu. Rev. Neurosci.* **24**, 653–675.

Rhodes, P.A. (1999). Functional implications of active currents in the dendrites of pyramidal neurons. In P.S. Ulinski, E.G. Jones, and A. Peters, eds., *Cerebral Cortex*, Volume 13, *Models of Cortical Circuits*. Kluwer, New York, pp. 139–200.

Rhodes, P.A. and Llinás, R.R. (2001). Apical tuft efficacy in layer 5 pyramidal cells from the rat visual cortex. *J. Physiol. (Lond.)* **536**, 167–187.

Richardson, T.L., Turner, R.W., and Miller, J.J. (1984). Extracellular fields influence transmembrane potentials and synchronization of hippocampal neuronal activity. *Brain Res.* **294**, 255–262.

Rieke, F., Warland, D., de Ruyter van Steveninck, R., and Bialek, W. (1998). *Spikes: Exploring the Neural Code.* MIT Press, Cambridge, MA.

Rigas, A. (1983). *Point Processes and Time Series Analysis: Theory and Applications to Complex Physiological Systems*. Ph.D. Dissertation, University of Glasgow, UK.

Rinzel, J. (1985). Excitation dynamics: insights from simplified membrane models. *Fed. Proc.* **44**, 2944–2946.

Rinzel, J. (1990). Mechanisms for nonuniform propagation along excitable cables. *Ann. N.Y. Acad. Sci.* **591**, 51–61.

Rinzel, J. and Miller, R.N. (1980). Numerical calculations of stable and unstable periodic solutions to the Hodgkin–Huxley equations. *Math. Biosci.* **49**, 27–59.

Rinzel, J. and Rall, W. (1974). Transient response in a dendritic neuron model for current injected at one branch. *Biophys. J.* **14**, 759–790.

Rinzel, J., Terman, D., Wang, X.-J., and Ermentrout, B. (1998). Propagating activity patterns in large-scale inhibitory neuronal networks. *Science* **279**, 1351–1355.

Rissman, P. (1977). The leading edge approximation to the nerve axon problem. *B. Math. Biol.* **39**, 43–58.

Roberts, W. (1994). Localization of Ca^{2+} signals by a mobile Ca^{2+} buffer in frog saccular hair cells. *Nature* **14**, 3246–3262.

Roelfsema, P.R., Engel, A.K., König, P., and Singer, W. (1997). Visuomotor integration is associated with zero time-lag synchronization among cortical areas. *Nature* **385**, 157–161.

Rolls, E.T. (1989). Functions of neuronal networks in the hippocampus and neocortex in memory. In J.H. Byrne and W.O. Berry, eds., *Neural Models of Plasticity*. Academic, San Diego, CA, pp. 240–265.

Rolls, E.T. and Stringer, S.M. (2000). On the design of neural networks in the brain by genetic evolution. *Prog. Neurobiol.* **61**, 557–579.

Rolls, E.T. and Treves, A. (1998). *Neural Networks and Brain Function*. Oxford University Press, Oxford.

Rose, P.K. and Dagum, A. (1988). Nonequivalent cylinder models of neurons: interpretation of voltage transients generated by somatic current injection. *J. Neurophysiol.* **60**, 125–148.

Rosenberg, J.R., Murray-Smith, D.J., and Rigas, A. (1982). An introduction to the application of system identification techniques to elements of the neuromuscular system. *Trans. Inst. Meas. Control* **4**, 187–201.

Rosenberg, J.R., Amjad, A.M., Breeze, P., Brillinger, D.R., and Halliday, D.M. (1989). The Fourier approach to the identification of functional coupling between neuronal spike trains. *Prog. Biophys. Mol. Biol.* **53**, 1–31.

Rosenberg, J.R., Halliday, D.M., Breeze, P., and Conway, B.A. (1998). Identification of patterns of neuronal activity–partial spectra, partial coherence, and neuronal interactions. *J. Neurosci. Meth.* **83**, 57–72.

Rosenblatt, M. (1983). Cumulants and cumulant spectra. In D.R. Brillinger and P.R. Krishnaiah, eds., *Handbook of Statistics*, Volume 3. North Holland, New York, pp. 369–382.

Rosenfalck, P. (1969). Intra- and extracellular potential fields of active nerve and muscle fibres. *Acta. Physiol. Scand.* (Suppl.) **321**, 1–168.

Rozsa, B., Zelles, T., Sylvester-Vizi, E., and Lendvai, B. (2004). Distance-dependent scaling of calcium transients evoked by backpropagating spikes and synaptic activity in dendrites of hippocampal interneurons. *J. Neurosci.* **24**, 661–670.

Rumelhart, D.E., McClelland, J.L., and The PDP Research Group (1986). *Parallel Distributed Processing: Explorations in the Microstructure of Cognition, Volume 1: Foundations*. MIT Press, Cambridge, MA.

Russell, A.R. (1992). Using tactile whiskers to measure surface contours. In *Proc. IEEE Int. Conf. Robot. Automat.*, pp. 1295–1299.

Russo, R.E. and Hounsgaard, J. (1994). Short-term plasticity in turtle dorsal horn neurons mediated by L-type Ca^{2+} channels. *Neuroscience* **61**, 191–197.

Russo, R.E. and Hounsgaard, J. (1999). Dynamics of intrinsic electrophysiological properties in spinal cord neurons. *Prog. Biophys. Mol. Biol.* **72**, 329–365.

Sabah, N.H. and Leibovic, K.N. (1969). Subthreshold oscillatory responses of the Hodgkin–Huxley cable model for the squid giant axon. *Biophys. J.* **9**, 1206–1222.

Sabah, N.H. and Leibovic, K.N. (1972). The effects of membrane parameters on the properties of the nerve impulse. *Biophys. J.* **12**, 1132–1144.

Sabah, N.H. and Spangler, R.A. (1970). Repetitive response of the Hodgkin–Huxley model for the squid giant axon. *J. Theor. Biol.* **29**, 155–171.

Sabatini, B.L. and Svoboda, K. (2000). Analysis of calcium channels in single spines determined by optical fluctuation analysis. *Nature* **408**, 589–593.

Sabatini, B.L., Maravell, M., and Svoboda, K. (2001). Ca^{2+} signaling in dendritic spines. *Curr. Opin. Neurobiol.* **11**, 349–356.

Sabatini, B.L., Oertner, T.G., and Svoboda, K. (2002). The life cycle of Ca^{2+} ions in dendritic spines. *Neuron* **33**, 439–452.

Safronov, B.V. (1999). Spatial distribution of Na^{+} and K^{+} channels in spinal dorsal horn neurons: role of the soma, axon and dendrites in spike generation. *Prog. Neurobiol.* **59**, 217–241.

Sakaba, T., Ishikane, H., and Tachibana, M. (1997). Ca^{2+}-activated K^{+} current at presynaptic terminals of goldfish retinal bipolar cells. *Neurosci. Res.* **27**, 219–228.

Sakatani, S. and Hirose, A. (2003). The influence of neuron shape changes on the firing characteristics. *Neurocomputing* **52**, 355–362.

Sala, F. and Hernandez-Cruz, A. (1990). Ca^{2+} diffusion modeling in a spherical neuron. Relevance of buffering properties. *Biophys. J.* **57**, 313–324.

Sanes, J.R. and Lichtman, J.W. (1999). Can molecules explain long-term potentiation? *Nat. Neurosci.* **2**, 597–604.

Sangrey, T.D., Friesen, W.O., and Levy, W.B. (2004). Analysis of the optimal channel density of the squid giant axon using a reparameterized Hodgkin–Huxley model. *J. Neurophysiol.* **91**, 2541–2550.

Saul, A.B. and Humphrey, A.L. (1992). Evidence of input from lagged cells in the lateral geniculate nucleus to simple cells in cortical area 17 of the cat. *J. Neurophysiol.* **68**, 1190–1208.

Savtchenko, L.P., Gogan, P., Korogod, S.M., and Tyc-Dumont, S. (2001). Imaging stochastic spatial variability of active channel clusters during excitation of single neurons. *Neurosci. Res.* **39**, 431–446.

Sayer, R.J., Friedlander, M.J., and Redman, S.J. (1990). The time course and amplitude of EPSPs evoked at synapses between pairs of CA3/CA1 neurons in the hippocampal slice. *J. Neurosci.* **10**, 826–836.

Scannell, J.W., Burns, G.A.P.C., Hilgetag, C.C., O'Neil, M.A., and Young, M.P. (1999). The connectional organization of the cortico-thalamic system of the cat. *Cereb. Cortex* **9**, 277–299.

Schaefer, A.T., Larkum, M.E., Sakmann, B., and Roth, A. (2003). Coincidence detection in pyramidal neurons is tuned by their dendritic branching pattern. *J. Neurophysiol.* **89**, 3143–3154.

Scheier, C. and Lambrinos, D. (1996). Categorization in a real-world agent using haptic exploration and active perception. In P. Maes, M. Mataric, J.-A. Meyer, J. Pollack, and S.W. Wilson, eds., *From Animals to Animats: Proceedings of the Fourth International Conference on Simulation of Adaptive Behavior*. MIT Press, Cambridge, MA, pp. 65–75.

Schena, M., Shalon, D., Davis, R.W., and Brown P.O. (1995). Quantitative monitoring of gene expression patterns with a complementary DNA microarray. *Science* **270**, 467–470.

Schiegg, A., Gerstner, W., Ritz, R., and van Hemmen, J.L. (1995). Intracellular Ca^{2+} stores can account for the time course of LTP induction: a model of Ca^{2+} dynamics in dendritic spines. *J. Neurophysiol.* **74**, 1046–1055.

Schierwagen, A.K. (1986). Segmental cable modelling of electrotonic transfer properties of deep superior colliculus neurons in the cat. *J. Hirnforsch.* **27**, 679–690.

Schierwagen, A.K. (1990). Identification problems in distributed parameter neuron models. *Automatica* **26**, 739–755.

Schierwagen, A.K. and Grantyn, R. (1986). Quantitative morphological analysis of deep superior colliculus neurons stained intracellularly with HRP in the cat. *J. Hirnforsch.* **27**, 611–623.

Schiller, J., Schiller, Y., Stuart, G., and Sakmann, B. (1997). Calcium action potentials restricted to distal apical dendrites of rat neocortical pyramidal neurons. *J. Physiol. (Lond.)* **505**, 605–616.

Schiller, J., Schiller, Y., and Clapham, D.E. (1998). NMDA receptors amplify calcium influx into dendritic spines during associative pre- and postsynaptic activation. *Nat. Neurosci.* **1**, 114–118.

Schiller, J., Major, G., Koester, H.J., and Schiller, Y. (2000). NMDA spikes in basal dendrites of cortical pyramidal neurons. *Nature* **404**, 285–289.

Schlue, W.R., Richter, D.W., Mauritz, R.-H., and Nacimiento, A.C. (1974). Responses of cat spinal motoneuron somata and axons to linearly rising currents. *J. Neurophysiol.* **37**, 303–309.

Schmalbruch, H. and Jahnsen, H. (1981). Gap-junctions on CA3 pyramidal cells of guinea pig hippocampus shown by freeze-fracture. *Brain Res.* **217**, 175–178.

Schneidman, E., Freedman, B., and Segev, I. (1998). Ion channel stochasticity may be critical in determining the reliability and precision of spike timing. *Neural Comput.* **10**, 1679–1703.

Schreiber, T. (2000). Measuring information transfer. *Phys. Rev. Lett.* **85**, 461–464.

Schultz, W., Dayan, P., and Montague, P.R. (1997). A neural substrate of prediction and reward. *Science* **275**, 1593–1599.

Schuster, S., Marhl, M., and Hofer, T. (2002). Modelling of simple and complex calcium oscillations. From single-cell responses to intercellular signalling. *Eur. J. Biochem.* **269**, 1333–1355.

Schuster, T.H. (1992). Gap junctions and electrotonic coupling between hippocampal neurons: recent evidence and possible significance — a review. *Concept Neurosci.* **3**, 135–155.

Schwartz, D.R., Kardia, S.L.R., Shedden, K.A., Kuick, R., Michailidids, G., Taylor, J.M.G., Misek, D.E., Wu, R., Zhai, Y., Darrah, D.M., Reed, H., Ellenson, L.H., Giordano, T.J., Fearon, E.R., Hanash, S.M., and Cho, K.R. (2002). Gene expression in ovarian cancer reflects both morphology and biological behavior, distinguishing clear cell from other poor prognosis ovarian carcinomas. *Cancer Res.* **62**, 4722–4729.

Schwartzkroin, P.A. and Haglund, M.M. (1986). Spontaneous rhythmic synchronous activity in epileptic human and normal monkey temporal lobe. *Epilepsia* **27**, 523–533.

Schwindt, P.C. and Calvin, W.H. (1973). Nature of conductance underlying rhythmic firing in cat spinal motoneurons. *J. Neurophysiol.* **36**, 955–973.

Schwindt, P.C. and Crill, W.E. (1977). A persistent negative resistance in cat lumbar motoneurons. *Brain Res.* **120**, 173–178.

Schwindt, P.C. and Crill, W.E. (1980a). Role of persistent inward current in motoneuron bursting during spinal seizures. *J. Neurophysiol.* **43**, 1296–1318.

Schwindt, P.C. and Crill, W.E. (1980b). Properties of a persistent inward current in normal and TEA-injected motoneurons. *J. Neurophysiol.* **43**, 1700–1724.

Schwindt, P.C. and Crill, W.E. (1995). Amplification of synaptic current by persistent sodium conductance in apical dendrite of neocortical neurons. *J. Neurophysiol.* **74**, 2220–2224.

Schwindt, P.C. and Crill, W.E. (1996). Equivalence of amplified current flowing from dendrite to soma measured by alteration of repetitive firing and by voltage clamp in layer 5 pyramidal neurons. *J. Neurophysiol.* **76**, 3731–3739.

Schwindt, P.C. and Crill, W.E. (1997). Modification of current transmitted from apical dendrite to soma by blockade of voltage- and Ca^{2+}-dependent conductances in rat neocortical pyramidal neurons. *J. Neurophysiol.* **78**, 187–198.

Schwindt, P.C. and Crill, W.E. (1999). Mechanisms underlying burst and regular spiking evoked by dendritic depolarisation in layer 5 cortical pyramidal neurons. *J. Neurophysiol.* **81**, 1341–1354.

Scott, A.C. (2002). *Neuroscience: A Mathematical Primer*. Springer-Verlag, Berlin and New York.

Scott, A.C. and Luzader, S.D. (1979). Coupled solitary waves in neurophysics. *Phys. Scripta* **20**, 395–401.

Sculptoreanu, A., Figourov, A., and Degroat, W.C. (1995). Voltage-dependent potentiation of neuronal L-type calcium channels due to state-dependent phosphorylation. *Am. J. Physiol.* **38**, C725–C732.

Searle, J. (1984). *Minds, Brains, and Science*. Harvard University Press, Cambridge, MA.

Segal, M. (1995). Dendritic spines for neuroprotection: a hypothesis. *Trends Neurosci.* **18**, 468–471.

Segal, M. (2001). Rapid plasticity of dendritic spine: hints to possible functions? *Prog. Neurobiol.* **63**, 61–70.

Segev, I. and Burke, R.E. (1989). Compartmental models of complex neurons. In C. Koch and I. Segev, eds., *Methods in Neuronal Modelling: From Synapses to Networks*. MIT Press, Cambridge, MA, pp. 93–136.

Segev, I. and London, M. (2000). Untangling dendrites with quantitative models. *Science* **290**, 744–750.

Segev, I. and London, M. (2003). Dendritic processing. In M.A. Arbib, ed., *Handbook of Brain Theory and Neural Networks*, 2nd edition. MIT Press, Cambridge, MA.

Segev, I. and Rall, W. (1988). Computational study of an excitable dendritic spine. *J. Neurophysiol.* **2**, 499–523.

Segev, I. and Rall, W. (1998). Excitable dendrites and spines: earlier theoretical insights elucidate recent direct observations. *Trends Neurosci.* **21**, 453–460.

Segev, I., Rinzel, G.M., and Shepherd, G.M., eds. (1995). *The Theoretical Foundation of Dendritic Function: Selected Papers of Wilfrid Rall With Commentaries*. MIT Press, Cambridge, MA.

Seth, A.K., McKinstry, J.L., Edelman, G.M., and Krichmar, J.L. (2004a). Spatiotemporal processing of whisker input supports texture discrimination by a brain-based device. In S. Schaal, A. Ijspeert, A. Billard, S. Vijayakumar, J. Hallam, and J.A. Meyer, eds., *Animals to Animats*, Volume 8. MIT Press, Cambridge, MA, pp. 130–139.

Seth, A.K., McKinstry, J.L., Edelman, G.M., and Krichmar, J.L. (2004b). Texture discrimination by an autonomous mobile brain-based device with whiskers. In *IEEE Conference on Robotics and Automation*. New Orleans, LA, pp. 4925–4930.

Seth, A.K., McKinstry, J.L., Edelman, G.M., and Krichmar, J.L. (2004c). Visual binding through reentrant connectivity and dynamic synchronization in a brain-based device. *Cereb. Cortex* **14**, 1185–1199.

Seung, H.S., Lee, D.D., Reis, B.Y., and Tank, D.W. (2000). Stability of the memory of eye position in a recurrent network of conductance-based model neurons. *Neuron*, **26**, 259–271.

Shadlen, M.N. and Newsome, W.T. (1998). The variable discharge of cortical neurons: implications of connectivity, computation and information coding. *J. Neurosci.* **18**, 3870–3896.

Shadowitz, A. (1975). *The Electromagnetic Field.* McGraw-Hill, New York.

Shannon, C.E. and Weaver, W. (1949). *The Mathematical Theory of Communication.* University Illinois Press, Urbana, IL.

Shao, L.-R., Halvorsrud, R., Borg-Graham, L., and Storm, J.F. (1999). The role of BK-type Ca^{2+}-dependent K^+ channels in spike broadening during repetitive firing in rat hippocampal pyramidal cells. *J. Physiol. (Lond.)* **521**, 135–146.

Sheasby, B.W. and Fohlmeister, J.F. (1999). Impulse encoding across the dendritic morphologies of retinal ganglion cells. *J. Neurophysiol.* **81**, 1685–1698.

Shehadeh, L., Liebovitch, L.S., and Jirsa, V.K. (submitted). Relationships between the global structure of genetic networks and mRNA levels measured by cDNA microarrays. *J. Biol. Phys.*

Shelton, D.P. (1985). Membrane resistivity estimated for the Purkinje neuron by means of a passive computer model. *Neuroscience* **14**, 111–131.

Shen Y., Lu T., and Yang, X.L. (1999). Modulation of desensitization at glutamate receptors in isolated crucian carp horizontal cells by concanavalin A, cyclothiazide, aniracetam and PEPA. *Neuroscience* **89**, 979–890.

Shen-Orr, S., Milo, R., Mangan, S., and Alon, U. (2002). Network motifs in the transcriptional regulation network of *Escherichia coli. Nat. Genet.* **31**, 64–68.

Shepherd, G.M. (1979). *The Synaptic Organisation of the Brain.* Oxford University Press, NewYork.

Shepherd, G.M. (1991). *Foundations of the Neuron Doctrine*. Oxford University Press, New York.

Shepherd, G.M. (1994). *Neurobiology*, 3rd edition. Oxford University Press, New York.

Sherman, A. and Rinzel, J. (1991). Model for synchronization of pancreatic beta-cells by gap junction coupling. *Biophys. J.* **59**, 547–559.

Sherman, A., Keizer, J., and Rinzel, J. (1990). Domain model for Ca-inactivation of Ca^{2+} channels at low channel density. *Biophys. J.* **58**, 985–995.

Sherman, A., Li, Y., and Keizer, J. (2002). Whole-cell models. In C. Fall, E. Marland, J. Wagner, and J. Tyson, eds., *Computational Cell Biology*. Springer-Verlag, New York, pp. 101–139.

Sherwood, S.L. (1966). *The Nature of Psychology. A Selection of Papers, Essays and Other Writings by Kenneth J. Craik.* Cambridge University Press, Cambridge.

Shiells, R.A. and Falk, G. (1994). Responses of rod bipolar cells isolated from dogfish retinal slice to concentration-jumps of glutamate. *Vis. Neurosci.* **11**, 1175–1183.

Shingai, R. and Christensen, B.N. (1983). Sodium and calcium currents measurment in isolated catfish horizontal cells under voltage clamp. *Neuroscience* **10**, 893–897.

Shreve, R.L. (1966). Statistical law of stream numbers. *J. Geol.* **74**, 17–37.

Siegel, M., Marder, E., and Abbott, L.F. (1994). Activity-dependent current distributions in model neurons. *Proc. Natl. Acad. Sci. USA* **91**, 11308–11312.

Sigworth, F.J. and Neher, E. (1980). Single sodium channel currents observed in cultured rat muscle cells. *Nature* **287**, 497–499.

Silver, R.A. (2003). Estimation of the nonuniform quantal parameters with multiple-probability fluctuation analysis: theory, application and limitations. *J. Neurosci. Meth.* **139**, 127–141.

Simoncelli, E.P. and Olshausen, B.A. (2001). Natural image statistics and neural representation. *Annu. Rev. Neurosci.* **24**, 1193–1216.

Simpson, P.B., Challiss, R.A., and Nahorski, S.R. (1995). Neuronal calcium stores: activation and function. Trends Neurosci. **18**, 299–306.

Singer, W. (1993). Synchronization of cortical activity and its putative role in information processing and learning. *Annu. Rev. Physiol.* **55**, 349–374.

Singer, W. (1995). Development and plasticity of cortical processing architectures. *Science* **270**, 758–764.

Sirovich, L. and Knight, B.W. (1977). On subthreshold solutions of the Hodgkin–Huxley equations. *Proc. Natl. Acad. Sci. USA* **74**, 5199–5202.

Sivilotti, L.G., Thompson, S.W.N., and Woolf, C.J. (1993). Rate of rise of the cumulative depolarization evoked by repetitive stimulation of small-caliber afferents is a predictor of action potential windup in rat spinal neurons *in vitro*. *J. Neurophysiol.* **69**, 1621–1631.

Skrzypek, J. (1984). Electrical coupling between horizontal cell bodies in the tiger salamander retina. *Vision Res.* **24**, 701–711.

Smit, G.J., Uylings, H.B.M., and Veldmaat-Wansink, L. (1972). The branching pattern in dendrites of cortical neurons. *Acta Morphol. Neer. Sc.* **9**, 253–274.

Smith, A.F.M. and Roberts, G.O. (1993). Bayesian computation via the Gibbs sampler and related Markov chain Monte Carlo methods. *J. R. Stat. Soc. B* **55**, 3–23.

Smith, D.O. (1983). Extracellular potassium levels and axon excitability during repetitive action potentials in crayfish. *J. Physiol. (Lond.)* **336**, 143–157.

Smith, D.O. and Hatt, H. (1976). Axon conduction block in a region of dense connective tissue in crayfish. *J. Neurophysiol.* **39**, 794–801.

Smith, G.D. (1996). Analytical steady-state solution to the rapid buffering approximation near an open Ca^{2+} channel. *Biophys. J.* **71**, 3064–3072.

Smith, G.D. (2002a). An extended DeYoung-Keizer-like Ip_3R receptor model that accounts for domain Ca^{2+}-mediated inactivation. In C.A. Condat and A. Baruzzi, eds., *Recent Research Developments in Biophysical Chemistry*, Volume II. Research Signpost, Kerala, pp. 37–55.

Smith, G.D. (2002b). Modeling the stochastic gating of ion channels. In C. Fall, E. Marland, J. Wagner, and J. Tyson, eds., *Computational Cell Biology*. Springer-Verlag, Berlin, pp. 291–325.

Smith, G.D. and Keizer, J. (1998). Spark-to-wave transition: saltatory transmission of Ca^{2+} waves in cardiac myocytes. *Biophys. Chem.* **72**, 87–100.

Smith, G.D., Lee, R., Oliver, J., and Keizer, J. (1996a). Effect of Ca^{2+} influx on intracellular free Ca^{2+} responses in antigen-stimulated RBL-2H3 cells. *Am. J. Physiol.* **270**, C939–C952.

Smith, G.D., Wagner, J., and Keizer, J. (1996b). Validity of the rapid buffering approximation near a point source of Ca^{2+} ions. *Biophys. J.* **70**, 2527–2539.

Smith, G.D., Keizer, J., Stern, M., Lederer, W., and Cheng, H. (1998). A simple numerical model of Ca^{2+} spark formation and detection in cardiac myocytes. *Biophys. J.* **75**, 15–32.

Smith, G.D., Dai, L., Muira, R., and Sherman, A. (2001). Asymptotic analysis of equations for the buffered diffusion of intracellular Ca^{2+}. *SIAM J. Appl. Math.* **61**, 1816–1838.

Smith, G.D., Pearson, J., and Keizer, J. (2002). Modeling intracellular Ca^{2+} waves and sparks. In C. Fall, E. Marland, J. Wagner, and J. Tyson, eds., *Computational Cell Biology*. Springer-Verlag, Berlin, pp. 200–232.

Smith, H. (1987). Oscillations and multiple steady states in a cyclic gene model with repression. *J. Math. Biol.* **25**, 169–190.

Smolen, P., Baxter, D.A., and Byrne, J.H. (2000a). Mathematical modeling of gene networks. *Neuron* **26**, 567–580.

Smolen, P., Baxter, D.A., and Byrne, J.H. (2000b). Modeling transcriptional control in gene networks methods, recent results, and future directions. *B. Math. Biol.* **62**, 247–292.

Sneyd, J. and Dufour, J.F. (2002). A dynamic model of the type-2 inositol trisphosphate receptor. *Proc. Natl. Acad. Sci. USA* **99**, 2398–2403.

Sneyd, J., Dale, P., and Duffy, A. (1998). Traveling waves in buffered systems: applications to Ca^{2+} waves. *SIAM J. Appl. Math.* **58**, 1178–1192.

Snow, R.W. and Dudek, F.E. (1986). Evidence for neuronal interactions by electrical field effects in the CA3 and dentate regions of rat hippocampus slices. *Brain Res.* **367**, 292–295.

Sochivko, D., Pereverzev, A., Smyth, N., Gissel, C., Schneider, T., and Beck, H. (2002). The $Ca_v2.3$ Ca^{2+} channel subunit contributes to R-type Ca^{2+} currents in murine hippocampal and neocortical neurons. *J. Physiol. (Lond.)* **542**, 699–710.

Soderling, T.R. (1993). Calcium/calmodulin-dependent protein kinase II: role in learning and memory. *Mol. Cell. Biochem.* **127–128**, 93–101.

Softky, W.R. (1994). Sub-millisecond coincidence detection in active dendritic trees. *Neuroscience* **58**, 13–41.

Softky, W.R. and Koch, C. (1993). The highly irregular firing of cortical cells is inconsistent with temporal integration of random EPSPs. *J. Neurosci.* **13**, 334–350.

Sokolov, M.V., Rossokhin, A.V., Astrelin, A.V., Frey, J.U., and Voronin, L.L. (2002). Quantal analysis suggests strong involvement of presynaptic mechanisms during the initial 3 hr maintenance of long-term potentiation in rat hippocampal CA1 area in vitro. *Brain Res.* **957**, 61–75.

Somjen, G.G. and Muller, M. (2000). Potassium-induced enhancement of persistent inward current in hippocampal neurons in isolation and in tissue slices. *Brain Res.* **885**, 102–110.

Song, H.J. and Poo, M.-M. (2001). The cell biology of neuronal navigation. *Nat. Cell Biol.* **3**, E81-E88.

Soto-Trevino, C., Thoroughman, K.A., Marder, E., and Abbott, L.F. (2001). Activity-dependent modification of inhibitory synapses in models of rhythmic neural networks. *Nat. Neurosci.* **4**, 297–303.

Spacek, J. and Harris, K.M. (1997). Three-dimensional organization of smooth endoplasmic reticulum in hippocampal CA1 dendrites and dendritic spines of the immature and mature rat. *J. Neurosci.* **17**, 190–203.

Spieker, S., Boose, A., Breit, S., and Dichgans, J. (1998). Long-term measurement of tremor. *Movement Disord.* **13**(Suppl. 3), 81–84.

Spira, M.E., Yarom, Y., and Parnas, I. (1976). Modulation of spike frequency by regions of special axonal geometry and by synaptic inputs. *J. Neurophysiol.* **39**, 882–899.

Sporns, O. (1998). Biological variability and brain function. In J. Cornwell, ed., *Consciousness and Human Identity*. Oxford University Press, New York, pp. 38–56.

Sporns, O. (2002a). Graph theory methods for the analysis of neural connectivity patterns. In R. Kötter, ed., *Neuroscience Databases. A Practical Guide*. Kluwer, Boston, MA, pp. 171–186.

Sporns, O. (2002b). Embodied cognition. In M. Arbib, ed., *Handbook of Brain Theory and Neural Networks*. MIT Press, Cambridge, MA, pp. 395–398.

Sporns, O. (2004). Complex neural dynamics. In V.K. Jirsa and J.A.S. Kelso, eds., *Coordination Dynamics: Issues and Trends*. Springer-Verlag, Berlin, pp. 197–215.

Sporns, O. and Alexander, W.H. (2002). Neuromodulation and plasticity in an autonomous robot. *Neural Networks* **15**, 761–774.

Sporns, O. and Kötter, R. (2004). Motifs in brain networks. *PLoS Biology* **2**, 1910–1918.

Sporns, O. and Pegors, T.K. (2004). Information-theoretical aspects of embodied artificial intelligence. In F. Iida, R. Pfeifer, L. Steels, and Y. Kuniyoshi, eds., *Embodied Artificial Intelligence, Lecture Notes in Computer Science*, Volume 3139. Springer-Verlag, Berlin, pp. 74–85.

Sporns, O. and Tononi, G. (2002). Classes of network connectivity and dynamics. *Complexity* **7**, 28–38.

Sporns, O. and Zwi, J. (2004). The small world of the cerebral cortex. *Neuroinformatics* **2**, 145–162.

Sporns, O., Tononi, G., and Edelman, G.M. (2000a). Theoretical neuroanatomy: relating anatomical and functional connectivity in graphs and cortical connection matrices. *Cereb. Cortex* **10**, 127–141.

Sporns, O., Tononi, G., and Edelman, G. (2000b). Connectivity and complexity: the relationship between neuroanatomy and brain dynamics. *Neural Networks* **13**, 909–922.

Sporns, O., Chialvo, D.R., Kaiser, M., and Hilgetag, C.C. (2004). Organization, development and function of complex brain networks. *Trends Cogn. Sci.* **8**, 418–425.

Spruston, N., Jaffe, D.B., Williams, S.H., and Johnston, D. (1993). Voltage- and space-clamp errors associated with the measurement of electrotonically remote synaptic events. *J. Neurophysiol.* **70**, 781–802.

Spruston, N. and Kath, W.L. (2004). Dendritic arithmetic. *Nat. Neurosci.* **7**, 567–569.

Spruston, N., Schiller, Y., Stuart, G., and Sakmann, B. (1995). Activity-dependent action potential invasion and calcium influx into hippocampal CA1 dendrites. *Science* **268**, 297–300.

Spyers-Ashby, J.M., Bain, P.G., and Roberts, S.J. (1998). A comparison of fast Fourier transform (FFT) and autoregressive (AR) spectral estimation techniques for the analysis of tremor data. *J. Neurosci. Meth.* **83**, 35–43.

Srinivasan, S.K. (1974). *Stochastic Point Processes and their Applications. Monograph No. 34*. Griffin, London.

Srinivasan, R., Russell, D.P., Edelman, G.M., and Tononi, G. (1999). Increased synchronization of neuromagnetic responses during conscious perception. *J. Neurosci.* **19**, 5435–5448.

Staley, K.J. (1999). Quantal GABA release: noise or not? *Nat. Neurosci.* **2**, 494–495.

Stanley, E. (1995). Calcium entry and the functional organization of the presynaptic transmitter release site. In H. Wheal and A. Thomson, eds., *Excitatory Amino Acids and Synaptic Transmission*, 2nd edition. Academic, New York.

Steinberg, I.Z. (1996). On the analytic solution of electrotonic spread in branched passive dendritic trees. *J. Comput. Neurosci.* **3**, 301–311.

Stemmler, M. and Koch, C. (1999). How voltage-dependent conductances can adapt to maximize the information encoded by neuronal firing rate. *Nat. Neurosci.* **2**, 521–527.

Stern, M. (1992). Buffering of Ca^{2+} in the vicinity of a channel pore. *Cell Calcium* **13**, 183–192.

Stewart, A.E. and Froehring, R.C. (2001). Effects of spike parameters and neuromodulators on action potential waveform-induced calcium entry into Pyramidal neurons. *J. Neurophysiol.* **85**, 1412–1422.

Stewart, I. (1992). Numerical methods: warning-handle with care! *Nature* **355**, 16–17.

Stiles, J.R. and Bartol, T.M. (2001). Monte Carlo methods for simulating realistic synaptic microphysiology using MCell. In E. De Schutter, ed., *Computational Neuroscience: Realistic Modeling for Experimentalists*. CRC, Boca Raton, FL, pp. 87–128.

Stockbridge, N. (1988). Differential conduction at axonal bifurcations. II. Theoretical basis. *J. Neurophysiol.* **59**, 1286–1295.

Stoney, S.D., Jr. (1994). Signal integration in the axonal tree due to branch point filtering action. In L. Urban, ed., *Cellular Mechanisms of Sensory Processing*. Springer-Verlag, Heidelberg.

Strack, S., Barban, M.A., Wadzinski, B.E., and Colbran, R.J. (1997a). Differential inactivation of postsynaptic density-associated and soluble Ca^{2+}/calmodulin-dependent protein kinase II by protein phosphatases 1 and 2A. *J. Neurochem.* **68**, 2119–2128.

Strack, S., Choi, S., Lovinger, D.M., and Colbran, R.J. (1997b). Translocation of autophosphorylated calcium/calmodulin-dependent protein kinase II to the postsynaptic density. *J. Biol. Chem.* **272**, 13467–13470.

Strassberg, A.F. and DeFelice, L.J. (1993). Limitations of the Hodgkin–Huxley formalism: effects of single channel kinetics upon transmembrane voltage dynamics. *Neural Comput.* **5**, 843–855.

Strain, G.M. and Brockman, W.H. (1975). A modified cable model for neuron processes with non-constant diameters. *J. Theor. Biol.* **51**, 475–494.

Stratford, K.J., Mason, A.J.R., Larkman, A.U., Major, G., and Jack, J.J.B. (1989). The modelling of pyramidal neurons in the visual cortex. In R. Durbin, C. Miall, and G. Mitchison, eds., *The Computing Neuron*. Addison-Wesley, Wokingham, UK, pp. 296–321.

Streekstra, G.J. and Van Pelt, J. (2002). Analysis of tubular structures in 3D confocal images. *Network: Comput. Neural Syst.* **13**, 381–395.

Stricker, C. and Redman, S.J. (2003). Quantal analysis based on density estimation. *J. Neurosci. Meth.* **130**, 159–171.

Stricker, C., Field, A.C., and Redman, S.J. (1996a). Statistical analysis of amplitude fluctuations in EPSCs evoked in rat CA1 pyramidal neurons in vitro. *J. Physiol. (Lond.)* **490**, 419–441.

Stricker, C., Field, A.C., and Redman, S.J. (1996b). Changes in quantal parameters of EPSCs in rat CA1 neurons in vitro after the induction of long-term potentiation. *J. Physiol. (Lond.)* **490**, 443–454.

Strier, D., Ventura, A., and Dawson, S. (2003). Saltatory and continuous calcium waves and the rapid buffering approximation. *Biophys. J.* **85**, 3575–3586.

Strogatz, S.H. (2001). Exploring complex networks. *Nature* **410**, 268–277.

Stuart, G.J. and Häusser, M. (1994). Initiation and spread of sodium action potentials in cerebellar Purkinje cells. *Neuron* **13**, 703–712.

Stuart, G.J. and Häusser, M. (2001). Dendritic coincidence detection of EPSPs and action potentials. *Nat. Neurosci.* **4**, 63–71.

Stuart, G.J. and Sakmann, B. (1994). Active propagation of somatic action potentials into neocortical pyramidal cell dendrites. *Nature* **367**, 69–72.

Stuart, G.J. and Sakmann, B. (1995). Amplification of EPSPs by axosomatic sodium channels in neocortical pyramidal neurons. *Neuron* **15**, 1065–1076.

Stuart, G.J. and Spruston, N. (1998). Determinants of voltage attenuation in neocortical pyramidal neuron dendrites. *J. Neurosci.* **18**, 3501–3510.

Stuart, G., Schiller, I., and Sakmann, B. (1997a). Action potential initiation and propagation in rat neocortical pyramidal neurons. *J. Physiol. (Lond.)* **505**, 617–632.

Stuart, G., Spruston, N., and Hausser, M. (1999a). *Dendrites*. Oxford University Press, New York.

Stuart, G., Spruston, N., and Hausser, M. (1999b). Dendrites as biochemical compartments. In G. Stuart, N. Spruston, and M. Hausser, eds., *Dendrites*. Oxford University Press, New York, pp. 161–192.

Stuart, G.J., Spruston, N., Sakmann, B., and Häusser, M. (1997b). Action potential initiation and backpropagation in neurons of the mammalian CNS. *Trends Neurosci.* **20**, 125–131.

Stuart, J.M., Segal, E., Koller, D., and Kim, S.K. (2003). A gene-coexpression network for global discovery of conserved genetic modules. *Science* **302**, 249–255.

Stuhmer, W., Methfessel, B., Sakmann, B., Noda, M., and Numa, S. (1987). Patch clamp characterization of sodium channels expressed from rat brain cDNA. *Eur. Biophys. J.* **14**, 131–138.

Stundzia, A.B. and Lumsden, C.J. (1996). Stochastic simulation of coupled reaction-diffusion processes. *J. Comput. Phys.* **127**, 196–207.

Subba Rao, T. and Gabr, M.M. (1984). *An Introduction to Bispectral Analysis and Bilinear Time Series Models*. Springer-Verlag, Berlin.

Sudhof, T.C., Newton, C.L., Archer, B.T., Ushkaryov, Ya., and Mignery, G.A. (1991). Structure of a novel INSP3 receptor. *EMBO J.* **10**, 3199–3206.

Suilleabhain, P.E. and Matsumoto, J.Y. (1998). Time-frequency analysis of tremors. *Brain* **121**, 2127–2134.

Sullivan, J.D. and Lees, A.J. (2000). Nonparkinsonian tremors. *Clin. Neuropharmacol.* **23**, 233–238.

Sullivan, J.M. and Lasater, E.M. (1990). Sustained and transient potassium currents of cultured horizontal cells isolated from white bass retina. *J. Neurophysiol.* **64**, 1758–1766.

Surkis, A., Taylor, B., Peskin, C.S., and Leonard, C.S. (1996). Quantitative morphology of physiologically identified and intracellularly labeled neurons from the guinea-pig laterodorsal tegmental nucleus in vitro. *Neuroscience* **74**, 375–392.

Sutton, R.S. and Barto, A.G. (1990). Time derivative models of Pavlovian reinforcement. In M. Gabriel and J. Moore, eds., *Learning and Computational Neuroscience: Foundations of Adaptive Networks*. MIT Press, Cambridge, MA, pp. 497–538.

Sutton, R.S. and Barto, A.G. (1998). *Reinforcement Learning*. MIT Press, Cambridge, MA.

Svirskis, G. and Hounsgaard, J. (1997). Depolarization-induced facilitation of a plateau-generating current in ventral horn neurons in the turtle spinal cord. *J. Neurophysiol.* **78**, 1740–1742.

Svirskis, G., Svirskiene, N., and Gutman, A. (1998). Theoretical analysis of the multistability of pyramidal neurons. *Biophysics-USSR* **43**, 87–91.

Svoboda, K., Tank, D.W., and Denk, W. (1996). Direct measurement of coupling between dendritic spines and shafts. *Science* **272**, 716–719.

Svoboda, K., Denk, W., Kleinfeld, D., and Tank, D.A. (1997). In vivo dendritic calcium dynamics in neocortical pyramidal neurons. *Nature* **385**, 161–165.

Swadlow, H.A. and Waxman, S.G. (1978). Activity-dependent variations in conduction properties of central axons. In S.G. Waxman ed., *Physiology and Pathology of Axons*. Raven, New York.

Swadlow, H.A., Kocsis, J.D., and Waxman, S.G. (1980). Modulation of impulse conduction along the axonal tree. *Annu. Rev. Biophys. Bio.* **9**, 143–179.

Swillens, S., Champeil, P., Combettes, L., and Dupont, G. (1998). Stochastic simulation of a single inositol 1,4,5-trisphosphate-sensitive Ca^{2+} channel reveals repetitive openings during 'blip-like' Ca^{2+} transients. *Cell Calcium* **23**, 291–302.

Szilagyi, T. and De Schutter, E. (2004). Effects of variability in anatomical reconstruction techniques on models of synaptic integration by dendrites: a comparison of the Internet archives. *Eur. J. Neurosci.* **19**, 1257–1266.

Tachibana, M. (1981). Membrane properties of solitary horizontal cells isolated from goldfish retina. *J. Physiol. (Lond.)* **321**, 141–161.

Tachibana, M. (1983). Ionic currents of solitary horizontal cells isolated from goldfish retina. *J. Physiol. (Lond.)* **345**, 329–351.

Tachibana, M. (1985). Permeability changes induced by L-glutamate in solitary retinal horizontal cells isolated from *Carassius auratus*. *J. Physiol. (Lond.)* **358**, 153–167.

Tachibana, M. and Kaneko, A. (1988). Retinal bipolar cells receive negative feedback input from GABAergic amacrine cells. *Visual Neurosci.* **1**, 297–305.

Tachibana, M., Okada, T., Arimura, T., Kobayashi, K., and Piccolino, M. (1993). Dihydropyridine-sensitive calcium current mediates neurotransmitter release from bipolar cells of the goldfish retina. *J. Neurosci.* **13**, 2898–2909.

Tagamets, M.A. and Horwitz, B. (1998). Integrating electrophysiological and anatomical experimental data to create a large-scale model that simulates a delayed match-to-sample human brain imaging study. *Cereb. Cortex* **8**, 310–320.

Tagamets, M.A. and Horwitz, B. (2000). A model of working memory: bridging the gap between electrophysiology and human brain imaging. *Neural Netw.* **13**, 941–952.

Tamori, Y. (1993). Theory of dendritic morphology. *Phys. Rev. E* **148**, 3124–3129.

Tang, Y., Stephenson, J., and Othmer, H. (1995). Simplification and analysis of models of Ca^{2+} dynamics based on Ip-sensitive Ca^{2+} channel kinetics. *Biophys. J.* **70**, 246–263.

Tao, H.-Z.W., Zhang, L.I., Bi, G.-Q., and Poo, M.-M. (2000). Selective presynaptic propagation of long-term potentiation in defined neural networks. *J. Neurosci.* **20**, 3233–3243.

Tasaki, I. (1959). Demonstration of two stable states of the nerve membrane in potassium-rich media. *J. Physiol. (Lond.)* **148**, 306–331.

Tate, C. and Woolfson, M.M. (1971). On modelling neural networks in the retina. *Vision Res.* **11**, 617–633.

Tauc, L. and Hughes, G.M. (1963). Modes of initiation and propagation of spikes in the branching axons of molluscan central neurons. *J. Gen. Physiol.* **46**, 533–549.

Tawfik, B. and Durand, D.M. (1994). Nonlinear parameter estimation by linear association: application to a five-parameter passive neuron model. *IEEE T. Bio-Med. Eng.* **41**, 461–469.

Taylor, C.P. and Dudek, F.E. (1982). Synchronous neural afterdischarges in rat hippocampal slices without active chemical synapses. *Science* **218**, 810–812.

Taylor, C.P. and Dudek, F.E. (1984). Excitation of hippocampal cells by an electrical field effect. *J. Neurophysiol.* **52**, 126–142.

Taylor, D.M., Helms Tillery, S.I., and Schwartz, A.B. (2002). Direct cortical control of 3D neuroprosthetic devices. *Science* **296**, 1829–1832.

Taylor, G.C., Coles, J.A., and Eilbeck, J.C. (1995). Mathematical modeling of weakly nonlinear pulses in a retinal neuron. *Chaos Solition Fract.* **5**, 407–413.

Thomas, R. and d'Ari, R. (1990). *Biological Feedback*. CRC, Boca Raton, FL.

Thomas, R., Thieffry, D., and Kauffman, M. (1999). Dynamical behavior of biological regulatory networks — I. Biological role of feedback loops and practical use of the concept of the loop-characteristic state. *B. Math. Biol.* **57**, 247–276.

Thompson, J.M.T. and Stewart, H.B. (1993). A tutorial glossary of geometrical dynamics. *Int. J. Bifurcat. Chaos* **3**, 223–239.

Tierney, L.J. (1994). Markov chains for exploring posterior distributions. *Ann. Stat.* **22**, 1701–1728.

Timmer, J., Gantert, C., Deuscal, G., and Honerkamp, J. (1993). Characteristics of hand tremor time series. *Biol. Cybern.* **70**, 75–80.

Timmer, J., Lauk, M., and Deuschl, G. (1996). Quantitative analysis of tremor time series. *Electroen. Clin. Neuro.* **101**, 461–468.

Timmer, J., Lauk, M., Pfleger, W., and Deuschl, G. (1998). Cross-spectral analysis of physiological tremor and muscle activity. I. Theory and application to unsynchronized electromyogram. *Biol. Cybern.* **78**, 349–357.

Tononi, G. and Edelman, G.M. (1998). Consciousness and complexity. *Science* **282**, 1846–1851.

Tononi, G. and Sporns, O. (2003). Measuring information integration. *BMC Neuroscience* **4**, 31.

Tononi, G., Sporns, O., and Edelman, G.M. (1992). Reentry and the problem of integrating multiple cortical areas: simulation of dynamic integration in the visual system. *Cereb. Cortex* **2**, 310–335.

Tononi, G., Sporns, O., and Edelman, G.M. (1994). A measure for brain complexity: relating functional segregation and integration in the nervous system. *Proc. Natl. Acad. Sci. USA* **91**, 5033–5037.

Tononi, G., Edelman, G.M., and Sporns, O. (1998). Complexity and the integration of information in the brain. *Trends Cogn. Sci.* **2**, 44–52.

Tononi, G., Sporns, O., and Edelman, G.M. (1999). Measures of degeneracy and redundancy in biololgical networks. *Proc. Natl. Acad. Sci. USA* **96**, 3257–3262.

Toth, T.I. and Crunelli, V. (2002). Modelling large scale neuronal networks using 'average neurons'. *NeuroReport* **13**, 1785–1788.

Tottene, A., Moretti, A., and Pietrobon, D. (1996). Functional diversity of P-type and R-type calcium channels in rat cerebellar neurons. *J. Neurosci.* **16**, 6353–6363.

Tottene, A., Volsen, S., and Pietrobon, D. (2000). α_{1E} subunits from the pore of three cerebellar R-type calcium channels with different pharmacological and permeation properties. *J. Neurosci.* **20**, 171–178.

Traub, R.D. and Jefferys, J.G.R. (1994). Mechanisms responsible for epilepsy in hippocampal slices predispose the brain to collective oscillations. In F. Ventriglia, ed., *Neural Modeling and Neural Networks*. Pergamon, Oxford.

Traub, R.D. and Llinás, R. (1979). Hippocampal pyramidal cells: significance of dendritic ionic conductances for neuronal function and epileptogenesis. *J. Neurophysiol.* **42**, 476–496.

Traub, R.D., Dudek, F.E., Snow, R.W., and Knowles, W.D. (1985a). Computer simulations indicate that electrical field effects contribute to the shape of the epileptiform field potential. *Neuroscience* **15**, 947–958.

Traub, R.D., Dudek, F.E., Taylor, C.P., and Knowles, W.D.(1985b). Simulations of hippocampal afterdischarges synchronized by electrical interactions. *Neuroscience* **14**, 1033–1038.

Traub, R.D., Knowles, W.D., Miles, R., and Wong, R.K.S. (1987a). Models of the cellular mechanism underlying propagation of epileptiform activity in the CA2-CA3 region of the hippocampal slice. *Neuroscience* **21**, 457–470.

Traub, R.D., Miles, R., and Wong, R.K.S. (1987b). Models of synchronized hippocampal bursts in the presence of inhibition. I. Single population events. *J. Neurophysiol.* **58**, 739–751.

Traub, R.D., Miles, R., Wong, R.K.S., Schulman, L.S., and Schneiderman, J.H. (1987c). Models of synchronized hippocampal bursts in the presence of inhibition. II. Ongoing spontaneous population events. *J. Neurophysiol.* **58**, 752–764.

Traub, R.D., Wong, R., Miles, R., and Michelson, H. (1991). A model of CA3 hippocampal pyramidal neuron incorporating voltage-clamp data on intrinsic conductances. *J. Neurophysiol.* **66**, 635–650.

Traub, R.D., Miles, R., and Buzsaki, G. (1992). Computer simulation of carbachol-driven rhythmic population oscillations in the CA3 region of the in vitro rat hippocampus. *J. Physiol. (Lond.)* **451**, 653–672.

Treisman, A. (1998). Feature binding, attention and object perception. *Philos. T. R. Soc. Lond. B* **353**, 1295–1306.

Treves, A., Rolls, E.T., and Tovee, M.J. (1996). On the time required for recurrent processing in the brain. In V. Torre and F. Conti, eds., *Neurobiology: Ionic Channels, Neurons, and the Brain*. NATO ASI Series, Plenum, New York.

Troy, W.C. (1976). Large amplitude periodic solutions of a system of equations derived from the Hodgkin–Huxley equations. *Arch. Rat. Mech. Anal.* **65**, 227–247.

Troyer, T.W. and Miller, K.D. (1997). Physiological gain leads to high ISI variability in a simple model of a cortical regular spiking cell. *Neural Comput.* **9**, 971–983.

Tseng, G.C., Oh, M.K., Rohlin, L., Liao, J.C., and Wong, W.H. (2001). Issues in cDNA microarray analysis: quality filtering, channel normalization, models of variations and assessment of gene effects. *Nucleic Acids Res.* **29**, 2549–2557.

Tsetlin, M.L. (1969). *Research in Theory of Automata and Modeling of Biological Systems*. Nauka, Moscow (in Russian).

Tsien, R. (1980). New calcium indicators and buffers with high selectivity against magnesium and protons: design, synthesis, and properties of prototype structures. *Biochemistry-US* **19**, 2396–2404.

Tsutsui, H. and Oka, Y. (2002). Slow removal of Na channel inactivation underlies the temporal filtering property in the teleost thalamic neurons. *J. Physiol. (Lond.)* **539**, 743–753.

Tuckwell, H.C. (1985). Some aspects of cable theory with synaptic reversal potentials. *J. Theoret. Neurobiol.* **4**, 113–127.

Tuckwell, H.C. (1988a). *Introduction to Theoretical Neurobiology, Volume 1. Linear Cable Theory and Dendritic Structure*. Cambridge University Press, New York.

Tuckwell, H.C. (1988b). *Introduction to Theoretical Neurobiology, Volume 2. Nonlinear and Stochastic Theories*. Cambridge University Press, Cambridge.

Tuckwell, H.C. (1988c). Use of Green's function matrices for systems of diffusion equations. *Int. J. Systems Sci.* **19**, 1663–1666.

Tuckwell, H.C. (1989). *Stochastic Processes in the Neurosciences*. SIAM, Philadelphia, PA.

Tuckwell, H.C. and Feng, J. (2003). A theoretical overview. In J. Feng, ed., *Computational Neuroscience: A Comprehensive Approach*. Chapman & Hall/CRC, Boca Raton, FL, pp. 1–30.

Tuckwell, H.C., Rospars, J.-P., Vermeulen, A., and Lansky, P. (1996). Time-dependent solutions for a cable model of an olfactory receptor neuron. *J. Theor. Biol.* **181**, 25–31.

Tukker, J.J., Taylor, W.R., and Smith, R.G. (2004). Direction selectivity in a model of the starburst amacrine cell. *Vis. Neurosci.* **21**, 611–625.

Turing, A.M. (1950). Computing machinery and intelligence. *Mind* **59**, 433–460.

Turner, D.A. (1984a). Segmental cable evaluation of somatic transients in hippocampal neurons (CA1, CA3, and Dentate). *Biophys. J.* **46**, 73–84.

Turner, D.A. (1984b). Conductance transients onto dendritic spines in a segmental cable model of CA1 and dentate hippocampal neurons. *Biophys. J.* **46**, 85–96.

Turner, D.A. and Schwartzkroin, P.A. (1980). Steady-state electrotonic analysis of intracellularly stained hippocampal neurons. *J. Neurophysiol.* **44**, 184–199.

Turner, D.A. and Schwartzkroin, P.A. (1983). Electrical characteristics of dendrites and dendritic spines in intracellularly stained CA3 and dentate hippocampal neurons. *J. Neurosci.* **3**, 2381–2394.

Turner, D.A. and West, M. (1993). Bayesian analysis of mixtures applied to post-synaptic potential fluctuations. *J. Neurosci. Meth.* **47**, 1–21.

Turner, D.A., Chen, Y., Isaac, J., West, M., and Wheal, H.V. (1997). Synaptic site heterogeneity in paired-pulse plasticity in CA1 pyramidal cells. *J. Physiol. (Lond.)* **500**, 441–462.

Turner, R.W., Baimbridge, K.G., and Miller, J.J. (1982). Calcium-induced long-term potentiation in the hippocampus. *Neuroscience* **7**, 1411–1416.

Turner, R.W., Maler, L., Deerinik, T., Levison, S.R., and Ellisman, M.H. (1994). TTX-sensitive dendritic sodium channels underlie oscillatory discharges in a vertebrate sensory neuron. *J. Neurosci.* **14**, 6453–6471.

Turrigiano, G.G., Leslie, K.R., Desai, N.S., Rutherford, L.C., and Nelson, S.B. (1998). Activity-dependent sealing of quantal amplitude in neocortical neurons. *Nature* **391**, 892–896.

Tylstedt, S. and Rask-Andersen, H. (2001). A 3-D model of membrane specializations between human auditory spiral ganglion cells. *J. Neurocytol.* **30**, 465–473.

Ueda, Y., Kaneko, A., and Kaneda, M. (1992). Voltage-dependent ionic currents in solitary horizontal cells isolated from cat retina. *J. Neurophysiol.* **68**, 1143–1150.

Uemura, E., Carriquiry, A., Kliemann, W., and Goodwin, J. (1995). Mathematical modeling of dendritic growth in vitro. *Brain Res.* **671**, 187–194.

Ulfhake, B. and Kellerth, J.-O. (1983). A quantitative morphological study of HRP-labelled cat alpha-motoneurons supplying different hindlimb muscles. *Brain Res.* **264**, 1–19.

Ulfhake, B. and Kellerth, J.-O. (1984). Electrophysiological and morphological measurements in cat gastrocnemius and soleus alpha motoneurons. *Brain Res.* **307**, 167–179.

Ungerleider, L.G. and Mishkin, M. (1982). Two cortical visual systems. In D.J. Ingle, M.A. Goodale, and R.J.W. Mansfield, eds., *Analysis of Visual Behavior*. MIT Press, Cambridge, MA, pp. 549–586.

Urban, N.N., Henze, D.A., and Barriounuevo, G. (1998). Amplification of perforant-path EPSPs in CA3 pyramidal cells by LVA calcium and sodium channels. *J. Neurophysiol.* **80**, 1558–1561.

Usher, M., Stemmler, M., and Koch, C. (1994). Network amplification of local fluctuations causes high spike rate variability, fractal firing patterns and oscillatory local field potentials. *Neural Comput.* **6**, 795–836.

Usrey, W.M. (2002). The role of spike timing for thalamocortical processing. *Curr. Opin. Neurobiol.* **12**, 411–417.

Usui, S. (2003). Visiome: neuroinformatics research in vision project. *Neural Netw.* **16**, 1293–1300.

Usui, S., Ishii, H., and Kamiyama Y. (1991). An analysis of retinal L-type horizontal cell responses by the ionic current model. *Neurosci Res.* (Suppl.) **15**, S91–S105.

Usui, S., Ishihara, A., Kamiyama, Y., and Ishii, H. (1996a). Ionic current model of bipolar cells in the lower vertebrate retina. *Vision Res.* **36**, 4069–4076.

Usui, S., Kamiyama, Y., Ishii, H., and Ikeno, H. (1996b). Reconstruction of retinal horizontal cell responses by the ionic current model. *Vision Res.* **36**, 1711–1719.

Usui, S., Kamiyama, Y., Ogura, T., Kodama, I., and Toyama, J. (1996c). Effects of zatebradine (UL-FS 49) on the vertebrate retina. In J. Toyama, M. Hiraoka, and I. Kodama, eds., *Recent Progress in Electropharmacology of the Heart*. CRC, Boca Raton, FL, pp. 37–46.

Uteshev, V.V., Patlak, J.B., and Pennefather, P.S. (2000). Analysis and implications of equivalent uniform approximations of nonuniform unitary synaptic systems. *Biophys. J.* **79**, 2825–2839.

Uylings, H.B.M. and Van Pelt, J. (2002). Measures for quantifying dendritic arborizations. *Network: Comput. Neural Syst.* **13**, 397–414.

Uylings, H.B.M., Kuypers, K., Diamond, M.C., and Veltman, W.A.M. (1978a). Effects of differential environment on plasticity of dendrites of cortical pyramidal neurons in adult rats. *Exp. Neurol.* **62**, 658–677.

Uylings, H.B.M., Kuypers, K., and Veltman, W.A.M. (1978b). Environmental influences on the neocortex in later life. In M.A. Corner, ed., *Maturation of the nervous system, Progress in Brain Research*, Volume 48. Elsevier, Amsterdam, pp. 261–273.

Uylings, H.B.M., Ruiz-Marcos, A., and Van Pelt, J. (1986). The metric analysis of three-dimensional dendritic tree patterns: a methodological review. *J Neurosci. Meth.* **18**, 127–151.

Uylings, H.B.M., Van Pelt, J., and Verwer, R.W.H. (1989). Topological analysis of individual neurons. In J.J. Capowski, ed., *Computer Techniques in Neuroanatomy*. Plenum, New York, pp. 215–239.

Uylings, H.B.M., Van Eden, C.G., Parnavelas, J.G., and Kalsbeek, A. (1990). The prenatal and postnatal development of rat cerebral cortex. In B. Kolb and R.C. Tees, eds., *The Cerebral Cortex of the Rat*. MIT Press, Cambridge, MA, pp. 35–76.

Uylings, H.B.M., Van Pelt, J., Parnavelas, J.G., and Ruiz-Marcos, A. (1994). Geometrical and topological characteristics in the dendritic development of cortical pyramidal and non-pyramidal neurons. In J. van Pelt, M.A. Corner, H.B.M. Uylings, and F.H. Lopes da Silva, eds., *The Self-Organizing Brain: From Growth Cones to Functional Networks, Progress in Brain Research*, Volume 102. Elsevier, Amsterdam, pp. 109–123.

Vandenberg, C.A. and Bezanilla, F. (1991a). Single-channel, macroscopic, and gating currents from sodium channels in the squid giant axon. *Biophys. J.* **60**, 1499–1510.

Vandenberg, C.A. and Bezanilla, F. (1991b). A sodium channel gating model based on single channel, macroscopic ionic, and gating currents in the squid giant axon. *Biophys. J.* **60**, 1499–1510.

Van Essen, D.C. and Maunsell, J.H.R. (1983). Hierarchical organization and functional streams in the visual cortex. *Trends Neurosci.* **6**, 370–375.

Van Essen, D.C., Anderson, C.H., and Felleman, D.J. (1992). Information processing in the primate visual system: an integrated systems perspective. *Science* **255**, 419–423.

Vanier, M.C. and Bower, J.M. (1999). A comparative survey of automated parameter search methods for compartmental neural models. *J. Comput. Neurosci.* **7**, 149–171.

Van Ooyen, A. (1994). Activity-dependent neural network development. *Network: Comput. Neural Syst.* **5**, 401–423.

Van Ooyen, A. (2001). Competition in the development of nerve connections: a review of models. *Network: Comput. Neural Syst.* **12**, R1–R47.

Van Ooyen, A. (2003). *Modeling Neural Development*. MIT Press, Cambridge, MA.

Van Ooyen, A. and Van Pelt, J. (2002). Competition in neuronal morphogenesis and the development of nerve connections. In G.A. Ascoli, ed., *Computational Neuroanatomy: Principles and Methods*. Humana, Totowa, NJ, pp. 219–244.

Van Ooyen, A. and Willshaw, D.J. (1999). Competition for neurotrophic factor in the development of nerve connections. *Proc. R. Soc. Lond. B* **266**, 883–892.

Van Ooyen, A. and Willshaw, D.J. (2000). Development of nerve connections under the control of neurotrophic factors: parallels with consumer-resource systems in population biology. *J. Theor. Biol.* **206**, 195–210.

Van Ooyen, A., Duijnhouwer, J., Remme, M.W.H., and Van Pelt, J. (2002). The effect of dendritic topology on firing patterns in model neurons. *Network: Comput. Neural Syst.* **13**, 311–325.

Van Ooyen, A., Graham, B.P., and Ramakers, G.J.A. (2001). Competition for tubulin between growing neurites during development. *Neurocomputing* **38–40**, 73–78.

Van Pelt, J. (1992). A simple vector implementation of the Laplace-transformed cable equations in passive dendritic trees. *Biol. Cybern.* **68**, 15–21.

Van Pelt, J. (1997). Effect of pruning on dendritic tree topology. *J. Theor. Biol.* **186**, 17–32.

Van Pelt, J. and Schierwagen, A.K. (1994). Electrotonic properties of passive dendritic trees — effect of dendritic topology. In J. Van Pelt, M.A. Corner, H.B.M. Uylings, and F.H. Lopes da Silva, eds., *The Self-Organizing Brain: from Growth Cones to Functional Networks, Progress in Brain Research*, Volume 102. Elsevier, Amsterdam, pp. 127–149.

Van Pelt, J. and Schierwagen, A.K. (1995). Dendritic electrotonic extent and branching pattern topology. In J.M. Bower, ed., *The Neurobiology of Computation*. Kluwer, Boston, MA, pp. 153–158.

Van Pelt, J. and Schierwagen, A. (2004). Morphological analysis and modeling of neuronal dendrites. *Math. Biosci.* **188**, 147–155.

Van Pelt, J. and Uylings, H.B.M. (1999). Modeling the natural variability in the shape of dendritic trees: application to basal dendrites of small rat cortical layer 5 pyramidal neurons. *Neurocomputing* **26–27**, 305–311.

Van Pelt, J. and Uylings, H.B.M. (2002). Branching rates and growth functions in the outgrowth of dendritic branching patterns. *Network: Comput. Neural Syst.* **13**, 261–281.

Van Pelt, J. and Uylings, H.B.M. (2003). Growth functions in dendritic outgrowth. *Brain Mind* **4**, 51–65.

Van Pelt, J. and Verwer, R.W.H. (1983). The exact probabilities of branching patterns under segmental and terminal growth hypotheses. *B. Math. Biol.* **45**, 269–285.

Van Pelt, J. and Verwer, R.W.H. (1985). Growth models (including terminal and segmental branching) for topological binary trees. *B. Math. Biol.* **47**, 323–336.

Van Pelt, J. and Verwer, R.W.H. (1986). Topological properties of binary trees grown with order-dependent branching probabilities. *B. Math. Biol.* **48**, 197–211.

Van Pelt, J., Uylings, H.B.M., and Verwer, R.W.H. (1989). Distributional properties of measures of tree topology. *Acta Stereol.* **8**, 465–470.

Van Pelt, J., Uylings, H.B.M., Verwer, R.W.H., Pentney, R.J., and Woldenberg, M.J. (1992). Tree asymmetry — a sensitive and practical measure for binary topological trees. *B. Math. Biol.* **54**, 759–784.

Van Pelt, J., Dityatev, A.E., and Uylings, H.B.M. (1997). Natural variability in the number of dendritic segments: model-based inferences about branching during neurite outgrowth. *J. Comp. Neurol.* **387**, 325–340.

Van Pelt, J., Schierwagen, A.K., and Uylings, H.B.M. (2001a). Modeling dendritic morphological complexity of deep layer cat superior colliculus neurons. *Neurocomputing* **38–40**, 403–408.

Van Pelt, J., Van Ooyen A., and Uylings, H.B.M. (2001b). Modeling dendritic geometry and the development of nerve connections. In E. de Schutter, ed., *Computational Neuroscience: Realistic Modeling for Experimentalists*. CRC, Boca Raton, FL, pp. 179–208.

Van Pelt, J., Van Ooyen, A., and Uylings, H.B.M. (2001c). The need for integrating neuronal morphology databases and computational environments in exploring neuronal structure and function. *Anat. Embryol.* **204**, 255–265.

Van Pelt, J., Graham, B.P., and Uylings, H.B.M. (2003). Formation of dendritic branching patterns. In A. van Ooyen, ed., *Modeling Neural Development*. MIT Press, Cambridge, MA, pp. 75–94.

Van Veen, M.P. and Van Pelt, J. (1993). Terminal and intermediate segment lengths in neuronal trees with finite length. *B. Math. Biol.* **55**, 277–294.

Varela, F., Lachaux, J.P., Rodriguez, E., and Martinerie, J. (2001). The brainweb: phase synchronization and large-scale integration. *Nat. Rev. Neurosci.* **2**, 229–239.

Varela, F.J., Thompson, E., and Rosch, E. (1991). *The Embodied Mind.* MIT Press, Cambridge, MA.

Varghese, A. and Boland, L.M. (2002). Computing transient gating charge movement of voltage-dependent ion channels. *J. Comp. Neurosci.* **12**, 123–137.

Vermeulen, A., Rospars, J.-P., Lansky, P., and Tuckwell, H.C. (1996). Coding of stimulus intensity in an olfactory receptor neuron: role of neuron spatial extent and passive dendritic backpropagation of action potentials. *B. Math. Biol.* **58**, 493–512.

Verwer, R.W.H., Van Pelt, J., and Uylings, H.B.M. (1992). An introduction to topological analysis of neurons. In M.G. Stewart, ed., *Quantitative Methods in Neuroanatomy*. Wiley, New York, pp. 295–323.

Vetter, P., Roth, A., and Hausser, M. (2001). Propagation of action potentials in dendrites depends on dendritic morphology. *J. Neurophysiol.* **85**, 926–937.

Vinje, W.E. and Gallant, J.L. (2000). Sparse coding and decorrelation in primary visual cortex during natural vision. *Science* **287**, 1273–1275.

Vitek, J.L. (2002). Pathophysiology of dystonia: a neuronal model. *Movement Disord.* **17** (suppl. 3), S49–S62.

Viviani, P. and Flash, T. (1995). Minimum-jerk, two-thirds power law, and isochrony: converging approaches to movement planning. *J. Exp. Psychol. Human* **21**, 32–53.

Volfovsky, N., Parnas, H., Segal, M., and Korkotian, E. (1999). Geometry of dendritic spines affects calcium dynamics in hippocampal neurons: theory and experiments. *J. Neurophysiol.* **81**, 450–462.

Volkmann, J., Joliot, M., Mogilner, A., Ioannides, A.A., Lado, F., Fazzini, E., Ribary, U., and Llinás, R. (1996). Central motor loop oscillations in parkinsonian resting tremor revealed by magnetoencephalography. *Neurology* **46**, 1359–1370.

Von Stein, A., Rappelsberger, P., Sarntheim, J., and Petsche, H. (1999). Synchronization between temporal and parietal cortex during multimodal object processing in man. *Cereb. Cortex* **9**, 137–150.

Wagner, J. and Keizer, J. (1994). Effects of rapid buffers on Ca^{2+} diffusion and Ca^{2+} oscillations. *Biophys. J.* **67**, 447–456.

Wagner, J., Li, Y., Pearson, J., and Keizer, J. (1998). Simulation of the fertilization Ca^{2+} wave in *Xenopus laevis* eggs. *Biophys. J.* **75**, 2088–2097.

Wahde, M. and Hertz, J. (2000). Coarse-grained reverse engineering of genetic regulatory networks. *Biosystems* **55**, 129–136.

Wall, P.D. (1995). Do nerve impulses penetrate terminal arborizations? A pre-presynaptic control mechanism. *Trends Neurosci.* **18**, 99–103.

Walmsley, B. (1995). Interpretation of 'quantal' peaks in distributions of evoked synaptic transmission at central synapses. *Proc. R. Soc. Lond. B* **261**, 245–250.

Walter, W.G. (1953). *The Living Brain.* Duckworth, London.

Wang, X. (1998). Calcium coding and adaptive temporal computation in cortical pyramidal neurons. *J. Neurophysiol.* **79**, 1549–1566.

Washington, S.D., Ascoli, G.A., and Krichmar, J.L. (2000). A statistical analysis of dendritic morphology's effect on neuron electrophysiology of CA3 pyramidal cells. *Neurocomputing* **32–33**, 261–269.

Waters, J. and Helmchen, F. (2004). Boosting of action potential backpropagation by neocortical network activity in vivo. *J. Neurosci.* **24**, 11127–11136.

Watts, D.J. (1999). *Small Worlds.* Princeton University Press, Princeton, NJ.

Watts, D.J. (2003). *Six Degrees. The Science of a Connected Age.* Norton, New York.

Watts, D.J. and Strogatz, S.H. (1998). Collective dynamics of "small world" networks. *Nature* **393**, 440–442.

Wegner, D.M. (2003). *The Illusion of Conscious Will.* MIT Press, Cambridge, MA.

Weinberg, R.J., Pierce, J.P., and Rustioni, A. (1990). Single fiber studies of ascending inputs to the cuneate nucleus of cats: I. Morphometry of primary afferent fibers. *J. Comp. Neurol.* **300**, 113–133.

Wen, X., Fuhrman, S., Michaels, G.S., Carr, D.B., Smith, S., Barker, J.L., and Somogyi, R. (1998). Large-scale temporal gene expression mapping of central nervous system development. *Proc. Natl. Acad. Sci. USA* **95**, 334–339.

Weng, J., McClelland, J., Pentland, A., Sporns, O., Stockman, I., Sur, M., and Thelen, E. (2001). Autonomous mental development by robots and animals. *Science* **291**, 599–600.

Werbos, P. (1974). *Beyond Regression: New Tools for Prediction and Analysis in the Behavioral Sciences.* Ph.D. Thesis, Harvard University Press, Cambridge, MA.

West, M. (1997). Hierarchical mixture models in neurological transmission analysis. *J. Am. Stat. Assoc.* **92**, 587–606.

West, M. and Harrison, P.J. (1997). *Bayesian Forecasting and Dynamic Models*, 2nd edition. Springer-Verlag, New York.

West, M. and Turner, D.A. (1994). Deconvolution of mixtures in analysis of neural synaptic transmission. *Statistician* **43**, 31–43.

Westerfield, M., Joyner, R.W., and Moore, J.W. (1978). Temperature-sensitive conduction failure at axon branch points. *J. Neurophysiol.* **41**, 1–7.

White, J.A., Manis, P.B., and Young, E.D. (1992). The parameter identification problem for the somatic shunt model. *Biol. Cybern.* **66**, 307–318.

White, J.A., Rubinstein, J.T., and Kay, A.R. (2000). Channel noise in neurons. *Trends Neurosci.* **23**, 131–137.

Whitehead, R.R. and Rosenberg, J.R. (1993). On trees as equivalent cables. *Proc. R. Soc. Lond. B* **252**, 103–108.

Whitford, K.L., Marillat, V., Stein, E.W., Goodman, C.S., Tessier-Lavigne, M., Chedotal, A., and Ghosh, A. (2002). Regulation of cortical dendrite development by slit-robo interactions. *Neuron* **33**, 47–61.

Whittington, M.A. and Traub, R.D. (2003). Interneuron diversity series: Inhibitory interneurons and network oscillations in vitro. *Trends Neurosci.* **26**, 676–682.

Wichmann, T. and DeLong, M.R. (1996). Functional and pathophysiological models of the basal ganglia. *Curr. Opin. Neurobiol.* **6**, 751–758.

Wickens, J. (1988). Electrically coupled but chemically isolated synapses: Dendritic spines and calcium in a rule for synaptic modification. *Prog. Neurobiol.* **31**, 507–528.

Wiener, N. (1930). Generalized harmonic analysis. *Acta. Math-Djursholm* **55**, 117–258.

Wiener, N. (1948). *Cybernetics. Or Control and Communication in the Animal and the Machine.* MIT Press, Cambridge, Mass.

Wiener, N. (1961). *Cybernetics*, 2nd edition. MIT Press, Cambridge, MA.

Wigstrom, H., Swann, J.W., and Andersen, P. (1979). Calcium dependency of synaptic long-lasting potentiation in the hippocampal slice. *Acta Physiol. Scand.* **105**, 126–128.

Williams, S.R. and Stuart, G.J. (1999). Mechanisms and consequences of action potential burst firing in rat neocortical pyramidal neurons. *J. Physiol. (Lond.)* **521**, 467–482.

Williams, S.R. and Stuart, G.J. (2000). Action potential backpropagation and somatic-dendritic distribution of ion channels in thalamocortical neurons. *J. Neurosci.* **20**, 1307–1317.

Williams, S.R. and Stuart, G.J. (2002). Dependence of EPSP efficacy on synapse location in neocortical pyramidal cells. *Science* **295**, 1907–1910.

Williams, S.R. and Stuart, G.J. (2003). Role of dendritic synapse location in the control of action potential output. *Trends Neurosci.* **26**, 147–154.

Wilson, C.J. (1984). Passive cable properties of dendritic spines and spiny neurons. *J. Neurosci.* **4**, 281–297.

Wilson, H.R. (1999a). Simplified dynamics of human and mammalian neocortical neurons. *J. Theor. Biol.* **200**, 375–388.

Wilson, H.R. (1999b). *Spikes, Decisions, and Actions: The Dynamical Foundations of Neuroscience.* Oxford University Press, New York.

Wilson, H.R. and Cowan, J.D. (1972). Excitatory and inhibitory interactions in localized populations of model neurons. *Biophys. J.* **12**, 1–24.

Wilson, H.R. and Cowan, J.D. (1973). A mathematical theory of the functional dynamics of cortical and thalamic nervous tissue. *Kybernetik* **13**, 55–80.

Woldenberg, M.J., O'Neill, M.P., Quackenbush, L.J., and Pentney, R.J. (1993). Models for growth, decline and regrowth of the dendrites of rat Purkinje cells induced from magnitude and link-length analysis. *J. Theor. Biol.* **162**, 403–429.

Wolfe, J. and Palmer, L.A. (1998). Temporal diversity in the lateral geniculate nucleus of cat. *Visual Neurosci.* **15**, 653–675.

Wong, R.K.S. and Prince, D.A.J. (1981). Afterpotential generation in hippocampal pyramidal cells. *J. Neurophysiol.* **45**, 86–97.

Woolsey, T.A. and Van der Loos, H. (1970). The structural organization of layer IV in the somatosensory region (SI) of mouse cerebral cortex. The description of a cortical field composed of discrete cytoarchitectonic units. *Brain Res.* **17**, 205–242.

Workman, C., Jensen, L.J., Jarmer, H., Berka, R., Gautier, L., Nielser, H.B., Saxild, H.H., Nielsen, C., Brunak, S., and Knudsen, S. (2002). A new nonlinear normalization method for reducing variability in DNA microarray experiments. *Genome Biol.* **3**, research 0048.1–0048.16.

Wos, L. and McCune, W. (1991). Automated theorem-proving and logic programming–a natural symbiosis. *J. Logic Program* **11**, 1–53.

Yamada, W.M., Koch, C., and Adams, P.R. (1989). Multiple channels and calcium dynamics. In Koch, C. and Segev, I.,, eds., *Methods in Neuronal Modeling*. MIT Press, Cambridge, MA, pp. 97–133.

Yamada, Y., Koizumi, A., Iwasaki, E., Watanabe, S.-I., and Kaneko, A. (2002). Propagation of action potentials from the soma to individual dendrite of cultured rat amacrine cells is regulated by local GABA input. *J. Neurophysiol.* **87**, 2858–2866.

Yang, H., Haddad, H., Tomas, C., Alsaker, K., and Papoutsakis, E.T. (2003a). A segmental nearest neighbor normalization and gene identification method gives superior results for DNA-array analysis. *Proc. Natl. Acad. Sci. USA* **100**, 1122–1127.

Yang, S.X., Ogmen, H., and Maguire, G. (2003b). Neural computations in the tiger salamander and mudpuppy outer retinae and an analysis of GABA action from horizontal cells. *Biol. Cybern.* **88**, 450–458.

Yang, Y.H., Dudoit, S., Luu, P., Lin, D.M., Peng, V., Ngai, J., and Speed, T.P. (2002). Normalization for cDNA microarray data: a robust composite method addressing single and multiple slide systematic variation. *Nucleic Acids Res.* **30**, e15.

Yao, Y., Choi, J., and Parker, I. (1995). Quantal puffs of intracellular Ca^{2+} evoked by inositol trisphosphate in Xenopus oocytes. *J. Physiol. (Lond.)* **482**, 533–553.

Yau, K.W. and Baylor, D.A. (1989). Cyclic GMP-activated conductance of retinal photoreceptor cells. *Annu. Rev. Neurosci.* **12**, 289–327.

Yee, H.C., Sweby, P.K., and Griffiths, D.F. (1991). Dynamical approach study of spurious steady-state numerical solutions of non-linear differential equations. I. The dynamics of time discretization and its implications for algorithm development in computational fluid dynamics. *J. Comput. Phys.* **97**, 249–310.

Yeung, M.K.S., Tegner, J., and Collins, J.J. (2002). Reverse engineering gene networks using singular value decomposition and robust regression. *Proc. Natl. Acad. Sci. USA* **9**, 6163–6168.

Yim, C.C., Krnjevic, K., and Dalkara T. (1986). Ephaptically generated potentials in CA1 neurons of rat's hippocampus in situ. *J. Neurophysiol.* **56**, 99–122.

Young, M.P. (1992). Objective analysis of the topological organization of the primate cortical visual system. *Nature* **358**, 152–155.

Young, M.P. (1993). The organization of neural systems in the primate cerebral cortex. *Proc. R. Soc. Lond. B* **252**, 13–18.

Young, M.P., Scannell, J.W., O'Neill, M.A., Hilgetag, C.C., Burns, G., and Blakemore, C. (1995). Non-metric multidimensional scaling in the analysis of neuroanatomical connection data and the organization of the primate cortical visual system. *Philos. T. R. Soc. Lond. B* **348**, 281–308.

Yu, Y., Liu, F., and Wang, W. (2001). Frequency sensitivity in Hodgkin–Huxley systems. *Biol. Cybern.* **84**, 227–235.

Yuh, C.H., Bolouri, H., and Davidson, E.H. (1998). Genomic cis-regulatory logic, experimental and computational analysis of sea urchin gene. *Science* **279**, 1896–1902.

Yuste, R., Gutnick, M.J., Saar, D., Delaney, K.R., and Tank, D. (1994). Ca^{++} accumulations in dendrites of neocortical pyramidal neurons: an apical band and evidence of two functional compartments. *Neuron* **13**, 23–43.

Yuste, R., Majewska, A., Cash, S.S., and Denk, W. (1999). Mechanisms of calcium influx into hippocampal spines: heterogeneity among spines, coincidence detection by NMDA receptors, and optical quantal analysis. *J. Neurosci.* **19**, 1976–1987.

Yuste, R., Majewska, A., and Holthoff, K. (2000). From form to function: calcium compartmentalization in dendritic spines. *Nat. Neurosci.* **3**, 653–659.

Zador, A. and Koch, C. (1994). Linearized models of calcium dynamics: formal equivalence to the cable equation. *J. Neurosci.* **14**, 4705–4715.

Zador, A., Koch, C., and Brown, T.H. (1990). Biophysical model of a Hebbian synapse. *Proc. Natl. Acad. Sci. USA* **87**, 6718–6722.

Zecevic, D. (1996). Multiple spike-initiation zones in single neurons revealed by voltage-sensitive dyes. *Nature* **381**, 322–325.

Zeki, S. (1993). *A Vision of the Brain.* Blackwell, Oxford.

Zeki, S. and Shipp, S. (1988). The functional logic of cortical connections. *Nature* **335**, 311–317.

Zenisek, D. and Matthews, G. (1998). Calcium action potentials in retinal bipolar neurons. *Vis. Neurosci.* **15**, 69–75.

Zhabotinsky, A.M. (2000). Bistability in the Ca^{2+}/calmodulin-dependent protein kinase-phosphatase system. *Biophys. J.* **79**, 2211–2221.

Zhang, L.-I. and Poo, M.-M. (2001). Electrical activity and development of neural circuits. *Nat. Neurosci.* **4** (Suppl.), 1207–1214.

Zhou, L. and Chiu, S.-Y. (2001). Computer model for action potential propagation through branch point in myelinated nerves. *J. Neurophysiol.* **85**, 197–210.

Zhou, Y. and Bell, J. (1994). Study of propagation along nonuniform excitable fibers. *Math. Biosci.* **119**, 169–203.

Zhu, J.J. (2000). Maturation of layer 5 Neocortical pyramidal neurons: amplifying salient layer 1 and layer 4 inputs by Ca^{++} action potentials in adult rat tuft dendrites. *J. Physiol. (Lond.)* **526**, 571–587.

Zirh, T.A., Lenz, F.A., Reich, S.G., and Dougherty, P.M. (1998). Patterns of bursting occurring in thalamic cells during parkinsonian tremor. *Neuroscience* **83**, 107–121.

Zukerman, I. and Litman, D. (2001). Natural language processing and user modeling: synergies and limitations. *User Model. User-Adap.* **11**, 129–158.

Index